SOME ... ROP...

Air (dry, at 20°C and 1 atm)

Density	1.21 kg/m^3
Specific heat molar at constant pressure	$1010 \text{ J/kg} \cdot \text{K}$
Ratio of molar specific heats	1.40
Speed of sound	343 m/s
Electrical breakdown strength	$3 \times 10^6 \text{ V/m}$
Effective molar mass	0.0289 kg/mol

Water

Density	1000 kg/m^3
Speed of sound	1460 m/s
Specific heat at constant pressure	$4190 \text{ J/kg} \cdot \text{K}$
Heat of fusion (0°C)	333 kJ/kg
Heat of vaporization (100°C)	2260 kJ/kg
Index of refraction ($\lambda = 589$ nm)	1.33
Molar mass	0.0180 kg/mol

Earth

Mass	$5.98 \times 10^{24} \text{ kg}$
Mean radius	$6.37 \times 10^6 \text{ m}$
Free-fall acceleration at the Earth's surface	9.81 m/s^2
Standard atmosphere	$1.01 \times 10^5 \text{ Pa}$
Period of satellite at 100-km altitude	86.3 min
Radius of the geosynchronous orbit	42,200 km
Escape speed	11.2 km/s
Magnetic dipole moment	$8.0 \times 10^{22} \text{ A} \cdot \text{m}^2$
Mean electric field at surface	150 V/m, down

Distance to:

Moon	$3.82 \times 10^8 \text{ m}$
Sun	$1.50 \times 10^{11} \text{ m}$
Nearest star	$4.04 \times 10^{16} \text{ m}$
Galactic center	$2.2 \times 10^{20} \text{ m}$
Andromeda galaxy	$2.1 \times 10^{22} \text{ m}$
Edge of the observable universe	$\sim 10^{26} \text{ m}$

THE GREEK ALPHABET

Alpha	A	α	Iota	I	ι	Rho	P	ρ
Beta	B	β	Kappa	K	κ	Sigma	Σ	σ
Gamma	Γ	γ	Lambda	Λ	λ	Tau	T	τ
Delta	Δ	δ	Mu	M	μ	Upsilon	Υ	υ
Epsilon	E	ϵ	Nu	N	ν	Phi	Φ	ϕ, φ
Zeta	Z	ζ	Xi	Ξ	ξ	Chi	X	χ
Eta	H	η	Omicron	O	o	Psi	Ψ	ψ
Theta	Θ	θ	Pi	Π	π	Omega	Ω	ω

FOURTH EDITION

FUNDAMENTALS OF
PHYSICS

VOLUME 1

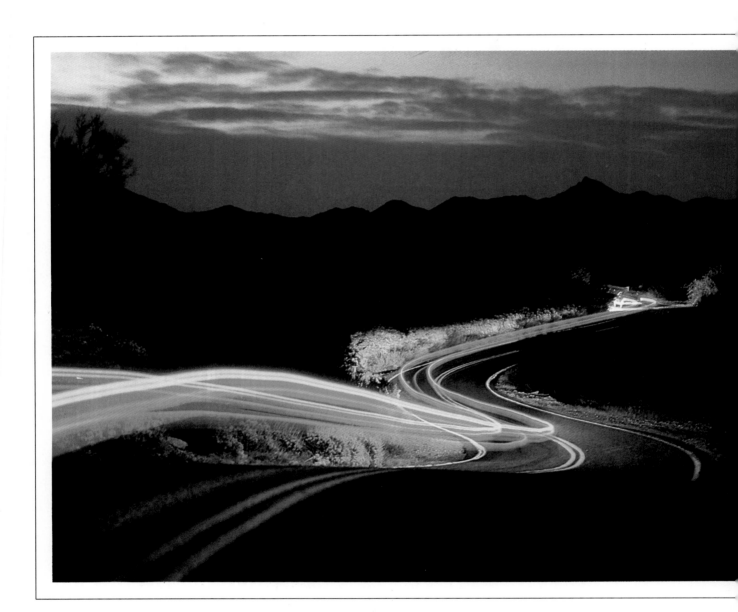

FUNDAMENTALS OF PHYSICS

VOLUME 1

DAVID HALLIDAY

University of Pittsburgh

ROBERT RESNICK

Rensselaer Polytechnic Institute

JEARL WALKER

Cleveland State University

JOHN WILEY & SONS, INC.

New York Chichester Brisbane Toronto Singapore

ACQUISITIONS EDITOR Cliff Mills
DEVELOPMENTAL EDITOR Barbara Heaney
MARKETING MANAGER Catherine Faduska
PRODUCTION SUPERVISOR Lucille Buonocore
INTERIOR DESIGN Dawn L. Stanley
COVER DESIGN Jeanette Jacobs Design
MANUFACTURING MANAGER Andrea Price
COPY EDITING SUPERVISOR Deborah Herbert
PHOTO RESEARCH DIRECTOR Stella Kupferberg
PHOTO RESEARCHERS Charles Hamilton
Hilary Newman
Pat Cadley
ILLUSTRATION Precision Graphics
ILLUSTRATION COORDINATOR Edward Starr
COVER PHOTO Courtesy FPG
International

Recognizing the importance of preserving what has been
written, it is a policy of John Wiley & Sons, Inc. to have books
of enduring value published in the United States printed on acid-
free paper, and we exert our best efforts to that end.

This book was set in 10/12 New Baskerville by Progressive Typog-
raphers and printed and bound by Von Hoffman Press. The cover
was printed by Phoenix Color.

Library of Congress Cataloging in Publication Data:

Halliday, David
   Fundamentals of physics / David Halliday, Robert Resnick,
Jearl Walker. — 4th ed.
      p.   cm.
   Includes index.
   ISBN 0-471-31041-7
   1. Physics. I. Resnick, Robert   II. Walker,
Jearl
   QC21.2.H35   1993
   530—dc20   92-32801
                                   CIP

Printed in the United States of America

10 9 8 7 6 5 4 3 2 1

PREFACE

Tremendous advances have taken place over the last few years in understanding the needs and preparation of physics students for their careers in science and engineering. In writing this fourth edition of *Fundamentals of Physics*, we have been guided by this ferment of activity. With the insights provided by a new coauthor, Jearl Walker, we have completely reexamined our approaches and coverage, and we hope that this new edition will contribute to the enhancement of physics education.

CHANGES IN THE FOURTH EDITION

Although we have retained the organizational framework of the third edition, we have rewritten many chapters and many sections of other chapters. Each chapter has been scrutinized to ensure clarity and currency of coverage, reflecting the needs of science and engineering students. In particular, changes have been made in the coverage of friction, work and energy, electrostatics, and optics.

We have completely reexamined current concepts and derivations to see whether there are better or clearer ways to treat them. In many instances, we have added more explanations or intermediate steps. We have also added more Sample Problems within each chapter, with the goal not only of providing more examples to students but also of tying these Sample Problems more closely to the end-of-chapter Exercises and Problems.

We have also analyzed all of the end-of-chapter Questions, Exercises, and Problems, adding many to their number and editing them for even greater clarity and interest. At the ends of most chapters we have added a new section of problems, entitled "Additional Problems," which are unreferenced to chapter sections.

We have devoted considerable attention to illustrating real-world applications of physics topics. A prime example is the "puzzler" that opens each chapter. These examples of curious phenomena, many of which are common, were chosen so as to intrigue a student. Explanations of the puzzlers are given within the chapters, either in a text discussion or within Sample Problems. Because students will likely see these or related phenomena well after the physics course is completed, the puzzlers should provide long-term reinforcement of the associated physics.

Because the diagrams that accompany discussions of physics are vital to understanding, we have reviewed every diagram in the book for its clarity and usefulness. Nearly all the diagrams in this edition have been changed in some way, and many new diagrams have been added.

In addition, we have used full color for all the diagrams and most of the photographs. In the diagrams, color allowed us to show the various parts more clearly, to emphasize the important aspects, and to give a sense of depth to three-dimensional situations.

CHAPTER FEATURES

The features of each chapter were carefully planned to motivate students and guide their reasoning processes.

Opening Puzzlers

Each chapter opens with a physics "puzzler," describing a curious phenomenon that is intended to entice a student. These puzzlers are carefully linked to the physics of the associated chapters, and the memorable photographs of the puzzlers have been chosen in order that the relevant physics also be memorable. The explanations are given within text, for qualitative explanations, or within Sample Problems, for quantitative explanations. When answered in the latter form, the puzzler is designed to prepare the students for some of the more challenging end-of-chapter problems.

Sample Problems

We have increased the number of Sample Problems in this edition to over 400, so as to provide problem-solving models for all aspects of the chapter. We have modified many of the Sample Problems from the previous edition to tie them more closely to the end-of-chapter Exercises and Problems. All Sample Problems have been carefully edited to make them even more helpful to students. Thus, more than 50% of the Sample Problems are in some way new.

These Sample Problems offer a student the chance to work through a problem with the authors, to see how to begin with a question and end with an answer. Thus the Sample Problems provide a bridge from the physics of the text to the end-of-chapter problems. They also provide an opportunity to sort out concepts, terminology, and symbolization, to strengthen mathematical skills, and to sharpen the ability to spot "dead-end" solution strategies.

Problem Solving Tactics

Careful attention to developing a student's problem-solving skills has been a hallmark of previous editions of this book. This feature is continued in the present edition with more than 70 sections titled Problem Solving Tactics, an increase of 50% over the third edition. In these sections, we emphasize techniques of skilled problem solvers, review the logic of Sample Problems, and discuss common misunderstandings of terminology and physics concepts. As in the third edition, most of these guideposts to learning fall within the first half of the book where students need the most help, but many now appear in the second half of the book when especially tricky situations arise.

Questions, Exercises, and Problems

The sets of end-of-chapter Questions, Exercises, and Problems are by far the largest and most varied of any introductory physics text. We have edited the highly praised sets of the earlier editions to achieve even greater clarity and interest and have added a substantial number of new applied and conceptual Questions, Exercises, and Problems. We have been careful to maintain the variation in level and breadth of scope that have characterized our texts.

At the same time, we have been careful not to discard the many tried-and-true problems that have survived the test of the classroom for many years. Long-time users of our text will not find their favorites missing.

A more generous use of figures and photographs serves better to illustrate the Questions, Exercises, and Problems than before.

Questions. These thought questions have always been a special feature of our books. They are used as sources of classroom discussion and for clarification of homework concepts. Now numbering approximately 1150, they relate even more to everyday phenomena, serve to arouse curiosity and interest, and stress conceptual aspects of physics.

Exercises and Problems. The total number of Exercises and Problems has been increased to over 3400 in this edition, up from 3160 in the third edition. Exercises, identified by an "E" after their number, typically involve one step or formula or represent a single application. They thus build student confidence for the problem sets. Problems are identified by a "P" after their number; among them are a small number of advanced problems, identified by asterisks (*) next to their number.

In addition to labeling Exercises "E" and Problems "P", we have organized them by difficulty for each section of the chapter. Our goal was to simplify the selection process for instructors in the face of the voluminous material now available. Hence, instructors can vary the content emphasis and the level of difficulty to suit their tastes and the preparation of the student body while still putting aside a significant number of Exercises and Problems for many years of instruction.

Additional Problems. At the request of many instructors, we have added a new section, entitled "Additional Problems," to the ends of most chapters. While working these problems, which are unreferenced to chapter sections, students must identify for themselves the relevant physics principles.

Applications and Guest Essays

To emphasize the relevance of what physicists do and to motivate the students further, we have included within the chapters numerous applications of physics in engineering, technology, medicine, and familiar everyday phenomena.

In addition, we feature 17 essays, written by distinguished scientists and distributed at appropriate locations within the text, on the applications of physics to special topics of student interest, such as dance, sports, the greenhouse effect, lasers, holography, and many more. (See the table of contents.)

Four of the essays are new, and the authors of the 13 essays carried over from the third edition have revised and updated their material. And most of the essays, although self-contained, make references to the material covered in the immediately preceding chapter(s) and contain questions to engage the student's thought process.

MODERN PHYSICS

Like the third edition, this edition is available in a single volume of 42 chapters, ending with relativity, and in an Extended Version of 49 chapters, containing in addition a development of quantum physics and its applications to atoms, solids, nuclei, and particles. The former is meant for introductory courses that treat quantum physics in a subsequent separate course or semester.

In the early chapters, we have sought to pave the way for the systematic study of quantum physics. We have done this in three ways. (1) In appropriate places we have called attention—by specific example—to the impact of quantum ideas on our daily lives. (2) We have stressed those concepts (conservation principles, symmetry arguments, reference frames, role of aesthetics, similarity of methods, use of models, field concepts, wave concepts, etc.) that are common to both classical and quantum physics. (3) Finally, we have included a number of short, optional sections in which selected quantum (and relativistic) ideas are presented in ways that lay the foundation for the detailed and systematic treatments of relativity, atomic, nuclear, solid state, and particle physics given in later chapters.

FLEXIBILITY

In addition to the quantum physics chapters and the optional sections on quantum topics, we have included numerous optional sections throughout the text that are of an advanced, historical, general, or specialized nature.

Thus, we have consciously made available much more material than any one course or instructor is expected to "cover." Just as a textbook alone is not a course, so a course does not include the entire textbook. Indeed, more can be "uncovered" by doing less. The process of physics and its essential unity can be revealed by judicious selective coverage of many

fewer chapters than are contained here and by coverage of only portions of many included chapters. Rather than give numerous examples of such coherent selections, we urge instructors to be guided by their own interests and circumstances and to plan ahead so that some topics in relativistic and quantum physics are always included.

SUPPLEMENTS

- *A Student's Companion* by J. RICHARD CHRISTMAN, U.S. Coast Guard Academy. Much more than a traditional study guide, this student manual is designed to be used in close conjunction with the text. The Student's Companion is divided into four parts, each of which corresponds to a major section of the text, beginning with an overview "chapter." These overviews are designed to help students understand how the important topics are integrated and how the text is organized. For each chapter of the text, the corresponding Companion chapter offers: Basic Concepts, Problem Solving, Notes, Mathematical Skills, Computer Projects and Notes.

- *Solutions Manual* by J. RICHARD CHRISTMAN, U.S. Coast Guard Academy and EDWARD DERRINGH, Wentworth Institute. This manual provides students with complete worked-out solutions to 30% of the exercises and problems found at the end of each chapter within the text.

- *Interactive Learningware*, by JAMES TANNER, Georgia Institute of Technology, with the assistance of Gary Lewis, Kennesaw State College. This software contains 200 problems from the end-of-chapter exercises and problems, presented in an interactive format, providing detailed feedback for the student. Problems from Chapters 1 to 22 are included in Part 1, from Chapters 23 to 42 in Part 2. The accompanying workbooks allow the student to keep a record of the worked-out problems. The Learningware is available in IBM 3.5'' and Macintosh formats.

- *Instructor's Manual* by J. RICHARD CHRISTMAN, U.S. Coast Guard Academy. This manual contains lecture notes outlining the most important topics of each chapter, as well as demonstration experiments, laboratory and computer exercises; film and video sources are also included. Separate sections contain articles that have appeared recently in the *American Journal of Physics* and *The Physics Teacher*.

- *Test Bank* by J. RICHARD CHRISTMAN, U.S. Coast Guard Academy. More than 2200 multiple-choice questions are included in the Test Bank for *Fundamentals of Physics*.
- *Computerized Test Bank.* IBM and Macintosh versions of the entire Test Bank are available with full editing features to help you customize tests.
- *Animated Illustrations.* Approximately 85 text illustrations are animated for enhanced lecture demonstrations.
- *Transparencies.* More than 200 four-color illustrations from the text are provided in a form suitable for projection in the classroom.

ACKNOWLEDGMENTS

A textbook contains far more contributions to the elucidation of a subject than those made by the authors alone. J. Richard Christman, of the U.S. Coast Guard Academy, has once again created many fine supplements for us; his knowledge of our book and his recommendations to students and faculty are invaluable. James Tanner, of Georgia Institute of Technology, has provided us with innovative software, closely tied to the text's exercises and problems. Albert Altman, of the University of Lowell, Massachusetts, and Harry Dulaney, of Georgia Institute of Technology, contributed many new problems to the text. We thank John Merrill, of Brigham Young University, and Edward Derringh, of the Wentworth Institute of Technology for their many contributions in the past.

Our guest essayists contributed their expertise in many areas of applied physics. We thank Charles Bean, Rensselaer Polytechnic Institute; Peter Brancazio, Brooklyn College of SUNY; Patricia Cladis, AT&T Bell Laboratories; Joseph Ford, Georgia Institute of Technology; Elsa Garmire, University of Southern California; Ivar Giaever, Rensselaer Polytechnic Institute; Tung H. Jeong, Lake Forest College; Barbara Levi, *Physics Today*; Kenneth Laws, Dickinson College; Peter Lindenfeld, State University of New Jersey–Rutgers; Suzanne Nagel, AT&T Laboratories; Sally K. Ride, University of California at San Diego; John Rigden, American Institute of Physics; Thomas D. Rossing, Northern Illinois University; and Raymond Turner, Clemson University.

A team of graduate students at Johns Hopkins University checked every exercise and problem, a truly formidable task. We thank Anton Andreev, Kevin Fournier, Jidong Jiang, John Kordomenos, Mark May, Jason McPhate, Patrick Morrissey, Mark Sincell, Olaf Vancura, John Q. Xiao, and Andrew Zwicker, our coordinator.

At John Wiley, publishers, we have been fortunate to receive strong coordination and support from Cliff Mills, our editor. He has guided our efforts and encouraged us along the way. Barbara Heaney has coordinated the developmental editing and multilayered preproduction process. Catherine Faduska, our marketing manager, has been tireless in her efforts on behalf of this edition, as well as the previous edition. Joan Kalkut has built a fine supporting package of ancillary materials. Anne Scargill edited the essays. Cathy Donovan and Julia Salsbury managed the review and administrative duties admirably.

We thank Lucille Buonocore, our able production manager, for pulling all the pieces together and guiding us through the complex production process. We also thank Dawn Stanley, for her design; Deborah Herbert, for supervising the detailed copy editing; Christina Della Bartolomea, for her copy editing; Edward Starr, for managing the line art program; Lilian Brady, for her proofreading; and all other members of the production team.

Stella Kupferberg and her team of photo researchers, particularly Charles Hamilton, Hilary Newman, and Pat Cadley, were inspired in their search for unusual and interesting photographs that communicate physics principles beautifully. We thank Edward Millman and Irene Nunes for their careful development of a full-color line art program, for which they scrutinized and suggested revisions of every piece. We also owe a debt of gratitude for the line art to the late John Balbalis, whose careful hand and understanding of physics can still be seen in every diagram.

Finally, we thank Edward Millman for his developmental work on the manuscript. With us, he has read every word, asking questions from the point of view of a student. Many of his questions and suggested changes have added to the clarity of this volume. Irene Nunes added a final, valuable developmental check in the last stages of the book.

Our external reviewers have been outstanding and we acknowledge here our debt to each member of that team:

Professor Maris A. Abolins
Michigan State University

Professor Barbara Andereck
Ohio Wesleyan University

Professor Albert Bartlett
University of Colorado

Professor Timothy J. Burns
Leeward Community College

Professor Joseph Buschi
Manhattan College

Professor Philip A. Casabella
Rensselaer Polytechnic Institute

Professor Randall Caton
Christopher Newport College

Professor Roger Clapp
University of South Florida

Professor W. R. Conkie
Queen's University

Professor Peter Crooker
University of Hawaii at Manoa

Professor William P. Crummett
Montana College of Mineral Science and Technology

Professor Robert Endorf
University of Cincinnati

Professor F. Paul Esposito
University of Cincinnati

Professor Jerry Finkelstein
San Jose State University

Professor Alexander Firestone
Iowa State University

Professor Alexander Gardner
Howard University

Professor Andrew L. Gardner
Brigham Young University

Professor John Gieniec
Central Missouri State University

Professor John B. Gruber
San Jose State University

Professor Ann Hanks
American River College

Professor Samuel Harris
Purdue University

Emily Haught
Georgia Institute of Technology

Professor Laurent Hodges
Iowa State University

Professor John Hubisz
College of the Mainland

Professor Joey Huston
Michigan State University

Professor Darrell Huwe
Ohio University

Professor Claude Kacser
University of Maryland

Professor Leonard Kleinman
University of Texas at Austin

Professor Arthur Z. Kovacs
Rochester Institute of Technology

Professor Kenneth Krane
Oregon State University

Professor Sol Krasner
University of Illinois at Chicago

Professor Robert R. Marchini
Memphis State University

Professor David Markowitz
University of Connecticut

Professor Howard C. McAllister
University of Hawaii at Manoa

Professor W. Scott McCullough
Oklahoma State University

Professor Roy Middleton
University of Pennsylvania

Professor Irvin A. Miller
Drexel University

Professor Eugene Mosca
United States Naval Academy

Professor Patrick Papin
San Diego State University

Professor Robert Pelcovits
Brown University

Professor Oren P. Quist
South Dakota State University

Professor Jonathan Reichert
SUNY–Buffalo

Professor Manuel Schwartz
University of Louisville

Professor John Spangler
St. Norbert College

Professor Ross L. Spencer
Brigham Young University

Professor Harold Stokes
Brigham Young University

Professor David Toot
Alfred University

Professor J. S. Turner
University of Texas at Austin

Professor T. S. Venkataraman
Drexel University

Professor Gianfranco Vidali
Syracuse University

Professor Fred Wang
Prairie View A & M

Professor George A. Williams
University of Utah

Professor David Wolfe
University of New Mexico

This new edition traces its origins to the text *Physics for Students of Science and Engineering* (John Wiley & Sons, Inc., 1960) by the authors of the third edition. Since that time it is estimated that well over 5 million students have been introduced to physics at the college or university level by this text and those that flowed from it, including among them translations into many languages. We dedicate this fourth edition to those students, and we hope that it will be as well received by those for whom it has been written.

DAVID HALLIDAY
5110 Kenilworth Place NE
Seattle WA 98105

ROBERT RESNICK
Rensselaer Polytechnic Institute
Troy NY 12181

JEARL WALKER
Cleveland State University
Cleveland OH 44115

FEATURED IN THIS BOOK

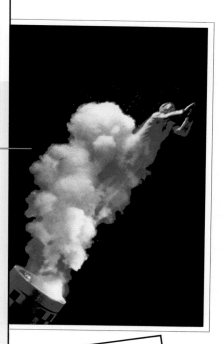

MOTION IN TWO AND THREE DIMENSIONS 4

When a "human cannonball" is shot from a cannon, the smoke and noise are theatrical, because the propulsion comes from a spring or compressed air rather than an explosion. But the danger is still real for two reasons. One is that the rapid propulsion in the muzzle usually results in a momentary blackout, from which the performer must awake if he or she is to land in the net without a broken neck. The other real danger is that the net may not be in the right place for the angle and speed of the launch.

CHAPTER OPENING PUZZLERS

Each chapter opens with an intriguing example of physics in action. By presenting high-interest applications of each chapter's concepts, the puzzlers motivate students to learn.

---

circle of radius r. Velocity vectors are drawn for two points, p and q, that are symmetric with respect to the y axis. These vectors, $\mathbf{v}_p$ and $\mathbf{v}_q$, have the same magnitude v but—because they point in different directions—they are different vectors. Their x and y components are

$$v_{px} = + v \cos \theta, \qquad v_{py} = + v \sin \theta$$

and

$$v_{qx} = + v \cos \theta, \qquad v_{qy} = - v \sin \theta.$$

The time required for the particle to move from p to q at constant speed v is

$$\Delta t = \frac{\text{arc}(pq)}{v} = \frac{r(2\theta)}{v}, \qquad (4\text{-}21)$$

in which arc(pq) is the length of the arc from p to q.

We now have enough information to calculate the components of the average acceleration $\overline{\mathbf{a}}$ experienced by the particle as it moves from p to q in Fig. 4-19. For the x component, we have

$$\overline{a}_x = \frac{v_{qx} - v_{px}}{\Delta t} = \frac{v \cos \theta - v \cos \theta}{\Delta t} = 0.$$

This result is not surprising because it is clear from symmetry in Fig. 4-19 that the x component of velocity has the same value at q and at p.

For the y component of the average acceleration, we find, making use of Eq. 4-21,

$$\sin \theta - v \sin \theta$$

FIGURE 4-20 Velocity and acceleration vectors for a particle in uniform circular motion. Both have constant magnitude but vary continuously in direction.

$(\sin \theta)/\theta$ approaches unity. In the relation given above for $\overline{a}_y$, we then have

$$a = \frac{v^2}{r} \qquad \text{(centripetal acceleration).} \qquad (4\text{-}22)$$

Our conclusion: when you see a particle moving at constant speed v in a circle (or a circular arc) of radius r, you may be sure that it has an acceleration, directed toward the center of the circle and having a magnitude v^2/r.

Figure 4-20 shows the relation between the velocity and acceleration vectors at various ing uniform circular m

---

BOXED EQUATIONS

Important equations are always boxed, to help students identify them.

FIGURE 4-17 Sample Problem 4-8. At this range, the harbor defense cannon can hit the pirate ship at two different elevation angles.

a. To what angle must the cannon be elevated to hit the pirate ship?

SOLUTION Solving Eq. 4-20 for $2\theta_0$ yields

$$2\theta_0 = \sin^{-1}\frac{gR}{v_0^2} = \sin^{-1}\frac{(9.8 \text{ m/s}^2)(560 \text{ m})}{(82 \text{ m/s})^2}$$

$$= \sin^{-1} 0.816.$$

There are two angles whose sine is 0.816, namely, 54.7° and 125.3°. Thus we find

$$\theta_0 = \tfrac{1}{2}(54.7°) \approx 27° \qquad \text{(Answer)}$$

and

$$\theta_0 = \tfrac{1}{2}(125.3°) \approx 63°. \qquad \text{(Answer)}$$

The commandant of the fort can elevate the guns to either of these two angles and (if only there were no intervening air!) hit the pirate ship.

b. What are the times of flight for the two elevation angles calculated above?

SOLUTION Solving Eq. 4-15 for t gives, for $\theta_0 = 27°$,

$$t = \frac{x - x_0}{v_0 \cos \theta_0} = \frac{560 \text{ m}}{(82 \text{ m/s})(\cos 27°)}$$

$$= 7.7 \text{ s.} \qquad \text{(Answer)}$$

Repeating the calculation for $\theta_0 = 63°$ yields $t = 15$ s. It is reasonable that the time of flight for the higher elevation angle should be larger.

c. How far should the pirate ship be from the fort if it is to be beyond range of the cannon?

SOLUTION We have seen that maximum range corresponds to an elevation angle θ_0 of 45°. Thus, from Eq. 4-20 with $\theta_0 = 45°$,

$$R = \frac{v_0^2}{g} \sin 2\theta_0 = \frac{(82 \text{ m/s})^2}{9.8 \text{ m/s}^2} \sin (2 \times 45°)$$

$$= 690 \text{ m.} \qquad \text{(Answer)}$$

As the pirate ship sails away, the two elevation angles at which the ship can be hit draw closer together, eventually merging at $\theta_0 = 45°$ when the ship is 690 m away. Beyond that point the ship is safe.

SAMPLE PROBLEM 4-9

Figure 4-18 illustrates the flight of Emanuel Zacchini over three Ferris wheels, located as shown and each 18 m high. He is launched with speed $v_0 = 26.5$ m/s, at an angle $\theta_0 = 53°$ from the horizontal and with an initial height of 3.0 m above the ground. The net in which he is to land is at the same height.

a. Does he clear the first Ferris wheel?

SOLUTION We place the origin on the cannon muzzle so that $x_0 = 0$ and $y_0 = 0$. To find his height y when $x = 23$ m, we use Eq. 4-19:

$$y = (\tan \theta_0)x - \frac{gx^2}{2(v_0 \cos \theta_0)^2}$$

$$= (\tan 53°)(23 \text{ m}) - \frac{(9.8 \text{ m/s}^2)(23 \text{ m})^2}{2(26.5 \text{ m/s})^2(\cos 53°)^2}$$

$$= 20.3 \text{ m.} \qquad \text{(Answer)}$$

Since he begins 3.0 m off the ground, he clears the Ferris wheel by about 5.3 m.

FIGURE 4-18 Sample Problem 4-9. The flight of a human cannonball over three Ferris wheels, and the desired placement of the net.

Eq. 4-16 and the fact that $y = 0$ when he lands. Then we have

$$y = (v_0 \sin \theta_0)t - \tfrac{1}{2}gt^2 = 0,$$

or

$$t = \frac{2v_0 \sin \theta_0}{g} = \frac{(2)(26.5 \text{ m/s})(\sin 53°)}{9.8 \text{ m/s}^2}$$

$$= 4.3 \text{ s.} \qquad \text{(Answer)}$$

d. How far from the cannon should the center of the net be positioned?

SOLUTION One way to answer is with Eq. 4-15, with $x_0 = 0$:

$$x = (v_0 \cos \theta_0)t$$

$$= (26.5 \text{ m/s})(\cos 53°)(4.3 \text{ s})$$

$$= 69 \text{ m,} \qquad \text{(Answer)}$$

which is the range R of the flight. Of course, Zacchini would use a wide net and bias it toward the cannon, because the air that he encounters will slow his flight, and he will not go as far as we have calculated.

PROBLEM SOLVING

TACTIC 5: SIGNIFICANT FIGURES
In some problems, you calculate a numerical value in one step of the problem and then use that number in a later step. In such cases, it is proper, strictly for the purposes of calculation, to keep more significant figures than you would accept in the final result.

In Sample Problem 4-6, for example, we calculated the time of flight of the capsule and found $t = 15.65$ s. If asked what the time of flight was, you would

4-7 UNIFORM CIRCULAR MOTION

A particle is in **uniform circular motion** if it travels in a circle or circular arc at constant speed. Although the speed does not vary, *the particle is accelerating.* That fact may be surprising because we usually think of acceleration as an increase in speed. But actually **v** is a vector, not a scalar. If **v** changes, even only in direction, there is an acceleration, and that is the case for uniform circular motion.

We use Fig. 4-19 to find the magnitude and direction of the acceleration. That figure represents a particle in uniform circular motion with speed v in a

FIGURE 4-19 A particle moves in uniform circular motion at constant speed v in a circle of radius r. Its velocities $\mathbf{v}_p$ and $\mathbf{v}_q$ at points p and q, equidistant from the y axis, are shown, along with its velocity components at those points. The instantaneous acceleration of the particle at any point is directed toward the center of the circle and has a mag-

---

SAMPLE PROBLEMS

The Sample Problems offer students the opportunity to work through the physics concepts just presented. Often built around real-world physics applications, they are closely coordinated with the end-of-chapter Questions, Exercises, and Problems.

ANSWERS TO PUZZLERS

All chapter-opening puzzlers are answered later in the chapter, either in text discussion or in a Sample Problem.

Careful attention to developing students' problem-solving skills is a hallmark of the text. These Tactics are closely related to the Sample Problems and can be found throughout the text, though most fall within the first half. They can help students in working assigned homework problems and in preparing for exams. Collectively, they represent the stock in trade of experienced problem solvers and of practicing scientists and engineers.

PROBLEM SOLVING TACTICS

FULL-COLOR LINE ART AND PHOTOGRAPHS

All line art has been redrawn in full color, carefully focusing on the important physics concepts. Photographs were chosen on the basis of whether they illuminated the concepts at hand.

Even if the effects of air could be neglected so that this motorcycle jump were simple projectile motion, it would still require careful planning if the motorcyclist is to land safely.

Sports involve many examples of projectile motion. However, the effects of air on the projectile nearly always increase the difficulty of determining the motion mathematically.

FIGURE 4-14

TABLE 4-1
TWO FLY BALLS[a]

	PATH I (AIR)	PATH II (VACUUM)
Range	323 ft	581 ft
Maximum height	174 ft	252 ft
Time of flight	6.6 s	7.9 s

[a]See Fig. 4-14. The launch angle is 60° and the launch speed is 100 mi/h.

fly ball that leaves the bat at an angle of 60° with the horizontal and an initial speed of 100 mi/h. (See "The Trajectory of a Fly Ball," by Peter J. Brancazio, *The Physics Teacher*, January 1985.) Path I (the baseball player's fly ball) is a calculated path that approximates normal conditions of play, in air. Path II (the physics professor's fly ball) is the path that the ball would follow in a vacuum. Table 4-1 gives some data for the two cases. Our assumption that air resistance can be neglected clearly does not apply to experiments carried out in such laboratories as Shea Stadium and Candlestick Park. We discuss the effect of the air on motion in Chapter 6.

Actually, *no* problem in physics can be solved *exactly*, no matter how many significant figures the calculated answer may contain: it is always necessary to make approximations. The physicist P. A. M. Dirac said that the trick is to divide the problem into two parts, one simple and the other small. You then solve the simple part exactly and do the best you can with the small part. Sometimes, the "small" part is small enough that you can neglect it entirely, as we do for air resistance at low speeds.

SAMPLE PROBLEM 4-6

A rescue plane is flying at a constant elevation of 1200 m with a speed of 430 km/h toward a point directly over a person struggling in the water (see Fig. 4-15). At what angle of sight ϕ should the pilot release a rescue capsule if it is to strike (very close to) the person in the water?

SOLUTION The initial velocity of the capsule is the same as that of the plane. That is, the initial velocity $\mathbf{v}_0$ is horizontal and has a magnitude of 430 km/h. We

REVIEW & SUMMARY

Review & Summary sections at the end of each chapter review the most important concepts and equations.

REVIEW & SUMMARY

Position Vector
The location of a particle relative to the origin of a coordinate system is given by a *position vector* **r**, which in unit-vector notation, is

$$\mathbf{r} = x\mathbf{i} + y\mathbf{j} + z\mathbf{k}, \tag{4-1}$$

where $x\mathbf{i}$, $y\mathbf{j}$, and $z\mathbf{k}$ are the *vector components* and x, y, and z are the *scalar components* of position vector **r**. A position vector might also be given by a magnitude and one or more angles for orientation.

Displacement
If a particle moves so that its position vector changes from being $\mathbf{r}_1$ to $\mathbf{r}_2$, then the particle's *displacement* $\Delta\mathbf{r}$ is

$$\Delta\mathbf{r} = \mathbf{r}_2 - \mathbf{r}_1. \tag{4-3}$$

Sample Problem 4-1 gives an example of displacement.

Average Velocity
If a particle undergoes a displacement in time Δt, its *average velocity* $\bar{\mathbf{v}}$ is

$$\bar{\mathbf{v}} = \frac{\Delta\mathbf{r}}{\Delta t}. \tag{4-4}$$

Velocity
As Δt in Eq. 4-4 is shrunk to 0, $\bar{\mathbf{v}}$ reaches a limit called the (instantaneous) *velocity*:

$$\mathbf{v} = \frac{d\mathbf{r}}{dt}, \tag{4-6}$$

which can be rewritten in unit-vector notation as

$$\mathbf{v} = v_x\mathbf{i} + v_y\mathbf{j} + v_z\mathbf{k}, \tag{4-7}$$

where $v_x = dx/dt$, $v_y = dy/dt$, and $v_z = dz/dt$. A two-dimensional example of **v** is given in Sample Problem 4-3.

When the position of a moving particle is plotted on a coordinate system, **v** is always tangent to the curve representing the particle's path.

Average Acceleration
If a particle's velocity changes from $\mathbf{v}_1$ to $\mathbf{v}_2$ in time Δt, its *average acceleration* during Δt is

$$\bar{\mathbf{a}} = \frac{\mathbf{v}_2 - \mathbf{v}_1}{\Delta t} = \frac{\Delta\mathbf{v}}{\Delta t}. \tag{4-9}$$

Acceleration
As Δt in Eq. 4-9 is shrunk to 0, $\bar{\mathbf{a}}$ reaches a limiting value called the (instantaneous) *acceleration*, which is

$$\mathbf{a} = \frac{d\mathbf{v}}{dt}. \tag{4-10}$$

In unit-vector notation,

$$\mathbf{a} = a_x\mathbf{i} + a_y\mathbf{j} + a_z\mathbf{k}, \tag{4-11}$$

where $a_x = dv_x/dt$, $a_y = dv_y/dt$, and $a_z = dv_z/dt$. An example of how Eq. 4-10 is used is in Sample Problem 4-4.

When **a** is constant, the components of **a**, **v**, and **r** along any axis can be treated as in the one-dimensional motion of Chapter 2. See Sample Problem 4-5.

Approximately 1,100 thought-provoking questions related to everyday phenomena emphasize the conceptual aspects of physics.

FIGURE 4-26 Question 4.

5. A shot is put (thrown) from above the athlete's shoulder level. The launch angle that will produce the longest range is less than 45°; that is, a flatter trajectory has a longer range. Explain why.

6. You are driving directly behind a pickup truck, going at the same speed as the truck. A crate falls from the bed of the truck to the road. Will your car hit the crate before the crate hits the road if you neither brake nor swerve? Explain your answer.

7. Trajectories are shown in Fig. 4-27 for three kicked footballs. Ignoring the effects of the air on the footballs,

10. In projectile motion when air resistance is negligible, is it ever necessary to consider three-dimensional motion rather than two-dimensional?

11. Is it possible to be accelerating if you are traveling at constant speed? Is it possible to round a curve with zero acceleration? With constant acceleration?

12. Show that, taking the Earth's rotation and revolution into account, a book resting on your table moves faster at night than it does during the daytime. In what reference frame is this statement true?

13. An aviator, pulling out of a dive, follows the arc of a circle and is said to have "experienced 3g's" in pulling out of the dive. Explain what this statement means.

14. A boy sitting in a railroad car moving at constant velocity throws a ball straight up into the air. Will the ball fall behind him? In front of him? Into his hands? What happens if the car accelerates forward or goes around a curve while the ball is in the air?

15. A woman on the observation platform of a train moving with constant velocity drops a coin while leaning over the rail. Describe the path of the coin as seen by (a) the woman on the train, (b) a person standing on the ground near the track, and (c) a person in a second train moving in the opposite direction to the first train on a parallel track.

EXERCISES AND PROBLEMS KEYED TO SECTIONS

A hallmark of this text, the exercises and problems have been increased in this edition to more than 3,400. Exercises, labeled with an "E", involve one step or formula or represent a single application. These exercises are designed to build student confidence. Problems, labeled with a "P", involve more than one step or formula or represent more than one application. Problems with an "*" are designed to challenge more advanced students.

EXERCISES & PROBLEMS

SECTION 4-2 POSITION AND DISPLACEMENT

1E. A watermelon has the following coordinates: $x = -5.0$ m, $y = 8.0$ m, and $z = 0$ m. Find its position vector (a) in unit-vector notation and (b) in terms of the vector's magnitude and orientation. (c) Sketch the vector on a right-handed coordinate system.

2E. The position vector for an electron is $\mathbf{r} = 5.0\mathbf{i} - 3.0\mathbf{j} + 2.0\mathbf{k}$, where the suppressed unit is meters. (a) Find the magnitude of $\mathbf{r}$. (b) Sketch the vector on a right-handed coordinate system.

3E. The position vector for a proton is initially $\mathbf{r} = 5.0\mathbf{i} - 6.0\mathbf{j} + 2.0\mathbf{k}$ and then later is $\mathbf{r} = -2.0\mathbf{i} + 6.0\mathbf{j} + 2.0\mathbf{k}$, where the suppressed unit is meters. (a) What is the proton's displacement vector, and (b) to what plane is it parallel?

4E. A positron undergoes a displacement $\Delta\mathbf{r} = 2.0\mathbf{i} - 3.0\mathbf{j} + 6.0\mathbf{k}$, ending with a position vector $\mathbf{r} = 3.0\mathbf{j} - 4.0\mathbf{k}$. (The suppressed unit is meters.) What was the positron's former position vector?

SECTION 4-3 VELOCITY AND AVERAGE VELOCITY

5E. A plane flies 300 mi east from city A to city B in 45.0 min and then 600 mi south from city B to city C in 1.50 h. (a) What displacement vector represents the total trip? What are (b) the average velocity vector and (c) the average speed for the trip?

6E. A train moving at a constant speed of 60.0 km/h moves east for 40.0 min, then in a direction 50.0° east of north for 20.0 min, and finally west for 50.0 min. What is the average velocity of the train during this run?

7E. In 3.50 h, a balloon drifts 21.5 km north, 9.70 k-

unit is m/s). (a) In unit-vector notation, what is the average acceleration $\bar{\mathbf{a}}$ over the 4.0 s? (b) What are the magnitude and orientation of $\bar{\mathbf{a}}$?

11E. A particle moves so that its position as a function of time in SI units is $\mathbf{r} = \mathbf{i} + 4t^2\mathbf{j} + t\mathbf{k}$. Write expressions for (a) its velocity and (b) its acceleration as functions of time.

12E. The position $\mathbf{r}$ of a particle moving in an xy plane is given by $\mathbf{r} = (2.00t^3 - 5.00t)\mathbf{i} + (6.00 - 7.00t^4)\mathbf{j}$. Here $\mathbf{r}$ is in meters and t in seconds. Calculate (a) $\mathbf{r}$, (b) $\mathbf{v}$, and (c) $\mathbf{a}$ when $t = 2.00$ s. (d) What is the orientation of a line that is tangent to the particle's path at $t = 2.00$ s?

13E. An ice boat sails across the surface of a frozen lake with constant acceleration produced by the wind. At a certain instant its velocity is $6.30\mathbf{i} - 8.4230\mathbf{j}$ in meters per second. Three seconds later, because of a shift in the wind, the boat is instantaneously at rest. What is its average acceleration during this 3-s interval?

14P. A particle A moves along the line $y = 30$ m with a constant velocity $\mathbf{v}$ ($v = 3.0$ m/s) directed parallel to the positive x axis (Fig. 4-28). A second particle B starts at the origin with zero speed and constant acceleration $\mathbf{a}$ ($a = 0.40$ m/s^2) at the same instant that particle A passes the y axis. What angle θ between $\mathbf{a}$ and the positive y axis would result in a collision between these two particles? (If your computation involves an equation with a term such as t^4, substitute $u = t^2$ and then consider solving the quadratic equation to get u.)

EXERCISES AND PROBLEMS KEYED TO SECTIONS *(duplicate handled)*

---

the river is 4.0 mi wide, how long will it take her to cross the river? (c) Suppose that instead of crossing the river she rows 2.0 mi *down* the river and then back to her starting point. How long will she take? (d) How long will she take to row 2.0 mi *up* the river and then back to her starting point? (e) In what direction should she head the boat if

92P. Galaxy Alpha is observed to be receding from us with a speed of 0.35c. Galaxy Beta, located in precisely the opposite direction, is also found to be receding from us at this same speed. What recessional speed would an observer on Galaxy Alpha find (a) for our galaxy and (b) for Galaxy Beta?

ADDITIONAL PROBLEMS

93. A baseball is hit at ground level. The ball reaches its maximum height above ground level 3.0 s after being hit. Then 2.5 s after reaching its maximum height, the ball barely clears a fence that is 97.5 m from where it was hit.

Assume the ground is level. (a) What maximum height above ground level is reached by the ball? (b) How high is the fence? (c) How far beyond the fence does the ball strike the ground?

ADDITIONAL PROBLEMS

These problems, not keyed to a particular section, challenge students to identify the concepts they involve.

Brief Contents

Contents

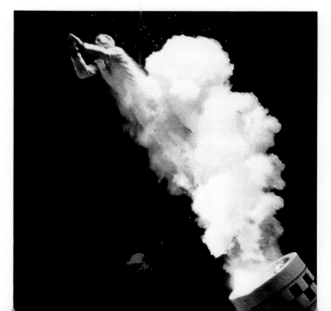

CHAPTER 5

FORCE AND MOTION—I *97*

*Can a man pull two railroad passenger cars
with his teeth?*

CHAPTER 6

FORCE AND MOTION—II *131*

*Why do cats survive long falls
better than shorter ones?*

CHAPTER 7

WORK AND KINETIC ENERGY *159*

*How much work is required
in lifting great weights?*

CHAPTER 8

CONSERVATION OF ENERGY *187*

How far down will a bungee-cord jumper fall?

CHAPTER 13

EQUILIBRIUM AND ELASTICITY *353*

Can you safely rest in a fissure during a chimney climb?

CHAPTER 14

OSCILLATIONS *381*

Why did only one section of the Nimitz Freeway collapse?

CHAPTER 15

GRAVITATION *411*

How can a black hole be detected?

CHAPTER 16

FLUIDS *443*

What factor occasionally kills novice scuba divers?

CHAPTER 17

WAVES—I *475*

How does a scorpion detect a beetle without using sight or sound?

CHAPTER 18

WAVES—II *503*

How does a bat detect a moth in total darkness?

CHAPTER 19

TEMPERATURE *533*

What feature of water allows aquatic life in cold climates?

APPENDICES *A1*

PROBLEM SOLVING TACTICS

FOURTH EDITION

FUNDAMENTALS OF
PHYSICS

VOLUME 1

MEASUREMENT 1

You can watch the sun set and disappear over a calm ocean once while you lie on a beach, and once again if you stand up. Surprisingly, by measuring the time between the two sunsets, you can approximate the radius of the Earth. How can such a simple observation be used to measure the Earth?

1-1 MEASURING THINGS

Physics is based on measurement. What is the time interval between two clicks of a counter? What is the temperature of the liquid helium in a vessel? What is the wavelength of the light from a certain laser? What is the electric current in a wire? The list goes on.

We start learning physics by learning how to measure the quantities that are involved in the laws of physics. Among these quantities are length, time, mass, temperature, pressure, and electrical resistance. We use many of these words in everyday speech. You might say, for example, "I will go to any *length* to help you as long as you don't *pressure* me." In physics, words like *length* and *pressure* have precise meanings, which we must not confuse with their (often vague) everyday meanings. In fact, the precise scientific meanings of length and pressure have nothing to do with their meanings in the quoted sentence. And that can be a problem: according to physicist Robert Oppenheimer, "Often the very fact that the words of science are the same as those of our common life and tongue can be more misleading than enlightening."

To describe a physical quantity we first define a **unit,** that is, a measure of the quantity that is defined to be exactly 1.0. Then we define a **standard,** that is, a reference to which all other examples of the quantity are compared. For example, the unit of length is the meter, and, as you will see, the standard for the meter is defined to be the length traveled by light in vacuum during a certain fraction of a second. We are free to define a unit and its standard in any way we care to. The important thing is to do so in such a way that scientists around the world will agree that our definitions are both sensible and practical.

Once we have set up a standard, for length, say, we must work out procedures by which any length whatever, be it the radius of a hydrogen atom, the wheelbase of a skateboard, or the distance to a star, can be expressed in terms of the standard. Many of our comparisons must be indirect. You cannot use a ruler, for example, to measure either the radius of an atom or the distance to a star.

There are so many physical quantities that it is a problem to organize them. Fortunately, they are not all independent. Speed, for example, is the ratio of a length to a time. So what we do is pick out—by international agreement—a small number of physical quantities, such as length and time, and assign standards to them alone. We then define all other physical quantities in terms of these *base quantities* and their standards. Speed, for example, is defined in terms of the base quantities length and time and the associated base standards.

The base standards must be both accessible and invariable. If we define the length standard as the distance between one's nose and the index finger on an outstretched arm, we certainly have an accessible standard—but it will, of course, vary from person to person. The demand for precision in science and engineering pushes us just the other way. We aim first for invariability, and we then exert great effort to make duplicates of the base standards accessible to those who need them.

1-2 THE INTERNATIONAL SYSTEM OF UNITS

In 1971, the 14th General Conference on Weights and Measures picked seven quantities as base quantities, thereby forming the basis of the International System of Units, abbreviated SI from its French name and popularly known as the **metric system.** Table 1-1 shows the units for the three base quantities—length, mass, and time—that we use in the early chapters of this book. The units for the quantities were chosen to be on a "human scale."

Many SI *derived units* are defined in terms of these base units. For example, the SI unit for power, called the **watt** (abbreviated as W), is defined in terms of the base units for mass, length, and time. Thus, as you will see in Chapter 7,

$$1 \text{ watt} = 1 \text{ W} = 1 \text{ kg} \cdot \text{m}^2/\text{s}^3. \quad (1\text{-}1)$$

To express the very large and the very small numbers that we often run into in physics, we use the so-called scientific notation, which employs powers of 10. In this notation,

$$3{,}560{,}000{,}000 \text{ m} = 3.56 \times 10^9 \text{ m} \quad (1\text{-}2)$$

and

$$0.000\ 000\ 492 \text{ s} = 4.92 \times 10^{-7} \text{ s}. \quad (1\text{-}3)$$

TABLE 1-1
SOME SI BASE UNITS

QUANTITY	UNIT NAME	UNIT SYMBOL
Length	meter	m
Time	second	s
Mass	kilogram	kg

TABLE 1-2
PREFIXES FOR SI UNITS[a]

FACTOR	PREFIX	SYMBOL	FACTOR	PREFIX	SYMBOL
10^{24}	yotta-	Y	10^{-24}	yocto-	y
10^{21}	zetta-	Z	10^{-21}	zepto-	z
10^{18}	exa-	E	10^{-18}	atto-	a
10^{15}	peta-	P	10^{-15}	femto-	f
10^{12}	tera-	T	**10^{-12}**	**pico-**	**p**
10^{9}	**giga-**	**G**	10^{-9}	nano-	n
10^{6}	**mega-**	**M**	10^{-6}	micro-	μ
10^{3}	**kilo-**	**k**	10^{-3}	milli-	m
10^{2}	hecto-	h	**10^{-2}**	**centi-**	**c**
10^{1}	deka-	da	10^{-1}	deci-	d

[a]The most commonly used prefixes are shown in bold type.

Since the advent of computers, scientific notation sometimes takes on an even briefer look, as in 3.56 E9 m and 4.92 E−7 s, where E stands for "exponent of ten." It is briefer still on some calculators, where E is replaced with an empty space.

As a further convenience when dealing with very large or very small measurements, we use the prefixes listed in Table 1-2. Attaching a prefix to a unit has the effect of multiplying by the associated factor. Thus we can express a particular electric power as

$$1.27 \times 10^9 \text{ watts} = 1.27 \text{ gigawatts} = 1.27 \text{ GW} \tag{1-4}$$

or a particular time interval as

$$2.35 \times 10^{-9} \text{ s} = 2.35 \text{ nanoseconds} = 2.35 \text{ ns.} \tag{1-5}$$

Some prefixes, as used in milliliter, centimeter, kilogram, and megabyte, are already familiar to you.

Appendix F and the inside back cover give conversion factors to systems of units other than SI. The United States is the only major country (in fact, almost the only country) that has not officially adopted the International System of Units.

1-3 CHANGING UNITS

We often need to change the units in which a physical quantity is expressed. We do so by a method called *chain-link conversion*. In this method, we multiply the original measurement by a **conversion factor** (a ratio of units that is equal to unity). For example, because 1 min and 60 s are identical time intervals,

we can write

$$\frac{1 \text{ min}}{60 \text{ s}} = 1 \quad \text{and} \quad \frac{60 \text{ s}}{1 \text{ min}} = 1.$$

This is *not* the same as writing $\frac{1}{60} = 1$ or $60 = 1$; the *number* and its *unit* must be treated together.

Because multiplying any quantity by unity leaves it unchanged, we can introduce such conversion factors wherever we find them useful. In chain-link conversion, we use the factors in such a way that the unwanted units cancel out. For example,

$$2 \text{ min} = (2 \text{ min})(1) = (2 \text{ min})\left(\frac{60 \text{ s}}{1 \text{ min}}\right)$$
$$= 120 \text{ s.} \tag{1-6}$$

If, by chance, you introduce the conversion factor in a way that the units do *not* cancel, simply invert the factor and try again. Note that units obey the same rules as do algebraic variables and numbers.

SAMPLE PROBLEM 1-1

The research submersible ALVIN is diving at a speed of 36.5 fathoms per minute.

a. Express this speed in meters per second. A *fathom* (fath) is precisely 6 ft.

SOLUTION To find the speed in meters per second, we write

$$36.5 \frac{\text{fath}}{\text{min}} = \left(36.5 \frac{\text{fath}}{\text{min}}\right)\left(\frac{1 \text{ min}}{60 \text{ s}}\right)\left(\frac{6 \text{ ft}}{1 \text{ fath}}\right)\left(\frac{1 \text{ m}}{3.28 \text{ ft}}\right)$$
$$= 1.11 \text{ m/s.} \tag{Answer}$$

b. What is this speed in miles per hour?

SOLUTION   As above, we have

$$36.5\ \frac{\text{fath}}{\text{min}} = \left(36.5\ \frac{\text{fath}}{\text{min}}\right)\left(\frac{60\ \text{min}}{1\ \text{h}}\right)\left(\frac{6\ \text{ft}}{1\ \text{fath}}\right)\left(\frac{1\ \text{mi}}{5280\ \text{ft}}\right)$$

$$= 2.49\ \text{mi/h.} \qquad \text{(Answer)}$$

c. What is this speed in light-years per year?

SOLUTION   A light-year (ly) is the distance that light travels in 1 year, 9.46×10^{12} km.

We start from the result of (a) above:

$$1.11\ \frac{\text{m}}{\text{s}} = \left(1.11\ \frac{\text{m}}{\text{s}}\right)\left(\frac{1\ \text{ly}}{9.46 \times 10^{12}\ \text{km}}\right)$$

$$\times \left(\frac{1\ \text{km}}{1000\ \text{m}}\right)\left(\frac{3.16 \times 10^7\ \text{s}}{1\ \text{y}}\right)$$

$$= 3.71 \times 10^{-9}\ \text{ly/y.} \qquad \text{(Answer)}$$

We can write this in the even more unlikely form of 3.71 nly/y, where "nly" is an abbreviation for nano-light-year.

   If you calculated the answer to part (a) above with your calculator "wide open," you might find 1.112804878 m/s appearing in the display; the precision implied by the nine decimal places in this result is totally meaningless. We have (properly) rounded this number to 1.11 m/s, which is the precision that is justified by the precision of the original data. The given speed of 36.5 fath/min consists of three digits, called **significant figures.** Any fourth digit that might appear to the right of the 5 is not known, and so the result of the conversion is not reliable beyond three digits, or three significant figures. Calculator results must always be modified to reflect such a limit of reliability.*

SAMPLE PROBLEM 1-2

How many square centimeters are in an area of 6.0 km²?

SOLUTION   Each kilometer in the original measure must be converted. The surest way is to separate them:

$$6.0\ \text{km}^2 = 6.0\ (\text{km})(\text{km}) = 6.0\ (\text{km})(\text{km})$$

$$\times \left(\frac{1000\ \text{m}}{1\ \text{km}}\right)\left(\frac{1000\ \text{m}}{1\ \text{km}}\right)\left(\frac{100\ \text{cm}}{1\ \text{m}}\right)$$

$$\times \left(\frac{100\ \text{cm}}{1\ \text{m}}\right)$$

$$= 6.0 \times 10^{10}\ \text{cm}^2. \qquad \text{(Answer)}$$

SAMPLE PROBLEM 1-3

Convert 60 mi/h to feet per second.

SOLUTION   To answer, you might convert miles to feet and hours to seconds. Or you might consult the inside back cover for a more direct conversion:

$$60\ \text{mi/h} = 60\ \text{mi/h}\left(\frac{3.28\ \text{ft/s}}{2.24\ \text{mi/h}}\right)$$

$$= 88\ \text{ft/s.} \qquad \text{(Answer)}$$

Note that, as in the previous examples, the conversion factor is equivalent to unity.

1-4 LENGTH

In 1792, the newborn Republic of France established a new system of weights and measures. As its cornerstone, the meter was defined to be one ten-millionth of the distance from the north pole to the equator. Eventually, for practical reasons, this Earth standard was abandoned and the meter came to be defined as the distance between two fine lines engraved near the ends of a platinum–iridium bar, the **standard meter bar,** which was kept at the International Bureau of Weights and Measures near Paris. Accurate copies of the bar were sent to standardizing laboratories throughout the world. These **secondary standards** were used to produce other, still more accessible standards so that ultimately every measuring device derived its authority from the standard meter bar through a complicated chain of comparisons.

   In 1959, the yard was legally defined to be

$$1\ \text{yard} = 0.9144\ \text{meter (exactly),} \qquad (1\text{-}7)$$

which is equivalent to

$$1\ \text{inch} = 2.54\ \text{centimeters (exactly).} \qquad (1\text{-}8)$$

---

*A fuller discussion of *significant figures* in problem solving appears in the Problem Solving Tactics of Chapter 4.

TABLE 1-3
SOME LENGTHS

LENGTH	METERS
Distance to the farthest observed quasar (1990)	2×10^{26}
Distance to the Andromeda galaxy	2×10^{22}
Distance to the nearest star (Proxima Centauri)	4×10^{16}
Distance to the farthest planet (Pluto)	6×10^{12}
Radius of the Earth	6×10^{6}
Height of Mt. Everest	9×10^{3}
Thickness of this page	1×10^{-4}
Wavelength of light	5×10^{-7}
Length of a typical virus	1×10^{-8}
Radius of the hydrogen atom	5×10^{-11}
Radius of a proton	$\sim 10^{-15}$

FIGURE 1-1 A "false-color" electron micrograph of influenza virus particles. The lipoproteins (colored yellow) taken from the host surround the core (colored green). The composite is less than 50 nm in diameter.

Table 1-3 lists some interesting lengths. One corresponds to a virus; an example is shown in Fig. 1-1.

Eventually, modern science and technology required a more precise standard than the distance between two fine scratches on a metal bar. In 1960, a new standard for the meter, based on the wavelength of light, was adopted. Specifically, the meter was redefined to be 1,650,763.73 wavelengths of a particular orange-red light emitted by atoms of krypton-86 in a gas discharge tube.* This awkward number of wavelengths was chosen so that the new standard would be as consistent as possible with the old meter-bar standard.

The krypton-86 atoms of the atomic length standard are available everywhere, are identical, and emit light of precisely the same wavelength. As Philip Morrison of MIT has pointed out, every atom is a storehouse of natural units, more secure than the International Bureau of Weights and Measures.

Figure 1-2 shows how the length of a master machinist's gauge block, used in industry as a precise secondary length standard, is compared with a reference standard at the National Institute of Standards and Technology (NIST). The dark *fringes* that cross the figure are formed by the cancellation of light waves with one another. If the fringe patterns on the two rectangular blocks match, the gauge blocks are of the same length. If the fringes fail to match by, say, one-tenth of a fringe, the blocks differ in length by one-twentieth of the wavelength of the light, or about 30 nm.

By 1983, the demand for higher precision had reached such a point that even the krypton-86 standard could not meet it, and in that year a bold step was taken. The meter was redefined as the distance traveled by a light wave in a specified time interval. In the words of the 17th General Conference on Weights and Measures:

FIGURE 1-2 A master machinist's gauge block (left) being compared with a reference standard (right) by means of light waves. If the bright and dark fringes match, the blocks are of equal length. These two blocks differ in length by about 25 nm, about the size of the virus in Fig. 1-1.

---

*The number 86 in the notation krypton-86 identifies a particular one of the five stable isotopes (or types) of this element. An equivalent notation is ^{86}Kr. The number is called the *mass number* of the isotope in question.

The meter is the length of the path traveled by light in vacuum during a time interval of 1/299,792,458 of a second.

This number was chosen so that the speed of light c could be exactly

$$c = 299,792,458 \text{ m/s.}$$

Measurements of the speed of light had become extremely precise, so it made sense to adopt the speed of light as a defined quantity and to use it to redefine the meter.

SAMPLE PROBLEM 1-4

Both 100 yd and 100 m are used as distances for dashes in track meets.

a. Which is longer?

SOLUTION From Eq. 1-7, 100 yd is equal to 91.44 m, so that 100 m is longer than 100 yd.

b. By how many meters is it longer?

SOLUTION We represent the difference by ΔL, where Δ is the capital Greek letter delta. Thus

$$\Delta L = 100 \text{ m} - 100 \text{ yd}$$

$$= 100 \text{ m} - 91.44 \text{ m} = 8.56 \text{ m.} \quad \text{(Answer)}$$

c. By how many feet is it longer?

SOLUTION We can express this difference by the method used in Sample Problem 1-1,

$$\Delta L = (8.56 \text{ m})\left(\frac{3.28 \text{ ft}}{1 \text{ m}}\right) = 28.1 \text{ ft.} \quad \text{(Answer)}$$

1-5 TIME

Time has two aspects. For civil and some scientific purposes, we want to know the time of day (see Fig. 1-3) so that we can order events in sequence. And in much scientific work, we want to know how long an event lasts. Thus any time standard must be able to answer two questions: "*When* did it happen?" and "What was its *duration*?" Table 1-4 shows some measured time intervals.

FIGURE 1-3 When the metric system was proposed in 1792, the hour was redefined to provide a 10-h day. The idea did not catch on. The maker of this 10-h watch wisely provided a small dial that kept conventional 12-h time. Do the two dials indicate the same time?

Any phenomenon that repeats itself is a possible time standard. The rotation of the Earth, which determines the length of the day, has been used in this way for centuries. A quartz clock, in which a quartz ring continuously vibrates, can be calibrated against the rotating Earth via astronomical observations and used to measure time intervals in the laboratory. However, the calibration cannot be carried out with

TABLE 1-4
SOME TIME INTERVALS

TIME INTERVAL	SECONDS
Lifetime of the proton (predicted)	$\sim 10^{39}$
Age of the universe	5×10^{17}
Age of the pyramid of Cheops	1×10^{11}
Human life expectancy (U.S.)	2×10^{9}
Length of a day	9×10^{4}
Time between human heartbeats	8×10^{-1}
Lifetime of the muon	2×10^{-6}
Shortest laboratory light pulse (1989)	6×10^{-15}
Lifetime of most unstable particle	$\sim 10^{-23}$
The Planck time[a]	$\sim 10^{-43}$

[a]This is the earliest time after the "Big Bang" at which the laws of physics as we know them can be applied.

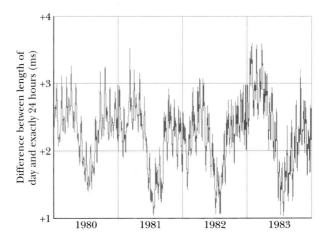FIGURE 1-4 The cesium atomic frequency standard at the National Institute of Standards and Technology in Boulder, Colorado. It is the primary standard for the unit of time in the United States. Dial (303) 499-7111 and set your watch by it. Dial (900) 410-8463 for Naval Observatory time signals.

the accuracy called for by modern scientific and engineering technology.

To meet the need for a better time standard, atomic clocks have been developed in several countries. Figure 1-4 shows such a clock, based on a characteristic frequency of the isotope cesium-133, at NIST. It forms the basis in this country for Coordinated Universal Time (UTC), for which time signals are available by shortwave radio (stations WWV and WWVH) and by telephone. (To set a clock accurately at your particular location, you would have to account for the travel time that is required for these signals to reach you.)

Figure 1-5 shows variations in the rate of rotation of the Earth over a 4-year period, as determined by comparison with a cesium clock.* Because of the seasonal variation displayed by Fig. 1-5, we suspect the rotating Earth when there is a difference between Earth and atom as timekeepers. The variation is probably due to tidal effects caused by the moon and to large-scale atmospheric winds.

The 13th General Conference on Weights and Measures in 1967 adopted a standard second based on the cesium clock:

---

*See ''The Earth's Inconstant Rotation,'' by John Wahr, *Sky and Telescope*, June 1986. See also ''Studying the Earth by Very-Long-Baseline Interferometry,'' by William E. Carter and Douglas S. Robertson, *Scientific American*, November 1986.

One second is the time taken by 9,192,631,770 vibrations of the light (of a specified wavelength) emitted by a cesium-133 atom.

In principle, two cesium clocks would have to run for 6000 years before their readings would differ by more than 1 s. Even such accuracy pales in comparison to that of clocks currently being developed; their precision may be as fine as 1 part in 10^{18}, that is, 1 s in 1×10^{18} s (about 3×10^{10} y).

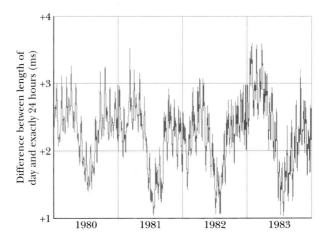

FIGURE 1-5 Variations in the length of the day over a 4-year period. Note that the entire vertical scale amounts to only 3 ms (= 0.003 s).

SAMPLE PROBLEM 1-5

Isaac Asimov proposed a unit of time based on the highest known speed and the smallest measurable distance. It is the *light-fermi,* the time taken by light to travel a distance of 1 fermi ($=1$ femtometer $= 1$ fm $= 10^{-15}$ m). How many seconds are there in a light-fermi?

SOLUTION   We find this time by dividing the distance by c, the speed of light ($\approx 3.00 \times 10^8$ m/s). Thus

$$1 \text{ light-fermi} = \frac{1 \text{ fm}}{\text{speed of light}} = \frac{10^{-15} \text{ m}}{3.00 \times 10^8 \text{ m/s}}$$

$$= 3.33 \times 10^{-24} \text{ s}. \qquad \text{(Answer)}$$

Table 1-4 tells us that the most unstable elementary particle known to date has a lifetime (on the average) of about 10^{-23} s before it disintegrates. We could say that its lifetime is about 3 light-fermis.

SAMPLE PROBLEM 1-6*

Suppose that you watch the sun set over a calm ocean while lying on the beach, starting a stopwatch just as the top of the sun disappears. You then stand, elevating your eyes by a height $h = 1.70$ m, and stop the watch when the top of the sun again disappears. If the elapsed time on the watch is $t = 11.1$ s, what is the radius r of the Earth?

SOLUTION   As shown in Fig. 1-6, your line of sight to the top of the sun, as the top first disappears, is tangent to the Earth's surface at your location, at point A. The figure also shows that your line of sight to the top of the sun as the top again disappears is tangent to the Earth's surface at point B. Let d represent the distance between point B and the location of your eyes when you are standing, and draw radii r as shown in Fig. 1-6. From the Pythagorean theorem, we then have

$$d^2 + r^2 = (r + h)^2 = r^2 + 2rh + h^2,$$

or

$$d^2 = 2rh + h^2. \qquad (1\text{-}9)$$

*Adapted from "Doubling Your Sunsets, or How Anyone Can Measure the Earth's Size with a Wristwatch and Meter Stick," by Dennis Rawlins, *American Journal of Physics,* Feb. 1979, Vol. 47, pp. 126–128. This works best at the equator.

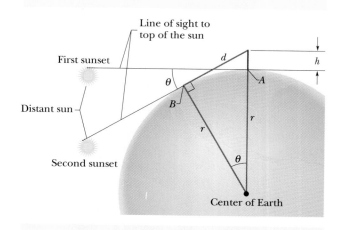

FIGURE 1-6   Sample Problem 1-6. Your line of sight to the top of the setting sun rotates through the angle θ when you move from a prone position at point A, elevating your eyes by a distance h. (Angle θ and distance h are exaggerated here for clarity.)

Because the height h is so much smaller than the Earth's radius r, the term h^2 is negligible compared to the term $2rh$, and we can rewrite Eq. 1-9 as

$$d^2 = 2rh. \qquad (1\text{-}10)$$

In Fig. 1-6, the angle between the two tangent points A and B is θ, which is also the angle through which the sun moves about the Earth during the measured time $t = 11.1$ s. During a full day, which is approximately 24 h, the sun moves through an angle of $360°$ about the Earth. This allows us to write

$$\frac{\theta}{360°} = \frac{t}{24 \text{ h}},$$

which, with $t = 11.1$ s, gives us

$$\theta = \frac{(360°)(11.1 \text{ s})}{(24 \text{ h})(60 \text{ min/h})(60 \text{ s/min})} = 0.04625°.$$

From Fig. 1-6 we see that $d = r \tan \theta$. Substituting this for d in Eq. 1-10 gives us

$$r^2 \tan^2\theta = 2rh$$

or

$$r = \frac{2h}{\tan^2\theta}.$$

Substituting $\theta = 0.04625°$ and $h = 1.70$ m, we find

$$r = \frac{(2)(1.70 \text{ m})}{\tan^2 0.04625°} = 5.22 \times 10^6 \text{ m}, \qquad \text{(Answer)}$$

which is within 20% of the accepted value (6.37×10^6 m) for the (mean) radius of the Earth.

1-6 MASS

The Standard Kilogram

The SI standard of mass is a platinum–iridium cylinder (Fig. 1-7) kept at the International Bureau of Weights and Measures near Paris and assigned, by international agreement, a mass of 1 kilogram. Accurate copies have been sent to standardizing laboratories in other countries, and the masses of other bodies can be determined by balancing them against a copy. Table 1-5 shows some measured masses expressed in kilograms.

The U.S. copy of the standard kilogram is housed in a vault at NIST. It is removed, no more than once a year, for the purpose of checking duplicate copies that are used elsewhere. Since 1889, it has been taken to France twice for recomparison with the primary standard. The procedure will

TABLE 1-5
SOME MASSES

OBJECT	KILOGRAMS
Known universe	$\sim 10^{53}$
Our galaxy	2×10^{41}
Sun	2×10^{30}
Moon	7×10^{22}
Asteroid Eros	5×10^{15}
Small mountain	1×10^{12}
Ocean liner	7×10^{7}
Elephant	5×10^{3}
Grape	3×10^{-3}
Speck of dust	7×10^{-10}
Penicillin molecule	5×10^{-17}
Uranium atom	4×10^{-25}
Proton	2×10^{-27}
Electron	9×10^{-31}

FIGURE 1-7 The international 1-kg standard of mass.

someday be replaced with a more reliable and accessible procedure involving a base unit of mass that is the mass of an atom.

A Second Mass Standard

The masses of atoms can be compared with each other more precisely than they can be compared with the standard kilogram. For this reason, we have a second mass standard. It is the carbon-12 atom, which, by international agreement, has been assigned a mass of 12 **atomic mass units** (u). The relation between the two standards is

$$1 \text{ u} = 1.6605402 \times 10^{-27} \text{ kg}, \qquad (1\text{-}11)$$

with an uncertainty of ± 10 in the last two decimal places. With a mass spectrometer, scientists can, with reasonable precision, determine the masses of other atoms relative to the mass of carbon-12. What we presently lack is a reliable means of extending that precision to more common units of mass, such as a kilogram.

REVIEW & SUMMARY

Measurement in Physics

Physics is based on measurement of physical quantities and the changes in physical quantities that take place in our universe. Certain physical quantities have been chosen as **base quantities** (such as length, time, and mass) and have been defined in terms of a **standard** and given a **unit** measure (such as meter, second, and kilogram). Other

physical quantities (such as speed) are defined in terms of the base quantities and their standards.

SI Units

The unit system emphasized in this book is the International System of Units (SI). The three physical quantities displayed in Table 1-1 serve as a base in the early chapters of this book. Standards, which must be both accessible and invariable, define units for these base quantities and are established by international agreement. These standards underlie all physical measurement, for both the base quantities and quantities derived from them. The prefixes of Table 1-2 simplify notation in many cases.

Unit Conversions

Conversion of units from one system to another (for example, from miles per hour to kilometers per second) may be performed by using **chain-link conversions** in which the units are treated as algebraic quantities and the original data are multiplied successively, by conversion factors written as unity, until the desired units are obtained. See Sample Problems 1-1 to 1-3.

The Meter

The unit of length—the meter—is defined as the distance traveled by light during a precisely specified time interval. The yard, together with its multiples and submultiples, is legally defined in this country in terms of the meter.

The Second

The unit of time—the second—was formerly defined in terms of the rotation of the Earth. It is now defined in terms of the vibrations of the light emitted by an atomic (cesium-133) source. Accurate time signals are disseminated worldwide by radio signals keyed to atomic clocks in standardizing laboratories.

The Kilogram

The unit of mass—the kilogram—is defined in terms of a particular platinum–iridium prototype kept near Paris, France. For measurements on atomic scale, the atomic mass unit, defined in terms of the atom carbon-12, is usually used.

QUESTIONS

1. How would you criticize this statement: "Once you have picked a standard, by the very meaning of 'standard' it is invariable"?

2. List characteristics other than accessibility and invariability that you would consider desirable for a physical standard.

3. Can you devise a system of base units (Table 1-1) in which time is not included? Explain.

4. Of the three base units listed in Table 1-1, only one—the kilogram—has a prefix (see Table 1-2). Would it be wise to redefine the mass of the standard platinum–iridium cylinder at the International Bureau of Weights and Measures as 1 g rather than 1 kg?

5. Why are there no SI base units for area and volume?

6. The meter was originally defined to be one ten-millionth of a meridian line that extends from the north pole to the equator and that passes through Paris. This definition disagrees with the standard meter bar by 0.023%. Does this mean that the standard meter bar is inaccurate to this extent? Explain.

7. In defining the meter bar as the standard of length, the temperature of the bar is specified. Can length be called a base quantity if another physical quantity, such as temperature, must be specified in choosing a standard?

8. In redefining the meter in terms of the speed of light, why did not the delegates to the 1983 General Conference on Weights and Measures simplify matters by defining the speed of light to be 3×10^8 m/s exactly? For that matter, why did they not define it to be 1 m/s exactly? Were both of these possibilities open to them? If so, why did they reject them?

9. What does the prefix "micro-" signify in the words "microwave oven"? It has been proposed that food that has been irradiated by gamma rays to lengthen its shelf life be marked "picowaved." What do you suppose that means?

10. Suggest a way to measure (a) the radius of the Earth, (b) the distance between the sun and the Earth, and (c) the radius of the sun.

11. Suggest a way to measure (a) the thickness of a sheet of paper, (b) the thickness of a soap bubble film, and (c) the diameter of an atom.

12. Name several repetitive phenomena occurring in nature which could serve as reasonable time standards.

13. You could define "1 s" to be one pulse beat of the current president of the American Physical Society. Galileo used his pulse as a timing device in some of his work. Why is a definition based on the atomic clock better?

14. What criteria should be satisfied by a good clock?

15. Cite the drawbacks to using the period of a pendulum, such as for a grandfather clock, as a time standard.

16. On June 30, 1981, the "minute" extending from 10:59 to 11:00 a.m. was arbitrarily lengthened to contain 61 s. The extra second—the *leap second*—was introduced to compensate for the fact that, as measured by our atomic time standard, the Earth's rotation rate is slowly decreasing. Why is it desirable to readjust our clocks in this way?

17. Why do we find it useful to have two standards of mass, the kilogram and the carbon-12 atom?

18. Is the current standard kilogram of mass accessible and invariable? Does it have simplicity for comparison purposes? Would an atomic standard be better in any respect?

19. Suggest objects whose masses would fall in the wide range in Table 1-5 between those of an ocean liner and a small mountain, and estimate their masses.

20. Opponents of a switch to the metric system often cloud the issue by saying things such as "Instead of buying one pound of butter you will have to ask for 0.454 kg of butter." The implication is that life would be more complicated. How might you refute this?

EXERCISES & PROBLEMS

SECTION 1-2 THE INTERNATIONAL SYSTEM OF UNITS

1E. Use the prefixes in Table 1-2 to express (a) 10^6 phones; (b) 10^{-6} phones; (c) 10^1 cards; (d) 10^9 lows; (e) 10^{12} bulls; (f) 10^{-1} mates; (g) 10^{-2} pedes; (h) 10^{-9} Nannettes; (i) 10^{-12} boos; (j) 10^{-18} boys; (k) 2×10^2 withits; (l) 2×10^3 mockingbirds. Now that you have the idea, invent a few similar expressions. (See, in this connection, p.

61 of *A Random Walk in Science,* compiled by R. L. Weber; Crane, Russak & Co., New York, 1974.)

2E. Some of the prefixes of the SI units have crept into everyday language. (a) What is the weekly equivalent of an annual salary of \$36 K ($= 36$ kilobucks)? (b) A lottery awards exactly 10 megabucks as the top prize, payable over 20 years. How much is in each monthly check? (c) The hard disk of a computer has a capacity of 30 MB ($= 30$ megabytes). At 8 bytes per word, how many words can it store? [In computerese, *kilo* means 1024 (= 2^{10}), not 1000.]

SECTION 1-4 LENGTH

3E. A space shuttle orbits the Earth at an altitude of 300 km. What is this distance in (a) miles and (b) millimeters?

4E. What is your height in meters?

5E. The micrometer (10^{-6} m = 1 μm) is often called the *micron*. (a) How many microns make up 1.0 km? (b) What fraction of a centimeter equals 1.0 μm? (c) How many microns are in 1.0 yd?

6E. The Earth is approximately a sphere of radius 6.37 $\times$ 10^6 m. (a) What is its circumference in kilometers? (b) What is its surface area in square kilometers? (c) What is its volume in cubic kilometers?

7E. Calculate the number of kilometers in 20.0 mi using only the following conversion factors: 1 mi = 5280 ft, 1 ft = 12 in., 1 in. = 2.54 cm, 1 m = 100 cm, and 1 km = 1000 m.

8E. Give the relation between (a) a square yard and a square foot; (b) a square inch and a square centimeter; (c) a square mile and a square kilometer; and (d) a cubic meter and a cubic centimeter.

9P. A unit of area, often used in measuring land areas, is the *hectare,* defined as 10^4 m^2. An open-pit coal mine consumes 75 hectares of land, down to a depth of 26 m, each year. What volume of earth, in cubic kilometers, is removed in this time?

10P. The *cord* is a volume of cut wood equal to a stack 8 ft long, 4 ft wide, and 4 ft high. How many cords of wood are in 1.0 m^3?

11P. A room is 20 ft, 2 in. long and 12 ft, 5 in. wide. What is the floor area in (a) square feet and (b) square meters? If the ceiling is 12 ft, $2\frac{1}{2}$ in. above the floor, what is the volume of the room in (c) cubic feet and (d) cubic meters?

12P. Antarctica is roughly semicircular in shape, with a radius of 2000 km. The average thickness of the ice cover is 3000 m. How many cubic centimeters of ice does Antarctica contain? (Ignore the curvature of the Earth.)

13P. A typical sugar cube has an edge length of 1 cm. If you had a cubical box that contained a mole of sugar cubes, what would its edge length be? (One mole = 6.02 $\times$ 10^{23} units.)

14P. Hydraulic engineers often use, as a unit of volume of water, the *acre-foot,* defined as the volume of water that will cover 1 acre of land to a depth of 1 ft. A severe thunderstorm dumps 2.0 in. of rain in 30 min on a town of area 26 km^2. What volume of water, in acre-feet, fell on the town?

15P. A certain brand of house paint claims a coverage of 460 ft^2/gal. (a) Express this quantity in square meters per liter. (b) Express this quantity in SI base units (see Appendixes A and F). (c) What is the inverse of the original quantity, and what is its physical significance?

16P. Astronomical distances are so large compared to terrestrial ones that much larger units of length are used for easy comprehension of the relative distances of astronomical objects. An *astronomical unit* (AU) is equal to the average distance from the Earth to the sun, about 92.9 $\times$ 10^6 mi. A *parsec* (pc) is the distance at which 1 AU would subtend an angle of exactly 1 second of arc (Fig. 1-8). A *light-year* (ly) is the distance that light, traveling through a vacuum with a speed of 186,000 mi/s, would cover in 1.0 year. (a) Express the distance from the Earth to the sun in parsecs and in light-years. (b) Express 1 ly and 1 pc in miles. Although "light-year" appears frequently in popular writing, the parsec is preferred by astronomers.

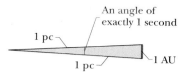

FIGURE 1-8 Problem 16.

17P. During a total eclipse, your view of the sun is almost exactly replaced by your view of the moon. Assuming that the distance from you to the sun is about 400 times the distance from you to the moon, (a) find the ratio of the sun's diameter to the moon's diameter. (b) What is the ratio of their volumes? (c) Hold up a dime (or another small coin) so that it would just eclipse the full moon, and measure the angle it subtends at the eye. From this experimental result and the given distance between the moon and the Earth (= 3.8 $\times$ 10^5 km), estimate the diameter of the moon.

18P*. The standard kilogram (see Fig. 1-7) is in the shape of a circular cylinder with its height equal to its diameter. Show that, for a circular cylinder of fixed volume, this equality gives the smallest surface area, thus minimizing the effects of surface contamination and wear.

19P*. The navigator of the oil tanker *Gulf Supernox* uses the satellites of what is called the Global Positioning System (GPS/NAVSTAR) to find the ship's latitude and longitude; see Fig. 1-9. If the values are 43°36′25.3″ N and

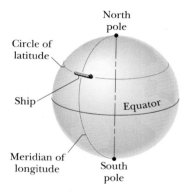

FIGURE 1-9 Problem 19.

77°31′48.2″ W with an accuracy of $\pm 0.5''$, what is the uncertainty in the tanker's position measured along (a) a north–south line and (b) an east–west line? (c) Where is the tanker?

SECTION 1-5 TIME

20E. Express the speed of light, 3.0×10^8 m/s, in (a) feet per nanosecond and (b) millimeters per picosecond.

21E. Enrico Fermi once pointed out that a standard lecture period (50 min) is close to 1 microcentury. How long is a microcentury in minutes, and what is the percent difference from Fermi's approximation?

22E. There are 365.25 days in 1 year. How many seconds is this?

23E. A certain pendulum clock (with a 12-h dial) happens to gain 1.0 min/day. After setting the clock to the correct time, how long must one wait until it again indicates the correct time?

24E. What is the age of the universe (see Table 1-4) in days?

25E. (a) A unit of time sometimes used in microscopic physics is the *shake*. One shake equals 10^{-8} s. Are there more shakes in a second than there are seconds in a year? (b) Humans have existed for about 10^6 years, whereas the universe is about 10^{10} years old. If the age of the universe is taken to be 1 "day," for how many "seconds" have humans existed?

26E. The maximum speeds of various animals are roughly as follows, in miles per hour: (a) snail, 3.0×10^{-2}; (b) spider, 1.2; (c) human, 23; and (d) cheetah, 70. Convert these data to meters per second. (All four calculations involve the identical conversion factor. You might want to determine that factor first and store it in the memory of your calculator, to use when needed.)

27P. An astronomical unit (AU) is the average distance of the Earth from the sun, approximately 1.50×10^8 km. The speed of light is about 3.0×10^8 m/s. Express the speed of light in terms of astronomical units per minute.

28P. Until 1883, every city and town in the United States kept its own local time. Today, travelers reset their watches only when the time change equals 1 h. How far, on the average, must you travel in degrees of longitude until your watch must be reset by 1.0 h? *Hint:* One rotation of the Earth requires 360° and about 24 h.

29P. On two *different* tracks, the winners of the mile race ran their races in 3 min, 58.05 s and 3 min, 58.20 s. In order to conclude that the runner with the shorter time was indeed faster, what is the maximum error, in feet, that can be permitted in laying out the mile distances?

30P. Five clocks are being tested in a laboratory. Exactly at noon, as determined by the WWV time signal, on successive days of a week the clocks read as in the following table. How would you arrange these five clocks in the order of their relative value as good timekeepers? Justify your choice.

CLOCK	SUN.	MON.	TUES.	WED.	THURS.	FRI.	SAT.
A	12:36:40	12:36:56	12:37:12	12:37:27	12:37:44	12:37:59	12:38:14
B	11:59:59	12:00:02	11:59:57	12:00:07	12:00:02	11:59:56	12:00:03
C	15:50:45	15:51:43	15:52:41	15:53:39	15:54:37	15:55:35	15:56:33
D	12:03:59	12:02:52	12:01:45	12:00:38	11:59:31	11:58:24	11:57:17
E	12:03:59	12:02:49	12:01:54	12:01:52	12:01:32	12:01:22	12:01:12

31P. Assuming the length of the day uniformly increases by 0.001 s per century, calculate the cumulative effect on the measure of time over 20 centuries. Such slowing of the Earth's rotation is indicated by observations of the occurrences of solar eclipses during this period.

32P*. The time it takes the moon to return to a given position as seen against the background of the fixed stars is called a *sidereal* month. The time interval between identical phases of the moon is called a *lunar* month. The lunar month is longer than the sidereal month. Why, and by how much?

SECTION 1-6 MASS

33E. Using conversions and data in the chapter, determine the number of hydrogen atoms required to obtain 1.0 kg of hydrogen. A hydrogen atom has a mass of 1.0 u.

34E. One molecule of water (H_2O) contains two atoms of hydrogen and one atom of oxygen. A hydrogen atom has a mass of 1.0 u and an atom of oxygen has a mass of 16 u, approximately. (a) What is the mass in kilograms of one molecule of water? (b) How many molecules of water are in the world's oceans, which have an estimated total mass of 1.4×10^{21} kg?

35E. The Earth has a mass of 5.98×10^{24} kg. The average mass of the atoms that make up the Earth is 40 u. How many atoms are in the Earth?

36P. What mass of water fell on the town in Problem 14 during the thunderstorm? One cubic meter of water has a mass of 10^3 kg.

37P. (a) Assuming that the density (mass/volume) of water is exactly 1 g/cm³, express the density of water in kilograms per cubic meter (kg/m³). (b) Suppose that it takes 10 h to drain a container of 5700 m³ of water. What is the "mass flow rate," in kilograms per second, of water from the container?

38P. A person on a diet might lose 2.3 kg (about 5 lb) per week. Express the mass loss rate in milligrams per second.

39P. The grains of fine California beach sand have an average radius of 50 μm and are made of silicon dioxide, 1 m³ of which has a mass of 2600 kg. What mass of sand grains would have a total surface area equal to the surface area of a cube 1 m on an edge?

40P. The density of iron is 7.87 g/cm³, and the mass of an iron atom is 9.27×10^{-26} kg. If the atoms are spherical and tightly packed, (a) what is the volume of an iron atom and (b) what is the distance between the centers of adjacent atoms?

MOTION ALONG A STRAIGHT LINE

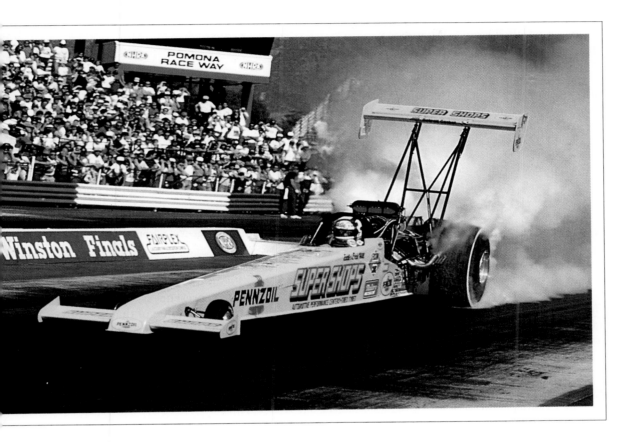

Blasting down a track in a dragster is an exhilarating example of straight-line motion, but what exactly, besides the noise, thrills the driver?

2-1 MOTION

The world, and everything in it, moves. Even seemingly stationary things, such as a roadway, move with the Earth's rotation, the Earth's orbit around the sun, the sun's orbit around the center of the Milky Way, and the galaxy's migration relative to other galaxies. The classification and comparison of motions (called **kinematics**) are often challenging. What exactly do you measure, and how do you compare?

Here are two examples of motion. In 1977, Kitty O'Neil set records for "terminal speed" and "elapsed time" for a dragster on a 440-yd run. From a standstill, she reached 392.54 mi/h (about 632.1 km/h) in a sizzling time of 3.72 s. In 1958, Eli Beeding, Jr. rode a rocket sled when it was shot along a track from a standstill to a speed of 72.5 mi/h (= 117 km/h) in the fantastic time of 0.04 s (less than the blink of an eye). How can we compare the two motions to see which must have been the more thrilling (or more frightening) — by final speeds, by elapsed time, or by some other quantity?

Before we attempt an answer, we first examine some general properties of motion that is restricted in three ways.

1. The motion is along a straight line only. The line may be vertical (a falling stone), horizontal (a car on a level highway), or slanted, but it must be straight.

2. The cause of the motion will not be specified until Chapter 5. In this chapter you study only the motion itself. Does the object speed up, slow down, stop, or reverse direction; and, if the motion does change, how is time involved in the change?

3. The moving object is either a **particle** (a pointlike object such as an electron) or an object that moves like a particle (every portion moves in the same direction and at the same rate). A pig slipping down a straight playground slide might be considered to be moving like a particle; however, a rotating playground merry-go-round would not because different points around its rim move in different directions.

2-2 POSITION AND DISPLACEMENT

To locate an object means to find its position relative to some reference point, often the **origin** (or zero point) of an axis such as the x axis in Fig. 2-1. The

FIGURE 2-1   Position is determined on an axis that is marked in units of length and that extends indefinitely in opposite directions.

positive direction of the axis is in the direction of increasing numbers, which is toward the right as the figure is drawn. The opposite direction is the **negative direction.**

For example, a particle might be located at $x = 5$ m, which means that it is 5 m in the positive direction from the origin. Were it at $x = -5$ m, it would be just as far from the origin but in the opposite direction.

A change from one position x_1 to another position x_2 is called **displacement** Δx, where

$$\Delta x = x_2 - x_1. \qquad (2\text{-}1)$$

(The symbol Δ, which represents a change in a quantity, means that the initial value of that quantity is to be subtracted from the final value.) When numbers are inserted for the position values, a displacement in the positive direction (toward the right in Fig. 2-1) always comes out positive, and one in the opposite direction (left in the figure) negative. For example, if the particle moves from $x_1 = 5$ m to $x_2 = 12$ m, then $\Delta x = (12$ m$) - (5$ m$) = +7$ m. The plus sign indicates that the motion is in the positive direction. If we ignore the sign (and thus the direction), we have the **magnitude** of Δx, which is 7 m. If the particle then returns to $x = 5$ m, the displacement for the full trip is zero. The actual number of meters covered is immaterial; displacement involves only the original and final positions.

Displacement is an example of a **vector quantity,** which is a quantity that has both a direction and a magnitude. We explore vectors more fully in Chapter 3 (in fact, some of you may have already read that chapter), but here all we need is the idea that displacement has two features: (1) its magnitude is the distance (such as the number of meters) between the original and final positions, and (2) its direction on an axis, from an original position to a final position, is represented by a plus or minus sign.

2-3 AVERAGE VELOCITY AND AVERAGE SPEED

A compact way to describe position is with a graph of position x plotted as a function of time t—a graph of $x(t)$. As a simple example, Fig. 2-2 shows $x(t)$ for a jack rabbit (which we treat as a particle) that is stationary at $x = -2$ m.

Figure 2-3a, also for a rabbit, is more interesting, because it involves motion. The rabbit is apparently first noticed at $t = 0$ when it is at the position $x = -5$ m. It moves toward $x = 0$, passes through

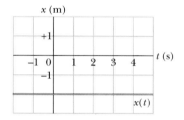

FIGURE 2-2 The graph of $x(t)$ for a jack rabbit that is stationary at $x = -2$ m. The value of x is -2 m for all times t.

(a)

(b)

FIGURE 2-3 (a) The graph of $x(t)$ for a moving jack rabbit. (b) The path associated with the graph. The scale below the x axis shows the times at which the rabbit reaches various x values.

If a member of this troupe were now to march back to where she started, her average velocity would be zero, because her net displacement would then be zero.

that point at $t = 3$ s, and then moves on to increasingly larger positive values of x.

Figure 2-3b depicts the actual motion of the rabbit and is something like what you would see. The graph is more abstract and quite unlike what you see, but it is richer in information. It also reveals how fast the rabbit moves. Several quantities are associated with the phrase "how fast." One of them is the **average velocity** $\bar{v}$, which is the ratio of the displacement Δx that occurs during a particular time interval Δt to that interval:*

$$\bar{v} = \frac{\Delta x}{\Delta t} = \frac{x_2 - x_1}{t_2 - t_1}. \qquad (2\text{-}2)$$

On a graph of x versus t, $\bar{v}$ is the **slope** of the straight line that connects two points on the $x(t)$ curve: one point corresponds to x_2 and t_2, and the other point corresponds to x_1 and t_1. Like displacement, $\bar{v}$ has both magnitude and direction. (Average velocity is another example of a vector quantity.) Its magnitude is the magnitude of the line's slope. A positive $\bar{v}$ (and slope) tells us that the line slants upward toward the right; a negative $\bar{v}$ (and slope), that the line slants upward to the left. The average velocity always has the same sign as the displacement because Δt is a positive number.

---

*In this book, a bar over a symbol usually means an average value of the quantity that the symbol represents.

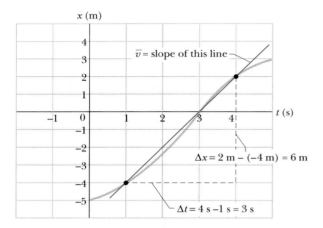

FIGURE 2-4   Calculations of average velocity between $t = 1$ s and $t = 4$ s.

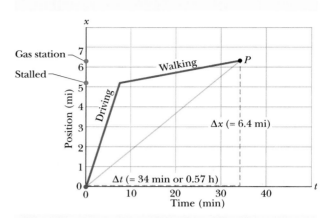

FIGURE 2-5   Sample Problem 2-1. The lines marked "Driving" and "Walking" are the position–time plots for the driver–walker in Sample Problem 2-1. The slope of the straight line joining the origin and point P is the average velocity for the trip.

Figure 2-4 shows the calculation of $\bar{v}$ for the rabbit of Fig. 2-3 for the time interval $t = 1$ s to $t = 4$ s. The average velocity during that time interval is $\bar{v} = +6$ m/3 s $= +2$ m/s, which is the slope of the straight line that connects the point on the curve at the beginning of the interval with the point on the curve at the end of the interval.

SAMPLE PROBLEM 2-1

You drive a beat-up pickup truck down a straight road for 5.2 mi at 43 mi/h, at which point you run out of fuel. You walk 1.2 mi farther, to the nearest gas station, in 27 min ($= 0.450$ h). What is your average velocity from the time that you started your truck to the time that you arrived at the station? Find the answer both numerically and graphically.

SOLUTION   To calculate $\bar{v}$ we need your displacement Δx, from start to finish, and the elapsed time Δt. Assume, for convenience, that your starting point is at the origin of an x axis (so $x_1 = 0$) and that you move in the positive direction. You end up at $x_2 = 5.2$ mi $+$ 1.2 mi $= +6.4$ mi, and so, $\Delta x = x_2 - x_1 = +6.4$ mi. To get the driving time, we rearrange Eq. 2-2 and insert the data about the driving:

$$\Delta t = \frac{\Delta x}{\bar{v}} = \frac{5.2 \text{ mi}}{43 \text{ mi/h}} = 0.121 \text{ h}.$$

So the total time, start to finish, is

$$\Delta t = 0.121 \text{ h} + 0.450 \text{ h} = 0.571 \text{ h}.$$

Finally, we insert Δx and Δt into Eq. 2-2:

$$\bar{v} = \frac{\Delta x}{\Delta t} = \frac{6.4 \text{ mi}}{0.571 \text{ h}} \approx +11 \text{ mi/h}. \quad \text{(Answer)}$$

To find $\bar{v}$ graphically, we must first plot $x(t)$, as in Fig. 2-5 where the start and finish points on the curve are the origin and P, respectively. Your average velocity is the slope of the straight line connecting those points. The dashed lines show that the slope also gives $\bar{v} = 6.4$ mi/0.57 h $= +11$ mi/h.

SAMPLE PROBLEM 2-2

Suppose that you next carry the fuel back to the truck, making the return trip in 35 min. What is your average velocity for the full journey, from the start of your driving to your arrival back at the truck with the fuel?

SOLUTION   As previously, we must find your displacement Δx from start to finish and then divide it by the time interval Δt between start and finish. In this problem, however, the finish is back at the truck. You started at $x_1 = 0$. Back at the truck you are at position $x_2 = 5.2$ mi. And so Δx is $5.2 - 0 = 5.2$ mi. The total time Δt you take in going from start to finish is

$$\Delta t = \frac{5.2 \text{ mi}}{43 \text{ mi/h}} + 27 \text{ min} + 35 \text{ min}$$

$$= 0.121 \text{ h} + 0.450 \text{ h} + 0.583 \text{ h} = 1.15 \text{ h}.$$

So

$$\overline{v} = \frac{\Delta x}{\Delta t} = \frac{5.2 \text{ mi}}{1.15 \text{ h}} \approx +4.5 \text{ mi/h}. \quad \text{(Answer)}$$

This is slower than the average velocity computed in Sample Problem 2-1 because here the displacement is smaller and the time interval longer.

Average speed $\overline{s}$ is a different way of describing "how fast" a particle moves. Whereas average velocity involves the particle's displacement Δx, the average speed involves the total distance covered (for example, the number of meters run), independent of direction. That is,

$$\overline{s} = \frac{\text{total distance}}{\Delta t}. \quad (2\text{-}3)$$

Average speed also differs from average velocity in that it does not include direction and thus lacks any algebraic sign. Sometimes $\overline{s}$ is the same (except for the absence of a sign) as $\overline{v}$. But, as demonstrated in Sample Problem 2-3, when an object doubles back on its path, the results can be quite different.

SAMPLE PROBLEM 2-3

In Sample Problem 2-2, what is your average speed?

SOLUTION From the beginning of your drive to your return to the truck with the fuel, you covered a total of 5.2 mi + 1.2 mi + 1.2 mi = 7.6 mi, taking 1.15 h, and so

$$\overline{s} = \frac{7.6 \text{ mi}}{1.15 \text{ h}} \approx 6.6 \text{ mi/h}. \quad \text{(Answer)}$$

PROBLEM SOLVING

TACTIC 1: READ THE PROBLEM CAREFULLY
For beginning problem solvers, no difficulty is more common than simply not understanding the problem. The best test of understanding is this: Can you explain the problem, in your own words, to a friend? Give it a try.

TACTIC 2: UNDERSTAND WHAT IS GIVEN
AND WHAT IS REQUESTED
Write down the given data, with units, using the symbols of the chapter. (In Sample Problems 2-1 and 2-2, the given data allow you to find your net displacement Δx and the corresponding time interval Δt.) Identify the unknown and its symbol. (In these sample problems, the unknown is your average velocity, symbol $\overline{v}$.) Then find the connection between the unknown and the data. (The connection is Eq. 2-2, the definition of average velocity.)

TACTIC 3: WATCH THE UNITS
Be sure to use a consistent set of units when putting numbers into the equations. In Sample Problems 2-1 and 2-2, which involve a truck, the logical units in terms of the given data are miles for distances, hours for time intervals, and miles per hour for velocities. You may need to make conversions.

TACTIC 4: THINK ABOUT YOUR ANSWER
Look at your answer and ask yourself whether it makes sense. Is it far too large or far too small? Is the sign correct? Are the units appropriate? In Sample Problem 2-1, for example, the correct answer is 11 mi/h. If you find 0.00011 mi/h, -11 mi/h, 11 mi/s, or 11,000 mi/h, you should realize at once that you have done something wrong. The error may lie in your method, in your algebra, or in your arithmetic. Check the problem carefully, being sure to start at the very beginning.

In Sample Problem 2-1, your answer must be greater than your speed of walking (2–3 mi/h) but less than the speed of the truck (43 mi/h). Finally, the answer to Sample Problem 2-2 must be less than that to Sample Problem 2-1 for two reasons: the displacement's magnitude is smaller in 2-2 and the time required is longer.

TACTIC 5: READING A GRAPH
Figures 2-2, 2-3a, 2-4, and 2-5 are examples of graphs that you should be able to read easily. In each graph, the variable on the horizontal axis is the time t, the direction of increasing time being to the right. In each, the variable on the vertical axis is the position x of the moving particle with respect to the origin, the direction of increasing x being upward.

Always note the units (seconds or minutes; meters, kilometers, or miles) in which the variables are expressed, and note whether the variables are positive or negative.

TACTIC 6: SIGNIFICANT FIGURES
If you were going to divide 137 jelly beans among 3 people, you would not think of giving each person *exactly* 137/3 or 45.66666666 · · · beans. You would give each person 45 beans and draw straws to see who would not get one of the remaining two. We need to develop that same kind of common sense in dealing with numerical calculations in physics.

In Sample Problem 2-1, for example, if you calculate the average velocity with your calculator wide open, you get $\overline{v} = 11.20840631$ mi/h. This number has 10 *significant figures.* The original data in the problem have only two significant figures.

In general, no final result should have more significant figures than the original data from which it was derived.

If multiple steps of calculation are involved, you should retain more significant figures than the original data have. However, when you come to the final result, you should round off according to the original data with the least significant figures. We did that in Sample Problem 2-1 to get $\bar{v} \approx 11$ mi/h. (Hereafter, the answer to a sample problem might be presented with the symbol = instead of $\approx$, but rounding off may still be involved.)

It is hard to escape the feeling that you are throwing away good data when you round off in this way but, in fact, you are doing the opposite; you are throwing away useless and misleading numbers. You may be able to set your calculator to do this rounding for you. Regardless of how you set it, your calculator continues to compute wide open internally, displaying only the rounded result that you ask it to show you.

When a number such as 3.15 or 3.15×10^3 is provided in a problem, the number of significant figures is apparent. But how about the number 3000? Is it known to only one significant figure (it could be written as 3×10^3)? Or is it known to as many as four significant figures (it could be written as 3.000×10^3)? In this book, we assume that all the zeros in such provided numbers as 3000 are significant, but you had better not make that assumption elsewhere.

TACTIC 7: SIGNIFICANT FIGURES AND DECIMAL PLACES

Don't confuse these. Consider the lengths 35.6 m, 3.56 m, 0.356 m, and 0.00356 m. They all have three significant figures but, in sequence, they have one, two, three, and five decimal places.

2-4 INSTANTANEOUS VELOCITY AND SPEED

You have now seen two ways to describe how fast something moves: average velocity and average speed, both of which are measured over a time interval Δt. But the phrase "how fast" more commonly refers to how fast a particle is moving at a given instant—its **instantaneous velocity** v (or simply **velocity**).

The velocity at any instant is obtained from the average velocity by shrinking the time interval Δt closer and closer to 0. As Δt dwindles, the average velocity approaches a limiting value, which is the velocity at that instant:

$$v = \lim_{\Delta t \to 0} \frac{\Delta x}{\Delta t} = \frac{dx}{dt}. \qquad (2\text{-}4)$$

Velocity is another vector and thus has an associated direction.

Table 2-1 shows an example of the limiting process. The first column gives the position x of a parti-

TABLE 2-1
THE LIMITING PROCESS

BEGINNING POINT		END POINT		INTERVALS		VELOCITY
x_1 (m)	t_1 (s)	x_2 (m)	t_2 (s)	Δx (m)	Δt (s)	$\Delta x/\Delta t$ (m/s)
5.00	1.00	9.00	3.00	4.00	2.00	+ 2.0
5.00	1.00	8.75	2.50	3.75	1.50	+ 2.5
5.00	1.00	8.00	2.00	3.00	1.00	+ 3.0
5.00	1.00	6.75	1.50	1.75	0.50	+ 3.5
5.00	1.00	5.760	1.200	0.760	0.200	+ 3.8
5.00	1.00	5.388	1.100	0.388	0.100	+ 3.9
5.00	1.00	5.196	1.050	0.196	0.050	+ 3.9
5.00	1.00	5.158	1.040	0.158	0.040	+ 4.0
5.00	1.00	5.119	1.030	0.119	0.030	+ 4.0

Δt shrinks ↓

a limiting value has been reached

cle at $t = 1$ s, which is when a time interval Δt begins. The third and fourth columns give the values of x and t at the end of Δt. And the fifth and sixth columns give the displacement Δx and the interval Δt (which we are shrinking). As Δt shrinks, $\bar{v}$ ($= \Delta x/ \Delta t$, in the last column) gradually changes until it reaches a limiting value of $+ 4.0$ m/s. That value is the instantaneous velocity v at $t = 1$ s.

In the language of calculus, the instantaneous velocity is the rate at which a particle's position x is changing with time at a given instant. According to Eq. 2-4, the velocity of a particle at any instant is the slope of its position curve at the point representing that instant.

Speed is the magnitude of a velocity; that is, speed is velocity that has been stripped of any indication of direction, either in words or via an algebraic sign.* A velocity of $+5$ m/s and one of -5 m/s both have an associated speed of 5 m/s. The speedometer in a car measures the speed, not the velocity, because it cannot ascertain anything about the direction of motion.

SAMPLE PROBLEM 2-4

Figure 2-6a is an $x(t)$ plot for an elevator cab that is initially stationary, then moves upward (which we take to be the positive direction), and then stops. Plot $v(t)$ as a function of time.

SOLUTION The slope, and so also the velocity, is zero in the intervals containing points a and d, when the cab is stationary. During the interval bc the cab moves with a constant velocity, and the slope of $x(t)$ is

$$v = \frac{\Delta x}{\Delta t} = \frac{24 \text{ m} - 4.0 \text{ m}}{8.0 \text{ s} - 3.0 \text{ s}} = +4.0 \text{ m/s}.$$

The plus sign indicates that the cab is moving in the positive x direction. These values are plotted in Fig. 2-6b. In addition, as the cab initially begins to move and then later slows to a stop, v varies as indicated in the intervals 1 s to 3 s and 8 s to 9 s. (Figure 2-6c is considered later.)

Given a $v(t)$ graph such as Fig. 2-6b, we could "work backward" to produce the shape of the associated $x(t)$ graph (Fig. 2-6a). However, without additional information, we would not know the actual

*Speed and average speed can be quite different, so you must be careful solving problems that involve either quantity.

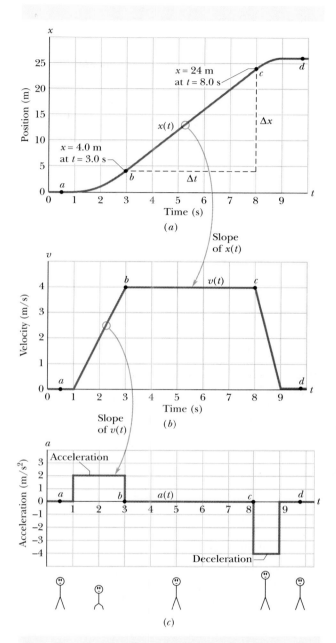

FIGURE 2-6 Sample Problem 2-4. (a) The $x(t)$ curve for an elevator cab that moves upward along an x axis. (b) The $v(t)$ curve for the cab. Note that it is the derivative of the $x(t)$ curve ($v = dx/dt$). (c) The $a(t)$ curve for the cab. It is the derivative of the $v(t)$ curve ($a = dv/dt$). The sketches suggest how a passenger's body might respond to the accelerations.

values for x, because the $v(t)$ graph indicates only *changes* in x. To find the change in x during any interval, we must, in the language of calculus, calculate the area "under the curve" on the $v(t)$ graph for the same interval. For example, during the interval in which the cab has a velocity of 4.0 m/s, the change in x is given by the "area" under the $v(t)$ curve:

$$\text{area} = (4.0 \text{ m/s})(8.0 \text{ s} - 3.0 \text{ s}) = +20 \text{ m}.$$

(This area is positive because the $v(t)$ curve is above the t axis.) Figure 2-6a shows that x does indeed increase by 20 m in the interval.

The instantaneous velocity of one of these speedboats is its velocity at the instant the photograph was taken. The boat may have had a different instantaneous velocity before or after that instant.

SAMPLE PROBLEM 2-5

The position of a particle moving on the x axis is given by

$$x = 7.8 + 9.2t - 2.1t^3. \qquad (2\text{-}5)$$

What is its velocity at $t = 3.5$ s? Is the velocity constant, or is it continuously changing?

SOLUTION For simplicity, the units have been omitted but you can insert them if you like by changing the coefficients to 7.8 m, 9.2 m/s, and -2.1 m/s^3. To solve the problem, we use Eq. 2-4 with the right side of Eq. 2-5 substituted for x:

$$v = \frac{dx}{dt} = \frac{d}{dt}(7.8 + 9.2t - 2.1t^3),$$

which becomes

$$v = 0 + 9.2 - (3)(2.1)t^2 = 9.2 - 6.3t^2. \qquad (2\text{-}6)$$

At $t = 3.5$ s,

$$v = 9.2 - (6.3)(3.5)^2 = -68 \text{ m/s}. \qquad \text{(Answer)}$$

At $t = 3.5$ s, the particle is moving toward decreasing x (note the minus sign) with a speed of 68 m/s. Since the quantity t appears in Eq. 2-6, the velocity v depends on t and so is continuously changing.

PROBLEM SOLVING

TACTIC 8: NEGATIVE NUMBERS

The line below is an x axis with its origin ($x = 0$) at the center. Using this scale, make sure you understand that, for example, -40 m is less than -10 m and that both are less than 20 m. Note also that 10 m is greater than -30 m.

The four arrows pointing to the right all represent increases in x, that is, positive values for Δx, the change in x. The four arrows pointing to the left represent decreases in x, that is, negative values for Δx.

Position (m)

TACTIC 9: DERIVATIVES AND SLOPES

Every derivative is the slope of a curve. In Sample Problem 2-4, for example, the velocity of the cab at any instant (a derivative; see Eq. 2-4) is the slope of the $x(t)$

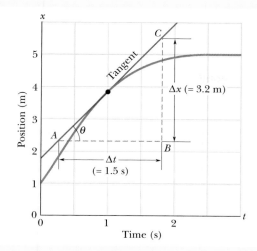

FIGURE 2-7 The derivative of a curve at any point is the slope of its tangent line at that point. At $t = 1.0$ s, the slope of the tangent line (and thus dx/dt, the instantaneous velocity) is $\Delta x/\Delta t = +2.1$ m/s.

curve of Fig. 2-6a at that instant. Here's how you can find a slope (and thus a derivative) graphically.

Figure 2-7 shows an $x(t)$ plot for a moving particle. To find the velocity of the particle at $t = 1$ s, put a dot on the curve at that point. Then draw a line tangent to the curve through the dot (*tangent* means *touching*; the tangent line touches the curve at a single point, the dot) judging carefully by eye. Then construct the right triangle ABC. (Although the slope is the same no matter what the size of this triangle, the larger the triangle the more precise will be your graphical measurement.) Find Δx and Δt, using the vertical and horizontal scales to provide the magnitude, the unit, and the sign. In Fig. 2-7 you find the slope (derivative) from the following equation:

$$\text{slope} = \frac{\Delta x}{\Delta t} = \frac{5.5 \text{ m} - 2.3 \text{ m}}{1.8 \text{ s} - 0.3 \text{ s}} = \frac{3.2 \text{ m}}{1.5 \text{ s}} = +2.1 \text{ m/s}.$$

As Eq. 2-4 tells you, this slope is the velocity of the particle at $t = 1$ s.

If you change the scale on the x or the t axis of Fig. 2-7, the appearance of the curve and the angle θ will change but the value you find for the velocity at $t = 1$ s will not.

If you have a mathematical expression for the function $x(t)$, as in Sample Problem 2-5, you can find the derivative dx/dt by the methods of calculus and avoid this graphical method.

2-5 ACCELERATION

When a particle's velocity changes, the particle is said to undergo **acceleration** (or to accelerate). The **average acceleration** $\overline{a}$ over an interval Δt is computed as

$$\overline{a} = \frac{v_2 - v_1}{t_2 - t_1} = \frac{\Delta v}{\Delta t}. \tag{2-7}$$

The **instantaneous acceleration** (or simply acceleration) is the derivative of the velocity:

$$a = \frac{dv}{dt}. \tag{2-8}$$

In words, the acceleration of a particle at any instant is the rate at which its velocity is changing at that instant. According to Eq. 2-8, the acceleration at any point is the slope of the curve of $v(t)$ at that point.

A common unit of acceleration is meter per second per second: m/(s·s) or m/s². You will see other units in the problems, but they will each be in the form of distance/(time·time) or distance/time². Acceleration has both magnitude and direction (it is yet another vector quantity). The algebraic sign represents the direction on an axis just as it does for displacement and velocity.

Figure 2-6c is a plot of the acceleration of the cab discussed in Sample Problem 2-4. Compare the curve with the $v(t)$ curve—each point on the $a(t)$ curve is the derivative (slope) of the corresponding point on the $v(t)$ curve. When v is constant (at either 0 or 4 m/s), the derivative is zero and so also is the acceleration. When the cab first begins to move, the $v(t)$ curve has a positive derivative (the slope is positive), which means that $a(t)$ is positive. When the cab slows to a stop, the derivative and slope of the $v(t)$ curve are negative; that is, $a(t)$ is negative.

Next compare the slopes of the $v(t)$ curve during the two acceleration periods. The one associated with the cab's stopping (commonly called "deceleration") is steeper, because the cab stops in half the time it took to get up to speed. The steeper slope means that the magnitude of the deceleration is larger than that of the acceleration, as indicated in Fig. 2-6c.

The sensations you would feel while riding in the cab of Fig. 2-6 are indicated by the sketched figures. When the car first accelerates, you feel as

FIGURE 2-8 Colonel J. P. Stapp in a rocket sled as it is brought up to high speed (acceleration out of the page) and then very rapidly braked (acceleration into the page).

though you are pressed downward; when later the cab is braked to a stop, you seem to be stretched upward. In between, you feel nothing special. Your body reacts to accelerations (it is an accelerometer) but not to velocities (it is not a speedometer). When you are in a car traveling at 60 mi/h or an airplane traveling at 600 mi/h, you have no bodily awareness of the motion. But if the car or plane quickly changes velocity, you may become keenly aware of the change, perhaps even frightened by it. Part of the thrill of an amusement park ride is due to the quick changes of velocity that you undergo. A more extreme example is shown in the photographs of Fig. 2-8, which were taken while a rocket sled was rapidly accelerated and then rapidly braked to a stop.

SAMPLE PROBLEM 2-6

a. When Kitty O'Neil set the dragster records for the greatest speed and least elapsed time, she reached 392.54 mi/h in 3.72 s. What was her average acceleration?

SOLUTION From Eq. 2-7, O'Neil's average acceleration was

$$\bar{a} = \frac{\Delta v}{\Delta t} = \frac{392.54 \text{ mi/h} - 0}{3.72 \text{ s} - 0}$$

$$= +106 \frac{\text{mi}}{\text{h} \cdot \text{s}}, \quad \text{(Answer)}$$

where the motion is taken to be in the positive x direction. In more conventional units, her acceleration was 47.1 m/s². Often, large accelerations are expressed in ''g'' units, where $1g = 9.8$ m/s² ($= 32$ ft/s²), as will be

explained in Section 2-8. O'Neil's average acceleration was 4.8g.

b. What was the average acceleration when Eli Beeding, Jr. reached 72.5 mi/h in 0.04 s on a rocket sled?

SOLUTION Again from Eq. 2-7,

$$\bar{a} = \frac{\Delta v}{\Delta t} = \frac{72.5 \text{ mi/h} - 0}{0.04 \text{ s} - 0}$$

$$= +1.8 \times 10^3 \frac{\text{mi}}{\text{h} \cdot \text{s}} \approx +800 \text{ m/s}^2, \quad \text{(Answer)}$$

or about 80g.

Recall our question in Section 2-1, where O'Neil and Beeding were introduced. How can we tell who had the more thrilling ride—by final speeds, by elapsed times, or by some other quantity? You now can answer that question. Because the human body senses acceleration rather than speed, you should compare accelerations, and so Beeding wins out, even though his final speed was considerably slower than O'Neil's. In fact, Beeding's acceleration could have been lethal had it continued for much longer.

PROBLEM SOLVING

TACTIC 10: AN ACCELERATION'S SIGN

Look again at the algebraic sign for the accelerations that are calculated in Sample Problem 2-6. In many common examples of acceleration, the sign has a common-sense meaning: positive acceleration means that the speed of an object (such as a car) is increasing, and negative acceleration means that the speed is decreasing (the object is undergoing deceleration).

Such meanings cannot be interpreted without some thought, however. For example, if a car with an initial velocity $v = -27$ m/s $(= -60$ mi/h) is braked to a stop in 5.0 s, $\bar{a} = +5.4$ m/s². The acceleration is *positive*, but the car has slowed. The reason is the difference in signs: the direction of the acceleration is opposite that of the velocity.

Here then is a better way to interpret the signs: if the signs of the velocity and acceleration are the same, a particle picks up speed; if the signs are opposite, the particle slows. The interpretation will have more meaning when we later explore the vector nature of velocity and acceleration.

SAMPLE PROBLEM 2-7

A particle's position is given by

$$x = 4 - 27t + t^3,$$

where the units of the coefficients are m, m/s, and m/s³, respectively, and the x axis is shown in Fig. 2-1.

a. Find $v(t)$ and $a(t)$.

SOLUTION To get $v(t)$, we differentiate $x(t)$ with respect to t:

$$v = -27 + 3t^2. \qquad \text{(Answer)}$$

To get $a(t)$, we differentiate $v(t)$ with respect to t:

$$a = +6t. \qquad \text{(Answer)}$$

b. Is there ever a time when $v = 0$?

SOLUTION Setting $v(t) = 0$ yields

$$0 = -27 + 3t^2,$$

which has the solution $t = \pm 3$ s. (Answer)

c. Describe the particle's motion for $t \geq 0$.

SOLUTION To answer, we examine the expressions for $x(t)$, $v(t)$, and $a(t)$.

At $t = 0$ the particle is at $x = +4$ m, is moving leftward with a velocity of -27 m/s, and is, at that instant, not accelerating.

For $0 < t < 3$ s, the particle continues to move to the left, but at decreasing speed, because it is accelerating to the right. (Check $v(t)$ and $a(t)$ for, say, $t = 2$ s.) The rate of the acceleration is increasing.

At $t = 3$ s, the particle stops momentarily ($v = 0$) and is as far to the left as it will ever get ($x = -50$ m). It continues to accelerate to the right at an increasing rate.

For $t > 3$ s, its acceleration to the right continues to increase, and its velocity, which is now also to the right, increases rapidly. (Note that the signs of v and a match.) The particle moves to the right without bound.

2-6 CONSTANT ACCELERATION: A SPECIAL CASE

In many common types of motion, the acceleration is either constant or approximately so. For example, you might accelerate a car at an approximately constant rate when a traffic light turns from red to green. (Graphs of your position, velocity, and acceleration would resemble those in Fig. 2-9.) If you later had to brake the car to a stop, the deceleration during the braking might also be approximately constant.

Such cases are so ubiquitous that a special set of equations has been derived for dealing with them. One approach to the derivation of the equations is given in this section. A second approach is given in the next section. Throughout both sections and later when you work on the homework problems, keep in mind that *the equations are valid only for con-*

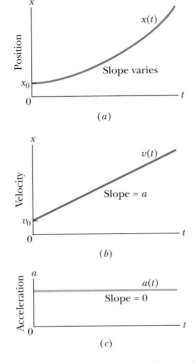

FIGURE 2-9 (*a*) The position $x(t)$ of a particle moving with constant acceleration. (*b*) Its velocity $v(t)$, given at each point by the slope of the curve in (*a*). (*c*) Its (constant) acceleration, equal to the (constant) slope of $v(t)$.

stant acceleration (or situations in which you can approximate the acceleration as being constant).

When the acceleration is constant, the distinction between average acceleration and instantaneous acceleration loses its meaning and we can write Eq. 2-7, with some changes in notation, as

$$a = \frac{v - v_0}{t - 0}.$$

Here v_0 is the velocity at time $t = 0$ and v is the velocity at any later time t. We can recast the equation as

$$v = v_0 + at. \qquad (2\text{-}9)$$

As a check, note that this equation reduces to $v = v_0$ for $t = 0$, as it must. As a further check, take the derivative of Eq. 2-9. Doing so yields $dv/dt = a$, which is the definition of a. Figure 2-9*b* shows a plot of Eq. 2-9, the $v(t)$ function.

In similar manner we can rewrite Eq. 2-2 (with a few changes in notation) as

$$x = x_0 + \overline{v}t, \qquad (2\text{-}10)$$

in which x_0 is the position of the particle at $t = 0$, and $\overline{v}$ is the average velocity between $t = 0$ and a later time t.

If you plot v against t using Eq. 2-9, a straight line results. Under these conditions, the *average* velocity over any time interval (say, $t = 0$ to a later time t) is the average of the velocity at the beginning of the interval ($= v_0$) and the velocity at the end of the interval ($= v$). For the interval $t = 0$ to the later time t then, the average velocity is

$$\overline{v} = \tfrac{1}{2}(v_0 + v). \qquad (2\text{-}11)$$

Substituting for v from Eq. 2-9 yields, after a little rearrangement,

$$\overline{v} = v_0 + \tfrac{1}{2}at. \qquad (2\text{-}12)$$

Finally, substituting Eq. 2-12 into Eq. 2-10 yields

$$x - x_0 = v_0 t + \tfrac{1}{2}at^2. \qquad (2\text{-}13)$$

As a check, note that putting $t = 0$ yields $x = x_0$, as it must. As a further check, taking the derivative of Eq. 2-13 yields Eq. 2-9, again as it must. Figure 2-9*a* is a plot of Eq. 2-13.

Five quantities can possibly be involved in any given problem regarding constant acceleration, namely, $x - x_0$, v, t, a, and v_0. Usually, one of these quantities is *not* involved in the problem, *either as a given or as an unknown*. We are then presented with three of the remaining quantities and asked to find the fourth.

Equations 2-9 and 2-13 each contain four of these quantities, but not the same four. In Eq. 2-9, the "missing ingredient" is the displacement, $x - x_0$. In Eq. 2-13, it is the velocity v. These two equations can also be combined in three ways to yield three additional equations, each of which involves a different "missing ingredient." Thus

$$v^2 = v_0^2 + 2a(x - x_0). \qquad (2\text{-}14)$$

This equation is useful if we do not know t and are not required to find it. We can, instead, eliminate the acceleration a between these same two equations to produce an equation in which a does not appear:

$$x - x_0 = \tfrac{1}{2}(v_0 + v)t. \qquad (2\text{-}15)$$

Finally, we can eliminate v_0, obtaining

$$x - x_0 = vt - \tfrac{1}{2}at^2. \qquad (2\text{-}16)$$

Note the subtle difference between this equation and Eq. 2-13. One involves the initial velocity v_0; the other involves the velocity v at time t.

TABLE 2-2
EQUATIONS FOR MOTION WITH CONSTANT ACCELERATION[a]

EQUATION NUMBER	EQUATION	MISSING QUANTITY
2-9	$v = v_0 + at$	$x - x_0$
2-13	$x - x_0 = v_0 t + \frac{1}{2}at^2$	v
2-14	$v^2 = v_0^2 + 2a(x - x_0)$	t
2-15	$x - x_0 = \frac{1}{2}(v_0 + v)t$	a
2-16	$x - x_0 = vt - \frac{1}{2}at^2$	v_0

[a]Make sure that the acceleration is indeed constant before using the equations in this table. Note that if you differentiate Eq. 2-13 you get Eq. 2-9. The other three equations are found by eliminating one or another of the variables between Eqs. 2-9 and 2-13.

Table 2-2 lists Eqs. 2-9, 2-13, 2-14, 2-15, and 2-16 and shows which one of the five possible quantities is missing from each. To solve a constant acceleration problem, you must decide which of the five quantities is *not* involved in the problem, either as a given or as an unknown. Select the correct equation from Table 2-2 and substitute for the three given quantities to find the unknown. Instead of using the table, you might find a solution more easily if you use only Eqs. 2-9 and 2-13, solving them as simultaneous equations when needed.

SAMPLE PROBLEM 2-8

Spotting a police car, you brake a Porsche from 75 km/h to 45 km/h over a displacement of 88 m.

a. What is the acceleration, assumed to be constant?

SOLUTION In this problem the time is not involved, being neither given nor requested. Table 2-2 then leads us to Eq. 2-14. Solving this equation for a yields

$$a = \frac{v^2 - v_0^2}{2(x - x_0)} = \frac{(45 \text{ km/h})^2 - (75 \text{ km/h})^2}{(2)(0.088 \text{ km})}$$

$$= -2.05 \times 10^4 \text{ km/h}^2 \approx -1.6 \text{ m/s}^2. \quad \text{(Answer)}$$

(In converting hours to seconds in the last step, we must convert *both* the hour units.) Note that the velocities are positive and the acceleration is negative, which is consistent with a slowing of the car.

b. What is the elapsed time?

SOLUTION Now time is not the missing ingredient, but the acceleration is. Table 2-2 suggests Eq. 2-15.

Solving that equation for t, we obtain

$$t = \frac{2(x - x_0)}{v_0 + v} = \frac{(2)(0.088 \text{ km})}{(75 + 45) \text{ km/h}}$$

$$= 1.5 \times 10^{-3} \text{ h} = 5.4 \text{ s}. \quad \text{(Answer)}$$

c. If you continue to slow down with the acceleration calculated in (a) above, how much time would elapse in bringing the car to rest from 75 km/h?

SOLUTION The quantity not given or asked for here is the displacement, $x - x_0$. Table 2-2 then suggests that we use Eq. 2-9. Solving for t gives

$$t = \frac{v - v_0}{a} = \frac{0 - (75 \text{ km/h})}{(-2.05 \times 10^4 \text{ km/h}^2)}$$

$$= 3.7 \times 10^{-3} \text{ h} = 13 \text{ s}. \quad \text{(Answer)}$$

d. In (c) above, what distance would be covered?

SOLUTION From Eq. 2-13, we have, for the displacement of the car,

$$x - x_0 = v_0 t + \frac{1}{2}at^2$$

$$= (75 \text{ km/h})(3.7 \times 10^{-3} \text{ h})$$

$$\quad + \frac{1}{2}(-2.05 \times 10^4 \text{ km/h}^2)(3.7 \times 10^{-3} \text{ h})^2$$

$$= 0.137 \text{ km} \approx 140 \text{ m}. \quad \text{(Answer)}$$

(Neglecting the sign on the acceleration would give a wrong result. When you work problems, you should always be alert to signs.)

e. Suppose that, on a second trial with the acceleration calculated in (a) above and a different initial velocity, you bring your car to rest after traversing 200 m. What was the total braking time?

SOLUTION The missing quantity here is the initial velocity, so we use Eq. 2-16. Noting that v (the final velocity) is zero and solving this equation for t, we obtain

$$t = \left(-\frac{(2)(x - x_0)}{a}\right)^{1/2} = \left(-\frac{(2)(200 \text{ m})}{-1.6 \text{ m/s}^2}\right)^{1/2}$$

$$= 16 \text{ s}. \quad \text{(Answer)}$$

PROBLEM SOLVING

TACTIC 11: CHECK THE DIMENSIONS

The dimension of a velocity is (L/T), that is, length (L) divided by time (T). The dimension of acceleration is (L/T²); and so on. In any physical equation, the dimensions of all terms must be the same. If you are in doubt about an equation, check its dimensions.

To check the dimensions of Eq. 2-13 ($x - x_0 = v_0 t + \frac{1}{2}at^2$), we note that every term must be a length, because that is the dimension of x and of x_0. The dimension of the term $v_0 t$ is $(L/T)(T)$, which is (L). The dimension of $\frac{1}{2}at^2$ is $(L/T^2)(T^2)$, which is also (L). This equation checks out. A pure number such as $\frac{1}{2}$ or π has no dimension.

2-7   ANOTHER LOOK AT CONSTANT ACCELERATION*

The first two equations in Table 2-2 are the basic equations from which the others are derived. Those two can be obtained by integration of the acceleration with the condition that a is constant. The definition of a (in Eq. 2-8) is

$$a = \frac{dv}{dt},$$

which can be rewritten as

$$dv = a\, dt.$$

If we take the *indefinite integral* (or *antiderivative*) of both sides, we get

$$\int dv = \int a\, dt,$$

which reduces to

$$v = \int a\, dt + C,$$

where C is a *constant of integration*. Since acceleration a is constant, it can be taken outside the integration. Then

$$v = a \int dt + C = at + C. \qquad (2\text{-}17)$$

To evaluate the constant C, we let $t = 0$, at which time $v = v_0$. Substituting these values into Eq. 2-17 (which must hold for all values of t, including $t = 0$) yields

$$v_0 = (a)(0) + C = C.$$

With this substitution, Eq. 2-17 takes the same form as Eq. 2-9.

To derive the other basic equation in Table 2-2, we rewrite the definition of velocity (Eq. 2-4) as

$$dx = v\, dt$$

and then take the indefinite integral of both sides to obtain

$$x = \int v\, dt + C',$$

where C' is another constant of integration. This time there is no reason to believe that v is constant, so we cannot move it outside the integration. But we can substitute for v with Eq. 2-9:

$$x = \int (v_0 + at)\, dt + C'.$$

Since v_0 is a constant, this can be rewritten as

$$x = v_0 \int dt + a \int t\, dt + C'.$$

Integration yields

$$x = v_0 t + \frac{1}{2}at^2 + C'. \qquad (2\text{-}18)$$

At time $t = 0$, we have $x = x_0$. Substituting these values in Eq. 2-18 yields $x_0 = C'$. Replacing C' with x_0 in Eq. 2-18 gives us Eq. 2-13.

2-8   FREE-FALL ACCELERATION

If you tossed an object either up or down and could somehow eliminate the effects of air on its flight, you

FIGURE 2-10   A feather and an apple, undergoing free fall in a vacuum, move downward at the same acceleration g. The acceleration causes the increase in distance between images during the fall.

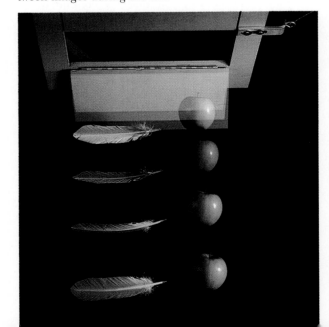

---

*This section is intended for those students who have had integral calculus.

TABLE 2-3
EQUATIONS FOR FREE FALL

EQUATION NUMBER	EQUATION	MISSING QUANTITY
2-19	$v = v_0 - gt$	$y - y_0$
2-20	$y - y_0 = v_0 t - \frac{1}{2}gt^2$	v
2-21	$v^2 = v_0^2 - 2g(y - y_0)$	t
2-22	$y - y_0 = \frac{1}{2}(v_0 + v)t$	g
2-23	$y - y_0 = vt + \frac{1}{2}gt^2$	v_0

would find that the object accelerates downward at a particular rate. That rate is called the **free-fall acceleration** g. The acceleration g is independent of the object's characteristics, such as mass, density, or shape.

Two examples of free-fall acceleration are shown in the photograph of Fig. 2-10, which is a stroboscopic series of photos of a feather and an apple. As these objects descend, they accelerate downward at the rate g, picking up speed. The value of g varies slightly with latitude and also with elevation. At sea level in the mid-latitudes the value is 9.8 m/s² (or 32 ft/s²), which is what you should use for the problems in this chapter.

The equations of motion in Table 2-2 for constant acceleration apply to free fall near the Earth's surface. That is, they apply to an object in vertical flight, either up or down, when the effects of the air can be neglected. However, we can make them simpler to use with two minor changes. (1) The directions of motion are along the vertical y axis instead of the x axis, with the positive direction of y upward. (This change will reduce confusion in later chapters when combined horizontal and vertical motions are examined.) (2) The free-fall acceleration is then negative, that is, downward on the y axis, and so we replace a with $-g$ in the equations.

With these small changes, the equations of Table 2-2 become, for free fall, the equations in Table 2-3.

SAMPLE PROBLEM 2-9

A worker drops a wrench down the elevator shaft of a tall building.

a. Where is the wrench 1.5 s later?

SOLUTION The missing ingredient is the velocity v, which is neither given nor requested. This suggests Eq. 2-20 of Table 2-3. Choose the release point of the wrench to be the origin of the y axis. Setting $y_0 = 0$, $v_0 = 0$, and $t = 1.5$ s in Eq. 2-20 gives

$$y = v_0 t - \tfrac{1}{2}gt^2$$
$$= (0)(1.5 \text{ s}) - \tfrac{1}{2}(9.8 \text{ m/s}^2)(1.5 \text{ s})^2$$
$$= -11 \text{ m}. \qquad \text{(Answer)}$$

The minus sign means that the wrench is below its release point, which we certainly expect.

b. How fast is the wrench falling just then?

SOLUTION The velocity of the wrench is given by Eq. 2-19:

$$v = v_0 - gt = 0 - (9.8 \text{ m/s}^2)(1.5 \text{ s})$$
$$= -15 \text{ m/s}. \qquad \text{(Answer)}$$

Here the minus sign means that the wrench is falling downward. Again, no great surprise. Figure 2-11 displays the important features of the motion up to $t = 4$ s.

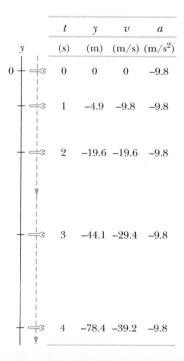

	t	y	v	a
	(s)	(m)	(m/s)	(m/s²)
0	0	0	0	-9.8
	1	-4.9	-9.8	-9.8
	2	-19.6	-19.6	-9.8
	3	-44.1	-29.4	-9.8
	4	-78.4	-39.2	-9.8

FIGURE 2-11 Sample Problem 2-9. The position, velocity, and acceleration of a freely falling object.

SAMPLE PROBLEM 2-10

In 1939, Joe Sprinz of the San Francisco Baseball Club attempted to break the record for catching a baseball dropped from the greatest height. Members of the Cleveland Indians had set the record the preceding year when they caught baseballs dropped about 700 ft from atop a building. Sprinz used a blimp at 800 ft. Ignore the effects of air on the ball and assume that the ball falls 800 ft.

a. Find its time of fall.

SOLUTION Mentally erect a vertical y axis with its origin at the point of the ball's release in the blimp, which means that $y_0 = 0$. The initial velocity v_0 is zero. The missing ingredient is v, and so Eq. 2-20 is required:

$$y - y_0 = v_0 t - \tfrac{1}{2} g t^2$$

$$-800 \text{ ft} = 0t - \tfrac{1}{2}(32 \text{ ft/s}^2) t^2$$

$$16 t^2 = 800$$

$$t = 7.1 \text{ s.} \qquad \text{(Answer)}$$

When taking a square root, we have the option of attaching a plus or minus sign to the square root. Here we choose the plus sign, since the ball reaches the ground *after* it is released.

b. What is the velocity of the ball just before it is caught?

SOLUTION To get the velocity from the original data, rather than from the result of (a), we use Eq. 2-21:

$$v^2 = v_0^2 - 2g(y - y_0)$$

$$= 0 - (2)(32 \text{ ft/s}^2)(-800 \text{ ft})$$

$$= 5.12 \times 10^4 \text{ ft}^2/\text{s}^2$$

$$v = -226 \text{ ft/s} \ (\approx -154 \text{ mi/h}). \quad \text{(Answer)}$$

Since the ball is moving downward, we choose the minus sign in our option of signs in taking the square root.

Neglecting the effects of air is actually unwarranted in such a fall. If you included them, you would find that the fall time was longer and the final speed was smaller than the values calculated above. Still, the speed must have been considerable, because when Sprinz finally managed to get a ball in his glove (on his fifth attempt) the impact slammed the glove and hand into his face, fracturing the upper jaw in 12 places, breaking five teeth, and knocking him unconscious. And he dropped the ball.

SAMPLE PROBLEM 2-11

A pitcher tosses a baseball straight up, with an initial speed of 12 m/s. See Fig. 2-12.

a. How long does it take to reach its highest point?

SOLUTION The ball is at its highest point when its velocity v becomes zero. From Eq. 2-19, we have

$$t = \frac{v_0 - v}{g} = \frac{12 \text{ m/s} - 0}{9.8 \text{ m/s}^2} = 1.2 \text{ s.} \quad \text{(Answer)}$$

b. How high does the ball rise above its release point?

SOLUTION We take the release point of the ball to be the origin of the y axis. Putting $y_0 = 0$ in Eq. 2-21 and solving for y, we obtain

$$y = \frac{v_0^2 - v^2}{2g} = \frac{(12 \text{ m/s})^2 - (0)^2}{(2)(9.8 \text{ m/s}^2)}$$

$$= 7.3 \text{ m.} \qquad \text{(Answer)}$$

If we wanted to take advantage of the fact that we also know the time of flight, having found it in (a), we could also calculate the height of rise from Eq. 2-23. Check it out.

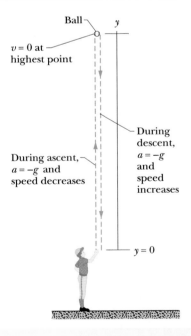

FIGURE 2-12 Sample Problem 2-11. A pitcher tosses a baseball straight up into the air. The equations of free fall apply for rising as well as for falling objects, provided that any effects from the air can be neglected.

c. How long will it take for the ball to reach a point 5.0 m above its release point?

SOLUTION Inspection of Eqs. 2-19 to 2-23 suggests that we try Eq. 2-20. With $y_0 = 0$, we have

$$y = v_0 t - \tfrac{1}{2}gt^2.$$

$$5.0 \text{ m} = (12 \text{ m/s})t - (\tfrac{1}{2})(9.8 \text{ m/s}^2)t^2.$$

If we temporarily omit the units (having noted that they are consistent), we can rewrite this as

$$4.9t^2 - 12t + 5.0 = 0.$$

Solving this quadratic equation for t yields*

$$t = 0.53 \text{ s} \quad \text{and} \quad t = 1.9 \text{ s}. \qquad \text{(Answer)}$$

There are two such times! This is not really surprising because the ball passes twice through $y = 5.0$ m, once on the way up and once on the way down.

We can check our findings because the time at which the ball reaches its maximum height should lie halfway between these two times, or at

$$t = \tfrac{1}{2}(0.53 \text{ s} + 1.9 \text{ s}) = 1.2 \text{ s}.$$

This is exactly what we found in (a) for the time to reach maximum height.

PROBLEM SOLVING

TACTIC 12: MINUS SIGNS

In Sample Problems 2-9, 2-10, and 2-11, many answers emerged automatically with minus signs. It is important to know what these signs mean. For falling body problems, we established a vertical axis (the y axis) and we chose—quite arbitrarily—its upward direction to be positive.

We then choose the origin of the y axis (that is, the $y = 0$ position) to suit the problem. In Sample Problem 2-9, the origin was the worker's hand; in Sample Problem 2-10 it was at the blimp; in Sample Problem 2-11 it was the pitcher's hand. A negative value of y means that the body is below the chosen origin.

A negative velocity means that the body is moving in the direction of decreasing y, that is, downward. This is true no matter where the body is located.

We have taken the acceleration $(= -9.8 \text{ m/s}^2)$ to be negative in all problems dealing with falling bodies. A negative acceleration means that, as time goes on, the velocity of the body becomes either less positive or

more negative. This is true no matter where the body is located and no matter how fast or in what direction it is moving. In Sample Problem 2-11, the acceleration of the ball is negative throughout its flight, whether the ball is rising or falling.

TACTIC 13: UNEXPECTED ANSWERS

Mathematics often generates answers that you might not have thought of as possibilities, as in Sample Problem 2-11c. If you get more answers than you expect, do not discard out of hand the ones that do not seem to fit. Examine them carefully for physical meaning; it is often there.

If time is your variable, even a negative value can mean something; negative time simply refers to time before $t = 0$, the (arbitrary) time at which you decided to start your stopwatch.

2-9 THE PARTICLES OF PHYSICS

As we progress through the book, we plan to step aside occasionally from the familiar world of large, tangible objects and look at nature on a much finer scale. The "particles" that we have dealt with in this chapter, for example, have included pigs, baseballs, and dragsters. In the spirit of our plan, we ask: "How small can a particle be? What are the *ultimate* particles of nature?" **Particle physics**—for so the field that relates to our inquiry is called—attracts the attention of many of the best and brightest of today's physicists.

The realization that matter, on its finest scale, is not continuous but is made up of atoms was the beginning of understanding for physics and chemistry. With the scanning tunneling microscope, we can now "photograph" these atoms, as Fig. 2-13 makes clear. It is also possible to keep single atoms in tiny electromagnetic "traps" and monitor them at leisure. A single electron was kept in such a trap at the University of Washington for 10 months before—by misadventure—it hit a wall and escaped.

We describe the "lumpiness" of matter by saying that matter is *quantized,* the word coming from the Latin word *quantus,* meaning "how much." Quantization is a central feature of nature, and you will see other physical quantities as we go along that are quantized when we look at them on a fine enough scale. This pervasiveness of quantization is reflected in the name we give to physics at the atomic and subatomic level—**quantum physics,** the

---

*See Appendix G for the formula used to solve a quadratic equation.

FIGURE 2-13   A hexagonal array of uranium atoms is revealed in this image from a scanning transmission electron microscope. The color has been added by a computer.

physics that deals with the ultimate particles of nature.

There is no sharp discontinuity between the quantum world and the world of large-scale objects. The quantum world and the laws that govern it are universal but, as we move from electrons and atoms to baseballs and automobiles, the fact of quantization becomes less noticeable and finally totally undetectable. The "graininess" effectively disappears, and the laws of **classical physics** that govern the motions of large objects emerge as special limiting forms of the more general laws of quantum physics.

The Structure of Atoms

An **atom** consists of a central, almost unimaginably compact and dense **nucleus** that is surrounded by one or more light-weight **electrons.** An atom is usually considered to be spherical; so is the nucleus. The radius of a typical atom is on the order of 10^{-10} m; the radius of a nucleus is 100,000 times smaller, about 10^{-15} m. An atom is held together by electrical attraction between the electrons, which are electrically negative, and **protons,** which are electrically positive and reside within the nucleus. The nature of that attraction is explored later in this book,

but for now you might realize that were it not present, atoms could not exist, and so neither could you.

The Structure of Nuclei

The simplest nucleus, that of common hydrogen, has a single proton. There are two other, rare, versions of hydrogen: they differ from the common version by the presence of one or two **neutrons** (electrically neutral particles) inside the nucleus. Hydrogen, in any of its versions, is an example of an **element;** each element is distinguished from all the others by the number of protons in the nucleus. When there is only one proton, the element is hydrogen. When, instead, there are six, the element is carbon. The various versions of each element are called **isotopes;** they are distinguished by the number of neutrons.

Roughly speaking, the purpose of the neutrons is to glue together the protons, which, being all electrically positive and closely packed, strongly repel one another. If the neutrons did not provide the glue, the only type of atom that could exist would be common hydrogen; all others would blow apart.

Such instability can be found in many isotopes of common elements, but thankfully not the elements on which your existence depends. For example, of the 17 isotopes of copper, all but two are unstable and undergo transformations to become other elements. The stable isotopes are the ones used in electronics and other technology.

The Structure of the Particles Within Atoms

The electron is simple but perplexingly so. It appears to be infinitesimal in size; that is, it has no size and no structure. It is a member of a family of other pointlike particles called **leptons;** there are six basic types, each with an **antiparticle** version.

Protons and neutrons are believed to be different from electrons and the other leptons, because each of the former appears to be a bundle of three simpler particles called **quarks,*** "up" or "down"

---

*On a quirk, the word "quark" was lifted from *Finnegans Wake* by James Joyce:

Three quarks for Muster Mark.
Sure he hasn't got much of a bark
And sure any he has it's all beside the mark.

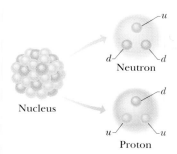

FIGURE 2-14 A representation of the nucleus of an atom, showing the neutrons and protons that make it up. These particles, in turn, are composed of "up" and "down" quarks.

quarks. A proton consists of two "up" quarks and one "down" quark, and a neutron the reverse (Fig. 2-14). Other, more exotic particles that were first thought to be fundamental appear to be similar bundles.

Provocatively, quarks come in six basic types* (each with its antiparticle) just as do leptons. Here physicists wonder: Is there a basic reason for the match in number of types? Or is the match simply coincidence? We do not know.

_____

*The other types are called charm, strange, top, and bottom (even physicists have their moments).

REVIEW & SUMMARY

Position

The *position* x of a particle on an axis locates it with respect to the **origin** of the axis. The position is either positive or negative, according to which side of the origin the particle is on, or zero if the particle is at the origin. The **positive direction** on an axis is the direction of increasing positive numbers; the opposite direction is the **negative direction.**

Displacement

The *displacement* Δx of a particle is the change in its position:

$$\Delta x = x_2 - x_1. \qquad (2\text{-}1)$$

Displacement is a vector quantity. It is positive if the particle has moved in the positive direction of the x axis, and negative if it has moved in the negative direction.

Average Velocity

When a particle has moved from position x_1 to x_2 during a time interval $\Delta t = t_2 - t_1$, its *average velocity* is

$$\bar{v} = \frac{\Delta x}{\Delta t}. \qquad (2\text{-}2)$$

The algebraic sign of $\bar{v}$ indicates the direction of motion ($\bar{v}$ is a vector quantity). Average velocity does not depend on the actual distance a particle covers, but instead depends on its original and final positions. Sample Problems 2-1 and 2-2 illustrate the calculation of average velocity.

On a graph of x versus t, the average velocity for a time interval Δt is the slope of the straight line connecting the points on the curve that represent the ends of the interval.

Average Speed

The *average speed* $\bar{s}$ of a particle depends on the full distance it covers in a time interval Δt:

$$\bar{s} = \frac{\text{total distance}}{\Delta t}. \qquad (2\text{-}3)$$

Instantaneous Velocity

If we allow Δt to approach zero in Eq. 2-2, then Δx will also approach zero; however, their ratio, which is $\bar{v}$, will approach a limiting value v, the *instantaneous velocity* (or simply **velocity**) of the particle at the time in question, or

$$v = \lim_{\Delta t \to 0} \frac{\Delta x}{\Delta t} = \frac{dx}{dt}. \qquad (2\text{-}4)$$

The instantaneous velocity (at a particular time) may be represented as the slope (at that particular time) of the graph of x versus t. See Sample Problem 2-3 and Fig. 2-9. Sample Problem 2-5 illustrates how we can find velocity by differentiating a function x(t). **Speed** is the magnitude of instantaneous velocity.

Average Acceleration

Average acceleration is the ratio of the change in velocity Δv that occurs within a time interval Δt to that time interval:

$$\bar{a} = \frac{\Delta v}{\Delta t}. \qquad (2\text{-}7)$$

The algebraic sign indicates the direction of $\bar{a}$. See Sample Problem 2-6 for examples.

Instantaneous Acceleration

Instantaneous acceleration (or simply **acceleration**) is the rate of change of velocity,

$$a = \lim_{\Delta t \to 0} \frac{\Delta v}{\Delta t} = \frac{dv}{dt}. \qquad (2\text{-}8)$$

Sample Problem 2-7 shows how to differentiate $v(t)$ to get $a(t)$. On a graph of v versus t, $a(t)$ is the slope of the curve.

Constant Acceleration

Figure 2-9 shows $x(t)$, $v(t)$, and $a(t)$ for the important case in which a is constant. In this circumstance, the five equations in Table 2-2 describe the motion:

$$v = v_0 + at, \tag{2-9}$$

$$x - x_0 = v_0 t + \tfrac{1}{2}at^2, \tag{2-13}$$

$$v^2 = v_0^2 + 2a(x - x_0), \tag{2-14}$$

$$x - x_0 = \tfrac{1}{2}(v_0 + v)t, \tag{2-15}$$

$$x - x_0 = vt - \tfrac{1}{2}at^2. \tag{2-16}$$

These equations are *not* valid when the acceleration is not constant. Sample Problem 2-8 illustrates the use of these equations.

Free-Fall Acceleration

An important example of straight-line motion with constant acceleration is that of an object rising or falling freely near the Earth's surface. The constant acceleration equations describe this motion, but we make two changes in notation: (1) we refer the motion to the vertical y axis with $+y$ vertically *up*; (2) we replace a with $-g$, where g is the magnitude of the free-fall acceleration. Near the Earth's surface, $g = 9.8 \text{ m/s}^2 \ (= 32 \text{ ft/s}^2)$. The free-fall equations, with these conventions, are shown as Eqs. 2-19 to 2-23. Sample Problems 2-9, 2-10, and 2-11 show how these equations can be used.

The Structure of Matter

All ordinary matter is composed of **atoms,** which, in a simple model, consist of **electrons** that surround a highly compact central core, the **nucleus. Neutrons** and **protons** reside inside nuclei. Each **element** is distinguished by the number of protons in its nucleus. Variations of an element, differing in the number of neutrons, are **isotopes** of the element.

Quarks and Leptons

Electrons appear to be pointlike particles with no size or internal structure, but protons and neutrons appear to have size and to contain more elementary particles, called **quarks.** There are six basic types of quarks, each with an antiparticle version. Electrons are members of a family of particles, **leptons,** that also come in six basic types, each with an antiparticle version.

QUESTIONS

1. What are some physical phenomena involving the Earth in which the Earth cannot be treated as a particle?

2. Can the speed of a particle ever be negative? If so, give an example; if not, explain why.

3. Each second a rabbit moves half the remaining distance from its nose to a head of lettuce. Does the rabbit ever get to the lettuce? What is the limiting value of the rabbit's velocity? Draw graphs showing the rabbit's position and average velocity versus time.

4. In this book, "average speed" means the ratio of the total distance covered to the elapsed time. However, sometimes the phrase is used to mean the magnitude of the average velocity. How, and in what cases, do the meanings differ?

5. In a qualifying two-lap heat, a racing car covers the first lap with an average speed of 90 mi/h. Can the driver take the second lap at a speed such that the average speed for the two laps is 180 mi/h? Explain.

6. Bob beats Judy by 10 m in a 100-m dash. Bob, claiming to want to give Judy an equal chance, agrees to race her again but to begin 10 m behind the starting line. Does this really give Judy an equal chance?

7. When the velocity is constant, can the average velocity over any time interval differ from the instantaneous velocity at any instant? If so, given an example; if not, explain why.

8. Can the average velocity of a particle moving along the x axis ever be $\frac{1}{2}(v_0 + v)$ if the acceleration is not uniform? Prove your answer with graphs.

9. (a) Can an object have zero velocity and still be accelerating? (b) Can an object have a constant velocity and still have a varying speed? In each case, give an example if your answer is yes; explain why if your answer is no.

10. Can the velocity of an object reverse direction when its acceleration is constant? If so, give an example; if not, explain why.

11. Can an object be increasing in speed as its acceleration decreases? If so, give an example; if not, explain why.

12. If a particle is accelerated from rest ($v_0 = 0$) at $x_0 = 0$, beginning at the time $t = 0$, Eq. 2-13 says that it is at position x at two different times, namely, $+\sqrt{2x/a}$ and $-\sqrt{2x/a}$. Does the negative root have meaning? If, instead, the particle had been moving before $t = 0$, does the negative root have more meaning?

13. What are some examples of falling objects for which it would be unreasonable to neglect the effects of air?

14. A person standing on the edge of a cliff at some height above the ground below throws one ball straight up with initial speed u and then throws another ball straight down with the same initial speed. Which ball, if either, has the larger speed when it hits the ground? Neglect air resistance.

15. Suppose that a hot-air balloonist drops an apple over the side while the balloon is accelerating upward at 4.0 m/s² during lift-off. What is the apple's acceleration once it is released? If the speed of the balloon is 2 m/s at the moment of release, what is the apple's speed just then?

16. On a planet where the value of g is one-half the value on Earth, an object is dropped from rest and falls to the ground. How is the time needed for it to reach the ground from rest related to the time required to fall the same distance on Earth?

17. A second ball is dropped down an elevator shaft 1 s after the first ball is dropped. (a) What happens to the distance between the balls as time goes on? (b) How does the ratio v_1/v_2 of the speed of the first ball to the speed of the second ball change as time goes on?

EXERCISES & PROBLEMS

In several of the problems that follow you are asked to graph position, velocity, and acceleration versus time. Usually a sketch will suffice, appropriately labeled and with straight and curved portions apparent. If you have a computer or programmable calculator, you might use it to produce the graph.

SECTION 2-3 AVERAGE VELOCITY AND AVERAGE SPEED

1E. Carl Lewis ran the 100-m dash in about 10 s, and Bill Rodgers ran the marathon (26 mi, 385 yd) in about 2 h 10 min. (a) What are their average speeds? (b) If Lewis could have maintained his sprint speed during a marathon, how long would he have taken to finish?

2E. During a hard sneeze, your eyes might shut for 0.50 s. If you are driving a car at 90 km/h, how far does it move during that time?

3E. On average, a blink lasts about 100 ms. How far does a MIG-25 "Foxbat" fighter travel during a pilot's blink if the plane's average velocity is 2110 mi/h?

4E. Boston Red Sox pitcher Roger Clemens could routinely throw a fastball at a horizontal speed of 160 km/h, as verified by radar gun. How long did the ball take to reach home plate 18.4 m away?

5E. Figure 2-15 is a plot of the age of ancient sediment, in millions of years, against the distance from a particular ocean ridge. Seafloor material extruded from this ridge moves away from it at approximately uniform speed. Find the average speed, in centimeters per year, at which this material recedes from the ridge.

FIGURE 2-15 Exercise 5.

6E. When the legal speed limit for the New York Thruway was increased from 55 mi/h (= 88.5 km/h) to 65 mi/h (= 105 km/h), how much time was saved by a motorist who drove the 435 mi (= 700 km) between the Buffalo entrance and the New York City exit at the legal speed limit?

7E. Using the tables in Appendix F and following the dimensions carefully, find the speed of light (= 3 × 10⁸ m/s) in miles per hour, feet per second, and light-years per year.

8E. An automobile travels on a straight road for 40 km at 30 km/h. It then continues in the same direction for another 40 km at 60 km/h. (a) What is the average velocity of the car during this 80-km trip? (Assume that it moves in the positive x direction.) (b) What is its average speed? (c) Graph x versus t and indicate how the average velocity is found on the graph.

9P. Compute your average velocity in the following two cases. (a) You walk 240 ft at a speed of 4.0 ft/s and then run 240 ft at a speed of 10 ft/s along a straight track. (b) You walk for 1.0 min at a speed of 4.0 ft/s and then run for 1.0 min at 10 ft/s along a straight track. (c) Graph x versus t for both cases and indicate how the average velocity is found on the graph.

10P. A car travels up a hill at a constant speed of 40 km/h and returns down the hill at a constant speed of 60 km/h. Calculate the average speed for the round trip.

11P. You drive on Interstate 10 from San Antonio to Houston, half the *time* at 35 mi/h (= 56 km/h) and the other half at 55 mi/h (= 89 km/h). On the way back you travel half the *distance* at 35 mi/h and the other half at 55 mi/h. What is your average speed (a) from San Antonio to Houston, (b) from Houston back to San Antonio, and (c) for the entire trip? (d) What is your average velocity for the entire trip? (e) Graph x versus t for part (a), assuming the motion is all in the positive x direction. Indicate how the average velocity can be found on the graph.

12P. The position of an object moving in a straight line is given by $x = 3t - 4t^2 + t^3$, where x is in meters and t in seconds. (a) What is the position of the object at t = 1, 2, 3, and 4 s? (b) What is the object's displacement between t = 0 and t = 4 s? (c) What is the average velocity for the time interval from t = 2 s to t = 4 s? (d) Graph x versus t for 0 ≤ t ≤ 4 s and indicate how the answer for (c) can be found on the graph.

13P. The position of a particle moving along the x axis is given in centimeters by $x = 9.75 + 1.50t^3$, where t is in seconds. Consider the time interval t = 2.00 s to t = 3.00 s and calculate (a) the average velocity; (b) the instantaneous velocity at t = 2.00 s; (c) the instantaneous velocity at t = 3.00 s; (d) the instantaneous velocity at t = 2.50 s; and (e) the instantaneous velocity when the particle is midway between its positions at t = 2.00 s and t = 3.00 s. (f) Graph x versus t and indicate your answers graphically.

14P. A high-performance jet plane, practicing radar avoidance maneuvers, is in horizontal flight 35 m above the level ground. Suddenly, the plane encounters terrain that slopes gently upward at 4.3°, an amount difficult to detect; see Fig. 2-16. How much time does the pilot have to make a correction if the plane is to avoid flying into the ground? The speed of the plane is 1300 km/h.

FIGURE 2-16 Problem 14.

15P. Two trains, each having a speed of 30 km/h, are headed at each other on the same straight track. A bird that can fly 60 km/h flies off the front of one train when they are 60 km apart and heads directly for the other train. On reaching the other train it flies directly back to the first train, and so forth. (We have no idea *why* a bird would behave in this way.) (a) How many trips can the bird make from one train to the other before they crash? (b) What is the total distance the bird travels?

SECTION 2-4 INSTANTANEOUS VELOCITY AND SPEED

16E. (a) If a particle's position is given by $x = 4 - 12t + 3t^2$ (where t is in seconds and x is in meters), what is its velocity at $t = 1$ s? (b) Is it moving toward increasing or decreasing x just then? (c) What is its speed just then? (d) Is the speed larger or smaller at later times? (Try answering the next two questions without further calculation.) (e) Is there ever an instant when the velocity is zero? (f) Is there a time after $t = 3$ s when the particle is moving leftward on the x axis?

17E. The graph in Fig. 2-17 pertains to an armadillo that scampers left (direction of decreasing x) and right along an x axis. (a) When, if ever, is the animal to the left of the origin on the axis? When, if ever, is its velocity (b) negative, (c) positive, or (d) zero?

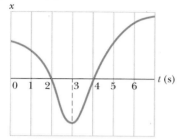

FIGURE 2-17 Exercise 17.

18E. Sketch a graph of x versus t for a mouse that is constrained in a narrow corridor (the x axis) and that scurries in the following sequence: (1) runs leftward (the direction of decreasing x) with a constant speed of 1.2 m/s, (2) gradually slows to 0.6 m/s toward the left, (3) gradually speeds up to 2.0 m/s toward the left, (4) gradually slows to a stop and then speeds up to 1.2 m/s toward the right. Where is the curve steepest? The least steep?

19P. How far does the runner whose velocity–time graph is shown in Fig. 2-18 travel in 16 s?

FIGURE 2-18 Problem 19.

SECTION 2-5 ACCELERATION

20E. A car accelerates at 9.2 km/h · s. What is its acceleration in m/s²?

21E. A particle had a velocity of 18 m/s and 2.4 s later its velocity was 30 m/s in the opposite direction. (a) What was the magnitude of the average acceleration of the particle during this 2.4-s interval? (b) Graph v versus t and indicate how the average velocity can be found on the graph.

22E. An object moves in a straight line as described by the velocity–time graph of Fig. 2-19. Sketch a graph that represents the acceleration of the object as a function of time.

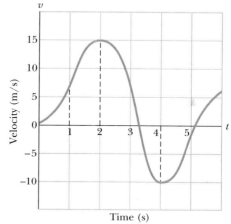

FIGURE 2-19 Exercise 22.

23E. The graph of x versus t in Fig. 2-20a is for a particle in straight-line motion. (a) State, for each of the intervals AB, BC, CD, and DE, whether the velocity v is positive, negative, or 0 and whether the acceleration a is positive, negative, or 0. (Ignore the end points of the intervals.) (b) From the curve, is there any interval over which the acceleration is obviously not constant? (c) If the axes are shifted upward together such that the time axis ends up running along the dashed line, do any of your answers change?

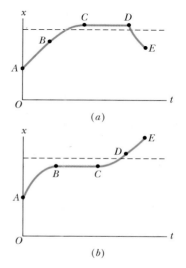

FIGURE 2-20 Exercises 23 and 24.

24E. Repeat Exercise 23 for the motion described by the graph of Fig. 2-20b.

25E. A particle moves along the x axis with $x(t)$ as shown in Fig. 2-21. Make rough sketches of velocity versus time and acceleration versus time for this motion.

26E. Sketch a graph that is a possible description of position as a function of time for a particle that moves along the x axis and, at $t = 1$ s, has (a) zero velocity and positive acceleration; (b) zero velocity and negative acceleration; (c) negative velocity and positive acceleration; (d) negative velocity and negative acceleration. (e) For which of these situations is the speed of the particle increasing at $t = 1$ s?

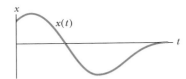

FIGURE 2-21 Exercise 25.

27E. Consider the two quantities $(dx/dt)^2$ and d^2x/dt^2. (a) Are these two equivalent expressions for the same thing? (b) What are the SI units of these two quantities?

28E. A particle moves along the x axis according to the equation $x = 50t + 10t^2$, where x is in meters and t is in seconds. Calculate (a) the average velocity of the particle during the first 3.0 s of its motion, (b) the instantaneous velocity of the particle at $t = 3.0$ s, and (c) the instantaneous acceleration of the particle at $t = 3.0$ s. (d) Graph x versus t and indicate how the answer to (a) can be obtained from the plot. (e) Indicate the answer to (b) on the graph. Plot v versus t and indicate on it the answer to (c).

29E. (a) If the position of a particle is given by $x = 20t - 5t^3$, where x is in meters and t is in seconds, when, if ever, is the particle's velocity zero? (b) When is its acceleration zero? (c) When is the acceleration negative? Positive? (d) Graph $x(t)$, $v(t)$, and $a(t)$.

30P. A man stands still from $t = 0$ to $t = 5.00$ min; from $t = 5.00$ min to $t = 10.0$ min he walks briskly in a straight line at a constant speed of 2.20 m/s. What are his average velocity and average acceleration during the time intervals (a) 2.00 min to 8.00 min and (b) 3.00 min to 9.00 min? (c) Sketch x versus t and v versus t, and indicate how the answers to (a) and (b) can be obtained from the graphs.

31P. If the position of an object is given by $x = 2.0t^3$, where x is measured in meters and t in seconds, find (a) the average velocity and the average acceleration between $t = 1.0$ s and $t = 2.0$ s and (b) the instantaneous velocities and the instantaneous accelerations at $t = 1.0$ s and $t = 2.0$ s. (c) Compare the average and instantaneous quantities and in each case explain why the larger one is larger. (d) Graph x versus t and v versus t, and indicate on the graphs your answers to (a) and (b).

32P. In an arcade video game, a spot is programmed to move across the screen according to $x = 9.00t - 0.750t^3$, where x is distance in centimeters measured from the left edge of the screen and t is time in seconds. When the spot reaches a screen edge, at either $x = 0$ or $x = 15.0$ cm, t is reset to 0 and the spot starts moving again according to $x(t)$. (a) At what time after starting is the spot instantaneously at rest? (b) Where does this occur? (c) What is its acceleration when this occurs? (d) In what direction is it moving just prior to coming to rest? (e) Just after? (f) When does it first reach an edge of the screen after $t = 0$?

33P. The position of a particle moving along the x axis depends on the time according to the equation $x = et^2 - bt^3$, where x is in feet and t in seconds. (a) What dimensions and units must e and b have? For the following, let their numerical values be 3.0 and 1.0, respectively. (b) At what time does the particle reach its maximum positive x position? (c) What distance does the particle cover in the first 4.0 s? (d) What is its displacement from $t = 0$ to $t = 4.0$ s? (e) What is its velocity at $t = 1.0$, 2.0, 3.0, and 4.0 s? (f) What is its acceleration at these times?

SECTION 2-6 CONSTANT ACCELERATION: A SPECIAL CASE

34E. The head of a rattlesnake can accelerate 50 m/s^2 in striking a victim. If a car could do as well, how long would it take for it to reach a speed of 100 km/h from rest?

35E. An object has a constant acceleration of $+3.2$ m/s^2. At a certain instant its velocity is $+9.6$ m/s. What is its velocity (a) 2.5 s earlier and (b) 2.5 s later?

36E. An automobile increases its speed uniformly from 25 to 55 km/h in 0.50 min. A bicycle rider uniformly speeds up to 30 km/h from rest in 0.50 min. Calculate their accelerations.

37E. Suppose a rocketship in deep space moves with constant acceleration equal to 9.8 m/s^2, which will give the illusion of normal gravity during the flight. (a) If it starts from rest, how long will it take to acquire a speed one-tenth that of light, which travels at 3.0×10^8 m/s? (b) How far will it travel in so doing?

38E. A jumbo jet must reach a speed of 360 km/h ($= 225$ mi/h) on the runway for takeoff. What is the least constant acceleration needed for takeoff from a 1.80-km runway?

39E. A muon (an elementary particle) enters an electric field with a speed of 5.00×10^6 m/s, whereupon the field slows it at the rate of 1.25×10^{14} m/s^2. (a) How far does the muon take to stop? (b) Graph x versus t and v versus t for the muon.

40E. An electron with initial velocity $v_0 = 1.50 \times 10^5$ m/s enters a region 1.0 cm long where it is electrically accelerated (Fig. 2-22). It emerges with a velocity $v = 5.70 \times 10^6$ m/s. What was its acceleration, assumed constant? (Such a process occurs in the electron gun in a cathode-ray tube, used in television receivers and oscilloscopes.)

Nonaccelerating   Accelerating
region   region

←1.0 cm→

Path of
electron

Source of
high voltage

FIGURE 2-22 Exercise 40.

41E. A car can be braked to a stop from 60 mi/h in 43 m. (a) What is the magnitude of the acceleration (really a deceleration) in SI units and "g units"? (Assume that a is constant.) (b) What is the stopping time? If your reaction time T for braking is 400 ms, to how many "reaction times" does the stopping time correspond?

42E. A world's land speed record was set by Colonel John P. Stapp when, on March 19, 1954, he rode a rocket-propelled sled that moved down a track at 1020 km/h. He and the sled were brought to a stop in 1.4 s. See Fig. 2-8. What acceleration did he experience? Express your answer in terms of the free-fall acceleration g.

43E. On a dry road a car with good tires may be able to brake with a deceleration of 4.92 m/s² (assume that it is constant). (a) How long does such a car, initially traveling at 24.6 m/s, take to come to rest? (b) How far does it travel in this time? (c) Graph x versus t and v versus t for the deceleration.

44E. A rocket-driven sled running on a straight level track is used to investigate the physiological effects of large accelerations on humans. One such sled can attain a speed of 1600 km/h in 1.8 s starting from rest. (a) Assume the acceleration is constant and find it in g units. (b) What is the distance traveled in this time?

45E. The brakes on your automobile are capable of creating a deceleration of 17 ft/s². (a) If you are going 85 mi/h and suddenly see a state trooper, what is the minimum time in which you can get your car under the 55-mi/h speed limit? (The answer reveals the futility of braking to keep your high speed from being detected with a radar gun.) (b) Graph x versus t and v versus t for such a deceleration.

46P. A certain drag racer can accelerate from 0 to 60 km/h in 5.4 s. (a) What is its average acceleration, in m/s², during this time? (b) How far will it travel during the 5.4 s, assuming its acceleration to be constant? (c) How much time would be required to go a distance of 0.25 km if the acceleration could be maintained at the same value?

47P. A train started from rest and moved with constant acceleration. At one time it was traveling 30 m/s, and

160 m farther on it was traveling 50 m/s. Calculate (a) the acceleration, (b) the time required to travel the 160 m mentioned, (c) the time required to attain the speed of 30 m/s, and (d) the distance moved from rest to the time the train had a speed of 30 m/s. (e) Graph x versus t and v versus t for the train, from rest.

48P. An automobile traveling 56.0 km/h is 24.0 m from a barrier when the driver slams on the brakes. The car hits the barrier 2.00 s later. (a) What was the automobile's constant deceleration before impact? (b) How fast was the car traveling at impact?

49P. A car moving with constant acceleration covers the distance between two points 60.0 m apart in 6.00 s. Its speed as it passes the second point is 15.0 m/s. (a) What is the speed at the first point? (b) What is the acceleration? (c) At what prior distance from the first point was the car at rest? (d) Graph x versus t and v versus t for the car, from rest.

50P. Two subway stops are separated by 1100 m. If the subway train accelerates at +1.2 m/s² from rest through the first half of the distance, and then decelerates at −1.2 m/s² through the second half, what are (a) its travel time and (b) its maximum speed? (c) Graph x, v, and a versus t for the trip.

51P. To stop a car, you require first a certain reaction time to begin braking; then the car slows under the constant braking deceleration. Suppose that the total distance moved by your car during these two phases is 186 ft when its initial speed is 50 mi/h, and 80 ft when the initial speed is 30 mi/h. What are (a) your reaction time and (b) the magnitude of the deceleration?

52P. When a driver brings a car to a stop by braking as hard as possible, the stopping distance can be regarded as the sum of a "reaction distance," which is initial speed times the driver's reaction time, and a "braking distance," which is the distance covered during braking. The following table gives typical values:

INITIAL SPEED (m/s)	REACTION DISTANCE (m)	BRAKING DISTANCE (m)	STOPPING DISTANCE (m)
10	7.5	5.0	12.5
20	15	20	35
30	22.5	45	67.5

(a) What reaction time is the driver assumed to have? (b) What is the car's stopping distance if the initial speed is 25 m/s?

53P. (a) If the maximum acceleration that is tolerable for passengers in a subway train is 1.34 m/s², and subway stations are located 806 m apart, what is the maximum speed a subway train can attain between stations? (b) What is the travel time between stations? (c) If the subway train stops for a 20-s interval at each station, what is the maximum average speed of a subway train? (d) Graph x, v, and a versus t.

54P. At the instant the traffic light turns green, an automobile starts with a constant acceleration a of 2.2 m/s². At the same instant a truck, traveling with a constant speed of 9.5 m/s, overtakes and passes the automobile. (a) How far beyond the traffic light will the automobile overtake the truck? (b) How fast will the car be traveling at that instant?

55P. An elevator cab in the New York Marquis Marriott has a total run of 624 ft. Its maximum speed is 1000 ft/min. Its acceleration and deceleration both have a magnitude of 4.0 ft/s². (a) How far does it move while accelerating to full speed from rest? (b) How long does it take to make the nonstop run, starting and ending at rest?

56P. As a high-speed passenger train traveling at 100 mi/h rounds a bend, the engineer is shocked to see that a locomotive has improperly entered onto the track from a siding 0.42 mi ahead; see Fig. 2-23. The locomotive is moving at 18 mi/h. The engineer of the passenger train immediately applies the brakes. (a) What must be the magnitude of the resulting constant acceleration if a collision is to be just avoided? (b) Assume that the engineer is at $x = 0$ when, at $t = 0$, he first spots the locomotive. Sketch the $x(t)$ curves representing the locomotive and train for the situation in which a collision is just avoided.

0.42 mi

FIGURE 2-23 Problem 56.

Add a curve to represent what happens if the braking rate is insufficient to avoid a crash.

57P. Two trains, one traveling at 72 km/h and the other at 144 km/h, are headed toward one another along a straight level track. When they are 950 m apart, each engineer sees the other's train and applies the brakes. If the brakes decelerate each train at the rate of 1.0 m/s², determine whether there is a collision.

58P. Sketch the $v(t)$ graph that would be associated with the $a(t)$ graph shown in Fig. 2-24.

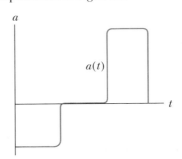

FIGURE 2-24 Problem 58.

SECTION 2-8 FREE-FALL ACCELERATION

59E. At a construction site a pipe wrench strikes the ground with a speed of 24 m/s. (a) From what height was it inadvertently dropped? (b) For how long was it falling? (c) Sketch graphs of y, v, and a versus t.

60E. (a) With what speed must a ball be thrown vertically up in order to rise to a maximum height of 50 m? (b) How long will it be in the air? (c) Sketch graphs of y, v, and a versus t. On the first two graphs, indicate the time at which 50 m is reached.

61E. Raindrops fall to Earth from a cloud 1700 m above the Earth's surface. If they were not slowed by air resistance, how fast would the drops be moving when they strike the ground? Would it be safe to walk outside during a rainstorm?

62E. The single cable supporting an unoccupied construction elevator breaks when the elevator is at rest at the top of a 120-m-high building. (a) With what speed does the elevator strike the ground? (b) For how long was it falling? (c) What was its speed when it passed the halfway point on the way down? (d) For how long had it been falling when it passed the halfway point?

63E. A hoodlum throws a stone vertically downward with an initial speed of 12 m/s from the roof of a building, 30.0 m above the ground. (a) How long does it take the stone to reach the ground? (b) What is the speed of the stone at impact?

64E. The Zero Gravity Research Facility at the NASA Lewis Research Center includes a 145-m drop tower. This is an evacuated vertical tower through which, among other possibilities, a 1-m-diameter sphere containing an experimental package can be dropped. (a) For how long is the sphere in free fall? (b) What is its speed just as it reaches the bottom of the tower? (c) When it hits the bottom of the tower, the sphere experiences an average deceleration of $25g$ as its speed is reduced to zero. Through what distance does its center travel during the deceleration?

65E. A model rocket, propelled by burning fuel, takes off vertically. Plot qualitatively (numbers not required) graphs of y, v, and a versus t for the rocket's flight. Indicate when the fuel is exhausted, when the rocket reaches maximum height, and when it returns to the ground.

66E. A rock is dropped from a 100-m-high cliff. How long does it take to fall (a) the first 50 m and (b) the second 50 m?

67P. A startled armadillo leaps upward (Fig. 2-25), rising 0.544 m in 0.200 s. (a) What was its initial speed? (b) What is its speed at this height? (c) How much higher does it go?

68P. A model rocket is fired vertically and ascends with a constant vertical acceleration of 4.00 m/s² for 6.00 s. Its fuel is then exhausted and it continues as a free-fall particle. (a) What is the maximum altitude reached? (b) What is the total time elapsed from takeoff until the rocket strikes the Earth?

69P. An object falls from a bridge that is 45 m above the water. It falls directly into a small boat moving with constant velocity that was 12 m from the point of impact when the object was released. What was the speed of the boat?

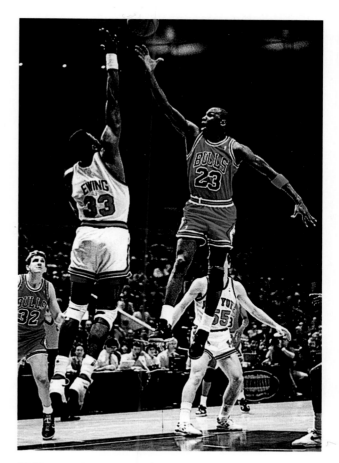

FIGURE 2-26 Problem 70.

FIGURE 2-25 Problem 67.

70P. A basketball player, standing near the basket to grab a rebound, jumps 76.0 cm vertically. How much time does the player spend (a) in the top 15.0 cm of this jump and (b) in the bottom 15.0 cm? Does this help explain why such players seem to hang in the air at the tops of their jumps? See Fig. 2-26.

71P. At the National Physical Laboratory in England, a measurement of the acceleration g was made by throwing a glass ball straight up in an evacuated tube and letting it return. Let ΔT_L in Fig. 2-27 be the time interval between the two passages of the ball across a lower level, ΔT_U the

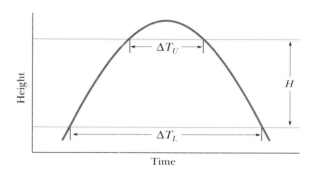

FIGURE 2-27 Problem 71.

41

time interval between the two passages across an upper level, and *H* the distance between the two levels. Show that

$$g = \frac{8H}{\Delta T_L^2 - \Delta T_U^2}.$$

72P. A ball of moist clay falls to the ground from a height of 15.0 m. It is in contact with the ground for 20.0 ms before coming to rest. What is the average acceleration of the clay during the time it is in contact with the ground? (Treat the ball as a particle.)

73P. A ball is thrown *down* vertically with an initial *speed* of v_0 from a height of *H*. (a) What will be its speed just before it strikes the ground? (b) How long will it take for the ball to reach the ground? (c) What would be the answers to (a) and (b) if the ball were thrown upward from the same height and with the same initial speed? Before solving any equations, decide if the answers here should be greater than, less than, or the same as in (a) and (b).

74P. Figure 2-28 shows a simple device for measuring your reaction time. It consists of a cardboard strip marked with a scale and two large dots. A friend holds the strip with his thumb and forefinger at the upper dot and you position your thumb and forefinger at the lower dot, being careful not to touch the strip. Your friend releases

FIGURE 2-28 Problem 74.

the strip, and you try to pinch it as soon as possible after you see it begin to fall. The mark at the place where you pinch the strip gives your reaction time. (a) How far from the lower dot should you place the 50.0-ms mark? (b) How much higher should the marks for 100, 150, 200, and 250 ms be? (For example, should the 100-ms marker be two times as far from the dot as the 50-ms marker? Can you find any pattern in the answers?)

75P. A juggler tosses balls vertically a certain distance into the air. How much higher must they be tossed if they are to spend twice as much time in the air?

76P. A stone is thrown vertically upward. On its way up it passes point *A* with speed *v*, and point *B*, 3.00 m higher than *A*, with speed $\frac{1}{2}v$. Calculate (a) the speed *v* and (b) the maximum height reached by the stone above point *B*.

77P. To test the quality of a tennis ball, you drop it onto the floor from a height of 4.00 m. It rebounds to a height of 3.00 m. If the ball was in contact with the floor for 10.0 ms, what was its average acceleration during contact?

78P. Water drips from the nozzle of a shower onto the floor 200 cm below. The drops fall at regular intervals of time, the first drop striking the floor at the instant the fourth drop begins to fall. Find the locations of the second and third drops when the first strikes the floor.

79P. A lead ball is dropped into a lake from a diving board 5.20 m above the water. It hits the water with a certain velocity and then sinks to the bottom with this same constant velocity. It reaches the bottom 4.80 s after it is dropped. (a) How deep is the lake? (b) What is the average velocity of the ball? (c) Suppose that all the water is drained from the lake. The ball is thrown from the diving board so that it again reaches the bottom in 4.80 s. What is the initial velocity of the ball?

80P. If an object travels half its total path in the last second of its fall from rest, find (a) the time and (b) the height of its fall. Explain the physically unacceptable solution of the quadratic time equation you obtain here.

81P. A woman fell 144 ft from the top of a building, landing on the top of a metal ventilator box, which crushed to a depth of 18 in. She survived without serious injury. What acceleration (assumed uniform) did she experience during the collision? Express your answer in terms of the free-fall acceleration *g*.

82P. A stone is dropped into a river from a bridge 144 ft above the water. Another stone is thrown vertically down 1.00 s after the first is dropped. Both stones strike the water at the same time. (a) What was the initial speed of the second stone? (b) Plot speed versus time on a graph for each stone, taking zero time as the instant the first stone was released.

83P. A parachutist bails out and freely falls 50 m. Then the parachute opens, and thereafter he decelerates at 2.0 m/s². He reaches the ground with a speed of 3.0 m/s. (a) How long is the parachutist in the air? (b) At what height did the fall begin?

84P. Two objects begin a free fall from rest from the same height 1.0 s apart. How long after the first object begins to fall will the two objects be 10 m apart?

85P. As Fig. 2-29 shows, Clara jumps from a bridge, followed closely by Jim. How long did Jim wait after Clara jumped? Assume that Jim is 170 cm tall and that the jumping-off level is at the top of the figure. Make scale measurements directly on the figure.

86P. A balloon is ascending at the rate of 12 m/s and is 80 m above the ground when a package is dropped. (a) How long does the package take to reach the ground? (b) With what speed does it hit the ground?

87P. An elevator without a ceiling is ascending with a constant speed of 10 m/s. A boy on the elevator throws a ball directly upward, from a height of 2.0 m above the elevator floor, just as the elevator floor is 28 m above the ground. The initial speed of the ball with respect to the elevator is 20 m/s. (a) What is the maximum height attained by the ball? (b) How long does it take for the ball to return to the elevator?

88P. A steel ball is dropped from a building's roof and passes a window, taking 0.125 s to fall from the top to the bottom of the window, a distance of 1.20 m. It then falls to a sidewalk and bounces "perfectly" back past the window, moving from bottom to top in 0.125 s. (The upward flight is a reverse of the fall.) The time spent below the bottom of the window is 2.00 s. How tall is the building?

FIGURE 2-29 Problem 85.

89P. A drowsy cat is startled by a flowerpot that sails first up and then down past an open window. The pot was in view for a total of 0.50 s, and the top-to-bottom height of the window is 2.00 m. How high above the top of the window did the flowerpot go?

ADDITIONAL PROBLEMS

90. An electric vehicle starts from rest and accelerates at a rate of 2.0 m/s² in a straight line until it reaches a speed of 20 m/s. The vehicle then slows at a constant rate of 1.0 m/s² until it stops. (a) How much time elapses from start to stop? (b) How far does the vehicle travel from start to stop?

91. A motorcycle is moving at 30 m/s when the rider applies the brakes, giving the motorcycle a constant deceleration. During the 3.0-s interval immediately after the brakes are applied, the speed decreases to 15 m/s. What total distance does the motorcycle travel from the instant the brakes are applied until it comes to rest?

92. The position of a particle as it moves along the x axis is given by $x = 15e^{-t}$ m, where t is in seconds. (a) What is the position of the particle at $t = 0$, 0.50, and 1.0 s? (b) What is the average velocity of the particle between $t = 0$ and $t = 1.0$ s? (c) What is the instantaneous velocity of the particle at $t = 0$, 0.50, and 1.0 s? (d) Plot x versus t for the particle for $0 \le t \le 1.0$ s, and estimate the instantaneous velocity at $t = 0.50$ s from the graph.

93. A ball is thrown vertically downward from the top of a 36.6-m-tall building. The ball passes the top of a window that is 12.2 m above the ground 2.00 s after being thrown. What is the speed of the ball as it passes the top of the window?

94. Figure 2-30 depicts the motion of a particle moving along the x axis with a constant acceleration. What is the value of the acceleration?

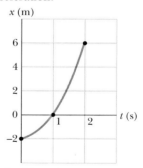

FIGURE 2-30 Problem 94.

95. A graph of x versus t for a particle in straight-line motion is shown in Fig. 2-31. (a) What is the average velocity of the particle between $t = 0.50$ s and $t = 4.5$ s? (b) What is the instantaneous velocity of the particle at $t = 4.5$ s? (c)

What is the average acceleration of the particle between $t = 0.50$ s and $t = 4.5$ s? (d) What is the instantaneous acceleration of the particle at $t = 4.5$ s?

FIGURE 2-31   Problem 95.

96. A rock is projected vertically upward from the edge of the top of a tall building. The rock reaches its maximum height above the top of the building 1.60 s after being launched. Then, after barely missing the edge of the building as it falls downward, the rock strikes the ground 6.00 s after it was launched. In SI units: (a) with what upward velocity was the rock initially projected, (b) what maximum height above the top of the building is achieved by the rock, (c) how tall is the building?

97. The position of a particle as it moves along the y axis is given by

$$y = 2.0 \sin\left(\frac{\pi}{4} t\right),$$

where t is in seconds and y is in centimeters. (a) What is the average velocity of the particle between $t = 0$ and $t = 2.0$ s? (b) What is the instantaneous velocity of the particle at $t = 0$, 1.0, and 2.0 s? (c) What is the average acceleration of the particle between $t = 0$ and $t = 2.0$ s? (d) What is the instantaneous acceleration of the particle at $t = 0$, 1.0, and 2.0 s? (e) Plot v versus t for $0 \leq t \leq 2.0$ s, and estimate the instantaneous acceleration at $t = 1.0$ s from the graph.

98. The speed of a bullet is measured to be 640 m/s as the bullet emerges from a 1.2-m-long barrel. Assuming constant acceleration, find the time that the bullet spends in the barrel after it is fired.

99. A particle moving along the x axis has a position given by $x = 16te^{-t}$ m, where t is in seconds. How far is the particle from the origin when it momentarily stops? (Do not consider its stop at $t = \infty$.)

RUSH-HOUR TRAFFIC FLOW

Jearl Walker
Cleveland State University

ESSAY 1

Traffic lights in a small town usually require no special sequencing. The flow of traffic through them may be haphazard, but the queues at red lights are seldom long. In contrast, traffic flow in a large city, especially during rush hours, requires careful regulation. Without it, the lines of cars lengthen until many intersections are blocked, throwing the area into gridlock. Since only the cars on the perimeter of the congested area can then move, hours may be needed to free the cars trapped in its interior.

Suppose you had to engineer the traffic light system for a one-way street that consists of several lanes along which rush-hour traffic flows. The lights are to be green for 50 s, yellow for 5 s, and red for 25 s (these values are typically employed for a heavily traveled route through a city). To promote the traffic flow, you might be tempted to increase the duration of the green light or decrease the duration of the red light. However, you must remember that the traffic on the perpendicular streets cannot be held in check for too long, or lengthy lines will develop there.

How should you time the onset of green lights at the various intersections? If you arrange for all the lights to turn green simultaneously, then the traffic can move for only 50 s. Upon each onset of the green lights, platoons of cars will move along the route until all the lights on the street turn red simultaneously. To maximize the distance traveled, drivers might race through the system. Large platoons of cars traveling at, say, 55 mi/h along a crowded city street would resemble a "grand prix," presenting an obvious danger.

A better and safer design is one in which the onset of the green lights is staggered so that the green at any particular intersection does not appear until the platoon leaders near the intersection. (The green should appear before they actually get there, or they will slow out of caution to avoid entering the intersection on a red.) Racing through the system is then futile; eventually the speeding driver is stopped by a light that has not yet changed from red to green.

Figure 1 shows part of the street that is to be controlled. Suppose that the leaders of a platoon of cars have just reached intersection 2, where the green appeared when they were a distance d from the intersection. They continue to travel at a certain speed

Jearl Walker is professor of physics at Cleveland State University. He received a B.S. in physics from M.I.T. and a Ph.D. in physics from the University of Maryland. From 1977 to 1990 he conducted "The Amateur Scientist" department of <u>Scientific American</u>. His book <u>The Flying Circus of Physics with Answers</u> is published in 10 languages.

A platoon of cars, almost in gridlock, works its way through a system of traffic lights.

FIGURE 1 The one-way street whose traffic flow is to be controlled.

v_p (the speed limit) to intersection 3, where the green appears when they are a distance d from it. The intersections are separated by distance D_{23}.

Question 1
What should be the time delay of the onset of green at intersection 3 relative to that at intersection 2 to keep the platoon moving smoothly? (In this question and the others here, always answer in terms of the given symbols.)

The situation (and the answer) changes if the platoon had been stopped by a red light at the previous intersection. For example, in Fig. 1 a platoon is stopped at intersection 1. When the green comes on there, the leaders require a certain time t_r to respond to the change and an additional time to accelerate at some rate a to the cruising speed v_p. During the acceleration, the leaders do move a certain distance, but less than if they had been moving at v_p.

Question 2
If the separation between intersections 1 and 2 is D_{12} and the green at intersection 2 is to appear when the leaders are a distance d from that intersection, how long after the light at intersection 1 turns green should the light at intersection 2 turn green?

With a system of delayed lights, the traffic can still freeze into place. The problem lies in the fact that once a platoon of drivers is stopped and then given another green light, they cannot all accelerate simultaneously. Instead, a "start-up wave" travels from the leader along the length of

the platoon at a speed of v_s. Each driver begins to react only when the start-up wave reaches that driver. Drivers behind the leaders also have a longer distance to travel to the next intersection.

Question 3
Suppose that a driver is a distance d_1 behind a platoon leader stopped at intersection 1, and that the duration of the green light at intersection 2 is t_{gr}. If the green at intersection 2 is to turn off when the driver is a distance d from that intersection (allowing the driver to just pass through on the yellow), then what should be the delay of the onset of green light between the intersections?

All these points are illustrated in Fig. 2, which shows the street plan on the left and a graph of the progress of the platoon (with the traffic light cycles) on the right. A length d_1 of a platoon, which was initially stopped at intersection 1, travels through the entire light system. The initial periods of acceleration are represented by curved lines, with cars farther back in the platoon accelerating later. The light at each intersection turns green a few seconds before the leaders reach it.

The figure also indicates that not all of the platoon gets through intersection 1 before the light there turns red again. If such failure is repeated for several cycles of the light, the length of the "abandoned" section grows, perhaps back to a preceding intersection where it will block cross traffic. Once that happens, gridlock sets in.

Question 4
What, on the graph, represents (a) v_p and (b) v_s? (c) What is the duration of the acceleration period?

A gridlock can develop even if the traffic light system is well designed. I was once trapped in a gridlock when a sudden, heavy snowfall abruptly caught the afternoon rush-hour traffic of Cleveland. Since the street I was on was slippery, the platoon leaders proceeded cautiously. The start-up waves also slowed. Within about 20 min, abandoned sections of platoons stretched back to previously crossed intersections, blocking them. For 2 mi along my route and along five parallel arteries out of the city, traffic came to almost a standstill. I made progress only because cars on the outward end of the route gradually escaped into the suburbs. As they left the pack, a start-up wave moved sedately through the 2-mi platoon, allowing me to creep forward by a few car lengths at a time. The problem worsened as the snow deepened and stalled cars blocked lanes. Although the drive normally took only 5 min, my escape from the gridlock that miserable day required over 2 h.

Answers to Questions

1. $t = D_{23}/v_p$.
2. $t = t_r + v_p/2a + (D_{12} - d)/v_p$.
3. $t = t_r + v_p/2a + d_1/v_s - t_{gr} + (D_{12} - d + d_1)/v_p$.
4. (a) Slope of straight section of $x(t)$ for moving drivers. (b) Slope of the $x(t)$ for start-up wave. (c) v_p/a.

FIGURE 2 Graphical representation of the progress of a platoon of cars initially stopped at intersection 1. The color bars show the actual durations of traffic lights.

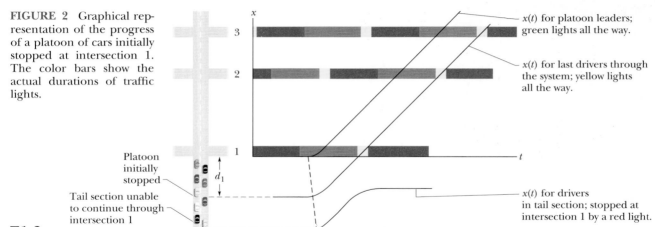

Platoon initially stopped

Tail section unable to continue through intersection 1

$x(t)$ for platoon leaders; green lights all the way.

$x(t)$ for last drivers through the system; yellow lights all the way.

$x(t)$ for drivers in tail section; stopped at intersection 1 by a red light.

VECTORS | 3

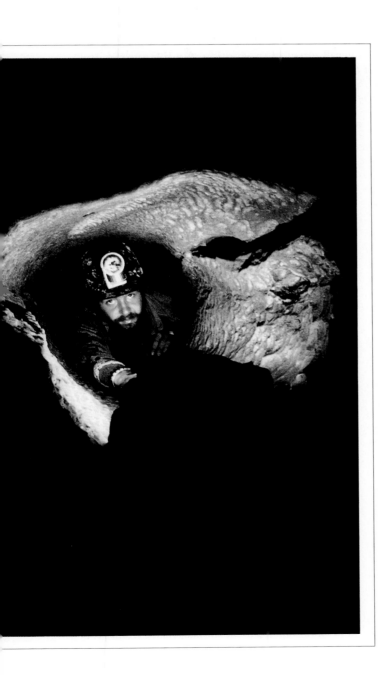

For two decades spelunking teams crawled, climbed, and squirmed through 200 km of Mammoth Cave and the Flint Ridge cave system, seeking a connection. The photograph shows Richard Zopf pushing his pack through the Tight Tube, far inside the Flint Ridge system. After 12 hours of "caving" along a labyrinthine route, Zopf and six others waded through a stretch of chilling water and found themselves in Mammoth Cave. Their breakthrough established the Mammoth–Flint cave system as being the longest cave in the world. How can their final point be related to their initial point other than in terms of the actual route they covered?

3-1 VECTORS AND SCALARS

A particle confined to a straight line can move in only two directions. We can take its motion to be positive in one of these directions and negative in the other. For a particle moving in three dimensions, however, a plus or minus sign is no longer enough to indicate the direction of the motion. Instead, we need an arrow to point out the direction. Such an arrow is a **vector.**

A vector has magnitude as well as direction and follows certain rules of combination, which we examine below. A **vector quantity** is a quantity that can be represented by a vector; that is, it is a quantity that has associated with it a magnitude and direction. Some physical quantities that can be represented by vectors are displacement, velocity, acceleration, force, and magnetic field.

Not all physical quantities involve a direction. Some examples are temperature, pressure, energy, mass, and time, none of which "points" in the spatial sense. We call such quantities **scalars** and we deal with them by the rules of ordinary algebra.

The simplest of all vectors is the **displacement vector,** a displacement being a change of position. If a particle changes its position by moving from A to B in Fig. 3-1a, we say that it undergoes a displacement from A to B, which we represent with an arrow, the symbol for a vector, pointing from A to B. To distinguish vector symbols from other kinds of arrows, we use the outline of a triangle as the arrowhead.

The arrows from A to B, from A' to B', and from A'' to B'' in Fig. 3-1a represent the same *change of position* for the particle and we make no distinction among them. All three arrows have the same magnitude and direction and hence are identical displacement vectors.

The displacement vector tells us nothing about the actual path that the particle takes. In Fig. 3-1b, for example, all three paths connecting points A and B correspond to the same displacement vector, that of Fig. 3-1a. Displacement vectors represent only the overall effect of the motion, not the motion itself.

SAMPLE PROBLEM 3-1

The 1972 team that connected the Mammoth–Flint cave system went from the Austin Entrance in the Flint Ridge system to Echo River in Mammoth Cave (see Fig. 3-2a), traveling a net 2.6 km westward, 3.9 km southward, and 25 m upward. What was their displacement vector?

SOLUTION   We first take an overhead view (Fig. 3-2b) to find their horizontal displacement d_h. The magnitude of d_h is given by the Pythagorean theorem:

$$d_h = \sqrt{(2.6 \text{ km})^2 + (3.9 \text{ km})^2} = 4.69 \text{ km}.$$

The angle θ relative to the west is given by

$$\tan \theta = \frac{3.9 \text{ km}}{2.6 \text{ km}} = 1.5,$$

or

$$\theta = \tan^{-1} 1.5 = 56°.$$

We next take a side view, Fig. 3-2c, to find the magnitude of the overall displacement d,

$$d = \sqrt{(4.69 \text{ km})^2 + (0.025 \text{ km})^2} = 4.69 \text{ km} \approx 4.7 \text{ km},$$

and the angle ϕ,

$$\phi = \tan^{-1} \frac{0.025 \text{ km}}{4.69 \text{ km}} = 0.3°.$$

So the team moved 4.7 km at an angle of 56° south of west and at an angle of 0.3° upward. The net vertical motion was, of course, insignificant compared to the horizontal motion, but that fact would have been no comfort to the team as they climbed up and down countless times. The route they actually covered was quite different from the displacement vector, which merely points from start to finish.

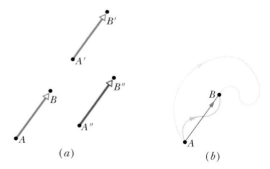

FIGURE 3-1   (a) All three arrows represent the same displacement. (b) All three paths connecting the two points correspond to the same displacement vector.

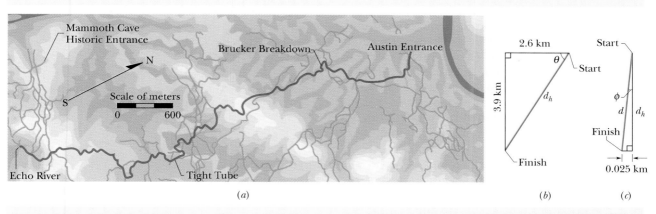

(a)

(b)   (c)

FIGURE 3-2 Sample Problem 3-1. (*a*) Part of the Mammoth–Flint cave system, with the spelunking team's route from the Austin Entrance to Echo River indicated. (*b*) An overhead view of the team's displacement.

(*c*) A side view. (Adapted from map by Cave Research Foundation.)

3-2 ADDING VECTORS: GRAPHICAL METHOD

Suppose that, as in Fig. 3-3*a*, a particle moves from *A* to *B* and then later from *B* to *C*. We can represent its overall displacement (no matter what its actual path) with two successive displacement vectors, *AB* and *BC*. The net effect of these two displacements is a single displacement from *A* to *C*. We call *AC* the

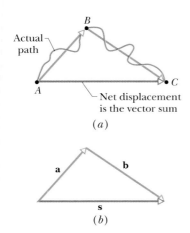

(*a*)

(*b*)

FIGURE 3-3 (*a*) *AC* is the vector sum of the vectors *AB* and *BC*. (*b*) The same vectors relabeled.

vector sum of the vectors *AB* and *BC*. This sum is not the usual algebraic sum, and we need more than simple numbers to specify it.

In Fig. 3-3*b*, we redraw the vectors of Fig. 3-3*a* and relabel them in the way that we shall use from now on, namely, with boldface symbols such as **a**, **b**, or **s**. In handwriting, you can place an arrow over the symbol, as in $\vec{a}$. If we want to indicate only the magnitude of the vector (a quantity that is always positive), we shall use the italic symbol, such as *a*, *b*, or *s*. (You can use just a handwritten symbol.) The boldface symbol always implies both properties of the vector, magnitude and direction.

We can represent the relation among the three vectors in Fig. 3-3*b* by

$$\mathbf{s} = \mathbf{a} + \mathbf{b}, \qquad (3\text{-}1)$$

in which we say that the vector **s** is the **vector sum** of vectors **a** and **b**. The procedure for adding vectors in this way (that is, graphically) is as follows: (1) On a sheet of paper, lay out the vector **a** to some convenient scale and at the proper angle. (2) Lay out vector **b** to the same scale, with its tail at the head of vector **a**, again at the proper angle. (3) Construct the vector sum **s** by drawing an arrow from the tail of **a** to the head of **b**. You can easily generalize this procedure to add more than two vectors.

Since vectors are new quantities, we must expect some new mathematical procedures. The symbol "+" in Eq. 3-1 and the words "add" and "sum" have different meanings than they have in arithmetic or in the usual algebra. They tell us to carry out quite a different operation, one that takes into account the *directions* associated with vectors as well as their magnitudes.

Vector addition, defined in this way, has two important properties. First, the order of addition does not matter. That is,

$$\mathbf{a} + \mathbf{b} = \mathbf{b} + \mathbf{a} \quad \text{(commutative law).} \quad (3\text{-}2)$$

Figure 3-4 should convince you that this is the case.

Second, if there are more than two vectors, it does not matter how we group them as we add them. Thus if we want to add vectors **a**, **b**, and **c**, we can add **a** and **b** first and then add their vector sum to **c**. On the other hand, we can add **b** and **c** first, and then add *that* sum to **a**. We get the same result either way. In equation form,

$$(\mathbf{a} + \mathbf{b}) + \mathbf{c} = \mathbf{a} + (\mathbf{b} + \mathbf{c}) \quad \begin{array}{l}\text{(associative}\\ \text{law).}\end{array} \quad (3\text{-}3)$$

A little quiet contemplation of Fig 3-5 will convince you that Eq. 3-3 is correct.

The vector $-\mathbf{b}$ is a vector with the same magnitude as **b** but the opposite direction (see Fig. 3-6). If you try to add the two vectors in Fig. 3-6, you will see that

$$\mathbf{b} + (-\mathbf{b}) = 0.$$

Adding $-\mathbf{b}$ has the effect of subtracting **b**! We use this property to define the difference between two vectors: Let $\mathbf{d} = \mathbf{a} - \mathbf{b}$. Then

$$\mathbf{d} = \mathbf{a} - \mathbf{b} = \mathbf{a} + (-\mathbf{b}) \quad \text{(subtraction).} \quad (3\text{-}4)$$

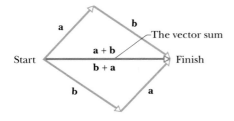

FIGURE 3-4 The two vectors **a** and **b** can be added in either order; see Eq. 3-2.

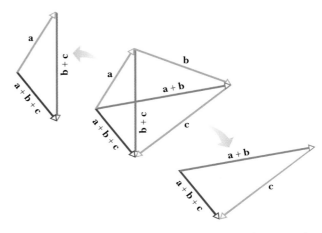

FIGURE 3-5 The three vectors **a**, **b**, and **c** can be grouped in any way as they are added; see Eq. 3-3.

FIGURE 3-6 The vectors **b** and $-\mathbf{b}$.

That is, we find the difference vector **d** by adding the vector $-\mathbf{b}$ to the vector **a**. Figure 3-7 shows how this is done graphically.

Remember, although we have used displacement vectors as a prototype, the rules for addition and subtraction hold for vectors of all kinds, whether they represent forces, velocities, or anything

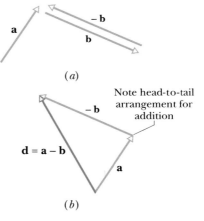

FIGURE 3-7 (*a*) Vectors **a**, **b**, and $-\mathbf{b}$. (*b*) Subtracting vector **b** from vector **a** is accomplished by adding vector $-\mathbf{b}$ to vector **a**.

else. However, as in ordinary arithmetic, it is still true that we can add only vectors of the same kind. We can add two displacements, for example, or two velocities, but it makes no sense to add a displacement and a velocity. In the world of scalars, that would be like trying to add 21 s and 12 m.

3-3 VECTORS AND THEIR COMPONENTS

Adding vectors graphically can be tedious. A neater and easier technique involves algebra but requires that the vectors be placed on a rectangular coordinate system. The x and y axes are usually drawn in the plane of the page, as in Fig. 3-8a. The z axis, which we ignore for now, comes directly out of the page at the origin.

Vector **a** in Fig. 3-8 is in the xy plane. If we drop perpendicular lines from the ends of **a** to the coordinate axes, the quantities a_x and a_y so formed are called the **components** of the vector **a** along the x and y directions. The process of forming them is called **resolving the vector.** In general, a vector will have three components, although for the case of Fig. 3-8a the component along the z axis happens to be

zero. As Fig. 3-8b shows, if you move a vector in such a way that it remains parallel to its original direction at all times, the values of its components remain unchanged. The directions of the components are consistent with that of the vector.

We can easily find the components of **a** in Fig. 3-8a from the right triangle there:

$$a_x = a \cos \theta \quad \text{and} \quad a_y = a \sin \theta, \quad (3\text{-}5)$$

where θ is the angle that the vector **a** makes with the direction of increasing x. Figure 3-8c shows that the vector and its x and y components form a right triangle. Depending on the value of θ, the components of a vector may be positive, negative, or zero. In a figure, we use smaller, filled-in triangles as arrowheads to indicate the signs of the components, according to the usual convention: positive in the direction of increasing coordinate values, and negative in the opposite direction. Figure 3-9 shows a vector **b** for which b_y is negative and b_x is positive.

Once a vector has been resolved into its components, the components themselves can be used in place of the vector. Instead of the two numbers a and θ, we can specify the vector with two other numbers, a_x and a_y. Both sets of numbers contain exactly the same information, and we can pass back and forth readily between the two descriptions. To obtain a and θ if we are given a_x and a_y, we note (see Fig. 3-8a) that

$$a = \sqrt{a_x^2 + a_y^2} \quad \text{and} \quad \tan \theta = \frac{a_y}{a_x}. \quad (3\text{-}6)$$

In solving problems, you may use *either* the a_x, a_y notation *or* the a, θ notation.

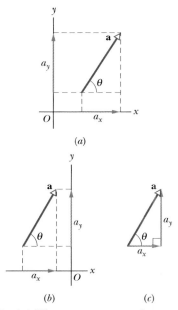

(a)

(b)   (c)

FIGURE 3-8 (a) The components of vector **a**. (b) The components are unchanged if the vector is shifted, as long as the magnitude and orientation are maintained. (c) The components form the legs of a right triangle whose hypotenuse is the magnitude of the vector.

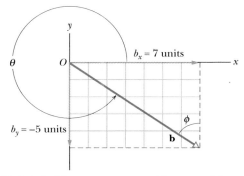

FIGURE 3-9 The components of **b** are positive on the x axis and negative on the y axis.

SAMPLE PROBLEM 3-2

A small airplane leaves an airport on an overcast day and is later sighted 215 km away, in a direction making an angle of 22° east of north. How far east and north is the airplane from the airport when sighted?

SOLUTION On an xy coordinate system the situation is as shown in Fig. 3-10, where for convenience the origin of the system has been placed at the airport. The airplane's displacement vector **d** points from the origin to where the airplane is sighted.

To answer the question, we find the components of **d**. With Eq. 3-5 and the angle $\theta = 68°$ ($= 90° - 22°$) we have

$$d_x = d \cos \theta = (215 \text{ km})(\cos 68°)$$

$$= 81 \text{ km} \qquad \text{(Answer)}$$

and

$$d_y = d \sin \theta = (215 \text{ km})(\sin 68°)$$

$$= 199 \text{ km}. \qquad \text{(Answer)}$$

So the airplane was spotted 199 km north and 81 km east of the airport.

PROBLEM SOLVING

~~~\M/~~~

TACTIC 1: ANGLES—DEGREES AND RADIANS

Angles that are measured relative to the positive direction of the x axis are positive if they are measured in the counterclockwise direction, and negative if clock-

wise. For example, 210° and −150° are the same angle. Most calculators (try yours) will accept angles in either form when a trig function is taken.

Angles may be measured in degrees or radians (rad). You can relate the two measures by remembering that one full circle is equivalent to 360° and to 2π rad. So if you needed to convert, say, 40° to radians, you would write

$$40° \frac{2\pi \text{ rad}}{360°} = 0.70 \text{ rad}.$$

Is the answer reasonable? Quickly check by realizing that 40° is $\frac{1}{9}$ of a full circle, and with a full circle equivalent to 2π rad or about 6.3 rad, the angle should be $\frac{1}{9}$ of 6.3, which it is. Or, check by remembering that 1 rad ≈ 57°.

Most calculators are in the degree mode when they are turned on, so that angles must be entered in degrees. However, you may be able to change yours to the radian mode. Check it out before your first exam.

TACTIC 2: TRIG FUNCTIONS

You need to know the definitions of the common trig functions—sine, cosine, and tangent—because they are part of the language of science and engineering. They are given in Fig. 3-11 in a form that does not depend on how the triangle is labeled.

You should also be able to sketch how the trig functions vary with angle, as in Fig. 3-12, in order to be able to judge whether a calculator result is reasonable. Even knowing the signs of the functions in the various quadrants can be of help.

TACTIC 3: INVERSE TRIG FUNCTIONS

The most important of the inverse trig functions are $\sin^{-1}$, $\cos^{-1}$, and $\tan^{-1}$. When they are taken on a calculator, you must consider the reasonableness of the answer it gives, because there is usually another possible answer that it does not give. The range of operation for a calculator in taking each inverse trig function is indicated in Fig. 3-12. As an example, $\sin^{-1}(0.5)$ has associated angles of 30° (which is displayed by the cal-

FIGURE 3-10 Sample Problem 3-2. A plane takes off from an airport at the origin and is later sighted at P.

$$\sin \theta = \frac{\text{leg opposite } \theta}{\text{hypotenuse}}$$

$$\cos \theta = \frac{\text{leg adjacent to } \theta}{\text{hypotenuse}}$$

$$\tan \theta = \frac{\text{leg opposite } \theta}{\text{leg adjacent to } \theta}$$

FIGURE 3-11 A triangle used to define the trigonometric functions. See Appendix G.

culator since 30° falls within its range of operation) and 150°. To see both angles, draw a horizontal line through 0.5 in Fig. 3-12a and note where it cuts the sine curve.

How do you distinguish a correct answer? As an example, reconsider the calculation of θ in Sample Problem 3-1, where $\tan \theta = 1.5$. Taking $\tan^{-1} 1.5$ on your calculator tells you that $\theta = 56°$, but $\theta = 236°$ ($= 180° + 56°$) also has a tangent of 1.5. Which is correct? From the physical situation (Fig. 3-2b), 56° is reasonable and 236° is clearly not.

TACTIC 4: MEASURING VECTOR ANGLES
Equation 3-5 and the second portion of Eq. 3-6 are valid only if the angle is measured relative to the positive direction of the x axis. If it is measured relative to some other direction, then the trig functions in Eq. 3-5 may have to be interchanged, and the ratio in Eq. 3-6 may have to be inverted. A safer method is to convert the given angle into one that is measured from the positive direction of the x axis, as shown in Sample Problem 3-2.

3-4 UNIT VECTORS

A **unit vector** is a vector that has a magnitude of exactly 1 and that points in a particular direction. It lacks both dimension and unit. Its sole purpose is to point, that is, to specify a direction. The unit vectors are in the positive directions of the x, y, and z axes and are labeled **i**, **j**, and **k**, as shown in Fig. 3-13.* The arrangement of axes in Fig. 3-13 is said to be a **right-handed coordinate system.** We use such coordinate systems exclusively in this text.

Unit vectors are very useful for expressing other vectors; for example, we can express **a** and **b** of Figs. 3-8 and 3-9 as

$$\mathbf{a} = a_x \mathbf{i} + a_y \mathbf{j} \tag{3-7}$$

and

$$\mathbf{b} = b_x \mathbf{i} + b_y \mathbf{j}, \tag{3-8}$$

(a)

(b)

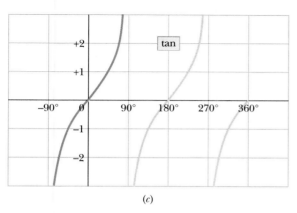

(c)

FIGURE 3-12 Three useful curves to remember. A calculator's range of operation for taking *inverse* trig functions is indicated by the darker portions of the colored curves.

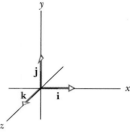

FIGURE 3-13 Unit vectors **i**, **j**, and **k** define the directions of a right-handed rectangular coordinate system. The system remains right-handed if it is rotated rigidly to a new orientation.

---

*To distinguish handwritten unit vectors from other symbols, including other vectors, you might top them with a "hat": $\hat{\imath}, \hat{\jmath}, \hat{k}$.

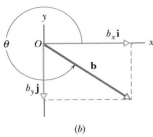

(a)

(b)

FIGURE 3-14 (a) The vector components of vector **a**. (b) The vector components of vector **b**.

which are redrawn in Fig. 3-14. The quantities $a_x\mathbf{i}$ and $a_y\mathbf{j}$ are the **vector components** of **a**, in contrast to a_x and a_y, which are its **scalar components** (or, as before, simply its **components**).

Look back for a moment at the displacement vector of Sample Problem 3-1. If you superimpose the coordinate system of Fig. 3-13 at the Austin Entrance of Fig. 3-2a, with **i** eastward, **j** northward, and **k** upward, then the displacement **d** to Echo River is neatly expressed as

$$\mathbf{d} = -(2.6 \text{ km})\mathbf{i} - (3.9 \text{ km})\mathbf{j} + (0.025 \text{ km})\mathbf{k}.$$

3-5 ADDING VECTORS BY COMPONENTS

Adding vectors with pencil, ruler, and protractor is tedious, has limited accuracy, and is challenging in three dimensions. In this section, we find a more direct technique—adding vectors by combining their components, axis by axis.

To start, consider the statement

$$\mathbf{r} = \mathbf{a} + \mathbf{b}, \tag{3-9}$$

which says that the vector **r** is the same as the vector (**a** + **b**). If that is so, then each component of **r** must be the same as the corresponding component of (**a** + **b**):

$$r_x = a_x + b_x, \tag{3-10}$$

$$r_y = a_y + b_y, \tag{3-11}$$

$$r_z = a_z + b_z. \tag{3-12}$$

In other words, two vectors are equal only if their corresponding components are equal. Equations 3-10 to 3-12 tell us that to add vectors **a** and **b**, we must (1) resolve the vectors into their components; (2) axis by axis, combine these components to get the components of the sum **r**; and (3) if necessary, combine the components of **r** to get **r** itself. (We have a choice here. We can express **r** in unit-vector notation, or we can give the magnitude of **r** and its orientation, using Eq. 3-6 in two dimensions or the method of Sample Problem 3-1 for three dimensions.)

SAMPLE PROBLEM 3-3

In a road rally, you are given the following instructions: from the starting point use available roads to drive 36 km due east to checkpoint "Able," then 45 km due north to "Baker," and then 25 km northwest to "Charlie." (The roadway and checkpoints are shown in Fig. 3-15.) At "Charlie," what are the magnitude and orientation of your displacement **d** from the starting point?

SOLUTION Figure 3-15 also shows a convenient orientation for an xy coordinate system, as well as vectors representing the three displacements you have undergone. The scalar components of **d** are

$$d_x = a_x + b_x + c_x = 36 \text{ km} + 0 + (25 \text{ km})(\cos 135°)$$

$$= (36 + 0 - 17.7) \text{ km} = 18.3 \text{ km}$$

and

$$d_y = a_y + b_y + c_y = 0 + 45 \text{ km} + (25 \text{ km})(\sin 135°)$$

$$= (0 + 45 + 17.7) \text{ km} = 62.7 \text{ km}.$$

Next, we use Eq. 3-6 to find the magnitude and direction of **d**:

$$d = \sqrt{d_x^2 + d_y^2} = \sqrt{(18.3 \text{ km})^2 + (62.7 \text{ km})^2}$$

$$= 65 \text{ km} \tag{Answer}$$

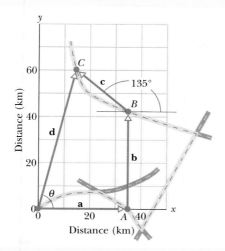

FIGURE 3-15 Sample Problem 3-3. A rally route, showing the origin, checkpoints Able, Baker, and Charlie, and the road network.

and

$$\theta = \tan^{-1} \frac{d_y}{d_x} = \tan^{-1} \frac{62.7 \text{ km}}{18.3 \text{ km}} = 74°, \quad \text{(Answer)}$$

where θ is the angle shown in the figure.

SAMPLE PROBLEM 3-4

Here are three vectors, each expressed in unit-vector notation:

$$\mathbf{a} = 4.2\mathbf{i} - 1.6\mathbf{j},$$

$$\mathbf{b} = -1.6\mathbf{i} + 2.9\mathbf{j},$$

$$\mathbf{c} = -3.7\mathbf{j}.$$

All three lie in the xy plane, and none of them has a z component. Find the vector $\mathbf{r}$ that is the sum of these three vectors. For convenience, the units have been omitted from these vector expressions; you may take them to be meters.

SOLUTION   From Eqs. 3-10 and 3-11 we have

$$r_x = a_x + b_x + c_x = 4.2 - 1.6 + 0 = 2.6$$

and

$$r_y = a_y + b_y + c_y = -1.6 + 2.9 - 3.7 = -2.4.$$

(a)

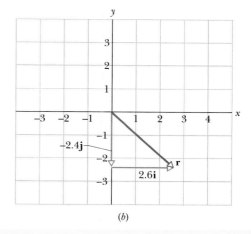

(b)

FIGURE 3-16   Sample Problem 3-4. Vector $\mathbf{r}$ is the vector sum of the other three vectors.

Thus

$$\mathbf{r} = 2.6\mathbf{i} - 2.4\mathbf{j}. \quad \text{(Answer)}$$

Figure 3-16a shows the three vectors and their sum. Figure 3-16b shows $\mathbf{r}$ and its vector components.

3-6 VECTORS AND THE LAWS OF PHYSICS

So far, in every illustration with a coordinate system, the x and y axes are parallel to the edges of the page. And so, when a vector $\mathbf{a}$ is included, its components a_x and a_y are also parallel to the edges (as in Fig.

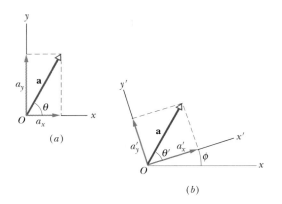

FIGURE 3-17 (*a*) The vector **a** and its components. (*b*) The same vector, with the axes of the coordinate system rotated through an angle ϕ.

3-17*a*). The only reason for that orientation of the axes is that it looks proper: there is no deeper reason. We could, instead, rotate the axes (but not the vector **a**) through an angle ϕ as in Fig. 3-17*b*, in which case the components would have new values, call them a'_x and a'_y. Since there are an infinite number of choices of ϕ, there are an infinite number of different pairs of component values for **a**.

Which then is the "right" pair of components? The answer is that they are all equally valid because each pair (with its axes) just gives us a different way of describing the same vector **a**; all produce the same magnitude and direction for the same vector. In Fig. 3-17 we have

$$a = \sqrt{a_x^2 + a_y^2} = \sqrt{a_x'^2 + a_y'^2} \qquad (3\text{-}13)$$

and

$$\theta = \theta' + \phi. \qquad (3\text{-}14)$$

The point is that we have great freedom in choosing a coordinate system, because the relations among vectors (including, for example, the vector addition of Eq. 3-1) do not depend on the location of the origin of the coordinate system or on the orientation of the axes. This is also true of the relations of physics; they are all independent of the choice of coordinate system. Add to that the simplicity and richness of the language of vectors and you can see why the laws of physics are almost always presented in that language: One equation, like Eq. 3-9, can represent three (or even more) relations, like Eqs. 3-10, 3-11, and 3-12.

3-7 MULTIPLYING VECTORS*

There are three ways in which vectors can be multiplied. None of them is exactly like the usual algebraic multiplication.

Multiplying a Vector by a Scalar

If we multiply a vector **a** by a scalar s, we get a new vector. Its magnitude is the product of the magnitude of **a** and the absolute value of s. Its direction is the direction of **a** if s is positive, but the opposite direction if s is negative. To divide **a** by s, we multiply **a** by $1/s$.

In either multiplication or division, the scalar may be a pure number or a physical quantity; in the latter case, the physical nature of the product differs from that of the original vector **a**.

A Glimpse Ahead

As an example, consider an equation from Chapter 5:

$$\mathbf{F} = m\mathbf{a},$$

where **a** is an acceleration vector, m is a mass (which is a positive scalar), and **F** is a force vector. Although you may not understand the significance of the equation, you should realize two things. First, since m is positive, **F** and **a** have the same direction. Second, since m is a physical quantity (in Chapter 1 its units are expressed as kilograms), **F** differs physically from the original vector **a**.

The Scalar Product

There are two ways to multiply a vector by a vector: one way produces a scalar and the other produces a new vector. Students commonly confuse the two ways, and so starting now, you should carefully distinguish between them.

The **scalar product** of the vectors **a** and **b** in Fig. 3-18*a* is written as $\mathbf{a} \cdot \mathbf{b}$ and defined to be

$$\mathbf{a} \cdot \mathbf{b} = ab \cos \phi, \qquad (3\text{-}15)$$

---

*This material will not be employed until later (Chapter 7 for scalar products and Chapter 12 for vector products), and so your instructor may wish to postpone assignment of the section.

(a)

Component of **b**
along direction of
a is $b \cos \phi$

Component of **a**
along direction of
b is $a \cos \phi$

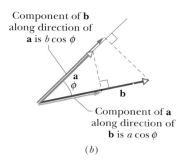

(b)

FIGURE 3-18 (a) Two vectors **a** and **b**, with an angle ϕ between them. (b) Each vector has a component along the direction of the other vector.

where a is the magnitude of **a**, b is the magnitude of **b**, and ϕ is the angle* between **a** and **b**. Note that there are only scalars on the right side (including the value of $\cos \phi$); thus the multiplication on the left side results in the *scalar* product. Also, because of the notation, the **a·b** is known as a **dot product** and is spoken as "a dot b."

A dot product can be regarded as the product of two quantities: (1) the magnitude of one of the vectors and (2) the scalar component of the second vector along the direction of the first vector. For example, in Fig. 3-18b, **a** has a scalar component $a \cos \phi$ along the direction of **b**; note that a perpendicular dropped from the head of **a** to **b** determines that component. Similarly, **b** has a scalar component $b \cos \phi$ along the direction of **a**. If ϕ is 0°, the component of one vector along the other is maximum, and so also is the dot product. If, instead, ϕ is 90°, the component of one vector along the other is zero, and so is the dot product.

Equation 3-15 can be rewritten as follows to emphasize the components and show that the order of

*In Fig. 3-18a, there are actually two angles between the vectors: ϕ and 360° − ϕ. Either can be used in Eq. 3-15, because their cosines are the same.

the multiplication is irrelevant:

$$\mathbf{a \cdot b} = \mathbf{b \cdot a} = (a \cos \phi)(b)$$
$$= (a)(b \cos \phi). \qquad (3\text{-}16)$$

In other words, the commutative law applies to a scalar product. When two vectors are in unit-vector notation, we write their dot product as

$$\mathbf{a \cdot b} = (a_x\mathbf{i} + a_y\mathbf{j} + a_z\mathbf{k}) \cdot (b_x\mathbf{i} + b_y\mathbf{j} + b_z\mathbf{k}), \qquad (3\text{-}17)$$

which obeys the **distributive law,** as demonstrated in Sample Problem 3-5.

A Glimpse Ahead

As one example of a scalar product, we select the definition of the work W (a scalar) that is done by a force **F** as its point of application moves through a displacement **d**. If ϕ is the angle between the vectors **F** and **d**, the work W is defined as

$$W = \mathbf{F \cdot d} = Fd \cos \phi.$$

We explore this definition more in Chapter 7.

SAMPLE PROBLEM 3-5

What is the angle ϕ between $\mathbf{a} = 3.0\mathbf{i} - 4.0\mathbf{j}$ and $\mathbf{b} = -2.0\mathbf{i} + 3.0\mathbf{k}$?

SOLUTION With Eq. 3-15, the dot product is

$$\mathbf{a \cdot b} = ab \cos \phi = \sqrt{3.0^2 + 4.0^2}\,\sqrt{2.0^2 + 3.0^2}\,\cos \phi$$
$$= 18.0 \cos \phi. \qquad (3\text{-}18)$$

We next find the dot product with Eq. 3-17:

$$\mathbf{a \cdot b} = (3.0\mathbf{i} - 4.0\mathbf{j}) \cdot (-2.0\mathbf{i} + 3.0\mathbf{k}).$$

With the distributive law, this yields

$$\mathbf{a \cdot b} = (3.0\mathbf{i}) \cdot (-2.0\mathbf{i}) + (3.0\mathbf{i}) \cdot (3.0\mathbf{k})$$
$$+ (-4.0\mathbf{j}) \cdot (-2.0\mathbf{i}) + (-4.0\mathbf{j}) \cdot (3.0\mathbf{k}).$$

We next apply Eq. 3-15 to each term. The angle for the first term is 0° and the angle for the other three terms is 90°. We then have

$$\mathbf{a \cdot b} = -(6.0)(1) + (9.0)(0) + (8.0)(0) - (12)(0)$$
$$= -6.0. \qquad (3\text{-}19)$$

Setting the results of Eqs. 3-18 and 3-19 equal to each other, we find

$$18.0 \cos \phi = -6.0,$$

or

$$\phi = \cos^{-1} \frac{-6.0}{18.0} = 109° \approx 110°. \quad \text{(Answer)}$$

The Vector Product

The **vector product** of **a** and **b**, written **a** × **b**, produces a third vector **c** whose magnitude is

$$c = ab \sin \phi, \quad (3\text{-}20)$$

where ϕ is the *smaller* of the two angles between **a** and **b**.* Because of the notation, **a** × **b** is also known as the **cross product** and in speech it is "a cross b." If **a** and **b** are parallel or antiparallel, **a** × **b** = 0. The magnitude of **a** × **b** is maximum when **a** and **b** are perpendicular to each other.

The direction of **c** is perpendicular to the plane that contains **a** and **b**. Figure 3-19a shows how to determine the direction of **c** with what is known as the **right-hand rule.** Place the vectors **a** and **b** tail to tail without altering their orientations and imagine a line that is perpendicular to their plane where they meet. Pretend to grasp the line with your *right* hand in such a way that your fingers would sweep **a** into **b** through the smaller angle between them. Your outstretched thumb points in the direction of **c**.

The order of the vector multiplication is important. In Fig. 3-19b, we are determining the direction of **c**' = **b** × **a**, so the fingers are placed to sweep **b** into **a** through the smaller angle. The thumb ends up in the opposite direction from previously, and so it must be that **c**' = −**c**; that is,

$$\mathbf{b} \times \mathbf{a} = - \mathbf{a} \times \mathbf{b}. \quad (3\text{-}21)$$

In other words, the commutative law does not apply to a vector product.

In unit-vector notation, a cross product is written

$$\mathbf{a} \times \mathbf{b} = (a_x \mathbf{i} + a_y \mathbf{j} + a_z \mathbf{k})$$
$$\times (b_x \mathbf{i} + b_y \mathbf{j} + b_z \mathbf{k}), \quad (3\text{-}22)$$

---

*In this case, you must use the smaller of the two angles between the vectors because $\sin \phi$ and $\sin(360° - \phi)$ differ in algebraic sign.

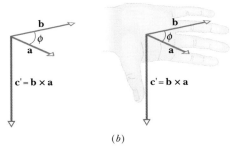

FIGURE 3-19 Illustration of the right-hand rule for vector products. (*a*) Swing vector **a** into vector **b** with the fingers of your right hand. Your thumb shows the direction of vector **c** = **a** × **b**. (*b*) Showing that (**a** × **b**) = − (**b** × **a**).

for which the distributive law applies, as demonstrated in Sample Problem 3-7.

A Glimpse Ahead

We first run into the vector product in Chapter 12 when we discuss a force **F** whose point of application is a distance **r** from a certain origin. The torque $\boldsymbol{\tau}$ (a turning effect) that this force exerts about the origin is defined to be

$$\boldsymbol{\tau} = \mathbf{r} \times \mathbf{F}.$$

SAMPLE PROBLEM 3-6

Vector **a** lies in the *xy* plane in Fig. 3-20. It has a magnitude of 18 units and points in a direction 250° from the direction of increasing *x*. Vector **b** has a magnitude of 12 units and points along the direction of increasing *z*.

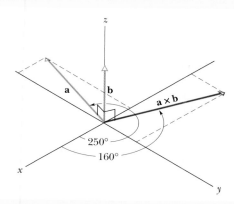

FIGURE 3-20 Sample Problem 3-6. A problem in vector multiplication.

a. What is the scalar product of these two vectors?

SOLUTION The angle ϕ between these two vectors is 90° so that, from Eq. 3-15,

$$\mathbf{a} \cdot \mathbf{b} = ab \cos \phi = (18)(12)(\cos 90°) = 0. \quad \text{(Answer)}$$

The scalar product of any two vectors that are at right angles to each other is zero. This is consistent with the fact that neither of these vectors has a component in the direction of the other vector.

b. What is the vector product $\mathbf{c}$ of vectors $\mathbf{a}$ and $\mathbf{b}$?

SOLUTION The magnitude of the vector product is, from Eq. 3-20,

$$ab \sin \phi = (18)(12)(\sin 90°) = 216. \quad \text{(Answer)}$$

The direction of $\mathbf{c}$ is at right angles to the plane formed by $\mathbf{a}$ and $\mathbf{b}$. It must then be at right angles to $\mathbf{b}$, which means that it is at right angles to the z axis. This in turn means that $\mathbf{c}$ must lie in the xy plane. The right-hand rule of Fig. 3-19 shows that $\mathbf{c}$ points as shown in Fig. 3-20. Because $\mathbf{c}$ is also perpendicular to $\mathbf{a}$, the direction of $\mathbf{c}$ makes an angle of 250° − 90° = 160° with the direction of increasing x.

SAMPLE PROBLEM 3-7

If $\mathbf{c} = \mathbf{a} \times \mathbf{b}$, where $\mathbf{a} = 3\mathbf{i} - 4\mathbf{j}$ and $\mathbf{b} = -2\mathbf{i} + 3\mathbf{k}$, what is $\mathbf{c}$?

SOLUTION From Eq. 3-22 we have

$$\mathbf{c} = (3\mathbf{i} - 4\mathbf{j}) \times (-2\mathbf{i} + 3\mathbf{k}),$$

which, with the distributive law, becomes

$$\mathbf{c} = -(3\mathbf{i} \times 2\mathbf{i}) + (3\mathbf{i} \times 3\mathbf{k}) + (4\mathbf{j} \times 2\mathbf{i}) - (4\mathbf{j} \times 3\mathbf{k}).$$

We next evaluate each term with Eq. 3-20, determining the direction with the right-hand rule. We find

$$\mathbf{c} = 0 - 9\mathbf{j} - 8\mathbf{k} - 12\mathbf{i} = -12\mathbf{i} - 9\mathbf{j} - 8\mathbf{k}. \quad \text{(Answer)}$$

The vector $\mathbf{c}$ is perpendicular to both $\mathbf{a}$ and $\mathbf{b}$, a fact that you can check by showing $\mathbf{c} \cdot \mathbf{a} = 0$ and $\mathbf{c} \cdot \mathbf{b} = 0$; that is, there is no component of $\mathbf{c}$ along the direction of either $\mathbf{a}$ or $\mathbf{b}$.

PROBLEM SOLVING
〰〰〰

TACTIC 5: COMMON ERRORS WITH CROSS PRODUCTS

Several errors are common in finding a cross product. (1) Failure to arrange them tail to tail is tempting when an illustration presents the vectors head to tail: you must mentally shift (or better, redraw) one of them to the proper arrangement without changing that vector's orientation. (2) Failing to use the right hand in applying the right-hand rule is easy when the right hand is occupied with a calculator or pencil. (3) Failure to sweep the first vector of the product into the second vector can occur when the orientations of the vectors require an awkward twisting of your hand to apply the right-hand rule. And sometimes it happens when you try to make the sweep mentally rather than actually using your hand. (4) Failure to work with a right-handed coordinate system results when you forget how to draw such a system (see Fig. 3-13).

REVIEW & SUMMARY

Scalars and Vectors

Scalars, such as temperature, have magnitude only. They are specified by a number with a unit (82°F) and obey the rules of arithmetic and ordinary algebra. *Vectors,* such as displacement, have both magnitude and direction (5 m, north) and obey the special rules of vector algebra.

Adding Vectors Geometrically

Two vectors **a** and **b** may be added geometrically by drawing them to a common scale and placing them head to tail. The vector connecting the tail of the first to the head of the second is the sum vector **s**, as Fig. 3-3 shows. To subtract **b** from **a**, reverse the direction of **b** to get $-\mathbf{b}$; then add $-\mathbf{b}$ to **a**: see Fig. 3-7. Vector addition and subtraction are commutative and obey the associative law.

Components of a Vector

The *components* a_x and a_y of any vector **a** are found by dropping perpendicular lines from the ends of **a** onto the coordinate axes, as Fig. 3-8 and Sample Problem 3-2 show. The components are given by

$$a_x = a \cos \theta \quad \text{and} \quad a_y = a \sin \theta, \quad (3\text{-}5)$$

where θ is measured from the positive direction of the x axis. The algebraic sign of a component indicates its direction along the associated axis. Given the components, we can reconstruct the vector from

$$a = \sqrt{a_x^2 + a_y^2} \quad \text{and} \quad \tan \theta = \frac{a_y}{a_x}, \quad (3\text{-}6)$$

where again θ is measured from the positive direction of the x axis.

Unit-Vector Notation

Often we find it useful to introduce *unit vectors* **i**, **j**, and **k**, whose magnitudes are unity and whose directions are the x, y, and z axes, respectively, of a right-handed coordinate system, such as shown in Fig. 3-13. We can write a vector **a** in terms of unit vectors as

$$\mathbf{a} = a_x \mathbf{i} + a_y \mathbf{j} + a_z \mathbf{k}, \quad (3\text{-}7)$$

in which $a_x \mathbf{i}$, $a_y \mathbf{j}$, and $a_z \mathbf{k}$ are the **vector components** and a_x, a_y, and a_z are the **scalar components** of **a**. Sample Problem 3-4 shows how to add vectors using unit vectors.

Adding Vectors in Component Form

To add vectors in component form, we use the rules

$$r_x = a_x + b_x; \quad r_y = a_y + b_y; \quad r_z = a_z + b_z. \quad (3\text{-}10 \text{ to } 3\text{-}12)$$

See Sample Problem 3-3.

Vectors and Physical Laws

Any physical situation involving vectors can be described using many possible coordinate systems. We usually choose the one that simplifies the work. However, the relation between the vector quantities does not depend on our choice. The laws of physics too are independent of our choice of coordinates.

Scalar Times Vector

The product of a scalar s and a vector **v** is a new vector whose magnitude is sv and whose direction is the same as that of **v** if s is positive, and opposite that of **v** if s is negative. To divide **v** by s, multiply **v** by $(1/s)$.

The Scalar Product

The scalar (or **dot**) product of two vectors is written as $\mathbf{a} \cdot \mathbf{b}$ and is the *scalar* quantity given by

$$\mathbf{a} \cdot \mathbf{b} = ab \cos \phi, \quad (3\text{-}15)$$

in which ϕ is the angle between the directions of **a** and **b**; see Fig. 3-18a. The scalar product may be positive, zero, or negative, depending on the value of ϕ. Figure 3-18b indicates that a scalar product is the product of the magnitude of one vector and the component of the second vector along the direction of the first vector. In unit-vector notation, we have

$$\mathbf{a} \cdot \mathbf{b} = (a_x \mathbf{i} + a_y \mathbf{j} + a_z \mathbf{k}) \cdot (b_x \mathbf{i} + b_y \mathbf{j} + b_z \mathbf{k}), \quad (3\text{-}17)$$

which obeys the distributive law, as demonstrated in Sample Problem 3-5. Note that $\mathbf{a} \cdot \mathbf{b} = \mathbf{b} \cdot \mathbf{a}$.

The Vector Product

The vector (or **cross**) product is written as $\mathbf{a} \times \mathbf{b}$ and is a *vector* **c** whose magnitude c is given by

$$c = ab \sin \phi, \quad (3\text{-}20)$$

in which ϕ is the smaller of the angles between the directions of **a** and **b**. The direction of **c** is at right angles to the plane defined by **a** and **b** and is given by a right-hand rule described in Fig. 3-19. Note that $\mathbf{a} \times \mathbf{b} = -\mathbf{b} \times \mathbf{a}$. In unit-vector notation we have

$$\mathbf{a} \times \mathbf{b} = (a_x \mathbf{i} + a_y \mathbf{j} + a_z \mathbf{k}) \times (b_x \mathbf{i} + b_y \mathbf{j} + b_z \mathbf{k}), \quad (3\text{-}22)$$

for which the distributive law applies. Sample Problems 3-6 and 3-7 illustrate vector products.

QUESTIONS

1. In 1969, three Apollo astronauts left Cape Canaveral, went to the moon and back, and splashed down in the Pacific Ocean. An admiral bid them good-bye at the Cape and then sailed to the Pacific Ocean in an aircraft carrier where he picked them up. Compare the displacements of the astronauts and the admiral.

2. Can two vectors having different magnitudes be combined to give a zero resultant? Can three vectors?

3. Can a vector have zero magnitude if one of its components is not zero?

4. Can the sum of the magnitudes of two vectors ever be equal to the magnitude of the sum of these two vectors?

5. Can the magnitude of the difference between two vectors ever be greater than the magnitude of either vector? Can it be greater than the magnitude of their sum? Give examples.

6. If three vectors add up to zero, must they all be in the same plane?

7. Explain in what sense a vector equation contains more information than a scalar equation.

8. Why do the unit vectors **i**, **j**, and **k** have no units?

9. Name several scalar quantities. Does the value of a scalar quantity depend on the coordinate system you choose?

10. You can order events in time. For example, event b may precede event c but follow event a, giving us a time order of events a, b, c. Hence there is a sense of time, distinguishing past, present, and future. Is time a vector therefore? If not, why not?

11. Do the commutative and associative laws apply to vector subtraction?

12. Can a scalar product be a negative quantity?

13. (a) If $\mathbf{a} \cdot \mathbf{b} = 0$, does it follow that **a** and **b** are perpendicular to one another? (b) If $\mathbf{a} \cdot \mathbf{b} = \mathbf{a} \cdot \mathbf{c}$, does it follow that **b** equals **c**?

14. If $\mathbf{a} \times \mathbf{b} = 0$, must **a** and **b** be parallel to each other? Is the converse true?

15. Must you specify a coordinate system when you (a) add two vectors, (b) form their scalar product, (c) form their vector product, or (d) find their components?

EXERCISES & PROBLEMS

SECTION 3-2 ADDING VECTORS: GRAPHICAL METHOD

1E. Consider two displacements, one of magnitude 3 m and another of magnitude 4 m. Show how the displacement vectors may be combined to get a resultant displacement of magnitude (a) 7 m, (b) 1 m, and (c) 5 m.

2E. What are the properties of two vectors **a** and **b** such that

    (a) $\mathbf{a} + \mathbf{b} = \mathbf{c}$ and $a + b = c$;

    (b) $\mathbf{a} + \mathbf{b} = \mathbf{a} - \mathbf{b}$;

    (c) $\mathbf{a} + \mathbf{b} = \mathbf{c}$ and $a^2 + b^2 = c^2$?

3E. A woman walks 250 m in the direction 30° east of north, then 175 m directly east. (a) Using graphical methods, find her final displacement from the starting point. (b) Compare the magnitude of her displacement with the distance she walked.

4E. A person walks in the following pattern: 3.1 km north, then 2.4 km west, and finally 5.2 km south. (a) Construct the vector diagram that represents this motion.

(b) How far and in what direction would a bird fly in a straight line to arrive at the same final point?

5E. A car is driven east for a distance of 50 km, then north for 30 km, and then in a direction 30° east of north for 25 km. Draw the vector diagram and determine the total displacement of the car from its starting point.

6P. Vector **a** has a magnitude of 5.0 units and is directed east. Vector **b** is directed 35° west of north and has a magnitude of 4.0 units. Construct vector diagrams for calculating $\mathbf{a} + \mathbf{b}$ and $\mathbf{b} - \mathbf{a}$. Estimate the magnitudes and directions of $\mathbf{a} + \mathbf{b}$ and $\mathbf{b} - \mathbf{a}$ from your diagrams.

7P. Three vectors **a**, **b**, and **c**, each having a magnitude of 50 units, lie in the xy plane and make angles of 30°, 195°, and 315° with the positive x axis, respectively. Find graphically the magnitudes and directions of the vectors (a) $\mathbf{a} + \mathbf{b} + \mathbf{c}$, (b) $\mathbf{a} - \mathbf{b} + \mathbf{c}$, and (c) a vector **d** such that $(\mathbf{a} + \mathbf{b}) - (\mathbf{c} + \mathbf{d}) = 0$.

8P. A bank in downtown Boston is robbed (see the map in Fig. 3-21). To elude police, the thieves escape by helicopter, making three successive flights described by the following displacements: 20 miles, 45° south of east; 33 miles, 26° north of west; 16 miles, 18° east of south. At the

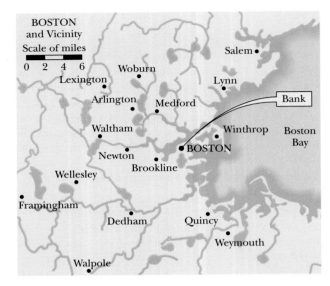

FIGURE 3-21   Problem 8.

end of the third flight they are captured. In what town are they apprehended? (Use the geometric method to add these displacements on the map.)

SECTION 3-3   VECTORS AND THEIR COMPONENTS

9E.   What are the x and y components of a vector **a** in the xy plane if its direction is 250° counterclockwise from the positive x axis and its magnitude is 7.3 units?

10E.   The x component of a certain vector is -25.0 units and the y component is $+40.0$ units. (a) What is the magnitude of the vector? (b) What is the angle between the direction of the vector and the positive x axis?

11E.   A displacement vector **r** in the xy plane is 15 m long and directed as shown in Fig. 3-22. Determine the x and y components of the vector.

FIGURE 3-22   Exercise 11.

12E.   A heavy piece of machinery is raised by sliding it 12.5 m along a plank oriented at 20.0° to the horizontal, as

FIGURE 3-23   Exercise 12.

shown in Fig. 3-23. (a) How high above its original position is it raised? (b) How far is it moved horizontally?

13E.   The minute hand of a wall clock measures 10 cm from axis to tip. What is the displacement vector of its tip (a) from a quarter after the hour to half past, (b) in the next half hour, and (c) in the next hour?

14E.   A ship sets out to sail to a point 120 km due north. An unexpected storm blows the ship to a point 100 km due east of its starting point. How far, and in what direction, must it now sail to reach its original destination?

15P.   A person desires to reach a point that is 3.40 km from her present location and in a direction that is 35.0° north of east. However, she must travel along streets that are oriented either north–south or east–west. What is the minimum distance she could travel to reach her destination?

16P.   Rock *faults* are ruptures along which opposite faces of rock have slid past each other. In Fig. 3-24, points A and B coincided before the rock in the foreground slid down to the right. The net displacement AB is along the plane of the fault. The horizontal component of AB is the *strike-slip* AC. The component of AB that is directly down the plane of the fault is the *dip-slip* AD. (a) What is the net displacement AB if the strike-slip is 22.0 m and the dip-slip is 17.0 m? (b) If the plane of the fault is inclined 52.0° to the horizontal, what is the vertical component of AB?

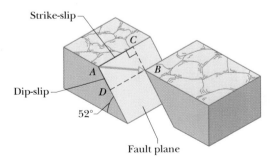

FIGURE 3-24   Problem 16.

17P.   A wheel with a radius of 45.0 cm rolls without slipping along a horizontal floor, as shown in Fig. 3-25. P is a dot painted on the rim of the wheel. At time t_1, P is at the point of contact between the wheel and the floor. At a

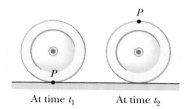

At time t_1 At time t_2

FIGURE 3-25 Problem 17.

later time t_2, the wheel has rolled through one-half of a revolution. What is the displacement of P during this interval?

18P. Two places A and B on the surface of the Earth in South America differ by 1° in latitude and 1° in longitude. Show that the magnitude of the displacement vector from A to B is approximately $d(1 + \cos^2 \lambda)^{1/2}$, where λ is the latitude of A and $d = 111$ km.

19P. A room has dimensions 10.0 ft × 12.0 ft × 14.0 ft. A fly starting at one corner flies around, ending up at the diagonally opposite corner. (a) What is the magnitude of its displacement? (b) Could the length of its path be less than this distance? Greater than this distance? Equal to this distance? (c) Choose a suitable coordinate system and find the components of the displacement vector in that system. (d) If the fly walks rather than flies, what is the length of the shortest path it can take?

SECTION 3-5 ADDING VECTORS BY COMPONENTS

20E. (a) Express the following angles in radians: 20.0°, 50.0°, 100°. (b) Convert the following angles to degrees: 0.330 rad, 2.10 rad, 7.70 rad.

21E. Find the vector components of the sum $\mathbf{r}$ of the vector displacements $\mathbf{c}$ and $\mathbf{d}$ whose components in meters along three perpendicular directions are $c_x = 7.4$, $c_y = -3.8$, $c_z = -6.1$; $d_x = 4.4$, $d_y = -2.0$, $d_z = 3.3$.

22E. (a) What is the sum in unit-vector notation of the two vectors $\mathbf{a} = 4.0\mathbf{i} + 3.0\mathbf{j}$ and $\mathbf{b} = -13\mathbf{i} + 7.0\mathbf{j}$? (b) What are the magnitude and direction of $\mathbf{a} + \mathbf{b}$?

23E. Calculate the x and y components, magnitudes, and directions of (a) $\mathbf{a} + \mathbf{b}$ and (b) $\mathbf{b} - \mathbf{a}$ if $\mathbf{a} = 3.0\mathbf{i} + 4.0\mathbf{j}$ and $\mathbf{b} = 5.0\mathbf{i} - 2.0\mathbf{j}$.

24E. Two vectors are given by $\mathbf{a} = 4\mathbf{i} - 3\mathbf{j} + \mathbf{k}$ and $\mathbf{b} = -\mathbf{i} + \mathbf{j} + 4\mathbf{k}$. Find (a) $\mathbf{a} + \mathbf{b}$, (b) $\mathbf{a} - \mathbf{b}$, and (c) a vector $\mathbf{c}$ such that $\mathbf{a} - \mathbf{b} + \mathbf{c} = 0$.

25E. Given two vectors $\mathbf{a} = 4.0\mathbf{i} - 3.0\mathbf{j}$ and $\mathbf{b} = 6.0\mathbf{i} + 8.0\mathbf{j}$, find the magnitudes and directions of (a) $\mathbf{a}$, (b) $\mathbf{b}$, (c) $\mathbf{a} + \mathbf{b}$, (d) $\mathbf{b} - \mathbf{a}$, and (e) $\mathbf{a} - \mathbf{b}$. How do the orientations of the last two compare?

26P. If $\mathbf{a} - \mathbf{b} = 2\mathbf{c}$, $\mathbf{a} + \mathbf{b} = 4\mathbf{c}$, and $\mathbf{c} = 3\mathbf{i} + 4\mathbf{j}$, then what are $\mathbf{a}$ and $\mathbf{b}$?

27P. Two vectors $\mathbf{a}$ and $\mathbf{b}$ have equal magnitudes of 10.0 units. They are oriented as shown in Fig. 3-26 and their vector sum is $\mathbf{r}$. Find (a) the x and y components of $\mathbf{r}$, (b) the magnitude of $\mathbf{r}$, and (c) the angle $\mathbf{r}$ makes with the positive x axis.

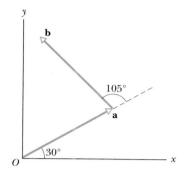

FIGURE 3-26 Problem 27.

28P. A golfer takes three putts to get the ball into the hole once it is on the green. The first putt displaces the ball 12 ft north, the second 6.0 ft southeast, and the third 3.0 ft southwest. What displacement was needed to get the ball into the hole on the first putt?

29P. A radar station detects an airplane approaching directly from the east. At first observation, the range to the plane is 1200 ft at 40° above the horizon. The plane is tracked for another 123° in the vertical east–west plane, the range at final contact being 2580 ft. See Fig. 3-27. Find the displacement of the plane during the period of observation.

FIGURE 3-27 Problem 29.

30P. (a) A man leaves his front door, walks 1000 m east, 2000 m north, and then takes a penny from his pocket and drops it from a cliff 500 m high. Set up a coordinate system and write down an expression, using unit vectors, for the displacement of the penny, from the house to its landing point. (b) The man then returns to his front door, following a different path on the return trip. What is his resultant displacement for the round trip?

31P. A particle undergoes three successive displacements in a plane, as follows: 4.00 m southwest, 5.00 m east, and 6.00 m in a direction 60.0° north of east. Choose the y axis pointing north and the x axis pointing east and find (a) the components of each displacement, (b) the components of the resultant displacement, (c) the magnitude and direction of the resultant displacement, and (d) the displacement that would be required to bring the particle back to the starting point.

32P. Prove that two vectors must have equal magnitudes if their sum is perpendicular to their difference.

33P. Two vectors of lengths a and b make an angle θ with each other when placed tail to tail. Prove, by taking components along two perpendicular axes, that the length of their sum is

$$r = \sqrt{a^2 + b^2 + 2ab \cos \theta}.$$

34P. (a) Using unit vectors, express the diagonals (the straight lines from one corner to another through the center) of a cube in terms of its edges, which have length a. (b) Determine the angles made by the diagonals with the adjacent edges. (c) Determine the length of the diagonals.

35P*. A person flies from Washington, DC, to Manila. (a) Describe the displacement vector. (b) What is its magnitude if the latitude and longitude of the two cities are 39° N, 77° W and 15° N, 121° E?

SECTION 3-6 VECTORS AND THE LAWS OF PHYSICS

36E. A vector $\mathbf{a}$ with a magnitude of 17.0 m is directed 56.0° counterclockwise from the $+ x$ axis, as shown in Fig. 3-28. (a) What are the components a_x and a_y of the vector? (b) A second coordinate system is inclined by 18.0° with respect to the first. What are the components a'_x and a'_y in this primed coordinate system?

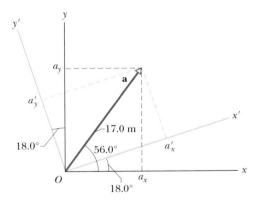

FIGURE 3-28 Exercise 36.

SECTION 3-7 MULTIPLYING VECTORS

37E. A vector $\mathbf{d}$ has a magnitude of 2.5 m and points north. What are the magnitudes and directions of the vectors (a) 4.0$\mathbf{d}$ and (b) 3.0$\mathbf{d}$?

38E. Consider $\mathbf{a}$ in the positive direction of x, $\mathbf{b}$ in the positive direction of y, and a scalar d. What is the direction of $\mathbf{b}/d$ if d is (a) positive and (b) negative? What is the magnitude of (c) $\mathbf{a} \cdot \mathbf{b}$ and (d) $\mathbf{a} \cdot \mathbf{b}/d$? What is the direction of (e) $\mathbf{a} \times \mathbf{b}$ and (f) $\mathbf{b} \times \mathbf{a}$? (g) What are the magnitudes of the cross products in (e) and (f)? (h) What are the magnitude and direction of $\mathbf{a} \times \mathbf{b}/d$?

39E. In a right-handed coordinate system show that

$$\mathbf{i} \cdot \mathbf{i} = \mathbf{j} \cdot \mathbf{j} = \mathbf{k} \cdot \mathbf{k} = 1$$

and

$$\mathbf{i} \cdot \mathbf{j} = \mathbf{j} \cdot \mathbf{k} = \mathbf{k} \cdot \mathbf{i} = 0.$$

If the coordinate system is rectangular but not right-handed, do the results change?

40E. In a right-handed coordinate system show that

$$\mathbf{i} \times \mathbf{i} = \mathbf{j} \times \mathbf{j} = \mathbf{k} \times \mathbf{k} = 0$$

and

$$\mathbf{i} \times \mathbf{j} = \mathbf{k}; \quad \mathbf{k} \times \mathbf{i} = \mathbf{j}; \quad \mathbf{j} \times \mathbf{k} = \mathbf{i}.$$

If the coordinate system is rectangular but not right-handed, do the results change?

41E. Show for any vector $\mathbf{a}$ that $\mathbf{a} \cdot \mathbf{a} = a^2$ and that $\mathbf{a} \times \mathbf{a} = 0$.

42E. Find (a) "north cross west," (b) "down dot south," (c) "east cross up," (d) "west dot west," and (e) "south cross south." Let each vector have unit magnitude.

43E. A vector $\mathbf{a}$ of magnitude 10 units and another vector $\mathbf{b}$ of magnitude 6.0 units point in directions differing by 60°. Find (a) the scalar product of the two vectors and (b) the magnitude of the vector product $\mathbf{a} \times \mathbf{b}$.

44E. Two vectors, $\mathbf{r}$ and $\mathbf{s}$, lie in the xy plane. Their magnitudes are 4.50 and 7.30 units, respectively, and their directions are 320° and 85.0°, respectively, as measured counterclockwise from the positive x axis. What are the values of (a) $\mathbf{r} \cdot \mathbf{s}$ and (b) $\mathbf{r} \times \mathbf{s}$?

45E. For the vectors in Fig. 3-29, calculate (a) $\mathbf{a} \cdot \mathbf{b}$, (b) $\mathbf{a} \cdot \mathbf{c}$, and (c) $\mathbf{b} \cdot \mathbf{c}$.

46E. For the vectors in Fig. 3-29, calculate (a) $\mathbf{a} \times \mathbf{b}$, (b) $\mathbf{a} \times \mathbf{c}$, and (c) $\mathbf{b} \times \mathbf{c}$.

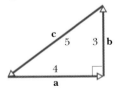

FIGURE 3-29 Exercises 45 and 46.

47P. *Scalar Product in Unit-Vector Notation.* Let two vectors be represented in terms of their coordinates as

$$\mathbf{a} = a_x\mathbf{i} + a_y\mathbf{j} + a_z\mathbf{k}$$

and

$$\mathbf{b} = b_x\mathbf{i} + b_y\mathbf{j} + b_z\mathbf{k}.$$

Show that

$$\mathbf{a} \cdot \mathbf{b} = a_x b_x + a_y b_y + a_z b_z.$$

48P. Use the definition of scalar product, $\mathbf{a} \cdot \mathbf{b} = ab \cos \theta$, and the fact that $\mathbf{a} \cdot \mathbf{b} = a_x b_x + a_y b_y + a_z b_z$ (see Problem 47) to calculate the angle between the two vectors given by $\mathbf{a} = 3.0\mathbf{i} + 3.0\mathbf{j} + 3.0\mathbf{k}$ and $\mathbf{b} = 2.0\mathbf{i} + 1.0\mathbf{j} + 3.0\mathbf{k}$.

49P. (a) Determine the components and magnitude of $\mathbf{r} = \mathbf{a} - \mathbf{b} + \mathbf{c}$ if $\mathbf{a} = 5.0\mathbf{i} + 4.0\mathbf{j} - 6.0\mathbf{k}$, $\mathbf{b} = -2.0\mathbf{i} + 2.0\mathbf{j} + 3.0\mathbf{k}$, and $\mathbf{c} = 4.0\mathbf{i} + 3.0\mathbf{j} + 2.0\mathbf{k}$. (b) Calculate the angle between $\mathbf{r}$ and the positive z axis.

50P. *Vector Product in Unit-Vector Notation.* Show that for the vectors $\mathbf{a}$ and $\mathbf{b}$ of Problem 47 $\mathbf{a} \times \mathbf{b} = \mathbf{i}(a_y b_z - a_z b_y) + \mathbf{j}(a_z b_x - a_x b_z) + \mathbf{k}(a_x b_y - a_y b_x)$.

51P. Two vectors are given by $\mathbf{a} = 3.0\mathbf{i} + 5.0\mathbf{j}$ and $\mathbf{b} = 2.0\mathbf{i} + 4.0\mathbf{j}$. Find (a) $\mathbf{a} \times \mathbf{b}$, (b) $\mathbf{a} \cdot \mathbf{b}$, and (c) $(\mathbf{a} + \mathbf{b}) \cdot \mathbf{b}$.

52P. Two vectors $\mathbf{a}$ and $\mathbf{b}$ have the components, in arbitrary units, $a_x = 3.2$, $a_y = 1.6$, $b_x = 0.50$, $b_y = 4.5$. (a) Find the angle between $\mathbf{a}$ and $\mathbf{b}$. (b) Find the components of a vector $\mathbf{c}$ that is perpendicular to $\mathbf{a}$, is in the xy plane, and has a magnitude of 5.0 units.

53P. Vector $\mathbf{a}$ lies in the yz plane $63°$ from the $+y$ axis, has a positive z component, and has magnitude 3.20 units. Vector $\mathbf{b}$ lies in the xz plane $48°$ from the $+x$ axis, has a positive z component, and has magnitude 1.40 units. Find (a) $\mathbf{a} \cdot \mathbf{b}$, (b) $\mathbf{a} \times \mathbf{b}$, and (c) the angle between $\mathbf{a}$ and $\mathbf{b}$.

54P. Three vectors are given by $\mathbf{a} = 3.0\mathbf{i} + 3.0\mathbf{j} - 2.0\mathbf{k}$, $\mathbf{b} = -1.0\mathbf{i} - 4.0\mathbf{j} + 2.0\mathbf{k}$, and $\mathbf{c} = 2.0\mathbf{i} + 2.0\mathbf{j} + 1.0\mathbf{k}$. Find (a) $\mathbf{a} \cdot (\mathbf{b} \times \mathbf{c})$, (b) $\mathbf{a} \cdot (\mathbf{b} + \mathbf{c})$, and (c) $\mathbf{a} \times (\mathbf{b} + \mathbf{c})$.

55P. Find the angles between the body diagonals of a cube with edge length a. See Problem 34.

56P. Show that the area of the triangle contained between the vectors $\mathbf{a}$ and $\mathbf{b}$ in Fig. 3-30 is $\frac{1}{2}|\mathbf{a} \times \mathbf{b}|$, where the vertical bars signify magnitude.

FIGURE 3-30 Problem 56.

57P. (a) Show that $\mathbf{a} \cdot (\mathbf{b} \times \mathbf{a})$ is zero for all vectors $\mathbf{a}$ and $\mathbf{b}$. (b) What is the value of $\mathbf{a} \times (\mathbf{b} \times \mathbf{a})$ if there is an angle ϕ between the directions of $\mathbf{a}$ and $\mathbf{b}$?

58P. The three vectors shown in Fig. 3-31 have magnitudes $a = 3.00$, $b = 4.00$, and $c = 10.0$. (a) Calculate the x and y components of these vectors. (b) Find the numbers p and q such that $\mathbf{c} = p\mathbf{a} + q\mathbf{b}$.

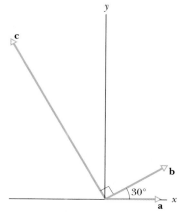

FIGURE 3-31 Problem 58.

59P. Show that $\mathbf{a} \cdot (\mathbf{b} \times \mathbf{c})$ is equal in magnitude to the volume of the parallelepiped formed on the three vectors $\mathbf{a}$, $\mathbf{b}$, and $\mathbf{c}$ as shown in Fig. 3-32.

FIGURE 3-32 Problem 59.

ADDITIONAL PROBLEMS

60. Oasis B is 25 km due east of oasis A. Starting from oasis A, a camel walks 24 km in a direction 15° south of east and then walks 8.0 km due north. How far is the camel then from oasis B?

61. A vector **B**, which has a magnitude of 8.0, is added to a vector **A**, which lies along the x axis. The sum of these two vectors is a third vector, which lies along the y axis and has a magnitude that is twice the magnitude of **A**. What is the magnitude of **A**?

62. If vector **B** is added to vector **A**, the result is $6.0\mathbf{i} + 1.0\mathbf{j}$. If **B** is subtracted from **A**, the result is $-4.0\mathbf{i} + 7.0\mathbf{j}$. What is the magnitude of **A**?

63. A vector **B**, when added to the vector $\mathbf{C} = 3.0\mathbf{i} + 4.0\mathbf{j}$, yields a resultant vector that is in the positive y direction and has a magnitude equal to that of **C**. What is the magnitude of **B**?

MOTION IN TWO AND THREE DIMENSIONS

4

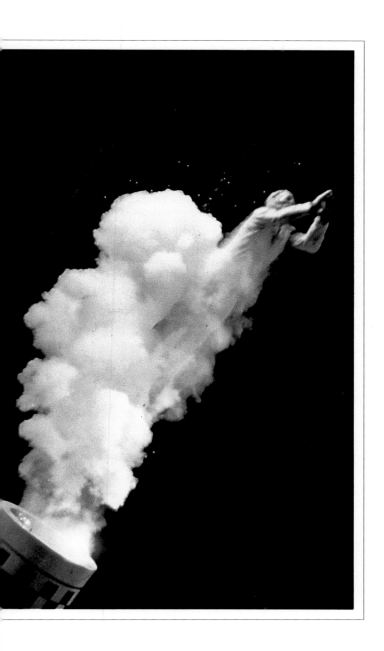

When a "human cannonball" is shot from a cannon, the smoke and noise are theatrical, because the propulsion comes from a spring or compressed air rather than an explosion.   But the danger is still real for two reasons.   One is that the rapid propulsion in the muzzle usually results in a momentary blackout, from which the performer must awake if he or she is to land in the net without a broken neck.   The other real danger is that the net may not be in the right place for the angle and speed of the launch.

4-1 MOVING IN TWO OR THREE DIMENSIONS

This chapter extends the material of the preceding two chapters to two and three dimensions. Many of the ideas of Chapter 2, such as position, velocity, and acceleration, are used here, but they are now a little more complex because of the extra dimensions. To keep the notation manageable, we use the vector algebra of Chapter 3. As you read this chapter, you might want to thumb back to those previous chapters to refresh your memory.

An example of two-dimensional motion is the flight of a human cannonball. The modern version of the stunt dates back to 1922 when the Zacchinis, a famous family of circus performers, first shot one of their troupe from a cannon to a net across an arena. To increase the excitement, the family gradually increased both the height and the length of the flight. By 1939 or 1940 the limit of reasonable safety was reached when Emanuel Zacchini soared over three Ferris wheels and through a horizontal distance of 225 feet.

How could Zacchini know where to place the net? And how could he be certain he would clear the Ferris wheels? You can be sure he did not want to answer either question through trial and error.

4-2 POSITION AND DISPLACEMENT

One general way of locating a particle-like object is with a **position vector r**, which is a vector that extends from a reference point (usually the origin of the coordinate system) to the object. In the unit-vector notation of Section 3-4, **r** can be written

$$\mathbf{r} = x\mathbf{i} + y\mathbf{j} + z\mathbf{k}, \tag{4-1}$$

where $x\mathbf{i}$, $y\mathbf{j}$, and $z\mathbf{k}$ are the vector components of **r**, and the coefficients x, y, and z are its scalar components. (This notation is slightly different from the notation we used in Chapter 3. Take a minute to convince yourself that the two are comparable.)

The coefficients x, y, and z give the object's location along the axes and relative to the origin. For instance, Fig. 4-1 shows an object P whose position vector at the moment is

$$\mathbf{r} = -3\mathbf{i} + 2\mathbf{j} + 5\mathbf{k}. \tag{4-2}$$

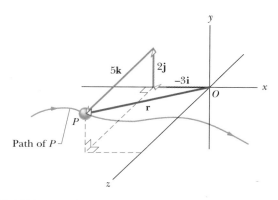

FIGURE 4-1   The position vector **r** for object P is the vector sum of its vector components, which are parallel to the coordinate axes.

Along the x axis P is 3 units from the origin, in the $-\mathbf{i}$ direction. Along the y axis, it is 2 units from the origin, in the $+\mathbf{j}$ direction. And along the z axis, it is 5 units from the origin, in the $+\mathbf{k}$ direction.

As an object moves, its position vector changes in such a way that the vector always extends to the object from the origin. If the object has position vector $\mathbf{r}_1$ at time t_1, and position vector $\mathbf{r}_2$ at a later time $t_1 + \Delta t$, then its *displacement* $\Delta\mathbf{r}$ during the time interval Δt is

$$\Delta\mathbf{r} = \mathbf{r}_2 - \mathbf{r}_1. \tag{4-3}$$

SAMPLE PROBLEM 4-1

The position vector for a particle is initially

$$\mathbf{r}_1 = -3\mathbf{i} + 2\mathbf{j} + 5\mathbf{k}$$

and then later is

$$\mathbf{r}_2 = 9\mathbf{i} + 2\mathbf{j} + 8\mathbf{k}$$

(see Fig. 4-2). What is the displacement from $\mathbf{r}_1$ to $\mathbf{r}_2$?

SOLUTION   Recall from Chapter 3 that we subtract two vectors in unit-vector notation by combining the components, axis by axis. So Eq. 4-3 becomes

$$\Delta\mathbf{r} = (9\mathbf{i} + 2\mathbf{j} + 8\mathbf{k}) - (-3\mathbf{i} + 2\mathbf{j} + 5\mathbf{k})$$

$$= 12\mathbf{i} + 3\mathbf{k}. \qquad \text{(Answer)}$$

The displacement vector is parallel to the xz plane, because it lacks any y component, a fact that is easier to pick out in the numerical result than in Fig. 4-2.

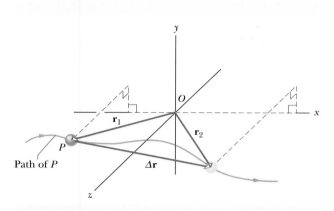

FIGURE 4-2 The displacement $\Delta \mathbf{r} = \mathbf{r}_2 - \mathbf{r}_1$ extends from the head of $\mathbf{r}_1$ to the head of $\mathbf{r}_2$.

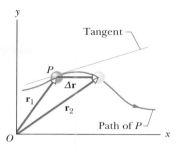

FIGURE 4-3 The position of particle P along its path is shown, both at time t_1 and at a later time $t_1 + \Delta t$. The vector $\Delta \mathbf{r}$ is the displacement of the particle during Δt. The tangent to the path at t_1 is shown.

The coefficients are the scalar components of $\mathbf{v}$:

$$v_x = \frac{dx}{dt}, \quad v_y = \frac{dy}{dt}, \quad \text{and} \quad v_z = \frac{dz}{dt}. \quad (4\text{-}8)$$

4-3 VELOCITY AND AVERAGE VELOCITY

If a particle moves through a displacement $\Delta \mathbf{r}$ in a time interval Δt, then its *average velocity* is

$$\bar{\mathbf{v}} = \frac{\Delta \mathbf{r}}{\Delta t}, \quad (4\text{-}4)$$

which can be written in expanded form as

$$\bar{\mathbf{v}} = \frac{\Delta x \mathbf{i} + \Delta y \mathbf{j} + \Delta z \mathbf{k}}{\Delta t}$$

$$= \frac{\Delta x}{\Delta t} \mathbf{i} + \frac{\Delta y}{\Delta t} \mathbf{j} + \frac{\Delta z}{\Delta t} \mathbf{k}. \quad (4\text{-}5)$$

The (instantaneous) *velocity* $\mathbf{v}$ is the value that $\bar{\mathbf{v}}$ approaches in the limit as we shrink Δt to 0. It can be written as the derivative

$$\mathbf{v} = \frac{d\mathbf{r}}{dt}. \quad (4\text{-}6)$$

Substituting for $\mathbf{r}$ from Eq. 4-1 yields

$$\mathbf{v} = \frac{d}{dt}(x\mathbf{i} + y\mathbf{j} + z\mathbf{k}) = \frac{dx}{dt}\mathbf{i} + \frac{dy}{dt}\mathbf{j} + \frac{dz}{dt}\mathbf{k},$$

which can be rewritten as

$$\mathbf{v} = v_x\mathbf{i} + v_y\mathbf{j} + v_z\mathbf{k}. \quad (4\text{-}7)$$

Figure 4-3 shows the path of a particle P that is restricted to the xy plane. As the particle travels to the right along the curve, its position vector sweeps to the right to keep up. At t_1 the position vector is $\mathbf{r}_1$, and at an arbitrary later time $t_1 + \Delta t$ the position vector is $\mathbf{r}_2$. The particle's displacement during Δt is $\Delta \mathbf{r}$. The particle's average velocity $\bar{\mathbf{v}}$ during Δt is, by Eq. 4-4, in the same direction as $\Delta \mathbf{r}$.

Three things happen as we shrink interval Δt toward zero: (1) the vector $\mathbf{r}_2$ in Fig. 4-3 moves toward $\mathbf{r}_1$ so that $\Delta \mathbf{r}$ shrinks toward zero; (2) the direction of $\Delta \mathbf{r}$ (and thus the direction of $\bar{\mathbf{v}}$) approaches the direction of the tangent line in Fig. 4-3; and (3) the average velocity $\bar{\mathbf{v}}$ approaches the instantaneous velocity $\mathbf{v}$.

In the limit as $\Delta t \rightarrow 0$, we have $\bar{\mathbf{v}} \rightarrow \mathbf{v}$ and, most important here, $\bar{\mathbf{v}}$ takes on the direction of the tangent line. Hence $\mathbf{v}$ has that direction as well. That is, the instantaneous velocity $\mathbf{v}$ of a particle is always tangent to the curve representing the path of the particle. This is shown in Fig. 4-4, where both $\mathbf{v}$ and its scalar components are included. The result is the same in three dimensions: $\mathbf{v}$ is always tangent to the particle's path.

Figure 4-5 represents a hockey puck constrained to move in a circle over very slippery ice, by a string fixed at O. The puck's position vector $\mathbf{r}$ changes in direction only; its magnitude (equal to the length of the string) remains constant. Again, the velocity vector at any instant is tangent to the path. If the string were to break, the puck would move off in a straight

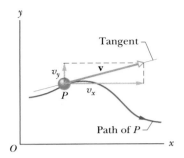

FIGURE 4-4 The velocity **v** of particle P along with its scalar components. Note that **v** lies along the tangent to the path.

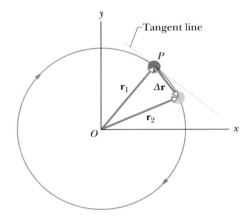

FIGURE 4-5 The particle is moving in a circular path about O. As Δt approaches zero, the vector $\Delta \mathbf{r}$ lines up with the tangent line. Hence **v** is along the tangent line.

line in the direction **v** had at the instant of the break. (The puck would not spiral outward, as if it had some lingering memory of its earlier circular path.)

4-4 ACCELERATION AND AVERAGE ACCELERATION

When a particle's velocity changes from $\mathbf{v}_1$ to $\mathbf{v}_2$ in a time period Δt, its average acceleration $\overline{\mathbf{a}}$ during Δt is

$$\overline{\mathbf{a}} = \frac{\mathbf{v}_2 - \mathbf{v}_1}{\Delta t} = \frac{\Delta \mathbf{v}}{\Delta t}. \qquad (4\text{-}9)$$

If we shrink Δt to 0, then in the limit $\overline{\mathbf{a}}$ approaches the (instantaneous) *acceleration* **a**; that is,

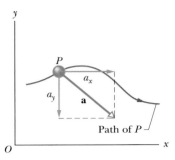

FIGURE 4-6 The acceleration **a** of particle P along with its scalar components.

$$\mathbf{a} = \frac{d\mathbf{v}}{dt}. \qquad (4\text{-}10)$$

If the velocity changes in *either* magnitude *or* direction (or both), there is an acceleration.

Substituting **v** from Eq. 4-7 into Eq. 4-10 yields

$$\mathbf{a} = \frac{d}{dt}\ (v_x\mathbf{i} + v_y\mathbf{j} + v_z\mathbf{k})$$

$$= \frac{dv_x}{dt}\,\mathbf{i} + \frac{dv_y}{dt}\,\mathbf{j} + \frac{dv_z}{dt}\,\mathbf{k}$$

or

$$\mathbf{a} = a_x\mathbf{i} + a_y\mathbf{j} + a_z\mathbf{k}, \qquad (4\text{-}11)$$

in which the three scalar components of the acceleration vector are given by

$$a_x = \frac{dv_x}{dt}, \quad a_y = \frac{dv_y}{dt}, \quad \text{and} \quad a_z = \frac{dv_z}{dt}. \qquad (4\text{-}12)$$

Figure 4-6 shows an acceleration vector **a** and its scalar components for the motion of a particle P in two dimensions.

SAMPLE PROBLEM 4-2

A rabbit runs across a parking lot on which a set of coordinate axes has, strangely enough, been drawn. The path is such that the components of the rabbit's position with respect to an origin of coordinates are given as functions of time by

$$x = -0.31t^2 + 7.2t + 28$$

and
$$y = 0.22t^2 - 9.1t + 30.$$

The units of the numerical coefficients in these equations are such that, if you substitute t in seconds, x and y are in meters.

a. Calculate the rabbit's position vector $\mathbf{r}$ (magnitude and direction) at $t = 15$ s.

SOLUTION At $t = 15$ s, the components of $\mathbf{r}$ are
$$x = (-0.31)(15)^2 + (7.2)(15) + 28 = 66 \text{ m}$$
and
$$y = (0.22)(15)^2 - (9.1)(15) + 30 = -57 \text{ m}.$$

The components and $\mathbf{r}$ itself are shown in Fig. 4-7a. The magnitude of $\mathbf{r}$ is given by
$$r = \sqrt{x^2 + y^2} = \sqrt{(66 \text{ m})^2 + (-57 \text{ m})^2}$$
$$= 87 \text{ m}. \qquad \text{(Answer)}$$

The angle θ between $\mathbf{r}$ and the direction of increasing x is
$$\theta = \tan^{-1}\frac{y}{x} = \tan^{-1}\left(\frac{-57 \text{ m}}{66 \text{ m}}\right) = -41°. \quad \text{(Answer)}$$

(Although $\theta = 139°$ has the same tangent as $-41°$, study of the signs of the components of $\mathbf{r}$ rules out $139°$.)

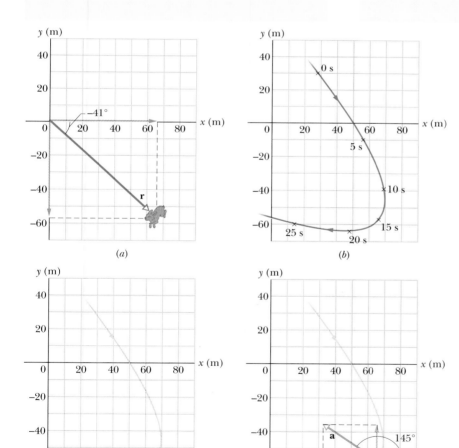

FIGURE 4-7 Sample Problems 4-2, 4-3, and 4-4. (*a*) The vector $\mathbf{r}$ and its components at $t = 15$ s. The magnitude of $\mathbf{r}$ is 87 m. (*b*) The path of a rabbit across a parking lot, showing the rabbit's position at the indicated times. (*c*) The velocity $\mathbf{v}$ of the rabbit at $t = 15$ s. Note that $\mathbf{v}$ is tangent to the path at the position of the rabbit at $t = 15$ s. (*d*) The acceleration $\mathbf{a}$ of the rabbit at $t = 15$ s. As it happens, the rabbit has this same acceleration for all points of its path.

b. Also calculate the position of the rabbit at $t = 0$, 5, 10, 20, and 25 s and sketch the rabbit's path.

SOLUTION Proceeding as in (a) leads to the following values of r and θ.

t (s)	x (m)	y (m)	r (m)	θ
0	28	30	41	$+47°$
5	56	-10	57	$-10°$
10	69	-39	79	$-29°$
15	66	-57	87	$-41°$
20	48	-64	80	$-53°$
25	14	-60	62	$-77°$

Figure 4-7b shows a plot of the rabbit's path.

SAMPLE PROBLEM 4-3

In Sample Problem 4-2, find the magnitude and the direction of the rabbit's velocity vector at $t = 15$ s.

SOLUTION The velocity component in the x direction (see Eq. 4-8) is

$$v_x = \frac{dx}{dt} = \frac{d}{dt}(-0.31t^2 + 7.2t + 28) = -0.62t + 7.2.$$

At $t = 15$ s, this becomes

$$v_x = (-0.62)(15) + 7.2 = -2.1 \text{ m/s}.$$

Similarly,

$$v_y = \frac{dy}{dt} = \frac{d}{dt}(0.22t^2 - 9.1t + 30) = 0.44t - 9.1.$$

At $t = 15$ s, this becomes

$$v_y = (0.44)(15) - 9.1 = -2.5 \text{ m/s}.$$

The vector $\mathbf{v}$ and its components are shown in Fig. 4-7c.
The magnitude and the direction of $\mathbf{v}$ are given by

$$v = \sqrt{v_x^2 + v_y^2} = \sqrt{(-2.1 \text{ m/s})^2 + (-2.5 \text{ m/s})^2}$$

$$= 3.3 \text{ m/s} \qquad \text{(Answer)}$$

and

$$\theta = \tan^{-1}\frac{v_y}{v_x} = \tan^{-1}\left(\frac{-2.5 \text{ m/s}}{-2.1 \text{ m/s}}\right)$$

$$= \tan^{-1} 1.19 = -130°. \qquad \text{(Answer)}$$

(Although 50° has the same tangent, inspection of the signs of the velocity components indicates that the de-

sired angle is in the third quadrant, given by $50° - 180° = -130°$.) The velocity vector in Fig. 4-7c is tangent to the path of the rabbit and points in the direction in which the rabbit is running at $t = 15$ s.

SAMPLE PROBLEM 4-4

In Sample Problem 4-2, determine both the magnitude and the direction of the rabbit's acceleration vector $\mathbf{a}$ at $t = 15$ s.

SOLUTION The acceleration components (see Eq. 4-12) are given by

$$a_x = \frac{dv_x}{dt} = \frac{d}{dt}(-0.62t + 7.2) = -0.62 \text{ m/s}^2$$

and

$$a_y = \frac{dv_y}{dt} = \frac{d}{dt}(0.44t - 9.1) = 0.44 \text{ m/s}^2.$$

We see that the acceleration does not vary with time; it is a constant. In fact, we have differentiated the time variable completely away! The components of $\mathbf{a}$, along with the vector itself, are shown in Fig. 4-7d.
The magnitude and direction of $\mathbf{a}$ are given by

$$a = \sqrt{a_x^2 + a_y^2} = \sqrt{(-0.62 \text{ m/s}^2)^2 + (0.44 \text{ m/s}^2)^2}$$

$$= 0.76 \text{ m/s}^2 \qquad \text{(Answer)}$$

and

$$\theta = \tan^{-1}\frac{a_y}{a_x} = \tan^{-1}\left(\frac{0.44 \text{ m/s}^2}{-0.62 \text{ m/s}^2}\right)$$

$$= 145°. \qquad \text{(Answer)}$$

The acceleration vector has the same magnitude and direction for all parts of the rabbit's path. Perhaps a strong southeast wind was blowing across the parking lot.

SAMPLE PROBLEM 4-5

A particle with velocity $\mathbf{v}_0 = -2.0\mathbf{i} + 4.0\mathbf{j}$ (in units of m/s) at $t = 0$ undergoes a constant acceleration $\mathbf{a}$ of magnitude $a = 3.0 \text{ m/s}^2$ at an angle $\theta = 130°$ from the positive direction of the x axis. What is the particle's velocity $\mathbf{v}$ at $t = 2.0$ s, in unit-vector notation and as a magnitude and direction (with respect to the positive direction of the x axis)?

SOLUTION Since **a** is constant, Eq. 2-9 applies; it should, however, be used separately to find v_x and v_y (the x and y components of velocity **v**) because their variations are independent of each other. We find

$$v_x = v_{0x} + a_x t$$

and

$$v_y = v_{0y} + a_y t.$$

Here $v_{0x} (= -2.0 \text{ m/s})$ and $v_{0y} (= 4.0 \text{ m/s})$ are the x and y components of $\mathbf{v}_0$, and a_x and a_y are the x and y components of **a**. To find a_x and a_y, we resolve **a** with Eq. 3-5:

$$a_x = a \cos \theta = (3.0 \text{ m/s}^2)(\cos 130°) = -1.93 \text{ m/s}^2,$$

$$a_y = a \sin \theta = (3.0 \text{ m/s}^2)(\sin 130°) = +2.30 \text{ m/s}^2.$$

When these values are inserted into the equations for v_x and v_y, we find that

$$v_x = -2.0 \text{ m/s} + (-1.93 \text{ m/s}^2)(2.0 \text{ s}) = -5.9 \text{ m/s},$$

$$v_y = 4.0 \text{ m/s} + (2.30 \text{ m/s}^2)(2.0 \text{ s}) = 8.6 \text{ m/s}.$$

So at $t = 2.0$ s, we have

$$\mathbf{v} = (-5.9 \text{ m/s})\mathbf{i} + (8.6 \text{ m/s})\mathbf{j}. \quad \text{(Answer)}$$

The magnitude of **v** is

$$v = \sqrt{(-5.9 \text{ m/s})^2 + (8.6 \text{ m/s})^2}$$

$$= 10 \text{ m/s}. \quad \text{(Answer)}$$

The angle of **v** is

$$\theta = \tan^{-1} \frac{8.6 \text{ m/s}}{-5.9 \text{ m/s}} = 124° \approx 120°. \quad \text{(Answer)}$$

Check the last line with your calculator. Does 124° appear on the display, or does $-55.5°$ appear? Now sketch the vector **v** with its components to see which angle is reasonable. To see why the calculator gives a mathematically possible answer but unreasonable result here, reread Tactic 3 in Chapter 3.

PROBLEM SOLVING

TACTIC 1: DRAWING A GRAPH

Figure 4-7 brings out many of the elements that you need to consider in drawing a graph. In Sample Problem 4-2b the quantities to be plotted are the position variables x and y, both expressed in meters. In this problem it makes sense to choose the same scale for both axes. You may have to experiment with the scale to get a graph that will fill the space well and be easy to

read. The problem asks you to calculate six points. If you find that you need more points to plot a smooth curve, calculate them from the formulas given in the problem statement. If one of your points falls distinctly off a smooth curve, you have probably made an error in calculating or in plotting.

The crosses on the graph of Fig. 4-7b show the time t at which the rabbit was at each position. By this device, you can display a third variable on your graph.

TACTIC 2: TRIG FUNCTIONS AND ANGLES

In Sample Problem 4-3, we were given $\theta = \tan^{-1} 1.19$ and asked to find θ. Your calculator will tell you $\theta = 50°$. However, Fig. 3-12c shows that $\theta = 230° (= 50° + 180°)$ has the same tangent. Inspection of the signs of the velocity components v_x and v_y in Fig. 4-7c tells us that this latter angle is the correct one.

There is still another decision to make. We can stick with 230° or we can relabel it as $-130°$. They are exactly the same angle, as is pointed out in Tactic 1 of Chapter 3. We chose $\theta = -130°$, purely as a matter of taste.

TACTIC 3: DRAWING VECTORS—DIRECTION

The vectors in Fig. 4-7 were oriented in the following way: (1) Choose the point at which you wish the tail of the vector to be. (2) From that point, draw a line in the direction of increasing x. (3) Using a protractor, mark off the appropriate angle θ counterclockwise from this line (if θ is positive) or clockwise (if θ is negative).

TACTIC 4: DRAWING VECTORS—LENGTH

The vector **r** in Fig. 4-7a should be drawn to the same scale as the two axes because it is a length. The velocity vector **v** in Fig. 4-7c and the acceleration vector **a** in Fig. 4-7d, however, have no established scale in this problem and you may make them as long or as short as you wish.

It makes no sense to ask whether, for example, a velocity vector should be longer or shorter than a displacement vector. They are different physical quantities, expressed in different units, and they have no common scale.

4-5 PROJECTILE MOTION

Here we consider a particle—that is, a **projectile**—that moves in two dimensions during free fall, with the free-fall acceleration **g**, which is directed downward. The projectile might be a golf ball (as in Fig. 4-8), a baseball, or any of a variety of other objects. Throughout, we shall assume that the air has no effect on the motion of the projectile.

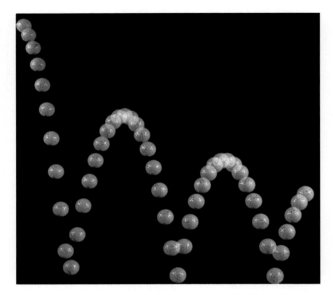

FIGURE 4-8 A stroboscopic photo of an orange golf ball bouncing off a hard surface. Between impacts, the ball exhibits projectile motion.

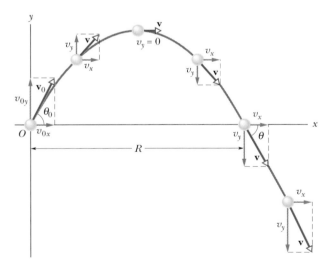

FIGURE 4-9 The path of a projectile that is launched at $x_0 = 0$ and $y_0 = 0$, with an initial velocity $\mathbf{v}_0$. The initial velocity and the velocities at various points along its path are shown, along with their components. Note that the horizontal velocity component remains constant but the vertical velocity component changes continuously. The *range R* is the horizontal distance the projectile has traveled when it returns to its launch height.

Figure 4-9, which is analyzed in the next section, shows the path followed by a projectile under such ideal conditions.

The projectile is launched with some initial velocity $\mathbf{v}_0$, which can be written

$$\mathbf{v}_0 = v_{0x}\mathbf{i} + v_{0y}\mathbf{j}. \qquad (4\text{-}13)$$

The components v_{0x} and v_{0y} can then be found if we know the angle θ_0 between $\mathbf{v}_0$ and the positive direction of the x axis:

$$v_{0x} = v_0 \cos \theta_0 \quad \text{and} \quad v_{0y} = v_0 \sin \theta_0. \quad (4\text{-}14)$$

During its two-dimensional motion, the projectile accelerates downward, and its position vector $\mathbf{r}$ and velocity vector $\mathbf{v}$ change continuously. However, the horizontal motion and the vertical motion are independent of one another (except for sharing a common time variable). That fact allows us to break up a problem about the two-dimensional motion into two separate and easier one-dimensional problems, one for the horizontal motion and one for the vertical motion. Here are two experiments that show that the horizontal motion and the vertical motion are independent.

Two Golf Balls

Figure 4-10 is a stroboscopic photograph of two golf balls, one released from rest and the other fired horizontally from a spring gun. They have the same vertical motion, each ball falling through the same vertical distance in the same interval of time. *The fact that one ball is moving horizontally while it is falling has no effect on its vertical motion.* To push the experiment to a limit, we say that, if you were to fire a rifle horizontally and at the same time drop a bullet, in the absence of air resistance the two bullets would reach the level ground at the same time.

A Great Student Rouser

Figure 4-11 shows a demonstration that has enlivened many a physics lecture. It involves a blow gun G, using a ball as a projectile. The target is a tin can C, suspended from a magnet M, and the tube of the blow gun is bore-sighted directly at the can. The experiment is arranged so that the magnet releases the can just as the ball leaves the blow gun.

If g (the free-fall acceleration) were zero, the ball would follow the straight line shown in Fig. 4-11

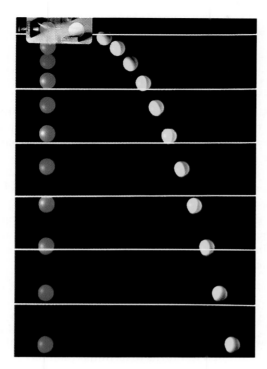

FIGURE 4-10 One ball is released from rest at the same instant that another ball is shot horizontally to the right. Their vertical motions are identical.

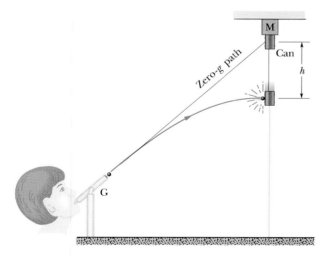

FIGURE 4-11 The ball always hits the falling can. Each falls a distance h from where it would be were there no free-fall acceleration.

and the can would float in place after the magnet released it. The ball would certainly hit the can.

However, g is *not* zero. The ball *still* hits the can! As Fig. 4-11 shows, during the time of flight of the ball, both ball and can fall the same distance h from their zero-g locations. The harder the demonstrator blows, the greater the ball's initial speed, the shorter the time of flight, and the smaller the value of h.

4-6 PROJECTILE MOTION ANALYZED

Now we are ready to analyze projectile motion, part by part.

The Horizontal Motion

Because there is *no* acceleration in the horizontal direction, the horizontal component of the velocity remains unchanged throughout the motion, as demonstrated in Fig. 4-12. The horizontal displacement $x - x_0$ from an initial position x_0 is given by Eq. 2-13, in which we put $a = 0$ and substitute v_{0x} ($= v_0 \cos \theta_0$) for v_0. Thus

$$x - x_0 = (v_0 \cos \theta_0)t. \qquad (4\text{-}15)$$

FIGURE 4-12 The vertical component of this skateboarder's velocity is changing, but not the horizontal component, which matches the skateboard's velocity. As a result, the skateboard stays underneath him, allowing him to land on it.

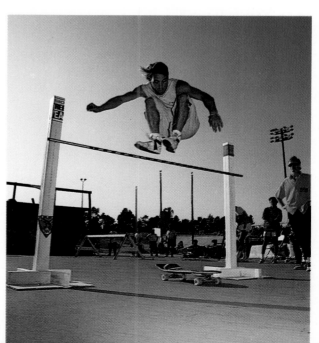

The Vertical Motion

The vertical motion is the motion we discussed in Section 2-8 for a particle in free fall. Equations 2-19 to 2-23 apply. Equation 2-20, for example, becomes

$$y - y_0 = (v_0 \sin \theta_0)t - \tfrac{1}{2}gt^2, \qquad (4\text{-}16)$$

where v_0 has been replaced with the vertical velocity component $v_0 \sin \theta_0$. As is illustrated in Fig. 4-9, the vertical velocity component behaves just as for a ball thrown vertically upward. It is directed upward initially, its magnitude steadily decreasing to zero, which marks the maximum height of the path. The vertical component then reverses direction, its magnitude becoming larger and larger as time goes on.

Equations 2-19 and 2-21 are also useful in analyzing projectile motion. Adapted to our purpose, they are

$$v_y = v_0 \sin \theta_0 - gt \qquad (4\text{-}17)$$

and

$$v_y^2 = (v_0 \sin \theta_0)^2 - 2g(y - y_0). \qquad (4\text{-}18)$$

The Equation of the Path

We can find the equation of the path (the **trajectory**) of the projectile by eliminating t between Eqs. 4-15 and 4-16. Solving Eq. 4-15 for t and substituting into Eq. 4-16, we obtain, after a little rearrangement,

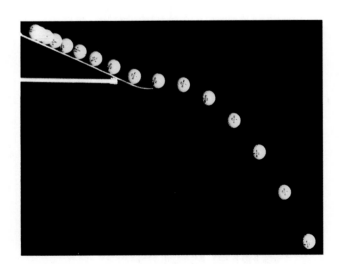

FIGURE 4-13 A stroboscopic photo of a ball rolling down an inclined plane and following a parabolic path after it is projected horizontally from the bottom of the plane.

$$y = (\tan \theta_0)x - \left(\frac{g}{2(v_0 \cos \theta_0)^2}\right)x^2 \quad \text{(trajectory)}.$$
$$(4\text{-}19)$$

This is the equation of the path shown in Fig. 4-9. For simplicity, we have here let $x_0 = 0$ and $y_0 = 0$ in Eqs. 4-15 and 4-16, respectively. Because g, θ_0, and v_0 are constants, Eq. 4-19 is of the form $y = ax + bx^2$, in which a and b are constants. This is the equation of a parabola.

Figure 4-13 illustrates a modern reenactment of an experiment done by Galileo in 1608, in which he first proved that a projectile follows a parabolic path. He did so by rolling a ball down a groove in an inclined plane, starting from various heights and measuring the corresponding positions at which the ball struck the floor.

The Horizontal Range

The *horizontal range* R of the projectile, as Fig. 4-9 shows, is the *horizontal* distance the projectile has traveled when it returns to its initial (launch) height. To find it, let us put $x - x_0 = R$ in Eq. 4-15 and $y - y_0 = 0$ in Eq. 4-16, obtaining

$$x - x_0 = (v_0 \cos \theta_0)t = R$$

and

$$y - y_0 = (v_0 \sin \theta_0)t - \tfrac{1}{2}gt^2 = 0.$$

Eliminating t between these two equations yields

$$R = \frac{2v_0^2}{g} \sin \theta_0 \cos \theta_0.$$

Using the identity $\sin 2\theta_0 = 2 \sin \theta_0 \cos \theta_0$ (see Appendix G), we obtain

$$R = \frac{v_0^2}{g} \sin 2\theta_0. \qquad (4\text{-}20)$$

Note that R has its maximum value when $\sin 2\theta_0 = 1$, which corresponds to $2\theta_0 = 90°$ or $\theta_0 = 45°$.

The Effects of the Air

We have assumed that the air through which the projectile moves has no effect on its motion, a reasonable assumption at low speeds. However, for greater speeds, the disagreement between our calculations and the actual motion of the projectile can be large because the air resists (or opposes) the motion. Figure 4-14, for example, shows two paths for a

Even if the effects of air could be neglected so that this motorcycle jump were simple projectile motion, it would still require careful planning if the motorcyclist is to land safely.

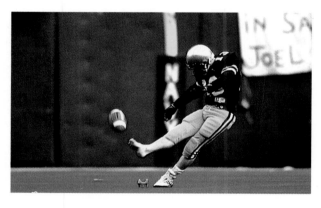

Sports involve many examples of projectile motion. However, the effects of air on the projectile nearly always increase the difficulty of determining the motion mathematically.

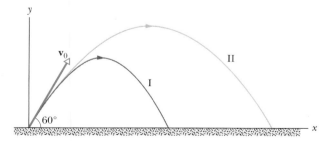

FIGURE 4-14 (I) The path of a fly ball, calculated (using a computer) by taking air resistance into account. (II) The path the ball would follow in a vacuum, calculated by the methods of this chapter. Also see Table 4-1.

TABLE 4-1
TWO FLY BALLS[a]

	PATH I (AIR)	PATH II (VACUUM)
Range	323 ft	581 ft
Maximum height	174 ft	252 ft
Time of flight	6.6 s	7.9 s

[a]See Fig. 4-14. The launch angle is 60° and the launch speed is 100 mi/h.

fly ball that leaves the bat at an angle of 60° with the horizontal and an initial speed of 100 mi/h. (See "The Trajectory of a Fly Ball," by Peter J. Brancazio, *The Physics Teacher,* January 1985.) Path I (the baseball player's fly ball) is a calculated path that approximates normal conditions of play, in air. Path II (the physics professor's fly ball) is the path that the ball would follow in a vacuum. Table 4-1 gives some data for the two cases. Our assumption that air resistance can be neglected clearly does not apply to experiments carried out in such laboratories as Shea Stadium and Candlestick Park. We discuss the effect of the air on motion in Chapter 6.

Actually, *no* problem in physics can be solved *exactly*, no matter how many significant figures the calculated answer may contain: it is always necessary to make approximations. The physicist P. A. M. Dirac said that the trick is to divide the problem into two parts, one simple and the other small. You then solve the simple part exactly and do the best you can with the small part. Sometimes, the "small" part is small enough that you can neglect it entirely, as we do for air resistance at low speeds.

SAMPLE PROBLEM 4-6

A rescue plane is flying at a constant elevation of 1200 m with a speed of 430 km/h toward a point directly over a person struggling in the water (see Fig. 4-15). At what angle of sight ϕ should the pilot release a rescue capsule if it is to strike (very close to) the person in the water?

SOLUTION The initial velocity of the capsule is the same as that of the plane. That is, the initial velocity $\mathbf{v}_0$ is horizontal and has a magnitude of 430 km/h. We

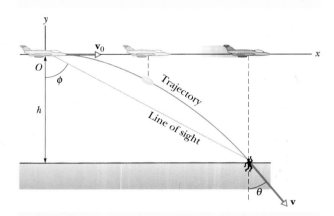

FIGURE 4-15   Sample Problem 4-6. A plane drops a rescue capsule and then continues in level flight. While the capsule is falling, its horizontal velocity component remains equal to the velocity of the plane. The capsule hits the water with velocity **v**, at angle θ from the vertical.

can find the time of flight of the capsule from Eq. 4-16,

$$y - y_0 = (v_0 \sin \theta_0) t - \tfrac{1}{2} g t^2.$$

Putting $y - y_0 = -1200$ m (the minus sign because the person is below the origin) and $\theta_0 = 0$, we obtain

$$-1200 \text{ m} = 0 - \tfrac{1}{2}(9.8 \text{ m/s}^2) t^2.$$

Solving for t yields*

$$t = \sqrt{\frac{(2)(1200 \text{ m})}{9.8 \text{ m/s}^2}} = 15.65 \text{ s}.$$

The horizontal distance covered by the capsule (and by the plane) in that time is given by Eq. 4-15:

$$x - x_0 = (v_0 \cos \theta_0) t$$
$$= (430 \text{ km/h})(\cos 0°)(15.65 \text{ s})(1 \text{ h}/3600 \text{ s})$$
$$= 1.869 \text{ km} = 1869 \text{ m}.$$

If $x_0 = 0$, then $x = 1869$ m. The angle of sight is then (see Fig. 4-15)

$$\phi = \tan^{-1} \frac{x}{h} = \tan^{-1} \frac{1869 \text{ m}}{1200 \text{ m}} = 57°. \quad \text{(Answer)}$$

Because the plane and the capsule have the same horizontal velocity, the plane remains vertically over the capsule while the capsule is in flight.

---

*We explain in Tactic 5 why we keep (temporarily) more significant figures for some quantities than the input data justify.

FIGURE 4-16   Sample Problem 4-7. Can he make the jump?

SAMPLE PROBLEM 4-7

A movie stuntman is to run across a rooftop and then horizontally off it, to land on the roof of the next building, as shown in Fig. 4-16. Before he attempts the jump, he wisely asks you to determine whether it is possible. Can he make the jump if his maximum rooftop speed is 4.5 m/s?

SOLUTION   The fall of 4.8 m will take a time t, which can be obtained from Eq. 4-16. Letting $y - y_0 = -4.8$ m (note the sign) and $\theta_0 = 0$, you rearrange Eq. 4-16 to obtain

$$t = \sqrt{-\frac{2(y - y_0)}{g}} = \sqrt{-\frac{(2)(-4.8 \text{ m})}{9.8 \text{ m/s}^2}}$$
$$= 0.990 \text{ s}.$$

You now ask: "How far would he move horizontally in this time?" The answer, from Eq. 4-15, is

$$x - x_0 = (v_0 \cos \theta_0) t$$
$$= (4.5 \text{ m/s})(\cos 0°)(0.990 \text{ s}) = 4.5 \text{ m}.$$

To reach the next building, the stuntman has to move 6.2 m horizontally. Your advice: "Don't jump."

SAMPLE PROBLEM 4-8

Figure 4-17 shows a pirate ship, moored 560 m from a fort defending the harbor entrance of an island. The harbor defense cannon, located at sea level, has a muzzle velocity of 82 m/s.

FIGURE 4-17 Sample Problem 4-8. At this range, the harbor defense cannon can hit the pirate ship at two different elevation angles.

a. To what angle must the cannon be elevated to hit the pirate ship?

SOLUTION Solving Eq. 4-20 for $2\theta_0$ yields

$$2\theta_0 = \sin^{-1}\frac{gR}{v_0^2} = \sin^{-1}\frac{(9.8 \text{ m/s}^2)(560 \text{ m})}{(82 \text{ m/s})^2}$$

$$= \sin^{-1} 0.816.$$

There are two angles whose sine is 0.816, namely, 54.7° and 125.3°. Thus we find

$$\theta_0 = \tfrac{1}{2}(54.7°) \approx 27° \qquad \text{(Answer)}$$

and

$$\theta_0 = \tfrac{1}{2}(125.3°) \approx 63°. \qquad \text{(Answer)}$$

The commandant of the fort can elevate the guns to either of these two angles and (if only there were no intervening air!) hit the pirate ship.

b. What are the times of flight for the two elevation angles calculated above?

SOLUTION Solving Eq. 4-15 for t gives, for $\theta_0 = 27°$,

$$t = \frac{x - x_0}{v_0 \cos \theta_0} = \frac{560 \text{ m}}{(82 \text{ m/s})(\cos 27°)}$$

$$= 7.7 \text{ s.} \qquad \text{(Answer)}$$

Repeating the calculation for $\theta_0 = 63°$ yields $t = 15$ s. It is reasonable that the time of flight for the higher elevation angle should be larger.

c. How far should the pirate ship be from the fort if it is to be beyond range of the cannon?

SOLUTION We have seen that maximum range corresponds to an elevation angle θ_0 of 45°. Thus, from Eq. 4-20 with $\theta_0 = 45°$,

$$R = \frac{v_0^2}{g} \sin 2\theta_0 = \frac{(82 \text{ m/s})^2}{9.8 \text{ m/s}^2} \sin(2 \times 45°)$$

$$= 690 \text{ m.} \qquad \text{(Answer)}$$

As the pirate ship sails away, the two elevation angles at which the ship can be hit draw closer together, eventually merging at $\theta_0 = 45°$ when the ship is 690 m away. Beyond that point the ship is safe.

SAMPLE PROBLEM 4-9

Figure 4-18 illustrates the flight of Emanuel Zacchini over three Ferris wheels, located as shown and each 18 m high. He is launched with speed $v_0 = 26.5$ m/s, at an angle $\theta_0 = 53°$ from the horizontal and with an initial height of 3.0 m above the ground. The net in which he is to land is at the same height.

a. Does he clear the first Ferris wheel?

SOLUTION We place the origin on the cannon muzzle so that $x_0 = 0$ and $y_0 = 0$. To find his height y when $x = 23$ m, we use Eq. 4-19:

$$y = (\tan \theta_0)x - \frac{gx^2}{2(v_0 \cos \theta_0)^2}$$

$$= (\tan 53°)(23 \text{ m}) - \frac{(9.8 \text{ m/s}^2)(23 \text{ m})^2}{2(26.5 \text{ m/s})^2(\cos 53°)^2}$$

$$= 20.3 \text{ m.} \qquad \text{(Answer)}$$

Since he begins 3.0 m off the ground, he clears the Ferris wheel by about 5.3 m.

FIGURE 4-18 Sample Problem 4-9. The flight of a human cannonball over three Ferris wheels, and the desired placement of the net.

b. If he reaches his maximum height when he is over the middle Ferris wheel, what is his clearance above it?

SOLUTION At maximum height, $v_y = 0$ and Eq. 4-18 becomes

$$v_y^2 = (v_0 \sin \theta_0)^2 - 2gy = 0.$$

Solving for y gives

$$y = \frac{(v_0 \sin \theta_0)^2}{2g} = \frac{(26.5 \text{ m/s})^2 (\sin 53°)^2}{(2)(9.8 \text{ m/s}^2)} = 22.9 \text{ m},$$

which means that he clears the middle Ferris wheel by 7.9 m.

c. What is his time of flight t?

SOLUTION Of the several ways to find t, we could use Eq. 4-16 and the fact that $y = 0$ when he lands. Then we have

$$y = (v_0 \sin \theta_0)t - \tfrac{1}{2}gt^2 = 0,$$

or

$$t = \frac{2v_0 \sin \theta_0}{g} = \frac{(2)(26.5 \text{ m/s})(\sin 53°)}{9.8 \text{ m/s}^2}$$

$$= 4.3 \text{ s.} \qquad \text{(Answer)}$$

d. How far from the cannon should the center of the net be positioned?

SOLUTION One way to answer is with Eq. 4-15, with $x_0 = 0$:

$$x = (v_0 \cos \theta_0)t$$

$$= (26.5 \text{ m/s})(\cos 53°)(4.3 \text{ s})$$

$$= 69 \text{ m,} \qquad \text{(Answer)}$$

which is the range R of the flight. Of course, Zacchini would use a wide net and bias it toward the cannon, because the air that he encounters will slow his flight, and he will not go as far as we have calculated.

PROBLEM SOLVING

TACTIC 5: SIGNIFICANT FIGURES

In some problems, you calculate a numerical value in one step of the problem and then use that number in a later step. In such cases, it is proper, strictly for the purposes of calculation, to keep more significant figures than you would accept in the final result.

In Sample Problem 4-6, for example, we calculated the time of flight of the capsule and found $t = 15.65$ s. If asked what the time of flight was, you would

have to round this and say 16 s. However, it is proper to use $t = 15.65$ s in further calculations, *provided* that you round the final result. If you round at every step, you will reduce the precision of your answer.

TACTIC 6: NUMBERS VERSUS ALGEBRA

One way to avoid numerical rounding errors is to solve problems algebraically, substituting numbers only in the final step. That is easy to do in the sample problems of this set and is the way experienced problem solvers operate. In these early chapters, however, we prefer to solve most problems in pieces, to give you a firmer numerical grasp of what is going on. As the text progresses, however, we shall stick more and more to the algebra and shall point out the advantages of doing so.

4-7 UNIFORM CIRCULAR MOTION

A particle is in **uniform circular motion** if it travels in a circle or circular arc at constant speed. Although the speed does not vary, *the particle is accelerating.* That fact may be surprising because we usually think of acceleration as an increase in speed. But actually $\mathbf{v}$ is a vector, not a scalar. If $\mathbf{v}$ changes, even only in direction, there is an acceleration, and that is the case for uniform circular motion.

We use Fig. 4-19 to find the magnitude and direction of the acceleration. That figure represents a particle in uniform circular motion with speed v in a

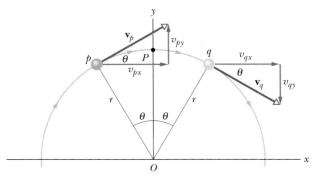

FIGURE 4-19 A particle moves in uniform circular motion at constant speed v in a circle of radius r. Its velocities $\mathbf{v}_p$ and $\mathbf{v}_q$ at points p and q, equidistant from the y axis, are shown, along with its velocity components at those points. The instantaneous acceleration of the particle at any point is directed toward the center of the circle and has a magnitude v^2/r.

circle of radius r. Velocity vectors are drawn for two points, p and q, that are symmetric with respect to the y axis. These vectors, $\mathbf{v}_p$ and $\mathbf{v}_q$, have the same magnitude v but—because they point in different directions—they are different vectors. Their x and y components are

$$v_{px} = +v \cos \theta, \qquad v_{py} = +v \sin \theta$$

and

$$v_{qx} = +v \cos \theta, \qquad v_{qy} = -v \sin \theta.$$

The time required for the particle to move from p to q at constant speed v is

$$\Delta t = \frac{\text{arc}(pq)}{v} = \frac{r(2\theta)}{v}, \qquad (4\text{-}21)$$

in which arc(pq) is the length of the arc from p to q.

We now have enough information to calculate the components of the average acceleration $\overline{\mathbf{a}}$ experienced by the particle as it moves from p to q in Fig. 4-19. For the x component, we have

$$\overline{a}_x = \frac{v_{qx} - v_{px}}{\Delta t} = \frac{v \cos \theta - v \cos \theta}{\Delta t} = 0.$$

This result is not surprising because it is clear from symmetry in Fig. 4-19 that the x component of velocity has the same value at q and at p.

For the y component of the average acceleration, we find, making use of Eq. 4-21,

$$\overline{a}_y = \frac{v_{qy} - v_{py}}{\Delta t} = \frac{-v \sin \theta - v \sin \theta}{\Delta t}$$

$$= -\frac{2v \sin \theta}{2r\theta/v} = -\left(\frac{v^2}{r}\right)\left(\frac{\sin \theta}{\theta}\right).$$

The minus sign tells us that this acceleration component points vertically downward in Fig. 4-19.

Now let us allow the angle θ in Fig. 4-19 to shrink, approaching zero as a limit. This requires that points p and q in that figure approach their midpoint, shown as point P at the top of the circle. The average acceleration $\overline{\mathbf{a}}$, whose components we have just found, then approaches the instantaneous acceleration $\mathbf{a}$ at point P.

The *direction* of this instantaneous acceleration vector at point P in Fig. 4-19 is downward, toward the center of the circle at O, because the direction of the average acceleration does not change as θ becomes smaller. To find the *magnitude* a of the instantaneous acceleration vector, we need only the mathematical fact that, as θ becomes smaller and smaller, the ratio

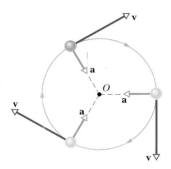

FIGURE 4-20 Velocity and acceleration vectors for a particle in uniform circular motion. Both have constant magnitude but vary continuously in direction.

$(\sin \theta)/\theta$ approaches unity. In the relation given above for $\overline{a}_y$, we then have

$$a = \frac{v^2}{r} \qquad \text{(centripetal acceleration)}. \qquad (4\text{-}22)$$

Our conclusion: when you see a particle moving at constant speed v in a circle (or a circular arc) of radius r, you may be sure that it has an acceleration, directed toward the center of the circle and having a magnitude v^2/r.

Figure 4-20 shows the relation between the velocity and acceleration vectors at various stages during uniform circular motion. Both vectors have constant magnitude as the motion progresses, but their directions change continuously. The velocity is always tangent to the circle in the direction of motion; the acceleration is always directed radially inward. Because of this, the acceleration associated with uniform circular motion is called a **centripetal** (meaning "center seeking") **acceleration,** a term coined by Isaac Newton.

The acceleration resulting from a change in the direction of a velocity is just as real as one resulting from a change in the magnitude of a velocity. In Fig. 2-8, for example, we showed Colonel John P. Stapp while his rocket sled was braked to rest. He was experiencing a velocity that was constant in direction but changing rapidly in magnitude. On the other hand, an astronaut whirling in the human centrifuge at the NASA Manned Spacecraft Center in Houston experiences a velocity that is constant in magnitude but changing rapidly in direction. The accelerations that these two people feel are indistinguishable.

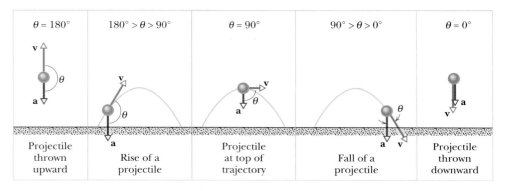

FIGURE 4-21 The velocity and acceleration vectors of a projectile for various motions. Note that the acceleration and velocity vectors do not have any fixed directional relation to each other.

There is no fixed relation between the direction of the velocity vector and that of the acceleration vector for a moving particle. Figure 4-21 shows examples in which the angle between these two vectors varies from 0° to 180°. In only one case do the two vectors happen to point in the same direction.

SAMPLE PROBLEM 4-10

A satellite is in circular Earth orbit, at an altitude $h = 200$ km above the Earth's surface. At that altitude, the free-fall acceleration g is 9.20 m/s². What is the orbital speed v of the satellite?

SOLUTION We have uniform circular motion around the Earth. We can find v using Eq. 4-22, with $a = g$ and with $r = R_E + h$, where R_E is the Earth's radius (see inside front cover or Appendix C):

$$g = \frac{v^2}{R_E + h}.$$

Solving for v gives

$$v = \sqrt{g(R_E + h)}$$
$$= \sqrt{(9.20 \text{ m/s}^2)(6.37 \times 10^6 \text{ m} + 200 \times 10^3 \text{ m})}$$
$$= 7770 \text{ m/s} = 7.77 \text{ km/s}. \qquad \text{(Answer)}$$

You can show that this is equivalent to 17,400 mi/h and that the satellite would take 1.47 h to complete one orbital revolution.

It may puzzle you that, although $g = 9.20$ m/s² at the position of the satellite orbit, an astronaut there would experience what is called "weightlessness," as if weight had somehow been eliminated. The explanation is that both the astronaut *and the spacecraft* are ac-

celerating toward the center of the Earth at 9.20 m/s². They are both in free fall, just like a passenger in a freely falling elevator cab. Hence the astronaut (and the elevator passenger) seem to float, as if without weight. See the essay "Physics in Weightlessness," by Sally Ride, following Chapter 15.

4-8 RELATIVE MOTION IN ONE DIMENSION

Suppose you see a duck flying north at, say, 20 mi/h. To another duck flying alongside, the first duck is at rest. In other words, the velocity of a particle depends on the **reference frame** of the person who is doing the measuring. For our purposes, a reference frame is the physical object to which you attach your coordinate system.

The reference frame that seems most natural to us in our daily comings and goings is the ground beneath our feet.* When a traffic officer tells you that you have been driving at 70 mi/h, the unspoken qualification, "in a coordinate system attached to the ground," is always taken for granted by each of you.

If you are traveling in an airplane or a spaceship, the reference frame of the Earth may not be the most convenient one. You are free to choose any reference frame that you wish. Once having made it, however, you must always be aware of your choice and you must be careful to make all your measurements with respect to the chosen reference frame.

---

*Even Shakespeare seemed to think so. He had Hamlet say, "this goodly frame, the earth. . . ."

FIGURE 4-22 Alex (frame A) and Barbara (frame B) watch car P. All motion is along the common x axis of the two frames. The vector $\mathbf{v}_{BA}$ shows the relative separation velocity of the two frames. The three position measurements shown are all made at the same instant of time.

Suppose that Alex (frame A) is parked by the side of a highway, watching car P (the "particle") speed past. Barbara (frame B), driving along the highway at constant speed, is also watching car P. Suppose that, as in Fig. 4-22, they both measure the position of the car at a given moment. From the figure we see that

$$x_{PA} = x_{PB} + x_{BA}. \qquad (4\text{-}23)$$

The terms in Eq. 4-23 are scalars and may be of either sign. The equation is read: "The position of P as measured by A *is equal to* the position of P as measured by B *plus* the position of B as measured by A." Note how this reading is supported by the sequence of the subscripts.

Taking the time derivative of Eq. 4-23, we obtain

$$\frac{d}{dt}(x_{PA}) = \frac{d}{dt}(x_{PB}) + \frac{d}{dt}(x_{BA}),$$

or (because $v = dx/dt$)

$$v_{PA} = v_{PB} + v_{BA}. \qquad (4\text{-}24)$$

This scalar equation is the relation between the velocities of the same object (car P) as measured in the two frames;* those measured velocities are different. In words, Eq. 4-24 says: "The velocity of P as measured by A *is equal to* the velocity of P as measured by B *plus* the velocity of B as measured by A." The term v_{BA} is the separation velocity of frame B with respect to frame A; see Fig. 4-22.

---

*It may help to note that, on the right-hand side of this equation, the two inner subscripts (B, B) are the same; the two outer ones (P, A) are the same as those on the left and appear in the same sequence.

We consider only frames that move at constant velocity *with respect to each other:* such frames are called **inertial reference frames.** In our example, this means that Barbara (frame B) will drive always at constant speed with respect to Alex (frame A). Car P (the moving particle), however, may speed up, slow down, come to rest, or reverse direction.

The time derivative of the velocity equation (Eq. 4-24) yields the acceleration equation,

$$a_{PA} = a_{PB}. \qquad (4\text{-}25)$$

Note that, because v_{BA} is a constant, its time derivative is zero. Equation 4-25 tells us that *observers on different inertial frames of reference (their velocity of separation is constant) will measure the same acceleration for the moving particle.*

SAMPLE PROBLEM 4-11

Alex, parked by the side of an east–west road, is watching car P, which is moving in a westerly direction. Barbara, driving east at a speed $v_{BA} = 52$ km/h, watches the same car. Take the easterly direction as positive.

a. If Alex measures a speed of 78 km/h for car P, what velocity will Barbara measure?

SOLUTION Equation 4-24 may be rearranged to yield

$$v_{PB} = v_{PA} - v_{BA}.$$

We have $v_{PA} = -78$ km/h, the minus sign telling us that car P is moving west, in the negative direction. We also have $v_{BA} = 52$ km/h, so that

$$v_{PB} = (-78 \text{ km/h}) - (52 \text{ km/h})$$
$$= -130 \text{ km/h}. \qquad \text{(Answer)}$$

If car P were connected to Barbara's car by a string wound up on a spool, the string would be unwinding at this speed as the two cars separated.

b. If Alex sees car P brake to a halt in 10 s, what acceleration (assumed constant) will he measure for it?

SOLUTION From Eq. 2-9 ($v = v_0 + at$) we have

$$a = \frac{v - v_0}{t} = \frac{0 - (-78 \text{ km/h})}{10 \text{ s}}$$

$$= \left(\frac{78 \text{ km/h}}{10 \text{ s}}\right)\left(\frac{1 \text{ m/s}}{3.6 \text{ km/h}}\right)$$

$$= 2.2 \text{ m/s}^2. \qquad \text{(Answer)}$$

c. What acceleration would Barbara measure for the braking car?

SOLUTION   Barbara sees the initial speed of the car as -130 km/h, as we calculated in (a) above. Although the car has braked to rest, it is at rest only in Alex's reference frame. To Barbara, car P is not at rest at all but is receding from her at 52 km/h so that its final velocity in her reference frame is -52 km/h. Thus, from the relation $v = v_0 + at$,

$$a = \frac{v - v_0}{t} = \frac{(-52 \text{ km/h}) - (-130 \text{ km/h})}{10 \text{ s}}$$

$$= 2.2 \text{ m/s}^2. \qquad \text{(Answer)}$$

This is exactly the same acceleration that Alex measured, which reassures us that we made no mistakes.

4-9 RELATIVE MOTION IN TWO DIMENSIONS

Now we move from the scalar world of relative motion in one dimension to the vector world of relative motion in two (and, by extension, in three) dimensions. Figure 4-23 shows reference frames A and B, now two dimensional. Our two observers are again watching a moving particle P. We once more assume that the two frames are separating at a constant velocity $\mathbf{v}_{BA}$ (the frames are inertial) and, for simplicity, we further assume that their x and y axes remain parallel to each other as they do so.

Let observers on frames A and B each measure the position of particle P at a certain instant. From

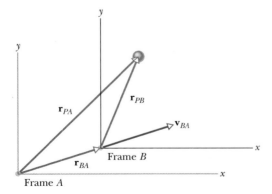

FIGURE 4-23   Reference frames in two dimensions. The vectors $\mathbf{r}_{PA}$ and $\mathbf{r}_{PB}$ show the positions of particle P in frames A and B, respectively. Vector $\mathbf{r}_{BA}$ shows the position of frame B with respect to frame A. Vector $\mathbf{v}_{BA}$ shows the (constant) separation velocity of the two frames.

the vector triangle in Fig. 4-23, we have the vector equation

$$\mathbf{r}_{PA} = \mathbf{r}_{PB} + \mathbf{r}_{BA}. \qquad (4\text{-}26)$$

This relation is the vector equivalent of the scalar Eq. 4-23.

If we take the time derivative of Eq. 4-26, we find a connection between the (vector) velocities of the particle as measured by our two observers; namely,

$$\mathbf{v}_{PA} = \mathbf{v}_{PB} + \mathbf{v}_{BA} \qquad (4\text{-}27)$$

This is the vector equivalent of the scalar Eq. 4-24. Note that the order of the subscripts is the same as in that equation, and that again $\mathbf{v}_{BA}$ is the constant relative velocity of frame B as observed by the observer on frame A.

If we take the time derivative of Eq. 4-27, we obtain a connection between the two measured accelerations; namely,

$$\mathbf{a}_{PA} = \mathbf{a}_{PB}. \qquad (4\text{-}28)$$

It remains true for motion in three dimensions that all observers on inertial reference frames will measure the same acceleration for a moving particle.

SAMPLE PROBLEM 4-12

A bat detects an insect (lunch) while the two are flying with velocities $\mathbf{v}_{BG}$ and $\mathbf{v}_{IG}$, respectively, measured with respect to the ground. See Fig. 4-24a. What is the velocity $\mathbf{v}_{IB}$ of the insect with respect to the bat, in unit-vector notation?

SOLUTION   From Fig. 4-24a, the velocities of the insect and the bat, relative to the ground, are given by

$$\mathbf{v}_{IG} = (5.0 \text{ m/s})(\cos 50°)\mathbf{i} + (5.0 \text{ m/s})(\sin 50°)\mathbf{j}$$

and

$$\mathbf{v}_{BG} = (4.0 \text{ m/s})(\cos 150°)\mathbf{i} + (4.0 \text{ m/s})(\sin 150°)\mathbf{j},$$

where in each term, the angle is relative to the positive direction of the x axis. The velocity $\mathbf{v}_{IB}$ of the *insect relative to the bat* is given by the vector sum of the velocity $\mathbf{v}_{IG}$ of the *insect relative to the ground* and the velocity $\mathbf{v}_{GB}$ of the *ground relative to the bat*; that is,

$$\mathbf{v}_{IB} = \mathbf{v}_{IG} + \mathbf{v}_{GB},$$

as shown in Fig. 4-24b. The vector $\mathbf{v}_{GB}$ is in the opposite direction of the vector $\mathbf{v}_{BG}$. Hence $\mathbf{v}_{BG} = -\mathbf{v}_{BG}$, and so

$$\mathbf{v}_{IB} = \mathbf{v}_{IG} + (-\mathbf{v}_{BG}).$$

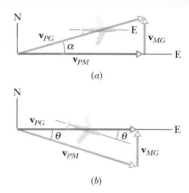

FIGURE 4-25 Sample Problem 4-13. (*a*) A plane, heading due east, is blown to the north. (*b*) To travel due east, the plane must head into the wind.

FIGURE 4-24 Sample Problem 4-12. (*a*) A bat detects an insect. (*b*) Velocity vectors of insect and bat.

Substituting the unit-vector expressions for $\mathbf{v}_{IG}$ and $\mathbf{v}_{BG}$ into this expression, we find

$$\mathbf{v}_{IB} = (5.0 \text{ m/s})(\cos 50°)\mathbf{i} + (5.0 \text{ m/s})(\sin 50°)\mathbf{j}$$

$$- (4.0 \text{ m/s})(\cos 150°)\mathbf{i} - (4.0 \text{ m/s})(\sin 150°)\mathbf{j}$$

$$= 3.21\mathbf{i} + 3.83\mathbf{j} + 3.46\mathbf{i} - 2.0\mathbf{j}$$

$$\approx (6.7 \text{ m/s})\mathbf{i} + (1.8 \text{ m/s})\mathbf{j}. \qquad \text{(Answer)}$$

SAMPLE PROBLEM 4-13

The compass in a plane indicates that is is headed due east; its airspeed indicator reads 215 km/h. (Airspeed is speed relative to the air.) A steady wind of 65.0 km/h is blowing due north.

a. What is the velocity of the plane with respect to ground?

SOLUTION The moving "particle" in this problem is the plane P. There are two reference frames, the ground (G) and the air mass (M). By a simple change of notation, we can rewrite Eq. 4-27 as

$$\mathbf{v}_{PG} = \mathbf{v}_{PM} + \mathbf{v}_{MG}. \qquad (4\text{-}29)$$

Figure 4-25*a* shows these vectors, which form a right triangle. The terms in Eq. 4-29 are, in sequence, the

velocity of the plane with respect to the ground, the velocity of the plane with respect to the air, and the velocity of the air with respect to the ground (that is, the wind velocity). Note the orientation of the plane, which is consistent with a due east reading on its compass. The plane is pointed due east but may not actually be moving in that direction.

The magnitude of the plane's velocity relative to the ground is found from

$$v_{PG} = \sqrt{v_{PM}^2 + v_{MG}^2}$$

$$= \sqrt{(215 \text{ km/h})^2 + (65.0 \text{ km/h})^2}$$

$$= 225 \text{ km/h}. \qquad \text{(Answer)}$$

The angle α in Fig. 4-25*a* follows from

$$\alpha = \tan^{-1} \frac{v_{MG}}{v_{PM}} = \tan^{-1} \frac{65.0 \text{ km/h}}{215 \text{ km/h}}$$

$$= 16.8°. \qquad \text{(Answer)}$$

Thus, with respect to the ground, the plane is flying at 225 km/h in a direction 16.8° north of east. Note that its speed relative to the ground (the "ground speed") is greater than its airspeed.

b. If the pilot wishes to fly due east, what must be the heading? That is, what must the compass read?

SOLUTION In this case the pilot must head into the wind so that the velocity of the plane with respect to the ground points east. The wind velocity is unchanged, and the vector diagram representing the new situation is as shown in Fig. 4-25*b*. Note that the three vectors still form a right triangle, as they did in Fig. 4-25*a*, and Eq. 4-29 still holds.

The pilot's ground speed is now

$$v_{PG} = \sqrt{v_{PM}^2 - v_{MG}^2}$$

$$= \sqrt{(215 \text{ km/h})^2 - (65.0 \text{ km/h})^2} = 205 \text{ km/h}.$$

As the orientation of the plane in Fig. 4-25b indicates, the pilot must head into the wind by an angle θ given by

$$\theta = \sin^{-1} \frac{v_{MG}}{v_{PM}} = \sin^{-1} \frac{65.0 \text{ km/h}}{215 \text{ km/h}} = 17.6°. \quad \text{(Answer)}$$

Note that for this orientation the ground speed is less than the airspeed.

4-10 RELATIVE MOTION AT HIGH SPEEDS (OPTIONAL)

An orbiting satellite has a speed of 17,000 mi/h. Before you call that a high speed you must answer this question: "High compared to what?" Nature has given us a standard; it is the speed of light c, where

$$c = 299,792,458 \text{ m/s} \quad \text{(speed of light).} \quad (4\text{-}30)$$

As we shall examine later, no entity—be it particle or wave—can move faster than the speed of light, no matter what reference frame is used for observation. By this standard, all tangible large-scale objects—no matter how high their speeds seem by ordinary standards—are very slow indeed. The speed of the orbiting satellite, for example, is only 0.0025% of the speed of light. Subatomic particles such as electrons or protons, however, can acquire speeds very close to (but never equal to or greater than) the speed of light. Experiment shows, for example, that an electron accelerated through 10 million volts acquires a speed of 0.9988c; if you accelerate it through 20 million volts, its speed increases, but only to 0.9997c. The speed of light is a barrier that objects can approach but never reach. (Alas, the ultra-light-speeds utilized in science fiction, such as during "warp drive" in *Star Trek*, when the speed is $c2^n$, where n is the warp number, are only fictional.)

We now ask: "How can we be sure that the kinematics we have examined so far, which was developed by studying very slow objects, also holds for very fast objects, such as highly energetic electrons or protons?" The answer, which we can find only from experiment, is that the kinematics for slow objects does *not* hold at speeds that approach the speed of light. Einstein's **theory of special relativity**, how-

ever, agrees with experiment at *all* speeds. Here we get a foretaste of that theory, which is introduced formally in Chapter 42.

At "slow" speeds—all speeds that can be acquired by common tangible objects—the kinematic equations of Einstein's theory reduce to those of the kinematics we have examined. The failure of the "slow kinematics" is gradual, its predictions agreeing less and less well with experiment as the speed increases. Here is an example: Equation 4-24,

$$v_{PA} = v_{PB} + v_{BA} \quad \text{(slow speeds),} \quad (4\text{-}31)$$

gives the relation of the speed of particle P as seen by an observer in frame B to that seen by an observer in frame A. The corresponding equation in Einstein's theory is

$$v_{PA} = \frac{v_{PB} + v_{BA}}{1 + v_{PB}v_{BA}/c^2} \quad \text{(all speeds).} \quad (4\text{-}32)$$

If $v_{PB} \ll c$ and $v_{BA} \ll c$ (which is always the case for ordinary objects), then the denominator in Eq. 4-32 is very close to unity and Eq. 4-32 reduces to Eq. 4-31, as we know it must.

The speed of light c is the central constant of Einstein's theory, and every relativistic equation contains it. A way to test such equations is to allow c to become infinitely large. Under those conditions, *all* speeds would be "slow" and "slow kinematics" would never fail. Putting $c \to \infty$ in Eq. 4-32 does indeed reduce that equation to Eq. 4-31.

SAMPLE PROBLEM 4-14

(Slow speeds) For the case where $v_{PB} = v_{BA} = 0.0001c$ ($= 67,000$ mi/h!), what do Eqs. 4-31 and 4-32 predict for v_{PA}?

SOLUTION From Eq. 4-31,

$$v_{PA} = v_{PB} + v_{BA}$$

$$= 0.0001c + 0.0001c$$

$$= 0.0002c. \quad \text{(Answer)}$$

From Eq. 4-32,

$$v_{PA} = \frac{v_{PB} + v_{BA}}{1 + v_{PB}v_{BA}/c^2}$$

$$= \frac{0.0001c + 0.0001c}{1 + (0.0001c)^2/c^2} = \frac{0.0002c}{1.00000001}$$

$$\approx 0.0002c. \quad \text{(Answer)}$$

Conclusion: At any speed acquired by ordinary tangible objects, Eqs. 4-31 and 4-32 yield essentially the same answer. We can use Eq. 4-31 ("slow kinematics") without a second thought.

SAMPLE PROBLEM 4-15

(High speeds) For $v_{PB} = v_{BA} = 0.65c$, what do Eqs. 4-31 and 4-32 predict for v_{PA}?

SOLUTION From Eq. 4-31,

$$v_{PA} = v_{PB} + v_{BA}$$
$$= 0.65c + 0.65c = 1.30c. \qquad \text{(Answer)}$$

From Eq. 4-32,

$$v_{PA} = \frac{v_{PB} + v_{BA}}{1 + v_{PB}v_{BA}/c^2}$$
$$= \frac{0.65c + 0.65c}{1 + (0.65c)(0.65c)/c^2}$$
$$= \frac{1.30c}{1.423} = 0.91c. \qquad \text{(Answer)}$$

Conclusion: At high speeds, "slow kinematics" and special relativity predict very different results. "Slow kinematics" involves no upper limit on speed and can easily (as in this case) predict a speed greater than the speed of light. Special relativity, on the other hand, *never* predicts a speed that exceeds c, no matter how high the combining speeds. Experiment agrees with special relativity on all counts.

REVIEW & SUMMARY

Position Vector

The location of a particle relative to the origin of a coordinate system is given by a *position vector* **r**, which in unit-vector notation is

$$\mathbf{r} = x\mathbf{i} + y\mathbf{j} + z\mathbf{k}, \qquad (4\text{-}1)$$

where $x\mathbf{i}$, $y\mathbf{j}$, and $z\mathbf{k}$ are the *vector components* and x, y, and z are the *scalar components* of position vector **r**. A position vector might also be given by a magnitude and one or more angles for orientation.

Displacement

If a particle moves so that its position vector changes from being $\mathbf{r}_1$ to $\mathbf{r}_2$, then the particle's *displacement* $\Delta \mathbf{r}$ is

$$\Delta \mathbf{r} = \mathbf{r}_2 - \mathbf{r}_1. \qquad (4\text{-}3)$$

Sample Problem 4-1 gives an example of displacement.

Average Velocity

If a particle undergoes a displacement in time Δt, its *average velocity* $\bar{\mathbf{v}}$ is

$$\bar{\mathbf{v}} = \frac{\Delta \mathbf{r}}{\Delta t}. \qquad (4\text{-}4)$$

Velocity

As Δt in Eq. 4-4 is shrunk to 0, $\bar{\mathbf{v}}$ reaches a limit called the (instantaneous) *velocity:*

$$\mathbf{v} = \frac{d\mathbf{r}}{dt}, \qquad (4\text{-}6)$$

which can be rewritten in unit-vector notation as

$$\mathbf{v} = v_x\mathbf{i} + v_y\mathbf{j} + v_z\mathbf{k}, \qquad (4\text{-}7)$$

where $v_x = dx/dt$, $v_y = dy/dt$, and $v_z = dz/dt$. A two-dimensional example of **v** is given in Sample Problem 4-3.

When the position of a moving particle is plotted on a coordinate system, **v** is always tangent to the curve representing the particle's path.

Average Acceleration

If a particle's velocity changes from $\mathbf{v}_1$ to $\mathbf{v}_2$ in time Δt, its *average acceleration* during Δt is

$$\bar{\mathbf{a}} = \frac{\mathbf{v}_2 - \mathbf{v}_1}{\Delta t} = \frac{\Delta \mathbf{v}}{\Delta t}. \qquad (4\text{-}9)$$

Acceleration

As Δt in Eq. 4-9 is shrunk to 0, $\bar{\mathbf{a}}$ reaches a limiting value called the (instantaneous) *acceleration*, which is

$$\mathbf{a} = \frac{d\mathbf{v}}{dt}. \qquad (4\text{-}10)$$

In unit-vector notation,

$$\mathbf{a} = a_x\mathbf{i} + a_y\mathbf{j} + a_z\mathbf{k}, \qquad (4\text{-}11)$$

where $a_x = dv_x/dt$, $a_y = dv_y/dt$, and $a_z = dv_z/dt$. An example of how Eq. 4-10 is used is in Sample Problem 4-4.

When **a** is constant, the components of **a**, **v**, and **r** along any axis can be treated as in the one-dimensional motion of Chapter 2. See Sample Problem 4-5.

Projectile Motion

Projectile motion is the motion of a particle that is launched with an initial velocity $\mathbf{v}_0$ and that then has the free-fall acceleration g. If $\mathbf{v}_0$ is expressed as a magnitude (the speed v_0) and an angle θ_0, the equations of motion along the horizontal x axis and vertical y axis are

$$x - x_0 = (v_0 \cos \theta_0)t, \tag{4-15}$$

$$y - y_0 = (v_0 \sin \theta_0)t - \tfrac{1}{2}gt^2, \tag{4-16}$$

$$v_y = v_0 \sin \theta_0 - gt, \tag{4-17}$$

and

$$v_y^2 = (v_0 \sin \theta_0)^2 - 2g(y - y_0). \tag{4-18}$$

The path of a particle in projectile motion is parabolic and given by

$$y = (\tan \theta_0)x - \frac{gx^2}{2(v_0 \cos \theta_0)^2}, \tag{4-19}$$

where the origin has been chosen so that x_0 and y_0 are both zero. The particle's **range** R, which is the horizontal distance from the launch point to the point at which the particle returns to the launch height, is

$$R = \frac{v_0^2 \sin 2\theta_0}{g}. \tag{4-20}$$

Sample Problems 4-6 to 4-9 involve projectile motion.

Uniform Circular Motion

If a particle travels along a circle or circular arc with radius r at constant speed v, it is in **uniform circular motion** and has an acceleration $\mathbf{a}$ of magnitude

$$a = \frac{v^2}{r}. \tag{4-22}$$

The direction of $\mathbf{a}$ is toward the center of the circle or circular arc, and $\mathbf{a}$ is said to be **centripetal.** An example of how to use Eq. 4-22 is given in Sample Problem 4-10.

Relative Motion

When two frames of reference A and B are moving relative to each other at constant velocity, they are said to be **inertial reference frames.** The velocity of a moving particle as measured by an observer in frame A, in general, differs from that measured from frame B. The two measured velocities are related by

$$\mathbf{v}_{PA} = \mathbf{v}_{PB} + \mathbf{v}_{BA}, \tag{4-27}$$

in which $\mathbf{v}_{BA}$ is the velocity of B with respect to A. Both observers measure the same acceleration for the particle; that is,

$$\mathbf{a}_{PA} = \mathbf{a}_{PB}. \tag{4-28}$$

Sample Problem 4-11 illustrates the use of these equations for one-dimensional motion; Sample Problems 4-12 and 4-13 treat two-dimensional motion.

If speeds near the speed of light are involved, Eq. 4-24 must be replaced with an equation derived using the **special theory of relativity.** For one-dimensional motion, the correct result is

$$v_{PA} = \frac{v_{PB} + v_{BA}}{1 + v_{PB}v_{BA}/c^2}, \tag{4-32}$$

which becomes identical to Eq. 4-24 if the speeds are all negligible compared to the speed of light c. Sample Problems 4-14 and 4-15 illustrate the use of this equation.

QUESTIONS

1. Can the acceleration of a body change direction (a) without the displacement immediately changing direction also and (b) without the velocity immediately changing direction? If so, give an example.

2. In broad jumping, sometimes called long jumping, what factors determine the length of the jump?

3. At what point in the path of a projectile is the speed a minimum? A maximum?

4. Figure 4-26 shows the path followed by a NASA Learjet in a run designed to simulate low-gravity conditions for a short period of time. Make an argument to show that, if the plane follows a particular parabolic path, the passengers will experience weightlessness.

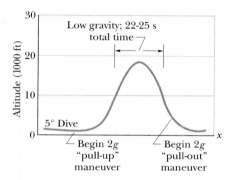

FIGURE 4-26 Question 4.

5. A shot is put (thrown) from above the athlete's shoulder level. The launch angle that will produce the longest range is less than 45°; that is, a flatter trajectory has a longer range. Explain why.

6. You are driving directly behind a pickup truck, going at the same speed as the truck. A crate falls from the bed of the truck to the road. Will your car hit the crate before the crate hits the road if you neither brake nor swerve? Explain your answer.

7. Trajectories are shown in Fig. 4-27 for three kicked footballs. Ignoring the effects of the air on the footballs, order the trajectories according to (a) time of flight, (b) initial vertical velocity component, (c) initial horizontal velocity component, and (d) initial speed. Order from largest to smallest and indicate any ties.

8. A rifle is "sighted-in" against a target, with both at the same elevation. Show that, at the same range, the rifle will shoot too high when shooting either uphill or downhill. (See "A Puzzle in Elementary Ballistics," by Ole Anton Haugland, *The Physics Teacher,* April 1983.)

9. When the Germans shelled Paris from 70 mi away with the WWI long-range artillery piece nicknamed Big Bertha, the shells were fired at an angle greater than 45°; the Germans had discovered that a greater angle gives a greater range, possibly even twice as long as with a 45° angle. Considering that the density of air decreases with altitude, explain their discovery.

10. In projectile motion when air resistance is negligible, is it ever necessary to consider three-dimensional motion rather than two-dimensional?

11. Is it possible to be accelerating if you are traveling at constant speed? Is it possible to round a curve with zero acceleration? With constant acceleration?

12. Show that, taking the Earth's rotation and revolution into account, a book resting on your table moves faster at night than it does during the daytime. In what reference frame is this statement true?

13. An aviator, pulling out of a dive, follows the arc of a circle and is said to have "experienced 3g's" in pulling out of the dive. Explain what this statement means.

14. A boy sitting in a railroad car moving at constant velocity throws a ball straight up into the air. Will the ball fall behind him? In front of him? Into his hands? What happens if the car accelerates forward or goes around a curve while the ball is in the air?

15. A woman on the observation platform of a train moving with constant velocity drops a coin while leaning over the rail. Describe the path of the coin as seen by (a) the woman on the train, (b) a person standing on the ground near the track, and (c) a person in a second train moving in the opposite direction to the first train on a parallel track.

16. If the acceleration of a body is constant as measured from a given reference frame, is it necessarily constant when measured from all other reference frames?

17. Equation 4-31 is so instinctively familiar from everyday experience that it is sometimes claimed to be "obviously correct, requiring no proof." Many so-called refutations of relativity theory turn out to be based on this claim. How would you refute someone who made this claim?

18. An elevator is descending at a constant speed. A passenger drops a coin to the floor. What acceleration would (a) the passenger and (b) a person at rest with respect to the elevator shaft observe for the falling coin?

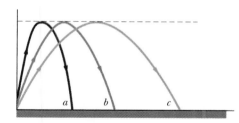

FIGURE 4-27 Question 7.

EXERCISES & PROBLEMS

SECTION 4-2 POSITION AND DISPLACEMENT

1E. A watermelon has the following coordinates: $x = -5.0$ m, $y = 8.0$ m, and $z = 0$ m. Find its position vector (a) in unit-vector notation and (b) in terms of the vector's magnitude and orientation. (c) Sketch the vector on a right-handed coordinate system.

2E. The position vector for an electron is $\mathbf{r} = 5.0\mathbf{i} - 3.0\mathbf{j} + 2.0\mathbf{k}$, where the suppressed unit is meters. (a) Find the magnitude of $\mathbf{r}$. (b) Sketch the vector on a right-handed coordinate system.

3E. The position vector for a proton is initially $\mathbf{r} = 5.0\mathbf{i} - 6.0\mathbf{j} + 2.0\mathbf{k}$ and then later is $\mathbf{r} = -2.0\mathbf{i} + 6.0\mathbf{j} + 2.0\mathbf{k}$, where the suppressed unit is meters. (a) What is the proton's displacement vector, and (b) to what plane is it parallel?

4E. A positron undergoes a displacement $\Delta\mathbf{r} = 2.0\mathbf{i} - 3.0\mathbf{j} + 6.0\mathbf{k}$, ending with a position vector $\mathbf{r} = 3.0\mathbf{j} - 4.0\mathbf{k}$. (The suppressed unit is meters.) What was the positron's former position vector?

SECTION 4-3 VELOCITY AND AVERAGE VELOCITY

5E. A plane flies 300 mi east from city A to city B in 45.0 min and then 600 mi south from city B to city C in 1.50 h. (a) What displacement vector represents the total trip? What are (b) the average velocity vector and (c) the average speed for the trip?

6E. A train moving at a constant speed of 60.0 km/h moves east for 40.0 min, then in a direction 50.0° east of north for 20.0 min, and finally west for 50.0 min. What is the average velocity of the train during this run?

7E. In 3.50 h, a balloon drifts 21.5 km north, 9.70 km east, and 2.88 km in upward elevation from its release point on the ground. Find (a) the magnitude of its average velocity and (b) the angle its average velocity makes with the horizontal.

8E. An ion's position vector is initially $\mathbf{r} = 5.0\mathbf{i} - 6.0\mathbf{j} + 2.0\mathbf{k}$, and 10 s later it is $\mathbf{r} = -2.0\mathbf{i} + 8.0\mathbf{j} - 2.0\mathbf{k}$. (The suppressed unit is meters.) What was its average velocity during the 10 s?

9E. The position of an electron is given by $\mathbf{r} = 3.0t\mathbf{i} - 4.0t^2\mathbf{j} + 2.0\mathbf{k}$ (where t is in seconds and the coefficients have the proper units for $\mathbf{r}$ to be in meters). (a) What is $\mathbf{v}(t)$ for the electron? (b) In unit-vector notation, what is $\mathbf{v}$ at $t = 2.0$ s? (c) What are the magnitude and direction of $\mathbf{v}$ just then?

SECTION 4-4 ACCELERATION AND AVERAGE ACCELERATION

10E. A proton initially has $\mathbf{v} = 4.0\mathbf{i} - 2.0\mathbf{j} + 3.0\mathbf{k}$ and then 4.0 s later has $\mathbf{v} = -2.0\mathbf{i} - 2.0\mathbf{j} + 5.0\mathbf{k}$ (the unshown unit is m/s). (a) In unit-vector notation, what is the average acceleration $\bar{\mathbf{a}}$ over the 4.0 s? (b) What are the magnitude and orientation of $\bar{\mathbf{a}}$?

11E. A particle moves so that its position as a function of time in SI units is $\mathbf{r} = \mathbf{i} + 4t^2\mathbf{j} + t\mathbf{k}$. Write expressions for (a) its velocity and (b) its acceleration as functions of time.

12E. The position $\mathbf{r}$ of a particle moving in an xy plane is given by $\mathbf{r} = (2.00t^3 - 5.00t)\mathbf{i} + (6.00 - 7.00t^4)\mathbf{j}$. Here $\mathbf{r}$ is in meters and t in seconds. Calculate (a) $\mathbf{r}$, (b) $\mathbf{v}$, and (c) $\mathbf{a}$ when $t = 2.00$ s. (d) What is the orientation of a line that is tangent to the particle's path at $t = 2.00$ s?

13E. An ice boat sails across the surface of a frozen lake with constant acceleration produced by the wind. At a certain instant its velocity is $6.30\mathbf{i} - 8.42\mathbf{j}$ in meters per second. Three seconds later, because of a shift in the wind, the boat is instantaneously at rest. What is its average acceleration during this 3-s interval?

14P. A particle A moves along the line $y = 30$ m with a constant velocity $\mathbf{v}$ ($v = 3.0$ m/s) directed parallel to the positive x axis (Fig. 4-28). A second particle B starts at the origin with zero speed and constant acceleration $\mathbf{a}$ ($a = 0.40$ m/s^2) at the same instant that particle A passes the y axis. What angle θ between $\mathbf{a}$ and the positive y axis would result in a collision between these two particles? (If your computation involves an equation with a term such as t^4, substitute $u = t^2$ and then consider solving the quadratic equation to get u.)

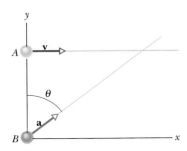

FIGURE 4-28   Problem 14.

15P. A particle leaves the origin with an initial velocity $\mathbf{v} = 3.00\mathbf{i}$, in meters per second. It experiences a constant acceleration $\mathbf{a} = -1.00\mathbf{i} - 0.500\mathbf{j}$, in meters per second squared. (a) What is the velocity of the particle when it reaches its maximum x coordinate? (b) Where is the particle at this time?

16P. The velocity $\mathbf{v}$ of a particle moving in the xy plane is given by $\mathbf{v} = (6.0t - 4.0t^2)\mathbf{i} + 8.0\mathbf{j}$. Here $\mathbf{v}$ is in meters per second and t (> 0) is in seconds. (a) What is the acceleration when $t = 3.0$ s? (b) When (if ever) is the acceleration zero? (c) When (if ever) is the velocity zero? (d) When (if ever) does the speed equal 10 m/s?

SECTION 4-6 PROJECTILE MOTION ANALYZED

> *In some of these problems, exclusion of the effects of the air is unwarranted but helps simplify the calculations.*

17E. A dart is thrown horizontally toward the bull's-eye, point *P* on the dart board of Fig. 4-29, with an initial speed of 10 m/s. It hits at point *Q* on the rim, vertically below *P*, 0.19 s later. (a) What is the distance *PQ*? (b) How far away from the dart board did the dart thrower stand?

FIGURE 4-29 Exercise 17.

18E. A rifle is aimed horizontally at a target 100 ft away. The bullet hits the target 0.75 in. below the aiming point. (a) What is the bullet's time of flight? (b) What is the muzzle velocity of the rifle?

19E. Electrons, like all other forms of matter, can undergo free fall. (a) If an electron is projected horizontally with a speed of 3.0×10^6 m/s, how far will it fall in traversing 1.0 m of horizontal distance? (b) Does the answer increase or decrease if the initial speed is increased?

20E. In a cathode-ray tube, a beam of electrons is projected horizontally with a speed of 1.0×10^9 cm/s into the region between a pair of horizontal plates 2.0 cm square. An electric field between the plates causes a constant downward acceleration of the electrons of magnitude 1.0×10^{17} cm/s^2. Find (a) the time required for the electrons to pass through the plates, (b) the vertical displacement of the beam in passing through the plates (it does not run into a plate), and (c) the velocity of the beam as it emerges from the plates.

21E. A ball rolls horizontally off the edge of a tabletop that is 4.0 ft high. It strikes the floor at a point 5.0 ft horizontally away from the edge of the table. (a) For how long was the ball in the air? (b) What was its speed at the instant it left the table?

22E. A projectile is fired horizontally from a gun that is 45.0 m above flat ground. The muzzle speed is 250 m/s. (a) How long does the projectile remain in the air? (b) At what horizontal distance from the firing point does it strike the ground? (c) What is the magnitude of the vertical component of its velocity as it strikes the ground?

23E. A baseball leaves a pitcher's hand horizontally at a speed of 100 mi/h. The distance to the batter is 60 ft. (a) How long does it take for the ball to travel the first 30 ft horizontally? The second 30 ft? (b) How far does the ball fall under gravity during the first 30 ft of its horizontal travel? (c) During the second 30 ft? (d) Why aren't the quantities in (b) and (c) equal? (Ignore the effect of air resistance.)

24E. A projectile is launched with an initial speed of 30 m/s at an angle of 60° above the horizontal. Calculate the magnitude and direction of its velocity (a) 2.0 s and (b) 5.0 s after launch.

25E. A stone is catapulted rightward with an initial velocity of 20.0 m/s at an angle of 40.0° above level ground. Find its horizontal and vertical displacements (a) 1.10 s, (b) 1.80 s, and (c) 5.00 s after launch.

26E. You throw a ball from a cliff with an initial velocity of 15.0 m/s at an angle of 20.0° below the horizontal. Find (a) its horizontal displacement and (b) its vertical displacement 2.30 s later.

27E. You throw a ball with a speed of 25.0 m/s at an angle of 40.0° above the horizontal directly toward a wall as shown in Fig. 4-30. The wall is 22.0 m from the release point of the ball. (a) How long is the ball in the air before it hits the wall? (b) How far above the release point does the ball hit the wall? (c) What are the horizontal and vertical components of its velocity as it hits the wall? (d) Has it passed the highest point on its trajectory when it hits?

FIGURE 4-30 Exercise 27.

28E. (a) Prove that for a projectile fired from level ground at an angle θ_0 above the horizontal, the ratio of the maximum height H to the range R is given by $H/R = \frac{1}{4} \tan \theta_0$. See Fig. 4-31. (b) For what angle θ_0 does $H = R$?

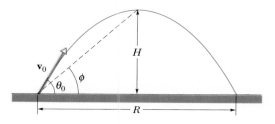

FIGURE 4-31 Exercises 28 and 29.

29E. A projectile is fired from level ground at an angle θ_0 above the horizontal. (a) Show that the elevation angle ϕ of the highest point as seen from the launch point is related to θ_0, the elevation angle of projection, by $\tan \phi = \frac{1}{2} \tan \theta_0$. See Fig. 4-31 and Exercise 28. (b) Calculate ϕ for $\theta_0 = 45°$.

30E. A stone is projected at a cliff of height h with an initial speed of 42.0 m/s directed 60.0° above the horizontal, as shown in Fig. 4-32. The stone strikes at A, 5.50 s after launching. Find (a) the height h of the cliff; (b) the speed of the stone just before impact at A; and (c) the maximum height H reached above the ground.

31P. In Sample Problem 4-6, find (a) the speed of the capsule when it hits the water and (b) the impact angle θ shown in Fig. 4-15.

FIGURE 4-32 Exercise 30.

32P. In the 1968 Olympics in Mexico City, Bob Beamon shattered the record for the long jump with a jump of 8.90 m. (See Fig. 4-33.) Assume that his speed on takeoff was 9.5 m/s, about equal to that of a sprinter. How close did this world-class athlete come to the maximum possible range in the absence of air resistance? The value of g in Mexico City is 9.78 m/s².

FIGURE 4-33 Problem 32. Bob Beamon's jump.

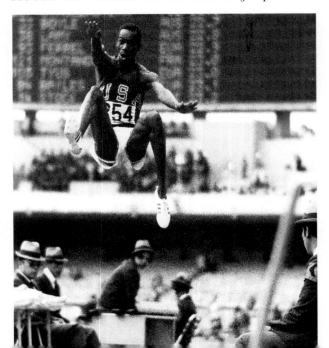

33P. A rifle with a muzzle velocity of 1500 ft/s shoots a bullet at a target 150 ft away. How high above the target must the rifle barrel be pointed so that the bullet will hit the target?

34P. Show that the maximum height reached by a projectile is $y_{max} = (v_0 \sin \theta_0)^2/2g$.

35P. In a detective story, a body is found 15 ft from the base of a building and beneath an open window 80 ft above. Would you guess the death to be accidental or not? Explain your answer.

36P. In Galileo's *Two New Sciences*, the author states that "for elevations [angles of projection] which exceed or fall short of 45° by equal amounts, the ranges are equal. . . ." Prove this statement. (See Fig. 4-34.)

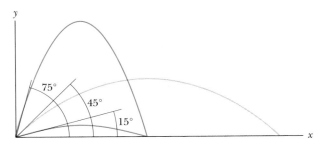

FIGURE 4-34 Problem 36.

37P. A ball is shot from the ground into the air. At a height of 9.1 m, the velocity is observed to be $\mathbf{v} = 7.6\mathbf{i} + 6.1\mathbf{j}$ in meters per second ($\mathbf{i}$ horizontal, $\mathbf{j}$ upward). (a) To what maximum height will the ball rise? (b) What will be the total horizontal distance traveled by the ball? (c) What is the velocity of the ball (magnitude and direction) the instant before it hits the ground?

38P. The range of a projectile depends not only on v_0 and θ_0 but also on the value g of the free-fall acceleration, which varies from place to place. In 1936, Jesse Owens established a world's running broad jump record of 8.09 m at the Olympic Games at Berlin ($g = 9.8128$ m/s²). Assuming the same values of v_0 and θ_0, by how much would his record have differed if he had competed instead in 1956 at Melbourne ($g = 9.7999$ m/s²)?

39P. A third baseman wishes to throw to first base, 127 ft distant. His best throwing speed is 85 mi/h. (a) If he throws the ball horizontally 3.0 ft above the ground, how far from first base will it hit the ground? (b) At what upward angle must the third baseman throw the ball if the first baseman is to catch it 3.0 ft above the ground? (c) What will be the time of flight in that case?

40P. During volcanic eruptions, chunks of solid rock can be blasted out of the volcano; these projectiles are called *volcanic bombs*. Figure 4-35 shows a cross section of Mt. Fuji, in Japan. (a) At what initial speed would a bomb have to be ejected, at 35° to the horizontal, from the vent at A in order to fall at the foot of the volcano at B? Ignore, for

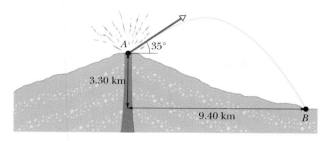

FIGURE 4-35 Problem 40.

the moment, the effects of air on the bomb's travel. (b) What would be the time of flight? (c) Would the effect of the air increase or decrease your answer in (a)?

41P. At what initial speed must the basketball player throw the ball, at 55° above the horizontal, to make the foul shot, as shown in Fig. 4-36?

FIGURE 4-36 Problem 41.

42P. A football player punts the football so that it will have a "hang time" (time of flight) of 4.5 s and land 50 yd away. If the ball leaves the player's foot 5.0 ft above the ground, what initial velocity (magnitude and direction) must the ball have?

43P. The B-52 shown in Fig. 4-37 is 49 m long and is traveling at an airspeed of 820 km/h over a bombing range. How far apart will the bomb craters be? Make any measurements you need directly from the figure. Assume that there is no wind, and ignore air resistance. How would air resistance affect your answer?

FIGURE 4-37 Problem 43.

44P. A projectile is fired with an initial speed $v_0 = 30.0$ m/s from the level ground at a target on the ground a distance $R = 20.0$ m away (Fig. 4-38). Find the two projection angles.

FIGURE 4-38 Problem 44.

45P. What is the maximum vertical height to which a baseball player can throw a ball if his maximum throwing *range* is 60 m?

46P. A football is kicked off with an initial speed of 64 ft/s at a projection angle of 45°. A receiver on the goal line, 60 yd away in the direction of the kick, starts running to meet the ball at that instant. What must be his average speed if he is to catch the ball just before it hits the ground? Neglect air resistance.

47P. A ball rolls horizontally off the top of a stairway with a speed of 5.0 ft/s. The steps are 8.0 in. high and 8.0 in. wide. Which step will the ball hit first?

48P. A certain airplane has a speed of 180 mi/h and is diving at an angle of 30.0° below the horizontal when a radar decoy is released. (See Fig. 4-39.) The horizontal distance between the release point and the point where the decoy strikes the ground is 2300 ft. (a) How high was the plane when the decoy was released? (b) How long was the decoy in the air?

FIGURE 4-39 Problem 48.

49P. A plane, diving at an angle of 53.0° with the vertical, releases a projectile at an altitude of 730 m. The projectile hits the ground 5.00 s after being released. (a) What is the speed of the plane? (b) How far did the projectile travel horizontally during its flight? (c) What were the horizontal and vertical components of its velocity just before striking the ground?

50P. A ball is thrown horizontally from a height of 20 m and hits the ground with a speed that is three times its initial speed. What was the initial speed?

51P. The launching speed of a certain projectile is five times the speed it has at its maximum height. Calculate the elevation angle at launching.

52P. (a) During a tennis match, a player serves at 23.6 m/s (as recorded by a radar gun), the ball leaving the racquet horizontally 2.37 m above the court surface. By how much does the ball clear the net, which is 12 m away and 0.90 m high? (b) Suppose the player serves the ball as before except that the ball leaves the racquet at 5.00° below the horizontal. Does the ball clear the net now?

53P. In Sample Problem 4-8, suppose that a second identical harbor defense cannon is emplaced 30 m above sea level, rather that at sea level. How much longer is the horizontal distance from launch to impact of the second cannon than that of the first, which was found to be 690 m, if the elevation angle of fire is 45°?

54P. A batter hits a pitched ball whose center is 4.0 ft above the ground so that its angle of projection is 45° and its *range* is 350 ft. The ball will be a home run if it clears a 24-ft-high fence that is 320 ft from home plate. Will the ball clear the fence? If so, by how much?

55P*. The kicker on a football team can give the ball an initial speed of 25 m/s. Within what two elevation angles must he kick the ball if he is to score a field goal from a point 50 m in front of goalposts whose horizontal bar is 3.44 m above the ground? (You might want to use $\sin^2 \theta + \cos^2 \theta = 1$ to get a relation between $\tan^2 \theta$ and $1/\cos^2 \theta$, and then solve the resulting quadratic equation.)

56P*. An antitank gun is located on the edge of a plateau that is 60 m above the surrounding plain (Fig. 4-40). The gun crew sights an enemy tank stationary on the plain at a horizontal distance of 2.2 km from the gun. At the same moment, the tank crew sees the gun and starts to move directly away from it with an acceleration of 0.90 m/s². If the antitank gun fires a shell with a muzzle speed of 240 m/s at an elevation angle of 10° above the horizontal, how long should the gun crew wait before firing if they are to hit the tank?

FIGURE 4-40   Problem 56.

57P*. A rocket is launched from rest and moves in a straight line at 70.0° above the horizontal with an acceleration of 46.0 m/s². After 30.0 s of the linear powered flight, the engines shut off and the rocket follows a para-

FIGURE 4-41   Problem 57.

bolic path back to Earth (see Fig. 4-41). Assume that the free-fall acceleration is 9.8 m/s² throughout and that effects of the air can be ignored. (a) Find the time of flight from launch to impact. (b) What is the maximum altitude reached? (c) What is the distance from launchpad to impact point?

SECTION 4-7 UNIFORM CIRCULAR MOTION

58E. In Bohr's model of the hydrogen atom, an electron orbits a proton in a circle of radius 5.28×10^{-11} m with a speed of 2.18×10^6 m/s. What is the acceleration of the electron in this model?

59E. (a) What is the acceleration of a sprinter running at 10 m/s when rounding a bend with a turn radius of 25 m? (b) To what does the acceleration **a** point?

60E. A magnetic field can force a charged particle to move in a circular path. Suppose that an electron experiences a radial acceleration of 3.0×10^{14} m/s² in a particular magnetic field. What is the speed of the electron if the radius of its circular path is 15 cm?

61E. A sprinter runs at 9.2 m/s around a circular track with a centripetal acceleration of 3.8 m/s². (a) What is the track radius? (b) How long does it take to go completely around the track at this speed?

62E. An Earth satellite moves in a circular orbit 640 km above the Earth's surface. The time for one revolution is 98.0 min. (a) What is the speed of the satellite? (b) What is the free-fall acceleration at the orbit?

63E. If a remote-controlled space probe can withstand the stresses of a 20*g* acceleration, (a) what is the minimum turning radius of such a craft moving at a speed of one-tenth the speed of light? (b) How long would it take to complete a 90° turn at this speed?

64E. A rotating fan completes 1200 revolutions every minute. Consider a point on the tip of the blade, which has a radius of 0.15 m. (a) Through what distance does the point move in one revolution? (b) What is the speed of the point? (c) What is its acceleration?

65E. The fast train known as the TGV (Train à Grande Vitesse) that runs south from Paris in France has a scheduled average speed of 216 km/h. (a) If the train goes around a curve at that speed and the acceleration experienced by the passengers is to be limited to 0.050g, what is the smallest radius of curvature for the track that can be tolerated? (b) If there is a curve with a 1.00-km radius, to what speed must the train be slowed to keep the acceleration below the limit?

66E. When a large star goes *supernova*, its core may be compressed so tightly that it becomes a *neutron star,* with a radius of about 20 km (about the size of the San Francisco area!). If a neutron star rotates once every second, (a) what is the speed of a particle on the star's equator, and (b) what is the particle's centripetal acceleration in m/s² and in *g*-units (where *g* is our earthly 9.8 m/s²)? (c) If the neutron star rotates even faster, what happens to the answers to (a) and (b)?

67E. An astronaut is rotated in a centrifuge at a radius of 5.0 m. (a) What is the astronaut's speed if the acceleration is 7.0*g*? (b) How many revolutions per minute are required to produce this acceleration?

68P. A carnival Ferris wheel has a 15-m radius and completes five turns about its horizontal axis every minute. (a) What is the acceleration of a passenger at the highest point? (b) What is the acceleration at the lowest point?

69P. (a) What is the centripetal acceleration of an object on the Earth's equator owing to the rotation of the Earth? (b) What would the period of rotation of the Earth have to be in order that objects on the equator have a centripetal acceleration equal to 9.8 m/s²?

70P. Calculate the acceleration of a person at latitude 40° owing to the rotation of the Earth. (See Fig. 4-42.)

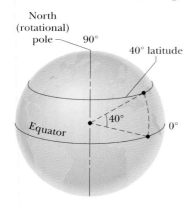

FIGURE 4-42 Problem 70.

71P. A particle *P* travels with constant speed on a circle of radius 3.00 m (Fig. 4-43) and completes one revolution in 20.0 s. The particle passes through *O* at *t* = 0. Find the magnitude and direction of each of the following vectors. (a) With respect to *O*, find the particle's position vector at *t* = 5.00 s, 7.50 s, and 10.0 s. For the 5.00-s interval from the end of the fifth second to the end of the tenth second, find the particle's (b) displacement and (c) average velocity. Find its (d) velocity and (e) acceleration at the beginning and end of that 5.00-s interval.

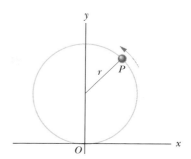

FIGURE 4-43 Problem 71.

72P. A boy whirls a stone in a horizontal circle 2.0 m above the ground by means of a string 1.5 m long. The string breaks, and the stone flies off horizontally and strikes the ground 10 m away. What was the centripetal acceleration of the stone while in circular motion?

SECTION 4-8 RELATIVE MOTION IN ONE DIMENSION

73E. A boat is traveling upstream at 14 km/h with respect to the water of a river. The water itself is flowing at 9 km/h with respect to the ground. (a) What is the velocity of the boat with respect to the ground? (b) A child on the boat walks from front to rear at 6 km/h with respect to the boat. What is the child's velocity with respect to the ground?

74E. A person walks up a stalled 15-m-long escalator in 90 s. When standing on the same escalator, now moving, the person is carried up in 60 s. How much time would it take that person to walk up the moving escalator? Does the answer depend on the length of the escalator?

75E. A transcontinental flight of 2700 mi is scheduled to take 50 min longer westward than eastward. The airspeed of the airplane is 600 mi/h and the jetstream it will fly through is presumed to be moving either due east or due west. What assumptions about the jetstream wind velocity are made in preparing the schedule?

76E. A cameraman on a pickup truck is traveling westward at 40 mi/h while he videotapes a cheetah that is moving westward 30 mi/h faster than the truck. Suddenly, the cheetah stops, turns, and then runs at 60 mi/h eastward, as measured by a suddenly nervous crew member who stands alongside the cheetah's path. The change in the animal's velocity took 2.0 s. What was its acceleration from the perspective of the cameraman? From the perspective of the nervous crew member?

77P. The airport terminal in Geneva, Switzerland, has a "moving sidewalk" to speed passengers through a long corridor. Peter, who walks through the corridor but does not use the moving sidewalk, takes 150 s to do so. Paul, who simply stands on the moving sidewalk, covers the same distance in 70 s. Mary not only uses the sidewalk but walks along it. How long does Mary take? Assume that Peter and Mary walk at the same speed.

SECTION 4-9   RELATIVE MOTION IN TWO DIMENSIONS

78E. In rugby (Fig. 4-44) a player can legally pass the ball to a teammate as long as the pass is not "forward" (it must not have a velocity component parallel to the length of the field and directed toward the other team's goal). Suppose a player runs parallel to the field's length with a speed of 4.0 m/s while he passes the ball with a speed of 6.0 m/s relative to himself. What is the smallest angle from the forward direction that keeps the pass legal?

FIGURE 4-44   Problem 78.

79E. Two highways intersect as shown in Fig. 4-45. At the instant shown, a police car *P* is 800 m from the intersection and moving at 80 km/h. Motorist *M* is 600 m from

the intersection and moving at 60 km/h. (a) In unit-vector notation, what is the velocity of the motorist with respect to the police car? (b) For the instant shown in the figure, how does the direction of the velocity found in (a) compare to the line of sight between the two cars? (c) If the cars maintain their velocities, do the answers to (a) and (b) change as the cars move nearer the intersection?

80E. Snow is falling vertically at a constant speed of 8.0 m/s. At what angle from the vertical do the snowflakes appear to be falling as viewed by the driver of a car traveling on a straight road with a speed of 50 km/h?

81E. In a large department store, a shopper is standing on the "up" escalator, which is traveling at an angle of 40° above the horizontal and at a speed of 0.75 m/s. He passes his daughter, who is standing on the identical, adjacent "down" escalator. (See Fig. 4-46.) Find the velocity of the shopper relative to his daughter in unit-vector notation.

FIGURE 4-46   Exercise 81.

FIGURE 4-45   Exercise 79.

82P. A helicopter is flying in a straight line over a level field at a constant speed of 6.2 m/s and at a constant altitude of 9.5 m. A package is ejected horizontally from the helicopter with an initial velocity of 12 m/s relative to the helicopter, and in a direction opposite the helicopter's motion. (a) Find the initial speed of the package relative to the ground. (b) What is the horizontal distance between the helicopter and the package at the instant the package strikes the ground? (c) What angle does the velocity vector of the package make with the ground at the instant before impact, as seen from the ground?

83P. A train travels due south at 30 m/s (relative to ground) in a rain that is blown toward the south by the wind. The path of each raindrop makes an angle of 22° with the vertical, as measured by an observer stationary on the Earth. An observer on the train, however, sees the drops fall perfectly vertically. Determine the speed of the raindrops relative to the Earth.

84P. Two ships, A and B, leave port at the same time. Ship A travels northwest at 24 knots and ship B travels at 28 knots in a direction 40° west of south. (1 knot = 1 nautical mile per hour; see Appendix F.) (a) What are the magnitude and direction of the velocity of ship A relative to B? (b) After what time will they be 160 nautical miles apart? (c) What will be the bearing of B from A at that time?

85P. A light plane attains an airspeed of 500 km/h. The pilot sets out for a destination 800 km to the north but discovers that the plane must be headed 20.0° east of north to fly there directly. The plane arrives in 2.00 h. What was the wind velocity vector?

86P. The New Hampshire State Police use aircraft to enforce highway speed limits. Suppose that one of the airplanes has a speed of 135 mi/h in still air. It is flying straight north so that it is at all times directly above a north–south highway. A ground observer tells the pilot by radio that a 70.0-mi/h wind is blowing, but neglects to give the wind direction. The pilot observes that in spite of the wind the plane can travel 135 mi along the highway in 1.00 h. In other words, the ground speed is the same as if there were no wind. (a) What is the direction of the wind? (b) What is the heading of the plane, that is, the angle between an axis along its length and the highway?

87P. A wooden boxcar is moving along a straight railroad track at speed v_1. A sniper fires a bullet (initial speed v_2) at it from a high-powered rifle. The bullet passes through both walls of the car, its entrance and exit holes being exactly opposite each other as viewed from within the car. From what direction, relative to the track, was the bullet fired? Assume that the bullet was not deflected upon entering the car, but that its speed decreased by 20%. Take $v_1 = 85$ km/h and $v_2 = 650$ m/s. (Why don't you need to know the width of the boxcar?)

88P. A woman can row a boat at 4.0 mi/h in still water. (a) If she is crossing a river where the current is 2.0 mi/h, in what direction must her boat be headed if she wants to reach a point directly opposite her starting point? (b) If the river is 4.0 mi wide, how long will it take her to cross the river? (c) Suppose that instead of crossing the river she rows 2.0 mi *down* the river and then back to her starting point. How long will she take? (d) How long will she take to row 2.0 mi *up* the river and then back to her starting point? (e) In what direction should she head the boat if

she wants to cross in the shortest possible time, and what is that time?

89P*. A battleship steams due east at 24 km/h. A submarine 4.0 km away fires a torpedo that has a speed of 50 km/h; see Fig. 4-47. If the bearing of the ship as seen from the submarine is 20° east of north, (a) in what direction should the torpedo be fired to hit the ship, and (b) what will be the running time for the torpedo to reach the battleship?

FIGURE 4-47 Problem 89.

90P*. A man wants to cross a river 500 m wide. His rowing speed (relative to the water) is 3000 m/h. The river flows at a speed of 2000 m/h. The man's walking speed on shore is 5000 m/h. (a) Find the path (combined rowing and walking) he should take to get to the point directly opposite his starting point in the shortest time. (b) How long does it take?

SECTION 4-10 RELATIVE MOTION AT HIGH SPEEDS

91E. An electron moves at speed $0.42c$ with respect to observer B. Observer B moves at speed $0.63c$ with respect to observer A, in the same direction as the electron. What does observer A measure for the speed of the electron?

92P. Galaxy Alpha is observed to be receding from us with a speed of $0.35c$. Galaxy Beta, located in precisely the opposite direction, is also found to be receding from us at this same speed. What recessional speed would an observer on Galaxy Alpha find (a) for our galaxy and (b) for Galaxy Beta?

ADDITIONAL PROBLEMS

93. A baseball is hit at ground level. The ball reaches its maximum height above ground level 3.0 s after being hit. Then 2.5 s after reaching its maximum height, the ball barely clears a fence that is 97.5 m from where it was hit.

Assume the ground is level. (a) What maximum height above ground level is reached by the ball? (b) How high is the fence? (c) How far beyond the fence does the ball strike the ground?

94. The position **r** of a particle moving in the xy plane is given by $\mathbf{r} = 2t\mathbf{i} + [2 \sin(\pi/4)(\text{rad/s})t]\mathbf{j}$, where **r** is in meters and t is in seconds. (a) Calculate the x and y components of the particle's position at $t = 0, 1.0, 2.0, 3.0$, and 4.0 s and sketch the particle's path in the xy plane for the interval $0 \leq t \leq 4.0$ s. (b) Calculate the components of the particle's velocity at $t = 1.0, 2.0$, and 3.0 s. Show that the velocity is tangent to the path of the particle and in the direction the particle is moving at each time by drawing the velocity vectors on the plot of the particle's path in part (a). (c) Calculate the components of the particle's acceleration at $t = 1.0, 2.0$, and 3.0 s.

95. A 200-m-wide river flows due east at a uniform speed of 2.0 m/s. A boat with a speed of 8.0 m/s relative to the water leaves the south bank pointed in a direction of 30° west of north. (a) What is the velocity of the boat relative to the Earth? (b) How long does the boat take to cross the river?

96. Two seconds after being projected from ground level, a projectile is displaced 40 m horizontally and 53 m vertically above its point of projection. (a) What are the horizontal and vertical components of the initial velocity of the projectile? (b) At the instant the projectile achieves its maximum height above ground level, how far is it displaced horizontally from its point of projection?

97. A particle starts from the origin at $t = 0$ with a velocity of 8.0$\mathbf{j}$ m/s and moves in the xy plane with a constant acceleration of $(4.0\mathbf{i} + 2.0\mathbf{j})$ m/s². (a) At the instant the x coordinate of the particle is 29 m, what is its y coordinate? (b) What is the speed of the particle at this time?

98. A car travels around a flat circle on the ground, at a constant speed of 12 m/s. At a certain instant the car has an acceleration of 3 m/s², toward the east. What are its distance and direction from the center of the circle at that instant if it is traveling (a) clockwise around the circle and (b) counterclockwise around the circle?

99. A golfer tees off from the top of a rise, giving the golf ball an initial velocity of 43 m/s at an angle of 30° above the horizontal. The ball strikes the fairway a horizontal distance of 180 m from the tee. Assume the fairway is level. (a) How high is the rise above the fairway? (b) What is the speed of the ball as it strikes the fairway?

100. After flying for 15 min in a wind blowing 42 km/h at an angle of 20° south of east, an airplane pilot finds himself over a town that is 55 km due north of his starting point. What is the speed of the airplane relative to the air?

FORCE AND MOTION—I

5

On April 4, 1974, John Massis of Belgium managed to move two passenger cars belonging to New York's Long Island Railroad. He did so by clamping his teeth down on a bit that was attached to the cars with a rope and then leaning backward while pressing his feet against the railway ties. The cars weighed about 80 tons. Did Massis have to pull with superhuman force to accelerate them? We shall answer the question shortly.

5-1 WHY DOES A PARTICLE CHANGE ITS VELOCITY?

Sometimes an object that we are watching—perhaps an automobile, a baseball, or a cat—will change its velocity. It will accelerate. Observation has taught us that when this happens we can always find one or more nearby objects that seem to be associated with that change. We relate the acceleration of a particle then to some interaction between the particle and its surroundings. We are so used to this that, when we see an object change its velocity without apparent cause, we suspect a trick. If a rolling ball suddenly changes direction, we look for a hidden magnet or an air jet.

Our central problem for this chapter is this: (1) We are given a particle (usually referred to as a *body,* from now on) whose characteristics (for example, mass, shape, volume, electric charge) we know. (2) We also know the locations and properties of all significant nearby objects. That is, we are fully informed about the body's environment. (3) We want to know how the body will move.

Isaac Newton (1642–1727), in putting forward his laws of motion and his theory of gravitation, first solved this problem. Here is our plan for following in his footsteps: (1) We introduce the concept of **force** (a push or pull), in terms of the acceleration given to a selected standard body. (2) We define **mass** and show how to assign a mass to a body, so that we may understand how different bodies, subject to the same environment, have different accelerations. (3) We find ways to calculate the force acting on a body from the properties of the body and its environment. That is, we look for **force laws.** (4) We show how multiple forces on a body can be combined to give a **net force.**

Figure 5-1 shows the relations among these quantities, a study known as **mechanics.** Force appears in both the force laws (which tell us how to calculate the force that will act on a body in a particular environment) and the laws of motion (which

tell us what acceleration a body will experience when a force acts on it). It is the crowning glory of Newton's mechanics that, for a fantastic range of phenomena, it predicts results that agree with experiment.

There are some important problems to which Newtonian mechanics does not give correct answers. As discussed in Chapter 4, if the speeds of the particles involved are an appreciable fraction of the speed of light, we must replace Newtonian mechanics with Einstein's special theory of relativity. For problems on the scale of atomic structure (for example, the motions of electrons within atoms), we must replace Newtonian mechanics with quantum mechanics. Physicists now view Newton's mechanics as a special case of these two, more comprehensive, theories. It is a very important special case, however, encompassing as it does the motions of objects that range in size from molecules to galaxies. Within this broad range it is highly accurate, as the successful maneuvering of space probes reminds us.

5-2 NEWTON'S FIRST LAW

Before Newton formulated his mechanics, it was thought that some influence or "force" was needed to keep a body moving at constant velocity. A body was then thought to be in its "natural state" when it was at rest. For it to move with constant velocity, it seemingly had to be propelled in some way. Otherwise, it would "naturally" stop moving.

This is not unreasonable. If you send a book sliding across a carpet it does indeed come to rest. If you want to make it move across the carpet with constant velocity you might, for example, tie a string to it and pull it across.

Slide the book over the ice of a skating rink, however, and it goes a lot farther. You can imagine smoother and longer surfaces, over which the book would slide farther and farther. In the limit you can think of a long, extremely smooth surface (said to be a **frictionless surface**), over which the book would show little sign at all of slowing down. We can in fact come close to this in the laboratory, by propelling a book over a horizontal air table, across which it moves on a film of air.

We are led to conclude that you do *not* need a force to keep a body moving with constant velocity. This fits in nicely with what we discussed in Section 4-8 about reference frames: a body at rest in one frame may be moving at constant velocity with re-

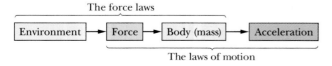

The force laws

| Environment | → | Force | → | Body (mass) | → | Acceleration |

The laws of motion

FIGURE 5-1 The relations of mechanics. The three left-hand boxes suggest that force is an interaction between a body and its environment. The three right-hand boxes suggest that a force acting on a body will accelerate it.

spect to another. *Rest* and *moving with constant velocity* are not all that different. And that leads us to the first of Newton's three laws of motion:

> **NEWTON'S FIRST LAW:** Consider a body on which no net force acts. If the body is at rest, it will remain at rest. If the body is moving with constant velocity, it will continue to do so.

Newton's first law is really a statement about reference frames, in that it defines the kinds of reference frames in which the laws of Newtonian mechanics hold. From this point of view the first law is expressed as follows:

> **NEWTON'S FIRST LAW:** If the net force acting on a body is zero, it is possible to find reference frames in which that body has no acceleration.

Newton's first law is sometimes called the *law of inertia* and the reference frames that it defines are called *inertial reference frames.*

Figure 5-2 shows how you can test a particular frame to see whether or not it is an inertial frame. With the railroad car at rest, mark the position of the stationary pendulum bob on the table. With the car in motion, the bob remains over the mark *only* if the car is moving in a straight line at constant speed. If the car is gaining or losing speed or is rounding a bend, the bob moves from its mark, and the car is a noninertial reference frame.

If you put a bowling ball at rest on a rotating merry-go-round, no identifiable force acts on the ball but it does not remain at rest. If you roll the ball out along a radial line it will veer away from that line. Rotating reference frames are *not* inertial frames. Strictly speaking, the Earth is not an inertial frame either, because of its rotation. However, unless we consider large-scale motions such as wind and ocean currents, we can usually make the approximation that the Earth is an inertial frame. In all that follows, unless stated otherwise, we shall make that approximation.*

5-3 FORCE

We now wish to define **force** carefully, in terms of the acceleration that it gives to a standard reference body. As a standard body, we use (or rather we imagine that we use) the standard kilogram of Fig. 1-7. This body has been assigned, exactly and by definition, a mass of 1 kg. Later, we shall show how to assign masses to other bodies.

We put the standard body on a horizontal frictionless table and pull the body to the right (Fig. 5-3) so that, by trial and error, it experiences a measured acceleration of 1 m/s². We then declare, as a matter of definition, that we are exerting a force on the standard body whose magnitude is 1 newton (abbreviated N).

We can exert a 2-N force on our standard body by pulling it so that its measured acceleration is 2 m/s², and so on. In general, we see that if our standard body has an acceleration a, we know that a force F must be acting on it and that the magnitude of the force (in newtons) is equal to the magnitude of the acceleration (in m/s²).

Acceleration is a vector. Is force a vector? We can easily enough assign a direction to force, namely, that of the acceleration that it produces in our standard body. That, however, is not enough to

FIGURE 5-3 A force **F** on the standard body gives that body an acceleration **a**.

FIGURE 5-2 Testing a railroad car to see whether or not it is an inertial reference frame.

---

*The Earth's noninertial character shows up in that a falling object does not fall straight down but veers to the east a little. At latitude 45°, for example, an object dropped from a height of 50 m would (neglecting air resistance) land 5 mm east of the spot it would have hit if the Earth were not rotating.

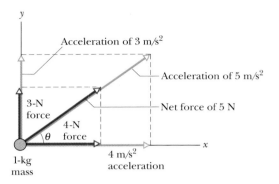

FIGURE 5-4 An overhead view of a 3-N force and a 4-N force acting simultaneously on the standard body, whose mass is exactly 1 kg. The body's acceleration turns out to be the same as if it had been acted on by a single force equal to the vector sum of the two actual forces. Forces add like vectors.

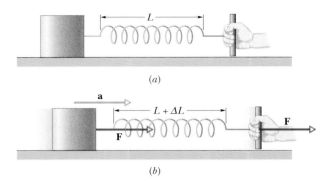

FIGURE 5-5 (*a*) A spring with length *L* is attached to the standard body, whose mass is 1 kg. (*b*) The standard body is given an acceleration **a** by pulling on the spring with force **F**, which stretches the spring by an amount ΔL. The surface is frictionless.

prove that force is a vector. We must also show that force obeys the laws of vector addition, and we can find out whether or not it does only by experiment.

Let us arrange to exert on our standard body either a 4-N force along the *x* axis or a 3-N force along the *y* axis, with both axes horizontal on the frictionless table. The first of these forces, acting alone, would produce an acceleration of 4 m/s² along the *x* axis. The second, acting alone, would produce an acceleration of 3 m/s² along the *y* axis.

What if the forces act simultaneously, as in Fig. 5-4? We would find by experiment (and only by experiment) that the acceleration of the standard body would be 5 m/s², directed as shown in Fig. 5-4. This is exactly the same acceleration that you would find if the standard body were acted on by a single force, equal to the vector sum (or *resultant*) of the two actual forces. This vector sum is the net force, with a magnitude of 5 N and directed as in Fig. 5-4. Experiments like this show beyond doubt that forces are vectors. They have magnitude; they have direction; they add according to the vector rules. Henceforth, we symbolize them with boldface letters, most often the letter **F**.

5-4 MASS

Everyday experience tells us that a given force will produce different accelerations in different bodies. Put a baseball and a bowling ball on the floor and give each one the same sharp kick; the acceleration of the baseball will be much greater. The difference

in the two accelerations is due to the difference between the mass of the baseball and the mass of the bowling ball. But what, exactly, is mass?

To get at this quantitatively, let us attach a spring to our standard body as in Fig. 5-5 and let us give it an acceleration a_0 of, say, 1 m/s². The force that we exert on the spring and that the spring exerts on the body is 1 N. We note carefully the extension ΔL of the spring that corresponds to this 1-N force. The mass m_0 of our standard body is, by definition, exactly 1 kg.

Now let us replace the standard body with an arbitrary body (body *X*), and let us apply the same 1-N force to it. We can do this by pulling on it in such a way that the spring is stretched by the same amount ΔL that it was in the experiment above with the standard body. Suppose that the acceleration a_X of body *X* turns out to be 0.25 m/s². We can use this experimental result to assign a mass m_X to body *X* by declaring that if the same force acts on two different bodies, we define the ratio of their masses to be the inverse ratio of their accelerations. Thus

$$\frac{m_X}{m_0} = \frac{a_0}{a_X},$$

or

$$m_X = m_0 \frac{a_0}{a_X} = (1 \text{ kg}) \frac{1 \text{ m/s}^2}{0.25 \text{ m/s}^2} = 4 \text{ kg}.$$

Thus body *X*, which receives only one-fourth the acceleration of the standard body when the same force acts on it, has, by this definition, four times the mass.

In this way, we can assign masses to bodies other than the standard body. Before we accept this method, however, let us test it, in two ways.

The First Test

Let us repeat the comparison to the standard body but with a different common force acting. Suppose, for example, that we stretch the spring farther, so that the acceleration a_0' of the standard body is now 5 m/s². That is, we decide to use a 5-N force rather than a 1-N force to make our mass comparisons.

We find by experiment that, if we let this same 5-N force act on body X, its acceleration a_X' is 1.25 m/s². The mass that we find for body X is then

$$m_X = m_0 \frac{a_0'}{a_X'} = (1 \text{ kg}) \frac{5 \text{ m/s}^2}{1.25 \text{ m/s}^2} = 4 \text{ kg},$$

exactly as before.

The Second Test

Suppose that—using the spring method—we have already compared a second body, body Y, with our standard body and found $m_Y = 6$ kg. Now let us compare body X and body Y directly. That is, we exert the same force **F** (of any convenient magnitude) on each of them and measure the accelerations a_X'' and a_Y'' that result. Let us say that we find $a_X'' = 2.4$ m/s² and $a_Y'' = 1.6$ m/s².

Now let us find the mass of body Y by comparing it—not with the standard body—but directly with body X, whose mass we already know. We find

$$m_Y = m_X \frac{a_X''}{a_Y''} = (4 \text{ kg}) \frac{2.4 \text{ m/s}^2}{1.6 \text{ m/s}^2} = 6 \text{ kg},$$

the same result as from the comparison with the standard body.

What Is Mass?

Thus our method of assigning masses to arbitrary bodies yields consistent results, no matter what force we use in making the comparison and no matter what body we use for a comparison standard. Mass seems to be truly an intrinsic characteristic of a body.

Since the word *mass* is used in everyday English, we should have some intuitive understanding of it, maybe something that we can physically sense. Is it a body's size, weight, or density? The answer is no, al-though those characteristics are sometimes confused with mass. *The **mass** of a body is the characteristic that relates the force on the body to the resulting acceleration.* Mass has no more familiar definition than that; the only time you can have a physical sensation of mass is in a situation where you attempt to accelerate a body. If, for example, you shove first a baseball and then a bowling ball away from you, you will notice that they have different masses.

5-5 NEWTON'S SECOND LAW

It is a tribute to Newton's genius that all of the detailed definitions, experiments, and observations that we have described so far can be summarized in a simple vector equation, which is called Newton's second law of motion:

$$\sum \mathbf{F} = m\mathbf{a} \qquad \text{(Newton's second law).} \qquad (5\text{-}1)$$

In using Eq. 5-1, we must first be quite certain what body we are applying it to. Then $\sum \mathbf{F}$ in Eq. 5-1 is the *vector* sum, or **net force,** of *all* the forces that act *on* that body. If you miss any forces (or count any of them twice), you will be in trouble. Only forces that act *on* the body are to be included. There may be many forces in a given problem and you must be sure to pick only those that act on the body with which you are dealing. Finally, $\sum \mathbf{F}$ includes only *external* forces, that is, forces exerted on the body by other bodies. We do not count internal forces, in which one part of the body exerts a force on another part.

To solve problems with Eq. 5-1, we often draw a **free-body diagram.** In it, the body is represented by a dot, and each external force (or the net force $\sum \mathbf{F}$) acting on the body is represented by a vector with its tail on the dot.

Like all vector equations, Eq. 5-1 is equivalent to three scalar equations:

$$\sum F_x = ma_x, \quad \sum F_y = ma_y, \quad \sum F_z = ma_z. \qquad (5\text{-}2)$$

These equations relate the three components of the net force acting on a body to the three components of the acceleration of that body.

Finally, we note that Newton's second law includes the formal statement of Newton's first law as a special case. That is, if no force acts on a body, Eq.

TABLE 5-1
UNITS IN NEWTON'S SECOND LAW
(Eqs. 5-1 and 5-2)

SYSTEM	FORCE	MASS	ACCELERATION
SI	newton (N)	kilogram (kg)	m/s^2
CGS	dyne	gram (g)	cm/s^2
British[a]	pound (lb)	slug	ft/s^2

[a]$1 \text{ lb} = 1 \text{ slug} \cdot \text{ft/s}^2$.

5-1 tells us that the body will not be accelerated. This does not trivialize Newton's first law; its role in defining the set of reference frames in which Newton's mechanics holds justifies its status as a separate law.

With Eq. 5-2 in SI units, we find

$$1 \text{ N} = (1 \text{ kg})(1 \text{ m/s}^2) = 1 \text{ kg} \cdot \text{m/s}^2, \quad (5\text{-}3)$$

consistent with our discussion in Section 5-3. Although we shall use SI units almost exclusively from now on, other systems of units are still in use. Chief among these are the British system and the CGS (centimeter–gram–second) system. Table 5-1 shows the units in which Eqs. 5-1 and 5-2 are expressed in these systems. (See also Appendix F.)

If we could neglect the rubbing between the puck and the ice, the velocity of the sliding puck would not change until one of the players applies a force to it or it runs into the wall.

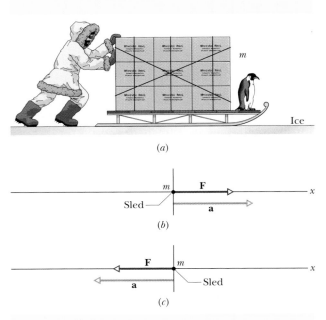

(a)

(b)

(c)

FIGURE 5-6 Sample Problem 5-1. (*a*) A student pushes a loaded sled over a frictionless surface. (*b*) A "free-body diagram," showing the net force acting on the sled and the acceleration it produces. (*c*) A free-body diagram for Sample Problem 5-2. The student now pushes in the opposite direction on the sled, reversing its acceleration.

SAMPLE PROBLEM 5-1

A student pushes a loaded sled whose mass m is 240 kg for a distance d of 2.3 m over the frictionless surface of a frozen lake. He exerts a constant horizontal force **F**, with magnitude $F = 130$ N, as he does so (see Fig. 5-6*a*). If the sled starts from rest, what is its final velocity?

SOLUTION Figure 5-6*b* is a free-body diagram for the situation. We lay out a horizontal x axis, we take the direction of increasing x to be to the right, and we treat the sled as a particle, represented by a dot. We assume that the x component F_x of the force **F** exerted by the student is the only horizontal force acting on the sled. We can then find the magnitude of the acceleration a_x of the sled from Newton's second law:

$$a_x = \frac{F_x}{m} = \frac{130 \text{ N}}{240 \text{ kg}} = 0.542 \text{ m/s}^2.$$

Because the acceleration is constant, we can use Eq. 2-14, $v^2 = v_0^2 + 2a(x - x_0)$, to find the final velocity.

Putting $v_0 = 0$ and $x - x_0 = d$, and identifying a_x as a, we solve for v:

$$v = \sqrt{2ad}$$

$$= \sqrt{(2)(0.542 \text{ m/s}^2)(2.3 \text{ m})} = 1.6 \text{ m/s}. \quad \text{(Answer)}$$

The force, the acceleration, the displacement, and the final velocity of the sled are all positive, which means that they all point to the right in Fig. 5-6b.

SAMPLE PROBLEM 5-2

The student in Sample Problem 5-1 wants to reverse the direction of the velocity of the sled in 4.5 s. With what constant force must he push on the sled to do so?

SOLUTION Let us first find the constant acceleration required to reverse the sled's velocity in 4.5 s, using Eq. 2-9, $v = v_0 + at$. Solving for a gives

$$a = \frac{v - v_0}{t} = \frac{(-1.6 \text{ m/s}) - (1.6 \text{ m/s})}{4.5 \text{ s}}$$

$$= -0.711 \text{ m/s}^2.$$

This is larger in magnitude than the acceleration in Sample Problem 5-1 (0.542 m/s^2), so it stands to reason that the student must push with a greater force this time. We find this greater force from Eq. 5-2, with a_x being a:

$$F_x = ma_x = (240 \text{ kg})(-0.711 \text{ m/s}^2)$$

$$= -171 \text{ N}. \quad \text{(Answer)}$$

The minus sign shows that the student is pushing the sled in the direction of decreasing x, that is, to the left in Fig. 5-6c, the free-body diagram for this situation.

SAMPLE PROBLEM 5-3

A crate whose mass m is 360 kg rests on the bed of a truck that is moving at a speed v_0 of 120 km/h, as in Fig. 5-7a. The driver applies the brakes and slows to a speed v of 62 km/h in 17 s. What force (assumed constant) acts on the crate during this time? Assume that the crate does not slide on the truck bed.

SOLUTION We first find the constant acceleration of the crate, using Eq. 2-9, $v = v_0 + at$:

$$(a)$$

$$(b)$$

FIGURE 5-7 Sample Problem 5-3. (a) A crate on a truck that is slowing down. (b) The free-body diagram for the crate. The force **F** causes the crate to have acceleration (or deceleration) **a**.

$$a = \frac{v - v_0}{t} = \frac{(62 \text{ km/h}) - (120 \text{ km/h})}{17 \text{ s}}$$

$$= \left(-3.41 \, \frac{\text{km}}{\text{h} \cdot \text{s}}\right)\left(\frac{1 \text{ h}}{3600 \text{ s}}\right)\left(\frac{1000 \text{ m}}{1 \text{ km}}\right)$$

$$= -0.947 \text{ m/s}^2.$$

As Fig. 5-7 shows, the velocity vector of the crate points to the right, its acceleration vector to the left.

The force on the crate follows from Newton's second law:

$$F = ma$$

$$= (360 \text{ kg})(-0.947 \text{ m/s}^2)$$

$$= -340 \text{ N}. \quad \text{(Answer)}$$

This force, which could be from straps securing the crate, acts in the same direction as the acceleration, namely, to the left in Fig. 5-7b.

SAMPLE PROBLEM 5-4

In a two-dimensional tug-of-war, Alex, Betty, and Charles pull on an automobile tire, at angles as shown in Fig. 5-8a, which is an overhead view. Alex pulls with a force F_A (220 N) and Charles with a force F_C (170 N). With what force F_B does Betty pull? The tire

FIGURE 5-8 Sample Problem 5-4. (*a*) An overhead view of three people pulling on a tire. (*b*) A free-body diagram for the tire.

remains stationary and the direction of Charles' pull is not given.

SOLUTION Figure 5-8*b* shows a free-body diagram for the tire. The acceleration of the tire is zero so, from Eq. 5-1, the net force on the tire must also be zero. That is,

$$\sum \mathbf{F} = \mathbf{F}_A + \mathbf{F}_B + \mathbf{F}_C = 0.$$

This vector equation is equivalent to the two scalar equations

$$\sum F_x = F_C \cos \phi - F_A \cos 47.0° = 0 \qquad (5\text{-}4)$$

and

$$\sum F_y = F_C \sin \phi + F_A \sin 47.0° - F_B = 0. \qquad (5\text{-}5)$$

The signs of the various terms in Eqs. 5-4 and 5-5 correspond to the directions of the corresponding force components in Fig. 5-8*b*. Substituting known values gives, for Eq. 5-4,

$$(170 \text{ N})(\cos \phi) = (220 \text{ N})(\cos 47.0°)$$

or

$$\phi = \cos^{-1} \frac{(220 \text{ N})(0.682)}{170 \text{ N}} = 28.0°.$$

Substituting into Eq. 5-5 gives us

$$F_B = F_C \sin \phi + F_A \sin 47.0°$$

$$= (170 \text{ N})(\sin 28.0°) + (220 \text{ N})(\sin 47.0°)$$

$$= 241 \text{ N}. \qquad \text{(Answer)}$$

Convince yourself that the three force vectors in Fig. 5-8*b*, if suitably shifted, form a closed triangle. That is, they add up to zero.

PROBLEM SOLVING

TACTIC 1: READ THE PROBLEM CAREFULLY

Read the problem statement several times until you have a good mental picture of what the situation is, what data are given, and what is requested. In Sample Problems 5-1 and 5-2, for example, you should tell yourself: "Someone is pushing a sled. Its speed changes, so acceleration is involved. The motion is along a straight line. A force is given in one problem and asked for in the other, and so the situation looks like Newton's second law applied to one-dimensional motion."

TACTIC 2: REREAD THE TEXT

If you know what the problem is about but don't know what to do next, put the problem aside and reread the text. If you are hazy about Newton's second law, reread that section. Study the sample problems. The one-dimensional-motion part of Sample Problems 5-1 and 5-2 should send you back to Chapter 2 and especially to Table 2-2, which displays all the equations you are likely to need.

TACTIC 3: DRAW A FIGURE

You may need two figures. One is a rough sketch of the actual real-world situation. When you draw the forces on it, place the tail of each force vector either on the boundary of or within the body feeling that force. The other figure is a free-body diagram in which the forces on a *single* body are drawn, with the body represented as a dot. Place the tail of each force vector on the dot.

TACTIC 4: WHAT IS YOUR SYSTEM?

If you are using Newton's second law, you must know to what body or system you are applying it. In Sample Problems 5-1 and 5-2, it is the sled (not the student or the ice). In Sample Problem 5-3, it is the crate (not the truck). In Sample Problem 5-4, it is the tire (not the ropes or the people).

TACTIC 5: WHAT IS YOUR REFERENCE FRAME?

Be clear as to what reference frame you are using. We used a reference frame attached to the Earth in all the sample problems so far. In Sample Problem 5-3, you have to make sure that you do not think it is the truck. In that problem, the truck is accelerating. A frame attached to it would be a noninertial frame.

TACTIC 6: CHOOSE YOUR AXES WISELY

In Sample Problem 5-4, we saved a lot of work by choosing one of our coordinate axes to coincide with one of the forces (the y axis with $\mathbf{F}_B$). Try working this problem with a coordinate system in which no axis coincides with a force.

5-6 SOME PARTICULAR FORCES

Weight

The **weight W** of a body is a force that pulls it directly toward a nearby astronomical body; in everyday circumstances that astronomical body is the Earth. The force is primarily due to an attraction—a **gravitational attraction**—between the masses of the two bodies, but further description of the force is delayed until Chapter 15. For now we consider only situations in which a body with mass m is located at a point where the magnitude of the free-fall acceleration is g. In that situation the *magnitude* of the weight (force) vector is

$$W = mg. \qquad (5\text{-}6)$$

The *vector* itself can be written either as

$$\mathbf{W} = -mg\mathbf{j} = -W\mathbf{j} \qquad (5\text{-}7)$$

On the moon, the diminished weight of equipment makes lifting the equipment much easier than on Earth.

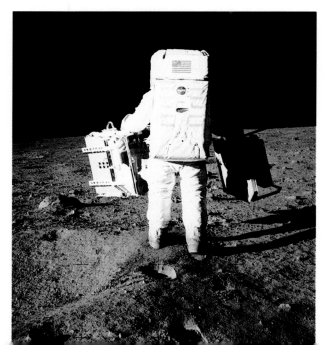

(where $+\mathbf{j}$ points upward, away from the Earth), or as

$$\mathbf{W} = m\mathbf{g}, \qquad (5\text{-}8)$$

where $\mathbf{g}$ represents the free-fall acceleration vector. In many cases, the choice of notation is up to you, but you need to understand what you mean by it and not get tripped up by, say, writing Eq. 5-6 when you mean Eq. 5-8.

Since weight is a force, its SI unit is the newton. *It is not mass,* and its magnitude at any given location depends on the value of g there. A bowling ball might weigh 71 N on the Earth, but only 12 N on the moon owing to the different free-fall acceleration there. The ball's mass, 7.2 kg, is the same in either place, because it is an intrinsic property of the ball. (If you want to lose weight, climb a mountain. Not only will the exercise reduce your mass, but the increased elevation means you are farther from the center of the Earth, and that means the value of g is less. So your weight will be less.)

Normally we assume that weight is measured from an inertial frame. If it is, instead, measured from a noninertial frame (an example comes up in Sample Problem 5-12), the measurement gives an **apparent weight** instead of the actual weight.

We can *weigh* a body by placing it on one of the pans of an equal-arm balance (Fig. 5-9) and then adding reference bodies (whose masses are known) on the other pan until we strike a balance. The masses on the pans then match and (assuming that g has the same value at both pans) so do the weights on the pans. We then know the mass m of the body. If we know the value of g for the location of the balance, we can find the weight of the body with Eq. 5-6.

FIGURE 5-9 An equal-arm balance. When the device is in balance, the masses on the left (L) and right (R) pans are equal.

$$\mathbf{W} = -mg\mathbf{j} = m\mathbf{g}$$

FIGURE 5-10   A spring scale. The reading is proportional to the *weight* of the object placed on the pan, and the scale gives that weight if marked in weight units. If, instead, it is marked in mass units, the reading is accurate only if the free-fall acceleration g is the same as where the scale was calibrated.

We can also weigh a body with a spring scale (Fig. 5-10). The body stretches a spring, moving a pointer along a scale that has previously been calibrated and marked in either mass or weight units. (Most bathroom scales in the United States work this way and read in pounds.) If marked in mass units, the scale is accurate only if the value of g is the same as where the scale was calibrated.

FIGURE 5-11   A normal force, directed upward, supports each elephant.

Normal Force

When a body is pressed against a surface, the body experiences a force that is perpendicular to the surface. The force is called the **normal force N**, the name coming from the mathematical term *normal*, meaning "perpendicular."

If a body rests on a horizontal surface as in Figs. 5-11 and 5-12*a*, **N** is directed upward and the body's weight $\mathbf{W} = m\mathbf{g}$ is directed downward. We find the magnitude of **N** from Eq. 5-2:

$$\sum F_y = N - mg = ma_y, \qquad (5\text{-}9)$$

and so, with $a_y = 0$,

$$N = mg. \qquad (5\text{-}10)$$

PROBLEM SOLVING

TACTIC 7: NORMAL FORCE

Equation 5-10 for the normal force holds only when **N** is directed upward and the vertical acceleration is zero. So you should not apply it for other orientations of **N** or when the vertical acceleration is not zero. Instead, learn the procedure for finding **N**: we use Newton's second law in its component form.

We are free to move **N** around in a figure as long as we maintain its orientation. For example, in Fig. 5-12*a* we can slide it downward so that its head is at the boundary of the body and the tabletop. However, **N** is least likely to be misinterpreted if its tail is at the boundary or somewhere within the body (as shown). An even better technique is to draw a free-body diagram such as in Fig. 5-12*b*, with the tail of **N** directly on the dot representing the body.

FIGURE 5-12   (*a*) The body resting on a tabletop experiences a normal force **N** perpendicular to the tabletop. (*b*) The corresponding free-body diagram.

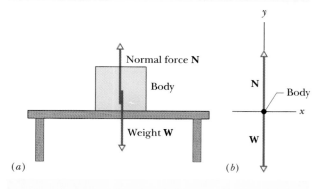

Friction

If we slide or attempt to slide a body over a surface, the motion is resisted by a bonding between the body and surface. (We discuss this more in the next chapter.) The resistance is considered to be a single force **f**, called the **frictional force,** or simply **friction.** The force runs parallel to the surface, opposite the direction of the intended motion (Fig. 5-13). Sometimes, to simplify a situation, friction is assumed to be negligible, and the surface is said to be *frictionless.*

Tension

When a cord (or a rope, cable, or other such object) is attached to a body and pulled taut, the cord is said to be under **tension.** It pulls on the body with a force **T**, whose direction is away from the body and along the cord at the point of attachment (Fig. 5-14*a*).

A cord is often regarded to be *massless* (meaning its mass is negligible compared to the body's mass)

FIGURE 5-13 A frictional force **f** opposes the attempted slide of a body over a surface.

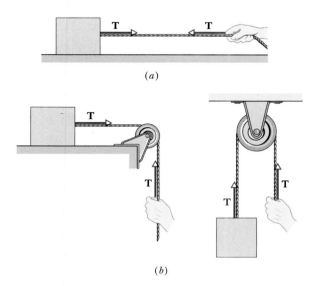

(a)

(b)

FIGURE 5-14 (*a*) The cord, pulled taut, is under tension. It pulls on the body at either end with force **T**. (*b*) The same is true if the cord runs around a massless, frictionless pulley.

and unstretchable. The cord exists only as a connection between two bodies. It pulls on the body at each end with the same magnitude *T*, even if the bodies and the cord are accelerating and even if the cord runs around a *massless, frictionless pulley* (Fig. 5-14*b*). Such a pulley has negligible mass compared to the bodies and has negligible friction on its axle opposing its rotation.

SAMPLE PROBLEM 5-5

Let us return to John Massis and the railroad cars, and assume that Massis pulled (with his teeth) on his end of the rope with a constant force that was 2.5 times his body weight, at an angle θ of 30° from the horizontal. His mass *m* was 80 kg. The weight *W* of the passenger cars was 7.0×10^5 N (about 80 tons), and he moved them 1.0 m along the rails. Assume the wheels encountered no retarding force from the rails as they rolled. What was the speed of the train just at the end of the pull?

SOLUTION Figure 5-15 is a free-body diagram for the cars, which are represented by a dot. The *x* axis runs along the rails. From Eq. 5-2, we have

$$\sum F_x = T \cos \theta = Ma_x, \qquad (5\text{-}11)$$

in which *M* is the mass of the cars.

With our assumptions, the pull from Massis is

$$T = 2.5mg = (2.5)(80 \text{ kg})(9.8 \text{ m/s}^2) = 1960 \text{ N}$$

(440 lb), which is about what a good middle-weight weight lifter can lift.

FIGURE 5-15 Sample Problem 5-5. Free-body diagram for the passenger cars pulled by Massis. The vectors are not drawn to scale; the tension in the rope is *much* smaller than the weight and normal force.

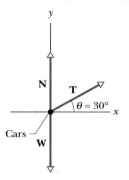

The weight W of the cars is

$$W = Mg,$$

where their mass M must be

$$M = \frac{W}{g} = \frac{7.0 \times 10^5 \text{ N}}{9.8 \text{ m/s}^2} = 7.143 \times 10^4 \text{ kg}.$$

With Eq. 5-11, we find their acceleration to be

$$a_x = \frac{T \cos \theta}{M} = \frac{(1960 \text{ N})(\cos 30°)}{7.143 \times 10^4 \text{ kg}}$$

$$= 2.376 \times 10^{-2} \text{ m/s}^2.$$

To solve for their speed at the end of the pull, we use Eq. 2-14, with subscripts for the x axis and with $v_0 = 0$ and $x - x_0 = 1.0$ m:

$$v_x^2 = v_{0x}^2 + 2a_x(x - x_0),$$

or

$$v_x = \sqrt{0 + (2)(2.376 \times 10^{-2} \text{ m/s}^2)(1.0 \text{ m})}$$

$$= 0.22 \text{ m/s}. \qquad \text{(Answer)}$$

Massis would have done better if the rope had been attached higher on the car, so that it was horizontal. Can you see why?

5-7 NEWTON'S THIRD LAW

Forces come in pairs. If a hammer exerts a force on a nail, the nail exerts an equal but oppositely directed force on the hammer. If you lean against a brick wall, the wall pushes back on you (Fig. 5-16). The situation has been summed up with the gentle words: "You cannot touch without being touched."

Let body A in Fig. 5-17 exert a force $\mathbf{F}_{BA}$ on body B; experiment shows that body B then exerts a force $\mathbf{F}_{AB}$ on body A. These two forces are equal in magnitude and oppositely directed. That is,

$$\mathbf{F}_{AB} = -\mathbf{F}_{BA} \qquad \text{(Newton's third law).} \qquad (5\text{-}12)$$

Note the order of the subscripts. $\mathbf{F}_{AB}$, for example, is the force exerted *on* body A *by* body B. Equation 5-12 holds regardless of whether the bodies move or remain stationary.

Equation 5-12 sums up Newton's third law of motion. Commonly, one of these forces (it does not matter which) is called the **action force.** The other member of the pair is then called the **reaction force.** Every time you find a force, a good question is: "Where is its reaction force?"

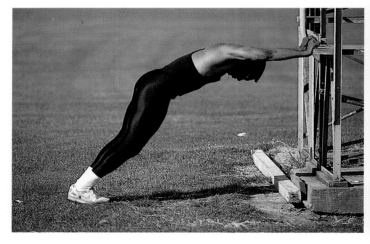

FIGURE 5-16 The man exerts a force to the right on the wall. The wall exerts a force to the left on the man. The magnitudes of the forces are equal.

FIGURE 5-17 Newton's third law. Body A exerts a force $\mathbf{F}_{BA}$ on body B, while body B exerts a force $\mathbf{F}_{AB}$ on body A, where $\mathbf{F}_{AB} = -\mathbf{F}_{BA}$.

The words, "To every action there is always an equal and opposite reaction," have become enshrined in the popular language and mean various things to various speakers. In physics, however, these words mean Eq. 5-12 and nothing else. In particular, cause and effect are not involved; either force can be the action force.

You may think: "If every force has an equal and opposite force associated with it, why don't they cancel each other? How can anything ever get moving?" The answer is simple. As Fig. 5-17 shows, the two members of an action–reaction pair *always* act on different bodies so that they cannot possibly cancel each other. If two forces act on the *same* body they are *not* an action–reaction pair, even though they may be equal and opposite. Let's identify the action–reaction pairs in two examples.

An Orbiting Satellite

Figure 5-18 shows a satellite orbiting the Earth. The only force that acts on it is $\mathbf{F}_{SE}$, the force exerted *on*

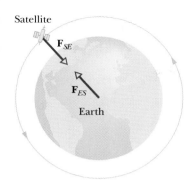

FIGURE 5-18 A satellite in Earth orbit. The forces shown are an action–reaction pair. Note that they act on different bodies.

the satellite *by* the gravitational pull of the Earth. Where is the corresponding reaction force? It is $\mathbf{F}_{ES}$, the force acting on the Earth owing to the gravitational pull of the satellite; its effective point of application is taken to be at the center of the Earth.

You may think that the tiny satellite cannot exert much of a gravitational pull on the Earth but it does, exactly as Newton's third law requires. That is, considering magnitudes only, $F_{ES} = F_{SE}$. The force $\mathbf{F}_{ES}$ causes the Earth to accelerate, but, because of the Earth's large mass, its acceleration is too small to be detected.

A Cantaloupe Resting on a Table*

Figure 5-19a shows a cantaloupe at rest on a table. The Earth pulls downward on the cantaloupe with a

---

*We ignore small complications caused by the rotation of the Earth.

force $\mathbf{F}_{CE}$, the cantaloupe's weight. The cantaloupe does not accelerate because this force is canceled by an equal and opposite normal force $\mathbf{F}_{CT}$ exerted on the cantaloupe by the table. (See Fig. 5-19b.) However, $\mathbf{F}_{CE}$ and $\mathbf{F}_{CT}$ do *not* form an action–reaction pair *because they act on the same body, the cantaloupe.*

The reaction force to $\mathbf{F}_{CE}$ is $\mathbf{F}_{EC}$, the (gravitational) force with which the cantaloupe attracts the Earth. This action–reaction pair is shown in Fig. 5-19c.

The reaction force to $\mathbf{F}_{CT}$ is $\mathbf{F}_{TC}$, the force on the table by the cantaloupe. This action–reaction pair is shown in Fig. 5-19d. The action–reaction pairs in this problem, and the bodies on which they act, are then

First pair: $\mathbf{F}_{CE} = -\mathbf{F}_{EC}$ (cantaloupe and Earth)

and

Second pair: $\mathbf{F}_{CT} = -\mathbf{F}_{TC}$ (cantaloupe and table).

We can use an accelerating elevator cab to sort the four forces shown in Fig. 5-19 properly into action–reaction pairs. Suppose that the cab is accelerating upward. The cantaloupe and the table, inside the cab, would then press against each other with a greater force. The contact forces $\mathbf{F}_{TC}$ and $\mathbf{F}_{CT}$ (see Fig. 5-19d) would increase in magnitude but would remain equal and opposite. However, the gravitational forces $\mathbf{F}_{CE}$ and $\mathbf{F}_{EC}$ (see Fig. 5-19c)

FIGURE 5-19 (a) A cantaloupe rests on a tabletop that rests on the Earth. (b) The forces *on the cantaloupe*, $\mathbf{F}_{CT}$ and $\mathbf{F}_{CE}$. The cantaloupe is stationary because these forces balance. (c) The action–reaction pair for the cantaloupe–Earth forces. (d) The action–reaction pair for the cantaloupe–table forces.

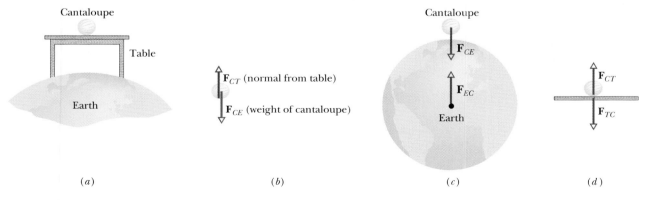

would remain unchanged. Both pairs of forces would continue to obey Newton's third law. The cantaloupe would accelerate because $\mathbf{F}_{CE}$ and $\mathbf{F}_{CT}$ (which are not an action–reaction pair) no longer cancel each other.

Would you have guessed that an orbiting satellite is simpler to analyze—in the context of Newton's third law—than a cantaloupe at rest on a table?

5-8 APPLYING NEWTON'S LAWS

The rest of the chapter consists of Sample Problems 5-6 to 5-12. You should pore over those samples, learning not just their particular answers but, instead, the procedure for attacking a problem. Especially important is knowing how to translate a sketch of a situation into a free-body diagram, with appropriate axes, so that Newton's laws can be applied. We begin with Sample Problem 5-6, which is worked out in exhaustive detail, using a question-and-answer format.

SAMPLE PROBLEM 5-6

Figure 5-20 shows a block (the *sliding block*) whose mass M is 3.3 kg. It is free to move along a horizontal frictionless surface such as an air table. The sliding block is connected by a cord that goes around a massless, frictionless pulley to a second block (the *hanging block*), whose mass m is 2.1 kg. The hanging block will fall and the sliding block will accelerate to the right.

Find (a) the acceleration of the sliding block, (b) the acceleration of the hanging block, and (c) the tension in the cord.

Q *What is this problem all about?*

You are given two massive objects, the sliding block and the hanging block. It might not occur to you but you are also given the Earth, which pulls on each of these objects; without the Earth, nothing would happen. There is a total of five forces on the blocks, as shown in Fig. 5-21:

1. The cord pulls to the right on the sliding block with a force of magnitude T.

2. The cord pulls upward on the hanging block with a force of the same magnitude T. This keeps the hanging block from falling freely, which it would otherwise do. We assume that the cord has the same tension through-

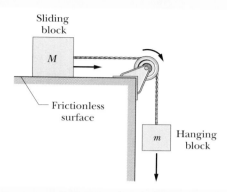

FIGURE 5-20 Sample Problem 5-6. A block of mass M on a horizontal frictionless surface is connected to a block of mass m by a cord that wraps over a pulley. Both cord and pulley are massless; that is, their masses are negligible compared to M and m. The pulley is frictionless; that is, the frictional force on its axle that opposes rotation is negligible. The arrows indicate the motion when the system is released from rest.

out its length; the pulley just serves to change the direction of this force, without changing its magnitude.

3. The Earth pulls down on the sliding block with a force $M\mathbf{g}$, the weight of the sliding block.

4. The Earth pulls down on the hanging block with a force $m\mathbf{g}$, the weight of the hanging block.

5. The table pushes up on the sliding block with a normal force $\mathbf{N}$.

There is another thing that you should note. We assume that the cord does not stretch, so that if the hanging block falls 1 mm in a certain time, the sliding block moves 1 mm to the right in that same interval. The blocks move together and their accelerations have the same magnitude a.

Q *How do I classify this problem? Should it suggest a particular law of physics to me?*

Yes, it should. Forces, masses, and accelerations are involved and this should suggest Newton's second law of motion, $\Sigma \mathbf{F} = m\mathbf{a}$.

Q *If I apply this law to this problem, to what body should I apply it?*

We focus on two bodies in this problem, the sliding block and the hanging block. Although they are extended objects, we can treat each block as a particle because every small part of it (every atom, say) moves in exactly the same way. Apply Newton's second law separately to each block.

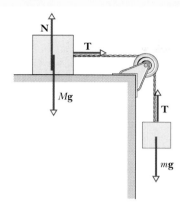

FIGURE 5-21 The forces acting on the two blocks of Fig. 5-20.

Q *What about the pulley?*

We cannot represent the pulley as a particle because different parts of it move in different ways. When we discuss rotation, we shall deal with pulleys in detail. Meanwhile, we get around the problem in a practical way, by using a pulley whose mass is negligible compared with the masses of the two blocks. The only function of the pulley (which we assume has no friction on its axle to retard rotation) is to change the direction of the cord that joins the two blocks.

Q *O.K. Now how do I apply $\Sigma\,\mathbf{F} = m\mathbf{a}$ to the sliding block?*

Represent the sliding block as a particle of mass M and draw *all* the forces that act *on* it, as in Fig. 5-22. This is the block's *free-body diagram*. There are three forces. Next, locate a horizontal axis (an x axis). It makes sense to draw this axis parallel to the table, in the direction in which the block moves.

Q *Thanks, but you still haven't told me how to apply $\Sigma\,\mathbf{F} = m\mathbf{a}$ to the sliding block. All you have done is to tell me how to draw a free-body diagram.*

Right you are. $\Sigma\,\mathbf{F} = m\mathbf{a}$ is a vector equation and you can write it as three scalar equations. Thus

$$\sum F_x = Ma_x, \quad \sum F_y = Ma_y, \quad \sum F_z = Ma_z, \quad (5\text{-}13)$$

in which $\Sigma\,F_x$, $\Sigma\,F_y$, and $\Sigma\,F_z$ are the components of the net force. From Eq. 5-10 we know there is no net force in the y direction: the weight $\mathbf{W}$ of the sliding block is balanced by the upward-acting normal force $\mathbf{N}$ on the block. No force acts in the z direction, which is perpendicular to the page. Thus only the first of Eqs. 5-13 is useful.

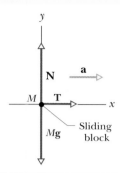

FIGURE 5-22 A free-body diagram for the sliding block of Fig. 5-20.

In the x direction, there is only one force component, so $\Sigma\,F_x = Ma_x$ becomes

$$T = Ma. \quad (5\text{-}14)$$

This contains two unknowns, T and a, so we cannot yet solve it. Recall, however, that we have not yet said anything about the hanging block.

Q *I agree. How do I apply $\Sigma\,\mathbf{F} = m\mathbf{a}$ to the hanging block?*

Draw a free-body diagram for the block, as in Fig. 5-23. This time, we use the second of Eqs. 5-13, finding

$$\sum F_y = T - mg = -ma, \quad (5\text{-}15)$$

where the minus sign on the right side of the equation indicates that the block accelerates downward, in the negative direction of the y axis. Equation 5-15 yields

$$mg - T = ma. \quad (5\text{-}16)$$

This contains the same two unknowns as Eq. 5-14 does. If you add these equations, T will cancel out. Solving

FIGURE 5-23 A free-body diagram for the hanging block of Fig. 5-20.

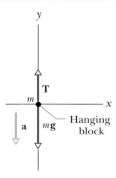

for *a* then yields

$$a = \frac{m}{M + m} g. \qquad \text{(Answer)} \qquad (5\text{-}17)$$

Substituting this result into Eq. 5-14 yields

$$T = \frac{Mm}{M + m} g. \qquad \text{(Answer)} \qquad (5\text{-}18)$$

Putting in the numbers gives, for these two quantities,

$$a = \frac{m}{M + m} g = \frac{2.1 \text{ kg}}{3.3 \text{ kg} + 2.1 \text{ kg}} (9.8 \text{ m/s}^2)$$

$$= 3.8 \text{ m/s}^2 \qquad \text{(Answer)}$$

and

$$T = \frac{Mm}{M + m} g = \frac{(3.3 \text{ kg})(2.1 \text{ kg})}{3.3 \text{ kg} + 2.1 \text{ kg}} (9.8 \text{ m/s}^2)$$

$$= 13 \text{ N}. \qquad \text{(Answer)}$$

Q   *The problem is now solved, right?*

That's a fair question, but we are here not only to solve problems but chiefly to learn physics. This problem is not really finished until we have studied the results to see if they make sense. This is often a much more confidence-building experience than simply getting the right answer.

Look first at Eq. 5-17. Note that it is dimensionally correct and also that the acceleration *a* will always be less than *g*. This is as it must be, because the hanging block is not in free fall. The cord pulls upward on it.

Look now at Eq. 5-18, which we rewrite in the form

$$T = \frac{M}{M + m} mg. \qquad (5\text{-}19)$$

In this form, it is easier to see that this equation is also dimensionally correct, because both *T* and *mg* are forces. Equation 5-19 also lets us see that the tension in the cord is always less than *mg*, the weight of the hanging block. That is a comforting thought because, if *T* were *greater* than *mg*, the hanging block would accelerate upward!

We can also check the results by studying special cases, in which we can guess what the answers must be. A simple example is to put *g* = 0, as if the experiment were carried out in interstellar space. We know that, in that case, the blocks would not move from rest and there would be no tension in the cord. Do the formulas predict this? Yes, they do. If you put *g* = 0 in Eqs. 5-17 and 5-18, you find *a* = 0 and *T* = 0. Two more special cases are *M* = 0 and *m* → ∞.

▬▬▬▬▬▬

SAMPLE PROBLEM 5-6—ANOTHER WAY

The acceleration *a* of the blocks of Fig. 5-20 can be found in two lines of algebra if we (a) employ an unconventional axis, call it *u*, that runs through *both* blocks and along the cord as shown in Fig. 5-24*a*, and then (b) mentally straighten out the *u* axis as in Fig. 5-24*b* and treat the blocks as being portions of a single composite body with mass *M* + *m*. A free-body diagram for the two-block system is shown in Fig. 5-24*c*.

SOLUTION   Note that there is only one force acting on the composite body along the *u* axis, and that is the force *m***g** in the positive direction of the axis. The tension **T** of Fig. 5-21 is now internal to the composite body and so does not enter into Newton's second law. The force exerted by the pulley on the cord is perpendicular to the *u* axis; so it too does not enter.

Using Eq. 5-2 as a guide, we write a component equation for the acceleration along the *u* axis:

$$\sum F_u = (M + m) a_u,$$

FIGURE 5-24   (*a*) An "axis" *u* runs through the system of Fig. 5-20. (*b*) The blocks are rearranged to straighten *u* and then are treated as a single body with mass *M* + *m*. (*c*) The associated free-body diagram, considering only forces along *u*. There is one such force.

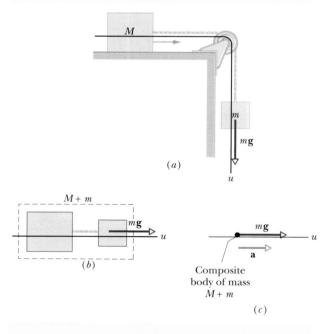

where the mass of the body is $M + m$. The acceleration of the composite along the u axis (and of the individual blocks, since they are connected) has magnitude a. The only force on the composite along the u axis has a magnitude of mg. So our equation becomes

$$mg = (M + m)a,$$

or

$$a = \frac{m}{M + m} g, \qquad (5\text{-}20)$$

which matches Eq. 5-17.

To find T, we apply Newton's second law to either block individually, obtaining either Eq. 5-14 or Eq. 5-16. We then substitute for a from Eq. 5-20 and solve for T, getting Eq. 5-18.

SAMPLE PROBLEM 5-7

A block whose mass M is 33 kg is pushed across a frictionless surface by a stick whose mass m is 3.2 kg, as in Fig. 5-25a. The block is moved (from rest) a distance $d = 77$ cm in 1.7 s at constant acceleration.

a. Identify all horizontal action–reaction force pairs in this problem.

SOLUTION As the exploded view of Fig. 5-25b shows, there are two action–reaction pairs:

First pair: $\mathbf{F}_{HS} = -\mathbf{F}_{SH}$ (hand and stick)

Second pair: $\mathbf{F}_{SB} = -\mathbf{F}_{BS}$ (stick and block).

The force $\mathbf{F}_{HS}$ on the hand by the stick is the force that you would feel if the hand in Fig. 5-25 were yours.

b. What force must the hand apply to the stick?

SOLUTION This is the force that accelerates the block and stick. To find it, we must first find the constant acceleration a, using Eq. 2-13:

$$x - x_0 = v_0 t + \tfrac{1}{2}at^2.$$

Setting $v_0 = 0$ and $x - x_0 = d$, and solving for a, give

$$a = \frac{2d}{t^2} = \frac{(2)(0.77 \text{ m})}{(1.7 \text{ s})^2} = 0.533 \text{ m/s}^2.$$

To find the force that the hand exerts, we apply Newton's second law to a system consisting of the stick and

the block taken together. Thus,

$$F_{SH} = (M + m)a = (33 \text{ kg} + 3.2 \text{ kg})(0.533 \text{ m/s}^2)$$

$$= 19.3 \text{ N} \approx 19 \text{ N}. \qquad \text{(Answer)}$$

c. With what force does the stick push on the block?

SOLUTION To find this force, we apply Newton's second law to the block alone:

$$F_{BS} = Ma = (33 \text{ kg})(0.533 \text{ m/s}^2)$$

$$= 17.6 \text{ N} \approx 18 \text{ N}. \qquad \text{(Answer)}$$

d. What is the net force on the stick?

SOLUTION We can find the magnitude F of this force in two ways. First, using results from (b) and (c) above, we have

$$F = F_{SH} - F_{SB} = 19.3 \text{ N} - 17.6 \text{ N}$$

$$= 1.7 \text{ N}. \qquad \text{(Answer)}$$

Note that we have used Newton's third law here, in assuming that $\mathbf{F}_{SB}$, the force on the stick by the block, has the same magnitude (17.6 N before rounding) as $\mathbf{F}_{BS}$.

The second way to arrive at an answer is to apply

FIGURE 5-25 Sample Problem 5-7. (a) A block of mass M is pushed over a frictionless surface by a stick of mass m. (b) An exploded view, showing the action–reaction pairs between the hand and the stick (first pair) and between the stick and the block (second pair).

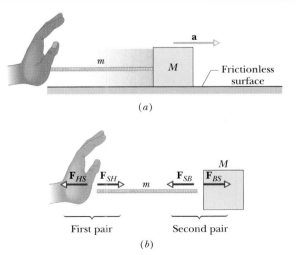

Newton's second law to the stick directly. We have

$$F = ma = (3.2 \text{ kg})(0.533 \text{ m/s}^2) = 1.7 \text{ N}, \quad \text{(Answer)}$$

in agreement with our first result. This is as it must be because the two methods are algebraically identical; check it out.

SAMPLE PROBLEM 5-8

Figure 5-26*a* shows a block of mass $m = 15$ kg hanging from three cords. What are the tensions in the cords?

SOLUTION   The free-body diagram for the block is shown in Fig. 5-26*b*: tension $\mathbf{T}_C$ from cord C pulls upward while the block's weight $m\mathbf{g}$ is directed downward.

FIGURE 5-26   Sample Problem 5-8. (*a*) A block of mass m hangs from three cords. (*b*) A free-body diagram for the block. (*c*) A free-body diagram for the knot at the intersection of the three cords.

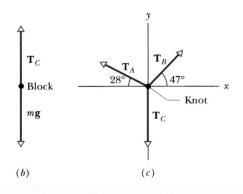

Since the system is at rest, Newton's second law for the block yields

$$\sum F_y = T_C - mg = 0,$$

or

$$T_C = mg = (15 \text{ kg})(9.8 \text{ m/s}^2)$$
$$= 147 \text{ N} \approx 150 \text{ N}. \quad \text{(Answer)}$$

The clue to our next step is to realize that the knot where the three cords join is the only point at which all three forces act, and it is this knot to which we should apply Newton's second law. Figure 5-26*c* shows the free-body diagram for the knot. Since the knot is not accelerated, the net force acting on it must be zero. Thus

$$\sum \mathbf{F} = \mathbf{T}_A + \mathbf{T}_B + \mathbf{T}_C = 0.$$

This vector equation is equivalent to the two scalar equations

$$\sum F_y = T_A \sin 28° + T_B \sin 47° - T_C = 0 \quad (5\text{-}21)$$

and

$$\sum F_x = -T_A \cos 28° + T_B \cos 47° = 0. \quad (5\text{-}22)$$

Note carefully that, when we write the x component of $\mathbf{T}_A$ as $T_A \cos 28°$, we must include a minus sign to show that it extends in the negative direction of the x axis.

Substituting numerical values into Eqs. 5-21 and 5-22 leads to

$$T_A(0.469) + T_B(0.731) = 147 \text{ N} \quad (5\text{-}23)$$

and

$$T_B(0.682) = T_A(0.883). \quad (5\text{-}24)$$

From Eq. 5-24, we have

$$T_B = \frac{0.883}{0.682} T_A = 1.29 \ T_A.$$

Substituting this quantity into Eq. 5-23 and solving for T_A, we obtain

$$T_A = \frac{147 \text{ N}}{0.469 + (1.29)(0.731)}$$
$$= 104 \text{ N} \approx 100 \text{ N}. \quad \text{(Answer)}$$

Finally, T_B is found from

$$T_B = 1.29 T_A = (1.29)(104 \text{ N})$$
$$= 134 \text{ N} \approx 130 \text{ N}. \quad \text{(Answer)}$$

SAMPLE PROBLEM 5-9

Figure 5-27a shows a block of mass $m = 15$ kg held by a cord on a frictionless inclined plane. What is the tension in the cord if $\theta = 27°$? What force does the plane exert on the block?

SOLUTION Figure 5-27b is the free-body diagram for the block. The following forces act on it: (1) a normal force **N**, exerted outward on the block by the plane on which it rests; (2) the tension **T** in the cord; and (3) the weight **W** ($= m\mathbf{g}$) of the block. Because the acceleration of the block is zero, the net force acting on the block must also be zero by Newton's second law:

$$\sum \mathbf{F} = \mathbf{T} + \mathbf{N} + m\mathbf{g} = 0. \qquad (5\text{-}25)$$

We choose a coordinate system with the x axis parallel to the plane. With this choice, not one but two forces (**N** and **T**) line up with coordinate axes (a bonus). Note that the angle between the weight vector and the negative direction of the y axis equals the slant angle of the plane. The x and y components of that vector are found with the triangle of Fig. 5-27c.

FIGURE 5-27 Sample Problems 5-9 and 5-10. (a) A block of mass m rests on a smooth plane, held there by a cord. (b) A free-body diagram for the block. Note how the coordinate axes are placed. (c) Finding the x and y components of mg.

(a)

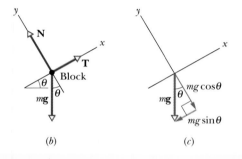

(b)   (c)

The component versions of Eq. 5-25 are

$$\sum F_x = T - mg \sin \theta = 0$$

and

$$\sum F_y = N - mg \cos \theta = 0.$$

Thus

$$T = mg \sin \theta$$
$$= (15 \text{ kg})(9.8 \text{ m/s}^2)(\sin 27°)$$
$$= 67 \text{ N} \qquad \text{(Answer)}$$

and

$$N = mg \cos \theta$$
$$= (15 \text{ kg})(9.8 \text{ m/s}^2)(\cos 27°)$$
$$= 131 \text{ N} \approx 130 \text{ N}. \qquad \text{(Answer)}$$

SAMPLE PROBLEM 5-10

Suppose you cut the cord holding the block on the plane in Fig. 5-27a. With what acceleration will the block move?

SOLUTION Cutting the cord removes the tension **T** in Fig. 5-27b. The two remaining forces do not cancel and, indeed, they cannot, because they do not act along the same line. Applying Newton's second law to the x components of the forces **N** and $m\mathbf{g}$ in Fig. 5-27b now yields

$$\sum F_x = 0 - mg \sin \theta = ma,$$

so that

$$a = -g \sin \theta. \qquad (5\text{-}26)$$

Note that the normal force **N** plays no role in producing the acceleration, its x component being zero.

Equation 5-26 yields

$$a = -(9.8 \text{ m/s}^2)(\sin 27°) = -4.4 \text{ m/s}^2. \qquad \text{(Answer)}$$

The minus sign indicates that the acceleration is in the direction of decreasing x, that is, down the plane.

Equation 5-26 reveals that the acceleration of the block is independent of its mass, just as the acceleration of a freely falling body is independent of the mass of the falling body. Indeed, Eq. 5-26 shows that an inclined plane can be used to "dilute" gravitation— "slow it down"—so that the effects of "falling" can be studied more easily. For $\theta = 90°$, Eq. 5-26 yields $a = -g$; for $\theta = 0°$, it yields $a = 0$. Both are expected results.

SAMPLE PROBLEM 5-11

Figure 5-28a shows two blocks connected by a cord that passes over a massless, frictionless pulley. Let $m = 1.3$ kg and $M = 2.8$ kg. Find the tension in the cord and the (common) magnitude of the acceleration of the two blocks.

SOLUTION   Figures 5-28b and 5-28c are free-body diagrams for the blocks. We are given $M > m$, so we expect M to fall and m to rise. That information allows us to assign the proper algebraic signs to the accelerations of the blocks.

Before we begin the calculations, note that the tension in the cord must be less than the weight of block M (otherwise, that block would not fall from rest) and greater than the weight of block m (otherwise, that block would not rise). The vectors in the two free-body

FIGURE 5-28   Sample Problem 5-11. (a) A block of mass M and one of mass m are connected by a cord that passes over a pulley. The directions in which the system will accelerate from rest are shown by the light arrows. (b) A free-body diagram for block m. (c) A free-body diagram for block M.

(a)

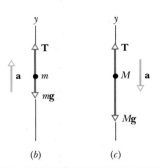

(b)   (c)

diagrams of Fig. 5-28 are drawn to represent these facts.

Applying Newton's second law to the block with mass m, which has acceleration a in the positive direction of the y axis, we find

$$T - mg = ma. \tag{5-27}$$

For the block with mass M, which has an acceleration of $-a$, we have

$$T - Mg = -Ma, \tag{5-28}$$

or

$$-T + Mg = Ma. \tag{5-29}$$

By adding Eqs. 5-27 and 5-29 (or eliminating T through substitution), we obtain

$$a = \frac{M - m}{M + m} g. \tag{5-30}$$

Substituting this result in either Eq. 5-27 or Eq. 5-29 and solving for T, we obtain

$$T = \frac{2mM}{M + m} g. \tag{5-31}$$

Equation 5-31 can be rewritten in the equivalent forms

$$T = \frac{M + M}{M + m} mg \quad \text{and} \quad T = \frac{m + m}{M + m} Mg. \tag{5-32}$$

The first of these formulations shows that $T > mg$ and the second shows that $T < Mg$. That is, the tension T is intermediate between the weights of the two bodies, as anticipated above. Furthermore, if $M = m$, Eqs. 5-30 and 5-31 reduce to $a = 0$ and $T = mg = Mg$, as we would expect. That is, if the blocks have equal masses, their acceleration is zero (the blocks do not move from rest) and the tension is equal to the weight of either of the two identical blocks. (Note that it is *not* twice the weight of one block.)

Inserting the given data, we obtain

$$a = \frac{M - m}{M + m} g = \frac{2.8 \text{ kg} - 1.3 \text{ kg}}{2.8 \text{ kg} + 1.3 \text{ kg}} (9.8 \text{ m/s}^2)$$

$$= 3.6 \text{ m/s}^2 \qquad \text{(Answer)}$$

and

$$T = \frac{2Mm}{M + m} g = \frac{(2)(2.8 \text{ kg})(1.3 \text{ kg})}{2.8 \text{ kg} + 1.3 \text{ kg}} (9.8 \text{ m/s}^2)$$

$$= 17 \text{ N.} \qquad \text{(Answer)}$$

You can show that the weights of the two blocks are 13 N ($= mg$) and 27 N ($= Mg$). So the tension ($= 17$ N) does indeed lie between these two values.

SAMPLE PROBLEM 5-11—ANOTHER WAY

Just as we redid Sample Problem 5-6 with an unconventional axis u, we can redo Sample Problem 5-11 with u.

SOLUTION Run the axis through the system as shown in Fig. 5-29a. Straighten out the axis as in Fig. 5-29b, and treat the blocks as a single body with a mass of $M + m$. Then draw a free-body diagram as in Fig. 5-29c. Note that along the u axis there are two forces on the composite of the blocks: mg in the negative direction of the u axis and Mg in the positive direction. (The force exerted on the cord by the pulley is perpendicular to the u axis.) The two forces along the u axis give the composite (and each block) an acceleration $\mathbf{a}$. Newton's second law in component form for motion along u is

$$\sum F_u = Mg - mg = (M + m)a, \qquad (5\text{-}33)$$

which yields

$$a = \frac{M - m}{M + m}\, g,$$

as previously. To get T we apply Newton's second law to either block, using a conventional axis y as in the original solution. For the block with mass m, we obtain Eq. 5-27. With the above result for a substituted into Eq. 5-27, we obtain Eq. 5-31.

FIGURE 5-29 (a) An "axis" u runs through the system of Fig. 5-28. (b) The blocks are rearranged to straighten u and then treated as a single body with mass $M + m$. (c) The associated free-body diagram, considering only forces along u. There are two such forces.

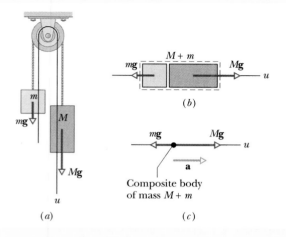

SAMPLE PROBLEM 5-12

A passenger of mass $m = 72.2$ kg stands on a platform scale in an elevator cab (Fig. 5-30). What is the scale reading for the acceleration values given in the figure?

SOLUTION We consider this problem from the point of view of an observer in an (inertial) reference frame fixed with respect to the Earth. Let this observer apply Newton's second law to the accelerating passenger. Figure 5-30a-e presents free-body diagrams for the passenger, treated as a particle (some particle!), for various accelerations of the cab.

Regardless of the acceleration of the cab, the Earth pulls downward on the passenger with a force having magnitude mg, in which $g = 9.80$ m/s^2 is the free-fall acceleration in the reference frame of the Earth. The scale platform pushes upward on the passenger with a normal force whose magnitude N equals the reading of the scale. The weight that the accelerating passenger would judge himself to have is what he reads on the scale. This quantity is often called the *ap-*

Your apparent weight differs from your true weight when the velocity of the elevator changes at the start and end of a ride, not during the rest of the ride when that velocity is constant.

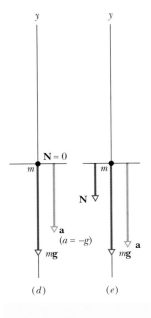

FIGURE 5-30 Sample Problem 5-12. A passenger with mass m in an elevator, standing on a spring scale which indicates his weight or apparent weight. (*a*) A free-body diagram for the case in which the acceleration of the elevator cab is zero. (*b*) For $a = +3.20$ m/s². (*c*) For $a = -3.20$ m/s². (*d*) For $a = -g = -9.80$ m/s². (*e*) For $a = -12.0$ m/s².

Reading of weight or apparent weight

parent weight, the term *weight* (or *true weight*) being reserved for the quantity *mg.*

Newton's second law yields

$$N - mg = ma,$$

or

$$N = m(g + a). \qquad (5\text{-}34)$$

a. What does the scale read if the cab is at rest or moving with constant speed? (See Fig. 5-30*a*.)

SOLUTION Here $a = 0$ and we have

$$N = m(g + a) = (72.2 \text{ kg})(9.80 \text{ m/s}^2 + 0)$$

$$= 708 \text{ N}. \qquad \text{(Answer)}$$

b. What does the scale read if the cab has an upward acceleration of magnitude 3.20 m/s²? (See Fig. 5-30*b*.)

SOLUTION An upward acceleration means that the cab is either moving up with increasing speed or down with decreasing speed. In either case, Eq. 5-34 yields

$$N = m(g + a) = (72.2 \text{ kg})(9.80 \text{ m/s}^2 + 3.20 \text{ m/s}^2)$$

$$= 939 \text{ N}. \qquad \text{(Answer)}$$

The passenger presses down on the scale with a greater force than if he were at rest. The passenger—reading the scale—might conclude that he has gained 231 N!

c. What does the scale read if the cab has a downward acceleration of magnitude 3.20 m/s²? (See Fig. 5-30*c*.)

SOLUTION A downward acceleration means that the cab is either moving up with decreasing speed or down with increasing speed. Equation 5-34 yields

$$N = m(g + a) = (72.2 \text{ kg})(9.80 \text{ m/s}^2 - 3.20 \text{ m/s}^2)$$

$$= 477 \text{ N}. \qquad \text{(Answer)}$$

The passenger presses down on the scale with less force than if the cab were at rest. He seems to have lost 231 N.

d. What does the scale read if the cable breaks, so that the cab falls freely? (See Fig. 5-30*d*.)

SOLUTION In this case, the passenger (and the scale) are in free fall, with $a = -g$. Equation 5-34 then yields

$$N = m(g + a) = m(g - g) = 0. \qquad \text{(Answer)}$$

Thus, in free fall, the scale reads zero and the passenger, in his accelerating frame, concludes that he is weightless. This is the same sense of weightlessness that astronauts experience in Earth orbit. In each case (elevator passenger or astronaut), the feeling of weightlessness arises *not* because the gravitational force has ceased to act—it hasn't—but because the vehicle (elevator cab or spacecraft) and its occupant are each in free fall, with the *same* acceleration.

e. What would happen if the cab were pulled (downward) with an acceleration of -12.0 m/s²? (See Fig. 5-30*e*.)

SOLUTION This is an acceleration whose magnitude exceeds that of free fall. Equation 5-34 yields

$$N = m(g + a) = (72.2 \text{ kg})(9.80 \text{ m/s}^2 - 12.0 \text{ m/s}^2)$$

$$= -159 \text{ N}. \qquad \text{(Answer)}$$

If the scale were screwed to the floor of the cab and the passenger's shoes glued to the scale platform, the scale would read backward, the reading being -159 N. If the passenger slipped out of his shoes, he would rise with respect to the cab until his head touched the ceiling, pushing against it with a force of 159 N. As seen from an inertial frame, the passenger would free-fall until his head touched the ceiling.

REVIEW & SUMMARY

Mechanics

The velocity of a particle or a particle-like body changes—that is, the particle accelerates—because the particle is acted on by one or more **forces**—pushes or pulls—exerted by other objects. **Mechanics** is the study of the relations between accelerations and forces. In the study we look for **force laws,** with which we can calculate the force acting on a body from the properties of the body and its environment.

Force

The magnitudes of forces are defined in terms of the acceleration they give the standard kilogram. A force that accelerates that standard body by exactly 1 m/s² is defined to have a magnitude of 1 N. The direction of the force is the direction of the acceleration. Forces are experimentally found to be vector quantities, so they are combined according to the rules of vector algebra. The **net force** on a body is the vector sum of all the forces acting on that body.

Mass

The **mass** m of a body is the characteristic of that body that relates the body's acceleration to the force (or net force) causing the acceleration. Masses are scalar quantities.

Newton's First Law

If there is no net force on a body, the body must remain at rest if it is initially at rest, or move in a straight line at constant speed if it is in motion. For such a body, there are reference frames, called *inertial frames,* from which the body's acceleration **a** will be measured as being zero. Measurements of **a** from other, noninertial, frames will indicate a nonexistent force on the body.

Newton's Second Law

The net force $\Sigma \mathbf{F}$ on a body with mass m is related to the body's acceleration **a** by

$$\sum \mathbf{F} = m\mathbf{a}, \qquad (5\text{-}1)$$

which may be written in its scalar component version:

$$\sum F_x = ma_x, \quad \sum F_y = ma_y, \quad \text{and} \quad \sum F_z = ma_z. \qquad (5\text{-}2)$$

The second law indicates that in SI units

$$1 \text{ N} = 1 \text{ kg} \cdot \text{m/s}^2. \qquad (5\text{-}3)$$

A **free-body diagram** is instrumental in solving problems with the second law: it is a stripped-down diagram in which only *one* body is considered. That body is represented by a dot. The external forces on the body are drawn as vectors, and a coordinate system is superimposed, oriented so as to simplify the solution.

Some Particular Forces

A body's **weight W** is the force on the body from a nearby astronomical body:

$$\mathbf{W} = m\mathbf{g}, \qquad (5\text{-}8)$$

where **g** is the free-fall acceleration vector. Usually the astronomical body is the Earth.

A **normal force N** is the force exerted on a body by a surface against which it is pressed. The normal force is always perpendicular to the surface.

A **frictional force f** is the force on a body when the body slides or attempts to slide along a surface. The force is parallel to the surface and directed so as to oppose the motion of the body. A **frictionless surface** is one where the frictional force is negligible.

A **tension T** is the force on a body from a taut cord at its point of attachment. The force points along the cord, away from the body. For **massless** cords (their mass is negligible) the pulls at both ends of the cord have the same magnitude T, even if the cord runs around a **massless, frictionless pulley** (the pulley's mass is negligible and it has negligible friction on its axle opposing its rotation).

Newton's Third Law

If body A exerts a force $\mathbf{F}_{BA}$ on body B, then B must exert a force $\mathbf{F}_{AB}$ on body A. The forces are equal in magnitude and opposite in direction:

$$\mathbf{F}_{AB} = -\mathbf{F}_{BA}. \qquad (5\text{-}12)$$

These forces act on *different* bodies.

QUESTIONS

1. If you stand facing forward during a bus or subway ride, why does a quick deceleration topple you forward and a quick increase in speed throw you backward? Why do you have better balance if you face toward the side of the bus or subway train?

2. Using Newton's first law, explain what happens to a child sitting in the front seat of a car if the seat belt is not used and the driver suddenly slams on the brakes. Suppose, instead, that the child is held by an adult who neglects the seat belt; if the car suddenly stops, is the child safer in the arms of the adult, or is the child actually in more danger? What happens to a person who rides at the back of a pickup truck if the truck suddenly stops?

3. A block with mass m is supported by a cord C from the ceiling, and a similar cord D is attached to the bottom of the block (Fig. 5-31). Explain this: if you give a sudden jerk to D, it will break; but if you pull on D steadily, C will break.

FIGURE 5-31 Question 3.

4. If two forces act on a moving body, is there any way that the body can move at (a) a constant speed or (b) a constant velocity? Is there any way that the velocity could be zero (c) for an instant or (d) continuously?

5. The owner's manual for a certain car suggests that your seat belt should be adjusted "to fit snugly" and that the front seat headrest should be adjusted *not* so that it fits comfortably at the back of your neck but so that "the top of the headrest is level with the top of your ears." Explain the wisdom of these instructions in terms of Newton's laws.

6. A Frenchman, filling out a form, writes "78 kg" in the space marked Poids (weight). However, weight is a force and the kilogram is a mass unit. What do Frenchmen

(among others) have in mind when they use a mass unit to report their weight? Why don't they report their weight in newtons? How many newtons does this Frenchman weigh? How many pounds?

7. What is your mass in slugs? Your weight in newtons?

8. Using force, length, and time as fundamental quantities, find the dimensions of mass.

9. A horse is urged to pull a wagon. The horse refuses to try, citing Newton's third law as a defense: the pull of the horse on the wagon is equal to but opposite the pull of the wagon on the horse. "If I can never exert a greater force on the wagon than it exerts on me, how can I ever start the wagon moving?" asks the horse. How would you reply?

10. Comment on whether the following pairs of forces are examples of action–reaction pairs: (a) The Earth attracts a brick; the brick attracts the Earth. (b) An airplane propeller pushes air in toward the tail; the air pushes the plane forward. (c) A horse pulls forward on a cart, accelerating it; the cart pulls backward on the horse. (d) A horse pulls forward on a cart without moving it; the cart pulls back on the horse. (e) A horse pulls forward on a cart without moving it; the Earth exerts an equal and opposite force on the cart. (f) The Earth pulls down on the cart; the ground pushes up on the cart with an equal and opposite force.

11. Comment on the following statements about mass and weight taken from examination papers. (a) Mass and weight are the same physical quantities expressed in different units. (b) Mass is a property of one object alone, whereas weight results from the interaction of two objects. (c) The weight of an object is proportional to its mass. (d) The mass of a body varies with changes in its local weight.

12. Describe several ways in which you could, even briefly, experience weightlessness.

13. The mechanical arm on a space shuttle can handle a 2200-kg satellite when extended to 12 m. Yet, on the ground, this remote manipulator system (RMS) cannot support its own weight. In the weightlessness of an orbiting shuttle, why would the RMS have to be able to exert any force at all?

14. In Fig. 5-32, there are four forces that are equal in magnitude. Do any three of them look as though they would leave a body's velocity unchanged when all three act on the body?

FIGURE 5-32 Question 14.

15. An elevator is supported by a single cable. There is no counterweight. The elevator receives passengers at the ground floor and takes them to the top floor, where they disembark. New passengers enter and are taken down to the ground floor. During this round-trip, when is the tension in the cable equal to the weight of the elevator plus passengers? When greater? When less?

16. You are on the flight deck of the orbiting space shuttle *Discovery* and someone hands you two wooden balls, outwardly identical. One, however, has a lead core but the other does not. Describe several ways of telling them apart.

17. You are an astronaut in the lounge of an orbiting space station and you remove the cover from a long thin jar containing a single olive. Describe several ways—all taking advantage of the mass of either the olive or the jar —to remove the olive from the jar.

18. A horizontal force acts on a body that is free to move. Can the force produce an acceleration if it is less than the weight of that body?

19. Why does the acceleration of a freely falling object not depend on the weight of the object?

20. What's the relation—if any—between the force acting on an object and the direction in which the object is moving?

21. A bird alights on a stretched telegraph wire. Does this change the tension in the wire? If so, by an amount less than, equal to, or greater than the weight of the bird?

22. In November 1984, astronauts Joe Allen and Dale Gardner salvaged a Westar-6 communications satellite in space and placed it into the cargo bay of the space shuttle *Discovery;* see Fig. 5-33. Describing the experience, Joe Allen said of the satellite, "It's not heavy; it's massive." What did he mean?

23. In a tug-of-war, three men pull on a rope to the left at *A* and three men pull to the right at *B* with forces of equal magnitude. Then a 5-lb weight is hung from the center of the rope. (a) Can the men get the rope *AB* to be horizontal? (b) If not, explain. If so, determine the magnitudes of the forces at *A* and *B* required to do this.

24. The following statement is true; explain it. Two teams are having a tug-of-war; the team that pushes harder (horizontally) against the ground wins.

25. A massless rope is strung over a frictionless pulley. A monkey holds onto one end of the rope, and a mirror, having the same weight as the monkey, is attached to the other end of the rope at the monkey's level. Can the monkey get away from its image seen in the mirror (a) by climbing up the rope, (b) by climbing down the rope, or (c) by releasing the rope?

26. You stand on the large platform of a spring scale and note your weight. You then take a step on this platform and note that the scale reads less than your weight at the

FIGURE 5-33 Question 22. Astronauts Joe Allen and Dale Gardner.

beginning of the step, and more than your weight at the end of the step. Explain.

27. Could you weigh yourself on a scale whose maximum reading is less than your weight? If so, how?

28. A weight is hung by a cord from the ceiling of an elevator. Order the following situations according to the tension they produce in the cord, with the situation producing the greatest tension listed first: (a) elevator at rest; (b) elevator rising with uniform speed; (c) elevator descending with decreasing speed; (d) elevator descending with increasing speed.

29. A woman stands on a spring scale in an elevator. Order the following situations according to the reading they produce on the scale, with the situation producing

FIGURE 5-34 Question 31.

the greatest reading listed first: (a) elevator stationary; (b) elevator cable breaks (free fall); (c) elevator accelerating upward; (d) elevator accelerating downward; (e) elevator moving at constant velocity.

30. Under what conditions could unequal masses be strung over a pulley without the pulley having any tendency to turn?

31. In Fig. 5-34, a needle has been placed in each end of a broomstick, the tips of the needles resting on the edges of filled wine glasses. The experimenter strikes the broomstick a swift and sturdy blow with a stout rod. The broomstick breaks and falls to the floor but the wine glasses remain in place and no wine is spilled. This impressive parlor stunt was popular at the end of the last century. What is the physics behind it? (If you try it, practice first with empty soft drink cans.)

EXERCISES & PROBLEMS

SECTION 5-3 FORCE

1E. If the 1-kg standard body has an acceleration of 2.00 m/s^2 at 20° to the positive direction of the x axis, then (a) what are the x and y components of the net force on it, and (b) what is the net force in unit-vector notation?

2E. If the 1-kg standard body is accelerated by $\mathbf{F}_1 = (3.0 \text{ N})\mathbf{i} + (4.0 \text{ N})\mathbf{j}$ and $\mathbf{F}_2 = (-2.0 \text{ N})\mathbf{i} + (-6.0 \text{ N})\mathbf{j}$, then (a) what is the net force in unit-vector notation, and what are the magnitude and direction of (b) the net force and (c) the acceleration?

3P. Suppose that the 1-kg standard body accelerates at 4.00 m/s^2 at 160° from the positive direction of the x axis, owing to two forces, one of which is $\mathbf{F}_1 = (2.50 \text{ N})\mathbf{i} + (4.60 \text{ N})\mathbf{j}$. What is the other force (a) in unit-vector notation and (b) as a magnitude and direction?

SECTION 5-5 NEWTON'S SECOND LAW

4E. While two forces act on it, a particle is to move continuously with $\mathbf{v} = (3 \text{ m/s})\mathbf{i} - (4 \text{ m/s})\mathbf{j}$. One of the forces is $\mathbf{F}_1 = (2 \text{ N})\mathbf{i} + (-6 \text{ N})\mathbf{j}$. What is the other force?

5E. Three forces act on a particle that moves with an unchanging velocity of $\mathbf{v} = (2 \text{ m/s})\mathbf{i} - (7 \text{ m/s})\mathbf{j}$. Two of the forces are $\mathbf{F}_1 = (2 \text{ N})\mathbf{i} + (3 \text{ N})\mathbf{j} + (-2 \text{ N})\mathbf{k}$ and $\mathbf{F}_2 = (-5 \text{ N})\mathbf{i} + (8 \text{ N})\mathbf{j} + (-2 \text{ N})\mathbf{k}$. What is the third force?

6E. There are two forces on the 2.0-kg box in Fig. 5-35 but only one is shown. The box moves strictly along the x axis. For each of the following values for the acceleration a_x of the box, find the second force: (a) 10 m/s^2, (b) 20 m/s^2, (c) 0, (d) -10 m/s^2, and (e) -20 m/s^2.

FIGURE 5-35 Exercise 6.

7E. There are two forces on the 2.0-kg box in Fig. 5-36 but only one is shown. The figure also shows the accelera-

tion of the box. Find the second force (a) in unit-vector notation and (b) as a magnitude and direction.

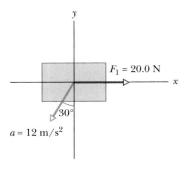

FIGURE 5-36 Exercise 7.

8E. Five forces pull on the 4.0-kg box in Fig. 5-37. Find the box's acceleration (a) in unit-vector notation and (b) as a magnitude and direction.

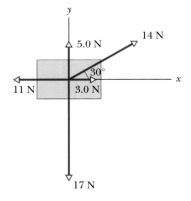

FIGURE 5-37 Exercise 8.

9P. Three astronauts, propelled by jet backpacks, push and guide a 120-kg asteroid toward a processing dock, exerting the forces shown in Fig. 5-38. What is the asteroid's

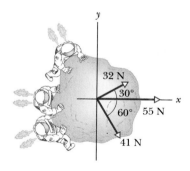

FIGURE 5-38 Problem 9.

acceleration (a) in unit-vector notation and (b) as a magnitude and direction?

10P. Figure 5-39 is an overhead view of a 12-kg tire that is to be pulled by three ropes. One force (F_1, with magnitude 50 N) is indicated. Orient the other two forces F_2 and F_3 so that the magnitude of the resulting acceleration of the tire is least, and find that magnitude if (a) $F_2 = 30$ N, $F_3 = 20$ N; (b) $F_2 = 30$ N, $F_3 = 10$ N; and (c) $F_2 = F_3 = 30$ N.

FIGURE 5-39 Problem 10.

SECTION 5-6 SOME PARTICULAR FORCES

11E. What are the mass and weight of (a) a 1400-lb snowmobile and (b) a 421-kg heat pump?

12E. What are the weight in newtons and the mass in kilograms of (a) a 5.0-lb bag of sugar, (b) a 240-lb fullback, and (c) a 1.8-ton automobile? (1 ton = 2000 lb.)

13E. A space traveler whose mass is 75 kg leaves the Earth. Compute his weight (a) on Earth, (b) on Mars, where $g = 3.8$ m/s², and (c) in interplanetary space, where $g = 0$. (d) What is his mass at each of these locations?

14E. A certain particle has a weight of 22 N at a point where the free-fall acceleration is 9.8 m/s². (a) What are the weight and mass of the particle at a point where the free-fall acceleration is 4.9 m/s²? (b) What are the weight and mass of the particle if it is moved to a point in space where the free-fall acceleration is zero?

15E. A weight-conscious penguin with a mass of 15.0 kg rests on a bathroom scale (Fig. 5-40). What are (a) the penguin's weight **W** and (b) the normal force **N** on the penguin? (c) What is the reading on the scale, assuming it is calibrated in weight units?

FIGURE 5-40 Exercise 15.

16E. A crude mobile hangs from the ceiling, with two metal pieces strung by massless cord, as shown in Fig. 5-41. The masses of the pieces are given. What is the tension in (a) the bottom cord and (b) the top cord?

3.5 kg

4.5 kg

FIGURE 5-41 Exercise 16.

17E. A three-piece mobile, strung by massless cord, is shown in Fig. 5-42. The masses of the highest and lowest pieces are given. The tension in the top cord is 199 N. What is the tension in (a) the lowest cord and (b) the middle cord?

4.8 kg

5.5 kg

FIGURE 5-42 Exercise 17.

18E. (a) An 11.0-kg salami is supported by a cord that runs to a spring scale, which is supported by another cord from the ceiling (Fig. 5-43*a*). What is the reading on the scale? (b) In Fig. 5-43*b* the salami is supported by a cord that runs around a pulley and to a scale. The opposite end of the scale is attached by a cord to a wall. What is the reading on the scale? (c) In Fig. 5-43*c* the wall has been replaced with a second 11.0-kg salami on the left, and the assembly is stationary. What is the reading on the scale now?

FIGURE 5-43 Exercise 18.

SECTION 5-8 APPLYING NEWTON'S LAWS

19E. When an airplane is in level flight, its weight is balanced by a vertical "lift," which is a force exerted by the air. How large is the lift on an airplane in such flight if its mass is 1.20×10^3 kg?

20E. What is the magnitude of the net force acting on a 3800-lb automobile accelerating at 12 ft/s²?

21E. An experimental rocket sled can be accelerated at a constant rate from rest to 1600 km/h in 1.8 s. What is the magnitude of the required average force if the sled has a mass of 500 kg?

22E. A car traveling at 53 km/h hits a bridge abutment. A passenger in the car moves forward a distance of 65 cm (with respect to the road) while being brought to rest by an inflated air bag. What magnitude of force (assumed constant) acts on the passenger's upper torso, which has a mass of 41 kg?

23E. If a nucleus captures a stray neutron, it must bring the neutron to a stop within the diameter of the nucleus by means of the *strong force*. That force, which "glues" the nucleus together, is essentially zero outside the nucleus. Suppose that a stray neutron with an initial speed of 1.4×10^7 m/s is just barely captured by a nucleus with diameter $d = 1.0 \times 10^{-14}$ m. Assuming that the force on

the neutron is constant, find its magnitude. The neutron's mass is 1.67×10^{-27} kg.

24E. A light beam from a satellite-based laser strikes an object ejected from an accidentally launched ballistic missile (Fig. 5-44). The beam exerts a force of 2.5×10^{-5} N on the object. If the "dwell time" of the beam on the object is 2.3 s, by how much is the object displaced from its flight path if it is (a) a 280-kg warhead and (b) a 2.1-kg decoy? (These displacements can be measured by observing the reflected beam.)

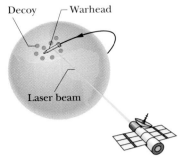

FIGURE 5-44 Exercise 24.

25E. In a modified tug-of-war game, two people pull in opposite directions, not on a rope, but on a 25-kg sled resting on an icy road. If the participants exert forces of 90 N and 92 N, what will be the magnitude of the sled's acceleration?

26E. A 450-lb motorcycle accelerates from rest to 55 mi/h in 6.0 s. (a) What is the magnitude of the motorcycle's acceleration? (b) What is the magnitude of the net force (assumed constant) that acts on it?

27E. Refer to Fig. 5-20 and suppose the two masses are $m = 2.0$ kg and $M = 4.0$ kg. (a) Decide without any calculations which of them should be hanging if the magnitude of the acceleration is to be largest. What are (b) the magnitude and (c) the associated tension in the cord?

28E. Refer to Fig. 5-27. Let the mass of the block be 8.5 kg and the angle θ equal 30°. Find (a) the tension in the cord and (b) the normal force acting on the block. (c) If the cord is cut, find the magnitude of the block's acceleration.

29E. A jet plane starts from rest on the runway and accelerates for takeoff at 2.3 m/s². It has two jet engines, each of which exerts a force on the plane (thrust) of 1.4×10^5 N. What is the weight of the plane?

30E. *Sunjamming.* A "sun yacht" is a spacecraft with a large sail that is pushed by sunlight. Although such a push is tiny in everyday circumstances, it can be large enough to send the spacecraft outward from the sun in a cost-free but slow propulsion. Suppose that the spacecraft has a mass of 900 kg and receives a push of 20 N. (a) What is the magnitude of the resulting acceleration? If the craft starts from rest, (b) how far will it travel in 1 day and (c) how fast will it then be moving?

31E. The tension at which a fishing line snaps is commonly called the line's "strength." What minimum strength is needed for a line that is to stop a 19-lb salmon in 4.4 in. if the fish is initially drifting at 9.2 ft/s? Assume a constant deceleration.

32E. An electron travels in a straight line from a "cathode" of a "vacuum tube" to its "anode," which is exactly 1.5 cm away. It starts with zero speed and reaches the anode with a speed of 6.0×10^6 m/s. (a) Assume constant acceleration and compute the magnitude of the force on the electron. (The force is electrical, but you do not need that fact.) The electron's mass is 9.11×10^{-31} kg. (b) Calculate the weight of the electron.

33E. An electron is projected horizontally at a speed of 1.2×10^7 m/s into an electric field that exerts a constant vertical force of 4.5×10^{-16} N on it. The mass of the electron is 9.11×10^{-31} kg. Determine the vertical distance the electron is deflected during the time it has moved 30 mm horizontally.

34E. A car that weighs 1.30×10^4 N is initially moving at a speed of 40 km/h when the brakes are applied and the car is brought to a stop in 15 m. Assuming that the force which stops the car is constant, find (a) the magnitude of that force and (b) the time required for the change in speed. If the initial speed is, instead, twice as great, and the car experiences the same force during the braking, how are (c) the stopping distance and (d) the stopping time changed? (There could be a lesson here about the danger of driving at high speeds.)

35E. Compute the initial upward acceleration of a rocket of mass 1.3×10^4 kg if the initial upward force produced by its engine (the thrust) is 2.6×10^5 N. Do not neglect the weight of the rocket.

36E. A rocket and its payload have a total mass of 5.0×10^4 kg. How large is the force produced by the engine (the thrust) when (a) the rocket is "hovering" over the launchpad just after ignition, and (b) when the rocket is accelerating upward at 20 m/s²?

37P. A 160-lb firefighter slides down a vertical pole with an acceleration of 10 ft/s², directed downward. What are the magnitudes and directions of the vertical forces (a) exerted by the pole on the firefighter and (b) exerted by the firefighter on the pole?

38P. A sphere of mass 3.0×10^{-4} kg is suspended from a cord. A steady horizontal breeze pushes the sphere so that the cord makes an angle of 37° with the vertical when at rest. Find (a) the magnitude of that push and (b) the tension in the cord.

39P. A 40-kg girl and an 8.4-kg sled are on the surface of a frozen lake, 15 m apart. By means of a rope, the girl exerts a horizontal 5.2-N force on the sled, pulling it toward her. (a) What is the acceleration of the sled? (b) What is the acceleration of the girl? (c) How far from the girl's initial position do they meet, assuming that no frictional forces act?

40P. Two blocks are in contact on a frictionless table. A horizontal force is applied to one block, as shown in Fig. 5-45. (a) If $m_1 = 2.3$ kg, $m_2 = 1.2$ kg, and $F = 3.2$ N, find the force of contact between the two blocks. (b) Show that if the same force F is applied to m_2 rather than to m_1, the force of contact between the blocks is 2.1 N, which is not the same value derived in (a). Explain the difference.

FIGURE 5-45 Problem 40.

41P. A worker drags a crate across a factory floor by pulling on a rope tied to the crate (Fig. 5-46). The worker exerts a force of 450 N on the rope, which is inclined at 38° to the horizontal, and the floor exerts a horizontal force of 125 N that opposes the motion. Calculate the acceleration of the crate (a) if its mass is 310 kg and (b) if its weight is 310 N.

FIGURE 5-46 Problem 41.

42P. You pull a short refrigerator with a constant force **F** across a greased (frictionless) floor, either with **F** horizontal (case 1) or with **F** tilted upward at an angle θ (case 2). (a) What is the ratio of the refrigerator's speed in case 2 to its speed in case 1 if you pull for a certain time t? (b) What is this ratio if you pull for a certain distance d?

43P. For sport, a 12-kg armadillo runs onto a large pond of level, frictionless ice with an initial velocity of 5.0 m/s along the positive direction of the x axis. Take its initial position on the ice as being the origin. It slips over the ice while being pushed by a wind with a force of 17 N in the positive direction of the y axis. In unit-vector notation, what are its (a) velocity and (b) position vectors when it has slid for 3.0 s?

44P. An elevator and its load have a combined mass of 1600 kg. Find the tension in the supporting cable when the elevator, originally moving downward at 12 m/s, is brought to rest with constant acceleration in a distance of 42 m.

45P. An object is hung from a spring balance attached to the ceiling of an elevator. The balance reads 65 N when the elevator is standing still. (a) What is the reading when the elevator is moving upward with a constant speed of 7.6 m/s? (b) What is the reading of the balance when the elevator is moving upward with a speed of 7.6 m/s while decelerating at a rate of 2.4 m/s²?

46P. A 1400-kg jet engine is fastened to the fuselage of a passenger jet by just three bolts (this is the usual practice). Assume that each bolt supports one-third of the load. (a) Calculate the force on each bolt as the plane waits in line for clearance to take off. (b) During flight, the plane encounters turbulence, which suddenly imparts an upward vertical acceleration of 2.6 m/s² to the plane. Calculate the force on each bolt now.

47P. In Fig. 5-47 a 15,000-kg helicopter is lifting a 4500-kg truck with an upward acceleration of 1.4 m/s². Calculate (a) the force the air exerts on the helicopter blades and (b) the tension in the upper supporting cable.

FIGURE 5-47 Problem 47.

48P. An 80-kg man jumps down to a concrete patio from a window ledge only 0.50 m above the ground. He neglects to bend his knees on landing, so that his motion is arrested in a distance of 2.0 cm. (a) What is the average acceleration of the man from the time his feet first touch the patio to the time he is brought fully to rest? (b) With what force does this jump jar his bone structure?

49P. Three blocks are connected, as shown in Fig. 5-48, on a horizontal frictionless table and pulled to the right

FIGURE 5-48 Problem 49.

with a force $T_3 = 65.0$ N. If $m_1 = 12.0$ kg, $m_2 = 24.0$ kg, and $m_3 = 31.0$ kg, calculate (a) the acceleration of the system and (b) the tensions T_1 and T_2.

50P. Figure 5-49 shows four penguins that are being playfully pulled along very slippery (frictionless) ice by a curator. The masses of three penguins and the tension in two of the cords are given. Find the mass that is not given.

FIGURE 5-49 Problem 50.

51P. An elevator weighing 6240 lb is pulled upward by a cable with an acceleration of 4.00 ft/s². (a) Calculate the tension in the cable. (b) What is the tension when the elevator is decelerating at the rate of 4.00 ft/s² but is still moving upward?

52P. An 80-kg person is parachuting and experiencing a downward acceleration of 2.5 m/s². The mass of the parachute is 5.0 kg. (a) What upward force is exerted on the open parachute by the air? (b) What downward force is exerted by the person on the parachute?

53P. An 85-kg man lowers himself to the ground from a height of 10.0 m by holding onto a rope that runs over a frictionless pulley to a 65-kg sandbag. (a) With what speed does the man hit the ground if he started from rest? (b) Is there anything he could do to reduce the speed with which he hits the ground?

54P. A new 26-ton Navy jet (Fig. 5-50) requires an airspeed of 280 ft/s for lift-off. Its engine develops a maximum force of 24,000 lb, but that is insufficient for reaching takeoff speed in the 300 ft available on an aircraft carrier. What minimum force (assumed constant) is needed from the catapult that is used to help launch the jet? Assume that the catapult and the jet's engine each exert a constant force over the 300-ft takeoff distance. (1 ton = 2000 lb.)

FIGURE 5-50 Problem 54.

55P. Imagine a landing craft approaching the surface of Callisto, one of Jupiter's moons. If the engine provides an upward force (thrust) of 3260 N, the craft descends at constant speed; if the engine provides only 2200 N, the craft accelerates downward at 0.39 m/s². (a) What is the weight of the landing craft in the vicinity of Callisto's surface? (b) What is the mass of the craft? (c) What is the free-fall acceleration near the surface of Callisto?

56P. A 52-kg circus performer is to slide down a rope that will snap if the tension exceeds 425 N. (a) What happens if the performer hangs stationary on the rope? (b) What is the magnitude of the least acceleration the performer can have so as to avoid breaking the rope?

57P. A chain consisting of five links, each of mass 0.100 kg, is lifted vertically with a constant acceleration of 2.50 m/s², as shown in Fig. 5-51. Find (a) the forces acting between adjacent links, (b) the force **F** exerted on the top link by the person lifting the chain, and (c) the *net* force accelerating each link.

FIGURE 5-51 Problem 57.

58P. A block of mass $m_1 = 3.70$ kg on a frictionless inclined plane of angle 30.0° is connected by a cord over a massless, frictionless pulley to a second block of mass $m_2 = 2.30$ kg hanging vertically (Fig. 5-52). What are (a) the magnitude of the acceleration of each block and (b) the direction of the acceleration of m_2? (c) What is the tension in the cord?

FIGURE 5-52 Problem 58.

59P. You need to lower a bundle of old roofing material weighing 100 lb to the ground with a rope that will snap if the tension in it exceeds 87 lb. (a) How can you avoid snapping the rope during the descent? (b) If the descent is 20 ft and you just barely avoid snapping the rope, with what speed will the bundle hit the ground?

60P. A block is projected up a frictionless inclined plane with initial speed v_0. The angle of incline is θ. (a) How far

up the plane does it go? (b) How long does it take to get there? (c) What is its speed when it gets back to the bottom? Find numerical answers for $\theta = 32.0°$ and $v_0 = 3.50$ m/s.

61P. A lamp hangs vertically from a cord in a descending elevator that decelerates at 2.4 m/s². (a) If the tension in the cord is 89 N, what is the mass of the lamp? (b) What is the tension in the cord when the elevator ascends with an upward acceleration of 2.4 m/s²?

62P. A 100-kg crate is pushed at constant speed up the frictionless 30.0° ramp shown in Fig. 5-53. (a) What horizontal force **F** is required? (b) What force is exerted by the ramp on the crate?

FIGURE 5-53 Problem 62.

63P. A 10-kg monkey climbs up a massless rope that runs over a frictionless tree limb and back down to a 15-kg package on the ground (Fig. 5-54). (a) What is the magnitude of the least acceleration the monkey must have if it is to lift the package off the ground? If, after the package is lifted, the monkey stops its climb and holds onto the rope, what are (b) its acceleration and (c) the tension in the rope?

FIGURE 5-54 Problem 63.

64P. Figure 5-55 shows a section of an alpine cable-car system. The maximum permissible mass of each car with occupants is 2800 kg. The cars, riding on a support cable, are pulled by a second cable attached to each pylon. What

Support cable

Pull cable

35°

FIGURE 5-55 Problem 64.

is the difference in tension between adjacent sections of pull cable if the cars are at maximum mass and are being accelerated up the 35° incline at 0.81 m/s²?

65P. An interstellar ship has a mass of 1.20×10^6 kg and is initially at rest relative to a star system. (a) What constant acceleration is needed to bring the ship up to a speed of $0.10c$ (where c is the speed of light) relative to the star system in 3.0 days? (You need not consider Einstein's special relativity.) (b) What is that acceleration in g units? (c) What force is required for the acceleration? (d) If the engines are shut down when $0.10c$ is reached, how long does the ship take (start to finish) to journey 5.0 light-months, the distance that light travels in 5.0 months?

66P. The elevator in Fig. 5-56 consists of a 1150-kg cage (A), a 1400-kg counterweight (B), a driving mechanism (C), a cable, and two pulleys. In operation, mechanism C grabs the cable, either forcing it along or retarding its motion. This process means that tension T_1 in the cable on one side of C differs from tension T_2 on the other side. Suppose that the upward acceleration of A and the downward acceleration of B have the magnitude $a = 2.0$ m/s². Neglecting the pulleys and the mass of the cable, find (a) T_1, (b) T_2, and (c) the force on the cable produced by C.

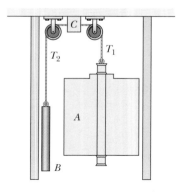

C

T_2 T_1

A

B

FIGURE 5-56 Problem 66.

67P. A 5.00-kg block is pulled along a horizontal frictionless floor by a cord that exerts a force $F = 12.0$ N at an angle $\theta = 25.0°$ above the horizontal, as shown in Fig. 5-57. (a) What is the acceleration of the block? (b) The force F is slowly increased. What is its value just before the block is lifted off the floor? (c) What is the acceleration of the block just before it is lifted off the floor?

5.00 kg   25.0°   **F**

FIGURE 5-57 Problem 67.

68P. In earlier days, horses pulled barges down canals in the manner shown in Fig. 5-58. Suppose that the horse pulls on the rope with a force of 7900 N at an angle of 18° to the direction of motion of the barge, which is headed straight along the canal. The mass of the barge is 9500 kg, and its acceleration is 0.12 m/s². Calculate the force exerted by the water on the barge.

7900 N

18°

FIGURE 5-58 Problem 68.

69P. A certain force gives mass m_1 an acceleration of 12.0 m/s² and mass m_2 an acceleration of 3.30 m/s². What acceleration would the force give to an object with a mass of (a) $m_2 - m_1$, or (b) $m_2 + m_1$?

70P. A hot-air balloon of mass M is descending vertically with downward acceleration a (Fig. 5-59). How much mass must be thrown out to give the balloon an upward acceleration a (same magnitude but opposite direction)? Assume that the upward force from the air (the lift) does not change because of the mass (the ballast) that is lost.

Ballast

FIGURE 5-59 Problem 70.

71P. A rocket with mass 3000 kg is fired from the ground at an angle of elevation of 60°. The motor creates a force on the rocket (thrust) of 6.0×10^4 N at a constant angle of 60° to the horizontal for 50 s and then cuts out. As a rough approximation, ignore the mass of fuel consumed and neglect forces from the air. Calculate (a) the altitude of the rocket at motor cutout and (b) the total horizontal distance from firing point to eventual impact with the ground (assumed to be level).

72P. A block of mass M is pulled along a horizontal frictionless surface by a rope of mass m, as shown in Fig. 5-60. A horizontal force **F** is applied to one end of the rope. (a) Show that the rope *must* sag, even if only by an imperceptible amount. Then, assuming that the sag is negligible, find (b) the acceleration of rope and block, (c) the force that the rope exerts on the block, and (d) the tension in the rope at its midpoint.

FIGURE 5-60 Problem 72.

73P. Figure 5-61 shows a man sitting in a bosun's chair that dangles from a massless rope, which runs over a mass-less, frictionless pulley and back down to the man's hand. The combined mass of the man and chair is 95.0 kg. (a) With what force must the man pull on the rope for him to rise at constant speed? (b) What force is needed for an upward acceleration of 1.30 m/s²? (c) Suppose, instead, that the rope on the right is held by a person on the ground. Repeat parts (a) and (b) for this new situation. (d) In each of the four cases, what is the force exerted on the ceiling by the pulley system?

FIGURE 5-61 Problem 73.

ADDITIONAL PROBLEMS

74. A motorcycle and 60.0-kg rider accelerate at 3.0 m/s² up a ramp inclined 10° above the horizontal. (a) What is the magnitude of the net force acting on the rider? (b) What is the magnitude of the force exerted by the motor-cycle on the rider?

75. A block weighing 3.0 N is at rest on a horizontal surface. A 1.0-N upward force is applied to the block by means of an attached vertical string. What are the magnitude and the direction of the force of the block on the horizontal surface?

76. A 1.0-kg mass on a frictionless inclined surface is connected to a 2.0-kg mass as shown in Fig. 5-62. The pulley is massless and frictionless. The 2.0-kg mass is acted on by an upward force **F** = 6.0 N and has a downward acceleration of 5.5 m/s². (a) What is the tension in the connecting cord? (b) What is the angle β?

77. Only two forces act on a 3.0-kg object that moves with an acceleration of 3.0 m/s² in the positive y direction. If one of the forces acts in the positive x direction and has a magnitude of 8.0 N, what is the magnitude of the other force?

78. A spaceship lifts off vertically from the moon, where the free-fall acceleration is 1.6 m/s². If the spaceship has

FIGURE 5-62 Problem 76.

an upward acceleration of 1.0 m/s² as it lifts off, what is the force of the spaceship on an astronaut who weighs 735 N on Earth?

79. A 0.20-kg hockey puck has a velocity of 2.0 m/s toward the east as it slides over the frictionless surface of a frozen lake. What are the magnitude and direction of the average force that must act on the puck during a 0.50-s interval to change its velocity to (a) 5.0 m/s, due west, and (b) 5.0 m/s, due south?

80. A 1.0-kg mass on a 37° incline is connected to a 3.0-kg mass on a horizontal surface (Fig. 5-63). The surfaces and the pulley are frictionless. If **F** = 12 N, what is the tension in the connecting cord?

FIGURE 5-63 Problem 80.

81. Only two forces act on a 3.0-kg mass. One force is 9.0 N, acting due east, and the other is 8.0 N, acting in a direction 62° north of west. What is the magnitude of the acceleration of the mass?

82. When the system in Fig. 5-64 is released from rest, the 3.0-kg mass has an acceleration of 1.0 m/s^2 to the right. The surfaces and the pulley are frictionless. (a) What is the tension in the connecting cord? (b) What is the value of M?

FIGURE 5-64 Problem 82.

FORCE AND MOTION—II

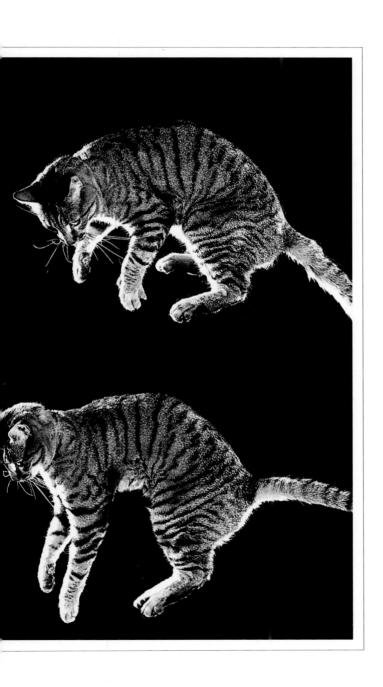

Cats, who enjoy sleeping on window sills, are often kept in apartment buildings. When a cat accidentally falls out of a window and onto a sidewalk, the extent of injury (such as the number of fractured bones or the certainty of death) <u>decreases</u> with height if the fall is more than seven or eight floors. (There is even a record of one cat who fell 32 floors and suffered only slight damage to its thorax and one tooth.) You'll see why shortly.

6-1 FRICTION

Frictional forces are unavoidable in our daily lives. Left to act alone, they would stop every rolling wheel and bring to a halt every rotating shaft. In an automobile, about 20% of the gasoline is used to counteract friction in the engine and in the drive train. On the other hand, if friction were totally absent, we could not walk or ride a bicycle. We could not hold a pencil and, if we could, it would not write. Nails and screws would be useless, woven cloth would fall apart, and knots would come undone.

In this chapter, we deal largely with the frictional forces that exist between dry solid surfaces, moving across each other at relatively slow speeds. Consider two simple experiments:

1. *First experiment.* Send a book sliding across a tabletop. A frictional force, exerted by the tabletop on the bottom of the sliding book, slows the book and eventually stops it. If you want to make the book move across the table with constant velocity, you must push or pull it with a steady force, with a magnitude matching that of the frictional force, which opposes the motion.

2. *Second experiment.* A heavy crate is resting on the floor of a warehouse. You push on it horizontally with a steady force but it does not move. That is because the force that you apply is balanced by a frictional force, exerted horizontally on the bottom of the crate by the floor and opposite the direction of your push. Remarkably, this frictional force automatically adjusts itself, both in magnitude and direction, to cancel exactly whatever force you decide to apply. Of course, if you can push hard enough, you will be able to move the crate (see the first experiment).

Figure 6-1 shows a similar situation in detail. In Fig. 6.1*a*, a block rests on a tabletop, its weight **W**

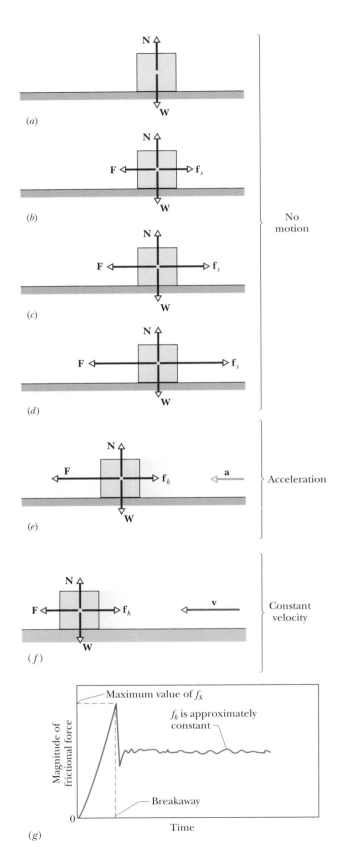

FIGURE 6-1 (*a*) The forces on a stationary block. (*b–d*) An external force **F**, applied to the block, is counterbalanced by an equal but opposite static frictional force **f**_s. As **F** is increased, **f**_s also increases, until **f**_s reaches a certain maximum value. (*e*) The block then "breaks away," accelerating suddenly to the left. (*f*) If the block is now to move with constant velocity, the applied force **F** must be reduced from the maximum value it had just before the block broke away. (*g*) Some experimental results for the sequence (*a*) through (*f*).

A "low rider," scraping the road surface, causes this luminous display. There is continuous stick-and-slip that rips off tiny pieces, which then briefly burn.

FIGURE 6-2 A magnified section of a highly polished steel surface. The height of the surface irregularities is several thousand atomic diameters.

balanced by an equal but opposite normal force **N**. In Fig. 6-1*b*, you exert a force **F** on the block, attempting to pull it to the left. In response, a frictional force $\mathbf{f}_s$ arises, pointing to the right, exactly matching the force that you have applied. The force $\mathbf{f}_s$ is called the **static frictional force.**

Figures 6-1*c* and 6-1*d* show that, as you increase your applied force, the static frictional force $\mathbf{f}_s$ increases also and the block remains at rest. At a certain value of the applied force, however, the block "breaks away" from its intimate contact with the tabletop and accelerates, moving to the left as in Fig. 6-1*e*. The frictional force that then opposes the motion is called the **kinetic frictional force** $\mathbf{f}_k$.

Usually, the kinetic frictional force, which acts when there is motion, is less than the maximum value of the static frictional force, which acts when there is no motion. So, if you wish the block to move across the surface with a constant speed, you must usually decrease the applied force once the block begins to move, as in Fig. 6-1*f*.

Figure 6-1*g* shows the results of an experiment in which the force on a 400-g block was slowly increased until breakaway occurred. Note the reduced force needed to keep the block moving at constant speed after breakaway.*

The transition from the static to the kinetic frictional force, though it may seem abrupt, is not instantaneous. In fact, kinetic friction for dry surfaces

at slow speeds proceeds by a "stick-and-slip" process, the slip being much like the initial breakaway from the static condition. It is these repetitive stick-and-slip events that cause dry surfaces moving across each other to squeak. Thus we have the squealing of skidding tires on a dry pavement, the squeaking of rusty hinges, and the painful sound made by fingernails dragged across a chalkboard. Sometimes, dry friction leads to pleasant sounds, as when a bow undergoes stick-and-slip as it is drawn over a violin string.

Basically, the frictional force is a force acting between the surface atoms of one body and those of another. If two highly polished and carefully cleaned metal surfaces are brought together in a very good vacuum, they cannot be made to slide over each other. Instead, they *cold-weld* together instantly, forming a single piece of metal. If machinist's polished gage blocks are brought together in air, they stick almost as firmly to each other and can be separated only by means of a wrenching motion. Under ordinary circumstances, however, such intimate atom-to-atom contact is not possible. Even a highly polished metal surface, as Fig. 6-2 shows, is far from being flat on the atomic scale. Moreover, the surfaces of everyday objects have layers of oxides and other contaminants that reduce cold-welding.

When two surfaces are placed together, only the high points touch each other. (It is like having the Alps of Switzerland turned over and placed down on the Alps of Austria.) The actual *micro*scopic area of contact is much less than the apparent *macro*scopic contact area, perhaps by a factor of 10^4. Many contact points cold-weld together. When the sur-

---

*See "Undergraduate Computer-Interfacing Projects," by Joseph Priest and John Snider, *The Physics Teacher*, May 1987.

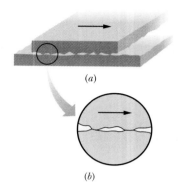

(a)

(b)

FIGURE 6-3 The mechanism of sliding friction. (a) The upper surface is sliding to the right over the lower surface in this enlarged view. (b) A detail, showing two spots where cold-welding has occurred. Force is required to break these welds and maintain the motion.

faces are pulled across each other, there is a continuous rupturing and reforming of welds as additional chance contacts are made (Fig. 6-3).

Experiments with radioactive tracers show that, when one dry metal surface is dragged across another, tiny metal fragments are indeed torn from each surface. The wear of piston rings has been tested by using rings that have been made radioactive by exposure in a nuclear reactor. Material torn from the rings is carried away by the lubricating oil, where it can be detected by its radioactivity.

6-2 PROPERTIES OF FRICTION

Experiment shows that when a body is pressed against a surface (both are dry and unlubricated) and a force **F** attempts to slide the body along the surface, the resulting frictional force has three properties:

PROPERTY 1. If the body does not move, then the static frictional force $\mathbf{f}_s$ and the component of **F** that is parallel to the surface are equal in magnitude, and $\mathbf{f}_s$ is directed opposite that component of **F**.

PROPERTY 2. The magnitude of $\mathbf{f}_s$ has a maximum value $f_{s,\max}$ that is given by

$$f_{s,\max} = \mu_s N, \qquad (6\text{-}1)$$

where μ_s is the **coefficient of static friction,** and N is the magnitude of the normal force. If the magnitude of the component of **F** that is parallel to the

surface exceeds $f_{s,\max}$, then the body begins to slide along the surface.

PROPERTY 3. If the body begins to slide along the surface, the magnitude of the frictional force rapidly decreases to a value f_k given by

$$f_k = \mu_k N, \qquad (6\text{-}2)$$

where μ_k is the **coefficient of kinetic friction.** Thereafter during the sliding, the kinetic frictional force $\mathbf{f}_k$ has a magnitude given by Eq. 6-2.

Properties 1 and 2 are worded in terms of a single force **F**, but they also hold for the net (or resultant) force due to several forces acting on the body. Equations 6-1 and 6-2 are *not* vector equations; the direction of $\mathbf{f}_s$ or $\mathbf{f}_k$ is always parallel to the surface and opposite the intended motion, and **N** is perpendicular to the surface.

The coefficients μ_s and μ_k are dimensionless and must be determined experimentally. Since their values depend on both the body and the surface, they are usually referred to with the preposition "between," as in "the value of μ_s *between* a sled and asphalt is 0.5." In this book we assume that the value of μ_k does not depend on the speed at which the body slides along the surface.

SAMPLE PROBLEM 6-1

Figure 6-4a shows a coin resting on a book that has been tilted at an angle θ with the horizontal. By trial and error you find that when θ is increased to 13°, the coin begins to slide down the book. What is the coefficient of static friction μ_s between the coin and the book?

SOLUTION Figure 6-4b is a free-body diagram for the coin when it is on the verge of sliding. The forces on the coin are the normal force **N**, pushing outward from the plane of the book, the weight **W** of the coin, and the frictional force $\mathbf{f}_s$, which points up the plane because the impending motion is down the plane. Since the coin is in equilibrium, the net force acting on it must be zero. From Newton's second law, we have

$$\sum \mathbf{F} = \mathbf{f}_s + \mathbf{W} + \mathbf{N} = 0. \qquad (6\text{-}3)$$

For the x components, this vector equation gives us

$$\sum F_x = f_s - W \sin \theta = 0, \quad \text{or} \quad f_s = W \sin \theta. \qquad (6\text{-}4)$$

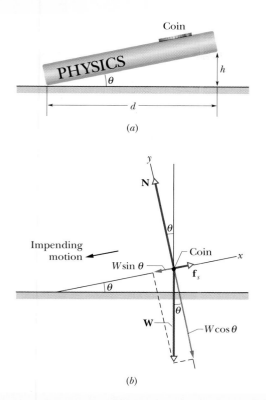

(a)

(b)

FIGURE 6-4 Sample Problem 6-1. (*a*) A coin is about to slide from rest down the cover of a book. (*b*) A free-body diagram for the coin, showing the three forces (drawn to scale) that act on it. The weight **W** is shown resolved into its components along the *x* and the *y* axes, whose orientations are chosen to simplify the problem.

For the *y* components, we have

$$\sum F_y = N - W \cos \theta = 0, \quad \text{or} \quad N = W \cos \theta. \quad (6\text{-}5)$$

When the coin is on the verge of sliding, the magnitude of the static frictional force acting on it has its maximum value $\mu_s N$. Substituting this into Eq. 6-4 and dividing by Eq. 6-5, we obtain

$$\frac{f_s}{N} = \frac{\mu_s N}{N} = \frac{W \sin \theta}{W \cos \theta} = \tan \theta,$$

or

$$\mu_s = \tan \theta = \tan 13° = 0.23. \quad \text{(Answer)} \quad (6\text{-}6)$$

Why not try to measure μ_s for a coin and this book? You do not need a protractor. Tan θ is the ratio h/d of the two lengths shown in Fig. 6-4*a*, and you can measure these lengths with a ruler.

(a)

(b)

FIGURE 6-5 Sample Problem 6-2. (*a*) A car, sliding to the right and finally stopping after a displacement *d*. (*b*) A free-body diagram for the decelerating car, the three force vectors being drawn to scale. The acceleration vector points to the left, in the direction of the frictional force f_k.

SAMPLE PROBLEM 6-2

If a car's wheels are "locked" (kept from rolling) during emergency braking, the car slides along the road. Ripped-off bits of tire and small melted sections of road form the "skid marks" that reveal the cold-welding during the slide. The record for the longest skid marks on a public road was reportedly set in 1960 by a Jaguar on the M1 highway in England—the marks were 290 m (about 950 ft) long! Assuming that $\mu_k = 0.60$, how fast was the car going when the wheels were locked?

SOLUTION Figure 6-5*a* depicts the car's travel; Fig. 6-5*b* is the car's free-body diagram during deceleration, showing the car's weight **W**, the normal force **N**, and the kinetic frictional force f_k acting on the car. We can use Eq. 2-14,

$$v^2 = v_0^2 + 2a_x(x - x_0),$$

with $v = 0$ and $x - x_0 = d$, to find the car's initial speed v_0. Substituting these values and rearranging yield

$$v_0 = \sqrt{-2a_x d}. \quad (6\text{-}7)$$

To find a_x, we apply Newton's second law along the x axis. If we ignore the effects of the air on the car, the only force along the x axis is $\mathbf{f}_k$. So, for the magnitude f_k, we have

$$f_k = -ma_x, \quad \text{or} \quad a_x = -\frac{f_k}{m} = -\frac{\mu_k N}{m}, \quad (6\text{-}8)$$

where m is the car's mass, and from Eq. 6-2, $f_k = \mu_k N$.

The normal force $\mathbf{N}$ has magnitude $N = W = mg$. Substituting this result into Eq. 6-8 yields

$$a_x = -\frac{\mu_k mg}{m} = -\mu_k g. \quad (6\text{-}9)$$

Substituting Eq. 6-9 into Eq. 6-7, we find

$$v_0 = \sqrt{2\mu_k gd} = \sqrt{(2)(0.60)(9.8 \text{ m/s}^2)(290 \text{ m})}$$

$$= 58 \text{ m/s} = 210 \text{ km/h} \qquad \text{(Answer)}$$

(about 130 mi/h). In obtaining the answer, we implicitly assumed that $v = 0$ at the far end of the skid marks. Actually, the marks ended only because the Jaguar left the road after 290 m. So v_0 was at least 210 km/h, and possibly much more.

SAMPLE PROBLEM 6-3

A woman pulls a loaded sled of mass $m = 75$ kg along a horizontal surface at constant velocity. The coefficient of kinetic friction μ_k between the runners and the snow is 0.10 and the angle ϕ in Fig. 6-6 is 42°.

a. What is the tension T in the rope?

SOLUTION Figure 6-6b is the free-body diagram for the sled. Applying Newton's second law in the horizontal direction yields

$$T \cos \phi - f_k = ma_x = 0, \quad (6\text{-}10)$$

where a_x is zero because the velocity is constant. In the vertical direction, we have

$$T \sin \phi + N - mg = ma_y = 0, \quad (6\text{-}11)$$

in which mg is the weight of the sled. From Eq. 6-2,

$$f_k = \mu_k N. \quad (6\text{-}12)$$

These three equations contain T, N, and f_k as unknowns. Eliminating N and f_k will allow us to find the remaining variable T.

(a)

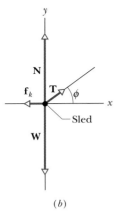

(b)

FIGURE 6-6 Sample Problem 6-3. (a) A woman exerts a force $\mathbf{T}$ on a sled, pulling it at a constant velocity. (b) A free-body diagram for the sled and its load, not to scale.

We start by adding Eqs. 6-10 and 6-12, obtaining

$$T \cos \phi = \mu_k N,$$

or

$$N = \frac{T \cos \phi}{\mu_k}. \quad (6\text{-}13)$$

Substituting this value of N into Eq. 6-11 and solving for T, we obtain

$$T = \frac{\mu_k mg}{\cos \phi + \mu_k \sin \phi} \quad (6\text{-}14)$$

$$= \frac{(0.10)(75 \text{ kg})(9.8 \text{ m/s}^2)}{\cos 42° + (0.10)(\sin 42°)}$$

$$= 91 \text{ N}, \qquad \text{(Answer)}$$

which is considerably less than the weight of the sled.

b. What is the normal force with which the snow pushes vertically upward on the sled?

SOLUTION Substituting T from Eq. 6-14 into Eq. 6-13 yields

$$N = \frac{\cos \phi}{\cos \phi + \mu \sin \phi} mg \qquad (6\text{-}15)$$

$$= \frac{\cos 42°}{\cos 42° + (0.10)(\sin 42°)} mg$$

$$= 0.917mg = 670 \text{ N}. \qquad \text{(Answer)}$$

Thus the upward pull on the sled reduces the normal force to 92% of the weight of the sled.

SAMPLE PROBLEM 6-4

In Fig. 6-7a, a crate of dilled pickles with mass $m_1 = 14$ kg moves along a plane that makes an angle of $\theta = 30°$ with the horizontal. That crate is connected to a crate of pickled dills with mass $m_2 = 14$ kg by a taut, massless cord that runs around a frictionless, massless pulley. The hanging crate of dills descends with constant velocity.

a. What are the magnitude and direction of the frictional force exerted on m_1 by the plane?

SOLUTION The fact that m_2 descends indicates that m_1 moves up the plane, and so a *kinetic* frictional force $\mathbf{f}_k$ must point down the plane.

We cannot use Eq. 6-2 to find the magnitude of $\mathbf{f}_k$, because we do not know the coefficient of kinetic friction μ_k between m_1 and the plane. However, we can use the techniques of Chapter 5. To start, we draw the free-body diagrams for m_1 and m_2 in Figs. 6-7b and 6-7c, where $\mathbf{T}$ is the pull from the tension in the cord and the weight vectors are $\mathbf{W}_1 = m_1\mathbf{g}$ and $\mathbf{W}_2 = m_2\mathbf{g}$.

With $\mathbf{W}_1$ resolved into x and y components we have, from Newton's second law applied to the x axis in Fig. 6-7b,

$$\sum F_x = T - f_k - m_1 g \sin \theta = m_1 a_x$$

$$= (m_1)(0) = 0, \qquad (6\text{-}16)$$

where $a_x = 0$, since m_1 must move at constant velocity. Next, for m_2 we apply Newton's second law along the y axis in Fig. 6-7c and use the fact that m_2 moves at constant velocity. We find

$$\sum F_y = T - m_2 g = m_2 a_y = (m_2)(0) = 0,$$

or

$$T = m_2 g. \qquad (6\text{-}17)$$

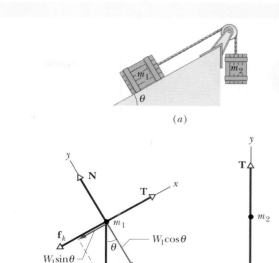

FIGURE 6-7 Sample Problem 6-4. (a) Mass m_1 moves up the plane while mass m_2 descends at constant velocity. (b) A free-body diagram for mass m_1. (c) A free-body diagram for mass m_2.

We then substitute T from Eq. 6-17 into Eq. 6-16 and solve for f_k, obtaining

$$f_k = m_2 g - m_1 g \sin \theta$$

$$= (14 \text{ kg})(9.8 \text{ m/s}^2) - (14 \text{ kg})(9.8 \text{ m/s}^2)(\sin 30°)$$

$$= 68.6 \text{ N} \approx 69 \text{ N}. \qquad \text{(Answer)}$$

b. What is μ_k?

SOLUTION We can use Eq. 6-2 to find μ_k, but first we need the magnitude of the normal force $\mathbf{N}$ acting on m_1. To find N, we apply Newton's second law for m_1 along the y axis in Fig. 6-7b:

$$\sum F_y = N - m_1 g \cos \theta = m_1 a_y = 0,$$

or

$$N = m_1 g \cos \theta.$$

From Eq. 6-2 we now have

$$\mu_k = \frac{f_k}{N} = \frac{f_k}{m_1 g \cos \theta}$$

$$= \frac{68.6 \text{ N}}{(14 \text{ kg})(9.8 \text{ m/s}^2)(\cos 30°)} = 0.58. \qquad \text{(Answer)}$$

The flow of an avalanche is an example of fluid flow, and any object caught in the flow would experience a drag force.

6-3 THE DRAG FORCE AND TERMINAL SPEED

A **fluid** is anything that can flow—generally either a gas or a liquid. When there is a relative velocity between a fluid and a body (either because the body moves through the fluid or because the fluid moves past the body), the body experiences a **drag force D** that opposes the relative motion and points in the direction in which the fluid flows relative to the body.

Here we examine only cases in which air is the fluid, the body is blunt (like a baseball or a falling cat) rather than slender (like a javelin), and the relative motion is fast enough so that the air becomes turbulent (breaks up into swirls) behind the body. In such cases, the magnitude of the drag force **D** is related to the relative speed v by an experimentally determined **drag coefficient** C according to

FIGURE 6-8 This skier crouches in an "egg position" so as to minimize her effective cross-sectional area and thus the air drag acting on her.

$$D = \tfrac{1}{2}C\rho Av^2, \qquad (6\text{-}18)$$

where ρ is the air density (mass per volume) and A is the **effective cross-sectional area** of the body (the area of a cross section taken perpendicularly to the velocity **v**). The drag coefficient C (typical values range from 0.4 to 1.0) is not truly a constant for a given body, because if v varies significantly, the value of C can vary as well. Here, we ignore such complications.

Downhill speed skiers know well that drag depends on A and v^2. To reach high speeds a skier must reduce D as much as possible by, for example, riding the skis in the "egg position" (Fig. 6-8) to minimize A.

Equation 6-18 indicates that when a blunt object falls from rest through air, D gradually increases from zero as the speed of the body increases. As suggested in Fig. 6-9, if the body falls far enough, D eventually equals the body's weight W (= mg), and the net vertical force on the body is then zero. By Newton's second law, the acceleration must also be zero then, and so the body's speed no longer increases. The body then falls at a constant **terminal speed** v_t, which we find by setting $D = mg$ in Eq. 6-18, obtaining

$$\tfrac{1}{2}C\rho Av_t^2 = mg,$$

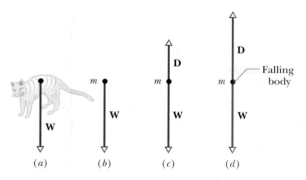

FIGURE 6-9 The forces that act on a body falling through air: (*a*) the body when it has just begun to fall, (*b*) the free-body diagram just then, and (*c*) the free-body diagram a little later, after a drag force has developed. (*d*) The drag force has increased until it balances the weight of the body. The body now falls at its constant terminal speed.

which gives

$$v_t = \sqrt{\frac{2mg}{C\rho A}}. \qquad (6\text{-}19)$$

Table 6-1 gives values of v_t for some common objects.

A cat must fall about six floors to reach terminal speed. Until it does so, $W > D$ and the cat accelerates downward because of the net downward force. Recall from Chapter 2 that your body is an accelero-

TABLE 6-1
SOME TERMINAL SPEEDS IN AIR

OBJECT	TERMINAL SPEED (m/s)	95% DISTANCE[a] (m)
16-lb Shot	145	2500
Sky diver (typical)	60	430
Baseball	42	210
Tennis ball	31	115
Basketball	20	47
Ping-Pong ball	9	10
Raindrop (radius = 1.5 mm)	7	6
Parachutist (typical)	5	3

[a]This is the distance through which the body must fall from rest to reach 95% of its terminal speed.

Source: Adapted from Peter J. Brancazio, *Sport Science*, Simon & Schuster, New York, 1984.

meter, not a speedometer. Because the cat too senses the acceleration, it is frightened and keeps its feet underneath its body, its head tucked in, and its spine bent upward, making A small, v_t large, and injury on landing likely.

However, if the cat does reach v_t, the acceleration vanishes and the cat relaxes somewhat, stretching its legs and neck horizontally outward and straightening its spine (it then resembles a flying squirrel). These actions increase A and D, and the cat begins to slow because now $D > W$—the net force is upward—until a new, smaller v_t is reached. The decrease in v_t reduces the possibility of injury on landing. Just before the end of the fall, when it nears the ground, the cat pulls its legs back beneath its body to prepare for the landing.

SAMPLE PROBLEM 6-5

If a falling cat reaches a first terminal speed of 60 mi/h while it is tucked in and then stretches out, doubling A, how fast is it falling when it reaches a new terminal speed?

SOLUTION We let v_{to} and v_{tn} represent the original and new terminal speeds, and A_o and A_n the original and new areas. We then use Eq. 6-19 to set up a ratio of speeds:

$$\frac{v_{tn}}{v_{to}} = \frac{\sqrt{2mg/C\rho A_n}}{\sqrt{2mg/C\rho A_o}} = \sqrt{\frac{A_o}{A_n}} = \sqrt{\frac{A_o}{2A_o}} = \sqrt{0.5} \approx 0.7,$$

which means that $v_{tn} \approx 0.7v_{to}$, or about 40 mi/h.

In April 1987, during a jump, parachutist Gregory Robertson noticed that fellow parachutist Debbie Williams had been knocked unconscious in a collision with a third sky diver and was unable to open her parachute. Robertson, who was well above Williams at the time and who had not yet opened his parachute for the 13,500-ft plunge, reoriented his body head-down so as to minimize A and maximize his downward speed. Reaching an estimated v_t of 200 mi/h, he caught up with Williams and then went into a horizontal "spread eagle" (as in Fig. 6-10) to increase D so that he could grab her. He opened her parachute and then, after releasing her, his own, with a scant 10 s before impact. Williams received extensive internal injuries due to her lack of control on landing but survived.

FIGURE 6-10   A parachutist in a horizontal "spread eagle" maximizes the air drag.

SAMPLE PROBLEM 6-6

A raindrop with radius $R = 1.5$ mm falls from a cloud that is at height $h = 1200$ m above the ground. The drag coefficient C for the drop is 0.60. Assume that the drop is spherical throughout its fall. The density of water ρ_w is 1000 kg/m³, and the density of air ρ_a is 1.2 kg/m³.

a. What is the terminal speed of the drop?

SOLUTION The volume of a sphere is $\frac{4}{3}\pi R^3$, and its effective area A is that of a circle with radius R. So, for the drop, we have

$$m = \tfrac{4}{3}\pi R^3 \rho_w \quad \text{and} \quad A = \pi R^2.$$

Then, from Eq. 6-19, we find

$$v_t = \sqrt{\frac{2mg}{C\rho_a A}} = \sqrt{\frac{8\pi R^3 \rho_w g}{3C\rho_a \pi R^2}} = \sqrt{\frac{8R\rho_w g}{3C\rho_a}}$$

$$= \sqrt{\frac{(8)(1.5 \times 10^{-3}\text{ m})(1000\text{ kg/m}^3)(9.8\text{ m/s}^2)}{(3)(0.60)(1.2\text{ kg/m}^3)}}$$

$$= 7.4\text{ m/s} \ (= 17\text{ mi/h}). \qquad \text{(Answer)}$$

Note that the height of the cloud does not enter into the calculation. The raindrop (see Table 6-1) reaches terminal speed after falling just a few meters.

b. What would have been the speed just before impact if there had been no drag force?

SOLUTION From Eq. 2-21, with $h = -(y - y_0)$ and $v_0 = 0$, we find

$$v = \sqrt{2gh} = \sqrt{(2)(9.8\text{ m/s}^2)(1200\text{ m})}$$

$$= 150\text{ m/s} \ (= 340\text{ mi/h}). \qquad \text{(Answer)}$$

Under these conditions, Shakespeare would scarcely have written, "it falleth like the gentle rain from heaven, upon the place beneath."

6-4   UNIFORM CIRCULAR MOTION

If a particle moves in a circle or a circular arc of radius r with a constant speed v, it is accelerated toward the center of that circle, and it is said to have uniform circular motion. The magnitude of this **centripetal acceleration** is constant and given by Eq. 4-22:

$$a = \frac{v^2}{r}. \qquad (6\text{-}20)$$

To account for the centripetal acceleration, a **centripetal force,** directed toward the center of the circle, must act on the body. The magnitude of this force is constant and is given by Newton's second law:

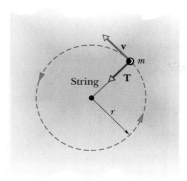

FIGURE 6-11 A hockey puck of mass m moves with constant speed v in a circular path on a horizontal frictionless surface. The centripetal force acting on the puck is **T**, the pull from the string.

$$F = ma = \frac{mv^2}{r}. \qquad (6\text{-}21)$$

If this force is not present, the body cannot undergo uniform circular motion. Both the centripetal acceleration and the centripetal force are vectors whose magnitudes are constant but whose directions are changing continuously so as always to point toward the center of the circle.

If the body in uniform circular motion is, say, a hockey puck whirled around on the end of a string

Curved flight at high speed requires a large centripetal force that makes this stunt dangerous even if the airplanes are not so close.

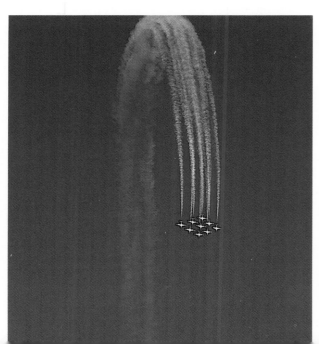

as in Fig. 6-11, the centripetal force is provided by the tension in the string. For the moon in its (nearly) uniform circular motion around the Earth, the centripetal force is the gravitational attraction of the Earth. Thus a centripetal force is not a new kind of force; it can be a tension force, a gravitational force, or any other force.

Let us compare and contrast two familiar examples of uniform circular motion:

1. *Rounding a curve in a car.* You are sitting in the center of the rear seat of a car moving at high speed over flat road. When the driver suddenly turns left, rounding a corner in a circular arc, you are slid across the seat toward the right and jammed against the interior wall of the car. What is going on?

The car, while it is moving in the circular arc, is in uniform circular motion. The centripetal force that causes the motion is a frictional force exerted by the roadway on the tires. This force points radially inward and—spread over the four tires—must have a magnitude given by Eq. 6-21.

You would have been in uniform circular motion in the center of the seat if the frictional force exerted on you by the seat had been great enough. However, it was not, and so you slid across the seat. Viewed from the reference frame of the ground, you actually continued moving in a straight line while the seat slid beneath you, until the wall of the car reached you. Its push on you was a centripetal force, and you then joined the car's uniform circular motion.

2. *Orbiting the Earth.* This time you are a passenger in the space shuttle *Atlantis,* orbiting the Earth and experiencing "weightlessness." What is going on in this case?

The centripetal force that keeps you and the shuttle in uniform circular motion is the gravitational attraction of the Earth for you and the shuttle. This force is directed radially inward, toward the center of the Earth, and has a magnitude given by Eq. 6-21.

In both car and shuttle you are in uniform circular motion, acted on by a centripetal force. Yet your experiences in the two situations are quite different. In the car, jammed up against the wall, you are aware of being compressed by the wall. In the orbiting shuttle, on the other hand, you are floating around with no sensation of any force acting on you. Why this great difference?

The difference is due to the nature of the two centripetal forces. In the car, the centripetal force is a *contact force*, exerted by the wall externally on the part of your body touching the wall. In the shuttle, the centripetal force is a *volume force*, due to the Earth's gravitational pull on every atom of your body and on every atom of the shuttle, in proportion to the mass of that atom. Thus there is no compression on any one part of your body, nor even any sensation of a force acting on you.

SAMPLE PROBLEM 6-7

Igor is a cosmonaut-engineer in the spacecraft *Vostok II*, orbiting the Earth at an altitude h of 520 km with a speed v of 7.6 km/s. Igor's mass m is 79 kg.

a. What is his acceleration?

SOLUTION Igor is in uniform circular motion in a circle of radius $R_E + h$, where R_E is the radius of the Earth. His centripetal acceleration is given by Eq. 6-20:

$$a = \frac{v^2}{r} = \frac{v^2}{R_E + h}$$

$$= \frac{(7.6 \times 10^3 \text{ m/s})^2}{6.37 \times 10^6 \text{ m} + 0.52 \times 10^6 \text{ m}}$$

$$= 8.38 \text{ m/s}^2 \approx 8.4 \text{ m/s}^2, \qquad \text{(Answer)}$$

which is the value of the free-fall acceleration at Igor's altitude. If he were lifted to that altitude and released, instead of being put into orbit there, he would fall toward the Earth's center, starting out with that value for the acceleration. The difference in the two situations is that when he orbits the Earth, he always has a "sideways" motion as well: as he falls, he also moves to the side, so that he ends up moving along a curved path around the Earth.

b. What (centripetal) gravitational force does the Earth exert on Igor?

SOLUTION The centripetal force is

$$F = ma = (79 \text{ kg})(8.38 \text{ m/s}^2)$$

$$= 660 \text{ N} \approx 150 \text{ lb.} \qquad \text{(Answer)}$$

If Igor were to stand on a scale placed on the top of a tower with height $h = 520$ km, the scale would read 660 N or 150 lb. In orbit, the scale (if Igor could "stand" on it) would read zero because he and the scale are in free fall together, so his feet do not actually press against it.

PROBLEM SOLVING

TACTIC 1: LOOKING THINGS UP

In Sample Problem 6-7, we had to know the radius of the Earth, which was not given in the problem statement. You need to become familiar with sources of this kind of information, starting with this book. Many useful data are given in the inside covers, in the various appendixes, and in tables (see the List of Tables given in the back inside covers). The *Handbook of Chemistry and Physics*, published every year by the Chemical Rubber Company (CRC), is an invaluable resource.

For practice, see if you can track down the density of iron, the series expansion of e^x, the number of centimeters in a mile, the mean distance of Saturn from the sun, the mass of the proton, the speed of light, and the atomic number of samarium. You can find them all in this book.

FIGURE 6-12 Sample Problem 6-8. (*a*) Advertisement and (*b*) free-body diagram for Diavolo.

SAMPLE PROBLEM 6-8

In a 1901 circus performance, Allo "Dare Devil" Diavolo introduced the stunt of riding a bicycle in a loop-the-loop (Fig. 6-12a). Assuming the loop is a circle with radius $R = 2.7$ m, what is the least speed v he could have at the top of the loop if he is to remain in contact with it there?

SOLUTION Figure 6-12b is a free-body diagram for Diavolo and the bicycle (taken together as being a single particle) at the top of the loop, showing the normal force **N** exerted on them by the loop, and their weight **W** $= m\mathbf{g}$. Their acceleration **a** points downward, toward the center of the loop, and, according to Eq. 6-20, has magnitude $a = v^2/R$. Applying Newton's second law along the y axis, we have

$$\sum F = -N - mg = -ma = -m\frac{v^2}{R}.$$

If he is on the verge of losing contact with the loop, $N = 0$, and so we then have

$$mg = m\frac{v^2}{R},$$

or

$$v = \sqrt{gR} = \sqrt{(9.8 \text{ m/s}^2)(2.7 \text{ m})}$$

$$= 5.1 \text{ m/s}. \qquad \text{(Answer)}$$

To avoid losing contact, Diavolo made certain that he had a faster speed at the top of the loop, in which case $N > 0$.

SAMPLE PROBLEM 6-9

Figure 6-13a shows a *conical pendulum*. Its bob, whose mass m is 1.5 kg, whirls around in a horizontal circle at constant speed v at the end of a cord whose length L, measured to the center of the bob, is 1.7 m. The cord makes an angle θ of 37° with the vertical. As the bob swings around in a circle, the cord sweeps out the surface of a cone. Find the period of the pendulum, that is, the time τ for the bob to complete one circle.

SOLUTION Figure 6-13b, the free-body diagram for the bob, shows the forces on the bob: the pull **T** from the cord due to the cord's tension, and the bob's weight **W** ($= m\mathbf{g}$). We place the origin of axes at the center of the bob, as shown. Instead of the usual x axis (which is stationary), we use a radial axis r that always points from the bob toward the center of the circle.

The y and r components of **T** are $T \cos \theta$ and $T \sin \theta$, respectively. Since $a_y = 0$, Newton's second law gives

$$T \cos \theta - mg = ma_y = 0, \quad \text{or} \quad T \cos \theta = mg. \quad (6\text{-}22)$$

There must be a net force along the r axis to provide the centripetal acceleration for the bob. The only force component in that direction is $T \sin \theta$. So for the r axis, Newton's second law gives

$$T \sin \theta = ma_r = \frac{mv^2}{R}, \qquad (6\text{-}23)$$

where R is the radius of the circular path. Dividing Eq. 6-23 by Eq. 6-22 and solving for v, we obtain

$$v = \sqrt{\frac{gR \sin \theta}{\cos \theta}}.$$

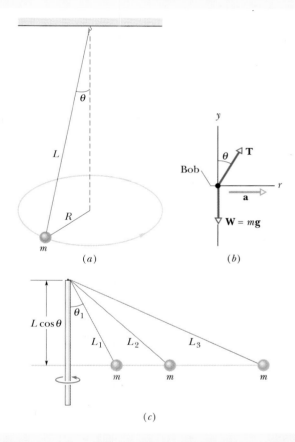

FIGURE 6-13 Sample Problem 6-9. (a) A conical pendulum, its cord making an angle θ with the vertical. (b) A free-body diagram for the pendulum bob. The axes point in the vertical and the radial directions. The resultant force (and thus the acceleration) points radially inward toward the center of the circle. (c) Three pendulums, of different lengths, are whirled around by a rotating shaft; their bobs circulate in the same horizontal plane, as Eq. 6-25 predicts.

We can substitute $2\pi R/\tau$ (the distance around the circle divided by the time to complete one circle) for the speed v of the bob. Doing so and solving for τ, we obtain

$$\tau = 2\pi\sqrt{\frac{R\cos\theta}{g\sin\theta}}. \qquad (6\text{-}24)$$

However, from Fig. 6-13a we see that $R = L\sin\theta$. Making this substitution in Eq. 6-24 yields

$$\tau = 2\pi\sqrt{\frac{L\cos\theta}{g}} \qquad (6\text{-}25)$$

$$= 2\pi\sqrt{\frac{(1.7\text{ m})(\cos 37°)}{9.8\text{ m/s}^2}} = 2.3\text{ s.} \qquad \text{(Answer)}$$

From Eq. 6-25 we see that the period τ does not depend on the mass of the bob, but only on $L\cos\theta$, the vertical distance of the bob from its point of support. Thus, as shown in Fig. 6-13c, if several conical pendulums of different lengths are swung from the same support with the same period, their bobs will all lie in the same horizontal plane.

SAMPLE PROBLEM 6-10

Figure 6-14a represents a car of mass $m = 1600$ kg traveling at a constant speed $v = 20$ m/s along a flat, circular road of radius $R = 190$ m. What is the minimum value of μ_s between the tires of the car and the road that will prevent the car from slipping?

FIGURE 6-14 Sample Problem 6-10. (a) A car moves around a flat curved road at constant speed. The frictional force $\mathbf{f}_s$ provides the necessary centripetal force. (b) A free-body diagram (not to scale) for the car.

SOLUTION The centripetal force that causes the car to travel in a circle is the radial frictional force $\mathbf{f}_s$ exerted on the tires by the road. (Although the car is moving, it is not sliding radially, and so the frictional force is $\mathbf{f}_s$ and not $\mathbf{f}_k$.)

Figure 6-14b, the free-body diagram for the car, shows the forces on the car: $\mathbf{f}_s$, $\mathbf{N}$, and $\mathbf{W} = m\mathbf{g}$. Since the car is not accelerating vertically, $a_y = 0$, and Newton's second law leads to the now familiar result $N = W = mg$.

In the radial direction, however, there must be a net force $\Sigma\,\mathbf{F}_r$ to give the car a centripetal acceleration $\mathbf{a}_r$. According to Newton's second law, $\Sigma\,F_r = ma_r$. Because the centripetal acceleration is v^2/R, and the only radial force we have is f_s, we find

$$f_s = \frac{mv^2}{R}. \qquad (6\text{-}26)$$

Next, recall that a body is on the verge of slipping when f_s reaches its maximum value, $\mu_s N$. Since the problem concerns just such a critical situation, we set $f_s = \mu_s N$ in Eq. 6-26 and then substitute mg for N. We get

$$\mu_s mg = \frac{mv^2}{R},$$

Turning left requires that frictional forces, pointing toward the center of rotation, act on the tires. For the cyclist to remain stable, he must lean toward that center.

(a)

(b)

or

$$\mu_s = \frac{v^2}{gR} \tag{6-27}$$

$$= \frac{(20 \text{ m/s})^2}{(9.8 \text{ m/s}^2)(190 \text{ m})} = 0.21. \quad \text{(Answer)}$$

If $\mu_s \geq 0.21$, the car will be held in a circle by $\mathbf{f}_s$. But if $\mu_s < 0.21$, the car will slide out of the circle.

Note two additional features of Eq. 6-27. First, the value of μ_s depends on the square of v. That means that much more friction is required as the turning speed increases. You may have noted this effect if you have ever taken a flat turn too fast and suddenly felt the tires slip. Second, the mass m does not appear in Eq. 6-27. That means Eq. 6-27 holds for a vehicle of any mass, from a kiddy car to a bicycle to a heavy truck.

SAMPLE PROBLEM 6-11

You cannot always count on friction to get your car around a curve, especially if the road is icy or wet. That is why highways are banked. As in Sample Problem 6-10, suppose that a car of mass m is moving at a constant speed v of 20 m/s around a curve, now banked,

FIGURE 6-15 Sample Problem 6-11. (*a*) A car moving around a curved banked roadway at a constant speed. The radially inward component of the normal force provides the necessary centripetal force. The bank angle is exaggerated for clarity. (*b*) A free-body diagram for the car. The resultant force (and thus the acceleration) points radially inward.

(a)

(b)

Banked tracks are needed for turns that are taken so quickly that friction alone cannot provide enough centripetal force.

whose radius R is 190 m (Fig. 6-15*a*). What bank angle θ would make reliance on friction unnecessary?

SOLUTION The centripetal acceleration and the required centripetal force are the same as in the previous problem. The effect of banking is to tilt the normal force $\mathbf{N}$ toward the center of curvature of the road, so that its inward radial component N_r can supply the required centripetal force.

Since there is no acceleration in the vertical direction,

$$N_y = N \cos \theta = W = mg. \tag{6-28}$$

In the radial direction the only force component is N_r (we have assumed a frictional force is unnecessary). So, by Eq. 6-21,

$$N_r = N \sin \theta = \frac{mv^2}{R}. \tag{6-29}$$

Dividing Eq. 6-29 by Eq. 6-28 gives

$$\tan \theta = \frac{v^2}{gR}.$$

Thus we have

$$\theta = \tan^{-1} \frac{v^2}{gR} \tag{6-30}$$

$$= \tan^{-1} \frac{(20 \text{ m/s})^2}{(9.8 \text{ m/s}^2)(190 \text{ m})} = 12°. \quad \text{(Answer)}$$

Equations 6-27 and 6-30 tell us that the critical coefficient of friction for an unbanked road is the same as the tangent of the bank angle for a banked road. The road must produce a certain centripetal force one way or the other—either with friction or by being banked.

SAMPLE PROBLEM 6-12

Even some seasoned roller coaster riders blanch at the thought of riding the Rotor, which is essentially a large hollow cylinder that is rotated rapidly around its central axis (Fig. 6-16). Before the ride begins, a rider enters the cylinder through a door on the side and stands on a floor against a canvas-covered wall. The door is closed, and as the cylinder begins to turn, the rider, wall, and floor move in unison. When the rider's speed reaches some predetermined value, the floor abruptly and alarmingly falls away. The rider does not fall with it but instead is pinned to the wall while the cylinder rotates, as if an unseen (and somewhat unfriendly) agent is pressing the body to the wall. Later, the floor is eased back to the rider's feet, the cylinder slows, and the rider sinks a few centimeters to regain footing on the floor. (Some riders consider all this to be fun.)

Suppose that the coefficient of static friction μ_s between the rider's clothing and the canvas is 0.40 and that the cylinder's radius R is 2.1 m.

a. What minimum speed v must the cylinder and rider have if the rider is not to fall when the floor drops?

SOLUTION The rider will not be pulled down by her weight **W** provided the magnitudes of **W** and the frictional force $\mathbf{f}_s$, which is exerted upward on her by the wall, are equal. At the minimum required speed, she is on the verge of slipping, which means that f_s must be at its maximum value of $\mu_s N$. So in this critical situation

$$\mu_s N = mg, \tag{6-31}$$

where m is her mass.

The normal force **N** is, as usual, perpendicular to the surface against which the body (here, the woman) is pressed, but note that now it points horizontally toward the central axis. That force accounts for the centripetal force that provides the woman's centripetal acceleration a_r and keeps her moving in a circle. Thus, by Eq. 6-21,

$$N = m \frac{v^2}{R}. \tag{6-32}$$

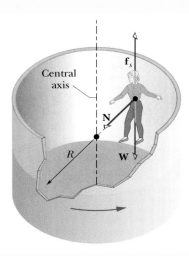

FIGURE 6-16 Sample Problem 6-12. A Rotor in an amusement park. The centripetal force is the normal force with which the wall pushes radially inward on the rider.

Substituting this expression for N in Eq. 6-31 and solving for v yield

$$v = \sqrt{\frac{gR}{\mu_s}} = \sqrt{\frac{(9.8 \text{ m/s}^2)(2.1 \text{ m})}{0.40}}$$

$$= 7.17 \text{ m/s} \approx 7.2 \text{ m/s}. \tag{Answer}$$

Note that the result is independent of the rider's mass; it holds for anyone riding the Rotor, from a child to a heavy adult.

b. If the rider's mass is 49 kg, what is the magnitude of the centripetal force on her?

SOLUTION According to Eq. 6-32,

$$N = m \frac{v^2}{R} = (49 \text{ kg}) \frac{(7.17 \text{ m/s})^2}{2.1 \text{ m}}$$

$$\approx 1200 \text{ N}. \tag{Answer}$$

Although this force is directed toward the central axis, the rider has an overwhelming sensation that the force pinning her against the wall is directed radially outward. Her sensation stems from the fact that she is in a noninertial frame (she and it are accelerating). As measured from such frames, forces can be illusionary. The illusion is part of the Rotor's attraction.

6-5 THE FORCES OF NATURE*

We have used **F** as a generic symbol for force. We have also used other symbols: **W** for the weight of a body, **T** for the pull from a cord under tension, **f** for a frictional force, **N** for a normal force, and **D** for the drag force exerted, for example, by air on a sky diver. At the fundamental level, all these forces fall into two types: (1) the **gravitational force,** of which weight is our only example, and (2) the **electromagnetic force,** which includes—without exception—all the others. The force that makes an electrically charged balloon stick to a wall and the force with which a magnet picks up an iron nail are other examples of the electromagnetic force. In fact, aside from the gravitational force, *all* forces that we can experience directly as a push or pull are electromagnetic when examined closely. That is, all such forces, including frictional forces, normal forces, contact forces, drag forces, and tension forces, involve, fundamentally, electromagnetic forces exerted by one atom on another. The tension in a taut rope, for example, is maintained only because the atoms of the rope attract one another.

*See P. C. W. Davies, *The Forces of Nature,* 2nd ed., Cambridge University Press, New York, 1986, for a very readable account.

The only other fundamental forces known to occur both act over such short distances that we cannot experience them directly through our sensory perceptions. They are the **weak force,** which is involved in certain kinds of radioactive decay, and the **strong force,** which binds together the constituents of protons and neutrons and which is the "glue" that holds together an atomic nucleus.

Physicists have long believed that nature has an underlying simplicity and that the number of fundamental forces can be reduced. Einstein spent most of his working life trying to interpret these forces as different aspects of a single *superforce.* He failed, but in the 1960s and 1970s, other physicists showed that the weak force and the electromagnetic force are different aspects of a single **electroweak force.** The quest for further reduction continues today, at the very forefront of physics. Table 6-2 lists the progress that has been made toward **unification** (as the goal is called) and gives some hints about the future.

The forces and the particles of nature are intimately connected with the origin and development of the universe. With a force unification theory in hand, it might be possible to answer questions such as: "How did the universe evolve to its present state?" and "What will the universe be like in the future?" In Chapter 49 of the extended version of this book, we shall explore such questions.

TABLE 6-2
THE QUEST FOR THE SUPERFORCE—A PROGRESS REPORT

DATE	RESEARCHER	ACHIEVEMENT
1687	Newton	Showed that the same laws apply to astronomical bodies and to objects on Earth. Unified celestial and terrestrial mechanics.
1820 1830s	Oersted Faraday	Showed, by brilliant experiments, that the then separate sciences of electricity and magnetism are intimately linked.
1873	Maxwell	Unified the sciences of electricity, magnetism, and optics into the single subject of electromagnetism.
1979	Glashow Salam Weinberg	Received the Nobel prize for showing that the weak force and the electromagnetic force could be viewed as different aspects of a single *electroweak force.* This reduced the number of fundamental forces from four to three.
1984	Rubbia van der Meer	Received the Nobel prize for verifying experimentally the predictions of the theory of the electroweak force.

Work in Progress

Grand unification theories: Called GUTs, these theories seek to unify the electroweak force and the strong force.
Supersymmetry theories: These theories seek to unify all forces, including the gravitational force, within a single framework.
Superstring theories: These theories interpret pointlike particles, such as electrons, as being unimaginably tiny, closed loops. Strangely, extra dimensions beyond the familiar four dimensions of spacetime appear to be required.

REVIEW & SUMMARY

Friction

When a force **F** attempts to slide a body along a surface, a **frictional force** is exerted on the body by the surface. The frictional force is parallel to the surface and directed so as to oppose the sliding. It is due to the bonding between the body and the surface.

If the body does not slide, the frictional force is a **static frictional force** $\mathbf{f}_s$. If there is sliding, the frictional force is a **kinetic frictional force** $\mathbf{f}_k$.

Properties of Friction

Property 1. If the body does not move, then the static frictional force $\mathbf{f}_s$ and the component of **F** that is parallel to the surface are equal in magnitude, and $\mathbf{f}_s$ is directed opposite that component.

Property 2. The magnitude of $\mathbf{f}_s$ has a maximum value $f_{s,max}$ that is given by

$$f_{s,max} = \mu_s N, \tag{6-1}$$

where μ_s is the **coefficient of static friction,** and N is the magnitude of the normal force. If the component of **F** that is parallel to the surface exceeds $f_{s,max}$, then the body begins to slide along the surface.

Property 3. If the body begins to slide along the surface, the magnitude of the frictional force rapidly decreases to a value f_k given by

$$f_k = \mu_k N, \tag{6-2}$$

where μ_k is the **coefficient of kinetic friction.** Thereafter during the sliding, the kinetic frictional force $\mathbf{f}_k$ has a magnitude given by Eq. 6-2.

Drag Force

When there is a relative velocity between air and a body, the body experiences a **drag force D** that opposes the relative motion and points in the direction in which the air flows relative to the body. The magnitude of **D** is related to the relative speed v by an experimentally determined **drag coefficient** C according to

$$D = \tfrac{1}{2}C\rho A v^2, \tag{6-18}$$

where ρ is the air density (mass per volume) and A is the **effective cross-sectional area** of the body (the area of a cross section taken perpendicularly to the velocity **v**).

Terminal Speed

When a blunt object falls far enough through air, the magnitudes of the drag force and the object's weight are equal. The body then falls at a constant **terminal speed** v_t given by

$$v_t = \sqrt{\frac{2mg}{C\rho A}}, \tag{6-19}$$

where m is the body's mass.

Uniform Circular Motion

If a particle moves in a circle or a circular arc with radius r at constant speed v, it is said to be in **uniform circular motion.** It has a **centripetal acceleration** with a magnitude given by

$$a = \frac{v^2}{r}, \tag{6-20}$$

which is due to a **centripetal force** with a magnitude given by

$$F = m\frac{v^2}{r}, \tag{6-21}$$

where m is the particle's mass. The vectors **a** and **F** point toward the center of curvature of the particle's path.

Fundamental Forces

The myriad examples of forces can be reduced to three fundamental types: **gravitational, electroweak** (a combination of the historic grouping of **electric** and **magnetic** forces with the **weak** force), and **strong.** Only the gravitational, electric, and magnetic forces are readily apparent in the everyday world. Physicists hope to reduce the list of three fundamental forces to a single force, the elusive *superforce* that would include all others.

QUESTIONS

1. There is a limit beyond which further polishing of a surface *increases* rather than decreases frictional resistance. Explain why.

2. Can the coefficient of static friction have a value greater than 1? What about the coefficient of kinetic friction?

3. A crate, heavier than you are, rests on a rough floor. The coefficient of static friction between the crate and the floor is the same as that between the soles of your shoes and the floor. Can you push the crate across the floor?

4. How could a person who is at rest on completely frictionless ice covering a pond reach shore? Could she do this by walking, rolling, swinging her arms, or kicking her feet? How could a person be placed in such a location in the first place?

5. Why do tires grip the road better on level ground than they do when going uphill or downhill?

6. What is the purpose of the curved surfaces, called spoilers, that are placed on the rear of race cars? They are designed so that air flowing past them exerts a downward force.

7. Which raindrops, if either, fall faster: small ones or large ones?

8. The terminal speed of a baseball is 95 mi/h. However, the measured speeds of pitched balls often exceed this, sometimes exceeding 100 mi/h. How can this be?

9. A log is floating downstream. How would you calculate the drag force acting on it?

10. What happens to a baseball that is fired downward

through air at twice its terminal speed—does it speed up, slow down, or continue to move with its initial speed?

11. Consider a ball thrown vertically up. Taking air resistance into account, would you expect the time during which the ball rises to be longer or shorter than the time during which it falls? Why? Make a qualitative graph of speed v versus time t for the ball.

12. Why are train roadbeds and highways banked on curves?

13. You are flying a plane at constant altitude and wish to make a 90° turn. Why do you bank the plane?

14. In the conical pendulum of Sample Problem 6-9, what happens to the period τ and the speed v when $\theta = 90°$? Why is this angle not achievable physically? Discuss the case for $\theta = 0°$.

15. A coin is put on a phonograph turntable. The motor is started but, before the final speed of rotation is reached, the coin flies off. Explain why.

16. A car is moving at constant speed along a country road that resembles a roller coaster track. Compare the force the car exerts on a horizontal section of the road to the force it exerts on the road at the top of a hill and at the bottom of a valley. Explain.

17. A passenger in the front seat of a car finds himself sliding toward the door as the driver makes a sudden left turn. Describe the forces on the passenger and on the car at this instant if the motion is viewed from a reference frame (a) attached to the Earth and (b) attached to the car.

EXERCISES & PROBLEMS

SECTION 6-2 PROPERTIES OF FRICTION

1E. A bedroom bureau with a mass of 45 kg, including drawers and clothing, rests on the floor. (a) If the coefficient of static friction between the bureau and the floor is 0.45, what is the minimum horizontal force a person must apply to start the bureau moving? (b) If the drawers and clothing, with 17-kg mass, are removed before the bureau is pushed, what is the new minimum magnitude?

2E. A baseball player with mass $m = 79$ kg, sliding into second base, is retarded by a force of friction $f = 470$ N. What is the coefficient of kinetic friction μ_k between the player and the ground?

3E. A worker pushes horizontally on a 35-kg crate with a 110-N force. The coefficient of static friction between the

crate and the floor is 0.37. (a) What is the frictional force exerted on the crate by the floor? (b) What is the maximum magnitude $f_{s,\text{max}}$ of the static frictional force under the circumstances? (c) Does the crate move? (d) Suppose, next, that a second worker pulls directly upward on the crate to help out. What is the least pull she can exert that will allow the first worker's 110-N push to move the crate? (e) If, instead, the second worker pulls horizontally to help out, what is the least pull she can exert to get the crate moving?

4E. The coefficient of static friction between Teflon and scrambled eggs is about 0.04. What is the smallest angle from the horizontal that will cause the eggs to slide across the bottom of a Teflon-coated skillet?

5E. A 100-N force is applied at an angle θ above the horizontal to a 25.0-kg chair sitting on the floor. (a) For each of the following angles θ, calculate the magnitude of the normal force of the floor on the chair and the horizontal component of the applied force: (i) 0°, (ii) 30.0°, (iii) 60.0°. (b) Take the coefficient of static friction between the chair and the floor to be 0.420 and, for each of the values of θ, decide if the chair remains at rest or slides.

6E. In Nevada and southern California, stones leave trails in the hard-baked desert floor, as if they had been migrating (Fig. 6-17). For years curiosity mounted about the unseen motion that caused the trails. The answer finally came in the 1970s: when an occasional storm hits the desert, a thin layer of mud may form over a still-firm base, greatly reducing the coefficient of friction between the stones and ground. If a strong wind accompanies the storm, it pushes the stones, leaving trails that are later baked hard by the sun. Suppose a stone's mass is 300 kg (about the greatest stone mass that has left a trail) and the coefficient of static friction is reduced to 0.15. Of what magnitude is the force from a horizontal gust that is needed to move the stone?

FIGURE 6-17 Exercise 6. Stone trail.

7E. What is the greatest acceleration that can be generated by a runner if the coefficient of static friction between shoes and track is 0.95? (Only one foot is on the track during the acceleration.)

8E. A person pushes horizontally with a force of 220 N on a 55-kg crate to move it across a level floor. The coefficient of kinetic friction is 0.35. (a) What is the magnitude of the frictional force? (b) What is the acceleration of the crate?

9E. A trunk with a weight of 220 N rests on the floor. The coefficient of static friction between the trunk and the floor is 0.41, while the coefficient of kinetic friction is 0.32. (a) What is the minimum magnitude for a horizontal force with which a person must push on the trunk to start it moving? (b) Once moving, what magnitude of horizontal force must the person apply to keep the trunk moving with constant velocity? (c) If the person continued to push with the force used to start the motion, what would be the acceleration of the trunk?

10E. A filing cabinet with a weight of 556 N rests on the floor. The coefficient of static friction between it and the floor is 0.68, and the coefficient of kinetic friction is 0.56. In four different attempts to move it, it is pushed with horizontal forces of (a) 222 N, (b) 334 N, (c) 445 N, and (d) 556 N. For each attempt, determine if the cabinet moves and calculate the magnitude of the frictional force the floor exerts on it. The cabinet is initially at rest for each attempt.

11E. A horizontal force F of 12 N pushes a block weighing 5.0 N against a vertical wall (Fig. 6-18). The coefficient of static friction between the wall and the block is 0.60, and the coefficient of kinetic friction is 0.40. Assume that the block is not moving initially. (a) Will the block start moving? (b) In unit-vector notation, what is the force exerted on the block by the wall?

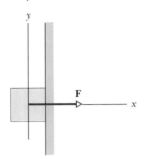

FIGURE 6-18 Exercise 11.

12E. A 49-kg rock climber is climbing a "chimney" between two rock slabs as shown in Fig. 6-19. The static coefficient of friction between her shoes and the rock is 1.2; between her back and the rock it is 0.80. She has reduced her push against the rock until her back and her shoes are on the verge of slipping. (a) What is her push against the rock? (b) What fraction of her weight is supported by the frictional force on her shoes?

FIGURE 6-19 Exercise 12.

13E. A house is built on the top of a hill with a nearby 45° slope (Fig. 6-20). An engineering study indicates that the slope angle should be reduced because the top layers of soil along the slope might slip past the lower layers. If the static coefficient of friction between two such layers is 0.5, what is the least angle ϕ through which the present slope should be reduced to prevent slippage?

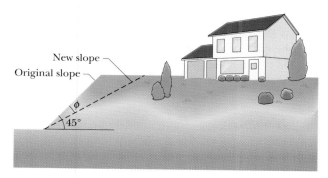

FIGURE 6-20   Exercise 13.

14E. The coefficient of kinetic friction in Fig. 6-21 is 0.20. What is the acceleration of the block if (a) it is sliding down the slope, and (b) it has been given an upward shove and is still sliding up the slope?

FIGURE 6-21   Exercise 14.

15E. A 110-g hockey puck slides on the ice for 15 m before it stops. (a) If its initial speed was 6.0 m/s, what was the magnitude of the frictional force on the puck during the sliding? (b) What was the coefficient of friction between the puck and the ice?

16P. A student wants to determine the coefficients of static friction and kinetic friction between a box and a plank. She places the box on the plank and gradually raises one end of the plank. When the angle of inclination with the horizontal reaches 30°, the box starts to slip and slides 2.5 m down the plank in 4.0 s. What are the coefficients of friction?

17P. A worker wishes to pile a cone of sand onto a circular area in his yard. The radius of the circle is R, and no sand is to spill onto the surrounding area (Fig. 6-22). If μ_s is the static coefficient of friction between each layer of

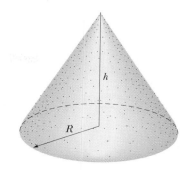

FIGURE 6-22   Problem 17.

sand along the slope and the sand beneath it (along which it might slip), show that the greatest volume of sand that can be stored in this manner is $\pi\mu_s R^3/3$. (The volume of a cone is $Ah/3$, where A is the base area, and h is the cone's height.)

18P. A ski that is placed on snow will stick to the snow. However, when the ski is moved along the snow, the rubbing warms and partially melts the snow, reducing the coefficient of friction and promoting sliding. Waxing the ski makes it water repellent and reduces friction with the resulting layer of water. A magazine reports that a new type of plastic ski is especially water repellent and that, on a gentle 200-m slope in the Alps, a skier reduced his top-to-bottom time from 61 to 42 s with the new skis. (a) Determine the magnitudes of his average acceleration with each pair of skis. (b) Assuming a 3.0° slope, compute the coefficient of kinetic friction for each case.

19P. An 11-kg block of steel is at rest on a horizontal table. The coefficient of static friction between block and table is 0.52. (a) What is the magnitude of the horizontal force that will just start the block moving? (b) What is the magnitude of a force acting upward 60° from the horizontal that will just start the block moving? (c) If the force acts down at 60° from the horizontal, how large can its magnitude be without causing the block to move?

20P. A railroad flatcar is loaded with crates having a coefficient of static friction of 0.25 with the floor. If the train is moving at 48 km/h, in how short a distance can the train be stopped at a constant deceleration without causing the crates to slide?

21P. A block slides down an inclined plane of slope angle θ with constant velocity. It is then projected up the same plane with an initial speed v_0. (a) How far up the incline will it move before coming to rest? (b) Will it slide down again? Give an argument to back your answer.

22P. A 68-kg crate is dragged across a floor by pulling on a rope inclined 15° above the horizontal. (a) If the coefficient of static friction is 0.50, what minimum tension in the rope is required to start the crate moving? (b) If $\mu_k = 0.35$, what is the magnitude of the initial acceleration of the crate?

23P. A pig slides down a 35° incline (Fig. 6-23) in twice the time it would take to slide down a frictionless 35° in-

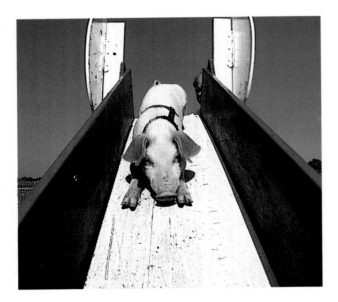

FIGURE 6-23 Problem 23.

cline. What is the coefficient of kinetic friction between the pig and the incline?

24P. In Fig. 6-24, *A* and *B* are blocks with weights of 44 N and 22 N, respectively. (a) Determine the minimum weight (block *C*) that must be placed on *A* to keep it from sliding, if μ_s between *A* and the table is 0.20. (b) Block *C* suddenly is lifted off *A*. What is the acceleration of block *A*, if μ_k between *A* and the table is 0.15?

Frictionless, massless pulley

FIGURE 6-24 Problem 24.

25P. A 3.5-kg block is pushed along a horizontal floor by a force $F = 15$ N that makes an angle $\theta = 40°$ with the horizontal (Fig. 6-25). The coefficient of kinetic friction between the block and floor is 0.25. Calculate (a) the mag-

FIGURE 6-25 Problem 25.

nitude of the frictional force exerted on the block and (b) the acceleration of the block.

26P. In Fig. 6-26 a fastidious worker pushes directly along the handle of a mop with a force **F**. The handle is at an angle θ with the vertical, and μ_s and μ_k are the coefficients of static and kinetic friction between the head of the mop and the floor. Ignore the mass of the handle and assume that all the mop's mass *m* is in its head. (a) If the mop head moves along the floor with a constant velocity, then what is *F*? (b) Show that if θ is less than a certain value θ_0, then **F** (still directed along the handle) is unable to move the mop head. Find θ_0.

FIGURE 6-26 Problem 26.

27P. A 5.0-kg block on an inclined plane is acted on by a horizontal force with magnitude 50 N (Fig. 6-27). The coefficient of kinetic friction between block and plane is 0.30. The coefficient of static friction is not given (but you might still know something about it). (a) What is the acceleration of the block if it is moving up the plane? (b) With the horizontal force still acting, how far up the plane will the block go if it has an initial upward speed of 4.0 m/s? (c) What happens to the block after it reaches the highest point? Give an argument to back your answer.

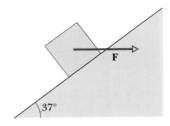

FIGURE 6-27 Problem 27.

28P. Figure 6-28 shows the cross section of a road cut into the side of a mountain. The solid line *AA'* represents a weak bedding plane along which sliding is possible. Block *B* directly above the highway is separated from uphill rock by a large crack (called a *joint*), so that only the force of friction between the block and the bedding plane prevents sliding. The mass of the block is 1.8×10^7 kg, the *dip angle* θ of the failure plane is 24°, and the coefficient of static friction between block and plane is 0.63. (a) Show that the

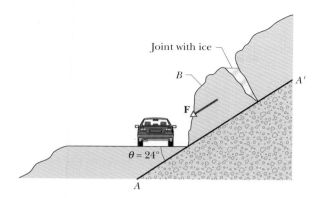

FIGURE 6-28 Problem 28.

block will not slide. (b) Water seeps into the joint and expands upon freezing, exerting on the block a force **F**
parallel to *AA'*. What minimum value of *F* will trigger a
slide?

29P. A block weighing 80 N rests on a plane inclined at
20° to the horizontal (Fig. 6-29). The coefficient of static
friction is 0.25, and the coefficient of kinetic friction is
0.15. (a) What is the minimum magnitude of the force **F**,
parallel to the plane, that will prevent the block from slipping down the plane? (b) What is the minimum magnitude *F* that will start the block moving up the plane? (c)
What value of *F* is required to move the block up the plane
at constant velocity?

FIGURE 6-29 Problem 29.

30P. Block *B* in Fig. 6-30 weighs 711 N. The coefficient
of static friction between block and horizontal surface is
0.25. Find the maximum weight of block *A* for which the
system will be stable.

FIGURE 6-30 Problem 30.

31P. Body *B* in Fig. 6-31 weighs 102 N, and body *A*
weighs 32 N. The coefficients of friction between *B* and
the incline are $\mu_s = 0.56$ and $\mu_k = 0.25$. Find the acceleration of the system if (a) *B* is initially at rest, (b) *B* is moving
up the incline, and (c) *B* is moving down the incline.

FIGURE 6-31 Problem 31.

32P. Two blocks are connected over a pulley as shown in
Fig. 6-32. The mass of block *A* is 10 kg and the coefficient
of kinetic friction is 0.20. Block *A* slides down the incline
at constant speed. What is the mass of block *B*?

FIGURE 6-32 Problem 32.

33P. Block m_1 in Fig. 6-33 has a mass of 4.0 kg and m_2 has
a mass of 2.0 kg. The coefficient of friction between m_2
and the horizontal plane is 0.50. The inclined plane is frictionless. Find (a) the tension in the cord and (b) the acceleration of the blocks.

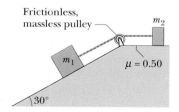

FIGURE 6-33 Problem 33.

34P. Two blocks, of weights 8.0 lb and 16 lb, are connected by a massless string and slide down a 30° inclined
plane. The coefficient of kinetic friction between the 8.0-
lb block and the plane is 0.10; that between the 16-lb
block and the plane is 0.20. Assuming that the 8.0-lb block

leads, find (a) the acceleration of the blocks and (b) the tension in the string. (c) Describe the motion if the blocks are reversed.

35P. Two masses, $m_1 = 1.65$ kg and $m_2 = 3.30$ kg, attached by a massless rod parallel to the inclined plane on which both slide (Fig. 6-34), travel down along the plane with m_1 trailing m_2. The angle of incline is $\theta = 30°$. The coefficient of kinetic friction between m_1 and the incline is $\mu_1 = 0.226$; that between m_2 and the incline is $\mu_2 = 0.113$. Compute (a) the tension in the rod and (b) the common acceleration of the two masses. (c) How would the answers to (a) and (b) change if m_2 trailed m_1?

FIGURE 6-34 Problem 35.

36P. A 4.0-kg block is put on top of a 5.0-kg block. To cause the top block to slip on the bottom one, which is held fixed, a horizontal force of at least 12 N must be applied to the top block. The assembly of blocks is now placed on a horizontal, frictionless table (Fig. 6-35). Find (a) the magnitude of the maximum horizontal force **F** that can be applied to the lower block so that the blocks will move together and (b) the resulting acceleration of the blocks.

FIGURE 6-35 Problem 36.

37P. A 40-kg slab rests on a frictionless floor. A 10-kg block rests on top of the slab (Fig. 6-36). The coefficient of static friction μ_s between the block and the slab is 0.60, whereas the kinetic coefficient μ_k is 0.40. The 10-kg block is pulled by a horizontal force with a magnitude of 100 N. What are the resulting acceleration magnitudes of (a) the block and (b) the slab?

FIGURE 6-36 Problem 37.

38P. The two blocks (with $m = 16$ kg and $M = 88$ kg) shown in Fig. 6-37 are not attached. The coefficient of

static friction between the blocks is $\mu_s = 0.38$, but the surface beneath M is frictionless. What is the minimum magnitude of the horizontal force **F** required to hold m against M?

FIGURE 6-37 Problem 38.

39P. A crate slides down an inclined right-angled trough as in Fig. 6-38. The coefficient of kinetic friction between the crate and the trough is μ_k. What is the acceleration of the crate in terms of μ_k, θ, and g?

FIGURE 6-38 Problem 39.

40P. A locomotive accelerates a 25-car train along a level track. Every car has a mass of 50 metric tons and is subject to a frictional force $f = 250v$, where the speed v is in meters per second and the force f is in newtons. At the instant when the speed of the train is 30 km/h, the acceleration is 0.20 m/s². (a) What is the tension in the coupling between the first car and the locomotive? (b) If this tension is the maximum force the locomotive can exert on the train, what is the steepest grade up which the locomotive can pull the train at 30 km/h?

41P. An initially stationary box of sand is to be pulled across a floor by means of a cord in which the tension should not exceed 1100 N. The coefficient of static friction between the box and floor is 0.35. (a) What should be the angle between the cord and the horizontal in order to pull the greatest possible amount of sand, and (b) what is the weight of the sand and box in that situation?

42P*. A 1000-kg boat is traveling at 90 km/h when its engine is shut off. The magnitude of the frictional force $\mathbf{f}_k$ between boat and water is proportional to the speed v of the boat: $f_k = 70v$, where v is in meters per second and f_k is in newtons. Find the time required for the boat to slow down to 45 km/h.

SECTION 6-3 THE DRAG FORCE AND TERMINAL SPEED

43E. Calculate the drag force on a missile 53 cm in diameter cruising with a speed of 250 m/s at low altitude, where the density of air is 1.2 kg/m³. Assume $C = 0.75$.

44E. The terminal speed of a sky diver in the spread-eagle position is 160 km/h. In the nose-dive position, the terminal speed is 310 km/h. Assuming that C does not change from one position to the other, find the ratio of the effective cross-sectional area A in the slower position to that in the faster position.

45E. Calculate the ratio of the drag force on a passenger jet flying with a speed of 1000 km/h at an altitude of 10 km to the drag force on a prop-driven transport flying at half the speed and half the altitude of the jet. At 10 km the density of air is 0.38 kg/m³ and at 5.0 km it is 0.67 kg/m³. Assume the planes have the same effective cross-sectional area and the same drag coefficient C.

46P. From the data in Table 6-1, deduce the diameter of the 16-lb shot. Assume $C = 0.49$.

SECTION 6-4 UNIFORM CIRCULAR MOTION

47E. If the coefficient of static friction for tires on a road is 0.25, at what maximum speed can a car round a level curve of 47.5-m radius without slipping?

48E. During an Olympic bobsled run, a European team takes a turn of radius 25 ft at a speed of 60 mi/h. How many g's do the riders experience during the turn?

49E. What is the smallest radius of an unbanked curve around which a bicyclist can travel if her speed is 18 mi/h and the coefficient of static friction between the tires and the road is 0.32?

50E. A car weighing 10.7 kN and traveling at 13.4 m/s attempts to round an unbanked curve with a radius of 61.0 m. (a) What force of friction is required to keep the car on its circular path? (b) If the coefficient of static friction between the tires and road is 0.35, is the attempt at taking the curve successful?

51E. A circular curve of highway is designed for traffic moving at 60 km/h. (a) If the radius of the curve is 150 m, what is the correct angle of banking of the road? (b) If the curve were not banked, what would be the minimum coefficient of friction between tires and road that would keep traffic from skidding at this speed?

52E. A banked circular highway curve is designed for traffic moving at 60 km/h. The radius of the curve is 200 m. Traffic is moving along the highway at 40 km/h on a stormy day. What is the minimum coefficient of friction between tires and road that will allow cars to negotiate the turn without sliding off the road?

53E. A child places a picnic basket on the outer rim of a merry-go-round that has a radius of 4.6 m and revolves once every 30 s. (a) What is the speed of a point on the rim of the merry-go-round? (b) How large must the coefficient of static friction between the basket and the merry-go-round be for the basket to stay on the ride?

54E. A conical pendulum is formed by attaching a 50-g mass to a 1.2-m string. The mass swings around a horizontal circle of radius 25 cm. (a) What is the speed of the

mass? (b) What is the acceleration of the mass? (c) What is the tension in the string?

55E. In the Bohr model of the hydrogen atom, the electron revolves in a circular orbit around the nucleus. If the radius is 5.3×10^{-11} m and the electron circles 6.6×10^{15} times per second, find (a) the speed of the electron, (b) the acceleration (magnitude and direction) of the electron, and (c) the centripetal force acting on the electron. (This force is the result of the attraction between the positively charged nucleus and the negatively charged electron.) The electron's mass is 9.11×10^{-31} kg.

56E. A mass m on a frictionless table is attached to a hanging mass M by a cord through a hole in the table (Fig. 6-39). Find the speed with which m must move for M to stay at rest.

FIGURE 6-39 Exercise 56.

57E. A stuntman drives a car over the top of a hill, the cross section of which can be approximated by a circle of radius 250 m, as in Fig. 6-40. What is the greatest speed at which he can drive without the car leaving the road at the top of the hill?

250 m

FIGURE 6-40 Exercise 57.

58P. A small coin is placed on a flat, horizontal turntable. The turntable is observed to make three revolutions in 3.14 s. (a) What is the speed of the coin when it rides without slipping at a distance 5.0 cm from the center of the turntable? (b) What is the acceleration (magnitude and direction) of the coin? (c) What is the magnitude of the frictional force acting on the coin if the coin has a mass of 2.0 g? (d) What is the coefficient of static friction between the coin and the turntable if the coin is observed to slide off the turntable when it is more than 10 cm from the center of the turntable?

59P. A small object is placed 10 cm from the center of a phonograph turntable. It is observed to remain on the table when it rotates at $33\frac{1}{3}$ rev/min (revolutions per minute) but slides off when it rotates at 45 rev/min. Between what limits must the coefficient of static friction between the object and the surface of the turntable lie?

60P. A bicyclist travels in a circle of radius 25.0 m at a constant speed of 9.00 m/s. The combined mass of the bicycle and rider is 85.0 kg. Calculate the magnitudes of (a) the force of friction exerted by the road on the bicycle and (b) the total force exerted by the road.

61P. A car is rounding a flat curve of radius $R = 220$ m at the curve's design speed $v = 94.0$ km/h. What force does a passenger with mass $m = 85.0$ kg exert on the seat cushion?

62P. A 150-lb student on a steadily rotating Ferris wheel has an apparent weight of 125 lb at the highest point. (a) What is the student's apparent weight at the lowest point? (b) What is the student's apparent weight at the highest point if the wheel's speed is doubled?

63P. A stone tied to the end of a string is whirled around in a vertical circle of radius R. Find the critical speed below which the string would become slack at the highest point.

64P. A certain string can withstand a maximum tension of 9.0 lb without breaking. A child ties a 0.82-lb stone to one end and, holding the other end, whirls the stone in a vertical circle of radius 3.0 ft, slowly increasing the speed until the string breaks. (a) Where is the stone on its path when the string breaks? (b) What is the speed of the stone as the string breaks?

65P. An airplane is flying in a horizontal circle at a speed of 480 km/h. If the wings of the plane are tilted 40° to the horizontal, what is the radius of the circle in which the plane is flying? See Fig. 6-41. Assume that the required force is provided entirely by an "aerodynamic lift" that is perpendicular to the wing surface.

FIGURE 6-41 Problem 65.

66P. A frigate bird is soaring in a circular path. Its bank angle (relative to the horizontal) is estimated to be 25° and the bird takes 13 s to complete one circle. (a) How fast is the bird flying? (b) What is the radius of the circle?

67P. A model airplane of mass 0.75 kg is flying at constant speed in a horizontal circle at one end of a 30-m cord and at a height of 18 m. The other end of the cord is tethered to the ground. The airplane circles 4.4 times per minute, and has its wings horizontal so that the air is pushing vertically upward. (a) What is the acceleration of the plane? (b) What is the tension in the cord? (c) What is the total upward force (lift) on the plane's wings?

68P. An old streetcar rounds a corner on unbanked tracks. If the radius of the tracks is 30 ft and the car's speed is 10 mi/h, what angle with the vertical will be made by the loosely hanging hand straps?

69P. Assume that the standard kilogram mass would weigh exactly 9.80 N at sea level on the Earth's equator if the Earth did not rotate. Then take into account the fact that the Earth does rotate, so that this mass moves in a circle of radius 6.40×10^6 m (the Earth's radius) at a constant speed of 465 m/s. (a) Determine the centripetal force needed to keep the standard mass moving in its circular path. (b) Determine the force exerted by the standard mass on a spring balance from which it is suspended at the equator (that force is its "apparent weight").

70P. As shown in Fig. 6-42, a 1.34-kg ball is connected by means of two massless strings to a vertical, rotating rod. The strings are tied to the rod, are taut, and form two sides of an equilateral triangle. The tension in the upper string is 35 N. (a) Draw the free-body diagram for the ball. (b) What is the tension in the lower string? (c) What is the net force on the ball at the instant shown in the figure? (d) What is the speed of the ball?

String length = 1.70 m each

1.70 m

Rotating rod

FIGURE 6-42 Problem 70.

71P. Because of the rotation of the Earth, a plumb bob may not point exactly toward the Earth's center but may deviate slightly from that direction. Calculate the deviation (a) at 40° latitude, (b) at the poles, and (c) at the equator.

ADDITIONAL PROBLEMS

72. A force **P**, parallel to a surface inclined 15° above the horizontal, acts on a 45-N block, as shown in Fig. 6-43. The coefficients of friction for the block and surface are $\mu_s = 0.50$ and $\mu_k = 0.34$. If the block is initially at rest, determine the magnitude and direction of the frictional force acting on the block for magnitudes of **P** of (a) 5.0 N, (b) 8.0 N, and (c) 15 N.

FIGURE 6-43 Problem 72.

73. A force **P** of magnitude 80 N is used to push a 5.0-kg block across the ceiling of a room as shown in Fig. 6-44. If the coefficient of kinetic friction between the block and surface is 0.40, what is the magnitude of the acceleration of the block?

FIGURE 6-44 Problem 73.

74. An amusement park ride consists of a car moving in a vertical circle on the end of a rigid boom of negligible mass. The combined weight of the car and riders is 5.0 kN, and the radius of the circle is 10 m. What are the magnitude and direction of the force of the boom on the car at the top of the circle if the car's speed there is (a) 5.0 m/s and (b) 12 m/s?

75. Two blocks are accelerated across a horizontal surface by a horizontal force applied to one of the blocks as shown in Fig. 6-45. The magnitude of the frictional force on the smaller block is 2.0 N, and the magnitude of the frictional force on the larger block is 4.0 N. If the magnitude of **F** is 12 N, what is the magnitude of the force exerted on the larger block by the smaller block?

FIGURE 6-45 Problem 75.

76. A 2.5-kg block is initially at rest on a horizontal surface. A 6.0-N horizontal force and a vertical force **P** are

applied to the block as shown in Fig. 6-46. The coefficients of friction for the block and surface are $\mu_s = 0.40$ and $\mu_k = 0.25$. Determine the magnitude and direction of the frictional force acting on the block if the magnitude of **P** is (a) 8.0 N, (b) 10 N, and (c) 12 N.

FIGURE 6-46 Problem 76.

77. As a 40-N block slides down a plane that is inclined at 25° to the horizontal, its acceleration is 0.80 m/s², directed up the plane. What is the coefficient of kinetic friction between the block and the plane?

78. A 45-kg skier skis over a frictionless circular hill of radius 15 m and then down the hill to a frictionless circular dip of 25 m, as shown in Fig. 6-47. At the top of the hill the ground exerts an upward force of 320 N on the skier, and at the bottom of the dip the ground exerts an upward force of 1.1 kN on her. What is the speed of the skier (a) at the top of the hill and (b) at the bottom of the dip?

FIGURE 6-47 Problem 78.

79. The three blocks in Fig. 6-48 are released from rest and accelerate at the rate of 1.5 m/s². If $M = 2.0$ kg, what is the magnitude of the frictional force on the block that slides horizontally?

FIGURE 6-48 Problem 79.

WORK AND KINETIC ENERGY | 7

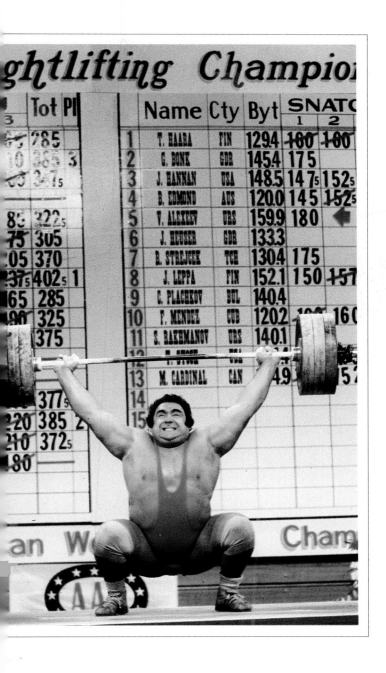

In the weight-lifting competition of the 1976 Olympics, Vasili Alexeev (on the left) astounded the world by lifting a record-breaking 562 lb (2500 N) from the floor to over his head (about 2 m). In 1957 Paul Anderson stooped beneath a reinforced wood platform, placed his hands on a short stool to brace himself, and then pushed upward on the platform with his back, lifting the platform and its load about a centimeter. On the platform were auto parts and a safe filled with lead; the composite weight of the load was 6270 lb (27,900 N)! Who, Alexeev or Anderson, did more work?

7-1 A WALK AROUND NEWTONIAN MECHANICS

Mountaineers are never satisfied with one view of a mountain, no matter how impressive that view may be. They always walk around it to study it from as many angles as possible. In that way they gain new insights and learn how to traverse familiar routes more easily. Sometimes they also see higher peaks, previously hidden from view, with the possibility of new conquests. We shall do the same with Newtonian mechanics.

In this chapter we discuss two new concepts: kinetic energy, which is a property that is associated with the state of motion of a body, and work, which gives rise to changes in kinetic energy. The ideas of work and kinetic energy provide a new angle from which to view Newtonian mechanics; and the rewards for doing so are new insights and new methods for solving some kinds of problems with remarkable ease.

More important, in Chapter 8 work and energy will provide your first glimpse of another peak, the law of conservation of energy. This law is consistent with Newton's laws, but it applies even where they fail—at speeds approaching the speed of light, for instance, and within the atom. It is independent of Newtonian mechanics and, at least so far, has no exceptions. It is, then, an even higher peak, and we begin the climb to its summit right here.

7-2 WORK: MOTION IN ONE DIMENSION WITH A CONSTANT FORCE

Imagine (see Fig. 7-1*a,b*) that, in an intramural bed race, you push on a wheeled bed with a steady horizontal force **F** and that the bed moves through a horizontal displacement **d**. In doing so, you do an amount of **work** W given by

$$W = Fd. \qquad (7-1)$$

Here F is the magnitude of the force that you apply and d is the magnitude of the displacement of the point on the bed where you apply that force. And W is said to be the work done on the bed by you, but more precisely it is the work done on the bed by the force **F** that you exert.

Figure 7-1*c* shows a free-body diagram, with the bed represented as a particle and the displacement

(a)

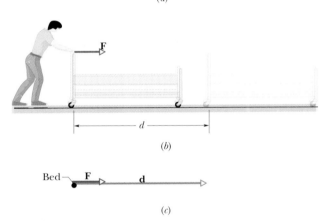

(b)

(c)

FIGURE 7-1 (*a*) A bed race. (*b*) A constant force **F** is exerted on a bed (with wheels), moving the bed through a distance *d*. (*c*) A free-body diagram for the bed showing the force **F**; the displacement **d** is also shown. The angle between **F** and **d** is zero.

also shown. Note that the angle ϕ between the force vector **F** and the displacement vector **d** is zero.

Figure 7-2 is a generalization of Fig. 7-1*c*, in which the angle ϕ between the force vector **F** and the displacement vector **d** is not zero. We define the work W done by the force **F** on the particle in this more general situation to be

$$W = Fd \cos \phi. \qquad (7-2)$$

Note that if $\phi = 0$, $\cos \phi = 1$; Eq. 7-2 then reduces to Eq. 7-1, as it must.

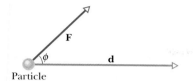

FIGURE 7-2 A constant force **F** acts on a particle that undergoes a displacement **d**. The two vectors make a constant angle ϕ with each other.

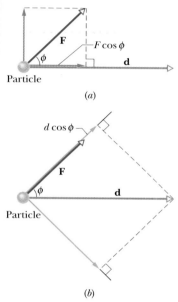

(a)

(b)

FIGURE 7-3 The vectors of Fig. 7-2: (a) the component of **F** in the direction of **d** is $F \cos \phi$; (b) the component of **d** in the direction of **F** is $d \cos \phi$.

We can rewrite Eq. 7-2 as

$$W = (d)(F \cos \phi) = (F)(d \cos \phi), \quad (7\text{-}3)$$

to show that there are two ways to calculate work: multiply the magnitude d of the displacement by the component of force in the direction of the displacement (Fig. 7-3a), *or* multiply the magnitude F of the force by the component of the displacement in the direction of the force (Fig. 7-3b). The two methods always give the same result.

If the force and the displacement are in the same direction, Eq. 7-2 tells us that the work done by the force has a positive value. If they are in opposite directions, then $\phi = 180°$, and Eq. 7-2 (with $\cos 180° = -1$) tells us that the work done by the force has a negative value. And if they are perpendicular, then $\phi = 90°$, and Eq. 7-2 (with $\cos 90° = 0$) tells us that the work done by the force is zero.

Suppose you pick up a cat, carry it across a room at constant speed, and then lower it back to the floor (Fig. 7-4). During the lift, your force **F** on the cat is in the same direction as the cat's upward displacement, and so your force does positive work on the cat. During the walk, the force with which you support the cat is perpendicular to the cat's displacement across the room, and so your force now does zero work on the cat. When you lower the cat, your force on it is upward, opposite its downward displacement, and so your force now does negative work on the cat.

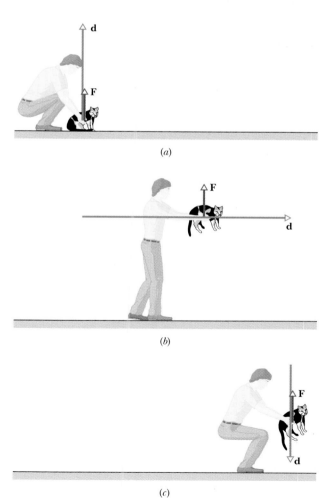

(a)

(b)

(c)

FIGURE 7-4 (a) While lifting a cat, you do positive work on it. (b) While carrying it across a room, you do no work on it. (c) While lowering it, you do negative work on it.

Using a harness, Paul Anderson lifts 30 people with a combined weight of about 2400 lb.

Note that we are not concerned with the common meaning of the word "work." Certainly, supporting and carrying the cat are tiring,* but neither involves work as defined by Eq. 7-2. Moreover, although we may say that *you* do work on the cat in raising and lowering it, what is meant is that the *force* you apply to the cat does the work.

Work is a scalar quantity, although the two quantities involved in its definition, force and displacement, are vectors. We can write Eq. 7-2 more compactly in vector form, as a **scalar** (or **dot**) **product**.† Thus

$$W = \mathbf{F} \cdot \mathbf{d} \qquad \text{(work: constant force).} \qquad (7\text{-}4)$$

This equation is identical to Eq. 7-2.

---

*Supporting an object and carrying an object require repeated contractions of muscles, which are tiring.

†This is our first application of the dot (or scalar) product notation in this text. You may wish to review Section 3-7, where the dot product of two vectors is defined and discussed and Eq. 7-4 is given as an example.

The SI unit of work is the **newton-meter**. The unit is used so often that it has been given a special name, the **joule** (J) after James Prescott Joule, an English scientist of the 1800s. In the British system of units, the unit of work is the **foot-pound** (ft · lb). The relations are

$$1 \text{ joule} = 1 \text{ J} = 1 \text{ N} \cdot \text{m} = 1 \text{ kg} \cdot \text{m}^2/\text{s}^2$$
$$= 0.738 \text{ ft} \cdot \text{lb}. \qquad (7\text{-}5)$$

Appendix F gives conversion factors to other units in which work is measured.

A convenient unit of work when dealing with atoms or with subatomic particles is the **electron-volt** (eV); its common multiples are the kiloelectron-volt (1 keV = 10^3 eV), the megaelectron-volt (1 MeV = 10^6 eV), and the gigaelectron-volt (1 GeV = 10^9 eV). We shall define the electron-volt precisely in Chapter 26, but we give here its relation to the joule; namely,

$$1 \text{ electron-volt} = 1 \text{ eV} = 1.60 \times 10^{-19} \text{ J}. \qquad (7\text{-}6)$$

So far, we have discussed situations where a single force acts on a particle and does work on the particle. Additional forces may also act on the particle, and they can also do work on it. The work done by each force must be calculated separately. To find the *total* work done on the particle, we can add up the separate work contributions of all the forces that act on it. Or we can find the net force, and then use it as the force in Eqs. 7-2 and 7-4.

SAMPLE PROBLEM 7-1

Let us return to the weight-lifting feats of Vasili Alexeev and Paul Anderson.

a. How much work was done by Alexeev (or, more precisely, by his applied force) in lifting a weight of 2500 N a distance of 2.0 m?

SOLUTION We ignore the brief acceleration of the weights during the beginning and end of the lift and assume that they are lifted at constant speed. (To assume otherwise would complicate the calculations but would not alter the result.) The magnitude of the upward force $\mathbf{F}_{VA}$ exerted by Alexeev then matches the weight he lifted:

$$F_{VA} = mg = 2500 \text{ N}.$$

(a)   (b)   (c)   (d)

FIGURE 7-5 Sample Problem 7-1. Free-body diagrams, with displacements also shown. (a) Force $\mathbf{F}_{VA}$ exerted by Vasili Alexeev on the weights he displaced by **d**. (b) Weight $m\mathbf{g}$ of the weights. (c) Both forces acting on the weights. (d) Force $\mathbf{F}_{PA}$ exerted by Paul Anderson on the weights he displaced by $\mathbf{d'}$.

The angle ϕ between the force $\mathbf{F}_{VA}$ and the displacement **d** of the weights is zero (Fig. 7-5a). From Eq. 7-2, the work done by $\mathbf{F}_{VA}$ is

$$W = F_{VA}d \cos \phi$$

$$= (2500 \text{ N})(2.0 \text{ m})(\cos 0°)$$

$$= 5000 \text{ J}. \qquad \text{(Answer)}$$

b. How much work was done on the weights by their weight $m\mathbf{g}$?

SOLUTION The magnitude of the weight is mg. The angle ϕ between the force vector $m\mathbf{g}$ and the displacement **d** is 180° (Fig. 7-5b). From Eq. 7-2, the work done by $m\mathbf{g}$ is

$$W = mgd \cos \phi$$

$$= (2500 \text{ N})(2.0 \text{ m})(\cos 180°)$$

$$= -5000 \text{ J}. \qquad \text{(Answer)}$$

c. How much work was done by the net force on the weights during the lift?

SOLUTION The net force on the weights during the lift is the sum of the forces shown in Fig. 7-5c, which is zero and, by Eq. 7-2, so is the work done by the net force. We get the same result by adding the answers to (a) and (b) to find the sum of the work done by Alexeev and the weight $m\mathbf{g}$.

d. While Alexeev held the weights stationary above his head, how much work did he do?

SOLUTION When he supports the weights, there is no displacement, and according to Eq. 7-2, no work is done.

e. How much work was done by Paul Anderson in lifting a weight of 27,900 N by $d' = 1.0$ cm?

SOLUTION The free-body diagram for the lift is shown in Fig. 7-5d. The magnitude of Anderson's force was $F_{PA} = 27,900$ N; we find the work he did as in (a):

$$W = F_{PA}d' = (27,900 \text{ N})(0.01 \text{ m})$$

$$= 279 \text{ J} \approx 300 \text{ J}. \qquad \text{(Answer)}$$

Anderson's lift required a tremendous upward force but comparatively little work because of the short displacement involved.

SAMPLE PROBLEM 7-2

Figure 7-6a shows two industrial spies sliding a floor safe 8.5 m along a straight line toward their truck. The push $\mathbf{F}_1$ of Spy 001 is 320 N, directed at an angle of 30° downward from the horizontal; the pull $\mathbf{F}_2$ of Spy 002 is 250 N, directed at 40° above the horizontal.

a. What is the total work done on the safe by the spies?

SOLUTION Figure 7-6b is a free-body diagram for the safe, considered to be a particle. We can find the total work done by the spies by finding the work done by each spy and then adding the results. From Eq. 7-2, the

FIGURE 7-6 Sample Problem 7-2. (a) Two spies move a floor safe. (b) A free-body diagram for the safe, with the displacement **d** of the safe included.

(a)

(b)

work done by Spy 001 is

$$W_1 = F_1 d \cos \phi_1 = (320 \text{ N})(8.5 \text{ m})(\cos 30°)$$

$$= 2356 \text{ J},$$

and the work done by Spy 002 is

$$W_2 = F_2 d \cos \phi_2 = (250 \text{ N})(8.5 \text{ m})(\cos 40°)$$

$$= 1628 \text{ J}.$$

So the total work W is

$$W = W_1 + W_2 = 2356 \text{ J} + 1628 \text{ J}$$

$$\approx 4000 \text{ J}. \qquad \text{(Answer)}$$

b. What is the work done on the safe by its weight $m\mathbf{g}$ and by the normal force $\mathbf{N}$ due to the floor?

SOLUTION Both of these forces are perpendicular to the displacement, so they each do no work on the safe.

SAMPLE PROBLEM 7-3

A 15-kg crate is pulled at constant speed a distance $d = 5.7$ m up a frictionless ramp, to a height h of 2.5 m above its starting point; see Fig. 7-7a.

a. What force $\mathbf{F}$ must the cable exert on the crate?

SOLUTION Figure 7-7b shows the free-body diagram for the crate. The crate is in equilibrium (because $a = 0$) so that, when we apply Newton's second law parallel to the ramp, we obtain

$$F = mg \sin \theta = (15 \text{ kg})(9.8 \text{ m/s}^2)(2.5 \text{ m}/5.7 \text{ m})$$

$$= 64.5 \text{ N} \approx 65 \text{ N}. \qquad \text{(Answer)}$$

Note that we have calculated $\sin \theta$ directly from the given values of h and d, with no need to find θ itself.

b. How much work is done on the crate by the force $\mathbf{F}$?

SOLUTION From Eq. 7-2, we have

$$W_F = Fd \cos \phi = (64.5 \text{ N})(5.7 \text{ m})(\cos 0°)$$

$$= 368 \text{ J} \approx 370 \text{ J}. \qquad \text{(Answer)}$$

Do not confuse the angle ϕ (which is the angle between the vectors $\mathbf{F}$ and $\mathbf{d}$ in Fig. 7-7) with the angle θ (which is the angle of the ramp).

c. If we raise the crate by the same height h on another ramp at a different slope θ, how much work is then done by $\mathbf{F}$?

(a)

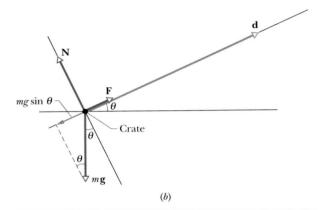

(b)

FIGURE 7-7 Sample Problem 7-3. (a) A crate is pulled up a frictionless ramp by a force parallel to the ramp. (b) A free-body diagram for the crate, showing all the forces that act on it. Its displacement $\mathbf{d}$ is also shown.

SOLUTION In (a), we have $F = mg \sin \theta$. In (b), $W_F = Fd \cos \phi = Fd \cos 0 = Fd$. Combining these equations, we have

$$W_F = mgd \sin \theta.$$

But $d \sin \theta = h$, so

$$W_F = mgh.$$

That is, the work done in raising the crate is independent of the angle of the ramp. If h is 2.5 m, then the work done is

$$W_F = (15 \text{ kg})(9.8 \text{ m/s}^2)(2.5 \text{ m})$$

$$= 368 \text{ J} \approx 370 \text{ J}. \qquad \text{(Answer)}$$

d. How much work would be required to lift the crate vertically upward, through a height h?

SOLUTION A force equal to the weight of the crate would be needed, and the angle ϕ between that force and the displacement would be zero. Thus

$$W_h = Fh \cos \phi = mgh \cos \phi$$

$$= (15 \text{ kg})(9.8 \text{ m/s}^2)(2.5 \text{ m})(\cos 0°)$$

$$= 368 \text{ J} \approx 370 \text{ J}. \qquad \text{(Answer)}$$

This is the same answer that we found in (b) and (c). The difference is that in (b) and (c) we applied smaller forces through larger distances. In other words, we used the ramps to lift the crate by applying a force smaller than the weight of the crate. That is what a ramp is for; it allows us to do the same amount of work with a reduced force.

e. What is the work done by the weight $m\mathbf{g}$ of the crate in (b), (c), and (d)?

SOLUTION Consider the general case of (c) in which the angle θ of the ramp is left as a symbol. From Eq. 7-4,

$$W_g = m\mathbf{g} \cdot \mathbf{d}.$$

From Fig. 7-7b the angle between $m\mathbf{g}$ and $\mathbf{d}$ is $\theta + 90°$, so we have

$$W_g = m\mathbf{g} \cdot \mathbf{d} = mgd \cos(\theta + 90°) = mgd(-\sin \theta).$$

From (c) we know that $d \sin \theta = h$, and so

$$W_g = -mgh = -368 \text{ J} \approx -370 \text{ J}. \qquad \text{(Answer)}$$

The result is the same whether the crate is lifted directly upward or along a ramp with any given angle.

SAMPLE PROBLEM 7-4

In Fig. 7-8a, a cord runs around a massless, frictionless pulley to a block with mass m. The pulley is fixed to the ceiling, and you pull downward on the free end of the cord.

a. What is the magnitude of the force $\mathbf{F}$ that you must exert on the cord to lift the block?

SOLUTION Assuming that the block is lifted at constant speed, the force $\mathbf{T}$ exerted on it by the cord must have magnitude $T = mg$. The force exerted by the cord on your hand has the same magnitude. Thus you must pull downward with a force of magnitude $F = mg$.

(a)

(b)

(c)

FIGURE 7-8 Sample Problem 7-4. (a) You lift a block a distance d by pulling on the free end of a cord that runs around a pulley. (b) You lift the same block the same distance by pulling on a cord that runs around two pulleys. (c) A free-body diagram for the lower pulley in (b).

b. Through what distance must your hand move to lift the block by a distance d?

SOLUTION You must move your hand downward by a distance d to lift the block by the same distance.

c. How much work is done on the block during that lift?

SOLUTION   The work done on the block by the cord is, from Eq. 7-1,

$$W = Td = mgd.$$

The work you do on the free end of the cord is the same:

$$W = Fd = mgd.$$

Thus you are said to do work *on the block by means of the cord.*

d. Another pulley arrangement is shown in Fig. 7-8b. The cord that loops around the lower pulley pulls upward on that pulley with a net force that is twice the tension T_L in that cord. What, now, is the magnitude of the force **F** that you must apply to the cord to lift the block?

SOLUTION   The free-body diagram for the lower pulley is Fig. 7-8c, where T_S is the tension in the short cord attached to the block and T_L is the tension in the long cord you are holding. Because the tension must be the same throughout the long cord, the force you exert has a magnitude equal to T_L. If the block is lifted at constant speed, Newton's second law gives us $2T_L = T_S$; then your force is given by

$$F = T_L = \frac{T_S}{2} = \frac{mg}{2},$$

which is half the force required of you in (a).

e. Through what distance must your hand move to lift the block a distance d?

SOLUTION   Because the long cord is wrapped around the lower pulley, that pulley will move only *half* as far as your hand. So to move the block up by d, you must pull down through a distance $2d$, which is twice as far as in (b).

f. How much work is done on the block in (e)?

SOLUTION   From Eq. 7-1, the work done on the block by the short cord is

$$W = T_S d = mgd.$$

This is the same as the work you do on the long cord,

$$W = F(2d) = \left(\frac{mg}{2}\right)(2d) = mgd.$$

So, again, you may be said to do work *on the block by means of the cord.*

g. What is the advantage of the pulley system in Fig. 7-8b over that in Fig. 7-8a?

SOLUTION   To lift the block by distance d with either system, you must do work equal to mgd. But the force you must exert with the double-pulley system in Fig. 7-8b is half that required with the single pulley in Fig. 7-8a.

SAMPLE PROBLEM 7-5

A runaway crate of prunes slides over a floor toward you. To slow the crate you push against it with a force **F** = (2.0 N)**i** + (− 6.0 N)**j**, while running backward (Fig. 7-9). During your pushing, the crate goes through a displacement **d** = (− 3.0 m)**i**. How much work have you done on the crate?

SOLUTION   From Eq. 7-4, your work is

$$W = \mathbf{F} \cdot \mathbf{d}$$

$$= [(2.0 \text{ N})\mathbf{i} + (− 6.0 \text{ N})\mathbf{j}] \cdot [(− 3.0 \text{ m})\mathbf{i}].$$

Recall that, of the possible unit-vector dot products, only **i**·**i**, **j**·**j**, and **k**·**k** are nonzero (see Section 3-7). Here we have

$$W = (2.0 \text{ N})(− 3.0 \text{ m})\mathbf{i} \cdot \mathbf{i} + (− 6.0 \text{ N})(− 3.0 \text{ m})\mathbf{j} \cdot \mathbf{i}$$

$$= (− 6.0 \text{ J})(1) + 0 = − 6.0 \text{ J}. \qquad \text{(Answer)}$$

PROBLEM SOLVING

TACTIC 1: FINDING THE WORK
In problems in which work is involved, draw a free-body diagram and make sure that you can answer these questions: With what particle am I dealing? What forces act on it? With which of these forces am I concerned? Is it perhaps the *net* force? What is the displacement of

FIGURE 7-9   Sample Problem 7-5. Slowing a runaway prune crate.

the particle? What is the angle between the displacement and each force? What is the sign of the work? Is the sign of the work physically reasonable? (You might want to review Sample Problems 7-1 to 7-5 with these questions in mind.)

7-3 WORK DONE BY A VARIABLE FORCE

One-Dimensional Analysis

Here we assume that the force acting on a particle and the displacement of that particle lie along the same straight line, which we may take as an x axis. We assume further that the magnitude of the force is *not constant* but depends on the particle's position.

Figure 7-10a shows a plot of such a one-dimensional, variable force. What work is done on the particle by this force as the particle moves from an initial point x_i to a final point x_f? We cannot use Eq. 7-2, because it applies only for a constant force **F**. To develop a new approach, let us divide the total displacement of the particle into a number of intervals of width Δx. We choose Δx small enough so that we can take the force $F(x)$ as being reasonably constant over that interval. We let $\overline{F(x)}$ be the average value of $F(x)$ within the interval.

The increment (small amount) of work ΔW done by the force over any particular interval is given by Eq. 7-2, or

$$\Delta W = \overline{F(x)} \, \Delta x. \qquad (7\text{-}7)$$

On the graph of Fig. 7-10b, ΔW is equal in magnitude to the area of the vertical strip to which it refers; $\overline{F(x)}$ is the height of the strip and Δx is its width.

To approximate the total work W done by the force as the particle moves from x_i to x_f, we add the areas of all the strips between x_i and x_f in Fig. 7-10b. That is,

$$W = \sum \Delta W = \sum \overline{F(x)} \, \Delta x. \qquad (7\text{-}8)$$

Equation 7-8 is an approximation because the broken "skyline" formed by the tops of the rectangular strips in Fig. 7-10b only approximates the actual curve in that figure.

We can make the approximation better by reducing the strip width Δx and using more strips, as in Fig. 7-10c. In the limit, we let the strip width approach zero; the number of strips then becomes infinitely large. We then have, as an exact result,

$$W = \lim_{\Delta x \to 0} \sum \overline{F(x)} \, \Delta x. \qquad (7\text{-}9)$$

This limit is exactly what we mean by the integral of the function $F(x)$ between the limits x_i and x_f. Thus Eq. 7-9 becomes

$$W = \int_{x_i}^{x_f} F(x) \, dx \qquad \text{(work: variable force).} \qquad (7\text{-}10)$$

If we know the function $F(x)$, we can substitute it into Eq. 7-10, introduce the proper limits of integration, carry out the integration, and thus find the work. (See Appendix G for a list of common integrals.) Geometrically, the work is equal to the area that lies under the $F(x)$ curve between the limits x_i and x_f, as in Fig. 7-10d.

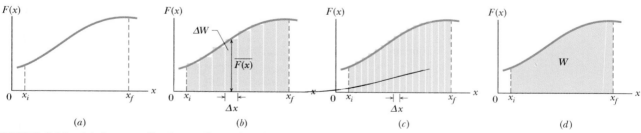

FIGURE 7-10 (*a*) A generalized one-dimensional force plotted against the displacement of a particle on which it is exerted. The particle moves from x_i to x_f. (*b*) The same as (*a*) but with the area under the curve divided into narrow strips. (*c*) The same as (*b*) but with the area divided into narrower strips. (*d*) The limiting case. The work done by the force is given by Eq. 7-10 and is represented geometrically by the shaded area under the curve.

Three-Dimensional Analysis

Consider a particle that is acted on by a three-dimensional force

$$\mathbf{F} = F_x\mathbf{i} + F_y\mathbf{j} + F_z\mathbf{k}, \qquad (7\text{-}11)$$

and let some or all of the components, F_x, F_y, and F_z, depend on the position of the particle. Furthermore, let the particle move through an incremental displacement

$$d\mathbf{r} = dx\mathbf{i} + dy\mathbf{j} + dz\mathbf{k}. \qquad (7\text{-}12)$$

The increment of work dW done on the particle by $\mathbf{F}$ during the displacement $d\mathbf{r}$ is, by Eq. 7-4,

$$dW = \mathbf{F} \cdot d\mathbf{r} = F_x\,dx + F_y\,dy + F_z\,dz. \qquad (7\text{-}13)$$

The work W done by $\mathbf{F}$ while the particle moves from an initial position r_i with coordinates (x_i, y_i, z_i) to a final position r_f with coordinates (x_f, y_f, z_f) is then

$$W = \int_{r_i}^{r_f} dW = \int_{x_i}^{x_f} F_x\,dx + \int_{y_i}^{y_f} F_y\,dy + \int_{z_i}^{z_f} F_z\,dz. \qquad (7\text{-}14)$$

If $\mathbf{F}$ has only an x component, then Eq. 7-14 reduces to Eq. 7-10.

SAMPLE PROBLEM 7-6

How much work is done by a force $\mathbf{F} = (3x\text{ N})\mathbf{i} + (4\text{ N})\mathbf{j}$, with x in meters, that acts on a particle as it moves from coordinates (2 m, 3 m) to (3 m, 0 m)?

SOLUTION From Eq. 7-14 we have

$$W = \int_2^3 3x\,dx + \int_3^0 4\,dy = 3\int_2^3 x\,dx + 4\int_3^0 dy.$$

Using the list of integrals in Appendix G, we obtain

$$W = 3\left[\frac{1}{2}x^2\right]_2^3 + 4\Big[y\Big]_3^0$$

$$= \tfrac{3}{2}[3^2 - 2^2] + 4[0 - 3]$$

$$= -4.5\text{ J} \approx -5\text{ J}. \qquad \text{(Answer)}$$

7-4 WORK DONE BY A SPRING

As an important example of a variable force, we consider the force exerted by a spring. Figure 7-11a shows a spring in its **relaxed state**; that is, it is neither

(a)

(b)

(c)

FIGURE 7-11 (a) A spring in its relaxed state. The origin of the x axis has been placed at the end of the spring. (b) The block is displaced by **d**, and the spring is stretched by an amount x. Note the restoring force **F** exerted by the spring. (c) The spring is compressed by an amount x. Again, note the restoring force.

compressed nor extended. One end is fixed, and a particle-like object, say, a block, is attached to the other, free end. In Fig. 7-11b, we have stretched the spring by pulling the block to the right. In reaction, the spring pulls on the block toward the left, so as to restore the relaxed state. (The spring's force is sometimes said to be a **restoring force**.) In Fig. 7-11c, we have compressed the spring by pushing the block to the left. The spring now pushes on the block toward the right, again so as to restore the relaxed state.

To a good approximation for many springs, the force **F** exerted by the spring is proportional to the displacement **d** of the free end from its position when the spring is in the relaxed state. The spring's force is given by

$$\mathbf{F} = -k\mathbf{d} \qquad \text{(Hooke's law)}, \qquad (7\text{-}15)$$

which is known as **Hooke's law** after Robert Hooke, an English scientist of the late 1600s. The minus sign in Eq. 7-15 indicates that the spring's force is always

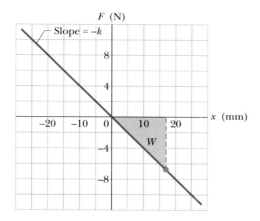

FIGURE 7-12 The force–distance plot for the spring of Sample Problems 7-7 and 7-8. The spring obeys Hooke's law (Eqs. 7-15 and 7-16) and has a spring constant $k = 410$ N/m. See Sample Problem 7-7 for the significance of the dot, and Sample Problem 7-8 for that of the shaded area marked W.

opposite in direction from the displacement of its free end. The constant k is called the **spring constant** and is a measure of the stiffness of the spring. The larger the k, the stiffer the spring; that is, the stronger the spring will pull or push for a given displacement. The SI unit for k is the newton per meter.

In Fig. 7-11 an x axis has been placed parallel to the length of the spring, with the origin ($x = 0$) at the position of the free end when the spring is in its relaxed state. For this common arrangement, Eq. 7-15 becomes

$$F = -kx \qquad \text{(Hooke's law).} \qquad (7\text{-}16)$$

Note that the spring's force is a variable force because it depends on the position of the free end: F can be symbolized as $F(x)$, as in Section 7-3. Hooke's law is a linear relationship; a possible plot of F is that in Fig. 7-12.

If we move the block from an initial position x_i to a final position x_f, we do work on the block and the spring does opposing work on the block. The work W done by the spring on the block is found by substituting F from Eq. 7-16 into Eq. 7-10 and integrating:

$$W = \int_{x_i}^{x_f} F(x) \, dx = \int_{x_i}^{x_f} (-kx) \, dx = -k \int_{x_i}^{x_f} x \, dx.$$

$$= (-\tfrac{1}{2}k) \left[x^2 \right]_{x_i}^{x_f} = (-\tfrac{1}{2}k)(x_f^2 - x_i^2) \qquad (7\text{-}17)$$

or

$$W = \tfrac{1}{2}kx_i^2 - \tfrac{1}{2}kx_f^2 \qquad \text{(work } \textit{by} \text{ a spring).} \qquad (7\text{-}18)$$

The work is positive if $x_i^2 > x_f^2$ and negative if $x_i^2 < x_f^2$. If $x_i = 0$ and if we call the final position x, then Eq. 7-18 becomes

$$W = -\tfrac{1}{2}kx^2. \qquad (7\text{-}19)$$

Equation 7-18 and its special case, Eq. 7-19, give the work done *by* the spring on an object. The work done by us (or by whatever stretches or compresses the spring) is the negative of this quantity.

Note that the length of the spring does not appear explicitly in the expressions for the force exerted by the spring (Eqs. 7-15 and 7-16) and the work done by the spring (Eqs. 7-18 and 7-19). The length of the spring is one of a number of factors that contribute to the value of the spring constant k, others being the spring geometry and the elastic properties of the material of the spring.

SAMPLE PROBLEM 7-7

You apply a 4.9-N force F to a block attached to the free end of a spring, stretching the spring from its relaxed length by 12 mm, as in Fig. 7-11b.

a. What is the spring constant of the spring?

SOLUTION The stretched spring pulls with a force of -4.9 N. From Eq. 7-16, with $x = 12$ mm, we have

$$k = -\frac{F}{x} = -\frac{-4.9 \text{ N}}{12 \times 10^{-3} \text{ m}}$$

$$= 408 \text{ N/m} \approx 410 \text{ N/m.} \qquad \text{(Answer)}$$

Here, x is positive (a stretching), and the force F exerted *by* the spring is negative. Note that we do not need to know the length of the spring. The plot of Eq. 7-16 in Fig. 7-12 refers to this spring. The slope of the plot is -410 N/m.

b. What force does the spring exert if you stretch it by 17 mm?

SOLUTION From Eq. 7-16 we have

$$F = -kx = -(408 \text{ N/m})(17 \times 10^{-3} \text{ m})$$

$$= -6.9 \text{ N.} \qquad \text{(Answer)}$$

The dot on the curve of Fig. 7-12 represents this force and the corresponding displacement. Note that x is positive and F is negative, as in Fig. 7-11b.

SAMPLE PROBLEM 7-8

You stretch the spring of Sample Problem 7-7 by 17 mm from its relaxed length; see Fig. 7-11b. How much work does the spring force do on the block?

SOLUTION   Because the spring is initially in its relaxed state, we can use Eq. 7-19:

$$W = -\tfrac{1}{2}kx^2 = -(\tfrac{1}{2})(408 \text{ N/m})(17 \times 10^{-3} \text{ m})^2$$

$$= -5.9 \times 10^{-2} \text{ J} = -59 \text{ mJ.} \qquad \text{(Answer)}$$

The shaded area in Fig. 7-12 represents this work. The work done by the spring is negative because the displacement of the block and the force exerted by the spring are in opposite directions. Note that the amount of work done by the spring would be the same if it had been compressed (rather than stretched) by 17 mm.

SAMPLE PROBLEM 7-9

The spring of Fig. 7-11b is initially stretched by 17 mm. You allow it to return slowly to its relaxed state and then compress it by 12 mm. How much work does the spring do during the total displacement of the block?

SOLUTION   For this situation, we have $x_i = +17$ mm (a stretching) and $x_f = -12$ mm (a compression). Equation 7-18 becomes

$$W = \tfrac{1}{2}kx_i^2 - \tfrac{1}{2}kx_f^2 = \tfrac{1}{2}k(x_i^2 - x_f^2)$$

$$= (\tfrac{1}{2})(408 \text{ N/m})[(17 \times 10^{-3} \text{ m})^2 - (12 \times 10^{-3} \text{ m})^2]$$

$$= 0.030 \text{ J} = 30 \text{ mJ.} \qquad \text{(Answer)}$$

In this case, the spring did more positive work (in moving from its initial stretched state to its relaxed state) than negative work (in moving farther from its relaxed state to its final compressed state). The net work done by the spring is thus positive.

PROBLEM SOLVING

TACTIC 2: DERIVATIVES AND INTEGRALS; SLOPES AND AREAS

If you know a function $y = F(x)$, you can find its derivative (for any value of x) or its integral (between any two values of x) from the rules of calculus. If you do not know the function analytically but have a plot of it, you can find both its derivative and integral graphically. You saw how to find the derivative graphically in Tactic 9 of Chapter 2. Here you'll see how to find an integral graphically.

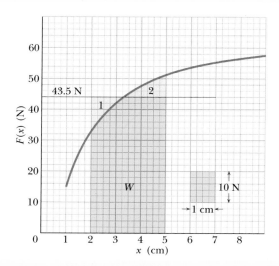

FIGURE 7-13   A generalized plot of $F(x)$. The shaded area (which represents the work) is approximated by a rectangle. The small rectangle at the right serves to calibrate the small squares in work units. Its 20 small squares are equivalent to 10 N·m.

Figure 7-13 is a plot of a particular force function $F(x)$. Let us find graphically the work W done by this force as the particle on which it acts moves from $x_i = 2.0$ cm to $x_f = 5.0$ cm. The work is the shaded area under the curve between these two points.

You can approximate this area with a rectangle formed by drawing a horizontal line across the figure. Draw it at a level such that the two areas marked "1" and "2" appear to be equal. A line at $F = 43.5$ N is about right, and the area of the equivalent rectangle $(= W)$ is then

$$W = \text{height} \times \text{base} = (43.5 \text{ N})(5.0 \text{ cm} - 2.0 \text{ cm})$$

$$= 130 \text{ N·cm} = 1.3 \text{ N·m} = 1.3 \text{ J.}$$

You can also find the area by counting the squares underneath the curve. The rectangle on the right of Fig. 7-13 can be used to calibrate the squares because it shows that 20 squares = 10 N·cm. By counting large blocks of squares where you can, you quickly find that the shaded area contains about 260 squares. The work is then

$$W = (260 \text{ squares})\left(\frac{10 \text{ N·cm}}{20 \text{ squares}}\right) = 130 \text{ N·cm}$$

$$= 1.3 \text{ J,}$$

just as above. *Remember:* on a two-dimensional graph, every derivative is a slope, and every integral is an area.

7-5 KINETIC ENERGY

If you see a hockey puck at rest on the ice and later hurtling toward the goal, you will probably conclude: "Someone hit it with a hockey stick." A physicist might say: "Someone did work on the puck, exerting a force on it over a small distance." Indeed, when we see a moving object, its very motion signals us that work has been done on it to establish that motion. What is there about the motion of a particle that can be related quantitatively to the work that must have been done on the particle?

The property that we are seeking is the **kinetic energy** of the particle, which is defined as

$$K = \tfrac{1}{2}mv^2 \qquad \text{(kinetic energy)}, \qquad (7\text{-}20)$$

where m is the particle's mass and v is its speed. Note that the kinetic energy depends on the square of the speed and thus can never be negative. Its units are the same as those for work (for example, the joule is the SI unit). Kinetic energy is a scalar quantity; it does not depend on the particle's direction of travel. Table 7-1 lists some kinetic energies.

If a single force **F** does work W on a particle, changing the particle's speed, then the kinetic energy of the particle changes from an initial value K_i to a final value K_f. The change in kinetic energy is numerically equal to the work done:

$$W = K_f - K_i = \Delta K, \quad \text{or} \quad K_f = K_i + W \qquad (7\text{-}21)$$
$$\text{(work–kinetic energy theorem)}.$$

Equation 7-21, in either form, is known as the

FIGURE 7-14 (*a*) The weight mg of a ball does positive work when the ball falls freely downward. (*b*) The weight does negative work on a ball moving vertically upward. (*c*) No net work is done on a falling ball that has reached its terminal speed, because the net force acting on the ball is zero, the weight mg being canceled by the drag force **D** exerted by the air.

work–kinetic energy theorem (or sometimes simply as the work–energy theorem). If several forces act on the particle, then the work W in Eq. 7-21 is the *total work done by all the forces* or, equivalently, the work done by the net force acting on the particle.

Equation 7-21 is useful in that it gives us a new way to look at familiar problems and makes the solution of certain kinds of problems much easier than with just Newton's laws of motion. Before proving the work–kinetic energy theorem, let us see how to apply it to some familiar situations.

A Particle in Free Fall

If you drop a baseball (Fig. 7-14a), the force acting on it is its weight mg. This force points in the direction in which the ball moves, so the work done by the weight on the falling ball is positive. According

TABLE 7-1
SOME KINETIC ENERGIES

ITEM	REMARKS	KINETIC ENERGY (J)
Meteor Crater meteor	5×10^{10} kg at 7200 m/s	1.3×10^{18}
Aircraft carrier *Nimitz*	91,400 tons at 30 knots	9.9×10^9
Orbiting satellite	100 kg at 300-km altitude	3.0×10^9
Trailer truck	18-wheeler at 60 mi/h	2.2×10^6
Football linebacker	110 kg at 9 m/s	4.5×10^3
NATO SS 109 bullet	4 g at 950 m/s	1.8×10^3
Pitched baseball	100 mi/h	1.5×10^2
Falling penny	3.2 g after falling 50 m	1.6
Bee in flight	1 g at 2 m/s	2×10^{-3}
Snail	5 g at 0.03 mi/h	4.5×10^{-7}
Electron in TV tube	20 keV	3.2×10^{-15}
Electron in copper	At absolute zero	6.7×10^{-19}

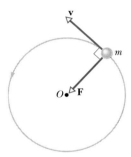

FIGURE 7-15 A particle in uniform circular motion. The centripetal force **F** does no work on the particle because it is always perpendicular to the direction in which the particle is moving.

to Eq. 7-21, the kinetic energy of the falling ball should increase. We know that it does because the speed increases.

If you throw a baseball vertically upward (Fig. 7-14b), during its rise the weight vector is directed opposite the motion, so the work done by this force on the ball is negative. Equation 7-21 tells us that the kinetic energy of the ball should decrease during the ascent. Again, we know that it does, since the speed decreases.

Suppose that a baseball is dropped from a great height and, eventually, falls through air at a constant terminal speed (Fig. 7-14c). Its weight then is exactly balanced by an upward-pointing drag force, and the net force acting on the ball is zero. So the total work being done on the ball is also zero. Equation 7-21 tells us that the kinetic energy of the ball remains constant ($\Delta K = 0$) as the ball falls with terminal speed.

A Particle in Uniform Circular Motion

The speed of a particle in uniform circular motion is constant, so its kinetic energy does not change. Equation 7-21 tells us that the centripetal force acting on the particle does no work on it. We know this to be true from Eqs. 7-2 and 7-4, because that force is always perpendicular to the direction in which the particle is moving (Fig. 7-15).

Proof of the Work–Kinetic Energy Theorem

This theorem, which holds quite generally for a particle moving in three dimensions, is a direct conse-

quence of Newton's second law. We shall prove it, however, only for the special case of a particle moving in one dimension. We assume that the net force acting on that particle may vary in magnitude.

Consider a particle of mass m, moving along the x axis, and acted on by a net force $F(x)$ that points along that axis. The work done on the particle by this force as the particle moves from an initial position x_i to a final position x_f is given by Eq. 7-10 as

$$W = \int_{x_i}^{x_f} F(x) \, dx = \int_{x_i}^{x_f} ma \, dx, \qquad (7\text{-}22)$$

in which we use Newton's second law to replace $F(x)$ by ma. We can recast the quantity $ma \, dx$ in Eq. 7-22 as

$$ma \, dx = m \frac{dv}{dt} dx. \qquad (7\text{-}23)$$

From the "chain rule" of calculus, we have

$$\frac{dv}{dt} = \frac{dv}{dx} \frac{dx}{dt} = \frac{dv}{dx} v, \qquad (7\text{-}24)$$

and Eq. 7-23 becomes

$$ma \, dx = m \frac{dv}{dx} v \, dx = mv \, dv. \qquad (7\text{-}25)$$

Substituting Eq. 7-25 into Eq. 7-22 yields

$$W = \int_{v_i}^{v_f} mv \, dv = m \int_{v_i}^{v_f} v \, dv$$
$$= \tfrac{1}{2}mv_f^2 - \tfrac{1}{2}mv_i^2. \qquad (7\text{-}26)$$

Note that when we change the variable from x to v we are required to express the limits on the integral in terms of the new variable. Note also that, because the mass m is a constant, we were able to move it outside the integral.

Recognizing the terms on the right of Eq. 7-26 as kinetic energies allows us to write this equation as

$$W = K_f - K_i = \Delta K,$$

which is the work–kinetic energy theorem.

SAMPLE PROBLEM 7-10

In 1896 in Waco, Texas, William Crush of the "Katy" railroad parked two locomotives at opposite ends of a 6.4-km track, fired them up, tied their throttles open, and then allowed them to crash head-on at full speed

(Fig. 7-16), in front of 30,000 spectators. Hundreds were hurt by flying debris; several were killed. Assuming the weight of each locomotive was 1.2×10^6 N and its acceleration prior to the collision was a constant 0.26 m/s^2, what was the total kinetic energy of the two locomotives just before the collision?

SOLUTION To find the kinetic energy of one locomotive, we need its speed just before the collision and its mass. To find the speed, we use Eq. 2-14,

$$v^2 = v_0^2 + 2a(x - x_0),$$

which, with $v_0 = 0$ and $x - x_0 = 3.2 \times 10^3$ m (half the initial separation), yields

$$v^2 = 0 + 2(0.26 \text{ m/s}^2)(3.2 \times 10^3 \text{ m}),$$

or

$$v = 40.8 \text{ m/s},$$

(about 90 mi/h). To find the mass of each locomotive, we divide its weight by g;

$$m = \frac{1.2 \times 10^6 \text{ N}}{9.8 \text{ m/s}^2} = 1.22 \times 10^5 \text{ kg}.$$

We now find their total kinetic energy just before the collision with

$$K = 2(\tfrac{1}{2}mv^2) = (1.22 \times 10^5 \text{ kg})(40.8 \text{ m/s})^2$$

$$= 2.0 \times 10^8 \text{ J}. \qquad \text{(Answer)}$$

This is equivalent to a detonation of about 100 lb of TNT. It is no wonder that nearby spectators were injured or killed by flying debris.

SAMPLE PROBLEM 7-11

A 500-kg elevator cab is descending with speed $v_i = 4.0$ m/s when the winch system that lowers it begins to slip,

FIGURE 7-16 Sample Problem 7-10. An 1896 crash of two locomotives.

allowing it to fall with a constant acceleration $\mathbf{a} = \mathbf{g}/5$ (Fig. 7-17a).

a. During its fall through a distance $d = 12$ m, what is the work W_1 done on the cab by its weight $m\mathbf{g}$?

SOLUTION The cab's free-body diagram during the 12-m fall is shown in Fig. 7-17b. Note that the angle between the cab's displacement $\mathbf{d}$ and its weight $m\mathbf{g}$ is $0°$. From Eq. 7-2 we find

$$W_1 = mgd \cos 0° = (500 \text{ kg})(9.8 \text{ m/s}^2)(12 \text{ m})(1)$$

$$= 5.88 \times 10^4 \text{ J} \approx 5.9 \times 10^4 \text{ J}. \qquad \text{(Answer)}$$

b. During the fall, what is the work W_2 done on the cab by the upward pull $\mathbf{T}$ exerted by the elevator cable?

SOLUTION To find W_2 with Eq. 7-2, we need $\mathbf{T}$ in Fig. 7-17b. Applying Newton's second law to the cab yields

$$\sum F = T - mg = ma,$$

or

$$T = m(g + a) = m(g - g/5)$$

$$= (500 \text{ kg})(\tfrac{4}{5})(9.8 \text{ m/s}^2)$$

$$= 3920 \text{ N}.$$

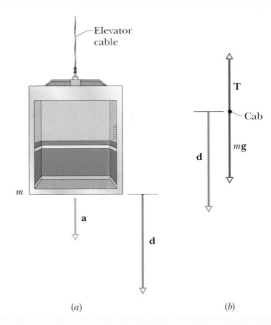

(a)   (b)

FIGURE 7-17 Sample Problem 7-11. An elevator cab, descending with speed v_i, suddenly begins to accelerate downward. (a) It moves through a displacement $\mathbf{d}$ with a constant acceleration $\mathbf{a} = \mathbf{g}/5$. (b) A free-body diagram for the cab, with the displacement included.

The angle between **T** and the cab's displacement **d** is 180°. We can now use Eq. 7-2 to find the work done by the pull **T**:

$$W_2 = Td \cos 180° = (3920 \text{ N})(12 \text{ m})(-1)$$

$$= -4.7 \times 10^4 \text{ J}. \qquad \text{(Answer)}$$

c. What is the total work W done on the cab during the 12-m fall?

SOLUTION The total work is the sum of the answers to (a) and (b):

$$W = W_1 + W_2 = 5.88 \times 10^4 \text{ J} - 4.7 \times 10^4 \text{ J}$$

$$= 1.18 \times 10^4 \text{ J} \approx 1.2 \times 10^4 \text{ J}. \qquad \text{(Answer)}$$

We can also find W a different way. We first find the net force on the cab, using Newton's second law:

$$\sum F = ma = (500 \text{ kg})\left(-\frac{9.8 \text{ m/s}^2}{5}\right) = -980 \text{ N}.$$

We next find the work done on the cab by that net force, which acts downward and thus at an angle of 0° with **d**:

$$W = (980 \text{ N})(12 \text{ m})\cos 0°$$

$$= 1.18 \times 10^4 \text{ J} \approx 1.2 \times 10^4 \text{ J}. \quad \text{(Answer)}$$

d. What is the kinetic energy of the cab at the end of the 12-m fall?

SOLUTION The kinetic energy K_i at the start of the fall, when the speed is $v_i = 4.0$ m/s, is given by

$$K_i = \tfrac{1}{2}mv_i^2 = \tfrac{1}{2}(500 \text{ kg})(4.0 \text{ m/s})^2 = 4000 \text{ J}.$$

The kinetic energy K_f at the end of the fall is given by Eq. 7-21,

$$K_f = K_i + W = 4000 \text{ J} + 1.18 \times 10^4 \text{ J}$$

$$= 1.58 \times 10^4 \text{ J} \approx 1.6 \times 10^4 \text{ J}. \qquad \text{(Answer)}$$

e. What is the speed v_f of the cab at the end of the 12-m fall?

SOLUTION From Eq. 7-20, we have

$$K_f = \tfrac{1}{2}mv_f^2,$$

which we solve for v_f:

$$v_f = \sqrt{\frac{2K_f}{m}} = \sqrt{\frac{(2)(1.58 \times 10^4 \text{ J})}{500 \text{ kg}}}$$

$$= 7.9 \text{ m/s}. \qquad \text{(Answer)}$$

FIGURE 7-18 Sample Problem 7-12. A block moves toward a spring. It will compress the spring momentarily by a maximum distance d.

SAMPLE PROBLEM 7-12

A block whose mass m is 5.7 kg slides on a horizontal frictionless tabletop with a constant speed v of 1.2 m/s. It is brought momentarily to rest by compressing a spring in its path (Fig. 7-18). By what maximum distance d is the spring compressed? The spring constant k is 1500 N/m.

SOLUTION From Eq. 7-19, the work done *by* the spring force *on* the block as the spring is compressed a distance d from its rest state is given by

$$W = -\tfrac{1}{2}kd^2.$$

The change in the kinetic energy of the block as it is stopped is

$$\Delta K = K_f - K_i = 0 - \tfrac{1}{2}mv^2.$$

The work–kinetic energy theorem (Eq. 7-21) requires that these two quantities be equal. Setting them so and solving for d, we obtain

$$d = v\sqrt{\frac{m}{k}} = (1.2 \text{ m/s})\sqrt{\frac{5.7 \text{ kg}}{1500 \text{ N/m}}}$$

$$= 7.4 \times 10^{-2} \text{ m} = 7.4 \text{ cm}. \qquad \text{(Answer)}$$

7-6 POWER

A contractor wishes to lift a load of bricks from the sidewalk to the top of a building. It is easy to calculate how much work is required to do this. The contractor, however, is much more interested in the rate at which the winch can do this work. Will the job take 5 min (acceptable) or 1 week (unacceptable)?

People who are considering buying an outboard motor are not concerned with how much work the motor can do. They want to know how rapidly it can do the work, for that determines how fast the boat can move.

No doubt you can think of many more situations that involve the rate at which work is done. That rate is called **power,** which is a scalar quantity. If an amount of work W is done in an amount of time Δt, the **average power** for that amount of time is defined to be

$$\overline{P} = \frac{W}{\Delta t} \qquad \text{(average power)}. \qquad (7\text{-}27)$$

The **instantaneous power** P is the instantaneous rate of doing work, which we can write as

$$P = \frac{dW}{dt} \qquad \text{(instantaneous power)}. \qquad (7\text{-}28)$$

The SI unit of power is the joule per second. This unit is used so often that it has a special name, the **watt** (W), after James Watt, who greatly improved the rate at which steam engines could do work. In the British system, the unit of power is the foot-pound per second. Often the horsepower is used. Some relations among these units are

$$1 \text{ watt} = 1 \text{ W} = 1 \text{ J/s} = 0.738 \text{ ft}\cdot\text{lb/s} \qquad (7\text{-}29)$$

and

$$1 \text{ horsepower} = 1 \text{ hp} = 550 \text{ ft}\cdot\text{lb/s}$$
$$= 746 \text{ W}. \qquad (7\text{-}30)$$

Inspection of Eq. 7-27 shows that work can be expressed as power multiplied by time, such as the common unit, the kilowatt-hour. Thus

$$1 \text{ kilowatt-hour} = 1 \text{ kW}\cdot\text{h} = (10^3 \text{ W})(3600 \text{ s})$$
$$= 3.60 \times 10^6 \text{ J}$$
$$= 3.60 \text{ MJ}. \qquad (7\text{-}31)$$

Perhaps because of our utility bills, the watt and the kilowatt-hour have become identified as electrical units. They can be used equally well as units for other examples of power and energy. Thus if you pick up this book from the floor and put it on a tabletop, you are free to report the work that you have done as $4 \times 10^{-6} \text{ kW}\cdot\text{h}$ (or more conveniently as $4 \text{ mW}\cdot\text{h}$).

We can also express the rate at which a force does work on a body in terms of that force and the body's velocity. For a particle moving in one dimen-

sion and acted on by a constant force $\mathbf{F}$, Eq. 7-28 becomes

$$P = \frac{dW}{dt} = \frac{F\,dx}{dt} = F\left(\frac{dx}{dt}\right).$$

We can write this more simply as

$$P = Fv, \qquad (7\text{-}32)$$

in which P is the instantaneous power and v is the particle's speed. In the more general case of motion in two or three dimensions, we can extend Eq. 7-32 to

$$P = \mathbf{F}\cdot\mathbf{v} \qquad \text{(instantaneous power)}. \qquad (7\text{-}33)$$

As an example, the truck in Fig. 7-19 is exerting a force $\mathbf{F}$ on its load, which has a velocity $\mathbf{v}$. The power of the truck at any instant is given by Eqs. 7-32 and 7-33.

SAMPLE PROBLEM 7-13

A load of bricks whose mass m is 420 kg is to be lifted by a winch a height h of 120 m in 5.0 min. What must be the minimum power of the winch motor?

SOLUTION   The work to be done is

$$W = mgh = (420 \text{ kg})(9.8 \text{ m/s}^2)(120 \text{ m})$$
$$= 4.94 \times 10^5 \text{ J}.$$

FIGURE 7-19   A truck pulls a heavy load, providing power.

From Eq. 7-27, the average power is

$$\overline{P} = \frac{W}{\Delta t} = \frac{4.94 \times 10^5 \, \text{J}}{5 \times 60 \, \text{s}}$$

$$= 1650 \, \text{W} = 1.65 \, \text{kW} = 2.2 \, \text{hp}. \quad \text{(Answer)}$$

The actual power must be greater than this because friction and other retarding forces work against the lift. We have assumed that the weights of the platform (which holds the bricks) and the cable are negligible compared with the weight of the bricks, and that the load will be lifted without appreciable acceleration.

SAMPLE PROBLEM 7-14

An 80-hp outboard motor, operating at full speed, can drive a speedboat at 22 knots (= 25 mi/h = 11 m/s). What is the forward thrust (force) of the motor?

SOLUTION From Eq. 7-32 we have

$$F = \frac{P}{v} = \frac{(80 \, \text{hp})(746 \, \text{W/hp})}{11 \, \text{m/s}}$$

$$= 5400 \, \text{N} \, (= 1200 \, \text{lb}). \quad \text{(Answer)}$$

Note that since the speed is constant, the thrust of the motor must be balanced by the drag force from the water.

7-7 KINETIC ENERGY AT HIGH SPEEDS (OPTIONAL)

For particles moving at speeds near the speed of light, Newtonian mechanics fails and must be replaced by Einstein's special theory of relativity.* One consequence is that we can no longer use the expression $K = \frac{1}{2}mv^2$ for the kinetic energy of a particle. Instead, we must use

$$K = mc^2\left(\frac{1}{\sqrt{1 - (v/c)^2}} - 1\right), \quad (7\text{-}34)$$

in which c is the speed of light.†

---

*Our first encounter with this fact was in Section 4-10 (Relative Motion at High Speeds).

†In relativistic equations as we present them throughout this book, the mass m will always be understood to be the mass measured when the particle is at rest or essentially at rest.

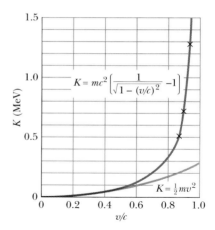

FIGURE 7-20 The relativistic (Eq. 7-34) and the classical (Eq. 7-20) expressions for the kinetic energy of an electron, plotted as a function of v/c, where v is the speed of the electron and c is the speed of light. Note that at low speeds, the two curves blend together, and at high speeds, they diverge widely. The crosses mark experimentally determined points, showing that, at high speeds, the relativistic curve agrees with experiment but the classical curve does not.

Figure 7-20 shows that these two formulas, which look so different, do indeed give widely different results at high speeds. Experiment shows that—beyond any doubt—the relativistic expression (Eq. 7-34) is correct and the classical expression (Eq. 7-20) is not. At low speeds, however, the two formulas merge, yielding the same result. In particular, both formulas yield $K = 0$ for $v = 0$.

All relativistic formulas must reduce to their classical counterparts at low speeds. To see how this comes about for Eq. 7-34, let us write that equation in the form

$$K = mc^2[(1 - \beta^2)^{-1/2} - 1]. \quad (7\text{-}35)$$

Here, for convenience, we have substituted the dimensionless *speed parameter* β for the speed ratio v/c.

At very low speeds, $v \ll c$ and therefore $\beta \ll 1$. At low speeds, we can then expand the quantity $(1 - \beta^2)^{-1/2}$ by the binomial theorem, obtaining (see Tactic 3)

$$(1 - \beta^2)^{-1/2} = 1 + \tfrac{1}{2}\beta^2 + \cdots. \quad (7\text{-}36)$$

Substituting Eq. 7-36 into Eq. 7-35 leads to

$$K = mc^2[(1 + \tfrac{1}{2}\beta^2 + \cdots) - 1]. \quad (7\text{-}37)$$

For very small β, the terms represented by the dots in Eq. 7-37 decrease rapidly in size. We can then, with little error, replace the sum with its first two terms, obtaining

$$K \approx (mc^2)[(1 + \tfrac{1}{2}\beta^2) - 1]$$

or, with $\beta = v/c$,

$$K \approx (mc^2)(\tfrac{1}{2}\beta^2) = \tfrac{1}{2}mv^2,$$

which is exactly what we set out to prove.

PROBLEM SOLVING

TACTIC 3: APPROXIMATIONS

Often we have a quantity written in the form $(a + b)^n$ and we want to find its approximate value for the case in which $b \ll a$. It is simplest to recast it in the form $(1 + x)^n$, where x is dimensionless and is much less than unity. Thus we can put

$$(a + b)^n = a^n(1 + b/a)^n = (a^n)(1 + x)^n.$$

This is of the desired form, with x ($= b/a$) being dimensionless. We can then evaluate $(1 + x)^n$ with the binomial theorem, keeping only as many terms as are appropriate to the problem. (That decision takes some experience.)

The binomial theorem (given in Appendix G) can be written

$$(1 + x)^n = 1 + \frac{n}{1!}x + \frac{n(n-1)}{2!}x^2 + \cdots. \quad (7\text{-}38)$$

In applying Eq. 7-38 in connection with Eq. 7-36, note that $x = -\beta^2$ and $n = -\tfrac{1}{2}$. The exclamation marks in Eq. 7-38 identify factorials, or products of all the whole numbers from the given number down to 1. As an example, $4! = 4 \times 3 \times 2 \times 1 = 24$; you probably have a key for factorials on your calculator.

As an exercise, evaluate $(1 + 0.045)^{-2.3}$ both on your calculator and by expansion, using Eq. 7-38 with $x = 0.045$ and $n = -2.3$. Check the various terms in the binomial sum to see how rapidly they decrease.

7-8 REFERENCE FRAMES

We apply Newton's laws of mechanics only in inertial reference frames. Recall that these frames move with respect to each other at constant velocity.

For some physical quantities, observers in different inertial reference frames will measure the exact same values. In Newtonian mechanics, these *invariant* quantities (as they are called) are force, mass, acceleration, and time. As an example, if an observer in one inertial frame finds that a certain particle has a mass of 3.15 kg, observers in all other inertial frames will measure the same mass for the particle. For other physical quantities, such as the displacement and velocity of a particle, observers in different inertial frames will measure different values; these quantities are *not* invariant.

If the displacement of a particle depends on the reference frame of the observer, so must the work done on the particle because work ($W = \mathbf{F} \cdot \mathbf{d}$) is defined in terms of displacement. If the displacement of a particle during a given interval is measured to be $+2.47$ m in one reference frame, it might be measured to be zero in another and -3.64 m in a third. Because the force $\mathbf{F}$ does not change (it is invariant), work that is positive in one reference frame can be zero in another and negative in a third.

What about the kinetic energy of the particle? If its velocity depends on the reference frame of the observer, then so must its kinetic energy because kinetic energy ($K = \tfrac{1}{2}mv^2$) is defined in terms of velocity. Does that mean that the work–kinetic energy theorem is in trouble?

From Galileo to Einstein, physicists have come to believe in the **principle of invariance:**

The laws of physics must have the same form in all inertial reference frames.

That is, even though some *physical quantities* have different values in different reference frames, the *laws of physics* must hold true in all frames. Behind this formal statement of invariance is a feeling that —in some deep sense—if different observers look at the same event, they must see nature working in the same way.

Among the laws to which this invariance principle applies is the work–kinetic energy theorem. Thus even though different observers watching the same moving particle might measure different values for work and for kinetic energy, they would all find that the work–kinetic energy theorem holds in their respective frames. Let us look at a simple example.

In Fig. 7-21, Sally is riding up in an elevator cab at constant speed and holding a book. Steve, stand-

FIGURE 7-21 Sally rides in an elevator, holding a book. Steve watches her. Both check the work–kinetic energy theorem in their respective reference frames, as it applies to the motion of the book.

ing on a facing balcony, observes her as the cab rises through a height h. What are the work–kinetic energy relations that apply to the book as viewed from these two reference frames?

1. *Sally's report.* "My reference frame is the elevator cab. I am exerting an upward force on the book but this force does no work because the book is not moving in my reference frame. The weight of the book, acting downward, does no work, for the same reason. Thus the total work done on the book by all the forces that act on it is zero. According to the work–kinetic energy theorem, the kinetic energy of the book should not change. That is what I observe; the kinetic energy of the book is zero in my reference frame and remains so. Everything fits together."

2. *Steve's report.* "My reference frame is the balcony. I see that Sally exerts a force $\mathbf{F}$ on her book. In my frame, the point of application of $\mathbf{F}$ is moving and the work $\mathbf{F}$ does as the book moves upward through a height h is $+ mgh$. I also know that the book's weight vector does work on the book, in the amount $- mgh$. Thus the total work done on the book during its rise is zero. According to the work–kinetic energy theorem, the kinetic energy of the book should not change. That is what I observe. In my frame, the kinetic energy is $\frac{1}{2}mv^2$ and remains so. Everything fits together."

Although Steve and Sally do not agree about the displacement of the book and its kinetic energy, they both agree that the work–kinetic energy theorem holds in their respective reference frames.

It does not matter what (inertial) reference frame you pick in which to solve a problem, provided that you (1) make sure you know what that frame is and (2) use that frame consistently throughout the problem.

REVIEW & SUMMARY

Work Done by a Constant Force

When a constant force $\mathbf{F}$ acts on a particle-like object while the object moves through a displacement $\mathbf{d}$, the force is said to do **work** W on the object. When $\mathbf{F}$ and $\mathbf{d}$ are along a single axis,

$$W = Fd. \tag{7-1}$$

When $\mathbf{F}$ and $\mathbf{d}$ are at a constant angle to each other,

$$W = \mathbf{F} \cdot \mathbf{d}. \tag{7-4}$$

This **scalar** (or **dot**) **product** can be evaluated as

$$\mathbf{F} \cdot \mathbf{d} = Fd \cos \phi, \tag{7-2}$$

where ϕ is the angle between $\mathbf{F}$ and $\mathbf{d}$. When more than one force acts on the object, the force in Eqs. 7-1, 7-2, and 7-4 is the **net force** on the object.

Work and Energy Units

The SI unit for work and energy is the **joule** (J), and the British unit is the **foot-pound** (ft · lb):

$$1 \text{ J} = 1 \text{ N} \cdot \text{m} = 1 \text{ kg} \cdot \text{m}^2/\text{s}^2$$
$$= 0.738 \text{ ft} \cdot \text{lb}. \tag{7-5}$$

The **electron-volt** (eV) is commonly used in atomic and nuclear physics:

$$1 \text{ eV} = 1.60 \times 10^{-19} \text{ J}. \tag{7-6}$$

The **kilowatt-hour** (kW · h) is commonly used in engineering:

$$1 \text{ kW} \cdot \text{h} = 3.6 \times 10^6 \text{ J}. \tag{7-31}$$

Work Done by a Variable Force

When the force $\mathbf{F}$ on a particle-like object depends on the position of the object, the work done by $\mathbf{F}$ on the object while the object moves from an initial position r_i with coordinates (x_i, y_i, z_i) to a final position r_f with coordinates (x_f, y_f, z_f) is

$$W = \int_{x_i}^{x_f} F_x \, dx + \int_{y_i}^{y_f} F_y \, dy + \int_{z_i}^{z_f} F_z \, dz. \quad (7\text{-}14)$$

If $\mathbf{F}$ has only an x component, then Eq. 7-14 reduces to

$$W = \int_{x_i}^{x_f} F \, dx. \quad (7\text{-}10)$$

Springs
The force $\mathbf{F}$ exerted by a spring is

$$\mathbf{F} = -k\mathbf{d} \qquad \text{(Hooke's law)}, \quad (7\text{-}15)$$

where $\mathbf{d}$ is the displacement of its free end from the position when the spring is in its **relaxed state** (neither compressed or extended), and k is the **spring constant** (a measure of the spring's stiffness). If an x axis is positioned with the origin at the location of the free end when the spring is in its relaxed state, Eq. 7-15 can be written as

$$F = -kx \qquad \text{(Hooke's law)}. \quad (7\text{-}16)$$

Work Done by a Spring
If an object is attached to the spring's free end, the work W done on the object by the spring when the object is moved from an initial position x_i to a final position x_f is

$$W = \tfrac{1}{2}kx_i^2 - \tfrac{1}{2}kx_f^2. \quad (7\text{-}18)$$

If $x_i = 0$ and $x_f = x$, then Eq. 7-18 becomes

$$W = -\tfrac{1}{2}kx^2. \quad (7\text{-}19)$$

Kinetic Energy
Kinetic energy is a scalar property that is associated with the state of motion of an object and is defined to be

$$K = \tfrac{1}{2}mv^2. \quad (7\text{-}20)$$

The units of kinetic energy are the same as those for work.

Work – Kinetic Energy Theorem
We can rewrite Newton's second law, $\mathbf{F} = m\mathbf{a}$, to relate the net work W done on a body and the change ΔK in the body's kinetic energy:

$$W = K_f - K_i = \Delta K, \quad \text{or} \quad K_f = K_i + W, \quad (7\text{-}21)$$

where K_i is the body's initial kinetic energy and K_f is its final value. Equation 7-21 (in either form) is known as the **work–kinetic energy theorem**.

Power
Power is the *rate* at which work is done. If a force does work W during a time interval Δt, the average **power** is

$$\overline{P} = \frac{W}{\Delta t}. \quad (7\text{-}27)$$

Instantaneous power is the instantaneous rate of doing work:

$$P = \frac{dW}{dt}. \quad (7\text{-}28)$$

If a force $\mathbf{F}$ acts on an object moving along a straight line with velocity $\mathbf{v}$, the instantaneous power is

$$P = \mathbf{F} \cdot \mathbf{v}. \quad (7\text{-}33)$$

Like work, power is a scalar. Its SI unit is the watt (W), and its British units are the foot-pound per second (ft·lb/s) and the horsepower (hp):

$$1 \text{ W} = 1 \text{ J/s} = 0.738 \text{ ft·lb/s}, \quad (7\text{-}29)$$

$$1 \text{ hp} = 550 \text{ ft·lb/s} = 746 \text{ W}. \quad (7\text{-}30)$$

Relativistic Kinetic Energy
The kinetic energy for objects moving at speeds near the speed of light c must be calculated using the relativistic expression

$$K = mc^2 \left(\frac{1}{\sqrt{1 - (v/c)^2}} - 1 \right). \quad (7\text{-}34)$$

This equation reduces to Eq. 7-20 when v is much less than c.

The Principle of Invariance
Some quantities (such as mass, force, acceleration, and time in Newtonian mechanics) are *invariant;* that is, they have the same numerical values when measured in different inertial reference frames. Others (for example, velocity, kinetic energy, and work) have different values in different frames. However, the *laws* of physics have the same *form* in all inertial reference frames. This is called the **principle of invariance**.

QUESTIONS

1. What other words are like "work" in that their colloquial meanings differ from their scientific meanings?

2. Why is it tiring to hold a heavy weight even though no work is done?

3. The inclined plane (Sample Problem 7-3) is a simple "machine" that enables us to do work with the application of a smaller force than is otherwise necessary. The same statement applies to a wedge, a lever, a screw, a gear wheel, and a set of pulleys (such as in Sample Problem 7-4). Far from saving us work, such machines in practice require that we do a little more work with them than without them. Why is this so? Why do we use such machines?

4. In a tug-of-war, one team is slowly giving way to the other. Is work being done on the losing team? How about the winning team?

5. Give a situation in which positive work is done by a static frictional force.

6. Suppose that the Earth orbits the sun in a perfect circle. Does the sun do any work on the Earth?

7. If you slowly lift a bowling ball from the floor, two forces act on the ball: its weight $m\mathbf{g}$ and your upward force $\mathbf{F} = -m\mathbf{g}$. These two forces cancel each other so that it would seem that no work is done. On the other hand, you know that you have done some work. What is wrong?

8. Figure 7-22 shows six situations in which two forces act simultaneously on a box *after* the box has been sent sliding over a frictionless surface, either to the left or right. The forces are either 1 N or 2 N in magnitude, as indicated by the vector lengths in the figure. For each situation, determine whether the work done on the box by the net force during the indicated displacement **d** is positive, negative, or zero.

FIGURE 7-22   Question 8.

9. In overhead views, Fig. 7-23 shows three situations in which a box is acted on by two forces of equal magnitude. As the box moves, the forces maintain their orientation relative to the velocity **v**. For each situation, determine whether the work done on the box by the net force during the motion is positive, negative, or zero.

10. An ant must carry a food morsel to the top of a cone (Fig. 7-24). Compare the work it does on the food if it follows a spiraling path around the cone to the work it does by climbing directly up the side of the cone.

FIGURE 7-23   Question 9.

FIGURE 7-24   Question 10.

11. A greased pig has a choice of three frictionless slides (Fig. 7-25) along which to slide to the ground. Compare the slides according to how much work the pig's weight $m\mathbf{g}$ does on the pig during the descent.

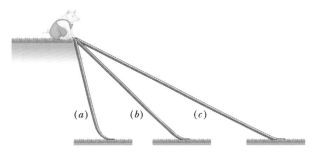

FIGURE 7-25   Question 11.

12. You cut a spring in half. What is the relation of the spring constant k for the original spring to that for either of the resulting half-springs? (*Hint:* Consider the amount by which each coil in a spring stretches for a given value of force.)

13. Springs A and B are identical except that A is stiffer than B; that is, $k_A > k_B$. Which spring does more work if they are stretched (a) by the same amount and (b) by the same force?

14. In picking up a book from the floor and putting it on a table, you do work. However, the initial and final values of the book's kinetic energy are zero. Is there a violation of the work–kinetic energy theorem here? Explain why or why not.

15. You throw a ball vertically into the air and catch it when it returns. What happens to the ball's kinetic energy during the flight? First ignore air resistance, and then take it into account.

16. Does the power needed to raise a box onto a platform depend on how fast it is raised?

17. You lift some library books from a lower to a higher shelf in time Δt. Does the work that you do depend on (a) the mass of the books, (b) the weight of the books, (c) the height of the upper shelf above the floor, (d) the time Δt, and (e) whether you move the books sideways or directly upward?

18. We have heard a lot about the "energy crisis." Would it be more accurate to speak of a "power crisis"?

19. We say that a 1-keV electron (meaning its kinetic energy is 1 keV) is a "classical" particle, a 1-MeV electron is a "relativistic" particle, and a 1-GeV electron is an "extremely relativistic" particle. What do these terms mean?

20. The displacement of a body depends on the reference frame of the observer who measures it. It follows that the work done on a body should also depend on the observer's reference frame. Suppose you drag a crate across a rough floor by pulling on it with a rope. Identify reference frames in which the work done on the crate by the rope would be (a) positive, (b) zero, and (c) negative.

21. Sally and Yuri are flying jet planes at the same speed on parallel low-altitude paths. Suddenly, Sally lowers her flaps and slows to a new speed. Consider how this looks from the reference frame of Yuri, who keeps on flying at the original speed. (a) Would he say that Sally's plane had gained or lost kinetic energy? (b) Would he say that the work done *on* her plane is positive or negative? (c) Would he conclude that the work–kinetic energy theorem holds? (d) Answer these same questions from the reference frame of Chang, who watches from the ground.

EXERCISES & PROBLEMS

SECTION 7-2 WORK: MOTION IN ONE DIMENSION WITH A CONSTANT FORCE

1E. (a) In 1975 the roof of Montreal's Velodrome, with a weight of 41,000 tons, was lifted by 4.0 in. so that it could be centered. How much work was done by the machines making the lift? (b) In 1960 Mrs. Maxwell Rogers of Tampa, Florida, reportedly raised one end of a car that weighed 3600 lb. The car had fallen onto her son when a jack failed. If her panic lift effectively raised 900 lb of car by 2 in., how much work did she do?

2E. To push a 50-kg crate across a frictionless floor, a worker applies a force of 210 N, directed 20° above the horizontal. As the crate moves 3.0 m, what work is done on the crate by (a) the worker, (b) the weight of the crate, and (c) the normal force exerted by the floor on the crate? (d) What is the total work done on the crate?

3E. To push a 25.0-kg crate up a frictionless incline, angled at 25° to the horizontal, a worker exerts a force of 209 N, parallel to the incline. As the crate slides 1.5 m, what work is done on the crate by (a) the worker, (b) the weight of the crate, and (c) the normal force exerted by the incline on the crate? (d) What is the total work done on the crate?

4E. A 102-kg object is initially moving in a straight line with a speed of 53 m/s. If it is stopped with a deceleration of 2.0 m/s², (a) what magnitude of force is required, (b) what distance does it travel while decelerating, and (c) what work is done by the decelerating force? (d) Answer questions (a)–(c) for a deceleration of 4.0 m/s².

5E. A 45-kg block of ice slides down a frictionless incline 1.5 m long and 0.91 m high. A worker pushes up on the ice parallel to the incline so that it slides down at constant speed. Find (a) the force exerted by the worker. Find the work done on the block by (b) the worker, (c) the weight of the block, (d) the normal force exerted by the surface of the incline on the block, and (e) the resultant (or net) force on the block.

6E. A floating ice block is pushed through a displacement $\mathbf{d} = (15 \text{ m})\mathbf{i} - (12 \text{ m})\mathbf{j}$ along a straight embankment by rushing water, which exerts a force $\mathbf{F} = (210 \text{ N})\mathbf{i} - (150 \text{ N})\mathbf{j}$ on the block. How much work does the water do on the block during the displacement?

7E. A particle moves along a straight path through a displacement $\mathbf{d} = (8 \text{ m})\mathbf{i} + c\mathbf{j}$ while a force of $\mathbf{F} = (2 \text{ N})\mathbf{i} - (4 \text{ N})\mathbf{j}$ acts on it. (Other forces also act on the particle.) What is the value of c if the work done by $\mathbf{F}$ on the particle is (a) zero, (b) positive, and (c) negative?

8E. In Fig. 7-26, a cord runs around two massless, frictionless pulleys; a canister with mass $m = 20$ kg hangs from one pulley; and you exert a force $\mathbf{F}$ on the free end of the cord. (a) What must be the magnitude of $\mathbf{F}$ if you are to lift the canister at a constant speed? (b) To lift the canister by 2.0 cm, how far must you pull the free end of the cord? During that lift, what is the work done on the canister by (c) you and (d) the weight $m\mathbf{g}$ of the canister?

FIGURE 7-26 Exercise 8.

9P. Figure 7-27 shows an arrangement of pulleys designed to facilitate the lifting of a heavy load *L*. Assume that friction can be ignored everywhere and that the lower two pulleys to which the load is attached weigh a total of 20.0 lb. An 840-lb load is to be raised 12.0 ft. (a) What is the minimum applied force *F* that can lift the load? (b) How much work must be done to lift the load 12.0 ft? (c) Through what distance must the free end of the rope be moved to lift the load that far? (d) How much work must be done by *F* to accomplish this task?

FIGURE 7-27 Problem 9.

10P. A worker pushed a 27-kg block 9.2 m along a level floor at constant speed with a force directed 32° below the horizontal. If the coefficient of kinetic friction is 0.20, how much work did the worker do on the block?

11P. A 50-kg trunk is pushed 6.0 m at constant speed up a 30° incline by a constant horizontal force. The coefficient of sliding friction between the trunk and the incline is 0.20. Calculate the work done by (a) the applied force and (b) the weight of the trunk.

12P. A 3.57-kg block is drawn at constant speed 4.06 m along a horizontal floor by a rope exerting a 7.68-N force at an angle of 15.0° above the horizontal. Compute (a) the work done by the rope on the block, and (b) the coefficient of kinetic friction between block and floor.

SECTION 7-3 WORK DONE BY A VARIABLE FORCE

13E. A 5.0-kg block moves in a straight line on a horizontal frictionless surface under the influence of a force that varies with position as shown in Fig. 7-28. How much work is done by the force as the block moves from the origin to *x* = 8.0 m?

FIGURE 7-28 Exercise 13.

14E. A 10-kg mass moves along the *x* axis. Its acceleration as a function of its position is shown in Fig. 7-29. What is the net work performed on the mass as it moves from *x* = 0 to *x* = 8.0 m?

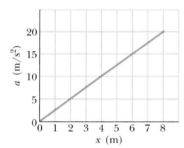

FIGURE 7-29 Exercise 14.

15P. (a) Estimate the work done by the force represented by the graph of Fig. 7-30 in displacing a particle from *x* =

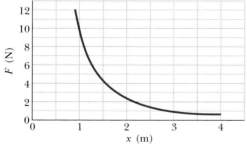

FIGURE 7-30 Problem 15.

1 m to $x = 3$ m. Refine your method to see how close you can come to the exact answer of 6 J. (b) The curve is given analytically by $F = a/x^2$, where $a = 9$ N·m². Show how to calculate the work by the rules of integration.

16P. The force exerted on an object is $F = F_0(x/x_0 - 1)$. Find the work done in moving the object from $x = 0$ to $x = 2x_0$ (a) by plotting $F(x)$ and finding the area under the curve and (b) by evaluating the integral analytically.

17P. What work is done by a force $\mathbf{F} = (2x\ \text{N})\mathbf{i} + (3\ \text{N})\mathbf{j}$, where x is in meters, that is exerted on a particle while it moves from a position of $\mathbf{r}_i = (2\ \text{m})\mathbf{i} + (3\ \text{m})\mathbf{j}$ to a position of $\mathbf{r}_f = -(4\ \text{m})\mathbf{i} - (3\ \text{m})\mathbf{j}$?

SECTION 7-4 WORK DONE BY A SPRING

18E. A spring with a spring constant of 15 N/cm has a cage attached to one end, as in Fig. 7-31. (a) How much work does the spring do on the cage if the spring is stretched from its relaxed length by 7.6 mm? (b) How much additional work is done by the spring if it is stretched by an additional 7.6 mm?

19E. During spring semester at MIT, residents of the parallel buildings of the East Campus dorms battle one another with large catapults that are made with surgical hose mounted on a window frame. A balloon filled with dyed

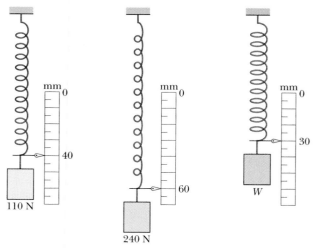

FIGURE 7-31 Exercise 18.

FIGURE 7-32 Problem 20.

water is placed in a pouch attached to the hose, and the hose is then stretched through the width of the room. Assume that the stretching of the hose obeys Hooke's law and has a spring constant of 100 N/m. If the hose is stretched by 5.00 m and then released, how much work does it do on the balloon in the pouch by the time it reaches its relaxed length?

20P. Figure 7-32 shows a spring with a pointer attached, hanging next to a scale graduated in millimeters. Three different weights are hung from the spring, in turn, as shown. (a) If all weight is removed from the spring, which mark on the scale will the pointer indicate? (b) What is the weight W?

SECTION 7-5 KINETIC ENERGY

21E. If a Saturn V rocket with an Apollo spacecraft attached has a combined mass of 2.9×10^5 kg and is to reach a speed of 11.2 km/s, how much kinetic energy will it then have?

22E. A conduction electron (mass $m = 9.11 \times 10^{-31}$ kg) in copper near the absolute zero of temperature has a kinetic energy of 6.7×10^{-19} J. What is the speed of the electron?

23E. Calculate the kinetic energies of the following objects moving at the given speeds: (a) a 110-kg football linebacker running at 8.1 m/s; (b) a 4.2-g bullet at 950 m/s; (c) the aircraft carrier *Nimitz*, 91,400 tons at 32 knots.

24E. A proton (mass $m = 1.67 \times 10^{-27}$ kg) is being accelerated in a linear accelerator. In each stage of the accelerator, the proton is accelerated along a straight line at 3.6×10^{15} m/s². If a proton enters such a stage moving initially with a speed of 2.4×10^7 m/s and the stage is 3.5 cm long, compute (a) its speed at the end of the stage and (b) the gain in kinetic energy resulting from the acceleration. The mass of the proton is 1.67×10^{-27} kg. Express the energy in electron-volts.

25E. A proton starting from rest is accelerated in a cyclotron to a final speed of 3.0×10^6 m/s (about 1% of the speed of light). How much work, in electron-volts, is done on the proton by the electrical force of the cyclotron that accelerated it?

26E. A single force acts on a body that moves along a straight line. A plot of velocity versus time for the body is shown in Fig. 7-33. Find the sign (plus or minus) of the

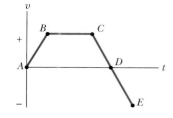

FIGURE 7-33 Exercise 26.

work done *by* the force *on* the body in each of the intervals *AB*, *BC*, *CD*, and *DE*.

27E. A firehose (Fig. 7-34) is uncoiled by pulling the end of the hose horizontally along a frictionless surface at the steady speed of 2.3 m/s. The mass of 1.0 m of the hose is 0.25 kg. How much kinetic energy is imparted in uncoiling 12 m of hose?

FIGURE 7-34   Exercise 27.

28E. From what height would an automobile weighing 12,000 N have to fall to gain the kinetic energy equivalent to what it would have when going 89 km/h? Why does the answer not depend on the weight of the car?

29E. A 1000-kg car is traveling at 60 km/h on a level road. The brakes are applied long enough to reduce the car's kinetic energy by 50 kJ. (a) What is the final speed of the car? (b) By how much more must the brakes reduce the kinetic energy to stop the car?

30E. On August 10, 1972, a large meteorite skipped across the atmosphere above western United States and Canada, much like a stone can be skipped across water. The accompanying fireball was so bright that it could be seen in the daytime sky (Fig. 7-35). The meteorite's mass was about 4×10^6 kg; its speed was about 15 km/s. Had it entered the atmosphere vertically, it would have hit the Earth's surface with about the same speed. (a) Calculate the meteorite's loss of kinetic energy (in joules) that would have been associated with the vertical impact. (b) Express the energy as a multiple of the explosive energy of 1 megaton of TNT, which is 4.2×10^{15} J. (c) The energy associated with the atomic bomb explosion over Hiroshima was equivalent to 13 kilotons of TNT. To how many "Hiroshima bombs" would the meteorite impact have been equivalent?

31E. An explosion at ground level leaves a crater with a diameter that is proportional to the energy of the explosion raised to the 1/3 power; an explosion of 1 megaton of TNT leaves a crater with a 1-km diameter. Below Lake Huron in Michigan there appears to be an ancient impact crater with a 50-km diameter. What was the kinetic energy associated with that impact, in terms of (a) megatons of TNT and (b) Hiroshima bomb equivalents (see preceding problem)? (Such meteorite or comet impacts may have significantly altered the Earth's climate and contributed to the extinction of the dinosaurs and other life forms.)

32P. A man racing his son has half the kinetic energy of the son, who has half the mass of the father. The man speeds up by 1.0 m/s and then has the same kinetic energy as the son. What were the original speeds of man and son?

33P. A force acts on a 3.0-kg particle in such a way that the position of the particle as a function of time is given by $x = 3.0t - 4.0t^2 + 1.0t^3$, where x is in meters and t is in seconds. Find the work done by the force from $t = 0$ to $t = 4.0$ s.

34P. The Earth circles the sun once a year. How much work would have to be done on the Earth to bring it to rest relative to the sun? See Appendix C for numerical data, and ignore the rotation of the Earth about its own axis.

35P. A helicopter lifts a 72-kg astronaut 15 m vertically from the ocean by means of a cable. The acceleration of the astronaut is $g/10$. How much work is done on the astronaut by (a) the helicopter and (b) her weight? What are (c) the kinetic energy and (d) the speed of the astronaut just before she reaches the helicopter?

36P. A cord is used to vertically lower an initially stationary block of mass *M* at a constant downward acceleration

FIGURE 7-35 A large meteorite skips across the atmosphere in the sky above the mountains.

of $g/4$. When the block has fallen a distance d, find (a) the work done by the cord on the block, (b) the work done by the weight of the block, (c) the kinetic energy of the block, and (d) the speed of the block.

37P. A crate with a mass of 230 kg hangs from the end of a 12.0-m rope. You push horizontally on the crate with a varying force **F** to move it 4.00 m to the side (Fig. 7-36). (a) What is the magnitude of **F** when the crate is in this final position? During the crate's displacement, what are (b) the total work done on it, (c) the work done by the weight of the crate, and (d) the work done by the pull on the crate from the rope? (e) From the answers to (b), (c), and (d) and the fact that the crate is motionless before and after its displacement, find the work you do on the crate. (f) Why is your work not equal to the product of the horizontal displacement and the answer to (a)?

12.0 m

F

←—4.00 m—→

FIGURE 7-36 Problem 37.

38P. A 250-g block is dropped onto a vertical spring with spring constant $k = 2.5$ N/cm (Fig. 7-37). The block becomes attached to the spring, and the spring compresses 12 cm before momentarily stopping. While the spring is being compressed, what work is done (a) by the block's

FIGURE 7-37 Problem 38.

weight and (b) by the spring? (c) What was the speed of the block just before it hits the spring? (d) If the speed at impact is doubled, what is the maximum compression of the spring? Assume the friction is negligible.

SECTION 7-6 POWER

39E. The loaded cab of an elevator has a mass of 3.0×10^3 kg and moves 210 m up the shaft in 23 s at constant speed. At what average rate does the cable do work on the cab?

40E. If a ski lift raises 100 passengers averaging 150 lb a height of 500 ft in 60 s, at constant speed, what average power is required of the lift?

41E. An elevator cab in the New York Marriott Marquis has a mass of 4500 kg and can carry a maximum load of 1800 kg. The cab is moving upward at full load at 3.8 m/s. What power is required to maintain that speed?

42E. (a) At a certain instant, a particle experiences a force of $\mathbf{F} = (4.0 \text{ N})\mathbf{i} - (2.0 \text{ N})\mathbf{j} + (9.0 \text{ N})\mathbf{k}$ while having a velocity of $\mathbf{v} = -(2.0 \text{ m/s})\mathbf{i} + (4.0 \text{ m/s})\mathbf{k}$. What is the instantaneous rate at which the force does work on the particle? (b) At some other time, the velocity consists of only a **j** component. If the force is unchanged, and the instantaneous power is -12 W, what is the velocity of the particle just then?

43P. A 1400-kg block of granite is pulled up an incline at a constant speed of 1.34 m/s by a steam winch (Fig. 7-38). The coefficient of kinetic friction between the block and incline is 0.40. What is the power of the winch?

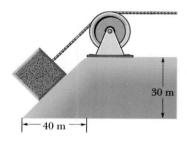

30 m

←—40 m—→

FIGURE 7-38 Problem 43.

44P. A 100-kg block is pulled at a constant speed of 5.0 m/s across a horizontal floor by an applied force of 122 N at 37° above the horizontal. At what rate is the applied force doing work?

45P. A horse pulls a cart with a force of 40 lb at an angle of 30° with the horizontal and moves along at a speed of 6.0 mi/h. (a) How much work does the horse do in 10 min? (b) What is the average power (in horsepower) of the horse?

46P. A 2.0-kg object accelerates uniformly from rest to a speed of 10 m/s in 3.0 s. (a) How much work is done on the object during the 3.0-s interval? What is the rate at which work is done on the object (b) at the end of the interval, and (c) at the end of the first half of the interval?

47P. A force of 5.0 N acts on a 15-kg body initially at rest. Compute (a) the work done by the force in the first, second, and third seconds and (b) the instantaneous power of the force at the end of the third second.

48P. A fully loaded, slow-moving freight elevator has a cab with a total mass of 1200 kg, which is required to travel upward 54 m in 3.0 min. The elevator's counterweight has a mass of 950 kg. Find the power (in horsepower) required of the elevator motor as the cable pulls upward on the cab. Ignore the work required to start and stop the elevator; that is, assume it travels at constant speed.

49P. The force (but not the power) required to tow a boat at constant velocity is proportional to the speed. If a speed of 2.5 mi/h requires 10 hp, how much horsepower does a speed of 7.5 mi/h require?

SECTION 7-7 KINETIC ENERGY AT HIGH SPEEDS

50E. An electron moves through 5.1 cm in 0.25 ns. (a) What is its speed in terms of the speed of light? (b) What is its kinetic energy in electron-volts? (c) What percentage error do you make if you use the classical formula to compute the kinetic energy?

51E. The work–kinetic energy theorem applies to particles at all speeds. How much work, in keV, must be done to increase the speed of an electron from rest (a) to $0.500c$, (b) to $0.990c$, and (c) to $0.999c$?

52P. An electron has a speed of $0.999c$. (a) What is its kinetic energy? (b) If its speed is increased by 0.05%, by what percentage will its kinetic energy increase?

ADDITIONAL PROBLEMS

53. The only force acting on a 2.0-kg body as it moves along the positive x axis has an x component $F_x = -6x$ N, where x is in meters. The velocity of the body at $x = 3.0$ m is 8.0 m/s. (a) What is the velocity of the body at $x = 4.0$ m? (b) At what positive value of x will the body have a velocity of 5.0 m/s?

54. A force **F** in the direction of increasing x acts on an object moving along the x axis. If the magnitude of the force is $F = 10e^{-x/2.0}$ N, where x is in meters, find the work done by **F** as the object moves from $x = 0$ to $x = 2.0$ m (a) by plotting $F(x)$ and finding the area under the curve and (b) by integrating to find the work analytically.

55. The only force acting on a 2.0-kg body as it moves along the x axis varies as shown in Fig. 7-39. The velocity of the body at $x = 0$ is 4.0 m/s. (a) What is the kinetic energy of the body at $x = 3.0$ m? (b) At what value of x will the body have a kinetic energy of 8.0 J? (c) What is the maximum kinetic energy attained by the body between $x = 0$ and $x = 5.0$ m?

56. A constant force of magnitude 10 N makes an angle of 150° (measured counterclockwise) with the direction of increasing x as it acts on a 2.0-kg object. How much work is done on the object by the force as the object moves from the origin to the point with position vector $(2.0$ m$)\mathbf{i} - (4.0$ m$)\mathbf{j}$?

57. A 0.30-kg mass sliding on a horizontal frictionless surface is attached to one end of a horizontal spring (with $k = 500$ N/m) whose other end is fixed. The mass has a kinetic energy of 10 J as it passes through its equilibrium position. (a) At what rate is the spring doing work on the mass as the mass passes through its equilibrium position? (b) At what rate is the spring doing work on the mass when the spring is compressed 0.10 m and the mass is moving away from the equilibrium position?

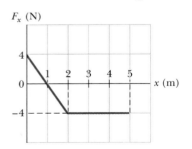

FIGURE 7-39 Problem 55.

CONSERVATION OF ENERGY

8

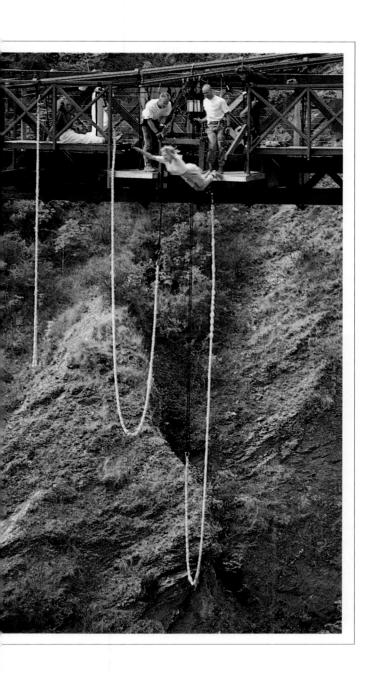

In the dangerous "sport" of bungee-jumping, a participant attaches a specially designed elastic cord to the ankles and leaps from a tall structure. How can we tell just how far down a jumper will go? The answer is, of course, important to the jumper, who is keenly aware of what lies below.

8-1 WORK AND POTENTIAL ENERGY

We begin this chapter with a definition: **energy** is a measure that is associated with a state (or condition) of one or more bodies. In Chapter 7 we considered one kind of energy—kinetic energy K, which is associated with the state of motion of a body. In this chapter we are concerned with another kind—potential energy U, which is the energy associated with the configuration of one or more bodies. (A third kind, thermal energy, which is associated with the random motions of atoms and molecules in a body, will also enter our discussions. A change in thermal energy is demonstrated by a change in the temperature of a body.)

When Vasili Alexeev lifted 562 lb above his head, he did not increase the weights' kinetic energy. But he did increase the separation between the weights and the Earth, which attract each other via the gravitational force. His work increased the gravitational potential energy of the weights–Earth system by changing its configuration, that is, by shifting the relative locations of the Earth and the weights. **Gravitational potential energy** is the energy associated with the state of separation between bodies that attract each other via the gravitational force.

If you compress or extend a spring, you change the relative locations of the coils within the spring. They resist that reconfiguration, and the result of your work is an increase in the elastic potential energy of the spring. **Elastic potential energy** is the energy associated with the state of compression or extension of an elastic (springlike) object.

In the language that describes energy, potential energy is often said to be ''stored'' in a system, in the sense that it could later result in motion. For example, the pole-vaulter in Fig. 8-1 is said to store potential energy in the pole when he bends it. When the pole straightens, it sends him soaring.

When each stone of the Rheims cathedral (Fig. 8-2) was lifted from a quarry to its present height, potential energy was ''stored'' in the stone–Earth

FIGURE 8-1 A pole-vaulter does work in bending his fiberglass vaulting pole, thus storing elastic potential energy. As he rises, the energy is returned to him in the form of gravitational potential energy. The record jump with a stiff pole (aluminum, steel, or bamboo) was 15 ft 7.75 in., which was set in 1942 and which stood for 17 years. When flexible fiberglass poles were allowed in competition, the record almost immediately rose to 17 ft.

FIGURE 8-2 Rheims cathedral, built about A. D. 1240. Each stone in the structure is a repository of gravitational potential energy. Where, do you think, are the stones that store the most potential energy?

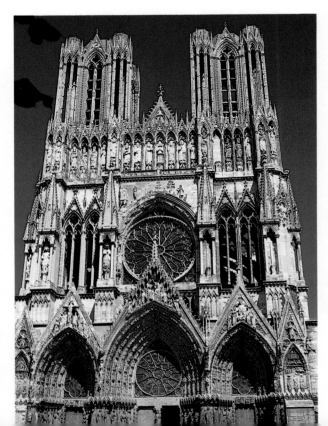

system. However, using less precise language, you might say that the energy is stored in the stone itself, and you might refer to "the potential energy of the stone" or even that of the cathedral as a whole. Such language is often acceptable, but you should keep in mind what is actually meant: the Earth is an important part of the energy-storing system.

8-2 MECHANICAL ENERGY

We now examine how three different forces change the state of a system to see if a potential energy can be assigned to the system's configuration. Our test (which is justified in Section 8-4) is this: the system will begin in an initial state, with a body in the system having a certain amount of kinetic energy, and the force will then do work on that body, changing the kinetic energy. If the system returns to its initial state, we will look to see whether the body then has its original amount of kinetic energy. If it does, then we can define a potential energy—a stored energy —for the system. You will see that this can be done when the force is a spring force or a gravitational force. But when the force is a kinetic frictional force, we cannot define a potential energy.

We will also examine the **mechanical energy** E of a system, which is the sum of its kinetic energy and its potential energy. Our goal is to see what happens to the value of the mechanical energy when a gravitational, frictional, or spring force acts within the system. Does E change? Or does it remain constant, in which case it is said to be **conserved?**

The Spring Force

Figure 8-3a shows a spring in its relaxed state, with one end fastened to a wall. A block of mass m, sliding with kinetic energy K, is just about to make contact with the free end of the spring. We assume that the horizontal plane is frictionless, that the spring is massless and *ideal* (it has no internal friction when compressed or extended), and that the spring obeys Hooke's law (Eq. 7-16):

$$F(x) = -kx. \qquad (8\text{-}1)$$

Here $F(x)$, the *spring force,* is the force exerted by the spring on the block, and x is the amount by which the spring is stretched or compressed.

Once the block of Fig. 8-3a reaches and then begins to compress the spring (Fig. 8-3b), it slows, losing kinetic energy until it momentarily stops against the compressed spring (Fig. 8-3c). Then, as

the spring expands, the block reverses its motion (Fig. 8-3d). When the block is on the verge of leaving the spring, the block has the same kinetic energy (Fig. 8-3e) as in Fig. 8-3a.

Since the kinetic energy is fully restored to the block when the block has made a "round-trip" and the system has returned to its original state (that is, the block is at its original location and the spring is at its original length), we can assign a potential energy to the block–spring system. That potential energy is associated with the state of compression of the spring.

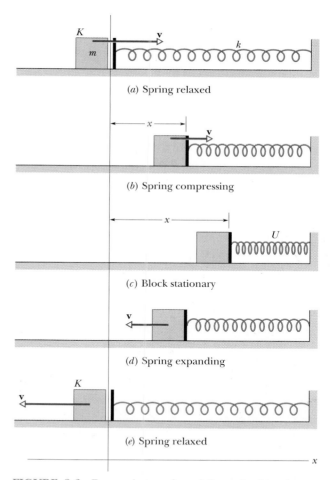

FIGURE 8-3 Energy is transferred from the kinetic energy K of a moving block to the potential energy U of a spring, and is returned as kinetic energy. (a) The block is about to strike the end of the spring, which is in its relaxed state. (b) The block slows and the spring compresses. (c) The block comes to rest, having transferred all its kinetic energy to the (compressed) spring. (d) The spring expands and the block moves. (e) The block has recovered its initial kinetic energy.

When the block is compressing the spring, the block is losing kinetic energy, but the spring is gaining potential energy. During this process we can say that energy is being transferred from the block to the spring. When the spring is expanding, it is losing potential energy but the kinetic energy of the block is increasing. During this process we can say that energy is being transferred from the spring to the block.

The mechanical energy E of the block–spring system is the sum of the block's kinetic energy and the spring's potential energy at any instant. The point of Fig. 8-3 is that E remains constant during the various stages of energy transfer. That is, the mechanical energy of the block–spring system is conserved. If E were not conserved, the block would not end up in its original state with its original kinetic energy.

The energy transfer between the block and spring is like a money transfer between a savings account and a checking account. If you transfer money from the savings account into the checking account, the value of the savings account decreases by the same amount that the value of the checking account increases, but the sum of the two accounts does not change; that is, it is conserved.

The conservation of mechanical energy for the block–spring system can be written as

$$E = U_1 + K_1 = U_2 + K_2 = \cdots$$
$$= U_n + K_n = \text{constant}, \tag{8-2}$$

By pulling back the string, Kevin Costner does work on the bow, storing energy in it. When he releases the string and arrow, much of the energy is transferred to the arrow as kinetic energy.

where the subscripts refer to different instants during the energy-transfer process. An equivalent way of writing Eq. 8-2, this time in terms of energy *changes*, is

$$\Delta K + \Delta U = 0, \tag{8-3}$$

which tells us that every change in the kinetic energy of the block is accompanied by an equal but opposite change in the elastic potential energy of the spring.

The Gravitational Force

Figure 8-4 shows a ball of mass m moving vertically upward near the Earth's surface and acted on by only its weight $m\mathbf{g}$, which is primarily due to the gravitational force exerted on the ball by the Earth. As the ball rises, the weight does work on the ball, decreasing its speed and thus its kinetic energy, and slowing it until it momentarily stops at c. Then the motion is reversed, and the weight again does work on the ball, but this time increasing its speed and kinetic energy.

Figures 8-3 and 8-4 are similar, and equivalent states are labeled with the same letters. In each fig-

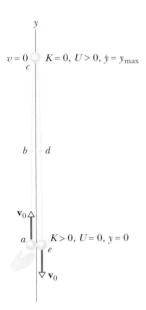

FIGURE 8-4 A ball of mass m is thrown upward. During the ascent, energy is transferred from the ball's kinetic energy to potential energy of the ball–Earth system, until the ball momentarily comes to rest at c. As the ball then falls, it regains its kinetic energy, drawing energy from the store of potential energy in the ball–Earth system. The labeled points correspond to parts (*a*) through (*e*) of Fig. 8-3.

ure, a particle-like body makes a round-trip during which it loses its kinetic energy, only to gain it all back by the time it returns to its starting point.

As the ball rises in Fig. 8-4, losing kinetic energy, we can say that energy is being transferred from the ball to the ball–Earth system, where it is stored as gravitational potential energy. As the ball falls and its kinetic energy increases, the potential energy of the ball–Earth system decreases. In this process we can say that the energy is being transferred from the system to the ball, where it is then kinetic energy. During the rise and fall of the ball, the mechanical energy of the system is conserved.

The Kinetic Frictional Force

Consider a block of mass m sliding along a floor and brought to rest by the frictional force exerted on it by the floor. This situation differs from the situations in Figs. 8-3 and 8-4, because there is no way the block can regain its original kinetic energy with a round-trip.

The reason is that energy is transferred from the block into thermal energy of the block and floor. The transfer is one-way (reversing it would be like unscrambling an egg). Thus, in this example, we cannot say that energy is stored as a potential energy. We say, instead, that energy is **dissipated** (its transfer cannot be reversed). And we must conclude that the mechanical energy of the block–floor system is not conserved but reduced.

8-3 DETERMINING THE POTENTIAL ENERGY

Now comes the quantitative part. Suppose that a single force **F**, either a weight or a spring force, acts on a particle and does an amount of work W. Combining Eq. 8-3 with the work–kinetic energy theorem,

$$W = \Delta K, \qquad (8\text{-}4)$$

we get

$$\Delta U = -W \qquad \text{(definition of } \Delta U\text{)}. \qquad (8\text{-}5)$$

Thus if a force changes the potential energy of a system by changing the configuration of the system, then the change in the potential energy is the negative of the work done by that force. We see also that the SI unit for potential energy is the same as that for work, namely, the joule.

For one-dimensional motion, Eq. 8-5 becomes

$$\Delta U = -W = -\int_{x_i}^{x} F(x)\, dx, \qquad (8\text{-}6)$$

where x_i represents the initial configuration of the system, and x any other configuration.

Only *changes* in potential energy are physically important. That fact allows us to simplify a situation involving potential energy by (1) choosing some arbitrary *reference configuration* x_0, (2) assigning it a potential energy $U(x_0)$ (which is usually set equal to zero), and (3) then calculating the potential energy of any other configuration relative to the reference configuration. If we want the potential energy $U(x)$ of some configuration x, we can find it with

$$U(x) = U(x_0) + \Delta U$$
$$= U(x_0) - \int_{x_0}^{x} F(x)\, dx. \qquad (8\text{-}7)$$

Let us clarify these ideas by applying them to two specific cases.

Elastic Potential Energy

Let us choose the reference configuration x_0 of a spring to be its relaxed state with its free end at $x = 0$. Let us further declare that its potential energy in this state is zero: $U(x_0) = 0$. We also assume that no force acts other than the spring force. Substituting $F(x)$ from Eq. 8-1 into Eq. 8-7 yields

$$U(x) = 0 - \int_{0}^{x} (-kx)\, dx,$$

or

$$U(x) = \tfrac{1}{2}kx^2 \qquad (8\text{-}8)$$

for the potential energy of the spring at any elongation or compression x. Because x appears as a square in Eq. 8-8, the potential energy of a spring has the same positive value when the spring is stretched or compressed by the same amount.

We now consider a system in which a block is attached to a spring and the block lies on a frictionless surface. We first push the block to compress or stretch the spring, and then we release the block. It moves back and forth. With Eq. 8-8 and $K = \tfrac{1}{2}mv^2$, Eq. 8-2 becomes, for the block–spring system,

$\frac{1}{2}kx^2 + \frac{1}{2}mv^2 = E$ (block–spring system),   (8-9)

in which the mechanical energy E of the system is a constant. Equation 8-9 is a statement of the law of conservation of mechanical energy for a block–spring system.

Figure 8-5 shows how energy shuttles back and forth between the spring and the block during a *cycle* of the motion, in which the block makes one round-trip. When the block is at $x = 0$, the energy is all kinetic. When the block is at $\pm x_{max}$, the energy is all potential. At intermediate points, the energy is shared between these two forms. The mechanical energy E, the sum of the energies in the two forms, remains constant at all times.

Note that no changes in potential energy are due to the weight of the block or the normal force exerted on the block by the surface over which it slides. These forces always act perpendicular to the motion of the block and thus do no work on it.

Gravitational Potential Energy

Consider a particle moving either up or down along a vertical y axis near the Earth's surface, acted on only by the gravitational force. That force is $F(y) = -mg$, where the minus sign indicates that the force points downward, in the direction of decreasing y. Let $y = 0$ be the reference position y_0 at which the potential energy of the particle–Earth system is zero.

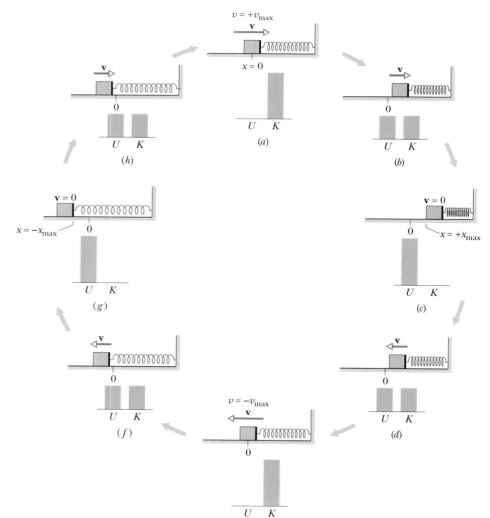

FIGURE 8-5 A block attached to a spring oscillates back and forth on a horizontal, friction-less surface. One full cycle of the motion is shown. We might consider the cycle to begin when the block is in any of the stages. The cycle is completed when the block returns to that stage. During the cycle the values of the potential and ki-netic energies of this block–spring system vary, as indicated by the bar graphs, but their total, which is the mechanical energy E of the system, does not. The energy E can be de-scribed as continuously shifting between the kinetic and poten-tial forms. In stages (*a*) and (*e*), all the energy is kinetic energy. The block then has its greatest speed, and the spring is neither stretched nor compressed. In stages (*c*) and (*g*), all the energy is potential energy. The block then has zero speed, and the spring is at its maximum stretch or compression. In stages (*b*), (*d*), (*f*), and (*h*), half the energy is kinetic energy and half is po-tential energy.

With these substitutions, Eq. 8-7 becomes

$$U(y) = 0 - \int_0^y (-mg)\, dy,$$

or

$$U(y) = mgy. \tag{8-10}$$

With the aid of Eq. 8-10 and $K = \frac{1}{2}mv^2$, Eq. 8-2 becomes, for the particle–Earth system,

$$mgy + \tfrac{1}{2}mv^2 = E \quad \text{(particle–Earth system)}, \tag{8-11}$$

in which E is the mechanical energy of the system. Although the position y and the speed v of the parti-cle may vary, they will always do so in such a way that E in Eq. 8-11 remains constant.

Equations 8-10 and 8-11 hold not only for a particle moving vertically but also for one that moves at an angle with the vertical in a vertical plane, as in projectile motion (such as when a ball is thrown in from the outfield). To show this, we calculate the work done on a particle moving in more than one dimension, using Eq. 7-14:

$$W = \int_{x_0}^x F_x\, dx + \int_{y_0}^y F_y\, dy + \int_{z_0}^z F_z\, dz, \tag{8-12}$$

where the subscripts on the limits have been modified to apply to the situation at hand. The only force doing work on the particle is its weight $m\mathbf{g}$, so $F_x =$

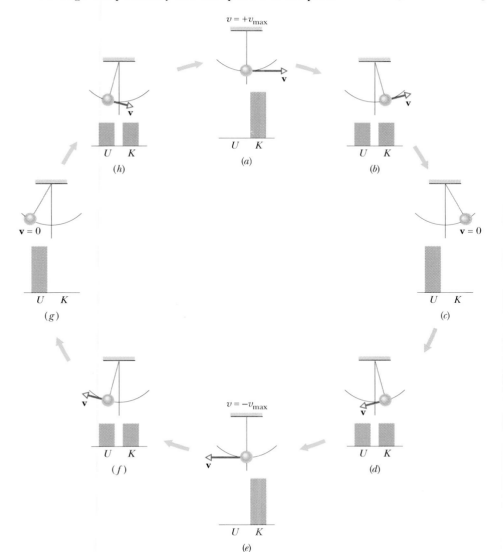

FIGURE 8-6 A pendulum, with its mass concentrated in a bob at the lower end, swings back and forth. One full cycle of the motion is shown. During the cycle the values of the potential and kinetic energies of the pendulum–Earth system vary as the bob rises and falls, but the mechanical energy E of the system does not. The energy E can be described as continuously shifting between the kinetic and potential forms. In stages (a) and (e), all the energy is kinetic energy. The bob then has its greatest speed and is at its lowest point. In stages (c) and (g), all the energy is potential energy. The bob then has zero speed and is at its highest point. In stages (b), (d), (f), and (h), half the energy is kinetic energy and half is potential energy. If the swinging involved a frictional force at the point where the pendulum is attached to the ceiling, or a drag force due to the air, then E would be dissipated, and eventually the pendulum would stop.

$F_z = 0$, and $F_y = -mg$. Thus the first and third terms on the right of Eq. 8-12 are zero. With $y_0 = 0$, the second term on the right is mgy. By substituting $-\Delta U$ for W, we then reduce Eq. 8-12 to Eq. 8-10. The result means (1) that the work done by the weight $m\mathbf{g}$ depends only on the initial and final values of y and not on the actual path taken between the initial and final positions, and (2) that the potential energy U of a particle acted on only by its weight depends on its vertical position y but not on its horizontal position x or z.

An example of this argument can be seen with a swinging pendulum consisting of a bob and a massless cord. Figure 8-6 shows the variation with time of the kinetic energy and the gravitational potential energy of such a pendulum. We take the potential energy of the system to be zero when the bob is at the bottom of its swing. In this configuration (Fig. 8-6a,e), the bob is moving with its greatest speed and the system energy is all kinetic. At the ends of its swing (Fig. 8-6c,g), the bob is momentarily at rest and the system energy is all potential. In other configurations, the energy is partly potential and partly kinetic, the sum E always remaining constant (in the

In olden days, a native Alaskan would be tossed via a blanket so as to see farther over the flat terrain. Nowadays, it is done just for fun. During the ascent of the child in the photograph, energy is transferred to gravitational potential energy. The maximum height is reached when that transfer is complete.

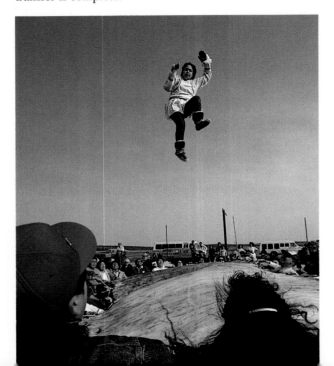

absence of friction). Note that no changes in potential energy are due to the force exerted on the bob by the pendulum cord. This force always acts perpendicular to the motion of the bob and thus does no work on it.

SAMPLE PROBLEM 8-1

A 2.0-kg sloth steps off a limb and drops 5.0 m to the ground (Fig. 8-7).

a. What was the sloth's initial potential energy U if $y_0 = 0$ is taken to be (1) at the ground, (2) at the bottom of a balcony that is 3.0 m above the ground, (3) at the limb, and (4) 1.0 m above the limb?

SOLUTION With Eq. 8-10, we can calculate U for each choice of $y_0 = 0$. For example, for (1) the sloth is initially at $y = 5.0$ m, and

$$U = mgy = (2.0 \text{ kg})(9.8 \text{ m/s}^2)(5.0 \text{ m})$$
$$= 98 \text{ J}. \qquad \text{(Answer)}$$

For the other choices, the values of U are

$$(2) \quad U = mgy = mg(2.0 \text{ m}) = 39 \text{ J},$$
$$(3) \quad U = mgy = mg(0) = 0 \text{ J},$$
$$(4) \quad U = mgy = mg(-1.0 \text{ m}) = -19.6 \text{ J}$$
$$\approx -20 \text{ J}. \qquad \text{(Answer)}$$

b. For each choice of reference configuration, what is the change in the potential energy of the sloth–Earth system due to the fall?

SOLUTION The change in the potential energy is given by

$$\Delta U = mgy_f - mgy_i = mg(y_f - y_i) = mg\,\Delta y.$$

For all four choices for y_0, we have $\Delta y = -5.0$ m. So for (1) through (4),

$$\Delta U = (2.0 \text{ kg})(9.8 \text{ m/s}^2)(-5.0 \text{ m})$$
$$= -98 \text{ J}. \qquad \text{(Answer)}$$

Thus although the value of U depends on the choice for $y_0 = 0$, the *change* in the potential energy does not.

c. What is the sloth's speed v just before landing?

SOLUTION Just before landing, all the sloth's potential energy has been transferred to kinetic energy. From Eq. 8-3 we have

$$\Delta K = -\Delta U. \qquad (8\text{-}13)$$

FIGURE 8-7 Sample Problem 8-1. The four choices for a *y* axis, each marked in units of meters.

Here, because the sloth's initial speed was zero,

$$\Delta K = K_f - K_i = \tfrac{1}{2}mv_f^2 - 0 = \tfrac{1}{2}(2.0 \text{ kg}) v_f^2,$$

and

$$\Delta U = -98 \text{ J}.$$

So

$$\tfrac{1}{2}(2.0 \text{ kg}) v_f^2 = 98 \text{ J},$$

and

$$v_f = \sqrt{\frac{98 \text{ J}}{1.0 \text{ kg}}} \approx 9.9 \text{ m/s}. \qquad \text{(Answer)}$$

d. If the sloth repeats the fall after stuffing itself with leaves and fruits, how does the answer to (c) change?

SOLUTION In part (b) we found that $\Delta U = mg\,\Delta y$. Substituting this into Eq. 8-13 yields

$$\Delta K = -\Delta U = -mg\,\Delta y,$$

which can be rewritten as

$$\tfrac{1}{2}mv_f^2 - 0 = -mg\,\Delta y$$

or

$$v_f = \sqrt{-2g\,\Delta y}.$$

Since *m* does not enter the calculation, *v* is independent of the sloth's mass. The speed just before landing is still

$$v_f = \sqrt{(-2)(9.8 \text{ m/s}^2)(-5.0 \text{ m})}$$

$$\approx 9.9 \text{ m/s}. \qquad \text{(Answer)}$$

SAMPLE PROBLEM 8-2

The spring of a spring gun is compressed a distance *d* of 3.2 cm from its relaxed state, and a ball of mass $m = 12$ g is put in the barrel. With what speed will the ball leave the barrel once the gun is fired? The spring constant *k* is 7.5 N/cm. Assume no friction and a horizontal gun barrel.

SOLUTION Let E_i be the mechanical energy of the gun–ball system in the initial state (before the gun has been fired), and E_f be the mechanical energy in the final state (just after the gun has been fired). Initially, the mechanical energy is the spring's potential energy $U_i = \tfrac{1}{2}kd^2$. In the final state, the mechanical energy is the ball's kinetic energy $K_f = \tfrac{1}{2}mv^2$. Because mechanical energy is conserved, we have

$$E_i = E_f,$$

$$U_i + K_i = U_f + K_f,$$

$$\tfrac{1}{2}kd^2 + 0 = 0 + \tfrac{1}{2}mv^2.$$

Solving for *v* yields

$$v = d\sqrt{\frac{k}{m}} = (0.032 \text{ m})\sqrt{\frac{750 \text{ N/m}}{12 \times 10^{-3} \text{ kg}}}$$

$$= 8.0 \text{ m/s}. \qquad \text{(Answer)}$$

SAMPLE PROBLEM 8-3

In Fig. 8-3, the block, whose mass *m* is 1.7 kg, has an initial speed *v* of 2.3 m/s. The spring constant *k* is 320 N/m.

a. By what distance *x* is the spring compressed in the configuration of Fig. 8-3*c*?

SOLUTION The mechanical energy *E* of the block–spring system must be the same in the configurations of

Figs. 8-3a and 8-3c. Thus

$$E_a = E_c,$$

$$U_a + K_a = U_c + K_c,$$

$$0 + \tfrac{1}{2}mv^2 = \tfrac{1}{2}kx^2 + 0.$$

Solving for x yields

$$x = v \sqrt{\frac{m}{k}} = (2.3 \text{ m/s}) \sqrt{\frac{1.7 \text{ kg}}{320 \text{ N/m}}}$$

$$= 0.17 \text{ m} = 17 \text{ cm}. \qquad \text{(Answer)}$$

b. For what value of x is the energy equally divided between potential and kinetic forms?

SOLUTION In Fig. 8-3a, the mechanical energy E is all kinetic and is

$$E = K = \tfrac{1}{2}mv^2 = (\tfrac{1}{2})(1.7 \text{ kg})(2.3 \text{ m/s})^2 = 4.50 \text{ J}.$$

The question is: For what compression x of the spring will its potential energy be half this value? That is, for what value of x is

$$U(x) = \tfrac{1}{2}kx^2 = \tfrac{1}{2}E$$

true? Solving this equation for x yields

$$x = \sqrt{\frac{E}{k}} = \sqrt{\frac{4.50 \text{ J}}{320 \text{ N/m}}}$$

$$= 0.12 \text{ m} = 12 \text{ cm}. \qquad \text{(Answer)}$$

Note that this is *not* half the maximum compression.

SAMPLE PROBLEM 8-4

In Fig. 8-8, a child of mass m is released from rest at the top of a three-dimensional water slide, a height $h = 8.5$ m above the level of the pool. How fast is the child moving when she reaches the pool? Assume that the slide is frictionless because of the water on it.

SOLUTION At first glance, this problem may seem to be impossible to solve because we are given no information about the child's mass or the shape of the slide. However, it is easy if we note that, in the absence of friction, the only force exerted on the child by the slide is a normal force, which is always perpendicular to the surface of the slide. Since the child always moves *along* the slide, this force can do no work on the child. The only force that does work is the weight $m\mathbf{g}$ of the child. The mechanical energy E is thus conserved throughout the motion and we can write, from Eqs. 8-2 and 8-11,

$$E_b = E_t$$

FIGURE 8-8 Sample Problem 8-4. A child slides down a water slide into a pool of water. The three-dimensional slide is shown flattened onto a plane. If friction can be neglected, the speed of the child at the bottom of the slide does not depend on the shape of the slide or the mass of the child.

or

$$\tfrac{1}{2}mv_b^2 + mgy_b = \tfrac{1}{2}mv_t^2 + mgy_t.$$

Here the subscript b refers to the bottom of the slide and t to the top. Dividing by m, the mass of the child, we find

$$v_b^2 = v_t^2 + 2g(y_t - y_b).$$

Putting $v_t = 0$ and $y_t - y_b = h$ leads to

$$v_b = \sqrt{2gh} = \sqrt{(2)(9.8 \text{ m/s}^2)(8.5 \text{ m})}$$

$$= 13 \text{ m/s}. \qquad \text{(Answer)}$$

This is the same speed that the child would reach if she fell 8.5 m. On an actual slide, some frictional forces would act and the child would not be moving quite so fast.

This problem is hard to solve directly with Newton's laws. Using conservation of mechanical energy makes the solution much easier. However, if you were asked to find the time taken for the child to reach the bottom of the slide, energy methods would be of no use; you would need to know the shape of the slide and you would have a difficult problem on your hands.

SAMPLE PROBLEM 8-5

A 61.0-kg bungee-cord jumper is on a bridge 45.0 m above a river. In its relaxed state, the elastic bungee cord has length $L = 25.0$ m. Assume that the cord obeys Hooke's law, with a spring constant of 160 N/m.

FIGURE 8-9 Sample Problem 8-5. A bungee-cord jumper at the lowest point of the jump.

a. If the jumper stops before reaching the water, what is the height h of her feet above the water at her lowest point?

SOLUTION As shown in Fig. 8-9, let d be the amount by which the cord has stretched when she momentarily stops at her lowest point. As a result of her jump, the change ΔU_g in her gravitational potential energy is

$$\Delta U_g = mg \, \Delta y = -mg(L + d),$$

where m is her mass. The change in the elastic potential energy U_e of the cord is

$$\Delta U_e = \tfrac{1}{2}kd^2.$$

Her kinetic energy K is zero both initially and when she stops.

From Eq. 8-3 and the above, we have

$$\Delta U_e + \Delta U_g + \Delta K = 0$$

$$\tfrac{1}{2}kd^2 - mg(L + d) + 0 = 0$$

$$\tfrac{1}{2}kd^2 - mgd - mgL = 0.$$

Inserting the data, we obtain

$$\tfrac{1}{2}(160 \text{ N/m})\,d^2 - (61.0 \text{ kg})\,(9.8 \text{ m/s}^2)\,d$$

$$- (61.0 \text{ kg})\,(9.8 \text{ m/s}^2)\,(25.0 \text{ m}) = 0,$$

which we solve for d with the quadratic formula, getting

$$d = 17.9 \text{ m}.$$

(The quadratic formula also gives a negative value for d, which has no meaning here.) Her feet are then a distance of $(L + d) = 42.9$ m below their initial height. Thus

$$h = 2.1 \text{ m}. \qquad \text{(Answer)}$$

If she happens to be especially tall, she could be in for a dunking.

b. What is the net force on her at her lowest point (in particular, is it zero)?

SOLUTION Her weight $m\mathbf{g}$ acts downward and has magnitude $mg = 597.8$ N. The upward force exerted on her by the cord at the moment of stopping is given by Hooke's law, $F = -kx$, where x is the displacement of the free end of the cord. Here, the displacement is downward and $x = -d$, so

$$F = -kx = -(160 \text{ N/m})\,(-17.9 \text{ m}) = 2864 \text{ N}.$$

The net force on the jumper is then

$$2864 \text{ N} - 597.8 \text{ N} \approx 2270 \text{ N}. \qquad \text{(Answer)}$$

Thus at the lowest point, where she momentarily stops, there is a net upward force of 2270 N acting on her. This force will yank her back upward.

PROBLEM SOLVING

TACTIC 1: CONSERVATION OF
MECHANICAL ENERGY

The following questions will help you to solve problems involving the conservation of mechanical energy. Review Sample Problems 8-1 to 8-5 with these questions in mind.

For what system is mechanical energy conserved? You should be able to draw a closed surface such that whatever is inside is your system and whatever is outside is the environment of that system. In Sample Problem 8-1 the system is the sloth + Earth. In Sample Problem 8-2, it is the ball + gun, taken together. In Sample Problem 8-3, it is the block + spring. In Sample Problem 8-4, it is the child + Earth. In Sample Problem 8-5, it is the jumper + Earth.

Is friction present? If friction and drag are present, mechanical energy will not be conserved.

Is your system isolated? Conservation of mechanical energy applies only to isolated systems. That means that no external forces should do work on the objects in the system. In Sample Problem 8-2, if you chose the ball alone as your system you would find that it is not isolated; the spring does work on it across the system boundary. Thus you cannot conserve mechanical energy for the ball alone (or the spring alone).

What are the initial and final states of your system? The system changes from some initial state to some final state. You apply the conservation of mechanical energy by saying that E has the same value in both these states. That is, $E_i = E_f$. You must be very clear about what these two states are.

8-4 CONSERVATIVE AND NONCONSERVATIVE FORCES

When a force changes the state of a system, if a change in potential energy can be associated with that change in state, the force is said to be **conservative**. Otherwise, the force is said to be **nonconservative**. A spring force and a weight (or gravitational) force are conservative; air drag and a frictional force are nonconservative.

There are two tests that we can apply to a force to find out whether it is conservative. The tests are totally equivalent so that, if the force meets either one, it will automatically meet the other.

First Test

Suppose that you throw a ball in the air and then catch it when it returns. Since the ball comes back to its original location, it is said to follow a *closed path*; that is, it makes a round-trip. We say too that potential energy of the ball–Earth system is stored during part of the round-trip (during the ball's ascent). But the idea of stored energy is meaningful only if the same amount of energy is taken back out of storage during the rest of the round-trip (during the ball's descent). That is, the net change in the potential energy for the system over a closed path must be zero.

That is indeed true for the ball–Earth system because when the ball completes the round-trip, all of the energy stored in the system during the ascent has been transferred back to kinetic energy of the ball.

From Eq. 8-5, this requirement that $\Delta U = 0$ for a round-trip means that the work W done by the force in question over such a round-trip must also be zero. Thus our first test can be stated as follows:

A force is conservative if the work it does on a particle that moves through a closed path is zero; otherwise, the force is nonconservative.

The gravitational force, by this criterion, is conservative. It did negative work on the ball while the ball was rising and an equal amount of positive work on the return trip. Its work for the round-trip was zero.

The requirement of zero work for a round-trip is not met by the frictional force. If you push an eraser for a certain distance across a chalkboard, the frictional force does negative work on the eraser. However, as you drag the eraser back to its starting point, the direction of the frictional force reverses automatically and that force *still* does negative work on the eraser. The total work done by the frictional force for a round-trip is not zero. The force of friction is nonconservative.

Second Test

Suppose that, as in Fig. 8-10a, a particle moves from a to b along path 1 and then back to a along path 2. If the force acting on the particle is conservative, then from our first test above, the work done on the particle for this round-trip must be zero. We can write

$$W_{ab,1} + W_{ba,2} = 0$$

or

$$W_{ab,1} = -W_{ba,2}. \qquad (8\text{-}14)$$

That is, the work in going from a to b along path 1 is the negative of the work in going from b to a along path 2. Now let us cause the particle to go from a to b along path 2, as in Fig. 8-10b. This merely reverses the direction of motion along this path, so that

$$W_{ab,2} = -W_{ba,2}. \qquad (8\text{-}15)$$

From Eqs. 8-14 and 8-15, we can then write

$$W_{ab,1} = W_{ab,2}, \qquad (8\text{-}16)$$

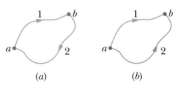

(a)   (b)

FIGURE 8-10 (*a*) A particle, acted on by a conservative force, moves in a round-trip, starting from point a, passing through point b, and returning to point a. (*b*) A particle starts from point a and proceeds to point b, following either of two available paths.

which tells us that the work done on a particle by a conservative force does not depend on the path followed between two points. Thus our second (equivalent) test can be stated as follows:

A force is conservative if the work it does on a particle that moves between two points is the same for all paths connecting those points; otherwise, the force is nonconservative.

Let us test a weight force by applying this criterion to it. Pick up a stone of mass m at point i in Fig. 8-11a, move it horizontally to point a, and then raise it through a vertical height h to point f. We label the path iaf as path 1. Along the horizontal path ia, the work done on the stone by its weight $m\mathbf{g}$ is zero because the force and the displacement are perpendicular. For the vertical path af, the work is $-mgh$. Thus the total work W done by $m\mathbf{g}$ on the stone along path 1 is $-mgh$.

Now consider path 2 in Fig. 8-11a, a straight line connecting points i and f. To find the work W done

by $m\mathbf{g}$ on the stone along this path, we note first that the angle between the force $m\mathbf{g}$ and the displacement $\mathbf{d}$ is $180° - \phi$, where the angle ϕ is defined in the figure. Thus we have

$$W = Fd \cos (180° - \phi) = -Fd \cos \phi.$$

But since $d \cos \phi = h$, we find that $W = -mgh$, just as for path 1.

We can also show that the work done by $m\mathbf{g}$ on the stone along a completely arbitrary path, such as path X in Fig. 8-11b, is also equal to $-mgh$. To do so, let us approximate path X by a series of vertical and horizontal steps. We can have as many steps as we need to make this stepped path arbitrarily close to path X. If the particle follows this stepped path in moving from i to f in Fig. 8-11b, no work will be done by $m\mathbf{g}$ along the horizontal segments. The sum of the vertical segments is just h, so that the work done by $m\mathbf{g}$ along the stepped path is again $-mgh$.

8-5 USING A POTENTIAL ENERGY CURVE

Suppose that a particle is constrained to move along the x axis. We can learn a lot from a plot of its potential energy $U(x)$ as a function of position (if one is available). Before we discuss the plot, though, we need one more relationship.

Finding the Force Analytically

Equation 8-7 tells us how to find the potential energy $U(x)$ in a one-dimensional situation if we know the force $F(x)$. But now we want to go the other way. That is, we know the potential energy function $U(x)$ and want to find the force.

For one-dimensional motion, the work W done by a force that acts on a particle as the particle moves through a distance Δx is $F(x) \, \Delta x$. We can then write Eq. 8-5 as

$$\Delta U(x) = -W = -F(x) \, \Delta x.$$

Solving for $F(x)$ and passing to the differential limit yield

$$F(x) = -\frac{dU(x)}{dx} \qquad \text{(one-dimensional motion).} \qquad (8\text{-}17)$$

(a)

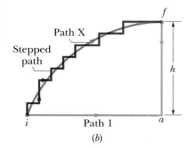

(b)

FIGURE 8-11 (*a*) A stone moves from point i to point f. The work done on the stone by its weight $m\mathbf{g}$ is the same for paths 1 and 2. (*b*) The work is also the same for path X and the stepped path that roughly follows it. In fact, the work is the same for *any* path connecting points i and f.

We can check this result by putting $U(x) = \frac{1}{2}kx^2$, which is the elastic potential energy function for a spring force. Equation 8-17 then yields, as expected, $F(x) = -kx$, which is Hooke's law. Similarly, we can substitute $U(x) = mgx$, which is the gravitational potential energy function for a particle of mass m at height x above the Earth's surface. Equation 8-17 then yields $F = -mg$, which is the weight force that acts on the particle.

The Potential Energy Curve

Figure 8-12a is a plot of a potential energy function $U(x)$ of a particle in one-dimensional motion. We now easily find the force $F(x)$ that acts on the particle by (graphically) taking the slope of the $U(x)$ curve at various points, as Eq. 8-17 instructs us to do. Figure 8-12b is a plot of $F(x)$ found in this way.

Turning Points

In the absence of friction, the mechanical energy E of the particle is a constant value given by

$$U(x) + K(x) = E. \qquad (8\text{-}18)$$

Here $K(x)$ is the *kinetic energy function* of the particle (this $K(x)$ gives the kinetic energy as a function of the particle's location x). We may rewrite Eq. 8-18 as

$$K(x) = E - U(x). \qquad (8\text{-}19)$$

Suppose that E (a constant value, remember) happens to be 5.0 J. It would be represented in Fig. 8-12 by a horizontal line that runs through the value 5.0 J on the energy axis. (It is shown in Fig. 8-12a.) Equation 8-19 tells us how to determine the kinetic energy K for any location x of the particle: on the $U(x)$ curve, find U for that location x and then subtract U from E. For example, if the particle is at any point to the right of x_5, then $K = 1.0$ J. The value of K is greatest (5.0 J) when the particle is at x_2, and least (0 J) when the particle is at x_1.

Since K can never be negative (because v^2 is always positive), the particle can never move to the left of x_1, where $E - U$ is negative. As the particle moves toward x_1 from x_2, K decreases (the particle slows) until $K = 0$ at x_1 (the particle stops). Note that when the particle reaches x_1, the force on the particle, given by Eq. 8-17, is positive. This means that the particle does not remain at x_1 but instead begins to move to the right, opposite its previous motion. Hence x_1 is called a **turning point.** There is

no turning point (where $K = 0$) on the right side of the graph. Once the particle heads toward increasing values of x, it will continue indefinitely.

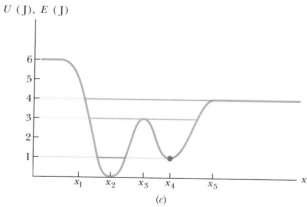

FIGURE 8-12 (a) A plot of $U(x)$, the potential energy function for a particle confined to move along the x axis. There is no friction, so mechanical energy is conserved. (b) A plot of the force $F(x)$ acting on the particle, derived from the potential energy plot by taking its slope at various points. (c) The $U(x)$ plot of (a) with three possible values of E shown.

Equilibrium Points

Figure 8-12c shows three additional values for E superposed on the plot of the potential energy $U(x)$. Let us see how they would change the situation. If $E = 4.0$ J (purple line), the turning point shifts from x_1 to a point between x_1 and x_2. Also, at any point to the right of x_5, the particle's mechanical energy is equal to its potential energy; it has no kinetic energy and no force acts on it, and so it must be stationary there. A particle at such a position is said to be in **neutral equilibrium.** (A marble placed on a horizontal tabletop is in that state.)

If $E = 3.0$ J (pink line), there are two turning points: one is between x_1 and x_2, and the other is between x_4 and x_5. In addition, x_3 is a point at which $K = 0$. If the particle is located exactly there, the force on it is also zero, and the particle remains stationary. However, if it is displaced even slightly in either direction, a nonzero force pushes it farther in the same direction, and the particle continues to move. A particle at such a position is said to be in **unstable equilibrium.** (A marble balanced on top of a bowling ball is an example.)

Next consider the particle's behavior if $E = 1.0$ J (green line). If we place it at x_4, it is stuck there. It cannot move left or right on its own because to do so would require a negative kinetic energy. If we push it slightly left or right, a restoring force appears that moves it back to x_4. A particle at such a position is said to be in **stable equilibrium.** (A marble placed at the bottom of a hemispherical bowl is an example.) If we place the particle in the cuplike *potential valley* centered at x_2, it can still move left and right somewhat, but only part way to x_1 or x_3.

8-6 CONSERVATION OF ENERGY

Let us assume that a kinetic frictional force **f** acts on the oscillating block of the block–spring system of Fig. 8-5. The action of this force causes the oscillations to decrease in amplitude until the system eventually comes to rest. From experiment we would find that this decrease in mechanical energy is accompanied by an increase in the *thermal energy* of the block and the floor over which it slides; they get warmer. Thermal energy is a form of internal energy, since it is associated with the random motions of the atoms and molecules within an object. We represent the change in thermal energy as ΔE_{int}.

Because ΔE_{int} is the change in the thermal energy of *both* the block *and* the floor over which it slides, we can account for the transfer of mechanical energy to thermal energy only if we consider a system that includes the block, the spring, *and the floor.* If we *isolate* the block–spring–floor system (so that no object that is outside the system can change any of the energy values of the objects inside the system), then the mechanical energy lost by the block and the spring is not lost by the system but is transferred within the system to thermal energy.

We are led to postulate that, for such an isolated system,

$$\Delta K + \Delta U + \Delta E_{int} = 0. \qquad (8\text{-}20)$$

The values of K, U, and E_{int} might change for one or more bodies within the isolated system, but their total for the system does not change. Note that Eq. 8-20 is an extension of Eq. 8-3 (the conservation of mechanical energy) in which the presence of the frictional force is taken into account.

It turns out that in all physical situations (including those involving, for example, electrical and magnetic phenomena) we can always identify additional energy quantities like E_{int} that permit us to expand the scope of our definition of energy and to retain, in a more generalized form, the law of conservation of energy. That is, we can always write, for an isolated system,

$$\Delta K + \Delta U + \Delta E_{int} + \left(\begin{array}{c}\text{changes in other}\\\text{forms of energy}\end{array}\right) = 0. \quad (8\text{-}21)$$

This further generalized law of conservation of energy can be expressed as follows:

> In an isolated system, energy can be transferred from one form to another, but the total energy of the system remains constant.

This statement is a generalization backed up by experiment. So far, it has never been contradicted by any experiment or observation of nature.

If forces act through the system boundary and do work W on bodies within the system, the system is not isolated, and Eq. 8-20 does not apply. In such cases we must modify Eq. 8-20 to

$$W = \Delta K + \Delta U + \Delta E_{int}. \qquad (8\text{-}22)$$

This tells us that if work W is done *on* a system by external forces, the total store of energy in the system, including all forms, is increased by an amount equal to W. If W is negative, which means that work is done *by* the system on objects in its surroundings, the energy store is correspondingly decreased.

Although it is sometimes convenient to consider a system that is *not* isolated from its surroundings, we are never forced to do so. We can always enlarge a system so that the external objects that act to change its energy content are then included in the enlarged system. This enlarged system *is* then an isolated system, and Eq. 8-20 applies. The forces in question continue to act but they do so within the enlarged system; the work that they do is internal to that system and is not included in the W of Eq. 8-22, which is the work done only by forces acting through the system boundary.

Figure 8-13 is an example of energy conservation. As the climber ascends the rock face in Fig. 8-13*a*, she is gaining gravitational potential energy at the expense of her own store of biochemical internal energy. In Fig. 8-13*b*, a climber descends, at a roughly constant speed, sliding down a rope which passes through a metal brake bar. The gravitational potential energy that she loses in descending appears now as internal or thermal energy in the rope and the brake bar. They get hot. (The purpose of the brake bar is to prevent the gravitational potential energy from being transferred into kinetic energy of a falling climber!)

Often in the history of physics, the law of conservation of energy has seemed to fail. Its apparent failure, however, has always stimulated a search for the reason for the failure. So far, the reason has always been found and the law of conservation of energy has always been maintained, perhaps in a more generalized form. As you will see in later chapters, it has become one of the grand unifying ideas of physical science.

Power

Now that you have seen how energy can be transferred from one form to another, and have been given a hint that additional forms of energy will be

FIGURE 8-13 (*Left*) Going up! Internal biochemical energy is transferred into gravitational potential energy. (*Right*) Coming down! Gravitational potential energy is transferred into internal energy of the ropes and the brake bar. They get hot.

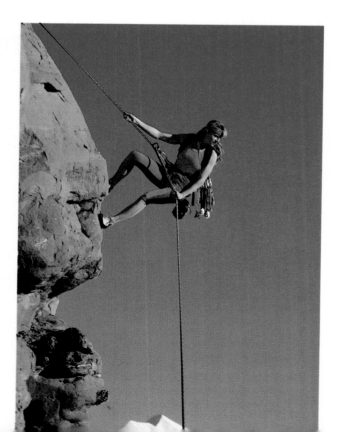

introduced in this book, we can expand the definition of power given in Section 7-6. There it was the rate at which work is done. In a more general sense, power is the rate at which an amount of energy is transferred from one form to another.

For example, as the rock climber ascends in Fig. 8-13a, her average power is the average rate at which she transfers biochemical internal energy to gravitational potential energy and to her own thermal energy. And her instantaneous power is the rate at which that transfer takes place at any given instant.

8-7 WORK DONE BY FRICTIONAL FORCES

Consider a block of mass m sliding along a horizontal floor and acted on by a constant (nonconservative) kinetic frictional force **f** and a constant (conservative) force **F**. For simplicity, let us assume that these two forces act in the same direction, opposite

The slide is lubricated with water so as to reduce the friction acting on someone sliding down it. However, there is still some friction involved, and the total mechanical energy is reduced during the slide, decreasing the person's speed at the bottom.

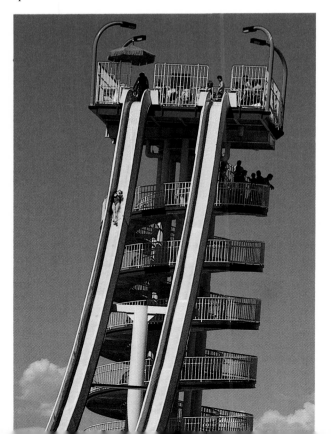

the (positive) direction of motion of the block. Let us choose the block (not the block–floor) as our system. The block alone is *not* an isolated system because forces **F** and **f** act on it through the system boundary and do work on it.

Let us now apply Newton's second law to the block. As we have used it so far, this law applies only to particles. However, if we assume that every point of the block moves in the same way, we can apply the law to the block as if it were a single particle. Thus along the block's line of motion,

$$\sum F = -F - f = ma. \qquad (8\text{-}23)$$

Since the forces are constant, the deceleration a of the block is also constant. Thus we can use Eq. 2-14 to relate a to the initial speed v_i and the final speed v_f of the block as it slides a distance d over the horizontal floor:

$$v_f^2 = v_i^2 + 2ad$$

or

$$a = \frac{v_f^2 - v_i^2}{2d}. \qquad (8\text{-}24)$$

Substituting a from Eq. 8-24 into Eq. 8-23 and rearranging, we find

$$-Fd - fd = \tfrac{1}{2}mv_f^2 - \tfrac{1}{2}mv_i^2 = \Delta K, \qquad (8\text{-}25)$$

where ΔK is the change in the kinetic energy of the block. The product Fd is the work done on the system by the conservative force. From Eq. 8-6, we can replace $-Fd$ with $-\Delta U$, where ΔU is the change in the potential energy of the system. Equation 8-25 then becomes

$$-fd = \Delta K + \Delta U = \Delta E. \qquad (8\text{-}26)$$

This equation tells us the following:

The product $-fd$, where f is the kinetic frictional force, is equal to the change ΔE in the mechanical energy of the system.

The product $-fd$ is negative because the frictional force points in the direction opposite that of the displacement. Hence ΔE is negative; that is, the mechanical energy of the system decreases because of the action of friction, just as we observe.

The left side of Eq. 8-26 looks like an expression for the work done on the block by the frictional force. However, this is not the case—and it is not the case because of the complex nature of the kinetic frictional force. As we discussed in Section 6-1, f is the average of a large number of highly complicated time-varying forces that act at the cold-welded points of contact between the block and the floor. The sum of the work done by all those complicated forces is *not* equal to $-fd$.

To gain more insight into the work done by the frictional force, let us apply Eqs. 8-22 and 8-26 to a block that slides to rest over a horizontal floor because of a kinetic frictional force exerted on it by the floor. (Our system consists of the block alone.) We have $\Delta U = 0$ in each of these equations because no changes in potential energy are involved. Equation 8-22, with W replaced by W_f, the work done by the frictional force, then becomes

$$W_f = \Delta K + \Delta E_{\text{int}}, \qquad (8\text{-}27)$$

where ΔE_{int} is the change in the internal energy *of only the block*. Equation 8-26 becomes

$$-fd = \Delta K. \qquad (8\text{-}28)$$

Comparison of these two equations supports our earlier assertion that $-fd$ is not equal to W_f, the work done by friction. Instead, $-fd$ is the loss of mechanical energy (dissipated by the frictional force), and W_f is the portion of that energy that crosses the system boundary, leaving the block. The remainder is transferred to internal energy of the block.

For example, suppose the sliding block initially has 100 J of kinetic energy and the action of the frictional force in stopping the block results in a 40-J increase in the block's internal energy. Then the energy dissipated by the frictional force is -100 J, and the work done by that force is, by Eq. 8-27, -60 J; that is, 60 J is transferred from the block to the floor.

We also see explicitly from Eq. 8-27 that the work–kinetic energy theorem (which would have to be $W_f = \Delta K$) does not hold for the sliding block. The reason is that the work–kinetic energy theorem is derived on the assumption that the system behaves like a single particle. However, *where energy transfers are concerned*, we cannot treat the sliding block as a particle, because a particle has no internal structure and thus cannot have internal energy.

You may object by pointing out that we treated the block as a particle in applying Newton's second law to derive Eq. 8-26. In that context, however, all that was required was that every point of the block move in the same way; energy considerations did not enter. It is not unusual to be able to treat a body as a particle in one situation but not in others.

SAMPLE PROBLEM 8-6

In Fig. 8-14, a circus beagle of mass 6.0 kg runs onto the left end of a curved ramp, with speed $v_0 = 7.8$ m/s at height $y_0 = 8.50$ m above the floor. It then slides to the right and comes to a momentary stop when it reaches a height $y = 11.1$ m from the floor. What is the increase in thermal energy of the beagle and ramp due to the sliding?

SOLUTION The isolated system to which Eq. 8-20 applies is the beagle–Earth–ramp system, because once the beagle begins to slide, the only forces acting on it are its weight $m\mathbf{g}$ (due to the Earth) and the forces (frictional force and normal force) exerted on it by the ramp. Since the normal force is always perpendicular to the beagle's direction of travel, it does no work on the beagle and thus cannot change any of the energy values in the beagle–Earth–ramp system. However, the frictional force does do work on the beagle, dissipating mechanical energy and increasing the thermal energy of the beagle and ramp by an amount ΔE_{int}.

At the stopping point, the speed of the beagle is zero, and thus so is its kinetic energy. Applying Eq. 8-20 to the beagle–Earth–ramp system, we find that

$$\Delta K + \Delta U + \Delta E_{\text{int}} = 0,$$

FIGURE 8-14 Sample Problem 8-6. A beagle, performing at a circus, slides along a curved ramp, starting with speed v_0 at height y_0, and reaching a height y where it comes to a momentary stop.

or

$$(0 - \tfrac{1}{2}mv_0^2) + mg(y - y_0) + \Delta E_{\text{int}} = 0.$$

Solving for ΔE_{int}, we get

$$\Delta E_{\text{int}} = \tfrac{1}{2}mv_0^2 - mg(y - y_0),$$

$$\Delta E_{\text{int}} = \tfrac{1}{2}(6.0 \text{ kg})(7.8 \text{ m/s})^2$$

$$- (6.0 \text{ kg})(9.8 \text{ m/s}^2)(11.1 \text{ m} - 8.5 \text{ m})$$

$$\approx 30 \text{ J}. \qquad \text{(Answer)}$$

SAMPLE PROBLEM 8-7

A steel ball whose mass m is 5.2 g is fired vertically downward from a height h_1 of 18 m with an initial speed v_0 of 14 m/s (Fig. 18-15a). It buries itself in sand to a depth h_2 of 21 cm.

FIGURE 8-15 Sample Problem 8-7. (a) A ball is fired downward, coming to rest in sand. Mechanical energy is conserved over path h_1 but not over path h_2, where a (nonconservative) frictional force acts on the ball. (b) Force **F** is shown acting on the ball.

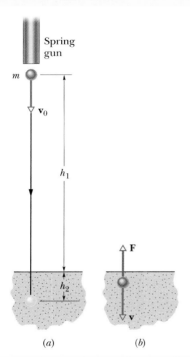

a. What is the change in the mechanical energy of the ball?

SOLUTION At the stopping depth h_2, the speed of the ball is zero, and thus so is its kinetic energy. The change in the mechanical energy of the ball is given by

$$\Delta E = \Delta K + \Delta U, \qquad (8\text{-}29)$$

or

$$\Delta E = (0 - \tfrac{1}{2}mv_0^2) - mg(h_1 + h_2),$$

where $-(h_1 + h_2)$ is the total *downward* displacement of the ball. Inserting the given data, we find

$$\Delta E = -\tfrac{1}{2}(5.2 \times 10^{-3} \text{ kg})(14 \text{ m/s})^2$$

$$- (5.2 \times 10^{-3} \text{ kg})(9.8 \text{ m/s}^2)(18 \text{ m} + 0.21 \text{ m})$$

$$= -1.437 \text{ J} \approx -1.4 \text{ J}. \qquad \text{(Answer)}$$

b. What is the change in the internal energy of the ball–Earth–sand system?

SOLUTION This system is an isolated system, because once the ball is fired, the only forces that act on it are its weight $m\mathbf{g}$ (due to the Earth) and the average upward force **F** exerted on it by the sand (Fig. 8-15b). Substituting Eq. 8-29 into Eq. 8-20, we find that for the ball–Earth–sand system

$$\Delta E + \Delta E_{\text{int}} = 0,$$

or

$$\Delta E_{\text{int}} = -\Delta E = -(-1.437 \text{ J}) \approx 1.4 \text{ J}. \qquad \text{(Answer)}$$

This tells us that as the ball moves through the sand, **F** dissipates all of the ball's mechanical energy, transferring that energy to internal energy (primarily thermal energy) of the ball and sand.

c. What is the magnitude of the average force **F** exerted on the ball by the sand?

SOLUTION The mechanical energy of the ball is conserved until it reaches the sand. Then, as the ball moves through a distance h_2 in the sand, its mechanical energy is changed by ΔE. So Eq. 8-26 ($-fd = \Delta E$) can be rewritten as

$$-Fh_2 = \Delta E.$$

Solving this for F, we find

$$F = \frac{\Delta E}{-h_2} = \frac{-1.437 \text{ J}}{-0.21 \text{ m}} = 6.84 \text{ N} \approx 6.8 \text{ N}. \qquad \text{(Answer)}$$

We could also find F by first using the techniques of Chapter 2 to find the ball's speed at the surface of the sand and its deceleration within the sand. Then, using Newton's second law, we would have F. Obviously, more algebraic steps would be required.

8-8 MASS AND ENERGY (OPTIONAL)

The science of chemistry was built up by assuming that in chemical reactions, energy and mass are each conserved.* In 1905, Einstein showed that, as a consequence of his special theory of relativity, mass can be considered to be another form of energy. Thus the law of conservation of energy is really the law of conservation of mass–energy.

In a chemical reaction, the amount of mass that is transferred into other forms of energy (or vice versa) is such a tiny fraction of the total mass involved that there is no hope of measuring the mass change with even the best laboratory balances. Mass and energy truly *seem* to be separately conserved. In a nuclear reaction, however, the energy released is often about a million times greater than in a chemical reaction, and the change in mass can easily be measured. Taking mass–energy transfers into account where nuclear reactions are involved becomes a matter of necessary laboratory routine.

Mass and energy are related by what is certainly the best-known equation in physics (see Fig. 8-16), namely,

$$E = mc^2, \qquad (8\text{-}30)$$

in which E is the energy equivalent (called the **mass energy**) of mass m, and c is the speed of light. Table 8-1 shows the energy equivalents of the masses of a few objects.

The amount of energy lying dormant in ordinary objects is enormous. The energy equivalent of the mass of a penny, for example, would cost over a million dollars to purchase from your local utility company. The mass equivalents of some energies are equally striking. The entire annual U.S. electrical energy production, for example, corresponds to a mass of only a few hundred kilograms of matter (stones, potatoes, library books, anything!).

In applying Eq. 8-30 to nuclear or chemical reactions between particles, we can rewrite it as

$$Q = -\Delta mc^2, \qquad (8\text{-}31)$$

FIGURE 8-16   Students of Shenandoah Junior High School, in Miami, Florida, honor Einstein on the 100th anniversary of his birth by spelling out his famous formula with their bodies. Courtesy of Rocky Raisen, physics teacher.

in which Q (called simply the *Q of the reaction*) is the energy released or absorbed in the reaction, and Δm is the corresponding decrease or increase in the mass of the particles as a result of the reaction. If the reaction is nuclear fission, less than 0.1% of the mass initially present is transformed into other forms of energy. If the reaction is a chemical one, the percentage is much less, typically a million times less. In the matter of extracting energy from matter, there is a long way to go.

In practice, SI units are rarely used in connection with Eq. 8-31, being too large for convenience.

TABLE 8-1
THE ENERGY EQUIVALENTS OF A FEW OBJECTS

OBJECT	MASS (kg)	ENERGY EQUIVALENT	
Electron	9.1×10^{-31}	8.2×10^{-14} J	$(= 511 \text{ keV})$
Proton	1.7×10^{-27}	1.5×10^{-10} J	$(= 938 \text{ MeV})$
Uranium atom	4.0×10^{-25}	3.6×10^{-8} J	$(= 225 \text{ GeV})$
Dust particle	1×10^{-13}	1×10^{4} J	$(= 2 \text{ kcal})$
Penny	3.1×10^{-3}	2.8×10^{14} J	$(= 78 \text{ GW·h})$

---

*In this section, the word *mass* and the symbol m always refer to the mass of an object as ordinarily measured, that is, while the object is at rest. In Chapter 42, we will further discuss mass and energy in Einstein's special theory of relativity.

Masses are usually entered in atomic mass units (abbreviated u; see Section 1-6), where

$$1 \text{ u} = 1.66 \times 10^{-27} \text{ kg}, \qquad (8\text{-}32)$$

and energies in electron-volts or multiples thereof. Equation 7-6 tells us that

$$1 \text{ eV} = 1.60 \times 10^{-19} \text{ J}. \qquad (8\text{-}33)$$

In the units of Eqs. 8-32 and 8-33, the multiplying constant c^2 has the values

$$c^2 = 9.32 \times 10^8 \text{ eV/u} = 9.32 \times 10^5 \text{ keV/u}$$

$$= 932 \text{ MeV/u}. \qquad (8\text{-}34)$$

SAMPLE PROBLEM 8-8

Suppose that 1.0 mol of (diatomic) oxygen interacts with 2.0 mol of (diatomic) hydrogen to produce 2.0 mol of water vapor, according to the reaction

$$2H_2 + O_2 \rightarrow 2H_2O.$$

The energy Q released is 4.85×10^5 J. What fraction of the mass of the reactants vanishes in order to generate this energy?

SOLUTION The mass decrease needed to supply the released energy follows from Eq. 8-31 as

$$\Delta m = \frac{-Q}{c^2} = \frac{-4.85 \times 10^5 \text{ J}}{(3.00 \times 10^8 \text{ m/s})^2} = -5.39 \times 10^{-12} \text{ kg}.$$

The mass M of the reactants is twice the molar mass (mass of 1 mole) of H_2 plus the molar mass of O_2. Using Appendix D, we find

$$M = 2(2.02 \text{ g}) + 32.0 \text{ g} = 36.0 \text{ g} = 0.036 \text{ kg}.$$

The desired fraction is

$$\frac{|\Delta m|}{M} = \frac{5.39 \times 10^{-12} \text{ kg}}{0.036 \text{ kg}} = 1.5 \times 10^{-10}. \quad \text{(Answer)}$$

Such a tiny fractional mass loss, typical of chemical reactions, cannot be detected by a laboratory balance, but the energy equivalent (4.85×10^5 J per mole of O_2) is easily detected.

SAMPLE PROBLEM 8-9

A typical nuclear fission reaction is

$$^{235}\text{U} + \text{n} \rightarrow {}^{140}\text{Ce} + {}^{94}\text{Zr} + 2\text{n},$$

in which n represents a neutron. The masses involved are

$$\text{mass}(^{235}\text{U}) = 235.04 \text{ u} \qquad \text{mass}(^{94}\text{Zr}) = 93.91 \text{ u}$$

$$\text{mass}(^{140}\text{Ce}) = 139.91 \text{ u} \qquad \text{mass}(\text{n}) = 1.00867 \text{ u}$$

a. What is the fractional change in the mass of the two interacting particles?

SOLUTION To find the mass change Δm we subtract the mass of the reacting particles from the mass of the product particles:

$$\Delta m = (139.91 + 93.91 + 2 \times 1.00867)$$

$$- (235.04 + 1.00867)$$

$$= -0.211 \text{ u}.$$

The mass of the reacting particles is

$$M = 235.04 + 1.00867 = 236.05 \text{ u},$$

and the fractional decrease in mass is

$$\frac{|\Delta m|}{M} = \frac{0.211 \text{ u}}{236.05 \text{ u}}$$

$$= 0.00089, \text{ or about } 0.1\%. \quad \text{(Answer)}$$

Although this is small, it is easily measurable and is much greater than the fractional decrease calculated for the chemical reaction in Sample Problem 8-8.

b. How much energy is released during each fission reaction?

SOLUTION From Eq. 8-31 we have

$$Q = -\Delta mc^2 = (-0.211 \text{ u})(932 \text{ MeV/u})$$

$$= 197 \text{ MeV}, \qquad \text{(Answer)}$$

where the value for c^2 is taken from Eq. 8-34. The energy release of 197 MeV per reaction is much larger than the energy releases of a few electron-volts per reaction that are typical of chemical reactions.

SAMPLE PROBLEM 8-10

The nucleus of an atom of deuterium (or heavy hydrogen) is called a **deuteron**. It is composed of a proton and a neutron. If a deuteron is torn apart, how much energy is involved in the process? Is energy absorbed or released by the reaction?

SOLUTION   The masses involved are

$$\text{deuteron: } m_d = 2.01355 \text{ u}$$

$$\left.\begin{array}{l}\text{proton: } m_p = 1.00728 \text{ u}\\\text{neutron: } m_n = 1.00867 \text{ u}\end{array}\right\} 2.01595 \text{ u}$$

Since the total mass of the separated proton and neutron is greater than the deuteron mass, energy must be added to the deuteron to cause the reaction. The resulting increase in mass is

$$\Delta m = (m_p + m_n) - m_d$$

$$= (1.00728 + 1.00867) - (2.01355) = 0.00240 \text{ u}.$$

The corresponding energy is then, from Eq. 8-31,

$$Q = -\Delta mc^2 = -(0.00240 \text{ u})(932 \text{ MeV/u})$$

$$= -2.24 \text{ MeV}. \qquad \text{(Answer)}$$

The quantity 2.24 MeV is known as the **binding energy** of the deuteron. By contrast, the energy required to pull the electron out of the hydrogen atom is only about 13.6 eV, which is about 6×10^{-6} times smaller.

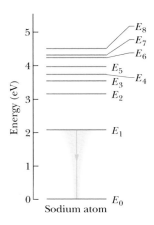

FIGURE 8-17   Some of the energy levels of a sodium atom, corresponding to the various quantum states in which the atom may exist, shown in an energy level diagram. The lowest state, marked E_0, is called the ground state. The atom emits the characteristic yellow sodium light when it transfers from the state with energy E_1 to its ground state, as the vertical arrow suggests. The atom cannot have an energy between those two levels, or between any two of the other levels shown.

8-9   ENERGY IS QUANTIZED (OPTIONAL)

The air, as we wave a hand through it, seems to be perfectly continuous. Yet on a fine enough scale, air is not continuous at all but comes in "lumps," that is, in particles of specific masses—mainly oxygen and nitrogen molecules. We say that mass is **quantized.** When we examine the atomic and subatomic worlds, in which we have no direct "fingertip" experience, we find that many other physical quantities are quantized, that is, restricted to specific (discrete) values. Energy is one of those quantities.

We would all agree that a swinging pendulum can have any reasonable energy that we choose to give it. Things are different in the atomic world, however. An atom can exist in only certain characteristic states—called **quantum states**—each associated with a discrete energy E.

Figure 8-17 shows the allowed energy values (or *energy levels*) for an isolated sodium atom, each value corresponding to a different quantum state. The lowest energy, labeled E_0 in Fig. 8-17 and usually assigned the arbitrary value of zero, is the **ground state** of the sodium atom. It is the state in which an isolated sodium atom usually exists, just as a marble in a bowl usually rests at the bottom of the bowl. To at-

tain a higher **excited state,** the sodium atom must absorb energy from some external source, perhaps by colliding with electrons in a sodium vapor lamp. When it returns to its ground state, it must decrease its energy, perhaps by radiating light.

Actually, the energy of *anything* consisting of atoms—including a swinging pendulum—is quantized. However, for large objects the allowed energy values are so close together that they cannot be distinguished and appear to be continuous. For such objects, we can ignore quantization totally because we cannot possibly detect it (or its effects).

Quantization and the Emission of Light

Let us see how neatly the concepts of quantization and conservation of energy join to explain the emission of light by single isolated atoms. Consider the yellow light emitted by sodium atoms, easily seen by throwing common salt (sodium chloride) into an open flame. That particular color of light is emitted as the sodium atoms transfer from the excited state with energy E_1 in Fig. 8-17 to the ground state with energy E_0; the transition is marked by an arrow in the figure. According to the **wave theory of light,** that particular color corresponds to a certain fre-

FIGURE 8-18 When an atom transfers from an excited state to a state of lower energy, it can decrease its energy by radiating a wave of light.

quency f. (The **frequency** of a wave is the number of times per second the wave vibrates as it moves past a given point; f has the unit of inverse second.)

In general, radiation such as light is emitted from *any* atom when the atom changes from *any* state with energy E_x to *any* state with lower energy E_y, as Fig. 8-18 suggests. The law of conservation of energy for light emission then takes the form

$$E_x - E_y = hf, \qquad (8\text{-}35)$$

in which h is a constant and f is the frequency of the emitted light. This equation, first put forward by the Danish physicist Niels Bohr, tells us that $E_x - E_y$, the energy lost by the atom, is carried away by the emitted light. Equation 8-35 governs the emission (and also the absorption) of radiation of all kinds—not only from atoms but from nuclei, molecules, or heated solids.

The constant h in Eq. 8-35 is called the **Planck constant** and has the value

$$\begin{aligned} h &= 6.63 \times 10^{-34}\,\text{J}\cdot\text{s} \\ &= 4.14 \times 10^{-15}\,\text{eV}\cdot\text{s} \end{aligned} \quad \begin{array}{l}\text{(Planck}\\ \text{constant).}\end{array} \quad (8\text{-}36)$$

This constant, named after the German physicist Max Planck who introduced it in 1900, is the central constant of quantum physics. It pops up in almost every quantum equation, just as the speed of light c pops up in every equation of the theory of relativity. We shall meet it again.

REVIEW & SUMMARY

Energy

Energy is a measure that is associated with a state of one or more bodies. **Kinetic energy** K is associated with the state of *motion* of a body. **Thermal energy** is associated with the random motions of atoms and molecules in a body. **Potential energy** is associated with the *configuration* of one or more bodies. In particular, we consider two types: **gravitational potential energy** is the energy associated with the state of separation between bodies that attract each other via the gravitational force; **elastic potential energy** is the energy associated with the state of compression or extension of an elastic (springlike) object.

Mechanical Energy

The *mechanical energy* E of a system is the sum of its kinetic energy K and its potential energy U. If only a gravitational force or a spring force acts within the system, changing the state of the system, kinetic energy can be transferred to potential energy, and vice versa. However, the value of E stays constant; that is, E is conserved. This **conservation of mechanical energy** can be written as

$$E = U_1 + K_1 = U_2 + K_2 = \text{constant}, \qquad (8\text{-}2)$$

where the subscripts refer to different instants during the energy-transfer process. An equivalent way of writing Eq. 8-2 is

$$\Delta K + \Delta U = 0. \qquad (8\text{-}3)$$

If a kinetic frictional force acts within the system, E is not conserved but is reduced, and the force is said to **dissipate** mechanical energy: it transfers the dissipated energy to **internal energy,** primarily thermal energy.

Elastic Potential Energy

For a spring that obey's Hooke's law, $F = -kx$, the **elastic potential energy** of the spring is given by

$$U = \tfrac{1}{2}kx^2. \qquad (8\text{-}8)$$

Gravitational Potential Energy

When an object near the Earth moves relative to the Earth, the change in the **gravitational potential energy** of the object–Earth system is given by

$$\Delta U = mg\,\Delta y,$$

where Δy is the change in the object's position along a vertical y axis. Usually U is set to zero at $y = 0$, and the object is said to have a potential energy given by

$$U = mgy. \qquad (8\text{-}10)$$

Conservative and Nonconservative Forces

A force is **conservative** if the work it does on a particle that moves through a closed path is zero; otherwise the force is **nonconservative.** Equivalently, a force is conservative if the work it does on a particle that moves between two points is the same for all paths connecting those points; otherwise the force is nonconservative.

Potential Energy Curves

If the **potential energy function** $U(x)$ of a particle is known, then the force that is responsible for changes in $U(x)$ is given by

$$F(x) = -\frac{dU(x)}{dx}. \qquad (8-17)$$

If $U(x)$ is given on a graph, then F at any value of x is the negative of the slope of the curve at that value of x.

The behavior of a particle moving along the x axis can be inferred from a plot of $U(x)$. The kinetic energy of the particle at any point x is given by

$$K(x) = E - U(x), \qquad (8-19)$$

where E is the mechanical energy of the particle. A **turning point** is where the particle reverses its motion (there, $K = 0$). The particle is in **equilibrium** at points where the slope of the $U(x)$ curve is zero.

Law of Conservation of Energy

In an isolated system, energy may be transferred from one form to another, but the total energy of the system always remains constant. This conservation law is expressed as

$$\Delta K + \Delta U + \Delta E_{\text{int}} + \left(\begin{array}{c}\text{changes in other}\\\text{forms of energy}\end{array}\right) = 0, \qquad (8-21)$$

where ΔE_{int} is the change in the internal energy of the bodies within the system.

Work Done on a System

If forces act through the boundary of a system and do work W on the system, the system is not isolated and the energy of the system is not conserved but changes. The work is related to the changes in energy by

$$W = \Delta K + \Delta U + \Delta E_{\text{int}}. \qquad (8-22)$$

Mass and Energy

The **mass energy** of an object is the energy equivalent of its mass, as given by

$$E = mc^2, \qquad (8-30)$$

where m is the mass of the object, and c is the speed of light.

Energy Is Quantized

The energy of a system, such as an atom, is quantized, which means that it can have only certain values. If the system is to change from one state with energy E_x to a lower state with energy E_y, the system must release energy, perhaps as a radiated wave such as light. The radiated wave will have a frequency f given by

$$E_x - E_y = hf, \qquad (8-35)$$

where h is the Planck constant,

$$h = 6.63 \times 10^{-34}\ \text{J} \cdot \text{s} = 4.14 \times 10^{-15}\ \text{eV} \cdot \text{s}. \qquad (8-36)$$

QUESTIONS

1. An automobile is moving along a highway. The driver jams on the brakes and the car skids to a halt, its kinetic energy decreasing to zero. What type of energy increases as a result of the action?

2. In Question 1, assume that the driver operates the brakes in such a way that there is no skidding or sliding. In this case, what type of energy increases?

3. You drop an object and observe that it bounces to half its original height. What conclusions can you draw? What if it bounces to 1.5 times its original height?

4. When an elevator descends from the top of a building and stops at the ground floor, what becomes of the energy that had been potential energy?

5. Why do mountain roads rarely go straight up the slope but wind up gradually?

6. Air bags greatly reduce the chance of injury in a car accident. Explain how they do so, in terms of energy transfers.

7. You see a duck flying by and declare it to have a certain amount of kinetic energy. However, another duck, who flies alongside the first one and who knows a bit of physics, declares it to have no kinetic energy at all. Who is right, you or the second duck? How does the law of conservation of energy fit into this situation?

8. An earthquake can release enough energy to devastate a city. Where does this energy "reside" an instant before the earthquake takes place?

9. Figure 8-19 shows a circular glass tube that is fastened to a vertical wall. The tube is filled with water except for an air bubble that is temporarily at rest at the bottom of the tube. Discuss the subsequent motion of the bubble in terms of energy transfers. Do so first while neglecting retarding forces and then while taking them into account.

10. Give physical examples of unstable equilibrium, neutral equilibrium, and stable equilibrium.

Bubble

FIGURE 8-19 Question 9.

11. In an article "Energy and the Automobile," which appeared in the October 1980 issue of *The Physics Teacher*, author Gene Waring states: "It is interesting to note that *all* the fuel input energy is eventually transformed to thermal energy and strung out along the car's path." Analyze the various mechanisms by which this might come about. Consider, for example, road friction, air resistance, braking, the car radio, the headlamps, the battery, internal engine and drive train losses, the horn, and so on. Assume a straight and level roadway.

12. Trace back to the sun as many of our present energy sources as you can. Can you think of any that cannot be so traced?

13. Explain, using work and energy ideas, how you can pump a swing to make it go higher. If the swing is initially at rest, can you get it going by pumping it?

14. Two disks are connected by a stiff spring (Fig. 8-20). Can you press the upper disk down far enough so that when it is released it will spring back and raise the lower disk off the table? Can mechanical energy be conserved in such a case?

FIGURE 8-20 Question 14.

15. Discuss the words "conservation of energy" as used (a) in this chapter and (b) in connection with an "energy crisis." How do these two usages differ?

16. The electric power for a small town is provided by a hydroelectric plant at a nearby river. If you turn off a light bulb in this system, conservation of energy requires that an equal amount of energy, perhaps in another form, appears somewhere else in the system. Where and in what form does this energy appear?

17. A spring is compressed by tying its ends together tightly. It is then placed in acid and dissolves. What happens to its stored potential energy?

18. The expression $E = mc^2$ tells us that ordinary objects such as coins or pebbles contain enormous amounts of energy. Why did these large stores of energy go unnoticed for so long?

19. "Nuclear explosions—mass for mass—release about a million times more energy than do chemical explosions because nuclear explosions are based on Einstein's $E = mc^2$ relation." What do you think of this statement?

20. How can mass and energy be equivalent in view of the fact that they are totally different physical quantities, defined in different ways and measured in different units?

EXERCISES & PROBLEMS

SECTION 8-3 DETERMINING THE POTENTIAL ENERGY

1E. A certain spring stores 25 J of potential energy when it is compressed by 7.5 cm. What is the spring constant?

2E. To disable ballistic missiles during their boost-phase (when they are blasting upward), an "electromagnetic rail gun," to be carried in low-orbit Earth satellites, is being developed. The gun (a type of kinetic-energy weapon) might fire a 2.4-kg projectile at 10 km/s by accelerating it with electromagnetic forces. Imagine using a spring instead. What must the spring constant be in order to achieve the desired speed if the spring is to be compressed 1.5 m from its relaxed state?

3E. You drop a 2.0-kg textbook to a friend who stands on the ground, which is 10 m below (Fig. 8-21). (a) If the potential energy is taken as being zero at ground level, then what is the potential energy of the book when you release it? (b) What is its kinetic energy just before your friend catches it in her outstretched hands, which are 1.5 m above the ground level? (c) How fast is the book moving as it is caught?

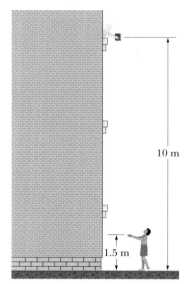

FIGURE 8-21 Exercise 3.

4E. A 200-lb man jumps out a window into a fire rescue net 30 ft below. The net stretches 9.0 ft before bringing him to rest and tossing him back into the air. What is the potential energy of the stretched net if mechanical energy is conserved?

5E. An 8.0-kg mortar shell is fired straight up with an initial speed of 100 m/s. (a) What is the kinetic energy of the shell as it leaves the mortar? (b) What is the change in the shell's potential energy during its ascent if we assume that air drag can be ignored?

6E. An ice flake is released from the edge of a hemispherical frictionless bowl whose radius is 22 cm (Fig. 8-22). How fast is the ice moving at the bottom of the bowl?

FIGURE 8-22 Exercise 6.

7E. A frictionless roller-coaster car tops the first hill in Fig. 8-23 with speed v_0. What is its speed at (a) point A, (b) point B, and (c) point C? (d) How high will it go on the last hill, which is too high for it to cross?

FIGURE 8-23 Exercise 7.

8E. A runaway truck with failed brakes is moving downgrade at 80 mi/h. Fortunately, there is an emergency escape ramp near the bottom of the hill. The inclination of the ramp is 15° (see Fig. 8-24). What minimum length must it have to make certain of bringing the truck to rest, at least momentarily? Real escape ramps are often covered with a thick layer of sand or gravel. Why?

FIGURE 8-24 Exercise 8.

9E. A volcanic ash flow is moving across horizontal ground when it encounters a 10° upslope. It is observed to travel 920 m on the upslope before coming to rest. The volcanic ash contains trapped gas, so the frictional force exerted on it by the ground is very small and can be ignored. At what speed was the ash flow moving just before encountering the upslope? (*Hint:* Use the conservation of mechanical energy for a particle at the front of the flow.)

10E. A projectile with a mass of 2.40 kg is fired from a cliff 125 m high with an initial velocity of 150 m/s, directed 41.0° above the horizontal. What are (a) the kinetic energy of the projectile just after firing and (b) its potential energy? Use the ground as $y = 0$. (c) Find the speed of the projectile just before it strikes the ground. As long as air drag can be ignored, do the answers depend on the mass of the projectile?

11E. Figure 8-25 shows an 8.00-kg stone resting on a spring. The spring is compressed 10.0 cm by the stone. (a) What is the spring constant? (b) The stone is pushed down an additional 30.0 cm and released. What is the potential energy of the compressed spring just before that release? (c) How high above the release position will the stone rise?

FIGURE 8-25 Exercise 11.

12E. A 5.0-g marble is fired vertically upward using a spring gun. The spring must be compressed 8.0 cm if the marble is to just reach a target 20 m above it. (a) What is the change in gravitational potential energy of the marble during its ascent? (b) What is the spring constant?

13E. A ball with mass m is attached to the end of a very light rod with length L and negligible mass. The other end of the rod is pivoted so that the ball can move in a vertical circle. The rod is held in the horizontal position shown in Fig. 8-26 and then given just enough of a downward push so that the ball swings down and around and just reaches the vertically upward position, having zero speed there. (a) What is the change in potential energy of the ball? (b) What initial speed was given to the ball?

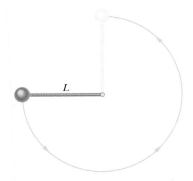

FIGURE 8-26 Exercise 13.

14E. A thin rod whose length is $L = 2.00$ m and whose mass is negligible is pivoted at one end so that it can rotate

FIGURE 8-27 Exercise 14.

in a vertical circle. A heavy ball with mass m is attached to the lower end. The rod is pulled aside through an angle $\theta = 30.0°$, as shown in Fig. 8-27, and then released. How fast is the ball moving at its lowest point?

15P. Figure 8-28a applies to the spring in a cork gun (Fig. 8-28b); it shows the spring force as a function of the stretch or compression of the spring. The spring is compressed by 5.5 cm and used to propel a 3.8-g cork from the gun. (a) What is the speed of the cork if it is released as the spring passes through its relaxed position? (b) Suppose, instead, that the cork sticks to the spring and stretches it 1.5 cm before separation occurs. What is the speed of the cork at the time of release in this case?

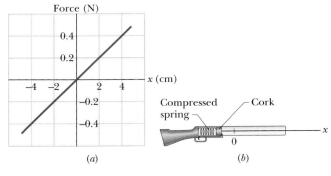

FIGURE 8-28 Problem 15.

16P. A 2.0-kg block is placed against a spring on a frictionless 30° incline (Fig. 8-29). The spring, whose spring constant is 19.6 N/cm, is compressed 20 cm and then released. How far along the incline does it send the block?

FIGURE 8-29 Problem 16.

17P. A spring can be compressed 2.0 cm by a force of 270 N. A block whose mass is 12 kg is released from rest at the top of a frictionless incline as shown in Fig. 8-30, the angle of the incline being $\theta = 30°$. The block comes to rest momentarily after it has compressed the spring by 5.5 cm. (a) How far has the block moved down the incline at this moment? (b) What is the speed of the block just as it touches the spring?

FIGURE 8-30 Problem 17.

18P. A 0.55-kg projectile is launched from the edge of a cliff with an initial kinetic energy of 1550 J and at its highest point is 140 m above the launch point. (a) What is the horizontal component of its velocity? (b) What was the vertical component of its velocity just after launch? (c) At one instant during its flight the vertical component of its velocity is 65 m/s. At that time, how far is it above or below the launch point?

19P. A 50-g ball is thrown from a window with an initial velocity of 8.0 m/s at an angle of 30° above the horizontal. Using energy methods determine (a) the kinetic energy of the ball at the top of its flight and (b) its speed when it is 3.0 m below the window. Does the answer to (b) depend on either (c) the mass of the ball or (d) the initial angle?

20P. The spring of a spring gun has a spring constant of 4.0 lb/in. When the gun is inclined upward by 30° to the horizontal, a 2.0-oz ball is shot to a height of 6.0 ft above the muzzle of the gun. (a) What was the muzzle speed of the ball? (b) By how much must the spring have been compressed initially?

21P. A 5.0-kg mortar shell is fired upward at an angle of 34° to the horizontal and with a muzzle speed of 100 m/s. (a) What is its initial kinetic energy? (b) What is the change in its potential energy by the time it reaches the top of its trajectory? (c) How high does the shell go?

22P. A pendulum consists of a 2.0-kg stone swinging on a 4.0-m string. The stone has a speed of 8.0 m/s when it passes its lowest point. (a) What is the speed when the string is at 60° to the vertical? (b) What is the largest angle with the vertical that the string will reach during the stone's motion? (c) Using the lowest point of the swing as the zero of gravitational potential energy, what is the total mechanical energy of the system?

23P. The string in Fig. 8-31 is $L = 120$ cm long, and the distance d to the fixed peg P is 75 cm. When the ball is released from rest in the position shown, it will swing along the dashed arc. How fast will it be going (a) when it

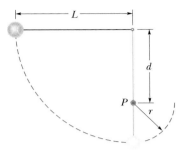

FIGURE 8-31 Problems 23 and 32.

reaches the lowest point in its swing and (b) when it reaches its highest point, after the string catches on the peg?

24P. One end of a vertical spring is fastened to the ceiling. An object is attached to the other end and slowly lowered to its equilibrium position, where the upward force exerted by the spring on the object balances the weight of the object. Show that the loss of gravitational potential energy of the object equals twice the gain in the spring's potential energy. (Why are these two quantities not equal?)

25P. A 2.0-kg block is dropped from a height of 40 cm onto a spring of spring constant $k = 1960$ N/m (Fig. 8-32). Find the maximum distance the spring is compressed.

FIGURE 8-32 Problem 25.

26P. Figure 8-33 shows a pendulum of length L. Its bob (which effectively has all the mass) has speed v_0 when the cord makes an angle θ_0 with the vertical. (a) Derive an

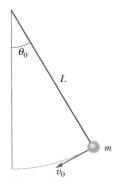

FIGURE 8-33 Problem 26.

expression for the speed of the bob when it is in its lowest position. What is the least value that v_0 can have if the cord is to swing up (b) to a horizontal position, and (c) to a vertical position with the cord remaining straight?

27P. Two children are playing a game in which they try to hit a small box on the floor with a marble fired from a spring-loaded gun that is mounted on a table. The target box is 2.20 m horizontally from the edge of the table; see Fig. 8-34. Bobby compresses the spring 1.10 cm, but the center of the marble falls 27.0 cm short of the center of the box. How far should Rhoda compress the spring to score a direct hit?

FIGURE 8-34 Problem 27.

28P. The magnitude of the gravitational force of attraction between a particle of mass m_1 and one of mass m_2 is given by

$$F(x) = G \frac{m_1 m_2}{x^2},$$

where G is a constant and x is the distance between the particles. (a) What is the potential energy function $U(x)$? Assume that $U(x) \to 0$ as $x \to \infty$. (b) How much work is required to increase the separation of the particles from $x = x_1$ to $x = x_1 + d$?

29P. A 20-kg object is acted on by a force given by $F = -3.0x - 5.0x^2$, where F is in newtons if x is in meters. As the force causes the object to move, assume that the object's mechanical energy is conserved. Also assume that the potential energy U of the object is zero at $x = 0$. (a) What is the potential energy of the object at $x = 2.0$ m? (b) If the object has a velocity of 4.0 m/s in the negative x direction when it is at $x = 5.0$ m, what is its speed as it passes through the origin? (c) Next assume that the potential energy of the object is -8.0 J at $x = 0$, and reanswer parts (a) and (b).

30P. A small block of mass m can slide along the frictionless loop-the-loop track shown in Fig. 8-35. (a) The block is released from rest at point P. What is the net force acting on it at point Q? (b) At what height above the bottom of the loop should the block be released so that it is on the verge of losing contact with the track at the top of the loop?

31P. Tarzan, who weighs 688 N, swings from a cliff at the end of a convenient vine that is 18 m long (Fig. 8-36). From the top of the cliff to the bottom of the swing, he descends by 3.2 m. The vine will break if the force on it exceeds 950 N. Does the vine break?

FIGURE 8-35 Problem 30.

FIGURE 8-36 Problem 31.

32P. In Fig. 8-31 show that, if the ball is to swing completely around the fixed peg, then $d > 3L/5$. (*Hint:* The ball must still be moving at the top of its swing. Do you see why?)

33P. An effectively massless rigid rod of length L has a ball with mass m attached to its end, forming a pendulum. The pendulum is inverted, with the rod straight up, and then released. What are (a) the ball's speed at the lowest point, and (b) the tension in the rod at that point? (c) The same pendulum is next put in a horizontal position and released from rest. At what angle from the vertical do the magnitudes of the tension in the rod and the weight of the ball match?

34P. An escalator joins one floor with another one 8.0 m above. The escalator is 12 m long and moves along its length at 60 cm/s. (a) What power must its motor deliver if the escalator is required to carry a maximum of 100 persons per minute, of average mass 75 kg? (b) An 80-kg man walks up the escalator in 10 s. How much work does the escalator do on him? (c) If this man turned around between floors and walked down the escalator so as to stay at the same level in space, would the escalator do work on him? If so, what power is required of it? (d) Is there any (other?) way the man could walk along the escalator without it doing work on him?

35P*. A chain is held on a frictionless table with one-fourth of its length hanging over the edge, as shown in Fig. 8-37. If the chain has length L and mass m, how much work is required to pull the hanging part back onto the table?

FIGURE 8-37   Problem 35.

36P*. A 3.20-kg block starts at rest and slides a distance d down a frictionless 30.0° incline, where it runs into a spring (Fig. 8-38). The block slides an additional 21.0 cm before it is brought to rest momentarily by compressing the spring, whose spring constant is 431 N/m. (a) What is the value of d? (b) What is the distance between the point of first contact and the point where the block's speed is greatest?

FIGURE 8-38   Problem 36.

FIGURE 8-39   Problem 37.

37P*. A boy is seated on the top of a hemispherical mound of ice (Fig. 8-39). He is given a very small push and starts sliding down the ice. Show that he leaves the ice at a point whose height is $2R/3$ if the ice is frictionless. (*Hint:* The normal force vanishes as he leaves the ice.)

SECTION 8-5   USING A POTENTIAL ENERGY CURVE

38E. A particle moves along the x axis through a region in which its potential energy $U(x)$ varies as in Fig. 8-40. (a) Plot the force $F(x)$ that acts on the particle, using the same x axis scale as in Fig. 8-40. (b) The particle has a (constant) mechanical energy E of 4.0 J. Plot its kinetic energy $K(x)$ directly on Fig. 8-40.

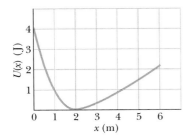

FIGURE 8-40   Exercise 38.

39P. The potential energy of a diatomic molecule (for example, H_2 or O_2) is given by

$$U = \frac{A}{r^{12}} - \frac{B}{r^6},$$

where r is the separation between the atoms that make up the molecule and A and B are positive constants. This potential energy is due to the force that binds the atoms together. (a) Find the equilibrium separation, that is, the distance between the atoms at which the force on each atom is zero. Is the force repulsive (the atoms are pushed apart) or attractive (they are pulled together) if their separation is (b) smaller and (c) larger than the equilibrium separation?

40P. A particle of mass 2.0 kg moves along the x axis through a region in which its potential energy $U(x)$ varies as shown in Fig. 8-41. When the particle is at $x = 2.0$ m, its velocity is -2.0 m/s. (a) What force acts on it at this position? (b) Between what limits of x does the particle move? (c) What is its speed at $x = 7.0$ m?

41P. Figure 8-42a shows two atoms with masses m and M ($m \ll M$), separated by distance r. Figure 8-42b shows the potential energy $U(r)$ of the two-atom *system* as a function of r. Describe the motion of the atoms (a) if the total mechanical energy E is greater than zero (as is E_1), and (b) if E is less than zero (as is E_2). For $E_1 = 1 \times 10^{-19}$ J and $r = 0.3$ nm, find (c) the potential energy of the system,

FIGURE 8-41 Problem 40.

(a)

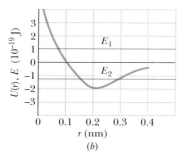

(b)

FIGURE 8-42 Problem 41.

(d) the total kinetic energy of the atoms, and (e) the force (magnitude and direction) acting on each atom. For what values of r is the force (f) repulsive, (g) attractive, and (h) zero?

SECTION 8-6 CONSERVATION OF ENERGY
SECTION 8-7 WORK DONE BY FRICTIONAL FORCES

42E. When a space shuttle (mass = 79,000 kg) returns to Earth from orbit, it enters the atmosphere at a speed of

FIGURE 8-43 Exercise 42. Shuttle touchdown.

18,000 mi/h, which is gradually reduced to a touchdown speed of 220 mi/h. What is its kinetic energy (a) at atmospheric entry and (b) at touchdown? See Fig. 8-43. (c) What happens to all the "missing" energy?

43E. It is claimed that large trees can evaporate as much as 900 kg of water per day. (a) Evaporation takes place from the leaves. To get there, the water must be raised from the roots of the tree. Assuming the average rise of water to be 9.0 m from the ground, how much energy must be supplied per day to raise the water? (b) At what rate is the energy supplied if the evaporation is assumed to occur during 12 h of the day?

44E. The summit of Mount Everest is 8850 m above sea level. (a) How much energy would a 90-kg climber expend against the gravitational force (his weight) in climbing to the summit from sea level? (b) How many candy bars, at 300 kcal per bar, would supply an energy equivalent to this? Your answer should suggest that work done against the gravitational force is a very small part of the energy expended in climbing a mountain.

45E. Approximately 5.5×10^6 kg of water drops 50 m over Niagara Falls every second. (a) How much potential energy is lost every second by the falling water? (b) What would be the power generated by an electric generating plant that could convert *all* of the water's potential energy to electrical energy? (c) If the utility company sold this energy at an industrial rate of 1 cent/kW·h, what would be its yearly income from this source?

46E. The area of the continental United States is about 8×10^6 km², and the average elevation of its land surface is about 500 m. The average yearly rainfall is 75 cm. Two-thirds of this rainwater returns to the atmosphere by evaporation, but the rest eventually flows into the oceans. If all this water could be used to generate electricity in hydroelectric power plants, what average power would be produced?

47E. A 75-g frisbee is thrown from a height of 1.1 m above the ground with a speed of 12 m/s. When it has reached a height of 2.1 m, its speed is 10.5 m/s. (a) How much work was done on the frisbee by its weight? (b) How much of the frisbee's mechanical energy was dissipated by air drag?

48E. An outfielder throws a baseball with an initial speed of 120 ft/s. Just before an infielder catches the ball at the same level, its speed is reduced to 110 ft/s. By how much did air drag reduce the mechanical energy of the ball? The weight of a baseball is 9.0 oz.

49E. Your car averages 30 mi/gal of gasoline, which provides 140 MJ/gal. (a) How far could your car travel on 1.0 kW·h of energy consumed? (b) If you are driving at 55 mi/h, at what rate are you expending energy?

50E. A 51-kg boy climbs, with constant speed, a vertical rope 6.0 m long in 10 s. (a) What is the increase in his gravitational potential energy? (b) What is the boy's average power during the climb?

51E. A 55-kg woman runs up a flight of stairs with a

height of 4.5 m in 3.5 s. What average power is required of her?

52E. A sprinter who weighs 670 N runs the first 7.0 m of a race in 1.6 s, starting from rest and accelerating uniformly. What are the sprinter's (a) speed and (b) kinetic energy at the end of that 1.6 s? (c) What average power does the sprinter generate during the 1.6-s interval?

53E. The luxury liner *Queen Elizabeth 2* (Fig. 8-44) is powered by a diesel-electric powerplant with a maximum power of 92 MW at a cruising speed of 32.5 knots. What forward force is exerted on the ship at this highest attainable speed? (1 knot = 6076 ft/h.)

54E. What power, in horsepower, is required of the engine of a 1600-kg car moving at 25.1 m/s on a level road if the forces of resistance total 703 N?

55E. A swimmer moves through the water at an average speed of 0.22 m/s. The average drag force opposing this motion is 110 N. What average power is required of the swimmer?

56E. The energy required for a person to run is about 335 J/m, regardless of speed. What is a runner's average power during (a) a 100-m dash (time = 10 s) and (b) a marathon (distance = 26.2 mi; time = 2 h 10 min)?

57E. (a) Show that the power required of an airplane cruising at constant speed v in level flight is proportional to v^3. Assume that the aerodynamic drag force is given by Eq. 6-18. (b) By what factor must the engines' power be increased to increase the airspeed by 50%?

58E. Each second, 1200 m^3 of water passes over a waterfall 100 m high. Assuming that three-fourths of the kinetic energy gained by the water in falling is transferred to electrical energy by a hydroelectric generator, at what rate does the generator produce electrical energy? (Note that 1 m^3 of water has a mass of 10^3 kg.)

59E. A 30-g bullet, initially traveling 500 m/s, penetrates 12 cm into a solid wall before it stops. (a) What is the reduction in the bullet's mechanical energy? (b) Assume that the force of the wall on the bullet is constant, and calculate its value.

60E. A 68-kg sky diver falls at a constant terminal speed of 59 m/s. At what rate is the gravitational potential energy of the Earth–sky diver system being reduced? What happens to the removed energy?

61E. A river descends 15 m through rapids. The speed of the water is 3.2 m/s upon entering the rapids, and 13 m/s as it leaves. What percentage of the potential energy lost by 10 kg of water in traversing the rapids is transferred to kinetic energy of water downstream? (Does the answer depend on the mass of water considered?) What happens to the rest of the energy?

62E. A projectile whose mass is 9.4 kg is fired vertically upward. On its upward flight, an energy of 68 kJ is dissipated because of air drag. How much higher would it have gone if the air drag had been made negligible (for example, by streamlining the projectile)?

63E. A 0.63-kg ball is thrown straight up into the air with an initial speed of 14 m/s. It reaches a height of 8.1 m, then falls back down. Assume that the only forces acting on the ball are air drag and the ball's weight, and calculate the work done during the ascent by the air drag.

64E. A 25-kg bear slides, from rest, 12 m down a lodgepole pine tree, moving with a speed of 5.6 m/s just before hitting the ground. (a) What change occurs in the potential energy of the bear? (b) What is the kinetic energy of the bear just before hitting the ground? (c) What is the average frictional force that acts on the bear?

65E. During a rockslide, a 520-kg rock slides from rest down a hillslope that is 500 m long and 300 m high. The coefficient of kinetic friction between the rock and the hill surface is 0.25. (a) What is the potential energy U of the rock just before the slide? (Let $U = 0$ at the bottom of the hill.) (b) How much mechanical energy is dissipated by frictional forces during the slide? (c) What is the kinetic energy of the rock as it reaches the bottom of the hill? (d) What is its speed then?

66P. As Fig. 8-45 shows, a 3.5-kg block is released from a compressed spring whose spring constant is 640 N/m. After leaving the spring at the spring's relaxed length, the

FIGURE 8-44 Exercise 53.

FIGURE 8-45 Problem 66.

block travels over a horizontal surface, with a coefficient of kinetic friction of 0.25, for a distance of 7.8 m before coming to rest. (a) How much mechanical energy was dissipated by the frictional force in bringing the block to rest? (b) What was the maximum kinetic energy of the block? (c) How far was the spring compressed before the block was released?

67P. You push a 2.0-kg block against a horizontal spring, compressing the spring by 15 cm. When you release the block, the spring forces it to slide across a table top. It stops 75 cm from where you released it. The spring constant is 200 N/m. What is the coefficient of kinetic friction between the block and the table?

68P. A moving 2.5-kg block (Fig. 8-46) collides with a horizontal spring whose spring constant is 320 N/m. The block compresses the spring a maximum distance of 7.5 cm from its rest position. The coefficient of kinetic friction between the block and the horizontal surface is 0.25. (a) How much work is done by the spring in bringing the block to rest? (b) How much mechanical energy is dissipated by the force of friction while the block is being brought to rest by the spring? (c) What was the speed of the block when it hit the spring?

FIGURE 8-46 Problem 68.

69P. Two snowy peaks are 850 m and 750 m above the valley between them. A ski run extends down from the top of the higher peak and then back up to the top of the lower one, with a total length of 3.2 km and an average slope of 30° (Fig. 8-47). (a) A skier starts from rest on the higher peak. At what speed will he arrive at the top of the lower peak if he just coasts without using the poles? Ignore friction. (b) Approximately what coefficient of kinetic friction between snow and skis would make him stop just at the top of the lower peak?

FIGURE 8-47 Problem 69.

70P. A factory worker accidentally releases a 400-lb crate that was being held at rest at the top of a 12-ft-long ramp inclined at 39° to the horizontal. The coefficient of kinetic friction between the crate and the ramp, and between the crate and the horizontal factory floor, is 0.28. (a) How fast is the crate moving as it reaches the bottom of the ramp? (b) How far will it subsequently slide across the factory floor? (Assume that the crate's kinetic energy does not change as it moves from the ramp onto the floor.) (c) Why don't the answers to (a) and (b) depend on the crate's mass?

71P. Two blocks are connected by a string, as shown in Fig. 8-48. They are released from rest. Show that, after they have moved a distance L, their common speed is given by

$$v = \sqrt{\frac{2(m_2 - \mu m_1)gL}{m_1 + m_2}},$$

in which μ is the coefficient of kinetic friction between the upper block and the surface. Assume that the pulley is massless and frictionless.

FIGURE 8-48 Problem 71.

72P. A 4.0-kg bundle starts up a 30° incline with 128 J of kinetic energy. How far will it slide up the plane if the coefficient of friction is 0.30?

73P. A cookie jar is moving up a 40° incline. At a point 1.8 ft from the bottom of the incline (measured along the incline), it has a speed of 4.5 ft/s. The coefficient of kinetic friction between jar and incline is 0.15. (a) How much farther up the incline will the jar move? (b) How fast will it be going when it slides back to the bottom of the incline?

74P. A certain spring is found *not* to conform to Hooke's law. The force (in newtons) it exerts when stretched a distance x (in meters) is found to have magnitude $52.8x + 38.4x^2$ in the direction opposing the stretch. (a) Compute the work required to stretch the spring from $x = 0.500$ m to $x = 1.00$ m. (b) With one end of the spring fixed, a particle of mass 2.17 kg is attached to the other end of the spring when it is extended by an amount $x = 1.00$ m. If the particle is then released from rest, compute its speed at the instant the spring has returned to the configuration in which the extension is $x = 0.500$ m. (c) Is the force exerted by the spring conservative or nonconservative? Explain your answer.

75P. A girl whose weight is 267 N slides down a 6.1-m playground slide that makes an angle of 20° with the horizontal. The coefficient of kinetic friction is 0.10. (a) Find the work done on her by her weight. (b) Find the amount of energy dissipated by the frictional force. (c) If she starts at the top with a speed of 0.457 m/s, what is her speed at the bottom?

76P. A 1500-kg automobile starts from rest on a horizontal road and gains a speed of 72 km/h in 30 s. (a) What is the kinetic energy of the auto at the end of the 30 s? (b) What is the average power required of the car during the 30-s interval? (c) What is the instantaneous power at the end of the 30-s interval, assuming that the acceleration was constant?

77P. In Fig. 8-49, a block slides along a track from one level to a higher level, by moving through an intermediate valley. The track is frictionless until the block reaches the higher level. There a frictional force stops the block in a distance d. The block's initial speed v_0 is 6.0 m/s; the height difference h is 1.1 m; and the coefficient of kinetic friction μ is 0.60. Find d.

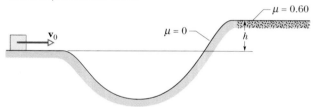

FIGURE 8-49   Problem 77.

78P. In the hydrogen atom, the magnitude of the force of attraction between the positively charged nucleus (a proton) and the negatively charged electron is given by

$$F = k \frac{e^2}{r^2},$$

where e is the magnitude of the charge of the electron and the proton, k is a constant, and r is the separation between electron and nucleus. Assume that the nucleus is fixed in place. Imagine that the electron, which is initially moving in a circle of radius r_1 about the nucleus, suddenly "jumps" into a circular orbit of smaller radius r_2 (Fig. 8-50). (a) Calculate the change in kinetic energy of the electron, using Newton's second law. (b) Using the relation between force and potential energy, calculate the change in potential energy of the atom. (c) By how much has the atom's total energy decreased in this process? (The missing energy is the energy of the light that the atom emits because of the electron's jump.)

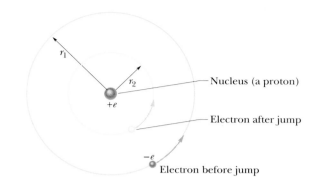

FIGURE 8-50   Problem 78.

79P. A stone with weight w is thrown vertically upward into the air with initial speed v_0. If a constant force f due to air drag acts on the stone throughout its flight, (a) show that the maximum height reached by the stone is

$$h = \frac{v_0^2}{2g(1 + f/w)}.$$

(b) Show that the speed of the stone upon impact with the ground is

$$v = v_0 \left(\frac{w - f}{w + f} \right)^{1/2}.$$

80P. A child's playground slide is in the form of an arc of a circle with a maximum height of 4.0 m, with a radius of 12 m, and with the ground tangent to the circle (Fig. 8-51). A 25-kg child starts from rest at the top of the slide and is observed to have a speed of 6.2 m/s at the bottom. (a) What is the length of the slide? (b) What average frictional force acts on the child over this distance? If, instead of the ground, a vertical line through the *top of the slide* is tangent to the circle, what are (c) the length of the slide and (d) the average frictional force on the child?

4.0 m

FIGURE 8-51   Problem 80.

81P. A small particle slides along a track with elevated ends and a flat central part, as shown in Fig. 8-52. The flat part has length L. The curved portions of the track are frictionless, but for the flat part the coefficient of kinetic friction is $\mu_k = 0.20$. The particle is released at point A,

FIGURE 8-52 Problem 81.

which is a height $h = L/2$ above the flat part of the track. Where does the particle finally come to rest?

82P. A massless rigid rod whose length is L has a ball of mass m attached to one end (Fig. 8-53). The other end is pivoted in such a way that the ball will move in a vertical circle. The system is launched from the horizontal position A with initial downward speed v_0. The ball just reaches point D and then stops. (a) Derive an expression for v_0 in terms of L, m, and g. (b) What is the tension in the rod when the ball is at B? (c) A little grit is placed on the pivot, after which the ball just reaches C when launched from A with the same speed as before. How much mechanical energy is dissipated by friction during this motion? (d) How much total mechanical energy is dissipated by friction before the ball finally comes to rest at B after oscillating back and forth several times?

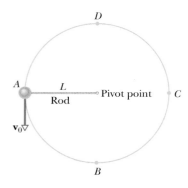

FIGURE 8-53 Problem 82.

83P. The cable of the 4000-lb elevator in Fig. 8-54 snaps when the elevator is at rest at the first floor, where the bottom is a distance $d = 12$ ft above a cushioning spring whose spring constant is $k = 10,000$ lb/ft. A safety device clamps the elevator against guide rails so that a constant frictional force of 1000 lb opposes the motion of the elevator. (a) Find the speed of the elevator just before it hits the spring. (b) Find the maximum distance x that the spring is compressed. (c) Find the distance that the elevator will bounce back up the shaft. (d) Using conservation of energy, find the approximate total distance that the elevator will move before coming to rest. Why is the answer not exact?

84P. While a 1710-kg automobile is moving at a constant speed of 15.0 m/s, the motor supplies 16.0 kW of power to

FIGURE 8-54 Problem 83.

overcome friction, wind resistance, and so on. (a) What is the effective retarding force associated with all the frictional forces combined? (b) What power must the motor supply if the car is to move up an 8.00% grade (8.00 m vertically for each 100 m horizontally) at 15.0 m/s? (c) At what downgrade, expressed as a percentage, would the car coast at 15.0 m/s?

85P. What power is required of a grinding machine, whose wheel has a radius of 20 cm and runs at 2.5 rev/s, when the tool to be sharpened is held against the wheel with a force of 180 N? The coefficient of kinetic friction between the tool and the wheel is 0.32.

86P. A truck can move up a hill that rises 1.0 ft every 50 ft with a speed of 15 mi/h. A resisting force equal to 1/25 the weight of the truck acts on the truck. How fast will the truck move down the hill if the truck's power is the same as it was moving up the hill? Assume that the resisting force also remains unchanged.

87P. If a locomotive with a power capability of 1.5 MW can accelerate a train from a speed of 10 m/s to 25 m/s in 6.0 min, (a) calculate the mass of the train. Find (b) the speed of the train, and (c) the force accelerating the train; give each answer as a function of time in seconds during the 6.0-min interval. (d) Find the distance moved by the train during the interval.

88P. The resistance to motion of an automobile depends on road friction, which is almost independent of speed, and on air drag, which is proportional to speed squared. For a car with a weight of 12,000 N, the total resistant force F is given by $F = 300 + 1.8v^2$, where F is in newtons and v is in meters per second. Calculate the power required to accelerate the car at 0.92 m/s^2 when the speed is 80 km/h.

89P*. A *governor* consists of two 200-g masses attached by light, rigid 10-cm rods to a vertical rotating axle (Fig.

FIGURE 8-55 Problem 89.

8-55). The rods are hinged so that the masses swing out from the axle as they rotate with it. However, when the angle θ is 45°, the masses encounter the wall of the cylinder in which the governor is rotating. (a) What is the minimum rate of rotation, in revolutions per minute, required for the masses to touch the wall? (b) If the coefficient of kinetic friction between the masses and the wall is 0.35, at what rate does the frictional force dissipate mechanical energy as a result of the masses rubbing against the wall when the mechanism is rotating at 300 rev/min?

90P*. A dragster with mass m races through a distance d from an initial dead stop. Assume that the engine provides a constant instantaneous power P for the entire race, and that the dragster can be treated as a particle. Find the elapsed time for the race.

91P*. At a factory, a 300-kg crate is dropped vertically from a packing machine onto a conveyor belt moving at 1.20 m/s (Fig. 8-56). (A motor maintains the belt's constant speed.) The coefficient of kinetic friction between belt and crate is 0.400. After a short time, slipping between the belt and the crate ceases and the crate then moves along with the belt. For the period of time during which the crate is being brought to rest relative to the belt, calculate, for a coordinate system at rest in the factory, (a) the kinetic energy supplied to the crate, (b) the magnitude of the kinetic frictional force acting on the crate, and (c) the energy supplied by the motor. (d) Why are the answers to (a) and (c) different?

SECTION 8-8 MASS AND ENERGY

92E. (a) How much energy in joules is represented by a mass of 102 g? (b) For how many years would this supply the energy needs of a one-family home consuming energy at the average rate of 1.00 kW?

93E. The magnitude M of an earthquake on the Richter scale is related to the released energy E in joules by the equation

$$\log E = 5.24 + 1.44M.$$

(a) The 1906 San Francisco earthquake was of magnitude 8.2 (Fig. 8-57). How much energy was released? (b) How much mass is equivalent to this amount of energy?

94E. The nucleus of an atom of gold contains 79 protons and 118 neutrons and has a mass of 196.9232 u. Calculate the binding energy of the nucleus. See Sample Problem 8-10 for other needed numerical data.

95P. A nuclear power plant in Oregon produces energy at the rate of 5980 MW, including 1030 MW of useful power and 2100 MW of power in thermal energy lost to

FIGURE 8-57 Exercise 93. Destruction on Nob Hill in San Francisco due to the earthquake of 1906. Over 400 km of the San Andreas fault line ruptured during the earthquake.

FIGURE 8-56 Problem 91.

the Columbia River. What is the mass equivalent of the energy lost in other forms at this plant in a year?

96P. The United States generated about 2.31×10^{12} kW·h of electrical energy in 1983. What is the mass equivalent of this energy?

97P. An aspirin tablet has a mass of 320 mg. For how many miles would the energy equivalent of this mass, in the form of gasoline, power a car? Assume 30 mi/gal and that gasoline provides 130 MJ/gal. Express your answer in terms of the circumference of the Earth.

98P. How much mass must be transferred into kinetic energy to accelerate a spaceship with a mass of 1.66×10^6 kg from rest to one-tenth the speed of light? Use the nonrelativistic formula for kinetic energy.

99P. The sun radiates energy at the rate of 4×10^{26} W. How many kilograms of sunlight does the Earth intercept in 1 day?

SECTION 8-9 ENERGY IS QUANTIZED

100E. By how much must the energy of an atom change in order to emit light of frequency 4.3×10^{14} s^{-1}?

101P. (a) A hydrogen atom has an energy of -3.40 eV. If its energy changes to -13.6 eV, what is the frequency of the light involved in the change? (b) Is the light emitted or absorbed?

ADDITIONAL PROBLEMS

102. A spring with spring constant $k = 200$ N/m is suspended vertically with its upper end fixed to the ceiling and its lower end at position $y = 0$ (Fig. 8-58). A block of weight 20 N is attached to the lower end, held still for a moment, and then released. What are the kinetic energy and the changes in the gravitational potential energy and the elastic potential energy of the spring–block system when the block is at y values of (a) -5.0 cm, (b) -10 cm, (c) -15 cm, and (d) -20 cm?

FIGURE 8-58 Problem 102.

103. A 60-kg skier leaves the end of a ski-jump ramp with a velocity of 24 m/s directed 25° above the horizontal. Suppose that as a result of air resistance the skier returns to the ground with a speed of 22 m/s and lands at a point that is 14 m vertically below the end of the ramp. How much energy is dissipated because of air resistance from the time the skier leaves the ramp until she returns to the ground?

104. A 0.40-kg particle moves under the influence of a single conservative force. At point A, where the particle has a speed of 10 m/s, the associated potential energy is 40 J. As the particle moves from A to B, the force does 25 J

of work on the particle. What is the value of the potential energy at point B?

105. A spring ($k = 200$ N/m) is fixed at the top of a frictionless plane that makes an angle of 40° with the horizontal, as shown in Fig. 8-59. A 1.0-kg mass is projected up the plane, from an initial position that is 0.60 m from the end of the uncompressed spring, with an initial kinetic energy of 16 J. (a) What is the kinetic energy of the mass at the instant it has compressed the spring 0.20 m? (b) With what kinetic energy must the mass be projected up the plane if it is to stop momentarily when it has compressed the spring by 0.40 m?

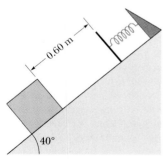

FIGURE 8-59 Problem 105.

106. A block is moved down an incline a distance of 5.0 m from point A to point B in Fig. 8-60 by a force $\mathbf{F} = 2.0$ N

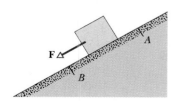

FIGURE 8-60 Problem 106.

acting parallel to the incline. The magnitude of the frictional force acting on the block between A and B is 10 N. If the kinetic energy of the block increases by 35 J between A and B, how much work is done on the block by its weight as the block moves from A to B?

107. At $t = 0$ a 1.0-kg ball is thrown from the top of a tall tower with velocity $\mathbf{v} = (18$ m/s$)\mathbf{i} + (24$ m/s$)\mathbf{j}$. What is the change in the ball's potential energy between $t = 0$ and $t = 6.0$ s?

108. In Fig. 8-61, the pulley is massless, and both it and the inclined plane are frictionless. If the system is released from rest, what is the combined kinetic energy of the two masses after the 2.0-kg mass has fallen 25 cm?

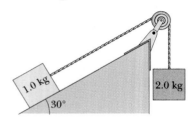

FIGURE 8-61 Problem 108.

109. Two blocks, of masses M and 2M where $M = 2.0$ kg, are connected to a spring of spring constant $k = 200$ N/m that has one end fixed, as shown in Fig. 8-62. The horizontal surface and the pulley are frictionless, and the pulley is massless. The system is released from rest with the spring

FIGURE 8-62 Problem 109.

unstretched. (a) What is the combined kinetic energy of the two masses after the hanging mass has fallen 0.090 m? (b) What is the kinetic energy of the hanging mass after it has fallen 0.090 m? (c) What maximum distance does the hanging mass fall before momentarily stopping?

110. A 20-kg block on a horizontal surface is attached to a horizontal spring of spring constant $k = 4.0$ kN/m. The block is pulled to the right so that the spring is extended 10 cm beyond its unstretched length, and the block is then released from rest. The frictional force between the sliding block and the surface has a magnitude of 80 N. (a) What is the kinetic energy of the block when it has moved 2.0 cm from its point of release? (b) What is the kinetic energy of the block when it first slides back through the point where the spring is unstretched? (c) What is the maximum kinetic energy attained by the block as it slides from its point of release to the point where the spring is unstretched?

SYSTEMS OF PARTICLES

*If you leap forward, chances are that your head and torso
will follow a parabolic path, like a baseball thrown in from
the outfield.   But when a skilled ballet dancer leaps across
the stage in a <u>grand jeté</u> as shown in the photograph, the
path taken by her head and torso is nearly horizontal during
much of the jump.   She seems to be floating across the stage.
The audience may not know Newton's laws of motion, but
they still sense that something unusual has happened.   How
does the ballerina seemingly "turn off" Newton's laws?*

9-1 A SPECIAL POINT

Physicists love to look at something complicated and find in it something simple and familiar. Here is an example. If you toss a baseball bat into the air, its motion as it turns is clearly more complicated than that of, say, a tossed ball (Fig. 9-1a). Every part of the bat moves in a different way from every other part, so you cannot represent the bat as a single particle; instead, it is a system of particles. However, if you look closely, you will find that one special point of the bat moves in a simple parabolic path, much as a tossed ball does. This point, called the **center of mass,** lies along the axis of the bat. You can locate it by balancing the bat on an outstretched finger: the center of mass is on the bat's axis just above your finger.

Every body has a center of mass. Figure 9-1b shows an ax thrown between two jugglers. The center of mass of the ax (and no other point) moves like a free particle, following a parabolic path. If the jugglers were doing their stunt in a dark room (!) and if a small light bulb were fastened to the center of mass of the ax, the simplicity of its motion would be clear.

9-2 THE CENTER OF MASS

In discussing center of mass, our plan is to start with simple systems of particles and then work our way up to extended objects like baseball bats.

Systems of Particles

Figure 9-2a shows two particles of masses m_1 and m_2 separated by a distance d. We have arbitrarily chosen the origin of the x axis to coincide with m_1. We *define* the position of the center of mass of this two-particle system to be

$$x_{cm} = \frac{m_2}{m_1 + m_2}\, d. \qquad (9\text{-}1)$$

Suppose, as an example, that $m_2 = 0$. Then there is only one particle (m_1), and the center of mass must lie at the position of that particle; Eq. 9-1 dutifully reduces to $x_{cm} = 0$. If $m_1 = 0$, there is again only one particle (m_2), and we have, as we expect, $x_{cm} = d$. If $m_1 = m_2$, the masses of the particles are equal and the center of mass should be halfway between them; Eq. 9-1 reduces to $x_{cm} = \frac{1}{2}d$, again as we

FIGURE 9-1 (*a*) The center of mass of a ball, at its center, follows a parabolic path when the ball is tossed. (*b*) The center of mass (the dot) of an ax that is tossed between two performers does also, but all other points of the ax follow more complicated paths.

(a)

(b)

expect. Finally, Eq. 9-1 tells us that, if neither m_1 nor m_2 is zero, x_{cm} can have only values that lie between zero and d; that is, the center of mass must lie somewhere between the two particles.

Figure 9-2b shows a more generalized situation, in which the coordinate system has been shifted leftward. The position of the center of mass is now defined as

$$x_{cm} = \frac{m_1 x_1 + m_2 x_2}{m_1 + m_2}. \qquad (9\text{-}2)$$

Note that if we put $x_1 = 0$, then x_2 becomes d and Eq. 9-2 reduces to Eq. 9-1, as it must. Note also that in spite of the shift of the coordinate system, the center of mass is still the same distance from each particle.

We can rewrite Eq. 9-2 as

$$x_{cm} = \frac{m_1 x_1 + m_2 x_2}{M}, \qquad (9\text{-}3)$$

in which M is the total mass of the system. Here, $M = m_1 + m_2$. We can extend this definition to a more general situation in which n particles are strung out along the x axis. Then the total mass is $M = m_1 + m_2 + \cdots + m_n$, and the location of the center of mass is

$$x_{cm} = \frac{m_1 x_1 + m_2 x_2 + m_3 x_3 + \cdots + m_n x_n}{M}$$

$$= \frac{1}{M} \sum_{i=1}^{n} m_i x_i. \qquad (9\text{-}4)$$

Here the subscript i is a running number, or index, that takes on all integer values from 1 to n. It identifies the various particles and their coordinates, in order.

If the particles are distributed in three dimensions, the center of mass must be identified by three coordinates. By extension of Eq. 9-4, they are

$$x_{cm} = \frac{1}{M} \sum_{i=1}^{n} m_i x_i,$$

$$y_{cm} = \frac{1}{M} \sum_{i=1}^{n} m_i y_i, \qquad (9\text{-}5)$$

$$z_{cm} = \frac{1}{M} \sum_{i=1}^{n} m_i z_i.$$

We can also define the center of mass with the language of vectors. The position of a particle whose coordinates are x_i, y_i, and z_i is given by a position vector:

$$\mathbf{r}_i = x_i \mathbf{i} + y_i \mathbf{j} + z_i \mathbf{k}. \qquad (9\text{-}6)$$

Here the index identifies the particle, and $\mathbf{i}$, $\mathbf{j}$, and $\mathbf{k}$ are unit vectors pointing, respectively, in the direction of the x, y, and z axes. Similarly, the position of the center of mass of a system of particles is given by a position vector:

$$\mathbf{r}_{cm} = x_{cm} \mathbf{i} + y_{cm} \mathbf{j} + z_{cm} \mathbf{k}. \qquad (9\text{-}7)$$

The three scalar equations of Eq. 9-5 can now be replaced by a single vector equation,

$$\mathbf{r}_{cm} = \frac{1}{M} \sum_{i=1}^{n} m_i \mathbf{r}_i, \qquad (9\text{-}8)$$

(a)

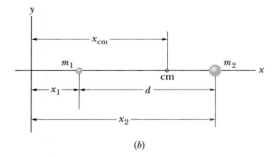

(b)

FIGURE 9-2 (a) Two particles of masses m_1 and m_2 are separated by a distance d. The dot labeled cm shows the position of the center of mass, calculated from Eq. 9-1. (b) The same as (a) except that the origin is located farther from the particles. The position of the center of mass is calculated from Eq. 9-3. The relative location of the center of mass (with respect to the particles) is the same in both cases.

where again M is the total mass of the system. You can check that this equation is correct by substituting Eqs. 9-6 and 9-7 into it, and then separating out the x, y, and z components. The scalar relations of Eq. 9-5 result.

Rigid Bodies

An ordinary object, such as a bat or an ax, contains so many particles (atoms) that we can best treat it as a continuous distribution of matter. The "particles" then become differential mass elements dm, the sums of Eq. 9-5 become integrals, and the coordinates of the center of mass are defined as

$$x_{cm} = \frac{1}{M} \int x \, dm,$$

$$y_{cm} = \frac{1}{M} \int y \, dm, \qquad (9\text{-}9)$$

$$z_{cm} = \frac{1}{M} \int z \, dm.$$

In principle, the integrals are to be evaluated for all the mass elements in the object. In practice, however, we rewrite them in terms of the coordinates of the mass elements. If the object has uniform density (mass per volume), then we can write

$$\frac{dm}{dV} = \frac{M}{V}, \qquad (9\text{-}10)$$

where dV is the volume occupied by a mass element, and V is the total volume of the object. We next substitute dm from Eq. 9-10 into Eq. 9-9, finding

$$x_{cm} = \frac{1}{V} \int x \, dV,$$

$$y_{cm} = \frac{1}{V} \int y \, dV, \qquad (9\text{-}11)$$

$$z_{cm} = \frac{1}{V} \int z \, dV.$$

These integrals are evaluated over the volume of the object. An example is given in Sample Problem 9-4.

Many objects have a point, a line, or a plane of symmetry. The center of mass of such an object then lies at that point, on that line, or in that plane. For example, the center of mass of a homogeneous sphere (which has a point of symmetry) is at the center of the sphere. The center of mass of a homogeneous cone (whose axis is a line of symmetry) lies on the axis of the cone. The center of mass of a banana (which has a plane of symmetry that splits it into two equal parts) lies somewhere in that plane.

The center of mass of an object need not lie within the object. There is no rubber at the center of

mass of an automobile tire, and no iron at the center of mass of a horseshoe.

SAMPLE PROBLEM 9-1

Figure 9-3 shows three particles of masses $m_1 = 1.2$ kg, $m_2 = 2.5$ kg, and $m_3 = 3.4$ kg located at the corners of an equilateral triangle of edge $a = 140$ cm. Where is the center of mass?

SOLUTION We choose our x and y coordinate axes so that one of the particles is located at the origin and so that the x axis coincides with one of the sides of the triangle. The coordinates of the three particles can be shown to have the following values:

PARTICLE	MASS (kg)	x (cm)	y (cm)
m_1	1.2	0	0
m_2	2.5	140	0
m_3	3.4	70	121

Due to our wise choice of coordinate axes, three of the coordinates in the above table are zero, simplifying the calculations. The total mass M of the system is 7.1 kg.

From Eq. 9-5, the coordinates of the center of mass are

$$x_{cm} = \frac{1}{M} \sum_{i=1}^{3} m_i x_i = \frac{m_1 x_1 + m_2 x_2 + m_3 x_3}{M}$$

$$= \frac{(1.2 \text{ kg})(0) + (2.5 \text{ kg})(140 \text{ cm}) + (3.4 \text{ kg})(70 \text{ cm})}{7.1 \text{ kg}}$$

$$= 83 \text{ cm} \qquad \text{(Answer)}$$

FIGURE 9-3 Sample Problem 9-1. Three particles having different masses form an equilateral triangle of side a. The center of mass is located by the position vector $\mathbf{r}_{cm}$.

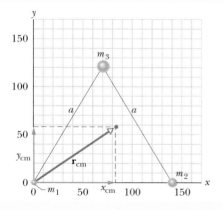

and

$$y_{cm} = \frac{1}{M} \sum_{i=1}^{3} m_i y_i = \frac{m_1 y_1 + m_2 y_2 + m_3 y_3}{M}$$

$$= \frac{(1.2 \text{ kg})(0) + (2.5 \text{ kg})(0) + (3.4 \text{ kg})(121 \text{ cm})}{7.1 \text{ kg}}$$

$$= 58 \text{ cm}. \qquad \text{(Answer)}$$

The center of mass is located by the position vector $\mathbf{r}_{cm}$ in Fig. 9-3.

SAMPLE PROBLEM 9-2

Find the center of mass of the uniform triangular plate that is shown in each part of Fig. 9-4.

SOLUTION Figure 9-4a shows the plate divided into thin slats, parallel to one side of the triangle. From symmetry, the center of mass of a thin, uniform slat is at its midpoint. The center of mass of the triangular plate must then lie somewhere along the line that connects the midpoint of all the slats. That bisecting line also connects the upper vertex with the midpoint of the opposite side. The plate would balance if it were placed on a knife-edge coinciding with this line of symmetry.

In Figs. 9-4b and 9-4c, we subdivide the plate into slats parallel to the other two sides of the triangle. Again, the center of mass must lie somewhere along each of the bisecting lines shown. Hence the center of mass of the plate must lie at the intersection of these three symmetry lines, as Fig. 9-4d shows. It is the only point that the three lines have in common.

You can check this conclusion experimentally by taking advantage of the (correct) intuitive notion that an object suspended from a point will orient itself so that its center of mass lies vertically below that point. Suspend the triangle from each vertex in turn, and draw a line vertically downward from the suspension point, as in Fig. 9-4e. The center of mass of the triangle will be at the intersection of the three lines.

FIGURE 9-4 Sample Problem 9-2. In (a), (b), and (c), the triangle is divided into thin slats, parallel to each of the three sides. The center of mass must lie along the bisecting lines shown. (d) The dot, the only point common to all three lines, is the position of the center of mass. (e) Finding the center of mass by suspending the triangle from each vertex in turn.

(a) (b) (c)

(d)

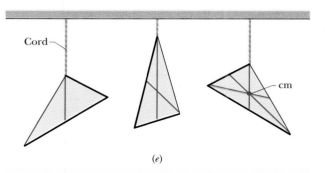

Cord

cm

(e)

SAMPLE PROBLEM 9-3

Figure 9-5a shows a circular metal plate of radius $2R$ from which a disk of radius R has been removed. Let us call this plate with a hole object X. Its center of mass is shown as a dot on the x axis. Locate this point.

SOLUTION Figure 9-5b shows object X before the disk was removed. Call the disk object D, and the original composite plate object C. From symmetry, the center of mass of object C is at the origin of a coordinate system placed as shown.

In finding the center of mass of a composite object, we can assume that the masses of its components are concentrated at their individual centers of mass. Thus object C can be treated as equivalent to two point masses, representing objects X and D. Figure 9-5c shows the positions of the centers of mass of these three objects.

The position of the center of mass of object C is given, from Eq. 9-2, as

$$x_C = \frac{m_D x_D + m_X x_X}{m_D + m_X},$$

in which x_D and x_X are the positions of the centers of mass of objects D and X, respectively. Noting that $x_C =$

(a)

(b)

(c)

FIGURE 9-5 Sample Problem 9-3. (a) Object X is a metal disk of radius $2R$ with a hole of radius R cut in it; its center of mass is at x_X. (b) Object D is a metal disk that fills the hole in object X; its center of mass is at $x_D = -R$. Object C is the composite object made up of objects X and D; its center of mass is at the origin of coordinates. (c) The centers of mass of the three objects.

0 and solving for x_X, we obtain

$$x_X = -\frac{x_D m_D}{m_X}. \qquad (9\text{-}12)$$

If ρ is the density (mass per volume) of the plate material and t is the thickness of the plate, we have

$$m_D = \pi R^2 \rho t \quad \text{and} \quad m_X = \pi(2R)^2 \rho t - \pi R^2 \rho t.$$

With these substitutions and with $x_D = -R$, Eq. 9-12 becomes

$$x_X = -\frac{(-R)(\pi R^2 \rho t)}{\pi(2R)^2 \rho t - \pi R^2 \rho t} = \tfrac{1}{3}R. \quad \text{(Answer)}$$

Note that the density and the thickness of the uniform plate cancel out.

SAMPLE PROBLEM 9-4

Silbury Hill (Fig. 9-6a), a mound on the plains near Stonehenge, was built 4600 years ago for unknown reasons, perhaps as a burial site. It is an incomplete right-circular cone (see Fig. 9-6b), with a flattened top of radius $r_2 = 16$ m, a base radius $r_1 = 88$ m, and a height $h = 40$ m. The sides of the cone make an angle $\theta = 30°$ with the horizontal.

a. Where is the center of mass of the mound?

SOLUTION   Because of the circular symmetry of the mound, the center of mass lies on the central vertical axis of the cone, at height z_{cm} above the base. To find z_{cm}, we use the last part of Eq. 9-11. We can simplify the integral by using the symmetry of the mound. To do this, we consider a thin, horizontal "wafer," as shown in Fig. 9-6b. The wafer has radius r, thickness dz, and cross-sectional area πr^2 and is at height z from the base. Its volume dV is

$$dV = \pi r^2\, dz. \qquad (9\text{-}13)$$

The mound consists of a stack of such wafers, with radii ranging from r_1 at the bottom of the stack to r_2 at the top. If the cone were complete, it would have a height that we call H in Fig. 9-6b. The radius r of each wafer is related to H by

$$\tan \theta = \frac{H}{r_1} = \frac{H - z}{r},$$

or

$$r = (H - z)\frac{r_1}{H}. \qquad (9\text{-}14)$$

Substituting Eqs. 9-13 and 9-14 into the last part of Eq. 9-11, we have

$$z_{cm} = \frac{1}{V}\int z\,dV = \frac{\pi r_1^2}{VH^2}\int_0^h z(H-z)^2\,dz$$

$$= \frac{\pi r_1^2}{VH^2}\int_0^h (z^3 - 2z^2H + zH^2)\,dz$$

$$= \frac{\pi r_1^2}{VH^2}\left[\frac{z^4}{4} - \frac{2z^3H}{3} + \frac{z^2H^2}{2}\right]_0^h$$

$$= \frac{\pi r_1^2 h^4}{VH^2}\left[\frac{1}{4} - \frac{2H}{3h} + \frac{H^2}{2h^2}\right]. \qquad (9\text{-}15)$$

The volume of a right-circular cone is $\frac{1}{3}\pi R^2 Z$, where R is the base radius, and Z is the height. The completed cone of Fig. 9-6b has height $H = r_1 \tan 30° = 50.8$ m. To find the volume of the mound, we subtract the volume of a cone with base radius r_2 and height $(H - h)$ from the volume of a cone with base radius r_1 and height H:

$$V = \tfrac{1}{3}\pi r_1^2 H - \tfrac{1}{3}\pi r_2^2(H - h)$$

$$= \tfrac{1}{3}\pi[(88\text{ m})^2(50.8\text{ m}) - (16\text{ m})^2(50.8\text{ m} - 40\text{ m})]$$

$$= 4.091 \times 10^5\text{ m}^3.$$

Substituting this and other known values into Eq. 9-15, we find

$$z_{cm} = \frac{\pi(88\text{ m})^2(40\text{ m})^4}{(4.091 \times 10^5\text{ m}^3)(50.8\text{ m})^2}$$

$$\times \left[\frac{1}{4} - \frac{2(50.8\text{ m})}{3(40\text{ m})} + \frac{(50.8\text{ m})^2}{2(40\text{ m})^2}\right]$$

$$= 12.37\text{ m} \approx 12\text{ m}. \qquad \text{(Answer)}$$

(a)

(b)

FIGURE 9-6 Sample Problem 9-4. (a) Silbury Hill in England, built by Neolithic people, required an estimated 1.8×10^7 hours of labor. Its purpose is not known. (b) An incomplete, right-circular cone that resembles Silbury Hill. A "wafer" of radius r and thickness dz is shown at height z from the base.

b. If Silbury Hill has density $\rho = 1.5 \times 10^3$ kg/m^3, then how much work was required to lift the dirt from the level of the base to build the mound?

SOLUTION The required work equals the gravitational potential energy U associated with the hill, which we can calculate by assuming that all the mass is concentrated at the center of mass. The mass of the hill is given by $m = \rho V$. We then have

$$W = U = mgz_{cm} = \rho V g z_{cm}$$

$$= (1.5 \times 10^3\text{ kg/m}^3)(4.091 \times 10^5\text{ m}^3)$$

$$\times (9.8\text{ m/s}^2)(12.37\text{ m})$$

$$= 7.4 \times 10^{10}\text{ J}. \qquad \text{(Answer)}$$

PROBLEM SOLVING

———————✺/\/\/✺———————

TACTIC 1: CENTER-OF-MASS PROBLEMS

Sample Problems 9-1 to 9-3 provide three strategies for simplifying center-of-mass problems. (1) Make full use of the symmetry of the object, be it point, line, or plane. (2) If the object can be divided into several parts, treat each of these parts as a particle, located at its own center of mass. (3) Choose your axes wisely: if your system is a group of particles, choose one of the particles as your origin. If your system is a body with a line of symmetry, there is your *x* axis! The choice of origin is completely arbitrary; the location of the center of mass is the same regardless of the origin from which it is measured.

9-3 NEWTON'S SECOND LAW FOR A SYSTEM OF PARTICLES

If you roll a cue ball at a second billiard ball that is at rest, you expect that the two-ball system will continue to have some forward motion after impact. You would be surprised, for example, if both balls came back toward you or if both moved to the right or to the left.

What continues to move forward, its steady motion completely unaffected by the collision, is the center of mass of the two balls. If you focus on this point—which is always halfway between these balls of identical mass—you can easily convince yourself by trial at a pool table that this is so. No matter whether the collision is glancing, head on, or somewhere in between, the center of mass moves majestically forward, as if the collision had never occurred. Let us look into this more closely.

We replace the pair of billiard balls with an assemblage of *n* particles of (possibly) different masses. We are interested not in the individual motions of these particles but *only* in the motion of their center of mass. Although the center of mass is just a point, it moves like a particle whose mass is equal to the total mass of the system; we can assign a position, a velocity, and an acceleration to it. We state (and shall prove below) that the (vector) equation that governs the motion of the center of mass of such a system of particles is

$$\sum \mathbf{F}_{\text{ext}} = M\mathbf{a}_{\text{cm}} \qquad \text{(system of particles).} \qquad (9\text{-}16)$$

Equation 9-16 is Newton's second law, governing the motion of the center of mass of a system of particles. It is remarkable that it retains the same form ($\sum \mathbf{F} = m\mathbf{a}$) that holds for the motion of a single particle. In using Eq. 9-16, the three quantities that appear in it must be evaluated with some care:

1. $\sum \mathbf{F}_{\text{ext}}$ is the *vector* sum of *all* the *external* forces that act on the system. Forces exerted by one part of the system on the other are called *internal forces*, and you must be careful not to include them in using Eq. 9-16.

2. *M* is the *total mass* of the system. We assume that no mass enters or leaves the system as it moves, so that *M* remains constant. The system is said to be **closed.**

3. $\mathbf{a}_{\text{cm}}$ is the acceleration of the *center of mass* of the system. Equation 9-16 gives no information about the acceleration of any other point of the system.

Equation 9-16, like all vector equations, is equivalent to three scalar equations involving the components of $\sum \mathbf{F}_{\text{ext}}$ and $\mathbf{a}_{\text{cm}}$ along the three coordinate axes. These equations are

$$\sum F_{\text{ext},x} = Ma_{\text{cm},x},$$
$$\sum F_{\text{ext},y} = Ma_{\text{cm},y}, \qquad (9\text{-}17)$$
$$\sum F_{\text{ext},z} = Ma_{\text{cm},z}.$$

Now we can examine the behavior of the billiard balls. Once the cue ball is rolling, no net external force acts on the (two-ball) system. And thus, because $\sum \mathbf{F}_{\text{ext}} = 0$, Eq. 9-16 tells us that $\mathbf{a}_{\text{cm}} = 0$ also. Because acceleration is the rate of change of velocity, we conclude that the velocity of the center of mass of the system of two balls does not change. When the two balls collide, the forces that come into play are *internal* forces, exerted by one ball on the other. Such forces do not contribute to $\sum \mathbf{F}_{\text{ext}}$, which remains zero. Thus the center of mass of the system, which was moving forward before the collision, must continue to move forward after the collision, with the same speed and in the same direction.

Consider now two other particles that interact with each other, the Earth (mass m_E) and the moon (mass m_M). In contrast with the billiard balls of our example above, an external force *does* act on this system, the gravitational attraction of the sun. According to Eq. 9-16, the center of mass of the Earth–

FIGURE 9-7 The sun's gravitational force is an external force that acts on the Earth–moon system. The center of mass of this system orbits the sun smoothly, as if all the mass of the system were concentrated at this point. The Earth and the moon follow wavy paths about this orbit. Here, the dots represent the centers of the Earth and moon, and the X's represent the center of mass of the system. (Actually, the latter point lies *within* the Earth, so the Earth "wobbles" much less than the figure suggests.)

FIGURE 9-8 A fireworks rocket explodes in flight. In the absence of air drag, the center of mass of the fragments would continue to follow the original parabolic path, until fragments began to hit the ground.

moon system moves under the influence of the sun's gravitational force as if a single particle of mass $m_E + m_M$ were concentrated at that point and as if the sun's gravitational force acted there. As the Earth and moon move around the moving center of mass, they travel along wavy paths, as Fig. 9-7 shows.

Equation 9-16 applies not only to a system of particles but also to a solid body, such as the ax of Fig. 9-1*b*. In that case, *M* in Eq. 9-16 is the mass of the ax and $\Sigma \mathbf{F}_{ext}$ is the weight $M\mathbf{g}$ of the ax. Equation 9-16 then tells us that $\mathbf{a}_{cm} = \mathbf{g}$. In other words,

the center of mass of the ax moves as if it were a single particle of mass *M*.

Figure 9-8 shows another interesting case. Suppose that at a fireworks display, a rocket is launched on a parabolic path. At a certain point, it explodes into fragments. If the explosion had not occurred, the rocket would have continued along the trajectory shown in the figure. The forces of the explosion are *internal* to the system (the rocket or its fragments); that is, they are forces exerted by one part of the system on another part. If we ignore air resistance, the total *external* force $\Sigma \mathbf{F}_{ext}$ acting on the system is the weight $M\mathbf{g}$ of the system, whether the

FIGURE 9-9 A grand jeté. (Adapted from *The Physics of Dance*, by Kenneth Laws, Schirmer Books, 1984.)

rocket explodes or not. Thus, from Eq. 9-16, the acceleration of the center of mass of the fragments (while they are in flight) remains equal to **g**, and the center of mass of the fragments follows the same parabolic trajectory that the unexploded rocket would have followed.

When a ballet dancer leaps across the stage in a grand jeté, she raises her arms and stretches her legs out horizontally as soon as her feet leave the stage (Fig. 9-9). These actions shift her center of mass upward through her body. Although the shifting center of mass faithfully follows a parabolic path across the stage, its movement relative to the body decreases the height that would be attained by the head and torso in a normal jump. The result is that the head and torso follow a nearly horizontal path.

Proof of Equation 9-16

Now let us prove this important equation. From Eq. 9-8 we have, for a system of n particles,

$$M\mathbf{r}_{cm} = m_1\mathbf{r}_1 + m_2\mathbf{r}_2 + m_3\mathbf{r}_3 + \cdots + m_n\mathbf{r}_n, \quad (9\text{-}18)$$

in which M is the total mass of the system and $\mathbf{r}_{cm}$ is the vector that locates the position of the center of mass, just as if the center of mass were a real particle.

Differentiating Eq. 9-18 with respect to time gives

$$M\mathbf{v}_{cm} = m_1\mathbf{v}_1 + m_2\mathbf{v}_2 + m_3\mathbf{v}_3 + \cdots + m_n\mathbf{v}_n. \quad (9\text{-}19)$$

Here $\mathbf{v}_1$ ($= d\mathbf{r}_1/dt$) is the velocity of the first particle, and so on, and $\mathbf{v}_{cm}$ ($= d\mathbf{r}_{cm}/dt$) is the velocity of the center of mass.

Differentiating Eq. 9-19 with respect to time leads to

$$M\mathbf{a}_{cm} = m_1\mathbf{a}_1 + m_2\mathbf{a}_2 + m_3\mathbf{a}_3 + \cdots + m_n\mathbf{a}_n. \quad (9\text{-}20)$$

Here $a_1(= d\mathbf{v}_1/dt)$ is the acceleration of the first particle, and so on, and $\mathbf{a}_{cm}(= d\mathbf{v}_{cm}/dt)$ is the acceleration of the center of mass. Although the center of mass is just a geometrical point, it has a position, a velocity, and an acceleration, as if it were a particle.

From Newton's second law, $m_1\mathbf{a}_1$ is the resultant force $\mathbf{F}_1$ that acts on the first particle, and so on. Thus we can rewrite Eq. 9-20 as

$$M\mathbf{a}_{cm} = \mathbf{F}_1 + \mathbf{F}_2 + \mathbf{F}_3 + \cdots + \mathbf{F}_n. \quad (9\text{-}21)$$

Among the forces that contribute to the right side of Eq. 9-21 will be forces that the particles of the system exert on each other (internal forces) and forces exerted on the particles from outside the system (ex-

ternal forces). By Newton's third law, the internal forces form action–reaction pairs and cancel out in the sum that appears on the right side of Eq. 9-21. What remains is the vector sum of all the *external* forces that act on the system. Equation 9-21 then reduces to Eq. 9-16, the relation that we set out to prove.

SAMPLE PROBLEM 9-5

A ball of mass m and radius R is placed inside a spherical shell of the same mass m and inner radius $2R$. The combination is at rest on a tabletop in the position shown in Fig. 9-10a. The ball is released, rolls back and forth inside, and finally comes to rest at the bottom of the shell, as in Fig. 9-10c. What is the displacement d of the shell during this process?

SOLUTION The only external forces acting on the ball–shell system are the downward weight of the system and the normal force exerted vertically upward by the table. Neither force has a horizontal component, so $\Sigma F_{ext,x} = 0$. From Eq. 9-17, the acceleration component $a_{cm,x}$ of the center of mass of the system must also be zero. Since there is no initial motion, the horizontal position of the center of mass of the system must be stationary throughout the rolling motion of the ball and the shell.

We can represent both ball and shell as single particles of mass m, located at their respective centers. Figure 9-10b shows the system before the ball is released, and Fig. 9-10d after the ball has come to rest at the bottom of the shell. We choose our origin to coincide with the initial position of the center of the shell. Figure 9-10b shows that, with respect to this origin, the center of mass of the ball–shell system is originally located a distance $\frac{1}{2}R$ to the left, halfway between the two particles (both have mass m). When the ball and shell finally come to rest, their centers of mass are at the same value of x as the center of mass of the ball–shell system. Thus, as Fig. 9-10d shows, the shell must move to the left through a displacement

$$d = \tfrac{1}{2}R \qquad \text{(Answer)}$$

to keep the center of mass stationary.

SAMPLE PROBLEM 9-6

Figure 9-11a shows a system of three particles, each particle acted on by a different external force and all initially at rest. What is the acceleration of the center of mass of this system?

(a)

(b)

(c)

(d)

FIGURE 9-10   Sample Problem 9-5. (*a*) A ball of radius *R* is released from this initial position, free to roll inside a spherical shell of inner radius 2*R*. (*b*) The centers of mass of the ball, the shell, and the combination of the two. (*c*) The final state, after the ball has come to rest. The shell has moved so that the center of mass of the system remains in place. (*d*) The centers of mass of the ball, the shell, and the combination of the two are now at the same value of *x*.

(a)

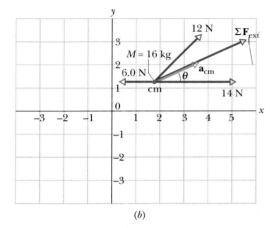

(b)

FIGURE 9-11   Sample Problem 9-6. (*a*) Three particles, placed at rest in the positions shown, are acted on by the external forces shown. The center of mass of the system is marked. (*b*) The forces are transferred to the center of mass of the system, which behaves like a particle whose mass *M* is equal to the total mass of the system. The net force and the acceleration of the center of mass are shown.

SOLUTION   The position of the center of mass, calculated by the method of Sample Problem 9-1, is marked by a dot in the figure. As Fig. 9-11*b* suggests, we treat this point as if it were a real particle, assigning to it a mass *M* equal to the total mass of the system (16 kg) and assuming that all external forces are applied at that point.

The *x* component of the net external force $\Sigma \mathbf{F}_{\text{ext}}$ acting on the center of mass is

$$\sum F_{\text{ext},x} = 14 \text{ N} - 6.0 \text{ N} + (12 \text{ N})(\cos 45°)$$

$$= 16.5 \text{ N},$$

and the y component is

$$\sum F_{\text{ext},y} = (12 \text{ N})(\sin 45°) = 8.49 \text{ N}.$$

The net external force thus has the magnitude

$$\sum F_{\text{ext}} = \sqrt{(16.5 \text{ N})^2 + (8.49 \text{ N})^2} = 18.6 \text{ N}$$

and makes an angle with the x axis given by

$$\theta = \tan^{-1} \frac{8.49 \text{ N}}{16.5 \text{ N}} = \tan^{-1} 0.515$$

$$= 27°. \qquad \text{(Answer)}$$

This is also the direction of the acceleration vector $\mathbf{a}_{\text{cm}}$ of the center of mass. From Eq. 9-16, the magnitude of $\mathbf{a}_{\text{cm}}$ is given by

$$a_{\text{cm}} = \frac{\sum F_{\text{ext}}}{M} = \frac{18.6 \text{ N}}{16 \text{ kg}} = 1.16 \text{ m/s}^2$$

$$\approx 1.2 \text{ m/s}^2. \qquad \text{(Answer)}$$

The three particles of Fig. 9-11a and their center of mass move with (different) constant accelerations. Since the particles start from rest, each will move, with ever-increasing speed, along a straight line in the direction of the force acting *on it*. The center of mass will move in the direction of $\mathbf{a}_{\text{cm}}$.

9-4 LINEAR MOMENTUM

Momentum is another of those words that have several meanings in everyday language but only a single precise meaning in physics. The **linear momentum** of a particle is a vector $\mathbf{p}$, defined as

$$\mathbf{p} = m\mathbf{v} \qquad \begin{array}{l}\text{(linear momentum} \\ \text{of a particle),}\end{array} \qquad (9\text{-}22)$$

in which m is the mass of the particle, and $\mathbf{v}$ is its velocity. (The adjective *linear* is often dropped, but it serves to distinguish $\mathbf{p}$ from *angular* momentum, which is introduced in Chapter 12.) Note that since m is always a positive scalar quantity, Eq. 9-22 tells us that $\mathbf{p}$ and $\mathbf{v}$ are in the same direction.

Newton actually expressed his second law of motion in terms of momentum:

The rate of change of the momentum of a particle is proportional to the net force acting on the particle and is in the direction of that force.

In equation form this becomes

$$\sum \mathbf{F} = \frac{d\mathbf{p}}{dt}. \qquad (9\text{-}23)$$

Substituting for $\mathbf{p}$ from Eq. 9-22 gives

$$\sum \mathbf{F} = \frac{d\mathbf{p}}{dt} = \frac{d}{dt}(m\mathbf{v}) = m\frac{d\mathbf{v}}{dt} = m\mathbf{a}.$$

Thus the relations $\sum \mathbf{F} = d\mathbf{p}/dt$ and $\sum \mathbf{F} = m\mathbf{a}$ are completely equivalent expressions of Newton's second law of motion as it applies to the motion of single particles in classical mechanics.

Momentum at Very High Speeds

For particles moving with speeds that are near the speed of light, Newtonian mechanics predicts results that do not agree with experiment. In such cases, we must use Einstein's theory of special relativity. In relativity, the formulation $\mathbf{F} = d\mathbf{p}/dt$ holds good, *provided* that we define the momentum of a particle not as $m\mathbf{v}$ but as

$$\mathbf{p} = \frac{m\mathbf{v}}{\sqrt{1 - (v/c)^2}}, \qquad (9\text{-}24)$$

in which c, the speed of light, is a sure indicator of a relativistic equation.

The speeds of common macroscopic objects such as baseballs, bullets, or space probes are so much less than the speed of light that the quantity $(v/c)^2$ in Eq. 9-24 is very much less than unity. Under these conditions, Eq. 9-24 reduces to Eq. 9-22 and Einstein's relativity theory reduces to Newtonian mechanics. For electrons and other subatomic particles, however, speeds very close to that of light are easily obtained and the definition in Eq. 9-24 *must* be used, often as a matter of routine engineering practice.

9-5 THE LINEAR MOMENTUM OF A SYSTEM OF PARTICLES

Consider now a system of n particles, each with its own mass, velocity, and linear momentum. The particles may interact with each other, and external forces may act on them as well. The system as a whole has a total linear momentum $\mathbf{P}$, which is defined to be the vector sum of the individual particle

TABLE 9-1
SOME DEFINITIONS AND LAWS IN CLASSICAL MECHANICS

LAW OR DEFINITION	SINGLE PARTICLE	SYSTEM OF PARTICLES
Newton's second law	$\Sigma\,\mathbf{F} = m\mathbf{a}$	$\Sigma\,\mathbf{F}_{\text{ext}} = M\mathbf{a}_{\text{cm}}$ (9-16)
Linear momentum	$\mathbf{p} = m\mathbf{v}$ (9-22)	$\mathbf{P} = M\mathbf{v}_{\text{cm}}$ (9-26)
Newton's second law	$\Sigma\,\mathbf{F} = d\mathbf{p}/dt$ (9-23)	$\Sigma\,\mathbf{F}_{\text{ext}} = d\mathbf{P}/dt$ (9-28)
Work–kinetic energy theorem	$W = \Delta K$	

linear momenta. Thus

$$\mathbf{P} = \mathbf{p}_1 + \mathbf{p}_2 + \mathbf{p}_3 + \cdots + \mathbf{p}_n$$

$$= m_1\mathbf{v}_1 + m_2\mathbf{v}_2 + m_3\mathbf{v}_3 + \cdots + m_n\mathbf{v}_n. \quad (9\text{-}25)$$

If we compare this equation with Eq. 9-19, we see that

$$\mathbf{P} = M\mathbf{v}_{\text{cm}} \qquad \begin{array}{l}\text{(linear momentum,}\\ \text{system of particles),}\end{array} \quad (9\text{-}26)$$

which gives us another way to define the linear momentum of a system of particles:

The linear momentum of a system of particles is equal to the product of the total mass M of the system and the velocity of the center of mass.

If we take the time derivative of Eq. 9-26, we find

$$\frac{d\mathbf{P}}{dt} = M\,\frac{d\mathbf{v}_{\text{cm}}}{dt} = M\mathbf{a}_{\text{cm}}. \quad (9\text{-}27)$$

Comparing Eqs. 9-16 and 9-27 allows us to write Newton's second law for a system of particles in the equivalent form

$$\sum \mathbf{F}_{\text{ext}} = \frac{d\mathbf{P}}{dt}. \quad (9\text{-}28)$$

This equation is the generalization of the single-particle equation $\Sigma\,\mathbf{F} = d\mathbf{p}/dt$ to a system of many particles. Table 9-1 displays the important relations that we have derived for single particles and for systems of particles.

9-6 CONSERVATION OF LINEAR MOMENTUM

Suppose that the sum of the external forces acting on a system of particles is zero (the system is iso-

lated), and that no particles leave or enter the system (the system is closed). Putting $\Sigma\,\mathbf{F}_{\text{ext}} = 0$ in Eq. 9-28 then yields $d\mathbf{P}/dt = 0$, or

$$\mathbf{P} = \text{constant} \qquad \text{(closed, isolated system).} \quad (9\text{-}29)$$

This important result, called the **law of conservation of linear momentum,** can also be written as

$$\mathbf{P}_i = \mathbf{P}_f \qquad \text{(closed, isolated system),} \quad (9\text{-}30)$$

where the subscripts refer to the values of $\mathbf{P}$ at some initial time and at a later time. Equations 9-29 and 9-30 tell us that, if no net external force acts on a system of particles, the total linear momentum of the system remains constant.

Like the law of conservation of energy that you met in Chapter 8, the law of conservation of linear momentum is a more general law than Newtonian mechanics itself. It holds in the subatomic realm, where Newton's laws fail. It holds for the highest particle speeds, where Einstein's relativity theory prevails, that is, where Eq. 9-24, rather than Eq. 9-22, must be used to find the linear momentum.

From Eq. 9-26 ($\mathbf{P} = M\mathbf{v}_{\text{cm}}$) we see that, if $\mathbf{P}$ is constant, then $\mathbf{v}_{\text{cm}}$, the velocity of the center of mass, is also constant. This in turn means that $\mathbf{a}_{\text{cm}}$, the acceleration of the center of mass, must be zero, which is consistent with Eq. 9-16 ($\Sigma\,\mathbf{F}_{\text{ext}} = M\mathbf{a}_{\text{cm}}$) when $\Sigma\,\mathbf{F}_{\text{ext}} = 0$. Thus the law of conservation of linear momentum is consistent with both forms of Newton's second law for systems of particles, as displayed in Table 9-1.

Equations 9-29 and 9-30 are vector equations and, as such, both are equivalent to three scalar equations corresponding to the conservation of linear momentum in three mutually perpendicular directions. Depending on the forces acting on a system, linear momentum might be conserved in only one or two directions but not in all directions.

As an example, suppose that you throw a ball across a field. During the ball's flight, the only external force acting on it is its weight $m\mathbf{g}$, which is directed vertically downward. In that situation, the vertical linear momentum of the ball (which we take as the system) changes, but since there is no horizontal external force, the horizontal linear momentum is conserved.

SAMPLE PROBLEM 9-7

Bullets, each of mass $m = 3.8$ g, are fired horizontally with speed $v = 1100$ m/s into a large wood block of mass $M = 12$ kg that is initially at rest on a horizontal table (Fig. 9-12). If the block is free to slide without friction across the table, what speed will it have acquired after it has absorbed eight bullets?

SOLUTION So that our system is closed, it must include both the block and the eight bullets, taken together, as indicated in Fig. 9-12.

The system is not isolated, because several external forces act on it: the bullets and the block all have weight, and the surface supporting the block exerts a normal force on it. However, these external forces are strictly vertical and cannot change the horizontal momenta of the bullets and block, and thus not the total horizontal momentum of the system either. Horizontally, no external forces act on the system. The forces involved in the collisions between the bullets and block are *internal forces* (within the system) and cannot change the horizontal momentum of the system.

Thus, since there is no external force acting to change the horizontal momentum of the system, we can apply the law of conservation of linear momentum (Eq. 9-30) in the direction of flight of the bullets. The initial horizontal linear momentum, measured while the bullets are still in flight* and the block is at rest, is

$$P_i = n(mv),$$

in which mv is the linear momentum of an individual bullet and $n = 8$. The final linear momentum, measured when all the bullets are in the block and the block is sliding over the table with speed V, is

$$P_f = (M + nm)V.$$

Conservation of linear momentum requires that

$$P_i = P_f$$

*For simplicity, we assume a rapid rate of fire, so that all the bullets are in flight before the first bullet strikes. (Actually, it doesn't make any difference.)

FIGURE 9-12 Sample Problem 9-7. A gun fires a stream of bullets toward a block of wood. The system to which we apply the law of conservation of linear momentum is indicated. This system consists of the block and eight bullets.

or

$$n(mv) = (M + nm)V.$$

Solving for V yields

$$V = \frac{nm}{M + nm}v$$

$$= \frac{(8)(3.8 \times 10^{-3} \text{ kg})}{12 \text{ kg} + (8)(3.8 \times 10^{-3} \text{ kg})}(1100 \text{ m/s})$$

$$= 2.8 \text{ m/s.} \qquad \text{(Answer)}$$

With the choice of system that we made, we did not have to consider the details of how the bullets hit the block.

SAMPLE PROBLEM 9-8

As Fig. 9-13 shows, a cannon whose mass M is 1300 kg fires a ball whose mass m is 72 kg in a horizontal direction with a velocity $\mathbf{v}$ relative to the cannon, which recoils (freely) with velocity $\mathbf{V}$ relative to the Earth. The magnitude of $\mathbf{v}$ is 55 m/s.

a. What is $\mathbf{V}$?

SOLUTION We choose the cannon plus the ball as our system. By doing so, we ensure that the forces involved in the firing of the cannon are internal to the system, and we do not have to deal with them. The external forces acting on the system have no components in the horizontal direction. Thus the horizontal component of the total linear momentum of the system must remain unchanged as the cannon is fired.

We choose the Earth as a reference frame, and we assume that all velocities are positive if they point to the right in Fig. 9-13. Because all velocities in the figure are

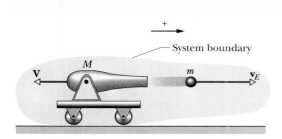

FIGURE 9-13 Sample Problem 9-8. A cannon of mass M fires a ball of mass m. The ball has velocity $\mathbf{v}_E$ relative to the Earth and velocity $\mathbf{v}$ relative to the cannon. The recoiling cannon has velocity $\mathbf{V}$ relative to the Earth. The arrow shows the direction taken as positive for all velocities.

horizontal, either to the left or right, they can be represented by the corresponding speeds, with an implied plus sign or an explicit minus sign. Thus velocity $\mathbf{v}$ is represented by v. Although Fig. 9-13 shows $\mathbf{V}$ pointing to the left, we actually do not yet know the direction of $\mathbf{V}$, and so it is represented by V, which can be either positive or negative.

The velocity of the ball relative to the Earth is $\mathbf{v}_E$, represented by v_E. The ball's velocity relative to the cannon is the difference between its velocity relative to the Earth and the cannon's velocity relative to the Earth. That is,

$$v = v_E - V,$$

or

$$v_E = v + V. \tag{9-31}$$

Before the cannon is fired, the system has an initial linear momentum P_i of zero. While the ball is in flight, the system has a horizontal linear momentum P_f which, with the aid of Eq. 9-31, is given by

$$P_f = MV + mv_E = MV + m(v + V),$$

in which the first term on the right is the linear momentum of the recoiling cannon and the second term that of the speeding ball.

Conservation of linear momentum in the horizontal direction requires that $P_i = P_f$, or

$$0 = MV + m(v + V).$$

Solving for V yields

$$V = -\frac{mv}{M + m} = -\frac{(72 \text{ kg})(55 \text{ m/s})}{1300 \text{ kg} + 72 \text{ kg}}$$

$$= -2.9 \text{ m/s}. \tag{Answer}$$

The minus sign tells us that the cannon recoils to the left in Fig. 9-13, as we know it does.

b. What is v_E?

SOLUTION From Eq. 9-31, we find

$$v_E = v + V = 55 \text{ m/s} + (-2.9 \text{ m/s})$$

$$= 52 \text{ m/s}. \tag{Answer}$$

Because of the recoil, the ball is moving a little slower relative to the Earth than it otherwise would.

Note the importance in this problem of choosing the system (cannon + ball) wisely and of being absolutely clear about the reference frame (Earth or recoiling cannon) to which the various measurements are referenced.

SAMPLE PROBLEM 9-9

A spaceship with mass M is traveling in deep space with velocity $v_i = 2100$ km/h relative to the sun. It ejects a rear stage of mass $0.20M$ with a relative speed $u = 500$ km/h. What then is the velocity of the ship?

SOLUTION We take the *ship + rear stage* as our system. We also take a velocity to be positive if it points in the ship's direction of travel. Because the system is closed and isolated, the linear momentum of the system is conserved; that is,

$$P_i = P_f, \tag{9-32}$$

where the subscripts i and f refer to values before and after the ejection, respectively. Before the ejection, we have

$$P_i = Mv_i. \tag{9-33}$$

Let U be the velocity of the ejected rear stage and v_f be the velocity of the ship after the ejection, both velocities measured relative to the sun. The total linear momentum of the system after the ejection is then

$$P_f = 0.20MU + 0.80Mv_f, \tag{9-34}$$

where the first term on the right is the linear momentum of the stage and the second term is that of the ship.

The relative speed u of the ejected stage is the difference in the velocities of the ship and the stage:

$$u = v_f - U,$$

or

$$U = v_f - u.$$

Substituting this expression for U into Eq. 9-34, and

then substituting Eqs. 9-33 and 9-34 into Eq. 9-32, we find

$$Mv_i = 0.20M(v_f - u) + 0.80Mv_f,$$

which gives us

$$v_f = v_i + 0.2u,$$

or

$$v_f = 2100 \text{ km/h} + (0.2)(500 \text{ km/h})$$

$$= 2200 \text{ km/h}. \qquad \text{(Answer)}$$

SAMPLE PROBLEM 9-10

Figure 9-14 shows two blocks that are connected by a spring and free to slide on a frictionless horizontal surface. The blocks, whose masses are m_1 and m_2, are pulled apart and then released from rest. What fraction of the total kinetic energy of the system will each block have at any later time?

SOLUTION   We take the two blocks and the spring as our system, and the horizontal surface on which they slide as our reference frame. We assume that velocities are positive if they point to the right in Fig. 9-14.

   The initial momentum P_i of the system before the blocks are released is zero. The final momentum, at any time after the blocks are released, is

$$P_f = m_1v_1 + m_2v_2,$$

in which v_1 and v_2 are the velocities of the blocks. Conservation of momentum requires that $P_i = P_f$, or

$$0 = m_1v_1 + m_2v_2.$$

Thus we have

$$\frac{v_1}{v_2} = -\frac{m_2}{m_1}. \qquad (9\text{-}35)$$

The minus sign tells us that the two velocities always have opposite signs. Equation 9-35 holds at every instant after release, no matter what the actual speeds of the blocks are.

   The kinetic energies of the blocks are $K_1 = \frac{1}{2}m_1v_1^2$ and $K_2 = \frac{1}{2}m_2v_2^2$, and the fraction we seek, for the block of mass m_1, is

$$\text{frac}_1 = \frac{K_1}{K_1 + K_2} = \frac{\frac{1}{2}m_1v_1^2}{\frac{1}{2}m_1v_1^2 + \frac{1}{2}m_2v_2^2}. \qquad (9\text{-}36)$$

From Eq. 9-35, $v_2 = -v_1(m_1/m_2)$. Making this substitu-

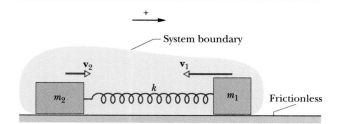

FIGURE 9-14   Sample Problem 9-10. Two blocks, resting on a frictionless surface and connected by a spring, have been pulled apart and then released from rest. The vector sum of their linear momenta remains zero during their subsequent motions. The system boundary is shown. The arrow shows the positive direction for all velocities.

tion into Eq. 9-36 leads, after a little algebra, to

$$\text{frac}_1 = \frac{m_2}{m_1 + m_2}. \qquad \text{(Answer)} \qquad (9\text{-}37)$$

Similarly, for the block of mass m_2,

$$\text{frac}_2 = \frac{m_1}{m_1 + m_2}. \qquad \text{(Answer)} \qquad (9\text{-}38)$$

Thus the kinetic-energy split is a constant, independent of time, the less massive block receiving the larger share of the available kinetic energy. If, for example, $m_2 = 10m_1$, then

$$\text{frac}_1 = \frac{10m_1}{m_1 + 10m_1} = 0.91$$

and

$$\text{frac}_2 = \frac{m_1}{m_1 + 10m_1} = 0.09.$$

If $m_2 \gg m_1$, we see from Eqs. 9-37 and 9-38 that $\text{frac}_1 \approx 100\%$ and $\text{frac}_2 \approx$ zero; that is, the lighter block gets essentially all the kinetic energy.

   Equations 9-37 and 9-38 apply equally well to a falling stone. Here m_2 is the mass of the Earth and m_1 that of the stone, and the spring of Fig. 9-14 is equivalent to the (attractive) gravitational force between the Earth and the stone. The best reference frame to adopt for this problem is a frame in which the center of mass of the stone–Earth system is at rest. In this frame, the stone has essentially all the kinetic energy ($\text{frac}_1 \approx 1$), and the Earth very little ($\text{frac}_2 \approx 0$). From Eq. 9-35, however, we see that the magnitudes of the linear momenta of the stone and the Earth remain equal at all times, the tiny speed v_2 of the Earth being compensated by m_2, its enormous mass.

SAMPLE PROBLEM 9-11

A firecracker placed inside a coconut of mass M, initially at rest on a frictionless floor, blows the fruit into three pieces and sends them sliding across the floor. An overhead view is shown in Fig. 9-15a. Piece C, with mass $0.30M$, has speed $v_{fC} = 5.0$ m/s. (The subscript f refers to the final speed of the piece.)

a. What is the speed of piece B, with mass $0.20M$?

SOLUTION We superimpose an xy coordinate system as shown in Fig. 9-15b, with the direction of decreasing x coinciding with $\mathbf{v}_{fA}$. The x axis makes an angle of $80°$ with $\mathbf{v}_{fC}$ and an angle of $50°$ with $\mathbf{v}_{fB}$.

The linear momentum of the coconut (and its pieces) is conserved along both the x and y axes, because the forces involved in the explosion are internal forces, and because there is no external force acting on the coconut along the x or y axis. Along the y axis the conservation of linear momentum is written as

$$P_{iy} = P_{fy}, \qquad (9\text{-}39)$$

where the subscript i refers to the initial value (before the explosion), and the subscript y refers to the y component of the vector $\mathbf{P}_i$ or $\mathbf{P}_f$.

The component P_{iy} of the initial linear momentum is zero, because the coconut is initially at rest. To get an expression for P_{fy}, we find the y component of the final linear momentum of each piece:

$$p_{fA,y} = 0,$$

$$p_{fB,y} = -0.20Mv_{fB,y} = -0.20Mv_{fB} \sin 50°,$$

$$p_{fC,y} = 0.30Mv_{fC,y} = 0.30Mv_{fC} \sin 80°.$$

(Note that $p_{fA,y} = 0$ because of our choice of axes.) Equation 9-39 can now be written as

$$P_{iy} = P_{fy} = p_{fA,y} + p_{fB,y} + p_{fC,y}.$$

Then, with $v_{fC} = 5.0$ m/s, we have

$$0 = 0 - 0.20Mv_{fB} \sin 50° + (0.30M)(5.0 \text{ m/s}) \sin 80°,$$

from which we find

$$v_{fB} = 9.64 \text{ m/s} \approx 9.6 \text{ m/s}. \qquad \text{(Answer)}$$

b. What is the speed of piece A?

SOLUTION Because linear momentum is also conserved along the x axis, we have

$$P_{ix} = P_{fx}, \qquad (9\text{-}40)$$

where $P_{ix} = 0$ because the coconut is initially at rest.

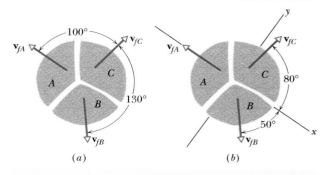

(a)   (b)

FIGURE 9-15 Sample Problem 9-11. Three pieces of an exploded coconut move off in three directions along a frictionless floor. (a) An overhead view of the event. (b) The same with a two-dimensional axis system imposed.

To get P_{fx}, we find the x components of the final momenta, using the fact that piece A must have a mass of $0.50M$:

$$p_{fA,x} = -0.50Mv_{fA},$$

$$p_{fB,x} = 0.20Mv_{fB,x} = 0.20Mv_{fB} \cos 50°,$$

$$p_{fC,x} = 0.30Mv_{fC,x} = 0.30Mv_{fC} \cos 80°.$$

Equation 9-40 can now be written as

$$P_{ix} = P_{fx} = p_{fA,x} + p_{fB,x} + p_{fC,x}.$$

Then, with $v_{fC} = 5.0$ m/s and $v_{fB} = 9.64$ m/s, we have

$$0 = -0.50Mv_{fA} + 0.20M(9.64 \text{ m/s}) \cos 50°$$

$$+ 0.30M(5.0 \text{ m/s}) \cos 80°,$$

from which we find

$$v_{fA} = 3.0 \text{ m/s}. \qquad \text{(Answer)}$$

PROBLEM SOLVING

TACTIC 2: CONSERVATION OF LINEAR MOMENTUM This is a good time to reread Tactic 1 of Chapter 8, which deals with problems involving the conservation of mechanical energy. The same broad suggestions hold here for problems involving the conservation of linear momentum.

First, make sure that you have chosen a closed, isolated system. *Closed* means that no matter (no particles) passes through the system boundary in any direction. *Isolated* means that the net external force acting on the system is zero. If the system is *not* isolated and closed, then Eqs. 9-29 and 9-30 do not hold.

Remember that linear momentum is a vector so that each component can be conserved separately, provided only that the corresponding component of the net external force is zero. In Sample Problem 9-8, there is no net external force acting horizontally on the cannon + ball system, so the horizontal component of linear momentum is conserved. However, the net external force acting vertically on this system is not zero; weight acts on the cannon ball while it is in flight. Thus the vertical component of linear momentum for this system is not conserved.

Select two appropriate states of the system (which you may choose to call the initial state and the final state) and write expressions for the linear momentum of the system in each of these two states. In writing these expressions, make sure that you know what inertial reference frame you are using, and make sure also that you include the entire system, not missing any part of it and not including objects that do not belong to your system. In Sample Problem 9-8 we must be particularly clear about the reference frame (Earth or recoiling cannon) to which the various measurements are referenced.

Finally, set your expressions for $\mathbf{P}_i$ and $\mathbf{P}_f$ equal to each other and solve for what is requested.

9-7 SYSTEMS WITH VARYING MASS: A ROCKET (OPTIONAL)

In the systems we have dealt with so far, we have assumed that the total mass of the system remains constant. Sometimes it does not, a rocket (Fig. 9-16) being a familiar example. Most of the mass of a rocket on its launching pad is fuel, all of which will eventually be burned and ejected from the nozzle of the rocket engine.

We handle the variation of the mass of the rocket as the rocket accelerates by applying Newton's second law, not to the rocket alone but to the rocket and its ejected combustion products taken together. The mass of *this* system does *not* change as the rocket accelerates.

Finding the Acceleration

Assume that you are in an inertial reference frame, watching a rocket accelerate through deep space with no gravitational or atmospheric drag forces act-

ing on it. At an arbitrary time t (see Fig. 9-17a), let M be the mass of the rocket and v its velocity.

Figure 9-17b shows how things stand a time interval dt later. The rocket now has velocity $v + dv$ and mass $M + dM$, where the change in mass dM is a *negative quantity*. The exhaust products released by the rocket during interval dt have mass $- dM$ and velocity U relative to our inertial reference frame.

Our system consists of the rocket and the exhaust products released during interval dt. The system is closed and isolated, so the linear momentum of the system must be conserved during dt; that is,

$$P_i = P_f, \qquad (9\text{-}41)$$

where the subscripts i and f indicate the values at the beginning and end of time interval dt. We can rewrite Eq. 9-41 as

$$Mv = - dM\,U + (M + dM)(v + dv), \qquad (9\text{-}42)$$

where the first term on the right is the linear momentum of the exhaust products released during interval dt, and the second term is the linear momentum of the rocket at the end of interval dt.

We can simplify Eq. 9-42 by using the speed u of the exhaust products relative to the rocket. We find this relative speed u by subtracting the velocity U of the exhaust products from the velocity $v + dv$ of the rocket at the end of interval dt:

$$u = (v + dv) - U.$$

FIGURE 9-16 Lift-off of Project Mercury spacecraft.

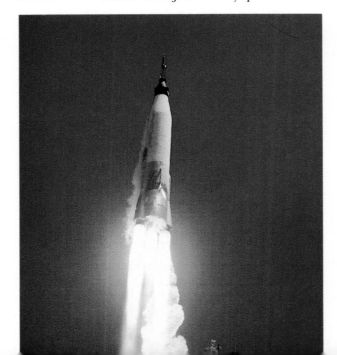

This gives us

$$U = v + dv - u. \qquad (9\text{-}43)$$

Substituting this result for U into Eq. 9-42 yields, with a little algebra,

$$-dM \, u = M \, dv. \qquad (9\text{-}44)$$

Dividing each side by dt yields

$$-\frac{dM}{dt} u = M \frac{dv}{dt}. \qquad (9\text{-}45)$$

We replace dM/dt (the rate at which the rocket loses mass) by $-R$, where R is the (positive) rate of fuel consumption. And we recognize that dv/dt is the acceleration of the rocket. With these changes, Eq. 9-45 becomes

$$Ru = Ma \qquad \text{(first rocket equation).} \qquad (9\text{-}46)$$

Equation 9-46 holds at any instant, the mass M, the fuel consumption rate R, and the acceleration a being evaluated at that instant.

The left side of Eq. 9-46 has the dimensions of a force ($\text{kg} \cdot \text{m/s}^2 = \text{N}$) and depends only on design characteristics of the rocket engine, namely, the rate R at which it ejects mass and the speed u with which that mass is ejected relative to the rocket. We call this term Ru the **thrust** of the rocket engine and represent it with T. Newton's second law emerges clearly if we write Eq. 9-46 as $T = Ma$, in which a is the acceleration of the rocket at the time that its mass is M.

Finding the Velocity

How will the velocity of a rocket change as it consumes its fuel? From Eq. 9-44 we have

$$dv = -u \frac{dM}{M}.$$

Integrating leads to

$$\int_{v_i}^{v_f} dv = -u \int_{M_i}^{M_f} \frac{dM}{M},$$

in which M_i is the initial mass of the rocket and M_f its final mass. Evaluating the integral then gives

$$v_f - v_i = u \ln \frac{M_i}{M_f} \qquad \text{(second rocket equation)} \qquad (9\text{-}47)$$

for the increase in the speed of the rocket during the change in mass from M_i to M_f.* We see here the advantage of multistage rockets, in which M_f is reduced by discarding successive stages when their fuel is depleted. An ideal rocket would reach its destination with only its payload remaining.

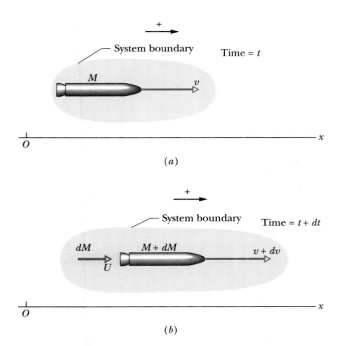

FIGURE 9-17 (a) An accelerating rocket of mass M at time t, as seen from an inertial reference frame. (b) The same but at time $t + dt$. The exhaust products released during interval dt are shown. The arrow shows the direction taken as positive for all velocities.

SAMPLE PROBLEM 9-12

A rocket whose initial mass M_i is 850 kg consumes fuel at the rate $R = 2.3$ kg/s. The speed u of the exhaust gases relative to the rocket engine is 2800 m/s.

a. What thrust does the rocket engine provide?

SOLUTION The thrust is

$$T = Ru = (2.3 \text{ kg/s})(2800 \text{ m/s})$$

$$= 6440 \text{ N} \approx 6400 \text{ N}. \qquad \text{(Answer)}$$

*The symbol "ln" in Eq. 9-47 means the *natural logarithm,* which is the logarithm taken to the base e ($= 2.718 \ldots$); punch "ln," not "log," on your calculator.

b. What is the initial acceleration of the rocket?

SOLUTION From Newton's second law, we have

$$a = \frac{T}{M_i} = \frac{6440 \text{ N}}{850 \text{ kg}} = 7.6 \text{ m/s}^2. \quad \text{(Answer)}$$

To be launched from the Earth's surface, a rocket must have an initial acceleration greater than $g = 9.8$ m/s². Put another way, the thrust T (= 6400 N) of the rocket engine must exceed the initial weight of the rocket, which is $M_i g = 850$ kg $\times$ 9.8 m/s² = 8300 N. Because the requirement of acceleration or thrust is not met, our rocket could not be launched from the Earth's surface. It would have to be launched into space by another, more powerful, rocket.

c. Suppose, instead, the rocket is launched from a spacecraft already in deep space, where we can neglect any gravitational force acting on it. The mass M_f of the rocket when its fuel is exhausted is 180 kg. What is its speed relative to the spacecraft at that time?

SOLUTION The rocket begins with a relative speed of $v_i = 0$. From Eq. 9-47 we have

$$v_f = u \ln \frac{M_i}{M_f}$$

$$= (2800 \text{ m/s}) \ln \frac{850 \text{ kg}}{180 \text{ kg}}$$

$$= (2800 \text{ m/s}) \ln 4.72 \approx 4300 \text{ m/s}. \quad \text{(Answer)}$$

Note that the ultimate speed of the rocket can exceed the exhaust speed u.

FIGURE 9-18 (a) A skater pushes herself away from a railing. The railing exerts a force $\mathbf{F}_{ext}$ on her. (b) The skater glides across the ice after push-off, with velocity $\mathbf{v}_{cm}$.

9-8 SYSTEMS OF PARTICLES: CHANGES IN KINETIC ENERGY (OPTIONAL)

Suppose a skater pushes herself away from a railing as in Fig. 9-18, picking up some kinetic energy in the process. Where does this energy come from? If you ask the skater, she will say, "From me!" and she will be right. The skater draws on her internal energy reserves in straightening her arm to push herself away from the railing and would be very aware of muscular exertion.

However, the skater's change in kinetic energy cannot be accounted for by applying the work–kinetic energy theorem of Section 7-5,

$$W = K_f - K_i = \Delta K \quad \text{(single particle)}, \quad (9\text{-}48)$$

which tells us that the increase in kinetic energy of a particle is equal to the work done by the net force $\mathbf{F}_{ext}$ acting on the particle. The reason is that the skater cannot be considered as a particle. The criterion for representing a body as a single particle is that every part of it moves in the same way. In the essential act of extending her arm to push herself away from the railing, the skater fails to meet this requirement. Thus, she must be treated as a system of particles, to which Eq. 9-48 simply does not apply.

This result leads us to another question, a formal one. We first met Newton's second law ($\Sigma \mathbf{F} = m\mathbf{a}$) in the context of a single particle. In Section 9-3 we saw that, by writing it in the form $\Sigma \mathbf{F}_{ext} = M\mathbf{a}_{cm}$, we could apply this law to a system of particles. To do so, we treat the system as a single particle of mass M, located at the center of mass of the system. We then assume that the net external force $\Sigma \mathbf{F}_{ext}$ acts on the system at that point. The question now is: "Can we use this form of Newton's second law to account for the changes in the kinetic energy of a system of particles?" We can.

We consider a system of particles, such as the skater, on which the net external force is the single force $\mathbf{F}_{ext}$, and we choose an x axis that points in the direction of this force. Suppose that the center of mass of the system moves a distance dx_{cm} along this axis in a time dt. For this one-dimensional motion Newton's second law is written as

$$F_{ext} = Ma_{cm}.$$

Multiplying each side of this equation by dx_{cm} yields

$$F_{ext}\, dx_{cm} = Ma_{cm}\, dx_{cm} = M\frac{dv_{cm}}{dt}\, dx_{cm}$$

$$= M\frac{dx_{cm}}{dt}\, dv_{cm}$$

or

$$F_{ext}\, dx_{cm} = Mv_{cm}\, dv_{cm}. \qquad (9\text{-}49)$$

As the force acts, let the center of mass of the system move from an initial position x_i to a final position x_f. Integrating Eq. 9-49 between these limits gives

$$\int_{x_i}^{x_f} F_{ext}\, dx_{cm} = (\tfrac{1}{2}Mv_{cm}^2)_f - (\tfrac{1}{2}Mv_{cm}^2)_i. \qquad (9\text{-}50)$$

This can be rewritten as

$$\int_{x_i}^{x_f} F_{ext}\, dx_{cm} = K_{cm,f} - K_{cm,i} = \Delta K_{cm}. \qquad (9\text{-}51)$$

If F_{ext} is a constant force, then we can simplify the left side of Eq. 9-51 by writing

$$\int_{x_i}^{x_f} F_{ext}\, dx_{cm} = F_{ext}\int_{x_i}^{x_f} dx_{cm} = F_{ext}(x_{cm,f} - x_{cm,i})$$

$$= F_{ext}d_{cm},$$

where d_{cm} is the displacement of the center of mass while the force F_{ext} acts. With this simplification, Eq. 9-51 becomes

$$F_{ext}d_{cm} = K_{cm,f} - K_{cm,i} = \Delta K_{cm}. \qquad (9\text{-}52)$$

Equations 9-51 and 9-52 tell us that we can account for the change in the kinetic energy of the center of mass of a system of particles by calculating the numerical value of the left side of Eq. 9-51 if the net

external force F_{ext} is a variable force, or the left side of Eq. 9-52 if F_{ext} is a constant force.

For example, if our skater pushes off from the railing with a constant force of 200 N while moving her center of mass through a displacement of $d_{cm} = 0.10$ m, the external force F_{ext} exerted on her by the railing is 200 N and, by Eq. 9-52, her kinetic energy increases by $(200\text{ N})(0.10\text{ m}) = 20$ J.

The ΔK_{cm} in Eqs. 9-51 and 9-52 is the change in the kinetic energy associated with only the *translational motion* of the system, that is, the motion of the system when it is treated as a single particle with all the mass concentrated at the center of mass. Any changes in energy associated with rotation of the system or with internal motions or vibrations are not included in ΔK_{cm}. (We return to this distinction in later chapters.)

In Section 8-7, we considered a block acted on by a kinetic frictional force f, as the block moves horizontally through a displacement d and undergoes a change ΔK in its kinetic energy. There we found that

$$-fd = \Delta K. \qquad (9\text{-}53)$$

Note that Eq. 9-53 is equivalent to Eq. 9-52: f is the constant external force acting on the system of particles (the block), d is the displacement of the block's center of mass, and ΔK is the change in kinetic energy of the center of mass. A minus sign is shown because the frictional force and the displacement are in opposite directions.

What Makes a Car Go?

A common situation in which an external force acts on a system of particles, changing the kinetic energy of the system's center of mass, is where a car is accelerated from rest, as in Fig. 9-19. In this situation, the

FIGURE 9-19 A car, initially at rest, accelerates to the right. The road exerts four frictional forces (two of them shown) on the bottoms of the tires. Taken together, these four forces constitute the net external force $\mathbf{F}_{ext}$ acting on the car.

external force $\mathbf{F}_{ext}$ is a frictional force exerted on the tires by the road. As the drive train of the car turns the wheels, the wheels push against the road, toward the rear. The force $\mathbf{F}_{ext}$ is the reaction to that push, and it points in the forward direction, that is, in the direction of the car's acceleration. As the car accelerates, its kinetic energy increases. Here we can ask the same question as with the skater: "Where does this energy come from?"

The kinetic energy acquired by the car (we would all agree) comes from the combustion of gasoline. Because of the (vitally necessary) internal motions of its engine, drive train, wheels, and tires, the car cannot be treated as a single particle, any more than could the skater. We must treat it as a system of particles with Eqs. 9-51 and 9-52 applying.

Equations 9-51 and 9-52 also apply to the braking of a car. In this situation, the frictional force exerted by the road on the tires points toward the rear of the car, and the kinetic energy of the car decreases. The left sides of Eqs. 9-51 and 9-52 are now negative, and the "lost" kinetic energy is transferred to the thermal energy of the braking surfaces. The brakes get hot.

SAMPLE PROBLEM 9-13

A car and its driver have mass $M = 1400$ kg. The car brakes to rest from a speed $v_{cm} = 24.0$ m/s, covering a distance $d_{cm} = 180$ m as it does so. What total frictional force $\mathbf{F}_{ext}$, assumed constant, does the road exert on the tires during this process? Assume that the driver manipulates the brakes so that the tires do not skid.

SOLUTION   Since the external force $\mathbf{F}_{ext}$ is assumed to be constant, we use Eq. 9-52 to account for the change in the kinetic energy of the center of mass of the car. In this situation $\mathbf{F}_{ext}$ points toward the rear of the car, opposite the displacement. Thus the left side of Eq. 9-52 is negative. With the final kinetic energy being zero and the initial kinetic energy written as $\frac{1}{2}Mv_{cm}^2$, we have

$$-F_{ext}d_{cm} = 0 - \tfrac{1}{2}Mv_{cm}^2,$$

which gives us

$$F_{ext} = \frac{Mv_{cm}^2}{2d_{cm}} = \frac{(1400 \text{ kg})(24.0 \text{ m/s})^2}{(2)(180 \text{ m})}$$

$$= 2240 \text{ N.} \qquad \text{(Answer)}$$

REVIEW & SUMMARY

Center of Mass

The **center of mass** of a system of discrete particles is defined to be the point whose coordinates are given by

$$x_{cm} = \frac{1}{M} \sum_{i=1}^{n} m_i x_i,$$

$$y_{cm} = \frac{1}{M} \sum_{i=1}^{n} m_i y_i, \qquad (9\text{-}5)$$

$$z_{cm} = \frac{1}{M} \sum_{i=1}^{n} m_i z_i,$$

or

$$\mathbf{r}_{cm} = \frac{1}{M} \sum_{i=1}^{n} m_i \mathbf{r}_i, \qquad (9\text{-}8)$$

where M is the total mass of the system. If the mass is continuously distributed, the center of mass is given by

$$x_{cm} = \frac{1}{M} \int x \, dm,$$

$$y_{cm} = \frac{1}{M} \int y \, dm, \qquad (9\text{-}9)$$

$$z_{cm} = \frac{1}{M} \int z \, dm.$$

If the density (mass per volume) is uniform, then Eq. 9-9 can be rewritten as

$$x_{cm} = \frac{1}{V} \int x \, dV,$$

$$y_{cm} = \frac{1}{V} \int y \, dV, \qquad (9\text{-}11)$$

$$z_{cm} = \frac{1}{V} \int z \, dV,$$

where V is the volume occupied by M.

Newton's Second Law for a System of Particles

The motion of the center of mass of any system of particles is governed by **Newton's second law for a system of particles**, which is written as

$$\sum \mathbf{F}_{ext} = M\mathbf{a}_{cm}. \qquad (9\text{-}16)$$

Here $\Sigma \mathbf{F}_{ext}$ is the resultant of the *external* forces acting on the system, M is the total mass of the system, and $\mathbf{a}_{cm}$ is the acceleration of the system's center of mass.

Linear Momentum and Newton's Second Law

For a single particle, we define a vector quantity $\mathbf{p}$ called the **linear momentum** as

$$\mathbf{p} = m\mathbf{v}, \qquad (9\text{-}22)$$

and write Newton's second law in terms of momentum:

$$\sum \mathbf{F} = \frac{d\mathbf{p}}{dt}. \qquad (9\text{-}23)$$

For a system of particles these relations become

$$\mathbf{P} = M\mathbf{v}_{cm} \quad \text{and} \quad \sum \mathbf{F}_{ext} = \frac{d\mathbf{P}}{dt}. \qquad (9\text{-}26, 9\text{-}28)$$

(See Table 9-1 for a summary of these important relations as applied to particles and to systems of particles.)

Relativistic Momentum

A more complete (relativistic) definition of linear momentum is

$$\mathbf{p} = \frac{m\mathbf{v}}{\sqrt{1 - (v/c)^2}}, \qquad (9\text{-}24)$$

a form that must be used whenever a particle's speed is near the speed of light c. Equation 9-24 is valid under any circumstances and is equivalent to Eq. 9-22 whenever $v \ll c$.

Conservation of Linear Momentum

If a system is isolated so that no net *external* force acts on the system, the linear momentum $\mathbf{P}$ of the system remains constant:

$$\mathbf{P} = \text{constant} \qquad \text{(closed, isolated system)}, \qquad (9\text{-}29)$$

which can also be written as

$$\mathbf{P}_i = \mathbf{P}_f \qquad \text{(closed, isolated system)}, \qquad (9\text{-}30)$$

where the subscripts refer to the values of $\mathbf{P}$ at some initial time and at a later time. Equations 9-29 and 9-30 are equivalent statements of the **law of conservation of linear momentum**.

Variable Mass Systems

Section 9-7 illustrates a strategy for dealing with systems of varying mass. We redefine the system, enlarging its boundaries until it encompasses a larger system whose mass *does* remain constant; then we apply the law of conservation of linear momentum. For a rocket, this means that the system includes both the rocket and its exhaust gases. Such analysis shows that in the absence of external forces a rocket accelerates at an instantaneous rate given by

$$Ru = Ma \qquad \text{(first rocket equation)} \qquad (9\text{-}46)$$

in which M is the rocket's instantaneous mass (including unexpended fuel), R is the fuel consumption rate, and u is the fuel's exhaust speed relative to the rocket. The term Ru is the **thrust** of the engine. For a rocket with constant R and u, whose speed changes from v_i to v_f when its mass changes from M_i to M_f,

$$v_f - v_i = u \ln \frac{M_i}{M_f} \qquad \begin{array}{l}\text{(second rocket}\\\text{equation)}.\end{array} \qquad (9\text{-}47)$$

Change in Kinetic Energy of a System of Particles

If a net external force F_{ext} acts on a system of particles, we can assume that the mass M of the system is concentrated at the center of mass, and that F_{ext} acts at that point. We can then account for a change ΔK_{cm} in the kinetic energy of the center of mass of the system. If F_{ext} is a variable force, we use

$$\int_{x_i}^{x_f} F_{ext}\, dx_{cm} = K_{cm,f} - K_{cm,i} = \Delta K_{cm}, \qquad (9\text{-}51)$$

where dx_{cm} is the incremental displacement of the center of mass, and where $K_{cm,i}$ and $K_{cm,f}$ are the initial and final values of the kinetic energy of the center of mass, respectively. If F_{ext} is a constant force, we use

$$F_{ext}d_{cm} = K_{cm,f} - K_{cm,i} = \Delta K_{cm}, \qquad (9\text{-}52)$$

where d_{cm} is the displacement of the center of mass. If the incremental displacement dx_{cm} in Eq. 9-51, or the displacement d_{cm} in Eq. 9-52, is in the same direction as F_{ext}, the left side of the applicable equation is positive, and the kinetic energy of the center of mass increases. If dx_{cm} or d_{cm} is in the opposite direction of F_{ext}, then the left side is negative, and the kinetic energy of the center of mass decreases.

QUESTIONS

1. Figure 9-20 shows (a) an isosceles triangle and (b) a right-circular cone whose diameter has the same length as the base of the triangle. The center of mass of the triangle is one-third of the way up from the base but that of the cone is only one-fourth of the way up. Can you explain this difference?

FIGURE 9-20 Question 1.

2. Where is the center of mass of the Earth's atmosphere?

3. An amateur sculptor decides to portray a bird (Fig. 9-21). Luckily, the final model is actually able to stand upright. The model is formed of a single sheet of metal of uniform thickness. Of the numbered points, which is most likely to be the center of mass?

FIGURE 9-21 Question 3.

4. Someone claims that when a skillful high jumper clears the bar her or his center of mass actually goes *under* the bar. Is this possible?

5. A bird is in a wire cage that hangs from a spring balance. Is the reading of the balance when the bird is flying greater than, less than, or the same as that when the bird sits in the cage?

6. Can a sailboat be propelled by air blown at the sails from a fan attached to the boat? Explain your answer.

7. A canoeist in a still pond can reach shore by jerking sharply on a rope attached to the bow of the canoe. How do you explain this? (It really can be done.)

8. How might a person sitting at rest on a frictionless horizontal surface get altogether off it?

9. A man stands still on a large sheet of slick ice; in his hand he holds a lighted firecracker. He throws the firecracker into the air. Describe briefly, but as exactly as you can, the motion of the center of mass of the firecracker and the motion of the center of mass of the system consisting of man and firecracker. It will be most convenient to describe each motion during each of the following periods: (a) after he throws the firecracker, but before it explodes; (b) between the explosion and the first piece of firecracker hitting the ice; (c) between the first fragment hitting the ice and the last fragment landing; and (d) during the time when all fragments have landed but none has reached the edge of the ice.

10. In 1920, a prominent newspaper editorialized as follows about the pioneering rocket experiments of Robert H. Goddard, dismissing the notion that a rocket could operate in a vacuum: "That Professor Goddard, with his 'chair' in Clark College and the countenancing of the Smithsonian Institution, does not know the relation of action to reaction, and of the need to have something better than a vacuum against which to react—to say that would be absurd. Of course, he only seems to lack the knowledge ladled out daily in high schools." What is wrong with this argument?

EXERCISES & PROBLEMS

SECTION 9-2 THE CENTER OF MASS

1E. (a) How far is the center of mass of the Earth–Moon system from the center of the Earth? (From Appendix C, obtain the masses of the Earth and the moon and the distance between the two.) (b) Express the answer to (a) as a fraction of the Earth's radius.

2E. The distance between the centers of the carbon (C) and oxygen (O) atoms in a carbon monoxide (CO) gas molecule is 1.131×10^{-10} m. Locate the center of mass of

a CO molecule relative to the carbon atom. (Find the mass of C and O in Appendix D.)

3E. (a) What are the coordinates of the center of mass of the three particles shown in Fig. 9-22? (b) What happens to the center of mass as the mass of the topmost particle is gradually increased?

4E. Three thin rods each of length L are arranged in an inverted U, as shown in Fig. 9-23. The two rods on the arms of the U each have mass M; the third rod has mass $3M$. Where is the center of mass of the assembly?

FIGURE 9-22 Exercise 3.

FIGURE 9-23 Exercise 4.

5E. A uniform square plate 6 m on a side has had a square piece 2 m on a side cut out of it (Fig. 9-24). The center of that piece is at $x = 2$ m, $y = 0$. The center of the square plate is at $x = y = 0$. Find the coordinates of the center of mass of the remaining piece.

FIGURE 9-24 Exercise 5.

6P. Show that the ratio of the distances of two particles from their center of mass is the inverse ratio of their masses.

7P. Figure 9-25 shows the dimensions of a composite slab; half the slab is made of aluminum (density = 2.70 g/cm³) and half of iron (density = 7.85 g/cm³). Where is the center of mass of the slab?

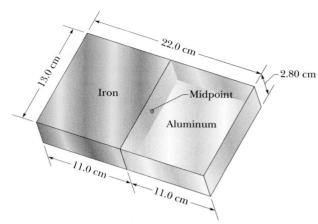

FIGURE 9-25 Problem 7.

8P. In the ammonia (NH_3) molecule (see Fig. 9-26), the three hydrogen (H) atoms form an equilateral triangle; the center of the triangle is 9.40×10^{-11} m from each hydrogen atom. The nitrogen (N) atom is at the apex of a pyramid, the three hydrogens forming the base. The nitrogen-to-hydrogen distance is 10.14×10^{-11} m, and the nitrogen-to-hydrogen atomic mass ratio is 13.9. Locate the center of mass relative to the nitrogen atom.

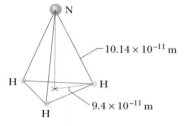

FIGURE 9-26 Problem 8.

9P. A cubical box, open at the top, with edge length 40 cm, is constructed from metal plate of uniform density and negligible thickness (Fig. 9-27). Find the coordinates of the center of mass of the box.

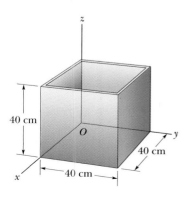

FIGURE 9-27 Problem 9.

10P. Two pieces of sheet metal, each in the shape of a right triangle with height $H = 2.0$ cm and length $L = 3.5$ cm, are shown in Fig. 9-28. (a) What are the coordinates of the center of mass of the composite structure? (b) If each piece is reversed left-for-right so that the 2.0-cm sides are against each other, what are the coordinates of the center of mass of the composite structure?

FIGURE 9-28   Problem 10.

11P*. The Great Pyramid of Cheops at El Gizeh, Egypt (Fig. 9-29a), had height $H = 147$ m before its topmost stone fell. Its base is a square with edge length $L = 230$ m (see Fig. 9-29b). Assuming that it has uniform density, find the original height of its center of mass above the base.

(a)

(b)

FIGURE 9-29   Problem 11.

12P*. A right cylindrical can with mass M, height H, and uniform density is initially filled with pop of mass m (Fig. 9-30). We punch small holes in the top and bottom to drain the pop; we then consider the height h of the center of mass of the can and any pop within it. What is h (a) initially, and (b) when all the pop has drained? (c) What happens to h during the draining of the pop? (d) If x is the height of the remaining pop at any given instant, find x (in terms of M, H, and m) when the center of mass reaches its lowest point.

FIGURE 9-30   Problem 12.

SECTION 9-3   NEWTON'S SECOND LAW FOR A SYSTEM OF PARTICLES

13E. Two skaters, one with mass 65 kg and the other with mass 40 kg, stand on an ice rink holding a pole with a length of 10 m and a negligible mass. Starting from the ends of the pole, the skaters pull themselves along the pole until they meet. How far will the 40-kg skater move?

14E. An old Chrysler with mass 2400 kg is moving along a straight stretch of road at 80 km/h. It is followed by a Ford with mass 1600 kg moving at 60 km/h. How fast is the center of mass of the two cars moving?

15E. A man of mass m clings to a rope ladder suspended below a balloon of mass M; see Fig. 9-31. The balloon is stationary with respect to the ground. (a) If the man

FIGURE 9-31   Exercise 15.

begins to climb the ladder at speed v (with respect to the ladder), in what direction and with what speed (with respect to the Earth) will the balloon move? (b) What is the state of the motion after the man stops climbing?

16E. Two particles P and Q are initially at rest 1.0 m apart. P has a mass of 0.10 kg and Q a mass of 0.30 kg. P and Q attract each other with a constant force of 1.0×10^{-2} N. No external forces act on the system. (a) Describe the motion of the center of mass. (b) At what distance from P's original position do the particles collide?

17E. A cannon and a supply of cannonballs are inside a sealed railroad car of length L, as in Fig. 9-32. The cannon fires to the right; the car recoils to the left. Fired cannonballs remain in the car after hitting the far wall and landing on the floor there. (a) After all the cannonballs have been fired, what is the greatest distance the car can have moved from its original position? (b) What is the speed of the car just after all the cannonballs have been fired?

FIGURE 9-32 Exercise 17.

18P. A stone is dropped at $t = 0$. A second stone, with twice the mass of the first, is dropped from the same point at $t = 100$ ms. (a) Where is the center of mass of the two stones at $t = 300$ ms? Neither stone has yet reached the ground. (b) How fast is the center of mass of the two-stone system moving at that time?

19P. Ricardo, mass 80 kg, and Carmelita, who is lighter, are enjoying Lake Merced at dusk in a 30-kg canoe. When the canoe is at rest in the placid water, they exchange seats, which are 3.0 m apart and symmetrically located with respect to the canoe's center. Ricardo notices that the canoe moved 40 cm relative to a submerged log during the exchange and calculates Carmelita's mass, which she has not told him. What is it?

20P. A shell is fired from a gun with a muzzle velocity of 20 m/s, at an angle of 60° with the horizontal. At the top of the trajectory, the shell explodes into two fragments of

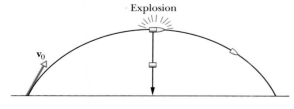

FIGURE 9-33 Problem 20.

equal mass (Fig. 9-33). One fragment, whose speed immediately after the explosion is zero, falls vertically. How far from the gun does the other fragment land, assuming level terrain and negligible air drag?

21P. Two identical containers of sugar are connected by a massless cord that passes over a massless, frictionless pulley with a diameter of 50 mm. The two containers are at the same level. Each originally has a mass of 500 g. (a) Locate the horizontal position of their center of mass. (b) Now 20 g of sugar is transferred from one container to the other, but the containers are prevented from moving. Locate the new horizontal position of their center of mass. (c) The two containers are now released. In what direction does the center of mass move? (d) What is its acceleration?

22P. A dog, weighing 10.0 lb, is standing on a flatboat so that he is 20.0 ft from the shore (to the left in Fig. 9-34a). He walks 8.00 ft on the boat toward shore and then halts. The boat weighs 40.0 lb, and one can assume there is no friction between it and the water. How far is the dog then from the shore? (*Hint:* See Fig. 9-34b. The dog moves leftward; the boat moves rightward; but does the center of mass of the *boat + dog* system move?)

(a)

(b)

FIGURE 9-34 Problem 22.

SECTION 9-4 LINEAR MOMENTUM

23E. What is the linear momentum of an automobile with weight 3200 lb traveling at 55 mi/h?

24E. Suppose that your mass is 80 kg. How fast would you have to run to have the same linear momentum as a 1600-kg car moving at 1.2 km/h?

25E. How fast must an 816-kg Volkswagen travel (a) to have the same linear momentum as a 2650-kg Cadillac going 16 km/h and (b) to have the same kinetic energy?

26E. What is the linear momentum of an electron of speed $0.99c$ ($= 2.97 \times 10^8$ m/s)?

27E. The linear momentum of a particle moving at 1.5×10^8 m/s is measured to be 2.9×10^{-19} kg·m/s. By finding its mass, identify the particle.

28P. An object is tracked by a radar station and found to have a position vector given by $\mathbf{r} = (3500 - 160t)\mathbf{i} + 2700\mathbf{j} + 300\mathbf{k}$, with $\mathbf{r}$ in meters and t in seconds. The radar station's x axis points east, its y axis north, and its z axis vertically up. If the object is a 250-kg missile warhead, what are (a) its linear momentum, (b) its direction of motion, and (c) the net force on it?

29P. A 0.165-kg cue ball with an initial speed of 2.00 m/s bounces off the rail in a game of pool, as shown from overhead in Fig. 9-35. For x and y axes located as shown, the bounce reverses the y component of the ball's velocity but does not alter the x component. (a) What is θ in Fig. 9-35? (b) What is the change in the ball's linear momentum in unit-vector notation? (The fact that the ball rolls is not relevant to either question.)

FIGURE 9-35 Problem 29.

30P. A 2100-kg truck traveling north at 41 km/h turns east and accelerates to 51 km/h. (a) What is the change in the kinetic energy of the truck? (b) What are the magnitude and direction of the change in the linear momentum of the truck?

31P. A 50-g ball is thrown from ground level into the air with an initial speed of 16 m/s at an angle of 30° above the horizontal. (a) What are the values of the kinetic energy of the ball initially and just before it hits the ground? (b) Find the corresponding values of the linear momentum (magnitude and direction). (c) Show that the change in linear momentum is just equal to the weight of the ball multiplied by the time of flight.

32P. A particle with mass m has linear momentum **p** of magnitude mc. What is its speed in terms of c, the speed of light?

SECTION 9-6 CONSERVATION OF LINEAR MOMENTUM

33E. A 200-lb man standing on a surface of negligible friction kicks forward a 0.15-lb stone lying at his feet, giving it a speed of 13 ft/s. What velocity does the man acquire as a result?

34E. Two blocks of masses 1.0 kg and 3.0 kg connected by a spring rest on a frictionless surface. They are given velocities toward each other such that the 1.0-kg block travels initially at 1.7 m/s toward the center of mass, which remains at rest. What is the initial velocity of the other block?

35E. A space vehicle is traveling at 4300 km/h relative to the Earth when the exhausted rocket motor is disengaged and sent backward with a speed of 82 km/h relative to the command module. The mass of the motor is four times the mass of the module. What is the speed of the command module relative to Earth after the separation?

36E. A 75-kg man is riding on a 39-kg cart traveling at a speed of 2.3 m/s. He jumps off with zero horizontal speed. What is the resulting change in the speed of the cart?

37E. A railroad flatcar of weight W can roll without friction along a straight horizontal track. Initially, a man of weight w is standing on the car, which is moving to the right with speed v_0; see Fig. 9-36. What is the change in velocity of the car if the man runs to the left so that his speed relative to the car is v_{rel}?

FIGURE 9-36 Exercise 37.

38P. The last stage of a rocket is traveling at a speed of 7600 m/s. This last stage is made up of two parts that are clamped together, namely, a rocket case with a mass of 290.0 kg and a payload capsule with a mass of 150.0 kg. When the clamp is released, a compressed spring causes the two parts to separate with a relative speed of 910.0 m/s. (a) What are the speeds of the two parts after they have separated? Assume that all velocities are along the same line. (b) Find the total kinetic energy of the two parts before and after they separate and account for the difference (if any).

39P. A vessel at rest explodes, breaking into three pieces. Two pieces, having equal mass, fly off perpendicular to one another with the same speed of 30 m/s. The third piece has three times the mass of each other piece. What are the direction and magnitude of its velocity immediately after the explosion?

40P. A radioactive nucleus, initially at rest, decays by emitting an electron and a neutrino perpendicular to each other. (A *neutrino* is one of the fundamental particles of physics.) The linear momentum of the electron is 1.2×10^{-22} kg·m/s and that of the neutrino is 6.4×10^{-23} kg·m/s. (a) Find the direction and magnitude of the linear momentum of the nucleus as it recoils from the decay. (b) The mass of the residual nucleus is 5.8×10^{-26} kg. What is its kinetic energy as it recoils?

41P. A 2140-kg railroad flatcar, which can move with negligible friction, is motionless next to a platform. A 242-kg sumo wrestler runs at 5.3 m/s along the platform (parallel to the track) and then jumps onto the flatcar. What is the speed of the flatcar if he then (a) stands on it, (b) runs at 5.3 m/s relative to it in his original direction, and (c) turns and runs at 5.3 m/s relative to the flatcar opposite his original direction?

42P. A rocket sled with a mass of 2900 kg moves at 250 m/s on a set of rails. At a certain point, a scoop on the sled dips into a trough of water located between the tracks and scoops water into an empty tank on the sled. By applying the conservation of linear momentum, determine the speed of the sled after 920 kg of water have been scooped up. Ignore any retarding force on the scoop.

43P. A 3.50-g bullet is fired horizontally at two blocks resting on a smooth tabletop, as shown in Fig. 9-37a. The bullet passes through the first block, with mass 1.20 kg, and embeds itself in the second, with mass 1.80 kg. Speeds of 0.630 m/s and 1.40 m/s, respectively, are thereby imparted to the blocks, as shown in Fig. 9-37b. Neglecting the mass removed from the first block by the bullet, find (a) the speed of the bullet immediately after it emerges from the first block and (b) the bullet's original speed.

(a)

(b)

FIGURE 9-37   Problem 43.

44P. A body of mass 8 kg is traveling at 2 m/s with no external force acting. At a certain instant an internal explosion occurs, splitting the body into two chunks of 4-kg mass each. The explosion gives the chunks an additional 16 J of kinetic energy. Neither chunk leaves the line of original motion. Determine the speed and direction of motion of each of the chunks after the explosion.

45P. You are on an iceboat (with combined mass M) on frictionless, flat ice. Along with you are two stones with masses m_1 and m_2 such that $M = 6.00m_1 = 12.0m_2$. To get the boat moving, you throw the stones rearward, either in succession or together, but in each case with a certain speed v_{rel} relative to the boat. What is the resulting speed of the boat if you throw (a) the stones simultaneously, (b) m_1 and then m_2, and (c) m_2 and then m_1?

46P. A 1400-kg cannon, which fires a 70.0-kg shell with a speed of 556 m/s relative to the muzzle, is set at an elevation angle of 39.0° above the horizontal. The cannon is mounted on frictionless rails, so that it recoils freely. (a) What is the speed of the shell with respect to the Earth? (b) At what angle with the ground is the shell projected? (*Hint:* The horizontal component of the linear momentum of the system remains unchanged as the gun is fired.)

SECTION 9-7   SYSTEMS WITH VARYING MASS: A ROCKET

47E. A rocket is moving away from the solar system at a speed of 6.0×10^3 m/s. It fires its engine, which ejects exhaust with a velocity of 3.0×10^3 m/s relative to the rocket. The mass of the rocket at this time is 4.0×10^4 kg, and it experiences an acceleration of 2.0 m/s². (a) What is the thrust of the engine? (b) At what rate, in kilograms per second, was exhaust ejected during the firing?

48E. A 6090-kg space probe, moving nose-first toward Jupiter at 105 m/s relative to the sun, fires its rocket engine, ejecting 80.0 kg of exhaust at a speed of 253 m/s relative to the space probe. What is the final velocity of the probe?

49E. A rocket at rest in space, where there is virtually no gravitational force, has a mass of 2.55×10^5 kg, of which 1.81×10^5 kg is fuel. The engine consumes fuel at the rate of 480 kg/s, and the exhaust speed is 3.27 km/s. The engine is fired for 250 s. (a) Find the thrust of the rocket engine. (b) What is the mass of the rocket after the engine burn? (c) What is the final speed attained?

50E. Consider a rocket at rest in deep space. What must be its *mass ratio* (ratio of initial to final mass) in order that, after firing its engine, the rocket's speed is (a) equal to the exhaust speed and (b) equal to twice the exhaust speed?

51E. During a lunar mission, it is necessary to increase the speed of the spacecraft by 2.2 m/s when it is moving at 400 m/s. The exhaust speed of the rocket engine is 1000 m/s. What fraction of the initial mass of the spacecraft must be burned and ejected for the increase?

52E. A railroad car moves at a constant speed of 3.20 m/s under a grain elevator. Grain drops into it at the rate of 540 kg/min. What force must be applied to the railroad car, in the absence of friction, to keep it moving at constant speed?

53P. A single-stage rocket, at rest in a certain inertial reference frame, has mass M when the rocket engine is ignited. Show that, when the mass has decreased to $0.368M$, the gases streaming out of the rocket engine at that time will be at rest in the original reference frame.

54P. A 6100-kg rocket is set for vertical firing from the Earth's surface. If the exhaust speed is 1200 m/s, how much gas must be ejected each second if the thrust (a) is to equal the weight of the rocket and (b) is to give the rocket an initial upward acceleration of 21 m/s²?

55P. A 5.4-kg toboggan carrying 35 kg of sand slides from rest down an icy 90-m-long slope inclined 30° below the horizontal. The sand leaks from the back of the toboggan at the rate of 2.3 kg/s. How long does it take the toboggan to reach the bottom of the slope? (*Hint:* Before you get too involved in calculations, consider how the acceleration of an object along an inclined plane depends on mass.)

56P. Two long barges are moving in the same direction in still water, one with a speed of 10 km/h and the other with a speed of 20 km/h. While they are passing each other, coal is shoveled from the slower to the faster one at a rate of 1000 kg/min; see Fig. 9-38. How much additional force must be provided by the driving engines of each barge if neither is to change speed? Assume that the shoveling is always perfectly sideways and that the frictional forces between the barges and the water do not depend on the weight of the barges.

57P. A jet airplane is traveling 180 m/s. The engine takes in 68 m³ of air, having a mass of 70 kg, each second. The air is used to burn 2.9 kg of fuel each second. The energy is used to compress the products of combustion and to eject them at the rear of the plane at 490 m/s relative to the plane. Find (a) the thrust of the jet engine and (b) its power in watts.

SECTION 9-8 SYSTEMS OF PARTICLES: CHANGES IN KINETIC ENERGY

58E. An automobile with passengers has weight 16,400 N and is moving at 113 km/h when the driver brakes to a stop. The road exerts a frictional force of 8230 N on the wheels. Using energy considerations, find the stopping distance.

59E. You crouch from a standing position, lowering your center of mass 18 cm in the process. Then you jump vertically. The average force exerted on you by the floor while you jump is three times your weight. What is your upward speed as you pass through your standing position in leaving the floor?

60E. A 55-kg woman leaps vertically from a crouching position in which her center of mass is 40 cm above the ground. As her feet leave the floor her center of mass is 90 cm above the ground and rises to 120 cm at the top of her leap. (a) What average force was exerted on her by the ground during the jump? (b) What maximum speed does she attain?

FIGURE 9-38 Problem 56.

61P. A 110-kg ice hockey player skates at 3.0 m/s toward a railing at the edge of the ice and then stops himself by grasping the railing with his outstretched arms. During this stopping process his center of mass moves 30 cm toward the railing. (a) What is the change in the kinetic energy of his center of mass during this process? (b) What average force must he exert on the railing?

62P. Many bicycles are braked by pressing a piece of hard rubber against each tire. Suppose that such a bicycle, initially traveling at 40.2 km/h, is stopped in 55 revolutions of its 35.6-cm-radius tires. The mass of bicycle plus rider is 51.1 kg. Treat the front and rear brakes as identical, and assume that the tires do not skid on the road. (a) What work is done by each brake? (The tire slides past the brake a certain distance while the brake rubs against it.) (b) Assume that the force of a brake against a tire is constant, and calculate its value. (c) What is the acceleration of the bicycle? (d) What is the horizontal force of the road on each tire?

ADDITIONAL PROBLEMS

63. A 0.70-kg mass is moving horizontally with a speed of 5.0 m/s when it strikes a vertical wall. The mass rebounds with a speed of 2.0 m/s. What is the magnitude of the change in linear momentum of the mass?

64. At the instant a 3.0-kg particle has a velocity of 6.0 m/s in the direction of decreasing y, a 4.0-kg particle has a velocity of 7.0 m/s in the direction of increasing x. What is the speed of the center of mass of the two-particle system?

65. A 2.0-kg mass sliding on a frictionless surface explodes into two 1.0-kg masses. After the explosion the ve-

locities of the 1.0-kg masses are 3.0 m/s, due north, and 5.0 m/s, 30° north of east. What was the original speed of the 2.0-kg mass?

66. A 1000-kg automobile is at rest at a traffic light. At the instant the traffic light turns green, the automobile starts to move with a constant acceleration of 4.0 m/s². At the same instant a 2000-kg truck, traveling at a constant speed of 8.0 m/s, overtakes and passes the automobile. (a) How far is the center of mass of the automobile–truck system from the traffic light at $t = 3.0$ s? (b) What is the speed of the center of mass of the automobile–truck system then?

COLLISIONS 10

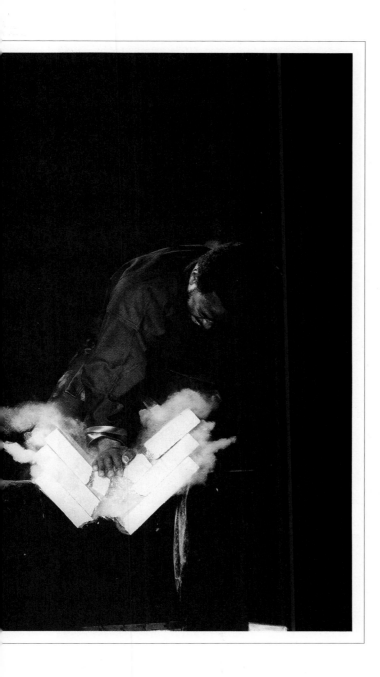

Ronald McNair, a physicist and one of the astronauts killed in the explosion of the <u>Challenger</u> space shuttle, was a black-belt in karate. Here he breaks several concrete slabs. In such karate demonstrations, a pine board or a concrete "patio block" is typically used. When struck, the board or block bends, storing energy like a stretched spring does, until a critical energy is reached. Then the object breaks. Surprisingly, the energy necessary to break the block is about one-third of that for the board, yet the board is considerably easier to break. Why?

10-1 WHAT IS A COLLISION?

In everyday language, a collision occurs when objects crash into each other. Although we will refine that definition, it conveys the meaning well enough. Within our daily experience, the things that collide might be billiard balls, a hammer and a nail, a baseball and a bat, and—all too often—automobiles. A stone thrown or dropped toward the Earth may be said to collide with it. Figure 10-1 shows the impressive evidence for one such event that occurred about 20,000 years ago. As Fig. 10-2 shows, collisions occur that are beyond our direct experience, ranging from the collisions of subatomic particles to the collisions of galaxies.

Let us formalize our definition:

A collision is an isolated event in which a relatively strong force acts on each of two or more colliding bodies for a relatively short time.

Moreover, it must be possible to make a clear distinction between times that are *before*, *during*, and *after* the collision. Figure 10-3 suggests the spirit of this definition. The system boundary surrounding the bodies in that figure suggests that, in an ideal collision, only internal forces (between the bodies) play a role.

When a racquet strikes a ball, the beginning and the end of the collision can be determined fairly

FIGURE 10-1   Meteor Crater in Arizona, the result of a collision about 20,000 years ago. The crater is about 1200 m wide and 200 m deep.

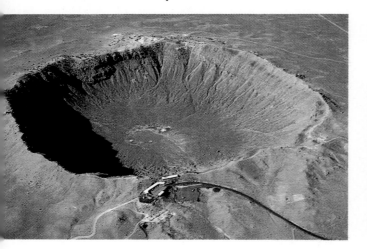

FIGURE 10-2   Collisions for a wide range of masses. (*a*) An alpha particle ($m \approx 10^{-26}$ kg) coming in from the left (whose trail is colored yellow in this false-color photograph) bounces off a nitrogen nucleus that had been stationary and that now moves toward the bottom right (red trail). (*b*) A cue ball ($m \approx 0.1$ kg) bounces off a colored ball. (*c*) A computer simulation of two galaxies ($m \approx 10^{40}$ kg) nearing collision.

(*a*)

(*b*)

(*c*)

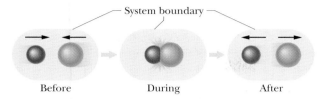

FIGURE 10-3 A flowchart showing the system in which a collision occurs.

precisely. The racquet–ball contact time (about 4 ms) is short compared with the time that the ball is in flight to and from the racquet. As Fig. 10-4 shows, the force exerted on the ball is strong enough to deform it temporarily. Similar deformation is shown in Fig. 10-5 for a somewhat longer lasting collision.

Note that our formal definition of collision does not require the "crash" of our informal definition. When a space probe approaches a large planet, swings around it, and then continues its course with increased speed (a *slingshot* encounter), that too is a collision. The probe and planet do not actually "touch," but, a collision does not require contact, and a collision force does not have to be a force of contact; it can just as easily be a gravitational force, as in this case.

FIGURE 10-4 In a tennis ball–tennis racquet collision, the ball is in contact with the racquet for about 4 ms. During a typical world-class tennis match, a ball is in contact with a racquet for a cumulative time of about 1 s per set.

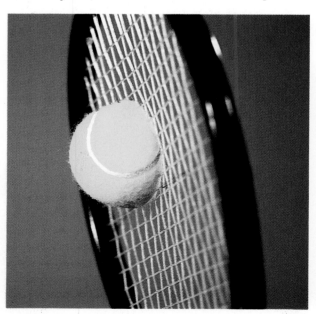

Many physicists today spend their time playing what we can call "the collision game." A principal goal of this game is to find out as much as possible about the forces that act during the collision, knowing the state of the particles both before and after the collision. Virtually all of our knowledge of the subatomic world—electrons, protons, neutrons, muons, quarks, and the like—comes from experiments of this kind. The rules of the collision game are the laws of conservation of momentum and of energy.

10-2 IMPULSE AND LINEAR MOMENTUM

Single Collision

Figure 10-6 shows the equal but opposite forces, $\mathbf{F}(t)$ and $-\mathbf{F}(t)$, that act during a simple head-on collision between two particle-like bodies of different masses. These forces will change the linear momentum of both bodies; the amount of the change will depend not only on the average values of the forces, but also on the time Δt during which they act. To see this quantitatively, let us apply Newton's second law in the form $\mathbf{F} = d\mathbf{p}/dt$ to body R on the right in Fig. 10-6. We have

$$d\mathbf{p} = \mathbf{F}(t)\ dt, \qquad (10\text{-}1)$$

FIGURE 10-5 The severity of the collision of the fist and body bag can be appreciated by the visible deformation of the bag and the shock felt by the fist and forearm.

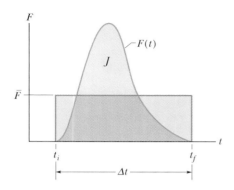

FIGURE 10-6 Two particle-like bodies L and R collide with each other. During the collision, body L exerts force $\mathbf{F}(t)$ on body R, and body R exerts force $-\mathbf{F}(t)$ on body L. Forces $\mathbf{F}(t)$ and $-\mathbf{F}(t)$ are an action–reaction pair. Their magnitudes vary with time during the collision but at any given instant, their magnitudes are equal.

in which $\mathbf{F}(t)$ is the time variation of the force, displayed as the curve in Fig. 10-7. Let us integrate Eq. 10-1 over the collision interval Δt, that is, from an initial time t_i (just before the collision) to a final time t_f (just after the collision). We obtain

$$\int_{\mathbf{p}_i}^{\mathbf{p}_f} d\mathbf{p} = \int_{t_i}^{t_f} \mathbf{F}(t)\ dt. \qquad (10\text{-}2)$$

The left side of this equation is $\mathbf{p}_f - \mathbf{p}_i$, the change in linear momentum of the body R. The right side, which is a measure of both the strength and the duration of the collision force, is called the collision **impulse J.** Thus

$$\mathbf{J} = \int_{t_i}^{t_f} \mathbf{F}(t)\ dt \qquad \text{(impulse defined).} \qquad (10\text{-}3)$$

Equation 10-3 tells us that the impulse is equal to the area under the $F(t)$ curve of Fig. 10-7.

In a pillow fight, the collisions are mild because the "give" of the pillow prolongs the collision time.

FIGURE 10-7 The curve shows the magnitude of the time-varying force $F(t)$ that acts on body R in Fig. 10-6. The height of the rectangle represents the average force $\overline{F}$ acting over the time interval Δt. The areas under the curve $F(t)$ and within the rectangle are the same and are equal to the magnitude of the impulse J in the collision of Fig. 10-6.

From Eqs. 10-2 and 10-3 we see that the change in linear momentum of each body in a collision is equal to the impulse that acts on that body:

$$\mathbf{p}_f - \mathbf{p}_i = \Delta\mathbf{p} = \mathbf{J} \qquad \begin{array}{l}\text{(impulse–linear}\\ \text{momentum theorem).}\end{array} \qquad (10\text{-}4)$$

Furthermore, from the conservation of linear momentum, we know that $\Delta\mathbf{p}$ for body R equals $-\Delta\mathbf{p}$ for body L. Equation 10-4 can also be written in component form as

$$p_{fx} - p_{ix} = \Delta p_x = J_x, \qquad (10\text{-}5)$$

$$p_{fy} - p_{iy} = \Delta p_y = J_y, \qquad (10\text{-}6)$$

The collisions in jousting are severe because they occur in a short time.

and

$$p_{fz} - p_{iz} = \Delta p_z = J_z. \qquad (10\text{-}7)$$

Impulse and linear momentum are both vectors, and they have the same units and dimensions. The **impulse–linear momentum theorem** of Eq. 10-4, like the work–kinetic energy theorem, is not a new and independent theorem but is a direct consequence of Newton's second law. Both theorems are special forms of this law, useful for special purposes.

If $\bar{F}$ is the average magnitude of the force in Fig. 10-7, we can write the magnitude of the impulse as

$$J = \bar{F} \, \Delta t, \qquad (10\text{-}8)$$

in which Δt is the duration of the collision. The value of $\bar{F}$ must be chosen so that the area within the rectangle of Fig. 10-7 is equal to the area under the $F(t)$ curve.

Series of Collisions

In Fig. 10-8, a steady stream of bodies, with identical linear momenta mv, collides with body R, which is fixed in place. In each collision of this one-dimensional situation, the impulse J acting on body R and the change in linear momentum Δp of a colliding body have the same magnitude but opposite directions. If n bodies collide in time interval Δt, then the total impulse J acting on body R during Δt is

$$J = -n \, \Delta p. \qquad (10\text{-}9)$$

Substituting Eq. 10-9 into Eq. 10-8, we find the average force $\bar{F}$ acting on body R during the collisions:

$$\bar{F} = -\frac{n}{\Delta t} \Delta p = -\frac{n}{\Delta t} m \, \Delta v. \qquad (10\text{-}10)$$

The ratio $n/\Delta t$ is the rate at which the bodies collide with body R.

FIGURE 10-8 A steady stream of bodies, with identical linear momenta, collide with body R, which is fixed in place. The average force $\bar{\mathbf{F}}$ on body R is to the right and has a magnitude that depends on the rate at which the bodies collide or, equivalently, the rate at which mass collides.

If the colliding bodies stop upon impact, then in Eq. 10-10 we substitute

$$\Delta v = v_f - v_i = 0 - v = -v, \qquad (10\text{-}11)$$

where $v_i \, (= v)$ and $v_f \, (= 0)$ are the velocities before and after the collision, respectively. If, instead, the colliding bodies bounce directly backward from body R with no change in speed, then $v_f = -v$ and we substitute

$$\Delta v = v_f - v_i = -v - v = -2v. \qquad (10\text{-}12)$$

In time interval Δt, an amount of mass $\Delta m = nm$ collides with body R. With this result, we can rewrite Eq. 10-10 as

$$\bar{F} = -\frac{\Delta m}{\Delta t} \Delta v, \qquad (10\text{-}13)$$

where $\Delta m/\Delta t$ is the rate at which mass collides with body R. Here again we can substitute Eq. 10-11 or 10-12 for Δv, depending on what the colliding bodies do.

SAMPLE PROBLEM 10-1

A pitched 140-g baseball, in horizontal flight with a speed v_i of 39 m/s, is struck by a batter. After leaving the bat, the ball travels in the opposite direction with speed v_f, also 39 m/s.

a. What impulse J acts on the ball while it is in contact with the bat?

SOLUTION We can calculate the impulse from the change it produces in the ball's linear momentum, using Eq. 10-4 for one-dimensional motion. Let us choose the direction in which the bat is moving to be positive. From Eq. 10-4 we have

$$J = p_f - p_i = mv_f - mv_i$$
$$= (0.14 \text{ kg})(39 \text{ m/s}) - (0.14 \text{ kg})(-39 \text{ m/s})$$
$$= 10.9 \text{ kg} \cdot \text{m/s} \approx 11 \text{ kg} \cdot \text{m/s}. \qquad \text{(Answer)}$$

With our sign convention, the initial velocity of the ball is negative and the final velocity is positive. The impulse turned out to be positive, which tells us that the direction of the impulse vector acting on the ball is the direction in which the bat was swinging, which makes sense.

b. The impact time Δt for the baseball–bat collision is 1.2 ms, a typical value. What average force acts on the baseball?

SOLUTION   From Eq. 10-8 we have

$$\overline{F} = \frac{J}{\Delta t} = \frac{10.9 \text{ kg} \cdot \text{m/s}}{0.0012 \text{ s}}$$

$$= 9100 \text{ N,} \qquad \text{(Answer)}$$

which is about a ton. The *maximum* force will be larger than this. The sign of the average force exerted on the ball is positive, which means that the direction of the force vector is the same as that of the impulse vector.

c. What is the average acceleration a of the baseball?

SOLUTION   We find this with

$$\overline{a} = \frac{\overline{F}}{m} = \frac{9100 \text{ N}}{0.14 \text{ kg}} = 6.5 \times 10^4 \text{ m/s}^2. \qquad \text{(Answer)}$$

This is about 6600 times the acceleration of a freely falling body.

In defining a collision, we assumed that no net external force acts on the colliding bodies. That is not true in this case, because the weight $m\mathbf{g}$ of the ball always acts on the ball, whether the ball is in flight or in contact with the bat. However, this force, with a magnitude of 1.4 N, is negligible compared to the average force exerted by the bat, which has a magnitude of 9100 N. We are quite safe in treating the collision as "isolated."

SAMPLE PROBLEM 10-2

As in Sample Problem 10-1, the baseball approaches the bat at a speed v_i of 39 m/s; but now the collision is not head-on, and the ball leaves the bat with a speed v_f of 45 m/s at an upward angle of 30° from the horizontal (Fig. 10-9). What is the average force $\overline{\mathbf{F}}$ exerted on the ball if the collision lasts 1.2 ms?

SOLUTION   We find the components J_x and J_y of the impulse from Eqs. 10-5 and 10-6:

$$J_x = p_{fx} - p_{ix} = mv_{fx} - mv_{ix}$$

$$= (0.14 \text{ kg}) [(45 \text{ m/s})(\cos 30°) - (-39 \text{ m/s})]$$

$$= 10.92 \text{ kg} \cdot \text{m/s,}$$

and

$$J_y = p_{fy} - p_{iy} = mv_{fy} - mv_{iy}$$

$$= (0.14 \text{ kg}) [(45 \text{ m/s})(\sin 30°) - 0]$$

$$= 3.150 \text{ kg} \cdot \text{m/s.}$$

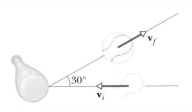

FIGURE 10-9   Sample Problem 10-2. A bat collides with a pitched baseball, sending it off at an angle of 30° from the horizontal.

The magnitude of the impulse $\mathbf{J}$ is given by

$$J = \sqrt{J_x^2 + J_y^2}$$

$$= \sqrt{(10.92 \text{ kg} \cdot \text{m/s})^2 + (3.150 \text{ kg} \cdot \text{m/s})^2}$$

$$= 11.37 \text{ kg} \cdot \text{m/s.}$$

From Eq. 10-8, the magnitude $\overline{F}$ of the average force is

$$\overline{F} = \frac{J}{\Delta t} = \frac{11.37 \text{ kg} \cdot \text{m/s}}{0.0012 \text{ s}}$$

$$= 9475 \text{ N} \approx 9500 \text{ N.} \qquad \text{(Answer)}$$

The impulse $\mathbf{J}$ is angled upward from the horizontal by θ, where θ is given by

$$\tan \theta = \frac{J_y}{J_x} = \frac{3.150 \text{ kg} \cdot \text{m/s}}{10.92 \text{ kg} \cdot \text{m/s}} = 0.288,$$

or

$$\theta = 16°. \qquad \text{(Answer)}$$

The average force $\overline{\mathbf{F}}$ is in the same direction as $\mathbf{J}$. Note that, unlike Sample Problem 10-1, $\overline{\mathbf{F}}$ and $\mathbf{J}$ in this two-dimensional situation are not in the same direction taken by the ball as it leaves the bat.

10-3 ELASTIC COLLISIONS IN ONE DIMENSION

Stationary Target

Consider a simple head-on collision of two bodies of different masses. For convenience, we take one of the bodies to be at rest before the collision; that body will be the "target," and the other body will be the "projectile."* We assume that this two-body sys-

---

*If the target body is moving with respect to our laboratory frame, we can always find another inertial reference frame in which the target body is initially at rest.

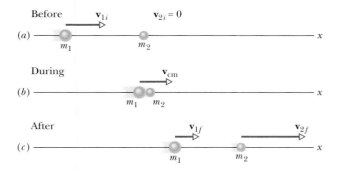

FIGURE 10-10 Two bodies undergo an elastic collision. One of them (the target body with mass m_2) is initially at rest before the collision. The velocities are shown (a) before, (b) during, and (c) after the collision. (The velocity during the collision is the velocity of the center of mass of the two bodies, which are momentarily touching.) The velocities are drawn to scale for the case in which $m_1 = 3m_2$.

tem is closed (no mass enters or leaves it) and isolated (no net external force acts on it). We assume also that the kinetic energy of the system is the same before and after the collision. Such collisions, in which kinetic energy is conserved, are of a special type called **elastic collisions.** The linear momentum of a closed, isolated system is *always* conserved in a collision, whether or not the collision is elastic, because the forces involved in the collision are all internal forces.

Applying the laws of conservation of linear momentum and of kinetic energy to the collision of Fig. 10-10 gives us

$$m_1 v_{1i} = m_1 v_{1f} + m_2 v_{2f} \qquad \text{(linear momentum),} \qquad (10\text{-}14)$$

and

$$\tfrac{1}{2} m_1 v_{1i}^2 = \tfrac{1}{2} m_1 v_{1f}^2 + \tfrac{1}{2} m_2 v_{2f}^2 \quad \text{(kinetic energy).} \quad (10\text{-}15)$$

In each of these equations, the subscript i identifies the initial velocities and the subscript f the final velocities of the bodies. If we know the masses of the bodies and if we also know v_{1i}, the initial velocity of body 1, the only unknown quantities are v_{1f} and v_{2f}, the final velocities of the two bodies. With two equations at our disposal, we should be able to find these two unknowns.

To do so, we rewrite Eq. 10-14 as

$$m_1(v_{1i} - v_{1f}) = m_2 v_{2f} \qquad (10\text{-}16)$$

and Eq. 10-15 as*

$$m_1(v_{1i} - v_{1f})(v_{1i} + v_{1f}) = m_2 v_{2f}^2. \qquad (10\text{-}17)$$

After dividing Eq. 10-17 by Eq. 10-16 and doing some more algebra, we obtain

$$v_{1f} = \frac{m_1 - m_2}{m_1 + m_2} v_{1i} \qquad (10\text{-}18)$$

and

$$v_{2f} = \frac{2m_1}{m_1 + m_2} v_{1i}. \qquad (10\text{-}19)$$

We note from Eq. 10-19 that v_{2f} is always positive (the target body with mass m_2 always moves forward). From Eq. 10-18 we see that v_{1f} may be of either sign (the projectile body with mass m_1 moves forward if $m_1 > m_2$ but rebounds if $m_1 < m_2$). Let us look at a few special situations.

Equal Masses

If $m_1 = m_2$, Eqs. 10-18 and 10-19 reduce to

$$v_{1f} = 0 \quad \text{and} \quad v_{2f} = v_{1i},$$

which we might call a pool player's result. It predicts that after a head-on collision of bodies with equal masses, body 1 (initially moving) stops dead in its tracks and body 2 (initially at rest) takes off with the initial speed of body 1. In head-on collisions, bodies of equal mass simply exchange velocities. This is true even if the target particle is not initially at rest.

A Massive Target

In terms of Fig. 10-10, a massive target means that $m_2 \gg m_1$. For example, we might fire a golf ball at a cannonball. Equations 10-18 and 10-19 then reduce to

$$v_{1f} \approx - v_{1i} \quad \text{and} \quad v_{2f} \approx \left(\frac{2m_1}{m_2}\right) v_{1i}. \qquad (10\text{-}20)$$

This tells us that body 1 (the golf ball) simply bounces back in the same direction from which it came, its speed essentially unchanged. Body 2 (the cannonball) moves forward at a low speed, because the quantity in parentheses in Eq. 10-20 is much less than unity. All this is what we should expect.

---

*In this step, we use the identity $a^2 - b^2 = (a - b)(a + b)$. It reduces the amount of algebra needed to solve the simultaneous equations, Eqs. 10-16 and 10-17.

A Massive Projectile

This is the opposite case; that is, $m_1 \gg m_2$. This time, we fire a cannonball at a golf ball. Equations 10-18 and 10-19 reduce to

$$v_{1f} \approx v_{1i} \quad \text{and} \quad v_{2f} \approx 2v_{1i}. \quad (10\text{-}21)$$

Equation 10-21 tells us that body 1 (the cannonball) simply keeps on going, scarcely slowed by the collision. Body 2 (the golf ball) charges ahead at twice the speed of the cannonball.

You may wonder: "Why twice?" As a starting point in thinking about the matter, recall the collision described by Eq. 10-20, in which the velocity of the incident light body (the golf ball) changed from $+v$ to $-v$, a velocity *change* of $2v$. The same *change* in velocity (from zero to $2v$) occurs in this example also.

Motion of the Center of Mass

The center of mass of two colliding bodies continues to move, totally uninfluenced by the collision. This follows from conservation of linear momentum and Eq. 9-26,

$$P = Mv_{cm} = (m_1 + m_2)v_{cm}, \quad (10\text{-}22)$$

which relates the linear momentum P of the system of two bodies to v_{cm}, the velocity of their center of mass. Because the momentum P is unchanged by the collision, v_{cm} must also be unchanged. So the center of mass continues to move in the same direction and at the same speed. From Eq. 10-22, the velocity of the center of mass for the collision shown in Fig. 10-10 (target initially at rest) is

$$v_{cm} = \frac{P}{m_1 + m_2} = \frac{m_1}{m_1 + m_2} v_{1i}. \quad (10\text{-}23)$$

Figure 10-11, a series of freeze-frames of a typical elastic collision, shows that the center of mass does indeed move steadily forward, unaffected in any way by the collision.

Moving Target

Now that we have examined the elastic collision of a projectile and a stationary target, let us examine the situation in which both bodies are moving before they undergo an elastic collision.

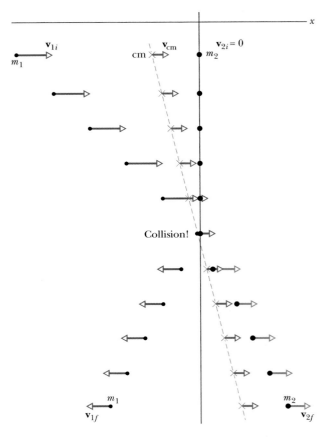

FIGURE 10-11 Some freeze-frames of two bodies undergoing an elastic collision. Body 2 is initially at rest, and $m_2 = 3m_1$. The velocity of the center of mass is also shown. Note that it is unaffected by the collision.

For the situation of Fig. 10-12, the conservation of linear momentum is written as

$$m_1 v_{1i} + m_2 v_{2i} = m_1 v_{1f} + m_2 v_{2f}, \quad (10\text{-}24)$$

and the conservation of kinetic energy is written as

$$\tfrac{1}{2} m_1 v_{1i}^2 + \tfrac{1}{2} m_2 v_{2i}^2 = \tfrac{1}{2} m_1 v_{1f}^2 + \tfrac{1}{2} m_2 v_{2f}^2. \quad (10\text{-}25)$$

To solve these simultaneous equations for v_{1f} and v_{2f}, we first rewrite Eq. 10-24 as

$$m_1(v_{1i} - v_{1f}) = -m_2(v_{2i} - v_{2f}), \quad (10\text{-}26)$$

and Eq. 10-25 as

$$m_1(v_{1i} - v_{1f})(v_{1i} + v_{1f})$$
$$= -m_2(v_{2i} - v_{2f})(v_{2i} + v_{2f}). \quad (10\text{-}27)$$

After dividing Eq. 10-27 by Eq. 10-26 and doing

FIGURE 10-12 Two bodies headed for an elastic collision.

some more algebra, we obtain

$$v_{1f} = \frac{m_1 - m_2}{m_1 + m_2} v_{1i} + \frac{2m_2}{m_1 + m_2} v_{2i} \quad (10\text{-}28)$$

and

$$v_{2f} = \frac{2m_1}{m_1 + m_2} v_{1i} + \frac{m_2 - m_1}{m_1 + m_2} v_{2i}. \quad (10\text{-}29)$$

Note that the assignment of subscripts 1 and 2 to the bodies is arbitrary. If we exchange those subscripts in Fig. 10-12 and in Eqs. 10-28 and 10-29, we end up with the same set of equations. Note also that if we set $v_{2i} = 0$, body 2 becomes a stationary target, and Eqs. 10-28 and 10-29 reduce to Eqs. 10-18 and 10-19, respectively.

SAMPLE PROBLEM 10-3

Two metal spheres, suspended by vertical cords, initially touch, as shown in Fig. 10-13. Sphere 1, with mass $m_1 = 30$ g, is pulled to the left to a height $h_1 = 8.0$ cm, and then released. After swinging down, it undergoes an elastic collision with sphere 2, whose mass $m_2 = 75$ g.

a. What is velocity v_{1f} of sphere 1 just after the collision?

SOLUTION We let v_{1i} represent the speed of sphere 1 just before the collision. The sphere begins its descent with no kinetic energy and with gravitational potential energy m_1gh_1. Just before the collision, sphere 1 has kinetic energy $\frac{1}{2}m_1v_{1i}^2$. During the descent, the conservation of mechanical energy gives us

$$\tfrac{1}{2}m_1v_{1i}^2 = m_1gh_1,$$

which we solve for the speed v_{1i} of sphere 1 just before the collision:

$$v_{1i} = \sqrt{2gh_1} = \sqrt{(2)(9.8\text{ m/s}^2)(0.080\text{ m})} = 1.252\text{ m/s}.$$

Although sphere 1 swings down in two dimensions, its collision with sphere 2 is horizontal and thus in one dimension, and we can represent the *velocity* of sphere 1 just before that collision as v_{1i}.

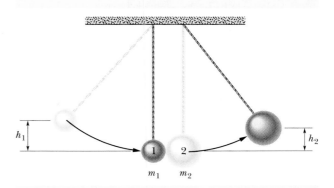

FIGURE 10-13 Sample Problem 10-3. Two metal spheres suspended by cords just touch when they are at rest. Sphere 1, with mass m_1, is pulled to the left to height h_1, and then released. The subsequent elastic collision with sphere 2 sends sphere 2 to height h_2.

To find velocity v_{1f} of sphere 1 just after the collision, we use Eq. 10-18:

$$v_{1f} = \frac{m_1 - m_2}{m_1 + m_2} v_{1i} = \frac{0.030\text{ kg} - 0.075\text{ kg}}{0.030\text{ kg} + 0.075\text{ kg}} (1.252\text{ m/s})$$

$$= -0.537\text{ m/s} \approx -0.54\text{ m/s}. \quad \text{(Answer)}$$

The minus sign tells us that sphere 1 moves to the left just after the collision.

b. To what height h_1' does sphere 1 swing to the left after the collision?

SOLUTION Sphere 1 begins its upward swing with kinetic energy $\frac{1}{2}m_1v_{1f}^2$. When it momentarily stops at height h_1', it has gravitational potential energy m_1gh_1'. Conserving mechanical energy during the upward swing, we find

$$m_1gh_1' = \tfrac{1}{2}m_1v_{1f}^2,$$

or

$$h_1' = \frac{v_{1f}^2}{2g} = \frac{(-0.537\text{ m/s})^2}{(2)(9.8\text{ m/s}^2)}$$

$$= 0.0147\text{ m} \approx 1.5\text{ cm}. \quad \text{(Answer)}$$

c. What is the velocity v_{2f} of sphere 2 just after the collision?

SOLUTION From Eq. 10-19, we have

$$v_{2f} = \frac{2m_1}{m_1 + m_2} v_{1i} = \frac{(2)(0.030\text{ kg})}{0.030\text{ kg} + 0.075\text{ kg}} (1.252\text{ m/s})$$

$$= 0.715\text{ m/s} \approx 0.72\text{ m/s}. \quad \text{(Answer)}$$

d. To what height h_2 does sphere 2 swing after the collision?

SOLUTION Sphere 2 begins its ascent with kinetic energy $\frac{1}{2}m_2 v_{2f}^2$. When it momentarily stops at height h_2, it has gravitational potential energy $m_2 g h_2$. Conservation of mechanical energy during the ascent gives us

$$m_2 g h_2 = \tfrac{1}{2} m_2 v_{2f}^2,$$

or

$$h_2 = \frac{v_{2f}^2}{2g} = \frac{(0.715 \text{ m/s})^2}{(2)(9.8 \text{ m/s}^2)}$$

$$= 0.0261 \text{ m} \approx 2.6 \text{ cm}. \qquad \text{(Answer)}$$

SAMPLE PROBLEM 10-4

In a nuclear reactor, newly produced fast neutrons must be slowed down before they can participate effectively in the chain-reaction process. This is done by allowing them to collide with the nuclei of atoms in a *moderator*.

a. By what fraction is the kinetic energy of a neutron (of mass m_1) reduced in a head-on elastic collision with a nucleus of mass m_2, initially at rest?

SOLUTION The initial and the final kinetic energies of the neutron are

$$K_i = \tfrac{1}{2} m_1 v_{1i}^2 \quad \text{and} \quad K_f = \tfrac{1}{2} m_1 v_{1f}^2.$$

The fraction we seek is then

$$\text{frac} = \frac{K_i - K_f}{K_i} = \frac{v_{1i}^2 - v_{1f}^2}{v_{1i}^2} = 1 - \frac{v_{1f}^2}{v_{1i}^2}. \quad (10\text{-}30)$$

For such a collision we have, from Eq. 10-18,

$$\frac{v_{1f}}{v_{1i}} = \frac{m_1 - m_2}{m_1 + m_2}. \qquad (10\text{-}31)$$

Substituting Eq. 10-31 into Eq. 10-30 yields, after a little algebra,

$$\text{frac} = \frac{4 m_1 m_2}{(m_1 + m_2)^2}. \quad \text{(Answer)} \quad (10\text{-}32)$$

b. Evaluate the fraction for lead, carbon, and hydrogen. The ratios of nuclear mass to neutron mass ($= m_2/m_1$) for these nuclei are 206 for lead, 12 for carbon, and about 1 for hydrogen.

SOLUTION The following values of the fraction can be calculated with Eq. 10-32: for lead ($m_2 = 206 m_1$),

$$\text{frac} = \frac{(4)(206)}{(1 + 206)^2} = 0.019 \text{ or } 1.9\%; \quad \text{(Answer)}$$

for carbon ($m_2 = 12 m_1$),

$$\text{frac} = \frac{(4)(12)}{(1 + 12)^2} = 0.28 \text{ or } 28\%; \quad \text{(Answer)}$$

and for hydrogen ($m_2 \approx m_1$),

$$\text{frac} = \frac{(4)(1)}{(1 + 1)^2} = 1 \text{ or } 100\%. \quad \text{(Answer)}$$

These results explain why water, which contains lots of hydrogen, is a much better moderator of neutrons than lead.

SAMPLE PROBLEM 10-5

A target glider, whose mass m_2 is 350 g, is at rest on an air track, a distance $d = 53$ cm from the end of the track. A projectile glider, whose mass m_1 is 590 g, approaches the target glider with velocity $v_{1i} = -75$ cm/s and collides elastically with it (Fig. 10-14a). The target glider rebounds elastically from a short spring at the end of the track and meets the projectile glider for a second time, as shown in Fig. 10-14b. How far from the end of the track does this second collision occur?

SOLUTION Let us use Eqs. 10-18 and 10-19 to find the velocities of the two gliders after they collide. From Eq. 10-18,

$$v_{1f} = v_{1i} \frac{m_1 - m_2}{m_1 + m_2} = (-75 \text{ cm/s}) \frac{590 \text{ g} - 350 \text{ g}}{590 \text{ g} + 350 \text{ g}}$$

$$= -19 \text{ cm/s}. \qquad (10\text{-}33)$$

From Eq. 10-19,

$$v_{2f} = v_{1i} \frac{2 m_1}{m_1 + m_2} = (-75 \text{ cm/s}) \frac{(2)(590 \text{ g})}{590 \text{ g} + 350 \text{ g}}$$

$$= -94 \text{ cm/s}. \qquad (10\text{-}34)$$

At the second collision, glider 1 will have traveled a distance $d - x$, and glider 2 a distance $d + x$. Their travel times t for these distances are equal, so that

$$t = \frac{d - x}{v_{1f}} = \frac{d + x}{v_{2f}}.$$

Substituting from Eqs. 10-33 and 10-34 and putting $d = 53$ cm, we obtain

$$\frac{53 \text{ cm} - x}{-19 \text{ cm/s}} = \frac{53 \text{ cm} + x}{-94 \text{ cm/s}}.$$

Solving for x yields, after a little algebra,

$$x = 35 \text{ cm}. \qquad \text{(Answer)}$$

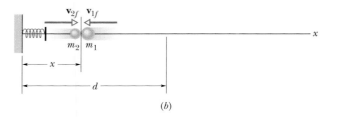

FIGURE 10-14 Sample Problem 10-5. (*a*) Two gliders on an air track are about to collide. Glider m_2 is initially at rest. (*b*) After they collide, glider m_2 rebounds elastically from a spring at the left end of the track and meets glider m_1 for a second time. Where does this second encounter occur?

See Tactic 1 to learn what is wrong with this method of solving the problem.

PROBLEM SOLVING

TACTIC 1: NUMBERS VERSUS ALGEBRA

Beginning problem solvers often substitute numbers too soon. This "rush to the numbers" may build a feeling of confidence, but it often gives a very restricted view of the overall problem. Sample Problem 10-5, in which we have deliberately used the numbers too soon, is a good example. If you solve this problem using symbols only, you get, by replacing the steps in the numerical solution above,

$$x = d\,\frac{m_1 + m_2}{3m_1 - m_2},$$

which shows us that the result is independent of the initial speed of the glider.

We can rewrite this in terms of the dimensionless mass ratio r as

$$x = d\,\frac{r + 1}{3r - 1} \quad \text{in which} \quad r = \frac{m_1}{m_2}.$$

Thus, the algebraic solution shows that the result depends not on the individual masses, *but only on their*

ratio. For the data of Sample Problem 10-5, $r = (590\text{ g})/(350\text{ g}) = 1.69$ and $x = 35$ cm, in agreement with the numerical solution.

The algebraic solution also allows us to check for special cases. For example, $r = 1$ corresponds to $x = d$, which we expect. (The second collision takes place exactly where the first took place.) We can ask questions such as: "For what value of r will $x = \frac{1}{2}d$?" or "For what values of r will $x = 2d$?" (The answers are $r = 3$ and $r = 0.6$.) We can also ask: "What is the smallest value of r for which a second collision will indeed take place?" The answer is $r = \frac{1}{3}$, corresponding to $x \rightarrow \infty$.

You may object that if certain input data are not needed, we should not have given them in the problem statement. But in the real world, problem solvers must supply their own input data and must decide for themselves what is needed and what is not. Best advice: Stick to the algebra as long as you can, and study the requirements and implications of the equation with which you end up.

10-4 INELASTIC COLLISIONS IN ONE DIMENSION

An **inelastic collision** is one in which the kinetic energy of the system of colliding bodies is not conserved. If you drop a Superball onto a hard floor, it loses very little of its kinetic energy on impact and rebounds to almost its original height. If the ball did regain the original height, its collision with the floor would be elastic. However, the small loss of kinetic energy in the collision lowers the rebound height, and therefore the collision is somewhat inelastic.

A dropped golf ball will lose more of its kinetic energy and will rebound to only 60% of its original height. This collision is noticeably inelastic. If you drop a ball of putty onto the floor, it sticks to the floor and does not rebound at all. Because the putty sticks and does not rebound, this collision is said to be a **completely inelastic collision.** (In any inelastic collision, the kinetic energy that is lost is transferred to some other form of energy, perhaps thermal energy.)

In this section we restrict ourselves largely to completely inelastic collisions. Although a system of colliding bodies always loses kinetic energy in any inelastic collision, the linear momentum of the system is always conserved (provided the system is isolated and closed). Since kinetic energy and linear

This ball–window collision is inelastic because some of the ball's mechanical energy is transferred to sound and to the breaking of chemical bonds where the window shatters.

where *V* represents the final velocity of the stuck-together bodies. Equation 10-36 tells us that the final speed is always less than that of the incoming body.

Figure 10-16 (compare it with Fig. 10-11) shows that the motion of the center of mass in a completely inelastic collision is unaffected by the collision. Here, even though the bodies stick together, the kinetic energy associated with the motion of the center of mass is still present. The only way in which kinetic energy can vanish *totally* in an inelastic collision is if the reference frame is fixed with respect to the center of mass of the colliding bodies. Then, all motion relative to the reference frame ceases when the bodies stick together. If the target body m_2 happens to be exceedingly massive, the center of mass of the system essentially coincides with that of the target. This is the situation in which a putty ball is dropped onto the floor, the target body being the Earth. As the ball hits, all the kinetic energy is dissi-

momentum both involve the speeds of the colliding bodies, the conservation of linear momentum limits just how much kinetic energy is lost by a system in an inelastic collision. When the bodies stick together in a completely inelastic collision, the amount of kinetic energy that is lost is the maximum allowed by the conservation of linear momentum. In some situations, the system may lose all its kinetic energy.

Figure 10-15 shows a one-dimensional inelastic collision in which one body is initially stationary. The law of conservation of linear momentum holds, so

$$m_1 v = (m_1 + m_2) V, \qquad (10\text{-}35)$$

or

$$V = v \frac{m_1}{m_1 + m_2}, \qquad (10\text{-}36)$$

FIGURE 10-15 An inelastic collision between two bodies. Before the collision, the body of mass m_2 is at rest. Afterward, the bodies stick together, which is the criterion for a *completely* inelastic collision. The velocities are drawn to scale for the case in which $m_1 = 3m_2$.

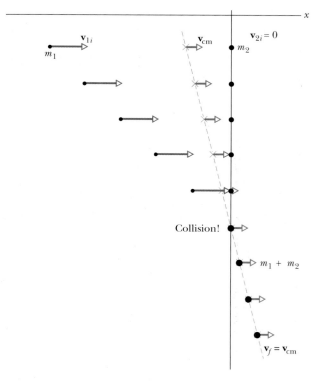

FIGURE 10-16 Some freeze-frames of two bodies undergoing a completely inelastic collision. Body 2 is initially at rest, and $m_2 = 3m_1$. The bodies stick together after the collision and move forward together. The velocity of the center of mass is also shown. Note that it is unaffected by the collision, and that it is equal to the final velocity of the stuck-together bodies.

pated, being transferred into some other form of energy.

If both bodies are moving prior to a collision in which they stick together, we replace Eq. 10-35 with

$$m_1 v_1 + m_2 v_2 = (m_1 + m_2) V, \qquad (10\text{-}37)$$

where $m_1 v_1$ is the initial linear momentum of one body, and $m_2 v_2$ is that of the other body. Here again, all the kinetic energy is dissipated only if the reference frame is fixed with respect to the center of mass of the bodies. Figure 10-17 shows an example: identical cars driven at identical speeds underwent an almost head-on and almost completely inelastic collision. Their center of mass was stationary relative to the Earth. So a bystander, who would have been stationary relative to the Earth, would have seen these cars stop because of the collision, rather than move in one direction or another afterward.

FIGURE 10-17 Two cars after an almost head-on, almost completely inelastic collision.

SAMPLE PROBLEM 10-6

A *ballistic pendulum* (Fig. 10-18) is a device that was used to measure the speeds of bullets before electronic timing devices were developed. The device consists of a large block of wood of mass $M = 5.4$ kg, hanging from two long cords. A bullet of mass $m = 9.5$ g is fired into the block, coming quickly to rest. The *block + bullet* then swing upward, their center of mass rising a vertical distance $h = 6.3$ cm before the pendulum comes momentarily to rest at the end of its arc.

a. What was the speed v of the bullet just prior to the collision?

SOLUTION Just after the collision, the *bullet + block* have speed V. Applying the conservation of linear momentum to the collision, we have

$$mv = (M + m) V.$$

Since the bullet and block stick together, the collision is completely inelastic, and kinetic energy is not conserved during it. However, *after* the collision, mechanical energy *is* conserved, because no force then acts to dissipate that energy. So the kinetic energy of the system when the block is at the bottom of its arc must equal the potential energy of the system when the

FIGURE 10-18 Sample Problem 10-6. A ballistic pendulum, formerly used to measure the speeds of rifle bullets.

block is at the top:

$$\tfrac{1}{2}(M + m)V^2 = (M + m)gh.$$

Eliminating V between these two equations leads to

$$v = \frac{M + m}{m}\sqrt{2gh}$$

$$= \left(\frac{5.4 \text{ kg} + 0.0095 \text{ kg}}{0.0095 \text{ kg}}\right)\sqrt{(2)(9.8 \text{ m/s}^2)(0.063 \text{ m})}$$

$$= 630 \text{ m/s}. \qquad \text{(Answer)}$$

The ballistic pendulum is a kind of "transformer," exchanging the high speed of a light object (the bullet) for the low—and thus more easily measurable—speed of a massive object (the block).

b. What is the initial kinetic energy of the bullet? How much of this energy remains as mechanical energy of the swinging pendulum?

SOLUTION   The kinetic energy of the bullet is

$$K_b = \tfrac{1}{2}mv^2 = (\tfrac{1}{2})(0.0095 \text{ kg})(630 \text{ m/s})^2$$

$$= 1900 \text{ J}. \qquad \text{(Answer)}$$

The mechanical energy of the swinging pendulum is equal to its potential energy when the block is at the top of its swing, or

$$E = (M + m)gh$$

$$= (5.4 \text{ kg} + 0.0095 \text{ kg})(9.8 \text{ m/s}^2)(0.063 \text{ m})$$

$$= 3.3 \text{ J}. \qquad \text{(Answer)}$$

Thus only 3.3/1900 or 0.2% of the original kinetic energy of the bullet is transferred to mechanical energy of the pendulum. The rest is transferred to thermal energy of the block and bullet, or goes into the breaking of wood fibers as the bullet bores into the wood.

SAMPLE PROBLEM 10-7

A karate expert strikes downward with his fist (of mass $m_1 = 0.70$ kg), breaking a 0.14-kg board (Fig. 10-19a). He then does the same to a 3.2-kg concrete block. The spring constants k for bending are 4.1×10^4 N/m for the board and 2.6×10^6 N/m for the block. Breaking occurs at a deflection d of 16 mm for the board and 1.1 mm for the block (Fig. 10-19c).*

*The data are taken from "The Physics of Karate," by S. R. Wilk, R. E. McNair, and M. S. Feld, *American Journal of Physics*, September 1983.

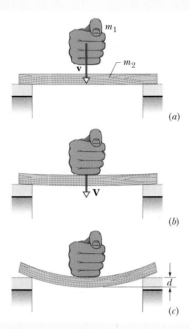

FIGURE 10-19   Sample Problem 10-7. (*a*) A karate expert strikes at a flat object with speed v. (*b*) The fist and object undergo a completely inelastic collision, and bending begins. The *fist* + *object* then have speed V. (*c*) The object breaks when its center has been deflected by an amount d. We assume that the fist and object stop just then.

a. Just before the board and block break, what is the energy stored in each?

SOLUTION   We treat the bending as the compression of a spring for which Hooke's law applies. The stored energy is then, from Eq. 7-19, $U = \tfrac{1}{2}kd^2$. For the board,

$$U = \tfrac{1}{2}(4.1 \times 10^4 \text{ N/m})(0.016 \text{ m})^2$$

$$= 5.248 \text{ J} \approx 5.2 \text{ J}. \qquad \text{(Answer)}$$

For the block,

$$U = \tfrac{1}{2}(2.6 \times 10^6 \text{ N/m})(0.0011 \text{ m})^2$$

$$= 1.573 \text{ J} \approx 1.6 \text{ J}. \qquad \text{(Answer)}$$

b. What fist speed v is required to break the board and the block? Assume that mechanical energy is conserved during the bending, that the fist and struck object stop just before the break, and that the fist–object collision at the onset of bending (Fig. 10-19b) is totally inelastic.

SOLUTION   If mechanical energy is conserved during the bending, then the kinetic energy K of the fist–

object system at the onset of bending must be equal to U just at breaking: 5.2 J for the board and 1.6 J for the block. The fist speed required to break the object is the speed required to produce that kinetic energy K. We may write K as

$$K = \tfrac{1}{2}(m_1 + m_2)V^2 = U,$$

so that

$$V = \sqrt{\frac{2U}{m_1 + m_2}},$$

where V is the speed of *fist + object* at the onset of bending, $m_1 = 0.70$ kg, and m_2 is 0.14 kg for the board or 3.2 kg for the block. For the board we have

$$V = \sqrt{\frac{2(5.248\text{ J})}{0.70\text{ kg} + 0.14\text{ kg}}} = 3.534\text{ m/s}.$$

For the block we have

$$V = \sqrt{\frac{2(1.573\text{ J})}{0.70\text{ kg} + 3.2\text{ kg}}} = 0.8981\text{ m/s}.$$

Just before hitting the board or block, the fist has the speed v of Eq. 10-36. So, by rearranging that equation, we get

$$v = \left(\frac{m_1 + m_2}{m_1}\right)V.$$

For the board we find

$$v = \left(\frac{0.70\text{ kg} + 0.14\text{ kg}}{0.70\text{ kg}}\right)(3.534\text{ m/s})$$

$$\approx 4.2\text{ m/s}, \qquad \text{(Answer)}$$

and for the block we find

$$v = \left(\frac{0.70\text{ kg} + 3.2\text{ kg}}{0.70\text{ kg}}\right)(0.8981\text{ m/s})$$

$$\approx 5.0\text{ m/s}. \qquad \text{(Answer)}$$

The fist speed must be about 20% faster for the fist to break the block, because the larger mass of the block makes the transfer of energy to the block more difficult.

10-5 COLLISIONS IN TWO DIMENSIONS

Here we consider a glancing collision between a projectile body and a target body at rest. In Figure 10-20 we see a typical situation. The distance b by which the collision fails to be head-on is called the *impact parameter*. It is a measure of the directness of the collision, $b = 0$ corresponding to head-on. After

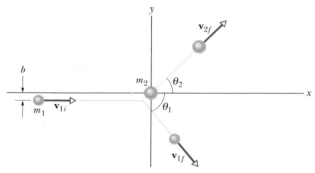

FIGURE 10-20 An elastic collision between two bodies in which the collision is not head-on. The body with mass m_2 (the target) is initially at rest. The distance b is the impact parameter.

the collision, the two bodies fly off at angles θ_1 and θ_2, as the figure shows.

From the conservation of linear momentum (a vector relation), we can write two scalar equations:

$$m_1 v_{1i} = m_1 v_{1f} \cos\theta_1 + m_2 v_{2f} \cos\theta_2 \quad (x\text{ component})$$
$$(10\text{-}38)$$

and

$$0 = -m_1 v_{1f} \sin\theta_1 + m_2 v_{2f} \sin\theta_2 \quad (y\text{ component}).$$
$$(10\text{-}39)$$

If the collision is elastic, kinetic energy is also conserved:

$$\tfrac{1}{2}m_1 v_{1i}^2 = \tfrac{1}{2}m_1 v_{1f}^2 + \tfrac{1}{2}m_2 v_{2f}^2 \quad (\text{kinetic energy}). \quad (10\text{-}40)$$

These three equations contain seven variables: two masses, m_1 and m_2; three speeds, v_{1i}, v_{1f}, and v_{2f}; and two angles, θ_1 and θ_2. If we know any four of these quantities, we can solve the three equations for the remaining three quantities. Often the known quantities are the two masses, the initial speed, and one of the angles. The unknowns to be solved for are then the two final speeds and the remaining angle.

SAMPLE PROBLEM 10-8

Two particles of equal masses have an elastic collision, the target particle being initially at rest. Show that (un-

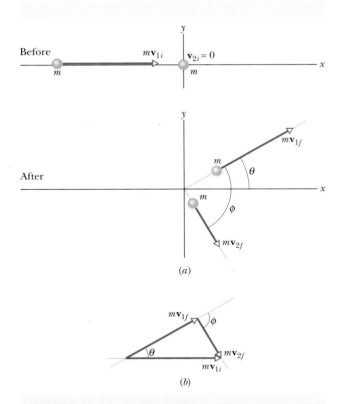

(a)

(b)

FIGURE 10-21 Sample Problem 10-8. A neat proof of the fact that, in an elastic collision between two particles of the same mass, the particles fly off at 90° to each other afterward. For this to hold true, the target particle must be initially at rest and the collision must not be head-on.

less the collision is head-on) the two particles will always move off perpendicular to each other after the collision.

SOLUTION There is a temptation to jump into the problem and solve it in a straightforward way, by applying Eqs. 10-38, 10-39, and 10-40. There is a neater way.

Figure 10-21a shows the situation both before and after the collision, each particle with its linear momentum vector attached. Because linear momentum is conserved in the collision, these vectors must form a closed triangle, as Fig. 10-21b shows. (The vector $m\mathbf{v}_{1i}$ must be the vector sum of $m\mathbf{v}_{1f}$ and $m\mathbf{v}_{2f}$.) Because the masses of the particles are equal, the closed linear-momentum triangle of Fig. 10-21b is also a closed velocity triangle, because dividing by the scalar m does not change the relation of the vectors. That is,

$$\mathbf{v}_{1i} = \mathbf{v}_{1f} + \mathbf{v}_{2f}. \qquad (10\text{-}41)$$

Because kinetic energy is conserved, Eq. 10-40 holds. With the equal mass m canceled out, this equation becomes

$$v_{1i}^2 = v_{1f}^2 + v_{2f}^2. \qquad (10\text{-}42)$$

Equation 10-42 applies to the lengths of the sides in the triangle of Fig. 10-21b. For it to hold, the triangle must be a right triangle (and Eq. 10-42 is then the Pythagorean theorem). Therefore the angle ϕ between the vectors $\mathbf{v}_{1f}$ and $\mathbf{v}_{2f}$ must be 90°, which is what we set out to prove.

SAMPLE PROBLEM 10-9

Two skaters collide and embrace, in a completely inelastic collision. That is, they stick together after impact, as suggested by Fig. 10-22, where the origin is placed at the point of collision. Alfred, whose mass m_A is 83 kg, is originally moving east with speed $v_A = 6.2$ km/h. Barbara, whose mass m_B is 55 kg, is originally moving north with speed $v_B = 7.8$ km/h.

a. What is the velocity $\mathbf{V}$ of the couple after impact?

SOLUTION Linear momentum is conserved during the collision. We can write, for the linear momentum components in the x and y directions,

$$m_A v_A = MV \cos \theta \qquad (x \text{ component}) \qquad (10\text{-}43)$$

and

$$m_B v_B = MV \sin \theta \qquad (y \text{ component}), \qquad (10\text{-}44)$$

in which $M = m_A + m_B$. Dividing Eq. 10-44 by Eq. 10-43 yields

$$\tan \theta = \frac{m_B v_B}{m_A v_A} = \frac{(55 \text{ kg})(7.8 \text{ km/h})}{(83 \text{ kg})(6.2 \text{ km/h})} = 0.834.$$

Thus

$$\theta = \tan^{-1} 0.834 = 39.8° \approx 40°. \qquad \text{(Answer)}$$

From Eq. 10-44 we then have

$$V = \frac{m_B v_B}{M \sin \theta} = \frac{(55 \text{ kg})(7.8 \text{ km/h})}{(83 \text{ kg} + 55 \text{ kg})(\sin 39.8°)}$$

$$= 4.86 \text{ km/h} \approx 4.9 \text{ km/h}. \qquad \text{(Answer)}$$

b. What is the velocity of the center of mass of the two skaters before and after the collision?

SOLUTION We can answer this without further calculation. After the collision, the velocity of the center of mass is the same as the velocity that we calculated in part (a), namely, 4.9 km/h at 40° north of east (Fig.

FIGURE 10-22 Sample Problem 10-9. Two skaters, Alfred (A) and Barbara (B), have a completely inelastic collision. Afterward, they move off together at angle θ, with speed V. The path of their center of mass is shown. The position of the center of mass for the indicated positions of the skaters before the collision is also shown.

10-22). Because the velocity of the center of mass is not changed by the collision, this same value must prevail before the collision.

c. What is the fractional change in the kinetic energy of the skaters because of the collision?

SOLUTION The initial kinetic energy is

$$K_i = \tfrac{1}{2}m_A v_A^2 + \tfrac{1}{2}m_B v_B^2$$

$$= (\tfrac{1}{2})(83 \text{ kg})(6.2 \text{ km/h})^2 + (\tfrac{1}{2})(55 \text{ kg})(7.8 \text{ km/h})^2$$

$$= 3270 \text{ kg} \cdot \text{km}^2/\text{h}^2.$$

The final kinetic energy is

$$K_f = \tfrac{1}{2}MV^2$$

$$= (\tfrac{1}{2})(83 \text{ kg} + 55 \text{ kg})(4.86 \text{ km/h})^2$$

$$= 1630 \text{ kg} \cdot \text{km}^2/\text{h}^2.$$

The fractional change is then

$$\text{frac} = \frac{K_f - K_i}{K_i}$$

$$= \frac{1630 \text{ kg} \cdot \text{km}^2/\text{h}^2 - 3270 \text{ kg} \cdot \text{km}^2/\text{h}^2}{3270 \text{ kg} \cdot \text{km}^2/\text{h}^2}$$

$$= -0.50. \qquad \text{(Answer)}$$

Thus 50% of the initial kinetic energy is lost as a result of the collision.

PROBLEM SOLVING

TACTIC 2: MORE ABOUT UNITS

More often than not, it is wise to express all physical quantities in their basic SI units; thus all speeds in meters per second, all masses in kilograms, and so on. Sometimes, however, it is not necessary to do this. In Sample Problem 10-9a, for example, the units cancel out when we calculate the angle θ. In Sample Problem 10-9c, they cancel when we calculate the dimensionless quantity frac. In this latter case, for example, there is no need to change the kinetic energy units to joules, the basic SI energy unit; we can leave them in the units $\text{kg} \cdot \text{km}^2/\text{h}^2$ because we can look ahead and see that they are going to cancel when we compute the dimensionless ratio frac.

10-6 REACTIONS AND DECAY PROCESSES (OPTIONAL)

Here we consider collisions (called **reactions**) in which the identity and even the number of interacting nuclear particles change because of the collisions. We consider too the *spontaneous decay* of unstable particles, in which one particle is transformed to two other particles. For both kinds of events, there is a clear distinction between times "before the event" and times "after the event," and the laws of conservation of linear momentum and of *total* energy apply. In short, we can treat these events by the methods we have already developed for collisions.

SAMPLE PROBLEM 10-10

A radioactive nucleus of uranium-235 decays spontaneously to thorium-231 by emitting an alpha particle*:

$$^{235}\text{U} \rightarrow \alpha + {}^{231}\text{Th}.$$

The alpha particle ($m_\alpha = 4.00$ u) has a kinetic energy K_α of 4.60 MeV. What is the kinetic energy of the recoiling thorium-231 nucleus ($m_{\text{Th}} = 231$ u)?

SOLUTION The ^{235}U nucleus is initially at rest in the laboratory. After decay, the alpha particle flies off and

---

*An alpha particle (symbolized as α, helium-4, or ⁴He) is the nucleus of a helium atom.

the ^{231}Th recoils in the opposite direction, with kinetic energies K_α and K_{Th}, respectively. Applying the law of conservation of linear momentum leads to

$$0 = m_{\text{Th}}v_{\text{Th}} + m_\alpha v_\alpha,$$

which we can recast as

$$m_{\text{Th}}v_{\text{Th}} = -m_\alpha v_\alpha. \qquad (10\text{-}45)$$

Because $K = \frac{1}{2}mv^2$, we can square each side of Eq. 10-45 and rewrite it as

$$m_{\text{Th}}K_{\text{Th}} = m_\alpha K_\alpha.$$

Thus

$$K_{\text{Th}} = K_\alpha \frac{m_\alpha}{m_{\text{Th}}} = (4.60 \text{ MeV}) \frac{4.00 \text{ u}}{231 \text{ u}}$$

$$= 7.97 \times 10^{-2} \text{ MeV} = 79.7 \text{ keV.} \quad \text{(Answer)}$$

We see that, of the total amount of kinetic energy made available during the decay (4.60 MeV + 0.0797 MeV = 4.68 MeV), the heavy recoiling nucleus receives only about 0.0797/4.68 or 1.7%.

SAMPLE PROBLEM 10-11

A nuclear reaction of great importance for the generation of energy by nuclear fusion is the so-called d-d reaction, one form of which is

$$d + d \rightarrow t + p. \qquad (10\text{-}46)$$

The particles represented by these letters are all isotopes of hydrogen, with the following properties:

SYMBOLS		NAME	MASS
p	^{1}H	Proton	$m_p = 1.00783$ u
d	^{2}H	Deuteron	$m_d = 2.01410$ u
t	^{3}H	Triton	$m_t = 3.01605$ u

a. How much kinetic energy appears because of the mass change Δm that occurs in this reaction?

SOLUTION From Einstein's relation $E = mc^2$, we know that the change in mass energy associated with Δm is $\Delta m \, c^2$. From Eq. 8-31, the Q of a reaction is defined as $Q = -\Delta m \, c^2$. So, in this problem, we find

$$Q = -\Delta m \, c^2 = (2m_d - m_p - m_t)c^2$$

$$= (2 \times 2.01410 \text{ u} - 1.00783 \text{ u} - 3.01605 \text{ u})$$

$$\times (932 \text{ MeV/u})$$

$$= (0.00432 \text{ u})(932 \text{ MeV/u})$$

$$= 4.03 \text{ MeV.} \qquad \text{(Answer)}$$

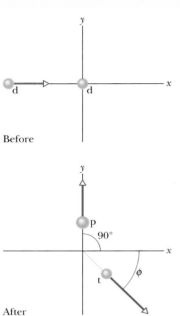

FIGURE 10-23   Sample Problem 10-11. A moving deuteron (d) strikes a stationary deuteron. The d-d reaction of Eq. 10-46 occurs and the product particles (p and t) move off as shown.

We have used the value of 932 MeV/u for c^2 given by Eq. 8-34.

A positive value for Q (as in this case) means that the reaction is **exothermic;** in this reaction mass energy is transferred into kinetic energy. Only 0.00432/(2 × 2.01410) or about 0.1% of the mass originally present has been so transferred. A negative Q signals an **endothermic** reaction, in which the transfer is from kinetic energy to mass energy. And $Q = 0$ means an *elastic encounter*, with no mass change and kinetic energy conserved.

b. A deuteron of kinetic energy $K_d = 1.50$ MeV strikes a stationary deuteron, initiating the reaction of Eq. 10-46. A proton is observed to move off at an angle of 90° with the incident direction, with a kinetic energy of 3.39 MeV; see Fig. 10-23. What is the kinetic energy of the triton?

SOLUTION The energy Q released due to the decrease in mass energy appears as an increase in the kinetic energies of the particles. Thus we can write

$$Q = K_p + K_t - K_d,$$

where Q is the energy computed for this reaction in part (a). Solving for K_t gives us

$$K_t = Q + K_d - K_p$$
$$= 4.03 \text{ MeV} + 1.50 \text{ MeV} - 3.39 \text{ MeV}$$
$$= 2.14 \text{ MeV}. \qquad \text{(Answer)}$$

c. At what angle ϕ with the incident direction (see Fig. 10-23) does the triton emerge?

SOLUTION We have not yet made use of the fact that linear momentum is conserved in the reaction of Eq. 10-46. The law of conservation of linear momentum yields two scalar equations:

$$m_d v_d = m_t v_t \cos \phi \qquad (x \text{ component}) \quad (10\text{-}47)$$

and

$$0 = m_p v_p + m_t v_t \sin \phi \qquad (y \text{ component}). \quad (10\text{-}48)$$

From Eq. 10-48 we have

$$\sin \phi = -\frac{m_p v_p}{m_t v_t}. \qquad (10\text{-}49)$$

Using the relation $K = \frac{1}{2}mv^2$, we can rewrite each linear momentum mv as $\sqrt{2mK}$ and thus recast Eq. 10-49 as

$$\phi = \sin^{-1}\left(-\sqrt{\frac{m_p K_p}{m_t K_t}}\right)$$
$$= \sin^{-1}\left(-\sqrt{\frac{(1.01 \text{ u}) (3.39 \text{ MeV})}{(3.02 \text{ u}) (2.14 \text{ MeV})}}\right)$$
$$= \sin^{-1}(-0.728) = -46.7°. \qquad \text{(Answer)}$$

REVIEW & SUMMARY

Collisions

In a **collision,** two bodies exert strong forces on each other for a relatively short time. These forces are internal to the two-body system and are significantly larger than any external force during the collision. The laws of conservation of energy and of linear momentum, applied immediately before and after a collision, allow us to predict the outcome of the collision and to understand the interactions between the colliding bodies.

Impulse and Linear Momentum

Applying Newton's second law in momentum form to a particle-like body involved in a collision leads to the **impulse–linear momentum theorem:**

$$\mathbf{p}_f - \mathbf{p}_i = \Delta \mathbf{p} = \mathbf{J}, \qquad (10\text{-}4)$$

where $\mathbf{p}_f - \mathbf{p}_i = \Delta \mathbf{p}$ is the change in the body's linear momentum, and $\mathbf{J}$ is the **impulse** due to the force exerted on the body by the other body in the collision:

$$\mathbf{J} = \int_{t_i}^{t_f} \mathbf{F}(t) \; dt. \qquad (10\text{-}3)$$

If $\overline{F}$ is the average of $\mathbf{F}(t)$ during the collision, and Δt is the duration of the collision, then for one-dimensional motion

$$J = \overline{F} \, \Delta t. \qquad (10\text{-}8)$$

When a steady stream of bodies, each with mass m and speed v, collide with a body fixed in position, the average

force on the fixed body is

$$\overline{F} = -\frac{n}{\Delta t} \Delta p = -\frac{n}{\Delta t} m \, \Delta v, \qquad (10\text{-}10)$$

where $n/\Delta t$ is the rate at which the bodies collide with the fixed body, and Δv is the change in velocity of each colliding body. This average force can also be written as

$$\overline{F} = -\frac{\Delta m}{\Delta t} \Delta v, \qquad (10\text{-}13)$$

where $\Delta m/\Delta t$ is the rate at which mass collides with the fixed body. In Eqs. 10-10 and 10-13, $\Delta v = -v$ if the bodies stop upon impact or $\Delta v = -2v$ if they bounce directly backward with no change in speed.

Elastic Collision — One Dimension

An *elastic collision* is one in which the kinetic energy of a system of two colliding bodies is conserved. For a one-dimensional situation in which one body (the target) is stationary and the other body (the projectile) is initially moving, conservation of kinetic energy and of linear momentum leads to the following relations:

$$v_{1f} = \frac{m_1 - m_2}{m_1 + m_2} v_{1i} \qquad (10\text{-}18)$$

and

$$v_{2f} = \frac{2m_1}{m_1 + m_2} v_{1i}. \qquad (10\text{-}19)$$

Here subscripts i and f refer to the velocities immediately before and after the collision, respectively. If both bodies are moving prior to the collision, their velocities immediately after the collision are given by

$$v_{1f} = \frac{m_1 - m_2}{m_1 + m_2} v_{1i} + \frac{2m_2}{m_1 + m_2} v_{2i} \qquad (10\text{-}28)$$

and

$$v_{2f} = \frac{2m_1}{m_1 + m_2} v_{1i} + \frac{m_2 - m_1}{m_1 + m_2} v_{2i}. \qquad (10\text{-}29)$$

Inelastic Collision

An *inelastic collision* is one in which the kinetic energy of a system of two colliding bodies is not conserved. The total linear momentum of the system must, however, still be conserved. If the colliding bodies stick together, the collision is **completely inelastic,** and the reduction in kinetic energy is maximum (but not necessarily zero). For one-dimensional motion, with one body initially stationary, and the other with initial velocity v, the velocity of the stuck-together bodies is found by applying conservation of linear momentum to the system:

$$m_1 v = (m_1 + m_2) V. \qquad (10\text{-}35)$$

If both bodies are moving prior to the collision, conserva-

tion of linear momentum is written as

$$m_1 v_1 + m_2 v_2 = (m_1 + m_2) V. \qquad (10\text{-}37)$$

Collisions in Two Dimensions

Collisions in two dimensions are governed by the conservation of vector linear momentum, a condition that leads to two component equations. These determine the final motion if the collision is completely inelastic. Otherwise, the laws of conservation of linear momentum and of energy generally lead to equations that cannot be solved completely unless other experimental data, such as the final direction of one of the velocities, are available.

Reactions and Decay

In a *reaction* or *decay* of nuclear particles, linear momentum and *total* energy are conserved. If the mass of a system of such particles changes by Δm, the mass energy of the system changes by $\Delta m\, c^2$. The Q of the reaction or decay is defined as

$$Q = -\Delta m\, c^2.$$

This process is said to be **exothermic,** and Q is a positive quantity, if mass energy is transferred to kinetic energy of particles in the system. It is said to be **endothermic,** and Q is a negative quantity, if kinetic energy of particles in the system is transferred to mass energy.

QUESTIONS

1. Explain how conservation of linear momentum applies to a handball bouncing off a wall.

2. Can the impulse of a force be zero, even if the force is not zero? Explain why or why not.

3. Figure 10-24 shows a popular carnival "strongarm" device, in which contestants try to see how high they can raise a weighted marker by hitting a target with a sledgehammer. What physical quantity does the device measure? Is it the average force, the maximum force, the work done, the impulse, the energy transferred, the linear momentum transferred, or something else? Discuss your answer.

4. Different batters swing bats differently. What features of the swing help determine the speed and trajectory of the ball?

5. Many features of cars, such as collapsible steering wheels and padded dashboards, are meant to protect passengers during accidents. Explain their usefulness, using the impulse concept.

6. Why does the use of gloves make modern boxing somewhat safer than bare-knuckle fighting? When stuntpeople fall from buildings, why does landing on an air bag help protect them from injury or death? Why can some victims falling from great heights survive the landing if the ground is covered with soft snow, if they crash through tree branches before reaching the ground, or if they land on the side of a ravine and then slide down it? In each case, argue your point from considerations of average force.

7. A parachutist carrying a pumpkin as a Halloween stunt finds that the pumpkin is ripped out of her grasp when she opens her parachute. Explain why.

8. It is said that, during a 30-mi/h collision, a 10-lb child can exert a 300-lb force against a parent's grip. How can such a large force come about?

9. The following statement was taken from an exam paper: "The collision between two helium atoms is perfectly elastic, so that momentum is conserved." Is the statement logically correct? Explain.

10. You are driving along a highway at 50 mi/h, followed by another car moving at the same speed. You slow to 40 mi/h but the other driver does not and there is a collision. What are the initial velocities of the colliding cars as seen from the reference frame of (a) yourself, (b) the other driver, and (c) a state trooper, who is in a patrol car parked by the roadside? (d) A judge asks whether you bumped into the other driver or the other driver bumped into you. As a physicist, how would you answer?

11. It is obvious from inspection of Eqs. 10-16 and 10-17 that a valid solution to the problem of finding the final velocities of two particles in a one-dimensional elastic collision is $v_{1f} = v_{1i}$ and $v_{2f} = v_{2i} = 0$. What does this mean physically?

12. Two identical cubical blocks, moving in the same direction with a common speed v, strike a third such block initially at rest on a horizontal frictionless surface. What is the motion of the blocks after the collision? Does it matter whether or not the two initially moving blocks were in contact? Does it matter whether these two blocks were glued together?

13. Drop, in succession, a baseball and a basketball from about shoulder height above a hard floor, and note how high each rebounds. Then align the baseball above the basketball (with a small separation as in Fig. 10-25a) and

FIGURE 10-24 Question 3.

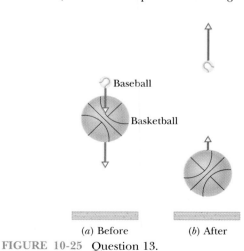

Baseball

Basketball

(a) Before   (b) After

FIGURE 10-25 Question 13.

drop them simultaneously. (Be prepared to duck, and guard your face.) Why does the basketball almost "go dead" on the floor as in Fig. 10-25b, while the baseball shoots up to the ceiling, going higher than the sum of the individual baseball and basketball rebounds? (See also Problem 38.)

14. Two clay balls of equal mass and speed strike each other head-on, stick together, and come to rest. Kinetic energy is certainly not conserved. What happened to the energy? How is linear momentum conserved?

15. A football player, momentarily at rest on the field, catches a football as he is tackled by a running player on the other team. This is certainly a collision (inelastic!), and linear momentum must be conserved. In the reference frame of the football field, there is linear momentum before the collision but there seems to be none after the collision. Is linear momentum really conserved? If so, explain how. If not, explain why not.

16. Consider a one-dimensional elastic collision between a moving object A and an object B initially at rest. How would you choose the mass of B, in comparison to the

mass of A, in order that B should recoil with (a) the greatest speed, (b) the greatest linear momentum, and (c) the greatest kinetic energy?

17. An inverted hourglass is weighed on a sensitive balance from the time the first grain moves to after the last grain has landed. How does the weight vary during that time? Why?

18. An evacuated box is at rest on a frictionless table. You punch a small hole in one face so that air can enter. (See Fig. 10-26.) How will the box move? What argument did you use to arrive at your answer?

FIGURE 10-26 Question 18.

EXERCISES & PROBLEMS

SECTION 10-2 IMPULSE AND LINEAR MOMENTUM

1E. A 95-kg athlete is running at 4.2 m/s. What impulse will stop him?

2E. The linear momentum of a 1500-kg car increases by 9.0×10^3 kg·m/s in 12 s. (a) What is the magnitude of the force that accelerated the car? (b) By how much did the speed of the car increase?

3E. A cue stick strikes a pool ball, exerting an average force of 50 N over a time of 10 ms. If the ball has mass 0.20 kg, what speed does it have after impact?

4E. The National Transportation Safety Board is testing the crash-worthiness of a new car. The 2300-kg vehicle, moving at 15 m/s, is allowed to collide with a bridge abutment, being brought to rest in a time of 0.56 s. What force, assumed constant, acted on the car during impact?

5E. A ball of mass m and speed v strikes a wall perpendicularly and rebounds in the opposite direction with undiminished speed. (a) If the time of collision is Δt, what is the average force exerted by the ball on the wall? (b) Evaluate this average force numerically for a rubber ball with mass 140 g moving at 7.8 m/s; the duration of the collision is 3.8 ms.

6E. A 150-g (weight ≈ 5.3 oz) baseball pitched at a speed of 40 m/s (≈ 130 ft/s) is hit straight back to the pitcher at a speed of 60 m/s (≈ 200 ft/s). What average force was exerted by the bat if it was in contact with the ball for 5.0 ms?

7E. Until he was in his seventies, Henri LaMothe excited audiences by belly-flopping from a height of 40 ft into 12 in. of water (Fig. 10-27). Assuming that he stops just as

FIGURE 10-27 Exercise 7.

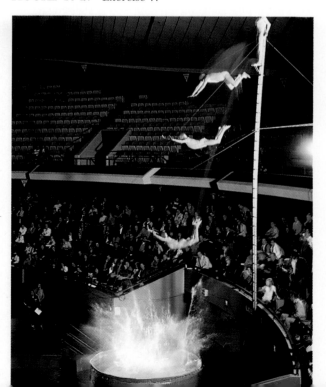

he reaches the bottom of the water, what is the average force on him from the water? Assume his weight is 160 lb.

8E. In February 1955, a paratrooper fell 1200 ft from an airplane without being able to open his chute. His impact in the snow on the ground resembled a mortar round exploding. Assume that his speed at impact was 56 m/s (terminal speed), that his mass (including gear) was 85 kg, and that the force on him from the snow was at the survivable limit of 1.2×10^5 N. What is the minimum depth of snow that would have stopped him safely?

9E. A force that averages 1200 N is applied to a 0.40-kg steel ball moving at 14 m/s in a collision lasting 27 ms. If the force is in a direction opposite the initial velocity of the ball, find the final speed of the ball.

10E. A 1.2-kg medicine ball drops vertically onto a floor, hitting with a speed of 25 m/s. It rebounds with an initial speed of 10 m/s. (a) What impulse acts on the ball during the contact? (b) If the ball is in contact with the floor for 0.020 s, what is the average force exerted on the floor?

11E. A golfer hits a golf ball, giving it an initial velocity of magnitude 50 m/s directed 30° above the horizontal. Assuming that the mass of the ball is 46 g and the club and ball are in contact for 1.7 ms, find (a) the impulse on the ball, (b) the impulse on the club, (c) the average force exerted on the ball by the club, and (d) the work done on the ball.

12P. A 1400-kg car moving at 5.3 m/s is initially traveling north in the positive y direction. After completing a 90° right-hand turn to the positive x direction in 4.6 s, the inattentive operator drives into a tree, which stops the car in 350 ms. In unit-vector notation, what is the impulse on the car (a) during the turn and (b) during the collision? What is the magnitude of the average force that acts on the car (c) during the turn and (d) during the collision? (e) What is the angle between the average force in (c) and the positive x direction?

13P. The force on a 10-kg object increases uniformly from zero to 50 N in 4.0 s. What is the object's final speed if it started from rest?

14P. A pellet gun fires ten 2.0-g pellets per second with a speed of 500 m/s. The pellets are stopped by a rigid wall. (a) What is the momentum of each pellet? (b) What is the kinetic energy of each pellet? (c) What is the average force exerted by the stream of pellets on the wall? (d) If each pellet is in contact with the wall for 0.6 ms, what is the average force exerted on the wall by each pellet while in contact? Why is this so different from the force in (c)?

15P. A machine gun fires 50-g bullets at a speed of 1000 m/s. The gunner, holding the machine gun in his hands, can exert an average force of 180 N against the gun. Determine the maximum number of bullets he can fire per minute while still holding the gun steady.

16P. It is well known that bullets and other missiles fired at Superman simply bounce off his chest (Fig. 10-28). Suppose that a gangster sprays Superman's chest with 3-g

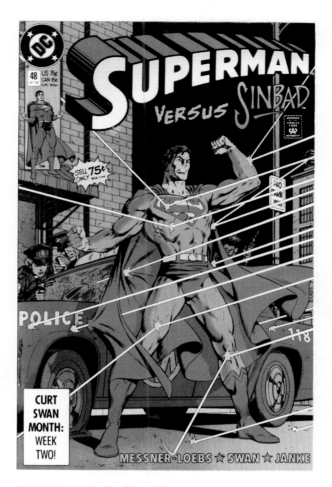

FIGURE 10-28 Problem 16.

bullets at the rate of 100 bullets/min, the speed of each bullet being 500 m/s. Suppose too that the bullets rebound straight back with no change in speed. What is the average force exerted by the stream of bullets on Superman's chest?

17P. Each minute, a special game warden's machine gun fires 220 10-g rubber bullets with a muzzle velocity of 1200 m/s. How many bullets must be fired at an 85-kg animal charging toward the warden at 4.0 m/s in order to stop the animal in its tracks? (Assume that the bullets travel horizontally and drop to the ground after striking the target, with no rebound.)

18P. During a violent thunderstorm, hail of diameter 1.0 cm falls at a speed of 25 m/s. There are estimated to be 120 hailstones per cubic meter of air. Ignore the bounce of the hail on impact. (a) What is the mass of each hailstone (density = 0.92 g/cm^3)? (b) What average force is exerted by hail on a 10 m × 20 m flat roof at the height of the storm?

19P. A stream of water impinges on a stationary "dished" turbine blade, as shown in Fig. 10-29. The speed of the water is v, both before and after it strikes the curved surface of the blade, and the mass of water striking the blade per unit time is constant at the value μ. Find the force exerted by the water on the blade.

FIGURE 10-29 Problem 19.

20P. A stream of water from a hose is sprayed on a wall. If the speed of the water is 5.0 m/s and the hose sprays 300 cm³/s, what is the average force exerted on the wall by the stream of water? Assume that the water does not spatter back appreciably. Each cubic centimeter of water has a mass of 1.0 g.

21P. Figure 10-30 shows an approximate plot of force versus time during the collision of a 58-g tennis ball with a wall. The initial velocity of the ball is 34 m/s perpendicular to the wall; it rebounds with the same speed, also perpendicular to the wall. What is F_{max}, the maximum value of the contact force during the collision?

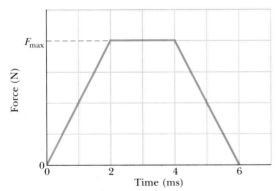

FIGURE 10-30 Problem 21.

22P. A ball having a mass of 150 g strikes a wall with a speed of 5.2 m/s and rebounds with only 50% of its initial kinetic energy. (a) What is the speed of the ball immediately after rebounding? (b) What was the magnitude of the impulse of the ball on the wall? (c) If the ball was in contact with the wall for 7.6 ms, what was the magnitude of the average force exerted by the wall on the ball during this time interval?

23P. A 300-g ball with a speed v of 6.0 m/s strikes a wall at an angle θ of 30° and then rebounds with the same

FIGURE 10-31 Problem 23.

speed and angle (Fig. 10-31). It is in contact with the wall for 10 ms. (a) What was the impulse on the ball? (b) What was the average force exerted by the ball on the wall?

24P. A 2500-kg unmanned space probe is moving in a straight line at a constant speed of 300 m/s. Control rockets on the space probe execute a burn in which a thrust of 3000 N acts for 65.0 s. (a) What is the change in linear momentum (magnitude only) of the probe if the thrust is backward, forward, or directly sideways? (b) What is the change in kinetic energy under the same three conditions? Assume that the mass of the ejected fuel is negligible compared to the mass of the space probe.

25P. A force exerts an impulse J on an object of mass m, changing its speed from v to u. The force and the object's motion are along the same straight line. Show that the work done by the force is $\frac{1}{2}J(u + v)$.

26P. A spacecraft is separated into two parts by detonating the explosive bolts that hold them together. The masses of the parts are 1200 kg and 1800 kg; the magnitude of the impulse on each part is 300 N·s. With what relative speed do the two parts separate?

27P. A croquet ball with mass 0.50 kg is struck by a mallet, receiving the impulse shown in the graph of Fig. 10-32. What is the ball's velocity just after the force has become zero?

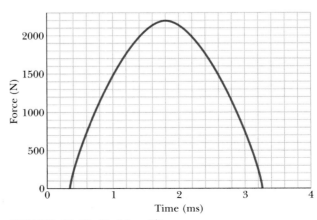

FIGURE 10-32 Problem 27.

28P. Spacecraft *Voyager 2* (of mass m and speed v relative to the sun) approaches the planet Jupiter (of mass M and speed V relative to the sun) as shown in Fig. 10-33. The spacecraft rounds the planet and departs in the opposite direction. What is its speed, relative to the sun, after this slingshot encounter? Assume $v = 12$ km/s and $V = 13$ km/s (the orbital speed of Jupiter). The mass of Jupiter is very much greater than the mass of the spacecraft; $M \gg m$. (For added information, see "The Slingshot Effect: Explanation and Analogies," by Albert A. Bartlett and Charles W. Hord, *The Physics Teacher*, November 1985.)

FIGURE 10-33 Problem 28.

SECTION 10-3 ELASTIC COLLISIONS IN ONE DIMENSION

29E. The blocks in Fig. 10-34 slide without friction. (a) What is the velocity **v** of the 1.6-kg block after the collision? (b) Is the collision elastic? (c) Suppose the initial velocity of the 2.4-kg block is the reverse of what is shown. Can the velocity **v** of the 1.6-kg block after the collision be in the direction shown?

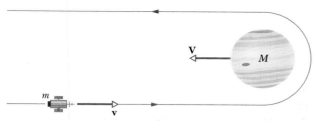

FIGURE 10-34 Exercise 29.

30E. A hovering fly is approached by an enraged elephant charging at 2.1 m/s. Assuming that the collision is elastic, at what speed does the fly rebound? Note that the projectile (the elephant) is much more massive than the stationary target (the fly).

31E. An electron collides elastically with a hydrogen atom initially at rest. (All motions are along the same straight line.) What percentage of the electron's initial kinetic energy is transferred to the hydrogen atom? The mass of the hydrogen atom is 1840 times the mass of the electron.

32E. An α particle (mass 4 u) experiences an elastic head-on collision with a gold nucleus (mass 197 u) that is originally at rest. What percentage of its original kinetic energy does the α particle lose?

33E. A cart with mass 340 g moving on a frictionless linear air track at an initial speed of 1.2 m/s strikes a second cart of unknown mass at rest. The collision between the carts is elastic. After the collision, the first cart continues in its original direction at 0.66 m/s. (a) What is the mass of the second cart? (b) What is its speed after impact?

34E. A body of 2.0-kg mass makes an elastic collision with another body at rest and continues to move in the original direction but with one-fourth of its original speed. What is the mass of the struck body?

35P. A steel ball of mass 0.500 kg is fastened to a cord 70.0 cm long and fixed at the far end, and is released when the cord is horizontal (Fig. 10-35). At the bottom of its path, the ball strikes a 2.50-kg steel block initially at rest on a frictionless surface. The collision is elastic. Find (a) the speed of the ball and (b) the speed of the block, both just after the collision.

FIGURE 10-35 Problem 35.

36P. A platform scale is calibrated to indicate the mass in kilograms of an object placed on it. Particles fall from a height of 3.5 m and collide with the balance pan of the scale. The collisions are elastic; the particles rebound upward with the same speed they had before hitting the pan. If each particle has a mass of 110 g and collisions occur at the rate of 42 s^{-1}, what is the average scale reading?

37P. Two titanium spheres approach each other head-on with the same speed and collide elastically. After the collision, one of the spheres, whose mass is 300 g, remains at rest. What is the mass of the other sphere?

38P. A ball of mass m is aligned above a ball of mass M (with a slight separation, as in Fig. 10-25a), and the two are dropped simultaneously from height h. (Assume the radius of each ball is negligible compared to h.) (a) If M rebounds elastically from the floor and then m rebounds elastically from M, what ratio m/M results in M stopping upon its collision with m? (The answer is approximately the mass ratio of a baseball to a basketball, as in Question 13.) (b) What height does m then reach?

39P. A block of mass m_1 is at rest on a long frictionless table, one end of which is terminated in a wall. Another

FIGURE 10-36 Problem 39.

block of mass m_2 is placed between the first block and the wall and set in motion to the left, toward m_1, with constant speed v_{2i}, as in Fig. 10-36. Assuming that all collisions are completely elastic, find the value of m_2 (in terms of m_1) for which both blocks move with the same velocity after m_2 has collided once with m_1 and once with the wall. Assume the wall to have infinite mass.

40P*. An elevator is moving up a shaft at 6.0 ft/s. At the instant the elevator is 60 ft from the top, a ball is dropped from the top of the shaft. The ball rebounds elastically from the elevator roof. (a) To what height can it rise relative to the top of the shaft? (b) Do the same problem assuming the elevator is moving down at 6.0 ft/s. (*Hint:* The velocity of the ball relative to the elevator is merely reversed by the collision.)

SECTION 10-4 INELASTIC COLLISIONS IN ONE DIMENSION

41E. Meteor Crater in Arizona (Fig. 10-1) is thought to have been formed by the impact of a meteor with the Earth some 20,000 years ago. The mass of the meteor is estimated at 5×10^{10} kg, and its speed at 7200 m/s. What speed would such a meteor impart to the Earth in a head-on collision?

42E. A 6.0-kg box sled is coasting across the ice at a speed of 9.0 m/s when a 12-kg package is dropped into it from above. What is the new speed of the sled?

43E. A 5.20-g bullet moving at 672 m/s strikes a 700-g wooden block at rest on a frictionless surface. The bullet emerges with its speed reduced to 428 m/s. Find the resulting speed of the block.

44E. Two 2.0-kg masses, A and B, collide. The velocities before the collision are $\mathbf{v}_A = 15\mathbf{i} + 30\mathbf{j}$ and $\mathbf{v}_B = -10\mathbf{i} + 5.0\mathbf{j}$. After the collision, $\mathbf{v}'_A = -5.0\mathbf{i} + 20\mathbf{j}$. All speeds are given in meters per second. (a) What is the final velocity of B? (b) How much kinetic energy was gained or lost in the collision?

45E. A bullet of mass 10 g strikes a ballistic pendulum of mass 2.0 kg. The center of mass of the pendulum rises a vertical distance of 12 cm. Assuming that the bullet remains embedded in the pendulum, calculate the bullet's initial speed.

46E. A 5.0-kg block with a speed of 3.0 m/s collides with a 10-kg block that has a speed of 2.0 m/s in the same direction. After the collision, the 10-kg block is observed to be traveling in the original direction with a speed of 2.5 m/s. (a) What is the speed of the 5.0-kg block immediately after the collision? (b) By how much does the total kinetic energy of the system of two blocks change because of the collision? (c) Suppose, instead, that the 10-kg block ends up with a speed of 4.0 m/s. What then is the change in the total kinetic energy? (d) Account for the result you obtained in (c).

47E. A bullet of mass 4.5 g is fired horizontally into a 2.4-kg wooden block at rest on a horizontal surface. The coefficient of kinetic friction between block and surface is 0.20. The bullet comes to rest in the block, which moves 1.8 m. (a) What is the speed of the block immediately after the bullet comes to rest within it? (b) At what speed is the bullet fired?

48P. Two cars A and B slide on an icy road as they attempt to stop at a traffic light. The mass of A is 1100 kg, and the mass of B is 1400 kg. The coefficient of kinetic friction between the locked wheels of both cars and the road is 0.13. Car A succeeds in coming to rest at the light, but car B cannot stop and rear-ends car A. After the collision, A comes to rest 8.2 m ahead of the impact point and B 6.1 m ahead; see Fig. 10-37. Both drivers had their brakes locked throughout the incident. (a) From the distance each car moved after the collision, find the speed of each car immediately after impact. (b) Use conservation of linear momentum to find the speed at which car B struck car A. On what grounds can the use of linear-momentum conservation be criticized here?

FIGURE 10-37 Problem 48.

49P. A 3.0-ton weight falling through a distance of 6.0 ft drives a 0.50-ton pile 1.0 in. into the ground. Assuming that the weight–pile collision is completely inelastic, find the average force of resistance exerted by the ground.

50P. Two particles, one having twice the mass of the other, are held together with a compressed spring between them. The energy stored in the spring is 60 J. How much kinetic energy does each particle have after they are released? Assume that all the stored energy is transferred to the particles and that neither particle is attached to the spring after they are released.

51P. An object with mass m and speed v explodes into two pieces, one three times as massive as the other; the explosion takes place in gravity-free space. The less massive piece comes to rest. How much kinetic energy was added to the system in the explosion?

52P. A box is put on a scale that is marked in units of mass and adjusted to read zero when the box is empty. A stream of marbles is then poured into the box from a height h above its bottom at a rate of R (marbles per second). Each marble has mass m. If the collisions between the marbles and the box are completely inelastic, find the scale reading at time t after the marbles begin to fill the box. Determine a numerical answer when $R = 100 \text{ s}^{-1}$, $h = 7.60$ m, $m = 4.50$ g, and $t = 10.0$ s.

53P. A 35-ton railroad freight car collides with a stationary caboose car. They couple together, and 27% of the initial kinetic energy is dissipated as heat, sound, vibrations, and so on. Find the weight of the caboose.

54P. A ball of mass m is projected with speed v_i into the barrel of a spring gun of mass M initially at rest on a frictionless surface; see Fig. 10-38. The ball sticks in the barrel at the point of maximum compression of the spring. No energy is lost in friction. (a) What is the speed of the spring gun after the ball comes to rest in the barrel? (b) What fraction of the initial kinetic energy of the ball is stored in the spring?

FIGURE 10-38 Problem 54.

55P. A block of mass $m_1 = 2.0$ kg slides along a frictionless table with a speed of 10 m/s. Directly in front of it, and moving in the same direction, is a block of mass $m_2 = 5.0$ kg moving at 3.0 m/s. A massless spring with spring constant $k = 1120$ N/m is attached to the near side of m_2, as shown in Fig. 10-39. When the blocks collide, what is the maximum compression of the spring? (*Hint:* At the moment of maximum compression of the spring, the two blocks move as one. Find the velocity by noting that the collision is completely inelastic to this point.)

FIGURE 10-39 Problem 55.

56P. Two 22.7-kg ice sleds are placed a short distance apart, one directly behind the other, as shown in Fig. 10-40. A 3.63-kg cat, standing on one sled, jumps across to

FIGURE 10-40 Problem 56.

the other and immediately back to the first. Both jumps are made at a speed of 3.05 m/s relative to the ice. Find the final speeds of the two sleds.

57P. The bumper of a 1200-kg car is designed so that it can just absorb all the energy when the car runs head-on into a solid wall at 5.00 km/h. The car is involved in a collision in which it runs at 70.0 km/h into the rear of a 900-kg car moving at 60.0 km/h in the same direction. The 900-kg car is accelerated to 70.0 km/h as a result of the collision. (a) What is the speed of the 1200-kg car immediately after impact? (b) What is the ratio of the kinetic energy absorbed in the collision to that which can be absorbed by the bumper of the 1200-kg car?

58P. A railroad freight car weighing 32 tons and traveling at 5.0 ft/s overtakes one weighing 24 tons and traveling at 3.0 ft/s in the same direction. Find (a) the speed of the cars after collision and (b) the loss of kinetic energy during collision if the cars couple together. (c) If instead, as is very unlikely, the collision is elastic, find the speeds of the cars after collision.

59P*. An electron of mass m collides head-on with an atom of mass M, initially at rest. As a result of the collision, a characteristic amount of energy E is stored internally in the atom. What is the minimum initial speed v_0 that the electron must have? (*Hint:* Conservation principles lead to a quadratic equation for the final electron speed v, and a quadratic equation for the final atom speed V. The minimum value v_0 follows from the requirement that the radical in the solutions for v and V be real.)

SECTION 10-5 COLLISIONS IN TWO DIMENSIONS

60E. An α particle collides with an oxygen nucleus, initially at rest. The α particle is scattered at an angle of 64.0° above its initial direction of motion, and the oxygen nucleus recoils at an angle of 51.0° below this initial direction. The final speed of the nucleus is 1.20×10^5 m/s. What are (a) the final speed and (b) the initial speed of the α particle? (The mass of an α particle is 4.0 u, and the mass of an oxygen nucleus is 16 u.)

61E. A proton (atomic mass 1 u) with a speed of 500 m/s collides elastically with another proton at rest. The original proton is scattered 60° from its initial direction. (a) What is the direction of the velocity of the target proton after the collision? (b) What are the speeds of the two protons after the collision?

62E. A certain nucleus, at rest, spontaneously disintegrates into three particles. Two of them are detected; their masses and velocities are as shown in Fig. 10-41. (a) In

y

16.7×10^{-27} kg

x

6.00×10^6 m/s

Original nucleus

8.35×10^{-27} kg

8.00×10^6 m/s

FIGURE 10-41 Exercise 62.

unit-vector notation, what is the linear momentum of the third particle, which is known to have a mass of 11.7×10^{-27} kg? (b) How much kinetic energy appears in the disintegration process?

63E. In a game of pool, the cue ball strikes another ball initially at rest. After the collision, the cue ball moves at 3.50 m/s along a line making an angle of 22.0° with its original direction of motion, and the second ball has a speed of 2.00 m/s. Find (a) the angle between the direction of motion of the second ball and the original direction of motion of the cue ball and (b) the original speed of the cue ball. (c) Is kinetic energy conserved?

64E. Two vehicles A and B are traveling west and south, respectively, toward the same intersection where they collide and lock together. Before the collision, A (total weight 2700 lb) is moving with a speed of 40 mi/h and B (total weight 3600 lb) has a speed of 60 mi/h. Find the magnitude and direction of the velocity of the (interlocked) vehicles immediately after the collision.

65E. In a game of billiards, the cue ball is given an initial speed V and strikes the pack of 15 stationary balls. All 16 balls then engage in numerous ball–ball and ball–cushion collisions. Some time later, it is observed that (by some accident) all 16 balls have the same speed v. Assuming that all collisions are elastic and ignoring the rotational aspect of the balls' motion, calculate v in terms of V.

66P. A 20.0-kg body is moving in the direction of the positive x axis with a speed of 200 m/s when, owing to an internal explosion, it breaks into three parts. One part, whose mass is 10.0 kg, moves away from the point of explosion with a speed of 100 m/s along the positive y axis. A second fragment, with a mass of 4.00 kg, moves along the negative x axis with a speed of 500 m/s. (a) What is the velocity of the third (6.00-kg) fragment? (b) How much energy was released in the explosion? Ignore effects due to gravity.

67P. Two balls A and B, having different but unknown masses, collide. A is initially at rest, and B has speed v. After collision, B has speed $v/2$ and moves perpendicularly to its original motion. (a) Find the direction in which ball A moves after collision. (b) Can you determine the speed of A from the information given? Explain.

68P. Show that if a neutron is scattered through 90° in an elastic collision with a deuteron that is initially at rest, the neutron loses two-thirds of its initial kinetic energy to the deuteron. (The mass of a neutron is 1.0 u; the mass of a deuteron is 2.0 u.)

69P. After a totally inelastic collision, two objects of the same mass and initial speed are found to move away together at half their initial speed. Find the angle between the initial velocities of the objects.

70P. Two pendulums, both of length l, are initially situated as in Fig. 10-42. The left pendulum is released and strikes the other. Assume that the collision is completely inelastic, and neglect the mass of the strings and any frictional effects. How high does the center of mass of the pendulum system rise after the collision?

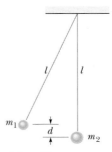

FIGURE 10-42 Problem 70.

71P. A billiard ball moving at a speed of 2.2 m/s strikes an identical stationary ball a glancing blow. After the collision, one ball is found to be moving at a speed of 1.1 m/s in a direction making a 60° angle with the original line of motion. (a) Find the velocity of the other ball. (b) Can the collision be inelastic, given these data?

72P. A ball with an initial speed of 10 m/s collides elastically with two identical balls whose centers are on a line perpendicular to the initial velocity and that are initially in contact with each other (Fig. 10-43). The first ball is aimed directly at the contact point and all motion is frictionless. Find the velocities of all three balls after the collision. (*Hint*: With friction absent, each impulse is directed along the line of centers of the colliding balls, normal to the colliding surfaces.)

FIGURE 10-43 Problem 72.

73P. A barge with mass 1.50×10^5 kg is proceeding down river at 6.2 m/s in heavy fog when it collides broadside with a barge heading directly across the river; see Fig. 10-44. The second barge has mass 2.78×10^5 kg and was moving at 4.3 m/s. Immediately after impact, the second barge finds its course deflected by 18° in the downriver direction and its speed increased to 5.1 m/s. The river current was practically zero at the time of the accident. (a)

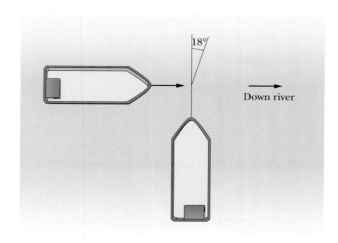

FIGURE 10-44 Problem 73.

What are the speed and direction of motion of the first barge immediately after the collision? (b) How much kinetic energy is lost in the collision?

SECTION 10-6 REACTIONS AND DECAY PROCESSES

74E. The precise masses in the reaction

$$p + {}^{19}F \rightarrow \alpha + {}^{16}O$$

have been determined to be

$$m_p = 1.007825 \text{ u}, \qquad m_\alpha = 4.002603 \text{ u},$$

$$m_F = 18.998405 \text{ u}, \qquad m_O = 15.994915 \text{ u}.$$

Calculate the Q of the reaction from these data.

75E. A particle called Σ^- (sigma minus) is initially at rest and decays spontaneously into two other particles according to

$$\Sigma^- \rightarrow \pi^- + n.$$

The masses are

$$m_\Sigma = 2340.5 m_e, \quad m_\pi = 273.2 m_e, \quad m_n = 1838.65 m_e,$$

where m_e (9.11×10^{-31} kg) is the electron mass. (a) How much kinetic energy is generated in this process? (b) How do the linear momenta of the decay products (π^- and n) compare? (c) Which product gets the larger share of the generated kinetic energy?

76P*. An α particle with kinetic energy 7.70 MeV strikes a ${}^{14}N$ nucleus at rest. An ${}^{17}O$ nucleus and a proton are produced; the proton is emitted at 90° to the direction of the incident α particle and has a kinetic energy of 4.44 MeV. The masses of the various particles are: α particle, 4.00260 u; ${}^{14}N$, 14.00307 u; proton, 1.007825 u; and ${}^{17}O$, 16.99914 u. (a) What is the kinetic energy of the oxygen nucleus? (b) What is the Q of the reaction?

77P*. Consider the α decay of radium (Ra) to radon (Rn), according to the reaction

$$^{226}Ra \rightarrow \alpha + {}^{222}Rn.$$

The masses of the various nuclei are: ${}^{226}Ra$, 226.0254 u; α, 4.0026 u; ${}^{222}Rn$, 222.0175 u. (a) Calculate the Q of the reaction. (b) What value of Q would be obtained if the accurate masses given above were rounded off to three significant figures? What is the kinetic energy of (c) the α particle and (d) the radon nucleus? (For this calculation the rounded-off values of the masses *can* be used; why?)

ADDITIONAL PROBLEMS

78. A 6.0-kg mass and a 4.0-kg mass are moving on a frictionless surface, as shown in Fig. 10-45. A spring of spring constant $k = 8000$ N/m is fixed to the 4.0-kg mass. The 6.0-kg mass has an initial velocity of 8.0 m/s toward the right, and the 4.0-kg mass has an initial velocity of 2.0 m/s toward the right. Eventually, the larger mass overtakes the smaller mass. (a) What is the velocity of the 4.0-kg mass at

the instant the 6.0-kg mass has a velocity of 6.4 m/s toward the right? (b) What is the elastic potential energy of the system just then?

79. A soccer player kicks a soccer ball of mass 0.45 kg that is initially at rest. The player's foot is in contact with the ball for 3.0×10^{-3} s, and the force of the kick is given by

$$F(t) = [(6.0 \times 10^6)t - (2.0 \times 10^9)t^2] \text{ N},$$

for $0 \leq t \leq 3.0 \times 10^{-3}$ s, where t is in seconds. Find the magnitudes of the following: (a) the impulse imparted to the ball, (b) the average force exerted by the player's foot on the ball during the period of contact, (c) the maximum force exerted by the player's foot on the ball during the period of contact, and (d) the ball's velocity immediately after it loses contact with the player's foot.

FIGURE 10-45 Problem 78.

80. A 1.0-kg block at rest on a horizontal frictionless surface is connected to an unstretched spring ($k = 200$ N/m) whose other end is fixed (Fig. 10-46). A 2.0-kg block whose speed is 4.0 m/s collides with the 1.0-kg block. If the two blocks stick together after the one-dimensional collision, what maximum compression of the spring occurs when the blocks momentarily stop?

FIGURE 10-46 Problem 80.

81. A completely inelastic collision occurs between a 3.0-kg mass moving upward at 20 m/s and a 2.0-kg mass moving downward at 12 m/s. How high does the combined mass rise above the point of collision?

82. A 5.0-kg mass with an initial velocity of 4.0 m/s, due east, collides with a 4.0-kg mass whose initial velocity is 3.0 m/s, due west. After the collision the 5.0-kg mass has a velocity of 1.2 m/s, due south. (a) What is the magnitude of the velocity of the 4.0-kg mass after the collision? (b) How much energy is dissipated in the collision?

83. A 0.30-kg softball has a velocity of 12 m/s at an angle

FIGURE 10-47 Problem 83.

of 35° below the horizontal just before making contact with a bat. The ball leaves the bat 2.0 ms later with a vertical velocity of 10 m/s as shown in Fig. 10-47. What is the magnitude of the average force acting on the ball while it is in contact with the bat?

84. A 3.0-kg object moving at 8.0 m/s in the direction of increasing x has a one-dimensional, completely elastic collision with an object of mass M, initially at rest. After the collision the object of mass M has a velocity of 6.0 m/s in the direction of increasing x. What is M?

85. A 60-kg man is ice skating due north with a velocity of 6.0 m/s when he collides with a 38-kg child. The man and child stay together and have a velocity of 3.0 m/s at an angle of 35° north of east immediately after the collision. What were the magnitude and direction of the child's velocity just before the collision?

ROTATION | 11

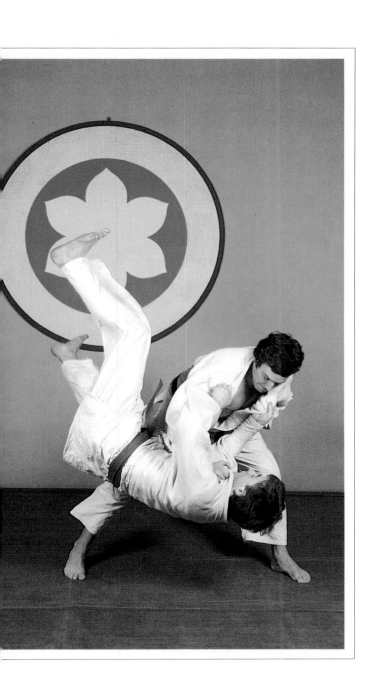

In judo, a weaker and smaller fighter who understands physics can defeat a stronger and larger fighter who does not. This fact is demonstrated by the basic "hip throw," in which a fighter rotates his opponent around his hip and— if the throw is successful—onto the mat. Without the proper use of physics, the throw requires considerable strength and can easily fail. But what is the advantage offered by physics?

11-1 A SKATER'S LIFE

The graceful movement of a figure skater can be used to illustrate, in an aesthetically pleasing way, two kinds of pure, or unmixed, motion. Figure 11-1*a* shows a skater gliding across the ice in a straight line with constant speed. Her motion is one of pure **translation.** Figure 11-1*b* shows her spinning at a constant rate about a fixed vertical axis, in a motion of pure **rotation.** It is this second kind of motion that we deal with in this chapter.

Translation is motion along a straight line, the motion we have discussed almost exclusively so far. Rotation is the motion of wheels, gears, motors, the hands of clocks, the rotors of jet engines, and the blades of helicopters. It is the motion of spinning electrons and atoms, hurricanes, and spinning planets, stars, and galaxies. It is the motion of acrobats, high divers, and orbiting astronauts. Rotation is all around us.

11-2 THE ROTATIONAL VARIABLES

In this chapter, we deal with the rotation of a *rigid* body about a *fixed* axis. The first of these restrictions means that we shall not examine the rotation of such objects as the sun, because the sun—a ball of gas—is not a rigid body. Our second restriction rules out objects like a bowling ball rolling down a bowling lane. Such a ball is in *rolling* motion, rotating about a *moving* axis.

Figure 11-2 shows a rigid body of arbitrary shape in pure *rotation around a fixed axis,* called the **axis of rotation** or the **rotation axis.** Every point of the body moves in a circle whose center lies on the axis of rotation, and every point moves through the same angle during a particular time interval. We contrast this with a body moving in pure *translation in a fixed direction.* In pure translation, every point of the body moves in a straight line, and every point moves through the same *linear distance* during a particular time interval. (Comparisons between linear and angular motion will be a constant part of what follows.)

We deal now—one at a time—with the angular equivalents of the linear quantities position, displacement, velocity, and acceleration.

Angular Position

Figure 11-2 also shows a reference line, fixed in the body, perpendicular to the rotation axis, and rotating with the body. We can describe the motion of

FIGURE 11-1 Figure skater Kristi Yamaguchi in motion of (*a*) pure translation and (*b*) pure rotation. In the first case, the motion is along a fixed direction. In the second case, it is about a fixed axis.

(*a*)

(*b*)

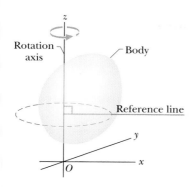

FIGURE 11-2 A rigid body of arbitrary shape in pure rotation about the z axis of a coordinate system. The position of the *reference line* with respect to the rigid body is arbitrary, but it is perpendicular to the rotation axis. It is fixed in the body and it rotates with the body.

the rotating body by specifying the **angular position** of this line, that is, the angle of the line relative to a fixed axis. In Fig. 11-3, the angular position θ is measured relative to the x axis, and θ is given by

$$\theta = \frac{s}{r} \quad \text{(radian measure)}. \quad (11\text{-}1)$$

Here s is the length of arc (or arc distance) of any circle cut off by the x axis and the reference line, and r is the radius of that circle.

An angle defined in this way is measured in **radians** (rad) rather than in revolutions (rev) or degrees. The radian, being the ratio of two lengths, is a pure number and thus has no dimension. Because the circumference of a circle of radius r is $2\pi r$, there

are 2π radians in a complete circle:

$$1 \text{ rev} = 360° = \frac{2\pi r}{r} = 2\pi \text{ rad} \quad (11\text{-}2)$$

or

$$1 \text{ rad} = 57.3° = 0.159 \text{ rev}. \quad (11\text{-}3)$$

We do *not* reset θ to zero with each complete rotation of the reference line about the rotation axis. If the reference line completes two revolutions, then $\theta = 4\pi$ rad.

For pure translational motion along the x direction, we know all there is to know about a moving body if we know $x(t)$, its position as a function of time. Similarly, for pure rotation, we know all there is to know about the rotation of a body if we know $\theta(t)$, the angular position of the body's reference line as a function of time.

Angular Displacement

If the body of Fig. 11-3 rotates about the rotation axis as in Fig. 11-4, changing the angular position of the reference line from θ_1 to θ_2, the body undergoes an **angular displacement** $\Delta\theta$ given by

$$\Delta\theta = \theta_2 - \theta_1. \quad (11\text{-}4)$$

This definition of angular displacement holds not only for the rigid body as a whole but also for *every particle of that body.*

If a body is in translational motion along an x axis, its displacement Δx is either positive or negative, depending on whether the body is moving in

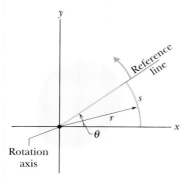

FIGURE 11-3 The rotating rigid body of Fig. 11-2 in cross section, viewed from above. The plane of the cross section is perpendicular to the rotation axis. In this position, the reference line makes an angle θ with the x axis.

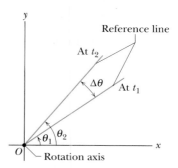

FIGURE 11-4 The reference line of the rigid body of Figs. 11-2 and 11-3 is at angular position θ_1 at time t_1 and at angular position θ_2 at a later time t_2. The quantity $\Delta\theta$ $(= \theta_2 - \theta_1)$ is the angular displacement that occurs during the interval Δt $(= t_2 - t_1)$. The body itself is not shown.

the direction of increasing x or decreasing x. Similarly, the angular displacement $\Delta\theta$ of a rotating body can be either positive or negative, depending on whether the body is rotating in the direction of increasing θ (counterclockwise, as in Figs. 11-3 and 11-4) or decreasing θ (clockwise).

Angular Velocity

Suppose (see Fig. 11-4) that our rotating body is at angular position θ_1 at time t_1 and at angular position θ_2 at time t_2. We define the **average angular velocity** of the body to be

$$\overline{\omega} = \frac{\theta_2 - \theta_1}{t_2 - t_1} = \frac{\Delta\theta}{\Delta t}, \qquad (11\text{-}5)$$

in which $\Delta\theta$ is the angular displacement that occurs during the time interval Δt. (The letter ω is the Greek omega.)

The **(instantaneous) angular velocity** ω, with which we shall be most concerned, is the limit of the ratio in Eq. 11-5 as Δt is made to approach zero. Thus

$$\omega = \lim_{\Delta t \to 0} \frac{\Delta\theta}{\Delta t} = \frac{d\theta}{dt}. \qquad (11\text{-}6)$$

If we know $\theta(t)$, we can find the angular velocity ω by differentiation. This definition of angular velocity holds not only for the rotating rigid body as a whole but also for *every particle of that body*. The unit of angular velocity is commonly the radian per second (rad/s) or revolution per second (rev/s).

If a particle moves in translation along an x axis, its linear velocity v can be either positive or negative, depending on whether the particle is moving in the direction of increasing x or decreasing x. Similarly, the angular velocity ω of a rotating rigid body can be either positive or negative, depending on whether the body is rotating in the direction of increasing θ (counterclockwise) or of decreasing θ (clockwise). The magnitude of an angular velocity is called the **angular speed,** which is also represented with ω.

Angular Acceleration

If the angular velocity of a rotating body is not constant, then the body has an angular acceleration. Let ω_2 and ω_1 be the angular velocities at times t_2 and t_1, respectively. The **average angular acceleration** of the rotating body is defined as

$$\overline{\alpha} = \frac{\omega_2 - \omega_1}{t_2 - t_1} = \frac{\Delta\omega}{\Delta t}, \qquad (11\text{-}7)$$

in which $\Delta\omega$ is the change in the angular velocity that occurs during the time interval Δt. The **(instantaneous) angular acceleration** α, with which we shall be most concerned, is the limit of this quantity as Δt is made to approach zero. Thus

$$\alpha = \lim_{\Delta t \to 0} \frac{\Delta\omega}{\Delta t} = \frac{d\omega}{dt}. \qquad (11\text{-}8)$$

This definition of angular acceleration holds not only for the rotating rigid body as a whole but also for *every particle of that body*. The unit of angular acceleration is commonly radians per second-squared (rad/s^2) or revolutions per second-squared (rev/s^2).

SAMPLE PROBLEM 11-1

The angular position of a reference line on a spinning wheel is given by

$$\theta = t^3 - 27t + 4,$$

where t is in seconds and θ is in radians.

a. Find $\omega(t)$ and $\alpha(t)$.

SOLUTION   To get $\omega(t)$, we differentiate $\theta(t)$ with respect to t:

$$\omega = \frac{d\theta(t)}{dt} = 3t^2 - 27. \qquad \text{(Answer)}$$

To get $\alpha(t)$, we differentiate $\omega(t)$ with respect to t:

$$\alpha = \frac{d\omega(t)}{dt} = \frac{d(3t^2 - 27)}{dt} = 6t. \qquad \text{(Answer)}$$

b. Is there ever a time when $\omega = 0$?

SOLUTION   Setting $\omega(t) = 0$ yields

$$0 = 3t^2 - 27,$$

which we solve, finding

$$t = \pm 3 \text{ s}. \qquad \text{(Answer)}$$

That is, the angular velocity is momentarily zero 3 s before and 3 s after our clock reads zero.

c. Describe the wheel's motion for $t \geq 0$.

SOLUTION To answer, we examine the expressions for $\theta(t)$, $\omega(t)$, and $\alpha(t)$.

At $t = 0$, the reference line on the wheel is at $\theta = 4$ rad, and the wheel is rotating with an angular velocity of -27 rad/s (that is, *clockwise* at an angular speed of 27 rad/s) and an angular acceleration of zero.

For $0 < t < 3$ s, the wheel continues to rotate clockwise, but at decreasing angular speed, because it now has a positive (counterclockwise) angular acceleration. (Check $\omega(t)$ and $\alpha(t)$ for, say, $t = 2$ s.)

At $t = 3$ s, the wheel stops momentarily ($\omega = 0$) and has rotated as far clockwise as it will ever get (the reference line is now at $\theta = -50$ rad).

For $t > 3$ s, the wheel's angular acceleration continues to increase. Its angular velocity, which is now also counterclockwise, increases rapidly because the signs of ω and α match.

SAMPLE PROBLEM 11-2

A top is spun on a floor with angular acceleration

$$\alpha = 5t^3 - 4t,$$

where the coefficients are in units compatible with seconds and radians. At $t = 0$, the top has angular velocity 5 rad/s, and a reference line on it is at angular position $\theta = 2$ rad.

a. Obtain an expression for the angular velocity $\omega(t)$ of the top.

SOLUTION From Eq. 11-8 we have

$$d\omega = \alpha\, dt,$$

which we integrate to get

$$\omega = \int \alpha\, dt = \int (5t^3 - 4t)\, dt$$
$$= \tfrac{5}{4}t^4 - \tfrac{4}{2}t^2 + C.$$

To evaluate the constant of integration C, we note that $\omega = 5$ rad/s at $t = 0$. Substituting these values in our expression for ω yields

$$5 \text{ rad/s} = 0 - 0 + C,$$

so $C = 5$ rad/s. Then

$$\omega = \tfrac{5}{4}t^4 - 2t^2 + 5. \qquad \text{(Answer)}$$

b. Obtain an expression for the angular position $\theta(t)$ of the top.

SOLUTION From Eq. 11-6 we have

$$d\theta = \omega\, dt,$$

which we integrate to get

$$\theta = \int \omega\, dt = \int (\tfrac{5}{4}t^4 - 2t^2 + 5)\, dt,$$
$$= \tfrac{1}{4}t^5 - \tfrac{2}{3}t^3 + 5t + C'$$
$$= \tfrac{1}{4}t^5 - \tfrac{2}{3}t^3 + 5t + 2, \qquad \text{(Answer)}$$

where C' is evaluated by noting that $\theta = 2$ rad at $t = 0$.

11-3 ANGULAR QUANTITIES AS VECTORS: AN ASIDE

We can describe the position, velocity, and acceleration of a single particle by means of vectors. If the particle is confined to a straight line, however, we do not really need the power of vectors. Such a particle has only two directions available to it, and we can designate these with plus and minus signs.

In the same way, a rigid body rotating about a fixed axis can rotate only clockwise or counterclockwise about this axis, and again we can select between them by means of plus and minus signs. The question arises: "Can we treat the angular displacement, velocity, and acceleration of a rotating body as vectors?" The answer (with a caution that we clarify below) is: "Yes."

Consider the angular velocity. Figure 11-5a shows a phonograph record rotating about a fixed spindle. The record has a fixed rotation rate ω ($= 33\tfrac{1}{3}$ rev/min) and a fixed direction of rotation (clockwise as viewed from above). By convention, we represent its angular velocity as a vector pointing along the axis of rotation, as in Fig. 11-5b. We choose the length of this vector according to some convenient scale, for example, with 10 rev/min equal to 1 cm.

We establish a direction for the vector ω by using a **right-hand rule,** as Fig. 11-5c shows. Curl your right hand about the rotating disk, your fingers pointing *in the direction of rotation.* Your extended thumb will then point in the direction of the angular velocity vector. If the disk were to rotate in the opposite sense, the right-hand rule would tell you that the angular velocity vector then points in the opposite direction.

It is not easy to get used to representing angular

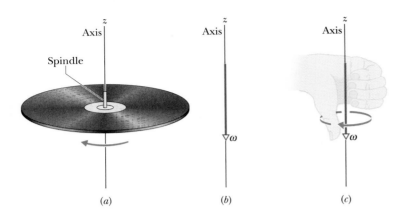

FIGURE 11-5 (*a*) An LP record rotating about a vertical axis that coincides with the axis of the spindle. (*b*) The angular velocity of the rotating record can be represented by the vector ω, lying along the axis and pointing down, as shown. (*c*) We establish the direction of the angular velocity vector as downward by using a right-hand rule. When the fingers of the right hand curl around the record and point the way it is moving, the extended thumb points in the direction of ω.

quantities as vectors. We instinctively expect that something should be moving *along* the direction of a vector. That is not the case here. Instead, something (the rigid body) is rotating *around* the direction of the vector. In the world of pure rotation, a vector defines an axis of rotation, not a direction in which something moves. Nonetheless, the vector also defines the motion. Furthermore, it obeys all the rules for vector manipulation discussed in Chapter 3. The angular acceleration α is another vector, and it too obeys those rules.

Now for the caution referred to above. Angular *displacements* (unless they are very small) *cannot* be treated as vectors. Why not? We can certainly give

them both magnitude and direction, just as we did for the angular velocity vector in Fig. 11-5. However, that is (as the mathematicians say) a necessary condition but not sufficient. To be represented as a vector, a quantity must *also* obey the rules of vector addition, one of which says that, if you add two vectors, the order in which you add them does not matter. Angular displacements fail this test.

To see this, place a book flat on the floor, as in Fig. 11-6*a*. Now give the book two successive 90° angular displacements, *first* about the (horizontal) *x* axis and *then* about the (vertical) *y* axis, using the right-hand rule as a guide to positive rotation in each case.

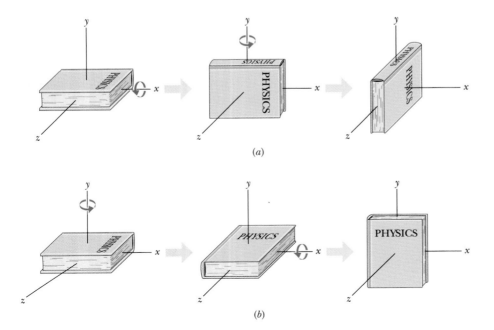

(a)

(b)

FIGURE 11-6 (*a*) From its initial position on the left, the book is given two successive 90° rotations, first about the (horizontal) *x* axis and then about the (vertical) *y* axis. (*b*) The book is given the same rotations, but in the reverse order. If angular displacement were truly a vector quantity, the order of these displacements would not matter. It clearly does matter, so (large) angular displacements are not vector quantities, even though we can assign magnitude and direction to them.

TABLE 11-1
**EQUATIONS OF MOTION FOR CONSTANT LINEAR
AND FOR CONSTANT ANGULAR ACCELERATION**

EQUATION NUMBER	LINEAR FORMULA	MISSING VARIABLE		ANGULAR FORMULA	EQUATION NUMBER
(2-9)	$v = v_0 + at$	x	θ	$\omega = \omega_0 + \alpha t$	(11-9)
(2-13)	$x = v_0 t + \frac{1}{2}at^2$	v	ω	$\theta = \omega_0 t + \frac{1}{2}\alpha t^2$	(11-10)
(2-14)	$v^2 = v_0^2 + 2ax$	t	t	$\omega^2 = \omega_0^2 + 2\alpha\theta$	(11-11)
(2-15)	$x = \frac{1}{2}(v_0 + v)t$	a	α	$\theta = \frac{1}{2}(\omega_0 + \omega)t$	(11-12)
(2-16)	$x = vt - \frac{1}{2}at^2$	v_0	ω_0	$\theta = \omega t - \frac{1}{2}\alpha t^2$	(11-13)

Now, with a book in the same initial position (Fig. 11-6*b*), carry out these two angular displacements in the reverse order (that is, *first* about the *y* axis and *then* about the *x* axis). As the figure shows, the book ends up in a very different orientation.

Thus the same two operations produce different results, depending on the order in which you carry them out. Addition of angular displacements is thus not commutative, so angular displacements are not vector quantities. With some practice, you should be able to show that the final positions of the book are much closer together if you use displacements much smaller than 90°. In the limiting case of differential angular displacements (such as $d\theta$ in Eq. 11-6), angular displacements *can* be treated as vectors.

11-4 ROTATION WITH CONSTANT ANGULAR ACCELERATION

In pure translation, motion with a *constant linear acceleration* (for example, that of a falling body) is an important special case. In Table 2-2, we displayed a series of equations that hold for such motion.

In pure rotation, the case of *constant angular acceleration* is also important, and a parallel set of equations holds for this case also. We shall not derive them here, but simply write them from the corresponding linear equations, substituting equivalent angular quantities for the linear ones. Table 11-1 displays both sets of equations. For simplicity, we let $x_0 = 0$ and $\theta_0 = 0$ in these equations. With those *initial conditions,* a linear displacement Δx ($= x - x_0$) is equal to x, and an angular displacement $\Delta \theta$ ($= \theta - \theta_0$) is equal to θ.

SAMPLE PROBLEM 11-3

A grindstone wheel (Fig. 11-7) has a constant angular acceleration $\alpha = 0.35$ rad/s². It starts from rest (that is, $\omega_0 = 0$) with an arbitrary reference line horizontal, at angular position $\theta_0 = 0$.

a. What is the angular displacement θ of the reference line (and hence of the wheel) at $t = 18$ s?

SOLUTION We use Eq. 11-10 of Table 11-1 to obtain

$\theta = \omega_0 t + \frac{1}{2}\alpha t^2$

$\quad = (0)(18\text{ s}) + (\frac{1}{2})(0.35\text{ rad/s}^2)(18\text{ s})^2$

$\quad = 56.7\text{ rad} \approx 57\text{ rad} \approx 3200° \approx 9.0\text{ rev.}$ (Answer)

b. What is the wheel's angular velocity at $t = 18$ s?

SOLUTION We now use Eq. 11-9 of Table 11-1 to get

$\omega = \omega_0 + \alpha t$

$\quad = 0 + (0.35\text{ rad/s}^2)(18\text{ s})$

$\quad = 6.3\text{ rad/s} = 360°/\text{s} = 1.0\text{ rev/s.}$ (Answer)

FIGURE 11-7 Sample Problems 11-3 and 11-4. A grindstone. At $t = 0$ the reference line (which we imagine to be marked on the stone) is horizontal.

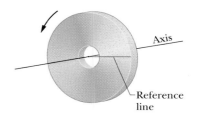

SAMPLE PROBLEM 11-4

For the grindstone of Sample Problem 11-3, let us assume the same angular acceleration ($\alpha = 0.35$ rad/s²), but let us now assume that the wheel does not start from rest but has an initial angular velocity ω_0 of $- 4.6$ rad/s; that is, the angular acceleration acts initially to slow the wheel.

a. At what time t will the grindstone momentarily stop?

SOLUTION   Solving Eq. 11-9 ($\omega = \omega_0 + \alpha t$) for t yields

$$t = \frac{\omega - \omega_0}{\alpha} = \frac{0 - (- 4.6 \text{ rad/s})}{0.35 \text{ rad/s}^2} = 13 \text{ s}. \quad \text{(Answer)}$$

b. At what time will the grindstone have rotated such that its angular displacement is five revolutions in the positive direction of rotation? (The angular displacement of the reference line will then be $\theta = 5$ rev.)

SOLUTION   The wheel is initially rotating in the negative direction with $\omega_0 = - 4.6$ rad/s, but its angular acceleration α is positive. This initial opposition of the signs of angular velocity and angular acceleration means that the wheel slows in its rotation in the negative direction, stops, and then reverses to rotate in the positive direction. After the reference line comes back through its initial orientation of $\omega = 0$, the wheel must turn an additional five revolutions to reach the angular

FIGURE 11-8   A plot of angular position versus time for the grindstone of Sample Problem 11-4. Negative times (that is, times before $t = 0$) have been included. The two roots of Eq. 11-14 are indicated by points d and e.

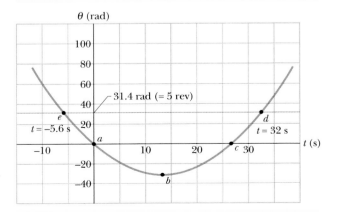

displacement we want. This is all "taken care of" if we use Eq. 11-10:

$$\theta = \omega_0 t + \tfrac{1}{2}\alpha t^2.$$

Substituting known values and setting $\theta = 5$ rev $= 10\pi$ rad give us

$$10\pi \text{ rad} = (- 4.6 \text{ rad/s})t + (\tfrac{1}{2})(0.35 \text{ rad/s}^2)\, t^2.$$

Note that, for t in seconds, the units in this equation are consistent. Dropping units (for convenience) and rearranging give

$$t^2 - 26.3t - 180 = 0. \quad (11\text{-}14)$$

Solving this quadratic equation for t and discarding the negative root, we obtain

$$t = 32 \text{ s}. \quad \text{(Answer)}$$

PROBLEM SOLVING

TACTIC 1: UNEXPECTED ANSWERS

Do not hasten to throw away one root of a quadratic equation as meaningless. Often, as in Sample Problem 11-4, a discarded root has physical meaning.

The two solutions to Eq. 11-14 are $t = 32$ s and $t = - 5.6$ s. We chose the first (positive) solution and ignored the second as perhaps meaningless. But is it? A negative time in this problem simply means a time before $t = 0$, that is, a time before you started to pay attention to what was going on.

Figure 11-8 is a plot of the angular position θ of the reference line on the grindstone of Sample Problem 11-4 as a function of time, for both positive and negative times. It is a plot of Eq. 11-10 ($\theta = \omega_0 t + \tfrac{1}{2}\alpha t^2$), using as constants $\omega_0 = - 4.6$ rad/s and $\alpha = + 0.35$ rad/s². Point a corresponds to $t = 0$, at which time we arbitrarily took the angular position of the reference line to be zero. The wheel was moving in the direction of decreasing θ at that time, and continues to do so until coming to rest at point b at $t = 13$ s. It then reverses, with the reference line returning to its $\theta = 0$ position at point c and going on for five additional revolutions ($= 31.4$ rad) to point d. This latter point ($t = 32$ s) is the root that we accepted as the answer to our problem.

Note, however, that the reference line was at this same angular position at $t = - 5.6$ s, before the "official start" of the problem. This root (point e) is just as valid as the root at point d. More important, by asking if this negative root can have any physical meaning, you learn a little more about the rotation of the wheel.

SAMPLE PROBLEM 11-5

During an analysis of a helicopter engine, you determine that the rotor's velocity changes from 320 rev/min to 225 rev/min in 1.50 min as the rotor is slowing to a stop.

a. What is the average angular acceleration of the rotor blades during this interval?

SOLUTION From Eq. 11-7,

$$\bar{\alpha} = \frac{\omega - \omega_0}{\Delta t} = \frac{225 \text{ rev/min} - 320 \text{ rev/min}}{1.50 \text{ min}}$$

$$= -63.3 \text{ rev/min}^2. \qquad \text{(Answer)}$$

The minus sign reminds us that the rotor blades are slowing.

b. With this same average acceleration, how long will the rotor blades take to stop from their initial angular velocity of 320 rev/min?

SOLUTION Solving Eq. 11-9 ($\omega = \omega_0 + \alpha t$) for t gives

$$t = \frac{\omega - \omega_0}{\alpha} = \frac{0 - 320 \text{ rev/min}}{-63.3 \text{ rev/min}^2}$$

$$= 5.1 \text{ min}. \qquad \text{(Answer)}$$

c. How many revolutions will the rotor blades make in coming to rest from their initial angular velocity of 320 rev/min?

SOLUTION Solving Eq. 11-11 ($\omega^2 = \omega_0^2 + 2\alpha\theta$) for θ gives us

$$\theta = \frac{\omega^2 - \omega_0^2}{2\alpha} = \frac{0 - (320 \text{ rev/min})^2}{(2)(-63.3 \text{ rev/min}^2)}$$

$$= 809 \text{ rev}. \qquad \text{(Answer)}$$

11-5 THE LINEAR AND ANGULAR VARIABLES

In Section 4-7, we discussed uniform circular motion, in which a particle travels at constant linear speed v along a circle and around an axis of rotation. When a rigid body, such as a merry-go-round (Fig. 11-9), turns around an axis of rotation, each particle in the body moves in its own circle around that axis. Since the body is rigid, all the particles

make one revolution in the same amount of time; that is, they all have the same angular speed ω.

However, the farther a particle is from the axis, the greater the circumference of its circle is, and so the faster its linear speed v must be. You can notice this on a merry-go-round. You turn with the same angular speed ω regardless of your distance from the center, but your linear speed v increases noticeably if you move to the outside edge of the merry-go-round.

We often need to relate the linear variables s, v, and a for a particular point in a rotating body to the angular variables θ, ω, and α for that body. The two sets of variables are related by r, the *perpendicular distance* of the point from the rotation axis. This perpendicular distance is the distance between the point and the rotation axis, measured along a perpendicular to the axis. The radius r of the circle traveled by the point around the axis of rotation gives the perpendicular distance.

The Position

If a reference line on a rigid body rotates through an angle θ, a point within the body moves along a circular arc by a distance s given by Eq. 11-1:

$$s = \theta r \qquad \text{(radian measure).} \qquad (11\text{-}15)$$

This is the first of our linear–angular relations. The angle θ must be measured in radians because Eq. 11-15 is itself the definition of angular measure in radians.

FIGURE 11-9 All of the riders on a merry-go-round rotate around the central axis at the same angular speed ω. But riders farther from the axis move with greater linear speed v.

The Speed

Differentiating Eq. 11-15 with respect to time—with r held constant—leads to

$$\frac{ds}{dt} = \frac{d\theta}{dt} r.$$

But ds/dt is the linear speed (the magnitude of the linear velocity) of the point in question and $d\theta/dt$ is the angular speed of the rotating body, so

$$v = \omega r \qquad \text{(radian measure).} \qquad (11\text{-}16)$$

Again, the angular speed ω must be expressed in radian measure. Equation 11-16 tells us that, since all

(a)

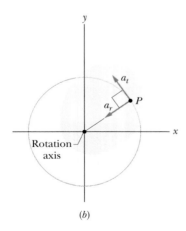

(b)

FIGURE 11-10 The rotating rigid body of Fig. 11-2, shown in cross section. Every point of the body (such as P) moves in a circle around the rotation axis. (a) The linear velocity $\mathbf{v}$ of every point is tangent to the circle in which the point moves. (b) The linear acceleration $\mathbf{a}$ of the point has (in general) two components: a tangential component a_t and a radial component a_r.

points within the rigid body have the same angular speed ω, the greater the radius r is, the greater the linear speed v. Figure 11-10a shows that the linear velocity is always tangent to the circular path of the point in question, for the same reason as in uniform circular motion (Section 4-7).

The Acceleration

Differentiating Eq. 11-16 with respect to time—again with r held constant—leads to

$$\frac{dv}{dt} = \frac{d\omega}{dt} r. \qquad (11\text{-}17)$$

Here we run up against a complication. In Eq. 11-17, dv/dt represents only the part of the linear acceleration that is responsible for changes in the *magnitude* v of the linear velocity $\mathbf{v}$. Like $\mathbf{v}$, that part of the linear acceleration is tangential to the path of the point in question. We call it the *tangential component* a_t of the linear acceleration of the point, and we write

$$a_t = \alpha r \qquad \text{(radian measure).} \qquad (11\text{-}18)$$

In addition, as Eq. 4-22 tells us, a particle (or point) moving in a circular path has a *radial component* of linear acceleration, $a_r = v^2/r$, that is responsible for changes in the *direction* of the linear velocity $\mathbf{v}$. By substituting for v from Eq. 11-16, we can write this component as

$$a_r = \frac{v^2}{r} = \omega^2 r \qquad \text{(radian measure).} \qquad (11\text{-}19)$$

Thus, as Fig. 11-10b shows, the linear acceleration of a point on a rotating rigid body has, in general, two components. The radial component a_r (given by Eq. 11-19) is always present as long as the angular velocity of the body is not zero. The tangential component a_t (given by Eq. 11-18) is present as long as the angular acceleration is not zero.

SAMPLE PROBLEM 11-6

Figure 11-11 shows a centrifuge used to subject astronaut trainees to high accelerations. The radius r of the circle traveled by an astronaut is 15 m.

a. At what constant angular velocity must the centrifuge rotate if the astronaut is to be subject to a linear acceleration that is 11 times that of free fall? This is about the maximum acceleration that a highly trained fighter pilot can tolerate—for a short time—without blacking out.

SOLUTION   Because the angular velocity is constant, the angular acceleration α $(= d\omega/dt)$ is zero and so (see Eq. 11-18) is the tangential component of the linear acceleration. This leaves only the radial component. From Eq. 11-19 $(a_r = \omega^2 r)$ we have

$$\omega = \sqrt{\frac{a_r}{r}} = \sqrt{\frac{(11)(9.8 \text{ m/s}^2)}{15 \text{ m}}}$$

$$= 2.68 \text{ rad/s} \approx 26 \text{ rev/min}. \qquad \text{(Answer)}$$

b. What is the linear speed of the astronaut under these conditions?

SOLUTION   From Eq. 11-16,

$$v = \omega r = (2.68 \text{ rad/s})(15 \text{ m}) = 40 \text{ m/s}. \qquad \text{(Answer)}$$

c. What is the tangential acceleration of the astronaut if the centrifuge accelerates uniformly from rest to the angular velocity of part (a) in 120 s?

SOLUTION   Since the angular acceleration is constant during the speed-up of the centrifuge, Eq. 11-9 applies:

$$\alpha = \frac{\omega - \omega_0}{t} = \frac{2.68 \text{ rad/s} - 0}{120 \text{ s}} = 0.0223 \text{ rad/s}^2.$$

With Eq. 11-18, we then find

$$a_t = \alpha r = (0.0223 \text{ rad/s}^2)(15 \text{ m})$$

$$= 0.33 \text{ m/s}^2. \qquad \text{(Answer)}$$

Although the final radial acceleration a_r is large (and alarming), the tangential acceleration during the speed-up is not.

PROBLEM SOLVING

TACTIC 2: UNITS FOR ANGULAR VARIABLES
In Eq. 11-1 $(\theta = s/r)$, we committed ourselves to the use of radian measure for all angular variables. That is, we must express angular displacements in radians, angular velocities in rad/s and rad/min, and angular accelerations in rad/s^2 and rad/min^2. The only exceptions to this rule are equations that involve *only* angular variables, such as the angular equations listed in Table 11-1. Here you are free to use any unit you wish for the angular variables. That is, you may use radians, degrees, or revolutions, as long as you use them consistently.

In equations where radian measure must be used, you need not keep track of the unit "radian" (rad) algebraically, as you must do for other units. You can add or delete it at will, to suit the context. In Sample Problem 11-6a the unit was added to the answer; in Sample Problem 11-6b and c it was deleted.

FIGURE 11-11   Sample Problem 11-6. A centrifuge in Cologne, Germany, is used to accustom astronauts to the large acceleration experienced during a takeoff.

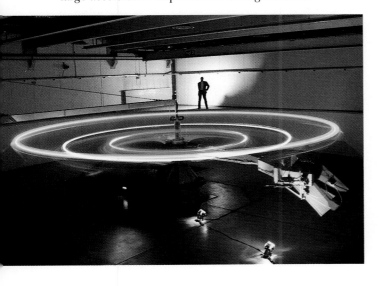

11-6 KINETIC ENERGY OF ROTATION

The rapidly rotating blade of a table saw certainly has kinetic energy. How can we express it? We cannot use the familiar formula $K = \frac{1}{2}mv^2$ directly because it applies only to particles, and we would not know what to use for v.

Instead, we shall treat the table saw (and any other rotating rigid body) as a collection of particles—all with different speeds. We can then add up the kinetic energies of these particles to find the kinetic energy of the body as a whole. In this way we obtain, for the kinetic energy of a rotating body,

$$K = \tfrac{1}{2}m_1 v_1^2 + \tfrac{1}{2}m_2 v_2^2 + \tfrac{1}{2}m_3 v_3^2 + \cdots$$

$$= \sum \tfrac{1}{2}m_i v_i^2, \qquad (11\text{-}20)$$

in which m_i is the mass of the ith particle and v_i is its speed. The sum is taken over all the particles that make up the body.

The problem with Eq. 11-20 is that v_i is not the same for all particles. We solve this problem by substituting for v from Eq. 11-16 ($v = \omega r$), so that we have

$$K = \sum \tfrac{1}{2} m_i (\omega r_i)^2 = \tfrac{1}{2} \left(\sum m_i r_i^2 \right) \omega^2, \quad (11\text{-}21)$$

in which ω *is* the same for all particles.

The quantity in parentheses on the right side of Eq. 11-21 tells us how the mass of the rotating body is distributed about its axis of rotation. We call that quantity the **rotational inertia*** I of the body with respect to that axis. Thus

$$I = \sum m_i r_i^2 \quad \text{(rotational inertia).} \quad (11\text{-}22)$$

Substituting into Eq. 11-21 yields, as the expression we seek,

$$K = \tfrac{1}{2} I \omega^2 \quad \text{(radian measure).} \quad (11\text{-}23)$$

Because we have used the relation $v = \omega r$ in deriving Eq. 11-23, ω must be expressed in radian measure. The SI unit for I is the kilogram-square meter ($\text{kg} \cdot \text{m}^2$).

Equation 11-23, which gives the kinetic energy of a rigid body in pure rotation, is the angular equivalent of the formula $K = \tfrac{1}{2} M v_{\text{cm}}^2$, which gives the kinetic energy of a rigid body in pure translation. In each case, there is a factor of $\tfrac{1}{2}$. Where mass M (which can be called the *translational inertia*) appears in one formula, I (the *rotational inertia*) appears in the other. Finally, each equation contains as a factor the square of a speed, translational or rotational as appropriate. The kinetic energies of translation and of rotation are not different kinds of energy. They are both kinetic energy, expressed in ways that are appropriate to the motion at hand.

The rotational inertia of a rotating body depends not only on its mass but also on how that mass is distributed with respect to the rotation axis. Figure 11-12 suggests a convincing way to develop a physical feeling for rotation inertia. Figure 11-12a shows the exterior of either of two plastic rods that outwardly appear to be identical. Both are about 1 m long. The dimensions and the weights of the two rods are the same and both rods balance at their midpoints. If you grasp each rod at its center and move them back

FIGURE 11-12 (*a*) Two plastic rods like this seem identical until you try to twist them back and forth rapidly about their midpoints. Rod (*c*) wiggles readily; rod (*b*) does not. The secret is in the internal distribution of the weight about the axis of rotation. Although both rods have the same mass, the rotational inertia of rod (*b*) about an axis through its midpoint is considerably greater than that of rod (*c*).

and forth rapidly in translational motion, you still cannot tell them apart.

However, a truly striking difference appears if, with a twisting wrist motion, you twist the rods (like a baton) rapidly in back-and-forth angular motion. One rod wiggles quite easily; the other does not. As Figs. 11-12*b* and 11-12*c* show, the "easy" rod has internal weights concentrated near its center and the "hard" rod has them at its ends. Although the masses of the rods are equal, their rotational inertias about a central axis are quite different because their mass distributions are different.

11-7 CALCULATING THE ROTATIONAL INERTIA

If a rigid body is made up of discrete particles, we can calculate its rotational inertia from Eq. 11-22. If the body is continuous, we can replace the sum in Eq. 11-22 with an integral, and the definition of rotational inertia becomes

$$I = \int r^2 \, dm \quad \text{(rotational inertia, continuous).} \quad (11\text{-}24)$$

In the sample problems that follow this section, we calculate I for both kinds of bodies. In general, the rotational inertia of any rigid body with respect to a

---

*Often called the *moment of inertia*.

TABLE 11-2
SOME ROTATIONAL INERTIAS

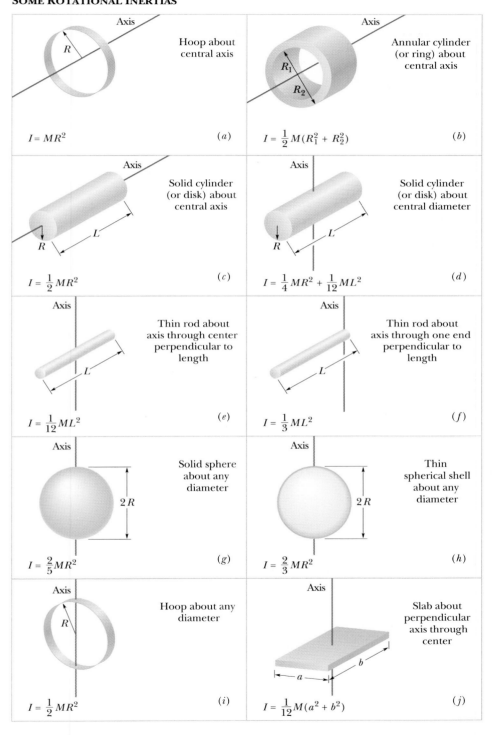

(a) Hoop about central axis

$$I = MR^2$$

(b) Annular cylinder (or ring) about central axis

$$I = \frac{1}{2}M(R_1^2 + R_2^2)$$

(c) Solid cylinder (or disk) about central axis

$$I = \frac{1}{2}MR^2$$

(d) Solid cylinder (or disk) about central diameter

$$I = \frac{1}{4}MR^2 + \frac{1}{12}ML^2$$

(e) Thin rod about axis through center perpendicular to length

$$I = \frac{1}{12}ML^2$$

(f) Thin rod about axis through one end perpendicular to length

$$I = \frac{1}{3}ML^2$$

(g) Solid sphere about any diameter

$$I = \frac{2}{5}MR^2$$

(h) Thin spherical shell about any diameter

$$I = \frac{2}{3}MR^2$$

(i) Hoop about any diameter

$$I = \frac{1}{2}MR^2$$

(j) Slab about perpendicular axis through center

$$I = \frac{1}{12}M(a^2 + b^2)$$

rotation axis depends on (1) the shape of the body, (2) the perpendicular distance from the axis to the body's center of mass, and (3) the orientation of the body with respect to the axis.

Table 11-2 gives the rotational inertias of several common bodies, about various axes. Study the table carefully to get a feeling for how the distribution of mass affects the rotational inertia.

The Parallel-Axis Theorem

If you know the rotational inertia of a body about any axis that passes through its center of mass, you can find its rotational inertia about any other axis parallel to this axis with the **parallel-axis theorem,** which is

$$I = I_{cm} + Mh^2 \qquad \text{(parallel-axis theorem).} \qquad (11\text{-}25)$$

Here M is the mass of the body, and h is the perpendicular distance between the two (parallel) axes. In words, this theorem can be stated as follows:

The rotational inertia of a body about any axis is equal to the rotational inertia ($= Mh^2$) it would have about that axis if all its mass were concentrated at its center of mass *plus* its rotational inertia ($= I_{cm}$) about a parallel axis through its center of mass.

Proof of the Parallel-Axis Theorem

Let O be the center of mass of the arbitrarily shaped body shown in cross section in Fig. 11-13. Place the origin of coordinates at O. Consider an axis through O perpendicular to the plane of the figure, and another axis through point P parallel to the first axis. Let the coordinates of P be a and b.

Let dm be a mass element with coordinates x and y. The rotational inertia of the body about the axis through P is then, from Eq. 11-24,

$$I = \int r^2 \, dm = \int [(x - a)^2 + (y - b)^2] \, dm,$$

which we can rearrange as

$$I = \int (x^2 + y^2) \, dm - 2a \int x \, dm$$
$$- 2b \int y \, dm + \int (a^2 + b^2) \, dm. \qquad (11\text{-}26)$$

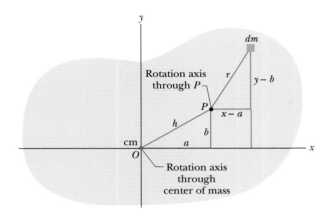

FIGURE 11-13 A rigid body in cross section, with its center of mass at O. The parallel-axis theorem (Eq. 11-25) relates the rotational inertia of the body about an axis through O to that about a parallel axis through a point such as P, a distance h from the center of mass. Both axes are perpendicular to the plane of the figure.

From the definition of the center of mass (Eq. 9-9), the middle two integrals of Eq. 11-26 give the coordinates of the center of mass (multiplied by a constant) and thus are zero. Because $x^2 + y^2$ is equal to R^2, where R is the distance from O to dm, the first integral is simply I_{cm}, the rotational inertia of the body about an axis through its center of mass. Inspection of Fig. 11-13 shows that the last term in Eq. 11-26 is Mh^2, where M is the total mass. Thus Eq. 11-26 reduces to Eq. 11-25, which is the relation that we set out to prove.

SAMPLE PROBLEM 11-7

Figure 11-14 shows a rigid body consisting of two particles of mass m connected by a rod of length L and negligible mass.

a. What is the rotational inertia of this body about an axis through its center, perpendicular to the rod (see Fig. 11-14a)?

SOLUTION From Eq. 11-22 we have

$$I = \sum m_i r_i^2 = (m)(\tfrac{1}{2}L)^2 + (m)(\tfrac{1}{2}L)^2$$
$$= \tfrac{1}{2}mL^2. \qquad \text{(Answer)}$$

b. What is the rotational inertia of the body about an

FIGURE 11-14 Sample Problem 11-7. A rigid body consists of two particles of mass m joined by a rod of negligible mass.

axis through one end of the rod and parallel to the first axis, as in Fig. 11-14b?

SOLUTION We can use the parallel-axis theorem of Eq. 11-25. We have just calculated I_{cm} in part (a), and the distance h between the parallel axes is half the length of the rod. Thus, from Eq. 11-25,

$$I = I_{cm} + Mh^2 = \tfrac{1}{2}mL^2 + (2m)(\tfrac{1}{2}L)^2$$

$$= mL^2. \qquad \text{(Answer)}$$

We can check this result by direct calculation, using Eq. 11-22:

$$I = \sum m_i r_i^2 = (m)(0)^2 + (m)(L)^2$$

$$= mL^2. \qquad \text{(Answer)}$$

SAMPLE PROBLEM 11-8

Figure 11-15 shows a thin, uniform rod of mass M and length L.

a. What is its rotational inertia about an axis perpendicular to the rod, through its center of mass?

SOLUTION We choose as a mass element a slice dx of the rod. The center of the slice is located at position x. The mass per unit length of the rod is M/L, so the mass dm of the element dx is

$$dm = \left(\frac{M}{L}\right) dx.$$

FIGURE 11-15 Sample Problem 11-8. A uniform rod of length L and mass M.

From Eq. 11-24 we have

$$I = \int x^2 \, dm = \int_{x=-L/2}^{x=+L/2} x^2 \left(\frac{M}{L}\right) dx$$

$$= \frac{M}{3L} \left[x^3 \right]_{-L/2}^{+L/2} = \frac{M}{3L} \left[\left(\frac{L}{2}\right)^3 - \left(-\frac{L}{2}\right)^3 \right]$$

$$= \tfrac{1}{12}ML^2. \qquad \text{(Answer)}$$

This agrees with the result given in Table 11-2(e).

b. What is the rotational inertia of the rod about an axis perpendicular to the rod through its end point?

SOLUTION We combine the result in part (a) with the parallel-axis theorem (Eq. 11-25), obtaining

$$I = I_{cm} + Mh^2$$

$$= \tfrac{1}{12}ML^2 + (M)(\tfrac{1}{2}L)^2 = \tfrac{1}{3}ML^2, \qquad \text{(Answer)}$$

which agrees with the result given in Table 11-2(f).

SAMPLE PROBLEM 11-9

A hydrogen chloride molecule consists of a hydrogen atom whose mass m_H is 1.01 u and a chlorine atom whose mass m_{Cl} is 35.0 u. The centers of the two atoms are a distance $d = 1.27 \times 10^{-10}$ m = 127 pm apart (Fig. 11-16). What is the rotational inertia of the molecule about an axis perpendicular to the line joining the two atoms and passing through the center of mass of the molecule?

SOLUTION Let x be the distance from the center of mass of the molecule to the chlorine atom. Then from Fig. 11-16 and Eq. 9-3, we see that

$$0 = \frac{-m_{Cl}x + m_H(d-x)}{m_{Cl} + m_H},$$

or

$$m_{Cl}x = m_H(d-x).$$

FIGURE 11-16 Sample Problem 11-9. A hydrogen chloride molecule, shown schematically. A rotation axis extends through its center of mass and is perpendicular to the line joining the two atomic centers.

This yields

$$x = \frac{m_H}{m_{Cl} + m_H} d. \qquad (11\text{-}27)$$

From Eq. 11-22, the rotational inertia about an axis through the center of mass is

$$I = \sum m_i r_i^2 = m_H(d - x)^2 + m_{Cl}x^2.$$

Substituting for x from Eq. 11-27 leads, after some algebra, to

$$I = d^2 \frac{m_H m_{Cl}}{m_{Cl} + m_H} = (127 \text{ pm})^2 \frac{(1.01 \text{ u})(35.0 \text{ u})}{35.0 \text{ u} + 1.01 \text{ u}}$$

$$= 15,800 \text{ u} \cdot \text{pm}^2. \qquad \text{(Answer)}$$

These units are convenient enough when dealing with the rotational inertia of molecules. If we use the ångstrom as a length unit ($1 \text{ Å} = 10^{-10} \text{ m}$), the answer above becomes $I = 1.58 \text{ u} \cdot \text{Å}^2$, an even more convenient unit.

SAMPLE PROBLEM 11-10

With modern technology, it is possible to construct a flywheel that would store enough energy to run an automobile. The energy is stored as rotational kinetic energy when the flywheel is initially made to spin by a machine. The stored energy is then gradually transferred to the automobile by a gear system as the automobile is being driven. Suppose that such a wheel is made in the form of a solid cylinder whose mass M is 75 kg and whose radius R is 25 cm. If the wheel is spun at 85,000 rev/min, how much rotational kinetic energy can it store?

SOLUTION The rotational inertia of the cylindrical wheel follows from Table 11-2(c):

$$I = \tfrac{1}{2}MR^2 = (\tfrac{1}{2})(75 \text{ kg})(0.25 \text{ m})^2 = 2.34 \text{ kg} \cdot \text{m}^2.$$

The angular velocity of the wheel is

$$\omega = (85{,}000 \text{ rev/min})(2\pi \text{ rad/rev})(1 \text{ min}/60 \text{ s})$$

$$= 8900 \text{ rad/s}.$$

From Eq. 11-23, the kinetic energy of rotation is then

$$K = \tfrac{1}{2}I\omega^2 = (\tfrac{1}{2})(2.34 \text{ kg} \cdot \text{m}^2)(8900 \text{ rad/s})^2$$

$$= 9.3 \times 10^7 \text{ J} = 26 \text{ kW} \cdot \text{h}. \qquad \text{(Answer)}$$

This amount of energy, used with the expected efficiency, would take a small car about 200 mi.

11-8 TORQUE

A door knob is located as far as possible from the door's hinge line for a good reason. If you want to open a heavy door you must certainly apply a force; that alone, however, is not enough. Where you apply that force and in what direction you push are also important. If you apply your force nearer to the hinge line than the knob, or at any angle other than 90° to the plane of the door, you must use a greater force to move the door than if you apply the force at the knob and perpendicular to the door's plane.

Figure 11-17a shows a cross section of a body that is free to rotate about an axis passing through O and perpendicular to the cross section. A force $\mathbf{F}$ is applied at point P, whose position relative to O is defined by a position vector $\mathbf{r}$. Vectors $\mathbf{F}$ and $\mathbf{r}$ make an angle ϕ with each other. (For simplicity, we consider only forces that have no component parallel to the rotation axis.)

To determine how $\mathbf{F}$ results in a rotation of the body around the rotation axis, we resolve $\mathbf{F}$ into two components (Fig. 11-17b). One component, called the *radial component* F_r, points along $\mathbf{r}$. This component does not cause rotation, because it acts along a line that extends through O. (If you pull on a door parallel to the plane of the door, you do not rotate the door.) The other component of $\mathbf{F}$, called the *tangential component* F_t, is perpendicular to $\mathbf{r}$ and has magnitude $F_t = F \sin \phi$. This component *does* cause rotation. (If you pull on a door perpendicular to its plane, you rotate the door.)

The ability of F_t to rotate the body depends not only on its magnitude, but also on just how far from O it, and thus the force $\mathbf{F}$, is applied. To include both these factors, we define a quantity called **torque** τ as the product of the two factors and write it as

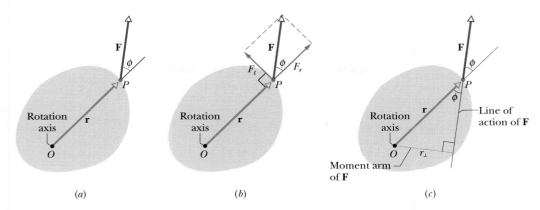

(a)　　　　　　　(b)　　　　　　　(c)

FIGURE 11-17 (a) A force **F** acts at point P on a rigid body that is free to rotate about an axis through O, perpendicular to the plane of the cross section shown. The torque exerted by this force is $Fr \sin \phi$. (b) The torque

can be written as $F_t r$, where F_t is the tangential component of **F**. (c) The torque can also be written as $Fr_\perp$, where $r_\perp$ is the moment arm of **F**.

$$\tau = (r)(F \sin \phi). \qquad (11\text{-}28)$$

Two equivalent ways of computing the torque are

$$\tau = (r)(F \sin \phi) = rF_t \qquad (11\text{-}29)$$

and

$$\tau = (r \sin \phi)(F) = r_\perp F, \qquad (11\text{-}30)$$

where $r_\perp$ is the perpendicular distance between the rotation axis at O and an extended line running through the vector **F** (Fig. 11-17c). This extended line is called the **line of action** of **F**, and $r_\perp$ is called the **moment arm** of **F**. In Fig. 11-17b, we can describe r, the magnitude of **r**, as being the moment arm of the force component F_t.

Torque, which comes from the Latin word meaning "to twist," may be loosely identified as the turning or twisting action of the force **F**. When you apply a force to an object—such as a screwdriver or pipe wrench—with the purpose of turning that object, you are applying a torque. The SI unit of torque is the newton-meter (N·m).* A torque τ is positive if it tends to rotate the body counterclockwise, in the direction of increasing θ, as in Fig. 11-17. It is negative if it tends to rotate the body clockwise.

The definition of torque in Eq. 11-28 can be rewritten as a *vector cross product:*

$$\boldsymbol{\tau} = \mathbf{r} \times \mathbf{F}. \qquad (11\text{-}31)$$

Thus torque $\boldsymbol{\tau}$ is a vector that is directed perpendicular to the plane containing **r** and **F**. The magnitude of $\boldsymbol{\tau}$ is given by Eqs. 11-28, 11-29, and 11-30, and its direction is given by the right-hand rule. We shall return to Eq. 11-31 in Chapter 12.

11-9 NEWTON'S SECOND LAW FOR ROTATION

Figure 11-18 shows a simple case of rotation about a fixed axis. The rotating rigid body consists of a single

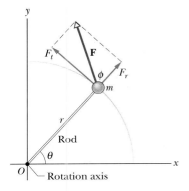

FIGURE 11-18 A simple rigid body, free to rotate about an axis through O, consists of a particle of mass m fastened to the end of a rod of length r and negligible mass. An applied force **F** causes the body to rotate.

---

*The newton-meter is also the unit of work. Torque and work, however, are quite different quantities and must not be confused. Work is often expressed in joules (1 J = 1 N·m), but torque never is.

particle of mass m fastened to the end of a massless rod of length r. A force **F** acts as shown, causing the particle to move in a circle. The particle has a tangential component of acceleration a_t governed by Newton's second law:

$$F_t = ma_t.$$

The torque acting on the particle is, from Eq. 11-29,

$$\tau = F_t r = ma_t r.$$

From Eq. 11-18 ($a_t = \alpha r$) we can write this as

$$\tau = m(\alpha r)r = (mr^2)\alpha. \qquad (11\text{-}32)$$

The quantity in parentheses on the right side of Eq. 11-32 is the rotational inertia of the particle about the rotation axis (see Eq. 11-22). So, Eq. 11-32 reduces to

$$\tau = I\alpha \qquad \text{(radian measure)}. \qquad (11\text{-}33)$$

For the situation in which more than one force is applied to the particle, we can extend Eq. 11-33 as

$$\sum \tau = I\alpha \qquad \text{(radian measure)}, \qquad (11\text{-}34)$$

where $\sum \tau$ is the net torque acting on the particle. Equation 11-34 is the angular (or rotational) form of Newton's second law.

Although we derived Eq. 11-33 for the special case of a single particle rotating about a fixed axis, it holds for any rigid body rotating about a fixed axis, because any such body can be thought of as an assembly of single particles.

SAMPLE PROBLEM 11-11

Figure 11-19a shows a uniform disk whose mass M is 2.5 kg and whose radius R is 20 cm mounted on a fixed horizontal axle. A block whose mass m is 1.2 kg hangs from a massless cord that is wrapped around the rim of the disk. Find the acceleration of the falling block (assuming that it does fall), the angular acceleration of the disk, and the tension in the cord. The cord does not slip, and there is no friction at the axle.

SOLUTION Figure 11-19b shows a free-body diagram for the block. The block accelerates downward, so its weight (= mg) must exceed the tension T in the cord.

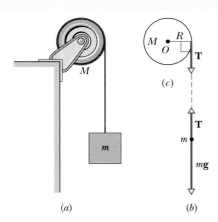

FIGURE 11-19 Sample Problems 11-11 and 11-13. (a) The falling block causes the disk to rotate. (b) A free-body diagram for the block. (c) A free-body diagram for the disk.

From Newton's second law, we have

$$T - mg = ma. \qquad (11\text{-}35)$$

Figure 11-19c shows a free-body diagram for the disk. The torque acting on the disk is $-TR$, negative because it rotates the disk clockwise. From Table 11-2(c), we know that the rotational inertia I of the disk is $\frac{1}{2}MR^2$. (Two other forces also act on the disk, its weight Mg and the normal force N exerted on the disk by its support. Both of these forces act at the axis of the disk, however, so that they exert no torque on the disk.) Applying Newton's second law in angular form, $\tau = I\alpha$, and Eq. 11-18 to the disk, we obtain

$$-TR = \frac{1}{2}MR^2 \left(\frac{a}{R}\right).$$

This reduces to

$$T = -\frac{1}{2}Ma. \qquad (11\text{-}36)$$

In replacing α with a/R, we have assumed—because the cord does not slip—that the linear acceleration of the block is equal to the (tangential) linear acceleration of the rim of the disk.

Combining Eqs. 11-35 and 11-36 leads to

$$a = -g\,\frac{2m}{M + 2m} = -(9.8 \text{ m/s}^2)\,\frac{(2)(1.2 \text{ kg})}{2.5 \text{ kg} + (2)(1.2 \text{ kg})}$$

$$= -4.8 \text{ m/s}^2. \qquad \text{(Answer)}$$

We then use Eq. 11-36 to find T:

$$T = -\frac{1}{2}Ma = -\frac{1}{2}(2.5 \text{ kg})(-4.8 \text{ m/s}^2)$$

$$= 6.0 \text{ N}. \qquad \text{(Answer)}$$

As we should expect, the acceleration of the falling block is less than g, and the tension in the cord ($= 6.0$ N) is less than the weight of the hanging block ($= mg = 11.8$ N). We see also that the acceleration of the block and the tension depend on the mass of the disk but not on its radius. As a check, we note that the formulas derived above predict $a = -g$ and $T = 0$ for the case of a massless disk ($M = 0$). This is what we would expect; the block simply falls as a free body, trailing the string behind it.

The angular acceleration of the disk follows from Eq. 11-18:

$$\alpha = \frac{a}{R} = \frac{-4.8 \text{ m/s}^2}{0.20 \text{ m}} = -24 \text{ rad/s}^2. \quad \text{(Answer)}$$

SAMPLE PROBLEM 11-12

To throw an 80-kg opponent with a basic judo hip throw, you intend to pull his uniform with a force **F** and a moment arm $d_1 = 0.30$ m from the pivot point (rotation axis) on your right hip, about which you wish to rotate him with an angular acceleration of -12 rad/s², that is, a clockwise acceleration (Fig. 11-20). Assume that his rotational inertia I relative to the pivot point is 15 kg·m².

a. What must the magnitude of **F** be if you initially bend him forward to bring his center of mass to your hip (Fig. 11-20a)?

SOLUTION If his center of mass is at the rotation axis, his weight vector produces no torque about that axis. The only torque on him then is due to your pull **F**. From Eqs. 11-29 and 11-33, we have, for the clockwise torque,

$$\tau = -d_1 F = I\alpha,$$

which gives

$$F = \frac{-I\alpha}{d_1} = \frac{-(15 \text{ kg·m}^2)(-12 \text{ rad/s}^2)}{0.30 \text{ m}}$$

$$= 600 \text{ N.} \quad \text{(Answer)}$$

b. What must the magnitude of **F** be if he remains upright and his weight vector has a moment arm $d_2 = 0.12$ m from the pivot point (Fig. 11-20b)?

SOLUTION In this situation, your opponent's weight provides a positive (counterclockwise) torque that counters your torque. From Eqs. 11-29 and 11-34, we have

$$\sum \tau = -d_1 F + d_2 mg = I\alpha,$$

FIGURE 11-20 Sample Problem 11-12. (a) Correctly executed and (b) incorrectly executed judo hip throw.

which gives

$$F = -\frac{I\alpha}{d_1} + \frac{d_2 mg}{d_1}.$$

From part (a), we know that the first term on the right is equal to 600 N. Substituting this and the known data, we have

$$F = 600 \text{ N} + \frac{(0.12 \text{ m})(80 \text{ kg})(9.8 \text{ m/s}^2)}{0.30 \text{ m}}$$

$$= 913.6 \text{ N} \approx 910 \text{ N.} \quad \text{(Answer)}$$

The results indicate that you will have to pull 50% harder if you do not initially bend your opponent to bring his center of mass to your hip. A good judo fighter knows this lesson from physics. (An analysis of judo and aikido is given in "The Amateur Scientist," *Scientific American*, July 1980.)

11-10 WORK, POWER, AND THE WORK–KINETIC ENERGY THEOREM

In Fig. 11-18, imagine that the rigid body (which consists of a single particle of mass m fastened to the end of a rod of negligible mass) rotates through an angle $d\theta$. The work done by the force **F** is

$$dW = \mathbf{F} \cdot d\mathbf{s} = (F_t)(r\, d\theta),$$

in which ds ($= r\, d\theta$) is the distance (length of arc) traveled by the particle. From Eq. 11-29 we see that

$F_t r$ is the torque τ, so that

$$dW = \tau \, d\theta. \tag{11-37}$$

The work done during a finite angular displacement from θ_i to θ_f is then

$$W = \int_{\theta_i}^{\theta_f} \tau \, d\theta. \tag{11-38}$$

Equation 11-38, which holds for any rigid body rotating about a fixed axis, is the rotational equivalent of Eq. 7-10, which is

$$W = \int_{x_i}^{x_f} F \, dx.$$

We can find the power P for rotational motion from Eq. 11-37:

$$P = \frac{dW}{dt} = \tau \frac{d\theta}{dt} = \tau \omega. \tag{11-39}$$

This is the rotational analog of Eq. 7-32, $P = Fv$, which gives the rate at which a force F does work on a particle moving in a fixed direction with speed v.

To find the angular equivalent of the work–kinetic energy theorem, we start from Eq. 11-38, substituting $I\alpha$ for the torque τ. Thus

$$W = \int_{\theta_i}^{\theta_f} I\alpha \, d\theta = \int_{\theta_i}^{\theta_f} I \left(\frac{d\omega}{dt} \right) d\theta$$

$$= \int_{\omega_i}^{\omega_f} I \left(\frac{d\theta}{dt} \right) d\omega = \int_{\omega_i}^{\omega_f} I\omega \, d\omega.$$

Carrying out the integration yields, with the help of Eq. 11-23 ($K = \frac{1}{2}I\omega^2$),

$$W = \tfrac{1}{2}I\omega_f^2 - \tfrac{1}{2}I\omega_i^2 = K_f - K_i = \Delta K. \tag{11-40}$$

This *work–kinetic energy theorem* tells us that the work done by the net torque acting on a rotating rigid body is equal to the change in the rotational kinetic energy of that body. This is the angular equivalent of the work–kinetic energy theorem for translational motion (Eq. 7-21), which we may restate as follows: the work done on a rigid body (or on a particle) by the net force that acts on it is equal to the change in the translational kinetic energy of that body.

Table 11-3 summarizes the equations that apply to the rotation of a rigid body about a fixed axis and the equivalent relations for translational motion.

SAMPLE PROBLEM 11-13

a. In the arrangement of Fig. 11-19, through what angle does the disk rotate in 2.5 s? Assume that the system starts from rest.

SOLUTION From Eq. 11-10, we have, putting $\omega_0 = 0$ and using the value of α calculated in Sample Problem 11-11,

$$\theta = \omega_0 t + \tfrac{1}{2}\alpha t^2$$

$$= 0 + (\tfrac{1}{2})(-24 \text{ rad/s}^2)(2.5 \text{ s})^2$$

$$= -75 \text{ rad.} \qquad \text{(Answer)}$$

b. What is the angular velocity of the disk at $t = 2.5$ s?

TABLE 11-3
SOME CORRESPONDING RELATIONS FOR TRANSLATIONAL AND ROTATIONAL MOTION

PURE TRANSLATION (FIXED DIRECTION)		PURE ROTATION (FIXED AXIS)	
Position	x	Angular position	θ
Velocity	$v = dx/dt$	Angular velocity	$\omega = d\theta/dt$
Acceleration	$a = dv/dt$	Angular acceleration	$\alpha = d\omega/dt$
Translational inertia (mass)	m	Rotational inertia	I
Newton's second law	$F = ma$	Newton's second law	$\tau = I\alpha$
Work	$W = \int F \, dx$	Work	$W = \int \tau \, d\theta$
Kinetic energy	$K = \tfrac{1}{2}mv^2$	Kinetic energy	$K = \tfrac{1}{2}I\omega^2$
Power	$P = Fv$	Power	$P = \tau\omega$
Work–kinetic energy theorem	$W = \Delta K$	Work–kinetic energy theorem	$W = \Delta K$

SOLUTION We can find this with Eq. 11-9. Putting $\omega_0 = 0$ and using the value of α calculated in Sample Problem 11-11, we obtain

$$\omega = \omega_0 + \alpha t$$

$$= 0 + (-24 \text{ rad/s}^2)(2.5 \text{ s})$$

$$= -60 \text{ rad/s}. \qquad \text{(Answer)}$$

c. What is the kinetic energy of the disk at $t = 2.5$ s?

SOLUTION From Eq. 11-23, the kinetic energy of the disk is $\frac{1}{2}I\omega^2$, in which $I = \frac{1}{2}MR^2$. Thus, using the value of ω found in (b),

$$K = \frac{1}{2}I\omega^2 = \frac{1}{2}(\frac{1}{2}MR^2)\omega^2$$

$$= (\frac{1}{4})(2.5 \text{ kg})(0.20 \text{ m})^2(-60 \text{ rad/s})^2$$

$$= 90 \text{ J}. \qquad \text{(Answer)}$$

Another approach to this problem is to calculate the change in the rotational kinetic energy using the work–kinetic energy theorem. To do so, we must calculate the work done on the disk by the torque that acts

on it. This torque, exerted by the tension T in the cord, is constant so that, from Eq. 11-38,

$$W = \int_{\theta_i}^{\theta_f} \tau \, d\theta = \tau \int_{\theta_i}^{\theta_f} d\theta = \tau(\theta_f - \theta_i).$$

For the torque τ we use $-TR$, in which T is the tension in the cord ($= 6.0$ N; see Sample Problem 11-11) and R ($= 0.20$ m) is the radius of the disk. The quantity $\theta_f - \theta_i$ is just the angular displacement that we calculated in (a). Thus

$$W = \tau(\theta_f - \theta_i) = -TR(\theta_f - \theta_i)$$

$$= -(6.0 \text{ N})(0.20 \text{ m})(-75 \text{ rad})$$

$$= 90 \text{ J}. \qquad \text{(Answer)}$$

Because K_i is zero (the system starts from rest), Eq. 11-40 tells us that this answer is equal to K.

SAMPLE PROBLEM 11-14

A rigid sculpture, consisting of a thin hoop (of mass m and radius $R = 0.15$ m) and two thin rods (each of mass m and length $L = 2.0R$), is arranged as shown in Fig. 11-21. The sculpture can pivot around a horizontal axis in the plane of the hoop, passing through its center.

a. In terms of m and R, what is the sculpture's rotational inertia I about the rotation axis?

SOLUTION From Table 11-2(i), the hoop has rotational inertia $I_{\text{hoop}} = \frac{1}{2}mR^2$ about its diameter. From

FIGURE 11-21 Sample Problem 11-14. A rigid sculpture consisting of a hoop and two rods can rotate around an axis.

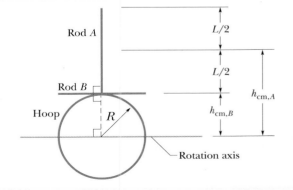

To spin a roulette wheel, you apply a torque and do work, giving the wheel kinetic energy for rotation.

Table 11-2(e), rod A has rotational inertia $I_{cm,A} = mL^2/12$ about an axis through its center of mass and parallel to the rotation axis. To find the rotational inertia I_A of rod A about the rotation axis, we use Eq. 11-25, the parallel-axis theorem:

$$I_A = I_{cm,A} + mh_{cm,A}^2 = \frac{mL^2}{12} + m\left(R + \frac{L}{2}\right)^2$$

$$= 4.33\,mR^2,$$

where we have used the fact that $L = 2.0R$, and where $h_{cm,A}$ ($= R + L/2$) is the perpendicular distance between the center of rod A and the rotation axis.

We treat rod B similarly: it has zero rotational inertia $I_{cm,B}$ about an axis along its length, and its rotational inertia I_B about the rotation axis is

$$I_B = I_{cm,B} + mh_{cm,B}^2 = 0 + mR^2 = mR^2.$$

Here $h_{cm,B}$ ($= R$) is the perpendicular distance between rod B and the rotation axis. Thus the rotational inertia I of the sculpture about the rotation axis is

$$I = I_{hoop} + I_A + I_B = \tfrac{1}{2}mR^2 + 4.33\,mR^2 + mR^2$$

$$= 5.83\,mR^2 \approx 5.8\,mR^2. \qquad \text{(Answer)}$$

b. Starting from rest, the sculpture rotates around the rotation axis from the initial upright orientation of Fig. 11-21. What is its angular speed ω about the axis when it is inverted?

SOLUTION Before the sculpture moves, its center of mass is a distance y_{cm} above the rotation axis, with y_{cm} given by Eq. 9-5:

$$y_{cm} = \frac{m(0) + mR + m(R + L/2)}{3m} = R,$$

where $3m$ is the mass of the sculpture. As the sculpture rotates, its center of mass descends to the same distance *below* the rotation axis. Thus the center of mass undergoes a vertical displacement of $\Delta y_{cm} = -2R$. During the descent, gravitational potential energy U of the sculpture is transferred to kinetic energy K of its rotation. The change ΔU in potential energy is the product of the sculpture's weight ($3mg$) and the vertical displacement Δy_{cm}. Thus

$$\Delta U = 3mg\,\Delta y_{cm} = 3mg(-2R) = -6mgR.$$

From Eq. 11-23, the associated change in kinetic energy is

$$\Delta K = \tfrac{1}{2}I\omega^2 - 0 = \tfrac{1}{2}I\omega^2.$$

So by conservation of mechanical energy

$$\Delta K + \Delta U = 0,$$

and

$$\tfrac{1}{2}I\omega^2 - 6mgR = 0.$$

Substituting $I = 5.83\,mR^2$ and solving for ω give

$$\omega = \sqrt{\frac{12g}{5.83R}} = \sqrt{\frac{(12)(9.8\ \text{m/s}^2)}{(5.83)(0.15\ \text{m})}}$$

$$= 12\ \text{rad/s}. \qquad \text{(Answer)}$$

As gymnast Kurt Thomas rotates around the bar to begin his dismount, the rotational kinetic energy he had at the bottom of his rotation is transferred (mostly) to gravitational potential energy. This transfer is due to the negative work done on him by the torque about the bar that is produced by his weight, which effectively acts at his center of mass.

REVIEW & SUMMARY

Angular Position

To describe the rotation of a rigid body about a fixed axis, called the **rotation axis,** we assume a **reference line** is fixed in the body, perpendicular to that axis and rotating with the body. We measure the **angular position** θ of this line relative to a fixed axis. When θ is measured in **radians,**

$$\theta = \frac{s}{r} \quad \text{(radian measure)}, \quad (11\text{-}1)$$

where s is the arc length of a circle of radius r that is swept out by θ. Radian measure is related to angle measure in revolutions and degrees by

$$1 \text{ rev} = 360° = 2\pi \text{ rad.} \quad (11\text{-}2)$$

Angular Displacement

A body that rotates about a rotation axis, changing its angular position from θ_1 to θ_2, undergoes an **angular displacement**

$$\Delta\theta = \theta_2 - \theta_1, \quad (11\text{-}4)$$

where $\Delta\theta$ is positive for counterclockwise rotation and negative for clockwise rotation.

Angular Velocity and Speed

If a body rotates through an angular displacement $\Delta\theta$ in a time interval Δt, its **average angular velocity** $\overline{\omega}$ is

$$\overline{\omega} = \frac{\Delta\theta}{\Delta t}. \quad (11\text{-}5)$$

The **(instantaneous) angular velocity** ω of the body is

$$\omega = \frac{d\theta}{dt}. \quad (11\text{-}6)$$

Both $\overline{\omega}$ and ω are vectors, with directions given by the **right-hand rule** of Fig. 11-5. They are positive for counterclockwise rotation and negative for clockwise rotation. The magnitude of the body's angular velocity is the **angular speed.**

Angular Acceleration

If the angular velocity of a body changes from ω_1 to ω_2 in a time interval $\Delta t = t_2 - t_1$, the **average angular acceleration** $\overline{\alpha}$ of the body is

$$\overline{\alpha} = \frac{\omega_2 - \omega_1}{t_2 - t_1} = \frac{\Delta\omega}{\Delta t}. \quad (11\text{-}7)$$

The **(instantaneous) angular acceleration** α of a body is given by

$$\alpha = \frac{d\omega}{dt}. \quad (11\text{-}8)$$

Both $\overline{\alpha}$ and α are vectors.

The Kinematic Equations for Constant Angular Acceleration

Constant angular acceleration ($\alpha = $ constant) is an important special case of rotational motion. The appropriate kinematic equations, given in Table 11-1, are

$$\omega = \omega_0 + \alpha t, \quad (11\text{-}9) \qquad \theta = \tfrac{1}{2}(\omega_0 + \omega)t, \quad (11\text{-}12)$$

$$\theta = \omega_0 t + \tfrac{1}{2}\alpha t^2, \quad (11\text{-}10) \qquad \theta = \omega t - \tfrac{1}{2}\alpha t^2. \quad (11\text{-}13)$$

$$\omega^2 = \omega_0^2 + 2\alpha\theta, \quad (11\text{-}11)$$

Linear and Angular Variables

A point in a rigid rotating body, at a *perpendicular distance r* from the rotation axis, moves in a circle with radius r. If the body rotates through an angle θ, the point moves along an arc with length s given by

$$s = \theta r \quad \text{(radian measure)}, \quad (11\text{-}15)$$

where θ is in radians.

The linear velocity **v** of the point is tangent to the circle; the point's linear speed v (the magnitude of its linear velocity) is given by

$$v = \omega r \quad \text{(radian measure)}, \quad (11\text{-}16)$$

where ω is the angular speed (in radians per second) of the body.

The linear acceleration **a** of the point has both *tangential* and *radial* components. The tangential component is

$$a_t = \alpha r \quad \text{(radian measure)}, \quad (11\text{-}18)$$

where α is the magnitude of the angular acceleration (in radians per second-squared) of the body. The radial component of **a** is

$$a_r = \frac{v^2}{r} = \omega^2 r \quad \text{(radian measure)}. \quad (11\text{-}19)$$

Rotational Kinetic Energy and Rotational Inertia

The kinetic energy K of a rigid body rotating about a fixed axis is given by

$$K = \tfrac{1}{2}I\omega^2 \quad \text{(radian measure)}, \quad (11\text{-}23)$$

in which I is the **rotational inertia** of the body, defined as

$$I = \sum m_i r_i^2 \quad (11\text{-}22)$$

for a system of discrete particles and as

$$I = \int r^2 \, dm \quad (11\text{-}24)$$

for a body with continuously distributed mass. The r and r_i in these expressions represent the perpendicular distance from the axis of rotation to each mass element in the body.

The Parallel-Axis Theorem

The *parallel-axis theorem* relates the rotational inertia of a body about any axis to that of the same body about a parallel axis through the center of mass:

$$I = I_{cm} + Mh^2. \tag{11-25}$$

Here h is the perpendicular distance between the two axes.

Torque

Torque is the turning or twisting action on a body about a rotation axis due to a force $\mathbf{F}$. If $\mathbf{F}$ is exerted at a point given by the position vector $\mathbf{r}$ relative to the axis, then the torque $\boldsymbol{\tau}$ (a vector quantity) is

$$\boldsymbol{\tau} = \mathbf{r} \times \mathbf{F}. \tag{11-31}$$

The magnitude of the torque is

$$\tau = rF_t = r_\perp F = rF \sin \phi, \tag{11-28, 11-29, 11-30}$$

where F_t is the component of $\mathbf{F}$ perpendicular to $\mathbf{r}$, and ϕ is the angle between $\mathbf{r}$ and $\mathbf{F}$. The quantity $r_\perp$ is the perpendicular distance between the rotation axis and an extension of a line running through the $\mathbf{F}$ vector. This extension is called the **line of action** of $\mathbf{F}$, and $r_\perp$ is called the **moment arm** of $\mathbf{F}$. Similarly, r is the moment arm of F_t.

The SI unit of torque is the newton-meter (N·m). A torque τ is positive if it tends to rotate a body counter-clockwise, and negative if it tends to rotate the body in the clockwise direction.

Newton's Second Law in Angular Form

The rotation analog of Newton's second law is

$$\sum \tau = I\alpha, \tag{11-34}$$

where $\sum \tau$ is the net torque acting on a particle or rigid body, I is the rotational inertia of the particle or body about the rotation axis, and α is the resulting angular acceleration about that axis.

Rotational Work and Energy

The equations used for calculating work and power in rotational motion are analogs of the corresponding equations governing translational motion and are

$$W = \int_{\theta_i}^{\theta_f} \tau \, d\theta \tag{11-38}$$

and

$$P = \frac{dW}{dt} = \tau \frac{d\theta}{dt} = \tau\omega. \tag{11-39}$$

The form of the work–kinetic energy theorem used for rotating bodies is

$$W = \tfrac{1}{2}I\omega_f^2 - \tfrac{1}{2}I\omega_i^2 = K_f - K_i = \Delta K. \tag{11-40}$$

QUESTIONS

1. Could the angular quantities θ, ω, and α be expressed in degrees instead of radians in the rotational equations of Table 11-1? Explain.

2. In what sense is the radian a "natural" measure of angle and the degree an "arbitrary" measure of that same quantity? (Consider the definitions of the two measures.)

3. Does the vector representing the angular velocity of a wheel rotating about a fixed axis necessarily have to lie along that axis?

4. Experiment by rotating a book after the fashion of Fig. 11-6, but this time using angular displacements of 180° rather than 90°. What do you conclude about the final positions of the book? Does this change your mind about whether (finite) angular displacements can be treated as vectors?

5. Hold your right arm downward, palm toward your thigh. Keeping your wrist rigid, (1) lift the arm until it is horizontal and forward, (2) move it horizontally until it is pointed toward the right, and (3) then bring it down to your side. Your palm faces forward. If you start over, but reverse the sequence of changes, why does your palm *not* face forward?

6. Suppose you hold one quarter flat against a table while you rotate a second quarter around it without slippage (Fig. 11-22). When the second quarter returns to its original position, through what angle has it turned?

FIGURE 11-22   Question 6.

7. The rotation of the sun can be monitored by tracking *sunspots*, magnetic storms on the sun that appear dark against the otherwise bright solar disk. Figure 11-23*a* shows the initial positions of five spots, and Fig. 11-23*b* the positions of these same spots one solar rotation later. What can we conclude about the structure of the sun from these two observations?

(a)   *(b)*

FIGURE 11-23   Question 7.

8. Why is it suitable to express α in revolutions per second-squared in the expression $\theta = \omega_0 t + \frac{1}{2}\alpha t^2$ but not in the expression $a_t = \alpha r$?

9. A rigid body is free to rotate about a fixed axis. Can the body have nonzero angular acceleration even if the angular velocity of the body is (perhaps instantaneously) zero? What is the linear equivalent of this question? Give examples to illustrate both the angular and linear situations.

10. A golfer swings a golf club, making a long drive from the tee. Do all points on the club have the same angular velocity ω at every instant, as long as the club is in motion?

11. When we say that a point on the equator has an angular speed of 2π rad/day, what reference frame do we have in mind?

12. Taking into account the Earth's rotation and orbiting (which are in the same direction), explain whether a tree moves faster during the day or during the night. With respect to what reference frame is your answer given?

13. Imagine a wheel rotating about its axle, and consider a point on its rim. When the wheel rotates with constant angular velocity, does the point have a radial acceleration? A tangential acceleration? When the wheel rotates with constant angular acceleration, does the point have a radial acceleration? A tangential acceleration? Do the magnitudes of these accelerations change with time?

14. Suppose that you were asked to determine the distance traveled by a needle in playing a phonograph record. What information would you need? Discuss the situation from the point of view of reference frames (a) fixed in the room, (b) fixed on the rotating record, and (c) fixed on the arm of the record player.

15. What is the relation between the angular velocities of a pair of coupled gears of different radii?

16. Can the mass of a body be considered as concentrated at the center of mass of the body for purposes of computing its rotational inertia? If yes, explain why. If no, offer a counterexample.

17. When you stand upright, about what axis is the rotational inertia of your body (a) the least and (b) the greatest? (c) For (a) and (b), how can you change the value of your rotational inertia?

18. If two circular disks of the same weight and thickness are made from metals having different densities, which disk, if either, will have the larger rotational inertia about its central axis?

19. Suggest ways in which the rotational inertia about a particular axis of a body with a very complicated shape could be measured experimentally.

20. Five solids with identical masses are shown in cross section in Fig. 11-24. The cross sections have equal heights

FIGURE 11-24 Question 20.

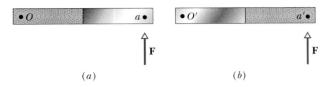

FIGURE 11-25 Question 21.

and equal maximum widths. Which one has the largest rotational inertia about an axis through its center of mass and perpendicular to its cross section? Which has the smallest?

21. Figure 11-25a shows a meter stick, half of which is wood and half of which is steel, that is pivoted at the wooden end at O. A force **F** is applied to the steel end at a. In Fig. 11-25b, the stick is pivoted at the steel end at O', and the same force is applied at the wooden end at a'. Does the same angular acceleration result in both cases? If not, in which case is the angular acceleration greater?

22. Comment on each of these assertions about skiing. (a)

In downhill racing, one wants skis that do not turn easily. (b) In slalom racing, one wants skis that turn easily. (c) Therefore the rotational inertia of downhill skis should be larger than that of slalom skis. (d) Considering that there is low friction between skis and snow and that the skier's center of mass is about over the center of the skis, how does a skier exert torques to turn or to stop a turn? (From "The Physics of Ski Turns," by J. I. Shonie and D. L. Mordick, *The Physics Teacher*, December 1972.)

23. Consider a straight stick standing on end on (frictionless) ice. What will be the path of its center of mass if it falls?

EXERCISES & PROBLEMS

SECTION 11-2 THE ROTATIONAL VARIABLES

1E. (a) What angle in radians is subtended from the center of a circle of radius 1.20 m by an arc of length 1.80 m? (b) Express the same angle in degrees. (c) The angle between two radii of a circle is 0.620 rad. What arc length is subtended if the radius is 2.40 m?

2E. During a time interval t the flywheel of a generator turns through the angle $\theta = at + bt^3 - ct^4$, where a, b, and c are constants. Write expressions for its (a) angular velocity and (b) angular acceleration.

3E. Our sun is 2.3×10^4 ly (light-years) from the center of our Milky Way galaxy and is moving in a circle around that center at a speed of 250 km/s. (a) How long does it take the sun to make one revolution about the galactic center? (b) How many revolutions has the sun completed since it was formed about 4.5×10^9 years ago?

4E. The angular position of a point on the rim of a rotating wheel is given by $\theta = 4.0t - 3.0t^2 + t^3$, where θ is in radians if t is given in seconds. (a) What are the angular velocities at $t = 2.0$ s and $t = 4.0$ s? (b) What is the average angular acceleration for the time interval that begins at $t = 2.0$ s and ends at $t = 4.0$ s? (c) What are the instantaneous angular accelerations at the beginning and end of this time interval?

5E. The angular position of a point on a rotating wheel is given by $\theta = 2 + 4.0t^2 + 2t^3$, where θ is in radians and t is in seconds. At $t = 0$, what are (a) the angular position and

(b) the angular velocity? (c) What is the angular velocity at $t = 4.0$ s? (d) Calculate the angular acceleration at $t = 2.0$ s. (e) Is the angular acceleration constant?

6P. A wheel rotates with an angular acceleration α given by $\alpha = 4at^3 - 3bt^2$, where t is the time and a and b are constants. If the wheel has an initial angular speed ω_0, write equations for (a) the angular speed and (b) the angular displacement as functions of time.

7P. What is the angular speed of (a) the second hand, (b) the minute hand, and (c) the hour hand of a watch?

8P. A good baseball pitcher can throw a baseball toward home plate at 85 mi/h with a spin of 1800 rev/min. How many revolutions does the baseball make on its way to home plate? For simplicity, assume that the 60-ft trajectory is a straight line.

9P. A diver makes 2.5 complete revolutions on the way from a 10-m-high platform to the water below. Assuming zero initial vertical velocity, calculate the average angular velocity during a dive.

10P. A wheel has eight spokes and a radius of 30 cm. It is mounted on a fixed axle and is spinning at 2.5 rev/s. You want to shoot a 20-cm-long arrow parallel to this axle and through the wheel without hitting any of the spokes. Assume that the arrow and the spokes are very thin; see Fig. 11-26. (a) What minimum speed must the arrow have? (b) Does it matter where between the axle and rim of the wheel you aim? If so, where is the best location?

FIGURE 11-26   Problem 10.

SECTION 11-4 ROTATION WITH CONSTANT ANGULAR ACCELERATION

11E. A disk, initially rotating at 120 rad/s, is slowed down with a constant angular acceleration of magnitude 4.0 rad/s². (a) How much time elapses before the disk stops? (b) Through what angle does the disk rotate in coming to rest?

12E. A phonograph turntable rotating at $33\frac{1}{3}$ rev/min slows down and stops in 30 s after the motor is turned off. (a) Find its (uniform) angular acceleration in units of rev/min². (b) How many revolutions did it make in this time?

13E. The angular speed of an automobile engine is increased from 1200 rev/min to 3000 rev/min in 12 s. (a) What is its angular acceleration in rev/min², assuming it to be uniform? (b) How many revolutions does the engine make during this time?

14E. A heavy flywheel rotating on its axis is slowing down because of friction in its bearings. At the end of the first minute, its angular velocity is 0.90 of its initial angular velocity of 250 rev/min. Assuming constant frictional forces, find its angular velocity at the end of the second minute.

15E. The flywheel of an engine is rotating at 25.0 rad/s. When the engine is turned off, the flywheel decelerates at a constant rate and comes to rest after 20.0 s. Calculate (a) the angular acceleration (in rad/s²) of the flywheel, (b) the angle (in rad) through which the flywheel rotates in coming to rest, and (c) the number of revolutions made by the flywheel in coming to rest.

16E. Starting from rest, a disk rotates about its axis with constant angular acceleration. After 5.0 s, it has rotated through 25 rad. (a) What was the angular acceleration during this time? (b) What was the average angular velocity? (c) What is the instantaneous angular velocity of the disk at the end of the 5.0 s? (d) Assuming that the acceleration does not change, through what additional angle will the disk turn during the next 5.0 s?

17E. A pulley wheel that is 8.0 cm in diameter has a 5.6-m long cord wrapped around its periphery. Starting from rest, the wheel is given a constant angular acceleration of 1.5 rad/s². (a) Through what angle must the wheel turn for the cord to unwind? (b) How long does it take?

18P. A wheel turns through 90 rev in 15 s, its angular speed at the end of the period being 10 rev/s. (a) What was its angular speed at the beginning of the 15-s interval, assuming constant angular acceleration? (b) How much time had elapsed between the time the wheel was at rest and the beginning of the 15-s interval?

19P. A wheel has a constant angular acceleration of 3.0 rad/s². In a 4.0-s interval, it turns through an angle of 120 rad. Assuming the wheel started from rest, how long had it been in motion at the start of this 4.0-s interval?

20P. A wheel, starting from rest, rotates with constant angular acceleration 2.00 rad/s². During a certain 3.00-s interval, it turns through 90.0 rad. (a) How long had the wheel been turning before the start of the 3.00-s interval? (b) What was the angular velocity of the wheel at the start of the 3.00-s interval?

21P. A flywheel completes 40 rev as it slows from an angular speed of 1.5 rad/s to a complete stop. (a) Assuming uniform acceleration, what is the time required for it to come to rest? (b) What is the angular acceleration? (c) How much time is required for it to complete the first 20 of the 40 rev?

22P. At $t = 0$, a flywheel has an angular velocity of 4.7 rad/s, an angular acceleration of -0.25 rad/s², and a reference line at $\theta_0 = 0$. (a) Through what maximum angle θ_{max} will the reference line turn in the positive direction? At what times t will the line be at (b) $\theta = \frac{1}{2}\theta_{max}$ and (c) $\theta = -10.5$ rad (consider both positive and negative values of t)? (d) Graph θ versus t, and indicate the answers to (a), (b), and (c) on the graph.

23P. A disk rotates about a fixed axis starting from rest and accelerates with constant angular acceleration. At one time it is rotating at 10 rev/s. After undergoing 60 more complete revolutions its angular speed is 15 rev/s. Calculate (a) the angular acceleration, (b) the time required to complete the 60 revolutions mentioned, (c) the time required to attain the 10-rev/s angular speed, and (d) the number of revolutions from rest until the time the disk attained the 10-rev/s angular speed.

24P. Starting from rest at $t = 0$, a wheel undergoes a constant angular acceleration. When $t = 2.0$ s, the angular velocity of the wheel is 5.0 rad/s. The acceleration continues until $t = 20$ s, when it abruptly ceases. Through what angle does the wheel rotate in the interval $t = 0$ to $t = 40$ s?

25P. A pulsar is a rapidly rotating neutron star that emits radio pulses with precise synchronization, one such pulse for each rotation of the star. The *period T* of rotation is the time for one rotation. It is found by measuring the time between pulses. At present, the pulsar in the central region of the Crab nebula (see Fig. 11-27) has a period of rotation of $T = 0.033$ s, and this is observed to be increasing at the rate of 1.26×10^{-5} s/y. (a) Show that the angular velocity ω of the star is related to T by $\omega = 2\pi/T$. (b) What is the value of the angular acceleration in rad/s²? (c)

If its angular acceleration is constant, how many years from now will the pulsar stop rotating? (d) The pulsar originated in a supernova explosion seen in the year A.D. 1054. What was T of the pulsar when it was born? (Assume constant angular acceleration since then.)

SECTION 11-5   THE LINEAR AND ANGULAR VARIABLES

26E. What is the acceleration of a point on the rim of a 12-in. ($= 30$ cm) diameter record rotating at $33\frac{1}{3}$ rev/min?

27E. A phonograph record on a turntable rotates at $33\frac{1}{3}$ rev/min. (a) What is the angular speed in rad/s? What is the linear speed of a point on the record at the needle at (b) the beginning and (c) the end of the recording? The distances of the needle from the turntable axis are 5.9 in. and 2.9 in., respectively, at these two positions.

28E. What is the angular speed of a car rounding a circular turn of radius 110 m at 50 km/h?

29E. A flywheel with a diameter of 1.20 m is rotating at 200 rev/min. (a) What is the angular velocity of the flywheel in rad/s? (b) What is the linear velocity of a point on the rim of the flywheel? (c) What constant angular acceleration (rev/min²) will increase the wheel's speed to 1000 rev/min in 60 s? (d) How many revolutions does the wheel make during this 60-s interval?

30E. A point on the rim of a 0.75-m diameter grinding wheel changes speed uniformly from 12 m/s to 25 m/s in 6.2 s. What is the average angular acceleration of the grinding wheel during this interval?

31E. The Earth's orbit about the sun is almost a circle. With respect to the sun, what are the Earth's (a) angular speed, (b) linear speed, and (c) acceleration?

32E. An astronaut is being tested in a centrifuge. The centrifuge has a radius of 10 m and, in starting, rotates according to $\theta = 0.30t^2$, where t in seconds gives θ in radians. When $t = 5.0$ s, what are the astronaut's (a) angular velocity, (b) linear speed, (c) tangential acceleration (magnitude only), and (d) radial acceleration (magnitude only)?

33E. What are (a) the angular speed, (b) the radial acceleration, and (c) the tangential acceleration of a spaceship negotiating a circular turn of radius 3220 km at a constant speed of 29,000 km/h?

34E. A coin of mass M is placed a distance R from the center of a phonograph turntable. The coefficient of static friction is μ_s. The angular speed of the turntable is slowly increased to a value ω_0 at which time the coin slides off. (a) Find ω_0 in terms of the quantities M, R, g, and μ_s. (b) Make a sketch showing the approximate path of the coin as it flies off the turntable.

35P. (a) What is the angular speed about the polar axis of a point on the Earth's surface at a latitude of 40° N? (b) What is the linear speed? (c) What are the corresponding values for a point at the equator?

36P. The flywheel of a steam engine runs with a constant angular speed of 150 rev/min. When steam is shut off, the friction of the bearings and of the air brings the wheel to rest in 2.2 h. (a) What is the constant angular acceleration, in rev/min², of the wheel during the slowdown? (b) How many rotations does the wheel make before coming to rest? (c) What is the tangential component of the linear acceleration of a particle that is 50 cm from the axis of rotation when the flywheel is turning at 75 rev/min? (d) What is the magnitude of the net linear acceleration of the particle in part (c)?

37P. A gyroscope flywheel of radius 2.83 cm is accelerated from rest at 14.2 rad/s² until its angular speed is 2760 rev/min. (a) What is the tangential acceleration of a point on the rim of the flywheel? (b) What is the radial acceleration of this point when the flywheel is spinning at full speed? (c) Through what distance does a point on the rim move during the acceleration?

38P. If an airplane propeller of radius 1.5 m rotates at 2000 rev/min while the airplane flies at a speed of 480 km/h relative to the ground, what is the speed of a point on the tip of the propeller, as seen by (a) the pilot and (b) an observer on the ground? Assume that the plane's velocity is parallel to the propeller's axis of rotation.

39P. An early method of measuring the speed of light makes use of a rotating slotted wheel. A beam of light passes through a slot at the outside edge of the wheel, as in

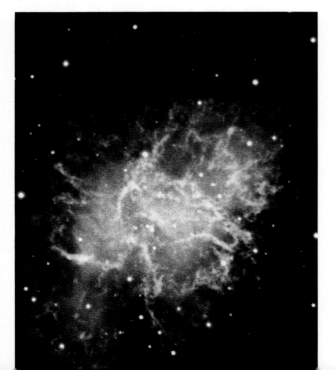

FIGURE 11-27   Problem 25. The Crab nebula resulted from a star whose explosion was seen in A.D. 1054. In addition to the gaseous debris seen here, the explosion left a spinning neutron star at its center.

FIGURE 11-28 Problem 39.

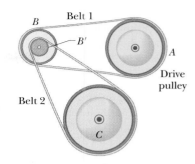

FIGURE 11-30 Problem 42.

Fig. 11-28, travels to a distant mirror, and returns to the wheel just in time to pass through the next slot in the wheel. One such slotted wheel has a radius of 5.0 cm and 500 slots at its edge. Measurements taken when the mirror was $l = 500$ m from the wheel indicated a speed of light of 3.0×10^5 km/s. (a) What was the (constant) angular speed of the wheel? (b) What was the linear speed of a point on its edge?

40P. A car starts from rest and moves around a circular track of radius 30.0 m. Its speed increases at the constant rate of 0.500 m/s². (a) What is the magnitude of its *net* linear acceleration 15.0 s later? (b) What angle does this net acceleration vector make with the car's velocity at this time?

41P. Wheel A of radius $r_A = 10$ cm is coupled by belt B to wheel C of radius $r_C = 25$ cm, as shown in Fig. 11-29. Wheel A increases its angular speed from rest at a uniform rate of 1.6 rad/s². Determine the time required for wheel C to reach a rotational speed of 100 rev/min, assuming the belt does not slip. (*Hint:* If the belt does not slip, the linear speeds at the rims of the two wheels must be equal.)

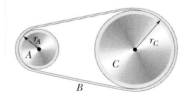

FIGURE 11-29 Problem 41.

42P. Four pulleys are connected by two belts as shown in Fig. 11-30. Pulley A (radius 15 cm) is the drive pulley, and it rotates at 10 rad/s. Pulley B (radius 10 cm) is connected by belt 1 to pulley A. Pulley B' (radius 5 cm) is concentric with pulley B and is rigidly attached to it. Pulley C (radius 25 cm) is connected by belt 2 to pulley B'. Calculate (a) the linear speed of a point on belt 1, (b) the angular speed of pulley B, (c) the angular speed of pulley B', (d) the linear speed of a point on belt 2, and (e) the angular speed of pulley C. (See hint to preceding problem.)

43P. A phonograph turntable is rotating at $33\frac{1}{3}$ rev/min. A small object is on the turntable 6.0 cm from the axis of rotation. (a) Calculate the acceleration of the object, assuming that it does not slip. (b) What is the minimum value of the coefficient of static friction between the object and the turntable? (c) Suppose that the turntable achieved its angular speed by starting from rest and undergoing a constant angular acceleration for 0.25 s. Calculate the minimum coefficient of static friction required for the object not to slip during the acceleration period.

44P. A compact disc has inner and outer radii for its recorded material of 2.50 cm and 5.80 cm. At playback, the disc is scanned at a constant linear speed of 130 cm/s, starting from its inner edge and moving outward. What are the angular speeds for scanning at the (a) inner radius and (b) outer radius? (c) Is the angular acceleration constant? (d) The spiraled scan lines are 1.60 μm apart; what is the total length of the scan? (e) What is the playing time?

SECTION 11-6 KINETIC ENERGY OF ROTATION

45E. Calculate the rotational inertia of a wheel that has a kinetic energy of 24,400 J when rotating at 602 rev/min.

46P. The oxygen molecule, O_2, has a total mass of 5.30×10^{-26} kg and a rotational inertia of 1.94×10^{-46} kg·m² about an axis through the center perpendicular to the line joining the atoms. Suppose that such a molecule in a gas has a speed of 500 m/s and that its rotational kinetic energy is two-thirds of its translational kinetic energy. Find its angular velocity.

SECTION 11-7 CALCULATING THE ROTATIONAL INERTIA

47E. Calculate the kinetic energies of two uniform solid cylinders, each rotating about its central axis. They have the same mass, 1.25 kg, and rotate with the same angular velocity, 235 rad/s, but the first has a radius of 0.25 m and the second a radius of 0.75 m.

48E. A molecule has a rotational inertia of 14,000 u·pm² and is spinning at an angular speed of 4.3×10^{12} rad/s. (a) Express the rotational inertia in kg·m². (b) Calculate the rotational kinetic energy in electron-volts.

49E. The masses and coordinates of four particles are as follows: 50 g, $x = 2.0$ cm, $y = 2.0$ cm; 25 g, $x = 0$, $y = 4.0$ cm; 25 g, $x = -3.0$ cm, $y = -3.0$ cm; 30 g, $x = -2.0$ cm, $y = 4.0$ cm. What is the rotational inertia of this collection with respect to the (a) x, (b) y, and (c) z axes? (d) If the answers to (a) and (b) are A and B, respectively, then what is the answer to (c) in terms of A and B?

50E. A communications satellite is a solid cylinder with mass 1210 kg, diameter 1.21 m, and length 1.75 m. Prior to launching from the shuttle cargo bay, it is set spinning at 1.52 rev/s about the cylinder axis (Fig. 11-31). Calculate the satellite's (a) rotational inertia about the rotation axis and (b) rotational kinetic energy.

FIGURE 11-31 Exercise 50.

51E. Two particles, each with mass m, are fastened to each other, and to a rotation axis at O, by two thin rods, each with length l and mass M as shown in Fig. 11-32. The combination rotates around the rotation axis with angular velocity ω. Obtain algebraic expressions for (a) the rotational inertia of the combination about O and (b) the kinetic energy of rotation about O.

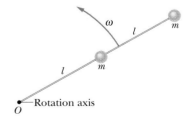

FIGURE 11-32 Exercise 51.

52E. Each of the three helicopter rotor blades shown in Fig15. 11-33 is 5.20 m long and has a mass of 240 kg. The rotor is rotating at 350 rev/min. (a) What is the rotational inertia of the rotor assembly about the axis of rotation?

FIGURE 11-33 Exercise 52.

(Each blade can be considered a thin rod; why?) (b) What is the kinetic energy of rotation?

53E. Assume the Earth to be a sphere of uniform density. Calculate (a) its rotational inertia and (b) its rotational kinetic energy. (c) Suppose that this energy could be harnessed for our use. For how long could the Earth supply 1.0 kW of power to each of the 4.2×10^9 persons on Earth?

54E. Calculate the rotational inertia of a meter stick, with mass 0.56 kg, about an axis perpendicular to the stick and located at the 20-cm mark. (Treat the stick as a thin rod.)

55P. Show that the axis about which a given rigid body has its smallest rotational inertia must pass through its center of mass.

56P. Derive the expression for the rotational inertia of a hoop of mass M and radius R about its central axis; see Table 11-2(a).

57P. Figure 11-34 shows a uniform solid block of mass M and edge lengths a, b, and c. Calculate its rotational inertia about an axis through one corner and perpendicular to the large faces of the block.

FIGURE 11-34 Problem 57.

58P. (a) Show that the rotational inertia of a solid cylinder of mass M and radius R about its central axis is equal to the rotational inertia of a thin hoop of mass M and radius $R/\sqrt{2}$ about its central axis. (b) Show that the rotational inertia I of any given body of mass M about any given axis is equal to the rotational inertia of an *equivalent hoop* about that axis, if the hoop has the same mass M and a radius k given by

$$k = \sqrt{\frac{I}{M}}.$$

The radius k of the equivalent hoop is called the *radius of gyration* of the given body.

59P. Delivery trucks that operate by making use of energy stored in a rotating flywheel have been used in Europe. The trucks are charged by using an electric motor to get the flywheel up to its top speed of 200π rad/s. One such flywheel is a solid, homogeneous cylinder with a mass of 500 kg and a radius of 1.0 m. (a) What is the kinetic energy of the flywheel after charging? (b) If the truck operates with an average power requirement of 8.0 kW, for how many minutes can it operate between chargings?

SECTION 11-8 TORQUE

60E. A small 0.75-kg ball is attached to one end of a 1.25-m-long massless rod, and the other end is hung from a pivot. When the resulting pendulum is 30° from the vertical, what is the magnitude of the torque about the pivot?

61E. A bicyclist of mass 70 kg puts all his weight on each downward-moving pedal as he pedals up a steep road. Take the diameter of the circle in which the pedals rotate to be 0.40 m, and determine the magnitude of the maximum torque he exerts in the process.

62E. The length of a bicycle pedal arm is 0.152 m, and a downward force of 111 N is applied by the foot. What is the magnitude of the torque about the pivot point when the arm makes an angle of (a) 30°, (b) 90°, and (c) 180° with the vertical?

63P. The body in Fig. 11-35 is pivoted at O, and two forces act on it as shown. (a) Find an expression for the magnitude of the net torque on the body about the pivot. (b) If $r_1 = 1.30$ m, $r_2 = 2.15$ m, $F_1 = 4.20$ N, $F_2 = 4.90$ N, $\theta_1 = 75.0°$, and $\theta_2 = 60.0°$, what is the net torque about the pivot?

FIGURE 11-35 Problem 63.

64P. The body in Fig. 11-36 is pivoted at O. Three forces act on it in the directions shown on the figure: $F_A = 10$ N at point A, 8.0 m from O; $F_B = 16$ N at point B, 4.0 m from O; and $F_C = 19$ N at point C, 3.0 m from O. What is the net torque about O?

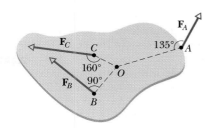

FIGURE 11-36 Problem 64.

SECTION 11-9 NEWTON'S SECOND LAW FOR ROTATION

65E. When a torque of 32.0 N·m is applied to a certain wheel, it acquires an angular acceleration of 25.0 rad/s². What is the rotational inertia of the wheel?

66E. Leaving the board, a diver changed her angular velocity from zero to 6.20 rad/s in 220 ms. Her rotational inertia is 12.0 kg·m². (a) What was the angular acceleration during the jump? (b) What external torque acted on the diver during the jump?

67E. A cylinder having a mass of 2.0 kg can rotate about an axis through its center O. Forces are applied as in Fig. 11-37: $F_1 = 6.0$ N, $F_2 = 4.0$ N, $F_3 = 2.0$ N, and $F_4 = 5.0$ N. Also, $R_1 = 5.0$ cm and $R_2 = 12$ cm. Find the magnitude and direction of the angular acceleration of the cylinder, assuming that during the rotation, the forces maintain their same angles relative to the cylinder.

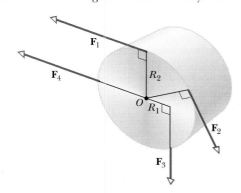

FIGURE 11-37 Exercise 67.

68E. A small object with mass 1.30 kg is mounted on one end of a rod 0.780 m long and of negligible mass. The system rotates in a horizontal circle about the other end of the rod at 5010 rev/min. (a) Calculate the rotational inertia of the system about the axis of rotation. (b) Air resistance exerts a force of 2.30×10^{-2} N on the object, directed opposite its direction of motion. What torque must be applied to the system to keep it rotating at constant speed?

69E. A thin spherical shell has a radius of 1.90 m. An applied torque of 960 N·m imparts to the shell an angular acceleration equal to 6.20 rad/s² about an axis through the center of the shell. (a) What is the rotational inertia of the shell about the axis of rotation? (b) Calculate the mass of the shell.

70P. A pulley, with a rotational inertia of 1.0×10^{-3} kg·m² about its axle and a radius of 10 cm, is acted on by a force, applied tangentially at its rim. The force magnitude varies in time as $F = 0.50t + 0.30t^2$, where F is in newtons if t is given in seconds. The pulley is initially at rest. At $t = 3.0$ s what are (a) its angular acceleration and (b) its angular velocity?

71P. Figure 11-38 shows the massive shield door at a neutron test facility at Lawrence Livermore Laboratory; this is the world's heaviest hinged door. The door has a mass of 44,000 kg, a rotational inertia about an axis through its hinges of 8.7×10^4 kg·m², and a width of 2.4 m. Neglecting friction, what steady force, applied at its outer edge and perpendicular to the plane of the door, can move it from rest through an angle of 90° in 30 s? Assume no friction acts on the hinges.

72P. Two uniform solid spheres have the same mass, 1.65 kg, but one has a radius of 0.226 m while the other has a radius of 0.854 m. (a) For each of the spheres, find the torque required to bring the sphere from rest to an angular velocity of 317 rad/s in 15.5 s. Each sphere rotates about an axis through its center. (b) For each sphere, what force applied tangentially at the equator would provide the needed torque?

73P. In an Atwood's machine (Fig. 5-28), one block has a mass of 500 g, and the other a mass of 460 g. The pulley, which is mounted in horizontal frictionless bearings, has a radius of 5.00 cm. When released from rest, the heavier block is observed to fall 75.0 cm in 5.00 s (without the string slipping on the pulley). (a) What is the acceleration of each block? What is the tension in the part of the cord that supports (b) the heavier block and (c) the lighter

block? (d) What is the angular acceleration of the pulley? (e) What is its rotational inertia?

74P. Figure 11-39 shows two blocks, each of mass m, suspended from the ends of a rigid weightless rod of length $l_1 + l_2$, with $l_1 = 20$ cm and $l_2 = 80$ cm. The rod is held in the horizontal position shown in the figure and then released. Calculate the accelerations of the two blocks as they start to move.

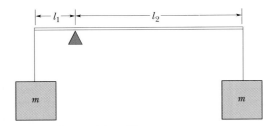

FIGURE 11-39 Problem 74.

75P. Two identical blocks, each of mass M, are connected by a massless string over a pulley of radius R and rotational inertia I (Fig. 11-40). The string does not slip on the pulley; it is not known whether or not there is friction between the table and the sliding block; the pulley's axis is frictionless. When this system is released, it is found that the pulley turns through an angle θ in time t and the acceleration of the blocks is constant. (a) What is the angular acceleration of the pulley? (b) What is the acceleration of the two blocks? (c) What are the tensions in the upper and lower sections of the string? All answers are to be expressed in terms of M, I, R, θ, g, and t.

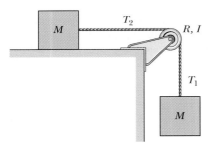

FIGURE 11-40 Problem 75.

76P. The length of the day is increasing at the rate of about 1 ms/century. This is primarily due to frictional forces generated by movement of water in the world's shallow seas as a response to the tidal forces exerted by the sun and moon. (a) At what rate is the Earth losing rotational kinetic energy? (b) What is its angular acceleration? (c) What tangential force, exerted at (average) latitudes 60° N and 60° S, is being applied by the oceans to the near-coastal seabed?

77P. A tall, cylinder-shaped chimney falls over when its base is ruptured. Treating the chimney as a thin rod with

FIGURE 11-38 Problem 71.

height *h*, express (a) the radial and (b) tangential components of the linear acceleration of the top of the chimney as a function of the angle θ made by the chimney with the vertical. (c) At what angle θ does the linear acceleration equal *g*?

SECTION 11-10   WORK, POWER, AND THE WORK–KINETIC ENERGY THEOREM

78E. (a) If $R = 12$ cm, $M = 400$ g, and $m = 50$ g in Fig. 11-19, find the speed of *m* after it has descended 50 cm starting from rest. Solve the problem using energy-conservation principles. (b) Repeat (a) with $R = 5.0$ cm.

79E. An automobile engine develops 100 hp ($= 74.6$ kW) when rotating at a speed of 1800 rev/min. What torque does it deliver?

80E. A 32.0-kg wheel, essentially a thin hoop with radius 1.20 m, is rotating at 280 rev/min. It must be brought to a stop in 15.0 s. (a) How much work must be done to stop it? (b) What is the required power?

81E. A thin rod of length *l* and mass *m* is suspended freely from one end. It is pulled aside and allowed to swing like a pendulum, passing through its lowest position with an angular speed ω. (a) Calculate its kinetic energy as it passes through its lowest position. (b) How high does its center of mass rise above its lowest position? Neglect friction and air resistance.

82P. A meter stick is held vertically with one end on the floor and is then allowed to fall. Find the speed of the other end when it hits the floor, assuming that the end on the floor does not slip. (*Hint:* Consider the stick to be a thin rod and use conservation of energy.)

83P. A rigid body is made of three identical thin rods, each with length *l*, fastened together in the form of a letter H (Fig. 11-41). The body is free to rotate about a horizontal axis that runs along the length of one of the legs of the H. The body is allowed to fall from rest from a position in which the plane of the H is horizontal. What is the angular speed of the body when the plane of the H is vertical?

FIGURE 11-41   Problem 83.

84P. Calculate (a) the torque, (b) the energy, and (c) the average power required to accelerate the Earth from rest to its present angular speed about its axis in 1 day.

85P. A uniform helicopter rotor blade (see Fig. 11-33) is 7.80 m long and has a mass of 110 kg. (a) What force is exerted on the bolt attaching the blade to the rotor axle when the rotor is turning at 320 rev/min? (*Hint:* For this

calculation the blade can be considered to be a point mass at the center of mass. Why?) (b) Calculate the torque that must be applied to the rotor to bring it to full speed from rest in 6.7 s. Ignore air resistance. (The blade cannot be considered to be a point mass for this calculation. Why not? Assume the mass distribution of a uniform, thin rod.) (c) How much work did the torque do on the blade for the blade to reach a speed of 320 rev/min?

86P. A uniform spherical shell of mass *M* and radius *R* rotates about a vertical axis on frictionless bearings (Fig. 11-42). A massless cord passes around the equator of the shell, over a pulley of rotational inertia *I* and radius *r*, and is attached to a small object of mass *m* that is otherwise free to fall under the influence of gravity. There is no friction on the pulley's axle; the cord does not slip on the pulley. What is the speed of the object after it has fallen a distance *h* from rest? Use work–energy considerations.

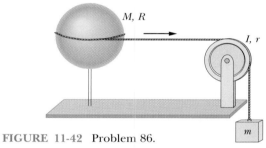

FIGURE 11-42   Problem 86.

87P. A thin steel rod of length 1.20 m and mass 6.40 kg has attached to each end a small ball of mass 1.06 kg. The rod is constrained to rotate in a horizontal plane about a vertical axis through its midpoint. At a certain instant, it is observed to be rotating with an angular velocity of 39.0 rev/s. Because of friction, it comes to rest 32.0 s later. Compute, assuming a constant frictional torque, (a) the angular acceleration, (b) the retarding torque exerted by the friction, (c) the total work done by the friction, and (d) the number of revolutions executed during the 32.0 s. (e) Now suppose that the frictional torque is known not to be constant. Which, if any, of the quantities (a), (b), (c), or (d) can still be computed without additional information? If such a quantity exists, give its value.

88P*. A car is fitted with an energy-conserving flywheel, which in operation is geared to the driveshaft so that the flywheel rotates at 240 rev/s when the car is traveling at 80 km/h. The total mass of the car is 800 kg; the flywheel weighs 200 N and is a uniform disk 1.1 m in diameter. The car descends a 1500-m-long, 5° slope, starting from rest, with the flywheel engaged and no power generated from the engine. Neglecting friction and the rotational inertia of the wheels, find (a) the speed of the car at the bottom of the slope, (b) the angular acceleration of the flywheel at the bottom of the slope, and (c) the rate at which energy is stored in the rotation of the flywheel by the driveshaft at the bottom of the slope.

ADDITIONAL PROBLEMS

89. Four identical particles of mass 0.50 kg each are placed at the vertices of a 2.0 m × 2.0 m square and held there by four massless rods, which form the sides of the square. What is the rotational inertia of this rigid body about an axis (a) that passes through the midpoints of opposite sides and lies in the plane of the square, (b) that passes through the midpoint of one of the sides and is perpendicular to the plane of the square, and (c) that lies in the plane of the square and passes through two diagonally opposite particles?

90. An object rotates about a fixed axis, and a reference line on the object is given by $\theta = 0.40e^{2t}$, where θ is in radians and t is in seconds. Consider a point on the object that is 4.0 cm from the axis of rotation. At $t = 0$, what is the magnitude of (a) the tangential component of the acceleration of the point, and (b) the radial component of the acceleration of the point?

91. A wheel of radius 0.20 m is mounted on a frictionless horizontal axis. A massless cord is wrapped around the wheel and attached to a 2.0-kg object that slides on a frictionless surface inclined at an angle of 20° with the horizontal, as shown in Fig. 11-43. The object accelerates down the incline at 2.0 m/s². What is the rotational inertia of the wheel about its axis of rotation?

FIGURE 11-43 Problem 91.

92. The turntable of a record player has an angular speed of 8.0 rad/s when it is switched off. Three seconds later, the turntable is observed to have an angular speed of 2.6 rad/s. Through how many radians does the turntable rotate from the time it is turned off until it stops? (Assume constant angular acceleration.)

93. Two thin disks, each of mass 4.0 kg and radius 0.40 m, are attached as shown in Fig. 11-44 to form a rigid body.

FIGURE 11-44 Problem 93.

What is the rotational inertia of this body about an axis A that is perpendicular to the plane of the disks and passes through the center of one of the disks?

94. The rigid body shown in Fig. 11-45 consists of three particles connected by massless rods. It is to be rotated about an axis perpendicular to its plane through point P. If $M = 0.40$ kg, $a = 30$ cm, and $b = 50$ cm, how much work is required to take the body from rest to an angular speed of 5.0 rad/s?

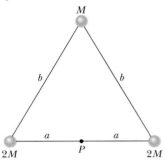

FIGURE 11-45 Problem 94.

95. A uniform rod of mass 1.5 kg is 2.0 m long (Fig. 11-46). The rod can pivot about a horizontal, frictionless pin through one end. It is released from rest at an angle of 40° above the horizontal. (a) What is the angular acceleration of the rod at the instant it is released? (The rotational inertia of the rod about the pin is 2.0 kg·m².) (b) Use the principle of conservation of energy to determine the angular speed of the rod as it passes through the horizontal position.

FIGURE 11-46 Problem 95.

96. A uniform cylinder of radius 10 cm and mass 20 kg is mounted so as to rotate freely about a horizontal axis that is parallel to and 5.0 cm from the longitudinal axis of the cylinder. (a) What is the rotational inertia of the cylinder about the axis of rotation? (b) If the cylinder is released from rest with its axis at the same height as the axis about which the cylinder rotates, what is the angular speed of the cylinder as it passes through its lowest position? (*Hint:* Use the principle of conservation of energy.)

ROLLING, TORQUE, AND ANGULAR MOMENTUM

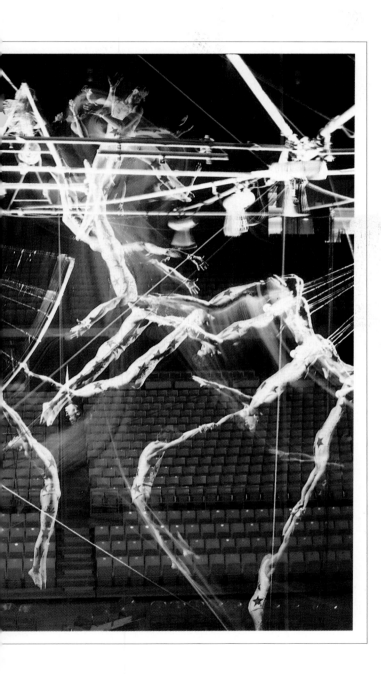

In 1897, a European "aerialist" made the first triple somersault during the flight from a swinging trapeze to the hands of a partner. For the next 85 years aerialists attempted to complete a <u>quadruple</u> somersault, but not until 1982 was it done before an audience: Miguel Vazquez of the Ringling Bros. and Barnum & Bailey Circus rotated his body in four complete circles in midair before his brother Juan caught him. Both were stunned by their success. Why was the feat so difficult, and what feature of physics made it (finally) possible?

12-1 ROLLING

When a bicycle moves along a straight track, the center of each wheel moves forward in pure translation. A point on the rim of the wheel, however, traces out a more complex path, as Fig. 12-1 shows. In what follows, we analyze the motion of a rolling wheel first by viewing it as a combination of pure translation and pure rotation, and then by viewing it as rotation alone.

Rolling as Rotation and Translation Combined

Imagine that you are watching the wheel of a bicycle, which passes you at constant speed while rolling smoothly, without slipping, along a street. As shown in Fig. 12-2, the center of mass O of the wheel moves forward at constant speed v_{cm}. The point P at which the wheel contacts the street also moves forward at speed v_{cm}, so that it always remains directly below O.

During a time interval t, you see both O and P move forward by a distance s. The bicycle rider sees the wheel rotate through an angle θ about the center of the wheel, with the point of the wheel that was touching the street at the beginning of t moving through arc length s. Equation 11-15 relates the arc length s to the rotation angle θ:

$$s = R\theta, \tag{12-1}$$

where R is the radius of the wheel. The linear speed v_{cm} of the center of the wheel (the center of mass of this uniform wheel) is ds/dt, and the angular speed ω of the wheel about its center is $d\theta/dt$. So, differentiating Eq. 12-1 with respect to time gives us

$$v_{cm} = \omega R \quad \text{(rolling motion)}. \tag{12-2}$$

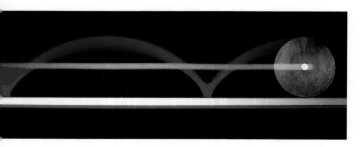

FIGURE 12-1   A time exposure photograph of a rolling disk. Small lights have been attached to the disk, one at its center and one at its edge. The latter traces out a curve called a *cycloid*.

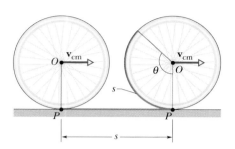

FIGURE 12-2   The center of mass O of a rolling wheel moves a distance s at velocity $\mathbf{v}_{cm}$, while the wheel rotates through angle θ. The contact point P between the wheel and the surface over which the wheel rolls also moves a distance s.

(Note that Eq. 12-2 holds only if the wheel rolls smoothly; that is, it does not slip over the street.)

Figure 12-3 shows that the rolling motion of a wheel is a combination of purely translational and purely rotational motions. Figure 12-3a shows the purely rotational motion (as if the rotation axis through the center were stationary): every point on the wheel rotates about the center with angular speed ω. (This is the type of motion we considered in Chapter 11.) Every point on the outside edge of the wheel has linear speed v_{cm} given by Eq. 12-2. Figure 12-3b shows the purely translational motion (as if the wheel did not rotate at all): every point on the wheel moves to the right with speed v_{cm}.

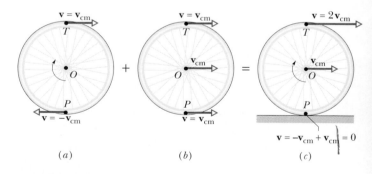

(a)   (b)   (c)

FIGURE 12-3   Rolling motion of a wheel as a combination of purely rotational motion and purely translational motion. (a) The purely rotational motion: all points on the wheel move with the same angular speed ω. Points on the outside edge of the wheel all move with the same linear speed $v = v_{cm}$. The linear velocities $\mathbf{v}$ of two such points, at top (T) and bottom (P) of the wheel, are shown. (b) The purely translational motion: all points on the wheel move to the right with the same linear velocity $\mathbf{v}_{cm}$ as the center of the wheel. (c) The rolling motion of the wheel is the combination of (a) and (b).

The combination of Figs. 12-3*a* and 12-3*b* yields the actual rolling motion of the wheel, Fig. 12-3*c*. Note that, in this combination of motions, the portion of the wheel at the bottom (at point *P*) is stationary, and the portion at the top (at point *T*) is moving at speed $2v_{cm}$, faster than any other portion of the wheel. These results are demonstrated in Fig. 12-4, which is a time exposure of a rolling bicycle wheel. The blurring of the spokes near the top of the wheel compared with the sharper images of the spokes near the bottom of the wheel shows that the wheel is moving faster near its top than near its bottom.

The motion of any round body rolling smoothly over a surface can be separated into purely translational and purely rotational motions, as in Figs. 12-3*a* and 12-3*b*.

Rolling as Pure Rotation

Figure 12-5 suggests another way to look at a rolling wheel, namely, as pure rotation about an axis that always extends through the point where the wheel contacts the street as the wheel moves, that is, about an axis passing through point *P* in Fig. 12-3*c* and perpendicular to the plane of the figure. The vectors in Fig. 12-5 represent the instantaneous velocities of various points on the rolling wheel.

Question. What angular speed about this new axis will a stationary observer assign to a rolling bicycle wheel?

Rotational axis at *P*

FIGURE 12-5 Rolling can be viewed as pure rotation, with angular speed ω, about an axis that always extends through *P*. The vectors show the instantaneous linear velocities of selected points on the rolling wheel. You can obtain the vectors by combining the translational and rotational motions as in Fig. 12-3.

Answer. The same angular speed ω that the rider assigns to the wheel as she or he observes it in pure rotation about an axis through its center of mass.

Let us use this answer to calculate the linear speed of the top of the rolling wheel from the point of view of a stationary observer. If we call the wheel's radius *R*, the top is a distance $2R$ from the axis through *P* in Fig. 12-5, so its linear speed should be (using Eq. 12-2)

$$v_{\text{top}} = (\omega)(2R) = 2(\omega R) = 2v_{\text{cm}},$$

in exact agreement with Fig. 12-3*c*. You can similarly verify the linear speeds shown for the portion of the wheel at points *O* and *P* in Fig. 12-3*c*.

The Kinetic Energy

Let us calculate the kinetic energy of the rolling wheel as measured by the stationary observer. If we view the rolling as pure rotation about an axis through *P* in Fig. 12-5, we have

$$K = \tfrac{1}{2}I_P\omega^2, \tag{12-3}$$

in which ω is the angular speed of the wheel, and I_P is the rotational inertia of the wheel about the axis through *P*. From the parallel-axis theorem (Eq. 11-25) we have

$$I_P = I_{\text{cm}} + MR^2, \tag{12-4}$$

in which *M* is the mass of the wheel, and I_{cm} is its rotational inertia about an axis through its center of mass. Substituting Eq. 12-4 into Eq. 12-3, we obtain

$$K = \tfrac{1}{2}I_{\text{cm}}\omega^2 + \tfrac{1}{2}MR^2\omega^2,$$

FIGURE 12-4 A photograph of a rolling bicycle wheel. The spokes near the top of the wheel are more blurred than those near the bottom of the wheel because they are moving faster, as Fig. 12-3*c* shows.

As this hoop rolls, does the dog walk as fast as the pony?

and using the relation $v_{cm} = \omega R$ (Eq. 12-2) yields

$$K = \tfrac{1}{2}I_{cm}\omega^2 + \tfrac{1}{2}Mv_{cm}^2. \qquad (12\text{-}5)$$

We can interpret the first of these terms ($\tfrac{1}{2}I_{cm}\omega^2$) as the kinetic energy associated with the rotation of the wheel about an axis through its center of mass (as in Fig. 12-3a), and the second term ($\tfrac{1}{2}Mv_{cm}^2$) as the kinetic energy associated with the translational motion of the wheel (as in Fig. 12-3b).

The Role of Friction

If the wheel rolls at constant speed, as in Fig. 12-2, there is no tendency for the wheel to slide at the point of contact P, and thus there is no frictional force acting on the wheel there. However, if a force acts on the wheel, changing the speed v_{cm} of the center of the wheel or the angular speed ω about the center, then there is a tendency for the wheel to slide at P, and a frictional force acts on the wheel, at the portion in contact, to oppose that tendency. Until the wheel actually begins to slide, the frictional

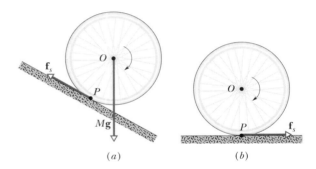

FIGURE 12-6 (*a*) A wheel rolls down an incline without sliding. A static frictional force $\mathbf{f}_s$ acts on the wheel at P, opposing the wheel's tendency to slide because of its weight $M\mathbf{g}$. (*b*) A wheel rolls horizontally without sliding while its angular speed is increased. A static frictional force $\mathbf{f}_s$ acts on the wheel at P, opposing the tendency to slide. If in (*a*) or (*b*) the wheel slides, the frictional force is a kinetic frictional force $\mathbf{f}_k$.

force is a *static* frictional force $\mathbf{f}_s$. If the wheel begins to slide, then the force is a *kinetic* frictional force $\mathbf{f}_k$.

In Fig. 12-6a, a wheel rolls down an incline. Its weight $M\mathbf{g}$ acts on the wheel at the wheel's center of mass. Since that force $M\mathbf{g}$ lacks any moment arm relative to the center of the wheel, it cannot produce a torque about the center and cause the wheel to rotate about the center. But since $M\mathbf{g}$ tends to slide the wheel down the incline, a frictional force acts on the wheel at P, the portion of the wheel in contact with the incline, to oppose the tendency to slide. This force, which points up along the incline, *does* have a moment arm relative to the center; the moment arm is the radius of the wheel. Thus this frictional force produces a torque about the center and causes the wheel to rotate about the center.

In Fig. 12-6b, a wheel is being made to rotate faster while rolling along a flat surface, as on an accelerating bicycle. The increase in ω tends to slide the bottom of the wheel toward the left. A frictional force acts on the wheel toward the right at P to oppose the tendency to slide. (This frictional force is the external force acting on an accelerating bicycle–rider system that causes the system to accelerate.)

SAMPLE PROBLEM 12-1

A uniform solid cylindrical disk, whose mass M is 1.4 kg and whose radius R is 8.5 cm, rolls across a horizontal table at a speed v of 15 cm/s.

a. What is the instantaneous velocity of the top of the rolling disk?

SOLUTION When we speak of the speed of a rolling object, we always mean the speed of its center. From Fig. 12-3c we see that the speed of the top of the disk is just twice this, or

$$v_{top} = 2v_{cm} = (2)(15 \text{ cm/s}) = 30 \text{ cm/s}. \quad \text{(Answer)}$$

b. What is the angular speed ω of the rolling disk?

SOLUTION From Eq. 12-2 we have

$$\omega = \frac{v_{cm}}{R} = \frac{15 \text{ cm/s}}{8.5 \text{ cm}} = 1.8 \text{ rad/s}$$

$$= 0.28 \text{ rev/s}. \quad \text{(Answer)}$$

The value applies whether the axis of rotation is taken to be an axis through point P in Fig. 12-5 or an axis through the center of mass.

c. What is the kinetic energy K of the rolling disk?

SOLUTION From Eq. 12-5 we have, putting $I_{cm} = \frac{1}{2}MR^2$ and using the relation $v_{cm} = \omega R$,

$$K = \frac{1}{2}I_{cm}\omega^2 + \frac{1}{2}Mv_{cm}^2$$

$$= (\frac{1}{2})(\frac{1}{2}MR^2)(v_{cm}/R)^2 + \frac{1}{2}Mv_{cm}^2 = \frac{3}{4}Mv_{cm}^2$$

$$= \frac{3}{4}(1.4 \text{ kg})(0.15 \text{ m/s})^2$$

$$= 0.024 \text{ J} = 24 \text{ mJ}. \quad \text{(Answer)}$$

d. What fraction of the kinetic energy is associated with the motion of translation and what fraction with the motion of rotation about an axis through the center of mass?

SOLUTION The kinetic energy associated with translation is the second term of Eq. 12-5, or $\frac{1}{2}Mv_{cm}^2$. The fraction we seek is then, using the expression derived in (c) above,

$$\text{frac} = \frac{\frac{1}{2}Mv_{cm}^2}{\frac{3}{4}Mv_{cm}^2} = \frac{2}{3} \text{ or } 67\%. \quad \text{(Answer)}$$

The remaining 33% is associated with rotation about an axis through the center of mass.

The relative split between translational and rotational energy depends on the rotational inertia of the rolling object. As Table 12-1 shows, the rolling object (the hoop) that has its mass farthest from the central axis of rotation (and so has the largest rotational inertia) has the largest share of its kinetic energy in rotational motion. The object (the sphere) that has its mass closest to the central axis of rotation (and so has the smallest rotational inertia) has the smallest share in that form.

TABLE 12-1

THE RELATIVE SPLITS BETWEEN ROTATIONAL AND TRANSLATION ENERGIES FOR ROLLING OBJECTS

OBJECT	ROTATIONAL INERTIA I_{cm}	PERCENTAGE OF ENERGY IN	
		TRANSLATION	ROTATION
Hoop	$1MR^2$	50%	50%
Disk	$\frac{1}{2}MR^2$	67%	33%
Sphere	$\frac{2}{5}MR^2$	71%	29%
General[a]	βMR^2	$100\dfrac{1}{1+\beta}$	$100\dfrac{\beta}{1+\beta}$

[a]β may be computed for any rolling object as I_{cm}/MR^2.

The formulas at the bottom of Table 12-1 apply to the generic rolling object with **rotational inertia parameter** β. This parameter has the value 1 for a hoop, $\frac{1}{2}$ for a disk, and $\frac{2}{5}$ for a sphere.

SAMPLE PROBLEM 12-2

A bowling ball, whose radius R is 11 cm and whose mass M is 7.2 kg, rolls from rest down a ramp whose length L is 2.1 m. The ramp is inclined at an angle θ of 34° to the horizontal; see the sphere in Fig. 12-7. How fast is the ball moving when it reaches the bottom of the ramp? Assume the ball is uniform in density.

SOLUTION The center of the ball falls a vertical distance $h = L \sin \theta$, so that the ball loses a gravitational potential energy of $MgL \sin \theta$. This loss of potential energy equals the gain in kinetic energy. Thus we can write (see Eq. 12-5)

$$MgL \sin \theta = \frac{1}{2}I_{cm}\omega^2 + \frac{1}{2}Mv_{cm}^2. \quad (12\text{-}6)$$

FIGURE 12-7 Sample Problems 12-2 and 12-3. A hoop, a disk, and a sphere roll from rest down a ramp of angle θ. Although released from rest at the same position and time, they arrive at the bottom in the order shown.

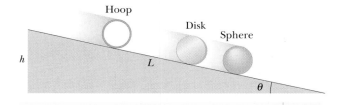

From Table 11-2(g) we see that, for a solid sphere, $I_{cm} = \frac{2}{5}MR^2$. We can also replace ω with its equal, v_{cm}/R. Substituting both these quantities into Eq. 12-6 yields

$$MgL \sin \theta = (\tfrac{1}{2})(\tfrac{2}{5})(MR^2)(v_{cm}/R)^2 + \tfrac{1}{2}Mv_{cm}^2.$$

Solving for v_{cm} yields

$$v_{cm} = \sqrt{(\tfrac{10}{7})gL \sin \theta}$$
$$= \sqrt{(\tfrac{10}{7})(9.8 \text{ m/s}^2)(2.1 \text{ m})(\sin 34°)}$$
$$= 4.1 \text{ m/s.} \qquad \text{(Answer)}$$

Note that the answer does not depend on the mass or radius of the ball.

SAMPLE PROBLEM 12-3

Here we generalize the result of Sample Problem 12-2. A uniform hoop, disk, and sphere, having the same mass M and the same radius R, are released simultaneously from rest at the top of a ramp whose length L is 2.5 m and whose ramp angle θ is 12° (Fig. 12-7).

a. Which object reaches the bottom first?

SOLUTION Table 12-1 gives us the answer. The sphere puts the largest share of its kinetic energy (71%) into translational motion, so it wins the race. Next comes the disk and then the hoop.

b. How fast are the objects moving at the bottom of the ramp?

SOLUTION In rolling down the ramp, the center of mass of each object falls the same vertical distance h. Like a body in free fall, the object loses potential energy in amount Mgh and thus gains this amount of kinetic energy. At the bottom of the ramp then, the total kinetic energies of all three objects are the same. How these kinetic energies are divided between the translational and rotational forms depends on each object's distribution of mass.

From Eq. 12-5 we can write (putting $\omega = v_{cm}/R$)

$$Mgh = \tfrac{1}{2}I_{cm}\omega^2 + \tfrac{1}{2}Mv_{cm}^2$$
$$= \tfrac{1}{2}I_{cm}(v_{cm}^2/R^2) + \tfrac{1}{2}Mv_{cm}^2$$
$$= \tfrac{1}{2}(I_{cm}/R^2)v_{cm}^2 + \tfrac{1}{2}Mv_{cm}^2. \qquad (12\text{-}7)$$

Putting $h = L \sin \theta$ and solving for v_{cm}, we obtain

$$v_{cm} = \sqrt{\frac{2gL \sin \theta}{1 + I_{cm}/MR^2}}, \qquad \text{(Answer)} \quad (12\text{-}8)$$

which is the symbolic answer to the question.

Note that the speed depends not on the mass or the radius of the rolling object, but only on the distribution of its mass about its central axis, which enters through the term I_{cm}/MR^2. A marble and a bowling ball will have the same speed at the bottom of the ramp and will thus roll down the ramp in the same time. A bowling ball will beat a disk of any mass or radius, and almost anything will beat a hoop. (An exception is shown in Question 8.)

For the rolling hoop (see Table 12-1) we have $I_{cm}/MR^2 = 1$, so Eq. 12-8 yields

$$v_{cm} = \sqrt{\frac{2gL \sin \theta}{1 + I_{cm}/MR^2}}$$
$$= \sqrt{\frac{(2)(9.8 \text{ m/s}^2)(2.5 \text{ m})(\sin 12°)}{1 + 1}}$$
$$= 2.3 \text{ m/s.} \qquad \text{(Answer)}$$

A similar calculation yields $v_{cm} = 2.6$ m/s for the disk ($I_{cm}/MR^2 = \frac{1}{2}$) and 2.7 m/s for the sphere ($I_{cm}/MR^2 = \frac{2}{5}$). This supports our prediction of (a) above that the win, place, and show sequence in this race will be sphere, disk, and hoop.

SAMPLE PROBLEM 12-4

Figure 12-8 shows a round uniform body of mass M and radius R rolling down a ramp that slopes at an angle θ. This time let us analyze the motion directly from Newton's laws, rather than by energy methods as we did in Sample Problem 12-3.

a. What is the linear acceleration of the rolling body?

SOLUTION Figure 12-8 also shows the forces that act on the body: its weight $M\mathbf{g}$, a normal force $\mathbf{N}$, and a static frictional force $\mathbf{f}_s$. The weight can be considered to act at the center of mass, which is at the center of this uniform body. The normal force and the frictional force act on the portion of the body in contact with the ramp at point P. The weight and normal force have zero moment arms about an axis through the center of the body. So they cannot cause the body to rotate about that center. Clockwise rotation of the body results from a negative torque due to the frictional force; that force has a moment arm of R about the center of the body.

We now apply the linear form of Newton's second law ($\Sigma F = Ma$) along the ramp, taking the positive direction to be up the ramp. We obtain

$$\Sigma F = f_s - Mg \sin \theta = Ma. \qquad (12\text{-}9)$$

This equation has two unknowns, f_s and a. To obtain another equation in these same two unknowns, we next apply the angular form of Newton's second law ($\Sigma \tau =$

$I\alpha$) about the rotation axis through the center of mass. (Although we derived the relation $\Sigma\ \tau = I\alpha$ in Chapter 11 for an axis fixed in an inertial frame, it holds also for a rotation axis through the center of mass of this or another accelerating body, provided that the axis does not change direction.) We get

$$\Sigma\ \tau = -f_s R = I_{cm}\alpha = I_{cm}\frac{a}{R}, \qquad (12\text{-}10)$$

where we employ the relation $\alpha = a/R$ (Eq. 11-18).

Solving Eq. 12-10 for the frictional force f_s gives

$$f_s = -\frac{I_{cm}}{R^2}a, \qquad (12\text{-}11)$$

where the minus sign reminds us that the frictional force $\mathbf{f}_s$ acts in the direction opposite that of the acceleration $\mathbf{a}$. Substituting Eq. 12-11 into Eq. 12-9 and solving for a yield

$$a = -\frac{g\sin\theta}{1 + I_{cm}/MR^2}. \qquad \text{(Answer)} \quad (12\text{-}12)$$

We could have, instead, found a second equation by summing torques and applying Newton's law in angular form about an axis through the *point of contact P*. This time, $\Sigma\ \tau$ would consist only of the torque due to the force component $Mg\sin\theta$, acting at the center of the body with moment arm R:

$$\Sigma\ \tau = -(Mg\sin\theta)(R) = I_P\alpha = I_P\frac{a}{R}, \qquad (12\text{-}13)$$

where I_P is the rotational inertia about an axis through P. To find I_P, we would use the parallel-axis theorem:

$$I_P = I_{cm} + MR^2. \qquad (12\text{-}14)$$

Substituting I_P from Eq. 12-14 in Eq. 12-13 and solving for a, we would again obtain Eq. 12-12.

b. What is the frictional force f_s?

SOLUTION Substituting Eq. 12-12 into Eq. 12-11 yields

$$f_s = Mg\frac{\sin\theta}{1 + MR^2/I_{cm}}. \qquad \text{(Answer)} \quad (12\text{-}15)$$

Study of Eq. 12-15 shows that the frictional force is less than $Mg\sin\theta$, the component of the weight that acts parallel to the ramp. This is necessarily true if the object is to accelerate down the ramp.

Table 12-1 shows that if the rolling body is a solid disk, $I_{cm}/MR^2 = \frac{1}{2}$. The acceleration and the frictional force then follow from Eqs. 12-12 and 12-15 and are

$$a = -\tfrac{2}{3}g\sin\theta \quad \text{and} \quad f_s = \tfrac{1}{3}Mg\sin\theta.$$

c. What is the speed of the rolling body at the bottom of the ramp if the ramp has length L?

SOLUTION The motion is one of constant acceleration, so we can use the relation

$$v^2 = v_0^2 + 2a(x - x_0). \qquad (12\text{-}16)$$

Putting $x - x_0 = -L$ and $v_0 = 0$ and introducing a from Eq. 12-12, we obtain Eq. 12-8—the result that we derived by energy methods. You should not be surprised because, after all, everything that we have said about mechanical energy is consistent with Newton's laws.

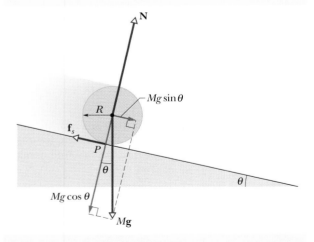

FIGURE 12-8 Sample Problem 12-4. A round uniform body of radius R rolls down a ramp. The forces that act on it are its weight $M\mathbf{g}$, a normal force $\mathbf{N}$, and a frictional force $\mathbf{f}_s$ pointing up the ramp. (For clarity, $\mathbf{N}$ has been shifted along its line of action until its tail is at the center of the body.)

12-2 THE YO-YO

A yo-yo is a physics lab that you can fit in your pocket. If a yo-yo rolls down its string for a distance h, it loses potential energy in amount mgh but gains kinetic energy in both translational ($\frac{1}{2}mv_{cm}^2$) and rotational ($\frac{1}{2}I_{cm}\omega^2$) forms. When it is climbing back up, it loses kinetic energy and regains potential energy.

In a modern yo-yo the string is not tied to the axle but is looped around it. When the yo-yo reaches the bottom of its string there is a slight bounce that removes the small remaining translational kinetic energy. The yo-yo then spins, axle inside loop, with only rotational kinetic energy. The yo-yo keeps spinning ("sleeping") until you "wake it" by jerking on the string, causing the string to catch on the axle and the yo-yo to climb back up. The rotational kinetic energy of the yo-yo at the bottom of its string (and thus the sleeping time) can be considerably increased by throwing the yo-yo downward so that it starts down the string with some initial speeds v_{cm} and ω instead of rolling down from rest.

Let us analyze the motion of the yo-yo directly from Newton's second law. Figure 12-9a shows an idealized yo-yo, in which the thickness of the string can be neglected.* Figure 12-9b shows a free-body diagram, in which only the yo-yo axle is shown. Applying Newton's second law in its linear form ($\Sigma F = ma$) yields

$$\sum F = T - Mg = Ma. \qquad (12\text{-}17)$$

Here M is the mass of the yo-yo and T is the tension in the cord.

Applying Newton's second law in angular form ($\Sigma \tau = I\alpha$) about an axis through the center of mass yields

$$\sum \tau = TR_0 = I\alpha, \qquad (12\text{-}18)$$

where R_0 is the radius of the yo-yo axle, and I is the rotational inertia of the yo-yo about its central axis. The linear acceleration a of the yo-yo is downward (and thus negative). From the perspective of Fig. 12-9, the angular acceleration α of the yo-yo is counterclockwise (and thus positive) because from that perspective the torque given by Eq. 12-18 is counterclockwise. We relate α to a with the relation $a = -\alpha R_0$. Solving this relation for α ($= -a/R_0$) and substituting that into Eq. 12-18, we find

$$TR_0 = -\frac{Ia}{R_0}.$$

After eliminating T between this and Eq. 12-17, we solve for a, obtaining

$$a = -g\,\frac{1}{1 + I/MR_0^2}. \qquad (12\text{-}19)$$

Thus an ideal yo-yo rolls down its string with constant acceleration. For a small acceleration, you need a light yo-yo with a large rotational inertia and a small axle radius.

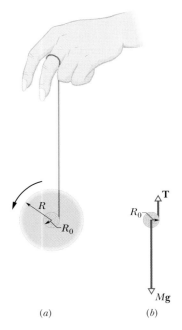

(a)   (b)

FIGURE 12-9 (a) A yo-yo, shown in cross section. The string, of assumed negligible thickness, is wound around an axle of radius R_0. (b) A free-body diagram for the falling yo-yo. Only the axle is shown.

SAMPLE PROBLEM 12-5

A yo-yo is constructed of two brass disks whose thickness b is 8.5 mm and whose radius R is 3.5 cm, joined by a short axle whose radius R_0 is 3.2 mm.

a. What is its rotational inertia about its central axis? Neglect the rotational inertia of the axle. The density ρ of brass is 8400 kg/m³.

SOLUTION The rotational inertia I of a disk about its central axis is $\frac{1}{2}MR^2$. In this problem, we can treat the two disks together, as a single disk. We first find its mass from

$$M = V\rho = (2)(\pi R^2)(b)(\rho)$$

$$= (2)(\pi)(0.035\text{ m})^2(0.0085\text{ m})(8400\text{ kg/m}^3)$$

$$= 0.550\text{ kg}.$$

The rotational inertia is then

$$I = \tfrac{1}{2}MR^2 = (\tfrac{1}{2})(0.550\text{ kg})(0.035\text{ m})^2$$

$$= 3.4 \times 10^{-4}\text{ kg}\cdot\text{m}^2. \qquad \text{(Answer)}$$

---

*In a real yo-yo, the thickness of the string cannot be neglected. It changes the effective radius of the yo-yo axle, which then varies with the amount of wound-up string.

b. A string of length $l = 1.1$ m and of negligible thickness is wound on the axle. What is the linear acceleration of the yo-yo as it rolls down the string from rest?

SOLUTION From Eq. 12-19,

$$a = -g \frac{1}{1 + I/MR_0^2}$$

$$= -\frac{9.8 \text{ m/s}^2}{1 + \frac{3.4 \times 10^{-4} \text{ kg} \cdot \text{m}^2}{(0.550 \text{ kg})(0.0032 \text{ m})^2}}$$

$$= -0.16 \text{ m/s}^2. \quad \text{(Answer)}$$

The acceleration points downward and has this value whether the yo-yo is rolling down the string or climbing it.

Note that the quantity I/MR_0^2 in Eq. 12-19 is simply the rotational inertia parameter β that was introduced in Table 12-1. For this yo-yo, we have $\beta = 60$, a much greater value than that for any of the objects listed in that table. The acceleration of our yo-yo is small, corresponding to that of a hoop rolling down a 1.9° ramp.

c. What is the tension in the string of the yo-yo?

SOLUTION We can find this by substituting a from Eq. 12-19 into Eq. 12-17 and solving for T. We find

$$T = \frac{Mg}{1 + MR_0^2/I}, \quad (12\text{-}20)$$

which tells us, as it must, that the tension in the string is smaller than the weight of the yo-yo. Numerically, we have

$$T = \frac{(0.550 \text{ kg})(9.8 \text{ m/s}^2)}{1 + (0.550 \text{ kg})(0.0032 \text{ m})^2/(3.4 \times 10^{-4} \text{ kg} \cdot \text{m}^2)}$$

$$= 5.3 \text{ N}. \quad \text{(Answer)}$$

This tension holds whether the yo-yo is falling or climbing.

12-3 TORQUE REVISITED

In Chapter 11 we defined torque τ for a rigid body that can rotate around a fixed axis, with each particle in the body constrained to move along a circle about that axis. We shall now generalize the definition of torque so that it may be applied to a *particle* (rather than just a rigid body) that moves relative to a fixed *point* (rather than a fixed axis) that is usually taken to be an origin. The path taken by the particle need not be a circle about that point, and need not even lie in a plane.

Figure 12-10 shows a particle at point P in the xy plane. Its position with respect to the origin O is specified by position vector $\mathbf{r}$. A single force $\mathbf{F}$ lying in the xy plane acts on the particle, vector $\mathbf{F}$ and an extension of vector $\mathbf{r}$ making an angle ϕ with each other.

The torque exerted on the particle by this force, with respect to the origin O, is a vector quantity defined as

$$\boldsymbol{\tau} = \mathbf{r} \times \mathbf{F} \quad \text{(torque defined).} \quad (12\text{-}21)$$

According to the rules for a vector product (see Fig. 3-19), $\boldsymbol{\tau}$ is perpendicular to the plane containing $\mathbf{r}$ and $\mathbf{F}$. Thus, the vector $\boldsymbol{\tau}$ in Fig. 12-10 is parallel to the z axis, in the direction of increasing z. In applying the right-hand rule to find this direction, it helps to slide the vector $\mathbf{F}$, without changing its direction, until its tail is at the origin O, as the pale purple vector in Fig. 12-10 shows. The tail of torque $\boldsymbol{\tau}$ is also at O.

Kayaking in "white water." If the kayaker rolls over, to avoid drowning he must extend his body and the paddle in the direction of the roll and thrust the paddle toward the bottom of the stream. Then the force exerted by the water on the paddle produces a torque that rights him.

(a)

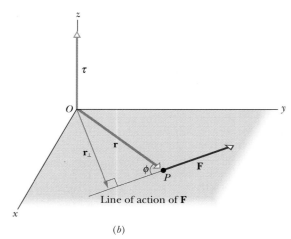

(b)

FIGURE 12-10 Defining torque. A force **F**, lying in the xy plane, acts on a particle at point P. This force exerts a torque $\boldsymbol{\tau}$ $(= \mathbf{r} \times \mathbf{F})$ on the particle with respect to the origin O. The torque vector points in the direction of increasing z. Its magnitude is given by $rF_\perp$ in (a) and by $r_\perp F$ in (b).

The magnitude of the vector $\boldsymbol{\tau}$ is given (see Eq. 3-20) by

$$\tau = rF \sin \phi. \tag{12-22}$$

Equation 12-22 can be rewritten as

$$\tau = rF_\perp, \tag{12-23}$$

where $F_\perp$ is the component of **F** perpendicular to **r** (Fig. 12-10a). Equation 12-22 can also be rewritten as

$$\tau = r_\perp F, \tag{12-24}$$

where $r_\perp$ (the moment arm of **F**) is the perpendicular distance between O and an extension (the line of action) of **F** (Fig. 12-10b).

Note that if the angle between the force **F** and the position vector **r** is 0 or 180°, then Eq. 12-22 tells us that the torque is zero. We get the same result with Eq. 12-23 (in which the perpendicular component $F_\perp = 0$) and with Eq. 12-24 (in which the moment arm $r_\perp = 0$).

Equations 12-22 to 12-24 agree with our earlier definition of torque, in which we treated only the special case of a force acting on a rigid body constrained to rotate about a fixed axis. (See Eqs. 11-28 to 11-30.) Recall that we identified the torque on a rigid body as being the *turning action* of an applied force: the torque tends to rotate (or turn) the body; that is, a position vector locating any portion of the body rotates about a fixed axis. Similarly, in Fig. 12-10, the torque acting on the particle tends to rotate the particle's position vector **r** about the origin.

SAMPLE PROBLEM 12-6

In Fig. 12-11a, three forces, each of magnitude 2.0 N, act on a particle. The particle is in the xz plane at point P given by position vector **r**, where $r = 3.0$ m and $\theta = 30°$. Force $\mathbf{F}_1$ is parallel to the x axis, force $\mathbf{F}_2$ is parallel to the z axis, and force $\mathbf{F}_3$ is parallel to the y axis. What is the torque, with respect to the origin O, due to each force?

SOLUTION Figures 12-11b and 12-11c are straight-on views of the xz plane, redrawn with vectors $\mathbf{F}_1$ and $\mathbf{F}_2$ shifted so their tails are at the origin, to better show the angles between those vectors and vector **r**. The angle between **r** and $\mathbf{F}_3$ is 90°. Applying Eq. 12-22 for each force, we find the magnitudes of the torques to be

$$\tau_1 = rF_1 \sin \phi_1 = (3.0 \text{ m})(2.0 \text{ N})(\sin 150°)$$

$$= 3.0 \text{ N·m},$$

$$\tau_2 = rF_2 \sin \phi_2 = (3.0 \text{ m})(2.0 \text{ N})(\sin 120°)$$

$$= 5.2 \text{ N·m},$$

and

$$\tau_3 = rF_3 \sin \phi_3 = (3.0 \text{ m})(2.0 \text{ N})(\sin 90°)$$

$$= 6.0 \text{ N·m}. \tag{Answer}$$

To find the directions of these torques, we use the right-hand rule, placing the fingers of the right hand so as to rotate **r** into **F** through the *smaller* of the two angles they make. Torque $\boldsymbol{\tau}_3$ is perpendicular to **r** and $\mathbf{F}_3$ (Fig. 12-11d) and at an angle of $\theta = 30°$ from the

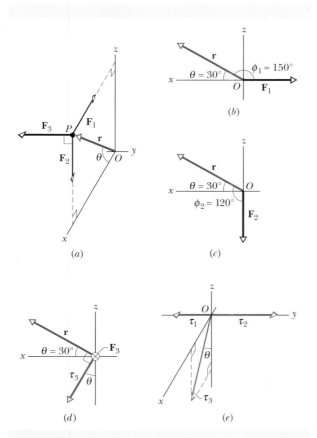

(a)

(b)

(c)

(d)

(e)

FIGURE 12-11 Sample Problem 12-6. (*a*) A particle at point *P* is acted on by three forces, each parallel to a coordinate axis. The angle ϕ (used in finding torque) is shown (*b*) for $\mathbf{F}_1$ and (*c*) for $\mathbf{F}_2$. (*d*) Torque τ_3 is perpendicular to both **r** and $\mathbf{F}_3$ (the $\otimes$ symbol indicates that $\mathbf{F}_3$ is into and perpendicular to the plane of the figure). (*e*) The torques (relative to the origin *O*) acting on the particle.

direction of decreasing *z*. In Fig. 12-11*d*, we represent $\mathbf{F}_3$ with a circled cross $\otimes$, suggesting the tail of an arrow. (Were it in the opposite direction, $\mathbf{F}_3$ would be represented by a circled dot $\odot$, suggesting the tip of an arrow.) All three of these torque vectors are shown in Fig. 12-11*e*.

12-4 ANGULAR MOMENTUM

Like all linear quantities, linear momentum has its angular counterpart. Figure 12-12 shows a particle with linear momentum **p** (= *m***v**) located at point *P* in the *xy* plane. The **angular momentum** ℓ of this particle with respect to the origin *O* is a vector quan-

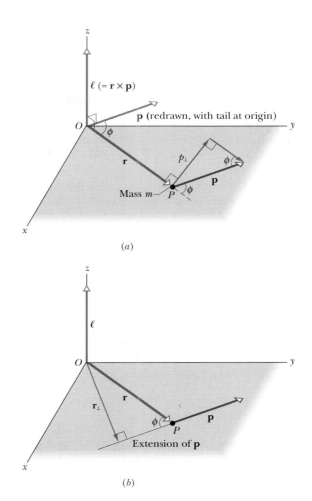

(a)

(b)

FIGURE 12-12 Defining angular momentum. A particle of mass *m* at point *P* has linear momentum **p** (= *m***v**), assumed to lie in the *xy* plane. The particle has angular momentum ℓ (= **r** × **p**) with respect to the origin *O*. The angular momentum vector points in the direction of increasing *z*. (*a*) The magnitude of ℓ is given by $\ell = rp_\perp = rmv_\perp$. (*b*) The magnitude of ℓ is also given by $\ell = r_\perp p = r_\perp mv$.

tity defined as

$$\ell = \mathbf{r} \times \mathbf{p} = m(\mathbf{r} \times \mathbf{v})$$
(angular momentum defined),   (12-25)

where **r** is the position vector of the particle with respect to *O*. As the particle moves relative to *O*, in the direction of its momentum **p** (= *m***v**), position vector **r** rotates around *O*. To have angular momentum, the particle does not itself have to rotate around *O*. Comparison of Eqs. 12-21 and 12-25 shows that angular momentum bears the same relation to linear momentum that torque does to force. The SI unit of angular momentum is the kilogram-meter-squared per second $(kg \cdot m^2/s)$, equivalent to the joule-second $(J \cdot s)$.

The angular momentum vector ℓ in Fig. 12-12 is parallel to the z axis, and it points in the direction of increasing z. Thus ℓ is positive, consistent with the counterclockwise rotation of the particle's position vector $\mathbf{r}$ around the z axis. (A negative ℓ, consistent with a clockwise rotation of $\mathbf{r}$ around the z axis, would point in the direction of decreasing z.)

The magnitude of ℓ is given by

$$\ell = rmv \sin \phi, \qquad (12\text{-}26)$$

where ϕ is the angle between $\mathbf{r}$ and $\mathbf{p}$. Equation 12-26 can be rewritten as

$$\ell = rp_\perp = rmv_\perp, \qquad (12\text{-}27)$$

where $p_\perp$ is the component of $\mathbf{p}$ perpendicular to $\mathbf{r}$ (as in Fig. 12-12a), and $p_\perp = mv_\perp$. Equation 12-26 can also be rewritten as

$$\ell = r_\perp p = r_\perp mv, \qquad (12\text{-}28)$$

where $r_\perp$ is the perpendicular distance between O and an extension of $\mathbf{p}$ (as in Fig. 12-12b). If a particle is moving directly away from the origin ($\phi = 0$) or directly toward it ($\phi = 180°$), Eq. 12-26 tells us that the particle has no angular momentum about that origin.

Just as is true for torque, angular momentum has meaning only with respect to a specified origin. Moreover, if the particle in Fig. 12-12 did not lie in the xy plane, or if the linear momentum $\mathbf{p}$ of the particle did not also lie in that plane, the angular momentum ℓ would not be parallel to the z axis. The direction of the angular momentum vector is always perpendicular to the plane formed by the vectors $\mathbf{r}$ and $\mathbf{p}$.

12-5 NEWTON'S SECOND LAW IN ANGULAR FORM

Newton's second law written in the form

$$\sum \mathbf{F} = \frac{d\mathbf{p}}{dt} \qquad \text{(single particle),} \qquad (12\text{-}29)$$

expresses the close relation between force and linear momentum for a single particle. We have seen enough of the parallelism between linear and angu-

lar quantities to be pretty sure that there is also a close relation between torque and angular momentum. Guided by Eq. 12-29, we can even guess that it must be

$$\sum \boldsymbol{\tau} = \frac{d\ell}{dt} \qquad \text{(single particle).} \qquad (12\text{-}30)$$

Equation 12-30 is indeed an angular form of Newton's second law for a single particle:

> The (vector) sum of all the torques acting on a particle is equal to the time rate of change of the angular momentum of that particle.

Equation 12-30 has no meaning unless the torques $\boldsymbol{\tau}$ and the angular momentum ℓ are defined with respect to the same origin.

Proof of Equation 12-30

We start with Eq. 12-25, the definition of angular momentum:

$$\ell = m(\mathbf{r} \times \mathbf{v}).$$

Differentiating* each side with respect to time t yields

$$\frac{d\ell}{dt} = m\left(\mathbf{r} \times \frac{d\mathbf{v}}{dt} + \frac{d\mathbf{r}}{dt} \times \mathbf{v}\right). \qquad (12\text{-}31)$$

But $d\mathbf{v}/dt$ is the acceleration $\mathbf{a}$ of the particle, and $d\mathbf{r}/dt$ is its velocity $\mathbf{v}$. Thus we can rewrite Eq. 12-31 as

$$\frac{d\ell}{dt} = m(\mathbf{r} \times \mathbf{a} + \mathbf{v} \times \mathbf{v}).$$

Now $\mathbf{v} \times \mathbf{v} = 0$ (the vector product of any vector with itself is zero because the angle between the two vectors is necessarily zero). This leads to

$$\frac{d\ell}{dt} = m(\mathbf{r} \times \mathbf{a}) = \mathbf{r} \times m\mathbf{a}.$$

We now use Newton's second law ($\sum \mathbf{F} = m\mathbf{a}$) to replace $m\mathbf{a}$ with its equal, the vector sum of the forces

---

*In differentiating a vector product, you must be sure not to change the order of the two quantities (here $\mathbf{r}$ and $\mathbf{v}$) that form that product. (See Eq. 3-21.)

that act on the particle, obtaining

$$\frac{d\boldsymbol{\ell}}{dt} = \mathbf{r} \times \left(\sum \mathbf{F} \right) = \sum (\mathbf{r} \times \mathbf{F}). \quad (12\text{-}32)$$

Finally, Eq. 12-21 shows us that $\mathbf{r} \times \mathbf{F}$ is the torque associated with the force $\mathbf{F}$, so Eq. 12-32 becomes

$$\sum \boldsymbol{\tau} = \frac{d\boldsymbol{\ell}}{dt}.$$

This is Eq. 12-30, the relation that we set out to prove.

SAMPLE PROBLEM 12-7

A penguin of mass m falls from rest at point A, a horizontal distance d from the origin O in Fig. 12-13.

a. What is the angular momentum of the falling penguin about O?

SOLUTION The angular momentum is given by Eq. 12-25 ($\boldsymbol{\ell} = \mathbf{r} \times \mathbf{p}$); its magnitude is (from Eq. 12-26)

$$\ell = rp \sin \phi.$$

Here, $r \sin \phi = d$ no matter how far the penguin falls, and $p = mv = m(gt)$. Thus ℓ has magnitude

$$\ell = mgtd. \quad \text{(Answer)} \quad (12\text{-}33)$$

The right-hand rule shows that the angular momentum vector $\boldsymbol{\ell}$ is directed into the plane of Fig. 12-13, in the direction of decreasing z. We represent $\boldsymbol{\ell}$ with a circled cross $\otimes$ at the origin. The vector $\boldsymbol{\ell}$ changes with time in magnitude only, its direction remaining unchanged.

b. What torque does the weight $m\mathbf{g}$ acting on the penguin exert about the origin O?

SOLUTION The torque is given by Eq. 12-21 ($\boldsymbol{\tau} = \mathbf{r} \times \mathbf{F}$); its magnitude is (from Eq. 12-22)

$$\tau = rF \sin \phi.$$

Again $r \sin \phi = d$, and $F = mg$. Thus

$$\tau = mgd = \text{a constant.} \quad \text{(Answer)} \quad (12\text{-}34)$$

Note that the torque is simply the product of the force mg and the moment arm d. The right-hand rule shows that the torque vector $\boldsymbol{\tau}$ is directed into the plane of Fig. 12-13, in the direction of decreasing z and hence parallel to $\boldsymbol{\ell}$. (Note that we can also derive Eq. 12-34 by differentiating Eq. 12-33 with respect to t and then substituting the result into Eq. 12-30.)

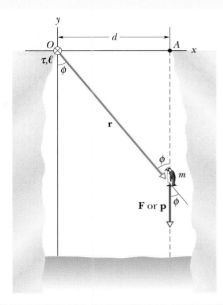

FIGURE 12-13 Sample Problem 12-7. A penguin of mass m falls vertically from point A. The torque $\boldsymbol{\tau}$ and the angular momentum $\boldsymbol{\ell}$ of the falling penguin with respect to the origin O are directed into the plane of the figure at O.

We see that $\boldsymbol{\tau}$ and $\boldsymbol{\ell}$ depend very much (through d) on the location of the origin. If the penguin falls from the origin, we have $d = 0$ and thus no torque or angular momentum.

12-6 THE ANGULAR MOMENTUM OF A SYSTEM OF PARTICLES

Now we turn our attention to the motion of a system of particles with respect to an origin. Note that "a system of particles" includes a rigid body as a special case. The total angular momentum $\mathbf{L}$ of a system of particles is the (vector) sum of the individual angular momenta $\boldsymbol{\ell}$ of the particles:

$$\mathbf{L} = \boldsymbol{\ell}_1 + \boldsymbol{\ell}_2 + \boldsymbol{\ell}_3 + \cdots + \boldsymbol{\ell}_n = \sum_{i=1}^{n} \boldsymbol{\ell}_i, \quad (12\text{-}35)$$

in which $i \, (= 1, 2, 3, \ldots)$ labels the particles.

With time, the angular momenta of individual particles may change, either because of interactions within the system (between the individual particles) or because of influences that may act on the system from the outside. We can find the change with time of the angular momentum **L** of the system as a whole as these changes take place by taking the time derivative of Eq. 12-35. Thus

$$\frac{d\mathbf{L}}{dt} = \sum_{i=1}^{n} \frac{d\boldsymbol{\ell}_i}{dt}. \qquad (12\text{-}36)$$

From Eq. 12-30, $d\boldsymbol{\ell}_i/dt$ is just $\Sigma\ \boldsymbol{\tau}_i$, the (vector) sum of the torques that act on the ith particle.

Some torques are *internal,* associated with forces that the particles within the system exert on one another; other torques are *external,* associated with forces that act from outside the system. The internal forces, because of Newton's law of action and reaction, cancel in pairs.* In adding up the torques then, we need consider only those torques associated with external forces. Equation 12-36 then becomes

$$\sum \boldsymbol{\tau}_{\text{ext}} = \frac{d\mathbf{L}}{dt} \qquad \text{(system of particles).} \quad (12\text{-}37)$$

Equation 12-37 is Newton's second law for a system of particles, expressed in terms of angular quantities, and is analogous to $\Sigma\ \mathbf{F}_{\text{ext}} = d\mathbf{P}/dt$ (Eq. 9-23). In words, Eq. 12-37 tells us that the (vector) sum of the *external torques* acting on a system of particles is equal to the time rate of change of the *angular momentum* of that system. Equation 12-37 has meaning only if the torque and angular momentum vectors are referred to the same origin. In an inertial reference frame, Eq. 12-37 can be applied with respect to any point. In an accelerating frame (such as a wheel rolling down a ramp), Eq. 12-37 can be applied *only* with respect to the *center of mass* of the system.

12-7 THE ANGULAR MOMENTUM OF A RIGID BODY ROTATING ABOUT A FIXED AXIS

We next consider angular momentum for the special case in which a system of particles forms a rotating rigid body. Figure 12-14a shows such a situation: the body is constrained to rotate about a fixed axis that is taken to be the z axis and that extends through the body. The body rotates with angular speed ω.

*We must further assume that both forces in each internal action–reaction force pair have the same line of action.

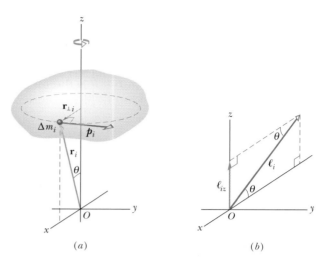

FIGURE 12-14 (*a*) A rigid body rotates about the z axis with angular speed ω. A mass element Δm_i within the body moves about the z axis in a circle with radius $r_{\perp i}$. It has linear momentum $\mathbf{p}_i$, and it is located relative to the origin O by position vector $\mathbf{r}_i$. Here Δm_i is shown when $r_{\perp i}$ is parallel to the x axis. (*b*) The angular momentum $\boldsymbol{\ell}_i$, with respect to O, of the mass element in (*a*). The z component ℓ_{iz} of $\boldsymbol{\ell}_i$ is also shown.

We can find the angular momentum of the rotating body by summing the z components of the angular momenta of the mass elements in the body. In Fig. 12-14a, a typical mass element Δm_i of the body moves around the z axis in a circular path. The position of the mass element is located relative to the

A strong breeze is blowing this performer forward and to his left. To keep from falling, he has shifted the heavy pole to his right, so that the combined center of mass of the performer-pole system remains over the rope. In that way, the moment arm of the system's weight about the rope is zero, and so is the torque due to the weight.

TABLE 12-2
**MORE CORRESPONDING RELATIONS FOR TRANSLATIONAL
AND ROTATIONAL MOTION**[a]

TRANSLATIONAL		ROTATIONAL	
Force	$\mathbf{F}$	Torque	$\boldsymbol{\tau}$ $(=\mathbf{r} \times \mathbf{F})$
Linear momentum	$\mathbf{p}$	Angular momentum	$\boldsymbol{\ell}$ $(=\mathbf{r} \times \mathbf{p})$
Linear momentum[b]	$\mathbf{P}$ $(= \Sigma\, \mathbf{p}_i)$	Angular momentum[b]	$\mathbf{L}$ $(= \Sigma\, \boldsymbol{\ell}_n)$
Linear momentum[b]	$\mathbf{P} = M\mathbf{v}_{cm}$	Angular momentum[c]	$L = I\omega$
Newton's law[b]	$\Sigma\, \mathbf{F}_{ext} = \dfrac{d\mathbf{P}}{dt}$	Newton's law[b]	$\Sigma\, \boldsymbol{\tau}_{ext} = \dfrac{d\mathbf{L}}{dt}$
Conservation law[d]	$\mathbf{P} = $ a constant	Conservation law[d]	$\mathbf{L} = $ a constant

[a] See also Table 11-3.   [b] For systems of particles, including rigid bodies.

[c] For a rigid body about a fixed axis, with L being the component along that axis.

[d] For an isolated system.

origin O by position vector $\mathbf{r}_i$. The radius of the circular path is $r_{\perp i}$, the perpendicular distance between the element and the z axis.

The magnitude of the angular momentum ℓ_i of this mass element, with respect to O, is given by Eq. 12-26:

$$\ell_i = (r_i)(p_i)(\sin 90°) = (r_i)(\Delta m_i\, v_i),$$

where p_i and v_i are the linear momentum and linear speed of the mass element, and 90° is the angle between $\mathbf{r}_i$ and $\mathbf{p}_i$. The angular momentum vector $\boldsymbol{\ell}_i$ for the mass element in Fig. 12-14a is shown in Fig. 12-14b.

We are interested in the component of $\boldsymbol{\ell}_i$ that is parallel to the rotation axis, here the z axis. That z component is

$$\ell_{iz} = \ell_i \sin \theta = (r_i \sin \theta)(\Delta m_i\, v_i) = r_{\perp i}\, \Delta m_i\, v_i.$$

The z component of the angular momentum for the rotating rigid body as a whole is found by adding up the contributions of all the mass elements that make up the body. Thus, because $v = \omega r_\perp$, we may write

$$L_z = \sum_{i=1}^{n} \ell_{iz} = \sum_{i=1}^{n} \Delta m_i\, v_i r_{\perp i} = \sum_{i=1}^{n} \Delta m_i(\omega r_{\perp i})\, r_{\perp i}$$

$$= \omega \left(\sum_{i=1}^{n} \Delta m_i\, r_{\perp i}^2 \right). \qquad (12\text{-}38)$$

We can remove ω from the summation here because it is a constant: it has the same value for all points of the rotating rigid body.

The quantity $\Sigma\, \Delta m_i\, r_\perp^2$ in Eq. 12-38 is the rotational inertia I of the body about the fixed axis (see Eq. 11-22). Thus Eq. 12-38 reduces to

$$L = I\omega \qquad \text{(rigid body, fixed axis).} \qquad (12\text{-}39)$$

We have dropped the subscript z, but you must remember that the angular momentum defined by Eq. 12-39 is only the angular momentum component that is parallel to the rotation axis. Also, I in that equation is the rotational inertia about that same axis.

Table 12-2, which supplements Table 11-3, extends our list of corresponding linear and angular relations.

SAMPLE PROBLEM 12-8

Figure 12-15 shows the Earth rotating on its axis as it orbits the sun.

FIGURE 12-15 Sample Problem 12-8. A perspective view of the Earth rotating on its axis as it moves in its orbit (assumed circular) around the sun. The angular momentum vectors are not drawn to scale; $\mathbf{L}_{orb}$ is actually about 4×10^6 times $\mathbf{L}_{rot}$.

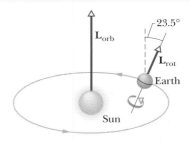

a. What angular momentum is associated with the Earth's rotation about its axis?

SOLUTION From Eq. 12-39 and Table 11-2(*g*) we have

$$L_{\text{rot}} = I\omega = \tfrac{2}{5}MR^2 \frac{2\pi \text{ rad}}{T},$$

in which M and R are the mass and the radius of the Earth and T (= 24 h = 8.64×10^4 s) is the time for one rotation of the Earth (T is the *period of rotation*). Then

$$L_{\text{rot}} = (\tfrac{2}{5})(5.98 \times 10^{24} \text{ kg})(6.37 \times 10^6 \text{ m})^2 \frac{2\pi}{8.64 \times 10^4 \text{ s}}$$

$$= 7.1 \times 10^{33} \text{ kg} \cdot \text{m}^2/\text{s}. \qquad \text{(Answer)}$$

The vector $\mathbf{L}_{\text{rot}}$ is parallel to the Earth's axis of rotation, pointing (as the right-hand rule will show) from the south pole to the north pole.

b. What angular momentum is associated with the Earth's orbital motion about the sun?

SOLUTION From Eq. 12-39, treating the Earth as a particle, we have

$$L_{\text{orb}} = I\omega = MR^2 \frac{2\pi \text{ rad}}{T},$$

in which R is now the mean Earth–sun distance and T (= 1 y = 3.16×10^7 s) is now the Earth's *period of revolution* around the sun. Then

$$L_{\text{orb}} = (5.98 \times 10^{24} \text{ kg})(1.50 \times 10^{11} \text{ m})^2 \frac{2\pi}{3.16 \times 10^7 \text{ s}}$$

$$= 2.7 \times 10^{40} \text{ kg} \cdot \text{m}^2/\text{s}. \qquad \text{(Answer)}$$

The vector $\mathbf{L}_{\text{orb}}$ is perpendicular to the plane of the Earth's orbit. Because of the inclination of the Earth's axis, the orbital angular momentum vector and the rotational angular momentum vector make an angle of 23.5° with each other. Both vectors remain constant in magnitude and direction as the Earth moves around its orbit during the year.

12-8 CONSERVATION OF ANGULAR MOMENTUM

So far we have discussed two powerful conservation laws, the conservation of energy and the conservation of linear momentum. Now we meet a third law of this type, the conservation of angular momentum. We start from Eq. 12-37 ($\Sigma \, \boldsymbol{\tau}_{\text{ext}} = d\mathbf{L}/dt$), which is Newton's second law in angular form. If no net external torque acts on the system, this equation be-

comes $d\mathbf{L}/dt = 0$, or

$$\mathbf{L} = \text{a constant} \qquad \text{(isolated system).} \qquad (12\text{-}40)$$

This equation represents the **law of conservation of angular momentum:**

───────

If no net external torque acts on a system, the (vector) angular momentum $\mathbf{L}$ of that system remains constant, no matter what changes take place within the system.

Equation 12-40 is a vector equation, and as such is equivalent to three scalar equations corresponding to the conservation of angular momentum in three mutually perpendicular directions.

Like the other two conservation laws that we have discussed, Eq. 12-40 holds beyond the limitations of Newtonian mechanics. It holds for particles whose speeds approach that of light (where the theory of relativity reigns), and it remains true in the world of subatomic particles (where quantum mechanics reigns). No exceptions to it have ever been found.

12-9 CONSERVATION OF ANGULAR MOMENTUM: SOME EXAMPLES

1. *The spinning volunteer.* Figure 12-16 shows a student seated on a stool that can rotate freely about a vertical axis. The student, who has been set into rotation at a modest initial angular speed ω_i, holds two dumbbells in his outstretched hands. His angular momentum vector $\mathbf{L}$ lies along the vertical axis, pointing upward in the figure.

The instructor now asks the student to pull in his arms; this reduces his rotational inertia from its initial value I_i to a smaller value I_f, because he moves mass closer to the rotation axis. His rate of rotation increases markedly, from ω_i to ω_f. If the student wishes to slow down, he has only to extend his arms once more.

No net external torque acts on the system consisting of the student, stool, and dumbbells. Thus the angular momentum of that system about the rotation axis must remain constant, no matter how the student maneuvers the weights. From Eqs. 12-39 and 12-40 we have

$$L = I\omega = \text{a constant},$$

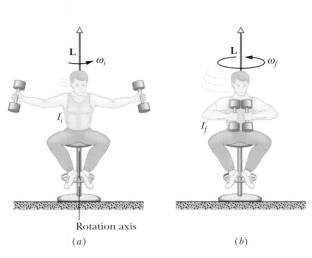

Rotation axis

(a)   (b)

FIGURE 12-16 (a) The student has a relatively large rotational inertia and a relatively small angular speed. (b) By decreasing his rotational inertia, the student automatically increases his angular speed. The angular momentum **L** of the rotating system remains unchanged.

or

$$I_i\omega_i = I_f\,\omega_f.$$

In Fig. 12-16a, the student's angular speed ω_i is relatively low and his rotational inertia I_i relatively high. In Fig. 12-16b, his angular speed must be greater to compensate for the decreased rotational inertia.

   2. *The springboard diver.* Figure 12-17 shows a diver doing a forward one-and-a-half somersault dive. As you should expect, her center of mass follows a parabolic path. She leaves the springboard with a definite angular momentum **L** about an axis through her center of mass, represented by a vector pointing into the plane of Fig. 12-17, perpendicular to the page. When she is in the air, the diver forms an isolated system and her angular momentum cannot change. By pulling her arms and legs into the closed *tuck position,* she can considerably reduce her rotational inertia about the same axis and thus considerably increase her angular velocity. Pulling out of the tuck position (into the *open layout position*) at the end of the dive increases her rotational inertia and thus slows her rotation rate so she can enter the water with little splash. Even in a more complicated dive involving twisting, the angular momentum of the diver must be conserved, in both magnitude *and* direction, throughout the dive.

   3. *Stabilizing a satellite (or a Frisbee).* Before a satellite is launched from the cargo bay of a space

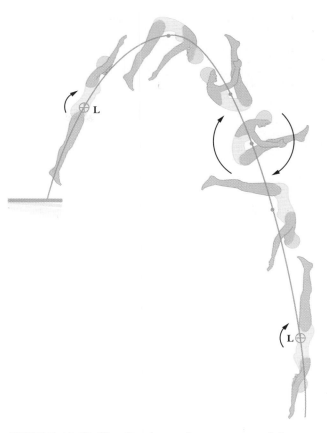

FIGURE 12-17 The diver's angular momentum **L** is constant throughout the dive, being represented by the tail $\otimes$ of an arrow that is perpendicular to the plane of the figure. Note also that her center of mass (see the dots) follows a parabolic path.

The diver begins a combination somersault and twist with one arm outstretched (so that the rotational inertia for the somersault is large) and with both arms near the body axis (so that the rotational inertia for the twist is small). By shifting arms and legs, she can control the rate of somersaulting and twisting.

FIGURE 12-18 Deployment of the Morelos-D satellite, a communications satellite for Mexico, from the bay of a space shuttle. The satellite is made to spin about its central axis to stabilize its orientation as it makes its way upward to its appointed orbit.

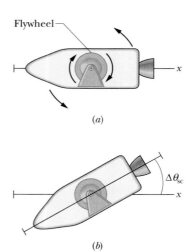

FIGURE 12-19 (*a*) An idealized spacecraft containing a flywheel. If the flywheel is made to rotate clockwise as shown, the spacecraft itself will rotate counterclockwise because the total angular momentum must remain zero. (*b*) When the flywheel is braked to rest, the spacecraft will also stop rotating but will have reoriented its axis by the angle $\Delta\theta_{sc}$.

shuttle (Fig. 12-18), it is made to spin about its central axis. Why?

The direction of the velocity of a moving particle is harder to change (by a given sidewise impulse) when the linear momentum of the particle is large than when it is small. In the same way, the orientation of a spinning object is harder to change (by a given external torque) when the object's angular momentum is large than when it is small. The orientation of a satellite that is *not* spinning might be changed by even a fairly small external torque, perhaps due to the thin residual atmosphere or to radiation pressure from sunlight. That of a spinning satellite will not.

In a more down-to-Earth example, a Frisbee is stabilized in flight in just the same way.

4. *Spacecraft orientation.* When an isolated system of particles has no angular momentum, can its orientation in space be changed by making internal changes in the system? If the system is not a rigid body, the answer is "Yes, under certain conditions."

Figure 12-19, which represents a spacecraft with a rigidly mounted flywheel, suggests a scheme for orientation control. The *spacecraft + flywheel* form an isolated system. If the total angular momentum **L** of the system is zero, it must remain so.

To change the orientation of the spacecraft, the flywheel is started up, as in Fig. 12-19*a*. The spacecraft will start to rotate in the opposite sense to maintain the system's angular momentum at zero. When the flywheel is then brought to rest, the spacecraft will also stop rotating but will have changed its orientation, as in Fig. 12-19*b*. At no time during this

The dog can predict the flight of a Frisbee because the rotation of the Frisbee disallows the buffeting by passing air that would disrupt the flight.

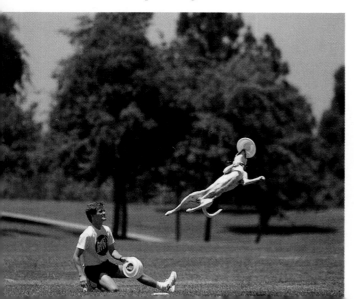

maneuver does the angular momentum of the system *spacecraft + flywheel* differ from zero.

Conservation of angular momentum requires that

$$I_{sc}\omega_{sc} + I_{fw}\omega_{fw} = 0 \qquad (12\text{-}41)$$

at all times. (Here the subscript sc refers to the spacecraft, and fw to the flywheel.) The two angular velocities have different signs, corresponding to the opposite directions of rotation of the spacecraft and the flywheel. Because $\omega = \Delta\theta/\Delta t$, we can write Eq. 12-41 as

$$I_{sc}\,\Delta\theta_{sc} = -I_{fw}\,\Delta\theta_{fw}$$

or

$$\Delta\theta_{sc} = -\frac{I_{fw}}{I_{sc}}\,\Delta\theta_{fw}.$$

Here $\Delta\theta_{sc}$ is the angle through which the spacecraft rotates in a given time, and $\Delta\theta_{fw}$ is the angle through which the flywheel rotates during that same time. The minus sign reminds us that these two angles are oppositely directed. Because $I_{fw} \ll I_{sc}$, the flywheel will have to make many revolutions to rotate the spacecraft through a modest angle. (Actually, engineering analysis favors thruster jets over flywheels as a means of reorienting spacecraft.)

Interestingly, the spacecraft *Voyager 2*, on its 1986 flyby of the planet Uranus, was set into unwanted rotation by this flywheel effect every time its tape recorder was turned on at high speed. The ground staff at the Jet Propulsion Laboratory had to program the on-board computer to turn on counteracting thruster jets every time the tape recorder was turned on or off.

5. *The incredible shrinking star.* When the nuclear fire in the core of a star burns low, the star may eventually begin to collapse, building up pressure in its interior. The collapse may go so far as to reduce the radius of the star from something like that of our sun to the incredibly small value of a few kilometers. The star then becomes a *neutron star,* so called because the material of which it is made has been compressed to an incredibly dense neutron gas.

During this shrinking process, the star is an isolated system and its angular momentum **L** cannot change. Because its rotational inertia is greatly reduced, its angular speed is correspondingly greatly increased, to as much as 600–800 revolutions per *second.* For comparison, our sun, a typical star, rotates about one revolution per month.

SAMPLE PROBLEM 12-9

Figure 12-20*a* shows a student, again sitting on a stool that can rotate freely about a vertical axis. The student, initially at rest, is holding a bicycle wheel whose rim is loaded with lead and whose rotational inertia I about its central axis is 1.2 kg·m². The wheel is rotating at an angular speed ω_i of 3.9 rev/s; from an overhead perspective, the rotation is counterclockwise. The axis of the wheel is vertical, and the angular momentum $\mathbf{L}_i$ of the wheel points vertically upward. The student now inverts the wheel (Fig. 12-20*b*); as a result, the student and stool rotate about the stool axis. With what angular speed and direction does the student then rotate? (The rotational inertia I_0 of the *student + stool + wheel* system about the stool axis is 6.8 kg·m².)

SOLUTION There is no net torque acting on the *student + stool + wheel* to change the angular momentum of that system about any vertical axis. The initial

FIGURE 12-20 Sample Problem 12-9. (*a*) A student holds a bicycle wheel rotating around the vertical. (*b*) The student inverts the wheel, setting himself into rotation. (*c*) The net angular momentum of the system must remain the same in spite of the inversion.

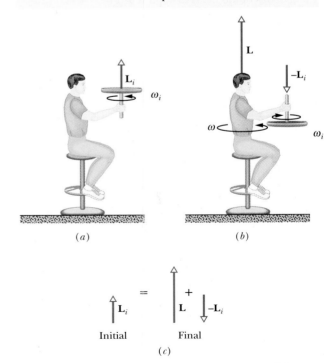

angular momentum $\mathbf{L}_i$ of the system is that of the bicycle wheel alone. After the wheel is inverted, the system must still have a net angular momentum of the same magnitude *and* direction.

After the inversion, the angular momentum of the wheel is $-\mathbf{L}_i$. In addition, the *student + stool* must acquire some angular momentum; call it $\mathbf{L}$. Then, as shown in Fig. 12-20*c*, we have

$$L_i = L + (-L_i),$$

or

$$L = 2L_i = I_0\omega,$$

in which ω is the angular speed acquired by the student after the wheel's inversion. This yields

$$\omega = \frac{2L_i}{I_0} = \frac{2I\omega_i}{I_0}$$

$$= \frac{(2)(1.2 \text{ kg} \cdot \text{m}^2)(3.9 \text{ rev/s})}{6.8 \text{ kg} \cdot \text{m}^2}$$

$$= 1.4 \text{ rev/s}. \qquad \text{(Answer)}$$

This positive result tells us that the student rotates counterclockwise about the stool axis as seen from an overhead perspective. If the student wishes to stop rotating, he has only to invert the wheel once more.

In inverting the wheel, the student will become well aware of the need to apply a torque. However, this torque is internal to the *student + stool + wheel* system and so cannot change the total angular momentum of this system.

We can, however, decide to adopt as our system *student + stool* alone, the wheel now being external to this new system. From this point of view, as the student exerts a torque on the wheel, the wheel exerts a reaction torque on him and this torque is now an external torque. It is the action of this external torque that changes the angular momentum of the *student + stool* system, setting it spinning. Whether a torque is internal or external depends only on how you choose to define your system.

SAMPLE PROBLEM 12-10

An aerialist is to make a triple somersault during a jump to his partner lasting a time $t = 1.87$ s. For the first and last quarter revolution, he is in the extended orientation shown in Fig. 12-21, with rotational inertia $I_1 = 19.9 \text{ kg} \cdot \text{m}^2$ around his center of mass. During the rest of the flight he is in a moderate tuck, with rotational inertia $I_2 = 5.50 \text{ kg} \cdot \text{m}^2$.

a. What must be his initial angular speed ω_1 around his center of mass?

SOLUTION He rotates in the extended position for a total angle $\theta_1 = 0.500$ rev in total time t_1, and in the tuck for an angle θ_2 in time t_2, as given by

$$t_1 = \frac{\theta_1}{\omega_1}, \qquad t_2 = \frac{\theta_2}{\omega_2}, \qquad (12\text{-}42)$$

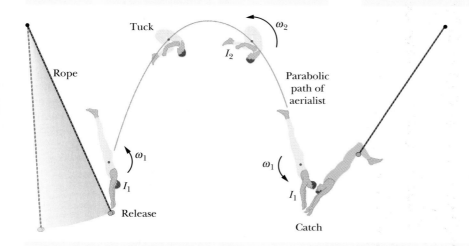

FIGURE 12-21 Sample Problem 12-10. The triple somersault.

Tuck

ω_2

I_2

Rope

Parabolic
path of
aerialist

ω_1

I_1

Release

ω_1

I_1

Catch

where ω_2 is his angular speed in the tuck. We find an expression for ω_2 by noting that his angular momentum is conserved throughout the flight:

$$I_2\omega_2 = I_1\omega_1,$$

from which

$$\omega_2 = \frac{I_1}{I_2}\omega_1. \qquad (12\text{-}43)$$

His total flight time is

$$t = t_1 + t_2,$$

which, with substitutions from Eqs. 12-42 and 12-43, becomes

$$t = \frac{\theta_1}{\omega_1} + \frac{\theta_2 I_2}{\omega_1 I_1} = \frac{1}{\omega_1}\left(\theta_1 + \theta_2\frac{I_2}{I_1}\right). \qquad (12\text{-}44)$$

Inserting the given data, we obtain

$$1.87\ \text{s} = \frac{1}{\omega_1}\left(0.500\ \text{rev} + 2.50\ \text{rev}\ \frac{5.50\ \text{kg}\cdot\text{m}^2}{19.9\ \text{kg}\cdot\text{m}^2}\right),$$

from which we find

$$\omega_1 = 0.6369\ \text{rev/s} \approx 0.637\ \text{rev/s}. \quad \text{(Answer)}$$

b. If he now attempts a quadruple somersault, with the same ω_1 and t, by using a tighter tuck, what must his rotational inertia I_2 be during the tuck?

The acrobat begins his flight with a certain amount of rotation and angular momentum. However, if he does not tuck during the flight so as to increase his rate of rotation, he would not land with the stilts beneath him.

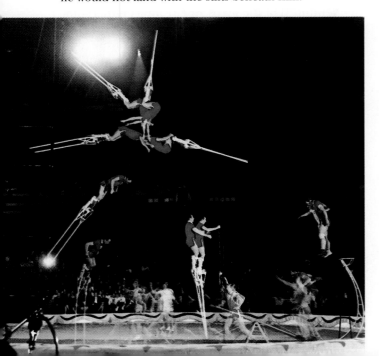

SOLUTION   The angle of rotation θ_2 during the tuck is now 3.50 rev (= 4.00 rev − 0.500 rev), and Eq. 12-44 becomes

$$1.87\ \text{s} = \frac{1}{0.6369\ \text{rev/s}}$$
$$\times\left(0.500\ \text{rev} + 3.50\ \text{rev}\ \frac{I_2}{19.9\ \text{kg}\cdot\text{m}^2}\right),$$

from which we find

$$I_2 = 3.929\ \text{kg}\cdot\text{m}^2 \approx 3.93\ \text{kg}\cdot\text{m}^2. \quad \text{(Answer)}$$

This smaller value for I_2 allows faster turning in the tuck and is about the smallest value possible for an aerialist. To make a four-and-a-half somersault, an aerialist would have to increase either the time of flight or the initial angular speed, but either change would make the catch by his partner more difficult. (Can you see why?)

c. For the quadruple, what is his rotation period T (the time for one rotation) during the tuck?

SOLUTION   We first find the angular speed ω_2 during the tuck from Eq. 12-43:

$$\omega_2 = \frac{I_1}{I_2}\omega_1 = \frac{19.9\ \text{kg}\cdot\text{m}^2}{3.929\ \text{kg}\cdot\text{m}^2}\ 0.6369\ \text{rev/s}$$
$$= 3.226\ \text{rev/s}.$$

Then we find the time T for one rotation from

$$T = \frac{1\ \text{rev}}{\omega_2} = \frac{1\ \text{rev}}{3.226\ \text{rev/s}} = 0.310\ \text{s}. \quad \text{(Answer)}$$

One reason the quadruple somersault is so difficult is that rotation occurs too quickly for the aerialist to see his surroundings clearly or to "fine-tune" the angular speed by adjusting his rotational inertia during flight.

SAMPLE PROBLEM 12-11

Four thin rods, each with mass M and length $d = 1.0$ m, are rigidly connected in the form of a plus sign; the entire assembly rotates in a horizontal plane around a vertical axle at the center, with initial (clockwise) angular velocity $\omega_i = -2.0$ rad/s (see Fig. 12-22). A mud ball with mass m and initial *speed* $v_i = 12$ m/s is thrown at and sticks to the end of one rod. Let $M = 3m$. What is the final angular velocity ω_f of the *plus sign + mud ball* system if the initial path of the mud ball is each of the four paths shown in Fig. 12-22: path 1 (contact is made when the ball's velocity is perpendicular to the

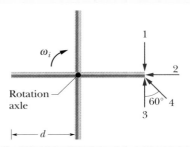

FIGURE 12-22 Sample Problem 12-11. An overhead view of four rigidly connected rods rotating around a central axle, and four paths a mud ball can take to stick onto one of the rods.

rod), path 2 (radial contact), path 3 (perpendicular contact), and path 4 (contact is made at 60° to the perpendicular)?

SOLUTION The total angular momentum L of the system about the axle is conserved during the collision:

$$L_f = L_i, \qquad (12\text{-}45)$$

where the subscripts f and i represent final and initial values. Let I_+ represent the rotational inertia of the plus sign about the axle. From Table 11-2(f), we have, for the four rods,

$$I_+ = 4\left(\frac{Md^2}{3}\right).$$

The rotational inertia of the mud ball about the axle as the ball rotates on the plus sign is $I_{mb} = md^2$. Let ℓ_i represent the initial (before contact) angular momentum of the mud ball about the axle, and ω_f represent the final angular velocity of the system. Using $L = I\omega$, we may rewrite Eq. 12-45 as

$$I_+\omega_f + I_{mb}\omega_f = I_+\omega_i + \ell_i,$$

and then as

$$(\tfrac{4}{3}Md^2)\omega_f + (md^2)\omega_f = (\tfrac{4}{3}Md^2)\omega_i + \ell_i. \quad (12\text{-}46)$$

Substituting $M = 3m$ and $\omega_i = -2.0$ rad/s and solving for ω_f, we find that

$$\omega_f = \frac{1}{5md^2}\left(4md^2(-2 \text{ rad/s}) + \ell_i\right). \quad (12\text{-}47)$$

We evaluate the magnitude of ℓ_i for paths 1 and 3 with Eq. 12-28, where $r_\perp = d$ and $v = v_i$. For path 2, we use Eq. 12-28 with $r_\perp = 0$. For path 4 we use Eq. 12-27, where $r = d$ and $v_\perp = v_i \cos 60°$. We then determine the sign of ℓ_i by seeing how a position vector lo-

cating the mud ball relative to the axle rotates around the axle as the mud ball approaches the plus sign: if it rotates clockwise, ℓ_i is negative; if it rotates counterclockwise, ℓ_i is positive. The results are:

path 1: $\ell_i = -mdv_i$;   path 2: $\ell_i = 0$;

path 3: $\ell_i = mdv_i$;   path 4: $\ell_i = mdv_i \cos 60°$.

Here speed $v_i = 12$ m/s. Substituting these values in turn into Eq. 12-47 along with $v_i = 12$ m/s, we find ω_f to be:

path 1: -4.0 rad/s; path 2: -1.6 rad/s;

path 3: 0.80 rad/s;   path 4: -0.40 rad/s.   (Answer)

12-10 PRECESSION OF A GYROSCOPE (OPTIONAL)

A simple gyroscope consists of a wheel that is fixed to a shaft and free to spin about the axis of the shaft. If the farther end of the shaft of a nonspinning gyroscope is placed on a support as in Fig. 12-23a and the gyroscope is released, the gyroscope falls by rotating downward around the support. Since the fall involves rotation, it is governed by Newton's second law in angular form, which, with Eq. 12-37, can be written as

$$\boldsymbol{\tau} = d\mathbf{L}/dt. \qquad (12\text{-}48)$$

This equation tells us that the torque causing the downward rotation (the fall) changes the angular momentum $\mathbf{L}$ of the gyroscope from its initial value of zero. The torque $\boldsymbol{\tau}$ is due to the gyroscope's weight $M\mathbf{g}$, acting at the center of mass of the gyroscope (which we take to be at the center of the wheel), with moment arm $\mathbf{r}$ relative to the support at O. The magnitude of the torque is

$$\tau = Mgr \sin 90° = Mgr \qquad (12\text{-}49)$$

(because the angle between $M\mathbf{g}$ and $\mathbf{r}$ is 90°), and its direction is as shown in Fig. 12-23a.

A rapidly spinning gyroscope that is released in the same manner behaves much differently. It first rotates downward slightly, but then it begins to rotate around a vertical axis through its support in a motion called **precession.** Why does the spinning gyroscope stay aloft, instead of falling over? The clue is that when the spinning gyroscope is released, the torque due to $M\mathbf{g}$ must change not an initial angular momentum of zero, but some already existing, finite

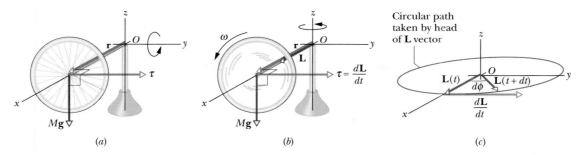

FIGURE 12-23 (*a*) A nonspinning gyroscope, with one end of its shaft on a support, falls by rotating around the *y* axis because of torque $\boldsymbol{\tau}$. (*b*) A rapidly spinning gyroscope, with angular momentum **L**, precesses around the *z* axis. (*c*) The change *d***L**/*dt* in angular momentum leads to a rotation of **L** about *O*.

angular momentum that is due to the spin of the gyroscope.

To see how this leads to precession, we first consider the angular momentum **L** of the gyroscope due to its spin. To simplify the situation, we assume the spin rate is so rapid that the angular momentum due to the gradual precession is negligible in comparison to **L**. We also assume that the shaft is horizontal when precession begins (see Fig. 12-23*b*). The magnitude of **L** is given by Eq. 12-39 as

$$L = I\omega, \quad (12\text{-}50)$$

where *I* is the rotational moment of the gyroscope around its shaft, and ω is the angular speed of the spin. The vector **L** points along the shaft, as shown in Fig. 12-23*b*. Since **L** is parallel to **r**, the torque $\boldsymbol{\tau}$ must be perpendicular to **L**.

According to Eq. 12-48, the torque $\boldsymbol{\tau}$ causes an incremental change *d***L** in the angular momentum of the gyroscope in an incremental time interval *dt*; that is,

$$d\mathbf{L} = \boldsymbol{\tau}\, dt. \quad (12\text{-}51)$$

However, for a *rapidly spinning* gyroscope, the magnitude of **L** is fixed by Eq. 12-50. Thus as the torque acts to change **L**, it can change only the direction of **L**, not the magnitude of **L**.

From Eq. 12-51 we see that the direction of *d***L** is in the direction of $\boldsymbol{\tau}$, perpendicular to **L**. The only way that **L** can be changed in the direction of $\boldsymbol{\tau}$ without its magnitude being changed is for **L** to rotate around the *z* axis as shown in Fig. 12-23*c*: **L** maintains its magnitude, the head of the **L** vector follows a circular path, and $\boldsymbol{\tau}$ is always tangent to that path. Since **L** must always point along the shaft of the gyroscope, the shaft must also rotate around the *z* axis

in the direction of $\boldsymbol{\tau}$. Thus we have the precession. Because the spinning gyroscope must obey Newton's law in angular form in response to any change in its initial angular momentum, it must precess instead of merely toppling over.

We can find the rate of precession Ω by first using Eqs. 12-49 and 12-51 to get the magnitude of *d***L**:

$$dL = \tau\, dt = Mgr\, dt. \quad (12\text{-}52)$$

As **L** changes by increment *d***L** in incremental time *dt*, the shaft and **L** precess around the *z* axis through incremental angle $d\phi$. (In Fig. 12-23*c*, angle $d\phi$ is exaggerated for clarity.) With the aid of Eqs. 12-50 and 12-52, we find that $d\phi$ is given by

$$d\phi = \frac{dL}{L} = \frac{Mgr\, dt}{I\omega}.$$

Dividing this expression by *dt* and setting the rate $\Omega = d\phi/dt$, we obtain

$$\Omega = \frac{Mgr}{I\omega} \quad \text{(precession rate).} \quad (12\text{-}53)$$

This result is valid under the assumption that the spin rate ω is rapid. Note that the precession rate Ω decreases if ω is increased. Note also that there would be no precession if a weight force Mg did not act on the gyroscope, but that since *I* is a function of *M*, the mass *M* cancels from Eq. 12-53; thus Ω is independent of the mass of the gyroscope.

Equation 12-53 also applies if the shaft of a spinning gyroscope is at an angle to the horizontal. It holds as well for a spinning top, which is essentially a spinning gyroscope at an angle to the horizontal.

12-11 ANGULAR MOMENTUM IS QUANTIZED (OPTIONAL)

A physical quantity is said to be **quantized** if it can exist with only certain discrete values, such that all intermediate values are prohibited. We have met so far with two examples: the quantization of mass (in Section 2-9) and the quantization of energy (in Section 8-9). Angular momentum is our third example.

Although quantization is universal, it makes itself known most strikingly only at the atomic and subatomic levels, and it is there that we must look for our evidence. All the particles of physics, such as the electron, the proton, and the pion, possess intrinsic, characteristic values of angular momentum, *as if* they spin like a top (however, they do not spin). This intrinsic angular momentum is given by the relation

$$\ell = s \frac{h}{2\pi}, \qquad (12\text{-}54)$$

in which s (called the *spin quantum number*) is an integer, a half-integer, or zero. The quantity h in Eq. 12-54 is the Planck constant, the basic constant of quantum physics.

The electron, for example, has a spin quantum number of $\frac{1}{2}$, so its intrinsic angular momentum is

$$\ell = (\tfrac{1}{2})(6.63 \times 10^{-34} \text{ J}\cdot\text{s}) \left(\frac{1}{2\pi}\right)$$

$$= 5.28 \times 10^{-35} \text{ J}\cdot\text{s}.$$

The smallness of h means that the quantization of angular momentum is not detectable in even the smallest of macroscopic objects. In the same way, the quantizations of mass and of energy are not directly apparent to our sensory perceptions.

The orbital motions of electrons in atoms—and of protons and neutrons in atomic nuclei—are also quantized. Angular momentum considerations lie at the heart of our understanding of the structure of matter at the atomic and subatomic levels. Confronted with a new particle, or with a quantum state of a nucleus, an atom, or a molecule, the first question a physicist is likely to ask is, "What is its angular momentum?" It is no surprise that the Planck constant —the foundation stone of the subatomic enterprise —has units of angular momentum.

REVIEW & SUMMARY

Rolling Bodies
For a wheel of radius R that rolls without slipping,

$$v_{\text{cm}} = \omega R, \qquad (12\text{-}2)$$

where v_{cm} is the speed of the wheel's center and ω is the angular velocity of the wheel about its center. The wheel may also be thought of as rotating instantaneously about a point P of the "road" that is in contact with the wheel. The angular velocity of the wheel about this point is the same as the angular velocity of the wheel about its center. With this in mind, we can show that the rolling wheel has kinetic energy

$$K = \tfrac{1}{2}I_{\text{cm}}\omega^2 + \tfrac{1}{2}Mv_{\text{cm}}^2, \qquad (12\text{-}5)$$

where I_{cm} is the rotational moment of the wheel about its center.

Analysis Using Newton's Second Law
Sample Problem 12-4 illustrates the use of Newton's second law, in forms appropriate for translation ($F = ma_{\text{cm}}$) and for rotation ($\tau = I\alpha$), for analyzing accelerated motion of rolling objects.

Torque as a Vector
In three dimensions, *torque* $\boldsymbol{\tau}$ is a vector quantity defined relative to a fixed point (usually an origin); it is

$$\boldsymbol{\tau} = \mathbf{r} \times \mathbf{F}, \qquad (12\text{-}21)$$

where $\mathbf{F}$ is a force applied to a particle, and $\mathbf{r}$ is a position vector locating the particle relative to the fixed point (or origin). The magnitude of $\boldsymbol{\tau}$ is given by

$$\tau = rF \sin\phi = rF_{\perp} = r_{\perp}F, \quad (12\text{-}22, 12\text{-}23, 12\text{-}24)$$

where ϕ is the angle between $\mathbf{F}$ and $\mathbf{r}$, $F_{\perp}$ is the component of $\mathbf{F}$ perpendicular to $\mathbf{r}$, and $r_{\perp}$ is the perpendicular distance between the fixed point and an extension (the line of action) of $\mathbf{F}$. The direction of $\boldsymbol{\tau}$ is given by the right-hand rule.

Angular Momentum of a Particle
The **angular momentum** ℓ of a particle with linear momentum $\mathbf{p}$, mass m, and linear velocity $\mathbf{v}$ is a vector quantity defined relative to a fixed point (usually an origin); it is

$$\ell = \mathbf{r} \times \mathbf{p} = m(\mathbf{r} \times \mathbf{v}). \qquad (12\text{-}25)$$

The magnitude of ℓ is given by

$$\ell = rmv \sin \phi \qquad (12\text{-}26)$$

$$= rp_\perp = rmv_\perp \qquad (12\text{-}27)$$

$$= r_\perp p = r_\perp mv, \qquad (12\text{-}28)$$

where ϕ is the angle between $\mathbf{r}$ and $\mathbf{p}$, $p_\perp$ and $v_\perp$ are the components of $\mathbf{p}$ and $\mathbf{v}$ perpendicular to $\mathbf{r}$, and $r_\perp$ is the perpendicular distance between the fixed point and an extension of $\mathbf{p}$. The direction of ℓ is given by the right-hand rule.

Newton's Second Law in Angular Form for a Particle

Newton's second law for a particle can be written in vector angular form as

$$\sum \boldsymbol{\tau} = \frac{d\ell}{dt}, \qquad (12\text{-}30)$$

where $\sum \boldsymbol{\tau}$ is the net torque acting on the particle, and ℓ is the angular momentum of the particle.

Angular Momentum of a System of Particles

The angular momentum $\mathbf{L}$ of a system of particles is the vector sum of the angular momenta of the individual particles:

$$\mathbf{L} = \ell_1 + \ell_2 + \cdots + \ell_n = \sum_{i=1}^{n} \ell_i. \qquad (12\text{-}35)$$

The time rate of change of this angular momentum is equal to the sum of the external torques on the system (the torques due to interactions of the particles of the system with particles external to the system). The exact relation is

$$\sum \boldsymbol{\tau}_{\text{ext}} = \frac{d\mathbf{L}}{dt} \qquad \text{(system of particles)}. \qquad (12\text{-}37)$$

Rigid Body Angular Momentum

For a rigid body rotating about a fixed axis, the component of its angular momentum parallel to the rotation axis is

$$L = I\omega \qquad \text{(rigid body, fixed axis)}. \qquad (12\text{-}39)$$

Conservation of Angular Momentum

The angular momentum $\mathbf{L}$ of a system remains constant if the net external torque acting on the system is zero:

$$\mathbf{L} = a \text{ constant} \qquad \text{(isolated system)}. \qquad (12\text{-}40)$$

This is the **law of conservation of angular momentum.** It is one of the fundamental conservation laws of nature, having been verified even in situations (involving high-speed particles or subatomic dimensions) in which Newton's laws are not applicable.

Precession of a Top

The **precession** of a top may be analyzed in terms of Eq. 12-37. The precession rate Ω of the top is given by

$$\Omega = \frac{Mgr}{I\omega}. \qquad (12\text{-}53)$$

Quantized Angular Momentum

Angular momentum is a **quantized** quantity, occurring in nature only in integral or half-integral multiples of $h/2\pi$, h being the Planck constant.

QUESTIONS

1. A cannon ball and a marble roll from rest down an incline. Which gets to the bottom first?

2. A cylindrical can filled with beef and an identical can filled with water both roll down an incline. Compare their angular and linear accelerations. Explain the difference.

3. A solid wooden cylinder rolls down two different inclined planes of the same height but with different angles of inclination. Will it reach the bottom with the same speed in each case? Will it take longer to roll down one incline than the other? Explain your answers.

4. A solid brass cylinder and a solid wooden cylinder have the same radius and mass, the wooden cylinder being longer. You release them together at the top of an incline. Which will reach the bottom first? Suppose that the cylinders are now made to be the same length (and radius) and that the masses are made to be equal by boring a hole along the axis of the brass cylinder. Which cylinder will win the race now? Explain your answers. Assume that the cylinders roll without slipping.

5. Ruth and Roger are cycling along a path at the same speed. The wheels of Ruth's bike are a little larger in diameter than the wheels of Roger's bike. How do the angular speeds of their wheels compare? What about the speeds of the top portion of their wheels?

6. If a car's speedometer is set to read at a speed proportional to the angular speed of its rear wheels, is it necessary to correct the reading when tires with a larger outside diameter (such as snow tires) are used?

7. A cylindrical drum is pushed along from above with a board. From the initial position shown in Fig. 12-24, it rolls forward on the ground a distance $L/2$, equal to half the length of the board. There is no slipping at any contact. Where, is the board then? How far has the woman walked?

FIGURE 12-24 Question 7.

8. Two heavy disks are connected by a short rod of much smaller radius. This system is placed on a ramp so that the disks hang over the sides as in Fig. 12-25. The system rolls down the ramp without slipping. (a) Near the bottom of the ramp the disks touch the horizontal table and the system takes off with greatly increased translational speed. Explain why. (b) If this system raced a hoop (of any radius) down the ramp, which would reach the bottom

FIGURE 12-25 Question 8.

first? (c) Show that the system has $\beta > 1$, where β is the rotational inertia parameter of Table 12-1.

9. A yo-yo falls to the bottom of its cord and then climbs back up. Does it reverse its direction of rotation at the bottom? Explain your answer.

10. A yo-yo is resting on a horizontal table and is free to roll (Fig. 12-26). If the string is pulled by a horizontal force such as $\mathbf{F}_1$, which way will the yo-yo roll? What happens when force $\mathbf{F}_2$ is applied (its line of action passes through the point of contact of the yo-yo and table)? If the string is pulled vertically with force $\mathbf{F}_3$, what happens?

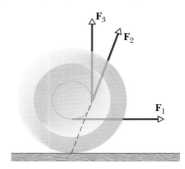

FIGURE 12-26 Question 10.

11. A rear-wheel-drive car accelerates quickly from rest, and the driver observes that the car "noses up." Why does it do that? Would a front-wheel-drive car do that?

12. The mounting bolts that fasten the engines of jet planes to the structural framework of the plane are arranged to snap apart if the (rapidly rotating) engine suddenly seizes up because of some mishap. Why are such "structural fuses" used?

13. Is there any advantage in setting the wheels of the landing gear of an airplane in rotation just before the plane lands? If so, how would you find the optimum speed and direction of rotation?

14. A disgruntled hockey player throws a hockey stick along the ice. It rotates about its center of mass as it slides and is eventually brought to rest by friction. Why must its rotational motion stop exactly when its center of mass comes to rest?

15. When the angular velocity ω of an object increases, its angular momentum may or may not also increase. Give an example in which it does and one in which it does not.

16. A student stands on a table rotating with an angular speed ω while holding two equal dumbbells at arm's length. Without moving anything else, the two dumbbells are dropped. What change, if any, is there in the student's angular speed? Is angular momentum conserved? Explain your answers.

17. A helicopter flies off, its rotor blades rotating. Why doesn't the body of the helicopter rotate in the opposite direction?

18. If the entire population of the world moved to Antarctica, would the length of the day be affected? If so, in what way?

19. A circular turntable rotates at constant angular velocity about a vertical axis. There is no friction and no driving torque. A circular pan rests on the turntable and rotates with it (see Fig. 12-27). The bottom of the pan is covered with a layer of ice of uniform thickness, which is, of course, also rotating with the pan. Suppose the ice melts, but none of the water escapes from the pan. Is the angular velocity now greater than, the same as, or less than the original velocity? Give reasons for your answer.

20. You can distinguish between a raw egg and a hard-boiled one by spinning each egg on the table. Explain how. Also, if you stop a spinning raw egg by very briefly touching the top, why will it resume spinning?

21. Figure 12-28*a* shows an acrobat propelled upward by a trampoline with zero angular momentum. Can the acrobat, by maneuvering the body, manage to land on his back as in Fig. 12-28*b*? Interestingly, 38% of questioned diving coaches and 34% of a sample of physicists gave the wrong answer. What do *you* think? (From ''Do Springboard Divers Violate Angular Momentum Conservation?'' by Cliff Frohlich, *American Journal of Physics,* July 1979.)

FIGURE 12-28 Question 21.

FIGURE 12-27 Question 19.

22. Can you ''pump'' a swing so that it turns in a complete circle, moving completely around its support? Assume (if you wish) that the seat of the swing is connected to its support by a rigid rod rather than by a rope or a chain. Explain your answer.

23. A massive spinning wheel can be used for a stabilizing effect on a ship. If the wheel is mounted with its axis of rotation at right angles to the ship deck, what is its effect when the ship tends to roll from side to side?

24. Why must a football quarterback put a lot of spin on the ball if it is to stay aligned with its path during a long pass?

EXERCISES & PROBLEMS

SECTION 12-1 ROLLING

1E. A thin-walled pipe rolls along the floor. What is the ratio of its translational kinetic energy to its rotational kinetic energy about an axis parallel to its length and through its center of mass?

2E. A 140-kg hoop rolls along a horizontal floor so that its center of mass has a speed of 0.150 m/s. How much work must be done on the hoop to stop it?

3E. An automobile traveling 80.0 km/h has tires of 75.0-cm diameter. (a) What is the angular speed of the tires about the axle? (b) If the car is brought to a stop uniformly in 30.0 turns of the tires (without skidding), what is the angular acceleration of the wheels? (c) How far does the car advance during this braking period?

4E. A 1000-kg car has four 10-kg wheels. When the car is moving, what fraction of the total kinetic energy of the car is due to rotation of the wheels about their axles? Assume that the wheels have the same rotational inertia as uniform disks of the same mass and size. Explain why you do not need to know the radius of the wheels.

5E. An automobile has a total mass of 1700 kg. It accelerates from rest to 40 km/h in 10 s. Assume each wheel is a uniform 32-kg disk. Find, for the end of the 10-s interval, (a) the rotational kinetic energy of each wheel about its axle, (b) the total kinetic energy of each wheel, and (c) the total kinetic energy of the automobile.

6E. A uniform sphere rolls down an incline. (a) What must be the incline angle if the linear acceleration of the center of the sphere is to be $0.10g$? (b) For this angle, what would be the acceleration of a frictionless block sliding down the incline?

7E. A solid sphere of weight 8.00 lb rolls up an incline with an inclination angle of 30.0°. At the bottom of the incline the center of mass of the sphere has a translational speed of 16.0 ft/s. (a) What is the kinetic energy of the sphere at the bottom of the incline? (b) How far does the sphere travel up the incline? (c) Does the answer to (b) depend on the weight of the sphere?

8E. A wheel of radius 0.250 m, which is moving initially at 43.0 m/s, rolls to a stop in 225 m. Calculate (a) its linear acceleration and (b) its angular acceleration. (c) The wheel's rotational inertia is 0.155 kg·m²; calculate the torque exerted by the friction on the wheel, about the center of the wheel.

9P. Consider a 66-cm-diameter tire on a car traveling at 80 km/h on a level road in the direction of increasing x. As seen by a passenger in the car, what are the linear velocity and the magnitude of the linear acceleration of (a) the center of the wheel, (b) a point at the top of the tire, and (c) a point at the bottom of the tire? (d) Repeat (a) through (c), in the same order, for a stationary observer alongside the road.

10P. A body of radius R and mass m is rolling horizontally without slipping with speed v. It then rolls up a hill to a maximum height h. (a) If $h = 3v^2/4g$, what is the body's rotational inertia? (b) What might the body be?

11P. A homogeneous sphere starts from rest at the upper end of the track shown in Fig. 12-29 and rolls without slipping until it rolls off the right-hand end. If $H = 60$ m and $h = 20$ m and the track is horizontal at the right-hand end, how far horizontally from point A does the sphere land on the floor?

12P. A small sphere, with radius r and mass m, rolls without slipping on the inside of a large fixed hemisphere with radius R and a vertical axis of symmetry. It starts at the top from rest. (a) What is its kinetic energy at the bottom? (b) What fraction of its kinetic energy at the bottom is associated with rotation about an axis through its center of

mass? (c) What normal force does the small sphere exert on the hemisphere at the bottom if $r \ll R$?

13P. A small solid marble of mass m and radius r rolls without slipping along the loop-the-loop track shown in Fig. 12-30, having been released from rest somewhere on the straight section of track. (a) From what minimum height h above the bottom of the track must the marble be released in order that it not leave the track at the top of the loop? (The radius of the loop-the-loop is R; assume $R \gg r$.) (b) If the marble is released from height $6R$ above the bottom of the track, what is the horizontal component of the force acting on it at point Q?

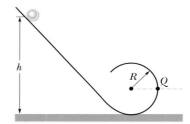

FIGURE 12-30 Problem 13.

14P. A solid cylinder of radius 10 cm and mass 12 kg starts from rest and rolls without slipping a distance of 6.0 m down a house roof that is inclined at 30°. (See Fig. 12-31.) (a) What is the angular speed of the cylinder about its center as it leaves the house roof? (b) The outside wall of the house is 5.0 m high. How far from the edge of the roof does the cylinder hit the level ground?

FIGURE 12-31 Problem 14.

15P*. An apparatus for testing the skid resistance of automobile tires is constructed as shown in Fig. 12-32. The tire is initially motionless and is held in a framework with negligible mass that is pivoted at point *B*. The rotational inertia of the wheel about its axis is 0.750 kg·m², its mass is 15.0 kg, and its radius is 0.300 m. The tire is free to rotate around point *A*. The tire is placed on the surface of a conveyor belt that is moving with a surface velocity of 12.0 m/s. (a) If the coefficient of kinetic friction between the tire and the conveyor belt is 0.60 and if the tire does not bounce, what time will be required for the wheel to reach its final angular velocity? (b) What will be the length of the skid mark on the conveyor surface?

FIGURE 12-32 Problem 15.

SECTION 12-2 THE YO-YO

16E. A yo-yo has a rotational inertia of 950 g·cm² and a mass of 120 g. Its axle radius is 3.2 mm, and its string is 120 cm long. The yo-yo rolls from rest down to the end of the string. (a) What is its acceleration? (b) How long does it take to reach the end of the string? As it reaches the end of the string, what are its (c) linear speed, (d) translational kinetic energy, (e) rotational kinetic energy, and (f) angular speed?

17P. Suppose that the yo-yo in Exercise 16, instead of rolling from rest, is thrown so that its initial speed down the string is 1.3 m/s. What is its angular speed at the bottom of the string?

SECTION 12-3 TORQUE REVISITED

18E. Given that $\mathbf{r} = x\mathbf{i} + y\mathbf{j} + z\mathbf{k}$ and $\mathbf{F} = F_x\mathbf{i} + F_y\mathbf{j} + F_z\mathbf{k}$, show that the torque $\boldsymbol{\tau} = \mathbf{r} \times \mathbf{F}$ is given by

$$\boldsymbol{\tau} = (yF_z - zF_y)\mathbf{i} + (zF_x - xF_z)\mathbf{j} + (xF_y - yF_x)\mathbf{k}.$$

19E. Show that, if $\mathbf{r}$ and $\mathbf{F}$ lie in a given plane, the torque $\boldsymbol{\tau} = \mathbf{r} \times \mathbf{F}$ has no component in that plane.

20E. What are the magnitude and direction of the torque about the origin on a plum at coordinates $(-2.0$ m, 0, 4.0 m) due to force $\mathbf{F}$ whose only component is (a) $F_x = 6.0$ N, (b) $F_x = -6.0$ N, (c) $F_z = 6.0$ N, and (d) $F_z = -6.0$ N?

21E. What are the magnitude and direction of the torque about the origin on a particle at coordinates $(0, -4.0$ m, 3.0 m) due to (a) force $\mathbf{F}_1$ with components $F_{1x} = 2.0$ N,

and $F_{1y} = F_{1z} = 0$, and (b) force $\mathbf{F}_2$ with components $F_{2x} = 0$, $F_{2y} = 2.0$ N, and $F_{2z} = 4.0$ N?

22P. Force $\mathbf{F} = (2.0$ N)$\mathbf{i} - (3.0$ N)$\mathbf{k}$ acts on a pebble with position vector $\mathbf{r} = (0.50$ m)$\mathbf{j} - (2.0$ m)$\mathbf{k}$, relative to the origin. What is the resulting torque acting on the pebble about (a) the origin and (b) a point with coordinates (2.0 m, 0, -3.0 m)?

23P. What is the torque about the origin on a sand grain at coordinates $(3.0$ m, -2.0 m, 4.0 m) due to (a) force $\mathbf{F}_1 = (3.0$ N)$\mathbf{i} - (4.0$ N)$\mathbf{j} + (5.0$ N)$\mathbf{k}$, (b) force $\mathbf{F}_2 = (-3.0$ N)$\mathbf{i} - (4.0$ N)$\mathbf{j} - (5.0$ N)$\mathbf{k}$, and (c) the net force of $\mathbf{F}_1$ and $\mathbf{F}_2$? (d) Repeat (c) for a point with coordinates $(3.0$ m, 2.0 m, 4.0 m) instead of the origin.

24P. What is the net torque about the origin on a flea at coordinates $(0, -4.0$ m, 5.0 m) when forces $\mathbf{F}_1 = (3.0$ N)$\mathbf{k}$ and $\mathbf{F}_2 = (-2.0$ N)$\mathbf{j}$ act on the flea?

25P. Force $\mathbf{F} = (-8.0$ N)$\mathbf{i} + (6.0$ N)$\mathbf{j}$ acts on a particle with position vector $\mathbf{r} = (3.0$ m)$\mathbf{i} + (4.0$ m)$\mathbf{j}$. What are (a) the torque on the particle about the origin and (b) the angle between $\mathbf{r}$ and $\mathbf{F}$?

SECTION 12-4 ANGULAR MOMENTUM

26E. A 1200-kg airplane is flying in a straight line at 80 m/s, 1.3 km above the ground. What is the magnitude of its angular momentum with respect to a point on the ground directly under the path of the plane?

27E. Two objects are moving as shown in Fig. 12-33. What is their total angular momentum about point *O*?

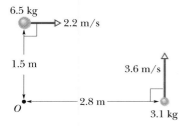

FIGURE 12-33 Exercise 27.

28E. A particle *P* with mass 2.0 kg has position vector $\mathbf{r}$ $(r = 3.0$ m) and velocity $\mathbf{v}$ $(v = 4.0$ m/s) as shown in Fig. 12-34. It is acted on by force $\mathbf{F}$ $(F = 2.0$ N). All three vec-

FIGURE 12-34 Exercise 28.

tors lie in the *xy* plane. About the origin, what are (a) the angular momentum of the particle and (b) the torque acting on the particle?

29E. If we are given *r*, *p*, and *φ*, we can calculate the angular momentum of a particle from Eq. 12-26. Sometimes, however, we are given the components (*x*, *y*, *z*) of **r** and (*v_x*, *v_y*, *v_z*) of **v** instead. (a) Show that the components of **ℓ** along the *x*, *y*, and *z* axes are then given by

$$\ell_x = m(yv_z - zv_y),$$

$$\ell_y = m(zv_x - xv_z),$$

$$\ell_z = m(xv_y - yv_x).$$

(b) Show that if the particle moves only in the *xy* plane, the resultant vector for the angular momentum has only a *z* component.

30E. At a certain time, the position vector in meters of a 0.25-kg object is $\mathbf{r} = 2.0\mathbf{i} - 2.0\mathbf{k}$. At that instant, its velocity in meters per second is $\mathbf{v} = -5.0\mathbf{i} + 5.0\mathbf{k}$, and the force in newtons acting on it is $\mathbf{F} = 4.0\mathbf{j}$. (a) What is the angular momentum of the object about the origin? (b) What torque acts on it? (*Hint:* See Exercises 18 and 29.)

31P. Calculate the angular momentum, about the Earth's center, of an 84-kg person on the equator of the rotating Earth.

32P. Show that the angular momentum, about any point, of a single particle moving with constant velocity remains constant throughout the motion.

33P. Two particles, each of mass *m* and speed *v*, travel in opposite directions along parallel lines separated by a distance *d*. Find an expression, in terms of *m*, *v*, and *d*, for the total angular momentum of the system about any origin.

34P. A 2.0-kg object moves in a plane with velocity components $v_x = 30$ m/s and $v_y = 60$ m/s as it passes through the point $(x, y) = (3.0, -4.0)$ m. (a) What is its angular momentum relative to the origin at this moment? (b) What is its angular momentum relative to the point $(-2.0, -2.0)$ m at this same moment?

35P. (a) Use the data given in the appendices to compute the total angular momentum of all the planets owing to their revolution about the sun. (b) What fraction of this angular momentum is associated with the planet Jupiter?

SECTION 12-5 NEWTON'S SECOND LAW IN ANGULAR FORM

36E. A 3.0-kg particle is at $x = 3.0$ m, $y = 8.0$ m with a velocity of $\mathbf{v} = (5.0 \text{ m/s})\mathbf{i} - (6.0 \text{ m/s})\mathbf{j}$. It is acted on by a 7.0-N force in the negative *x* direction. (a) What is the angular momentum of the particle? (b) What torque acts on the particle? (c) At what rate is the angular momentum of the particle changing with time?

37E. A particle is acted on by two torques about the origin: $\boldsymbol{\tau}_1$ has a magnitude of 2.0 N·m and points in the di-

rection of increasing *x*, and $\boldsymbol{\tau}_2$ has a magnitude of 4.0 N·m and points in the direction of decreasing *y*. What are the magnitude and direction of $d\boldsymbol{\ell}/dt$, where $\boldsymbol{\ell}$ is the angular momentum of the particle?

38E. What torque about the origin acts on a particle moving in the *xy* plane if the particle has the following values of angular momentum about the origin:

(a) -4.0 kg·m²/s,   (c) $-4.0\sqrt{t}$ kg·m²/s,
(b) $-4.0t^2$ kg·m²/s,   (d) $-4.0/t^2$ kg·m²/s?

39E. A 3.0-kg toy car on the *x* axis has velocity $\mathbf{v} = -2.0t^3$ m/s along that axis. About the origin and for $t > 0$, what are (a) the car's angular momentum and (b) the torque acting on the car? (c) Repeat (a) and (b) for a point with coordinates (2.0 m, 5.0 m, 0) instead of the origin. (d) Repeat (a) and (b) for a point with coordinates (2.0 m, -5.0 m, 0) instead of the origin.

40P. At $t = 0$, a 2.0-kg particle has position vector $\mathbf{r} = (4.0 \text{ m})\mathbf{i} - (2.0 \text{ m})\mathbf{j}$ relative to the origin. Its velocity is given by $\mathbf{v} = (-6.0t^4 \text{ m/s})\mathbf{i} + (3.0 \text{ m/s})\mathbf{j}$. About the origin and for $t > 0$, what are (a) the particle's angular momentum and (b) the torque acting on the particle? (c) Repeat (a) and (b) for a point with coordinates (-2.0 m, -3.0 m, 0) instead of the origin.

41P. A projectile of mass *m* is fired from the ground with an initial speed v_0 and an initial angle θ_0 above the horizontal. (a) Find an expression for its angular momentum about the firing point as a function of time. (b) Find the rate at which the angular momentum changes with time. (c) Evaluate $\mathbf{r} \times \mathbf{F}$ directly and compare the result with (b). Why should the results be identical?

SECTION 12-7 THE ANGULAR MOMENTUM OF A RIGID BODY ROTATING ABOUT A FIXED AXIS

42E. A sanding disk with rotational inertia 1.2×10^{-3} kg·m² is attached to an electric drill whose motor delivers a torque of 16 N·m. Find (a) the angular momentum and (b) the angular speed of the disk 33 ms after the motor is turned on.

43E. The angular momentum of a flywheel having a rotational inertia of 0.140 kg·m² decreases from 3.00 to 0.800 kg·m²/s in 1.50 s. (a) What is the average torque acting on the flywheel during this period? (b) Assuming a uniform angular acceleration, through what angle will the flywheel have turned? (c) How much work was done on the wheel? (d) What is the average power of the flywheel?

44E. Three particles, each of mass *m*, are fastened to each other and to a rotation axis by three massless strings, each with length *l* as shown in Fig. 12-35. The combination rotates around the rotational axis at *O* with angular velocity *ω* in such a way that the particles remain in a straight line. In terms of *m*, *l*, and *ω*, and relative to point *O*, what are

FIGURE 12-35 Exercise 44.

(a) the rotational inertia of the combination, (b) the angular momentum of the middle particle, and (c) the total angular momentum of the three particles?

45E. A uniform rod rotates in a horizontal plane about a vertical axis through one end. The rod is 6.00 m long, weighs 10.0 N, and rotates at 240 rev/min clockwise when seen from above. Calculate (a) the rotational inertia of the rod about the axis of rotation and (b) the angular momentum of the rod.

46P. Wheels A and B in Fig. 12-36 are connected by a belt that does not slip. The radius of wheel B is three times the radius of wheel A. What would be the ratio of the rotational inertias I_A/I_B if (a) both wheels had the same angular momenta and (b) both wheels had the same rotational kinetic energy?

FIGURE 12-36 Problem 46.

47P. An impulsive force $F(t)$ acts for a short time Δt on a rotating rigid body with rotational inertia I. Show that

$$\int \tau \, dt = \overline{F}R(\Delta t) = I(\omega_f - \omega_i),$$

where R is the moment arm of the force, $\overline{F}$ is the average value of the force during the time it acts on the body, and ω_i and ω_f are the angular velocities of the body just before and just after the force acts. [The quantity $\int \tau \, dt = \overline{F}R(\Delta t)$ is called the *angular impulse*, in analogy with $\overline{F}(\Delta t)$, the linear impulse.]

48P*. Two cylinders having radii R_1 and R_2 and rotational inertias I_1 and I_2, respectively, are supported by axes perpendicular to the plane of Fig. 12-37. The large

FIGURE 12-37 Problem 48.

cylinder is initially rotating with angular velocity ω_0. The small cylinder is moved to the right until it touches the large cylinder and is caused to rotate by the frictional force between the two. Eventually, slipping ceases, and the two cylinders rotate at constant rates in opposite directions. Find the final angular velocity ω_2 of the small cylinder in terms of I_1, I_2, R_1, R_2, and ω_0. (*Hint:* Neither angular momentum nor kinetic energy is conserved. Apply the angular impulse equation of Problem 47.

49P*. A beginning bowler throws a bowling ball of mass M and radius $R = 11$ cm down the lane with an initial speed $v_0 = 8.5$ m/s. The ball is thrown in such a way that it skids for a certain distance before it starts to roll. It is not rotating at all when it first hits the lane, its motion being pure translation. The coefficient of kinetic friction between the ball and the lane is 0.21. (a) For what length of time does the ball skid? (*Hint:* As the ball skids, its speed v decreases and its angular speed ω increases until skidding ceases and rolling begins. (b) How far down the lane does it skid? (c) How many revolutions does it make before it starts to roll? (d) How fast is it moving when it starts to roll? (e) Does the answer to (d) depend on M, R, or v_0? Explain.

SECTION 12-9 CONSERVATION OF ANGULAR MOMENTUM: SOME EXAMPLES

50E. The rotor of an electric motor has rotational inertia $I_m = 2.0 \times 10^{-3}$ kg·m² about its central axis. The motor is used to change the orientation of the space probe in which it is mounted. The motor axis is mounted parallel to the axis of the probe, which has rotational inertia $I_p = 12$ kg·m² about its axis. Calculate the number of revolutions of the rotor required to turn the probe through 30° about its axis.

51E. A man stands on a frictionless platform that is rotating with an angular speed of 1.2 rev/s; his arms are outstretched and he holds a weight in each hand. With his hands in this position the rotational inertia of the system of man, weights, and platform is 6.0 kg·m². If by moving the weights the man decreases the rotational inertia of the system to 2.0 kg·m², (a) what is the resulting angular speed of the platform and (b) what is the ratio of the new kinetic energy of the system to the original kinetic energy? (c) What provided the added kinetic energy?

52E. Two disks are mounted on low-friction bearings on the same axle and can be brought together so that they couple and rotate as one unit. (a) The first disk, with rotational inertia 3.3 kg·m², is set spinning at 450 rev/min. The second disk, with rotational inertia 6.6 kg·m², is set spinning at 900 rev/min in the same direction as the first. They then couple together. What is their angular speed after coupling? (b) If instead the second disk is set spinning at 900 rev/min in the direction opposite the first disk's rotation, what is the angular speed after coupling?

53E. A wheel is rotating freely with an angular speed of 800 rev/min on a shaft whose rotational inertia is negligible. A second wheel, initially at rest and with twice the rotational inertia of the first, is suddenly coupled to the same shaft. (a) What is the angular speed of the resultant combination of the shaft and two wheels? (b) What fraction of the original rotational kinetic energy is lost?

54E. The rotational inertia of a collapsing spinning star changes to one-third its initial value. What is the ratio of the new rotational kinetic energy to the initial rotational kinetic energy?

55E. Suppose that the sun runs out of nuclear fuel and suddenly collapses to form a so-called white dwarf star, with a diameter equal to that of the Earth. Assuming no mass loss, what would then be the sun's new rotation period (time for one rotation), which currently is about 25 days? Assume that the sun and the white dwarf are uniform, solid spheres.

56E. In a playground, there is a small merry-go-round of radius 1.20 m and mass 180 kg. Its radius of gyration (see Problem 58 of Chapter 11) is 91.0 cm. A child of mass 44.0 kg runs at a speed of 3.00 m/s along a path that is tangent to the rim of the initially stationary merry-go-round and then jumps on. Neglect friction between the bearings and the shaft of the merry-go-round. Calculate (a) the rotational inertia of the merry-go-round about its axis of rotation; (b) the angular momentum of the child, while running, about the axis of rotation of the merry-go-round; and (c) the angular velocity of the merry-go-round and child after the child has jumped on.

57P. With center and spokes of negligible mass, a certain bicycle wheel has a thin rim of radius 1.14 ft and weight 8.36 lb; it can turn on its axle with negligible friction. A man holds the wheel above his head with the axis vertical while he stands on a turntable free to rotate without friction; the wheel rotates clockwise, as seen from above, with an angular speed of 57.7 rad/s, and the turntable is initially at rest. The rotational inertia of wheel + man + turntable about the common axis of rotation is 1.54 slug·ft². The man's hand suddenly stops the rotation of the wheel (relative to the turntable). Determine the resulting angular velocity (magnitude and direction) of the system.

58P. Two skaters, each of mass 50 kg, approach each other along parallel paths separated by 3.0 m. They have equal and opposite velocities of 1.4 m/s. The first skater carries one end of a long pole with negligible mass, and the second skater grabs the other end of it as she passes; see Fig. 12-38. Assume frictionless ice. (a) Describe quantitatively the motion of the skaters after they are connected by the pole. (b) By pulling on the pole, the skaters reduce their separation to 1.0 m. What is their angular speed then? (c) Calculate the kinetic energy of the system in parts (a) and (b). (d) From where does the added kinetic energy come?

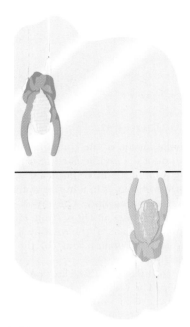

FIGURE 12-38 Problem 58.

59P. Two children, each with mass M, sit on opposite ends of a narrow board with length L and mass M (the same as the mass of each child). The board is pivoted at its center and is free to rotate in a horizontal circle without friction. (Treat it as a thin rod.) (a) What is the rotational inertia of the board plus the children about a vertical axis through the center of the board? (b) What is the angular momentum of the system if it is rotating with angular speed ω_0 in a clockwise direction as seen from above? What is the direction of the angular momentum? (c) While the system is rotating, the children pull themselves toward the center of the board until they are half as far from the center as before. What is the resulting angular speed in terms of ω_0? (d) What is the change in kinetic energy of the system as a result of the children changing their positions? (From where does the added kinetic energy come?)

60P. A toy train track is mounted on a large wheel that is free to turn with negligible friction about a vertical axis (Fig. 12-39). A toy train of mass m is placed on the track and, with the system initially at rest, the electrical power is

FIGURE 12-39 Problem 60.

turned on. The train reaches a steady speed v with respect to the track. What is the angular velocity ω of the wheel, if its mass is M and its radius R? (Treat the wheel as a hoop, and neglect the mass of the spokes and hub.)

61P. A cockroach of mass m runs counterclockwise around the rim of a lazy Susan (a circular dish mounted on a vertical axle) of radius R and rotational inertia I with frictionless bearings. The cockroach's speed (relative to the Earth) is v, whereas the lazy Susan turns clockwise with angular speed ω_0. The cockroach finds a bread crumb on the rim and, of course, stops. (a) What is the angular speed of the lazy Susan after the cockroach stops? (b) Is mechanical energy conserved?

62P. A girl of mass M stands on the rim of a frictionless merry-go-round of radius R and rotational inertia I that is not moving. She throws a rock of mass m horizontally in a direction that is tangent to the outer edge of the merry-go-round. The speed of the rock, relative to the ground, is v. Afterwards, what are (a) the angular speed of the merry-go-round and (b) the linear speed of the girl?

63P. If the polar ice caps of the Earth melted and the water returned to the oceans, the oceans would be made deeper by about 30 m. What effect would this have on the Earth's rotation? Make an estimate of the resulting change in the length of the day. (Concern has been expressed that warming of the atmosphere resulting from industrial pollution could cause the ice caps to melt.)

64P*. The particle of mass m in Fig. 12-40 slides down the

FIGURE 12-40 Problem 64.

frictionless surface and collides with the uniform vertical rod, sticking to it. The rod pivots about O through the angle θ before momentarily coming to rest. Find θ in terms of the other parameters given in the figure.

65P*. Two 2.00-kg balls are attached to the ends of a thin rod of negligible mass, 50.0 cm long. The rod is free to rotate in a vertical plane without friction about a horizontal axis through its center. While the rod is horizontal (Fig. 12-41), a 50.0-g putty wad drops onto one of the balls with a speed of 3.00 m/s and sticks to it. (a) What is the angular speed of the system just after the putty wad hits? (b) What is the ratio of the kinetic energy of the entire system after the collision to that of the putty wad just before? (c) Through what angle will the system rotate until it momentarily stops?

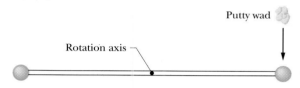

FIGURE 12-41 Problem 65.

SECTION 12-10 PRECESSION OF A GYROSCOPE

66E. A top is spinning at 30 rev/s about an axis making an angle of 30° with the vertical. Its mass is 0.50 kg, and its rotational inertia about its central axis is 5.0×10^{-4} kg·m². The center of mass is 4.0 cm from the pivot point. If the spin is clockwise as seen from above, what are the magnitude and direction of the rate of precession?

67P. A gyroscope consists of a rotating uniform disk with a 50-cm radius suitably mounted at the center of a 12-cm-long axle (of negligible mass) so that the gyroscope can spin and precess freely. Its spin rate is 1000 rev/min. Find the rate of precession (in rev/min) if the axle is supported at one end and is horizontal.

ADDITIONAL PROBLEMS

68. A uniform thin rod of length 0.50 m and mass 4.0 kg can rotate in a horizontal plane about a vertical axis through its center. The rod is at rest when a 3.0-g bullet traveling in the horizontal plane of the rod is fired into one end of the rod. As viewed from above, the direction of the bullet's velocity makes an angle of 60° with the rod (Fig. 12-42). If the bullet lodges in the rod and the angular velocity of the rod is 10 rad/s immediately after the

FIGURE 12-42 Problem 68. The rod viewed from above.

collision, what is the magnitude of the bullet's velocity just before impact?

69. A horizontal platform in the shape of a circular disk rotates on a frictionless bearing about a vertical axle through the center of the disk. The platform has a mass of 150 kg, a radius of 2.0 m, and a rotational inertia of 300 kg·m² about the axis of rotation. A 60-kg student walks slowly from the rim of the platform toward the center. If the angular speed of the system is 1.5 rad/s when the student starts at the rim, what is the angular speed when she is 0.50 m from the center? (*Hint:* Use the principle of conservation of angular momentum.)

70. A constant horizontal force of 10 N is applied to a wheel of mass 10 kg and radius 0.30 m as shown in Fig. 12-43. The wheel rolls without slipping on the horizontal surface, and the acceleration of its center of mass is 0.60 m/s². (a) What are the magnitude and direction of

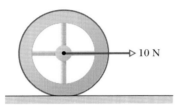

FIGURE 12-43 Problem 70.

the frictional force on the wheel? (b) What is the rotational inertia of the wheel about an axis through its center of mass and perpendicular to the plane of the wheel?

71. An 0.80-kg particle is located at the position $\mathbf{r} = (2.0 \text{ m})\mathbf{i} + (3.0 \text{ m})\mathbf{j}$. The momentum of the particle lies in the xy plane and has a magnitude of 2.4 kg·m/s and a direction of 115° measured counterclockwise from the direction of increasing x. What is the angular momentum of the particle about the origin, in unit-vector form?

72. A phonograph record of mass 0.10 kg and radius 0.10 m rotates about a vertical axis through its center with an angular speed of 4.7 rad/s. The rotational inertia of the record about its axis of rotation is 5.0×10^{-4} kg·m². A wad of putty of mass 0.020 kg drops vertically onto the record from above and sticks to the edge of the record. What is the angular speed of the record immediately after the putty sticks to it?

73. A hollow sphere of mass 3.0 kg and radius 0.15 m, with rotational inertia $I = 0.040$ kg·m² about its center of mass, rolls without slipping up a surface inclined at 30° to the horizontal. At a certain initial position, the sphere's total kinetic energy is 20 J. (a) How much of this initial kinetic energy is rotational? (b) What is the speed of the center of mass of the sphere at the initial position? What are (c) the total kinetic energy of the sphere and (d) the speed of its center of mass after it has moved 1.0 m up along the incline from its initial position?

THE MECHANICS OF TURNS IN DANCE

Kenneth Laws
Dickinson College

Dancers performing on stage move in an amazing variety of ways—some graceful in their simplicity, some dazzling in their athletic complexity. Some motions or positions may inspire the reaction, "Wow! That looks impossible!" Indeed, dancers often move in ways that cause us to marvel, sometimes even appearing to violate laws of physics. Now *that* observation calls for analysis!

The moving human body is not a rigid object whose dimensions and configuration are constant and easily measurable. But some movements in the "vocabulary" of dance can be described sufficiently rigorously to enable us to apply principles of classical mechanics to the body moving through space under the influence of gravitational and other forces.

A particularly interesting category of dance movement involves rotation—turns on the floor or in the air, rotations about vertical, horizontal, or inclined axes, and turns in which the illusion of performing the impossible is created. The basis for an analysis of turns is the relationship between torque and angular momentum. For instance, how does a dancer arrange for the torque on the body necessary to initiate a *pirouette* (a rotation about a vertical axis with one foot on the floor)? Or how is a leaping turn accomplished when it appears that the body starts rotating after it leaves the floor? How does a dancer (or ice skater) change the rate of rotation during a turning movement?

A dancer usually begins a *pirouette* with both feet on the floor, one behind the other (Fig. 1). Pushing sideways in one direction with one foot and the other direction with the other produces two equal and opposite forces against the floor, with some perpendicular distance *d* between them (Fig. 2). The corresponding forces exerted by the floor on the feet pro-

FIGURE 1 Preparation position for a *pirouette*.

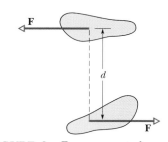

FIGURE 2 Forces exerted on the floor by the feet. Reaction forces exerted on the feet produce a torque that causes rotational acceleration for a *pirouette*.

FIGURE 3 The *pirouette* position for a rotating dancer.

Kenneth Laws is professor of physics at Dickinson College in Carlisle, PA, where he has been teaching since 1962. He earned his B.S., M.S., and Ph.D. degrees from Caltech, the University of Pennsylvania, and Bryn Mawr College, respectively. Since 1976 he has been studying classical ballet at the Central Pennsylvania Youth Ballet, and he has recently been applying the principles of physics to dance movement. This work has led to numerous lectures and classes around the country and has culminated in a book, The Physics of Dance, published in 1984 (paperback 1986) by Schirmer Books.

duce a torque on the body, which causes it to gain angular momentum. When the dancer rises onto one foot to the normal *pirouette* position (Fig. 3), a net rotation exists whose rate is determined by the magnitude of the torque, its duration, and the rotational inertia of the body in its rotating configuration. (See Eqs. 12-37 and 12-39.) Note that since there is no net horizontal force, there is no linear acceleration of the body.

The magnitude of the torque depends both on the magnitude of the forces and on the separation of the feet. That separation is typically a half meter, but may be as small as a few centimeters in a "fifth position" preparation, in which the feet are next to each other and antiparallel. In that case a dancer is often observed to "wind up" with the arms, causing them to start rotating before the rest of the body rises into its *pirouette* configuration. That wind-up serves the purpose of allowing angular momentum to be stored in a rotating part of the body—the arms—while the feet are still planted and able to exert forces against the floor. Extending the duration of the exerted torque compensates for the smaller torque arising from the small separation of the forces, allowing a significant final angular momentum to be acquired.

Once the magnitude of the torque and its duration have been established, the rate of rotation is determined by the rotational inertia of the body. In a normal *pirouette* position of Fig. 3, the rotational inertia is small, and substantial rates of turn are possible —sometimes greater than two revolutions per second. *Arabesque* turns or *grandes pirouettes*, in which one leg is extended to the rear or the side, respectively, are slower turns, since the body's rotational inertia in these configurations is significantly greater. By changing the distribution of body mass relative to the axis of rotation, the dancer can change the angular velocity during the turn. An ice skater, for instance, increases the rate of turn during a *pirouette* by bringing the

arms and gesture leg closer to the rotation axis.

Suppose two dancers perform a *pirouette* to the same music (same timing and hence equal angular accelerations), but one dancer is 15% larger in each linear dimension. Note that a dancer's volume, and hence mass, depends on the third power of the linear size, and the body's rotational inertia depends on the second power of the distance of each increment of mass from the rotation axis.

Question 1

How much more torque is required of the larger dancer than the smaller dancer to initiate the pirouette?

But now recognize that, for the same shape of preparation position before starting the turn, the larger dancer's feet will be 15% farther apart.

Question 2

How much greater horizontal force between the feet and the floor must the larger dancer exert in order to perform the identical pirouette?

An interesting situation arises when the torque for a *pirouette* is exerted by the hands of a partner rather than the floor. Such "supported" *pirouettes,* common in classical ballet, are often performed as shown in Fig. 4. Suppose the woman is standing in such a *pirouette* position, in preparation for a turn to the right. If her partner attempts to initiate the turn

FIGURE 4   A partnered, or "supported," *pirouette.*

by pulling back with his right hand and pushing forward with his left, she will indeed turn, but her body will rotate through a considerable angle before the torque has had a chance to build up much angular momentum. The resulting rate of turn is therefore limited.

Now suppose that, before the man exerts the torque on the woman's waist, she extends her right leg to the front and a little to her left (*croisé*). Now when he exerts the torque, her leg moves from front to her right side. The straight leg extended horizontally has a rotational inertia about four times that of the body rotating in the *pirouette* position, so when rotating it can have a substantial amount of angular momentum while the rest of her body remains facing the audience. Thus the length of time during which the man can exert the torque against his partner's waist is extended considerably, producing a significantly larger final angular momentum. When the woman does finally bring her right leg from the side into the *pirouette* position with the foot at her left knee, the large angular momentum is transferred from the rotating leg to the body as a whole, producing a far more rapid rate of turn than was possible without the use of the rotating leg.

Another movement demonstrates a similar process involving the transfer of angular momentum between different parts of the body. A series of *fouetté* turns, commonly seen in ballet, represents a movement in which the style of the dance form and the mechanical properties that allow for the smooth execution of the movement fit beautifully together. Figure 5 shows one turn in a *fouetté* turn series. This is a form of repeated *pirouette* in which once during each revolution, when the dancer is facing the audience, the right leg moves from its *pirouette* position at the left knee toward the front, rotates from front to side while straight, then returns to the left knee. During that time the angular momentum of the turn is stored in

(a)

(b)

FIGURE 5 Lisa de Ribère performing *fouetté* turns. Note that she rotates very little while the leg has most of the angular momentum in views (*a*), (*b*), and (*c*). Then the body turns rapidly in (*d*) and (*e*) while the leg is held close to the body, where it has less rotational inertia.

(c)

(d)

(e)

the rotating leg, allowing the rest of the body to pause in its rotation while facing the audience. This pause accomplishes two purposes. First, it makes the shape of the movement consistent with the style of classical ballet, in which the body is usually in a position "open" to the audience. In this case, a significant fraction of the total time for the turn is spent with the body facing the audience between successive turns. Second, the pause allows the dancer to descend from a pointed supporting left foot

down to a flat foot. From this position a twist against the floor with the flat left foot can produce the torque that replaces the angular momentum lost to friction during the preceding turn.

Let us calculate the ratio of the time spent facing the audience to the time spent turning during each cycle of the movement. Assume that the rotational inertia of the body rotating at angular velocity ω_b in the normal *pirouette* position is $\mathbf{I}_b = 0.62 \text{ K} \cdot \text{m}^2$, and for the horizontal leg rotating at ω_l around a vertical axis through the

hip joint is $\mathbf{I}_l = 2.55 \text{ K} \cdot \text{m}^2$. (These and other numerical data for a dancer's body can be found in *The Physics of Dance* by Kenneth Laws, Schirmer Books, 1984, p. 137.) If the angular momentum is approximately constant during one complete cycle of the movement, and the transitions between the two configurations are brief, the ratio ω_b/ω_l can be calculated. The leg by itself during its active phase rotates through an angle of about 90°, but the body as a whole must rotate through a complete 360° turn.

Question 3

Calculate the ratio of pause time (leg rotating) to turn time.

Jumps are common in dance, and jumps with turns in the air are particularly impressive. A *tour jeté* is a jump with a 180° turn around an almost-vertical axis, the legs crossing in the air so that the landing is to the foot opposite the take-off foot. (See Fig. 6.) The movement is most effective if the rotation apparently occurs only after the dancer leaves the floor. Can the body rotate so as to change its orientation in the air even if it has no rotational momentum?

Zero angular momentum turns are indeed possible. Note that as the dancer leaves the floor (Fig. 6a) the left leg is extended to the front where it has a large rotational inertia, being far from the axis of rotation. But the torso, head, right leg, and arms are close to the rotation axis. Therefore the left leg, with its large rotational inertia, can rotate through a small angle in one direction while the rest of the body rotates through a large angle—close to 180°—in the other direction, the angular momenta of the two rotations adding to zero throughout that rotation process. Then, after the turn has occurred, the legs reverse positions, the left leg descending for the landing, and the right rising to a similar position in space, now behind the body. The entire turn has taken place with no net angular momentum.

These turning movements—*pirouettes*, supported *pirouettes*, *fouetté* turns, *tours jetés*—are only a small part of the varied vocabulary of dance. When observing dance, we can appreciate these and many other movements more deeply when aesthetic appreciation is enriched by an understanding of the way the dancers work within the constraints of physical law.

An Experiment

A *demi-fouetté* is a zero angular momentum turn that can be performed quite simply and demonstrates the principle described in the analysis of the *tour jeté* above. Starting with the arms overhead and the left leg extended to the front, rise onto the ball of the supporting right foot, then quickly rotate the horizontal left leg to the left around to the rear of the body. This movement results in the torso, head, arms, and supporting leg rotating to the right. Through what angle has the left leg rotated? Through what angle has the rest of the body rotated? What do those angles tell you about the moment of inertia of the rotating leg, which is far from the axis of rotation, compared to the moment of inertia of the rest of the body, which is kept as close as possible to the rotation axis?

Since the friction between the supporting foot and the floor does perturb the process, a better way to perform the movement is to jump into the air from the initial position described above, *then* carry out the rotating movements, landing again on the same foot after the turn.

Answers to Questions

1. About twice as much!
2. 75%.
3. About 1!

FIGURE 6 The sequence of positions in a *tour jeté* performed by Sean Lavery.

(*a*)

(*b*)

(*c*)

(*d*)

EQUILIBRIUM AND ELASTICITY | 13

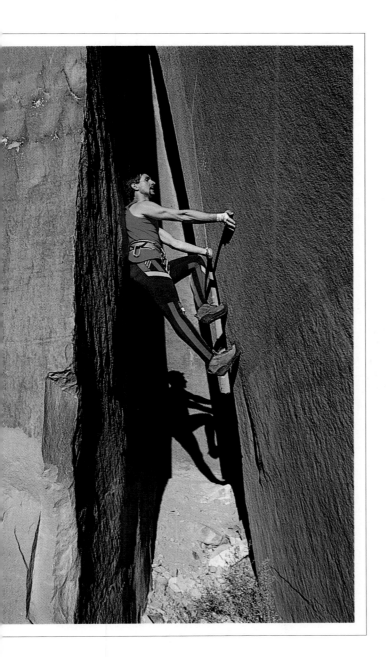

Rock climbing may be the ultimate physics exam. Failure can mean death, and even "partial credit" can mean severe injury. For example, in a long chimney climb, in which your shoulders are pressed against one wall of a wide, vertical fissure and your feet are pressed against the opposite wall, you need to rest occasionally or you will fall due to exhaustion. Here the exam consists of a single question: What can you do to relax your push on the walls in order to rest? If you relax without considering the physics, the walls will not hold you up. This life-and-death one-question exam is answered in this chapter.

13-1 EQUILIBRIUM

Consider these objects: (1) a book resting on a table, (2) a hockey puck sliding across the ice, (3) the rotating blades of a ceiling fan, and (4) the wheel of a bicycle that is traveling along a straight path at constant speed. For each of these four objects:

1. The linear momentum **P** of its center of mass is constant.

2. Its angular momentum **L** about its center of mass, or about any other point, is also constant.

We say that such objects are in **equilibrium.** The two requirements for equilibrium are then

$$\mathbf{P} = \text{a constant} \quad \text{and} \quad \mathbf{L} = \text{a constant.} \quad (13\text{-}1)$$

Our concern in this chapter is with situations in which the constants in Eq. 13-1 are in fact zero. That is, we are concerned largely with objects that are not moving in any way—either in translation or in rotation—in the reference frame from which we observe them. Such objects are in **static equilibrium.** Of the four objects mentioned at the beginning of this section, only one—the book resting on the table—is in static equilibrium.

The balancing rock of Fig. 13-1 is another example of an object that, for the present at least, is in static equilibrium. It shares this property with countless other structures, such as cathedrals, houses, filing cabinets, and taco stands, that remain stationary over time.

As we discussed in Section 8-5, if a body returns to a state of static equilibrium after having been displaced from it by a force, the body is said to be in *stable* static equilibrium. A marble placed at the bottom of a hemispherical bowl is an example. If, however, a small force can displace the body and disrupt the equilibrium, the body is in *unstable* static equilibrium. The construction worker in Fig. 13-2 is just one example.

The analysis of static equilibrium is very important in engineering practice. The design engineer must isolate and identify all the external forces and torques that act on a structure and, by good design and wise choice of materials, make sure that the

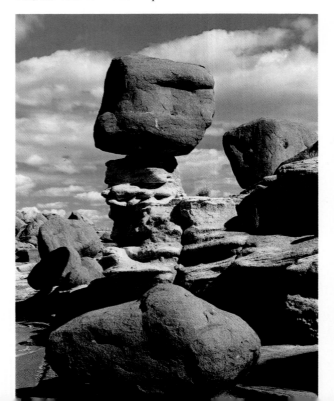

FIGURE 13-1 A balancing rock near Petrified Forest National Park in Arizona. Although its perch seems precarious, the rock is in static equilibrium.

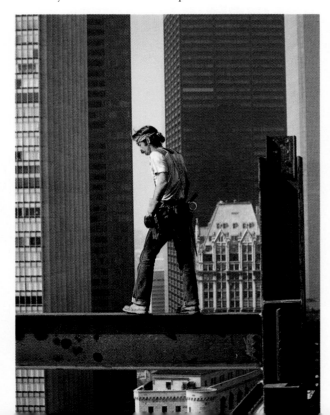

FIGURE 13-2 A balancing construction worker in New York City is in unstable static equilibrium.

structure will tolerate these loads. Such analyses are necessary to make sure, for example, that bridges do not collapse under their traffic and wind loads, and that the landing gear of aircraft will survive the shock of rough landings.

13-2 THE REQUIREMENTS OF EQUILIBRIUM

The translational motion of a body is governed by Newton's second law in its linear form, given by Eq. 9-28 as

$$\sum \mathbf{F}_{\text{ext}} = \frac{d\mathbf{P}}{dt}. \qquad (13\text{-}2)$$

If the body is in translational equilibrium, that is, if **P** is a constant, then $d\mathbf{P}/dt = 0$ and we must have

$$\sum \mathbf{F}_{\text{ext}} = 0 \qquad \text{(balance of forces).} \qquad (13\text{-}3)$$

The rotational motion of a body is governed by Newton's second law in its angular form, given by Eq. 12-37 as

$$\sum \boldsymbol{\tau}_{\text{ext}} = \frac{d\mathbf{L}}{dt}. \qquad (13\text{-}4)$$

If the body is in rotational equilibrium, that is, if **L** is a constant, then $d\mathbf{L}/dt = 0$ and we must have

$$\sum \boldsymbol{\tau}_{\text{ext}} = 0 \qquad \text{(balance of torques).} \qquad (13\text{-}5)$$

Thus the two requirements for a body to be in equilibrium are as follows:

1. The vector sum of all the external forces that act on the body must be zero.

2. The vector sum of all the external torques that act on the body, measured about *any* possible point, must also be zero.

These requirements obviously hold for *static* equilibrium as well as the more general equilibrium in which **P** and **L** are constant but not zero.

If you doubt that structures are held in equilibrium by a nice balance of external forces and torques, look at Fig. 13-3. The weight of the building had been balanced, perhaps for decades, by forces that were exerted upward on the structure by its foundation. When the foundation was blown out, sideways, by explosive charges, some of the forces

were suddenly removed, and the first requirement for equilibrium, Eq. 13-3, was suddenly no longer satisfied. At the instant the photograph was taken, the structure was accelerating downward and was certainly not in equilibrium, static or otherwise.

Equations 13-3 and 13-5, as vector equations, are each equivalent to three independent scalar equations, one for each direction of the coordinate axes:

Balance of forces	Balance of torques	
$\sum F_x = 0$	$\sum \tau_x = 0$	
$\sum F_y = 0$	$\sum \tau_y = 0$	(13-6)
$\sum F_z = 0$	$\sum \tau_z = 0.$	

For convenience, we have dropped the subscript ext.

We simplify matters by considering only situations in which the forces that act on the body lie in the *xy* plane. This means that the only torques that can act on the body must tend to cause rotation around an axis parallel to the *z* axis. With this assumption, we eliminate one force equation and two torque equations from Eq. 13-6, leaving

$$\sum F_x = 0 \qquad \text{(balance of forces),} \qquad (13\text{-}7)$$

$$\sum F_y = 0 \qquad \text{(balance of forces),} \qquad (13\text{-}8)$$

and

$$\sum \tau_z = 0 \qquad \text{(balance of torques).} \qquad (13\text{-}9)$$

Here, F_x and F_y are, respectively, the *x* and the *y* components of the external forces that act on the

FIGURE 13-3 If the forces that hold a structure in equilibrium are removed, the structure cannot remain in equilibrium.

body, and τ_z represents the torques that these forces exert about the z axis, or about *any* axis parallel to it.

A hockey puck sliding at constant velocity over ice satisfies Eqs. 13-7, 13-8, and 13-9 and is thus in equilibrium *but not in static equilibrium.* For static equilibrium, the linear momentum **P** of the puck must be not only constant but also zero; the puck must be resting on the ice. As the mathematicians say, Eqs. 13-7, 13-8, and 13-9 are *necessary* conditions for static equilibrium but they are not *sufficient.*

13-3 THE CENTER OF GRAVITY

One force that always acts on a body near the surface of the Earth is the body's weight, the gravitational force. Earlier, we did not hesitate to describe this with a single vector $M\mathbf{g}$ and to attach that vector to the center of mass of the body. Weight, however, is a *body force,* acting separately on every atom of the body. We can replace the vector sum of all these tiny forces with a single force acting at the center of mass only if the gravitational acceleration vector **g** has the same magnitude and direction for all points of space occupied by the body. Let us prove that this is so.

Consider a body of arbitrary shape, as in Fig. 13-4. There is a point, called the body's **center of gravity** (cg) where the vector sum of all the weight forces on the atoms is said to act. If the body is supported at this point, it is in static equilibrium for any orientation, and the required supporting force is equal to the weight Mg of the body. The center of gravity may or may not coincide with the center of mass.

To demonstrate these features of the center of gravity, and to locate it, we place the origin O of a coordinate system at the center of gravity and divide the body into small mass elements Δm, one of which is shown in Fig. 13-4. The weight $\Delta m\mathbf{g}$ of this element acts downward. We shall assume that **g** has the same value (magnitude and direction) for all points of the body. We put an upward force **F**′ at O and we hope to show that this single force will hold the body in equilibrium (1) if and only if the magnitude of **F**′ is Mg and (2) if O, which we have identified as the center of gravity, is in fact the center of mass.

If the body is to be in equilibrium, Eqs. 13-3 and 13-5 must be satisfied. We look first at Eq. 13-3, which deals with the balance of forces. We can write this equation as

$$\sum \mathbf{F}_{\text{ext}} = \mathbf{F}' + \sum \Delta m\, \mathbf{g} = \mathbf{F}' + \mathbf{g}\left(\sum \Delta m\right)$$
$$= \mathbf{F}' + M\mathbf{g} = 0, \qquad (13\text{-}10)$$

in which the sum is taken over all the mass elements that make up the body. Equation 13-10 tells us that Eq. 13-3, the balance of forces condition, will be satisfied if **F**′ is directed opposite $M\mathbf{g}$ and has magnitude Mg. Our first point is proved. Note that, in Eq. 13-10, we were able to remove **g** from the sum only because we have assumed it is the same for all points of the body.

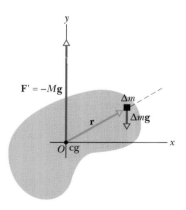

FIGURE 13-4 The weight of a body, though distributed throughout its volume, may be balanced by a single force of magnitude Mg acting at the center of gravity of the body. If **g** is the same throughout the body, the center of gravity coincides with the body's center of mass.

Michel Menin walks a taut wire 10,335 ft above French farmland, adjusting the position of a heavy pole to keep his center of mass and the pole over the wire in spite of wind gusts.

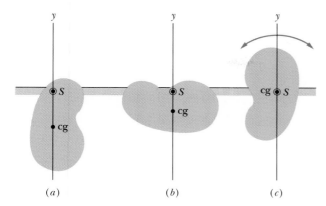

FIGURE 13-5   A body suspended from an arbitrary point *S*, as in (*a*) and (*b*), will be in (stable) equilibrium only if its center of gravity hangs vertically below its suspension point. (*c*) If a body is suspended from its center of gravity, that body is in equilibrium no matter what its orientation.

Now let us look at Eq. 13-5, the balance of torques. The vector $\mathbf{F}'$ in Fig. 13-4, which passes through O, has no torque about the z axis through that point. Thus Eq. 13-5 becomes

$$\sum \boldsymbol{\tau}_{\text{ext}} = \sum \mathbf{r} \times (\Delta m \, \mathbf{g})$$
$$= \left(\sum \Delta m \, \mathbf{r} \right) \times \mathbf{g} = 0. \quad (13\text{-}11)$$

The quantity $\sum \Delta m \, \mathbf{r}$ in this equation, which tells us how the mass of the body is distributed about point O, is zero if point O is the center of mass; see Eq. 9-8. Thus Eq. 13-11 tells us that Eq. 13-5, the balance of torques condition, will be satisfied if point O is indeed the center of mass.* Our second point is proved.

Figures 13-5*a* and 13-5*b* suggest that, if the force $\mathbf{F}'$, which is equal to $-M\mathbf{g}$, is applied to an arbitrary point S of a body, the body will be in equilibrium *only* if its center of gravity lies vertically below its suspension point. Otherwise $\sum \boldsymbol{\tau}_{\text{ext}}$ is not zero (Eq. 13-11 is not satisfied), and the body rotates until its center of gravity *is* below the suspension point. (Indeed, we saw in Sample Problem 9-2 how we can use this fact to find the center of mass of a body.) However, if $\mathbf{F}'$ is applied at the center of gravity of the body, as in Fig. 13-5*c*, the body will be in equilibrium no matter what its orientation. That is, you can turn

the body any way you wish about that point and it will remain in equilibrium.

If $\mathbf{g}$ does *not* have the same value for all points occupied by the body, we cannot remove it, as a constant, from the sum in Eq. 13-10 and our proof fails. The center of gravity and the center of mass are no longer guaranteed to coincide in such cases.

13-4 SOME EXAMPLES OF STATIC EQUILIBRIUM

In this section we examine six sample problems involving static equilibrium. In each, we select a system of one or more objects and apply the equations of equilibrium (Eqs. 13-7, 13-8, and 13-9). The forces involved in the equilibrium are all in the xy plane, which means that the torques involved are parallel to the z axis. Thus, in applying Eq. 13-9, the balance of torques, we select an axis parallel to the z axis for calculating the torques. Although Eq. 13-9 must be satisfied for *any* such choice of axis, we shall see that certain choices simplify the application of Eq. 13-9 by eliminating one or more unknown force terms.

SAMPLE PROBLEM 13-1

A uniform beam of length L whose mass m is 1.8 kg rests with its ends on two digital scales, as in Fig. 13-6*a*. A uniform block whose mass M is 2.7 kg rests on the beam, its center a distance $L/4$ from the beam's left end. What do the scales read?

We choose as our system the beam and the block, taken together. Figure 13-6*b* is a free-body diagram for this system, showing all the forces that act on it. The scales push upward at the ends of the beam with forces $\mathbf{F}_l$ and $\mathbf{F}_r$. The magnitudes of these two forces are the scale readings that we seek. The weight of the beam, $m\mathbf{g}$, acts downward at the beam's center of mass. Similarly, $M\mathbf{g}$, the weight of the block, acts downward at *its* center of mass. In Fig. 13-6*b*, the block is represented by a dot within the boundary of the beam, and the vector $M\mathbf{g}$ is drawn with its tail on that dot. (In drawing Fig. 13-6*b* from 13-6*a*, the vector $M\mathbf{g}$ is shifted in the direction it points, that is, along its line of action. The shift does not alter $M\mathbf{g}$, or a torque due to $M\mathbf{g}$ about any axis perpendicular to the figure.)

Our system is in static equilibrium so that the balance of forces equations (Eqs. 13-7 and 13-8) and the balance of torques equation (Eq. 13-9) apply. We solve this problem in two equivalent ways.

*In the cross product of Eq. 13-11, we were careful not to change the order in which the vectors $\mathbf{r}$ and $\mathbf{g}$ appear. We were free to move Δm to the left, however, because mass is a scalar quantity.

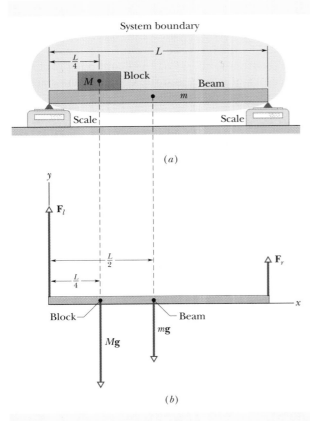

(a)

(b)

FIGURE 13-6 Sample Problem 13-1. (*a*) A beam of mass *m* supports a block of mass *M*. What are the readings on the digital scales that support the ends of the beam? The system boundary is marked. (*b*) A free-body diagram, showing the forces that act on the system *beam + block*.

FIRST SOLUTION The forces have no *x* components, so Eq. 13-7, which is $\sum F_x = 0$, provides no information. Equation 13-8 gives, for the magnitudes of the *y* components,

$$\sum F_y = F_l + F_r - Mg - mg = 0. \quad (13\text{-}12)$$

We have two unknown forces (F_l and F_r) but we cannot find them separately because we have only one equation. Fortunately, we have another equation at hand, namely, Eq. 13-9, the balance of torques equation.

We can apply Eq. 13-9 to *any* axis perpendicular to the plane of Fig. 13-6. Let us choose an axis through the left end of the beam. We take as positive torques that—acting alone—would produce a counterclockwise rotation of the beam about our axis. We then have,

from Eq. 13-9,

$$\sum \tau_z = (F_l)(0) + (F_r)(L) - (mg)(L/2) - (Mg)(L/4)$$
$$= 0,$$

or

$$F_r = (g/4)(2m + M)$$
$$= (\tfrac{1}{4})(9.8 \text{ m/s}^2)(2 \times 1.8 \text{ kg} + 2.7 \text{ kg})$$
$$= 15 \text{ N.} \qquad \text{(Answer)} \quad (13\text{-}13)$$

Note how choosing an axis that passes through the application point of one of the unknown forces, F_l, eliminates that force from Eq. 13-9 and allows us to solve directly for the other force.

If we solve Eq. 13-12 for F_l and substitute known quantities, we find

$$F_l = (M + m)g - F_r$$
$$= (2.7 \text{ kg} + 1.8 \text{ kg})(9.8 \text{ m/s}^2) - 15 \text{ N}$$
$$= 29 \text{ N.} \qquad \text{(Answer)}$$

SECOND SOLUTION As a check, let us solve this problem in a different way, applying the balance of torques equation about two different axes. Choosing first an

To be stable, the center of mass of this stack of performers and chairs must be on a vertical line that cuts through the stage somewhere between the legs of the lowest chair.

axis through the left end of the beam, as we did above, we find Eq. 13-13 and the solution $F_r = 15$ N.

For an axis passing through the right end of the beam, Eq. 13-9 yields

$$\sum \tau_z = (F_r)(0) - (F_l)(L) + (mg)(L/2)$$
$$+ (Mg)(3L/4)$$
$$= 0.$$

Solving for F_l, we find

$$F_l = (g/4)(2m + 3M)$$
$$= (\tfrac{1}{4})(9.8 \text{ m/s}^2)(2 \times 1.8 \text{ kg} + 3 \times 2.7 \text{ kg})$$
$$= 29 \text{ N}, \qquad \text{(Answer)}$$

in agreement with our earlier result. Note that the length of the beam does not enter this problem explicitly, but only as it affects the mass of the beam.

SAMPLE PROBLEM 13-2

A bowler holds a bowling ball whose mass M is 7.2 kg in the palm of his hand. As Fig. 13-7a shows, his upper arm is vertical and his lower arm is horizontal. What forces must the biceps muscle and the bony structure of the upper arm exert on the lower arm? The forearm and hand together have a mass m of 1.8 kg, and the needed dimensions are as shown in Fig. 13-7a.

SOLUTION Our system is the lower arm and the bowling ball, taken together. Figure 13-7b shows a free-body diagram of the system. (The ball is represented by a dot within the boundary of the lower arm; the ball's weight vector Mg has its tail on that dot. In drawing Fig. 13-7b from 13-7a, the vector Mg is shifted along its line of action; the shift does not alter Mg, or a torque due to Mg about any axis perpendicular to the figure.) The unknown forces are **T**, the force exerted by the biceps muscle, and **F**, the force exerted by the upper arm on the lower arm. The forces are all vertical.

From Eq. 13-8, which is $\Sigma F_y = 0$, we find, considering magnitudes only,

$$\sum F_y = T - F - mg - Mg = 0. \qquad (13\text{-}14)$$

Applying Eq. 13-9 about an axis through O at the elbow, and taking torques that would cause counterclockwise rotations as positive, we obtain

$$\sum \tau_z = (T)(d) + (F)(0) - (mg)(D) - (Mg)(L)$$
$$= 0. \qquad (13\text{-}15)$$

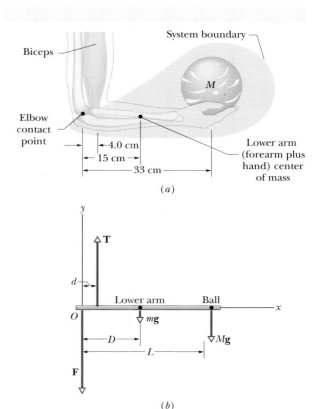

FIGURE 13-7 Sample Problem 13-2. (a) A hand holds a bowling ball. The system boundary is marked. (b) A free-body diagram, of the *lower arm + ball* system, showing the forces that act. The vectors are not to scale; the powerful forces exerted by the biceps muscle and the elbow joint are much larger than either weight shown.

By choosing our axis to pass through point O, we have eliminated the variable F from Eq. 13-15. Solved for T, it yields

$$T = g \, \frac{mD + ML}{d}$$
$$= (9.8 \text{ m/s}^2) \, \frac{(1.8 \text{ kg})(15 \text{ cm}) + (7.2 \text{ kg})(33 \text{ cm})}{4.0 \text{ cm}}$$
$$= 648 \text{ N} \approx 650 \text{ N}. \qquad \text{(Answer)}$$

Thus the biceps muscle must pull up on the forearm with a force that is about nine times larger than the weight of the bowling ball.

If we solve Eq. 13-14 for F and substitute known quantities into it, we find

$$F = T - g(M + m)$$

$$= 648 \text{ N} - (9.8 \text{ m/s}^2)(7.2 \text{ kg} + 1.8 \text{ kg})$$

$$= 560 \text{ N}, \qquad \text{(Answer)}$$

which is about eight times the weight of the bowling ball.

SAMPLE PROBLEM 13-3

A ladder whose length L is 12 m and whose mass m is 45 kg rests against a wall. Its upper end is a distance h of 9.3 m above the ground, as in Fig. 13-8a. The center of mass of the ladder is one-third of the way up the ladder. A fire fighter whose mass M is 72 kg climbs the ladder until her center of mass is halfway up. Assume that the wall, but not the ground, is frictionless. What forces are exerted on the ladder by the wall and by the ground?

SOLUTION Figure 13-8b shows a free-body diagram of the *fire fighter + ladder* system. (The fire fighter is represented by a dot within the boundary of the ladder; her weight vector $M\mathbf{g}$ has its tail on that dot. In drawing Fig. 13-8b from 13-8a, the vector $M\mathbf{g}$ is shifted along its line

of action; the shift does not alter $M\mathbf{g}$, or a torque due to $M\mathbf{g}$ about any axis perpendicular to the figure.) The wall exerts a horizontal force $\mathbf{F}_w$ on the ladder; it can exert no vertical force because the wall–ladder contact is assumed to be frictionless. The ground exerts a force $\mathbf{F}_g$ on the ladder with a horizontal component F_{gx} (due to friction) and a vertical component F_{gy} (the usual normal force). We choose coordinate axes as shown, with the origin O at the point where the ladder meets the ground. The distance a from the wall to the foot of the ladder is readily found from

$$a = \sqrt{L^2 - h^2} = \sqrt{(12 \text{ m})^2 - (9.3 \text{ m})^2} = 7.58 \text{ m}.$$

From Eqs. 13-7 and 13-8, the balance of forces equations, we have for the system, respectively,

$$\sum F_x = F_w - F_{gx} = 0 \qquad (13-16)$$

and

$$\sum F_y = F_{gy} - Mg - mg = 0. \qquad (13-17)$$

Equation 13-17 yields

$$F_{gy} = g(M + m) = (9.8 \text{ m/s}^2)(72 \text{ kg} + 45 \text{ kg})$$

$$= 1146.6 \text{ N} \approx 1100 \text{ N}. \qquad \text{(Answer)}$$

We next balance the torques acting on the system, choosing an axis through O, perpendicular to the plane of the figure. The moment arms about O for $\mathbf{F}_w$, $M\mathbf{g}$, $m\mathbf{g}$, and $\mathbf{F}_g$ are h, $a/2$, $a/3$, and zero, respectively. Note that the zero moment arm for $\mathbf{F}_g$ means this force

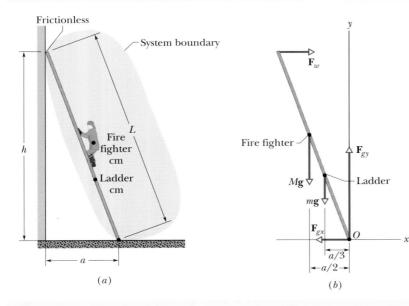

FIGURE 13-8 Sample Problems 13-3 and 13-4. (a) A fire fighter climbs halfway up a ladder that is leaning against a frictionless wall. The ground is not frictionless. (b) A free-body diagram, showing (to scale) the forces that act on the system.

produces zero torque about O. From Eq. 13-9, the balance of torques equation, we have

$$\sum \tau_z = -(F_w)(h) + (Mg)(a/2) + (mg)(a/3)$$
$$= 0. \tag{13-18}$$

Note that our choice of a location for an axis eliminated two variables, F_{gx} and F_{gy}, from the balance of torques equation. Solving Eq. 13-18 for F_w, we find that

$$F_w = \frac{ga(M/2 + m/3)}{h}$$

$$= \frac{(9.8 \text{ m/s}^2)(7.58 \text{ m})(72/2 \text{ kg} + 45/3 \text{ kg})}{9.3 \text{ m}}$$

$$= 407 \text{ N} \approx 410 \text{ N}. \qquad \text{(Answer)}$$

From Eq. 13-16 we then have

$$F_{gx} = F_w = 410 \text{ N}. \qquad \text{(Answer)}$$

SAMPLE PROBLEM 13-4

In Sample Problem 13-3, the coefficient of static friction μ_s between the ladder and the ground is 0.53. How far up the ladder can the fire fighter go before the ladder starts to slip?

SOLUTION The forces that act have the same labels as in Fig. 13-8. Let q be the fraction of the ladder's length the fire fighter climbs before slipping occurs (she is then a horizontal distance qa from O). At the onset of slipping, we have

$$F_{gx} = \mu_s F_{gy}, \tag{13-19}$$

in which F_{gx} is the static frictional force (usually symbolized as f_s) and F_{gy} is the normal force (usually symbolized as N).

If we apply Eq. 13-9, the balance of torques equation, about an axis through O we have, also at the onset of slipping,

$$\sum \tau_z = -(F_w)(h) + (mg)(a/3) + (Mg)(qa) = 0,$$

or

$$F_w = \frac{ga}{h}\left(\tfrac{1}{3}m + Mq\right). \tag{13-20}$$

This equation shows us that as the fire fighter climbs the ladder, that is, as q increases, the force F_w exerted by the wall must increase if equilibrium is to be maintained. To find q at the onset of slipping, we must first find F_w.

Equation 13-7, the balance of forces equation for the x direction, gives

$$\sum F_x = F_w - F_{gx} = 0.$$

If we combine this equation with Eq. 13-19 we find, for the ladder at the onset of slipping,

$$F_w = F_{gx} = \mu_s F_{gy}. \tag{13-21}$$

From Eq. 13-8, the balance of forces equation for the y direction, we have

$$\sum F_y = F_{gy} - Mg - mg,$$

or

$$F_{gy} = (M + m)g. \tag{13-22}$$

We find, combining Eqs. 13-21 and 13-22,

$$F_w = \mu_s g(M + m). \tag{13-23}$$

If, finally, we combine Eqs. 13-20 and 13-23 and solve for q, we have

$$q = \frac{\mu_s h}{a}\frac{(M + m)}{M} - \frac{m}{3M}$$

$$= \frac{(0.53)(9.3 \text{ m})}{7.6 \text{ m}}\frac{(72 \text{ kg} + 45 \text{ kg})}{72 \text{ kg}} - \frac{45 \text{ kg}}{(3)(72 \text{ kg})}$$

$$= 0.85. \qquad \text{(Answer)} \quad (13-24)$$

The fire fighter can climb 85% of the way up the ladder before it starts to slip.

You can show from Eq. 13-24 that the fire fighter can climb all the way up the ladder (which corresponds to $q = 1$) without slipping if $\mu_s > 0.61$. On the other hand, the ladder will slip when weight is put on the first rung (which corresponds to $q \approx 0$) if $\mu_s < 0.11$.

SAMPLE PROBLEM 13-5

Figure 13-9a shows a safe, whose mass M is 430 kg, hanging by a rope from a boom whose dimensions a and b are 1.9 m and 2.5 m, respectively. The boom's uniform beam has a mass m of 85 kg; the mass of the horizontal cable is negligible.

a. Find the tension T in the cable.

SOLUTION Figure 13-9b is a free-body diagram of the beam, which we take as our system. The forces acting on it are the tension **T** exerted by the cable, the weight M**g** of the safe, the weight m**g** of the beam itself, and the force components F_h and F_v exerted on the beam by the hinge that fastens it to the wall.

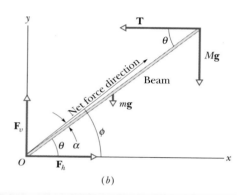

FIGURE 13-9 Sample Problem 13-5. (a) A heavy safe is hung from a boom consisting of a horizontal steel cable and a uniform beam. (b) A free-body diagram for the beam. Note that the net force acting on the beam due to $\mathbf{F}_v$ and $\mathbf{F}_h$ does not point directly along the beam axis.

Let us apply Eq. 13-9, the balance of torques equation, to an axis through the hinge perpendicular to the plane of the figure. Taking torques that would cause counterclockwise rotations as positive, we have

$$\sum \tau_z = (T)(a) - (Mg)(b) - (mg)(\tfrac{1}{2}b) = 0.$$

By our wise choice of an axis, we have eliminated the unknown forces F_h and F_v from this equation (since they create no torque about that axis), leaving only the unknown force T. Solving for T yields

$$T = \frac{gb(M + \tfrac{1}{2}m)}{a}$$

$$= \frac{(9.8 \text{ m/s}^2)(2.5 \text{ m})(430 \text{ kg} + 85/2 \text{ kg})}{1.9 \text{ m}}$$

$$= 6090 \text{ N} \approx 6100 \text{ N}. \qquad \text{(Answer)}$$

b. Find the forces F_h and F_v exerted on the beam by the hinge.

SOLUTION We now apply the balance of forces equations. From Eq. 13-7 we have

$$\sum F_x = F_h - T = 0,$$

and so

$$F_h = T = 6090 \text{ N} \approx 6100 \text{ N}. \qquad \text{(Answer)}$$

From Eq. 13-8 we have

$$\sum F_y = F_v - mg - Mg = 0,$$

and so

$$F_v = g(m + M) = (9.8 \text{ m/s}^2)(85 \text{ kg} + 430 \text{ kg})$$

$$= 5047 \text{ N} \approx 5000 \text{ N}. \qquad \text{(Answer)}$$

c. What is the magnitude F of the net force exerted by the hinge on the beam?

SOLUTION From the figure we see that

$$F = \sqrt{F_h^2 + F_v^2}$$

$$= \sqrt{(6090 \text{ N})^2 + (5047 \text{ N})^2} \approx 7900 \text{ N}. \qquad \text{(Answer)}$$

Note that F is substantially greater than either the combined weights of the safe and the beam, 5000 N, or the tension in the horizontal wire, 6100 N.

d. What is the angle α between the direction of the net force $\mathbf{F}$ exerted on the beam by the hinge and the center line of the beam?

SOLUTION From the figure we see that

$$\theta = \tan^{-1}\frac{a}{b} = \tan^{-1}\frac{1.9 \text{ m}}{2.5 \text{ m}} = 37.2°,$$

and

$$\phi = \tan^{-1}\frac{F_v}{F_h} = \tan^{-1}\frac{5047 \text{ N}}{6090 \text{ N}} = 39.6°.$$

So

$$\alpha = \phi - \theta = 39.6° - 37.2° = 2.4°. \qquad \text{(Answer)}$$

If the weight of the beam were small enough to neglect, you would find $\alpha = 0$; that is, the hinge force would point directly along the beam axis.

SAMPLE PROBLEM 13-6

In Fig. 13-10, a rock climber with mass $m = 55$ kg rests during a "chimney climb" in a fissure of width $w = 1.0$ m. Her center of mass is a horizontal distance $d = 0.20$ m from the wall against which her shoulders are pressed. The coefficient of static friction between her shoes and the wall is $\mu_1 = 1.1$, and between her shoulders and the wall it is $\mu_2 = 0.70$.

a. What minimum horizontal push must she exert on the walls to keep from falling?

SOLUTION Her horizontal push against the wall is the same at her shoulders and at her feet. That push must be equal in magnitude to the normal force $\mathbf{N}$ on her due to the wall at each of those places. Thus there is no net horizontal force on her and Eq. 13-7, $\Sigma F_x = 0$, is satisfied.

As her weight $m\mathbf{g}$ acts to slide her down the walls, static frictional forces $\mathbf{f}_1$ and $\mathbf{f}_2$ automatically act upward at feet and shoulders, respectively, to counter the tendency to slide. Thus Eq. 13-8, $\Sigma F_y = 0$, is satisfied and we have

$$f_1 + f_2 = mg. \qquad (13\text{-}25)$$

Let us assume that she initially pushes hard against the walls, and then she relaxes her push. As she decreases her push, the magnitude N of the normal force decreases, and so do the products $\mu_1 N$ and $\mu_2 N$ that limit the static friction at feet and shoulders, respectively (see Eq. 6-1).

Suppose N is reduced until the limit $\mu_1 N$ equals the magnitude f_1 of the frictional force at her feet, and the limit $\mu_2 N$ equals the magnitude f_2 of the frictional force at her shoulders. She is then on the verge of slipping at both places. If she were to relax her push any more, the resulting decrease in the limits $\mu_1 N$ and $\mu_2 N$ would leave $f_1 + f_2$ less than mg, and she would fall. Thus for the situation when she is on the verge of slipping, Eq. 13-25 becomes

$$\mu_1 N + \mu_2 N = mg, \qquad (13\text{-}26)$$

which yields

$$N = \frac{mg}{\mu_1 + \mu_2} = \frac{(55 \text{ kg})(9.8 \text{ m/s}^2)}{1.1 + 0.70} = 299 \text{ N} \approx 300 \text{ N}.$$

FIGURE 13-10 Sample Problem 13-6. The forces on a climber resting in a rock chimney. The push of the climber on the chimney walls gives rise to the normal forces $\mathbf{N}$ (which are equal in magnitude) and the frictional forces $\mathbf{f}_1$ and $\mathbf{f}_2$.

Thus her minimum horizontal push must be about 300 N.

b. For that push, what must be the vertical distance h between her feet and shoulders if she is to be stable?

SOLUTION To satisfy Eq. 13-9, $\Sigma \tau = 0$, the forces acting on her must not create a net torque about *any* rotation axis. Let us consider a rotational axis perpendicular to the page at her shoulders. The net torque about that axis is given by

$$\Sigma \tau = -f_1 w + Nh + mgd = 0. \qquad (13\text{-}27)$$

Solving this for h, setting $f_1 = \mu_1 N$, and substituting $N = 299$ N and other known values, we find

$$h = \frac{f_1 w - mgd}{N} = \frac{\mu_1 Nw - mgd}{N} = \mu_1 w - \frac{mgd}{N}$$

$$= (1.1)(1.0 \text{ m}) - \frac{(55 \text{ kg})(9.8 \text{ m/s}^2)(0.20 \text{ m})}{299 \text{ N}}$$

$$= 0.739 \text{ m} \approx 0.74 \text{ m}. \qquad \text{(Answer)}$$

We would find the same required value of h if we chose any other rotation axis perpendicular to the page, such as one at her feet.

c. What are the magnitudes of the frictional forces supporting her?

SOLUTION For $N = 299$ N, we have

$$f_1 = \mu_1 N = (1.1)(299 \text{ N})$$

$$= 328.9 \text{ N} \approx 330 \text{ N,} \qquad \text{(Answer)}$$

and from Eq. 13-25 we then have

$$f_2 = mg - f_1 = (55 \text{ kg})(9.8 \text{ m/s}^2) - 328.9 \text{ N}$$

$$= 210.1 \text{ N} \approx 210 \text{ N.} \qquad \text{(Answer)}$$

d. Is she stable if she exerts the same force (299 N) on the rock when her feet are higher, with $h = 0.37$ m?

SOLUTION Again $N = 299$ N, and the net torque on the climber about any rotation axis must be zero. For the axis at her shoulders, we obtain f_1 from Eq. 13-27:

$$f_1 = \frac{Nh + mgd}{w}$$

$$= \frac{(299 \text{ N})(0.37 \text{ m}) + (55 \text{ kg})(9.8 \text{ m/s}^2)(0.20 \text{ m})}{1.0 \text{ m}}$$

$$= 218 \text{ N.}$$

This is less than the limit $\mu_1 N$ (= 329 N) and thus is possible to attain.

Next we use Eq. 13-25 to find the value required of f_2 so that $\Sigma F_y = 0$:

$$f_2 = mg - f_1 = (55 \text{ kg})(9.8 \text{ m/s}^2) - 218 \text{ N} = 321 \text{ N.}$$

This exceeds the limit $\mu_2 N$ (= 209 N) and thus is impossible to attain with a push of 299 N. The only way the climber can avoid slipping when $h = 0.37$ m (or for any value of h less than 0.74 m) is by pushing harder than 299 N on the rock so as to increase the limit $\mu_2 N$.

Similarly, if $h > 0.74$ m, she must also exert a force greater than 299 N on the rock to be stable. Here, then, is the advantage of knowing the physics before you climb a chimney. When you need to rest, you will avoid the (dire) error of novice climbers who place their feet too high or too low. Instead, you will know that there is a "best" distance between shoulders and feet, requiring the least push, and giving you a good chance to rest.

PROBLEM SOLVING

TACTIC 1: STATIC EQUILIBRIUM PROBLEMS
We have consistently followed these procedures in solving Sample Problems 13-1 through 13-6:

1. Draw a *sketch* of the problem.

2. Select the *system* to which you will apply the laws of equilibrium, drawing a closed curve around it on your sketch to fix it clearly in your mind. In some situations you can select a single object as the system; it is the object you wish to be in equilibrium (such as the rock climber in Sample Problem 13-6). In other situations, you might include additional objects in the system *if* their inclusion simplifies the calculations for equilibrium. For example, suppose in Sample Problems 13-3 and 13-4 you select only the ladder as the system. Then in Fig. 13-8*b* you will have to account for additional unknown forces exerted on the ladder by the hands and feet of the fire fighter. These additional unknowns complicate the calculations for equilibrium. The system of Fig. 13-8 was chosen to include the fire fighter so that those unknown forces are *internal* to the system and thus are not required in the solutions of Sample Problems 13-3 and 13-4.

3. Draw a *free-body diagram* of the system. Show all the forces that act on the system, labeling them clearly and making sure that their points of application and lines of action are correctly shown.

4. Draw in the *x and y axes* of a coordinate system. Choose them so that at least one axis is parallel to one or more of the unknown forces. Resolve into

In order to be stable, the unicyclist must align the contact point of the tire on the road with his center of mass, but slight errors require that he continuously roll the wheel back and forth to adjust the alignment.

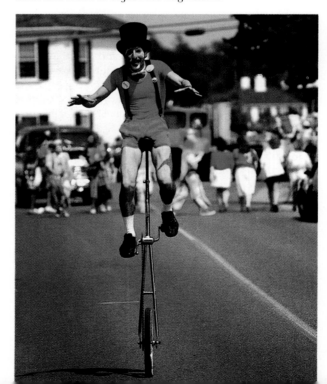

components those forces that do not lie along one of the axes. In all of our sample problems it made sense to choose the *x* axis horizontal and the *y* axis vertical.

5. Write the two *balance of forces equations,* using symbols throughout.

6. Choose one or more axes perpendicular to the plane of the figure and write the *balance of torques equation* for each axis. If you choose an axis that passes through the line of action of an unknown force, the equation will be simplified because that particular force will not appear in it.

7. *Solve* the equations that you have written *algebraically* for the unknowns. Some students feel more confident in substituting numbers with units in the independent equations at this stage, especially if the algebra is particularly involved. With experience, however, proceeding algebraically becomes preferable; see Tactic 6 of Chapter 4.

8. Finally, *substitute numbers* with units in your algebraic solutions, obtaining numerical values for the unknowns.

13-5 INDETERMINATE STRUCTURES

For the problems of this chapter, we have only three independent equations at our disposal, two balance of forces equations, and one balance of torques equation about a given axis. Thus if a problem has more than three unknowns, we cannot solve it.

It is easy to find such problems. In Sample Problems 13-3 and 13-4, for example, we could have assumed that there is friction between the wall and the ladder. Then there would have been a vertical frictional force acting where the ladder touches the wall, making a total of four unknown forces. With only three equations, we could not have solved this problem.

Consider also an unsymmetrically loaded car. What are the forces—all different—on the four tires? Again, we cannot find them because we have only three independent equations with which to work. Similarly, we can solve an equilibrium problem for a table with three legs but not one with four legs. Problems like these, in which there are more unknowns than equations, are called **indeterminate.**

And yet, solutions exist to indeterminate problems in the real world. If you rest the tires of the car on four platform scales, each scale will register a definite reading, the sum of the readings being the weight of the car. What is eluding us in our efforts to find the individual scale readings?

The problem is that we have assumed—without making a great point of it—that the bodies to which we apply the equations of static equilibrium are perfectly rigid. By this we mean that they do not deform when forces are applied to them. Strictly, there are no such bodies. The tires of the car, for example, deform easily under load until the car settles into a position of static equilibrium.

We have all had experience with a wobbly restaurant table, which we usually level by putting folded paper under one of the legs. If a big enough elephant sat on such a table, however, you may be sure that, if the table did not collapse, it would deform just like the tires of a car. Its legs would all touch the floor, the forces acting upward on the table legs would all assume definite values as in Fig. 13-11, and the table would no longer wobble. But how do we find the values of those forces acting upward on the legs?

To solve such indeterminate equilibrium problems, we must supplement equilibrium equations with some knowledge of *elasticity,* the branch of physics and engineering that describes how real bodies deform when forces are applied to them. The next section provides an introduction to this subject.

FIGURE 13-11 The table is an indeterminate structure. The four forces on the table legs are different in magnitude and cannot be found from the laws of static equilibrium alone.

13-6 ELASTICITY

When a large number of atoms come together to form a solid, such as an iron nail, they settle into equilibrium positions in a three-dimensional *lattice,* a repetitive arrangement in which each atom has a well-defined equilibrium distance from its nearest neighbors. The atoms are held together by interatomic forces that are represented by springs in Fig. 13-12.* The lattice is remarkably rigid, which is another way of saying that the "interatomic springs" are extremely stiff. It is for this reason that we perceive many ordinary objects such as ladders, tables, and spoons as perfectly rigid. Of course, some ordinary objects, such as garden hoses or rubber gloves, do not strike us as rigid at all. The atoms that make up these objects *do not* form a rigid lattice like that of Fig. 13-12 but are aligned in long flexible molecular chains, each chain being only loosely bound to its neighbors.

All real "rigid" bodies are to some extent **elastic,** which means that we can change their dimensions, slightly, by pulling, pushing, twisting, or compressing them. To get a feeling for the orders of magnitude involved, consider a steel rod, 1 m long and 1 cm in diameter. If you hang a subcompact car from the end of such a rod, the rod will stretch, but only by about 0.5 mm, or 0.05%. Furthermore, the rod will return to its original length when the car is removed.

If you hang two cars from the rod, the rod will be permanently stretched and will not recover its original length when you remove the load. If you hang three cars from the rod, the rod will break. Just before rupture, the elongation of the rod will be less than 0.2%. Although deformations of this size seem small, they are important in engineering practice. (Whether a wing under load will stay on an airplane is obviously important.)

Figure 13-13 shows three ways that a solid might change its dimensions when forces act on it. In Fig. 13-13a, a cylinder is stretched. In Fig. 13-13b, a cylinder is deformed by a force, much as one might deform a pack of cards or a book. In Fig. 13-13c, a

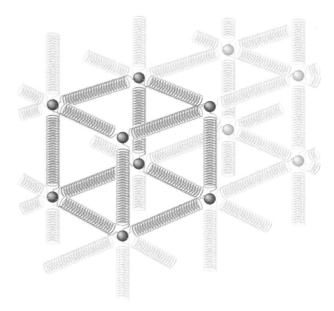

FIGURE 13-12 The atoms of a metallic solid are distributed on a repetitive three-dimensional lattice. The springs represent interatomic forces.

solid object, placed in a fluid under high pressure, is compressed uniformly on all sides. What the three modes have in common is that a **stress,** or deforming force per unit area, produces a **strain,** or unit deformation. In Fig. 13-13, there is *tensile stress* (associated with stretching) in (a), *shearing stress* in (b), and *hydraulic stress* in (c).

The stresses and the strains take different forms in the three cases of Fig. 13-13, but—over the range of engineering usefulness—stress and strain are proportional to each other. The constant of proportionality is called a **modulus of elasticity,** so that

$$\text{stress} = \text{modulus} \times \text{strain}. \qquad (13\text{-}28)$$

Figure 13-14 shows the relation between stress and strain for a steel test cylinder such as that of Fig. 13-15. In a standard test, the tensile stress on a test cylinder is slowly increased from zero to the point where the cylinder fractures, and the strain is carefully measured. For a substantial range of applied stresses, the stress–strain relation is linear, and it is here that Eq. 13-28 applies. If the stress is increased beyond the **yield strength** S_y of the specimen, the specimen becomes permanently deformed and does not recover its original dimensions when the stress is removed. If the stress continues to increase, the

---

*Ordinary metal objects, such as an iron nail, are made up of grains of iron, each grain formed as a more or less perfect lattice, such as that of Fig. 13-12. The forces between the grains are much weaker than the forces that hold the lattice together, so that rupture usually occurs along grain boundaries.

(a)   (b)   (c)

FIGURE 13-13 (*a*) A cylinder subject to *tensile stress* stretches by an amount ΔL. (*b*) A cylinder subject to *shearing stress* deforms like a pack of playing cards. (*c*) A solid

sphere subject to uniform *hydraulic stress* from a fluid shrinks in volume. All the deformations shown are greatly exaggerated.

specimen eventually ruptures, at a stress called the **ultimate strength** S_u.

Tension and Compression

For simple tension or compression, the stress is defined as F/A, the force divided by the area over which it acts (the force is perpendicular to the area, as you can see in Fig. 13-13*a*). The strain, or deformation, is then the dimensionless quantity $\Delta L/L$, the fractional (or sometimes percentage) change in length of the specimen. If the specimen is a long rod and the stress does not exceed the yield strength, then not only the entire rod but also any section of it experiences the same strain when a given stress is applied. Because the strain is dimensionless, the modulus in Eq. 13-28 has the same dimensions as the stress, namely, force per unit area.

The modulus for tensile and compressive stresses is called **Young's modulus,** which is represented in engineering practice by the symbol E. Equation 13-28 becomes

$$\frac{F}{A} = E\,\frac{\Delta L}{L}. \tag{13-29}$$

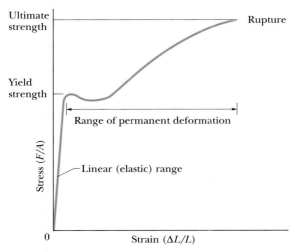

FIGURE 13-14 A stress–strain curve for a steel test specimen such as that of Fig. 13-15. The specimen deforms permanently when the stress is equal to the *yield strength* of the material. It ruptures when the stress is equal to the *ultimate strength* of the material.

FIGURE 13-15 A test specimen, used to determine a stress–strain curve such as that of Fig. 13-14. The change of a predetermined length L is measured in a tensile stress–strain test.

FIGURE 13-16 A strain gauge of overall dimensions 9.8 mm by 4.6 mm. The gauge is fastened with adhesive to the object whose strain is to be measured. The electrical resistance of the gauge varies with the strain, permitting strains up to about 3% to be measured.

The strain $\Delta L/L$ in a specimen can often be measured conveniently with a *strain gauge* (Fig. 13-16). These simple and useful devices, which can be attached directly to operating machinery with adhesives, are based on the principle that the electrical properties of the gauge are dependent on the strain it undergoes.

Although the modulus for an object may be almost the same for tension and compression, the ultimate strength may well be different for the two cases. Concrete, for example, is very strong in compression but is so weak in tension that it is almost never used in that manner. Table 13-1 shows the Young's modulus and other elastic properties for some materials of engineering interest.

Shearing

In the case of shearing, the stress is also a force per unit area but the force vector lies in the plane of the area rather than at right angles to it. The strain is the dimensionless ratio $\Delta x/L$, the quantities being defined as shown in Fig. 13-13b. The modulus, which is given the symbol G in engineering practice, is called the **shear modulus.** For shearing, Eq. 13-28 is written as

$$\frac{F}{A} = G\frac{\Delta x}{L}. \qquad (13\text{-}30)$$

Shearing stresses play a critical role in the buckling of shafts that rotate under load, and in bone fractures caused by bending.

Hydraulic Compression

In Fig. 13-13c, the stress is the same as the fluid pressure p on the object, once more a force per unit area, and the strain is $\Delta V/V$, where V is the volume of the specimen and ΔV is the absolute value of the change in volume. The modulus, with symbol B, is called the **bulk modulus** of the material, the object is said to be under *hydraulic compression,* and the pressure can be called the *hydraulic stress.*

For this situation, we write Eq. 13-28 as

$$p = B\frac{\Delta V}{V}. \qquad (13\text{-}31)$$

The bulk modulus of water is 2.2×10^9 N/m², and that of steel is 16×10^{10} N/m². The pressure at the

TABLE 13-1
SOME ELASTIC PROPERTIES OF SELECTED MATERIALS OF ENGINEERING INTEREST

MATERIAL	DENSITY ρ (kg/m³)	YOUNG'S MODULUS E (10^9 N/m²)	ULTIMATE STRENGTH S_u (10^6 N/m²)	YIELD STRENGTH S_y (10^6 N/m²)
Steel[a]	7860	200	400	250
Aluminum	2710	70	110	95
Glass	2190	65	50[b]	—
Concrete[c]	2320	30	40[b]	—
Wood[d]	525	13	50[b]	—
Bone	1900	9[b]	170[b]	—
Polystyrene	1050	3	48	—

[a] Structural steel (ASTM-A36).   [c] High strength.

[b] In compression.   [d] Douglas fir.

bottom of the Pacific Ocean, at its average depth of about 4000 m, is 4.0×10^7 N/m². The fractional compression $\Delta V/V$ of a volume of water at this depth, caused by pressure alone, is 1.8%; that for a steel object is only about 0.025%. In general, solids —with their rigid atomic lattices—are less compressible than liquids, in which the atoms or molecules are less tightly coupled to their neighbors.

SAMPLE PROBLEM 13-7

A structural steel rod has a radius R of 9.5 mm and a length L of 81 cm. A force F of 6.2×10^4 N (about 7 tons) stretches it axially.

a. What is the stress in the rod?

SOLUTION From its definition, the stress is

$$\text{stress} = \frac{F}{A} = \frac{F}{\pi R^2} = \frac{6.2 \times 10^4 \text{ N}}{(\pi)(9.5 \times 10^{-3} \text{ m})^2}$$

$$= 2.2 \times 10^8 \text{ N/m}^2. \qquad \text{(Answer)}$$

The yield strength for structural steel is 2.5×10^8 N/m², so that this rod is dangerously close to its yield strength.

b. What is the elongation of the rod under this load? What is the strain?

SOLUTION From Eq. 13-29, using the result we have just calculated and the value of E for steel (Table 13-1), we obtain

$$\Delta L = \frac{(F/A)L}{E} = \frac{(2.2 \times 10^8 \text{ N/m}^2)(0.81 \text{ m})}{2.0 \times 10^{11} \text{ N/m}^2}$$

$$= 8.9 \times 10^{-4} \text{ m} = 0.89 \text{ mm}. \qquad \text{(Answer)}$$

Thus the strain

$$\Delta L/L = (8.9 \times 10^{-4} \text{ m})/(0.81 \text{ m})$$

$$= 1.1 \times 10^{-3} = 0.11\%. \qquad \text{(Answer)}$$

SAMPLE PROBLEM 13-8

The femur, which is the principal bone of the thigh, has a minimum diameter in an adult male of about 2.8 cm, corresponding to a cross-sectional area A of 6×10^{-4} m². At what compressive load would it break?

SOLUTION From Table 13-1 we see that the ultimate strength S_u for bone in compression is 170×10^6 N/m². The compressive force at fracture is the force F that produces the stress S_u, or

$$F = S_u A = (170 \times 10^6 \text{ N/m}^2)(6 \times 10^{-4} \text{ m}^2)$$

$$= 1.0 \times 10^5 \text{ N}. \qquad \text{(Answer)}$$

This is 23,000 lb or 11 tons. Although this is a large force, it can be encountered during, for example, an unskillful parachute landing on hard ground. The force need not be sustained to break the bone; a few milliseconds will do it.

SAMPLE PROBLEM 13-9

A table has three legs that are 1.00 m in length and a fourth leg that is longer by a distance $d = 0.50$ mm, so that the table wobbles slightly. A heavy steel cylinder whose mass M is 290 kg is placed upright on the table so that all four legs compress and the table no longer wobbles. The legs are wooden cylinders whose cross-sectional area A is 1.0 cm². Young's modulus E for the wood is 1.3×10^{10} N/m². Assume that the tabletop remains level and that the legs do not buckle. With what force does the floor push upward on each leg?

SOLUTION We take the table plus cylinder as our system. The situation is like Fig. 13-11, except we now have a steel cylinder on the table. If the tabletop remains level, each of the three short legs must be compressed by the same amount ΔL_3 and thus by the same force F_3. The single long leg must be compressed by a larger amount ΔL_1, by a larger force F_1, and we must have

$$\Delta L_1 = \Delta L_3 + d. \qquad (13\text{-}32)$$

We can rewrite Eq. 13-29 as $\Delta L = FL/EA$. We then substitute this relationship into Eq. 13-32, letting L represent the initial length of the three short legs and the approximate length of the long leg. Equation 13-32 becomes

$$F_1 L = F_3 L + dAE. \qquad (13\text{-}33)$$

From Eq. 13-8, the balance of forces in the vertical direction, we have for our system

$$\sum F_y = 3F_3 + F_1 - Mg = 0. \qquad (13\text{-}34)$$

If we solve Eqs. 13-33 and 13-34 for the unknown force F_3, we find

$$F_3 = \frac{Mg}{4} - \frac{dAE}{4L}$$

$$= \frac{(290 \text{ kg})(9.8 \text{ m/s}^2)}{4}$$

$$- \frac{(5.0 \times 10^{-4} \text{ m})(10^{-4} \text{ m}^2)(1.3 \times 10^{10} \text{ N/m}^2)}{(4)(1.00 \text{ m})}$$

$$= 711 \text{ N} - 163 \text{ N} = 548 \text{ N}. \qquad \text{(Answer)}$$

Similarly, we find

$$F_1 = \frac{Mg}{4} + \frac{3dAE}{4L}$$

$$= 711 \text{ N} + 489 \text{ N} = 1200 \text{ N}. \qquad \text{(Answer)}$$

You can show that, to reach their equilibrium configuration, the three short legs were each compressed by 0.42 mm and the single long leg by 0.92 mm, the difference being 0.50 mm.

REVIEW & SUMMARY

Static Equilibrium

A rigid body at rest is said to be in **static equilibrium.** For such a body, the vector sum of the external forces acting on it is zero:

$$\sum \mathbf{F}_{\text{ext}} = 0 \qquad \text{(balance of forces).} \qquad (13\text{-}3)$$

If all the forces lie in the xy plane, this vector equation is equivalent to two scalar component equations:

$$\sum F_x = 0 \quad \text{and} \quad \sum F_y = 0 \qquad \begin{array}{c}\text{(balance of}\\ \text{forces).}\end{array} \qquad (13\text{-}7, 13\text{-}8)$$

Static equilibrium also implies that the vector sum of the external torques acting on the body about *any* point is zero, or

$$\sum \tau_{\text{ext}} = 0 \qquad \text{(balance of torques).} \qquad (13\text{-}5)$$

If the forces lie in the xy plane, any torque is parallel to the z axis, and Eq. 13-5 is equivalent to the single scalar component equation

$$\sum \tau_z = 0 \qquad \text{(balance of torques).} \qquad (13\text{-}9)$$

Center of Gravity

The gravitational force acts individually on all the particles of a body. The net effect of all the individual actions may be found by imagining an equivalent total gravitational force $M\mathbf{g}$ acting at a specific point called the **center of gravity.** If the gravitational acceleration $\mathbf{g}$ is the same for all the particles in a body, the center of gravity is at the center of mass.

Elastic Moduli

Three **elastic moduli** are used to describe the elastic behavior (deformations) of objects as they respond to the forces that act on them. The **strain** (fractional change in size) is linearly related to the applied **stress** (force per unit area) in each case. The general relation is

$$\text{stress} = \text{modulus} \times \text{strain.} \qquad (13\text{-}28)$$

Tension and Compression

When an object is under tension or compression, Eq. 13-28 is written as

$$\frac{F}{A} = E \frac{\Delta L}{L}, \qquad (13\text{-}29)$$

where $\Delta L/L$ is the strain of the object, F is the magnitude of the applied force $\mathbf{F}$ causing the strain, A is the cross-sectional area over which $\mathbf{F}$ is applied (perpendicular to A, as in Fig. 13-13a), and E is **Young's modulus** for the object.

Shearing

When an object is under a shearing stress, Eq. 13-28 is written as

$$\frac{F}{A} = G \frac{\Delta x}{L}, \qquad (13\text{-}30)$$

where $\Delta x/L$ is the strain of the object, Δx is the displacement of one end of the object in the direction of the applied force $\mathbf{F}$ (as in Fig. 13-13b), and G is the **shear modulus** of the object.

Hydraulic Compression

When an object undergoes *hydraulic compression* due to a stress exerted by a surrounding fluid, Eq. 13-28 is written as

$$p = B \frac{\Delta V}{V}, \qquad (13\text{-}31)$$

where p is the pressure (*hydraulic stress*) on the object due to the fluid, $\Delta V/V$ (strain) is the absolute value of the fractional change in the object's volume due to that pressure, and B is the **bulk modulus** of the object.

QUESTIONS

1. In a simple pendulum, is the bob in equilibrium at any point of its swing?

2. If a certain rigid body is thrown upward without spinning, it does not begin spinning during its flight, provided that air resistance can be neglected. What does this simple result imply about the location of the center of gravity?

3. A wheel rotating at constant angular velocity ω about a fixed axis is in equilibrium because no net external force or torque acts on it. However, the particles that make up the wheel undergo a centripetal acceleration **a** directed toward the axis. Since $\mathbf{a} \neq 0$, how can the wheel be said to be in equilibrium?

4. Give several examples of bodies that are not in equilibrium, even though the resultant of all the forces acting on them is zero.

5. Which is more likely to break in use, a hammock that is stretched tightly or one that sags quite a bit? Why?

6. A ladder is at rest with its upper end against a wall and the lower end on the ground. Is it more likely to slip when a person stands on it at the bottom or at the top? Explain.

7. A picture hangs on a wall by two wires. How should the wires be arranged to be under minimum tension? Explain why the tension is larger for other arrangements of the wires.

8. A framed rectangular picture hangs by one wire over a frictionless nail. Why is the picture in unstable equilibrium if the wire is "too short" (this is one reason why pictures become skewed), and in stable equilibrium if the wire is "long enough"? Experimentally define "too short" and "long enough" by considering the angle θ that the wire makes at the nail and the angle α between the diagonals of the picture, as shown in Fig. 13-17.

9. Stand facing the edge of an open door, one foot on each side of the door. You will find that you are not able to stand on your toes. Why?

10. Sit in a straight-backed chair and try to stand up without leaning forward. Why can't you do it?

11. How does a long balancing pole help a tightrope walker to maintain balance?

FIGURE 13-17   Question 8.

12. A composite block made up of wood and metal rests on a (rough) table top. In which of the two orientations shown in Fig. 13-18 can you tip it over with the least force **F** applied at the same height on the block?

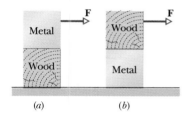

FIGURE 13-18   Question 12.

13. If you have a can of soda pop on your tray during a choppy flight, for what levels of fluid content is the can most likely and least likely to tip over?

14. A popular beer drinker's challenge is to balance a can on its edge. How is the stunt accomplished?

15. A horizontal beam supported at both ends is loaded in the middle. Show that the upper part of the beam is under compression whereas the lower part is under tension.

16. Why are reinforcing rods used in concrete structures? (Compare the tensile strength of concrete to its compressive strength.)

EXERCISES & PROBLEMS

SECTION 13-4   SOME EXAMPLES OF STATIC EQUILIBRIUM

1E. An eight-member family, whose weights in pounds are indicated in Fig. 13-19, is balanced on a see-saw. What

is the number of the person who causes the largest torque, about the rotation axis at *fulcrum f,* directed (a) out of the page and (b) into the page?

2E. A rigid square object of negligible weight is acted on by three forces that pull on its corners as shown, to scale,

FIGURE 13-19 Exercise 1.

in Fig. 13-20. (a) Is the first requirement of equilibrium satisfied? (b) Is the second requirement of equilibrium satisfied? (c) If the answer to either (a) or (b) is no, could a fourth force restore the equilibrium of the object? If so, specify the magnitude, direction, and point of application of the needed force.

FIGURE 13-20 Exercise 2.

3E. A certain nut is known to require forces of 40 N exerted on its shell from both sides to crack it. What forces F will be required when it is placed in the nutcracker shown in Fig. 13-21?

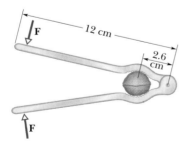

FIGURE 13-21 Exercise 3.

4E. The leaning Tower of Pisa (see Fig. 13-22) is 55 m high and 7.0 m in diameter. The top of the tower is displaced 4.5 m from the vertical. Treating the tower as a uniform, circular cylinder, (a) what additional displacement, measured at the top, will bring the tower to the verge of toppling? (b) What angle with the vertical will the tower make at that moment?

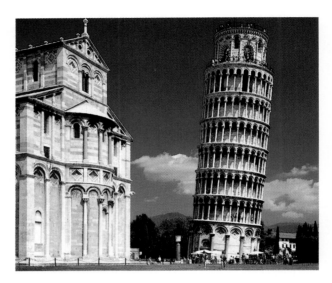

FIGURE 13-22 Exercise 4.

5E. A particle is acted on by forces given, in newtons, by $F_1 = 10\mathbf{i} - 4\mathbf{j}$ and $F_2 = 17\mathbf{i} + 2\mathbf{j}$. (a) What force F_3 balances these forces? (b) What direction does F_3 have relative to the x axis?

6E. A bow is drawn until the tension in the string is equal to the force exerted by the archer. What is the angle between the two parts of the string?

7E. A rope, assumed massless, is stretched horizontally between two supports that are 3.44 m apart. When an object of weight 3160 N is hung at the center of the rope, the rope is observed to sag by 35.0 cm. What is the tension in the rope?

8E. In Fig. 13-23, a man is trying to get his car out of mud on the shoulder of a road. He ties one end of a rope tightly around the front bumper and the other end tightly around a utility pole 60 ft away. He then pushes sideways on the rope at its midpoint with a force of 125 lb, displacing the center of the rope 1.0 ft from its previous position, and the car barely moves. What force does the rope exert on the car? (The rope stretches somewhat.)

FIGURE 13-23 Exercise 8.

9E. The system in Fig. 13-24 is in equilibrium, but it begins to slip if any additional mass is added to the 5.0-kg object. What is the coefficient of static friction between the 10-kg block and the plane on which it rests?

10E. A uniform sphere of weight W and radius r is held in place by a rope attached to a frictionless wall a distance L above the center of the sphere, as in Fig. 13-25. Find (a)

FIGURE 13-24 Exercise 9.

FIGURE 13-25 Exercise 10.

the tension in the rope and (b) the force exerted on the sphere by the wall.

11E. An automobile with a mass of 1360 kg has 3.05 m between the front and rear axles. Its center of gravity is located 1.78 m behind the front axle. Determine (a) the force exerted on each of the front wheels (assumed the same) and (b) the force exerted on each of the back wheels (assumed the same) by the level ground.

12E. A 160-lb man is walking across a level bridge and stops one-fourth of the way from one end. The bridge is uniform and weighs 600 lb. What are the vertical forces exerted on the bridge by its supports at (a) the far end and (b) the near end?

13E. A diver of weight 580 N stands at the end of a 4.5-m diving board of negligible weight. The board is attached to

FIGURE 13-26 Exercise 13.

two pedestals 1.5 m apart, as shown in Fig. 13-26. What are the magnitude and direction of the force on the board from (a) the left pedestal and (b) the right pedestal? (c) Which pedestal is being stretched, and which compressed?

14E. A meter stick balances horizontally on a knife-edge at the 50.0-cm mark. With two nickels stacked over the 12.0-cm mark, the stick is found to balance at the 45.5-cm mark. A nickel has a mass of 5.0 g. What is the mass of the meter stick?

15E. A beam is carried by three men, one man at one end and the other two supporting the beam between them on a crosspiece placed so that the load is equally divided among the three men. Where is the crosspiece placed? (Neglect the mass of the crosspiece.)

16E. A 75-kg window cleaner uses a 10-kg ladder that is 5.0 m long. He places one end down 2.5 m from a wall and rests the upper end against a cracked window and climbs the ladder. He climbs 3.0 m up the ladder when the window breaks. Neglecting friction between the ladder and window and assuming that the base of the ladder did not slip, find (a) the force exerted on the window by the ladder just before the window breaks and (b) the magnitude and direction of the force exerted on the ladder by the ground just before the window breaks.

17E. Figure 13-27 shows the anatomical structures in the lower leg and foot that are involved in standing tip-toe with the heel raised off the floor so that the foot effectively contacts the floor at only one point, shown as P in the figure. Calculate, in terms of a person's weight W, the forces that must be exerted on the foot by (a) the calf muscle (at A) and (b) the lower-leg bones (at B) when the person stands tip-toe on one foot. Assume that $a = 5.0$ cm and $b = 15$ cm.

FIGURE 13-27 Exercise 17.

18E. A uniform cubical crate is 0.750 m on each side and weighs 500 N. It rests on the floor with one edge against a very small, fixed obstruction. At what height above the floor must a horizontal force of 350 N be applied to the crate to just tip it?

19P. Two identical, uniform, frictionless spheres, each of weight W, rest in a rigid rectangular container as shown in Fig. 13-28. Find, in terms of W, the forces acting on the spheres due to (a) the container surfaces and (b) one another, if the line of centers of the spheres makes an angle of 45° with the horizontal.

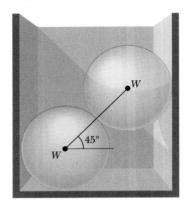

FIGURE 13-28 Problem 19.

20P. An 1800-lb construction bucket is suspended by a cable A that is attached at O to two other cables B and C, making angles of 51° and 66° with the horizontal (Fig. 13-29). Find the tension in (a) cable A, (b) cable B, and (c) cable C. (*Hint:* To avoid solving two equations in two unknowns, position the axes as shown in the figure.)

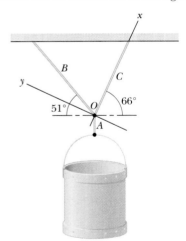

FIGURE 13-29 Problem 20.

21P. The system in Fig. 13-30 is in equilibrium with the string in the center exactly horizontal. Find (a) tension T_1, (b) tension T_2, (c) tension T_3, and (d) angle θ.
22P. The force $\mathbf{F}$ in Fig. 13-31 is just sufficient to hold the 14-lb block and weightless pulleys in equilibrium. There is no appreciable friction. Calculate the tension T in the upper cable.

FIGURE 13-30 Problem 21.

FIGURE 13-31 Problem 22.

23P. A balance is made up of a rigid rod with mass M, supported at and free to rotate about a point not at the center of the rod. It is balanced by unequal weights placed in the pans at each end of the rod. When an unknown mass m is placed in the left-hand pan, it is balanced by a mass m_1 placed in the right-hand pan; and when the mass m is placed in the right-hand pan, it is balanced by a mass m_2 in the left-hand pan. Show that $m = \sqrt{m_1 m_2}$.
24P. A 15-kg weight is being lifted by the pulley system shown in Fig. 13-32. The upper arm is vertical, whereas

FIGURE 13-32 Problem 24.

the forearm makes an angle of 30° with the horizontal. What forces are being exerted on the forearm by (a) the triceps muscle and (b) the upper-arm bone (the humerus)? The forearm and hand together have a mass of 2.0 kg with a center of mass 15 cm (measured along the arm) from the point where the forearm and upper-arm bones are in contact. The triceps muscle pulls vertically upward at a point 2.5 cm behind the contact point.

25P. A 50.0-kg uniform square sign, 2.00 m on a side, is hung from a 3.00-m rod of negligible mass. A cable is attached to the end of the rod and to a point on the wall 4.00 m above the point where the rod is fixed to the wall, as shown in Fig. 13-33. (a) What is the tension in the cable? What are the (b) horizontal and (c) vertical components of the force exerted by the wall on the rod?

FIGURE 13-33 Problem 25.

26P. In Fig. 13-34, a 55-kg rock climber is in a lie-back climb along a fissure, with hands pulling on one side of the fissure and feet pressed against the opposite side. The fissure has width $w = 0.20$ m, and the center of mass of the climber is a horizontal distance $d = 0.40$ m from the fissure. The coefficient of static friction between hands

FIGURE 13-34 Problem 26.

and rock is $\mu_1 = 0.40$, and between boots and rock it is $\mu_2 = 1.2$. (a) What is the least horizontal pull by the hands and push by the feet that will keep him stable? (b) For the horizontal pull of (a), what must be the vertical distance h between hands and feet? (c) If the climber encounters wet rock, so that μ_1 and μ_2 are reduced, what happens to the answer to (a) and (b), respectively?

27P. In Fig. 13-35, what magnitude of force **F** applied horizontally at the axle of the wheel is necessary to raise the wheel over an obstacle of height h? Take r as the radius of the wheel and W as its weight.

FIGURE 13-35 Problem 27.

28P. Forces $\mathbf{F}_1$, $\mathbf{F}_2$, and $\mathbf{F}_3$ act on the structure of Fig. 13-36 as shown in an overhead view. We wish to put the structure in equilibrium by applying a force, at a point such as P, whose vector components are $\mathbf{F}_h$ and $\mathbf{F}_v$. We are given that $a = 2.0$ m, $b = 3.0$ m, $c = 1.0$ m, $F_1 = 20$ N, $F_2 = 10$ N, and $F_3 = 5.0$ N. Find (a) F_h, (b) F_v, and (c) d.

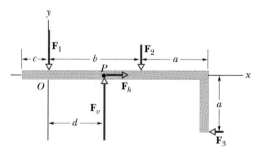

FIGURE 13-36 Problem 28.

29P. A trap door in a ceiling is 0.91 m square, has a mass of 11 kg, and is hinged along one side with a catch at the opposite side. If the center of gravity of the door is 10 cm toward the hinged side from the door's center, what forces must (a) the catch and (b) the hinge sustain?

30P. Four identical bricks, each of length L, are put on top of one another (Fig. 13-37) in such a way that part of each extends beyond the one beneath. Find, in terms of L, the maximum values of (a) a_1, (b) a_2, (c) a_3, (d) a_4, and (e) h, such that the stack is in equilibrium.

31P. One end of a uniform beam weighing 50.0 lb and 3.00 ft long is attached to a wall with a hinge. The other end is supported by a wire (see Fig. 13-38). (a) Find the tension in the wire. What are the (b) horizontal and (c) vertical components of the force of the hinge?

FIGURE 13-37 Problem 30.

FIGURE 13-38 Problem 31.

32P. A door 2.1 m high and 0.91 m wide has a mass of 27 kg. A hinge 0.30 m from the top and another 0.30 m from the bottom each support half the door's weight. Assume that the center of gravity is at the geometrical center of the door and determine (a) the vertical and (b) horizontal force components exerted by each hinge on the door.

33P. The system in Fig. 13-39 is in equilibrium. A mass of 225 kg hangs from the end of the strut; the strut has a mass of 45.0 kg. Find (a) the tension T in the cable and the (b) horizontal and (c) vertical force components exerted on the strut by the hinge.

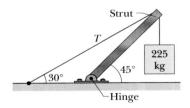

FIGURE 13-39 Problem 33.

34P. A nonuniform bar of weight W is suspended at rest in a horizontal position by two massless cords as shown in Fig. 13-40. One cord makes the angle $\theta = 36.9°$ with the vertical; the other makes the angle $\phi = 53.1°$ with the vertical. If the length L of the bar is 6.10 m, compute the

FIGURE 13-40 Problem 34.

distance x from the left-hand end of the bar to its center of gravity.

35P. In Fig. 13-41, a thin horizontal bar AB of negligible weight and length L is pinned to a vertical wall at A and supported at B by a thin wire BC that makes an angle θ with the horizontal. A weight W can be moved anywhere along the bar; its position is defined by the distance x from the wall of its center of mass. As a function of x, find (a) the tension in the wire, and the (b) horizontal and (c) vertical components of the force exerted on the bar by the pin at A.

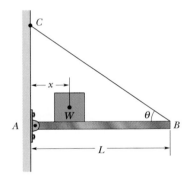

FIGURE 13-41 Problems 35 and 36.

36P. In Fig. 13-41, suppose the length L of the uniform bar is 3.0 m and its weight is 200 N. Also, let $W = 300$ N and $\theta = 30°$. The wire can withstand a maximum tension of 500 N. (a) What is the maximum possible distance x before the wire breaks? With W placed at this maximum x, what are the (b) horizontal and (c) vertical components of the force exerted on the bar by the pin at A?

37P. Two uniform beams, A and B, are attached to a wall with hinges and then loosely bolted together as in Fig. 13-42. Find the horizontal and vertical components, respectively, of the force on (a) beam A due to its hinge, (b) beam A due to the bolt, (c) beam B due to its hinge, and (d) beam B due to the bolt.

38P. Four identical, uniform bricks of length L are stacked on a table in two ways, as shown in Fig. 13-43 (compare with Problem 30). We seek to maximize the overhang distance h in both arrangements. Find the optimum distances a_1, a_2, b_1, and b_2, and calculate h for the two arrangements. (See "The Amateur Scientist," *Scientific American*, June 1985, for a discussion and an even better version of arrangement (b).)

FIGURE 13-42 Problem 37.

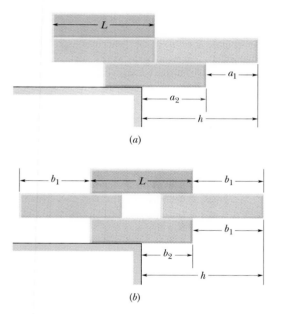

(a)

(b)

FIGURE 13-43 Problem 38.

39P. A uniform plank with length $L = 20$ ft and weight $W = 100$ lb, rests on the ground and against a frictionless roller at the top of a wall of height $h = 10$ ft (see Fig. 13-44). The plank remains in equilibrium for any value of

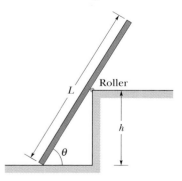

FIGURE 13-44 Problem 39.

$\theta \geq 70°$ but slips if $\theta < 70°$. Find the coefficient of static friction between the plank and the ground.

40P. For the stepladder shown in Fig. 13-45, sides AC and CE are each 8.0 ft long and hinged at C. Bar BD is a tie rod 2.5 ft long, halfway up. A man weighing 192 lb climbs 6.0 ft along the ladder. Assuming that the floor is frictionless and neglecting the weight of the ladder, find (a) the tension in the tie rod and the forces exerted on the ladder by the floor at (b) A and (c) E. (*Hint:* It will help to isolate parts of the ladder in applying the equilibrium conditions.)

FIGURE 13-45 Problem 40.

41P. By means of a turnbuckle G, bar AB of the square frame $ABCD$ in Fig. 13-46 is put in tension, as if its ends A and B were subject to the horizontal, outward forces **T** shown. Determine the forces on the other bars; identify those bars that are in tension and those that are in compression. The diagonals AC and BD pass each other freely at E. Symmetry considerations can lead to considerable simplification in this and similar problems.

FIGURE 13-46 Problem 41.

42P. A uniform cube of side length L rests on a horizontal floor. The coefficient of static friction between cube and floor is μ. A horizontal pull P is applied perpendicular to one of the vertical faces of the cube, at a distance h above the floor on the vertical midline of the cube face. As P is slowly increased, the cube will either (a) begin to slide or (b) begin to tip. What is the condition on μ for (a) to occur? For (b)? (*Hint:* At the onset of tipping, where is the normal force located?)

43P. A cubical box is filled with sand and weighs 890 N. We wish to "roll" the box by pushing horizontally on one of the upper edges. (a) What minimum force is required? (b) What minimum coefficient of static friction is required? (c) Is there a more efficient way to roll the box? If so, find the smallest possible force that would have to be applied directly to the box to roll it. (*Hint:* See the hint for Problem 42.)

44P. A crate, in the form of a cube with edge lengths of 4.0 ft, contains a piece of machinery whose design is such that the center of gravity of the crate and its contents is located 1.0 ft above its geometrical center. The crate rests on a ramp which makes an angle θ with the horizontal. As θ is increased from zero, an angle will be reached at which the crate will either start to slide down the ramp or tip over. Which event will occur (a) when the coefficient of static friction is 0.60 and (b) when it is 0.70? In each case, give the angle at which the event occurs. (*Hint:* See the hint for Problem 42.)

45P*. A car on a horizontal road makes an emergency stop by applying the brakes so that all four wheels lock and skid along the road. The coefficient of kinetic friction between tires and road is 0.40. The separation between the front and rear axles is 4.2 m, and the center of mass of the car is located 1.8 m behind the front axle and 0.75 m above the road; see Fig. 13-47. The car weighs 11 kN. Calculate (a) the braking deceleration of the car, (b) the normal force on each wheel, and (c) the braking force on each wheel. (*Hint:* Although the car is not in translational equilibrium, it *is* in rotational equilibrium.)

FIGURE 13-47   Problem 45.

SECTION 13-6   ELASTICITY

46E. Figure 13-48 shows the stress–strain curve for quartzite. What are (a) Young's modulus and (b) the approximate yield strength for this material?

47E. After a fall, a 95-kg rock climber finds himself dangling from the end of a rope that had been 15 m long and 9.6 mm in diameter but which has stretched by 2.8 cm. For the rope, calculate (a) the strain, (b) the stress, and (c) Young's modulus.

48E. A mine elevator is supported by a single steel cable 2.5 cm in diameter. The total mass of the elevator cage

FIGURE 13-48   Exercise 46.

plus occupants is 670 kg. By how much does the cable stretch when the elevator is (a) at the surface, 12 m below the elevator motor, and (b) at the bottom of the 350-m-deep shaft? (Neglect the mass of the cable.)

49E. Suppose the (square) beam in Fig. 13-9a is of Douglas fir. What must be its thickness to keep the compressive stress on it to $\frac{1}{6}$ of its ultimate strength? (See Sample Problem 13-5.)

50E. A horizontal aluminum rod 4.8 cm in diameter projects 5.3 cm from a wall. A 1200-kg object is suspended from the end of the rod. The shear modulus of aluminum is 3.0×10^{10} N/m². Neglecting the rod's weight, find (a) the shear stress on the rod, and (b) the vertical deflection of the end of the rod.

51E. A solid copper cube has an edge length of 85.5 cm. How much pressure must be applied to the cube to reduce the edge length to 85.0 cm? The bulk modulus of copper is 1.4×10^{11} N/m².

52P. A 150-m-long tunnel 7.2 m high and 5.8 m wide (with a flat roof) is to be constructed 60 m beneath the ground. (See Fig. 13-49.) The tunnel roof is to be supported entirely by square steel columns, each with a cross-sectional area of 960 cm². The density of the ground ma-

FIGURE 13-49   Problem 52.

terial is 2.8 g/cm³. (a) What is the total weight that the columns must support? (b) How many columns are needed to keep the compressive stress on each column at one-half of its ultimate strength?

53P. A rectangular slab of rock rests on a 26° incline; see Fig. 13-50. The slab has dimensions 43 m long, 2.5 m thick, and 12 m wide. Its density is 3.2 g/cm³. The coefficient of static friction between the slab and the underlying rock is 0.39. (a) Calculate the component of the slab's weight acting parallel to the incline. (b) Calculate the static force of friction. By comparing (a) and (b), you can see that the slab is in danger of sliding and is prevented from doing so only by chance protrusions between the two layers of rock. (c) To stabilize the slab, bolts are driven perpendicular to the incline. If each bolt has a cross-sectional area of 6.4 cm² and will snap under a shearing stress of 3.6 × 10⁸ N/m², what is the minimum number of bolts needed? Assume that the bolts do not affect the normal force.

FIGURE 13-50 Problem 53.

54P. In Fig. 13-51, a lead brick rests horizontally on cylinders A and B. The areas of the top faces of the cylinders

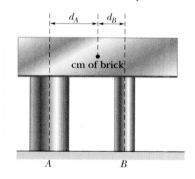

FIGURE 13-51 Problem 54.

are related by $A_A = 2A_B$; the Young's moduli of the cylinders are related by $E_A = 2E_B$. The cylinders had identical lengths before the brick was placed on them. (a) What fraction of the brick's weight is supported by cylinder A, and (b) what fraction by cylinder B? The horizontal distances between the center of mass of the brick and the center lines of the cylinders are d_A for cylinder A and d_B for cylinder B. (c) What is the ratio d_A/d_B?

55P. In Fig. 13-52, a 103-kg uniform log hangs by two steel wires, A and B, both of radius 1.20 mm. Initially, wire A was 2.50 m long and 2.00 mm shorter than wire B. The log is now horizontal. What forces are exerted on it by (a) wire A and (b) wire B? (c) What is the ratio d_A/d_B?

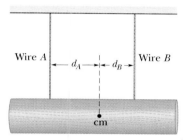

FIGURE 13-52 Problem 55.

56P. Figure 13-53 is an overhead view of a rigid rod that turns about a vertical axle until the identical rubber stoppers A and B are forced against rigid walls at distances r_A and r_B from the axle. Initially the stoppers touch the walls, without being compressed. Then force F is applied perpendicular to the rod at distance R from the axle. What forces compress (a) A and (b) B?

FIGURE 13-53 Problem 56.

ADDITIONAL PROBLEMS

57. A uniform beam of length 12 m is supported by a horizontal cable and pin as shown in Fig. 13-54. The tension in the cable is 400 N. What are (a) the weight of the beam and (b) the horizontal and vertical components of the force of the pin on the beam?

58. A scaffold of mass 60 kg and length 5.0 m is sup-

FIGURE 13-54 Problem 57.

ported in a horizontal position by one vertical cable at each end. A window washer of mass 80 kg stands at a point 1.5 m from one end. What is the tension in the cable (a) closest to the window washer and (b) farthest away from the window washer?

59. A uniform beam having a weight of 60 N and a length of 3.2 m is pinned at its lower end and acted on by a horizontal force **F** of magnitude 50 N at its upper end (Fig. 13-55). The beam is held vertical by a cable that makes an angle of 25° with the ground. What are (a) the tension in the cable and (b) the horizontal and vertical components of the force of the pin on the beam?

FIGURE 13-55 Problem 59.

60. A uniform 10-m ladder weighs 200 N. The ladder leans against a vertical, frictionless wall at a point 8.0 m

FIGURE 13-56 Problem 60.

above the ground, as shown in Fig. 13-56. A horizontal force F is applied to the ladder at a point 2.0 m from the base of the ladder (as measured along the ladder). (a) If $F = 50$ N, what is the force of the ground on the ladder, in terms of the unit vectors shown? (b) If $F = 150$ N, what is the force of the ground on the ladder, in unit-vector form? (c) Suppose the coefficient of static friction between the ladder and ground is 0.38; for what minimum value of F will the base of the ladder just start to move toward the wall?

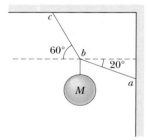

FIGURE 13-57 Problem 61.

61. The system shown in Fig. 13-57 is in equilibrium. If $M = 2.0$ kg, what is the tension in (a) string *ab* and (b) string *bc*?

OSCILLATIONS | 14

Just as the third game of the 1989 World Series was about to begin in Oakland, California, seismic waves from a magnitude 7.1 earthquake near Loma Prieta, 100 km distant, hit the area, causing extensive damage and killing 67 people. The photograph shows part of a 1.4-km stretch of the Nimitz Freeway, where an upper deck collapsed onto a lower deck, trapping motorists and causing dozens of deaths. Obviously, the collapse was due to violent shaking by the seismic waves. But why was that particular stretch of the freeway so severely damaged when the rest of the freeway, almost identical in construction, escaped collapse?

14-1 OSCILLATIONS

We are surrounded by oscillations—motions that repeat themselves. There are swinging chandeliers, boats bobbing at anchor, and the surging pistons in the engines of cars. There are oscillating guitar strings, drums, bells, diaphragms in telephones and speaker systems, and quartz crystals in wristwatches. Less evident are the oscillations of the molecules of the air that transmit the sensation of sound, of the atoms in a solid that convey the sensation of temperature, and of the electrons in the antennas of radio and TV transmitters.

Oscillations are not confined to material objects such as violin strings and electrons. Light, radio waves, x rays, and gamma rays are also oscillatory motions. You will study such oscillations in later chapters and will be helped greatly there by analogy with the mechanical oscillations that you are about to study here.

Oscillations in the real world are usually *damped*, by which we mean that the motion dies out gradually, transferring mechanical energy to thermal energy by the action of frictional forces. Although we cannot totally eliminate such loss of mechanical energy, we can replenish the energy from some source. The children in Fig. 14-1, for example, know that by swinging their legs or torsos they can "pump" the swing and maintain or enhance the oscillations. In doing this, they transfer biochemical energy to mechanical energy of the oscillating systems.

FIGURE 14-1 A child soon learns how to maintain the oscillations of a swing by transferring energy into its motion.

14-2 SIMPLE HARMONIC MOTION

Figure 14-2 shows a sequence of "snapshots" of a simple oscillating system, a particle moving repeatedly back and forth about the origin of the x axis. In this section we simply describe the motion. Later, we shall discuss how to attain such motion.

One important property of oscillatory motion is its **frequency,** or number of oscillations that are completed each second. The symbol for frequency is f, and its SI unit is the **hertz** (abbreviated Hz), where

$$1 \text{ hertz} = 1 \text{ Hz} = 1 \text{ oscillation per second}$$

$$= 1 \text{ s}^{-1}. \qquad (14\text{-}1)$$

Related to the frequency is the **period** T of the mo-

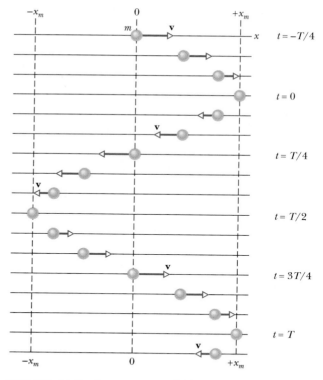

FIGURE 14-2 A sequence of "snapshots" (taken at equal time intervals) showing the position of a particle as it oscillates back and forth about the origin along the x axis, between the limits $+x_m$ and $-x_m$. The vector arrows are scaled to indicate the velocity of the particle. The velocity is maximum when the particle is at the origin, and zero when it is at $\pm x_m$. If the time t is chosen to be zero when the particle is at $+x_m$, then the particle returns to $+x_m$ at $t = T$, where T is the period of the motion. The motion is then repeated.

tion, which is the time for one complete oscillation (or **cycle**). That is,

$$T = \frac{1}{f}. \qquad (14\text{-}2)$$

Any motion that repeats itself at regular intervals is called **periodic motion** or **harmonic motion.** We are interested here in motion that repeats itself in a particular way, namely, like that in Fig. 14-2. It turns out that for such motion the displacement of the particle from the origin is given as a function of time by

$$x(t) = x_m \cos(\omega t + \phi) \quad \text{(displacement)}, \quad (14\text{-}3)$$

in which x_m, ω, and ϕ are constants. This motion is called **simple harmonic motion** (SHM), a term that means that the periodic motion is a sinusoidal function of time.

The quantity x_m in Eq. 14-3, a positive constant whose value depends on how the motion was started, is called the **amplitude** of the motion; the subscript m stands for *maximum* because the amplitude is the magnitude of the maximum displacement of the particle in either direction. The cosine function in Eq. 14-3 varies between the limits ± 1, so the displacement $x(t)$ varies between the limits $\pm x_m$, as Fig. 14-2 shows.

The time-varying quantity $(\omega t + \phi)$ in Eq. 14-3 is called the **phase** of the motion, and the constant ϕ is called the **phase constant** (or **phase angle**). The value of ϕ depends on the displacement and the velocity of the particle at $t = 0$. For the $x(t)$ plots of Fig. 14-3a, the phase constant ϕ is zero (compare the plot and Eq. 14-3 for $t = 0$).

It remains to interpret the constant ω. The displacement $x(t)$ must return to its initial value after one period T of the motion. That is, $x(t)$ must equal $x(t + T)$ for all t. To simplify our analysis, let us put $\phi = 0$ in Eq. 14-3. From that equation we then have

$$x_m \cos \omega t = x_m \cos \omega(t + T).$$

The cosine function first repeats itself when its argument (the phase) has increased by 2π radians so that we must have, in the above equation,

$$\omega(t + T) = \omega t + 2\pi$$

or

$$\omega T = 2\pi.$$

Thus from Eq. 14-2,

$$\omega = \frac{2\pi}{T} = 2\pi f. \qquad (14\text{-}4)$$

The quantity ω is called the **angular frequency** of the motion; its SI unit is the radian per second. (To be consistent, then, ϕ must be in radians.) Figure 14-3 compares $x(t)$ for two simple harmonic motions that differ either in amplitude, in period (and thus in frequency and angular frequency), or in phase constant.

SHM—The Velocity

By differentiating Eq. 14-3, we can find the velocity of a particle moving with simple harmonic motion.

(a)

(b)

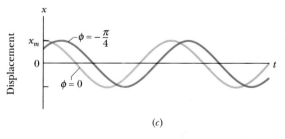

(c)

FIGURE 14-3 In all three cases, the blue curve is obtained from Eq. 14-3 with $\phi = 0$. (a) The red curve differs from the blue curve *only* in that its amplitude x'_m is greater. (b) The red curve differs from the blue curve *only* in that its period is $T' = T/2$. (c) The red curve differs from the blue curve *only* in that $\phi = -\pi/4$ rad, rather than zero.

That is,

$$v(t) = \frac{dx}{dt} = \frac{d}{dt} [x_m \cos(\omega t + \phi)]$$

or

$$v(t) = -\omega x_m \sin(\omega t + \phi) \quad \text{(velocity).} \quad (14\text{-}5)$$

Figure 14-4a is a plot of Eq. 14-3 with $\phi = 0$. Figure 14-4b shows Eq. 14-5, also with $\phi = 0$. Analogous to the amplitude x_m in Eq. 14-3, the positive quantity ωx_m in Eq. 14-5 is called the **velocity amplitude** v_m. As you can see in Fig. 14-4b, the velocity of the oscillating particle varies between the limits $\pm v_m = \pm \omega x_m$. Note also in that figure how the curve of $v(t)$ is *shifted* (to the left) from the curve of $x(t)$ by one-quarter period: when the magnitude of the displacement is greatest (that is, $x(t) = x_m$), the magnitude of the velocity is least (that is, $v(t) = 0$). And when the magnitude of the displacement is least (that is, zero), the magnitude of the velocity is greatest (that is, v_m).

FIGURE 14-4 (a) The displacement $x(t)$ of a particle oscillating in SHM with phase angle ϕ equal to zero (see Eq. 14-3). The period T marks one complete oscillation. (b) The velocity $v(t)$ of the particle (see Eq. 14-5). (c) The acceleration $a(t)$ of the particle (see Eq. 14-6). Note the relative shift (or phase difference) between the curves.

SHM—The Acceleration

Knowing the velocity $v(t)$ for simple harmonic motion, we can find the acceleration of the oscillating particle by differentiating once more. Thus we have, from Eq. 14-5,

$$a(t) = \frac{dv}{dt} = \frac{d}{dt} [-\omega x_m \sin(\omega t + \phi)]$$

or

$$a(t) = -\omega^2 x_m \cos(\omega t + \phi) \quad \text{(acceleration).} \quad (14\text{-}6)$$

Figure 14-4c is a plot of Eq. 14-6 for the case $\phi = 0$. The positive quantity $\omega^2 x_m$ in Eq. 14-6 is called the **acceleration amplitude** a_m. That is, the acceleration of the particle varies between the limits $\pm a_m = \pm \omega^2 x_m$, as Fig. 14-4c shows. Note also that the curve of $a(t)$ is shifted (to the left) by one-quarter period relative to the curve of $v(t)$.

We can combine Eqs. 14-3 and 14-6 to yield

$$a(t) = -\omega^2 x(t), \quad (14\text{-}7)$$

which is the hallmark of simple harmonic motion: the acceleration is proportional to the displacement but opposite in sign, and the two quantities are related by the square of the angular frequency. Thus when the displacement has its greatest positive value, the acceleration has its greatest negative value and conversely. When the displacement is zero, the acceleration is also zero.

PROBLEM SOLVING

TACTIC 1: PHASE ANGLES
Note the effect of the phase angle ϕ on a plot of $x(t)$. When $\phi = 0$, $x(t)$ has a graph like that in Fig. 14-4a, a typical cosine curve. A *negative* value for ϕ shifts the curve *rightward* along the t axis (as in Fig. 14-3c), while a *positive* value shifts it *leftward*.

Two plots of SHM with different phase angles are said to have a *phase difference,* or each is said to be *phase shifted* from the other or *out of phase* with the other. The curves in Fig. 14-3c, for example, have a phase difference of $\pi/4$ rad.

Because SHM repeats after each period T and the cosine function repeats after each 2π rad, one period T

represents a phase difference of 2π rad. In Fig. 14-4, $x(t)$ is phase shifted to the right from $v(t)$ by one-quarter period or $-\pi/2$ rad, and to the right from $a(t)$ by one-half period or $-\pi$ rad. A phase shift of 2π rad causes a curve of SHM to coincide with itself. The sine curve is shifted by $-\pi/2$ rad from the cosine curve.

14-3 SIMPLE HARMONIC MOTION: THE FORCE LAW

Once we know how the acceleration of a particle varies with time, we can use Newton's second law to learn what force must act on the particle to give it that acceleration. If we combine Newton's second law and Eq. 14-7, we find, for simple harmonic motion,

$$F = ma = -(m\omega^2)x. \qquad (14\text{-}8)$$

This result—a force proportional to the displacement but opposite in sign—is familiar. It is Hooke's law

$$F = -kx \qquad (14\text{-}9)$$

for a spring, the spring constant here being

$$k = m\omega^2. \qquad (14\text{-}10)$$

We can in fact take Eq. 14-9 as an alternative definition of simple harmonic motion, which states:

> Simple harmonic motion is the motion executed by a particle of mass m subject to a force that is proportional to the displacement of the particle but opposite in sign.

The block–spring system of Fig. 14-5 forms a **linear simple harmonic oscillator** (linear oscillator, for short), where "linear" indicates that F is proportional to x rather than to some other power of x. The angular frequency ω of the simple harmonic motion of the block is related to the spring constant k and the mass m of the block by Eq. 14-10, which yields

$$\omega = \sqrt{\frac{k}{m}} \qquad \text{(angular frequency).} \qquad (14\text{-}11)$$

By combining Eqs. 14-4 and 14-11, we can write, for the **period** of the linear oscillator of Fig. 14-5,

FIGURE 14-5 A simple harmonic oscillator. The block —like the particle of Fig. 14-2—moves in simple harmonic motion once it has been pulled to the side and released. Its displacement is then given by Eq. 14-3.

$$T = 2\pi \sqrt{\frac{m}{k}} \qquad \text{(period).} \qquad (14\text{-}12)$$

Equations 14-11 and 14-12 tell us that a large angular frequency (and thus a small period) goes with a stiff spring (large k) and a light block (small m).

Every oscillating system, be it the linear oscillator of Fig. 14-5, a diving board, or a violin string, has some element of "springiness" and some element of "inertia," or mass, and thus resembles a linear oscillator. In the linear oscillator of Fig. 14-5, these elements are located in separate parts of the system, the springiness being entirely in the spring, which we assume to be massless, and the inertia being entirely in the block, which we assume to be rigid. In a violin string, however, the two elements are both within the string itself, as you will see in Chapter 17.

SAMPLE PROBLEM 14-1

A block whose mass m is 680 g is fastened to a spring whose spring constant k is 65 N/m. The block is pulled a distance $x = 11$ cm from its equilibrium position at $x = 0$ on a frictionless surface and released from rest at $t = 0$ (see Fig. 14-5).

a. What force does the spring exert on the block just before the block is released?

SOLUTION From Hooke's law

$$F = -kx = -(65 \text{ N/m})(0.11 \text{ m})$$
$$= -7.2 \text{ N.} \qquad \text{(Answer)}$$

The minus sign reminds us that the spring force acting on the block, which points back toward the origin, is opposite the displacement of the block, which points away from the origin.

b. What are the angular frequency, the frequency, and the period of the resulting oscillation?

SOLUTION   From Eq. 14-11 we have

$$\omega = \sqrt{\frac{k}{m}} = \sqrt{\frac{65 \text{ N/m}}{0.68 \text{ kg}}} = 9.78 \text{ rad/s}$$

$$\approx 9.8 \text{ rad/s}. \qquad \text{(Answer)}$$

The frequency follows from Eq. 14-4, which yields

$$f = \frac{\omega}{2\pi} = \frac{9.78 \text{ rad/s}}{2\pi} = 1.56 \text{ Hz} \approx 1.6 \text{ Hz}, \qquad \text{(Answer)}$$

and the period from Eq. 14-2, which yields

$$T = \frac{1}{f} = \frac{1}{1.56 \text{ Hz}} = 0.64 \text{ s} = 640 \text{ ms}. \qquad \text{(Answer)}$$

c. What is the amplitude of the oscillation?

SOLUTION   As we discussed in Section 8-4, the mechanical energy of a spring–block system like that in Fig. 14-5 is conserved because friction is not involved. Since the block is released from rest 11 cm from its equilibrium point, it has kinetic energy of zero whenever it is again 11 cm from that point (see Fig. 8-5). Thus its maximum displacement is 11 cm; that is,

$$x_m = 11 \text{ cm}. \qquad \text{(Answer)}$$

d. What is the maximum speed of the oscillating block?

SOLUTION   From Eq. 14-5 we see that the velocity amplitude is

$$v_m = \omega x_m = (9.78 \text{ rad/s})(0.11 \text{ m})$$

$$= 1.1 \text{ m/s}. \qquad \text{(Answer)}$$

This maximum speed occurs when the oscillating block is rushing through the origin; compare Figs. 14-4a and 14-4b.

e. What is the magnitude of the maximum acceleration of the block?

SOLUTION   From Eq. 14-6 we see that the acceleration amplitude is

$$a_m = \omega^2 x_m = (9.78 \text{ rad/s})^2 (0.11 \text{ m})$$

$$= 11 \text{ m/s}^2. \qquad \text{(Answer)}$$

This maximum acceleration occurs when the block is at the ends of its path. At those points, the force acting on the block has its maximum magnitude; compare Figs. 14-4a and 14-4c.

f. What is the phase constant ϕ for the motion?

SOLUTION   At $t = 0$, the moment of release, the displacement of the block has its maximum value x_m and the velocity of the block is zero. If we put these *initial conditions*, as they are called, into Eqs. 14-3 and 14-5, we find

$$1 = \cos \phi \quad \text{and} \quad 0 = \sin \phi,$$

respectively. The smallest angle that satisfies both these requirements is $\phi = 0$. (Any angle that is an integer multiple of 2π rad also satisfies these requirements.)

SAMPLE PROBLEM 14-2

At $t = 0$, the displacement $x(0)$ of the block in a linear oscillator like that of Fig. 14-5 is -8.50 cm. Its velocity $v(0)$ then is -0.920 m/s and its acceleration $a(0)$ is $+47.0$ m/s^2.

a. What are the angular frequency ω and the frequency f of this system?

SOLUTION   If we put $t = 0$ in Eqs. 14-3, 14-5, and 14-6, we find

$$x(0) = x_m \cos \phi, \qquad (14\text{-}13)$$

$$v(0) = -\omega x_m \sin \phi, \qquad (14\text{-}14)$$

and

$$a(0) = -\omega^2 x_m \cos \phi. \qquad (14\text{-}15)$$

These three equations contain three unknowns, namely, x_m, ϕ, and ω. We should be able to find all three, but here we need only ω.

If we divide Eq. 14-15 by Eq. 14-13, the result is

$$\omega = \sqrt{-\frac{a(0)}{x(0)}} = \sqrt{-\frac{47.0 \text{ m/s}^2}{-0.0850 \text{ m}}}$$

$$= 23.5 \text{ rad/s}. \qquad \text{(Answer)}$$

The frequency f follows from Eq. 14-4 and is

$$f = \frac{\omega}{2\pi} = \frac{23.5 \text{ rad/s}}{2\pi} = 3.74 \text{ Hz}. \qquad \text{(Answer)}$$

b. What is the phase constant ϕ?

SOLUTION   If we divide Eq. 14-14 by Eq. 14-13, we find

$$\frac{v(0)}{x(0)} = \frac{-\omega x_m \sin \phi}{x_m \cos \phi} = -\omega \tan \phi.$$

Solving for $\tan \phi$, we find

$$\tan \phi = -\frac{v(0)}{\omega x(0)} = -\frac{-0.920 \text{ m/s}}{(23.5 \text{ rad/s})(-0.0850 \text{ m})}$$

$$= -0.461.$$

This equation has two solutions:

$$\phi = -25° \quad \text{and} \quad \phi = 155°.$$

You will see in part (c) how to choose between them.

c. What is the amplitude x_m of the motion?

SOLUTION From Eq. 14-13 we have, provisionally putting $\phi = 155°$,

$$x_m = \frac{x(0)}{\cos \phi} = \frac{-0.0850 \text{ m}}{\cos 155°} = 0.094 \text{ m}$$

$$= 9.4 \text{ cm.} \qquad \text{(Answer)}$$

If we had chosen $\phi = -25°$, we would have found $x_m = -9.4$ cm. However, the amplitude of the motion must always be a *positive* constant, so $-25°$ cannot be the correct phase constant. We must therefore have, in part (b),

$$\phi = 155°. \qquad \text{(Answer)}$$

SAMPLE PROBLEM 14-3

In Fig. 14-6a, a uniform bar with mass m lies symmetrically across two rapidly rotating, fixed rollers, A and B, with distance $L = 2.0$ cm between the bar's center of mass and each roller. The rollers, whose directions of rotation are shown in the figure, slip against the bar with coefficient of kinetic friction $\mu_k = 0.40$. Suppose the bar is displaced horizontally by a distance x, as in Fig. 14-6b, and then released.

a. What is the angular frequency ω of the resulting horizontal simple harmonic (back and forth) motion of the bar?

SOLUTION To find ω, we find an expression for the horizontal acceleration a of the bar as a function of x and compare it with Eq. 14-7. We do so by applying Newton's second law vertically and horizontally; then we use the law in angular form about the contact point between the bar and roller A.

The vertical forces acting on the bar are its weight $m\mathbf{g}$ and supporting forces $\mathbf{F}_A$ due to roller A and $\mathbf{F}_B$ due to roller B. Since there is no net vertical force acting

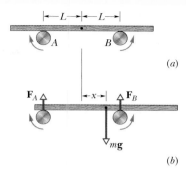

FIGURE 14-6 Sample Problem 14-3. (a) A bar is in equilibrium on two rotating rollers, A and B, that slip beneath it. (b) The bar is displaced by x and then released.

on the bar, Newton's second law gives us

$$\sum F_y = F_A + F_B - mg = 0. \qquad (14\text{-}16)$$

The horizontal forces acting on the bar are the kinetic frictional forces $f_{kA} = \mu_k F_A$ (toward the right) due to roller A, and $f_{kB} = -\mu_k F_B$ (toward the left) due to roller B. Horizontally, Newton's second law gives us

$$\mu_k F_A - \mu_k F_B = ma,$$

or

$$a = \frac{\mu_k F_A - \mu_k F_B}{m}. \qquad (14\text{-}17)$$

The bar experiences no net torque about an axis perpendicular to the plane of Fig. 14-6 through the contact point between the bar and roller A. So Newton's second law for torques about that axis gives us

$$\sum \tau = F_A(0) + F_B 2L - mg(L + x) + f_{kA}(0) + f_{kB}(0)$$

$$= 0, \qquad (14\text{-}18)$$

where the forces $\mathbf{F}_A$, $\mathbf{F}_B$, $m\mathbf{g}$, $\mathbf{f}_{kA}$, and $\mathbf{f}_{kB}$ have moment arms about that axis of 0, L, $L + x$, 0, and 0, respectively.

Solving Eqs. 14-16 and 14-18 for F_A and F_B, we find

$$F_A = \frac{mg(L - x)}{2L} \quad \text{and} \quad F_B = \frac{mg(L + x)}{2L}.$$

Substituting these results into Eq. 14-17 yields

$$a = -\frac{\mu_k g}{L} x. \qquad (14\text{-}19)$$

Comparison of Eq. 14-19 with Eq. 14-7 reveals that the bar is undergoing simple harmonic motion with angular frequency ω given by

$$\omega^2 = \frac{\mu_k g}{L}.$$

Thus

$$\omega = \sqrt{\frac{\mu_k g}{L}} = \sqrt{\frac{(0.40)\,(9.8 \text{ m/s}^2)}{0.020 \text{ m}}}$$

$$= 14 \text{ rad/s.} \qquad \text{(Answer)}$$

b. What is the period T of the bar's motion?

SOLUTION   From Eq. 14-4 we find

$$T = \frac{2\pi}{\omega} = \frac{2\pi}{14 \text{ rad/s}} = 0.45 \text{ s.} \qquad \text{(Answer)}$$

Note that, as long as the bar is able to slide over the rollers, ω and T are independent of the mass m.

14-4 SIMPLE HARMONIC MOTION: ENERGY CONSIDERATIONS

Figure 8-5 shows how the energy of a linear oscillator shuttles back and forth between kinetic and potential forms, while their sum—the mechanical energy E—remains constant. We now consider this situation quantitatively.

The potential energy of a linear oscillator like that of Fig. 14-5 is associated entirely with the spring. Its value depends on how much the spring is stretched or compressed, that is, on $x(t)$. We use Eq. 14-3 to find

$$U(t) = \tfrac{1}{2}kx^2 = \tfrac{1}{2}kx_m^2 \cos^2(\omega t + \phi). \quad (14\text{-}20)$$

The kinetic energy of the system is associated entirely with the block. Its value depends on how fast the block is moving, that is, on $v(t)$. We use Eq. 14-5 to find

$$K(t) = \tfrac{1}{2}mv^2 = \tfrac{1}{2}m(-\omega x_m)^2 \sin^2(\omega t + \phi). \quad (14\text{-}21)$$

If we use Eq. 14-11 to substitute k/m for ω^2, we can write Eq. 14-21 as

$$K(t) = \tfrac{1}{2}mv^2 = \tfrac{1}{2}kx_m^2 \sin^2(\omega t + \phi). \quad (14\text{-}22)$$

The mechanical energy follows from Eqs. 14-20 and 14-22 and is

$$E = U + K$$

$$= \tfrac{1}{2}kx_m^2 \cos^2(\omega t + \phi) + \tfrac{1}{2}kx_m^2 \sin^2(\omega t + \phi)$$

$$= \tfrac{1}{2}kx_m^2 \left[\cos^2(\omega t + \phi) + \sin^2(\omega t + \phi)\right].$$

However, for any angle α,

$$\cos^2 \alpha + \sin^2 \alpha = 1.$$

Thus the quantity in the square brackets above is unity and we have

$$E = U + K = \tfrac{1}{2}kx_m^2, \qquad (14\text{-}23)$$

which is indeed a constant, independent of time. The potential energy and the kinetic energy of the linear oscillator are shown as a function of time in

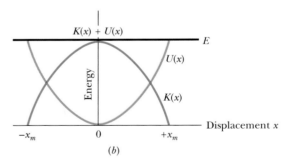

FIGURE 14-7 (a) The potential energy $U(t)$, the kinetic energy $K(t)$, and the mechanical energy E as functions of time, for a linear harmonic oscillator. Note that all energies are positive and that the potential energy and the kinetic energy peak twice during every period. (b) The potential energy $U(x)$, the kinetic energy $K(x)$, and the mechanical energy E as functions of position, for a linear harmonic oscillator with amplitude x_m. Note that for $x = 0$ the energy is all kinetic, and for $x = \pm x_m$ it is all potential.

Fig. 14-7*a*, and as a function of displacement in Fig. 14-7*b*.

You might now understand why an oscillating system normally contains an element of springiness and an element of inertia: it uses the former to store its potential energy and the latter to store its kinetic energy. The system of Sample Problem 14-3 is an exception, because no potential energy is involved. Instead, the energy transferred to the bar by the rollers actually varies over time.

SAMPLE PROBLEM 14-4

a. What is the mechanical energy of the linear oscillator of Sample Problem 14-1?

SOLUTION We find, substituting data from Sample Problem 14-1 into Eq. 14-23,

$$E = \tfrac{1}{2}kx_m^2 = (\tfrac{1}{2})(65 \text{ N/m})(0.11 \text{ m})^2$$

$$= 0.393 \text{ J} \approx 0.39 \text{ J}. \qquad \text{(Answer)}$$

This value remains constant throughout the motion.

b. What is the potential energy of this oscillator when the particle is halfway to its end point, that is, when $x = \pm \tfrac{1}{2}x_m$?

SOLUTION For either displacement, the potential energy is given by Eq. 14-20:

$$U = \tfrac{1}{2}kx^2 = \tfrac{1}{2}k(\tfrac{1}{2}x_m)^2 = \tfrac{1}{4}(\tfrac{1}{2}kx_m^2)$$

$$= \tfrac{1}{4}E = (\tfrac{1}{4})(0.393 \text{ J}) = 0.098 \text{ J}. \qquad \text{(Answer)}$$

c. What is the kinetic energy of the oscillator when $x = \tfrac{1}{2}x_m$?

SOLUTION We find this from

$$K = E - U$$

$$= 0.393 \text{ J} - 0.098 \text{ J} \approx 0.30 \text{ J}. \qquad \text{(Answer)}$$

Thus, at this point during the oscillation, 25% of the energy is in potential form and 75% in kinetic form.

14-5 AN ANGULAR SIMPLE HARMONIC OSCILLATOR

Figure 14-8 shows an angular version of a linear simple harmonic oscillator; the element of springiness or elasticity is associated with the twisting of a suspension wire rather than the extension and com-

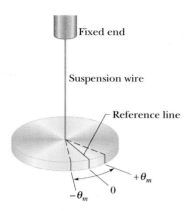

FIGURE 14-8 An angular simple harmonic oscillator, or torsion pendulum, is an angular version of the linear simple harmonic oscillator of Fig. 14-5. The disk oscillates in a horizontal plane; the reference line oscillates with angular amplitude θ_m. The twist in the suspension wire stores potential energy like a spring and provides the restoring torque.

pression of a spring. The device is called a **torsion pendulum,** with *torsion* referring to the twisting.

If we rotate the disk in Fig. 14-8 from its rest position (indicated by a reference line marked 0) and release it, it will oscillate about that position in **angular simple harmonic motion.** Rotating the disk through an angle θ, in either direction, introduces a restoring torque given by

$$\tau = -\kappa\theta. \qquad (14\text{-}24)$$

Here κ (Greek *kappa*) is a constant, called the **torsion constant,** that depends on the length, diameter, and material of the suspension wire.

Comparison of Eq. 14-24 with Eq. 14-9 leads us to suspect that Eq. 14-24 is the angular form of Hooke's law, and that we can transform Eq. 14-12, which gives the period of linear SHM, into an equation for the period of angular SHM: we replace the spring constant k in Eq. 14-12 with its equivalent, the constant κ of Eq. 14-24, and we replace the mass m in Eq. 14-12 with *its* equivalent, the rotational inertia I of the oscillating disk. These replacements lead to

$$T = 2\pi\sqrt{\frac{I}{\kappa}} \qquad \text{(torsion pendulum)} \qquad (14\text{-}25)$$

for the period of the angular simple harmonic oscillator, or torsion pendulum.

SAMPLE PROBLEM 14-5

As Fig. 14-9a shows, a thin rod whose length L is 12.4 cm and whose mass m is 135 g is suspended at its midpoint from a long wire. Its period T_a of angular SHM is measured to be 2.53 s. An irregular object, which we call object X, is then hung from the same wire, as in Fig. 14-9b, and its period T_b is found to be 4.76 s.

a. What is the rotational inertia of object X about its suspension axis?

SOLUTION   In Table 11-2(e), the rotational inertia of the thin rod about a perpendicular axis through its midpoint is given as $\frac{1}{12}mL^2$. Thus we have

$$I_a = \tfrac{1}{12}mL^2 = (\tfrac{1}{12})(0.135 \text{ kg})(0.124 \text{ m})^2$$

$$= 1.73 \times 10^{-4} \text{ kg} \cdot \text{m}^2.$$

Now let us write Eq. 14-25 twice, once for the rod and once for object X:

$$T_a = 2\pi \sqrt{\frac{I_a}{\kappa}} \quad \text{and} \quad T_b = 2\pi \sqrt{\frac{I_b}{\kappa}}.$$

Here the subscripts refer to Figs. 14-9a and 14-9b. The constant κ, which is a property of the wire, is the same for both figures; only the periods and the rotational inertias are different.

Let us square each of these two equations, divide the second by the first, and solve the resulting equation for I_b. The result is

$$I_b = I_a \frac{T_b^2}{T_a^2} = (1.73 \times 10^{-4} \text{ kg} \cdot \text{m}^2) \frac{(4.76 \text{ s})^2}{(2.53 \text{ s})^2}$$

$$= 6.12 \times 10^{-4} \text{ kg} \cdot \text{m}^2. \qquad \text{(Answer)}$$

b. What would be the period of oscillation if both objects were fastened together and hung from the wire, as in Fig. 14-9c?

We can again write Eq. 14-25 twice, this time as

$$T_a = 2\pi \sqrt{\frac{I_a}{\kappa}} \quad \text{and} \quad T_c = 2\pi \sqrt{\frac{I_c}{\kappa}}.$$

We find, after dividing again and putting $I_c = I_a + I_b$, that

$$T_c = T_a \sqrt{\frac{I_c}{I_a}} = T_a \sqrt{\frac{I_a + I_b}{I_a}} = T_a \sqrt{1 + \frac{I_b}{I_a}}$$

$$= (2.53) \sqrt{1 + \frac{6.12 \times 10^{-4} \text{ kg} \cdot \text{m}^2}{1.73 \times 10^{-4} \text{ kg} \cdot \text{m}^2}}$$

$$= 5.39 \text{ s.} \qquad \text{(Answer)}$$

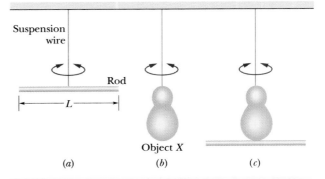

(a)   (b)   (c)

FIGURE 14-9   Sample Problem 14-5. Three torsion pendulums: (a) a rod, (b) an irregular object, and (c) the objects rigidly connected.

14-6 PENDULUMS

We turn now to a class of simple harmonic oscillators in which the springiness is associated with the gravitational force rather than with the elastic properties of a twisted wire or a compressed or stretched spring.

The Simple Pendulum

If you hang an apple at the end of a long thread fixed at its upper end, and then set the apple swinging by a small distance left and right, you easily see that the apple's motion is periodic. Is it, in fact, simple harmonic motion? To idealize this situation, we consider a **simple pendulum,** which consists of a particle of mass m suspended from an unstretchable, massless string of length L, as in Fig. 14-10. The mass is free to swing back and forth in a plane, to the left and right of a vertical line through the point at which the upper end of the string is fixed.

The element of inertia in this pendulum is the mass of the particle, and the element of springiness is in the gravitational attraction between the particle and the Earth. Potential energy can be associated with the varying vertical distance between the swinging particle and the Earth; we may view that varying distance as the varying length of a "gravitational spring."

The forces acting on the particle in Fig. 14-10 are its weight $m\mathbf{g}$ and the tension $\mathbf{T}$ in the string. We

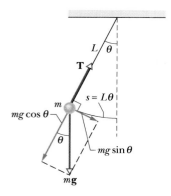

FIGURE 14-10   A simple pendulum. The forces acting on the bob are its weight mg and the tension **T** in the cord. The tangential component $mg \sin \theta$ of the weight is a restoring force that brings the pendulum back to the central position.

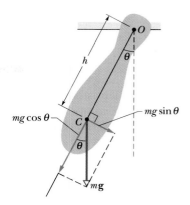

FIGURE 14-11   A physical pendulum. The restoring torque is $(mg \sin \theta)(h)$. When $\theta = 0$, the center of gravity C hangs directly below the point of suspension O.

resolve mg into a radial component $mg \cos \theta$ and a component $mg \sin \theta$ that is tangent to the path taken by the particle. This tangential component is a restoring force, because it always acts opposite the displacement of the particle so as to bring the particle back toward its central location, the *equilibrium position* ($\theta = 0$) where it would be at rest were it not swinging. We write the restoring force as

$$F = - mg \sin \theta, \qquad (14\text{-}26)$$

where the minus sign indicates that F acts opposite the displacement.

If we assume that the angle θ in Fig. 14-10 is small, then $\sin \theta$ is very nearly equal to θ in radians. For example, when $\theta = 5.00°\ (= 0.0873 \text{ rad})$, $\sin \theta = 0.0872$, a difference of only about 0.1%. The displacement s of the particle measured along its arc is equal to $L\theta$. Thus, for small θ, we have $\sin \theta \approx \theta$, and Eq. 14-26 becomes

$$F \approx - mg\theta = - mg\,\frac{s}{L} = -\left(\frac{mg}{L}\right) s. \quad (14\text{-}27)$$

A glance back at Eq. 14-9 shows that we again have Hooke's law, with the displacement now being arc length s instead of x. Thus *if a simple pendulum swings through a small angle,* it is a linear oscillator like the block–spring oscillator of Fig. 14-5; that is, it undergoes simple harmonic motion. Now the amplitude of the motion is measured as the **angular amplitude** θ_m, the maximum angle of swing. And the spring constant k is mg/L, the effective spring constant of the pendulum's gravitational spring.

By substituting mg/L for k in Eq. 14-12, we find, for the period of a simple pendulum,

$$T = 2\pi \sqrt{\frac{m}{k}} = 2\pi \sqrt{\frac{m}{mg/L}} \qquad (14\text{-}28)$$

or

$$T = 2\pi \sqrt{\frac{L}{g}} \qquad \text{(simple pendulum).} \quad (14\text{-}29)$$

Equation 14-29 holds only if angular amplitude θ_m is small (which is what is assumed in the exercises and problems for this chapter unless otherwise stated).

The element of inertia seems to be missing in Eq. 14-29 because the period is independent of the mass of the particle. This comes about because the element of springiness, which is the gravitational spring constant mg/L, is itself proportional to the mass of the particle, and the two masses cancel in Eq. 14-28. Figure 8-6 shows how energy shuttles back and forth between potential and kinetic forms during every oscillation of a simple pendulum.

The Physical Pendulum

Most pendulums in the real world are not even approximately "simple." Figure 14-11 shows a generalized **physical pendulum,** as we shall call it, with its weight mg acting at its center of mass C.

When the pendulum of Fig. 14-11 is displaced through an angle θ in either direction from its equilibrium position, a restoring torque appears. This torque acts about an axis through the suspension point O in Fig. 14-11 and has the magnitude

$$\tau = - (mg \sin \theta)(h). \qquad (14\text{-}30)$$

Here $mg \sin \theta$ is the tangential component of the weight $m\mathbf{g}$, and h (which is equal to OC) is the moment arm of this force component. The minus sign indicates that the torque is a restoring torque. That is, it is a torque that always acts to reduce the angle θ to zero.

We once more decide to limit our interest to small amplitudes, so that $\sin \theta \approx \theta$. Then Eq. 14-30 becomes

$$\tau \approx - (mgh)\, \theta. \qquad (14\text{-}31)$$

Comparison with Eq. 14-24 shows that we again have Hooke's law in angular form. Thus a physical pendulum undergoes simple harmonic motion *if the* angular amplitude θ_m of its motion is small. The term mgh of Eq. 14-31 is analogous to the torsion constant κ of Eq. 14-24. Substituting mgh for κ in Eq. 14-25, we find

$$T = 2\pi \sqrt{\frac{I}{mgh}} \qquad \text{(physical pendulum)} \qquad (14\text{-}32)$$

for the period of a physical pendulum when θ_m is small. Here I is the rotational inertia of the pendulum—about an axis through its point of support perpendicular to its plane of swing—and h is the distance between the point of support and the center of mass of the swinging pendulum. We know that a physical pendulum will not swing if we hang it from its center of mass. Formally, this corresponds to putting $h = 0$ in Eq. 14-32. That equation then predicts $T \to \infty$, which implies that such a pendulum will never complete one swing.

The physical pendulum of Fig. 14-11 includes the simple pendulum as a special case. For the latter, h would be the length L of the string and I would be mL^2. Making these substitutions in Eq. 14-32 leads to

$$T = 2\pi \sqrt{\frac{I}{mgh}} = 2\pi \sqrt{\frac{mL^2}{mgL}} = 2\pi \sqrt{\frac{L}{g}},$$

which is exactly Eq. 14-29, the expression for the period of a simple pendulum.

Measuring g

We can use a physical pendulum to measure the free-fall acceleration g. (Countless thousands of such measurements have been made in the course of geophysical prospecting.)

To analyze a simple case, take the pendulum to be a uniform rod of length L, suspended from one end. For such a pendulum, h in Eq. 14-32, the distance between the suspension point and the center of mass, is $\frac{1}{2}L$. Table 11-2(f) tells us that the rotational inertia of this pendulum about a perpendicular axis through one end is $\frac{1}{3}mL^2$. If we put $h = \frac{1}{2}L$ and $I = \frac{1}{3}mL^2$ in Eq. 14-32 and solve for g, we find

$$g = \frac{8\pi^2 L}{3T^2}. \qquad (14\text{-}33)$$

Thus by measuring L and the period T, we can find the value of g. (If precise measurements are to be made, a number of refinements are needed, such as swinging the pendulum in an evacuated chamber.)

SAMPLE PROBLEM 14-6

A meter stick, suspended from one end, swings as a physical pendulum; see Fig. 14-12a.

a. What is its period of oscillation?

SOLUTION From Table 11-2(f) we see that the rotational inertia of a rod or stick of length L about a perpendicular axis through one end is $\frac{1}{3}mL^2$. The distance h from the point of suspension to the center of mass, which is point C in Fig. 14-12a, is $\frac{1}{2}L$. If we substitute these two quantities into Eq. 14-32, we find

$$T = 2\pi \sqrt{\frac{I}{mgh}} = 2\pi \sqrt{\frac{\frac{1}{3}mL^2}{mg(\frac{1}{2}L)}} = 2\pi \sqrt{\frac{2L}{3g}} \qquad (14\text{-}34)$$

$$= 2\pi \sqrt{\frac{(2)(1.00 \text{ m})}{(3)(9.8 \text{ m/s}^2)}} = 1.64 \text{ s.} \qquad \text{(Answer)}$$

b. What would be the length L_0 of a simple pendulum that would have the same period (see Fig. 14-12b)?

SOLUTION Setting Eqs. 14-29 and 14-34 equal yields

$$T = 2\pi \sqrt{\frac{L_0}{g}} = 2\pi \sqrt{\frac{2L}{3g}}.$$

You can see by inspection that

$$L_0 = \tfrac{2}{3}L = (\tfrac{2}{3})(100 \text{ cm}) = 66.7 \text{ cm.} \qquad \text{(Answer)}$$

FIGURE 14-12 Sample Problem 14-6. (*a*) A meter stick suspended from one end as a physical pendulum. (*b*) A simple pendulum whose length L is chosen so that the periods of the two pendulums are equal. Point P on the pendulum of (*a*) marks the *center of oscillation*, a distance L_0 from the suspension point.

FIGURE 14-13 Sample Problem 14-7. (*a*) The pendulum of Fig. 14-12*a* reproduced. (*b*) The pendulum reversed, and suspended from its center of oscillation. Reversing the pendulum in this way does not change its period.

The length of the equivalent simple pendulum is marked as point P in Fig. 14-12*a*. This point, called the **center of oscillation,** has some interesting properties (see Sample Problem 14-7). The center of oscillation is not a fixed point on the pendulum but has meaning only with respect to a specified point of suspension.

SAMPLE PROBLEM 14-7

Suppose that the pendulum of Fig. 14-12*a* is inverted and suspended from point P, as in Fig. 14-13*b*. What will be its period of oscillation?

SOLUTION Let us first find the rotational inertia of the meter stick about a perpendicular axis through point P. Table 11-2(*e*) tells us that the rotational inertia about point C, the center of mass, is $\frac{1}{12}mL^2$. The distance h between points C and P in Fig. 14-13*b* can be found from Fig. 14-13*a* to be

$$h = \frac{2L}{3} - \frac{L}{2} = \frac{L}{6}.$$

From the parallel-axis theorem we have, for the rotational inertia of the pendulum about a perpendicular axis through P,

$$I = I_{cm} + mh^2 = \tfrac{1}{12}mL^2 + m(L/6)^2 = \tfrac{1}{9}mL^2.$$

If we substitute these values of h and I into Eq. 14-32, the result is

$$T = 2\pi \sqrt{\frac{I}{mgh}} = 2\pi \sqrt{\frac{\frac{1}{9}mL^2}{mg(L/6)}}$$

$$= 2\pi \sqrt{\frac{2L}{3g}}. \qquad \text{(Answer)}$$

This is exactly Eq. 14-34. Although we have shown it only for a specific case, the following theorem is true in general:

To every point of suspension O of a physical pendulum, there corresponds a *center of oscillation P*, distant from O by the length of the equivalent simple pendulum. The physical pendulum has the same period whether it is suspended from O, as in Fig. 14-13*a*, or from P, as in Fig. 14-13*b*.

SAMPLE PROBLEM 14-8

A disk whose radius R is 12.5 cm is suspended, as a physical pendulum, from a point halfway between its rim and its center C (see Fig. 14-14). Its period T is

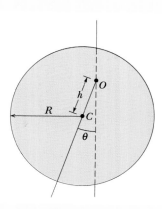

FIGURE 14-14 Sample Problem 14-8. A physical pendulum consisting of a uniform disk suspended from a point (O) that is halfway from the center of the disk to the rim.

measured to be 0.871 s. What is the free-fall acceleration g at the location of the pendulum?

SOLUTION The rotational inertia I_{cm} of a disk about its central axis is $\frac{1}{2}mR^2$. From the parallel-axis theorem, the rotational inertia about an axis parallel to the central axis and through the point of suspension O, as in Fig. 14-14, is

$$I = I_{cm} + mh^2 = \frac{1}{2}mR^2 + m(\tfrac{1}{2}R)^2 = \frac{3}{4}mR^2.$$

If we put $I = \frac{3}{4}mR^2$ and $h = \frac{1}{2}R$ in Eq. 14-32, we find

$$T = 2\pi\sqrt{\frac{I}{mgh}} = 2\pi\sqrt{\frac{\frac{3}{4}mR^2}{mg(\frac{1}{2}R)}} = 2\pi\sqrt{\frac{3R}{2g}}.$$

Solving for g then gives us

$$g = \frac{6\pi^2 R}{T^2} = \frac{(6\pi^2)(0.125\ \text{m})}{(0.871\ \text{s})^2} = 9.76\ \text{m/s}^2.\quad\text{(Answer)}$$

SAMPLE PROBLEM 14-9

In Fig. 14-15a, a marmoset of mass m_2 clutches a massless cord wrapped around a disk, of radius $R = 20$ cm and mass $M = 8m_2$, that pivots about a horizontal axis through the center of the disk at O. Mass m_1 ($=4m_2$) is attached to the disk at a distance $r = R/2$ from O.

a. When the *disk + marmoset + m_1* system is in equilibrium, what is angle ϕ between the vertical and a line from O to m_1 (Fig. 14-15b)?

SOLUTION In equilibrium, the net torque on the disk about O is zero. Let the counterclockwise direction be the positive angular direction. Then the torques about O due to m_1 and m_2 are $-m_1gr\sin\phi$ and m_2gR, respectively. So we have

$$\sum\tau = -m_1gr\sin\phi + m_2gR = 0,$$

or

$$\sin\phi = \frac{m_2R}{m_1r}.\quad(14\text{-}35)$$

With $m_1 = 4m_2$ and $r = R/2$, Eq. 14-35 yields

$$\sin\phi = \frac{m_2R}{(4m_2)(R/2)} = \tfrac{1}{2},\ \text{or}\ \phi = 30°.\quad\text{(Answer)}$$

(a)

(b)

(c)

FIGURE 14-15 Sample Problem 14-9. (*a*) An equilibrium system consisting of mass m_1, a marmoset of mass m_2, and a disk of radius R and mass M, pivoted at O. (*b*) Forces acting on the disk in the equilibrium orientation. (*c*) Same as (*b*) but for a different orientation.

b. In terms of m_2 and R, what is the rotational inertia I of the system about O?

SOLUTION The marmoset's mass m_2 acts as though it is located on the rim of the disk because its pull on the disk, via the cord, is there. By Eq. 11-22, it has rotational inertia m_2R^2 about O. The rotational inertias about O of m_1 and the disk are m_1r^2 and $\frac{1}{2}MR^2$, respectively, the latter by Table 11-2(c). So the rotational inertia of the system about O is

$$I = m_2R^2 + m_1r^2 + \tfrac{1}{2}MR^2.$$

With $m_1 = 4m_2$, $M = 8m_2$, and $r = R/2$ this yields

$$I = m_2R^2 + m_2R^2 + 4m_2R^2$$

$$= 6m_2R^2. \qquad \text{(Answer)} \qquad (14\text{-}36)$$

c. The disk is rotated counterclockwise from equilibrium through a small angle θ and released (Fig. 14-15c). What is the angular frequency ω of the resulting simple harmonic motion?

SOLUTION To find ω, we find an expression for the angular acceleration α (about O) in terms of θ and compare it to Eq. 14-7. About O, the torque due to m_1 is now

$$-m_1gr\sin(\phi + \theta) = -m_1gr(\sin\phi\cos\theta + \sin\theta\cos\phi)$$

$$= -m_1gr(\sin\phi + \theta\cos\phi).$$

Here, because θ is small, we have put $\sin\theta \approx \theta$ and $\cos\theta \approx 1$. Here also $\phi = 30°$. The angular displacement θ does not change the torque due to m_2, so the net torque about O is

$$\sum \tau = m_2gR - m_1gr(\sin\phi + \theta\cos\phi).$$

Using Newton's second law in angular form to set $\sum \tau = I\alpha$, substituting for $\sin\phi$ from Eq. 14-35, and substituting for I from Eq. 14-36 yield

$$m_2gR - m_1gr\left(\frac{m_2R}{m_1r} + \theta\cos\phi\right) = I\alpha = 6m_2R^2\alpha.$$

Solving for α in terms of θ, we obtain

$$\alpha = -\frac{m_1gr\cos\phi}{6m_2R^2}\theta.$$

With $m_1 = 4m_2$ and $r = R/2$, we obtain

$$\alpha = -\frac{g\cos\phi}{3R}\theta. \qquad (14\text{-}37)$$

Equation 14-37 is an angular version of Eq. 14-7. It implies simple harmonic motion with angular acceleration α and angular displacement θ. Comparing Eqs.

14-37 and 14-7 reveals that

$$\omega^2 = \frac{g\cos\phi}{3R},$$

so that

$$\omega = \sqrt{\frac{g\cos\phi}{3R}} = \sqrt{\frac{(9.8\ \text{m/s}^2)(\cos 30°)}{3(0.20\ \text{m})}}$$

$$= 3.8\ \text{rad/s}. \qquad \text{(Answer)}$$

(After "'Atwood's' Oscillator," by Thomas B. Greenslade, Jr., *American Journal of Physics*, December 1988.)

14-7 SIMPLE HARMONIC MOTION AND UNIFORM CIRCULAR MOTION

In 1610, Galileo, using his newly constructed telescope, discovered the four principal moons of Jupiter. Over weeks of observation, each moon seemed to him to be moving back and forth relative to the planet in what today we would call simple harmonic motion; the disk of the planet was the midpoint of the motion. The record of Galileo's observations, written in his own hand, is still available. A. P. French of MIT used Galileo's data to work out the position of the moon Callisto relative to Jupiter. In the results shown in Fig. 14-16, the circles are based on Galileo's observations and the curve is a best fit to

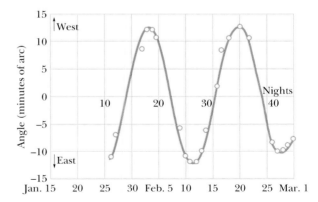

FIGURE 14-16 The angle between Jupiter and its moon Callisto as seen from Earth. The circles are based on Galileo's 1610 measurements. The curve is a best fit, strongly suggesting simple harmonic motion. At Jupiter's mean distance, 10 minutes of arc correspond to about 2×10^6 km. The time axis covers about 6 weeks of observations. (Adapted from A. P. French, *Newtonian Mechanics*, W. W. Norton & Company, New York, 1971, p. 288.)

the data. The curve strongly suggests Eq. 14-3, the displacement function for SHM. A period of about 16.8 days can be measured from the plot.

Actually, Callisto moves with essentially constant speed in an essentially circular orbit around Jupiter. Its true motion—far from being simple harmonic—is uniform circular motion. What Galileo saw—and what you can see with a good pair of binoculars and a little patience—is the projection of this uniform circular motion on a line in the plane of the motion. We are led by Galileo's remarkable observations to the conclusion that simple harmonic motion is uniform circular motion viewed edge on. In more formal language:

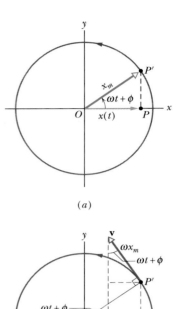

(a)

> Simple harmonic motion is the projection of uniform circular motion on a diameter of the circle in which the latter motion occurs.

Let us look at this conclusion more closely. Figure 14-17a shows a particle P'—the reference particle—in uniform circular motion with angular speed ω in a circle—the *reference circle*—whose radius is x_m. At any time t, the angular position of the particle is $\omega t + \phi$, where ϕ is its initial angular position. Let us project P' onto the x axis; its projection is a point P, which we imagine to be a second particle. The location of P as a function of time is then given by

$$x(t) = x_m \cos(\omega t + \phi),$$

which is precisely Eq. 14-3. Our conclusion is correct. If reference particle P' moves in uniform circular motion, its projection particle P will move in simple harmonic motion.

This relation throws new light on the angular frequency ω of simple harmonic motion, and you can see more clearly how the "angular" originates. The quantity ω is simply the constant angular speed with which the reference particle P' moves along its reference circle. The phase constant ϕ has a value determined by the position of reference particle P' along its reference circle at $t = 0$.

Figure 14-17b shows the velocity of the reference particle. The magnitude of the velocity vector is ωx_m and its projection on the x axis is

$$v(t) = -\omega x_m \sin(\omega t + \phi),$$

which is exactly Eq. 14-5. The minus sign appears because the velocity component of P in Fig. 14-17b points to the left, in the direction of decreasing x.

(b)

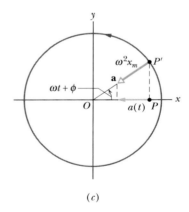

(c)

FIGURE 14-17 (a) A reference particle P' moving with uniform circular motion in a reference circle of radius x_m. Its projection P on the x axis executes simple harmonic motion. (b) The projection of the velocity **v** of the reference particle is the velocity of simple harmonic motion. (c) The projection of the acceleration **a** of the reference particle is the acceleration of simple harmonic motion.

Figure 14-17c shows the acceleration of the reference particle. The magnitude of the acceleration vector is $\omega^2 x_m$ and its projection on the x axis is

$$a(t) = -\omega^2 x_m \cos(\omega t + \phi),$$

which is exactly Eq. 14-6. Thus whether we look at the displacement, the velocity, or the acceleration, the projection of uniform circular motion is indeed simple harmonic motion.

14-8 DAMPED SIMPLE HARMONIC MOTION (OPTIONAL)

A pendulum will hardly swing under water, because the water exerts a drag force on the pendulum that quickly eliminates the motion. A pendulum swinging in air does better, but still the motion dies out because the air exerts a drag force on the pendulum (and friction acts at its support).

When the motion of an oscillator is reduced by an external force, the oscillator and its motion are said to be **damped.** An idealized example of a damped oscillator is shown in Fig. 14-18: a block with mass m oscillates on a spring with spring constant k. From the mass, a rod extends to a vane (both assumed massless) that is submerged in a liquid. As the vane moves up and down, the liquid exerts an inhibiting drag force on it and thus on the entire oscillating system. With time, the mechanical energy

FIGURE 14-18 An idealized damped simple harmonic oscillator. A vane immersed in a liquid exerts, on the oscillating block, a damping force given by $-b(dx/dt)$, where b is the damping constant.

of the block–spring system decreases, as energy is transferred to thermal energy of the liquid and vane.

Let us assume that the liquid exerts a **damping force F_d** that is proportional in magnitude to the velocity $\mathbf{v}$ of the vane and block (an assumption that is accurate if the vane moves slowly). Then

$$F_d = -bv, \tag{14-38}$$

where b is a **damping constant** that depends on the characteristics of the vane and the liquid, and the minus sign indicates that $\mathbf{F}_d$ opposes the motion. The total force acting on the block is then

$$\sum F = -kx - bv,$$

or, if we set $v = dx/dt$,

$$\sum F = -kx - b\frac{dx}{dt}. \tag{14-39}$$

If a particle is acted on by the force of Eq. 14-39, its displacement as a function of time turns out to be

$$x(t) = x_m e^{-bt/2m} \cos(\omega_d t + \phi), \tag{14-40}$$

where ω_d, the angular frequency of the damped oscillator, is given by

$$\omega_d = \sqrt{\frac{k}{m} - \frac{b^2}{4m^2}}. \tag{14-41}$$

If $b = 0$ (there is no damping), then Eq. 14-41 reduces to Eq. 14-11 ($\omega = \sqrt{k/m}$) for the angular frequency of an undamped oscillator, and Eq. 14-40 reduces to Eq. 14-3 for the displacement of an undamped oscillator. If the damping constant is not zero but small (so that $b \ll \sqrt{km}$), then $\omega_d \approx \omega$.

We can regard Eq. 14-40 as a cosine function whose amplitude, which is $x_m e^{-bt/2m}$, gradually decreases with time, as Fig. 14-19 suggests. For an undamped oscillator, the mechanical energy is constant and is given by Eq. 14-23, or $E = \frac{1}{2}kx_m^2$. If the oscillator is damped, the mechanical energy is not constant but decreases with time. If the damping is small, we can find $E(t)$ by replacing x_m in Eq. 14-23 with $x_m e^{-bt/2m}$, the amplitude of the damped oscillations. Doing so, we find

$$E(t) \approx \frac{1}{2}kx_m^2 e^{-bt/m}, \tag{14-42}$$

which tells us that the mechanical energy decreases exponentially with time.

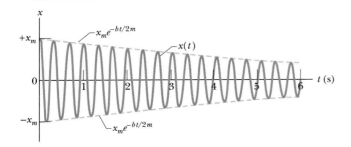

FIGURE 14-19 The displacement function $x(t)$ for the damped oscillator of Fig. 14-18. See Sample Problem 14-10. The amplitude, which is $x_m\, e^{-bt/2m}$, decreases exponentially with time.

SAMPLE PROBLEM 14-10

For the damped oscillator of Fig. 14-18, $m = 250$ g, $k = 85$ N/m, and $b = 70$ g/s.

a. What is the period of the motion?

SOLUTION From Eq. 14-12 we have

$$T = 2\pi \sqrt{\frac{m}{k}} = 2\pi \sqrt{\frac{0.25 \text{ kg}}{85 \text{ N/m}}} = 0.34 \text{ s}. \quad \text{(Answer)}$$

b. How long does it take for the amplitude of the damped oscillations to drop to half its initial value?

SOLUTION The amplitude, displayed in Eq. 14-40 as $x_m\, e^{-bt/2m}$, has the value x_m at $t = 0$. Thus we must find t in the expression

$$\tfrac{1}{2}x_m = x_m\, e^{-bt/2m}.$$

Let us divide both sides by x_m and take the natural logarithm of the equation that remains. We find

$$\ln \tfrac{1}{2} = \ln(e^{-bt/2m}) = -bt/2m,$$

or

$$t = \frac{-2m \ln \tfrac{1}{2}}{b} = \frac{-(2)(0.25 \text{ kg})(\ln \tfrac{1}{2})}{0.070 \text{ kg/s}}$$

$$= 5.0 \text{ s}. \quad \text{(Answer)}$$

Note, from the answer we found in (a), that this is about 15 periods of oscillation.

c. How long does it take for the mechanical energy to drop to one-half its initial value?

SOLUTION From Eq. 14-42 we see that the mechanical energy, which is $\tfrac{1}{2}kx_m^2\, e^{-bt/m}$, has the value $\tfrac{1}{2}kx_m^2$ at $t = 0$. Thus we must find t in

$$\tfrac{1}{2}(\tfrac{1}{2}kx_m^2) = \tfrac{1}{2}kx_m^2\, e^{-bt/m}.$$

If we divide by $\tfrac{1}{2}kx_m^2$ and solve for t, as we did above, we find

$$t = \frac{-m \ln \tfrac{1}{2}}{b} = \frac{-(0.25 \text{ kg})(\ln \tfrac{1}{2})}{0.070 \text{ kg/s}} = 2.5 \text{ s}. \quad \text{(Answer)}$$

This is exactly half the time calculated in (b), or about 7.5 periods of oscillation. Figure 14-19 was drawn to illustrate this sample problem.

14-9 FORCED OSCILLATIONS AND RESONANCE (OPTIONAL)

A person swinging passively in a swing is an example of *free oscillation*. If a kind friend pulls or pushes the swing periodically, as in Fig. 14-20, we have *forced oscillations*. There are now *two* angular frequencies with which to deal: (1) the *natural* angular frequency of the system, which is the angular frequency at which it would oscillate if it were disturbed and

FIGURE 14-20 Two frequencies are suggested in this painting by Nicholas Lancret: (1) the natural frequency at which the lady—left to herself—would swing, and (2) the frequency at which her friend tugs on the rope. If these two frequencies match, there is resonance.

then left to oscillate freely, and (2) the angular frequency of the external driving force. We have been representing the natural angular frequency with ω, but we will now switch to ω_0 and let ω represent the angular frequency of the driving force.

We can use Fig. 14-18 to represent an idealized forced simple harmonic oscillator if we allow the structure marked "rigid support" to move up and down at a variable angular frequency ω. A forced oscillator will *always* oscillate at the angular frequency ω of the driving force, its displacement $x(t)$ being given by

$$x(t) = x_m \cos(\omega t + \phi), \qquad (14\text{-}43)$$

where x_m is the amplitude of the oscillations.

How large the displacement amplitude x_m will be depends on a complicated function of ω and ω_0. The velocity amplitude v_m of the oscillations is easier to describe: it is greatest when

$$\omega = \omega_0 \qquad (\text{resonance}), \qquad (14\text{-}44)$$

a condition called **resonance**. Equation 14-44 is also *approximately* the condition at which the displacement amplitude x_m of the oscillations will be greatest. Thus if you push a swing at its natural frequency, the displacement and velocity amplitudes will increase to large values, a fact that children learn quickly by trial and error. If you push at other frequencies, either higher or lower, the displacement and velocity amplitudes will be smaller.

Figure 14-21 shows approximately how the displacement amplitude of an oscillator depends on the angular frequency ω of the driving force, for three different values of the damping coefficient b. Note that for all three the amplitude x_m is approximately greatest when $\omega/\omega_0 = 1$, that is, when the resonance condition of Eq. 14-44 is satisfied. The curves of Fig. 14-21 show that the smaller the damping, the taller and narrower is the *resonance peak*.

All mechanical structures have one or more natural frequencies. Care must be taken not to subject a structure to a strong external driving force that matches one of these frequencies, or the resulting oscillations of the structure may rupture it. The image of a soprano shattering a wine glass comes at once to mind. Aircraft designers make sure that none of the natural frequencies at which a wing can vibrate matches the angular frequency of the engines at cruising speed. It obviously would be dangerous for a wing to flap violently at certain engine speeds.

The collapse of a 1.4-km stretch of the Nimitz Freeway, shown in this chapter's opening photograph, is an example of the destruction that resonance can produce. Oscillations of the ground due to the earthquake's seismic waves had their greatest velocity amplitude at an angular frequency of about 9 rad/s, which almost exactly matched a natural angular frequency of the individual horizontal sections of the freeway. The collapse was confined to a particular 1.4-km stretch because that section of the

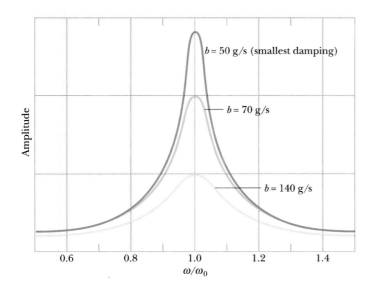

FIGURE 14-21 The displacement amplitude x_m of a forced oscillator varies as the angular frequency ω of the driving force is varied. These three curves are approximately symmetric about $\omega/\omega_0 = 1$, the resonance condition, at which the amplitude is approximately the greatest. The curves correspond to three values of the damping constant b, the smallest value giving the greatest amplitude at resonance.

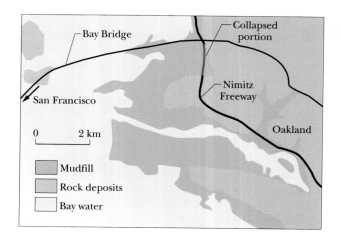

FIGURE 14-22 The geological structure in the Oakland portion of the San Francisco Bay area. The collapsed stretch of the Nimitz Freeway is indicated. (Adapted from "Sediment-Induced Amplification and the Collapse of the Nimitz Freeway," by S. E. Hough et al., *Nature*, April 26, 1990.)

freeway was built on a loosely structured mudfill whose velocity amplitude was *at least five times larger* than that of the rock deposits under the rest of the freeway (see Fig. 14-22). The Nimitz tragedy is a lesson for engineers and geologists who must plan for even more destructive earthquakes in California.

REVIEW & SUMMARY

Frequency

Every oscillatory, or periodic, motion has a *frequency f* (the number of oscillations per second) measured, in the SI system, in hertz:

$$1 \text{ hertz} = 1 \text{ Hz} = 1 \text{ oscillation per second}$$
$$= 1 \text{ s}^{-1}. \quad (14\text{-}1)$$

Period

The *period T*, the time required for one complete oscillation, or **cycle,** is related to frequency by

$$T = \frac{1}{f}. \quad (14\text{-}2)$$

Simple Harmonic Motion

In *simple harmonic motion* (SHM), the displacement $x(t)$ of a particle from its equilibrium position is described by the equation

$$x = x_m \cos(\omega t + \phi) \quad \text{(displacement)}, \quad (14\text{-}3)$$

in which x_m is the **amplitude** of the displacement, the quantity $(\omega t + \phi)$ is the **phase** of the motion, and ϕ is the **phase constant.** The **angular frequency** ω is related to the period and frequency of the motion by

$$\omega = \frac{2\pi}{T} = 2\pi f \quad \text{(angular frequency).} \quad (14\text{-}4)$$

Differentiating Eq. 14-3 leads to equations for velocity and acceleration SHM as functions of time:

$$v = -\omega x_m \sin(\omega t + \phi) \quad \text{(velocity)} \quad (14\text{-}5)$$

and

$$a = -\omega^2 x_m \cos(\omega t + \phi) \quad \text{(acceleration).} \quad (14\text{-}6)$$

In Eq. 14-5, the positive quantity ωx_m is the **velocity amplitude** v_m of the motion. In Eq. 14-6, the positive quantity $\omega^2 x_m$ is the **acceleration amplitude** a_m of the motion.

The Linear Oscillator

A particle with mass m that moves under the influence of a Hooke's law restoring force given by $F = -kx$ exhibits simple harmonic motion with

$$\omega = \sqrt{\frac{k}{m}} \quad \text{(angular frequency)} \quad (14\text{-}11)$$

and

$$T = 2\pi \sqrt{\frac{m}{k}} \quad \text{(period).} \quad (14\text{-}12)$$

Such a system is called a **linear simple harmonic oscillator.**

Energy

A particle in simple harmonic motion has, at any time, kinetic energy $K = \frac{1}{2}mv^2$ and potential energy $U = \frac{1}{2}kx^2$. If no friction is present, the mechanical energy $E = K + U$ remains constant even though K and U change.

Pendulums

Examples of simple harmonic motion are the **torsion pendulum** of Fig. 14-8, the **simple pendulum** of Fig. 14-10, and the **physical pendulum** of Fig. 14-11. Their periods of oscillation for small oscillations are, respectively,

$$T = 2\pi \sqrt{\frac{I}{\kappa}}, \quad T = 2\pi \sqrt{\frac{L}{g}}, \quad (14\text{-}25, 14\text{-}29)$$

and

$$T = 2\pi \sqrt{\frac{I}{mgh}}. \quad (14\text{-}32)$$

In each case, the period is determined by an inertial term divided by a "springiness" term that measures the strength of the oscillator's restoring force.

Simple Harmonic Motion and Uniform Circular Motion
Simple harmonic motion is the projection of uniform circular motion onto the diameter of the circle in which the latter motion occurs. Figure 14-17 shows that all parameters of circular motion (position, velocity, and acceleration) project to the corresponding values for simple harmonic motion.

Damped Harmonic Motion
The mechanical energy E in a real oscillating system decreases during the oscillations because external forces, such as a drag force, inhibit the oscillations and transfer mechanical energy to thermal energy. The oscillator and its motion are said to be **damped.** If a **damping force** is given by $F_d = -bv$, where v is the velocity of the oscillator and b is a **damping constant,** then the displacement of the oscillator is given by

$$x(t) = x_m e^{-bt/2m} \cos(\omega_d t + \phi), \qquad (14\text{-}40)$$

where ω_d, the angular frequency of the damped oscillator, is given by

$$\omega_d = \sqrt{\frac{k}{m} - \frac{b^2}{4m^2}}. \qquad (14\text{-}41)$$

If the damping constant is small ($b \ll \sqrt{km}$), then $\omega_d \approx \omega$, where ω is the angular frequency of the undamped oscillator. For small b, the mechanical energy E of the oscillator is given by

$$E(t) \approx \tfrac{1}{2} k x_m^2\, e^{-bt/m}. \qquad (14\text{-}42)$$

Forced Oscillations and Resonance
If an external driving force with angular frequency ω acts on an oscillating system with *natural* angular frequency ω_0, the system oscillates with angular frequency ω. The velocity amplitude v_m of the system is greatest when

$$\omega = \omega_0, \qquad (14\text{-}44)$$

a condition called **resonance.** The amplitude x_m of the system is (approximately) greatest under the same condition.

QUESTIONS

1. What are the phase differences of two linear oscillators, with identical mass and spring constant, in the three situations shown in Fig. 14-23?

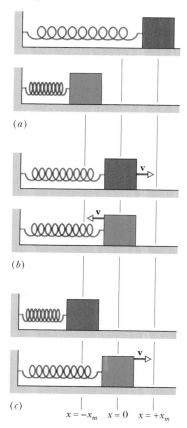

(a)

(b)

(c)

$x = -x_m \quad x = 0 \quad x = +x_m$

FIGURE 14-23 Question 1.

2. When mass m_1 is hung from spring A and a smaller mass m_2 is hung from spring B, the springs are stretched by the same distance. If the systems are then put into vertical simple harmonic motion with the same amplitude, which system will have more energy?

3. A spring has a force constant k, and a mass m is suspended from it. The spring is cut in half and the same mass is suspended from one of the halves. How are the frequencies of oscillation, before and after the spring is cut, related?

4. Suppose that a system consists of a block of unknown mass and a spring of unknown force constant. Show how we can predict the period of oscillation of this block–spring system simply by measuring the extension of the spring produced by attaching the block to it.

5. Any real spring has mass. If this mass is taken into account, explain qualitatively how this will affect the period of oscillation of a mass–spring system.

6. How are each of the following properties of a simple harmonic oscillator affected by doubling the amplitude: period, spring constant, total mechanical energy, maximum velocity, and maximum acceleration?

7. What changes could you make in a harmonic oscillator that would double the maximum speed of the oscillating mass?

8. What would happen to the motion of an oscillating system if the sign of the force term, $- kx$ in Eq. 14-9, were changed?

9. Will the frequency of oscillation of a torsion pendulum change if you take it to the moon? What about the frequencies of a simple pendulum, a mass–spring oscillator, and a physical pendulum, such as a wooden plank swinging from one end?

10. Predict by qualitative arguments whether the period of a pendulum will increase or decrease when its amplitude is increased.

11. A pendulum suspended from the ceiling of an elevator cab has period T when the cab is stationary. How is its period affected when the elevator moves (a) upward with constant speed, (b) downward with constant speed, (c) downward with constant upward acceleration, (d) upward with constant upward acceleration, (e) upward with constant downward acceleration $a < g$, and (f) downward with constant downward acceleration $a > g$? (g) In which cases, if any, does the pendulum swing upside down?

12. A pendulum mounted in a cart has period T when the cart is stationary and on a horizontal plane. How is the period affected if the cart is on a plane inclined at angle θ with the horizontal (Fig. 14-24) while (a) stationary, (b) moving down the plane with constant speed, (c) moving up the plane with constant speed, (d) moving up the plane with constant acceleration up the plane, (e) moving down the plane with constant acceleration up the plane,

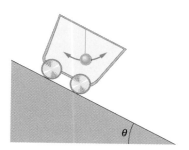

FIGURE 14-24 Questions 12 and 13.

(f) moving down the plane with constant acceleration $a < g \sin \theta$ down the plane, (g) moving down the plane with constant acceleration $a = g \sin \theta$ down the plane.

13. What is the angle between the vertical and the equilibrium position for the pendulum in the seven situations of Question 12?

14. How can a pendulum be used so that it traces out a sinusoidal curve?

15. Why are damping devices often used on machinery? Give an example.

16. A singer, holding a note of the right frequency, can shatter a glass if the glassware is of high quality. This cannot be done if the glassware quality is low. Explain why in terms of the damping constant of the glass.

EXERCISES & PROBLEMS

SECTION 14-3 SIMPLE HARMONIC MOTION: THE FORCE LAW

1E. An object undergoing simple harmonic motion takes 0.25 s to travel from one point of zero velocity to the next such point. The distance between those points is 36 cm. Calculate (a) the period, (b) the frequency, and (c) the amplitude of the motion.

2E. An oscillating mass–spring system takes 0.75 s to begin repeating its motion. Find (a) the period, (b) the frequency in hertz, and (c) the angular frequency of the system in radians per second.

3E. A 4.00-kg block hangs from a spring, extending it 16.0 cm from its unstretched position. (a) What is the spring constant? (b) The block is removed and a 0.500-kg body is hung from the same spring. If the spring is then stretched and released, what is its period of oscillation?

4E. An oscillator consists of a block of mass 0.500 kg connected to a spring. When set into oscillation with amplitude 35.0 cm, it is observed to repeat its motion every 0.500 s. Find (a) the period, (b) the frequency, (c) the angular frequency, (d) the spring constant, (e) the maximum speed, and (f) the maximum force exerted on the block.

5E. The vibration frequencies of atoms in solids at normal temperatures are of the order of 10^{13} Hz. Imagine the atoms to be connected to one another by "springs." Suppose that a single silver atom in a solid vibrates with this frequency and that all the other atoms are at rest. Compute the effective spring constant. One mole of silver has a mass of 108 g and contains 6.02×10^{23} atoms.

6E. What is the maximum acceleration of a platform that vibrates with an amplitude of 2.20 cm at a frequency of 6.60 Hz?

7E. A loudspeaker produces a musical sound by means of the oscillation of a diaphragm. If the amplitude of oscillation is limited to 1.0×10^{-3} mm, what frequencies will result in the acceleration of the diaphragm exceeding g?

8E. The scale of a spring balance which reads from 0 to 32.0 lb is 4.00 in. long. A package suspended from the balance is found to oscillate vertically with a frequency of 2.00 Hz. (a) What is the spring constant? (b) How much does the package weigh?

9E. A 20-N weight is hung from the bottom of a vertical spring, causing the spring to stretch 20 cm. (a) What is the spring constant? (b) This spring is now placed horizontally on a frictionless table. One end of it is held fixed and the other end is attached to a 5.0-N weight. The weight is then moved (stretching the spring) and released from rest. What is the period of oscillation?

10E. A 50.0-g mass is attached to the bottom of a vertical spring and set vibrating. If the maximum speed of the mass is 15.0 cm/s and the period is 0.500 s, find (a) the spring constant of the spring, (b) the amplitude of the motion, and (c) the frequency of oscillation.

11E. A 1.00×10^{-20}-kg particle is vibrating with simple harmonic motion with a period of 1.00×10^{-5} s and a maximum speed of 1.00×10^3 m/s. Calculate (a) the angular frequency and (b) the maximum displacement of the particle.

12E. A small body of mass 0.12 kg is undergoing simple harmonic motion of amplitude 8.5 cm and period 0.20 s. (a) What is the maximum value of the force acting on it? (b) If the oscillations are produced by a spring, what is the spring constant?

13E. In an electric shaver, the blade moves back and forth over a distance of 2.0 mm. The motion is simple harmonic, with frequency 120 Hz. Find (a) the amplitude, (b) the maximum blade speed, and (c) the maximum blade acceleration.

14E. A speaker diaphragm is vibrating in simple harmonic motion with a frequency of 440 Hz and a maximum displacement of 0.75 mm. What are (a) the angular frequency, (b) the maximum speed, and (c) the maximum acceleration of this diaphragm?

15E. An automobile can be considered to be mounted on four identical springs as far as vertical oscillations are concerned. The springs of a certain car are adjusted so that the vibrations have a frequency of 3.00 Hz. (a) What is the spring constant of each spring if the mass of the car is 1450 kg and the weight is evenly distributed over the springs? (b) What will be the vibration frequency if five passengers, averaging 73.0 kg each, ride in the car? (Again, consider an even distribution of weight.)

16E. A body oscillates with simple harmonic motion according to the equation

$$x = (6.0 \text{ m}) \cos[(3\pi \text{ rad/s})t + \pi/3 \text{ rad}].$$

At $t = 2.0$ s, what are (a) the displacement, (b) the velocity, (c) the acceleration, and (d) the phase of the motion? Also, what are (e) the frequency and (f) the period of the motion?

17E. A particle executes linear SHM with frequency 0.25 Hz about the point $x = 0$. At $t = 0$, it has displacement $x = 0.37$ cm and zero velocity. For the motion, determine (a) the period, (b) the angular frequency, (c) the amplitude, (d) the displacement at time t, (e) the velocity at time t, (f) the maximum speed, (g) the maximum acceleration, (h) the displacement at $t = 3.0$ s, and (i) the speed at $t = 3.0$ s.

18E. The piston in the cylinder head of a locomotive has a stroke (twice the amplitude) of 0.76 m. If the piston moves with simple harmonic motion with an angular frequency of 180 rev/min, what is its maximum speed?

19P. Figure 14-25 shows an astronaut on a Body Mass Measuring Device (BMMD). Designed for use on orbiting space vehicles, its purpose is to allow astronauts to measure their mass in the "weightless" conditions in Earth orbit. The BMMD is a spring-mounted chair; an astronaut measures his or her period of oscillation in the chair; the mass follows from the formula for the period of an oscillating block–spring system. (a) If M is the mass of the astronaut and m the "effective" mass of that part of the BMMD that also oscillates, show that

$$M = (k/4\pi^2)T^2 - m,$$

where T is the period of oscillation and k is the spring constant. (b) The spring constant was $k = 605.6$ N/m for the BMMD on Skylab Mission Two; the period of oscillation of the empty chair was 0.90149 s. Calculate the effective mass of the chair. (c) With an astronaut in the chair, the period of oscillation became 2.08832 s. Calculate the mass of the astronaut.

20P. A 2.00-kg block hangs from a spring. A 300-g body hung below the block stretches the spring 2.00 cm farther. (a) What is the spring constant? (b) If the 300-g body is removed and the block is set into oscillation, find the period of motion.

21P. The end point of a spring vibrates with a period of 2.0 s when a mass m is attached to it. When this mass is increased by 2.0 kg, the period is found to be 3.0 s. Find the value of m.

22P. The end of one of the prongs of a tuning fork that executes simple harmonic motion of frequency 1000 Hz has an amplitude of 0.40 mm. Find (a) the maximum acceleration and (b) the maximum speed of the end of the prong. Find (c) the acceleration and (d) the speed of the end of the prong when it has a displacement of 0.20 mm.

23P. A 0.10-kg block oscillates back and forth along a straight line on a frictionless horizontal surface. Its displacement from the origin is given by

$$x = (10 \text{ cm})\cos[(10 \text{ rad/s})t + \pi/2 \text{ rad})].$$

(a) What is the oscillation frequency? (b) What is the maximum speed acquired by the block? At what value of x does this occur? (c) What is the maximum acceleration of the block? At what value of x does this occur? (d) What force, applied to the block, results in the given oscillation?

24P. At a certain harbor, the tides cause the ocean surface to rise and fall a distance d in simple harmonic motion, with a period of 12.5 h. How long does it take for the water to fall a distance $d/4$ from its maximum height?

25P. Two blocks ($m = 1.0$ kg and $M = 10$ kg) and a spring ($k = 200$ N/m) are arranged on a horizontal, frictionless surface as shown in Fig. 14-26. The coefficient of static friction between the two blocks is 0.40. What is the maximum possible amplitude of the simple harmonic motion if no slippage is to occur between the blocks?

FIGURE 14-26   Problem 25.

26P. A block is on a horizontal surface (a shake table) that is moving horizontally with simple harmonic motion of frequency 2.0 Hz. The coefficient of static friction between block and surface is 0.50. How great can the amplitude of the SHM be if the block is not to slip along the surface?

FIGURE 14-25   Problem 19.

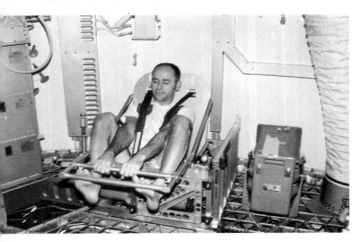

27P. A block is on a piston that is moving vertically with simple harmonic motion. (a) If the SHM has period 1.0 s, at what amplitude of motion will the block and piston separate? (b) If the piston has an amplitude of 5.0 cm, what is the maximum frequency for which the block and piston will be in contact continuously?

28P. An oscillator consists of a block attached to a spring ($k = 400$ N/m). At some time t, the position (measured from the system's equilibrium location), velocity, and acceleration of the block are $x = 0.100$ m, $v = -13.6$ m/s, and $a = -123$ m/s². Calculate (a) the frequency, (b) the mass of the block, and (c) the amplitude of oscillation for the motion.

29P. A simple harmonic oscillator consists of a block of mass 2.00 kg attached to a spring of spring constant 100 N/m. When $t = 1.00$ s, the position and velocity of the block are $x = 0.129$ m and $v = 3.415$ m/s. (a) What is the amplitude of the oscillations? What were the (b) position and (c) velocity of the mass at $t = 0$ s?

30P. A massless spring hangs from the ceiling with a small object attached to its lower end. The object is initially held at rest in a position y_i such that the spring is not stretched. The object is then released from y_i and oscillates up and down, with its lowest position being 10 cm below y_i. (a) What is the frequency of the oscillation? (b) What is the speed of the object when it is 8.0 cm below the initial position? (c) An object of mass 300 g is attached to the first object, after which the system oscillates with half the original frequency. What is the mass of the first object? (d) Relative to y_i, where is the new equilibrium (rest) position with both objects attached to the spring?

31P. Two particles oscillate in simple harmonic motion along a common straight line segment of length A. Each particle has a period of 1.5 s but they differ in phase by $\pi/6$ rad. (a) How far apart are they (in terms of A) 0.50 s after the lagging particle leaves one end of the path? (b) Are they then moving in the same direction, toward each other, or away from each other?

32P. Two particles execute simple harmonic motion of the same amplitude and frequency along the same straight line. They pass one another moving in opposite directions each time their displacement is half their amplitude. What is the phase difference between them?

33P. Two identical springs are attached to a block of mass m and to fixed supports as shown in Fig. 14-27. Show that the frequency of oscillation on the frictionless surface is

$$f = \frac{1}{2\pi}\sqrt{\frac{2k}{m}}.$$

34P. Suppose that the two springs in Fig. 14-27 have different spring constants k_1 and k_2. Show that the frequency f of oscillation of the block is then given by

$$f = \sqrt{f_1^2 + f_2^2},$$

where f_1 and f_2 are the frequencies at which the block

FIGURE 14-27 Problems 33 and 34.

would oscillate if connected only to spring 1 or only to spring 2.

35P. Two springs are joined and connected to a mass m as shown in Fig. 14-28. The surface is frictionless. If the springs both have force constant k, show that the frequency of oscillation of m is

$$f = \frac{1}{2\pi}\sqrt{\frac{k}{2m}}.$$

FIGURE 14-28 Problem 35.

36P. A block weighing 14.0 N, which slides without friction on a 40.0° incline, is connected to the top of the incline by a massless spring of unstretched length 0.450 m and spring constant 120 N/m, as shown in Fig. 14-29. (a) How far from the top of the incline does the block stop? (b) If the block is pulled slightly down the incline and released, what is the period of the ensuing oscillations?

Frictionless

40.0°

FIGURE 14-29 Problem 36.

37P. A uniform spring whose unstretched length is L has a force constant k. The spring is cut into two pieces of unstretched lengths L_1 and L_2, with $L_1 = nL_2$. (a) What are the corresponding force constants k_1 and k_2 in terms of n and k? (b) If a block is attached to the original spring, as in Fig. 14-5, it oscillates with frequency f. If the spring is replaced with the piece L_1 or L_2, the corresponding frequency is f_1 or f_2. Find f_1 and f_2 in terms of f.

38P. Three 10,000-kg ore cars are held at rest on a 30° incline on a mine railway using a cable that is parallel to the incline (Fig. 14-30). The cable is observed to stretch 15 cm just before the coupling between the lower cars breaks, detaching the lowest car. Assuming that the cable obeys Hooke's law, find (a) the frequency and (b) the am-

FIGURE 14-30   Problem 38.

plitude of the resulting oscillations of the remaining two cars.

SECTION 14-4 SIMPLE HARMONIC MOTION: ENERGY CONSIDERATIONS

39E. Find the mechanical energy of a block–spring system having a spring constant of 1.3 N/cm and an amplitude of 2.4 cm.

40E. An oscillating block–spring system has a mechanical energy of 1.00 J, an amplitude of 10.0 cm, and a maximum speed of 1.20 m/s. Find (a) the spring constant, (b) the mass, and (c) the frequency of oscillation.

41E. A vertical spring stretches 9.6 cm when a 1.3-kg block is hung from its end. (a) Calculate the spring constant. This block is then displaced an additional 5.0 cm downward and released from rest. Find (b) the period, (c) the frequency, (d) the amplitude, (e) the total energy, and (f) the maximum speed of the resulting SHM.

42E. A 5.00-kg object on a horizontal frictionless surface is attached to a spring with spring constant 1000 N/m. The object is displaced 50.0 cm horizontally and given an initial velocity of 10.0 m/s back toward the equilibrium position. (a) What is the frequency of the motion? What are (b) the initial potential energy of the block–spring system, (c) the initial kinetic energy, and (d) the amplitude of the oscillation?

43E. A (hypothetical) large slingshot is stretched 1.50 m to launch a 130-g projectile with speed sufficient to escape from the Earth (11.2 km/s). Assume the elastic sling obeys Hooke's law. (a) What is the spring constant of the device, if all the potential energy is converted to kinetic energy? (b) Assume that an average person can exert a force of 220 N. How many people would be required to stretch the slingshot?

44E. When the displacement in SHM is one-half the amplitude x_m, what fraction of the total energy is (a) kinetic and (b) potential? (c) At what displacement, in terms of the amplitude, is the energy of the system half kinetic and half potential?

45E. A block of mass M, at rest on a horizontal frictionless table, is attached to a rigid support by a spring of constant k. A bullet of mass m and velocity v strikes the block as shown in Fig. 14-31. The bullet remains embedded in the block. Determine (a) the velocity of the block immediately after the collision and (b) the amplitude of the resulting simple harmonic motion.

FIGURE 14-31   Exercise 45.

46P. A 3.0-kg particle is in simple harmonic motion in one dimension and moves according to the equation

$$x = (5.0 \text{ m})\cos[(\pi/3 \text{ rad/s})t - \pi/4 \text{ rad}].$$

(a) At what value of x is the potential energy of the particle equal to half the total energy? (b) How long does it take the particle to move to this position x from the equilibrium position?

47P. A 10-g particle is undergoing simple harmonic motion with an amplitude of 2.0×10^{-3} m. The maximum acceleration experienced by the particle is 8.0×10^3 m/s²; the phase constant is $-\pi/3$ rad. (a) Write an equation for the force on the particle as a function of time. (b) What is the period of the motion? (c) What is the maximum speed of the particle? (d) What is the total mechanical energy of this simple harmonic oscillator?

48P. A massless spring with spring constant 19 N/m hangs vertically. A body of mass 0.20 kg is attached to its free end and then released. Assume that the spring was unstretched before the body was released. Find (a) how far below the initial position the body descends, and (b) the frequency and (c) the amplitude of the resulting motion, assumed to be simple harmonic.

49P. A 4.0-kg block is suspended from a spring with a spring constant of 500 N/m. A 50-g bullet is fired into the block from directly below with a speed of 150 m/s and is embedded in the block. (a) Find the amplitude of the resulting simple harmonic motion. (b) What fraction of the original kinetic energy of the bullet appears as mechanical energy in the harmonic oscillator?

50P*. A solid cylinder is attached to a horizontal massless spring so that it can roll without slipping along a horizontal surface (Fig. 14-32). The spring constant k is 3.0 N/m. If the system is released from rest at a position in which the spring is stretched by 0.25 m, find (a) the translational kinetic energy and (b) the rotational kinetic energy of the

FIGURE 14-32   Problem 50.

cylinder as it passes through the equilibrium position. (c) Show that under these conditions the center of mass of the cylinder executes simple harmonic motion with period

$$T = 2\pi \sqrt{\frac{3M}{2k}},$$

where M is the mass of the cylinder. (*Hint:* Find the time derivative of the total mechanical energy.)

SECTION 14-5   AN ANGULAR SIMPLE HARMONIC OSCILLATOR

51E. A flat uniform circular disk has a mass of 3.00 kg and a radius of 70.0 cm. It is suspended in a horizontal plane by a vertical wire attached to its center. If the disk is rotated 2.50 rad about the wire, a torque of 0.0600 N·m is required to maintain the disk in this position. Calculate (a) the rotational inertia of the disk about the wire, (b) the torsion constant, and (c) the angular frequency of this torsion pendulum when it is set oscillating.

52P. A 95-kg solid sphere with a 15-cm radius is suspended by a vertical wire attached to the ceiling of a room. A torque of 0.20 N·m is required to twist the sphere through an angle of 0.85 rad. What is the period of oscillation when the sphere is released from this position?

53P. An engineer wants to find the rotational inertia of an odd-shaped object of mass 10 kg about an axis through its center of mass. The object is supported with a wire along the desired axis. The wire has a torsion constant $\kappa = 0.50$ N·m. If this torsion pendulum oscillates through 20 complete cycles in 50 s, what is the rotational inertia of the odd-shaped object?

54P. The balance wheel of a watch oscillates with an angular amplitude of π rad and a period of 0.500 s. Find (a) the maximum angular speed of the wheel, (b) the angular speed of the wheel when its displacement is $\pi/2$ rad, and (c) the angular acceleration of the wheel when its displacement is $\pi/4$ rad.

SECTION 14-6   PENDULUMS

55E. What is the length of a simple pendulum whose period is 1.00 s at a point where $g = 32.2$ ft/s^2?

56E. A 2500-kg demolition ball swings from the end of a crane, as shown in Fig. 14-33. The length of the swinging segment of cable is 17 m. (a) Find the period of swing, assuming that the system can be treated as a simple pendulum. (b) Does the period depend on the ball's mass?

57E. What is the length of a simple pendulum that marks seconds by completing a full swing to the left and right every 2.0 s?

58E. If a simple pendulum with length 1.50 m makes 72.0 oscillations in 180 s, what is the acceleration of gravity at its location?

FIGURE 14-33 Exercise 56.

59E. Two oscillating systems that you have studied are the block–spring and the simple pendulum. There is an interesting relation between them. Suppose that you hang a weight on the end of a spring, and when the weight is at rest, the spring is stretched a distance h. Show that the frequency of this block–spring system is the same as that of a simple pendulum whose length is h. See Fig. 14-34.

FIGURE 14-34 Exercise 59.

60E. A performer, seated on a trapeze, is swinging back and forth with a period of 8.85 s. If she stands up, thus raising the center of mass of the system *trapeze + performer* by 35.0 cm, what will be the new period of the trapeze? Treat *trapeze + performer* as a simple pendulum.

61E. A simple pendulum with length L is swinging freely with small angular amplitude. As the pendulum passes its central (or equilibrium) position, its cord is suddenly and rigidly clamped at its midpoint. In terms of the original period T of the pendulum, what will the new period be?

62E. A pendulum is formed by pivoting a long thin rod of length L and mass m about a point on the rod that is a distance d above the center of the rod. (a) Find the period of this pendulum in terms of d, L, m, and g, assuming that it swings with small amplitude. What happens to the period if (b) d is decreased, (c) L is increased, or (d) m is increased?

63E. A physical pendulum consists of a meter stick that is pivoted at a small hole drilled through the stick a distance x from the 50-cm mark. The period of oscillation is observed to be 2.5 s. Find the distance x.

64E. A physical pendulum consists of a uniform solid disk (of mass M and radius R) supported in a vertical plane by a pivot located a distance d from the center of the disk (Fig. 14-35). The disk is displaced by a small angle and released. Find an expression for the period of the resulting simple harmonic motion.

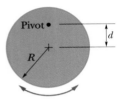

FIGURE 14-35   Exercise 64.

65E. A uniform circular disk whose radius R is 12.5 cm is suspended, as a physical pendulum, from a point on its rim. (a) What is its period of oscillation? (b) At what radial distance $r < R$ is there a point of suspension that gives the same period?

66E. A pendulum consists of a uniform disk with radius 10.0 cm and mass 500 g attached to a uniform rod with length 500 mm and mass 270 g; see Fig. 14-36. (a) Calculate the rotational inertia of the pendulum about the pivot. (b) What is the distance between the pivot and the center of mass of the pendulum? (c) Calculate the period of oscillation.

FIGURE 14-36   Exercise 66.

67E. In Sample Problem 14-6, we saw that a physical pendulum has a center of oscillation at distance $2L/3$ from its point of suspension. Show that the distance between the point of suspension and the center of oscillation for a physical pendulum of any form is I/mh, where I and h have the meanings assigned to them in Eq. 14-32, and m is the mass of the pendulum.

68E. A meter stick swinging from one end oscillates with a frequency f_0. What would be the frequency, in terms of f_0, if the bottom half of the stick were cut off?

69P. A stick with length L oscillates as a physical pendulum, pivoted about point O in Fig. 14-37. (a) Derive an expression for the period of the pendulum in terms of L and x, the distance from the point of support to the center of mass of the pendulum. (b) For what value of x/L is the period a minimum? (c) Show that if $L = 1.00$ m and $g = 9.80$ m/s^2, this minimum is 1.53 s.

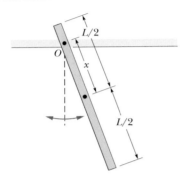

FIGURE 14-37   Problem 69.

70P. The center of oscillation of a physical pendulum has this interesting property: if an impulsive force (assumed horizontal and in the plane of oscillation) acts at the center of oscillation, no reaction is felt at the point of support. Baseball players (and players of many other sports) know that unless the ball hits the bat at this point (called the "sweet spot" by athletes), the reaction due to the impact will sting their hands. To prove this property, let the stick in Fig. 14-13a simulate a baseball bat. Suppose that a horizontal force **F** (due to impact with the ball) acts toward the right at P, the center of oscillation. The batter is assumed to hold the bat at O, the point of support of the stick. (a) What acceleration does point O undergo as a result of **F**? (b) What angular acceleration is produced by **F** about the center of mass of the stick? (c) As a result of the angular acceleration in (b), what linear acceleration does point O undergo? (d) Considering the magnitudes and directions of the accelerations in (a) and (c), convince yourself that P is indeed the "sweet spot."

71P. What is the frequency of a simple pendulum 2.0 m long (a) in a room, (b) in an elevator accelerating upward at a rate of 2.0 m/s^2, and (c) in free fall?

72P. A simple pendulum of length L and mass m is suspended in a car that is traveling with constant speed v around a circle of radius R. If the pendulum undergoes small oscillations in a radial direction about its equilibrium position, what will its frequency of oscillation be?

73P. For a simple pendulum, find the angular amplitude θ_m at which the restoring torque required for simple harmonic motion deviates from the actual restoring torque by 1.0%. (See "Trigonometric Expansions" in Appendix G.)

74P. The bob on a simple pendulum of length R moves in an arc of a circle. (a) By considering that the acceleration of the bob as it moves through its equilibrium position is that for uniform circular motion (mv^2/R), show that the tension in the string at that position is $mg(1 + \theta_m^2)$ if the angular amplitude θ_m is small. (See "Trigonometric Expansions" in Appendix G.) (b) Is the tension at other positions of the bob larger, smaller, or the same?

75P. A long uniform rod of length L and mass m is free to rotate in a horizontal plane about a vertical axis through its center. A spring with force constant k is connected horizontally between one end of the rod and a fixed wall. See Fig. 14-38. When the rod is in equilibrium it is parallel to the wall. What is the period of the small oscillations that result when the rod is rotated slightly and released?

FIGURE 14-38 Overhead view for Problem 75.

76P. A wheel is free to rotate about its fixed axle. A spring is attached to one of its spokes a distance r from the axle, as shown in Fig. 14-39. (a) Assuming that the wheel is a hoop of mass m and radius R, obtain the angular frequency of small oscillations of this system in terms of m, R, r, and the spring constant k. How does the result change if (b) $r = R$ and (c) $r = 0$?

FIGURE 14-39 Problem 76.

77P. A 2.5-kg disk, 42 cm in diameter, is supported by a massless rod, 76 cm long, which is pivoted at its end, as in Fig. 14-40. (a) The massless torsion spring is initially not connected. What is the period of oscillation? (b) The tor-

76 cm

42 cm

FIGURE 14-40 Problem 77.

sion spring is now connected so that, in equilibrium, the rod hangs vertically. What should be the torsional constant of the spring so that the new period of oscillation is 0.50 s shorter than before?

78P. A physical pendulum has two possible pivot points A and B; point A has a fixed position and B is adjustable

FIGURE 14-41 Problem 78.

along the length of the pendulum, as shown in Fig. 14-41. The period of the pendulum when suspended from A is found to be T. The pendulum is then reversed and suspended from B, which is moved until the pendulum again has period T. Show that the free-fall acceleration g is given by

$$g = \frac{4\pi^2 L}{T^2},$$

in which L is the distance between A and B for equal periods T. (Note that g can be measured in this way without knowing the rotational inertia of the pendulum or any of its dimensions except L.)

79P*. A uniform rod with length L swings from a pivot as a physical pendulum. How far, in terms of L, should the pivot be from the center of mass to minimize the period?

SECTION 14-8 DAMPED SIMPLE HARMONIC MOTION

80E. The amplitude of a lightly damped oscillator decreases by 3.0% during each cycle. What fraction of the energy of the oscillator is lost in each full oscillation?

81E. In Sample Problem 14-10, what is the ratio of the amplitude of the damped oscillations to the initial amplitude when 20 full oscillations have elapsed?

82E. For the system shown in Fig. 14-18, the block has a mass of 1.50 kg and the spring constant is 8.00 N/m. The damping force is given by $-b(dx/dt)$, where $b = 230$ g/s. Suppose that the block is pulled down a distance 12.0 cm and released. (a) Calculate the time required for the amplitude to fall to one-third of its initial value. (b) How many oscillations are made by the block in this time?

83P. A damped harmonic oscillator consists of a block ($m = 2.00$ kg), a spring ($k = 10.0$ N/m), and a damping force $F = -bv$. Initially, it oscillates with an amplitude of

25.0 cm; because of the damping, the amplitude falls to three-fourths of this initial value at the completion of four oscillations. (a) What is the value of b? (b) How much energy has been "lost" during these four oscillations?

84P. (a) In Eq. 14-39, find the ratio of the maximum damping force $(-b\, dx/dt)$ to the maximum spring force $(-kx)$ during the first oscillation for the data of Sample Problem 14-10. (b) Does this ratio change appreciably during later oscillations?

85P. Assume that you are examining the characteristics of the suspension system of a 2000-kg automobile. The suspension "sags" 10 cm when the weight of the entire automobile is placed on it. In addition, the amplitude of oscillation decreases by 50% during one complete oscillation. Estimate the values of k and b for the spring and shock absorber system of one wheel, assuming each wheel supports 500 kg.

SECTION 14-9 FORCED OSCILLATIONS AND RESONANCE

86E. For Eq. 14-43, suppose the amplitude x_m is given by

$$x_m = \frac{F_m}{[m^2(\omega^2 - \omega_0^2)^2 + b^2\omega^2]^{1/2}},$$

where F_m is the (constant) amplitude of the external oscillating force exerted on the spring by the rigid support. At resonance, what are (a) the amplitude and (b) the velocity amplitude of the oscillating object?

87P. A 2200-lb car carrying four 180-lb people travels over a rough "washboard" dirt road with corrugations 13 ft apart. The car bounces with maximum amplitude when its speed is 10 mi/h. The car now stops and the four people get out. By how much does the car body rise on its suspension owing to this decrease in weight?

ADDITIONAL PROBLEMS

88. A simple harmonic oscillator consists of a block attached to a spring of spring constant $k = 200$ N/m. The block slides back and forth along a straight line on a frictionless surface, with equilibrium point $x = 0$ and amplitude 0.20 m. A graph of the velocity v of the block as a function of time t is shown in Fig. 14-42. What are (a) the period of the simple harmonic motion, (b) the mass of the block, (c) the displacement of the block at $t = 0$, (d) the acceleration of the block at $t = 0.10$ s, and (e) the maximum kinetic energy attained by the block?

FIGURE 14-42 Problem 88.

89. A simple harmonic oscillator consists of a block of mass 0.50 kg attached to a spring. The block slides back and forth along a straight line on a frictionless surface with equilibrium point $x = 0$. At $t = 0$ the block is at its equilibrium point and is moving in the direction of increasing x. A graph of the magnitude of the net force **F** on the block as a function of its position is shown in Fig. 14-43. What are (a) the amplitude and (b) the period of the simple harmonic motion, (c) the magnitude of the maximum acceleration experienced by the block, and (d) the maximum kinetic energy attained by the block?

FIGURE 14-43 Problem 89.

90. A block weighing 20 N oscillates at one end of a vertical spring for which $k = 100$ N/m; the other end of the spring is attached to the ceiling. At a certain instant the spring is stretched 0.30 m beyond its unstretched length (the length when no weight is attached), and the block has zero velocity. (a) What is the net force on the block at this instant? What are (b) the amplitude and (c) the period of the resulting simple harmonic motion? (d) What is the maximum kinetic energy of the block as it oscillates?

91. A physical pendulum consists of two meter sticks that are joined together as shown in Fig. 14-44. What is its period of oscillation about a pin inserted through point A?

FIGURE 14-44 Problem 91.

GRAVITATION | 15

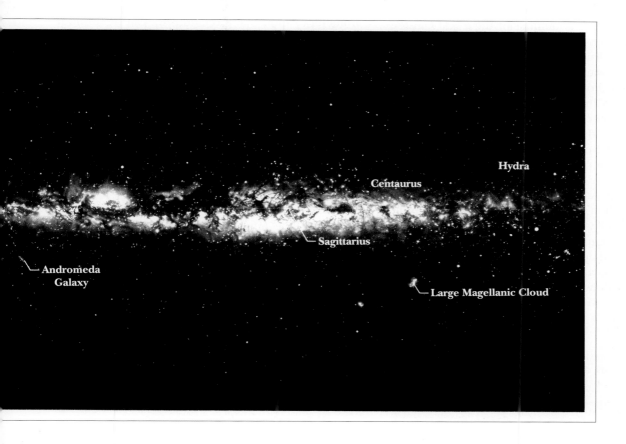

Hydra

Centaurus

Sagittarius

Andromeda
Galaxy

Large Magellanic Cloud

The Milky Way galaxy is a disk-shaped collection of dust, planets, and billions of stars, including our sun and solar system. The force that binds it or any other galaxy together is the same force that holds the moon in orbit and you to the Earth—the gravitational force. That force is also responsible for one of nature's strangest objects, the black hole, a star that has completely collapsed onto itself. The gravitational force near a black hole is so strong that not even light can escape it. But if that is the case, how can a black hole be detected?

15-1 THE WORLD AND THE GRAVITATIONAL FORCE

The drawing that opens this chapter shows our view of the Milky Way galaxy: we are near the edge of the disk of the galaxy, about 26,000 light-years (2.5×10^{20} m) from its center, which in the drawing lies in the direction of the star collection known as Sagittarius. Our galaxy is a member of the Local Cluster of galaxies, which includes the Andromeda galaxy (Fig. 15-1) at a distance of 2.3×10^6 light-years, and several closer dwarf galaxies, such as the Large Magellanic Cloud shown in the opening drawing.

The Local Cluster is one of about 100 clusters that form the Local Supercluster of galaxies. Measurements taken during and since the 1980s suggest that the Local Supercluster and the supercluster consisting of the clusters Hydra and Centaurus are all moving toward an exceptionally massive region called the Great Attractor. This region appears to be about 150 million light-years away, on the opposite side of the Milky Way from us, past the clusters Hydra and Centaurus.

The force that binds these progressively larger structures, from galaxy to supercluster, and which may be slowly drawing them all toward the Great Attractor, is the gravitational force. That force not only holds you to the Earth but also reaches out across the vastness of intergalactic space.

15-2 NEWTON'S LAW OF GRAVITATION

Physicists like to examine seemingly unrelated phenomena to show that a relationship can be found if they are examined closely enough. This search for unification has been going on for centuries. In 1665, the 23-year-old Isaac Newton made a basic contribution to physics when he showed that the force that holds the moon in its orbit is the same force that makes an apple fall. We take this so much for granted now that it is not easy for us to comprehend the ancient view that the motions of earthbound bodies and heavenly bodies were different in kind and were governed by different laws.

Newton concluded that not only does the Earth attract an apple and the moon but *every body in the universe attracts every other body*. This thought takes a little getting used to because the familiar attraction of the Earth for earthbound bodies is so great that it swamps the attractions that those earthbound bodies have for each other. For example, the Earth attracts an apple with a force of a few ounces. Two apples, placed 1 ft apart, also attract each other, but the force of attraction of each acting on the other is less than the weight of a speck of dust.

Quantitatively, Newton proposed what we call **Newton's law of gravitation:** every particle attracts any other particle with a **gravitational force** whose magnitude is given by

$$F = G \frac{m_1 m_2}{r^2}$$

(Newton's law of gravitation).   (15-1)

Here m_1 and m_2 are the masses of the particles, r is the distance between them, and G is the **gravitational constant** whose value is

$$G = 6.67 \times 10^{-11} \text{ N} \cdot \text{m}^2/\text{kg}^2$$
$$= 6.67 \times 10^{-11} \text{ m}^3/\text{kg} \cdot \text{s}^2. \quad (15\text{-}2)$$

As Fig. 15-2 shows, particle m_2 attracts particle m_1 with a gravitational force **F** that is directed toward particle m_2. And particle m_1 attracts particle m_2 with a gravitational force $-$**F** that is directed toward m_1. The forces **F** and $-$**F** form an action–reaction pair

FIGURE 15-1 The Andromeda galaxy. Located 2.3×10^6 light-years from us, and faintly visible to the naked eye, it is very similar to our home galaxy, the Milky Way.

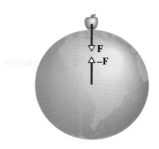

FIGURE 15-2 Two particles, of masses m_1 and m_2 and with a separation of r, attract each other according to Newton's law of gravitation, Eq. 15-1. The forces of attraction, **F** and $-$ **F**, are equal in magnitude and are in opposite directions.

FIGURE 15-3 The apple pulls up on the Earth just as hard as the Earth pulls down on the apple.

and are opposite in direction but equal in magnitude. They depend on the separation of the two particles, but not on the location of the two: the particles could be in a deep cave or in deep space. And **F** and $-$ **F** are not altered by the presence of other bodies, even if those bodies lie between the two particles we are considering.

The strength of the gravitational force, that is, how strongly two given masses at a given separation attract each other, depends on the value of the gravitational constant G. If G—by some miracle—were suddenly multiplied by a factor of 10, you would be crushed to the floor by the Earth's attraction. If it were divided by this factor, the Earth's attraction would be weak enough for you to jump over a tall building.

Although Newton's law of gravitation applies strictly to particles, we can also apply it to real objects as long as the sizes of the objects are small compared to the distance between them. The moon and the Earth are far enough apart so that, to a good approximation, we can treat them both as particles. But what about an apple and the Earth? From the point of view of the apple, the broad and level Earth, stretching out to the horizon beneath it, certainly does not look like a particle.

Newton solved the apple–Earth problem by proving an important theorem:

A uniform spherical shell of matter attracts a particle that is outside the shell as if all the shell's mass were concentrated at its center.

The Earth can be thought of as a nest of such shells, one within another, and each attracting a particle outside the Earth's surface as if the mass of that shell

were at the center of the shell. Thus, from the apple's point of view, the Earth *does* behave like a particle, located at the center of the Earth and having a mass equal to that of the Earth.

Suppose, as in Fig. 15-3, that the Earth pulls down on an apple with a force of 0.80 N. The apple must then pull up on the Earth with a force of 0.80 N, which we take to act at the center of the Earth. Although the forces are matched in magnitude, they produce different accelerations when the apple is released. For the apple, the acceleration is about 9.8 m/s², the familiar acceleration of a falling body near the Earth's surface. For the Earth, the acceleration measured in a reference frame attached to the center of mass of the apple–Earth system is only about 1×10^{-25} m/s².

15-3 GRAVITATION AND THE PRINCIPLE OF SUPERPOSITION

Given a group of particles, we find the net (or resultant) gravitational force exerted on any one of them by using the **principle of superposition.** This is a general principle that says a net effect is the sum of the individual effects. Here, the principle means that we first compute the gravitational force that acts on our selected particle due to each of the other particles in turn. We then find the net force by adding these forces vectorially, as we have previously added forces.

For n interacting particles, we can write the principle of superposition for gravitational forces as

$$\mathbf{F}_1 = \mathbf{F}_{12} + \mathbf{F}_{13} + \mathbf{F}_{14} + \mathbf{F}_{15} + \cdot \cdot \cdot + \mathbf{F}_{1n}. \quad (15\text{-}3)$$

Here $\mathbf{F}_1$ is the net force on particle 1 and, for example, $\mathbf{F}_{13}$ is the force exerted on particle 1 by particle

3. We can express this equation more compactly as a vector sum

$$\mathbf{F}_1 = \sum_{i=2}^{n} \mathbf{F}_{1i}, \qquad (15\text{-}4)$$

in which i is called an *index*.

What about the gravitational force exerted on a particle by a real extended object? It is found by dividing the object into units small enough to treat as particles and then using Eq. 15-4 to find the vector sum of the forces exerted on the particle by all the units. In the limiting case, we can divide the extended object into differential units of mass dm, each of which exerts only a differential force $d\mathbf{F}$ on the particle. In this limit, the sum of Eq. 15-4 becomes an integral and we have

$$\mathbf{F}_1 = \int d\mathbf{F}, \qquad (15\text{-}5)$$

in which the integral is taken over the entire extended object. If the object is a sphere or a spherical shell, we can avoid the integration of Eq. 15-5 by assuming the object's mass is concentrated at the object's center and using Eq. 15-1.

SAMPLE PROBLEM 15-1

Figure 15-4a shows an arrangement of five masses, with $m_1 = 8.0$ kg, $m_2 = m_3 = m_4 = m_5 = 2.0$ kg, $a = 2.0$ cm, and $\theta = 30°$. What is the net gravitational force $\mathbf{F}_1$ acting on m_1 due to the other masses?

SOLUTION From Eq. 15-4 we know that $\mathbf{F}_1$ is the vector sum of forces $\mathbf{F}_{12}$, $\mathbf{F}_{13}$, $\mathbf{F}_{14}$, and $\mathbf{F}_{15}$, which are the gravitational forces acting on m_1 due to the other masses. Because m_2 and m_4 are equal and they are both a distance $r = 2a$ from m_1, we have from Eq. 15-1

$$F_{12} = F_{14} = \frac{Gm_1 m_2}{(2a)^2}. \qquad (15\text{-}6)$$

Similarly, since m_3 and m_5 are equal and are both a distance $r = a$ from m_1, we have

$$F_{13} = F_{15} = \frac{Gm_1 m_3}{a^2}. \qquad (15\text{-}7)$$

Figure 15-4b is a free-body diagram for m_1. It and Eq. 15-6 show that $\mathbf{F}_{12}$ and $\mathbf{F}_{14}$ are equal in magnitude but opposite in direction; thus those forces cancel. Inspection of Fig. 15-4b and Eq. 15-7 reveals that the x com-

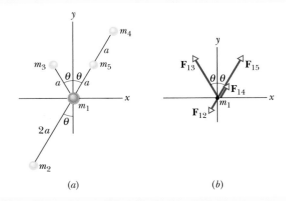

FIGURE 15-4 Sample Problem 15-1. (a) An arrangement of five masses. (b) The forces acting on m_1 due to the other four masses.

ponents of $\mathbf{F}_{13}$ and $\mathbf{F}_{15}$ also cancel, and that their y components are identical in magnitude and both point toward increasing y. Thus $\mathbf{F}_1$ points toward increasing y, and its magnitude is twice the y component of $\mathbf{F}_{13}$:

$$F_1 = 2F_{13} \cos\theta = 2\frac{Gm_1 m_3}{a^2} \cos\theta$$

$$= 2\frac{(6.67 \times 10^{-11} \text{ m}^3/\text{kg}\cdot\text{s}^2)(8.0 \text{ kg})(2.0 \text{ kg})}{(0.020 \text{ m})^2}$$

$$\times \cos 30°$$

$$= 4.6 \times 10^{-6} \text{ N}. \qquad \text{(Answer)}$$

Note that the presence of m_5 along the line between m_1 and m_4 does not in any way alter the gravitational force exerted by m_4 on m_1.

SAMPLE PROBLEM 15-2

In Fig. 15-5, a particle of mass $m_1 = 0.67$ kg is a distance $d = 23$ cm from one end of a uniform rod with length $L = 3.0$ m and mass $M = 5.0$ kg. What is the magnitude of the gravitational force $\mathbf{F}_1$ on the particle due to the rod?

SOLUTION We consider a differential mass dm of the rod, located a distance r from m_1 and occupying a length dr along the rod. From Eq. 15-1, we may write the differential gravitational force dF on m_1 due to dm as

$$dF = \frac{Gm_1}{r^2} dm. \qquad (15\text{-}8)$$

To find the net gravitational force on m_1 from all the differential masses along the length of the rod, we must

FIGURE 15-5 Sample Problem 15-2. A particle of mass m_1 is a distance d from one end of a rod of length L. A differential mass dm of the rod is a distance r from m_1.

integrate over that length. To do so, we use the fact that the rod is uniform in density to write

$$\frac{dm}{dr} = \frac{M}{L},$$ (15-9)

from which we substitute $dm = (M/L)\,dr$ in Eq. 15-8. Then to get F_1, we substitute dF into Eq. 15-5 and integrate:

$$F_1 = \int dF = \int_d^{L+d} \frac{Gm_1}{r^2}\frac{M}{L}\,dr = \frac{Gm_1 M}{L}\int_d^{L+d}\frac{dr}{r^2}$$

$$= -\frac{Gm_1 M}{L}\left[\frac{1}{r}\right]_d^{L+d} = -\frac{Gm_1 M}{L}\left[\frac{1}{L+d}-\frac{1}{d}\right]$$

$$= \frac{Gm_1 M}{d(L+d)}$$

$$= \frac{(6.67\times10^{-11}\ \mathrm{m^3/kg\cdot s^2})\,(0.67\ \mathrm{kg})\,(5.0\ \mathrm{kg})}{(0.23\ \mathrm{m})\,(3.0\ \mathrm{m}+0.23\ \mathrm{m})}$$

$$= 3.0\times10^{-10}\ \mathrm{N}.$$ (Answer)

PROBLEM SOLVING

TACTIC 1: SYMMETRY

In Sample Problem 15-1 we used the symmetry of the situation to reduce the time and amount of calculation involved in the solution. By realizing that m_2 and m_4 are positioned symmetrically about m_1, and thus that $\mathbf{F}_{12}$ and $\mathbf{F}_{14}$ cancel, we avoided calculating either force. And by realizing that the x components of $\mathbf{F}_{13}$ and $\mathbf{F}_{15}$ cancel and that their y components are identical and add, we saved even more effort.

15-4 GRAVITATION NEAR THE EARTH'S SURFACE

Let us ignore the rotation of the Earth for a moment and assume it is a nonrotating uniform sphere. The magnitude of the gravitational force acting on a par-

ticle of mass m, located outside the Earth a distance r from the Earth's center, is then given by Eq. 15-1 as

$$F = G\frac{Mm}{r^2},$$ (15-10)

in which M is the mass of the Earth. If the particle is released, this gravitational force causes it to fall toward the center of the Earth with an acceleration we shall call the **gravitational acceleration** a_g. Applying Newton's second law along the path of the falling particle, we find

$$F = ma_g.$$ (15-11)

Substituting F from Eq. 15-10 into Eq. 15-11 and solving for a_g, we find

$$a_g = \frac{GM}{r^2}.$$ (15-12)

Table 15-1 shows some values of a_g at various altitudes above the Earth's surface.

The gravitational acceleration a_g is due strictly to the gravitational force exerted on the particle by the Earth. It differs from the free-fall acceleration g that we would measure for a falling particle, because the real Earth is not uniform or spherical, and because the real Earth rotates. Similarly, the gravitational force exerted on the particle differs from the weight mg we would measure for the particle for the same three reasons. We now examine those reasons:

1. *The Earth is not uniform.* The density of the Earth varies radially as shown in Fig. 15-6, and the density

TABLE 15-1
VARIATION OF a_g WITH ALTITUDE

ALTITUDE (km)	a_g (m/s²)
0	9.83
5	9.81
10	9.80
50	9.68
100	9.53
400[a]	8.70
35,700[b]	0.225
380,000[c]	0.0027

[a] A typical space shuttle altitude.

[b] The altitude of communication satellites.

[c] The distance to the moon.

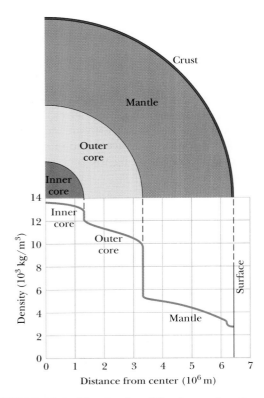

FIGURE 15-6 The density of Earth as a function of distance from the center. The limits of the solid inner core, the largely liquid outer core, and the solid mantle are shown, but the crust of the Earth is too thin to show clearly on this plot.

of the crust (or outer section) of the Earth varies from region to region over the Earth's surface.

2. The Earth is not a sphere. The Earth is approximately an ellipsoid, flattened at the poles and bulging at the equator. Its equatorial radius is greater than its polar radius by 21 km. Thus a point at the poles is closer to the dense core of the Earth than is a point on the equator. We would expect that the free-fall acceleration g would increase as one proceeds, at sea level, from the equator toward the poles. Figure 15-7 shows that this is indeed what happens. The measured values of g in this figure include both the equatorial bulge effect and effects resulting from the rotation of the Earth; see below.

3. The Earth is rotating. The rotation axis runs through the north and south poles of the Earth. An object located on the Earth's surface anywhere except at those poles must rotate in a circle about the rotation axis and thus must have a centripetal acceleration that points toward the center of the circle it

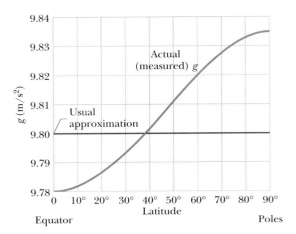

FIGURE 15-7 The variation of g with latitude at sea level. About 65% of the effect is due to the rotation of the Earth, the remaining 35% resulting from the Earth's shape as a flattened sphere.

travels. This centripetal acceleration requires a centripetal force that also points toward that center.

To see how this third requirement means that the free-fall acceleration g differs from the gravitational acceleration a_g, and that weight differs from the gravitational force, we analyze a simple situation in which a crate of mass m rests on a scale at the equator. Figure 15-8a shows this situation as viewed from a point in space above the north pole.

Figure 15-8b is a free-body diagram for the crate. The centripetal acceleration **a** of the crate points toward the center of the circle it travels, which is coincident with the center of the Earth (assumed spherical). The Earth exerts a gravitational force on the crate with magnitude ma_g (Eq. 15-11). The scale exerts an upward normal force **N** on the crate. Applying Newton's second law to the crate, with the positive direction being radially inward toward the Earth's center, we find

$$\sum F = ma_g - N = ma. \qquad (15\text{-}13)$$

The magnitude of **N** is the reading on the scale and thus is equal to the weight mg of the crate. Substituting mg for N in Eq. 15-13 yields

$$ma_g - mg = ma, \qquad (15\text{-}14)$$

which shows that the weight mg of the crate differs from the magnitude ma_g of the gravitational force acting on the crate. Dividing Eq. 15-14 by m shows also that g is different from a_g.

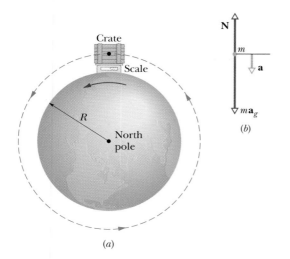

N

m

$\mathbf{a}$

$m\mathbf{a}_g$

(b)

(a)

FIGURE 15-8 (a) A crate on the rotating Earth, resting on a platform scale at the equator. The view is along the Earth's rotational axis, looking down on the north pole. (b) A free-body diagram for the crate. The crate is in uniform circular motion and is thus accelerated toward the center of the Earth. The gravitational force on it has magnitude ma_g. The normal force **N** from the scale has magnitude mg, where g is the free-fall acceleration.

For the centripetal acceleration a we can substitute $\omega^2 R$, where ω is the angular speed of the Earth's rotation, and R is the radius of the circular path taken by the crate. (R is essentially the radius of the Earth, as shown in Fig. 15-8.) For ω we can substitute $2\pi/T$, where $T = 24$ h is approximately the time for one rotation of the Earth. When we make these substitutions and divide by m, Eq. 15-14 gives us

$$a_g - g = \omega^2 R = \left(\frac{2\pi}{T}\right)^2 R \qquad (15\text{-}15)$$

$$= 0.034 \text{ m/s}^2.$$

Thus the free-fall acceleration g (9.78 m/s² ≈ 9.8 m/s²) measured on the equator of the real, rotating Earth is slightly less than the gravitational acceleration a_g due strictly to the gravitational force. The difference between g and a_g becomes progressively smaller as the crate is moved to progressively greater latitudes, because the crate then travels in a smaller circle, making R in Eq. 15-15 smaller. In many practical problems, we can approximate the free-fall acceleration g as being the gravitational acceleration a_g. And we can approximate the weight mg of an object as being the gravitational force (given by Eq. 15-10) acting on it.

SAMPLE PROBLEM 15-3

Consider a pulsar, a collapsed star of extremely high density, with a mass M equal to that of the sun (1.98 × 10³⁰ kg), a radius R of only 12 km, and a rotational period T of 0.041 s. At its equator, by what percentage does the free-fall acceleration g differ from the gravitational acceleration a_g?

SOLUTION To find a_g on the surface of the star, we use Eq. 15-12, with R replacing r and with M being the mass of the star. Substituting in the known values gives

$$a_g = \frac{GM}{R^2} = \frac{(6.67 \times 10^{-11} \text{ m}^3/\text{kg} \cdot \text{s}^2)(1.98 \times 10^{30} \text{ kg})}{(12{,}000 \text{ m})^2}$$

$$= 9.2 \times 10^{11} \text{ m/s}^2.$$

Dividing Eq. 15-15 (which applies to any rotating body) by a_g and substituting the known data, we now find

$$\frac{a_g - g}{a_g} = \left(\frac{2\pi}{T}\right)^2 \frac{R}{a_g} = \left(\frac{2\pi}{0.041 \text{ s}}\right)^2 \frac{12{,}000 \text{ m}}{9.2 \times 10^{11} \text{ m/s}^2}$$

$$= 3.1 \times 10^{-4} = 0.031\%. \qquad \text{(Answer)}$$

Even though a pulsar rotates extremely rapidly, its rotation reduces the free-fall acceleration from the gravitational acceleration only slightly, because its radius is so small.

SAMPLE PROBLEM 15-4

On a uniform, spherical, nonrotating planet with radius $R_s = 5.1 \times 10^3$ km, what is the fractional decrease in the value of the free-fall acceleration g for a particle when it is lifted from the surface to an elevation $h = 1.5$ km?

SOLUTION Let g_s and g_h be the values of g at the surface and at elevation h, respectively. Since the planet is uniform, spherical, and nonrotating, the free-fall acceleration g is equal to the gravitational acceleration a_g. So we can use Eq. 15-12 (with M as the planet's mass) to find either. At the surface, we have

$$g_s = \frac{GM}{R_s^2}. \qquad (15\text{-}16)$$

At elevation h, we have

$$g_h = \frac{GM}{(R_s + h)^2}. \qquad (15\text{-}17)$$

Since $h \ll R_s$, we should not evaluate Eq. 15-17 directly on a calculator. We can, however, recast Eq. 15-17 to use the binomial expansion (see Appendix G, and Tactic 3 in Chapter 7):

$$g_h = \frac{GM}{R_s^2 \left(1 + \frac{h}{R_s}\right)^2} = \frac{GM}{R_s^2}\left(1 + \frac{(-2)h}{R_s} + \cdots\right).$$

Retaining only the first two terms of the expansion and setting $g_s = GM/R_s^2$ (Eq. 15-16), we have

$$g_h \approx g_s\left(1 - \frac{2h}{R_s}\right),$$

which gives

$$\frac{g_h - g_s}{g_s} \approx -\frac{2h}{R_s} \tag{15-18}$$

$$= -\frac{2(1.5 \text{ km})}{5.1 \times 10^3 \text{ km}} = -5.9 \times 10^{-4}. \quad \text{(Answer)}$$

We can also answer by finding dg/g_s, where dg is the differential change in g due to a differential change dr in r. From Eq. 15-12, with g replacing a_g, we find

$$dg = -\frac{2GM}{r^3}\,dr. \tag{15-19}$$

Since $h \ll R_s$, the radius r here is approximately R_s throughout the lift, and we can treat h as being the differential change dr. With these substitutions and with $g_s = GM/R_s^2$, Eq. 15-19 yields

$$\frac{dg}{g_s} = -\frac{2h}{R_s},$$

which is equivalent to Eq. 15-18.

15-5 MEASURING THE GRAVITATIONAL CONSTANT G

The constant G can be found from Eq. 15-1, by measuring the gravitational force between two spheres of known mass, separated by a known distance. The first person to do this was Henry Cavendish, in 1798, more than a century after Newton advanced his law.

Figure 15-9 shows Cavendish's apparatus. Two small lead spheres, each of mass m, are fastened to the ends of a rod that is suspended from its midpoint by a fine fiber, forming a *torsion balance*. Two large lead spheres, each of mass M, can be placed in either configuration AA or configuration BB, as shown in the figure. In either configuration, the large spheres will attract the small ones, exerting a torque on the rod. The rod will then rotate to an equilibrium orientation in which this gravitational torque and a torque due to the twisting of the fiber are in

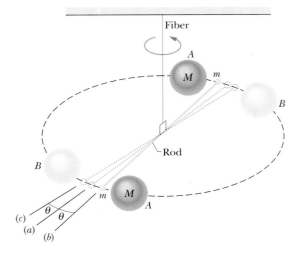

FIGURE 15-9 The apparatus used in 1798 by Henry Cavendish to measure the gravitational constant G. The large spheres of mass M, shown in configuration AA, can also be moved to configuration BB. (*a*) Orientation of the rod when the large spheres are absent and the fiber is not twisted. (*b*) Equilibrium orientation when large spheres are at AA. (*c*) Equilibrium orientation when large spheres are at BB. The angle θ is exaggerated.

balance. The two equilibrium orientations of the rod, corresponding to the two configurations of the large spheres, will differ by an angle 2θ. Sample Problem 15-5 shows how to calculate G from the measured data.

Knowing G, we can find the mass M of the Earth with Eq. 15-12. At the Earth's surface, r in Eq. 15-12 is the Earth's radius R. If we make the approximation $a_g \approx g \approx 9.8 \text{ m/s}^2$ and rearrange Eq. 15-12, we get

$$M = \frac{gR^2}{G} = \frac{(9.8 \text{ m/s}^2)(6.37 \times 10^6 \text{ m})^2}{6.67 \times 10^{-11} \text{ m}^3/\text{kg} \cdot \text{s}^2}$$

$$= 6.0 \times 10^{24} \text{ kg}. \tag{15-20}$$

Knowing the mass of the Earth, we can find its average density by dividing by its volume. The resulting value is about 5.5 times the density of water. However, the average density of the rocks that form the Earth's crust is only about 3 times the density of water. This tells us that the Earth's core must be much denser than its crust (see Fig. 15-6). Who would think that, by measuring the force between two lumps of lead, you could "weigh the Earth" and also find out something about its core, far beyond the depth of any drill hole?

SAMPLE PROBLEM 15-5

In a repetition of the Cavendish experiment of Fig. 15-9, let us take $M = 12.7$ kg, $m = 9.85$ g, and the length L of the rod connecting the two small spheres to be 52.4 cm. The angle 2θ between the rod's two equilibrium positions b and c in Fig. 15-9 is 0.516°. In either position, the distance R between the centers of m and M is 10.8 cm, and the twisted fiber exerts a torque τ of magnitude 3.75×10^{-10} N·m on the rod about a vertical axis through its center. What value of the gravitational constant G results from these data?

SOLUTION The torque exerted by the fiber is balanced by the torques associated with the gravitational forces that the large spheres exert on the small ones. The magnitude of force **F** on each small sphere is equal to GMm/R^2 and the moment arm for each force is $\frac{1}{2}L$, half the length of the rod. The torque caused by both forces is then

$$\tau = (2F)(\tfrac{1}{2}L) = FL = \frac{GMmL}{R^2}.$$

Solving this for G yields

$$G = \frac{\tau R^2}{MmL} = \frac{(3.75 \times 10^{-10}\text{ N·m})(0.108\text{ m})^2}{(12.7\text{ kg})(9.85 \times 10^{-3}\text{ kg})(0.524\text{ m})}$$

$$= 6.67 \times 10^{-11}\text{ N·m}^2/\text{kg}^2, \qquad \text{(Answer)}$$

in agreement with the value reported in Eq. 15-2.

15-6 GRAVITATION INSIDE THE EARTH

Newton's shell theorem states that a uniform spherical shell of matter exerts a gravitational force on a particle located outside the shell as if all the shell's mass were concentrated at its center. The theorem can also be applied to a situation in which a particle is located *inside* such a shell, to show the following:

A uniform shell of matter exerts no gravitational force on a particle located inside it.

If the density of the Earth were uniform, the gravitational force acting on a particle would be a maximum at the Earth's surface. It would, as we expect, decrease as we move outward. If we were to move inward, perhaps down a deep mine shaft, the gravitational force would change for two reasons. (1) It

would tend to increase because we are moving closer to the center of the Earth. (2) It would tend to decrease because the shell of the Earth's crust that lies outside the particle's radial position would not exert any force on the particle.

For the real Earth, the outer crust is so much less dense than the core that the second influence above is greatly weakened, and the first influence prevails. That is, the gravitational force on a particle would *increase* slightly as we lowered the particle down a deep mine shaft. Eventually, of course, the force would reach a maximum and then decrease to zero at the center of the Earth.

For an Earth of uniform density, however, the second influence would prevail, so that the gravitational force would decrease as we penetrated the Earth.

SAMPLE PROBLEM 15-6

Suppose a tunnel runs through the Earth from pole to pole, as in Fig. 15-10. Assume that the Earth is a nonrotating, uniform sphere.

a. Find the gravitational force on a particle of mass m dropped into the tunnel when it reaches a distance r from the Earth's center.

SOLUTION The force that acts on the particle is associated only with the mass M' of the Earth that lies within a sphere of radius r. The portion of the Earth

FIGURE 15-10 Sample Problem 15-6. A particle oscillates back and forth along a tunnel drilled through the Earth. A particle dropped in at one end of the tunnel would emerge at the other end about 42 min later.

that lies outside this sphere does not exert any net force on the particle.

Mass M' is given by

$$M' = \rho V' = \rho \frac{4\pi r^3}{3}, \qquad (15\text{-}21)$$

in which V' is the volume occupied by M' (which lies within the dashed line in Fig. 15-10), and ρ is the assumed uniform density of the Earth.

The force acting on the particle is then, using Eqs. 15-1 and 15-21,

$$F = -\frac{GmM'}{r^2} = -\frac{Gm\rho 4\pi r^3}{3r^2} = -\left(\frac{4\pi mG\rho}{3}\right) r$$

$$= -Kr, \qquad \text{(Answer)} \qquad (15\text{-}22)$$

in which K, a constant, is equal to $4\pi mG\rho/3$. We have inserted a minus sign to indicate that the force $\mathbf{F}$ and the displacement $\mathbf{r}$ are in opposite directions, the former being toward the center of the Earth and the latter away from that point. Thus Eq. 15-22 tells us that the force acting on the particle is proportional to the displacement of the particle but oppositely directed, precisely the criterion for simple harmonic motion (Section 14-3).

b. (Optional) If this tunnel were used to deliver mail, how long would the mail take to travel through the Earth? Assume $\rho = 5.5 \times 10^3$ kg/m³, the mean density of the Earth.

SOLUTION The force in Eq. 15-22 will cause the mail to move in simple harmonic motion about the center of the Earth, taking half a period to make the trip from pole to pole. From Eq. 14-12, this half period is given by

$$\tfrac{1}{2}T = \left(\frac{1}{2}\right) 2\pi \sqrt{\frac{m}{K}} = \pi \sqrt{\frac{3m}{4\pi mG\rho}} = \sqrt{\frac{3\pi}{4G\rho}}$$

$$= \sqrt{\frac{3\pi}{(4)(6.67 \times 10^{-11} \text{ m}^3/\text{kg} \cdot \text{s}^2)(5.5 \times 10^3 \text{ kg/m}^3)}}$$

$$= 2530 \text{ s} = 42 \text{ min.} \qquad \text{(Answer)}$$

Note that the delivery time is independent of the mass of the mail.

15-7 GRAVITATIONAL POTENTIAL ENERGY

In Section 8-3, we discussed the gravitational potential energy of a particle due to the Earth. We were careful to keep the particle near the Earth's surface, so that we could regard the gravitational force as

constant. We arbitrarily defined the potential energy of the particle–Earth system to be zero when the particle was on the Earth's surface.

Here we broaden our view and consider two particles, of masses m and M, separated by a distance r. For concreteness, we take M to be the Earth and m to be a baseball but our conclusions will apply generally, no matter what the relative masses of the particles. We also change our zero-potential-energy configuration to one in which the two particles are separated by a distance so large as to be taken as infinite.

The gravitational potential energy of the two-particle system, as we shall prove below, is

$$U(r) = -\frac{GMm}{r} \qquad \begin{array}{l}\text{(gravitational}\\ \text{potential energy).}\end{array} \qquad (15\text{-}23)$$

According to Eq. 15-23, $U(r)$ approaches zero as r approaches infinity. This is as it should be, because we made this a requirement at the outset.

The potential energy given by Eq. 15-23 is a property of the system of two particles rather than of either particle alone. There is no way to divide this energy and say that so much belongs to one particle and so much to the other. Nevertheless, if $M \gg m$, as is true for the Earth and a baseball, we often speak of "the potential energy of the baseball." We can get away with this because, when a baseball moves in the vicinity of the Earth, changes in the potential energy of the baseball–Earth system appear almost entirely as changes in the kinetic energy of the baseball, since changes in the kinetic energy of the Earth are too small to be measured. Similarly, in Section 15-9 we shall speak of "the potential energy of an artificial satellite" orbiting the Earth, because the satellite's mass is so much smaller than the Earth's mass. For bodies of comparable mass, however, we have to be careful to treat them as a system when we speak of their potential energy.

If our system contains more than two particles, we consider each pair of particles in turn, calculate the gravitational potential energy of that pair with Eq. 15-23 as if the other particles were not there, and then sum the results. Applying Eq. 15-23 to each of the three pairs of Fig. 15-11, for example, gives the potential energy of the system as

$$U = -\left(\frac{Gm_1 m_2}{r_{12}} + \frac{Gm_1 m_3}{r_{13}} + \frac{Gm_2 m_3}{r_{23}}\right). \qquad (15\text{-}24)$$

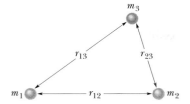

FIGURE 15-11 Three particles exert gravitational forces on each other. The gravitational potential energy of the system is the sum of the gravitational potential energies of each of the three possible pairs of the particles.

A *globular cluster* (Fig. 15-12) in the constellation Sagittarius is a good example of a naturally occurring system of particles. It contains about 70,000 stars and about 2.5×10^9 pairs of stars. Contemplation of this structure suggests the enormous amount of gravitational potential energy stored in the universe.

Proof of Eq. 15-23

Let a baseball, starting from rest at a great distance from the Earth, fall toward point *P*, as in Fig. 15-13. The potential energy of the baseball–Earth system when the baseball reaches *P* is the negative of the work *W* done by the gravitational force as the baseball moves to *P* from its distant position. Thus, from Eq. 8-6 (generalized to vector form), we have

$$U = -W = -\int_\infty^r \mathbf{F}(x) \cdot d\mathbf{x}. \quad (15\text{-}25)$$

The limits on the integral are the initial distance of

FIGURE 15-12 A globular cluster, like this one in the constellation Sagittarius, contains tens of thousands of stars in an almost spherical collection. There are many such clusters in our Milky Way galaxy, some of which are visible with a small telescope.

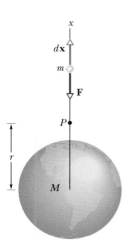

FIGURE 15-13 A baseball of mass *m* falls toward the Earth from infinity, along a radial line passing through point *P* at a distance *r* from the center of the Earth.

the baseball, which we take to be infinitely great, and its final distance *r*.

The vector $\mathbf{F}(x)$ in Eq. 15-25 points radially inward toward the center of the Earth in Fig. 15-13, and the vector $d\mathbf{x}$ points radially outward, the angle ϕ between them being 180°. Thus

$$\mathbf{F}(x) \cdot d\mathbf{x} = F(x)(\cos 180°)(dx)$$
$$= -F(x)\,dx. \quad (15\text{-}26)$$

For $\mathbf{F}(x)$ in Eq. 15-26, we now substitute from Newton's law of gravitation (Eq. 15-1) written as a function of *x*, obtaining

$$\mathbf{F}(x) \cdot d\mathbf{x} = -\frac{GMm}{x^2}\,dx.$$

Putting that result into Eq. 15-25, we obtain

$$U = \int_\infty^r \left(\frac{GMm}{x^2}\right) dx = -\left[\frac{GMm}{x}\right]_\infty^r = -\frac{GMm}{r},$$

which is Eq. 15-23.

Equation 15-23 holds no matter what path the baseball takes in moving in toward the Earth. Consider a path made up of small steps, as in Fig. 15-14. No work is done along circular steps like *AB*, because along them the force is perpendicular to the displacement. The total work done along all the radial steps, such as *BC*, is the same as the work done in moving along a single radial line, as in Fig. 15-13. Thus the work done on a particle by the gravitational force when the particle moves between any

FIGURE 15-14 The work done by the gravitational force as a baseball moves from A to E is independent of the path followed.

A space shuttle is accelerated to a speed of about 8.0 km/s, which is (as planned) insufficient for it to escape from Earth. In contrast, *Pioneer* 10 achieved a speed of about 14.4 km/s, which has allowed it to escape not only Earth, but also the solar system.

two points is independent of the path that the particle follows and depends only on the particle's initial and final positions. The work is computed as the negative of the change in potential energy between the two points:

$$W = -\Delta U = -(U_f - U_i), \qquad (15\text{-}27)$$

where U_f and U_i are the potential energies associated with the final and initial positions. This is what we mean when we say, as we did in Section 8-4, that the gravitational force is *conservative*. And, as discussed there, if the work *did* depend on the path, as it does for frictional forces, the whole concept of gravitational potential energy would be meaningless.

The Potential Energy and the Force

In the proof of Eq. 15-23, we derived the potential energy function U from the force function F. We should be able to go the other way, that is, to start from the potential energy function and derive the force. Guided by Eq. 8-17, we can write

$$F = -\frac{dU}{dr} = -\frac{d}{dr}\left(-\frac{GMm}{r}\right)$$
$$= -\frac{GMm}{r^2}. \qquad (15\text{-}28)$$

This is just Newton's law of gravitation, the minus sign reminding us that the force is attractive.

Escape Speed

If you fire a projectile upward, usually it will slow, stop momentarily, and return to Earth. There is, however, a certain initial speed that will cause it to move upward forever, theoretically coming to rest only at infinity. This initial speed is called the **escape speed**.

Consider a projectile of mass m, leaving the Earth's surface with escape speed v. It has a kinetic energy K given by $\frac{1}{2}mv^2$ and a potential energy U given by Eq. 15-23:

$$U = -\frac{GMm}{R},$$

in which M is the mass of the Earth and R its radius.

When the projectile reaches infinity, it stops and thus has no kinetic energy. It also has no potential energy because this is our zero-potential-energy configuration. Its total energy at infinity is therefore zero. From the principle of conservation of energy, its total energy at the Earth's surface must also have been zero, so

$$K + U = \tfrac{1}{2}mv^2 + \left(-\frac{GMm}{R}\right) = 0.$$

This yields

$$v = \sqrt{\frac{2GM}{R}}. \qquad (15\text{-}29)$$

Equation 15-29 can be applied to find the escape speed of a projectile from any astronomical body provided we substitute the mass of the body for M and the radius of the body for R. Table 15-2 shows escape speeds from some astronomical bodies.

SAMPLE PROBLEM 15-7

The masses of the Earth and the moon are 5.98×10^{24} kg and 7.36×10^{22} kg, respectively, and their mean separation distance d is 3.82×10^8 m. What is the gravitational potential energy of the moon–Earth system?

SOLUTION From Eq. 15-23,

$$U = -\frac{GMm}{d}$$

$$= -\frac{\left(6.67 \times 10^{-11}\ \dfrac{\text{N·m}^2}{\text{kg}^2}\right)(5.98 \times 10^{24}\ \text{kg})(7.36 \times 10^{22}\ \text{kg})}{3.82 \times 10^8\ \text{m}}$$

$$= -7.68 \times 10^{28}\ \text{J.} \qquad \text{(Answer)}$$

An energy of this magnitude is equal to world energy production, at its present rate, for about 10^8 years.

SAMPLE PROBLEM 15-8

Verify the escape speed from the Earth shown in Table 15-2. Ignore effects caused by air drag and the Earth's rotation.

SOLUTION Substituting values into Eq. 15-29 gives

$$v = \sqrt{\frac{2GM}{R}}$$

$$= \sqrt{\frac{2(6.67 \times 10^{-11}\ \text{m}^3/\text{kg·s}^2)(5.98 \times 10^{24}\ \text{kg})}{6.37 \times 10^6\ \text{m}}}$$

$$= 1.12 \times 10^4\ \text{m/s}$$

$$= 11.2\ \text{km/s or } 25,000\ \text{mi/h.} \qquad \text{(Answer)}$$

TABLE 15-2
SOME ESCAPE SPEEDS

BODY	MASS (kg)	RADIUS (m)	ESCAPE SPEED (km/s)
Ceres[a]	1.17×10^{21}	3.8×10^5	0.64
Moon	7.36×10^{22}	1.74×10^6	2.38
Earth	5.98×10^{24}	6.37×10^6	11.2
Jupiter	1.90×10^{27}	7.15×10^7	59.5
Sun	1.99×10^{30}	6.96×10^8	618
Sirius B[b]	2×10^{30}	1×10^7	5200
Neutron star[c]	2×10^{30}	1×10^4	2×10^5

[a] The most massive of the asteroids.

[b] A *white dwarf* (a star in a final stage of evolution) that is a companion of the bright star Sirius.

[c] The collapsed core of a star that remains after that star has exploded in a *supernova* event.

The escape speed does not depend on the direction in which the projectile is fired. However, attaining that speed is easier if the projectile is fired in the direction the launch site is moving as the Earth rotates about its axis. For example, rockets are launched eastward at Cape Canaveral in order to take advantage of the Cape's eastward speed of 1500 km/h.

15-8 PLANETS AND SATELLITES: KEPLER'S LAWS

The motions of the planets, as they seemingly wander against the background of the stars, have been a puzzle since the dawn of history. The "loop-the-loop" motion of Mars, shown in Fig. 15-15, was

FIGURE 15-15 The path of the planet Mars as it moved against a background of the constellation Capricorn during 1971. Its position on four selected days is marked. Both Mars and Earth are moving in orbits around the sun so that we see the position of Mars relative to us; this sometimes results in an apparent loop in the path of Mars.

particularly baffling. Johannes Kepler (1571–1630), after a lifetime of study, worked out the empirical laws that govern these motions. Tycho Brahe (1546–1601), the last of the great astronomers to make observations without the help of a telescope, compiled the extensive data from which Kepler was able to derive the three laws of planetary motion that now bear his name. Later, Isaac Newton (1642–1727) showed that Kepler's empirical laws followed from his law of gravitation.

We discuss each of Kepler's laws in turn. Although we apply the laws here to planets orbiting the sun, they hold equally well for satellites, either natural or artificial, orbiting the Earth or any other massive central body.

1. THE LAW OF ORBITS: All planets move in elliptical orbits, with the sun at one focus.

Figure 15-16 shows a planet, of mass m, moving in such an orbit around the sun, whose mass is M. We assume that $M \gg m$, so that the center of mass of the planet–sun system is virtually at the center of the sun.

The orbit in Fig. 15-16 is described by giving its **semimajor axis** a and its **eccentricity** e, the latter defined so that ea is the distance from the center of the ellipse to either focus F or F'. An eccentricity of zero corresponds to a circle, in which the two foci merge to a single central point. The eccentricities of the planetary orbits are not large, so—sketched on paper—they look circular. The eccentricity of the

ellipse of Fig. 15-16, which has been made large for clarity, is 0.74. The eccentricity of the Earth's orbit is only 0.0167.

2. THE LAW OF AREAS: A line that connects a planet to the sun sweeps out equal areas in equal times.

Qualitatively, this second law tells us that the planet will move most slowly when it is farthest from the sun and most rapidly when it is nearest to the sun. As it turns out, Kepler's second law is totally equivalent to the law of conservation of angular momentum. Let us prove it.

The area of the shaded wedge in Fig. 15-17a closely approximates the area swept out in time Δt. The area ΔA of the wedge is approximately the area of a triangle with base $r \Delta \theta$ and height r. Thus $\Delta A \approx \frac{1}{2} r^2 \Delta \theta$. This expression for ΔA becomes more exact as Δt (and, hence, $\Delta \theta$) approaches zero. The instantaneous rate at which area is being swept out is then

$$\frac{dA}{dt} = \frac{1}{2} r^2 \frac{d\theta}{dt} = \frac{1}{2} r^2 \omega, \qquad (15\text{-}30)$$

(a)

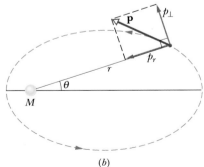

(b)

FIGURE 15-17 (a) In time Δt, the line r connecting the planet to the sun sweeps through an angle $\Delta \theta$, sweeping out an area ΔA. (b) The linear momentum $\mathbf{p}$ of the planet and its components.

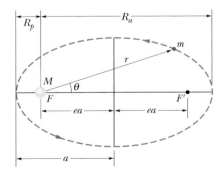

FIGURE 15-16 A planet of mass m moving in an elliptical orbit around the sun. The sun, of mass M, is at one focus F of the ellipse. The other, or "empty," focus is F'. Each focus is a distance ea from the center, e being the eccentricity of the orbit. The semimajor axis a of the ellipse, the perihelion (nearest the sun) distance R_p, and the aphelion (farthest from the sun) distance R_a are also shown.

in which ω is the angular speed of the rotating line, labeled r, connecting the sun and planet.

Figure 15-17b shows the linear momentum **p** of the planet, along with its components. From Eq. 12-27, the magnitude of the angular momentum **L** of the planet about the sun is given by the product of r and the component of **p** perpendicular to r, or

$$L = rp_\perp = (r)(mv_\perp) = (r)(m\omega r)$$

$$= mr^2\omega, \qquad (15\text{-}31)$$

where we have replaced $v_\perp$ with its equivalent ωr. Eliminating $r^2\omega$ between Eqs. 15-30 and 15-31 leads to

$$\frac{dA}{dt} = \frac{L}{2m}. \qquad (15\text{-}32)$$

If dA/dt is constant, as Kepler said it is, then Eq. 15-32 means that L must also be constant—angular momentum is conserved. So Kepler's second law is indeed equivalent to the law of conservation of angular momentum.

----

3. THE LAW OF PERIODS: The square of the period of any planet is proportional to the cube of the semimajor axis of its orbit.

On February 7, 1984, at a height of 102 km above Hawaii and with a speed of about 29,000 km/h, Bruce McCandless stepped into space from a space shuttle and became the first human satellite.

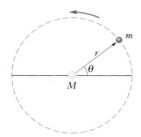

FIGURE 15-18 A planet of mass m moving around the sun in a circular orbit of radius r.

Let us consider a circular orbit with radius r (the radius is equivalent to the semimajor axis). Applying Newton's second law, $F = ma$, to the orbiting planet in Fig. 15-18 yields

$$\frac{GMm}{r^2} = (m)(\omega^2 r). \qquad (15\text{-}33)$$

Here we have substituted from Newton's law of gravitation for the force F, and substituted $\omega^2 r$ for the centripetal acceleration. If we replace ω with $2\pi/T$, where T is the period of the motion, we find

$$T^2 = \left(\frac{4\pi^2}{GM}\right)r^3 \qquad \text{(law of periods).} \qquad (15\text{-}34)$$

The quantity in parentheses is a constant, its value depending only on the mass of the central body.

Equation 15-34 holds also for elliptical orbits, provided we replace r with a, the semimajor axis of the ellipse. Table 15-3 shows how well this law holds for the orbits of the solar system: it predicts that the ratio T^2/a^3 has essentially the same value for every planetary orbit.

TABLE 15-3
KEPLER'S LAW OF PERIODS FOR THE SOLAR SYSTEM

PLANET	SEMIMAJOR AXIS a (10^{10} m)	PERIOD T (y)	T^2/a^3 (10^{-34} y^2/m^3)
Mercury	5.79	0.241	2.99
Venus	10.8	0.615	3.00
Earth	15.0	1.00	2.96
Mars	22.8	1.88	2.98
Jupiter	77.8	11.9	3.01
Saturn	143	29.5	2.98
Uranus	287	84.0	2.98
Neptune	450	165	2.99
Pluto	590	248	2.99

SAMPLE PROBLEM 15-9

An Earth satellite, in circular orbit at an altitude h of 230 km above the Earth's surface, has a period T of 89 min. What mass of the Earth follows from these data?

SOLUTION We apply Kepler's law of periods to the satellite–Earth system. Solving Eq. 15-34 for M yields

$$M = \frac{4\pi^2 r^3}{GT^2}. \qquad (15\text{-}35)$$

The radius r of the satellite orbit is

$$r = R + h = 6.37 \times 10^6 \text{ m} + 230 \times 10^3 \text{ m}$$

$$= 6.60 \times 10^6 \text{ m},$$

in which R is the radius of the Earth. Substituting this value and the period T into Eq. 15-35 yields

$$M = \frac{(4\pi^2)(6.60 \times 10^6 \text{ m})^3}{(6.67 \times 10^{-11} \text{ m}^3/\text{kg} \cdot \text{s}^2)(89 \times 60 \text{ s})^2}$$

$$= 6.0 \times 10^{24} \text{ kg.} \qquad \text{(Answer)}$$

In the same way, we could find the mass of the sun from the period and radius of the Earth's orbit (assumed circular), and the mass of Jupiter from the period and orbital radius of any one of its moons (without knowing the mass of that moon).

SAMPLE PROBLEM 15-10

Comet Halley has a period of 76 years and, in 1986, had a distance of closest approach to the sun, called the *perihelion distance* R_p, of 8.9×10^{10} m. Table 15-3 shows that this is between the orbits of Mercury and Venus.

a. What is R_a, the comet's farthest distance from the sun, called the *aphelion distance*?

SOLUTION We can find the semimajor axis of the orbit of comet Halley from Eq. 15-34. Substituting a for r in that equation and solving for a yields

$$a = \left(\frac{GMT^2}{4\pi^2}\right)^{1/3}. \qquad (15\text{-}36)$$

If we substitute the mass M of the sun, 1.99×10^{30} kg, and the period T of the comet, 76 years or 2.4×10^9 s, into Eq. 15-36 we find that $a = 2.7 \times 10^{12}$ m.
Study of Fig. 15-16 shows that

$$R_p = a - ea \qquad (15\text{-}37)$$

and

$$R_a = a + ea. \qquad (15\text{-}38)$$

Adding these two equations and solving for R_a yields

$$R_a = 2a - R_p$$

$$= (2)(2.7 \times 10^{12} \text{ m}) - 8.9 \times 10^{10} \text{ m}$$

$$= 5.3 \times 10^{12} \text{ m.} \qquad \text{(Answer)}$$

Table 15-3 shows that this is just a little less than the semimajor axis of Pluto.

b. What is the eccentricity of the orbit of comet Halley?

SOLUTION Subtracting Eq. 15-37 from Eq. 15-38 and solving for e, we obtain

$$e = \frac{R_a - R_p}{2a}$$

$$= \frac{5.3 \times 10^{12} \text{ m} - 8.9 \times 10^{10} \text{ m}}{(2)(2.7 \times 10^{12} \text{ m})}$$

$$= 0.96. \qquad \text{(Answer)}$$

This cometary orbit, with an eccentricity approaching unity, is a long thin ellipse.

SAMPLE PROBLEM 15-11

Observations of the light from a certain star indicate that it is part of a binary (two-star) system. This visible star has orbital speed $v = 270$ km/s, orbital period $T = 1.70$ days, and approximate mass $m_1 = 6M_s$, where M_s is the sun's mass, 1.99×10^{30} kg. Assuming that the visible star and its companion object, which is dark and unseen, are in circular orbits (see Fig. 15-19), determine the approximate mass m_2 of the dark object.

SOLUTION Like those of the two-particle systems in Section 9-2, the center of mass of this two-star system lies along a line connecting their centers. The position of the center of mass is labeled point O in Fig. 15-19, and the visible star and dark object orbit O with orbital radii r_1 and r_2, respectively, and separation $r = r_1 + r_2$. From Eq. 15-1, the gravitational force on the visible star from the dark object is

$$F = \frac{Gm_1 m_2}{r^2}.$$

Applying Newton's second law, $F = ma$, to the visible star then yields

$$\frac{Gm_1 m_2}{r^2} = m_1 a = (m_1)(\omega^2 r_1), \qquad (15\text{-}39)$$

where ω is the angular speed of the visible star, and $\omega^2 r_1$ is its centripetal acceleration toward O.

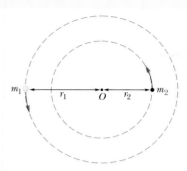

FIGURE 15-19 Sample Problem 15-11. A visible star with mass m_1 and a dark, unseen object with mass m_2 orbit around the center of mass of the two-star system at O.

We can obtain another equation in these same variables by locating the center of mass O. Because O is a distance r_1 from the visible star, we can use Eq. 9-1 to write

$$r_1 = \frac{m_2 r}{m_1 + m_2}.$$

This yields

$$r = r_1 \frac{m_1 + m_2}{m_2}. \qquad (15\text{-}40)$$

Now substituting Eq. 15-40 for r and $2\pi/T$ for ω into Eq. 15-39 gives us, after some algebraic manipulation,

$$\frac{m_2^3}{(m_1 + m_2)^2} = \frac{4\pi^2}{GT^2} r_1^3. \qquad (15\text{-}41)$$

We still have two unknowns, m_2 and r_1. We can, however, find r_1 from the orbital motion of the visible star: the orbital period T is equal to the circumference of the orbit $(2\pi r_1)$ divided by the speed v of the planet. This gives us

$$T = \frac{2\pi r_1}{v},$$

or

$$r_1 = \frac{vT}{2\pi}. \qquad (15\text{-}42)$$

If we substitute $m_1 = 6M_s$ and Eq. 15-42 for r_1, then Eq. 15-41 becomes

$$\frac{m_2^3}{(6M_s + m_2)^2} = \frac{v^3 T}{2\pi G}$$

$$= \frac{(2.7 \times 10^5 \text{ m/s})^3 (1.70 \text{ days}) (86{,}400 \text{ s/day})}{(2\pi)(6.67 \times 10^{-11} \text{ N} \cdot \text{m}^2/\text{kg}^2)}$$

$$= 6.90 \times 10^{30} \text{ kg},$$

or

$$\frac{m_2^3}{(6M_s + m_2)^2} = 3.47 M_s. \qquad (15\text{-}43)$$

We can solve this cubic equation for m_2, or, since we are working with approximate masses anyway, we can substitute integer multiples of M_s for m_2 until we find

$$m_2 \approx 9M_s. \qquad \text{(Answer)}$$

The data here approximate those for the binary system LMC X-3 in the Large Magellanic Cloud (shown in the figure that begins this chapter). From other data, the dark object is known to be especially compact: it may be a star that collapsed under its own gravitational pull to become a neutron star or a black hole. Since a neutron star cannot have a mass larger than about $2M_s$, the result $m_2 \approx 9M_s$ strongly suggests that the dark object is a black hole.

15-9 SATELLITES: ORBITS AND ENERGY (OPTIONAL)

As a satellite orbits the Earth on its elliptical path, both its speed, which fixes its kinetic energy K, and its distance from the center of the Earth, which fixes its potential energy U, fluctuate with fixed periods. However, the mechanical energy E of the satellite remains constant. (Since the satellite's mass is so much smaller than the Earth's mass, we assign U and E of the Earth–satellite system to the satellite alone.)

The potential energy is given by Eq. 15-23 and is

$$U = -\frac{GMm}{r}.$$

Here r is the radius of the orbit, assumed for the time being to be circular.

To find the kinetic energy of a satellite in a circular orbit, we write Newton's second law, $F = ma$, for the satellite,

$$\frac{GMm}{r^2} = m \frac{v^2}{r}, \qquad (15\text{-}44)$$

where v^2/r is the centripetal acceleration of the satellite. Then from Eq. 15-44, the kinetic energy is

$$K = \tfrac{1}{2}mv^2 = \frac{GMm}{2r}, \qquad (15\text{-}45)$$

which shows us that for a satellite in a circular orbit, $K = -U/2$.

The total mechanical energy is

$$E = K + U = \frac{GMm}{2r} - \frac{GMm}{r} = -\frac{GMm}{2r}, \quad (15\text{-}46)$$

which shows us that for a satellite in a circular orbit, the total energy E is the negative of the kinetic energy K. It can be shown that Eq. 15-46 continues to hold if we replace r with a, the semimajor axis of an elliptical orbit. Then we have, as the total mechanical energy for an elliptical orbit,

$$E = -\frac{GMm}{2a} \quad \text{(total mechanical energy).} \quad (15\text{-}47)$$

Equation 15-47 tells us that the total energy of an orbiting satellite depends only on the semimajor axis of its orbit and not on its eccentricity e. For example, four orbits with the same semimajor axis are shown in Fig. 15-20; the same satellite would have the same total mechanical energy E in all four orbits. Figure 15-21 shows the variation of K, U, and E for a satellite moving in a circular orbit about a massive central body.

Suppose that the space shuttle *Enterprise* is in a circular orbit of radius r and that, at point P in Fig. 15-22, the commander orders a short burn from a forward pointing thruster, so that the speed of the ship abruptly decreases. What path will *Enterprise* then follow?

The burn reduces the kinetic energy of the ship and thus reduces E; that is, E becomes more negative. This, from Eq. 15-47, requires a decrease in a, the semimajor axis of the orbit. Equation 15-34 then requires that T, the orbital period, must also de-

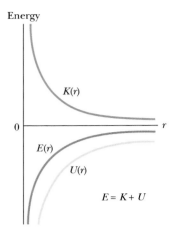

FIGURE 15-21 The variation of kinetic energy K, potential energy U, and total energy E with radius r for a satellite in a circular orbit. For any value of r, the values of U and E are negative, the value of K is positive, and $E = -K$. As r approaches infinity, all three energy curves approach a value of zero.

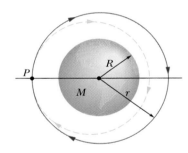

FIGURE 15-22 A spaceship is in a circular Earth orbit of radius r. At point P, the commander fires an instantaneous burst in the forward direction, slowing the ship. The ship then moves in the dashed elliptical orbit.

crease. *Enterprise* will follow the elliptical orbit shown dashed in Fig. 15-22. This maneuver is analyzed further in Sample Problem 15-13.

SAMPLE PROBLEM 15-12

A playful astronaut puts a bowling ball whose mass m is 7.20 kg into a circular orbit about the Earth at an altitude h of 350 km.

a. What is the kinetic energy of the ball?

SOLUTION The orbital radius r is given by

$$r = R + h = 6370 \text{ km} + 350 \text{ km} = 6.72 \times 10^6 \text{ m,}$$

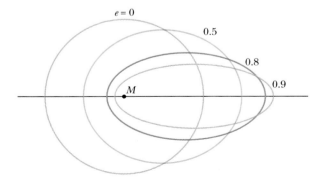

FIGURE 15-20 All four orbits have the same semimajor axis a and thus correspond to the same total mechanical energy E. Their eccentricities e are marked.

in which R is the radius of the Earth. From Eq. 15-45, the kinetic energy is

$$K = \frac{GMm}{2r}$$

$$= \frac{(6.67 \times 10^{-11} \, \text{N} \cdot \text{m}^2/\text{kg}^2)(5.98 \times 10^{24} \, \text{kg})(7.20 \, \text{kg})}{(2)(6.72 \times 10^6 \, \text{m})}$$

$$= 2.14 \times 10^8 \, \text{J} = 214 \, \text{MJ}. \qquad \text{(Answer)}$$

b. What is the ball's potential energy?

SOLUTION Since the ball is in a circular orbit, we have

$$U = -2K = -428 \, \text{MJ}. \qquad \text{(Answer)}$$

c. What is the ball's mechanical energy E?

SOLUTION Since the ball is in a circular orbit, we have

$$E = -K = -214 \, \text{MJ}. \qquad \text{(Answer)}$$

d. What was the mechanical energy E_0 of the ball on the launchpad at Cape Canaveral?

SOLUTION On the launchpad, the ball has some kinetic energy because of the rotation of the Earth but we can show that this is small enough to neglect. Thus the total energy E_0 is equal to the potential energy U_0, which is given by Eq. 15-23:

$$E_0 = U_0 = -\frac{GMm}{R}$$

$$= -\frac{(6.67 \times 10^{-11} \, \text{N} \cdot \text{m}^2/\text{kg}^2)(5.98 \times 10^{24} \, \text{kg})(7.20 \, \text{kg})}{6.37 \times 10^6 \, \text{m}}$$

$$= -4.51 \times 10^8 \, \text{J} = -451 \, \text{MJ}. \qquad \text{(Answer)}$$

You might be tempted to say that the potential energy of the ball on the Earth's surface is zero. Recall, however, that our zero-potential-energy configuration is one in which the ball is removed to a great distance from the Earth. You also might be tempted to use Eq. 15-47 to find E_0, but recall that Eq. 15-47 applies to a satellite *in orbit*.

The *increase* in the mechanical energy of the ball as it was launched was

$$\Delta E = E - E_0 = (-214 \, \text{MJ}) - (-451 \, \text{MJ})$$

$$= 237 \, \text{MJ}. \qquad \text{(Answer)}$$

You can buy this amount of energy from your utility company for a few dollars. Correspondingly, the increase in mechanical energy of the Hubble Space Telescope, from launchpad to deployment (Fig. 15-23) was worth only a few thousand dollars.

FIGURE 15-23 Deployment of the Hubble Space Telescope from the bay of a space shuttle.

SAMPLE PROBLEM 15-13

Two small spaceships, each with mass $m = 2000$ kg, are in the circular Earth orbit of Fig. 15-22, at an altitude h of 400 km. Igor, the commander of one of the ships, arrives at any fixed point in the orbit 90 s ahead of Sally, the commander of the other spaceship.

a. What are the period and the speed of the two ships in this circular orbit?

SOLUTION The radius of their circular orbit is

$$r = R + h = 6370 \, \text{km} + 400 \, \text{km}$$

$$= 6770 \, \text{km} = 6.77 \times 10^6 \, \text{m}.$$

The period then follows from Eq. 15-34 and is

$$T_0 = \sqrt{\frac{4\pi^2 r^3}{GM}}$$

$$= \sqrt{\frac{(4\pi^2)(6.77 \times 10^6 \, \text{m})^3}{(6.67 \times 10^{-11} \, \text{m}^3/\text{kg} \cdot \text{s}^2)(5.98 \times 10^{24} \, \text{kg})}}$$

$$= 5540 \, \text{s} \; (= 92.3 \, \text{min}). \qquad \text{(Answer)}$$

The orbital speed is obtained by dividing the circumference of the orbit by the period:

$$v_0 = \frac{2\pi r}{T_0} = \frac{(2\pi)(6.77 \times 10^6 \, \text{m})}{5540 \, \text{s}}$$

$$= 7680 \, \text{m/s}. \qquad \text{(Answer)}$$

b. At a point such as *P* in Fig. 15-22, Sally, wanting to get ahead of Igor, fires an instantaneous burst in the forward direction, *reducing* her speed by 1.00%. After she executes her burn, Sally will follow the elliptical orbit shown dashed in Fig. 15-22. What are the speed, kinetic energy, and potential energy of her ship immediately after the burn?

SOLUTION Sally's speed at point *P*, just after the burn, is

$$v = (0.99)v_0 = (0.99)(7680 \text{ m/s})$$

$$= 7600 \text{ m/s}. \quad \text{(Answer)}$$

So her new kinetic energy at point *P* is

$$K = \tfrac{1}{2}mv^2 = \tfrac{1}{2}(2000 \text{ kg})(7.60 \times 10^3 \text{ m/s})^2$$

$$= 5.78 \times 10^{10} \text{ J}. \quad \text{(Answer)}$$

Her potential energy at point *P* remains unchanged and is

$$U = -\frac{GMm}{r}$$

$$= -\frac{(6.67 \times 10^{-11} \text{ N} \cdot \text{m}^2/\text{kg}^2)(5.98 \times 10^{24} \text{ kg})(2000 \text{ kg})}{6.77 \times 10^6 \text{ m}}$$

$$= -11.8 \times 10^{10} \text{ J}. \quad \text{(Answer)}$$

(Note that since the burn puts Sally into an elliptical orbit, the relation $U = -2K$ for a circular orbit does not apply.)

c. In her new elliptical orbit, what are the total energy, the semimajor axis, and the orbital period?

SOLUTION Sally's new total energy, which is a constant for the new orbit, is the sum of her potential energy and new kinetic energy at point *P*:

$$E = K + U = 5.78 \times 10^{10} \text{ J} + (-11.8 \times 10^{10} \text{ J})$$

$$= -6.02 \times 10^{10} \text{ J}. \quad \text{(Answer)}$$

(Note that the relation $E = -K$ for a circular orbit does not apply to Sally's new elliptical orbit.)

From Eq. 15-47, the semimajor axis of Sally's new orbit is

$$a = -\frac{GMm}{2E}$$

$$= -\frac{(6.67 \times 10^{-11} \text{ m}^3/\text{kg} \cdot \text{s}^2)(5.98 \times 10^{24} \text{ kg})(2000 \text{ kg})}{(2)(-6.02 \times 10^{10} \text{ J})}$$

$$= 6.63 \times 10^6 \text{ m}. \quad \text{(Answer)}$$

Note that *a* is smaller, by 2.1%, than the radius *r* of the original circular orbit.

TABLE 15-4
PROPERTIES OF THE TWO ORBITS
IN SAMPLE PROBLEM 15-13

PROPERTY	CIRCULAR ORBIT	ELLIPTICAL ORBIT
Semimajor axis (a, 10^6 m)	6.77	6.63
Closest distance (R_p, 10^6 m)	6.77	6.49
Farthest distance (R_a, 10^6 m)	6.77	6.77
Eccentricity (e)	0	0.021
Period (T, s)	5540	5370
Energy (E, 10^{10} J)	−5.90	−6.02

The period in Sally's new orbit follows from Eq. 15-34, with *a* substituted for *r*:

$$T = \sqrt{\frac{4\pi^2 a^3}{GM}}$$

$$= \sqrt{\frac{(4\pi^2)(6.63 \times 10^6 \text{ m})^3}{(6.67 \times 10^{-11} \text{ m}^3/\text{kg} \cdot \text{s}^2)(5.98 \times 10^{24} \text{ kg})}}$$

$$= 5370 \text{ s} \ (= 89.5 \text{ min}). \quad \text{(Answer)}$$

This is shorter than the period of Igor's orbit by 170 s. Thus Sally will arrive back at point *P* ahead of Igor by 170 s − 90 s, or 80 s. By slowing down, Sally managed to get ahead of Igor! To get back into the original circular orbit, still remaining 80 s ahead of Igor, all Sally has to do is restore her original speed by executing the same burn, but in a backward direction, the next time she is at point *P*.

Sally was able to get ahead by slowing down for two reasons. (1) As Fig. 15-22 shows, the length of her new orbit is shorter. (2) By decreasing her speed at point *P*, she changed her orbit so that, until she returned to *P*, she dipped closer to the planet, thereby transferring potential energy into kinetic energy and increasing her speed. Table 15-4 compares the properties of the original circular orbit and the new elliptical orbit.

15-10 A CLOSER LOOK AT GRAVITATION (OPTIONAL)

Einstein once said: "I was sitting in a chair in the patent office at Bern when all of a sudden a thought occurred to me: 'If a person falls freely, he will not feel his own weight.' I was startled. This simple thought made a deep impression on me. It impelled me toward a theory of gravitation."

Thus Einstein tells us how he was led to formulate his *general theory of relativity,** which interprets gravitation not as a force but as a curvature in space and time. Although we cannot explore this important theory here, we can at least look at its fundamental postulate, the **principle of equivalence.**

Interpreted a little loosely, this principle says that gravitation and acceleration are equivalent. If a physicist were locked up in a small box as in Fig. 15-24, he would not be able to tell the difference between gravitation and acceleration. Suppose the physicist is standing on a platform scale. In Fig. 15-24a, the box is at rest on the Earth; in Fig. 15-24b, it is accelerating through interstellar space at 9.8 m/s². The physicist cannot tell the difference. Moreover, if he watches an object—say a cantaloupe of mass m —fall past him, he sees the same relative acceleration in both situations.

Figure 15-25 shows another aspect. In Fig. 15-25a, the physicist is floating around in a freely falling elevator cab. (Ignore the rotation of the Earth.) In Fig. 15-25b, he is floating around in interstellar space. Again, he cannot tell the difference; he feels "weightless" in both situations.

A familiar fact follows from the principle of equivalence; namely, in a vacuum, all bodies fall with the same acceleration. If, for example, pears fell with a larger acceleration than cantaloupes, the experiments of Figs. 15-24 and 15-25 would fail. In an orbiting space shuttle, pears would fall toward the Earth but cantaloupes would not.

We can express our claim about falling bodies by recalling that there are two different ways that we can assign a mass to a body. First, we can hang it from a spring scale (say, at the north pole to avoid the effect of the Earth's rotation). We may call the result of this measurement the **gravitational mass** m_g. This is the mass that appears in Newton's law of gravitation, which we can write as

$$F = G \frac{Mm_g}{R^2}. \qquad (15\text{-}48)$$

Alternatively, we can measure the mass by measuring its acceleration in response to a known force,

(a)   (b)

FIGURE 15-24 (a) A physicist in a box, resting on the Earth, sees a cantaloupe falling with acceleration $a = 9.8$ m/s². (b) If he and the box accelerate in deep space at 9.8 m/s², the cantaloupe has the same acceleration relative to him. It is not possible, by doing experiments within the box, for the physicist to tell which situation he is in. For example, the platform scale on which he stands reads the same in both situations.

obtaining what we may call the **inertial mass** m_i. This is the mass that appears in Newton's second law, which we can write as

$$F = m_i a. \qquad (15\text{-}49)$$

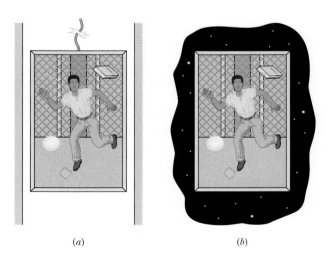

(a)   (b)

FIGURE 15-25 (a) A physicist in a freely falling elevator cab. (b) The elevator cab drifting in deep space. Again, it is not possible, by doing experiments within the cab, for the physicist to tell which situation he is in.

*Not to be confused with Einstein's *special* theory of relativity, treated in Chapter 42. For an overview of the general theory, which includes the special theory as a limiting case, see Robert Resnick and David Halliday, *Basic Concepts*, 2nd ed., John Wiley & Sons, New York, 1985, Supplementary Topic C.

These masses need not be equal to each other. The only way we can find out whether or not they are equal is to measure them and see. Experiment shows that, to a precision of perhaps 1 part in 10^{12}, *these two masses are identical.*

In Newtonian physics, the experimental fact that $m_g = m_i$ could be regarded as nothing but an astonishing coincidence. In Einstein's general theory of relativity, it enters in a natural way through the prin-

ciple of equivalence: if gravitation and acceleration are equivalent, then masses measured via gravitation and via acceleration must be equal.

Looking back, we see a direct line of thought about gravitation, from Galileo at the tower of Pisa (maybe, if he actually did the experiment), through Newton watching the falling apple, to Einstein daydreaming in his chair in the Swiss patent office. The story of gravitation is far from ended.

REVIEW & SUMMARY

The Law of Gravitation

Every particle in the universe attracts every other particle with a **gravitational force** whose magnitude is

$$F = G \frac{m_1 m_2}{r^2} \qquad \text{(Newton's law of gravitation).} \quad (15\text{-}1)$$

The **gravitational constant** G is a universal constant whose value is $6.67 \times 10^{-11} \ \text{N} \cdot \text{m}^2/\text{kg}^2$.

The Gravitational Behavior of Uniform Spherical Shells

Equation 15-1 holds only for particles. The gravitational force between extended bodies must generally be found by adding (integrating) the individual forces on individual particles within the bodies. However, if either of the bodies is a uniform spherical shell or a spherically symmetric solid, the net gravitational force it exerts on an *external* object may be computed as if the mass of the symmetric body were located at its center.

Superposition

Gravitational forces obey the **principle of superposition;** that is, the total force $\mathbf{F}_1$ on a particle labeled as particle 1 is the sum of the forces exerted on it by all other particles taken one at a time:

$$\mathbf{F}_1 = \sum_{i=2}^{n} \mathbf{F}_{1i}, \quad (15\text{-}4)$$

in which the sum is a vector sum of the forces $\mathbf{F}_{1i}$ exerted on particle 1 by particles 2, 3, . . . , n. The gravitational force $\mathbf{F}_1$ exerted on a particle by an extended body is found by dividing the body into units of differential mass dm, each of which exerts a differential force $d\mathbf{F}$ on the particle, and then integrating to find the net of those forces:

$$\mathbf{F}_1 = \int d\mathbf{F}. \quad (15\text{-}5)$$

Gravitational Acceleration

The *gravitational acceleration* a_g of a particle (of mass m) is due strictly to the gravitational force acting on it. When the particle is at distance r from the center of a uniform, spherical body of mass M, the magnitude of the gravitational force on the particle is given by Eq. 15-1. In addition, by Newton's second law,

$$F = ma_g, \quad (15\text{-}11)$$

and thus a_g is given by

$$a_g = \frac{GM}{r^2}. \quad (15\text{-}12)$$

Free-Fall Acceleration and Weight

The free-fall acceleration $\mathbf{g}$ of a particle near the Earth differs slightly from the gravitational acceleration $\mathbf{a}_g$, and the weight $m\mathbf{g}$ of the particle differs from the gravitational force (Eq. 15-1) acting on the particle, because the Earth is not uniform or spherical and because the Earth rotates.

Gravitation Within a Spherical Shell

A uniform shell of matter exerts no gravitational force on a particle located inside it. This means that if a particle is located inside a uniform sphere of matter at distance r from its center, the gravitational force exerted on the particle is due only to mass M' that lies within a sphere of radius r. This mass M' is given by

$$M' = \rho \frac{4\pi r^3}{3}, \quad (15\text{-}21)$$

where ρ is the density of the sphere.

Gravitational Potential Energy

The gravitational potential energy $U(r)$ of two particles, with masses M and m and separated by a distance r, is the negative of the work that would be done by the gravitational force of one particle acting on the other as the particles fell to the distance r from an infinite (very large) separation. This energy is

$$U = -\frac{GMm}{r} \qquad \text{(gravitational potential energy).} \qquad (15\text{-}23)$$

Potential Energy of a System

If a system contains more than two particles, the total gravitational potential energy is the sum of terms representing the potential energies of all the pairs; as an example, we have for three particles, of masses m_1, m_2, and m_3,

$$U = -\left(\frac{Gm_1m_2}{r_{12}} + \frac{Gm_1m_3}{r_{13}} + \frac{Gm_2m_3}{r_{23}} \right). \qquad (15\text{-}24)$$

Escape Speed

An object will escape the gravitational pull of an astronomical body with mass M and radius R if its speed near the body's surface is at least equal to the **escape speed** given by

$$v = \sqrt{\frac{2GM}{R}}. \qquad (15\text{-}29)$$

Kepler's Three Laws

Gravitational attraction holds the solar system together and makes possible orbiting Earth satellites, both natural and artificial. Such motions are governed by Kepler's three laws of planetary motion, all of which are direct consequences of Newton's laws of motion and gravitation:

1. *The law of orbits.* All planets move in elliptical orbits with the sun at one focus.

2. *The law of areas.* A line joining any planet to the sun sweeps out equal areas in equal times (a statement equivalent to conservation of angular momentum).

3. *The law of periods.* The square of the period T of any planet about the sun is proportional to the cube of the semimajor axis a of the orbit. For circular orbits with radius r, the semimajor axis a is replaced by r and the law is written as

$$T^2 = \left(\frac{4\pi^2}{GM} \right) r^3 \qquad \text{(law of periods),} \qquad (15\text{-}34)$$

where M is the mass of the attracting body—the sun in the case of the solar system. This result is generally valid for elliptical planetary orbits, when the semimajor axis a is inserted in place of the circular radius r.

Energy in Planetary Motion

When a planet or satellite with mass m moves in a circular orbit with radius r, its potential energy U and kinetic energy K are given by

$$U = -\frac{GMm}{r} \quad \text{and} \quad K = \frac{GMm}{2r}. \qquad (15\text{-}23, 15\text{-}45)$$

The total mechanical energy $E = K + U$ is

$$E = -\frac{GMm}{2a}, \qquad (15\text{-}47)$$

in which the radius r has been replaced with the semimajor axis a to indicate that this expression for total energy is also valid for more general elliptical orbits.

The Principle of Equivalence

Einstein pointed out that gravitation and acceleration are equivalent. This **principle of equivalence** led him to a theory of gravitation (the *general theory of relativity*) that views gravitational effects in terms of a curvature in space and time. One of the consequences of the theory is that **gravitational mass,** which determines the strength of an object's participation in the gravitational interaction, is equivalent to **inertial mass,** the mass that appears in Newton's second law of motion.

QUESTIONS

1. If the gravitational force acts on all bodies in proportion to their masses, why doesn't a heavy body fall correspondingly faster than a light body?

2. How does the net gravitational force on a space probe vary as it travels from the Earth to the moon? Would its mass change?

3. Would we have more sugar to the pound at the north pole or the equator? What about sugar to the kilogram?

4. Because the Earth bulges near the equator, the source of the Mississippi River, although high above sea level, is nearer to the center of the Earth than is its mouth. How can the river flow "uphill"?

5. One clock uses an oscillating spring; a second clock uses a pendulum. Both are taken to Mars. Will they keep the same time there that they kept on Earth? Will they agree with each other? Explain. (The mass of Mars is one-tenth that of Earth, and its radius is half that of Earth.)

6. Two identical cars travel at the same speed in opposite directions on an east–west highway. Which car presses down harder on the road?

7. In Sample Problem 15-6, the tunnel transit time was derived on the assumption of an Earth of uniform density. Would this time be larger or smaller if the actual density distribution of the Earth, with its dense inner core, were taken into account? Explain your answer.

8. The masses m in Figs. 15-26a and 15-26b are identical. Which system of masses has the smaller gravitational potential energy?

FIGURE 15-26 Question 8.

9. A satellite orbiting the Earth around the Arctic Circle would be very useful in maintaining surveillance of this strategically important part of the globe. Why don't we put one up?

10. As a car speeds around a curve, the passengers tend to be thrown radially outward. Why are astronauts in a space shuttle not similarly affected as their shuttle speeds in orbit around the Earth?

11. The gravitational force exerted by the sun on the moon is about twice as great as the gravitational force exerted by the Earth on the moon. Why then doesn't the moon escape from the Earth?

12. Explain why the following reasoning is wrong. "The sun attracts all bodies on the Earth. At midnight, when the sun is directly below, it pulls on an object in the same direction as the pull of the Earth on that object; at noon, when the sun is directly above, it pulls on an object in a direction opposite the pull of the Earth. Hence all objects should be heavier at midnight (or at night) than they are at noon (or during the day)."

13. If lunar tides slow down the rotation of the Earth (owing to friction), the angular momentum of the Earth decreases. What happens to the motion of the moon as a consequence of the conservation of angular momentum? Does the sun (and solar tides) play a role here?

14. A satellite in low Earth orbit experiences a small drag force from the Earth's atmosphere. What happens to its speed because of this drag force? (Compare this situation with that of Sample Problem 15-13c.)

15. Would you expect the total mechanical energy of the solar system to be constant? The total angular momentum? Explain your answers.

16. Objects at rest on the Earth's surface move in circular paths with a period of 24 h. Are they "in orbit" in the sense that an Earth satellite is in orbit? Why not? What would the period of the Earth have to be to put such objects in true orbit?

17. Can a satellite be in an orbit in a plane not passing through the Earth's center? Explain your answer.

18. As measured by an observer on Earth, would there be any difference in the periods of two satellites, each in a circular orbit near the Earth in an equatorial plane, but one moving eastward and the other westward?

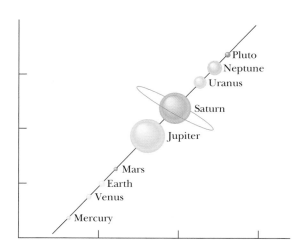

FIGURE 15-27 Question 25.

19. After *Sputnik I* was put into orbit, some scientists said that it would not return to Earth but would burn up in its *descent.* Considering the fact that it did not burn up in its *ascent,* how is this possible?

20. What happens to the angular momentum of an artificial Earth satellite as it descends through the atmosphere?

21. An artificial satellite is in a circular orbit about the Earth. How will its orbit change if one of its rockets is momentarily fired (a) toward the Earth, (b) away from the Earth, (c) in a forward direction, (d) in a backward direction, and (e) perpendicular to the plane of the orbit?

22. What advantage does Florida have over California for launching satellites?

23. For a flight to Mars, a rocket is fired in the direction the Earth is moving in its orbit. For a flight to Venus, it is fired backward along that orbit. Explain why.

24. Saturn is about six times farther from the sun than Mars. Which planet has (a) the greater period of revolution, (b) the greater orbital speed, and (c) the greater angular speed?

25. What is being plotted in Fig. 15-27? Put numbers with units on each axis and adjust the horizontal axis if needed.

26. You are a passenger on the SS *Arthur C. Clarke,* the first interstellar spaceship that rotates about a central axis to simulate Earth gravity. If you are in an enclosed cabin, how can you tell that you are not on Earth?

EXERCISES & PROBLEMS

SECTION 15-2 NEWTON'S LAW OF GRAVITATION

1E. What must the separation be between a 5.2-kg particle and a 2.4-kg particle for their gravitational attraction to be 2.3×10^{-12} N?

2E. Some believe that the positions of the planets at the time of birth influence the newborn. Others deride this belief and claim that the gravitational force exerted on a baby by the obstetrician is greater than that exerted by the planets. To check this claim, calculate and compare the gravitational force exerted on a 6-kg baby (a) by a 70-kg obstetrician who is 1 m away, (b) by the massive planet Jupiter ($m = 2 \times 10^{27}$ kg) at its closest approach to Earth ($= 6 \times 10^{11}$ m), and (c) by Jupiter at its greatest distance from Earth ($= 9 \times 10^{11}$ m). (d) Is the claim correct?

3E. The sun and Earth each exert a gravitational force on the moon. What is the ratio F_{sun}/F_{Earth} of these two forces? (The average sun–moon distance is equal to the sun–Earth distance.)

4E. One of the *Echo* satellites consisted of an inflated spherical aluminum balloon 30 m in diameter and of mass 20 kg. Suppose a meteor having a mass of 7.0 kg passes within 3.0 m of the surface of the satellite. What is the gravitational force on the meteor from the satellite at the closest approach?

5P. A mass M is split into two parts, m and $M - m$, which are then separated by a certain distance. What ratio m/M maximizes the gravitational force between the parts?

SECTION 15-3 GRAVITATION AND THE
PRINCIPLE OF SUPERPOSITION

6E. How far from the Earth must a space probe be along a line toward the sun so that the sun's gravitational pull balances the Earth's?

7E. A spaceship is on a straight-line path between the Earth and the moon. At what distance from the Earth is the net gravitational force on the spaceship zero?

8P. What is the percent change in the acceleration of the Earth toward the sun when the alignment of Earth, sun, and moon changes from an eclipse of the sun (with the moon between Earth and sun) to an eclipse of the moon (Earth between moon and sun)?

9P. Four spheres, with masses $m_1 = 400$ kg, $m_2 = 350$ kg, $m_3 = 2000$ kg, and $m_4 = 500$ kg, have (x, y) coordinates of (0, 50 cm), (0, 0), (-80 cm, 0), and (40 cm, 0), respectively. What is the net gravitational force $\mathbf{F}_2$ on m_2 due to the other masses?

10P. In Fig. 15-28, four spheres form the corners of a square whose side is 2.0 cm long. What are the magnitude and direction of the net gravitational force from them on a central sphere with mass $m_5 = 250$ kg?

11P. In Fig. 15-29, two spheres of mass m and a third sphere of mass M form an equilateral triangle, and a fourth sphere of mass m_4 is at the center of the triangle. If the net gravitational force on m_4 from the three other spheres is to be zero, what is M in terms of m?

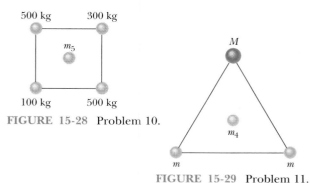

FIGURE 15-28 Problem 10.

FIGURE 15-29 Problem 11.

12P. Two spheres with masses $m_1 = 800$ kg and $m_2 = 600$ kg are separated by 0.25 m. What is the net gravitational force (in both magnitude and direction) from them on a 2.0-kg sphere located 0.20 m from m_1 and 0.15 m from m_2?

13P. The masses and coordinates of three spheres are as follows: 20 kg, $x = 0.50$ m, $y = 1.0$ m; 40 kg, $x = -1.0$ m, $y = -1.0$ m; 60 kg, $x = 0$ m, $y = -0.50$ m. What is the magnitude of the gravitational force on a 20-kg sphere located at the origin due to the other spheres?

14P*. A thin rod with mass M is bent in a semicircle of radius R, as in Fig. 15-30. (a) What is its gravitational force (both magnitude and direction) on a particle with mass m at P, the center of curvature? (b) What would be the force on m if the rod were a complete circle?

Rod of mass M

FIGURE 15-30 Problem 14.

15P. The following problem is from the 1946 "Olympic" examination of Moscow State University (see Fig. 15-31). A spherical hollow is made in a lead sphere of radius R, such that its surface touches the outside surface of the lead sphere and passes through its center. The mass of the sphere before hollowing was M. With what gravitational force will the hollowed-out lead sphere attract a small sphere of mass m, which lies at a distance d from the center of the lead sphere on the straight line connecting the centers of the spheres and of the hollow?

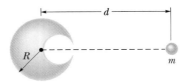

FIGURE 15-31 Problem 15.

SECTION 15-4 GRAVITATION NEAR THE EARTH'S SURFACE

16E. If a pendulum has a period of exactly 1.00 s at the equator, what would be its period at the south pole? Use Fig. 15-7.

17E. (a) Calculate the gravitational acceleration on the surface of the moon from values of the mass and radius of the moon found in Appendix C. (b) What would be the period of a "seconds pendulum" (period = 2.0 s on Earth) on the surface of the moon?

18E. At what altitude above the Earth's surface would the gravitational acceleration be 4.9 m/s²?

19E. You weigh 120 lb at the sidewalk level outside the World Trade Center in New York City. Suppose that you ride from this level to the top of one of its 1350-ft towers. Ignoring the Earth's rotation, how much less would you weigh there because you are slightly farther from the center of the Earth?

20E. A typical neutron star may have a mass equal to that of the sun but a radius of only 10 km. (a) What is the gravitational acceleration at the surface of such a star? (b) How fast would an object be moving if it fell from rest through a distance of 1.0 m on such a star? (Assume the star does not rotate.)

21E. An object lying on the Earth's equator is accelerated (a) toward the center of the Earth because the Earth rotates, (b) toward the sun because the Earth revolves around the sun in an almost circular orbit, and (c) toward the center of our galaxy. The period of the sun's revolution about the galactic center is 2.5×10^8 years, and the sun's distance from this center is 2.2×10^{20} m. Calculate these three accelerations in terms of $g = 9.8$ m/s².

22E. (a) What will an object weigh on the moon's surface if it weighs 100 N on the Earth's surface? (b) How many Earth radii must this same object be from the center of the Earth if it is to weigh the same as it does on the moon?

23P. The fact that g varies from place to place over the Earth's surface drew attention when Jean Richer in 1672 took a pendulum clock from Paris to Cayenne, French Guiana, and found that it lost 2.5 min/day. If $g = 9.81$ m/s² in Paris, what is g in Cayenne?

24P. (a) If g is to be determined by dropping an object through a distance of exactly 10 m, what percent error in the measurement of the time of fall results in a 0.1% error in the value of g? (b) How accurately (in seconds) would you have to measure the time for 100 oscillations of a 10-m-long pendulum to achieve the same percent error in the measurement of g?

25P. The fastest possible rate of rotation of a planet is that for which the gravitational force on material at the equator just barely provides the centripetal force needed for the rotation. (Why?) (a) Show that the corresponding shortest period of rotation is given by

$$T = \sqrt{\frac{3\pi}{G\rho}},$$

where ρ is the density of the planet, assumed to be homogeneous. (b) Evaluate the rotation period assuming a density of 3.0 g/cm³, typical of many planets, satellites, and asteroids. No astronomical object has ever been found to be spinning with a period shorter than that determined by this analysis.

26P. Certain neutron stars (extremely dense stars) are believed to be rotating at about 1 rev/s. If such a star has a radius of 20 km, what must be its minimum mass so that material on its surface remains in place during the rapid rotation?

27P. In Fig. 15-32, identical masses m hang from strings of different lengths on a balance at the surface of the Earth. The strings have negligible mass and differ in length by h. Assume the Earth is spherical with density $\rho = 5.5$ g/cm³. (a) Show that the difference in the weights, due to one mass being closer to the Earth than the other, is $8\pi G\rho mh/3$. (b) Find the difference in length that will give a weight difference of 1 part in a million.

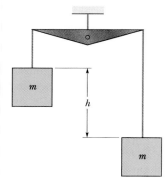

FIGURE 15-32 Problem 27.

28P. A scientist is making a precise measurement of g at a certain point in the Indian Ocean (on the equator) by timing the swings of a pendulum of accurately known construction. To provide a stable base the measurements are conducted in a submerged submarine. It is observed that a slightly different result for g is obtained when the submarine is moving eastward than when it is moving westward, the speed in each case being 16 km/h. Account for this difference and calculate the fractional error $\Delta g/g$ in g for either travel direction.

29P. A body is suspended on a spring balance in a ship sailing along the equator with a speed v. (a) Show that the scale reading will be very close to $W_0 (1 \pm 2\,\omega v/g)$, where ω is the angular speed of the Earth and W_0 is the scale reading when the ship is at rest. (b) Explain the plus or minus.

30P*. Sensitive meters that measure the local free-fall acceleration **g** can be used to detect the presence of deposits of near-surface rocks of density significantly greater or less than that of the surroundings. Cavities such as caves and abandoned mine shafts can also be located. (a) Show that the vertical component of **g** a distance x from a point directly above the center of a spherical cavern (see Fig. 15-33) is less than what would be expected assuming a uniform distribution of rock of density ρ, by the amount

$$\Delta g = \frac{4\pi}{3} R^3 G\rho \frac{d}{(d^2 + x^2)^{3/2}},$$

where R is the radius of the cavern and d is the depth of its center. (b) These values of Δg, called *anomalies,* are usually very small and are expressed in milligals, where 1 gal = 1 cm/s². Suppose oil prospectors doing a gravity survey

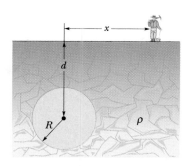

FIGURE 15-33 Problem 30.

over a straight-line distance of 300 m find Δg varying from 10 milligals to a maximum of 14 milligals at the midpoint of the 300-m distance. Assuming that the larger anomaly value was recorded directly over the center of a spherical cavern known to be in the region, find the radius of the cavern and the vertical depth to its roof. Nearby rocks have a density of 2.8 g/cm³. (c) Suppose that the cavern, instead of being empty, is completely flooded with water. What do the gravity readings in (b) now indicate for its radius and roof depth?

SECTION 15-6 GRAVITATION INSIDE THE EARTH

31E. Two concentric shells of uniform density having masses M_1 and M_2 are situated as shown in Fig. 15-34. Find the force on a particle of mass m when the particle is located at (a) $r = a$, (b) $r = b$, and (c) $r = c$. The distance r is measured from the center of the shells.

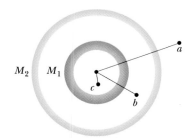

FIGURE 15-34 Exercise 31.

32E. With what speed would mail pass through the center of the Earth if it were dropped down the tunnel of Sample Problem 15-6?

33E. Show that, at the bottom of a vertical mine shaft dug to depth D, the measured value of g will be

$$g = g_s \left(1 - \frac{D}{R}\right),$$

g_s being the surface value. Assume that the Earth is a uniform sphere of radius R.

34P. Figure 15-35 shows, not to scale, a cross section through the interior of the Earth. Rather than being uniform throughout, the Earth is divided into three zones: an

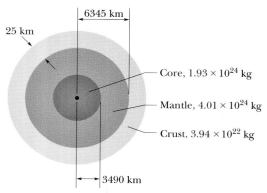

FIGURE 15-35 Problem 34. Not to scale.

outer *crust,* a *mantle,* and an inner *core.* The dimensions of these zones and the masses contained within them are shown on the figure. The Earth has total mass 5.98×10^{24} kg and radius 6370 km. Ignore rotation and assume that the Earth is spherical. (a) Calculate g at the surface. (b) Suppose that a bore hole (the *Moho*) is driven to the crust–mantle interface at a depth of 25 km; what would be the value of g at the bottom of the hole? (c) Suppose that the Earth was a uniform sphere with the same total mass and size. What would be the value of g at a depth of 25 km?

FIGURE 15-36 Problem 35.

(a)

(b)

(See Exercise 33.) (Precise measurements of g are sensitive probes of the interior structure of the Earth, although results can be clouded by local density variations.)

35P. (a) Figure 15-36 shows a planetary object of uniform density ρ and radius R. Show that the compressive stress S on the material near its center due to the weight pressing on that material is given by

$$S = \tfrac{2}{3}\pi G \rho^2 R^2.$$

(*Hint:* Construct a narrow column of cross-sectional area A extending from the center to the surface. The weight of the material in the column is mg_{av}, where m is the mass of material in the column and g_{av} is the value of g midway between center and surface.) (b) In our solar system, objects (for example, asteroids, small satellites, comets) with "diameters" less than 600 km can be very irregular in shape (see Fig. 15-36*b,* which shows Hyperion, a small satellite of Saturn), whereas those with larger diameters are spherical. Only if the rock has sufficient strength to resist gravitation can such an object maintain a nonspherical shape. Calculate the ultimate compressive strength of the rock making up asteroids. Assume a density of 4000 kg/m³. (c) What is the largest possible "diameter" of a nonspherical satellite made of concrete of density $\rho = 3000$ kg/m³. (See Table 13-1.)

36P. To alleviate the traffic congestion between two cities such as Boston and Washington, DC, engineers have proposed building a rail tunnel along a chord line connecting the cities (Fig. 15-37). A train, unpropelled by any engine and starting from rest, would fall through the first half of the tunnel and then move up the second half. Assuming the Earth is a uniform sphere and ignoring air drag and friction, (a) show that the travel between cities is half a cycle of simple harmonic motion, and (b) find the travel time.

FIGURE 15-37 Problem 36. Not to scale.

SECTION 15-7 GRAVITATIONAL POTENTIAL ENERGY

37E. (a) What is the gravitational potential energy of the two-particle system in Exercise 1? If you triple the separation between the particles, how much work is done (b) by the gravitational force between the particles and (c) by you?

38E. (a) In Problem 9, remove m_1 and calculate the gravitational potential energy of the remaining three-particle system. (b) If m_1 is then put back in place, is the potential energy of the four-particle system more or less than that of the system in (a)? (c) In (a), is the work done by you to remove m_1 positive or negative? (d) Is the work done by you to replace m_1 in (b) positive or negative?

39E. In Problem 5, what ratio of m/M gives the least gravitational potential energy for the system?

40E. The mean diameters of Mars and Earth are 6.9×10^3 km and 1.3×10^4 km, respectively. The mass of Mars is 0.11 times the Earth's mass. (a) What is the ratio of the mean density of Mars to that of Earth? (b) What is the value of g on Mars? (c) What is the escape speed on Mars?

41E. A spaceship is idling at the fringes of our galaxy, 80,000 light-years from the galactic center. What is the ship's escape speed from the galaxy? The mass of the galaxy is 1.4×10^{11} times that of our sun. Assume, for simplicity, that the matter forming the galaxy is distributed in a uniform sphere.

42E. Calculate the amount of energy required to escape from (a) the moon and (b) Jupiter relative to that required to escape from Earth.

43E. Show that the escape speed from the sun at the Earth's distance from the sun is $\sqrt{2}$ times the speed of the Earth in its orbit, assumed to be a circle. (This is a specific case of a general result for circular orbits: $v_{esc} = \sqrt{2}v_{orb}$.)

44E. A particle of comet dust with mass m is a distance R from the Earth's center and a distance r from the moon's center. If the Earth's mass is M_E and the moon's mass is M_m, what is the gravitational potential energy of the particle?

45E. Upon "burning out," a large star can collapse under its own gravitational force to become a *black hole* (*frozen star* in Russian). The star's surface then has a radius R_s such that the removal of a mass m from the surface to infinity would require work equal to the mass energy mc^2 of that mass. If M_s is the star's mass, show with Newton's law of gravitation that $R_s = GM_s/c^2$. (The correct value for R_s is actually twice this. To get it, we would have to apply Einstein's general theory of relativity rather than Newton's law.)

46P. The three spheres in Fig. 15-38, with masses $m_1 = 800$ g, $m_2 = 100$ g, and $m_3 = 200$ g, have their centers on a common line, with $L = 12$ cm and $d = 4.0$ cm. You move the middle sphere until its center-to-center separation from m_3 is $d = 4.0$ cm. How much work is done on m_2

FIGURE 15-38 Problem 46.

(a) by you and (b) by the net gravitational force on m_2 due to m_1 and m_3?

47P. A rocket is accelerated to a speed of $v = 2\sqrt{gR_e}$ near the Earth's surface (where the Earth's radius is R_e) and then coasts upward. (a) Show that it will escape from the Earth. (b) Show that very far from the Earth its speed is $v = \sqrt{2gR_e}$.

48P. (a) What is the escape speed on a spherical asteroid whose radius is 500 km and whose gravitational acceleration at the surface is 3.0 m/s^2? (b) How far from the surface will a particle go if it leaves the asteroid's surface with a radial speed of 1000 m/s? (c) With what speed will an object hit the asteroid if it is dropped from 1000 km above the surface?

49P. A projectile is fired vertically from the Earth's surface with an initial speed of 10 km/s. Neglecting air drag, how far above the surface of the Earth will it go?

50P. In a double star system, two stars of mass 3.0×10^{30} kg each rotate about the system's center of mass at a radius of 1.0×10^{11} m. (a) What is their common angular speed? (b) If a meteorite passes through this center of mass perpendicular to the orbital plane of the stars, what value must its speed exceed at that point if it is to escape to "infinity" from the star system?

51P. Two neutron stars are separated by a distance of 10^{10} m. They each have a mass of 10^{30} kg and a radius of 10^5 m. They are initially at rest with respect to one another. (a) How fast are they moving when their separation has decreased to one-half its initial value? (b) How fast are they moving just before they collide?

52P. A sphere of matter, of mass M and radius a, has a concentric cavity of radius b, as shown in cross section in Fig. 15-39. (a) Sketch a curve of the gravitational force F exerted by the sphere on a particle of mass m, located a distance r from the center of the sphere, as a function of r in the range $0 \leq r \leq \infty$. Consider $r = 0$, b, a, and ∞ in particular. (b) Sketch the corresponding curve for the potential energy $U(r)$ of the system.

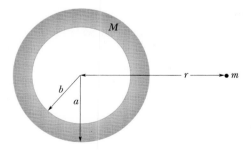

FIGURE 15-39 Problem 52.

53P*. Several planets (Jupiter, Saturn, Uranus) possess nearly circular surrounding rings, perhaps composed of material that failed to form a satellite. In addition, many galaxies contain ringlike structures. Consider a homoge-

FIGURE 15-40   Problem 53.

neous ring of mass M and radius R. (a) What gravitational attraction does it exert on a particle of mass m located a distance x from the center of the ring along its axis? See Fig. 15-40. (b) Suppose the particle falls from rest as a result of the attraction of the ring of matter. Find an expression for the speed with which it passes through the center of the ring.

54P*. The gravitational force between two particles with masses m and M, initially at rest at great separation, pulls them together. Show that at any instant the speed of either particle relative to the other is $\sqrt{2G(M + m)/d}$, where d is their separation at that instant. (*Hint:* Use conservation of energy and conservation of linear momentum.)

SECTION 15-8   PLANETS AND SATELLITES: KEPLER'S LAWS

55E. The mean distance of Mars from the sun is 1.52 times that of the Earth from the sun. From Kepler's law of periods, calculate the number of years required for Mars to make one revolution about the sun; compare your answer with the value given in Appendix C.

56E. The planet Mars has a satellite, Phobos, which travels in an orbit of radius 9.4×10^6 m with a period of 7 h 39 min. Calculate the mass of Mars from this information.

57E. Determine the mass of the Earth from the period T and the radius r of the moon's orbit about the Earth: $T = 27.3$ days and $r = 3.82 \times 10^5$ km. Assume the moon orbits the center of the Earth rather than the center of mass of the Earth–moon system.

58E. Our sun, with mass 2.0×10^{30} kg, revolves about the center of the Milky Way galaxy, which is 2.2×10^{20} m away, once every 2.5×10^8 years. Assuming that the stars in the galaxy each have a mass equal to that of our sun, that they are distributed uniformly in a sphere about the galactic center, and that our sun is essentially at the edge of that sphere, estimate roughly the number of stars in the galaxy.

59E. A satellite is placed in a circular orbit with a radius equal to one-half the radius of the moon's orbit. What is its period of revolution in lunar months? (A lunar month is the period of revolution of the moon.)

60E. (a) What linear speed must an Earth satellite have to be in a circular orbit at an altitude of 160 km? (b) What is the period of revolution?

61E. Most asteroids revolve around the sun between Mars and Jupiter. However, several "Apollo asteroids" with diameters of about 30 km move in orbits that cross the orbit of the Earth. The orbit of one of these is shown to scale in Fig. 15-41. By taking measurements directly from the figure, deduce the asteroid's period of revolution in years.

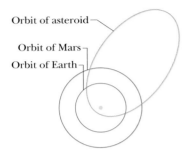

FIGURE 15-41   Exercise 61.

62E. An Earth satellite, moving in an elliptic orbit, is 360 km above the Earth's surface at its farthest point and 180 km above at its closest point. Calculate (a) the semimajor axis and (b) the eccentricity of the orbit. (*Hint:* See Sample Problem 15-10.)

63E. The sun's center is at one focus of the Earth's orbit. How far from this focus is the other focus? Express your answer in terms of the solar radius, 6.96×10^8 m. The eccentricity of the Earth's orbit is 0.0167 and the semimajor axis may be taken to be 1.50×10^{11} m. See Fig. 15-16.

64E. (a) Using Kepler's third law (Eq. 15-34), express the gravitational constant G in terms of the astronomical unit AU as a length unit, the solar mass M_s as a mass unit, and the year as a time unit. (One astronomical unit = 1 AU = 1.496×10^{11} m. One solar mass = $1M_s = 1.99 \times 10^{30}$ kg. One year = 1 y = 3.156×10^7 s.) (b) What form does Kepler's third law take in these units?

65E. A satellite hovers over a certain spot on the equator of the rotating Earth. What is the altitude of its orbit (called a *geosynchronous orbit*)?

66E. By colossal error, the satellite of Exercise 65 is inserted in orbit so that it moves in the direction *opposite* to the Earth's rotation. How often would such a "wrong-way" satellite pass over any given point on the equator?

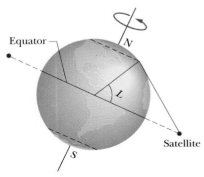

FIGURE 15-42   Exercise 67.

67E. What is the greatest latitude L "visible" to the satellite of Exercise 65? See Fig. 15-42.

68P. Assume that the satellite of Exercise 65 is in orbit at the longitude of Chicago. You are in Chicago (latitude 47.5°) and want to pick up the signals it broadcasts. In what direction should you point your antenna?

69P. In 1610, Galileo used his telescope to discover four prominent moons around Jupiter. Their mean orbital radii a and periods T are

NAME	a (10^8 m)	T (days)
Io	4.22	1.77
Europa	6.71	3.55
Ganymede	10.7	7.16
Callisto	18.8	16.7

(a) Plot log a (y axis) against log T (x axis) and show that you get a straight line. (b) Measure the slope of the line and compare it with the value that you expect from Kepler's third law. (c) Find the mass of Jupiter from the intercept of this line with the y axis. (*Note:* You can use log–log graph paper to avoid taking logarithms.)

70P. A binary star system consists of two stars, each with the same mass as the sun, revolving about their center of mass. The distance between them is the same as the distance between the Earth and the sun. What is their period of revolution in years?

71P. Show how, guided by Kepler's third law (Eq. 15-34), Newton could deduce that the force holding the moon in its orbit, assumed circular, depends on the inverse square of the distance from the center of the Earth.

72P. A certain triple-star system consists of two stars, each of mass m, revolving about a central star of mass M in the same circular orbit of radius r. The two stars are always at opposite ends of a diameter of the circular orbit (see Fig. 15-43). Derive an expression for the period of revolution of the stars.

FIGURE 15-43 Problem 72.

73P. (a) What is the escape speed from the sun for an object in the Earth's orbit (of orbital radius R) but far from the Earth? (b) If an object already has a speed equal to the Earth's orbital speed, what additional speed must it be given to escape as in (a)? (c) Suppose an object is launched from Earth in the direction of the Earth's orbital motion. What speed must it be given during the launch so that when it is far from Earth, but still at a distance of about R from the sun, it can escape from the sun? (This is the speed required for an Earth-launched object to escape from the sun.)

74P*. Three identical stars of mass M are located at the vertices of an equilateral triangle with side L. At what speed must they move if they all revolve under the influence of one another's gravitational force in a circular orbit circumscribing the triangle while still preserving the equilateral triangle?

75P*. A satellite is put into a circular orbit with the intention that it hover over a certain spot on the Earth's surface. However, the satellite's orbital radius is erroneously made 1.0 km too large for this to happen. At what rate and in what direction does the point directly below the satellite move across the Earth's surface?

SECTION 15-9 SATELLITES: ORBITS AND ENERGY

76E. An asteroid, whose mass is 2.0×10^{-4} times the mass of the Earth, revolves in a circular orbit around the sun at a distance that is twice the Earth's distance from the sun. (a) Calculate the period of revolution of the asteroid in years. (b) What is the ratio of the kinetic energy of the asteroid to that of the Earth?

77E. Consider two satellites, A and B, of equal mass m, moving in the same circular orbit of radius r around the Earth, of mass M_E, but in opposite senses of rotation and therefore on a collision course (see Fig. 15-44). (a) In terms of G, M_E, m, and r, find the total mechanical energy $E_A + E_B$ of the two-satellite-plus-Earth system before collision. (b) If the collision is completely inelastic so that wreckage remains as one piece of tangled material (mass = $2m$), find the total mechanical energy immediately after collision. (c) Describe the subsequent motion of the wreckage.

FIGURE 15-44 Exercise 77.

78P. Two Earth satellites, A and B, each of mass m, are to be launched into circular orbits about the Earth's center. Satellite A is to orbit at an altitude of 4000 mi. Satellite B is to orbit at an altitude of 12,000 mi. The radius of the Earth R_E is 4000 mi. (a) What is the ratio of the potential energy of satellite B to that of satellite A, in orbit? (b) What is the ratio of the kinetic energy of satellite B to that of satellite A, in orbit? (c) Which satellite has the greater total energy if each has a mass of 1.0 slug? By how much?

79P. Use the conservation of mechanical energy and Eq.

15-47 to show that if an object is in an elliptical orbit about a planet, then its distance r from the planet and speed v are related by

$$v^2 = GM \left(\frac{2}{r} - \frac{1}{a} \right).$$

80P. Use the result of Problem 79 and data contained in Sample Problem 15-10 to calculate (a) the speed v_p of comet Halley at perihelion and (b) its speed v_a at aphelion. (c) Using conservation of angular momentum relative to the sun, find the ratio of the comet's perihelion distance R_p to its aphelion distance R_a in terms of v_p and v_a.

81P. (a) Does it take more energy to get a satellite up to 1000 mi above the Earth than to put it in circular orbit once it is there? (Take the Earth's radius to be 4000 mi.) (b) What about 2000 mi? (c) What about 3000 mi?

82P. One way to attack a satellite in Earth orbit is to launch a swarm of pellets in the same orbit as the satellite but in the opposite direction. Consider a satellite in a circular orbit whose altitude above the Earth's surface is 500 km and which collides with a pellet having mass 4.0 g. (a) What is the kinetic energy of the pellet in the reference frame of the satellite? (b) What is the ratio of this

kinetic energy to the kinetic energy of a 4.0-g bullet from a modern army rifle with a muzzle velocity of 950 m/s?

83P. Consider a satellite in a circular orbit about the Earth. State how the following properties of the satellite depend on the radius r of its orbit: (a) period, (b) kinetic energy, (c) angular momentum, and (d) speed.

84P. What are (a) the speed and (b) the period of a 220-kg satellite in an approximately circular orbit 640 km above the surface of the Earth? Suppose the satellite loses mechanical energy at the average rate of 1.4×10^5 J per orbital revolution. Adopting the reasonable approximation that the trajectory is a "circle of slowly diminishing radius," determine the satellite's (c) altitude, (d) speed, and (e) period at the end of its 1500th revolution. (f) What is the magnitude of the average retarding force? (g) Is angular momentum around the Earth's center conserved for the satellite or the satellite–Earth system?

85P. The orbit of the Earth about the sun is *almost* circular: the closest and farthest distances are 1.47×10^8 km and 1.52×10^8 km, respectively. Determine the corresponding variations in (a) total energy, (b) potential energy, (c) kinetic energy, and (d) orbital speed. (*Hint:* Use conservation of energy and angular momentum.)

ADDITIONAL PROBLEMS

86. A projectile is launched from the surface of a planet with mass M and radius R; the launch speed is $(GM/R)^{1/2}$. Use the principle of conservation of energy to determine the maximum distance from the center of the planet achieved by the projectile. Express your result in terms of R.

87. Zero, a hypothetical planet, has a mass of 5.0×10^{23} kg, a radius of 3.0×10^6 m, and no atmosphere. A 10-kg space probe is to be launched vertically from its surface. (a) If the probe is launched with an initial energy of 5.0×10^7 J, what will be its kinetic energy when it is 4.0×10^6 m from the center of Zero? (b) If the probe is to achieve a maximum distance of 8.0×10^6 m from the center of Zero, with what initial kinetic energy must it be launched from the surface of Zero?

88. A satellite circles a planet of unknown mass in a circular orbit of radius 2.0×10^7 m. The magnitude of the gravitational force exerted on the satellite by the planet is 80 N. (a) What is the kinetic energy of the satellite in this orbit? (b) What would be the magnitude of the gravitational force exerted on the satellite by the planet if the radius of the orbit were increased to 3.0×10^7 m?

89. Three 5.0-kg masses are located in the xy plane as shown in Fig. 15-45. What is the magnitude of the net gravitational force on the mass at the origin due to the other two masses?

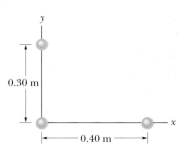

FIGURE 15-45 Problem 89.

90. A 20-kg mass is located at the origin, and a 10-kg mass is located on the x axis at $x = 0.80$ m. The 10-kg mass is released from rest while the 20-kg mass is held in place at the origin. (a) What is the gravitational potential energy of the two-mass system immediately after the 10-kg mass is released? (b) What is the kinetic energy of the 10-kg mass after it has moved 0.20 m toward the 20-kg mass?

91. A sphere of uniform density has a mass of 1.0×10^4 kg and a radius of 1.0 m. What is the gravitational force due to the sphere on a particle of mass m at a distance of (a) 1.5 m and (b) 0.50 m from the center of the sphere? (c) Write a general expression for the gravitational force on m at a distance $r \leq 1.0$ m from the center of the sphere.

PHYSICS IN WEIGHTLESSNESS

Sally Ride
University of California, San Diego

Feeling Gravity's Pull

Newspaper stories describing astronauts "away from the pull of Earth's gravity" prompted a high school teacher to complain that "spaceflight sure makes it hard on a science teacher." Her point was that terms like "zero-*g*" are misleading and that many people interpret that to mean "no gravity." Of course, the space shuttle is *not* free of the shackles of gravity. In fact, it is gravity that keeps the spacecraft, and everything inside it, in orbit around Earth. The confusion arises because space shuttle astronauts *are* "weightless"—they would float above any scales attached to the floor. They aren't weightless because they are "away from gravity," but because the shuttle and everything in it (astronauts and scales included) are in free fall. An astronaut could no more stand on scales in the space shuttle than an Earth-bound scientist could in the classic (though fortunately rare) freely falling elevator.

An orbiting spacecraft "falls" in that it falls away from the straight line it *would* follow into interplanetary space if there were no forces acting on it. It doesn't come crashing to Earth because it has sufficient horizontal velocity to travel "over the horizon." As it falls, the surface of the Earth curves away from it. It is theoretically possible to put a satellite (or a rock) into an orbit only meters above the Earth, but its energy would quickly be dissipated by air resistance (and also probably by buildings and hills!). To stay in orbit for more than just a few revolutions, a spacecraft has to be given enough energy to get it into an orbit above most of the Earth's atmosphere.

The space shuttle is boosted into orbit by the thrust from two solid rockets and three liquid-fueled engines; see Fig. 1. The solid rockets burn for the first 2 min, the launch engines for the first $8\frac{1}{2}$ min. This is enough integrated thrust or impulse

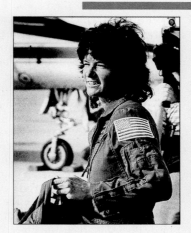

Sally K. Ride is a NASA space shuttle astronaut. She earned a B.S. in physics and a B.A. in English from Stanford University in 1973, and a Ph.D. in physics from Stanford in 1978. After graduate school, she was selected for the Astronaut Corps. She has flown in space twice: on the seventh space shuttle mission (STS-7, the second flight of the <u>Challenger</u>, launched in June 1983), and the thirteenth shuttle mission (STS-41G, launched in October 1984). In 1986, she was appointed to the Presidential Commission investigating the space shuttle <u>Challenger</u> accident. Since the completion of the investigation, she has acted as Special Assistant to the Administrator of NASA, helping to develop NASA's long-range plans for human exploration of space. She is currently a professor of physics and director of the California Space Institute at the University of California, San Diego.

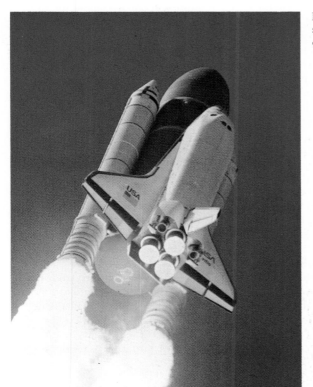

FIGURE 1 The space shuttle is boosted into orbit by five rockets.

to put the space shuttle into an elliptical low-Earth orbit (the precise orbit varies from flight to flight). At the apogee of that orbit (half a world away from the launchpad), the shuttle's small orbital engines are burned for a couple of minutes to add enough energy to circularize the orbit. These orbital engines shut off when the correct velocity is achieved. No engines are required to keep the shuttle in orbit. Gravity takes care of that. In a typical circular orbit 400 km above the surface of the Earth, the space shuttle has a velocity of 8 km/s and takes only 90 min to circle the Earth.

Once the space shuttle and everything in it have been given the velocity necessary to orbit the Earth, gravity does not accelerate objects toward the "floor" of the shuttle; it accelerates both the objects *and* the floor. All the objects inside are in the same orbit—all falling around the Earth together. That they fall together is a result of the equivalence principle, first demonstrated by Galileo. He showed that (neglecting air resistance) if a heavy object and a light object are dropped from the same height, they will hit the ground at the same time. This has been verified many times—often more precisely, but never more dramatically, than by astronaut Dave Scott on *Apollo 15*. He brought a hammer and a feather to the surface of the moon (which has no atmosphere, so no air resistance), stood outside the lunar module, held them at spacesuited arm's length, and let them fall. They hit the lunar surface together. The equivalence principle is demonstrated on every shuttle flight: since all things inside—astronauts, pencils, satellites, notebooks, extra socks—fall at the same rate, they don't develop motion relative to each other. They "float."

Physiological Effects of Gravity
There are also physiological effects associated with weightlessness. The human body evolved on Earth; it undergoes changes when it finds itself in this new environment. Perhaps the

most visible change is that astronauts' faces become puffy. On Earth, gravity is pulling the fluid in the body toward the feet. In orbit, the equilibrium distribution of fluid is different, and it tends to shift toward the upper body.

Another interesting effect is that astronauts grow about an inch in height while in orbit. Since there is no downward force on the spine, the spongy discs in the spinal column are no longer compressed. As the discs relax, astronauts "grow." The effect isn't permanent and astronauts shrink back to their "normal" height on return to Earth.

In a weightless environment, the cardiovascular system doesn't need to work very hard to pump blood around the body. It's easier to get the blood back up from the legs, or up to the brain, and the cardiovascular muscles become deconditioned. This isn't a problem as long as astronauts are in orbit, but when they return to Earth the cardiovascular system will once again be called on to pump blood against gravity and must be in condition to do so. This isn't a major problem if an astronaut is in orbit only for a week or so, but it is an important consideration for extended flights in a space station, and considerable research is required before astronauts can be sent off to Mars.

There is no "preferred direction" in weightlessness, no "upside down" or "right-side up." Physiologically, there's no way to distinguish up from down. The sensors that contribute to our balance and help us determine our orientation are located in the inner ear and are sensitive to gravity on Earth. When the head tilts, thin hairlike structures bend and send signals to the brain that the head isn't upright. In weightlessness, these sensors don't register differences in orientation, and there are no other familiar physiological indicators (such as fluid rushing to the head) to provide clues to the brain about the body's orientation. Astronauts feel the same whether their feet are pointed toward the Earth or toward the stars.

Adapting to Outer Space
Astronauts have to adapt to an environment that can't be simulated on Earth. Things in weightlessness *seem* to be subject to a different set of physical laws. The laws, of course, are the same, but sometimes the implications of those laws are much more apparent. For example, on Earth, frictional effects make it difficult to study Newton's laws of motion. Friction is hard to avoid because of gravity. Gravity holds things in contact with the ground, or table, or floor. When gravity no longer holds things in contact with each other, it's easy to avoid frictional effects. Weightlessness is a great improvement on an air table! In fact, an astronaut may seem to be an involuntary part of an elementary physics lab. Newton's laws of motion become very real . . . and this takes some getting used to. An astronaut put at rest in the middle of the cabin, unable to reach the floor, the ceiling, or any of the walls, will remain at rest —stranded in the middle of the cabin —until a friend comes along to supply an external force. A peanut set in motion will remain in motion until it hits a wall, a ceiling, or somebody's mouth. And a sharp tap on the shoulder can give a sufficient impulse to send an astronaut drifting across the room.

Human beings have learned to deal with the law of action and reaction on Earth, where they are anchored to the ground. Anyone who pulls open a drawer unconsciously reacts against the floor. When an unanchored astronaut pulls on a drawer the result is frustrating, but predictable; see Fig. 2. The drawer doesn't open, but the astronaut moves toward the drawer. And if that unanchored astronaut uses a screwdriver to apply a torque to a screw, the result will be a spinning astronaut, not a turning screw.

The effects of surface tension are very much in evidence on Earth: soap bubbles form, water drops hang on leaky faucets, and a meniscus forms on columns of water that rise up in glass tubes. Surface tension is a result

FIGURE 2   Even simple operations require an astronaut to be anchored.

of intermolecular forces. The molecules in a liquid feel some attraction for each other, so those molecules on the surface see a small net force holding them in. Similarly, if a liquid is in contact with a solid, the molecules in the liquid will feel some small attraction to those on the surface of the solid.

Surface tension tends to minimize the surface-to-volume ratio of a liquid. This is evident in weightlessness, where liquids do indeed coalesce into spheres. On Earth this isn't as obvious: spilled milk lies in a puddle on the floor; in weightlessness, the same milk doesn't splatter on the floor but forms a sphere floating in the middle of the room.

The residual forces between molecules at the interface of a solid and a liquid can cause the liquid to "cling" to the solid. It's because of surface tension that near-normal (that is, earthlike) dining is possible in space. Astronauts eat out of open cartons and use spoons to get the food to their mouths. The trick, of course, is to have "sticky" foods. Most of the food is dehydrated, and vacuum-packed in plastic cartons with thin plastic tops. It's rehydrated by poking the needle of a water gun through the plastic top, and injecting water. All these foods (e.g., macaroni and cheese, shrimp cocktail, tomato soup) are at least partly liquid once they are rehydrated, and surface tension will hold them inside the container—or on a spoon. Astronauts can snip open a container of soup and eat it with a spoon. The convenient difference is that if the spoon is tilted (or "dropped," or spun), the soup stays on it.

Surface tension keeps rehydrated food on spoons, but it also helps drinks escape from their containers. If a plastic straw is used to drink from a carton, the molecules of the liquid will feel some attraction to molecules

Astronauts at work in an orbiting space shuttle. Every action must be thought out and not taken for granted, owing to the continuous free fall of astronauts and shuttle.

of the straw, including those just above the surface of the liquid. The attraction is just enough that, in weightlessness, the liquid will crawl up a thin straw and collect in a large drop—a sphere—at the opening of the straw. Space shuttle straws all come with small clamps to pinch them closed and keep the drinks from climbing out.

In orbit, a column of liquid has no weight; there is no hydrostatic pressure and therefore no buoyant effects and no sedimentation. A cork would not bob, a bubble would not rise to the surface of a liquid (which means that dissolved gas stays in carbonated beverages, so they aren't very good to drink), and there would be no layer of chocolate at the bottom of a glass of chocolate milk.

This same principle—that denser material sinks and less dense material rises—also produces heat convection. On Earth, convection occurs when one part of a liquid or gas is heated or cooled. A heated blob expands, becomes less dense, and therefore (on Earth) rises; a cooled blob becomes more dense and falls. Convection doesn't occur in weightlessness. Again, since a column of (for example) air has no weight, hot air doesn't rise: it expands when heated but stays where it is.

An interesting demonstration (that for obvious safety reasons hasn't been performed yet) would be to light a match or candle in an orbiting spacecraft. As oxygen in the air is burned and depleted around a flame, the warmed gas rises and cooler air moves in to replace it and to deliver more oxygen to be consumed. Without convection, a candle would quickly burn itself out.

A weightless world is different from the world with which we are familiar. Some common physical effects are absent, while others are glaringly apparent. As you can imagine, it is an unusual living environment. It's also a unique laboratory environment: one that offers the opportunity to perform fundamental experiments in physics, chemistry, and physiology under new laboratory conditions.

FLUIDS 16

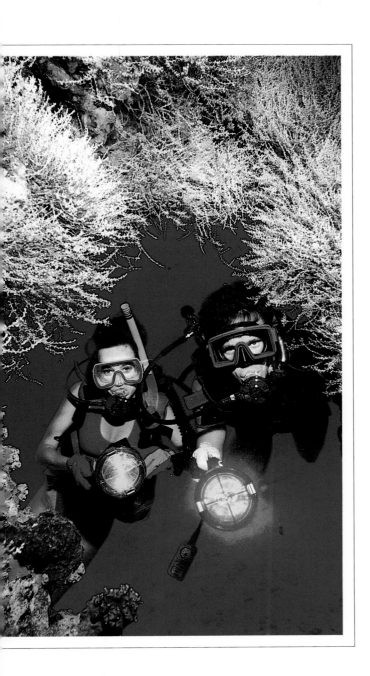

As a diver descends, the force exerted by the water on the diver's body increases noticeably, even for a relatively shallow descent to the bottom of a swimming pool. However, in 1975, using scuba gear with a special gas mixture for breathing, William Rhodes emerged from a chamber that had been lowered 1000 ft into the Gulf of Mexico, and he then swam to a record depth of 1148 ft.   Strangely, a novice scuba diver practicing in a swimming pool might be in more danger from the force exerted by the water on the body than was Rhodes.   And occasionally, novice scuba divers die when they neglect that danger.   What is this potentially lethal risk?

16-1 FLUIDS AND THE WORLD AROUND US

Fluids—which include both liquids and gases—play a central role in our daily lives. We breathe and drink them, and a rather vital fluid circulates in the human cardiovascular system. There is the fluid ocean, the fluid atmosphere, and—deep within the Earth—its fluid core.

In a car, there are fluids in the tires, the gas tank, the radiator, the combustion chambers of the engine, the exhaust manifold, the battery, the air conditioning system, the windshield wiper reservoir, the lubrication system, and the hydraulic system. (*Hydraulic* means operated via a liquid.) The next time you see a large piece of earth-moving machinery, count the hydraulic cylinders that permit the machine to do its work. Large jet planes have scores of them.

We use the kinetic energy of a moving fluid in windmills, and the potential energy of another fluid in hydroelectric power plants. Given time, fluids carve the landscape. We often travel great distances just to watch fluids move. Perhaps it is time to see what physics can tell us about fluids.

16-2 WHAT IS A FLUID?

A **fluid,** in contrast to a solid, is a substance that can flow. Fluids conform to the boundaries of any container in which we put them. They do so because a fluid cannot sustain a force that is tangential to its surface. (In the more formal language of Section 13-6, a fluid is a substance that cannot support a shearing stress. It can, however, exert a force in the direction perpendicular to its surface.) Some materials, such as pitch, take a long time to conform to the boundaries of a container, but they do so eventually; thus we classify them as fluids.

You may wonder why we lump liquids and gases together and call them fluids. After all (you may say), liquid water is as different from steam as it is from ice. Actually, it is not. Ice, like other crystalline solids, has its constituent atoms organized in a fairly rigid three-dimensional array called a crystalline lattice. In neither steam nor liquid water, however, is there any such orderly long-range arrangement; instead, intermolecular interactions are restricted to neighboring molecules.

16-3 DENSITY AND PRESSURE

When we discussed rigid bodies, we were concerned with particular lumps of matter, such as wooden blocks, baseballs, or metal rods. Physical quantities that we found useful, and in whose terms we expressed Newton's laws, were *mass* and *force*. We typically spoke, for example, of a 3.6-kg block acted on by a 25-N force.

With fluids, we are more interested in properties that vary from point to point in the extended substance than with properties of specific lumps of that substance. It is more useful to speak of **density** and **pressure** than of mass and force.

Density

To find the density ρ of a fluid at any point, we isolate a small volume element ΔV around that point and measure the mass Δm of the fluid contained within that element. The **density** is then

TABLE 16-1
SOME DENSITIES

MATERIAL OR OBJECT		DENSITY (kg/m³)
Interstellar space		10^{-20}
Best laboratory vacuum		10^{-17}
Air: 20°C and 1 atm		1.21
20°C and 50 atm		60.5
Styrofoam		1×10^2
Water: 20°C and 1 atm		0.998×10^3
20°C and 50 atm		1.000×10^3
Seawater: 20°C and 1 atm		1.024×10^3
Whole blood		1.060×10^3
Ice		0.917×10^3
Iron		7.9×10^3
Mercury		13.6×10^3
The Earth: average		5.5×10^3
	core	9.5×10^3
	crust	2.8×10^3
The sun: average		1.4×10^3
	core	1.6×10^5
White dwarf star (core)		10^{10}
Uranium nucleus		3×10^{17}
Neutron star (core)		10^{18}
Black hole (1 solar mass)		10^{19}

$$\rho = \frac{\Delta m}{\Delta V}. \qquad (16\text{-}1)$$

In theory, the density at any point in a fluid is the limit of this ratio as the volume element ΔV at that point is made smaller and smaller. In practice, we assume that a fluid sample is large compared to atomic dimensions and thus "smooth" (with uniform density), rather than "lumpy" with atoms. This assumption allows us to write Eq. 16-1 in the form $\rho = m/V$, where m and V are the mass and volume of the sample.

Density is a scalar property; its SI unit is the kilogram per cubic meter. Table 16-1 shows the densities of some substances and the average densities of some objects. Note that the density of gases (see Air in the table) varies considerably with pressure but that for liquids (see Water) does not. That is, gases are readily *compressible;* liquids are not.

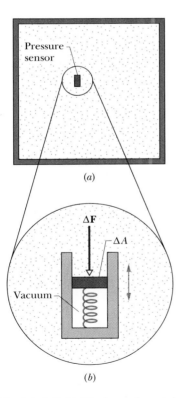

(a)

(b)

FIGURE 16-1 (a) A fluid-filled vessel containing a small pressure sensor, the details of which are shown in (b). The pressure is measured by the relative position of the piston in the sensor. At a given location, the pressure is independent of the orientation of the sensor.

Pressure

Let a small pressure-sensing device be suspended inside a fluid-filled vessel, as in Fig. 16-1a. The sensor (see Fig. 16-1b) consists of a piston of area ΔA riding in a close-fitting cylinder and resting against a spring. A readout arrangement allows us to record the amount by which the (calibrated) spring is compressed and thus the magnitude ΔF of the force that acts on the piston. We define the **pressure** exerted by the fluid on the piston as

$$p = \frac{\Delta F}{\Delta A}. \qquad (16\text{-}2)$$

In theory, the pressure at any point in the fluid is the limit of this ratio as the area ΔA of the piston, centered on that point, is made smaller and smaller. However, if the force is uniform over a flat area A, we can write Eq. 16-2 as $p = F/A$.

We find by experiment that the pressure p defined by Eq. 16-2 has the same value at a given point in a fluid at rest, no matter how the pressure sensor is oriented. Pressure is a scalar, having no directional properties. It is true that the force acting on the piston of our pressure sensor is a vector, but Eq. 16-2 involves only the *magnitude* of that force, a scalar quantity.

TABLE 16-2
SOME PRESSURES

	PRESSURE (Pa)
Center of the sun	2×10^{16}
Center of the Earth	4×10^{11}
Highest sustained laboratory pressure	1.5×10^{10}
Deepest ocean trench (bottom)	1.1×10^{8}
Spike heels on a dance floor	1×10^{6}
Automobile tire[a]	2×10^{5}
Atmosphere at sea level	1.0×10^{5}
Normal blood pressure[a,b]	1.6×10^{4}
Loudest tolerable sound[a,c]	30
Faintest detectable sound[a,c]	3×10^{-5}
Best laboratory vacuum	10^{-12}

[a] Pressure in excess of atmospheric pressure.

[b] The systolic pressure, corresponding to 120 torr on the physician's pressure gauge.

[c] Pressure at the ear drum, at 1000 Hz.

The SI unit of pressure is the newton per square meter, which is given a special name, the **pascal** (Pa). In metric countries, tire pressure gauges are calibrated in kilopascals. The pascal is related to some other common (non-SI) pressure units as follows:

$$1 \text{ atm} = 1.01 \times 10^5 \text{ Pa} = 760 \text{ torr} = 14.7 \text{ lb/in.}^2.$$

The *atmosphere* (atm) is, as the name suggests, the approximate average pressure of the atmosphere at sea level. The *torr* (named for Evangelista Torricelli, who invented the mercury barometer in 1674) was formerly called the *millimeter of mercury* (mm Hg). The pound per square inch is often abbreviated psi. Table 16-2 shows some pressures.

SAMPLE PROBLEM 16-1

A living room has floor dimensions of 3.5 m and 4.2 m and height of 2.4 m.

a. What does the air in the room weigh?

SOLUTION We have, where V is the volume of the room and ρ is the density of air at 1 atm (see Table 16-1),

$$W = mg = \rho V g$$
$$= (1.21 \text{ kg/m}^3)(3.5 \text{ m} \times 4.2 \text{ m} \times 2.4 \text{ m})(9.8 \text{ m/s}^2)$$
$$= 418 \text{ N} \approx 420 \text{ N.} \qquad \text{(Answer)}$$

This is about 94 lb. Would you have guessed that the air in a room could weigh so much?

b. What force does the atmosphere exert on the floor of the room?

SOLUTION The force is

$$F = pA = (1.0 \text{ atm})\left(\frac{1.01 \times 10^5 \text{ N/m}^2}{1 \text{ atm}}\right)$$
$$\times (3.5 \text{ m} \times 4.2 \text{ m})$$
$$= 1.5 \times 10^6 \text{ N.} \qquad \text{(Answer)}$$

This force ($\approx$ 170 tons) is the weight of a column of air covering the floor and extending all the way to the top of the atmosphere. It is equal to the force that would be exerted on the floor if (in the absence of the atmosphere) the room were filled with mercury to a depth of 30 in. Why doesn't this enormous force break the floor?

16-4 FLUIDS AT REST

Figure 16-2a shows a tank of water—or other liquid—open to the atmosphere. As every diver knows, the pressure *increases* with depth below the air–water interface. The diver's depth gauge, in fact, is a pressure sensor much like that of Fig. 16-1b. As every mountaineer knows, the pressure *decreases* with altitude as one ascends into the atmosphere. The pressures encountered by the diver and the mountaineer are usually called *hydrostatic pressures*, because they are due to fluids that are static (at rest).

Let us look first at the increase in pressure with depth below a water surface. We set up a vertical y axis, its origin being at the air–water interface and the direction of increasing y being up. Consider a water sample contained in a hypothetical right circular cylinder of base area A, and let y_1 and y_2 (both of which are *negative* numbers) be the depths of the upper and the lower cylinder faces, respectively, below the surface.

Figure 16-2b shows a free-body diagram for the water in the cylinder. The water sample is in equilibrium, its weight (downward) being exactly balanced by the difference between the force $F_2 = p_2 A$ acting upward on its lower face and the force $F_1 = p_1 A$ acting downward on its upper face. Thus

$$F_2 = F_1 + W. \qquad (16\text{-}3)$$

The volume V of the cylinder is $A(y_1 - y_2)$. Thus the mass m of the water in the cylinder is $\rho A(y_1 - y_2)$, in which ρ is the density of water. The weight W is then

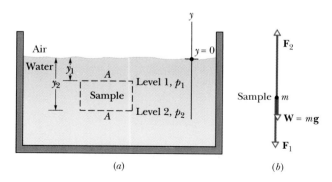

(a)   (b)

FIGURE 16-2   (a) A sample of water is contained in a hypothetical cylinder of base area A. (b) A free-body diagram of the water sample. The water in the sample is in static equilibrium, its weight being balanced by the net upward buoyant force that acts on it; see Eq. 16-3.

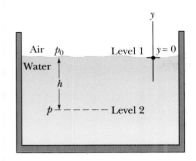

FIGURE 16-3 The pressure p increases with depth h below the water surface; see Eq. 16-5.

$\rho Ag(y_1 - y_2)$. Substituting this for W in Eq. 16-3 yields

$$p_2 A = p_1 A + \rho Ag(y_1 - y_2),$$

or

$$p_2 = p_1 + \rho g(y_1 - y_2). \qquad (16\text{-}4)$$

If we seek the pressure p at a depth h below the surface, we choose level 1 to be the surface and level 2 to be a distance h below it, as in Fig. 16-3. Representing the atmospheric pressure by p_0, we then substitute

$$y_1 = 0, \quad p_1 = p_0 \quad \text{and} \quad y_2 = -h, \quad p_2 = p$$

into Eq. 16-4, which becomes

$$p = p_0 + \rho g h \qquad \text{(pressure at depth } h). \qquad (16\text{-}5)$$

As we expect, Eq. 16-5 reduces to $p = p_0$ for $h = 0$. As you shall see in Section 16-5, the term $\rho g h$ in Eq. 16-5 is called the **gauge pressure;** it is the difference between the pressure p and the atmospheric pressure p_0.

The pressure at a given depth depends on that depth but not on any horizontal dimension. Thus the pressure on a dam at its base depends on the depth of the water there and is quite independent of the amount of water backed up behind the dam or the depth of the water elsewhere. Lake Mead, for example, extends for many miles behind Hoover Dam and is 700 ft deep at the dam face. The dam would have to be built every bit as strong to hold back just a few thousand gallons of water in, say, a pond of the same depth!

Equation 16-5 holds no matter what the shape of the containing vessel. Consider Fig. 16-4a, which shows a tube of irregular shape immersed in a tank of water. The stopcock connected to the tube is open so that the tube can fill freely. You would not doubt that Eq. 16-5 correctly gives the pressure at all points at any horizontal level such as AA, no matter whether those points are in the tube or in the tank.

Now close the stopcock, an action that causes no pressure changes at any point. With the stopcock closed, lower the tank, leaving the water-filled tube in place as in Fig. 16-4b. Again, this action causes no pressure changes for the water within the tube, which was removed from contact with the water in the tank when the stopcock was closed. In particular, the pressure of the water in the tube at level AA in Fig. 16-4b retains its value, namely, that given by Eq.

FIGURE 16-4 (a) A tube is immersed in a tank of water, the stopcock being open. The pressure at level AA is given by Eq. 16-5. (b) The stopcock is closed and the tank is removed. The pressure at AA is still given by Eq. 16-5.

16-5. All that counts is the vertical distance h below the free surface.

Pressure decreases as one moves vertically upward in the atmosphere above the liquid surface in Fig. 16-3. Equation 16-5 (with h negative, according to the sign convention of Fig. 16-2) continues to hold. However, because the densities of gases are typically three orders of magnitude less than those of liquids, the magnitude of the pressure change for a given change in elevation in a gas is correspondingly less than in a liquid. Nevertheless, with a pocket aneroid barometer of good quality, you can detect the atmospheric pressure difference (about 0.1%) encountered in climbing a few flights of stairs. Atmospheric pressure drops to half its sea-level value at an altitude of about 18,000 ft.

SAMPLE PROBLEM 16-2

a. An enterprising diver reasons that if a typical 20-cm-long snorkel tube works, a 6.0-m-long tube should also. If he foolishly uses such a tube (Fig. 16-5), what is the pressure difference Δp between the external pressure on him and the air pressure in his lungs? Why is he in danger?

SOLUTION First consider the diver at depth $L = 6.0$ m without the snorkel tube. The external pressure on him is given by Eq. 16-5 as

$$p = p_0 + \rho g L.$$

His body adjusts to that pressure by contracting slightly until the internal pressures are in equilibrium with the external pressure. In particular, the average blood pressure increases, and the average air pressure in his lungs matches p.

If he then foolishly uses the 6.0-m-long tube to breathe, the pressurized air in his lungs will be expelled upward through the tube to the atmosphere, and the air pressure in his lungs will rapidly drop to atmospheric pressure p_0. Assuming he is in fresh water, the pressure difference Δp acting on him will then be

$$\Delta p = p - p_0 = \rho g L$$

$$= (1000 \text{ kg/m}^3)(9.8 \text{ m/s}^2)(6.0 \text{ m})$$

$$= 5.9 \times 10^4 \text{ Pa.} \qquad \text{(Answer)}$$

This pressure difference, about 0.6 atm, is sufficient to collapse the lungs and force the still pressurized blood into them, a process known as lung squeeze.

FIGURE 16-5 Sample Problem 16-2. DON'T TRY THIS. Because the external (water) pressure on your chest can be so much greater than the internal (air) pressure, you may not be able to expand your lungs to inhale.

b. A novice scuba diver practicing in a swimming pool takes enough air from his tank to fully expand his lungs before abandoning the tank and swimming to the surface. He ignores instructions and fails to exhale during his ascent. When he reaches the surface, the pressure difference between the external pressure on him and the air in his lungs is 70 torr. From what depth did he start? What potentially lethal danger does he face?

SOLUTION When he fills his lungs at depth L, the external pressure on him (and the air pressure within his lungs) is again given by Eq. 16-5 as

$$p = p_0 + \rho g L.$$

As he ascends, the external pressure on him decreases, until it is atmospheric pressure p_0 at the surface. His blood pressure also decreases, until it is normal. But unless he exhales, the air pressure in his lungs does not change. At the surface, the pressure difference between the air in his lungs and the air outside his body is given by

$$\Delta p = p - p_0 = \rho g L,$$

from which we find

$$L = \frac{\Delta p}{\rho g}$$

$$= \frac{70 \text{ torr}}{(1000 \text{ kg/m}^3)(9.8 \text{ m/s}^2)} \left(\frac{1.01 \times 10^5 \text{ Pa}}{760 \text{ torr}} \right)$$

$$= 0.95 \text{ m.} \qquad \text{(Answer)}$$

The pressure difference of 70 torr (about 9% of atmospheric pressure) is sufficient to rupture the diver's lungs and force air from them into the depressurized blood, which then carries the air to the heart, killing the diver. If the diver follows instructions and gradually

FIGURE 16-6   Sample Problem 16-3. The oil in the left arm stands higher than the water in the right arm because the oil is less dense than the water. Both fluid columns produce the same pressure P_{int} at the level of the interface.

exhales as he ascends, he allows the pressure in his lungs to equalize with the external pressure, and then there is no danger.

SAMPLE PROBLEM 16-3

The U-tube in Fig. 16-6 contains two liquids in static equilibrium: water of density ρ_w is in the right arm and oil of unknown density ρ_x is in the left. Measurement gives $l = 135$ mm and $d = 12.3$ mm. What is the density of the oil?

SOLUTION   If the pressure at the oil–water interface in the left arm is p_{int}, then the pressure in the right arm at the level of that interface must also be p_{int}, because (like level AA in Fig. 16-4) the left and right arms are connected by water below the level of the interface. In the right arm, the interface is a distance l below the free surface of the *water* and we have, from Eq. 16-5,

$$p_{int} = p_0 + \rho_w g l \quad \text{(right arm)}.$$

In the left arm, the interface is a distance $l + d$ below the free surface of the *oil* and we have, again from Eq. 16-5,

$$p_{int} = p_0 + \rho_x g(l + d) \quad \text{(left arm)}.$$

Equating these two expressions and solving for the unknown density yield

$$\rho_x = \rho_w \frac{l}{l + d} = (1000 \text{ kg/m}^3) \frac{135 \text{ mm}}{135 \text{ mm} + 12.3 \text{ mm}}$$

$$= 916 \text{ kg/m}^3. \qquad \text{(Answer)}$$

Note that the answer does not depend on the atmospheric pressure p_0 or the free-fall acceleration g.

16-5 MEASURING PRESSURE

The Pressure of the Atmosphere

Figure 16-7*a* shows a very basic *mercury barometer*, a device used to measure the pressure of the atmosphere. To construct it, the long glass tube is filled with mercury and inverted with its open end in a dish of mercury, as the figure shows. The space above the mercury column contains only mercury vapor, whose pressure is so small at ordinary temperatures that it can be neglected.

We can use Eq. 16-4 to find the atmospheric pressure p_0 in terms of the height h of the mercury column. We choose level 1 of Fig. 16-2 to be that of the air–mercury interface and level 2 to be that of the top of the mercury column, as labeled in Fig. 16-7*a*. We then substitute

$$y_1 = 0 \quad p_1 = p_0 \quad \text{and} \quad y_2 = h, \quad p_2 = 0$$

into Eq. 16-4, finding that

$$p_0 = \rho g h. \qquad (16\text{-}6)$$

For a given pressure, the height h of the mercury column does not depend in any way on the cross-sectional area of the vertical tube. The fanciful mercury barometer of Fig. 16-7*b* gives the same reading as that of Fig. 16-7*a*; all that counts is the vertical distance h between the mercury levels.

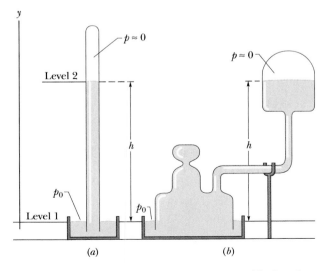

FIGURE 16-7   (*a*) A mercury barometer. (*b*) Another mercury barometer. The distance h is the same in both cases.

Equation 16-6 shows that, for a given pressure, the height of the column of mercury depends on the value of g at the location of the barometer and on the density of mercury, which varies with temperature. The column height (in millimeters) is numerically equal to the pressure (in torr) *only* if the barometer is at a place where g has its accepted standard value of 9.80665 m/s² *and* the temperature of the mercury is 0°C. If these conditions do not prevail (and they rarely do), small corrections must be made before the height of the mercury column can be transformed into a pressure.

The Open-Tube Manometer

Many times, as when we inflate the tires of an automobile or have our blood pressure measured, we do not want to know the *absolute pressure,* which is the actual or total pressure. Instead, we are interested only in the so-called *gauge pressure* p_g, which is the difference between the absolute pressure and the atmospheric pressure. In inflated tires or the human circulatory system, the (absolute) pressure is greater than atmospheric pressure, so the gauge pressure is a positive quantity, sometimes called the *overpressure.* If you suck on a straw to pull fluid up the straw, the (absolute) pressure in your lungs is actually less than atmospheric pressure. The gauge pressure in your lungs is then a negative quantity.

FIGURE 16-8 An open-tube manometer, connected so as to read the gauge pressure of the gas in the tank on the left. The right arm of the U-tube is open to the atmosphere.

An open-tube *manometer* (Fig. 16-8) measures the gauge pressure directly. It consists of a U-tube containing a liquid, with one end of the tube connected to the vessel whose gauge pressure we wish to measure and the other end open to the atmosphere. We can use Eq. 16-4 to find the gauge pressure in terms of the height h shown in Fig. 16-8. Let us choose levels 1 and 2 as shown in Fig. 16-8. We then substitute

$$y_1 = 0, \quad p_1 = p_0 \quad \text{and} \quad y_2 = -h, \quad p_2 = p$$

into Eq. 16-4, finding that

$$p_g = p - p_0 = \rho g h. \tag{16-7}$$

The gauge pressure p_g is directly proportional to h. The gauge pressure can be positive or negative, depending on whether $p > p_0$ or $p < p_0$.

SAMPLE PROBLEM 16-4

The column in a mercury barometer has a measured height h of 740.35 mm. The temperature is -5.0°C, at which temperature the density of mercury is 1.3608×10^4 kg/m³. The free-fall acceleration g at the site of the barometer is 9.7835 m/s². What is the atmospheric pressure in pascals and torr?

SOLUTION From Eq. 16-6 we have

$$p_0 = \rho g h$$

$$= (1.3608 \times 10^4 \text{ kg/m}^3)(9.7835 \text{ m/s}^2)(0.74035 \text{ m})$$

$$= 9.8566 \times 10^4 \text{ Pa}. \tag{Answer}$$

Barometer readings are usually expressed in torr, where 1 torr is the pressure exerted by a column of mercury 1 mm high at a place where g has an accepted standard value of 9.80665 m/s² and at a temperature (0.0°C) at which mercury has a density of 1.35955×10^4 kg/m³. Thus from Eq. 16-6,

$$1 \text{ torr} = (1.35955 \times 10^4 \text{ kg/m}^3)(9.80665 \text{ m/s}^2)$$

$$\times (1 \times 10^{-3} \text{ m})$$

$$= 133.326 \text{ Pa}.$$

Applying this conversion factor yields, for the atmospheric pressure recorded on the barometer,

$$p_0 = 9.8566 \times 10^4 \text{ Pa} = 739.29 \text{ torr}. \tag{Answer}$$

Note that the pressure in torr (739.29 torr) is numerically close to—but otherwise differs significantly from—the height h of the mercury column expressed in mm (740.35 mm).

16-6 PASCAL'S PRINCIPLE

When you squeeze one end of a tube of toothpaste, you are watching **Pascal's principle** in action. This principle is also the basis for the Heimlich maneuver, in which a sharp pressure increase properly applied to the abdomen is transmitted to the throat, forcefully ejecting a food particle that has become lodged there. The principle was first stated clearly in 1652 by Blaise Pascal (for whom the unit of pressure is named):

A change in the pressure applied to an enclosed fluid is transmitted undiminished to every portion of the fluid and to the walls of the containing vessel.

Demonstrating Pascal's Principle

Consider the case in which the fluid is an incompressible liquid contained in a tall cylinder, as in Fig. 16-9. The cylinder is fitted with a piston on which a container of lead shot rests. The atmosphere, container, and shot exert pressure p_{ext} on the piston and thus on the liquid. The pressure p at any point P in the liquid is then

$$p = p_{ext} + \rho g h. \tag{16-8}$$

Let us add a little more lead shot to the piston to increase p_{ext} by an amount Δp_{ext}. The quantities ρ, g, and h in Eq. 16-8 are unchanged, so the pressure change at P is

$$\Delta p = \Delta p_{ext}. \tag{16-9}$$

This pressure change is independent of h, so it must hold for all points within the liquid, as Pascal's principle states.

Pascal's Principle and the Hydraulic Lever

Figure 16-10 shows how Pascal's principle can be made the basis of a hydraulic lever. In operation, let an external force of magnitude F_i be exerted downward on the left-hand (or input) piston, whose area is A_i. An incompressible liquid in the device then exerts an upward force of magnitude F_o on the right-hand (or output) piston, whose area is A_o. To keep the system in equilibrium, an external load must exert a downward force of magnitude F_o on the output piston. The force F_i applied on the left and the force F_o exerted by the load on the right produce a change Δp in the pressure of the liquid that is given by

$$\Delta p = \frac{F_i}{A_i} = \frac{F_o}{A_o}.$$

Thus

$$F_o = F_i \frac{A_o}{A_i}. \tag{16-10}$$

FIGURE 16-9 Weights loaded onto the piston create a pressure p_{ext} at the top of the enclosed (incompressible) liquid. If p_{ext} is increased, by adding more weights, the pressure increases by the same amount at all points within the liquid.

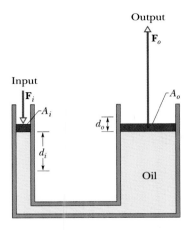

FIGURE 16-10 A hydraulic arrangement, used to magnify a force $\mathbf{F}_i$. The work done by $\mathbf{F}_i$, however, is not magnified, and is the same for both the input and output forces.

Equation 16-10 shows that the output force F_o exerted on the load must be larger than the input force F_i if $A_o > A_i$, as is the case in Fig. 16-10.

If we move the input piston downward a distance d_i, the output piston moves upward a distance d_o, such that the same volume V of the incompressible liquid is displaced at both pistons. Then

$$V = A_i d_i = A_o d_o,$$

which we can write as

$$d_o = d_i \frac{A_i}{A_o}. \qquad (16\text{-}11)$$

This shows that, if $A_o > A_i$ (as in Fig. 16-10), the output piston moves a smaller distance than the input piston moves.

From Eqs. 16-10 and 16-11 we can write the output work as

$$W = F_o d_o = \left(F_i \frac{A_o}{A_i} \right) \left(d_i \frac{A_i}{A_o} \right) = F_i d_i, \qquad (16\text{-}12)$$

which shows that the work W done *on* the input piston by the applied force is equal to the work W done *by* the output piston in lifting the load placed on it.

We see here that a given force, exerted over a given distance, can be transformed to a larger force, exerted over a smaller distance. The product of force and distance remains unchanged so that the same work is done. However, there is often tremendous advantage in being able to exert the larger force. Most of us, for example, cannot lift an automobile and welcome the availability of a hydraulic jack, even though we have to pump the handle farther than the automobile rises. In this device, the displacement d_i is accomplished not in a single stroke but over a series of small strokes.

16-7 ARCHIMEDES' PRINCIPLE

Figure 16-11 shows a student in a swimming pool, manipulating a very thin plastic sack filled with water. She will find that it is in static equilibrium, tending neither to rise nor to sink. Yet the water in the sack has weight and—for that reason—should sink. The weight must be balanced by an upward force whose magnitude is equal to the weight of the water in the sack.

This upward **buoyant force** $\mathbf{F}_b$ is exerted on water in the sack by the water that surrounds the sack. This buoyant force exists because—as you have already seen—the pressure in the water increases

FIGURE 16-11   A thin-walled plastic sack of water is in static equilibrium in the pool. Its weight must be balanced by an upward force exerted on the sack by the water surrounding it.

with depth below the surface, so the pressure near the bottom of the sack is greater than the pressure near the top.

Let us remove the sack of water. Figure 16-12a shows the forces acting at the hole in the water formerly occupied by the sack. The buoyant force, which points up, is the vector sum of all these forces.

Let us fill the hole in Fig. 16-12a with a stone of exactly the same dimensions, as in Fig. 16-12b. *The same upward buoyant force that acted on the water-filled sack will act on the stone.* However, this force is too small to balance the weight of the stone, so the stone will sink. Even though the stone sinks, the water's buoyant force reduces its apparent weight, making the stone easier to lift as long as it is under water.

If we now fill the hole of Fig. 16-12a with a block of wood of the same dimensions, as in Fig. 16-12c, the same upward buoyant force acts on the wood. This time, however, the upward force will be

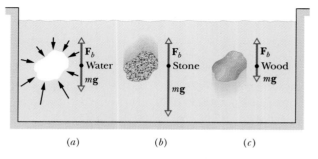

FIGURE 16-12   (*a*) The water surrounding the hole in the water exerts forces on the boundary of the hole, the resultant being an upward buoyant force acting on whatever fills the hole. (*b*) For a stone of the same volume as the hole, the weight exceeds the buoyant force. (*c*) For a lump of wood of the same volume, the weight is less than the buoyant force.

In the late evening of August 21, 1986, something (possibly a volcanic tremor) disturbed Cameroon's Lake Nyos, which has a high concentration of dissolved carbon dioxide. The disturbance caused that gas to form bubbles. Being lighter than the surrounding fluid (the water), those bubbles were buoyed to the surface, where they released the carbon dioxide. The gas, being heavier than the surrounding fluid (now the air), rushed down the mountainside like a river, asphyxiating 1700 persons and also the scores of hogs seen here.

greater than the weight of the wood, and the wood will rise toward the surface. We summarize these facts with *Archimedes' principle:*

> A body wholly or partially immersed in a fluid will be buoyed up by a force equal to the weight of the fluid that the body displaces.

Let us see how this principle can explain floating. When the piece of wood in Fig. 16-12*c* rises enough to break through the surface, it displaces less water than when it is submerged. According to Archimedes' principle, its buoyant force decreases. The wood will continue to rise out of the water until the buoyant force acting on it has decreased to exactly the weight of the wood. It is then in static equilibrium; it is floating.

The Equilibrium of Floating Objects

Occasionally, sailing vessels or warships are modified, by adding taller masts or heavier guns, in such a way that they become top heavy and tend to capsize in moderately rough seas. Icebergs often tumble as they melt. All this suggests that torques must play a role in the equilibrium of floating objects.

As we have seen, the weight of a floating object (acting downward) is exactly balanced by the buoyant force (acting upward). However, these two forces do not always act at the same point. The weight acts at the center of mass of the floating object; the buoyant force acts at the center of mass of the hole in the water, a point called the **center of buoyancy.**

If a floating body is tilted by a small angle from its equilibrium position, the shape of the hole in the water changes and so then does the location of the center of buoyancy. For such a floating object to be in stable equilibrium, its center of buoyancy must shift in such a way that the buoyant force (acting upward) and the weight (acting downward) provide a *restoring torque,* one that tends to return the body to its original upright position. If the torque acts in the opposite direction, the floating body will tilt farther and eventually tip over.

Figure 16-13 shows an unusual research vessel, the Floating Laboratory Instrument Platform (FLIP), that has *two* stable equilibrium positions. It can float "normally" on the surface or, by pumping water into its stern tanks, can "flip" to the position shown. In this position, it extends 55 ft above the surface and 300 ft below, providing a stable instrument platform for the study of ocean waves.

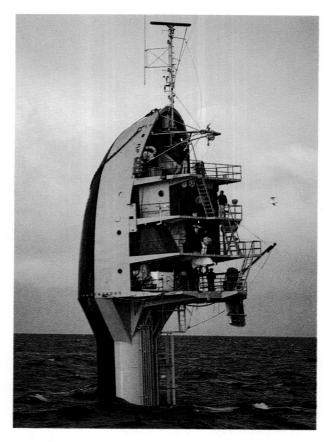

FIGURE 16-13 A research vessel (FLIP) used to study waves in deep water. It is towed to its work site floating horizontally and then pumps water into its stern tanks to flip to the position shown.

SAMPLE PROBLEM 16-5

The "tip of the iceberg" in popular speech has come to mean a small visible fraction of something that is mostly hidden. For real icebergs, what is this fraction?

SOLUTION The weight of an iceberg of total volume V_i is

$$W_i = \rho_i V_i g,$$

where $\rho_i = 917$ kg/m³ is the density of ice.

The weight of the displaced seawater, which is the buoyancy force F_b, is

$$W_w = F_b = \rho_w V_w g,$$

where $\rho_w = 1024$ kg/m³ is the density of seawater and V_w is the volume of the displaced water, that is, the sub-

---

merged volume of the iceberg. For the floating iceberg, these two forces are equal, or

$$\rho_i V_i g = \rho_w V_w g.$$

From this equation, we find that the fraction we seek is

$$\text{frac} = \frac{V_i - V_w}{V_i} = 1 - \frac{V_w}{V_i} = 1 - \frac{\rho_i}{\rho_w}$$

$$= 1 - \frac{917 \text{ kg/m}^3}{1024 \text{ kg/m}^3}$$

$$= 0.10 \text{ or } 10\%. \qquad \text{(Answer)}$$

SAMPLE PROBLEM 16-6

A spherical, helium-filled balloon has a radius R of 12.0 m. The balloon, support cables, and basket have a mass m of 196 kg. What maximum load M can the balloon carry? Take $\rho_{He} = 0.160$ kg/m³ and $\rho_{air} = 1.25$ kg/m³.

SOLUTION The weight of the displaced air, which is the buoyant force, and the weight of the helium in the balloon are

$$W_{air} = \rho_{air} V g \quad \text{and} \quad W_{He} = \rho_{He} V g,$$

in which $V\, (= 4\pi R^3/3)$ is the volume of the balloon.

At balance, from Archimedes' principle,

$$W_{air} = W_{He} + mg + Mg$$

or

$$M = (\tfrac{4}{3}\pi)(R^3)(\rho_{air} - \rho_{He}) - m$$

$$= (\tfrac{4}{3}\pi)(12.0 \text{ m})^3(1.25 \text{ kg/m}^3 - 0.160 \text{ kg/m}^3)$$

$$- 196 \text{ kg}$$

$$= 7690 \text{ kg}. \qquad \text{(Answer)}$$

A body with this mass would weigh 17,000 lb at sea level.

16-8 IDEAL FLUIDS IN MOTION

Previously, you were sometimes assigned a problem and told: "Neglect friction." That was a tacit admission that, if you had included friction, the problem would have been too difficult. That is the case here. The motion of *real* fluids is complicated and not yet fully understood. We discuss instead the motion of an **ideal fluid** that is simpler to handle mathematically. Although our results may not agree fully with the behavior of real fluids, they will be close enough to be useful. Here are four assumptions that we make about our ideal fluid:

1. *Steady flow.* In *steady* or *laminar flow* the velocity of the moving fluid at any fixed point does not change with time, either in magnitude or in direction. The gentle flow of water near the center of a quiet stream is steady; that in a chain of rapids is not. Figure 16-14, which shows the smoke rising from a cigarette, shows a transition from steady flow to *nonsteady,* or *turbulent,* flow. The speed of the smoke particles increases as they rise and, at a certain critical speed, the flow changes its character from steady to nonsteady.

2. *Incompressible flow.* We assume, as we have already done for fluids at rest, that our ideal fluid is incompressible. That is, its density has a constant value.

3. *Nonviscous flow.* Roughly speaking, the viscosity of a fluid is a measure of how resistive the fluid is to flow. For example, thick honey is more resistive to flow than water, and so honey is said to be more viscous than water. Viscosity is the analog of friction between solids. Both are mechanisms by which the kinetic energy of moving objects can be transferred

into thermal energy. In the absence of friction, a block could glide at constant speed along a horizontal surface. In the same way, an object moving through a nonviscous fluid would experience no *viscous drag force,* that is, no resistive force due to viscosity. Lord Rayleigh pointed out that, in an ideal fluid, a ship's propeller would not work but, on the other hand, a ship (once set into motion) would not need a propeller!

4. *Irrotational flow.* Although it need not concern us further, we also assume that the flow is *irrotational.* To test for this property, let a tiny grain of dust move with the fluid. Although this test body may (or may not) move in a circular path, in irrotational flow the test body will not rotate about an axis through its own center of mass. For a loose analogy, the motion of a Ferris wheel is rotational; that of its passengers is irrotational.

16-9 STREAMLINES AND THE EQUATION OF CONTINUITY

Figure 16-15 shows streamlines traced out by dye injected into a moving fluid; Fig. 16-16 shows similar streamlines revealed by smoke. A **streamline** is the path traced out by a tiny fluid element, which we may call a fluid "particle." As the fluid particle moves, its velocity may change, both in magnitude and in direction. As Fig. 16-17 shows, its velocity vector at any point will always be tangent to the streamline at that point. Streamlines never cross because, if they did, a fluid particle arriving at the intersection

FIGURE 16-14 At a certain point, the flow of heated gas rising from a cigarette changes from steady to turbulent.

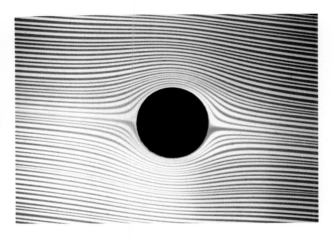

FIGURE 16-15 The steady flow of a fluid around a cylinder, as revealed by a dye tracer.

FIGURE 16-16 Smoke reveals streamlines in airflow past a car in a wind tunnel test.

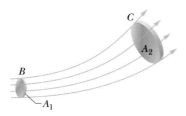

FIGURE 16-18 A tube of flow is defined by the streamlines that form its boundary. The flow rate of fluid must be the same for all cross sections of the tube of flow.

would have to assume two different velocities simultaneously, an impossibility.

In flows like that of Figs. 16-15 and 16-16, we can isolate a *tube of flow* whose boundary is made up of streamlines. Such a tube acts like a pipe because any fluid particle that enters it cannot escape through its walls; if it did, we would have a case of streamlines crossing each other.

Figure 16-18 shows two cross sections, of areas A_1 and A_2, along a thin tube of flow. Let us station ourselves at B and monitor the fluid, moving with speed v_1, for a short time interval Δt. During this interval, a fluid particle will move a small distance $v_1 \Delta t$ and a volume ΔV of fluid, given by

$$\Delta V = A_1 v_1 \Delta t,$$

will pass through area A_1.

The fluid is incompressible and cannot be created or destroyed. Thus in this same time interval, the same volume of fluid must pass point C, farther

down the tube of flow. If the speed there is v_2, this means that

$$\Delta V = A_1 v_1 \Delta t = A_2 v_2 \Delta t,$$

or,

$$A_1 v_1 = A_2 v_2.$$

Thus along the tube of flow we find

$$R = Av = \text{a constant}, \qquad (16\text{-}13)$$

in which R, whose SI unit is cubic meters per second, is the **volume flow rate.** Equation 16-13 is called the **equation of continuity** for fluid flow. It tells us that the flow is faster in the narrower parts of a tube of flow, where the streamlines are closer together, as in Fig. 16-19.

Equation 16-13 is actually an expression of the law of conservation of mass in a form useful in fluid mechanics. In fact, if we multiply R by the (constant) density of the fluid, we get the quantity $Av\rho$, which is the **mass flow rate,** whose SI unit is kilograms per second. Equation 16-13 effectively tells us that the mass that flows through point B in Fig. 16-18 each second must be equal to the mass that flows through point C each second.

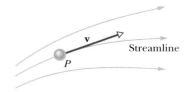

FIGURE 16-17 A fluid particle P traces out a streamline as it moves. The velocity of the particle is tangent to the streamline at every point.

FIGURE 16-19 When a channel, such as a pipe, constricts, the streamlines draw closer together, signaling an increase in the fluid velocity. The arrow shows the direction of flow.

SAMPLE PROBLEM 16-7*

The cross-sectional area A_0 of the aorta (the major blood vessel emerging from the heart) of a normal resting person is 3 cm² and the speed v_0 of the blood is 30 cm/s. A typical capillary (diameter $\approx$ 6 μm) has a cross-sectional area A of 3×10^{-7} cm² and a flow speed v of 0.05 cm/s. How many capillaries does such a person have?

SOLUTION All the blood that passes through the capillaries must have passed through the aorta so that, from Eq. 16-13,

$$A_0 v_0 = nAv,$$

where n is the number of capillaries. Solving for n yields

$$n = \frac{A_0 v_0}{Av} = \frac{(3 \text{ cm}^2)(30 \text{ cm/s})}{(3 \times 10^{-7} \text{ cm}^2)(0.05 \text{ cm/s})}$$

$$= 6 \times 10^9 \text{ or 6 billion.} \qquad \text{(Answer)}$$

You can easily show that the combined cross-sectional area of the capillaries is about 600 times the area of the aorta.

SAMPLE PROBLEM 16-8

Figure 16-20 shows how the stream of water emerging from a faucet "necks down" as it falls. The cross-sectional area A_0 is 1.2 cm², and A is 0.35 cm². The two levels are separated by a vertical distance h = 45 mm. At what rate does water flow from the tap?

SOLUTION From the equation of continuity (Eq. 16-13) we have

$$A_0 v_0 = Av, \qquad (16\text{-}14)$$

where v_0 and v are the water velocities at the corresponding levels. From Eq. 2-21 we can also write, because the water is falling freely with acceleration g,

$$v^2 = v_0^2 + 2gh. \qquad (16\text{-}15)$$

*See "Life in Moving Fluids," by Steven Vogel, Princeton University Press, 1981, for a fascinating account of the role of fluid flow in biology.

FIGURE 16-20 Sample Problem 16-8. As water falls from a tap, its speed increases. Because the flow rate must be the same at all cross sections, the stream must "neck down."

Eliminating v between Eqs. 16-14 and 16-15 and solving for v_0, we obtain

$$v_0 = \sqrt{\frac{2ghA^2}{A_0^2 - A^2}}$$

$$= \sqrt{\frac{(2)(9.8 \text{ m/s}^2)(0.045 \text{ m})(0.35 \text{ cm}^2)^2}{(1.2 \text{ cm}^2)^2 - (0.35 \text{ cm}^2)^2}}$$

$$= 0.286 \text{ m/s} = 28.6 \text{ cm/s}.$$

The volume flow rate R is then

$$R = A_0 v_0 = (1.2 \text{ cm}^2)(28.6 \text{ cm/s})$$

$$= 34 \text{ cm}^3/\text{s}. \qquad \text{(Answer)}$$

At this rate, it would take about 3 s to fill a 100-mL beaker.

16-10 BERNOULLI'S EQUATION

Figure 16-21 represents a tube of flow (or an actual pipe, for that matter) through which an ideal fluid is flowing at a steady rate. In a time interval Δt, suppose that a volume of fluid ΔV, colored dark blue in Fig. 16-21a, enters the tube at the left (or input) end and an identical volume, also colored dark blue in Fig. 16-21b, emerges at the right (or output) end. The emerging volume must be the same as the en-

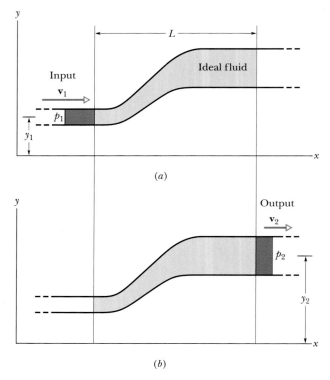

(a)

(b)

FIGURE 16-21 Fluid flows through a pipe at a steady rate. During a time interval Δt, the net effect is that the amount of fluid shown by the dark blue area in (a) is transferred from the input end to the output end, as shown in (b).

tering volume because the fluid is incompressible, with an assumed constant density ρ.

Let y_1, v_1, and p_1 be the elevation, the speed, and the pressure of the fluid as it enters at the left and y_2, v_2, and p_2 be the corresponding quantities for the fluid as it emerges at the right. By applying the law of conservation of energy to the fluid, we shall show that these quantities are related by

$$p_1 + \tfrac{1}{2}\rho v_1^2 + \rho g y_1 = p_2 + \tfrac{1}{2}\rho v_2^2 + \rho g y_2. \quad (16\text{-}16)$$

We can rewrite this as

$$p + \tfrac{1}{2}\rho v^2 + \rho g y = \text{a constant.} \quad (16\text{-}17)$$

Equations 16-16 and 16-17 are equivalent forms of **Bernoulli's equation,** after Daniel Bernoulli whose

suggestion in 1738 eventually led to this equation.* Like the equation of continuity (Eq. 16-13), Bernoulli's equation is not a new principle but simply the reformulation of a familiar principle (the conservation of mechanical energy) in a form more suitable to fluid mechanics. As a check, let us apply Bernoulli's equation to fluids at rest, by putting $v_1 = v_2 = 0$ in Eq. 16-16. The result is

$$p_2 = p_1 + \rho g(y_1 - y_2),$$

which is Eq. 16-4.

A major prediction of Bernoulli's equation emerges if we take y to be a constant ($y = 0$, say) so that the fluid does not change elevation as it flows. Equation 16-16 then becomes

$$p_1 + \tfrac{1}{2}\rho v_1^2 = p_2 + \tfrac{1}{2}\rho v_2^2, \quad (16\text{-}18)$$

which tells us that:

> If the speed of a fluid particle increases as it travels along a streamline, the pressure of the fluid must decrease, and conversely.

Put another way, where the streamlines are relatively close together (that is, where the velocity is relatively great), the pressure is relatively low, and conversely.

The link between a change in speed and a change in pressure makes sense if you consider a fluid particle. When the particle nears a narrow region, the higher pressure behind it accelerates it so that it then has a greater speed in the narrow region. And when it nears a wide region, the higher pressure ahead of it decelerates it so that it then has a lesser speed in the wide region.

This result is perhaps the opposite of what you might expect. For example, if you put your hand outside the window of a car, you sense an *increase* of pressure associated with the relative speed of the moving outside air, not a decrease. The difficulty is that in "sensing" the pressure in this way you interfere with the flow. The pressure must be measured in ways that do not interfere. If, instead, you open

---

*For irrotational flow (which we assume), the constant in Eq. 16-17 has the same value for all points within the tube of flow; the points do not have to lie along the same streamline. Similarly, the points 1 and 2 in Eq. 16-16 can lie anywhere within the tube of flow.

the car window a small amount (which will not disturb the flow of the outside air) you will note that smoke generated inside the car drifts outside, in response to the *lower* outside pressure.

Bernoulli's equation is strictly valid only to the extent that the fluid is ideal. If viscous forces are present, thermal energy will be involved. We take no account of this in the derivation that follows.

Proof of Bernoulli's Equation

Let us take as our system the entire volume of the (ideal) fluid shown in Fig. 16-21. We shall apply the law of energy conservation to this system as it moves from its initial state (Fig. 16-21*a*) to its final state (Fig. 16-21*b*). The part of the fluid lying between the two vertical planes separated by a distance L in Fig. 16-21 does not change its properties during this process; we need be concerned only with changes that take place at the input and the output ends.

We apply the law of energy conservation in the form of the work–kinetic energy theorem,

$$W = \Delta K, \qquad (16\text{-}19)$$

which tells us that the change in the kinetic energy of our system must equal the net work done on the system.

The change in kinetic energy depends on the change in speed between the ends of the pipe and is

$$\Delta K = \tfrac{1}{2}\Delta m \, v_2^2 - \tfrac{1}{2}\Delta m \, v_1^2$$
$$= \tfrac{1}{2}\rho \, \Delta V (v_2^2 - v_1^2), \qquad (16\text{-}20)$$

in which $\Delta m \; (= \rho \, \Delta V)$ is the mass of the fluid (colored dark blue in Fig. 16-21) that enters at the input end and leaves at the output end during a small time interval Δt.

The work done on the system arises from two sources. The work W_g done by the weight ($\Delta m \, \mathbf{g}$) of mass Δm during the vertical lift of the mass from input level to the output level is

$$W_g = -\Delta m \, g (y_2 - y_1)$$
$$= -\rho g \, \Delta V (y_2 - y_1). \qquad (16\text{-}21)$$

This work is negative because the upward displacement and the downward weight force point in opposite directions.

Work W_p must also be done *on* the system (at the input end) to push it through the tube and *by* the

system (at the output end) to push the fluid ahead of it forward. In general, the work done by a force of magnitude F, acting on a fluid sample contained in a tube of area A, to move the fluid through a distance Δx, is

$$F \, \Delta x = (pA)(\Delta x) = (p)(A \, \Delta x) = p \, \Delta V.$$

The work W_p is then

$$W_p = -p_2 \, \Delta V + p_1 \, \Delta V$$
$$= -(p_2 - p_1)\Delta V. \qquad (16\text{-}22)$$

The work–kinetic energy theorem of Eq. 16-19 now becomes

$$W = W_g + W_p = \Delta K.$$

Substituting from Eqs. 16-20, 16-21, and 16-22 yields

$$-\rho g \, \Delta V (y_2 - y_1) - \Delta V (p_2 - p_1) = \tfrac{1}{2}\rho \, \Delta V (v_2^2 - v_1^2).$$

This, after a slight rearrangement, matches Eq. 16-16, which we set out to prove.

SAMPLE PROBLEM 16-9

Ethanol of density $\rho = 791 \text{ kg/m}^3$ flows smoothly through a horizontal pipe that tapers (as in Fig. 16-19) in cross-sectional area from $A_1 = 1.20 \times 10^{-3} \text{ m}^2$ to $A_2 = A_1/2$. The pressure difference Δp between the wide and narrow sections of pipe is 4120 Pa. What is the volume flow rate R of the ethanol?

SOLUTION Rearranging Eq. 16-18 (Bernoulli's equation for level flow) yields

$$p_1 - p_2 = \tfrac{1}{2}\rho v_2^2 - \tfrac{1}{2}\rho v_1^2 = \tfrac{1}{2}\rho(v_2^2 - v_1^2), \qquad (16\text{-}23)$$

where subscripts 1 and 2 refer to the wide and narrow sections of pipe, respectively. Equation 16-13 (the continuity equation) tells us that in Fig. 16-19 the flow is faster in the narrower section. Here that means that $v_2 > v_1$. Equation 16-23 then tells us that $p_1 > p_2$. Thus we have $p_1 - p_2 = \Delta p = 4120$ Pa.

Equation 16-13 also tells us the volume flow rate R is the same in the wide and narrow sections. So

$$R = v_1 A_1 = v_2 A_2,$$

which, with $A_2 = A_1/2$, gives

$$v_1 = \frac{R}{A_1} \quad \text{and} \quad v_2 = \frac{R}{A_2} = \frac{2R}{A_1}.$$

Substituting these expressions into Eq. 16-23, setting $p_1 - p_2 = \Delta p$, and rearranging give us

$$\Delta p = \tfrac{1}{2}\rho \left(\frac{4R^2}{A_1^2} - \frac{R^2}{A_1^2} \right) = \frac{3\rho R^2}{2A_1^2}.$$

Solving for R, we find

$$R = A_1 \sqrt{\frac{2\Delta p}{3\rho}}$$

$$= 1.20 \times 10^{-3} \text{ m}^2 \sqrt{\frac{(2)(4120 \text{ Pa})}{(3)(791 \text{ kg/m}^3)}}$$

$$= 2.24 \times 10^{-3} \text{ m}^3/\text{s}. \qquad \text{(Answer)}$$

16-11 SOME APPLICATIONS OF BERNOULLI'S EQUATION

Popping Windows

If a high wind blows past a window, the pressure on the outside of the window is reduced and the window may break outward. This mechanism plays a role when flat roofs are blown off buildings in hurricanes; the roofs are, at least in part, pushed up by pressure from the stagnant air below. Although roofs are designed to withstand a relatively large *downward* pressure difference (due perhaps to snow loading), they are often not designed to withstand a large *upward* pressure difference.

The Venturi Meter

The Venturi meter is a device used to measure the flow speed of a fluid in a pipe. The meter is connected between two sections of the pipe, as in Fig. 16-22. The cross-sectional areas A of the entrance and exit of the meter match the pipe's cross-sectional area. Between the entrance and exit, the fluid flows through a narrow region of cross-sectional area a. A manometer connects the wider portion of the meter to the narrower portion.

As the fluid flows from the pipe with speed v and into the narrow region, its speed increases to V. According to Bernoulli's equation, this increase in speed is accompanied by a decrease in the fluid's pressure. The liquid in the manometer shifts to the right in response to the pressure difference Δp that then exists between the wider (at A) and narrower (at a) regions of the meter. The height difference h

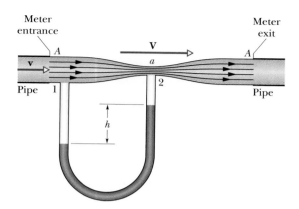

FIGURE 16-22 A Venturi meter is connected between two sections of pipe to measure the speed of the fluid flowing through the pipe.

between the surfaces of the liquid in the arms of the manometer is measured to find Δp.

Using Bernoulli's equation and the equation of continuity (Eq. 16-13), you can show that

$$v = \sqrt{\frac{2a^2 \, \Delta p}{\rho(A^2 - a^2)}}, \qquad (16\text{-}24)$$

where ρ is the density of the fluid. The volume flow rate R can be found as $R = Av$, and the device can be calibrated to read (and record) this flow rate directly.

The Hole in the Water Tank

In the old West, a desperado fires a bullet into an open water tank, creating a hole a distance h below the water surface, as Fig. 16-23 shows. How fast is the water moving as it emerges from the hole?

We take the level of the hole as our reference level for measuring elevations, and we note that the pressure at the top of the tank, and also at the hole, is atmospheric. Applying Bernoulli's equation (Eq. 16-16) gives us

$$p_0 + 0 + \rho g h = p_0 + \tfrac{1}{2}\rho v^2 + 0.$$

The 0 on the left denotes that the velocity of fluid at the top of the tank (that is, the speed with which the level is falling) is negligible. The 0 on the right reminds us that the level of the hole is our reference level for measuring gravitational potential energy. Thus we find

$$v = \sqrt{2gh}, \qquad (16\text{-}25)$$

FIGURE 16-23 The water flows from the hole with the same speed that it would have acquired had it fallen a vertical distance h.

which is the same speed that an object would acquire in falling from rest through a distance h.

An Airplane Wing

Figure 16-24 shows the streamlines about a moving airplane wing. We assume that the air approaches horizontally from the right with velocity $\mathbf{v}_{a1}$. The upward tilt of the wing, called the *attack angle,* causes a downward deflection of the airstream, which then has velocity $\mathbf{v}_{a2}$. Thus the wing exerts a force on the

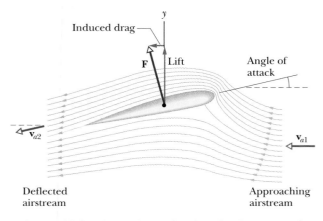

FIGURE 16-24 Streamlines showing the flow around a moving airplane wing. The air, approaching horizontally from the right, is deflected downward by the wing, which is tilted upward at an angle called the attack angle. In being deflected, the airstream exerts force $\mathbf{F}$ on the wing.

airstream to deflect it, and by Newton's third law, the airstream must exert an equal and opposite force on the wing. The vertical component of this force $\mathbf{F}$ on the wing is called *lift,* and the horizontal component is called *induced drag* (or simply *drag*).

The lift on the wing and the pattern of streamlines in Fig. 16-24 are consistent with Bernoulli's equation: the spacing of the streamlines is wider below the wing than above it, indicating that the speed of the air is smaller and the air pressure is greater below the wing than above it. The greater pressure below the wing is consistent with the existence of an upward force acting on the wing.

The lift that acts on an airplane wing (often called *dynamic lift*) must not be confused with the static or buoyant lift exerted, in accord with Archimedes' principle, on balloons or floating icebergs. Dynamic lift acts only when the object and the fluid stream are in relative motion.

16-12 THE FLOW OF "REAL" FLUIDS (OPTIONAL)

The preceding sections deal with the flow of an *ideal fluid,* its essential property being that it has zero viscosity. All real fluids are viscous, and that property has an important influence on their behavior. Let us look at some examples.

The Boundary Layer

One important effect of viscosity is that when there is relative motion between a fluid and a solid, the fluid molecules immediately next to the surface of the solid tend to stick to that surface. This tendency decreases rapidly with distance from the surface. The layer of fluid that exhibits this tendency to resist motion is called the *boundary layer.*

The existence of a boundary layer has many familiar consequences. For example, we might expect that dirt would not stick to a whirling fan blade but it does. The reason is that the air just next to the surface of the blade is not moving with respect to the blade, so that there is no mechanism for "blowing away" dirt particles. In the same way, you cannot blow all the fine dust from a tabletop; you must wipe it off. A dish cloth is far more effective in cleaning dishes than a simple rinse. Flowing water in a mountain stream is not the primary reason a stream bed is carved out of the rock. The boundary layer ensures

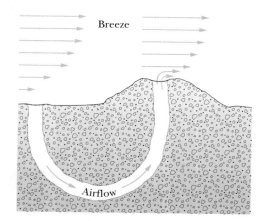

FIGURE 16-25   Prairie dogs build one entrance to a burrow through a mound, so that the entrance sticks up into air that is moving faster than the air flowing over a second entrance. Air molecules in the mound entrance are then entrained (removed) by the breeze, and air is drawn through the burrow.

that the water is mostly stagnant at its points of contact with the stream bed; it is the rock particles carried by the stream that do the major carving.

Entrainment

The prairie dog builds its burrow beneath the flat prairie floor. To ventilate its burrow, the animal builds a mound around one of the openings, as in Fig. 16-25. When a breeze blows across the prairie and over the mound, the breeze *entrains* (captures) and removes air molecules in the mounded opening. Owing to the boundary layer of the air over the flat terrain at the burrow's other (lower) opening, the breeze moves more slowly there, and entrainment is less. Because more air is removed at the mound, air is forced from the lower opening,

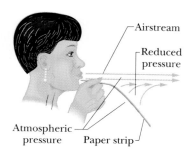

FIGURE 16-26   Entrainment and removal of air between an airstream and a paper strip.

through the burrow, and out through the mound, thus ventilating the burrow.

You can demonstrate entrainment by blowing a stream of air over a strip of paper held just below your lower lip (Fig. 16-26). The air in the stream is at atmospheric pressure (it has speed, not because its pressure has been reduced, but because your lungs gave it speed). As the air flows over the near section of the paper, it entrains and removes air molecules from between it and the paper, reducing the air pressure there. The atmospheric pressure below the paper then forces the paper to rise.

Dynamic Lift

The range of a properly hit golf ball is greatly extended because of the *dynamic lift* associated with its rotation. The boundary layer is intimately involved.

Figure 16-27*a* shows streamlines for the steady

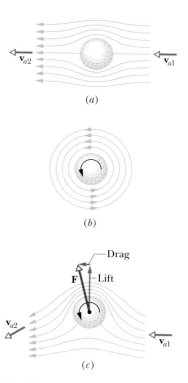

FIGURE 16-27   (*a*) Streamline flow around a (nonrotating) ball. The velocity of the air relative to the ball is initially $\mathbf{v}_{a1}$ and finally $\mathbf{v}_{a2}$. Here, $\mathbf{v}_{a2} = \mathbf{v}_{a1}$. (*b*) The circulation of air around a rotating ball due to viscosity. (*c*) The circulation of (*b*) deflects the oncoming airstream of (*a*), resulting in the stream exerting force **F** on the ball. The component of **F** parallel to $\mathbf{v}_{a1}$ is the induced drag. The component perpendicular to $\mathbf{v}_{a1}$ is the dynamic lift.

flow of air rushing past a stationary, nonrotating ball, at speeds low enough so that turbulence does not occur. Figure 16-27*b* shows streamlines for the air carried around by a stationary but rapidly rotating ball. This *circulation* (as it is called) is due to the viscosity of the air. Golf balls are systematically roughened by means of dimples in order to increase this circulation—and ·the dynamic lift that results from it.

Figure 16-27*c* shows the combination of circulation (resulting from the rotation of the ball) and steady flow (resulting from the translation of the ball through the air). The combined effect is a deflection of the oncoming airstream, similar to the deflection of the airstream around an airplane wing in Fig. 16-24. In Fig. 16-27*c*, the ball exerts a force on the airstream, and by Newton's third law, the airstream must exert an equal and opposite force **F** on the ball. The dynamic lift on the ball is the component of **F** perpendicular to the original velocity $\mathbf{v}_{a1}$ of the airstream.

The lift and the pattern of streamlines in Fig. 16-27*c* are consistent with Bernoulli's equation: the spacing of the streamlines is wider below the ball than above it, indicating that the air pressure is greater below the ball than above it. The greater pressure below the ball is consistent with there being an upward force acting on the ball.

To produce the lift of Fig. 16-27*c*, the golf club must strike the ball below the ball's center to give it "bottom spin" around a horizontal axis. If the ball is hit poorly so that the axis is not horizontal, the lift is not vertical and the result is the dreaded hook or

FIGURE 16-28 The *Baden-Baden*, known now as Flettner's ship after its inventor Anton Flettner, crossed the Atlantic in 1926, driven largely by the forces acting on the two vertical, rotating cylinders shown.

slice. A ball that is hit high and thus given "top spin" has downward lift, which greatly reduces its range.

Figure 16-27 also applies to a curve ball throw in baseball. In that situation, the rotation axis is vertical and the lift is horizontal, toward the pitcher's left or right, depending on the direction in which the ball rotates. Similar horizontal deflection is what drove the ship of Fig. 16-28 across the Atlantic. Motors rotated each of two tall cylinders around its central axis. When a wind blew past the cylinders, the airstreams were deflected, which resulted in the cylinders (and thus ship) being pushed horizontally.

REVIEW & SUMMARY

Density

The **density** of any material is defined as its mass per unit volume:

$$\rho = \frac{\Delta m}{\Delta V}. \qquad (16\text{-}1)$$

In most practical situations, where a material sample is large compared to atomic dimensions, we can write Eq. 16-1 as $\rho = m/V$.

Fluid Pressure

A **fluid** is a substance that can flow; it conforms to the boundaries of its container because it cannot sustain

shearing stress. It can, however, exert a force perpendicular to its surface. That force is described in terms of **pressure** *p*:

$$p = \frac{\Delta F}{\Delta A}, \qquad (16\text{-}2)$$

in which ΔF is the force acting on a surface element of area ΔA. If the force is uniform over a flat area, Eq. 16-2 can be written as $p = F/A$. The force resulting from fluid pressure at a particular point in a fluid has the same magnitude in all directions. **Gauge pressure** is the difference between actual pressure and atmospheric pressure.

Pressure Units

The SI unit for pressure is the **pascal** ($= 1$ N/m^2). Other units are 1 atm $= 1.01 \times 10^5$ Pa $= 760$ torr $= 14.7$ lb/in.^{2}.

Pressure Variation with Height and Depth

Pressure in a fluid at rest varies with vertical position y. For y measured positive up

$$p_2 = p_1 + \rho g(y_1 - y_2). \qquad (16\text{-}4)$$

The pressure is the same for all points at the same level. If h is the *depth* of a fluid sample below some reference level at which the pressure is p_0, Eq. 16-4 becomes

$$p = p_0 + \rho g h. \qquad (16\text{-}5)$$

Sample Problems 16-2 to 16-5 illustrate these ideas.

Pascal's Principle

Pascal's principle, which can be derived from Eq. 16-4, states that a change in the pressure applied to an enclosed fluid is transmitted undiminished to every portion of the fluid and to the walls of the containing vessel.

Archimedes' Principle

The surface of an immersed object is acted on by forces associated with the fluid pressure. The vector sum of those forces (called the **buoyant force**) acts vertically up through the center of mass of the displaced fluid (the **center of buoyancy**). Archimedes' principle states that the magnitude of the buoyant force is equal to the weight of the fluid displaced by the object. When an object floats, its weight is equal to the buoyant force acting on it.

Flow of Ideal Fluids

An *ideal fluid* is one that is incompressible and lacks viscosity, and whose flow is steady and irrotational. A **streamline** is the path followed by individual fluid particles. A *tube of flow* is a bundle of streamlines. The principle of conservation of mass shows that the flow within any tube of flow obeys the **equation of continuity:**

$$R = Av = \text{a constant}, \qquad (16\text{-}13)$$

in which R is the **volume flow rate,** A the cross-sectional area of the tube of flow at any point, and v the speed of the fluid, assumed to be constant across A. The **mass flow rate** $Av\rho$ is also constant.

Bernoulli's Equation

Applying the principle of conservation of mechanical energy to the flow of an ideal fluid leads to **Bernoulli's equation:**

$$p + \tfrac{1}{2}\rho v^2 + \rho g y = \text{a constant} \qquad (16\text{-}17)$$

along any tube of flow.

The pressure within real fluids is significantly affected by viscosity, which leads to boundary layer phenomena and dynamic lift.

QUESTIONS

1. Can you assign a coefficient of static friction between two surfaces, one of which is a fluid surface?

2. Make an estimate of the average density of your body. Explain a way in which you could get an accurate value using ideas in this chapter.

3. Explain the pressure variations in your blood as it circulates through your body.

4. Water is poured to the same level in each of the vessels shown in Fig. 16-29, all having the same base area. If the pressure is the same at the bottom of each vessel, the force experienced by the base of each vessel is the same. Why then do the three vessels have different weights when put on a scale? This apparently contradictory result is commonly known as the *hydrostatic paradox*.

FIGURE 16-29   Question 4.

5. Does Archimedes' principle hold in a vessel in free fall or in a satellite moving in a circular orbit?

6. A spherical bob made of cork floats half submerged in a pot of tea at rest on the Earth. Will the cork float or sink aboard a spaceship (a) coasting in free space and (b) on the surface of Jupiter?

7. How does a suction cup work?

8. What, if anything, happens to the buoyant force acting on a helium-filled balloon if you replace the helium with the same volume of hydrogen? (Hydrogen is less dense than helium.)

9. A block of wood floats in a pail of water in an elevator. When the elevator starts from rest and accelerates downward, does the block float higher above the water surface? What happens when the elevator accelerates upward?

10. Two identical buckets are filled to the brim with water, but one has a block of wood floating in the water. Which bucket, if either, is heavier?

11. Can you sink an iron ship by siphoning seawater into it? (See Problem 80.)

12. A beaker is exactly full of liquid water at its freezing point and has an ice cube floating in it, also at the freezing point. As the cube melts, what happens to the water level in these three cases: (a) the cube is solid ice; (b) the cube contains some grains of sand; (c) the cube contains some bubbles?

13. A ball floats on the surface of water in a container exposed to the atmosphere. Will the ball remain immersed at its former depth or will it sink or rise somewhat if (a) the container is covered and the air above the water is re-

moved and (b) the container is covered and the air is compressed?

14. Is the buoyant force acting on a submerged submarine the same at all depths? Explain why an inflated balloon will rise only to a certain height once it starts to rise, whereas a submarine will always sink to the very bottom of the ocean once it starts to sink, if no changes are made.

15. Why does a balloon weigh the same when deflated as when inflated with air at atmospheric pressure? Would the weights be the same if measured in a vacuum?

16. During World War II, a damaged freighter that was barely able to float in the salty water of the North Sea sank because it steamed up the Thames estuary toward the London docks. Why?

17. Design a lifeboat or sailboat that will right itself when inverted by rough water or strong wind. Explain the righting in terms of center of mass and center of buoyancy.

18. Why will a sinking ship often turn over as it becomes immersed in water?

19. A barge filled with scrap iron is in a closed, narrow canal lock. If the iron is thrown overboard into the water, what happens to the water level in the lock? What if it is thrown onto the land beside the canal?

20. A boat, floating in a swimming pool that is hardly wider than the boat, springs a small leak and gradually sinks until it is completely submerged. Explain what happens to the water level during the sinking.

21. A bucket of water is suspended from a spring balance. Does the balance reading change when a piece of iron suspended from string is immersed in the water? When a piece of cork is put in the water? (No water is spilled in the two situations.)

22. Why does a uniform wooden stick or log float horizontally? If enough iron is added to one end, it will float vertically. Explain this also.

23. Although there are practical difficulties, it is possible in principle to float an ocean liner in a few hundred gallons of water. Explain how.

24. Explain why a thin-walled pipe will burst more easily if, when there is a pressure differential between inside and outside, the excess pressure is on the outside.

25. Explain why the height of the liquid in the vertical standpipes in Fig. 16-30 indicates that the pressure drops

FIGURE 16-30   Question 25.

along the channel, even though the channel has a uniform cross section and even though the flowing liquid is incompressible.

26. In a lecture demonstration a Ping-Pong ball is kept in midair by a vertical jet of air. Is the equilibrium stable, unstable, or neutral? Explain.

FIGURE 16-31 Question 28.

Airflow

FIGURE 16-32 Question 30.

27. Two rowboats moving parallel to one another in the same direction tend to be pulled toward one another. Two automobiles moving parallel are also pulled together. Explain such phenomena using Bernoulli's equation.

28. Explain why you cannot remove the filter paper from the funnel of Fig. 16-31 by blowing into the narrow end.

29. Why can a discus be thrown farther *against* a 25-mi/h wind than with it?

30. On August 2, 1985, during a routine approach to the Dallas–Fort Worth airport, a Delta Airline L-1011 jet suddenly fell and crashed when it was unknowingly flown through a *microburst* (or *wind-shear*), whose flow is shown in Fig. 16-32. The crash killed 136 of the 167 people on board. The speed of the air in the microburst was 75 km/h relative to the ground. As the jet entered the microburst, its airspeed (speed relative to the air) was 300 km/h. The L-1011 jet must have an airspeed exceeding 210 km/h, or it *stalls* (loses lift). Why did the Delta jet crash?

31. Why does the shower curtain in a typical shower arrangement tend to flutter inward during a shower (Fig. 16-33)? (The flutter appears regardless of the water temperature, so circulation due to hot and cold air is not the primary answer.)

Curtain

Falling water

FIGURE 16-33 Question 31.

EXERCISES & PROBLEMS

SECTION 16-3 DENSITY AND PRESSURE

1E. Convert a density of 1.0 g/cm³ to kg/m³.

2E. Three liquids that will not mix are poured into a cylindrical container. The volumes and densities of the liquids are 0.50 L, 2.6 g/cm³; 0.25 L, 1.0 g/cm³; and 0.40 L, 0.80 g/cm³. What is the force acting on the bottom of the container due to these liquids? One liter = 1 L = 1000 cm³. (Ignore the contribution due to the atmosphere.)

3E. Find the pressure increase in the fluid in a syringe when a nurse applies a force of 42 N to the syringe's piston, of radius 1.1 cm.

4E. You inflate the front tires on your car to 28 psi. Later, you measure your blood pressure, obtaining a reading of 120/80, the readings being in mm Hg. In metric countries (which is to say, most of the rest of the world), these pressures are customarily reported in kilopascals (kPa). What are (a) your tire pressure and (b) your blood pressure in kilopascals?

5E. An office window has dimensions 3.4 m by 2.1 m. As a result of the passage of a storm, the outside air pressure drops to 0.96 atm, but inside the pressure is held at 1.0 atm. What net force pushes out on the window?

6E. A fish maintains its depth in fresh water by adjusting the air content of porous bone or air sacs to make its density the same as that of the water. Suppose that with its air sacs collapsed a fish has a density of 1.08 g/cm³. To what fraction of its expanded body volume must the fish inflate the air sacs to reduce its density to that of water?

7P. An airtight box having a lid with an area of 12 in.² is partially evacuated. If a force of 108 lb is required to pull the lid off the box and the outside atmospheric pressure is 15 lb/in.², what is the air pressure in the box?

8P. In 1654 Otto von Guericke, burgomaster of Magdeburg and inventor of the air pump, gave a demonstration before the Imperial Diet in which two teams of eight horses could not pull apart two evacuated brass hemispheres. (a) Assuming that the hemispheres have thin walls, so that R in Fig. 16-34 may be considered both the inside and outside radius, show that the force F required to pull apart the hemispheres is $F = \pi R^2 \, \Delta p$, where Δp is the difference between the pressures outside and inside the sphere. (b) Taking R equal to 1.0 ft and the inside pressure as 0.10 atm, find the force the teams of horses would have had to exert to pull apart the hemispheres. (c) Why are two teams of horses used? Would not one team prove the point just as well?

FIGURE 16-34 Problem 8.

SECTION 16-4 FLUIDS AT REST

9E. Calculate the hydrostatic difference in blood pressure between the brain and the foot in a person of height 1.83 m. The density of blood is 1.06×10^3 kg/m³.

10E. Find the pressure, in pascals, 150 m below the surface of the ocean. The density of seawater is 1.03 g/cm³ and the atmospheric pressure at sea level is 1.01×10^5 Pa.

11E. The sewer outlets of a house constructed on a slope are 8.2 m below street level. If the sewer is 2.1 m below street level, find the minimum pressure differential that

must be created by the sewage pump to transfer waste of average density 900 kg/m³.

12E. Figure 16-35 displays the *phase diagram* of carbon, showing the ranges of temperature and pressure in which carbon will crystallize either as diamond or graphite. What is the minimum depth at which diamonds can form if the local temperature is 1000°C and the subsurface rocks have density 3.1 g/cm³? Assume that, as in a fluid, the pressure is due to the weight of material lying above.

FIGURE 16-35 Exercise 12.

13E. The human lungs can operate against a pressure differential of up to about one-twentieth of an atmosphere. If a diver uses a snorkel for breathing, about how far below water level can she or he swim?

14E. A swimming pool has the dimensions 80 ft × 30 ft × 8.0 ft. (a) When it is filled with water, what is the force (resulting from the water alone) on the bottom, on the ends, and on the sides? (b) If you are concerned with whether or not the concrete walls and floor will collapse, is it appropriate to take the atmospheric pressure into account? Why?

15E. (a) Find the total weight of water on top of a nuclear submarine at a depth of 200 m, assuming that its (horizontal cross-sectional) hull area is 3000 m². (b) What water pressure would a diver experience at this depth? Express your answer in atmospheres. Do you think that occupants of a damaged submarine at this depth could escape without special equipment? Assume the density of seawater is 1.03 g/cm³.

16E. Crew members attempt to escape from a damaged submarine 100 m below the surface. What force must they apply to a pop-out hatch, which is 1.2 m by 0.60 m, to push it out? Assume that density of the ocean water is 1025 kg/m³.

17E. A simple open U-tube contains mercury. When 11.2 cm of water is poured into the right arm of the tube, how high does the mercury rise in the left arm from its initial level?

18P. Two identical cylindrical vessels with their bases at the same level each contain a liquid of density ρ. The area of either base is A, but in one vessel the liquid height is h_1

and in the other h_2. Find the work done by the gravitational force in equalizing the levels when the two vessels are connected.

19P. A cylindrical barrel has a narrow tube fixed to the top, as shown with dimensions in Fig. 16-36. The vessel is filled with water to the top of the tube. Calculate the ratio of the hydrostatic force exerted on the bottom of the barrel to the weight of the water contained inside. Why is the ratio not equal to one? (Ignore the presence of the atmosphere.)

FIGURE 16-36 Problem 19.

20P. (a) Consider a container of fluid subject to a *vertical upward* acceleration a. Show that the pressure variation with depth in the fluid is given by

$$p = \rho h(g + a),$$

where h is the depth and ρ is the density. (b) Show also that if the fluid as a whole undergoes a *vertical downward* acceleration a, the pressure at a depth h is given by

$$p = \rho h(g - a).$$

(c) What is the state of affairs in free fall?

21P. In analyzing certain geological features of the Earth, it is often appropriate to assume that the pressure at some horizontal *level of compensation,* deep in the Earth, is the same over a large region and is equal to that exerted by the weight of the overlying material. That is, the pressure on the level of compensation is given by the fluid pressure formula. This requires, for example, that mountains have low-density *roots;* see Fig. 16-37. Consider a mountain 6.0 km high. The continental rocks have a density of 2.9 g/cm³, and beneath the continent is the mantle, with a density of 3.3 g/cm³. Calculate the depth D of the root. (*Hint:* Set the pressure at points a and b equal; the depth y of the level of compensation will cancel out.)

22P. In Fig. 16-38, the ocean is about to overrun the continent. Find the depth h of the ocean using the level of compensation method shown in Problem 21.

FIGURE 16-37 Problem 21.

FIGURE 16-38 Problem 22.

23P. Water stands at a depth D behind the vertical upstream face of a dam, as shown in Fig. 16-39. Let W be the width of the dam. (a) Find the resultant horizontal force exerted on the dam by the gauge pressure of the water and (b) the net torque owing to that gauge pressure about a line through O parallel to the width of the dam. (c) Find the moment arm of the resultant horizontal force about the line through O.

FIGURE 16-39 Problem 23.

24P. Figure 16-40 shows a dam and part of the freshwater reservoir backed up behind it. The dam is made of concrete of density 3.2 g/cm³ and has the dimensions shown on the figure. (a) The force exerted by the water pushes horizontally on the dam face, and this is resisted by the force of static friction between the dam and the bedrock foundation on which it rests. The coefficient of friction is 0.47. Calculate the factor of safety against sliding, that is, the ratio of the maximum possible friction force to the

FIGURE 16-40 Problem 24.

force exerted by the water. (b) The water also tries to rotate the dam about a line running along the base of the dam through point A; see Problem 23. The torque resulting from the weight of the dam acts in the opposite sense. Calculate the factor of safety against rotation, that is, the ratio of the torque owing to the weight of the dam to the torque exerted by the water.

25P. A U-tube is filled with a single homogeneous liquid. The liquid is temporarily depressed in one side by a piston. The piston is removed and the level of the liquid in each side oscillates. Show that the period of oscillation is $\pi\sqrt{2L/g}$, where L is the total length of the liquid in the tube.

SECTION 16-5 MEASURING PRESSURE

26E. Calculate the height of a column of water that gives a pressure of 1 atm at the bottom. Assume that $g = 9.80$ m/s^2.

27E. If you can suck lemonade of density 1000 kg/m^3 up a straw a maximum height of 4.0 cm, what minimum gauge pressure (in atmospheres) can you produce in your lungs?

28P. What would be the height of the atmosphere if the air density (a) were constant and (b) decreased linearly to zero with height? Assume a sea-level density of 1.3 kg/m^3.

SECTION 16-6 PASCAL'S PRINCIPLE

29E. A piston of small cross-sectional area a is used in a hydraulic press to exert a small force $\mathbf{f}$ on the enclosed liquid. A connecting pipe leads to a larger piston of cross-sectional area A (Fig. 16-41). (a) What force $\mathbf{F}$ will the

larger piston sustain? (b) If the small piston has a diameter of 1.5 in. and the large piston one of 21 in., what weight on the small piston will support 2.0 tons on the large piston?

30E. In the hydraulic press of Exercise 29, through what distance must the large piston be moved to raise the small piston a distance of 3.5 ft?

SECTION 16-7 ARCHIMEDES' PRINCIPLE

31E. A tin can has a total volume of 1200 cm^3 and a mass of 130 g. How many grams of lead shot could it carry without sinking in water? The density of lead is 11.4 g/cm^3.

32E. A boat floating in fresh water displaces 8000 lb of water. (a) How many pounds of water would this boat displace if it were floating in salt water of density 68.6 lb/ft^3? (b) Would the volume of water displaced change? If so, by how much?

33E. About one-third of the body of a physicist swimming in the Dead Sea will be above the water line. Assuming that the human body density is 0.98 g/cm^3, find the density of the water in the Dead Sea. (Why is it so much greater than 1.0 g/cm^3?)

34E. An iron anchor appears 200 N lighter in water than in air. (a) What is the volume of the anchor? (b) How much does it weigh in air? The density of iron is 7870 kg/m^3.

35E. An object hangs from a spring balance. The balance registers 30 N in air, 20 N when this object is immersed in water, and 24 N when the object is immersed in another liquid of unknown density. What is the density of the other liquid?

36E. A cubical object of dimensions $L = 2.00$ ft on a side and weight $W = 1000$ lb in a vacuum is suspended by a rope in an open tank of liquid of density $\rho = 2.00$ slugs/ft^3 as in Fig. 16-42. (a) Find the total downward force exerted by the liquid and the atmosphere on the top of the object. (b) Find the total upward force on the bottom of the object. (c) Find the tension in the rope. (d) Calculate the buoyant force on the object using Archimedes' principle. What relation exists among all these quantities?

FIGURE 16-41 Exercise 29.

FIGURE 16-42 Exercise 36.

37E. A block of wood floats in water with two-thirds of its volume submerged. In oil the block floats with 0.90 of its volume submerged. Find the density of (a) the wood and (b) the oil.

38E. It has been proposed to move natural gas from the North Sea gas fields in huge dirigibles, using the gas itself to provide lift. Calculate the force required to tether such an airship to the ground for off-loading when it is fully loaded with 1.0×10^6 m³ of gas at a density of 0.80 kg/m³. (The weight of the airship is negligible by comparison.)

39E. The Goodyear blimp *Columbia* (see Fig. 16-43) is cruising slowly at low altitude, filled as usual with helium gas. Its maximum useful payload, including crew and cargo, is 1280 kg. How much more payload could the *Columbia* carry if you replaced the helium with hydrogen? Why not do it? The volume of the helium-filled interior space is 5000 m³. The density of helium gas is 0.16 kg/m³ and the density of hydrogen is 0.081 kg/m³.

40E. A helium balloon is used to lift a 40-kg payload to an altitude of 27 km, where the air density is 0.035 kg/m³. The balloon has a mass of 15 kg, and the density of the gas in the balloon is 0.0051 kg/m³. What is the volume of the balloon? Neglect the volume of the payload.

41P. A hollow sphere of inner radius 8.0 cm and outer radius 9.0 cm floats half submerged in a liquid of density 800 kg/m³. (a) What is the mass of the sphere? (b) Calculate the density of the material of which the sphere is made.

42P. A hollow spherical iron shell floats almost completely submerged in water. The outer diameter is 60.0 cm and the density of iron is 7.87 g/cm³. Find the inner diameter.

43P. An iron casting containing a number of cavities weighs 6000 N in air and 4000 N in water. What is the volume of the cavities in the casting? The density of iron is 7.87 g/cm³.

44P. (a) What is the minimum area of the top surface of a slab of ice 0.30 m thick floating on fresh water that will hold up an automobile of mass 1100 kg? (b) Does it matter where the car is placed on the block of ice?

45P. Three children, each of weight 80 lb, make a log raft by lashing together logs of diameter 1.0 ft and length 6.0 ft. How many logs will be needed to keep them afloat? Take the density of wood to be 50 lb/ft³.

46P. Assume the density of brass weights to be 8.0 g/cm³ and that of air to be 0.0012 g/cm³. What percent error arises from neglecting the buoyancy of air in weighing an object of mass m and density ρ on a beam balance?

47P. A car has a total mass of 1800 kg. The volume of air space in the passenger compartment is 5.00 m³. The volume of the motor and front wheels is 0.750 m³, and the volume of the rear wheels, gas tank, and trunk is 0.800 m³. Water cannot enter these areas. The car is parked on a hill; the handbrake cable snaps and the car rolls down the hill into a lake (see Fig. 16-44). (a) At first, no water enters the passenger compartment. How much of the car, in cubic meters, is below the water surface with the car floating as shown? (b) As water slowly enters, the car sinks. How many cubic meters of water are in the car as it disappears below the water surface? (The car remains horizontal, owing to a heavy load in the trunk.)

0.750 m³   5.00 m³   0.800 m³

FIGURE 16-44 Problem 47.

48P. A block of wood has a mass of 3.67 kg and a density of 600 kg/m³. It is to be loaded with lead so that it will float in water with 0.90 of its volume immersed. What mass of lead is needed (a) if the lead is on top of the wood and (b) if the lead is attached below the wood? The density of lead is 1.13×10^4 kg/m³.

49P. You place a glass beaker, partially filled with water, in a sink (Fig. 16-45). The beaker has a mass of 390 g and an interior volume of 500 cm³. You now start to fill the sink with water and you find, by experiment, that if the beaker is less than half full, it will float; but if it is more than half full, it remains on the bottom of the sink as the water rises to its rim. What is the density of the material of which the beaker is made?

50P. What is the acceleration of a rising balloon of hot air

FIGURE 16-43 Exercise 39.

FIGURE 16-45 Problem 49.

if the ratio of the air density outside the balloon to that inside is 1.39? Neglect the mass of the balloon fabric.

51P. A cylindrical wooden rod is loaded with lead at one end so that it floats upright in water as in Fig. 16-46. The length of the submerged portion is $l = 2.50$ m. The rod is set into vertical oscillation. (a) Show that the oscillation is simple harmonic. (b) Find the period of the oscillation. Neglect the fact that the water has a damping effect on the motion.

FIGURE 16-46 Problem 51.

52P*. The tension in a string holding a solid block below the surface of a liquid (of density greater than the solid) is T_0 when the containing vessel (Fig. 16-47) is at rest. Show that the tension T, when the vessel has an upward vertical acceleration a, is given by $T_0(1 + a/g)$.

FIGURE 16-47 Problem 52.

SECTION 16-9 STREAMLINES AND THE EQUATION OF CONTINUITY

53E. Figure 16-48 shows the merging of two streams to form a river. One stream has a width of 8.2 m, depth of 3.4 m, and current speed of 2.3 m/s. The other stream is 6.8 m wide, 3.2 m deep, and flows at 2.6 m/s. The width of the river is 10.5 m and the current speed is 2.9 m/s. What is its depth?

54E. A $\frac{3}{4}$-in. (inside diameter) water pipe is coupled to three $\frac{1}{2}$-in. pipes. (a) If the flow rates in the three smaller pipes are 7.0, 5.0, and 3.0 gal/min, what is the flow rate in the $\frac{3}{4}$-in. pipe? (b) What is the ratio of the speed of water in the $\frac{3}{4}$-in. pipe to that in the pipe carrying 7.0 gal/min?

55E. A garden hose having an internal diameter of 0.75 in. is connected to a lawn sprinkler that consists merely of an enclosure with 24 holes, each 0.050 in. in di-

FIGURE 16-48 Exercise 53.

ameter. If the water in the hose has a speed of 3.0 ft/s, at what speed does it leave the sprinkler holes?

56P. Water is pumped steadily out of a flooded basement at a speed of 5.0 m/s through a uniform hose of radius 1.0 cm. The hose passes out through a window 3.0 m above the water line. What is the power of the pump?

57P. A river 20 m wide and 4.0 m deep drains a 3000-km² land area in which the average precipitation is 48 cm/y. One-fourth of this rainfall returns to the atmosphere by evaporation, but the remainder ultimately drains into the river. What is the average speed of the river current?

SECTION 16-11 SOME APPLICATIONS OF BERNOULLI'S EQUATION

58E. Water is moving with a speed of 5.0 m/s through a pipe with a cross-sectional area of 4.0 cm². The water gradually descends 10 m as the pipe increases in area to 8.0 cm². (a) What is the speed of flow at the lower level? (b) If the pressure at the upper level is 1.5×10^5 Pa, what is the pressure at the lower level?

59E. Models of torpedoes are sometimes tested in a horizontal pipe of flowing water, much as a wind tunnel is used to test model airplanes. Consider a circular pipe of internal diameter 25.0 cm and a torpedo model, aligned along the axis of the pipe, with a diameter of 5.00 cm. The torpedo is to be tested with water flowing past it at 2.50 m/s. (a) With what speed must the water flow in the part of the pipe that is unconstricted by the model? (b) What will the pressure difference be between the constricted and unconstricted parts of the pipe?

60E. A water intake at a pump storage reservoir (see Fig. 16-49) has a cross-sectional area of 8.00 ft². The water flows in at a speed of 1.33 ft/s. At the generator building 600 ft below the intake point, the cross-sectional area is smaller than at the intake and the water flows out at

Reservoir

Generator
building

Intake
600 ft

Outlet

FIGURE 16-49 Exercise 60.

31.0 ft/s. What is the difference in pressure, in pounds per square inch, between inlet and outlet?

61E. A water pipe having a 1.0-in. inside diameter carries water into the basement of a house at a speed of 3.0 ft/s and a pressure of 25 lb/in.^{2}. If the pipe tapers to $\frac{1}{2}$ in. and rises to the second floor 25 ft above the input point, what are (a) the speed and (b) the water pressure at the second floor?

62E. How much work is done by pressure in forcing 1.4 m^3 of water through a 13-mm internal diameter pipe if the difference in pressure at the two ends of the pipe is 1.0 atm?

63E. In a horizontal oil pipeline of constant cross-sectional area, the pressure decrease between two points 1000 ft apart is 5.0 lb/in.^{2}. What is the energy loss per cubic foot of oil per foot?

64E. A tank of large area is filled with water to a depth $D = 1.0$ ft. A hole of cross section $A = 1.0$ in.^{2} in the bottom of the tank allows water to drain out. (a) What is the rate at which water flows out in cubic feet per second? (b) At what distance below the bottom of the tank is the cross-sectional area of the stream equal to one-half the area of the hole?

65E. Suppose that two tanks, 1 and 2, each with a large opening at the top, contain different liquids. A small hole is made in the side of each tank at the same depth h below the liquid surface, but the hole in tank 1 has half the cross-sectional area of the hole in tank 2. (a) What is the ratio ρ_1/ρ_2 of the densities of the fluids if it is observed that the mass flow rate is the same for the two holes? (b) What is the ratio of the volume flow rates from the two tanks? (c) To what height above the hole in the second tank should fluid be added or drained to equalize the volume flow rates?

66E. Air flows over the top of an airplane wing of area A with speed v_l, and past the underside of the wing (also of area A) with speed v_u. Show that in this simplified situation Bernoulli's equation predicts that the magnitude L of the upward lift force on the wing will be

$$L = \tfrac{1}{2}\rho A(v_l^2 - v_u^2),$$

where ρ is the density of the air.

67E. If the speed of flow past the lower surface of a wing is 110 m/s, what speed of flow over the upper surface will give a pressure difference of 900 Pa between upper and lower surfaces? Take the density of air to be 1.30×10^{-3} g/cm^3. (See Exercise 66.)

68E. An airplane has a wing area (each wing) of 10.0 m^2. At a certain airspeed, air flows over the upper wing surface at 48.0 m/s and over the lower wing surface at 40.0 m/s. What is the mass of the plane? Assume that the plane travels at constant velocity, that the air density is 1.20 kg/m^2, and that lift effects associated with the fuselage and tail assembly are small. Discuss the lift if the airplane, flying at the same airspeed, is (a) in level flight, (b) climbing at 15°, and (c) descending at 15°. (See Exercise 66.)

69E. A pitot tube (Fig. 16-50) is used to determine the airspeed of an airplane. It consists of an outer tube with a number of small holes B (four are shown); the tube is connected to one arm of a U-tube. The other arm of the U-tube is connected to a hole, A, at the front end of the device, which points in the direction the plane is headed. At A the air becomes stagnant so that $v_A = 0$. At B, however, the speed of the air presumably equals the airspeed of the aircraft. Use Bernoulli's equation to show that

$$v = \sqrt{\frac{2\rho g h}{\rho_{\text{air}}}},$$

where v is the airspeed of the plane and ρ is the density of the liquid in the U-tube.

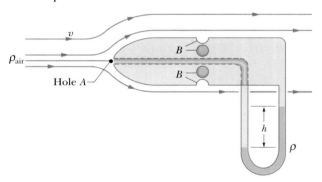

FIGURE 16-50 Exercise 69.

70E. A pitot tube (see Exercise 69) is mounted on an airplane wing to determine the speed of the plane relative to the air, which has a density of 1.03 kg/m^3. The tube contains alcohol and indicates a level difference h of 26.0 cm. What is the plane's speed relative to the air? The density of alcohol is 810 kg/m^3.

71P. A pitot tube (see Exercise 69) on a high-altitude aircraft measures a differential pressure of 180 Pa. What is the airspeed if the density of the air is 0.031 kg/m^3?

72P. In a hurricane, the air (density 1.2 kg/m^3) is blowing over the roof of a house at a speed of 110 km/h. (a) What is the pressure difference between inside and outside that tends to lift the roof? (b) What would be the lifting force on a roof of area 90 m^2?

73P. The windows in an office building are of dimensions 4.00 m by 5.00 m. On a stormy day, air is blowing at 30.0 m/s past a window on the 53rd floor. Calculate the net force on the window. The air density is 1.23 kg/m^3.

74P. A sniper fires a rifle bullet into a gasoline tank, making a hole 50 m below the surface of the gasoline. The tank was sealed and is under 3.0-atm absolute pressure, as shown in Fig. 16-51. The stored gasoline has a density of 660 kg/m³. At what speed v does the gasoline begin to shoot out of the hole?

FIGURE 16-51   Problem 74.

75P. A hollow tube has a disk DD attached to its end (Fig. 16-52). When air is blown through the tube, the disk attracts the card CC. Let the area of the card be A, and let v be the average airspeed between the card and the disk. Calculate the resultant upward force on CC. Neglect the card's weight; assume that $v_0 \ll v$, where v_0 is the airspeed in the hollow tube.

FIGURE 16-52   Problem 75.

76P. An 80-cm² plate of 500 g mass is hinged along one side. If air is blown over the upper surface only, what speed must the air have to hold the plate horizontal?

77P. If a person blows air with a speed of 15 m/s across the top of one side of a U-tube containing water, what will be the difference between the water levels on the two sides?

78P. A tank is filled with water to a height H. A hole is punched in one of the walls at a depth h below the water surface (Fig. 16-53). (a) Show that the distance x from the foot of the wall at which the resulting stream strikes the floor is given by $x = 2\sqrt{h(H - h)}$. (b) Could a hole be punched at another depth to produce a second stream

FIGURE 16-53   Problem 78.

that would have the same range? If so, at what depth? (c) At what depth should the hole be placed to make the emerging stream strike the ground at the maximum distance from the base of the tank?

79P. The fresh water behind a reservoir dam is 15 m deep. A horizontal pipe 4.0 cm in diameter passes through the dam 6.0 m below the water surface, as shown in Fig. 16-54. A plug secures the pipe opening. (a) Find the friction force between plug and pipe wall. (b) The plug is removed. What volume of water flows out of the pipe in 3.0 h?

FIGURE 16-54   Problem 79.

80P. A *siphon* is a device for removing liquid from a container. It operates as shown in Fig. 16-55. Tube ABC

FIGURE 16-55   Problem 80.

must initially be filled, but once this has been done the liquid will flow through the tube until the level in the container drops below the tube opening at *A*. The liquid has density ρ and negligible viscosity. (a) With what speed does the liquid emerge from the tube at *C*? (b) What is the pressure in the liquid at the topmost point *B*? (c) Theoretically, what is the greatest possible height h_1 that a siphon can lift water?

81P. By applying Bernoulli's equation and the equation of continuity to points 1 and 2 of Fig. 16-22, show that the speed of flow at the entrance (point 1) is

$$v = \sqrt{\frac{2a^2 \, \Delta p}{\rho(A^2 - a^2)}}.$$

82P. A Venturi meter has a pipe diameter of 10 in. and a throat diameter of 5.0 in. If the water pressure in the pipe is 8.0 lb/in.² and in the throat is 6.0 lb/in.², determine the rate of flow of water in cubic feet per second.

83P. Consider the Venturi tube of Fig. 16-22 without the manometer. Let *A* equal 5*a*. Suppose that the pressure p_1 at *A* is 2.0 atm. (a) Compute the values of *v* at *A* and *V* at *a* that would make the pressure p_2 at *a* equal to zero. (b) Compute the corresponding volume flow rate if the diameter at *A* is 5.0 cm. The phenomenon that occurs at *a* when p_2 falls to nearly zero is known as cavitation. The water vaporizes into small bubbles.

84P*. A jug contains 15 glasses of orange juice. When you open the tap at the bottom it takes 12.0 s to fill a glass with juice. If you leave the tap open, how long will it take to fill the remaining 14 glasses and thus empty the jug?

85P*. The total effective blade area of a small windmill is 4.6 m². (a) What is the maximum possible power of the windmill generator when a steady wind is blowing at 6.7 m/s? (b) What does this power become if the wind speed increases by 10%? The density of the air is 1.2 kg/m³.

ADDITIONAL PROBLEMS

86. Water flows through a horizontal pipe and is delivered into the atmosphere at a speed of 15 m/s as shown in Fig. 16-56. The diameters of the left and right sections of the pipe are 5.0 cm and 3.0 cm, respectively. (a) What volume of water is delivered into the atmosphere during a 10-min period? (b) What is the flow speed of the water in the left section of the pipe? (c) What is the gauge pressure in the left section of the pipe?

FIGURE 16-56 Problem 86.

87. A lead sinker of volume 0.40 cm³ and density 11.4 g/cm³ is used in fishing. The sinker is suspended from a vertical string whose other end is attached to the bottom of a spherical cork (of density 0.20 g/cm³) that is floating on the surface of a lake. Neglecting the effects of the line, hook, and bait, determine what the radius of the cork must be if it is to float with half its volume submerged.

88. A metal rod of length 80 cm and mass 1.6 kg has a uniform cross-sectional area of 6.0 cm². Due to a nonuniform density, the center of mass of the rod is 20 cm from one end of the rod. The rod is suspended in a horizontal position in water by ropes attached to both ends as shown in Fig. 16-57. (a) What is the tension in the rope closer to the center of mass? (b) What is the tension in the rope farther from the center of mass?

89. The L-shaped tank shown in Fig. 16-58 is filled with water and is open at the top. If *d* = 5.0 m, what are (a) the

FIGURE 16-57 Problem 88.

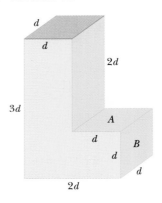

FIGURE 16-58 Problem 89.

force on face *A* and (b) the force on face *B* due to the water?

90. An opening of area 0.25 cm² in an otherwise closed beverage keg is 50 cm below the level of the liquid (of density 1.0 g/cm³) in the keg. What is the speed of the liquid flowing through the opening if the gauge pressure in the air space above the liquid is (a) zero and (b) 0.40 atm?

PHYSICS AND SPORTS: THE AERODYNAMICS OF PROJECTILES

Peter J. Brancazio
Brooklyn College

Peter J. Brancazio is a professor of physics at Brooklyn College, City University of New York. He received his Ph.D. in astrophysics from New York University in 1966. He is the author of two books: The Nature of Physics (Macmillan, 1975) and Sport Science (Simon & Schuster, 1984). His articles on the physics of baseball, football, and basketball have appeared in Discover, Physics Today, New Scientist, The Physics Teacher, and the American Journal of Physics. A lifelong athlete and sports fan, he is equally at home on the basketball court and in the classroom.

An object moving through a fluid always experiences some resistance to its movement. This force, exerted by the fluid on the object, necessarily alters the motion of the object to some degree. We would expect the retarding force to be fairly large if the object is moving through a liquid such as water; but if the fluid is a gas such as air, we might suppose that the force will be so small as to have virtually no effect on the motion of the object. However, as we shall see, the force that we commonly call *air resistance* cannot always be so easily ignored.

Air resistance is one manifestation of *aerodynamic force*—the force exerted by the air on a moving object. (When an object is moving through water, the force is said to be *hydrodynamic*.) Such forces are referred to as *dynamic* forces because they result from motion. Moreover, a force exists whether the object is at rest in a moving fluid or is moving in a stationary fluid; that is, the force is created by *relative* motion.

The study of fluid dynamics has considerable practical value in a wide range of areas, ranging from the flow of blood through capillaries to the design of boats, automobiles, and airplanes. It may come as a surprise, however, to learn that these same principles are also used by athletes to improve their performance in a variety of sports, for aerodynamic forces have a substantial effect on the motions of the various projectiles used in sports. These forces make it possible for a baseball pitcher to throw a curve ball and are responsible for the hook or slice of a poorly hit golf drive. They determine the proper technique for throwing a football or a javelin. They also represent the major force of resistance to the movement of a downhill skier or a cyclist.

In general, the fluid-dynamic force on an object will depend on the size, shape, and surface characteristics of the object as well as on its velocity relative to the fluid. The force, of course, will also depend on the properties of the fluid itself. A key property is the *viscosity* of the fluid, which is a measure of internal resistance to flow caused by interactions between the fluid molecules. Where the fluid comes into contact with the surface of an immersed object, its viscosity will create a frictional retarding force parallel to the surface. The viscous forces on an object are larger when the fluid is water rather than air; at room temperature, water is about 40 times more viscous than air—which explains why it takes more effort to wade across a swimming pool than to walk down the street. The nature of the surface of the moving object also plays a role; in general, the smoother the surface, the lower the viscous resistance.

An immersed object necessarily acts as an obstacle to the flow, forcing the fluid to change direction and accelerate around the object. The viscous friction between the fluid and the surface tends to remove energy from the fluid. These energy losses take place in a relatively thin layer of fluid, known as the *boundary layer,* that lies next to the surface. If the fluid is moving slowly, the frictional energy losses will be small; the fluid in the boundary layer will be able to accelerate and remain in contact with the surface of the object. However, at higher speeds the energy losses become great enough to prevent the fluid from following the surface contours. The result is that the boundary

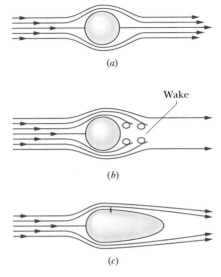

FIGURE 1 Fluid flow around a sphere: (*a*) low-velocity flow—no separation; (*b*) high-velocity flow—flow separates from surface, producing a low-pressure wake; and (*c*) flow around a streamlined body produces little or no wake.

layer tends to separate from the surface (see Fig. 1), creating a region behind the object known as the *wake* that is characterized by lower pressure and unsteady or turbulent motions. Under these conditions, the fluid pressure on the front or leading surface of the object will exceed the pressure on the rear surface, resulting in a net retarding force. This *form resistance* can be reduced by changing the shape of the object to make it more ''streamlined''—that is, by adjusting its contours so that the fluid will not separate from the surface.

The forces created by viscous and form resistance are distributed over the entire surface of an immersed object. However, it is common practice to sum and resolve them into two components: a *drag force*, which acts opposite the direction of the motion of the object relative to the fluid (that is, antiparallel to the velocity vector), and a *lift* force, which acts at right angles to the direction of motion. Despite its name, the

lift force should be thought of not as an upward (gravity-opposing) force, but rather as a lateral or sideways force that can deflect an object in any direction perpendicular to the velocity.

Aerodynamic Drag

The drag force on an object moving through a fluid always acts as a decelerating force, reducing the speed of the object relative to the fluid. In general, drag is produced by both viscous and form resistance. Viscous drag is significant for relatively small objects moving slowly through viscous fluids. For the sizes and speeds of the objects used as projectiles in sports—baseballs, tennis balls, and even the athletes themselves—viscous resistance is relatively small compared to form resistance, and the aerodynamic drag can be represented by a fairly simple equation. From energy and momentum considerations, it can be shown that the force per unit area exerted by a fluid on an immersed object should be proportional to the quantity $\frac{1}{2}\rho v^2$ (where ρ is fluid density and v is the relative velocity of the flow). That is, if D is the drag force,

$$D \propto \tfrac{1}{2}\rho A v^2.$$

where A is the effective frontal area of the object (its cross-sectional area perpendicular to the flow). This can be made into an equation by introducing as a constant of proportionality a dimensionless quantity known as the *drag coefficient, C_D*:

$$D = \tfrac{1}{2}\rho A C_D v^2.$$

The drag coefficient C_D takes into account the relative contributions of viscous and form resistances, and it depends on the nature of the object (size, shape, and irregularity and roughness of its surface) as well as on the characteristics of the flow. In general, the more streamlined an object is, the lower its drag coefficient will be—an important consideration in the design of objects that must move through fluids at high speeds.

As a rule, C_D must be determined by direct measurement. The standard

procedure is to place the object (or a suitably scaled model) in a wind tunnel, thereby taking advantage of the fact that the drag force depends on the relative velocity of object and fluid. For example, C_D for a baseball moving through still air at 90 mi/h can be measured by mounting a stationary baseball in a 90-mi/h wind.

The effect of aerodynamic drag on the motion of a falling body is described in Section 6-3. As a body falls through the air, both the speed and the drag increase until the drag force equals the weight of the body. At this point, the body has achieved its terminal speed (see Eq. 6-19).

One sport that makes use of the fact that falling bodies approach a terminal speed in air is skydiving. A person who jumps out of a plane falls with decreasing acceleration, approaching a terminal speed of as much as 200 mi/h (320 km/h). However, by changing the shape and orientation of their bodies as they fall, sky divers are able to increase or decrease the amount of aerodynamic drag, effectively selecting the terminal speed by changing the drag coefficient and frontal area. When spread-eagled with arms and legs extended, a sky diver feels the greatest drag and has the lowest terminal speed.

As a body falling through the air approaches its terminal speed, its motion deviates significantly from that of a freely falling body. Similarly, the motion of any object launched as a projectile diverges noticeably from the theoretical symmetrical parabolic path if its launching speed is comparable to its terminal speed. When an object is launched upward at an angle to the ground, aerodynamic drag retards both the vertical and horizontal components of its motion. Consequently, the maximum height of the trajectory as well as the horizontal range will be shortened. As the projectile rises, it decelerates more rapidly than it would in a vacuum; as it falls, it accelerates more slowly. The result is that the projectile will take less time to reach its peak than it does to descend, and the speed with which it reaches the ground will be less than the speed at which it was launched. Since the horizontal velocity component is continually decreasing, the projectile will travel a longer distance horizontally when it is rising than when it is falling. Consequently, the descent of the projectile is both steeper and slower than its rise.

A baseball has a terminal speed of about 95 mi/h (153 km/h). Under game conditions, pitched and batted balls are launched at speeds comparable to or even greater than the terminal speed. (Note that the terminal speed is not the maximum speed that a projectile can travel at in air, but rather the speed that the projectile attains when it is allowed to fall from rest.) Accordingly, it is clear that the trajectory of a baseball in flight must be altered substantially by aerodynamic drag. A typical fly ball has a range in air that is only about 60% of its range in a vacuum (see Fig. 4-14 for a graphic comparison).

Given its large size and asymmetrical shape, a football is particularly susceptible to the influence of aerodynamic drag. When it is traveling nose first, a football presents a smaller cross-sectional area and a more streamlined shape than when it is traveling broadside; the drag is roughly 10 times larger in the broadside orientation. This fact determines the best technique for throwing a football; namely, it should be "spiraled," or thrown nose first with a substantial spin about its long axis. (The spin gives the ball angular momentum, which stabilizes the orientation of the spin axis in flight.) Launching the football with its long axis angled laterally to its path will shorten its range considerably. The effect is especially noticeable when a football is kicked with little or no spin such that it rises in a nose-first orientation and descends broadside. The sudden increase in aerodynamic drag that occurs as the football passes the peak of its trajectory causes it to drop rather steeply and shortens the range of the kick. Good kickers try to get the ball to "turn over" (that is, turn its nose downward on the descent—a result of spiraling the kick) as a way to get more distance.

How Can Athletes Reduce Drag?

In sports such as swimming, cycling, or speed skating—where the energy for motion is provided entirely by the athlete—it is particularly important to keep the drag force as low as possible. In downhill skiing, where speeds as high as 80 mi/h (130 km/h) are attained, aerodynamic drag is virtually the only retarding force. The techniques and equipment used in these sports are designed in large measure to minimize drag. There are three basic approaches to achieving this purpose:

1. *Reduce the Effective Frontal Area.* Athletes are taught to adopt a stance in which their bodies collide with as little air as possible. By keeping the upper body parallel to the ground instead of upright, the effective frontal area of the body can be reduced by about 25%. Thus cyclists hunch down over the handlebars; speed skaters bend sharply at the waist and keep one arm tucked behind their backs; skiers crouch down over their skis in the "egg" position.

Swimmers move rather slowly through water (the top racing speeds are less than 5 mi/h), but they nevertheless experience substantial hydrodynamic resistance because of the relatively high density and viscosity of water. Swimmers reduce their effective frontal area in the water by kicking their feet as they swim. The major purpose of kicking the feet is *not* to propel the swimmer (the arm strokes provide approximately 75% of a swimmer's propulsive force) but to reduce drag by keeping the body horizontal in the water.

2. *Use Streamlined Equipment.* Wind tunnel studies of cyclists and skiers have led to numerous improvements in equipment design. It has been found that relatively small protrusions such as boot buckles, ski pole baskets, and cables and nuts on bicycles are sources of turbulence and measurable aerodynamic drag. Although these effects are relatively small, they are of some importance to competitors at the world-class level where fractions of a second make the difference between winning and losing. Accordingly, equipment has been modified so as to remove protruding parts or to cover them with smoothly contoured shields. The athlete's head is also a source of some turbulence, so some skiers and cyclists wear teardrop shaped helmets that produce a more streamlined flow of air around their heads.

3. *Wear Smooth, Skin-Tight Clothing.* Loose-fitting clothing increases the athlete's surface area and creates turbulence. "Aerodynamic clothing"— smooth, skin-tight bodysuits made of specially designed synthetic fabrics— has become increasingly popular with skiers, cyclists, and speed skaters. A smooth body surface is particularly important in competitive swimming; thus form-fitting swimsuits made of nonabsorbent material have become obligatory. Male swimmers routinely shave off exposed body hair—even on their heads—as a means of reducing hydrodynamic drag.

Aerodynamic Lift

As noted earlier, lift is that component of the interaction between an object and a fluid that is perpendicular to the direction of motion. In general, lift is generated by any effect that causes the fluid to change direction as it flows past the object. If the fluid acquires a component of velocity at right angles to its original direction as a result of its interaction with an immersed object, then the object must have exerted a force on the fluid giving it an acceleration in that direction. According to Newton's third law (the principle of action and reaction), the fluid must apply an equal and opposite force on the object. Thus if the object diverts the fluid to the left, the object will experience a lift force to the right. Among the effects that can create lift are (1) the object has an asymmetrical shape or orientation with respect to the flow; (2) the object is spinning; and (3) the lateral surface of the object is uneven (that is, one side is rougher than the other).

The Magnus Effect

The airflow around a sphere cannot generate lift because, by symmetry, it does not deflect the airflow one way or another. However, a *spinning* sphere does cause the airflow to change direction. This effect arises from the frictional losses in the boundary layer that cause the flow to separate from the surface and form a low-pressure wake. On a spinning sphere, one side is turning into the airflow, while the other is turning in the same direction as the airflow. On the latter side, the air is carried along by the surface, causing the separation to occur a little farther downstream. On the side turning into the airflow, the separation point is shifted toward the front. The result is that the wake is not symmetrical; the airflow is deflected to one side, and the sphere experiences a reaction force in the opposite direction (Fig. 2). The direction and strength of this force will depend on the rate and direction of spin. This phenomenon is known as the *Magnus effect* (after Gustav Magnus, who studied it in the 1850s).

The Magnus effect is familiar to anyone who has watched a ball curve in flight. It plays a key role in such sports as baseball, tennis, golf, and soccer. By applying an appropriate spin, an athlete can make a ball curve in any chosen direction. For example, if a soccer ball is kicked so that the impact point is right of center — giving the ball a counterclockwise spin as seen from above — it will curve from right to left; if the kick is aimed left of center, the ball will curve to the right.

The same principle is used by a baseball pitcher to throw a curve ball. The pitcher imparts spin to the ball by twisting his wrist as the pitch is delivered. However, the anatomy of the human arm is such that it twists more flexibly in one direction than the

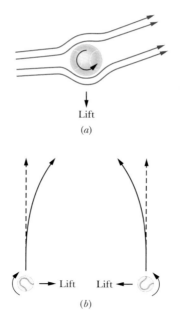

FIGURE 2 Lift on a spinning sphere (the Magnus effect): (*a*) the airflow around a spinning sphere and (*b*) paths of a curve ball thrown with different spin directions.

other. Thus it is easier for a right-handed pitcher to make a baseball spin counterclockwise than clockwise (as seen from above), whereas the reverse is true for a left-hander. A right-hander's curve ball moves from his right to his left (curving away from a right-handed batter). To make the pitch curve from left to right (a pitch known as a "screwball"), he must twist his wrist in the "wrong" direction. Few right-handed pitchers can throw this pitch effectively. Thus the Magnus effect, coupled with a quirk of human anatomy, underlies the maneuvering of the various combinations of right-handed and left-handed pitchers and batters that is so central to baseball strategy.

A baseball pitch must necessarily travel on a curved trajectory even if it is not spinning: because of gravity, it curves downward on a parabolic path. A baseball thrown horizontally at 85 mi/h (137 km/h) will drop about $3\frac{1}{2}$ ft (1.1 m) over the distance from the pitcher's hand to home

plate. A pitcher can add to or subtract from this vertical drop by using suitable spin. If the ball is thrown with a topspin (the top of the ball is turning toward the batter), then the lift force will act downward and the pitch will curve more sharply than normally (this pitch is called an overhand curve). If the ball is released with a backspin (by letting the ball roll off the fingertips with a downward snap of the wrist), the lift force will act upward, opposing the force of gravity. This is the "rising" fastball; however, the ball does not actually rise in flight—since it is not humanly possible to impart enough spin to make the lift exceed the weight of the baseball—but because the ball does not drop as much as it does under gravity alone, it gives the illusion of rising.

Topspin and backspin are also important in tennis: backspin (creating upward lift) tends to prolong the flight of a tennis ball, whereas topspin (producing downward lift) tends to shorten it. It is extremely difficult to hit a hard serve that will clear the net and land in the service box without giving it a topspin.

The fact that backspin prolongs the flight of a ball is particularly helpful to golfers. A well-hit drive should have a great deal of backspin on it: the horizontal grooves and sloped face of the clubhead serve to make this possible. Because of the lift produced by backspin, the trajectory of a golf drive is far from parabolic. The optimum launch angle tends to be much lower than 45° (it is generally between 20° and 30°); since gravity is being opposed by lift, the ball can be launched more horizontally for greater distance. However, if the clubface does not hit the ball squarely and is angled right or left, the ball will receive a side spin. This type of spin will make the ball hook (curve left) or slice (curve right) in flight. For the average golfer, the angle of the clubface at the moment of impact can sometimes make all the difference between a birdie and a bogie.

WAVES—I | 17

When a beetle moves along the sand within a few tens of centimeters of this sand scorpion, the scorpion immediately turns toward the beetle and dashes to its location to kill and eat it. The scorpion can do this without seeing (it is nocturnal) or hearing the beetle. How can it so precisely locate its prey?

17-1 WAVES AND PARTICLES

Two ways to get in touch with a friend in a distant city are to write a letter and to use the telephone.

The first choice (the letter) involves the concept of "particle." A material object moves from one point to another, carrying with it information and energy. Most of the previous chapters deal with particles or with systems of particles, ranging from electrons to baseballs to automobiles to planets.

The second choice (the telephone) involves the concept of "wave," the subject of this and the next chapter. In a wave, information and energy move from one point to another but no material object makes that journey. In your telephone call, a sound wave carries your message from your vocal cords to the telephone. From there, the wave is electromagnetic, passing along a copper wire or an optical fiber or through the atmosphere, possibly by way of a communications satellite. At the receiving end there is another sound wave, from a telephone to your friend's ear. Although the message is passed, nothing that you have touched reaches your friend. Leonardo da Vinci understood about waves when he wrote of water waves: "it often happens that the wave flees the place of its creation, while the water does not; like the waves made in a field of grain by the wind, where we see the waves running across the field while the grain remains in place."

Particle and *wave* are the two great concepts in classical physics, in the sense that we seem able to associate almost every branch of the subject with one or the other. The two concepts are quite different. The word *particle* suggests a tiny concentration of matter capable of transmitting energy. The word *wave* suggests just the opposite, namely, a broad distribution of energy, filling the space through which it passes. The job at hand is to put aside particles for a while and to learn something about waves.

17-2 WAVES

Mechanical Waves

A flag waving in the breeze is so familiar that when astronauts planted an American flag on the windless moon, they used a flag with built-in ripples so that it would look "natural." There are also water waves, in bodies of water ranging from an ocean to a wash basin. There are sound waves, in air and in water, and seismic waves, in the Earth's crust, mantle, and core. The central features of all these *mechanical waves* are that they are governed by Newton's laws and that they require a material medium, such as air, water, a flag, a stretched string, or a steel rod, to exist.

Electromagnetic Waves

The most familiar of electromagnetic waves is visible light, but almost as familiar are x rays, microwaves, and the waves that activate our radios and television sets. Many such waves pass right through you all the time.

Electromagnetic waves require no material medium to exist. Light from the stars, for example, travels to us through the near vacuum of deep space. All electromagnetic waves travel through vacuum with the same speed c, given by

$$c = 299{,}792{,}458 \text{ m/s} \qquad \text{(speed of light)}. \qquad (17\text{-}1)$$

We shall return to these waves in Chapter 38.

Matter Waves

Under certain experimental conditions, a beam of particles—electrons, for example—can exhibit wavelike properties. Such *matter waves* are governed by the laws of quantum physics. We discuss them more fully in Chapter 44 of the extended version of this book and refer to them here only so that our listing of the kinds of waves encountered in nature may be complete.

Much of what we discuss in this chapter will apply to waves of all kinds. When we seek specific illustrations, however, we shall draw them from the family of mechanical waves, focusing especially on waves traveling along a stretched string.

17-3 WAVES IN A STRETCHED STRING

Of all possible mechanical waves, a wave transmitted along a stretched string is perhaps the simplest. If you give one end of a stretched string a single up and down jerk, as in Fig. 17-1*a*, an *impulse* is passed along the string from particle to particle and thus a wave, in the form of a single *pulse*, travels along the string at velocity **v**. If you move your hand up and down in continuous simple harmonic motion (Fig.

17-1*b*), an extended sinusoidal wave travels along the string at velocity **v**.

We assume an "ideal" string, in which there are no frictionlike forces to cause the wave to die out as it travels along the string. We assume further that our string is so long that we do not have to be concerned about any "echos" that might rebound from its far end.

In studying the waves of Fig. 17-1, we can monitor the **wave form** (shape of the wave) as it moves to the right. Alternatively, we can monitor the motion of a specified element of the string, of length *dx* and corresponding mass *dm*, as it oscillates up and down as the wave passes through it. The displacement of any such oscillating string element is in the *y* direction, *perpendicular* to the direction of travel of the wave, which is the *x* direction. We call such wave motion **transverse.**

Figure 17-2 shows a sound wave, set up in a long, air-filled pipe by an oscillating piston. In this kind of mechanical wave, the displacements of small elements of air of mass *dm* are back and forth, *parallel*

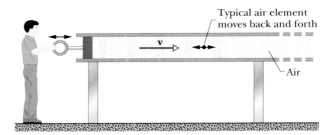

FIGURE 17-2 A sound wave is set up in an air-filled pipe by moving a piston back and forth. Because the oscillations of an element of the air (represented by the black dot) are parallel to the direction in which the wave travels, the wave is a *longitudinal wave*. The oscillations of elements of air are accompanied by a series of alternating compressions and expansions of the air.

to the direction in which the wave travels. We call this kind of wave motion **longitudinal;** we consider it further in Chapter 18.

The sand scorpion shown in the photograph opening this chapter uses waves of both transverse and longitudinal motion to locate its prey. When a beetle even slightly disturbs the sand, it sends pulses along the sand's surface (Fig. 17-3). One set of pulses is longitudinal, traveling with speed $v_l =$ 150 m/s. A second set is transverse, traveling with speed $v_t = $ 50 m/s.

The scorpion, with its eight legs spread roughly in a circle about 5 cm in diameter, intercepts the faster longitudinal pulses first and learns the direction of the beetle: it is in the direction of whichever

(*a*)

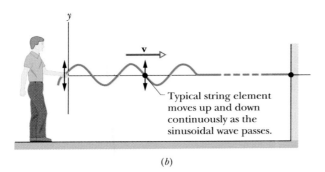

(*b*)

FIGURE 17-1 (*a*) Sending a single pulse down a long stretched string. (*b*) Sending a continuous sinusoidal wave down the string. Because the oscillations of any string element (represented here by a dot) are perpendicular to the direction in which the wave moves, the wave is a *transverse wave.*

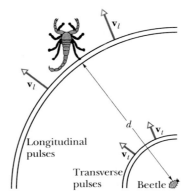

FIGURE 17-3 A beetle's motion sends fast longitudinal pulses and slower transverse pulses along the sand's surface. The sand scorpion first intercepts the longitudinal pulses; here, it is the right rear leg that senses the pulses earliest.

leg is disturbed earliest by the pulses. The scorpion then senses the time interval Δt between that first interception and the interception of the slower transverse waves and uses it to determine the distance d to the beetle. This distance is given by

$$\Delta t = \frac{d}{v_t} - \frac{d}{v_l},$$

and it turns out to be

$$d = (75 \text{ m/s}) \, \Delta t.$$

For example, if $\Delta t = 4.0$ ms, then $d = 30$ cm, which gives the scorpion a perfect fix on the beetle.

17-4 WAVELENGTH AND FREQUENCY

In general, a sinusoidal wave can be written with either a sine function or a cosine function. In this chapter we choose the former. We define a wave shape by giving a relation of the form $y = h(x, t)$, in which y is the transverse displacement of any string element as a function h of the position x of that element along the string and time t.

As a wave for particular study, we choose the sinusoidal wave of Fig. 17-1b, which is generated by moving the end of the string transversely in simple harmonic motion. Suppose that a transverse wave travels along a string, which we take to be stretched along an x axis. The wave can have many shapes, but fundamental to each is a **wavelength** λ and a **frequency** f. The wavelength is the distance along the x axis after which the shape of the wave begins to repeat itself. The frequency of the wave is the frequency at which any string element repeats its transverse oscillations due to the passage of the wave. We then write for the displacement y of the string element at position x and at time t

$$y(x, t) = y_m \sin(kx - \omega t), \qquad (17\text{-}2)$$

in which y_m is the **amplitude** of the wave; the subscript m stands for *maximum*, because the amplitude is the magnitude of the maximum displacement of the string element in either direction. The quantities k and ω are constants whose physical meaning we are about to discover.

You may wonder why, of the infinite variety of wave forms that are available, we pick the sinusoidal wave of Eq. 17-2 for detailed study. This is in fact a wise choice because, as you will see in Section 17-9,

all wave forms—including the pulse of Fig. 17-1a—can be constructed by adding up sinusoidal waves of carefully selected wavelengths and amplitudes. Understanding sinusoidal waves is thus the key to understanding waves of any shape.

There are no limits on the variables x and t in Eq. 17-2, so mathematically the equation describes a wave on a string of infinite length, existing for all time, from the remote past to the distant future. As a practical matter, we shall confine our attention to reasonably small ranges of each variable.

Because Eq. 17-2 has two independent variables (x and t), we cannot represent the dependent variable y on a single two-dimensional plot. We need something like a videotape to show it fully and in real time. We can, however, learn a lot from the two plots of Fig. 17-4.

Wavelength and Wave Number

Figure 17-4a shows how the transverse displacement y of Eq. 17-2 varies with position x at a fixed time, chosen to be $t = 0$. That is, the figure is a "snapshot" of the wave at that instant. With this restriction, Eq. 17-2 becomes

$$y(x, 0) = y_m \sin kx \qquad (t = 0). \qquad (17\text{-}3)$$

(a)

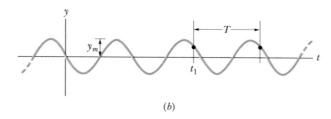

(b)

FIGURE 17-4 (a) A "snapshot" of a string at $t = 0$ as the sinusoidal wave of Eq. 17-2 travels along it. A typical wavelength λ is shown: it is the horizontal distance between two points on the string (represented by dots) with the same displacement. The amplitude y_m is also shown: it is the maximum displacement of the string. (b) A graph showing the displacement of a string element at $x = 0$ as a function of time as the sinusoidal wave passes through the element. A typical period is shown: it is the time between identical displacements of the string element (represented by dots).

We have defined the *wavelength* λ of a wave to be the distance at which the wave pattern begins to repeat itself. A typical wavelength is displayed in Fig. 17-4*a*. By definition, the displacement y is the same at both ends of this wavelength; that is, at $x = x_1$ and $x = x_1 + \lambda$. Hence, by Eq. 17-3,

$$y = y_m \sin kx_1 = y_m \sin k(x_1 + \lambda)$$

$$= y_m \sin(kx_1 + k\lambda). \qquad (17\text{-}4)$$

The sine function begins to repeat itself when its angle (or argument) is increased by 2π radians, so Eq. 17-4 will be true if $k\lambda = 2\pi$, or if

$$k = \frac{2\pi}{\lambda} \qquad \text{(angular wave number).} \qquad (17\text{-}5)$$

We call k the **angular wave number** of the wave; its SI unit is the radian per meter.

The **wave number,** symbolized by κ, is defined as $1/\lambda$ and is related to k by

$$\kappa = \frac{1}{\lambda} = \frac{k}{2\pi} \qquad \text{(wave number).} \qquad (17\text{-}6)$$

The wave number κ is the number of waves in a unit length of the wave pattern; its SI unit is the reciprocal meter (m^{-1}).

Frequency and Period

Figure 17-4*b* shows how the displacement y of Eq. 17-2 varies with time t at a fixed position, taken to be $x = 0$. If you were to monitor the string at that position, you would see that the motion of a single element of the string is up and down and given by

$$y(0, t) = y_m \sin(-\omega t)$$

$$= -y_m \sin \omega t \qquad (x = 0). \qquad (17\text{-}7)$$

Here we have made use of the fact that $\sin(-\alpha) = -\sin \alpha$, where α is any angle.

We define the **period** T of a wave to be the time interval which the motion of an oscillating string element (at any fixed position x) begins to repeat itself. A typical period is displayed in Fig. 17-4*b*. Applying Eq. 17-7 to each end of this time interval and equating the results yield

$$y = -y_m \sin \omega t = -y_m \sin \omega(t + T)$$

$$= -y_m \sin(\omega t + \omega T). \qquad (17\text{-}8)$$

This can be true only if $\omega T = 2\pi$, or if

$$\omega = \frac{2\pi}{T} \qquad \text{(angular frequency).} \qquad (17\text{-}9)$$

We call ω the **angular frequency** of the wave; its SI unit is the radian per second.

The *frequency* of the wave, symbolized by f, is defined as $1/T$ and is related to ω by

$$f = \frac{1}{T} = \frac{\omega}{2\pi} \qquad \text{(frequency).} \qquad (17\text{-}10)$$

The frequency f is the number of oscillations per unit time made by a given point on the string as the wave passes it. You saw in Chapter 14 that the frequency f is usually measured in units of the hertz or its multiples, where

$$1 \text{ hertz} = 1 \text{ Hz} = 1 \text{ oscillation/s.} \qquad (17\text{-}11)$$

17-5 THE SPEED OF A TRAVELING WAVE

Figure 17-5 shows two snapshots of the wave of Eq. 17-2, taken a small time interval Δt apart. The wave is traveling in the direction of increasing x (to the right in Fig. 17-5), the entire wave pattern moving a distance Δx in that direction during the interval Δt. The ratio $\Delta x/\Delta t$ (or, in the differential limit, dx/dt) is the **wave speed** v. How can we find its value?

Let us focus on a particular part of the wave pattern—perhaps a point of maximum displacement such as point A in Fig. 17-5. From Eq. 17-2, a given displacement y is defined by assigning a fixed value to the quantity $kx - \omega t$, which is called the **phase** of the wave. Thus putting

$$kx - \omega t = \text{a constant} \qquad (17\text{-}12)$$

FIGURE 17-5 A "snapshot" of the traveling wave of Eq. 17-2 at $t = 0$ and at a later time $t = \Delta t$. During the time interval Δt, the entire curve shifts distance Δx to the right.

defines a constant transverse displacement y, such as that for point A. Note that as t increases in Eq. 17-12, the location x of a certain transverse displacement increases. For example, as t increases, point A in Fig. 17-5 moves toward increasing x (toward the *right*). Thus the wave itself moves toward increasing x.

To find the wave speed v, we take the derivative of Eq. 17-12, getting

$$k\frac{dx}{dt} - \omega = 0$$

or

$$\frac{dx}{dt} = v = +\frac{\omega}{k}. \qquad (17\text{-}13)$$

This positive, constant result verifies that the wave is traveling in the direction of increasing x, that is, to the right in Fig. 17-5, at constant speed.

Using Eq. 17-5 ($k = 2\pi/\lambda$) and Eq. 17-9 ($\omega = 2\pi/T$), we can write the wave speed as

$$v = \frac{\omega}{k} = \frac{\lambda}{T} = \lambda f \qquad \text{(wave speed)}. \quad (17\text{-}14)$$

Equation 17-14 tells us that the wave moves through one wavelength in one period of oscillation.

Equation 17-2 describes a wave moving in the direction of increasing x. We can find the equation of a wave traveling in the opposite direction by replacing t in Eq. 17-2 by $-t$. This corresponds to keeping the quantity $kx + \omega t$ constant, which (compare Eq. 17-12) requires that x must *decrease* with time. Thus Eq. 17-2 holds for a wave traveling toward increasing x,

$$y(x, t) = y_m \sin(kx - \omega t) \quad \text{(increasing } x\text{)}, \quad (17\text{-}15)$$

whereas for a wave traveling toward decreasing x,

$$y(x, t) = y_m \sin(kx + \omega t) \quad \text{(decreasing } x\text{)}. \quad (17\text{-}16)$$

If you analyze the wave of Eq. 17-16 as we have just done for the wave of Eq. 17-2 (or Eq. 17-15) you will find for its velocity

$$\frac{dx}{dt} = -\frac{\omega}{k}. \qquad (17\text{-}17)$$

The minus sign (compare Eq. 17-13) verifies that the wave is indeed moving in the direction of decreasing x and justifies our device of switching the sign of the time variable.

Consider now a wave of generalized shape, given by

$$y(x, t) = h(kx \pm \omega t), \qquad (17\text{-}18)$$

where h represents *any* function, the sine function being one possibility. Our analysis above shows that all waves in which the variables x and t enter in the combination $kx \pm \omega t$ are traveling waves. Furthermore, all traveling waves *must* be of the form of Eq. 17-18. Thus $y(x, t) = \sqrt{ax + bt}$ represents a possible (though perhaps physically a little bizarre) traveling wave. The function $y(x, t) = \sin(ax^2 - bt)$, on the other hand, does *not* represent a traveling wave.

SAMPLE PROBLEM 17-1

A sinusoidal wave traveling along a string is described by

$$y(x, t) = 0.00327 \sin(72.1x - 2.72t), \quad (17\text{-}19)$$

in which the numerical constants are in SI units (0.00327 m, 72.1 rad/m, and 2.72 rad/s).

a. What is the amplitude of this wave?

SOLUTION By comparison with Eq. 17-15, we see that

$$y_m = 0.00327 \text{ m} = 3.27 \text{ mm}. \qquad \text{(Answer)}$$

b. What are the wavelength and the period of this wave?

SOLUTION By inspection of Eq. 17-19, we see that

$$k = 72.1 \text{ rad/m} \quad \text{and} \quad \omega = 2.72 \text{ rad/s}.$$

From Eq. 17-5 we have

$$\lambda = \frac{2\pi}{k} = \frac{2\pi \text{ rad}}{72.1 \text{ rad/m}} = 0.0871 \text{ m}$$

$$= 8.71 \text{ cm}. \qquad \text{(Answer)}$$

From Eq. 17-9 we have

$$T = \frac{2\pi}{\omega} = \frac{2\pi \text{ rad}}{2.72 \text{ rad/s}} = 2.31 \text{ s}. \qquad \text{(Answer)}$$

c. What are the wave number and the frequency of this wave?

SOLUTION From Eq. 17-6 we have

$$\kappa = \frac{1}{\lambda} = \frac{1}{0.0871 \text{ m}} = 11.5 \text{ m}^{-1}. \qquad \text{(Answer)}$$

From Eq. 17-10 we have

$$f = \frac{1}{T} = \frac{1}{2.31\ \text{s}} = 0.433\ \text{Hz}. \quad \text{(Answer)}$$

d. What is the speed of this wave?

SOLUTION From Eq. 17-14 we have

$$v = \frac{\omega}{k} = \frac{2.72\ \text{rad/s}}{72.1\ \text{rad/m}} = 0.0377\ \text{m/s}$$

$$= 3.77\ \text{cm/s}. \quad \text{(Answer)}$$

Note that all quantities calculated in (b), (c), and (d) are independent of the amplitude of the wave.

SAMPLE PROBLEM 17-2

For the string wave of Eq. 17-19, what is the displacement y at $x = 22.5$ cm ($= 0.225$ m) and $t = 18.9$ s? Take all numerical values to be exact.

SOLUTION From Eq. 17-19 we have

$$y = 0.00327 \sin(72.1 \times 0.225 - 2.72 \times 18.9)$$

$$= (0.00327\ \text{m}) \sin(-35.1855\ \text{rad})$$

$$= (0.00327\ \text{m})(0.588)$$

$$= 0.00192\ \text{m} = 1.92\ \text{mm}. \quad \text{(Answer)}$$

Thus the displacement is positive. (Be sure to change your calculator mode to radians before evaluating the sine.)

We need to assume that the numerical values in this problem are exact because we are asked to specify a displacement at a point several wavelengths from $x = 0$ and several periods after $t = 0$. It is as if you were asked to specify the location of a suitcase in a freight car near the end of a long train by making measurements backward from the engine. You had better be precise. If you make what seems like a modest error on a percentage basis, you could even end up in the wrong freight car!

SAMPLE PROBLEM 17-3

In Sample Problem 17-2, we showed that the transverse displacement of an element of the string for the wave of Eq. 17-19 at $x = 0.225$ m and $t = 18.9$ s is $y = 1.92$ mm.

a. What is u, the transverse speed of the same element of the string, at that place and at that time? (This speed, which is associated with the transverse oscillation of an element of the string, is in the y direction. Do not confuse it with v, the constant speed at which the wave form travels along the x axis.)

SOLUTION The generalized equation of the wave of Eq. 17-19 is

$$y(x, t) = y_m \sin(kx - \omega t). \quad (17\text{-}20)$$

In this expression, let us hold x constant but allow t (for the time being) to be a variable. We then take the derivative of this equation with respect to t, obtaining*

$$u = \frac{\partial y}{\partial t} = -\omega y_m \cos(kx - \omega t). \quad (17\text{-}21)$$

Substituting numerical values from Sample Problems 17-1 and 17-2, we obtain

$$u = (-2.72\ \text{rad/s})(3.27\ \text{mm}) \cos(-35.1855\ \text{rad})$$

$$= 7.20\ \text{mm/s}. \quad \text{(Answer)}$$

Thus, at $t = 18.9$ s, the element of string at $x = 22.5$ cm is moving in the direction of increasing y, with a speed of 7.20 mm/s.

b. What is the transverse acceleration a_y at that position and at that time?

SOLUTION From Eq. 17-21, treating x as a constant but allowing t (for the time being) to be a variable, we find

$$a_y = \frac{\partial u}{\partial t} = -\omega^2 y_m \sin(kx - \omega t).$$

Comparison with Eq. 17-20 shows that we can write this as

$$a_y = -\omega^2 y.$$

We see that the transverse acceleration of an oscillating string element is proportional to its transverse displacement but opposite in sign, completely consistent with the fact that the element is moving transversely in simple harmonic motion. Substituting numerical values yields

$$a_y = -(2.72\ \text{rad/s})^2(1.92\ \text{mm})$$

$$= -14.2\ \text{mm/s}^2. \quad \text{(Answer)}$$

Thus, at $t = 18.9$ s, the element of string at $x = 22.5$ cm is displaced by $+1.92$ mm and its acceleration is -14.2 mm/s².

---

*A derivative taken while one (or more) of the variables is treated as a constant is called a *partial derivative* and is represented by the symbol $\partial/\partial x$, rather than d/dx.

−35.1855 rad   +2.51361 rad

FIGURE 17-6   The two angles are different but all their trigonometric functions are identical.

PROBLEM SOLVING

TACTIC 1: VERY LARGE ANGLES

Sometimes, as in Sample Problems 17-2 and 17-3, an angle much greater than 2π rad (or 360°) will crop up and you are asked to find its sine or cosine. Adding or subtracting an integral multiple of 2π radians to such an angle does not change the value of any of its trigonometric functions. In Sample Problem 17-2, for example, the angle is -35.1855 rad. Adding $(6)(2\pi$ rad) to this angle yields

$$-35.1855 \text{ rad} + (6)(2\pi \text{ rad}) = 2.51361 \text{ rad},$$

an angle less than 2π rad that has the same trigonometric functions as -35.1855 rad (see Fig. 17-6). As an example, the sines of 2.51361 rad and -35.1855 rad are both 0.588.

Your calculator will reduce such large angles for you automatically. Caution: Do not round off large angles if you intend to take their sines or cosines. In taking the sine of a very large angle you are throwing away most of the angle and taking the sine of what is left over. If, for example, you were to round -35.1855 rad to -35 rad (a change of 0.5% and normally a reasonable step), you would be changing the sine of the angle by 27%. Also, if you change a large angle from degrees to radians, be sure to use an exact conversion factor (180° = π rad) rather than an approximate one (57.3° ≈ 1 rad).

17-6   WAVE SPEED ON A STRETCHED STRING

If a wave travels through a medium such as water, air, steel, or a stretched string, it must set the parti-cles of that medium into oscillation as it passes. For that to happen, the medium must possess both inertia (so that kinetic energy can be stored) and elasticity (so that potential energy can be stored). These two properties determine how fast the wave can travel in the medium. And conversely, it should be possible to calculate the speed of the wave through the medium in terms of these properties. We do so now for a stretched string, in two ways.

Dimensional Analysis

As its name suggests, the dimensional analysis of a situation requires that we examine carefully the dimensions of all the physical quantities that enter into the situation. In this case, we seek a speed v, which has the dimension of length divided by time, or LT^{-1}.

The inertia characteristic of a stretched string is the mass of a string element, which is represented by the mass m of the string divided by the length l of the string. We call this ratio the *linear density* μ of the string. Thus $\mu = m/l$, its dimension being ML^{-1}.

You cannot send a wave along a stretched string without further stretching it. The tension in the string does this stretching and must therefore represent the elastic characteristic of the string. The tension* τ is a force and has the dimension (think about $F = ma$) MLT^{-2}.

The problem is to combine μ (dimension ML^{-1}) and τ (dimension MLT^{-2}) in such a way as to generate v (dimension LT^{-1}). A little juggling of various combinations suggests

$$v = C\sqrt{\frac{\tau}{\mu}}, \qquad (17\text{-}22)$$

in which C is a dimensionless constant. The drawback of dimensional analysis is that it cannot give the value of such a dimensionless constant. In our second approach to determine the speed, you will see that Eq. 17-22 is indeed correct and that $C = 1$.

Derivation from Newton's Second Law

Instead of the sinusoidal wave of Fig. 17-1b, let us consider a single symmetrical pulse such as that of Fig. 17-7. For convenience, we choose a reference

---

*We would normally use the symbol T for tension but we want to avoid confusion with our use of this symbol for the period of oscillation.

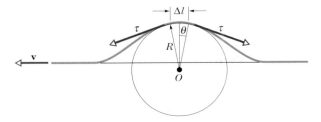

FIGURE 17-7 A symmetrical pulse, viewed from a reference frame in which the pulse is stationary and the string appears to move right to left with speed v. We find speed v by applying Newton's second law to a string element of length Δl, located at the top of the pulse.

frame in which the pulse remains stationary. That is, we run along with the pulse, keeping it constantly in view. In this frame, the string will appear to move past us, from right to left in Fig. 17-7, with speed v.

Consider a small segment of the pulse, of length Δl, forming an arc of a circle of radius R. A force equal in magnitude to the tension τ pulls tangentially on this segment at each end. The horizontal components of these forces cancel but the vertical components add to form a restoring force $\mathbf{F}$. In magnitude,

$$F = 2\tau \sin \theta \approx \tau (2\theta) = \tau \frac{\Delta l}{R} \quad \text{(force).} \quad (17\text{-}23)$$

We have used here the approximation that $\sin \theta \approx \theta$ for small angles, and we note that $2\theta = \Delta l / R$.

The mass of the segment is given by

$$\Delta m = \mu \, \Delta l \quad \text{(mass).} \quad (17\text{-}24)$$

At the moment shown in Fig. 17-7, the string element Δl is moving in an arc of a circle. Thus it has a centripetal acceleration toward the center of that circle, given by

$$a = \frac{v^2}{R} \quad \text{(acceleration).} \quad (17\text{-}25)$$

Equations 17-23, 17-24, and 17-25 contain the elements of Newton's second law. Combining them gives

$$\text{force} = \text{mass} \times \text{acceleration,}$$

or

$$\frac{\tau \, \Delta l}{R} = (\mu \, \Delta l) \frac{v^2}{R}.$$

Solving this equation for the speed v yields

$$v = \sqrt{\frac{\tau}{\mu}} \quad \text{(speed),} \quad (17\text{-}26)$$

in exact agreement with Eq. 17-22 if the constant C in that equation is given the value unity. Equation 17-26 gives the speed of the pulse in Fig. 17-7 and the speed of any other wave on the same string under the same tension.

Equation 17-26 tells us that the speed of a wave along a stretched ideal string depends only on the characteristics of the string and not on the frequency of the wave. The *frequency* of the wave is fixed entirely by whatever generates the wave (for example, the person in Fig. 17-1b). The *wavelength* of the wave is then fixed by Eq. 17-14 ($v = \lambda f$).

SAMPLE PROBLEM 17-4

In Fig. 17-8, a stranded climber has hooked himself onto a makeshift rope lowered by a rescuer. The rope consists of two sections: section 1 with length l_1 and linear density μ_1, and section 2 with length $l_2 = 2l_1$ and linear density $\mu_2 = 4\mu_1$. The climber happens to

FIGURE 17-8 Sample Problem 17-4. A stranded climber hangs from a rope consisting of two sections. His rescuer has secured the rope at top.

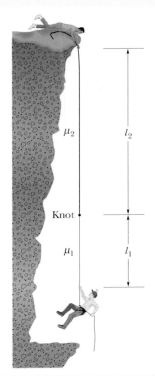

pluck the bottom end of the rope (as a "ready" signal) at the same time the rescuer plucks the top end.

a. What is the speed v_1 of the resulting pulses in section 1, in terms of their speed v_2 in section 2?

SOLUTION We assume that the mass of the rope sections is negligible relative to the mass of the climber. Thus the tension τ in the rope is equal to the weight of the climber and is the same in the two sections. From Eq. 17-26, speeds v_1 and v_2 are given by

$$v_1 = \sqrt{\frac{\tau}{\mu_1}} \quad \text{and} \quad v_2 = \sqrt{\frac{\tau}{\mu_2}}. \quad (17\text{-}27)$$

Dividing the first expression by the second, and substituting $\mu_2 = 4\mu_1$, we find

$$\frac{v_1}{v_2} = \sqrt{\frac{\tau}{\mu_1}}\sqrt{\frac{\mu_2}{\tau}} = \sqrt{\frac{\mu_2}{\mu_1}} = \sqrt{\frac{4\mu_1}{\mu_1}} = 2,$$

or

$$v_1 = 2v_2. \quad \text{(Answer)} \quad (17\text{-}28)$$

b. In terms of l_2, at what distance below the rescuer do the two pulses pass through each other?

SOLUTION We can simplify our calculations by first deciding if the pulses pass each other above or below the knot joining the two sections. Let t be the time the pulses take to reach each other. From Eq. 17-28, we know that the climber's pulse moves through section 1 twice as fast as the rescuer's pulse through section 2. Because $l_2 = 2l_1$, we also know that the climber's pulse must travel half as far as the rescuer's pulse to reach the knot. So the climber's pulse must reach the knot first, and the point of passing must be above the knot. Let us assume that the pulses pass at a distance d below the rescuer at time t.

To reach the point of passing, the rescuer's pulse must travel downward through a distance d at speed v_2 in time t. So

$$t = \frac{d}{v_2}. \quad (17\text{-}29)$$

To reach the point of passing, the climber's pulse must travel upward through distance l_1 at speed v_1 and then through distance $l_2 - d$ at speed v_2, all in time t. So

$$t = \frac{l_1}{v_1} + \frac{l_2 - d}{v_2}. \quad (17\text{-}30)$$

Substituting t from Eq. 17-29 into Eq. 17-30, and setting $l_1 = l_2/2$ and $v_1 = 2v_2$, we find

$$\frac{d}{v_2} = \frac{l_1}{v_1} + \frac{l_2 - d}{v_2} = \frac{l_2/2}{2v_2} + \frac{l_2 - d}{v_2}.$$

Multiplying through by v_2 and rearranging yield

$$d = \tfrac{5}{8}l_2. \quad \text{(Answer)}$$

17-7 THE SPEED OF LIGHT

A velocity has meaning only if you specify a reference frame. For the speed of a wave traveling along a stretched string, the logical reference frame is the undisturbed string itself. The situation is the same for sound waves in air, for ripples in water, for seismic waves in the Earth, and for all other mechanical waves. When we quote a speed, the implied reference frame is the material medium through which the wave travels.

However, visible light (like other electromagnetic waves) requires no medium for transmission. It travels freely through the vacuum of empty space. So when we say, "the speed of light c is 299,792,458 m/s," what reference frame do we have in mind?

The answer to this historic question, which was debated for decades, was given by Einstein in 1905, and it forms one of the two postulates on which his special theory of relativity is based:

▬▬▬

> The speed of light has the same value c in all directions and in all inertial reference frames.

In measuring the speed of light, the relative velocity of the source of light and the observer simply does not matter.

Think about what this says. Suppose that you have a pulsed laser light source in your laboratory and you measure the speed of the light in the beam of pulses that it emits, obtaining a value c. Suppose (see Fig. 17-9) that another experimenter, who is rushing away from you and your light source at a very high speed V along the direction of the beam, also measures the speed of the light in the beam. Einstein's postulate says that you and this second observer will measure the *same* value c for the speed of light. The speed V of the observer relative to you makes no difference.

Many of us, on first meeting this postulate, tend to reject it on the grounds that it violates "common sense." Before doing so, however, consider these facts:

Fact 1. The postulate does not apply to mechanical waves such as sound waves or to material particles such as baseballs—only to light. Light is different.

Fact 2. Our perception of "common sense" is developed from watching objects moving at speeds very much less than the speed of light. If the speed of light were very much lower (say 500 km/h) and thus

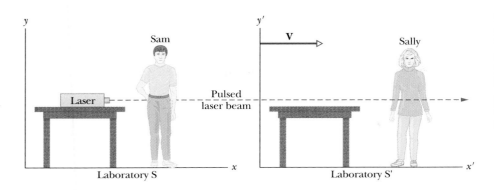

FIGURE 17-9 A laser in laboratory S emits a beam of light. The speed of this light is measured by two observers, one in laboratory S and the other in laboratory S′, which is moving at a high speed V with respect to S. Both observers measure the same speed c for the light.

closer to our daily experiences, none of us would likely have any conceptual problem with this postulate. It would then actually seem to be common sense.

Fact 3. The postulate itself and the predictions of the theory of relativity, for which it forms the foundation, have been tested exhaustively in the laboratory; the agreement with experiment is total. No exceptions have ever been confirmed.

We shall return to this postulate and explore its implications fully in Chapter 42, which deals with Einstein's theory of relativity.

17-8 ENERGY AND POWER IN A TRAVELING WAVE (OPTIONAL)

A wave moving along a stretched string transports both kinetic and potential energy. Let us consider each in turn.

Kinetic Energy

An element of the string of mass dm, oscillating transversely in simple harmonic motion as the wave passes through it, has kinetic energy associated with its transverse velocity **u**. When the element is rushing through its $y = 0$ position (see Fig. 17-10), its transverse velocity—and thus its kinetic energy—is a maximum. When the element is at its extreme position $y = y_m$, its transverse velocity—and thus again its kinetic energy—is zero.

Potential Energy

To send a sinusoidal wave along a previously straight string, the wave must necessarily stretch the string. As a string element of length dx oscillates transversely, its length must increase and decrease in a periodic way if the string element is to fit the sinusoidal wave form. Potential energy is associated with these length changes, just as for a spring.

When the string element is at its $y = y_m$ position (see Fig. 17-10), its length has its normal undisturbed value dx, so its stored potential energy is zero. However, when the element is rushing through its $y = 0$ position, it is stretched to its maximum extent, and its stored potential energy then has its maximum value. The oscillating string element thus has both its maximum kinetic energy and its maximum potential energy at $y = 0$.

The Transmitted Power

The kinetic energy dK associated with a string element of mass dm is given by

$$dK = \tfrac{1}{2}\, dm\, u^2, \qquad (17\text{-}31)$$

FIGURE 17-10 In string element ①, at position $y = y_m$, the stored kinetic energy and the stored potential energy are each zero. In string element ②, at position $y = 0$, these stored energies both have their maximum values. The kinetic energy depends on the transverse velocity of the string element. The potential energy depends on the amount by which the string element is stretched as the wave passes through it.

where u is the transverse speed of the oscillating string element, given by Eq. 17-21 as

$$u = \frac{\partial y}{\partial t} = -\omega y_m \cos(kx - \omega t). \quad (17\text{-}32)$$

Using this relation and putting $dm = \mu \, dx$, we rewrite Eq. 17-31 as

$$dK = \tfrac{1}{2}(\mu \, dx)(-\omega y_m)^2 \cos^2(kx - \omega t). \quad (17\text{-}33)$$

Dividing Eq. 17-33 by dt gives the rate at which the kinetic energy of a string element changes, and thus the rate at which kinetic energy is carried along by the wave. The ratio dx/dt that then appears on the right of Eq. 17-33 is the wave speed v, so we obtain

$$\frac{dK}{dt} = \tfrac{1}{2}\mu v \omega^2 y_m^2 \cos^2(kx - \omega t). \quad (17\text{-}34)$$

The *average* rate at which kinetic energy is transported is

$$\overline{\left(\frac{dK}{dt}\right)} = \tfrac{1}{2}\mu v \omega^2 y_m^2 \, \overline{\cos^2(kx - \omega t)}$$

$$= \tfrac{1}{4}\mu v \omega^2 y_m^2. \quad (17\text{-}35)$$

In Eq. 17-35 we have taken the average over an integer number of wavelengths and have used the fact that the average value of the square of a cosine function over an integer number of wavelengths is $\frac{1}{2}$.

Potential energy is also carried along with the wave, and at the same average rate given by Eq. 17-35. Although we will not examine the proof, you should recall that, in an oscillating system such as a pendulum or a spring–block system, the average kinetic energy and the average potential energy are indeed equal.

The **average power**, which is the average rate at which energy of both kinds is transmitted by the wave, is then

$$\overline{P} = 2\overline{\left(\frac{dK}{dt}\right)} \quad (17\text{-}36)$$

or, from Eq. 17-35,

$$\overline{P} = \tfrac{1}{2}\mu v \omega^2 y_m^2 \quad \text{(average power)}. \quad (17\text{-}37)$$

Factors μ and v in this equation depend on the material and tension of the string. Factors ω and y_m depend on the process that generates the wave. The fact that the average power transmitted by a wave depends on the square of its amplitude and also on the square of its angular frequency is a general result, true for waves of all types.

SAMPLE PROBLEM 17-5

A string has a linear density μ of 525 g/m and is stretched with a tension τ of 45 N. A wave whose frequency f and amplitude y_m are 120 Hz and 8.5 mm, respectively, is traveling along the string. At what average rate is the wave transporting energy along the string?

SOLUTION Before finding $\overline{P}$ from Eq. 17-37, we must calculate the angular frequency ω and the wave speed v. From Eq. 17-10,

$$\omega = 2\pi f = (2\pi)(120 \text{ Hz}) = 754 \text{ rad/s}.$$

From Eq. 17-26 we have

$$v = \sqrt{\frac{\tau}{\mu}} = \sqrt{\frac{45 \text{ N}}{0.525 \text{ kg/m}}} = 9.26 \text{ m/s}.$$

Equation 17-37 then yields

$$\overline{P} = \tfrac{1}{2}\mu v \omega^2 y_m^2$$

$$= (\tfrac{1}{2})(0.525 \text{ kg/m})(9.26 \text{ m/s})$$

$$\times (754 \text{ rad/s})^2 (0.0085 \text{ m})^2$$

$$= 100 \text{ W}. \quad \text{(Answer)}$$

17-9 THE PRINCIPLE OF SUPERPOSITION

It often happens that two or more waves pass simultaneously through the same region. When we listen to a concert, for example, sounds from many instruments fall simultaneously on our eardrums. The electrons in the antennas of our radio and TV sets are set in motion by a whole array of signals from different broadcasting centers. The water of a lake or harbor may be churned up by the wakes of many boats.

Suppose that two waves travel simultaneously along the same stretched string. Let $y_1(x, t)$ and $y_2(x, t)$ be the displacements that the string would experience if each wave acted alone. The displacement of the string when both waves act is then

$$y(x, t) = y_1(x, t) + y_2(x, t), \quad (17\text{-}38)$$

the sum being an algebraic sum. This is another example of the **principle of superposition,** which you met in Section 15-3. It says that when several effects occur simultaneously, their net effect is the sum of the individual effects. For waves on a stretched string, this means that the net displacement of the

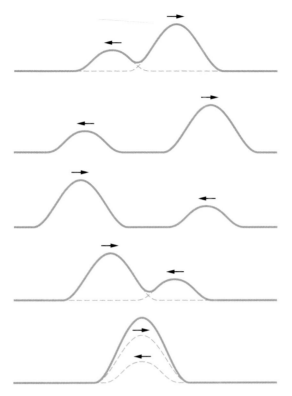

FIGURE 17-11 Two pulses travel in opposite directions along a stretched string. The superposition principle applies as they move through each other.

string at any point along its length is the sum of the displacements the waves would have produced individually. This result holds as long as the amplitudes of the waves are not too large, which we assume here.

Figure 17-11 shows a time sequence of "snapshots" of two pulses traveling in opposite directions in the same stretched string. When the pulses overlap, the displacement of the string is the algebraic sum of the individual displacements of the string, as Eq. 17-38 requires. Each pulse moves through the other, as if the other were not present.

Fourier Analysis

The French mathematician Jean Baptiste Fourier (1786–1830) explained how the principle of superposition can be used to analyze nonsinusoidal wave forms. He showed that any wave form can be represented as the sum of a large number of sinusoidal waves, of carefully chosen frequencies and amplitudes. The English physicist Sir James Jeans expressed it well:

[Fourier's] theorem tells us that every curve, no matter what its nature may be, or in what way it was originally obtained, can be exactly reproduced by superposing a sufficient number of simple harmonic curves—in brief, every curve can be built up by piling up waves.

Figure 17-12 shows an example of *Fourier series,* as such sums are called. The sawtooth curve in Fig. 17-12*a* shows the variation with time (at position $x = 0$) of the wave we wish to represent. The Fourier series that represents it can be shown to be

$$y(t) = -\frac{1}{\pi} \sin \omega t - \frac{1}{2\pi} \sin 2\omega t$$

$$-\frac{1}{3\pi} \sin 3\omega t \cdots , \qquad (17\text{-}39)$$

in which $\omega = 2\pi/T$, where T is the period of the sawtooth curve. The green curve of Fig. 17-12*a*, which represents the sum of the first six terms of Eq. 17-39, matches the sawtooth curve rather well. Figure 17-12*b* shows these six terms separately. By adding more terms, you could approximate the sawtooth curve as closely as you wish.

You can see now why we were justified in spending so much time analyzing the behavior of a sinusoidal wave. Once we understand that, Fourier's theorem opens the door to all other wave shapes.

17-10 DISPERSION (OPTIONAL)

A truly sinusoidal wave, such as that of Fig. 17-13*a*, has no beginning and no end, either in space or time, and all its intervals of length λ are identical. Such a wave is not only merely theoretical, but it would be useless in sending a signal from point to point because it is unchanging.

We *can* send a signal with a *pulse,* such as that of Fig. 17-13*b*, which might represent the quantity 1 in the binary arithmetic of computers. The pulse of Fig. 17-13*b* is *not* a pure sinusoidal wave, because it is finite in both space and time. From Fourier's analysis, this pulse can be built up by combining (infinitely extended) sinusoidal waves of appropriately chosen frequencies and amplitudes. It is impressive that an assembly of sinusoidal waves can be found such that they add up to the pulse of length l in Fig. 17-13*b*, canceling to zero everywhere beyond its limits.

If the combining sinusoidal waves all travel through a medium with the same speed, then their addition always yields a pulse with the same shape.

FIGURE 17-12 (a) The dashed sawtooth curve is approximated by the green curve, which is the result of keeping the first six terms in the sum of Eq. 17-39. (Keeping more terms would give closer approximations.) (b) Separate displays of the first six terms of that equation.

In such situations, the medium and the combining waves are said to be **dispersionless,** which means that the pulse retains its shape as it moves. If, however, the speed of a sinusoidal wave in a medium depends on its frequency (or, equivalently, its wavelength), the combining waves with different frequencies move with different speeds and do not add up to give a constant pulse shape. Instead, the pulse spreads (grows longer). In such a situation, the medium and the combining waves are said to exhibit **dispersion,** which means that the pulse changes shape as it moves.

For sinusoidal waves on a stretched string, the speed is given by Eq. 17-26 and we see that this speed does not depend on the wavelength. Sound waves behave in the same way, having only a single speed for all frequencies or wavelengths. So, sound waves traveling through air are dispersionless. If you shout toward a distant cliff, the returning echo retains the wave form of the shout; you can recognize your own words.

Light waves traveling through vacuum are also dispersionless. However, when light travels through a transparent medium such as glass or water, the speed then *does* depend on the wavelength; in such media, light waves exhibit dispersion. It is because of dispersion that sunlight spreads out into a spectrum of colors as it passes through a glass prism and forms a rainbow when it passes through falling raindrops.

The speed at which a pulse travels is called the

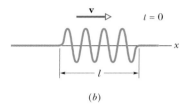

FIGURE 17-13   (a) A "snapshot" of the sinusoidal wave of Eq. 17-2, at $t = 0$. (b) A "snapshot" of a pulse of length l. This too can be represented by Fourier methods as the summation of many sinusoidal waves of different wavelengths and amplitudes.

group speed of the pulse. This is the speed at which signals or information can be transmitted by the pulse. If there is no dispersion, the group speed is the same as the common speed of the sinusoidal waves, given by Eq. 17-26 for the case of waves traveling along a stretched string. If there *is* dispersion, the group speed and the speeds of the sinusoidal waves differ, and the signal may be lost.

17-11 INTERFERENCE OF WAVES

Suppose that we send two sinusoidal waves of the same wavelength and amplitude in the same direction along a stretched string. The superposition principle applies. What resultant disturbance does it predict for the string?

Everything depends on the extent to which the waves are *in phase* (in step) with respect to each other. If they are exactly in phase, they will add up to double the displacement of either wave acting alone. If they are exactly out of phase, they will cancel everywhere, producing no disturbance at all. We call this phenomenon of cancellation and reinforcement **interference;** it applies to waves of all kinds.

Let two waves given by

$$y_1(x, t) = y_m \sin(kx - \omega t + \phi) \quad (17\text{-}40)$$

and

$$y_2(x, t) = y_m \sin(kx - \omega t) \quad (17\text{-}41)$$

travel along the same stretched string. These waves have the same angular frequency ω, the same angu-

lar wave number k, and the same amplitude y_m. They travel in the same direction, that of increasing x, with the same speed, given by Eq. 17-26. They differ only by a constant angle ϕ, which is called the **phase constant.**

From the principle of superposition (Eq. 17-38), the combined wave has displacement

$$y(x, t) = y_1(x, t) + y_2(x, t)$$
$$= y_m[\sin(kx - \omega t + \phi) + \sin(kx - \omega t)]. \quad (17\text{-}42)$$

From Appendix G we see that we can write the sum of the sines of two angles as

$$\sin \alpha + \sin \beta = 2 \sin \tfrac{1}{2}(\alpha + \beta)\cos \tfrac{1}{2}(\alpha - \beta). \quad (17\text{-}43)$$

Applying this relation to Eq. 17-42 yields

$$y(x, t) = [2y_m \cos \tfrac{1}{2}\phi]\sin(kx - \omega t + \tfrac{1}{2}\phi). \quad (17\text{-}44)$$

The resultant wave is thus also a sinusoidal wave, differing from the original waves only in its phase constant, which is $\tfrac{1}{2}\phi$, and in its amplitude, which is the quantity in the brackets, namely, $2y_m \cos \tfrac{1}{2}\phi$.

If $\phi = 0$, the two waves are exactly in phase. Then Eq. 17-44 reduces to

$$y(x, t) = 2y_m \sin(kx - \omega t) \qquad (\phi = 0). \quad (17\text{-}45)$$

The interference is fully *constructive*, and the resultant wave differs from the two combining waves only in having twice the amplitude. Figure 17-14a shows a case in which, although ϕ is not zero, it is very small.

If $\phi = \pi$ rad (or 180°), the two waves are exactly out of phase. Then $\cos \tfrac{1}{2}\phi$ becomes $\cos \pi/2 = 0$, and the amplitude of the resultant wave (Eq. 17-44) is zero. We have, then, for all values of x and t,

$$y(x, t) = 0 \qquad (\phi = \pi \text{ rad}). \quad (17\text{-}46)$$

So when the combining waves are exactly out of phase, their interference is totally *destructive*. Figure 17-14b shows a case in which, although ϕ is not π rad, it is close to that value. The two combining waves almost—but not quite—cancel.

SAMPLE PROBLEM 17-6

Two traveling waves, moving in the same direction along a stretched string, interfere with each other. The

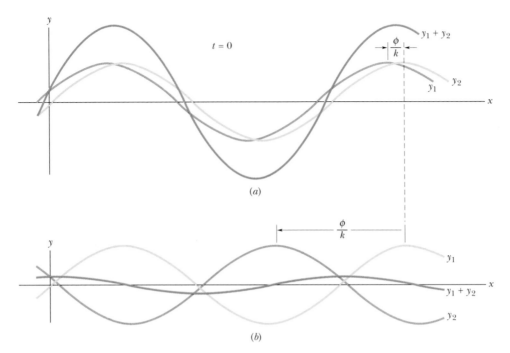

(a)

(b)

FIGURE 17-14 (a) Two waves, whose phases differ by only a small amount, reinforce each other, almost giving fully constructive interference. The quantity $\phi/k = (\phi/2\pi)\lambda$ gives the distance along the x axis between adjacent *crests* of the two waves. (b) Two waves, whose phases differ by almost π rad (180°), almost cancel each other.

amplitude y_m of each wave is 9.7 mm and the phase difference ϕ between them is 110°.

a. What is the amplitude y'_m of the wave formed by the interference of these two waves?

SOLUTION From Eq. 17-44 we have for the amplitude

$$y'_m = 2y_m \cos \tfrac{1}{2}\phi = (2)(9.7 \text{ mm})(\cos 110°/2)$$

$$= 11 \text{ mm.} \qquad \text{(Answer)}$$

b. What phase difference ϕ between the two combining waves would result in an amplitude for the combined wave that is the same as the identical amplitudes of the combining waves?

SOLUTION From Eq. 17-44 we have the requirement

$$2y_m \cos \tfrac{1}{2}\phi = y_m,$$

or

$$\phi = 2 \cos^{-1}\left(\tfrac{1}{2}\right) = 120° \text{ or } -120°$$

$$= 2.1 \text{ rad} \quad \text{or} \quad -2.1 \text{ rad.} \qquad \text{(Answer)}$$

Thus there are two solutions, one corresponding to the first wave leading the second in time, and the other corresponding to the second wave leading the first.

PROBLEM SOLVING

TACTIC 2: TRIGONOMETRIC IDENTITIES
In deriving Eq. 17-44 above (and in many other places), we have used trigonometric identities such as that of Eq. 17-43. Practicing engineers and scientists become familiar with these through constant use but, because there are so many of them, the beginner is often at a loss. Appendix G lists all the identities used in this book, plus a few more for completeness. It is helpful to study them, looking for patterns so that, for example, when you see something like $\sin \alpha + \sin \beta$ you will know that this can be expressed in another form.

Try to cast formulas into forms for which identities exist. For example, if you run across $3 \sin \alpha \cos \alpha$, you can write it as $\left(\tfrac{3}{2}\right)(2 \sin \alpha \cos \alpha)$, which (see Appendix G) is just $\tfrac{3}{2} \sin 2\alpha$. Rejoice when you see $\sin^2 \alpha + \cos^2 \alpha$ because it is equal to unity. Remember that, although the average value of a sine or a cosine function over one wavelength is zero, the average value over half a wavelength is $\tfrac{1}{2}\sqrt{2}$ and the average of the *square* of a sine or a cosine function over one wavelength is $\tfrac{1}{2}$.

17-12 STANDING WAVES

In the preceding section, we discussed two sinusoidal waves of the same wavelength and amplitude traveling *in the same direction* along a stretched string. What if they travel in opposite directions? We can see the result the two waves then give by applying the superposition principle.

Figure 17-15 suggests the situation graphically. It shows the two combining waves, one traveling to the left in Fig. 17-15a, the other to the right in Fig. 17-15b. Figure 17-15c shows their sum, obtained by applying the superposition principle graphically. The outstanding feature of the resultant wave is that there are places along the string, called **nodes**, where the string is permanently at rest. Four such nodes are marked by dots in Fig. 17-15c. Halfway between adjacent nodes are **antinodes,** where the amplitude of the resultant wave is a maximum. Wave patterns such as that of Fig. 17-15c are called **standing waves** because the patterns do not move; that is, the locations of the maxima and minima do not change. Let us analyze them mathematically.

We represent the two combining waves by

$$y_1(x, t) = y_m \sin(kx - \omega t) \qquad (17\text{-}47)$$

and

$$y_2(x, t) = y_m \sin(kx + \omega t). \qquad (17\text{-}48)$$

We introduce no phase constants here because it means nothing to speak of a phase difference between waves that travel in opposite directions. (As an analogy, the sweep-second hands of two clocks can

maintain a constant angular separation if the hands are rotating in the same direction, but they cannot do so if they are rotating in opposite directions.)

The principle of superposition gives, for the combined wave,

$$\begin{aligned} y(x, t) &= y_1(x, t) + y_2(x, t) \\ &= y_m \sin(kx - \omega t) + y_m \sin(kx + \omega t). \end{aligned}$$

Applying the trigonometric relation of Eq. 17-43 leads to

$$y(x, t) = [2y_m \sin kx] \cos \omega t. \qquad (17\text{-}49)$$

This is *not* a traveling wave because it is not of the form of Eq. 17-18. Equation 17-49 describes a standing wave.

The quantity $2y_m \sin kx$ in the brackets of Eq. 17-49 can be viewed as the amplitude of oscillation of the string element that is located at position x. However, since an amplitude is always positive, we take the absolute value of the quantity $2y_m \sin kx$ to be the amplitude.

In a traveling sinusoidal wave, the amplitude of the wave is the same for all string elements. That is not true for a standing wave, in which the amplitude *varies with position*. In the standing wave of Eq. 17-49, the amplitude varies as $\sin kx$. For example, the amplitude is zero for values of kx that give $\sin kx = 0$. Those values are

$$kx = n\pi, \qquad \text{for } n = 0, 1, 2, \ldots . \qquad (17\text{-}50)$$

Substituting $k = 2\pi/\lambda$ and rearranging, we get

FIGURE 17-15 How traveling waves produce standing waves. (a) and (b) represent "snapshots" of two waves of the same wavelength and amplitude, traveling in opposite directions, at five different instants during a single period of oscillation. (c) Their superposition at the five instants. Note the nodes and antinodes in (c), the nodes being represented by dots. There are no nodes or antinodes in the traveling waves (a) and (b).

$$x = n\frac{\lambda}{2}, \qquad \text{for } n = 0, 1, 2, \ldots$$

$$\text{(nodes)}, \quad (17\text{-}51)$$

as the positions of zero amplitude—the nodes—for the standing wave of Eq. 17-49. Note that adjacent nodes are separated by $\lambda/2$, half a wavelength.

The amplitude of the standing wave of Eq. 17-49 has a maximum value of $2y_m$, which occurs for values of kx that give $|\sin kx| = 1$. Those values are

$$kx = \tfrac{1}{2}\pi, \tfrac{3}{2}\pi, \tfrac{5}{2}\pi, \ldots = \frac{2n+1}{2}\pi$$

$$= (n + \tfrac{1}{2})\pi, \qquad \text{for } n = 0, 1, 2, \ldots. \quad (17\text{-}52)$$

Substituting $k = 2\pi/\lambda$ and rearranging, we get

$$x = (n + \tfrac{1}{2})\frac{\lambda}{2}, \qquad \text{for } n = 0, 1, 2, \ldots$$

$$\text{(antinodes)}, \quad (17\text{-}53)$$

as the positions of maximum amplitude—the antinodes—of the standing wave of Eq. 17-49. The antinodes are one-half wavelength apart and are located halfway between pairs of nodes.

Reflections at a Boundary

We can set up a standing wave in a stretched string by allowing a traveling wave to be reflected from the far end of the string. The incident (original) wave and the reflected wave can then be described by Eqs. 17-47 and 17-48, respectively, and they can combine to form a pattern of standing waves.

In Fig. 17-16, we use a single pulse to show how such reflections take place. In Fig. 17-16a, the string is fixed at its left end. When the pulse arrives at that end, it exerts an upward force on the support (the wall). By Newton's third law, the support exerts an equal but opposite force on the string. This reaction force generates a pulse at the support, which travels back along the string in a direction opposite that of the incident pulse. In a "hard" reflection of this kind, there must be a node at the support because the string is fixed there. The reflected and incident pulses must have opposite signs, so as to cancel each other at that point.

In Fig. 17-16b, the left end of the string is fastened to a light ring that is free to slide without friction along a rod. When the incident pulse arrives, the ring moves up the rod. As the ring moves, it pulls

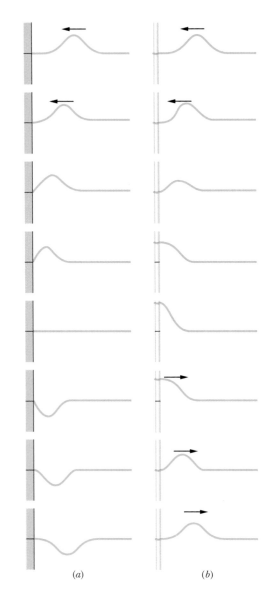

FIGURE 17-16 (a) A pulse incident from the right is reflected at a rigid wall. Note that the sign of the reflected pulse is reversed. (b) Here the termination is flexible, the loop being able to slide without friction up and down the rod. The pulse is reflected without a change of sign.

on the string, stretching the string and producing a reflected pulse with the same sign and amplitude as the incident pulse. Thus in such a "soft" reflection, the incident and reflected pulses reinforce each other, creating an antinode at the end of the string; the maximum displacement of the ring is twice the amplitude of either of these pulses.

FIGURE 17-17 Stroboscopic photographs reveal (imperfect) standing wave patterns on a string being made to oscillate by the vibrator at the left end. The patterns correspond to $n = 2, 3,$ and 4 in Eqs. 17-54 to 17-56.

17-13 STANDING WAVES AND RESONANCE

Figure 17-17 shows three standing wave patterns (or **oscillation modes**) that can be set up by vibrating one end of a stretched string at different frequencies, with the other end fixed. Note the nodes and antinodes. The oscillation modes of Fig. 17-17 occur only at sharply defined frequencies. We say that the system *resonates* at these frequencies. If the string is vibrated at some frequency other than these **resonant frequencies,** a standing wave is not set up, and the oscillations of the string are not substantial.

Consider a similar situation in which a string, such as a guitar string, is stretched between two clamps separated by a fixed distance l, and the string is somehow made to oscillate at a resonant frequency to set up a standing wave pattern. Since each end of the string is fixed, there must be a node at each end. The simplest pattern that meets this requirement is shown in Fig. 17-18a. There is an antinode at the center of the string, and the distance l is equal to $\lambda/2$, where λ is the wavelength required of the waves on the string to set up this standing wave pattern.

A second simple pattern meeting the requirement of fixed ends is shown in Fig. 17-18b. This pattern has three nodes and two antinodes; the distance l and the required wavelength are related by $l = \lambda$. A third pattern is shown in Fig. 17-18c. It has four nodes and three antinodes, and $l = 3\lambda/2$. If you like, you can continue the progression, drawing increasingly more complicated patterns. In each step of the progression, the pattern has one more node and one more antinode than the preceding step, and an additional $\lambda/2$ is fitted into distance l.

The relation between λ and l can be summarized as

$$l = \frac{n}{2}\lambda, \qquad \text{for } n = 1, 2, 3, \ldots . \quad (17\text{-}54)$$

Solving for λ gives us the wavelengths of the waves on the string for these successive patterns:

$$\lambda = \frac{2l}{n}, \qquad \text{for } n = 1, 2, 3, \ldots . \quad (17\text{-}55)$$

The resonant frequencies follow from Eq. 17-55 and Eq. 17-14:

$$f = \frac{v}{\lambda} = \frac{v}{2l}\,n, \qquad \text{for } n = 1, 2, 3, \ldots . \quad (17\text{-}56)$$

Equation 17-56 tells us that the resonant frequencies are integer multiples of the lowest resonant frequency, $f = v/2l$, which corresponds to $n = 1$. The oscillation mode with that lowest frequency is called

FIGURE 17-18 A string, stretched between two clamps, is made to oscillate in standing wave patterns. (a) The simplest possible pattern, the *fundamental* or *first harmonic*, is said to consist of one *loop*, which refers to the composite shape formed by the string in its extreme displacements (the solid and dashed lines). (b) The pattern of the *second harmonic* has two loops. (c) The pattern of the *third harmonic* has three loops.

FIGURE 17-19 Four of many possible standing wave patterns for the head of a kettledrum. They are made visible by sprinkling dark powder on the drumhead. As the drumhead is set into oscillation at a single frequency, by a mechanical vibrator at upper left in each photograph, the powder collects at the nodes, which are circles and straight lines (rather than points) in this two-dimensional example.

FIGURE 17-20 Hologram interferograms of three-dimensional standing waves on a vibrating c_5 handbell. The nodal regions are bright; the antinodal regions are patterned.

the *fundamental mode* or the *first harmonic*. The *second harmonic* is the oscillation mode with $n = 2$, the *third harmonic* is that with $n = 3$, and so on. The collection of all possible oscillation modes is called the **harmonic series,** and n is called the **harmonic number.**

If the string is wiggled at a frequency not given by Eq. 17-56, it will not be possible to transfer energy efficiently from the external oscillating agent to the string. During some time intervals, the external agent, say a mechanical vibrator, will do work on the string; during other intervals, the string will do work back on the vibrator. However, at *resonance*, that is, at a resonant frequency, the energy flow is entirely from the vibrator to the string. The oscillation amplitude and frictionlike losses build up until the oscillating string loses energy via those losses just as fast as it receives energy from the vibrator.

The phenomenon of resonance is common to all oscillating systems. Figure 17-19 shows four of the many (two-dimensional) oscillation modes of a kettledrum head, and Fig. 17-20 shows standing wave patterns on a handbell. In quantum physics, the states in which atoms may exist are interpreted as (three-dimensional) oscillation modes of the matter waves that represent the atomic electrons.

FIGURE 17-21 Sample Problem 17-7. A string under tension connected to a vibrator. For a fixed vibrator frequency, standing wave patterns will occur for discrete values of the tension in the string.

SAMPLE PROBLEM 17-7

In Fig. 17-21, a string, tied to a vibrator at P and running over a support at Q, is stretched by a block of mass m. The separation l between P and Q is 1.2 m, the linear density of the string is 1.6 g/m, and the frequency f of the vibrator is fixed at 120 Hz. The amplitude of the motion at P is small enough for that point to be considered a node. A node also exists at Q.

a. What mass m allows the vibrator to set up the fourth harmonic on the string?

SOLUTION The resonant frequencies are given by Eq. 17-56 as

$$f = \frac{v}{2l}\, n, \qquad \text{for } n = 1, 2, 3, \ldots \quad (17\text{-}57)$$

We need to set the tension τ in the string so that the vibrator frequency is equal to the frequency of the fourth harmonic as given by this equation.

The speed v of waves on the string is given by Eq. 17-26 as

$$v = \sqrt{\frac{\tau}{\mu}} = \sqrt{\frac{mg}{\mu}}, \qquad (17\text{-}58)$$

where the tension τ in the string is equal to the weight mg of the block. Substituting v from Eq. 17-58 into Eq. 17-57, setting $n = 4$ for the fourth harmonic, and solving for m give us

$$m = \frac{4l^2 f^2 \mu}{n^2 g} \qquad (17\text{-}59)$$

$$= \frac{(4)(1.2\ \text{m})^2 (120\ \text{Hz})^2 (0.0016\ \text{kg/m})}{(4)^2 (9.8\ \text{m/s}^2)}$$

$$= 0.846\ \text{kg} \approx 0.85\ \text{kg}. \qquad \text{(Answer)}$$

b. What standing wave mode is set up if $m = 1.00$ kg?

SOLUTION If we insert this value of m into Eq. 17-59 and solve for n, we find that $n = 3.7$, an impossibility since n must be an integer. Thus, with $m = 1.00$ kg, the vibrator cannot set up a standing wave on the string, and any oscillation of the string is small, perhaps even imperceptible.

REVIEW & SUMMARY

Transverse and Longitudinal Waves

Waves on a stretched string, the subject of this chapter, are **transverse** mechanical waves governed by Newton's laws. The particles of the medium (the string) oscillate perpendicular to the direction of motion of the wave. Waves in which particles of the medium oscillate parallel to the wave's direction of travel are **longitudinal** waves.

Sinusoidal Waves

A sinusoidal wave moving in the $+x$ direction has the mathematical form

$$y(x, t) = y_m \sin(kx - \omega t), \qquad (17\text{-}2)$$

where y_m is the **amplitude** of the wave, k the **angular wave number,** ω the **angular frequency,** and $kx - \omega t$ the **phase.** The **wavelength** λ and **wave number** κ (the number of waves per meter) are related to k by

$$\frac{k}{2\pi} = \kappa = \frac{1}{\lambda}. \qquad (17\text{-}6)$$

The **period** T and **frequency** f of the wave are related to ω by

$$\frac{\omega}{2\pi} = f = \frac{1}{T}. \qquad (17\text{-}10)$$

Finally, the **wave speed** v is related to these other parameters by

$$v = \frac{\omega}{k} = \frac{\lambda}{T} = \lambda f. \qquad (17\text{-}14)$$

Traveling Waves

In general, any function of the form

$$y(x, t) = h(kx \pm \omega t) \qquad (17\text{-}18)$$

can represent a **traveling wave** with a wave speed given by Eq. 17-14 and a wave shape given by the mathematical form of h. The plus (or minus) sign denotes a wave traveling in the $-x$ (or $+x$) direction.

Wave Speed on Stretched String

The speed of a wave on a stretched string with tension τ and linear density μ is

$$v = \sqrt{\frac{\tau}{\mu}}. \qquad (17\text{-}26)$$

The Speed of Light

The speed of light, in vacuum, has the same value c in all inertial reference frames. This invariance of the speed of light is one of the two postulates of Einstein's special theory of relativity and is now thought to be universally valid.

Power

The **average power,** the average rate at which energy is transmitted by a sinusoidal wave on a stretched string, is given by

$$\overline{P} = \tfrac{1}{2}\mu v \omega^2 y_m^2. \qquad (17\text{-}37)$$

Superposition

When two or more waves traverse the same medium, the displacement of any particle of the medium is the sum of the displacements that the individual waves would give it. This is called **superposition.**

Fourier Series

With a *Fourier series,* any wave can be constructed as the superposition of appropriate sinusoidal waves.

Dispersion

The transfer of information requires a **wave pulse,** which according to Fourier's analysis can be thought of as a superposition of sinusoidal waves. If the individual waves all have the same wave speed, the pulse shape is transmitted without change at the common speed of the waves; this situation is **dispersionless.** If the individual waves have different speeds, the wave is said to exhibit **dispersion** and the pulse changes shape as it travels. The pulse shape then travels at the **group speed.**

Interference of Waves

Two sinusoidal waves on the same string exhibit **interference,** adding or canceling according to the principle of superposition. If the two are traveling in the same direction and have the same amplitude y_m and frequency (and hence the same wavelength), but differ in phase by a **phase constant** ϕ, the result is a single wave with this same frequency:

$$y(x, t) = [2y_m \cos \tfrac{1}{2}\phi] \sin(kx - \omega t + \tfrac{1}{2}\phi). \qquad (17\text{-}44)$$

If $\phi = 0$, the waves are in phase and their interference is (fully) constructive; if $\phi = \pi/2$ rad, they are out of phase and their interference is destructive.

Standing Waves

The interference of two sinusoidal waves having the same frequency and amplitude but moving in opposite directions produces **standing waves** with the equation, for a string with fixed ends,

$$y(x, t) = [2y_m \sin kx]\cos \omega t. \qquad (17\text{-}49)$$

Standing waves are characterized by fixed positions of zero displacement called **nodes** and fixed positions of maximum displacement called **antinodes.**

Resonance

Standing waves on a string are typically induced by reflection of traveling waves from the ends of the string. If an end is fixed, it must be the position of a node; if it is free, it is the position of an antinode. These conditions limit the frequencies at which standing waves will occur on a given string. Each possible frequency is a **resonant frequency,** and the corresponding standing wave pattern is an **oscillation mode.** For a stretched string with fixed ends the resonant frequencies are

$$f = \frac{v}{\lambda} = \frac{v}{2l}\, n, \qquad \text{for } n = 1, 2, 3, \ldots. \qquad (17\text{-}56)$$

The oscillation mode corresponding to $n = 1$ is called the *fundamental mode* or the *first harmonic;* the mode corresponding to $n = 2$ is the *second harmonic;* and so on. The string will readily absorb energy if it is excited at one of its resonant frequencies; this is the phenomenon of **resonance.** Excited at other frequencies, it absorbs little energy.

QUESTIONS

1. How could you prove experimentally that energy can be transported by a wave?

2. Energy can be transferred by particles as well as by waves. How can we experimentally distinguish between these methods of energy transfer?

3. Can a wave motion be generated in which the particles of the medium vibrate with angular simple harmonic motion? If so, explain how and describe the wave.

4. The following functions, in which A is a constant, are of the form $h(x \pm vt)$:

$$y = A(x - vt), \qquad y = A(x + vt)^2,$$
$$y = A\sqrt{x - vt}, \qquad y = A\ln(x + vt).$$

Explain why these functions are not useful in wave motion.

5. Can one produce on a string a wave form that has a discontinuity in slope at a point, that is, one having a sharp corner? Explain.

6. Compare and contrast the behavior of (a) the mass of a mass–spring system, oscillating in simple harmonic motion, and (b) an element of a stretched string through which a traveling sinusoidal wave is passing. Discuss from the point of view of displacement, velocity, acceleration, and energy transfers.

7. A passing motorboat creates a wake that causes waves to wash ashore. As time goes on, the period of the arriving waves grows shorter and shorter. Why?

8. When two waves interfere, does one alter the progress of the other? Explain.

9. When waves interfere, is there a loss of energy? Explain your answer.

10. As Fig. 17-15 shows, twice during an oscillation the configuration of standing waves in a stretched string is a straight line, exactly what it would be if the string were not oscillating at all. Where is the energy of the standing wave at these times?

11. If two waves differ only in amplitude and travel in opposite directions through a medium, will they produce standing waves? Is energy transported? Are there any nodes?

12. In the discussion of transverse waves in a string, we have dealt only with displacements in a single plane, the xy plane. If all displacements lie in one plane, the wave is said to be *plane polarized*. Can there be displacements in a plane other than the single plane with which we dealt? If so, can

two differently plane-polarized waves be combined? What appearance would such a combined wave have?

13. A wave transmits energy. Does it transfer linear momentum? Can it transfer angular momentum?

14. In the Mexico City earthquake of 19 September 1985, areas with high damage alternated with areas of low damage. Also, buildings between 5 and 15 stories high sustained the most damage. Discuss these effects in terms of standing waves and resonance.

15. A string is stretched between two fixed supports that are separated by distance l. (a) For which harmonics is there a node at the point located $l/3$ from one of the supports? Is there a node, antinode, or some intermediate condition at a point located $2l/5$ from one support if (b) the fifth harmonic is set up and (c) the tenth harmonic is set up?

16. Strings A and B have identical lengths and linear densities, but string B is under greater tension than is string A. In Fig. 17-22, four situations, (a) through (d), are shown in which standing wave patterns exist on the strings. In which situation is there the possibility that strings A and B are oscillating at the same resonant frequency?

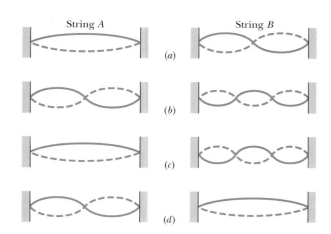

FIGURE 17-22   Question 16.

17. Guitar players know that prior to a concert, a guitar must be played and the strings tightened, because during the first few minutes of playing, the strings warm and loosen slightly. How does that loosening affect the resonant frequencies of the strings?

EXERCISES & PROBLEMS

SECTION 17-5 THE SPEED OF A TRAVELING WAVE

1E. A wave has a speed of 240 m/s and a wavelength of 3.2 m. What are the (a) frequency and (b) period of the wave?

2E. A wave has angular frequency 110 rad/s and wavelength 1.80 m. Calculate (a) the angular wave number and (b) the speed of the wave.

3E. By rocking a boat, a boy produces surface water waves on a previously quiet lake. He observes that the boat performs 12 oscillations in 20 s, each oscillation producing a wave crest 15 cm above the undisturbed surface of the lake. He further observes that a given wave crest reaches shore, 12 m away, in 6.0 s. What are (a) the period, (b) the speed, (c) the wavelength, and (d) the amplitude of this wave?

4E. The speed of electromagnetic waves in vacuum is 3.0×10^8 m/s. (a) Wavelengths of (visible) light waves range from about 400 nm in the violet to about 700 nm in the red. What is the range of frequencies of light waves? (b) The range of frequencies for shortwave radio (for example, FM radio and VHF television) is 1.5–300 MHz. What is the corresponding wavelength range? (c) X rays are also electromagnetic. Their wavelength range extends from about 5.0 nm to about 1.0×10^{-2} nm. What is the frequency range for x-rays?

5E. A sinusoidal wave travels along a string. The time for a particular point to move from maximum displacement to zero is 0.170 s. What are the (a) period and (b) frequency? (c) The wavelength is 1.40 m; what is the wave speed?

6E. Write the equation for a wave traveling in the negative direction along the x axis and having an amplitude of 0.010 m, a frequency of 550 Hz, and a speed of 330 m/s.

7E. A traveling wave on a string is described by

$$y = 2.0 \sin \left[2\pi \left(\frac{t}{0.40} + \frac{x}{80} \right) \right],$$

where x and y are in centimeters and t is in seconds. (a) For $t = 0$, plot y as a function of x for $0 \le x \le 160$ cm. (b) Repeat (a) for $t = 0.05$ s and $t = 0.10$ s. (c) From your graphs, what is the wave speed, and in which direction $(+x$ or $-x)$ is the wave traveling?

8E. Show that $y = y_m \sin(kx - \omega t)$ may be written in the alternative forms

$$y = y_m \sin k(x - vt), \qquad y = y_m \sin 2\pi \left(\frac{x}{\lambda} - ft \right),$$

$$y = y_m \sin \omega \left(\frac{x}{v} - t \right), \qquad y = y_m \sin 2\pi \left(\frac{x}{\lambda} - \frac{t}{T} \right).$$

9E. A single pulse, whose wave shape is given by the function $h(x - 5t)$, is shown in Fig. 17-23 for $t = 0$. Here x is in

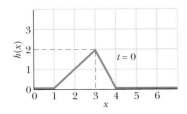

FIGURE 17-23 Exercise 9.

centimeters and t is in seconds. What are the (a) speed and (b) direction of travel of the pulse? (c) Plot $h(x - 5t)$ as a function of x for $t = 2$ s. (d) Plot $h(x - 5t)$ as a function of t for $x = 10$ cm.

10E. Show (a) that the maximum transverse speed of a particle in a string owing to a traveling wave is given by $u_{max} = \omega y_m = 2\pi f y_m$ and (b) that the maximum transverse acceleration is $a_{y,max} = \omega^2 y_m = 4\pi^2 f^2 y_m$.

11E. The equation of a transverse wave traveling in a string is given by

$$y = (2.0 \text{ mm}) \sin[(20 \text{ m}^{-1})x - (600 \text{ s}^{-1})t].$$

(a) Find the amplitude, frequency, velocity, and wavelength of the wave. (b) Find the maximum transverse speed of a particle in the string.

12E. (a) Write an expression describing a sinusoidal transverse wave traveling on a cord in the $+y$ direction with a wave number of 60 cm^{-1}, a period of 0.20 s, and an amplitude of 3.0 mm. Take the transverse direction to be the z direction. (b) What is the maximum transverse speed of a point on the cord?

13P. The equation of a transverse wave traveling along a very long string is given by $y = 6.0 \sin(0.020\pi x + 4.0\pi t)$, where x and y are expressed in centimeters and t is in seconds. Determine (a) the amplitude, (b) the wavelength, (c) the frequency, (d) the speed, (e) the direction of propagation of the wave, and (f) the maximum transverse speed of a particle in the string. (g) What is the transverse displacement at $x = 3.5$ cm when $t = 0.26$ s?

14P. (a) Write an equation describing a sinusoidal transverse wave traveling on a cord in the $+x$ direction with a wavelength of 10 cm, a frequency of 400 Hz, and an amplitude of 2.0 cm. (b) What is the maximum speed of a point on the cord? (c) What is the speed of the wave?

15P. Prove that if a transverse wave is traveling along a string, then the slope at any point of the string is numerically equal to the ratio of the particle speed to the wave speed at that point.

16P. A wave of frequency 500 Hz has a velocity of 350 m/s. (a) How far apart are two points that differ in phase by $\pi/3$ rad? (b) What is the phase difference between two displacements at a certain point at times 1.00 ms apart?

SECTION 17-6 WAVE SPEED ON A STRETCHED STRING

17E. What is the speed of a transverse wave in a rope of length 2.00 m and mass 60.0 g under a tension of 500 N?

18E. The heaviest and lightest strings in a certain violin have linear densities of 3.0 g/m and 0.29 g/m. What is the ratio, heaviest to lightest, of the diameters of these strings, assuming they are made of the same material?

19E. The speed of a wave on a string is 170 m/s when the tension is 120 N. To what value must the tension be increased in order to raise the wave speed to 180 m/s?

20E. The tension in a wire clamped at both ends is doubled without appreciably changing its length. What is the ratio of the new to the old wave speed for transverse waves in this wire?

21E. Show that, in terms of the tensile stress S and the volume density ρ, the speed v of transverse waves in a wire is given by

$$v = \sqrt{\frac{S}{\rho}}.$$

22E. The equation of a transverse wave on a string is

$$y = (2.0 \text{ mm}) \sin[(20 \text{ m}^{-1})x - (600 \text{ s}^{-1})t].$$

The tension in the string is 15 N. (a) What is the wave speed? (b) Find the linear density of this string in grams per meter.

23E. The linear density of a vibrating string is 1.6×10^{-4} kg/m. A transverse wave is propagating on the string and is described by the following equation: $y = (0.021 \text{ m}) \sin[(2.0 \text{ m}^{-1})x + (30 \text{ s}^{-1})t]$. (a) What is the wave speed? (b) What is the tension in the string?

24E. What is the fastest transverse wave that can be sent along a steel wire? Allowing for a reasonable safety factor, the maximum tensile stress to which steel wires should be subject is 7.0×10^8 N/m². The density of steel is 7800 kg/m³. Show that your answer does not depend on the diameter of the wire.

25P. A stretched string has a mass per unit length of 5.0 g/cm and a tension of 10 N. A sinusoidal wave on this string has an amplitude of 0.12 mm and a frequency of 100 Hz and is traveling toward decreasing x. Write an equation for this wave.

26P. For a wave on a stretched cord, find the ratio of the maximum particle speed (the speed with which a single particle in the cord moves transverse to the wave) to the wave speed. If a wave having a certain frequency and amplitude is sent along a cord, would this speed ratio depend on the material of which the cord is made, such as wire or nylon?

27P. A sinusoidal transverse wave is traveling along a string toward decreasing x. Figure 17-24 shows a plot of the displacement as a function of position at time $t = 0$. The string tension is 3.6 N and its linear density is 25 g/m. Calculate (a) the amplitude, (b) the wavelength, (c)

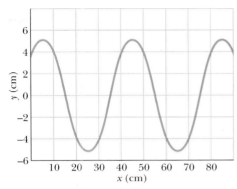

FIGURE 17-24 Problem 27.

the wave speed, and (d) the period of the wave. (e) Find the maximum speed of a particle in the string. (f) Write an equation describing the traveling wave.

28P. A sinusoidal wave is traveling on a string with velocity 40 cm/s. The displacement of the particles of the string at $x = 10$ cm is found to vary with time according to the equation $y = (5.0 \text{ cm}) \sin[1.0 - (4.0 \text{ s}^{-1})t]$. The linear density of the string is 4.0 g/cm. What are (a) the frequency and (b) the wavelength of the wave? (c) Write the general equation giving the transverse displacement of the particles of the string as a function of position and time. (d) Calculate the tension in the string.

29P. In Fig. 17-25a, string 1 has a linear density of 3.00 g/m, and string 2 has a linear density of 5.00 g/m. They are under tension owing to the hanging block of mass $M = 500$ g. (a) Calculate the wave speed in each string. (b) The block is now divided into two blocks (with $M_1 + M_2 = M$) and the apparatus rearranged as shown in Fig. 17-25b. Find M_1 and M_2 such that the wave speeds in the two strings are equal.

FIGURE 17-25 Problem 29.

30P. A wire 10.0 m long and having a mass of 100 g is stretched under a tension of 250 N. If two pulses, separated in time by 30.0 ms, are generated, one at each end of the wire, where will the pulses first meet?

31P. The type of rubber band used inside some baseballs and golf balls obeys Hooke's law over a wide range of elongation of the band. A segment of this material has an unstretched length l and a mass m. When a force F is applied, the band stretches an additional length Δl. (a) What is the speed (in terms of m, Δl, and the force constant k) of transverse waves on this rubber band? (b) Using your answer to (a), show that the time required for a transverse pulse to travel the length of the rubber band is proportional to $1/\sqrt{\Delta l}$ if $\Delta l \ll l$ and is constant if $\Delta l \gg l$.

32P*. A uniform rope of mass m and length l hangs from a ceiling. (a) Show that the speed of a transverse wave on the rope is a function of y, the distance from the lower end, and is given by $v = \sqrt{gy}$. (b) Show that the time a transverse wave takes to travel the length of the rope is given by $t = 2\sqrt{l/g}$.

SECTION 17-8 ENERGY AND POWER IN A TRAVELING WAVE

33E. Power P_1 is transmitted by a wave of frequency f_1 on a string with tension τ_1. What is the transmitted power P_2 in terms of P_1 (a) if the tension of the string is increased to $\tau_2 = 4\tau_1$, and (b) if, instead, the frequency is decreased to $f_2 = f_1/2$?

34E. A string 2.7 m long has a mass of 260 g. The tension in the string is 36 N. What must be the frequency of traveling waves of amplitude 7.7 mm in order that the average transmitted power be 85 W?

35P. A transverse sinusoidal wave is generated at one end of a long, horizontal string by a bar that moves up and down through a distance of 1.00 cm. The motion is continuous and is repeated regularly 120 times per second. The string has linear density 120 g/m and is kept under a tension of 90.0 N. Find (a) the maximum value of the transverse speed u and (b) the maximum value of the transverse component of the tension. (c) Show that the two maximum values calculated above occur at the same phase values for the wave. What is the transverse displacement y of the string at these phases? (d) What is the maximum power transferred along the string? (e) What is the transverse displacement y when this maximum power transfer occurs? (f) What is the minimum power transfer along the string? (g) What is the transverse displacement y when this minimum power transfer occurs?

SECTION 17-11 INTERFERENCE OF WAVES

36E. Two identical traveling waves, moving in the same direction, are out of phase by $\pi/2$ rad. What is the ampli-

tude of the combined wave in terms of the common amplitude y_m of the two combining waves?

37E. What phase difference between two otherwise identical traveling waves, moving in the same direction along a stretched string, will result in the combined wave having an amplitude 1.50 times that of the common amplitude of the two combining waves? Express your answer in both degrees and radians.

38P. A source S and a detector D of radio waves are a distance d apart on level ground (Fig. 17-26). Radio waves of wavelength λ reach D either along a straight path or by reflecting (bouncing) from a certain layer in the atmosphere. When the layer is at height H, the two waves reaching D are exactly in phase. If the layer gradually rises, the phase difference between the two waves gradually shifts, until they are exactly out of phase when the layer is at height $H + h$. Express λ in terms of d, h, and H.

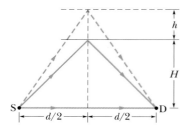

FIGURE 17-26 Problem 38.

39P. Three sinusoidal waves travel in the positive x direction along the same string. All three waves have the same frequency. Their amplitudes are in the ratio $1 : \frac{1}{2} : \frac{1}{3}$ and their phase constants are 0, $\pi/2$, and π, respectively. Plot the resultant wave form and discuss its behavior as t increases.

40P. Four sinusoidal waves travel in the positive x direction along the same string. Their frequencies are in the ratio $1 : 2 : 3 : 4$ and their amplitudes are in the ratio $1 : \frac{1}{2} : \frac{1}{3} : \frac{1}{4}$, respectively. When $t = 0$, at $x = 0$, the first and third waves are 180° out of phase with the second and fourth. Plot the resultant wave form when $t = 0$, and discuss its behavior as t increases.

41P*. Determine the amplitude of the resultant wave when two sinusoidal waves having the same frequency and traveling in the same direction are combined, if their amplitudes are 3.0 cm and 4.0 cm and they differ in phase by $\pi/2$ rad.

SECTION 17-13 STANDING WAVES AND RESONANCE

42E. A string under tension τ_i oscillates in the third harmonic at frequency f_3, and the waves on the string have wavelength λ_3. If the tension is increased to $\tau_f = 4\tau_i$ and the string is again made to oscillate in the third harmonic, what then is (a) the frequency of oscillation in terms of f_3 and (b) the wavelength of the waves in terms of λ_3?

43E. Two sinusoidal waves with identical wavelengths and amplitudes travel in opposite directions along a string with a speed of 10 cm/s. If the time interval between instants when the string is flat is 0.50 s, what is the wavelength of the waves?

44E. When played in a certain manner, the lowest resonant frequency of a certain violin string is concert A (440 Hz). What are the frequencies of the second and third harmonics of that string?

45E. A string fixed at both ends is 8.40 m long and has a mass of 0.120 kg. It is subjected to a tension of 96.0 N and set oscillating. (a) What is the speed of the waves on the string? (b) What is the longest possible wavelength for a standing wave? (c) Give the frequency of that wave.

46E. A nylon guitar string has a linear density of 7.2 g/m and is under a tension of 150 N. The fixed supports are 90 cm apart. The string is oscillating in the standing wave pattern shown in Fig. 17-27. Calculate the (a) speed, (b) wavelength, and (c) frequency of the waves whose superposition gives this standing wave.

FIGURE 17-27   Exercise 46.

47E. The equation of a transverse wave traveling along a string is given by

$$y = 0.15 \sin(0.79x - 13t),$$

in which x and y are expressed in meters and t is in seconds. (a) What is the displacement y at $x = 2.3$ m, $t = 0.16$ s? (b) Write the equation of a wave that, when added to the given one, would produce standing waves on the string. (c) What is the displacement of the resultant standing wave at $x = 2.3$ m, $t = 0.16$ s?

48E. A 120-cm length of string is stretched between fixed supports. What are the three longest possible wavelengths for standing waves in this string? Sketch the corresponding standing waves.

49E. A 125-cm length of string has a mass of 2.00 g. It is stretched with a tension of 7.00 N between fixed supports. (a) What is the wave speed for this string? (b) What is the lowest resonant frequency of this string?

50E. What are the three lowest frequencies for standing waves on a wire 10.0 m long having a mass of 100 g, which is stretched under a tension of 250 N?

51E. A 1.50-m wire has a mass of 8.70 g and is held under a tension of 120 N. The wire is held rigidly at both ends and set into vibration. Calculate (a) the velocity of waves on the wire, (b) the wavelengths of the waves that produce one- and two-loop standing waves on the string, and (c) the frequencies of the waves that produce one- and two-loop standing waves.

52E. One end of a 120-cm string is held fixed. The other end is attached to a weightless ring that can slide along a frictionless rod as shown in Fig. 17-28. What are the three longest possible wavelengths for standing waves in this string? Sketch the corresponding standing waves.

FIGURE 17-28   Exercise 52.

53E. String A is stretched between two clamps separated by distance l. String B, with the same linear density and under the same tension as string A, is stretched between two clamps separated by distance $4l$. Consider the first eight harmonics of string B. Which, if any, has a resonant frequency that matches a resonant frequency of string A?

54P. Two waves are propagating on the same very long string. A generator at the left end of the string creates a wave given by

$$y = (6.0 \text{ cm}) \cos \frac{\pi}{2} [(2.0 \text{ m}^{-1})x + (8.0 \text{ s}^{-1})t],$$

and one at the right end of the string creates the wave

$$y = (6.0 \text{ cm}) \cos \frac{\pi}{2} [2.0 \text{ m}^{-1})x - (8.0 \text{ s}^{-1})t].$$

(a) Calculate the frequency, wavelength, and speed of each wave. (b) Find the points at which there is no motion (the nodes). (c) At which points is the motion of the string a maximum?

55P. A string oscillates according to the equation

$$y = (0.50 \text{ cm}) \left[\sin \left(\frac{\pi}{3} \text{ cm}^{-1} \right) x \right] \cos[(40\pi \text{ s}^{-1}) t].$$

(a) What are the amplitude and speed of the waves whose superposition gives this oscillation? (b) What is the distance between nodes? (c) What is the speed of a particle of the string at the position $x = 1.5$ cm when $t = \frac{9}{8}$ s?

56P. A string is stretched between fixed supports separated by 75.0 cm. It is observed to have resonant frequencies of 420 and 315 Hz, and no other resonant frequencies between these two. (a) What is the lowest resonant frequency for this string? (b) What is the wave speed for this string?

57P. Two transverse sinusoidal waves travel in opposite directions along a string. Each wave has an amplitude of 0.30 cm and a wavelength of 6.0 cm. The speed of a transverse wave on the string is 1.5 m/s. Plot the shape of the string at times $t = 0$ (arbitrary), $t = 5.0$, $t = 10$, $t = 15$, and $t = 20$ ms.

58P. Two pulses travel along a string in opposite directions, as in Fig. 17-29. (a) If the wave speed v is 2.0 m/s and the pulses are 6.0 cm apart at $t = 0$, sketch the patterns when t is equal to 5.0, 10, 15, 20, and 25 ms. (b) What has happened to the energy at $t = 15$ ms?

FIGURE 17-29 Problem 58.

59P. Two waves on a string are described by the equations

$$y_1 = (0.10 \text{ m}) \sin 2\pi [(0.50 \text{ m}^{-1})x + (20 \text{ s}^{-1})t]$$

and

$$y_2 = (0.20 \text{ m}) \sin 2\pi [(0.50 \text{ m}^{-1})x - (20 \text{ s}^{-1})t].$$

Sketch the total response for the point on the string at $x = 3.0$ m; that is, plot y versus t for that value of x.

60P. A 3.0-m-long string is oscillating as a three-loop standing wave whose amplitude is 1.0 cm. The wave speed is 100 m/s. (a) What is the frequency? (b) Write equations for two waves that, when combined, will result in this standing wave.

61P. Vibration from a 600-Hz tuning fork sets up standing waves in a string clamped at both ends. The wave speed for the string is 400 m/s. The standing wave has four loops and an amplitude of 2.0 mm. (a) What is the length of the string? (b) Write an equation for the displacement of the string as a function of position and time.

62P. In an experiment on standing waves, a string 90 cm long is attached to the prong of an electrically driven tuning fork that oscillates perpendicular to the length of the string at a frequency of 60 Hz. The mass of the string is 0.044 kg. (a) What tension must the string be under (weights are attached to the other end) if it is to vibrate in four loops? (b) What would happen if the tuning fork were turned so as to vibrate parallel to the length of the string?

63P. Consider a standing wave that is the sum of two waves traveling in opposite directions but otherwise identical. Show that the maximum kinetic energy in each loop of the standing wave is $2\pi^2 \mu y_m^2 f v$.

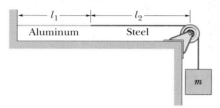

FIGURE 17-30 Problem 64.

64P. An aluminum wire, of length $l_1 = 60.0$ cm, cross-sectional area 1.00×10^{-2} cm², and density 2.60 g/cm³, is connected to a steel wire, of density 7.80 g/cm³ and the same cross-sectional area. The compound wire, loaded with a block of mass $m = 10.0$ kg, is arranged as in Fig. 17-30 so that the distance l_2 from the joint to the supporting pulley is 86.6 cm. Transverse waves are set up in the wire by using an external source of variable frequency. (a) Find the lowest frequency of excitation for which standing waves are observed such that the joint in the wire is a node. (b) How many nodes are observed at this frequency?

ADDITIONAL PROBLEMS

65. A rope, under a tension of 200 N and fixed at both ends, oscillates in a second-harmonic standing wave pattern. The displacement of the rope is given by

$$y = (0.10 \text{ m})(\sin \pi x/2)\sin 12\pi t,$$

where $x = 0$ at one end of the rope, x is in meters and t is in seconds. What are (a) the length of the rope, (b) the speed of the waves on the rope, and (c) the mass of the rope? (d) If the rope oscillates in a third-harmonic standing wave pattern, what will be the period of oscillation?

66. A transverse sinusoidal wave of wavelength 20 cm is moving to the right. The displacement from equilibrium of the particle at $x = 0$ as a function of time is shown in Fig. 17-31. (a) Make a rough sketch of one wavelength of the wave (the portion between $x = 0$ and $x = 20$ cm) at time $t = 0$. (b) What is the velocity of propagation of the wave? (c) Write the equation for the wave with all the constants evaluated. (d) What is the transverse velocity of the particle at $x = 0$ at $t = 5.0$ s?

67. A standing wave results from the sum of two transverse traveling waves given by

$$y_1 = 0.050 \cos(\pi x - 4\pi t),$$

$$y_2 = 0.050 \cos(\pi x + 4\pi t),$$

where x, y_1, and y_2 are in meters and t is in seconds. (a) What is the smallest positive value of x that corresponds to a node? (b) At what times during the interval $0 \le t \le 0.50$ s will the particle at $x = 0$ have zero velocity?

FIGURE 17-31 Problem 66.

WAVES—II | 18

This horseshoe bat can not only locate a moth flying in total darkness but can also determine the moth's relative speed, to "home in" on the insect. How does the detection system work? And how can a moth "jam" the system or otherwise reduce its effectiveness?

(a)

(b)

FIGURE 18-1 (a) An ultrasound image of a fetus search-
ing for a thumb to suck. The frequency of the sound waves
used was about 5 MHz. (b) The image of a transistor,
viewed in an acoustic microscope at a sound frequency of
4.2 GHz. The conducting strips at top and bottom are
about two micrometers wide.

18-1 SOUND WAVES

Sound waves, broadly defined, are mechanical waves
that can travel through a gas, liquid, or solid. Seismic
prospecting teams use such waves to probe the
Earth's crust for oil. Ships carry sound ranging gear
(sonar) to detect underwater obstacles. Submarines
use sound waves to stalk other submarines, largely by
listening for the characteristic acoustic signature of
propellers or reactor pumps or the occasional
dropped mess tray. Figure 18-1a, a computer-
processed image of a fetal head, shows how sound
waves can be used to explore the soft tissues of the
human body. Figure 18-1b shows how sound waves at
very high frequencies can be used to create images
of very tiny objects, with a resolution better than any
optical microscope.

In a solid there can be two types of waves. There
are *transverse* waves, in which the oscillations of small
elements of the solid are perpendicular to the direction in which the wave travels. There are also *longitudinal* waves, in which the oscillations are parallel to
the direction of travel.

A seismic exploration crew uses both types of
waves to determine rock formations below ground,
as in Fig. 18-2. The underground detonation of a
test charge produces transverse and longitudinal
waves that then reflect wherever the type of rock
changes.

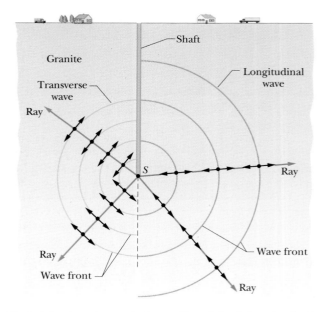

FIGURE 18-2 An explosion at S sends both longitudinal
and (slower) transverse waves traveling throughout the
granite in all directions. The waves are shown in separate
hemispheres and in cross section for convenience. *Wave
fronts* are surfaces (spherical in this situation) over which
the wave disturbance has the same value. *Rays* are lines
perpendicular to the wave fronts; they indicate the direction of travel of the wave fronts. The short, double arrows
superimposed on the rays here show the directions of the
oscillations of small elements of the medium.

In a gas or liquid, only longitudinal waves can be transmitted. To transmit transverse waves, a medium must be elastic when acted on by shearing stresses, so as to provide a restoring force. But fluids *flow* when acted on by shearing stresses; thus they cannot provide the restoring force and so cannot transmit transverse waves.

Sound waves are longitudinal waves. In this chapter, we shall mostly be concerned with sound waves defined in the usual sense, that is, as (longitudinal) mechanical waves traveling through air and having frequencies in the audible range.

18-2 THE SPEED OF SOUND

The speed of any mechanical wave, transverse or longitudinal, depends on both an inertial property of the medium (to store kinetic energy) and an elastic property of the medium (to store potential energy). Thus we can generalize Eq. 17-26, which gives the speed of a transverse wave along a taut string, by writing

$$v = \sqrt{\frac{\tau}{\mu}} = \sqrt{\frac{\text{elastic property}}{\text{inertial property}}}, \quad (18\text{-}1)$$

where τ is the tension in the string and μ is the string's linear density. If the medium is air, we can guess that the inertial property, corresponding to μ, is the volume density ρ of air. What shall we put for the elastic property?

In a taut string, potential energy is associated with the periodic stretching of the string elements as the wave passes through them. As a sound wave passes through air, potential energy is associated with periodic compressions and expansions of small volume elements of the air. The property that determines the extent to which an element of the medium changes its volume as the pressure (force per unit area) applied to it is increased or decreased is the **bulk modulus** B, which is defined as

$$B = -\frac{\Delta p}{\Delta V / V} \quad \text{(definition of } B). \quad (18\text{-}2)$$

Here $\Delta V / V$ is the fractional change in volume produced by a change in pressure Δp. As explained in Section 16-3, the SI unit for pressure is the newton per square meter, which is given a special name, the *pascal* (Pa). From Eq. 18-2 we see that the unit for B is also the pascal. The signs of Δp and ΔV are always opposite: when we increase the pressure on a fluid

TABLE 18-1
THE SPEED OF SOUND[a]

MEDIUM	SPEED (m/s)
Gases	
Air (0°C)	331
Air (20°C)	343
Helium	965
Hydrogen	1284
Liquids	
Water (0°C)	1402
Water (20°C)	1482
Seawater[b]	1522
Solids	
Aluminum	6420
Steel	5941
Granite	6000

[a]At 0°C and 1 atm pressure, except where noted.
[b]At 20°C and 3.5% salinity.

element (Δp positive), its volume decreases (ΔV negative). We include a minus sign in Eq. 18-2 so that B will always be a positive quantity. Substituting B for τ and ρ for μ in Eq. 18-1 yields

$$v = \sqrt{\frac{B}{\rho}} \quad \text{(speed of sound)} \quad (18\text{-}3)$$

for a medium with bulk modulus B and density ρ. Table 18-1 lists the speed of sound in various media.

The density of water is almost 1000 times greater than the density of air. If this were the only relevant factor, we would expect from Eq. 18-3 that the speed of sound in water would be considerably less than the speed of sound in air. However, Table 18-1 shows us that the reverse is true. We conclude (again from Eq. 18-3) that the bulk modulus of water must be more than 1000 times greater than that of air. This is indeed the case. Water is much more incompressible than air, which (see Eq. 18-2) is another way of saying that its bulk modulus is much greater.

Formal Derivation of Eq. 18-3

We now derive Eq. 18-3 by direct application of Newton's laws. Let a single compressional pulse travel (from right to left) with speed v through the air in a long tube. Let us run along with the pulse at that speed, so that the pulse appears to stand still in our reference frame. Figure 18-3a shows the situation as it is viewed from that frame. The pulse (labeled "compression zone") is standing still and air is moving at speed v through it from left to right.

Let the pressure of the undisturbed air be p and the pressure inside the pulse be $p + \Delta p$, where Δp is positive owing to the compression. Consider a slice of air of thickness Δx and area A, moving toward the pulse at speed v. As this fluid element enters the pulse, its leading face encounters a region of higher pressure and slows to speed $v + \Delta v$, in which Δv is negative. This slowing is complete when the rear face reaches the pulse, which requires time interval

$$\Delta t = \frac{\Delta x}{v}. \qquad (18\text{-}4)$$

Let us apply Newton's second law to the element. During Δt, the average force on the element's trailing face is pA toward the right, and the average force on the leading face is $(p + \Delta p)A$ toward the left (Fig. 18-3b). So the average net force on the element during Δt is

$$F = pA - (p + \Delta p)A$$

$$= - \Delta p\, A \qquad \text{(net force).} \qquad (18\text{-}5)$$

The minus sign indicates that the net force on the fluid element points to the left in Fig. 18-3b. The volume of the element is $A \Delta x$, so with the aid of Eq.

18-4, we can write its mass as

$$\Delta m = \rho A\, \Delta x = \rho A v\, \Delta t \qquad \text{(mass).} \qquad (18\text{-}6)$$

Finally, the average acceleration of the element during Δt is

$$a = \frac{\Delta v}{\Delta t} \qquad \text{(acceleration).} \qquad (18\text{-}7)$$

From Newton's second law ($F = ma$), we have, from Eqs. 18-5, 18-6, and 18-7,

$$- \Delta p\, A = (\rho A v\, \Delta t)\frac{\Delta v}{\Delta t},$$

which we can write as

$$\rho v^2 = - \frac{\Delta p}{\Delta v / v}. \qquad (18\text{-}8)$$

The air that occupies a volume $V (= Av\, \Delta t)$ outside the pulse is compressed by an amount ΔV ($= A\, \Delta v\, \Delta t$) as it enters the pulse. Thus

$$\frac{\Delta V}{V} = \frac{A\, \Delta v\, \Delta t}{Av\, \Delta t} = \frac{\Delta v}{v}. \qquad (18\text{-}9)$$

Substituting Eq. 18-9 and then Eq. 18-2 into Eq. 18-8 leads to

$$\rho v^2 = - \frac{\Delta p}{\Delta v / v} = - \frac{\Delta p}{\Delta V / V} = B.$$

Solving for v yields Eq. 18-3 for the speed of the air toward the right in Fig. 18-3, and thus for the actual speed of the pulse toward the left.

FIGURE 18-3 A pulse (compression) is sent down a long air-filled tube. The reference frame of the figure is chosen so that the pulse is at rest and the air moves from left to right. (a) A slice of air of width Δx is shown, moving toward the compression zone with speed v. (b) The leading face of the slice enters the zone. The forces acting on the leading and trailing faces (due to air pressure) are shown.

SAMPLE PROBLEM 18-1

One clue used by your brain to determine the direction of a source of sound is the time delay Δt between the arrival of the sound at the ear closer to the source and the arrival at the farther ear. Assume that the source is distant (so that a wave front from it is approximately straight instead of noticeably curved as in Fig. 18-2), and let D represent the separation between your ears.

a. Find an expression that gives Δt in terms of D and the angle θ between the direction of the source and the forward direction.

SOLUTION The situation is shown in Fig. 18-4 for a wave front approaching you from a source that is located in front of you and to your right. The time delay Δt is due to the distance d that the wave front must travel to reach your left ear (L) after it reaches your

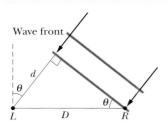

FIGURE 18-4 Sample Problem 18-1. A wave front travels distance $d \, (= D \sin \theta)$ to reach the left ear (L) after reaching the right ear (R).

right ear (R). From Fig. 18-4, we find

$$\Delta t = \frac{d}{v} = \frac{D \sin \theta}{v}, \quad \text{(Answer)} \quad (18\text{-}10)$$

where v is the speed of sound in air. From Eq. 18-10 and Fig. 18-4, we see that the time delay is least ($\Delta t = 0$) when $\theta = 0$ (the source is directly ahead), and it is maximum ($\Delta t = D/v$) when $\theta = 90°$ (the source is directly to the right). Based on a lifetime of experience, your brain correlates any detected value of Δt (from zero to the maximum value) with a value of θ (from zero to 90°) for the direction of the sound source.

b. Suppose that you are submerged in water at 20°C when a wave front arrives from directly to your right. Based on the time-delay clue, at what angle θ from the direction directly ahead will the source appear to be?

SOLUTION We find the time delay Δt_w in this situation with Eq. 18-10, substituting $\theta = 90°$ and using v_w, the speed of sound in water, instead of v, the speed of sound in air:

$$\Delta t_w = \frac{D \sin 90°}{v_w} = \frac{D}{v_w}. \quad (18\text{-}11)$$

Since v_w is about four times v, Δt_w is about one-fourth the maximum time delay in air. Based on experience, your brain will process the time delay as if it occurred in air. Thus the sound source will appear to be at an angle θ smaller than 90°. To find that apparent angle, we substitute the time delay D/v_w from Eq. 18-11 for Δt in Eq. 18-10, obtaining

$$\frac{D}{v_w} = \frac{D \sin \theta}{v}. \quad (18\text{-}12)$$

Substituting $v = 343$ m/s and $v_w = 1482$ m/s (from Table 18-1) into Eq. 18-12, we then find

$$\sin \theta = \frac{v}{v_w} = \frac{343 \text{ m/s}}{1482 \text{ m/s}} = 0.231$$

and thus

$$\theta = 13°. \quad \text{(Answer)}$$

18-3 TRAVELING SOUND WAVES

Here we look in some detail at the displacements and pressure variations that are associated with a sound wave passing through air. Figure 18-5a displays such a wave traveling toward the right through a long air-filled tube. Consider a thin slice of air of thickness Δx, located at a position x along the tube. As the wave passes through x, this element of air oscillates left and right in simple harmonic motion about its equilibrium position (Fig. 18-5b). Thus the oscillations of this air element due to a sound wave are like those of a string element due to a string wave, except that the air element oscillates *longitudinally* (parallel to the wave's direction of travel) and the string element oscillates *transversely* (perpendicular to the wave's direction of travel).

Owing to the simple harmonic motion of the air element, we can write a sinusoidal function, either a sine or cosine, for its displacement s due to the sound wave. In this chapter we use a cosine function:

$$s = s_m \cos(kx - \omega t). \quad (18\text{-}13)$$

Here s_m is the **displacement amplitude,** that is, the maximum displacement of the air element to either side of its equilibrium position (Fig. 18-5b).* Angular wave number k, angular frequency ω, frequency f, wavelength λ, speed v, and period T for a sound (longitudinal) wave are defined and interrelated exactly as for a transverse wave, except that λ is now the distance (again along the direction of travel) in which the pattern of compression and expansion due to the wave begins to repeat itself (Fig. 18-5a). (We assume s_m is much less than λ.)

As the wave moves, the air pressure at any position x in Fig. 18-5a rises and falls with time, the variation being given (as we shall prove below) by

$$\Delta p = \Delta p_m \sin(kx - \omega t). \quad (18\text{-}14)$$

A negative value of Δp in Eq. 18-14 corresponds to an expansion, and a positive value to a compression. We also prove below that the maximum pressure variation in the wave, Δp_m in Eq. 18-14, is related to

---

*For the transverse displacement of an element of a taut string, we used the symbol $y(x, t)$. We change the notation here to $s(x, t)$ to avoid writing $x(x, t)$ for the longitudinal displacement of an element of air.

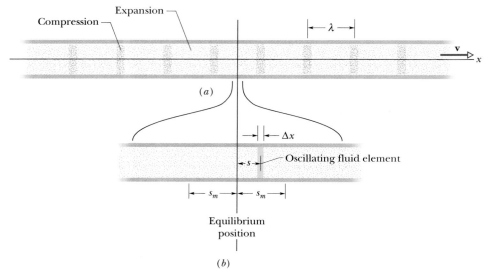

Expansion
Compression
λ
v
x
(a)

Δx
Oscillating fluid element
s
s_m   s_m
Equilibrium position

(b)

FIGURE 18-5 (*a*) A sound wave, traveling through a long air-filled tube with speed *v*, consists of a moving, periodic pattern of expansion and compression of the air. The wave is shown at an arbitrary instant. (*b*) A horizontally expanded view of a short piece of the tube. A fluid element of thickness Δx oscillates left and right in simple harmonic motion about its equilibrium position as the wave passes. At the instant shown in (*a*), the element happens to be displaced a distance *s* to the right of its equilibrium position. Its maximum displacement, either right or left, is s_m.

the maximum displacement, s_m in Eq. 18-13 and Fig. 18-5*b*, by

$$\Delta p_m = (v\rho\omega)s_m. \qquad (18\text{-}15)$$

The maximum pressure variation (or **pressure amplitude**) Δp_m is normally very much less than the pressure *p* that prevails when there is no wave.

Figure 18-6 shows plots of Eqs. 18-13 and 18-14 at $t = 0$. Note that the two waves are $\pi/2$ rad (or 90°) out of phase, the pressure variation being zero when the displacement is a maximum. Experimentally, it is usually easier to measure the pressure variation than the displacement.

Derivation of Eqs. 18-14 and 18-15

Figure 18-5*b* shows an oscillating element of air of area *A* and thickness Δx, with its center displaced from its equilibrium position by distance *s*.

From Eq. 18-2 we can write, for the pressure variation in the displaced element,

$$\Delta p = -B \frac{\Delta V}{V}. \qquad (18\text{-}16)$$

The quantity *V* in Eq. 18-16 is the volume of the element, given by

$$V = A \, \Delta x. \qquad (18\text{-}17)$$

The quantity ΔV in Eq. 18-16 is the change in volume that occurs when the element is displaced. This volume change comes about because the displacements of the two faces of the element are not quite the same, differing by some amount Δs. Thus we can write the change in volume as

$$\Delta V = A \, \Delta s. \qquad (18\text{-}18)$$

Substituting Eqs. 18-17 and 18-18 into Eq. 18-16 and passing to the differential limit yield

$$\Delta p = -B \frac{\Delta s}{\Delta x} = -B \frac{\partial s}{\partial x}. \qquad (18\text{-}19)$$

The symbols ∂ indicate that the derivative in Eq. 18-19 is a *partial derivative* which tells us how *s* changes with *x* when the time *t* is fixed. From Eq. 18-13 we then have, treating *t* as a constant,

$$\frac{\partial s}{\partial x} = \frac{\partial}{\partial x} \left[s_m \cos(kx - \omega t) \right] = -ks_m \sin(kx - \omega t).$$

Substituting this quantity for the partial derivative in Eq. 18-19 yields

$$\Delta p = Bks_m \sin(kx - \omega t).$$

Comparing this equation with Eq. 18-14 reveals that $Bks_m = \Delta p_m$.

Using Eq. 18-3, we can now write

$$\Delta p_m = (Bk)s_m = (v^2\rho k)s_m.$$

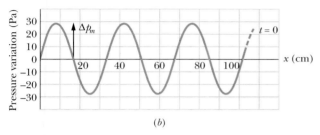

FIGURE 18-6 (a) A plot of the displacement function (Eq. 18-13) for $t = 0$. (b) A similar plot of the pressure variation function (Eq. 18-14). Both plots are for a 1000-Hz sound wave whose intensity is at the threshold of pain; see Sample Problem 18-2.

Equation 18-15, which we also promised to prove, follows at once if we eliminate k by using $v = \omega/k$ (Eq. 17-14).

SAMPLE PROBLEM 18-2

The maximum pressure variation Δp_m that the ear can tolerate in loud sounds is about 28 Pa. What is the displacement amplitude s_m for such a sound in air of density $\rho = 1.21$ kg/m^3, at a frequency of 1000 Hz?

SOLUTION From Eq. 18-15 we have

$$s_m = \frac{\Delta p_m}{v\rho\omega} = \frac{\Delta p_m}{v\rho 2\pi f}$$

$$= \frac{28 \text{ Pa}}{(343 \text{ m/s})(1.21 \text{ kg/m}^3)(2\pi)(1000 \text{ Hz})}$$

$$= 1.1 \times 10^{-5} \text{ m} = 11 \text{ }\mu\text{m}. \qquad \text{(Answer)}$$

We see that the displacement amplitude for even the loudest sound that the ear can tolerate is very small, being about one-seventh the thickness of this page. This amplitude is also very much smaller than the wavelength of a 1000-Hz sound wave in air, which is 34 cm.

The maximum tolerable pressure amplitude Δp_m ($= 28$ Pa) is correspondingly small, considering that

normal atmospheric pressure is about 10^5 Pa. Figure 18-6 is plotted for the conditions of this problem.

The pressure amplitude Δp_m for the *faintest* detectable sound at 1000 Hz is 2.8×10^{-5} Pa. Proceeding as above leads to $s_m = 1.1 \times 10^{-11}$ m or 11 pm, which is about 10 times smaller than the radius of a typical atom. The ear is indeed a sensitive detector of sound waves. The ear, in fact, can detect a pulse of sound whose total energy is as small as a few electron-volts, about the same as the energy required to remove an outer electron from a single atom.

Interference

Suppose sound waves of wavelength λ are emitted by two point sources that are in phase; that is, the emerging waves reach their maximum values at the same instants. If the waves then pass through a common point while traveling in (approximately) the same direction, they too will be in phase if they have traveled along paths with identical lengths to reach that point. However, if they have traveled along paths with a **path length difference** Δd, they may then not be in phase. We find their phase difference ϕ by recalling (see Section 17-4) that a phase difference of 2π rad corresponds to one wavelength. Thus we can write

$$\frac{\phi}{2\pi} = \frac{\Delta d}{\lambda}, \qquad (18\text{-}20)$$

or

$$\phi = \frac{\Delta d}{\lambda} 2\pi. \qquad (18\text{-}21)$$

As with transverse waves (see Section 17-11), sound waves undergo constructive interference and destructive interference. They undergo fully constructive interference when ϕ is zero or an integer multiple of 2π, that is, when

$$\phi = m2\pi, \qquad m = 0, 1, 2, \ldots$$
$$\text{(fully constructive interference).} \quad (18\text{-}22)$$

They undergo fully destructive interference when ϕ is an odd multiple of π, that is, when

$$\phi = (m + \tfrac{1}{2})2\pi, \qquad m = 0, 1, 2, \ldots$$
$$\text{(fully destructive interference).} \quad (18\text{-}23)$$

From Eq. 18-20, we see that these conditions correspond to

$$\Delta d = m\lambda \qquad \begin{array}{l}\text{(fully constructive} \\ \text{interference)}\end{array} \quad (18\text{-}24)$$

and

$$\Delta d = (m + \tfrac{1}{2})\lambda \qquad \text{(fully destructive interference)}, \qquad (18\text{-}25)$$

respectively.

SAMPLE PROBLEM 18-3

In Fig. 18-7, two point sources S_1 and S_2, which are in phase and separated by distance $D = 1.5\lambda$, emit identical sound waves of wavelength λ.

a. What is the phase difference of the waves from S_1 and S_2 at point P_1, along a line bisecting D, and what kind of interference exists there?

SOLUTION To reach P_1, waves emitted by S_1 and S_2 travel identical distances. Thus the path length difference Δd is 0, and from Eq. 18-21 we have

$$\phi = \frac{\Delta d}{\lambda}\, 2\pi = 0. \qquad \text{(Answer)}$$

From Eq. 18-22, we see that this situation corresponds to fully constructive interference, with $m = 0$.

b. What are the phase difference and kind of interference at P_2 in Fig. 18-7?

SOLUTION Point P_2 is on the line that extends through S_1 and S_2. So to reach P_2, waves emitted by S_1

FIGURE 18-7   Sample Problem 18-3. Two point sources S_1 and S_2 emit spherical sound waves that are in phase. The waves travel equal distances to reach point P_1. Point P_2 is on the line extending through S_1 and S_2.

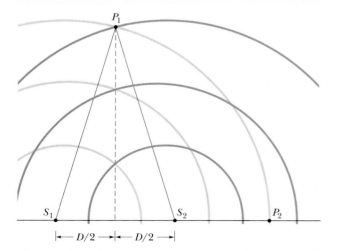

must travel a distance D farther than those emitted by S_2. From Eq. 18-21 with $\Delta d = D = 1.5\lambda$, we find

$$\phi = \frac{\Delta d}{\lambda}\, 2\pi = \frac{1.5\lambda}{\lambda}\, 2\pi = 3\pi \text{ rad}. \qquad \text{(Answer)}$$

From Eq. 18-23, we see that this situation corresponds to fully destructive interference, with $m = 1$. Note that the answer is independent of the distance between P_2 and the source S_2.

18-4 INTENSITY AND SOUND LEVEL

If you are trying to sleep when someone is operating a chain saw nearby, you are well aware that there is more to sound than frequency, wavelength, and speed. There is also intensity. The **intensity** I of a sound wave is defined to be the average rate per unit area at which energy is transmitted by the wave. The SI unit for intensity is thus watts per square meter (W/m^2). In a sound wave, the intensity I is related to the displacement amplitude s_m by

$$I = \tfrac{1}{2}\rho v \omega^2 s_m^2, \qquad (18\text{-}26)$$

a relation that we shall derive shortly.

You saw in Sample Problem 18-2 that the displacement amplitude at the human ear ranges from

Sound can cause the wall of a drinking glass to oscillate. If the sound produces a standing wave of those oscillations and if the intensity of the sound is large enough, the glass will shatter.

about 10^{-5} m for the loudest tolerable sound to about 10^{-11} m for the faintest detectable sound, a ratio of 10^6. From Eq. 18-26 we see that the intensity of a sound varies as the *square* of its amplitude, so the ratio of intensities at these two limits of the human auditory system is 10^{12}. Humans can hear over an enormous range of intensities.

The Decibel Scale

We deal with such an enormous range of values by using logarithms. Consider the relation

$$y = \log x,$$

in which x and y are variables. It is a property of this equation that if we *multiply* x by a factor of 10, y increases by $\log 10$ ($= 1$). Thus

$$y' = \log(10x) = \log 10 + \log x = 1 + y.$$

Similarly, if we multiply x by 10^{12}, y increases by only 12.

Thus instead of speaking of the intensity I of a sound wave, it is much more convenient to speak of a **sound level** β, defined as

$$\beta = (10 \text{ dB}) \log \frac{I}{I_0}. \qquad (18\text{-}27)$$

Here dB is the abbreviation for **decibel,** the unit of sound level, a name that was chosen to recognize the work of Alexander Graham Bell. I_0 in Eq. 18-27 is a standard reference intensity ($= 10^{-12}$ W/m^2), chosen because it is near the lower limit of the human range of hearing. For $I = I_0$, Eq. 18-27 gives $\beta = 10 \log 1 = 0$, so our standard reference level corre-

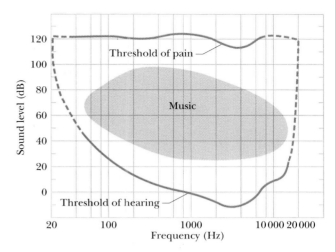

FIGURE 18-8 The average range of sound levels for human hearing. Both the threshold of pain and that of hearing depend on the frequency. The approximate ranges of frequencies and sound levels encountered in music are also shown.

sponds to zero decibels. Table 18-2 shows some values of β and the corresponding values of the intensity ratio I/I_0. The table shows us that β increases by the same *constant amount* every time the sound intensity I increases by a given *factor*. As it turns out, that is just the way the human hearing system operates, so that the ear can be uniformly sensitive over a wide range of sound intensities.

The response of the hearing system to sound is not the same at every frequency. Figure 18-8 shows how the thresholds of hearing and of pain vary through the acoustic spectrum for persons with average hearing. Table 18-3 shows the intensity levels of some sounds.

TABLE 18-2
VALUES FROM EQ. 18-27

β (dB)	I/I_0
0	$10^0 = 1$
10	$10^1 = 10$
20	$10^2 = 100$
30	$10^3 = 1000$
40	$10^4 = 10,000$
50	$10^5 = 100,000$
.	.
.	.
.	.
120	$10^{12} = 1,000,000,000,000$

TABLE 18-3
SOME SOUND LEVELS (dB)

Threshold of hearing	0
Rustle of leaves	10
Whisper (at 1 m)	20
City street, no traffic	30
Office, classroom	50
Normal conversation (at 1 m)	60
Jackhammer (at 1 m)	90
Rock group	110
Threshold of pain	120
Jet engine (at 50 m)	130
Saturn rocket (at 50 m)	200

Derivation of Eq. 18-26

Consider, in Fig. 18-5a, a thin slice of air of thickness dx, area A, and mass dm, oscillating back and forth as the sound wave of Eq. 18-13 passes through it. Its kinetic energy dK is

$$dK = \tfrac{1}{2}dm \; v_s^2. \qquad (18\text{-}28)$$

Here v_s is not the speed of the wave but the speed of the oscillating element of air, obtained from Eq. 18-13 as

$$v_s = \frac{\partial s}{\partial t} = -\omega s_m \sin(kx - \omega t).$$

Using this relation and putting $dm = \rho A \; dx$ allow us to rewrite Eq. 18-28 as

$$dK = \tfrac{1}{2}(\rho A \; dx)(-\omega s_m)^2 \sin^2(kx - \omega t). \quad (18\text{-}29)$$

Dividing Eq. 18-29 by dt gives the rate at which kinetic energy moves along with the wave. As we saw in Chapter 17 for transverse waves, the ratio dx/dt is the wave speed, so we have

$$\frac{dK}{dt} = \tfrac{1}{2}\rho A v \omega^2 s_m^2 \sin^2(kx - \omega t). \quad (18\text{-}30)$$

The *average* rate at which kinetic energy is transported is

$$\overline{\left(\frac{dK}{dt}\right)} = \tfrac{1}{2}\rho A v \omega^2 s_m^2 \; \overline{\sin^2(kx - \omega t)}$$

$$= \tfrac{1}{4}\rho A v \omega^2 s_m^2. \qquad (18\text{-}31)$$

To obtain this equation, we have used the fact that the average value of the square of a sine (or a cosine) function over one wavelength is $\tfrac{1}{2}$.

We assume that *potential* energy is carried along with the wave at this same average rate. The wave intensity I, which is the average rate per unit area at which energy of both kinds is transmitted by the wave, is then, from Eq. 18-31,

$$I = \frac{2\,\overline{(dK/dt)}}{A} = \tfrac{1}{2}\rho v \omega^2 s_m^2,$$

which is Eq. 18-26, the equation we set out to derive.

SAMPLE PROBLEM 18-4

a. Two sound waves have intensities I_1 and I_2. How do their sound levels compare?

SOLUTION   Let us write the ratio of the two intensities as

$$\frac{I_2}{I_1} = \frac{I_2/I_0}{I_1/I_0}.$$

Taking logarithms on each side and multiplying through by 10 dB, we have

$$(10 \text{ dB}) \log \frac{I_2}{I_1} = (10 \text{ dB}) \log \frac{I_2}{I_0} - (10 \text{ dB}) \log \frac{I_1}{I_0}.$$

From Eq. 18-27 we then see that the terms on the right are β_1 and β_2. So

$$\beta_2 - \beta_1 = (10 \text{ dB}) \log \frac{I_2}{I_1}. \quad \text{(Answer)} \quad (18\text{-}32)$$

Note that the *ratio* of two intensities corresponds to a *difference* in their sound levels.

b. If you multiply the intensity of a sound wave by 10, you add 10 dB to the sound level, as we have seen. If you multiply the intensity by a factor of 2.0, by how much do you raise its sound level?

SOLUTION   From Eq. 18-32,

$$\beta_2 - \beta_1 = (10 \text{ dB}) \log \frac{I_2}{I_1}$$

$$= (10 \text{ dB}) \log 2.0 = 3.0 \text{ dB}. \quad \text{(Answer)}$$

SAMPLE PROBLEM 18-5

a. In 1976, the Who set a record for the loudest concert: the sound level 46 m in front of the speaker systems was $\beta_2 = 120$ dB. What is the ratio of the intensity of the Who at that spot to the intensity of a jackhammer operating at sound level $\beta_1 = 92$ dB?

SOLUTION   From Eq. 18-32, we have

$$(10 \text{ dB}) \log \frac{I_2}{I_1} = 120 \text{ dB} - 92 \text{ dB} = 28 \text{ dB}.$$

Thus

$$\log \frac{I_2}{I_1} = \frac{28 \text{ dB}}{10 \text{ dB}} = 2.8.$$

Taking the antilog of both sides gives us

$$\frac{I_2}{I_1} = 630. \qquad \text{(Answer)}$$

Temporary exposure to sound intensities as great as those of a jackhammer and the 1976 Who concert results in a temporary reduction of hearing. Repeated or prolonged exposure can result in permanent reduction of hearing (Fig. 18-9). Loss of hearing is a clear risk for

FIGURE 18-9 Sample Problem 18-5. Peter Townsend of the Who, playing in front of a speaker system. His repeated and prolonged exposure to high-intensity sound, especially when playing into a speaker to produce feedback, resulted in a permanent reduction in his hearing.

anyone continually listening to, say, "heavy metal" at high volume.

b. How much more intense is an 80-dB shout than a 20-dB whisper?

SOLUTION Every 20-dB increase in sound level corresponds to multiplying the intensity by 100. We are dealing here with an increase in sound level of 60 (= 20 + 20 + 20) dB or three factors of 100. Thus the shout is $100 \times 100 \times 100$ or a million times stronger than the whisper. Would you have imagined that so much additional energy is required to raise your voice?

▬▬▬ **SAMPLE PROBLEM 18-6**

Sound waves are emitted uniformly in all directions from a point source S in Fig. 18-10.

a. What is the intensity of the sound waves at distance r from the source if the source emits energy at the rate

FIGURE 18-10 Sample Problem 18-6. A point source S sends out sound waves uniformly in all directions. The intensity I at a distance r from the source decreases as $1/r^2$.

P? Evaluate the intensity for $r = 2.50$ m and $P = 25.0$ W.

SOLUTION If we assume that the mechanical energy of the sound waves is conserved, then the rate at which energy is emitted by the point source (that is, the power P emitted by the source) must be equal to the rate at which energy passes through the surface area $4\pi r^2$ of a sphere of radius r centered on the source. Since intensity is the rate per unit area at which energy is transmitted,

$$I = \frac{P}{4\pi r^2}. \quad \text{(Answer)} \quad (18\text{-}33)$$

We see that the intensity of the sound drops off as the inverse square of the distance from the source. Numerically, we have

$$I = \frac{25.0 \text{ W}}{(4\pi)(2.50 \text{ m})^2} = 0.318 \text{ W/m}^2$$

$$= 318 \text{ mW/m}^2. \quad \text{(Answer)}$$

b. How does the amplitude of the sound waves change with distance r?

SOLUTION Substituting I from Eq. 18-26 into Eq. 18-33, we find

$$\tfrac{1}{2}\rho v \omega^2 s_m^2 = \frac{P}{4\pi r^2},$$

which yields

$$s_m = \sqrt{\frac{P}{2\pi\rho v\omega^2 r^2}} = \frac{1}{r}\sqrt{\frac{P}{2\pi\rho v\omega^2}}, \quad \text{(Answer)}$$

where all quantities within the square root sign are constant. Thus as the sound waves spread, their amplitude drops off as the inverse of the distance from the source.

18-5 SOURCES OF MUSICAL SOUND

Musical sounds can be set up by oscillating strings (guitar, piano, violin), membranes (kettledrum, snare drum), air columns (flute, oboe, pipe organ, and the fujara of Fig. 18-11), wooden blocks or steel bars (marimba, xylophone), and many other oscillating bodies. Most instruments involve more than a single oscillating part. In the violin, for example, not only the strings but also the body of the instrument participates in producing the sounds that we enjoy.

Recall from Chapter 17 that standing waves can be set up on a taut string that is fixed at both ends. They arise because waves traveling along the string are reflected back onto the string at each end. If the wavelength of the waves is suitably matched to the length of the string, the superposition of waves traveling in opposite directions produces a standing wave pattern (or oscillation mode). The wavelength required of the waves for such a match is one that corresponds to a *resonant frequency* of the string. The advantage of setting up standing waves is that the string then oscillates with a large, sustained amplitude, periodically pushing against the surrounding air and thus generating a noticeable sound wave

FIGURE 18-11 The air column within a fujara oscillates when that traditional Slovakian instrument is played.

with the same frequency as the oscillations of the string. This production of sound is of obvious importance to, say, a guitarist.

We can set up standing waves of sound in a pipe in a similar way. As sound waves travel through the air in a pipe, they reflect at each end and back through the pipe. (The reflection occurs even if an end is open, but the reflection is not as complete as when the end is closed.) If the wavelength of the sound waves is suitably matched to the length of the pipe, the superposition of waves traveling in opposite directions through the pipe sets up a standing wave pattern. The wavelength required of the sound waves for such a match is one that corresponds to a resonant frequency of the pipe. The advantage of such a standing wave is that the air in the pipe oscillates with a large, sustained amplitude, emitting a sound wave at any open end that has the same frequency as the oscillations in the pipe. This emission of sound is of obvious importance to, say, an organist.

Many other aspects of standing sound wave patterns are similar to those of string waves: the closed end of a pipe is like the fixed end of a string in that there must be a displacement node there. And the open end of a pipe is like the end of a string attached to a freely moving ring, as in Fig. 17-16b, in that there must be an antinode there. (Actually, the antinode for the open end of a pipe is located slightly beyond the end, but we shall not dwell on that fact here.)

The simplest standing wave pattern that can be set up in a pipe with two open ends is shown in Fig. 18-12a. There is a displacement antinode across each open end, as required. There is also a displacement node across the middle of the pipe. An easier way of representing this standing longitudinal sound wave is shown in Fig. 18-12b—by drawing it as a standing transverse string wave.

The standing wave pattern of Fig. 18-12a is called the *fundamental mode* or *first harmonic*. For it to be set up, the sound waves in a pipe of length L must have a wavelength given by $L = \lambda/2$, so that $\lambda = 2L$. Several more standing sound wave patterns for a pipe with two open ends are shown in Fig. 18-13a using string-wave representations. The *second harmonic* requires sound waves of wavelength $\lambda = L$. The *third harmonic* requires wavelength $\lambda = 2L/3$. And so on.

More generally, the resonant frequencies for a pipe of length L with two open ends correspond to

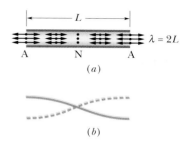

FIGURE 18-12 (*a*) The simplest standing wave pattern for (longitudinal) sound waves in a pipe with both ends open has a displacement antinode (A) across each end and a displacement node (N) across the middle. (The displacements represented by the double arrows are greatly exaggerated.) (*b*) The corresponding standing wave pattern for (transverse) string waves.

the wavelengths

$$\lambda = \frac{2L}{n}, \qquad n = 1, 2, 3, \ldots, \quad (18\text{-}34)$$

where *n* is called the *harmonic number*. The resonant frequencies are then given by

$$f = \frac{v}{\lambda} = \frac{nv}{2L}, \qquad n = 1, 2, 3, \ldots$$

$$\text{(pipe, two open ends),} \quad (18\text{-}35)$$

where *v* is the speed of sound.

Figure 18-13*b* shows (using string-wave representations) some of the standing sound wave patterns that can be set up in a pipe with only one open end. As required, across the open end there is a displacement antinode and across the closed end there is a displacement node. The simplest pattern requires sound waves having a wavelength given by $L = \lambda/4$, so that $\lambda = 4L$. The next simplest pattern requires a wavelength given by $L = 3\lambda/4$, so that $\lambda = 4L/3$. And so on.

More generally, the resonant frequencies for a pipe of length *L* with only one open end correspond to the wavelengths

$$\lambda = \frac{4L}{n}, \qquad n = 1, 3, 5, \ldots, \quad (18\text{-}36)$$

in which the harmonic number *n must be an odd number*. The resonant frequencies are then given by

$$f = \frac{v}{\lambda} = \frac{nv}{4L}, \qquad n = 1, 3, 5, \ldots$$

$$\text{(pipe, one open end).} \quad (18\text{-}37)$$

FIGURE 18-13 Standing wave patterns for string waves superimposed on pipes to represent standing sound wave patterns in the pipes. (*a*) Both ends of the pipe are open; any harmonic can exist. (*b*) Only one end is open; only odd harmonics can exist.

FIGURE 18-14 The families of saxophones and of violins, showing the relations between length and frequency range, the latter being shown by the horizontal bars. The frequency scale is suggested by the keyboard at the bottom, with the frequency increasing toward the right.

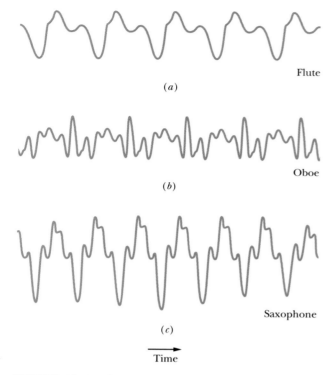

Flute

(a)

Oboe

(b)

Saxophone

(c)

→ Time

FIGURE 18-15 The wave form produced by (a) a flute, (b) an oboe, and (c) a saxophone when they all play the same note, with the same first harmonic frequency. The wave forms, and the resulting sound you would hear, differ because the instruments produce the higher harmonics at different intensities.

Note that only odd harmonics can exist in a pipe with only one open end. For example, the second harmonic, with $n = 2$, cannot be set up in such a pipe.

The length of a musical instrument reflects the range of frequencies over which the instrument is designed to function: smaller length implies higher frequency. Figure 18-14, for example, shows the saxophone and violin families, with their frequency ranges suggested by the piano keyboard. Note that, for every instrument, there is considerable overlap with its higher- and lower-frequency neighbors.

In any oscillating system that gives rise to a musical sound, whether it be a violin string or the air in an organ pipe, the fundamental and one or more of the higher harmonics are usually generated simultaneously. The resultant sound is the superposition of these components. For different instruments, the higher harmonics have different intensities, which accounts for the fact that a given note sounds differ-

ent on different instruments. Figure 18-15 shows, for example, the wave forms when the same note, with the same fundamental frequency, is sounded on three different instruments.

SAMPLE PROBLEM 18-7

Weak background noises from a room set up the fundamental standing wave in a cardboard tube of length $L = 67.0$ cm with two open ends. Assume that the speed of sound in the air within the tube is 343 m/s.

a. What frequency do you hear from the tube if you jam your ear against one end?

SOLUTION With your ear closing one end, the fundamental frequency is given by Eq. 18-37 with $n = 1$:

$$f = \frac{v}{4L} = \frac{343 \text{ m/s}}{(4)(0.670 \text{ m})} = 128 \text{ Hz}. \quad \text{(Answer)}$$

If the background noises set up any higher harmonics, such as the third harmonic, you may also hear frequencies that are *odd* multiples of 128 Hz.

b. What frequency do you hear from the tube if you move your head away enough so that the tube has two open ends?

SOLUTION With both ends open, the fundamental frequency is given by Eq. 18-35 with $n = 1$:

$$f = \frac{v}{2L} = \frac{343 \text{ m/s}}{(2)(0.670 \text{ m})} = 256 \text{ Hz}. \quad \text{(Answer)}$$

If the background noises also set up any higher harmonics, such as the second harmonic, you may also hear frequencies that are *integer* multiples of 256 Hz. However, sound with a frequency of 128 Hz is no longer emitted by the tube.

18-6 BEATS

If we listen, a few minutes apart, to two sounds whose frequencies are 552 and 564 Hz, most of us cannot tell one from the other. However, if the sounds reach our ears simultaneously, what we hear is a sound whose frequency is 558 Hz, the *average* of the two combining frequencies. We also hear a striking variation in the intensity of this sound: it increases and decreases in a slow wobbling *beat note*, whose frequency is 12 Hz, the *difference* between the two combining frequencies. Figure 18-16 shows this **beat** phenomenon.

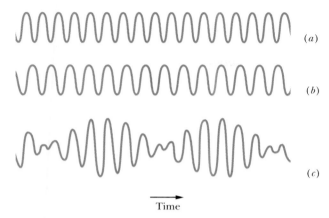

FIGURE 18-16 (*a*, *b*) The pressure variations Δp of two sound waves that are detected separately. The frequencies of the waves are nearly equal. (*c*) The resultant pressure variation if the two waves are detected simultaneously.

Let the time-dependent variations of the displacements due to two sound waves at a particular location be

$$s_1 = s_m \cos \omega_1 t \quad \text{and} \quad s_2 = s_m \cos \omega_2 t. \quad (18\text{-}38)$$

We have assumed, for simplicity, that the waves have the same amplitude. According to the superposition principle, the resultant displacement is

$$s = s_1 + s_2 = s_m(\cos \omega_1 t + \cos \omega_2 t).$$

Using the trigonometric identity (see Appendix G)

$$\cos \alpha + \cos \beta = 2 \cos \tfrac{1}{2}(\alpha - \beta) \cos \tfrac{1}{2}(\alpha + \beta)$$

allows us to write the resultant displacement as

$$s = 2s_m \cos \tfrac{1}{2}(\omega_1 - \omega_2)t \cos \tfrac{1}{2}(\omega_1 + \omega_2)t. \quad (18\text{-}39)$$

If we write

$$\omega' = \tfrac{1}{2}(\omega_1 - \omega_2) \quad \text{and} \quad \omega = \tfrac{1}{2}(\omega_1 + \omega_2), \quad (18\text{-}40)$$

we can then write Eq. 18-39 as

$$s(t) = [2s_m \cos \omega' t] \cos \omega t. \quad (18\text{-}41)$$

We now assume that the angular frequencies ω_1 and ω_2 of the combining waves are almost equal, which means that $\omega \gg \omega'$ in Eq. 18-40. We can then regard Eq. 18-41 as a cosine function whose angular frequency is ω and whose amplitude (which is not constant but varies with frequency ω') is the quantity in the brackets.

A beat, that is, a maximum of amplitude, will occur whenever $\cos \omega' t$ in Eq. 18-41 has the value $+1$ or -1, which happens twice in each repetition

of the cosine function. Because $\cos \omega' t$ has angular frequency ω', the angular frequency ω_{beat} at which beats occur is $\omega_{\text{beat}} = 2\omega'$. Then, with the aid of Eq. 18-40, we can write

$$\omega_{\text{beat}} = 2\omega' = (2)(\tfrac{1}{2})(\omega_1 - \omega_2) = \omega_1 - \omega_2.$$

Because $\omega = 2\pi f$, we can recast this as

$$f_{\text{beat}} = f_1 - f_2 \quad \text{(beat frequency)}. \quad (18\text{-}42)$$

Musicians use the beat phenomenon in tuning their instruments. If an instrument is sounded against a standard frequency (for example, the lead oboe's reference A) and tuned until the beat disappears, then the instrument is in tune with that standard. In musical Vienna, concert A (440 Hz) is available as a telephone service for the benefit of the city's many professional and amateur musicians.

SAMPLE PROBLEM 18-8

You wish to tune the note A_3 on a piano to its proper frequency of 220 Hz. You have available a tuning fork whose frequency is 440 Hz. How should you proceed?

SOLUTION These two frequencies are too far apart to produce beats. Recall that in our analysis of Eq. 18-41 we assumed that the two combining frequencies were reasonably close to each other. However, note that the second harmonic of A_3 is 2×220 or 440 Hz.

Let us say that the actual piano string is mistuned and that its fundamental frequency is not exactly 220 Hz. You listen for beats between the fundamental frequency of the tuning fork and the second harmonic of A_3 and you hear a beat frequency of 6 Hz. You then change the tension in the corresponding string until the beat note disappears. The string is then in tune.

Note that, from the observed beat frequency of 6 Hz, you cannot tell whether the fundamental frequency of the originally mistuned string is 223 Hz or 217 Hz. Both frequencies would produce the same beat frequency. You can easily experiment, however, by tightening the string slightly and noting whether the beat frequency increases or decreases. If it increases, you should be loosening the string.

18-7 THE DOPPLER EFFECT

A police car is parked by the side of the highway, sounding its 1000-Hz siren. If you are also parked by the highway, you will hear that same frequency. But if there is relative motion between you and the police car, either toward or away from each other, you will hear a different frequency. For example, if you are driving *toward* the police car at 120 km/h (about 75 mi/h), you will hear a *higher* frequency (1096 Hz, an *increase* of 96 Hz). If you are driving *away from* the police car at that same speed, you will hear a *lower* frequency (904 Hz, a *decrease* of 96 Hz).

These motion-related frequency changes are examples of the **Doppler effect.** The effect was proposed (although not fully worked out) in 1842 by the Austrian physicist Johann Christian Doppler. It was tested experimentally in 1845 by Buys Ballot in Holland, "using a locomotive drawing an open car with several trumpeters."

The Doppler effect holds not only for sound waves but also for electromagnetic waves, including microwaves, radio waves, and visible light. Police use the Doppler effect with microwaves to determine the speed of a car: a radar unit beams microwaves of a certain frequency f toward the oncoming car. The microwaves that reflect from the metal portions of the car back to the radar unit have a higher frequency f' owing to the motion of the car relative to the radar unit. The radar unit translates the difference between f' and f into the speed of the car, which is then displayed for the operator to see.

The displayed speed is the actual speed of the car only if the car is moving directly toward the radar unit. Any deviation in that alignment reduces f'. If the radar beam is perpendicular to the car's velocity, f' is equal to f, and the radar unit will display a speed of zero for the car. Similarly, a radar gun used in baseball to measure the speed of a pitched ball must be aimed toward an oncoming ball if the measurement is to be accurate.

The Doppler effect on visible light has allowed astronomers to determine the speeds of stars and galaxies relative to the Earth. The distant galaxies all turn out to be moving away from us, and the more distant a galaxy is, the faster it is moving. These facts were discovered by studying the frequencies of the light that our telescopes intercept from those galaxies: the frequencies are lower than they would be were the galaxies not moving.

In the analyses that follow, we restrict ourselves to sound waves, and we take as a reference frame the body of air through which these waves travel. We assume that there is no wind, so that this reference frame is identical to a frame fixed with respect to the Earth. Furthermore, we assume that the source S of the sound waves and the detector D of those waves move only along the line that joins them. We also assume that the speeds of S and D, which are measured *relative to the body of air*, are less than the speed of sound.

Detector Moving; Source Stationary

In Fig. 18-17, a detector D (represented by an ear) is moving at speed v_D toward a stationary source S that emits spherical wave fronts, of wavelength λ and frequency f, moving at the speed of sound v. The wave fronts are drawn one wavelength apart. The frequency detected by D is the rate at which the detector intercepts wave fronts (or individual wavelengths). If D were stationary, that rate would be f, but since D is moving into the waves, the rate of interception is greater, and thus the detected frequency f' is greater than f.

Let us for the moment consider the situation in which D is stationary (Fig. 18-18). In time t, the wave fronts move to the right a distance vt. The number of wavelengths in that distance vt is the number of wavelengths intercepted by D in time t, and that number is vt/λ. The rate at which D intercepts wavelengths, which is the frequency f detected by D, is

$$f = \frac{vt/\lambda}{t} = \frac{v}{\lambda}. \qquad (18\text{-}43)$$

In this situation, with D stationary, there is no Doppler effect: the frequency detected by D is the frequency emitted by S.

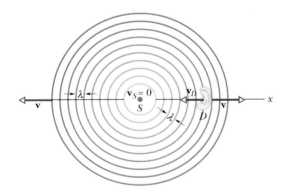

FIGURE 18-17 A stationary source of sound S emits spherical wave fronts, shown one wavelength apart, that expand outward at the speed of sound v. A sound detector D, represented by an ear, moves with velocity $\mathbf{v}_D$ toward the source. The detector senses a higher frequency because of its motion.

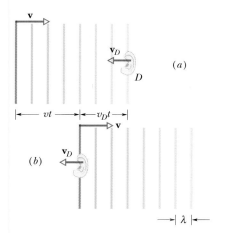

FIGURE 18-18 Wave fronts of Fig. 18-17 (a) reach and (b) pass a stationary detector D, moving distance vt to the right in time t.

Now let us again consider the situation in which D moves opposite the waves (Fig. 18-19). In time t, the wave fronts move to the right a distance vt as previously, but now D moves to the left a distance $v_D t$. Thus in this time t, the distance moved by the wave fronts relative to D is $vt + v_D t$. The number of wavelengths in this relative distance $vt + v_D t$ is the number of wavelengths intercepted by D in time t, and is $(vt + v_D t)/\lambda$. The *rate* at which D intercepts wavelengths in this situation is the frequency f', given by

$$f' = \frac{(vt + v_D t)/\lambda}{t} = \frac{v + v_D}{\lambda}. \qquad (18\text{-}44)$$

FIGURE 18-19 Wave fronts (a) reach and (b) pass detector D which moves opposite the waves. In time t, the waves move distance vt to the right and D moves distance $v_D t$ to the left.

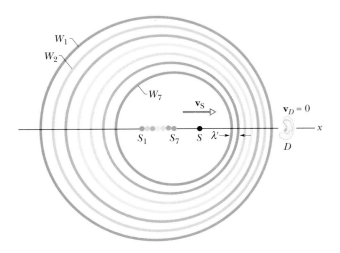

FIGURE 18-20 The detector D is stationary, with the source S moving toward it at speed v_S. Wave front W_1 was emitted when the source was at S_1, wave front W_7 when it was at S_7. At the moment depicted, the source is at S. The detector perceives a higher frequency because the moving source, chasing its own wave fronts, emits a reduced wavelength λ' in its direction of motion.

In the direction opposite that taken by S, the wavelength λ' of the waves is $vT + v_S T$. If D detects those waves, it detects frequency f' given by

$$f' = f\,\frac{v}{v + v_S}.\qquad (18\text{-}49)$$

Now f' must be less than f unless $v_S = 0$.

We can summarize Eqs. 18-48 and 18-49 with

$$f' = f\,\frac{v}{v \mp v_S}\qquad \begin{array}{l}\text{(source moving;}\\ \text{detector stationary).}\end{array}\qquad (18\text{-}50)$$

You can determine which sign applies in Eq. 18-50 by remembering the physical result: when the source moves toward the detector, the frequency is greater (*toward* means *greater*), which requires a minus sign in the denominator. Otherwise a plus sign should be used.

Source and Detector Both Moving

We can combine Eqs. 18-47 and 18-50 to produce the general Doppler effect equation, in which both the source and the detector are moving with respect to the air mass. Replacing the f in Eq. 18-50 (the frequency of the source) with the f' of Eq. 18-47

(the frequency associated with motion of the detector) leads to

$$f' = f\,\frac{v \pm v_D}{v \mp v_S}\qquad \begin{array}{l}\text{(detector moving;}\\ \text{source moving).}\end{array}\qquad (18\text{-}51)$$

Putting $v_S = 0$ in Eq. 18-51 reduces it to Eq. 18-47, and putting $v_D = 0$ reduces it to Eq. 18-50. The plus and minus signs are determined as in Eqs. 18-47 and 18-50 (*toward* means *greater*).

The Doppler Effect at Low Speeds

The Doppler effects for a moving detector (Eq. 18-47) and for a moving source (Eq. 18-50) are different, even though the detector and the source may be moving at the same speed. However, if the speeds are low enough (that is, if $v_D \ll v$ and $v_S \ll v$), the frequency changes produced by these two motions are essentially the same.

By using the binomial theorem (see Tactic 3 of Chapter 7), you can show that Eq. 18-51 can be written in the form

$$f' \approx f\left(1 \pm \frac{u}{v}\right)\qquad \text{(low speeds only),}\qquad (18\text{-}52)$$

in which $u\,(=|v_S \pm v_D|)$ is the *relative* speed of the source with respect to the detector. The rule for signs remains the same. If the source and the detector are moving *toward* each other, we anticipate a *greater* frequency; this requires that we choose the plus sign in Eq. 18-52. On the other hand, if the source and the detector are moving away from each other, we anticipate a frequency decrease and choose the minus sign in Eq. 18-52.

Supersonic Speeds

If a source is moving toward a stationary detector at a speed equal to the speed of sound, that is, if $v_S = v$, Eq. 18-50 predicts that the detected frequency f' will be infinitely great. This means that the source is moving so fast that it keeps pace with its own spherical wave fronts, as Fig. 18-21a suggests. What happens when the speed of the source *exceeds* the speed of sound?

For such *supersonic* speeds, Eq. 18-50 no longer applies. Figure 18-21b depicts the spherical wave fronts that originated at various positions of the source. The radius of any wave front in this figure is

FIGURE 18-22 A false-color, high-speed image of a 20-mm-caliber bullet traveling at about Mach 1.3. Note the prominent Mach cone produced by the nose of the bullet and the secondary cones produced by irregular features along the sides.

FIGURE 18-21 (*a*) A source of sound *S* moves at speed v_S very close to the speed of sound and thus almost as fast as the wave fronts it generates. (*b*) A source *S* moves at speed v_S faster than the speed of sound and thus faster than the wave fronts. When the source was at position S_1 it generated wave front W_1, and at position S_6 it generated W_6. All the spherical wave fronts expand at the speed of sound v and bunch along the surface of a cone called the Mach cone, forming a shock wave. The surface of the cone has half-angle θ and is tangent to all the wave fronts.

vt, where *v* is the speed of sound and *t* is the time that has elapsed since the source emitted that wave front. Note that all the wave fronts bunch along a V-shaped envelope in Fig. 18-21*b*, which in three dimensions is a cone called the *Mach cone*. A **shock wave** is said to exist along the surface of this cone, because the bunching of wave fronts causes an abrupt rise and fall of air pressure as the surface passes through any point. From Fig. 18-21*b*, we see that the half-angle θ of the cone, called the *Mach*

cone angle, is given by

$$\sin \theta = \frac{vt}{v_S t} = \frac{v}{v_S} \qquad \text{(Mach cone angle).} \qquad (18\text{-}53)$$

The ratio v_S/v is called the *Mach number*. When you hear that a particular plane has flown at Mach 2.3, it means that its speed was 2.3 times the speed of sound in the air through which the plane was flying.

The shock wave generated by a supersonic aircraft or projectile (see Fig. 18-22) produces a burst of sound called a *sonic boom*. A similar effect (called *Cerenkov radiation*) occurs for visible light when electrons travel through water or another transparent medium at speeds that are greater than the speed of light *in that medium*. The blue glow that emanates from the water in which highly radioactive reactor fuel rods are stored is caused by this effect. It is—if you wish—an "optical sonic boom," caused by the speeding electrons that are emitted by the radioactive atoms in the fuel rods.

SAMPLE PROBLEM 18-9

Bats navigate and search out prey by emitting, and then detecting reflections of, ultrasonic waves, which are sound waves with frequencies greater than can be heard by a human. Suppose a horseshoe bat flies toward a moth at speed $v_b = 9.0$ m/s, while the moth flies toward the bat with speed $v_m = 8.0$ m/s. From its nostrils, the bat emits ultrasonic waves of frequency f_{be} that reflect from the moth back to the bat with frequency f_{bd}. The bat adjusts the emitted frequency f_{be} until the returned frequency f_{bd} is 83 kHz, at which the bat's hearing is best.

a. What is f_m, the frequency of the waves heard and reflected by the moth?

SOLUTION The moth acts as both a detector (since it hears the waves) and an emitter (since it reflects the waves). To find f_m, we use Eq. 18-51, with the moth being the emitter (of the reflected waves with frequency f_m) and the bat being the detector (of the echo with frequency $f_{bd} = 83$ kHz). Since the detector moves toward the emitter (with speed v_b), we choose the plus sign in the numerator of Eq. 18-51. Since the emitter moves toward the detector (with speed v_m), we choose the minus sign in the denominator. We then have

$$f_{bd} = f_m \frac{v + v_b}{v - v_m},$$

or

$$83 \text{ kHz} = f_m \frac{343 \text{ m/s} + 9.0 \text{ m/s}}{343 \text{ m/s} - 8.0 \text{ m/s}},$$

from which

$$f_m = 78.99 \text{ kHz} \approx 79 \text{ kHz}. \quad \text{(Answer)}$$

b. What is the frequency f_{be} emitted by the bat?

SOLUTION We again use Eq. 18-51, but now with the bat as emitter (at frequency f_{be}) and the moth as detector (of frequency f_m). Since the detector moves toward the emitter (with speed v_m), we choose the plus sign in the numerator of Eq. 18-51. Since the emitter moves toward the detector (with speed v_b), we choose the minus sign in the denominator. We then have

$$f_m = f_{be} \frac{v + v_m}{v - v_b}$$

or

$$78.99 \text{ kHz} = f_{be} \frac{343 \text{ m/s} + 8.0 \text{ m/s}}{343 \text{ m/s} - 9.0 \text{ m/s}},$$

from which

$$f_{be} = 75 \text{ kHz}. \quad \text{(Answer)}$$

The bat determines the relative speed of the moth (17 m/s) from the 8 kHz (= 83 kHz − 75 kHz) it must lower its emitted frequency to hear an echo with a frequency of 83 kHz. Some moths evade capture by flying away from the direction in which they hear ultrasonic waves. That choice of flight reduces the frequency difference between what the bat emits and what it hears,

and then the bat may not notice the echo. Some moths avoid capture by clicking to produce their own ultrasonic waves, thus "jamming" the detection system and confusing the bat.

18-8 THE DOPPLER EFFECT FOR LIGHT (OPTIONAL)

It is tempting to apply to light waves the Doppler effect equation (Eq. 18-51) that we developed for sound waves in the preceding section, simply substituting c, the speed of light, for v, the speed of sound. We must avoid this temptation.

The reason is that sound waves—like all other mechanical waves—require a medium (the air in our examples) for their propagation but light waves, which can travel through a vacuum, do not. The speed of sound is always "with respect to the medium" but the speed of light cannot be because there may be no medium. As we discussed in Section 17-7, the speed of light always has the same value c, in all directions and in all inertial frames.

For sound waves "source in motion" and "detector in motion" refer separately to motion with respect to the medium. They represent physically different situations and lead to different equations for the Doppler effect. For light waves, however, all that can possibly matter is the *relative* motion of the source and the detector. "Source moving toward detector at speed u" and "detector moving toward source at speed u" are physically identical situations and there should be only one equation (not two) to describe the Doppler effect, an equation that must be derived on the basis of the theory of relativity.

Even though the Doppler equations for light and for sound are necessarily different, at low enough speeds they reduce to the same approximate result. This is not unexpected because *all* the predictions of relativity theory reduce to their classical counterparts at low enough speeds. Thus Eq. 18-52, with v replaced by c, holds for light waves if $u \ll c$, where u is the relative speed of the source and the detector. That is,

$$f' = f(1 \pm u/c) \quad \text{(light waves; } u \ll c\text{)}. \quad (18\text{-}54)$$

If the source and the detector are *approaching* each other, our rule for signs tells us to anticipate a frequency *increase*, which calls for the choice of a plus sign in Eq. 18-54.

In Doppler observations in astronomy, it is the

wavelength rather than the frequency that is more readily measured. This suggests that we replace f' and f in Eq. 18-54 with c/λ' and c/λ, respectively. Doing so, we find

$$\lambda' = \lambda(1 \pm u/c)^{-1} \approx \lambda(1 \mp u/c).$$

We can write this as

$$\frac{\lambda' - \lambda}{\lambda} = \mp\frac{u}{c}$$

or

$$u = \frac{\Delta\lambda}{\lambda} c \qquad \text{(light waves; } u \ll c\text{)}, \quad (18\text{-}55)$$

in which $\Delta\lambda$ is the magnitude of the Doppler wavelength shift. If the wavelength decreases (called a *blue shift* because the blue portion of the visible spectrum has the shortest wavelengths), the frequency necessarily increases and that—according to our rule for signs—means that the distance between the source and the detector is decreasing. If the wavelength increases (a *red shift*), the distance between the source and the detector is increasing.

SAMPLE PROBLEM 18-10

Figure 18-23 shows the measured intensity of light emitted by a distant galaxy known only by its catalog number, NGC 7319.* The most intense spectral line, whose wavelength λ is 513 nm, was emitted by oxygen atoms in the galaxy. The Doppler shift $\Delta\lambda$ ($= 12$ nm) shown on the figure indicates that this line has a wavelength 12 nm greater than would be measured for similar light emitted by a laboratory source. What is the speed of this galaxy with respect to Earth? Is it approaching or receding? (*All* lines in the spectrum of

---

*If you want to name a galaxy after a friend, feel free to do so. There are enough of them to allocate about 20 to every living person.

FIGURE 18-23 Sample Problem 18-10. The intensity of the light emitted by a galaxy about 3×10^8 light-years from Earth is shown here as a function of wavelength. The sharp intensity peaks can be associated with atoms such as hydrogen, nitrogen, and oxygen that are present in the galaxy. All of these lines are Doppler shifted toward longer wavelengths (a *red shift*) because the galaxy is receding from us at a rapid rate. The Doppler shift for the most intense line is shown. Courtesy of J. P. Huchra and his associates.

Fig. 18-23 are Doppler shifted; we discuss here only the most intense line.)

SOLUTION From Eq. 18-55 we find

$$u = \frac{\Delta\lambda}{\lambda} c = \frac{(12 \text{ nm})(3.00 \times 10^8 \text{ m/s})}{513 \text{ nm}}$$

$$= 7.0 \times 10^6 \text{ m/s}$$

$$= 7000 \text{ km/s.} \qquad \text{(Answer)}$$

More careful measurements by astronomers at the Center for Astrophysics, Harvard College Observatory, yield a value of 6764 km/s.

The observed wavelength is Doppler shifted toward longer wavelengths (a *red shift*) compared with light measured from a stationary laboratory source. Thus the observed frequency is *smaller* because of the motion of the galaxy. According to our rule for signs, this means that the galaxy must be *receding* from us. Measurements of this kind are the basis for the belief that the universe is expanding.

REVIEW & SUMMARY

Sound Waves

Sound waves are longitudinal mechanical waves that can travel through solids, liquids, or gases. The speed v of a sound wave in a medium having **bulk modulus** B and density ρ is

$$v = \sqrt{\frac{B}{\rho}} \qquad \text{(speed of sound).} \quad (18\text{-}3)$$

In air at 20°C, the speed of sound is 343 m/s.

Wave Equations for Displacement and Pressure

The longitudinal displacement s of a mass element in a medium, due to a sound wave, is

$$s = s_m \cos(kx - \omega t), \qquad (18\text{-}13)$$

where s_m is the **displacement amplitude** (maximum displacement) from equilibrium, $k = 2\pi/\lambda$, and $\omega = 2\pi f$, λ and f being the wavelength and frequency, respectively, of the sound wave. The pressure change Δp of the medium from the equilibrium pressure, due to the sound wave, is

$$\Delta p = \Delta p_m \sin(kx - \omega t), \qquad (18\text{-}14)$$

where the **pressure amplitude** is

$$\Delta p_m = (v\rho\omega) s_m. \qquad (18\text{-}15)$$

Inteference

The interference of two sound waves with identical wavelengths passing through a common point depends on their phase difference ϕ there. If the waves were emitted in phase, ϕ is given by

$$\phi = \frac{\Delta d}{\lambda} 2\pi, \qquad (18\text{-}21)$$

where Δd is their **path length difference** (the difference in the distances traveled by the waves to reach the common point). Conditions for total constructive and total destructive interference of the waves are given by

$$\phi = m\, 2\pi, \qquad m = 0, 1, 2, \ldots$$
$$\text{(fully constructive interference)} \qquad (18\text{-}22)$$

and

$$\phi = (m + \tfrac{1}{2})2\pi, \qquad m = 0, 1, 2, \ldots$$
$$\text{(fully destructive interference).} \qquad (18\text{-}23)$$

These conditions correspond to

$$\Delta d = m\lambda \qquad \begin{array}{l}\text{(fully constructive} \\ \text{interference)}\end{array} \qquad (18\text{-}24)$$

and

$$\Delta d = (m + \tfrac{1}{2})\lambda \qquad \begin{array}{l}\text{(fully destructive} \\ \text{interference).}\end{array} \qquad (18\text{-}25)$$

Sound Intensity

The **intensity** I of a sound wave is the average rate at which it transmits energy through a unit area and is found as

$$I = \tfrac{1}{2}\rho v \omega^2 s_m^2. \qquad (18\text{-}26)$$

Sound Level in Decibels

The *sound level* β in *decibels* (dB) is defined as

$$\beta = (10\ \text{dB}) \log \frac{I}{I_0}, \qquad (18\text{-}27)$$

where $I_0\ (= 10^{-12}\ \text{W/m}^2)$ is the reference intensity level to which any intensity I is compared. For every factor-of-10 increase in intensity, 10 dB is added to the sound level. The range of a human ear is approximately 0 to 120 dB.

Standing Wave Patterns in Pipes

Standing sound wave patterns can be set up in pipes. A pipe open at both ends will resonate at frequencies

$$f = \frac{v}{\lambda} = \frac{nv}{2L}, \qquad n = 1, 2, 3, \ldots$$
$$\text{(pipe, two open ends),} \qquad (18\text{-}35)$$

where v is the speed of sound in the air in the pipe. For a pipe closed at one end and open at the other, the frequencies are

$$f = \frac{v}{\lambda} = \frac{nv}{4L}, \qquad n = 1, 3, 5, \ldots$$
$$\text{(pipe, one open end).} \qquad (18\text{-}37)$$

Beats

Beats arise when two waves having slightly different angular frequencies, ω_1 and ω_2, are detected together. The equation giving the resultant particle displacement as a function of time is

$$s(t) = [2s_m \cos \omega' t] \cos \omega t. \qquad (18\text{-}41)$$

Here $\omega' = \tfrac{1}{2}(\omega_1 - \omega_2)$ and $\omega = \tfrac{1}{2}(\omega_1 + \omega_2)$. The beat frequency is

$$f_{\text{beat}} = f_1 - f_2. \qquad (18\text{-}42)$$

The Doppler Effect

The *Doppler effect* is a change in the observed frequency when the source or the detector moves relative to the medium. For sound the observed frequency f' is given in terms of the source frequency f by

$$f' = f\, \frac{v \pm v_D}{v \mp v_S} \qquad \begin{array}{l}\text{(detector and} \\ \text{source moving),}\end{array} \qquad (18\text{-}51)$$

where v_D is the speed of the detector relative to the medium, v_S is that of the source, and v is the speed of sound in the medium; the upper sign on v_S (or v_D) is used when the source (or detector) moves toward the detector (or source), while the lower sign is used when it moves away. When the speeds v_D and v_S are both much smaller than v, all four cases reduce to

$$f' \approx f\left(1 \pm \frac{u}{v}\right) \qquad \text{(low speeds only),} \qquad (18\text{-}52)$$

where u is the relative speed of the source with respect to the detector. The Doppler effect for light follows this same approximate expression, $f' = f(1 \pm u/c)$, if the relative velocity u is much less than the speed of light c.

Shock Wave

If the source speed relative to the medium exceeds the speed of sound in the medium, the Doppler equation no longer applies. In such a case, shock waves result. The half angle θ (see Fig. 18-21) of the wave front is given by

$$\sin \theta = \frac{v}{v_S} \qquad \text{(Mach cone angle).} \qquad (18\text{-}53)$$

QUESTIONS

1. In some science fiction movies, the explosion of a starship can apparently be heard in another starship, while both are in the vacuum of space. Is this possible? Is there any way the explosion can produce sound inside the second ship?

2. Ultrasonic waves can be used to reveal internal structures of the body. They can, for example, distinguish between liquid and soft human tissues far better than can x rays. Why?

3. What experimental evidence is there for assuming that the speed of sound in air is the same for all wavelengths?

4. In practice, the inverse square law does not apply exactly to the decrease in intensity of sounds with distance. Why not?

5. Could the reference intensity for audible sound be set so as to permit negative sound levels in decibels? If so, how?

6. What is the common purpose of the valves of a cornet and the slide of a trombone?

7. The bugle has no valves. How then can we sound different notes on it? To what notes is the bugler limited? Why?

8. The resonant frequencies of wind instruments rise as an orchestra warms up. Explain why.

9. When you strike one prong of a tuning fork, the other prong also oscillates, even if the bottom end of the fork is clamped firmly in a vise. How can this happen? That is, how does the second prong "get the word" that somebody has struck the first prong?

10. How can a sound wave travel down an organ pipe and be reflected at its open end? It seems that there is nothing there to reflect it.

11. How can we experimentally locate the positions of nodes and antinodes in a string, in an air column, and on a vibrating surface?

12. What physical properties of a sound wave correspond to the perceptions of pitch, loudness, and tone quality?

13. Does your singing really sound better in a shower? If so, what are the physical reasons?

14. Explain the audible tone produced by drawing a wet finger around the rim of a wine glass.

15. A lightning flash dissipates an enormous amount of energy and is essentially instantaneous. How is that energy transformed into the sound waves of thunder?

16. Sound waves can be used to measure the speed at which blood flows in arteries and veins. Explain how.

17. Suppose that George blows a whistle and Gloria hears it. She will hear an increased frequency whether she is running toward George or George is running toward her. Are the increases in frequency the same in each case? Assume the same running speeds.

18. Suppose that, in the Doppler effect for sound, the source and receiver are at rest in some reference frame, but the air is moving with respect to this frame. Will there be a change in the wavelength (or frequency) received?

19. Jenny, sitting on a bench, sees Lew, also sitting on a bench, across the campus. She blows a whistle to attract his attention. A steady wind is blowing from Jenny to Lew. How does the presence of the wind affect the frequency of the sound that Lew hears? How does the wind affect the sound travel time?

20. How might the Doppler effect be used in an instrument to detect the fetal heart beat? (Such measurements are routinely made.)

21. A satellite emits radio waves of constant frequency. These waves are picked up on the ground and made to beat against some standard frequency. The beat frequency is then sent through a loudspeaker and one "hears" the satellite signals. Describe how the sound changes as the satellite approaches, passes overhead, and recedes from the detector on the ground.

EXERCISES & PROBLEMS

SECTION 18-2 THE SPEED OF SOUND

Where needed in the problems, use

  speed of sound in air = 343 m/s = 1125 ft/s

and

  density of air = 1.21 kg/m³

unless otherwise specified.

1E. (a) A rule for finding your distance from a lightning flash is to count seconds from the time you see the flash until you hear the thunder and then divide the count by five. The result is supposed to give you the distance in miles. Explain this rule and determine the percent error in it at 20°C. (b) Devise a similar rule for obtaining the distance in kilometers.

2E. A column of soldiers, marching at 120 paces per minute, keep in step with the music of a band at the head of the column. It is observed that the soldiers in the rear end of the column are striding forward with the left foot when musicians are advancing with the right. What is the length of the column approximately?

3E. You are at a large outdoor concert, seated 300 m from the speaker system. The concert is also being broadcast live via satellite. Consider a listener 5000 km away. Who hears the music first, you or the listener, and by what time difference?

4E. Two spectators at a soccer game in Montjuic Stadium see, and a moment later hear, the ball being kicked on the playing field. The time delay for one spectator is 0.23 s and for the other 0.12 s. The lines through each spectator and the player kicking the ball meet at an angle of 90°. (a) How far is each spectator from the player? (b) How far are the spectators from each other?

5E. The average density of the Earth's crust 10 km beneath the continents is 2.7 g/cm³. The speed of longitudinal seismic waves at that depth, found by timing their arrival from distant earthquakes, is 5.4 km/s. Use this information to find the bulk modulus of the Earth's crust at that depth. For comparison, the bulk modulus of steel is about 16×10^{10} Pa.

6E. What is the value of the bulk modulus of oxygen at standard temperature and pressure if 1 mol (32.0 g) of oxygen occupies 22.4 L under these conditions and the speed of sound in oxygen is 317 m/s?

7P. An experimenter wishes to measure the speed of sound in an aluminum rod 10 cm long by measuring the time it takes for a sound pulse to travel the length of the rod. If results good to four significant figures are desired, how precisely must the length of the rod be known and how closely must she be able to resolve time intervals?

8P. The speed of sound in a certain metal is V. One end of a long pipe of that metal of length L is struck a hard blow. A listener at the other end hears two sounds, one from the wave that has traveled along the pipe and the other from the wave that has traveled through the air. (a) If v is the speed of sound in air, what time interval t elapses between the arrival of the two sounds? (b) Suppose that $t = 1.00$ s and the metal is steel. Find the length L.

9P. A man strikes a long aluminum rod at one end. Another man, at the other end with his ear close to the rod, hears the sound of the blow twice (once through air and once through the rod), with a 0.120-s interval between. How long is the rod?

10P. Earthquakes generate sound waves in the Earth. Unlike in a gas, there are both transverse (S) and longitudinal (P) sound waves in a solid. Typically, the speed of S waves is about 4.5 km/s and that of P waves 8.0 km/s. A seismograph records P and S waves from an earthquake. The first P waves arrive 3.0 min before the first S waves (Fig. 18-24). Assuming the waves traveled in a straight line, how far away did the earthquake occur?

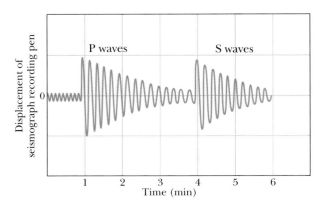

FIGURE 18-24 Problem 10.

11P. A stone is dropped into a well. The sound of the splash is heard 3.00 s later. What is the depth of the well?

SECTION 18-3 TRAVELING SOUND WAVES

12E. The audible frequency range for normal hearing is from about 20 Hz to 20 kHz. What are the wavelengths of sound waves at these frequencies?

13E. The shortest wavelength emitted by a bat is about 3.3 mm. What is the corresponding frequency?

14E. Diagnostic ultrasound of frequency 4.50 MHz is used to examine tumors in soft tissue. (a) What is the wavelength in air of such a sound wave? (b) If the speed of sound in tissue is 1500 m/s, what is the wavelength of this wave in tissue?

15E. (a) A conical loudspeaker has a diameter of 15.0 cm. At what frequency will the wavelength of the sound it emits in air be equal to its diameter? Be ten times its diameter? Be one-tenth its diameter? (b) Make the same calculations for a speaker of diameter 30.0 cm.

16E. Figure 18-1*b* shows a remarkably detailed image of a transistor in a microelectronic circuit, formed by an acoustic microscope. The sound waves have a frequency of 4.2 GHz. The speed of such waves in the liquid helium in which the specimen is immersed is 240 m/s. (a) What is the wavelength of these ultrahigh-frequency acoustic waves? (b) The ribbonlike conductors in the figure are about 2 μm wide. To how many wavelengths does this width correspond?

17P. (a) A continuous sinusoidal longitudinal wave is sent along a coiled spring from an oscillating source attached to it. The frequency of the source is 25 Hz, and the distance between successive points of maximum expansion in the spring is 24 cm. Find the wave speed. (b) If the maximum longitudinal displacement of a particle in the spring is 0.30 cm and the wave moves in the $-x$ direction, write the equation for the wave. Let the source be at $x = 0$ and the displacement at $x = 0$ when $t = 0$ be zero.

18P. The pressure in a traveling sound wave is given by the equation

$$\Delta p = (1.5 \text{ Pa}) \sin \pi[(1.00 \text{ m}^{-1})x - (330 \text{ s}^{-1})t].$$

Find (a) the pressure amplitude, (b) the frequency, (c) the wavelength, and (d) the speed of the wave.

19P. Two sound waves, from two different sources with the same frequency, 540 Hz, travel at a speed of 330 m/s. The sources are in phase. What is the phase difference of the waves at a point that is 4.40 m from one source and 4.00 m from the other? The waves are traveling in the same direction.

20P. At a certain point in space, two waves produce pressure variations given by

$$\Delta p_1 = \Delta p_m \sin \omega t,$$

$$\Delta p_2 = \Delta p_m \sin(\omega t - \phi).$$

What is the pressure amplitude of the resultant wave at this point when $\phi = 0$, $\phi = \pi/2$, $\phi = \pi/3$, and $\phi = \pi/4$?

21P. In Fig. 18-25, two speakers, separated by a distance of 2.00 m, are in phase. Assume the amplitudes of the sound from both speakers are approximately the same at the position of a listener, who is 3.75 m directly in front of one of the speakers. (a) For what frequencies in the audible range (20–20,000 Hz) is there a minimum signal? (b) For what frequencies is the sound a maximum?

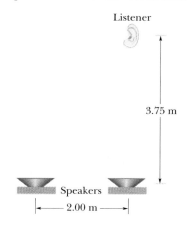

FIGURE 18-25 Problem 21.

22P. Two loudspeakers are located 11.0 ft apart on the stage of an auditorium. A listener is seated 60.0 ft from one and 64.0 ft from the other. A signal generator drives the two speakers in phase with the same amplitude and frequency. The frequency is swept through the audible range (20–20,000 Hz). (a) What are the three lowest frequencies at which the listener will hear minimum signal because of destructive interference? (b) What are the three lowest frequencies at which the listener will hear maximum signal?

23P. Two point sources of sound waves of identical wavelength λ and amplitude are separated by distance $D = 2.0\lambda$. The sources are in phase. (a) How many points of maximum signal (constructive interference) lie along a large circle around the sources? (b) How many points of minimum signal (destructive interference)?

24P. A sound wave of 40.0-cm wavelength enters the tube shown in Fig. 18-26. What must be the smallest radius r such that a minimum will be heard at the detector?

FIGURE 18-26 Problem 24.

25P. In Fig. 18-27, a point source S of sound waves lies near a reflecting wall AB. A sound detector D intercepts sound ray R_1 traveling directly from S. It also intercepts sound ray R_2 that reflects from the wall such that the *angle of incidence* θ_i is equal to the *angle of reflection* θ_r. Find two frequencies at which there is constructive interference of R_1 and R_2 at D. (Reflection of sound by the wall does not alter the phase of the sound wave.)

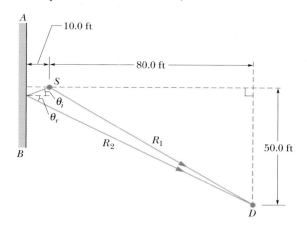

FIGURE 18-27 Problem 25.

26P*. Two point sources, separated by a distance of 5.00 m, emit sound waves at the same amplitude and frequency (300 Hz), but the sources are exactly out of phase. At what points along the line between the sources do the sound waves result in maximum oscillations of the air molecules? (*Hint:* One such point is midway between the sources. Can you see why?)

SECTION 18-4 INTENSITY AND SOUND LEVEL

27E. Sound with spherical wave fronts is emitted from a 1.0-W source. Assuming the energy of the waves is conserved, what is the intensity (a) 1.0 m from the source and (b) 2.5 m from the source?

28E. A source emits sound waves isotropically (that is, with equal intensity in all directions). The intensity of the waves 2.50 m from the source is 1.91×10^{-4} W/m². Assuming the energy of the waves is conserved, what is the power of the source?

29E. A note of frequency 300 Hz has an intensity of 1.00 μW/m². What is the amplitude of the air oscillations caused by this sound?

30E. Two sounds differ in sound levels by 1.00 dB. What is the ratio of the greater intensity to the smaller intensity?

31E. A certain sound level is increased by 30 dB. By what multiple is (a) its intensity increased and (b) its pressure amplitude increased?

32E. A salesperson claimed that a stereo system had a maximum audio power of 120 W. Testing the system with several speakers set up so as to simulate a point source, the consumer noted that she could get as close as 1.2 m with the volume full on before the sound hurt her ears. Should she report the firm to the Consumer Protection Agency?

33E. A certain loudspeaker produces a sound with a frequency of 2000 Hz and an intensity of 0.960 mW/m² at a distance of 6.10 m. Assume that there are no reflections and that the loudspeaker emits the same in all directions. (a) What is the intensity at 30.0 m? (b) What is the displacement amplitude at 6.10 m? (c) What is the pressure amplitude at 6.10 m?

34E. The source of a sound wave has a power of 1.00 μW. If it is a point source, (a) what is the intensity 3.00 m away and (b) what is the sound level in decibels at that distance?

35E. (a) If two sound waves, one in air and one in (fresh) water, are equal in intensity, what is the ratio of the pressure amplitude of the wave in water to that of the wave in air? Assume the water and the air are at 20°C. (See Table 16-1). (b) If the pressure amplitudes are equal instead, what is the ratio of the intensities of the waves?

36P. (a) Show that the intensity I of a wave is the product of the wave's energy per unit volume u and its speed v. (b) Radio waves travel at a speed of 3.00×10^8 m/s. Find u for a radio wave 480 km from a 50,000-W source, assuming the waves to be spherical.

37P. A *line source* of sound (for instance, a noisy freight train on a straight track) emits a cylindrical expanding sound wave. Assuming that the air absorbs no energy, find how (a) the intensity I and (b) the amplitude s_m of the wave depend on the perpendicular distance r from the source.

38P. A sound wave travels out uniformly in all directions from a point source. (a) Justify the following expression for the displacement s of the medium at any distance r from the source:

$$s = \frac{Y}{r} \sin k(r - vt).$$

Consider the speed, direction of propagation, periodicity, and intensity of the wave. (b) What are the dimensions of the constant Y?

39P. Find the ratios of (a) the intensities, (b) the pressure amplitudes, and (c) the particle displacement amplitudes for two sounds whose sound levels differ by 37 dB.

40P. At a distance of 10 km a 100-Hz horn, assumed to be a point source, is barely audible. At what distance would it begin to cause pain?

41P. You are standing at a distance D from a source that emits sound waves equally in all directions. You walk 50.0 m toward the source and observe that the intensity of these waves has doubled. Calculate the distance D.

42P. A certain loudspeaker (assumed to be a point source) emits 30.0 W of sound power. A small microphone of effective cross-sectional area 0.750 cm² is located 200 m from the loudspeaker. Calculate (a) the sound intensity at the microphone and (b) the power intercepted by the microphone.

43P. In a test, a subsonic jet flies overhead at an altitude of 100 m. The sound intensity on the ground as the jet passes overhead is 150 dB. At what altitude should the plane fly so that the ground noise is no greater than 120 dB, the threshold of pain? Ignore the finite time required for the sound to reach the ground.

44P. A hi-fi engineer has designed a speaker that is spherical in shape and emits sound isotropically (the same intensity in all directions). The speaker emits 10 W of acoustic power into a room with completely absorbent walls, floor, and ceiling (an *anechoic chamber*). (a) What is the intensity (W/m²) of the sound waves at 3.0 m from the center of the source? (b) How does the amplitude of the waves at 4.0 m compare with that at 3.0 m from the center of the source?

45P. Figure 18-28 shows an air-filled, acoustic interferometer, used to demonstrate the interference of sound waves. S is an oscillating diaphragm; D is a sound detector, such as the ear or a microphone. Path SBD can be varied in length, but path SAD is fixed. At D, the sound wave coming along path SBD interferes with that coming along path SAD. The sound intensity at D has a minimum value of 100 units at one position of B and continuously climbs to a maximum value of 900 units when B is shifted by 1.65 cm. Find (a) the frequency of the sound emitted by the source and (b) the ratio of the amplitude of the SAD wave to that of the SBD wave at D. (c) How can it happen that these waves have different amplitudes, considering that they originate at the same source?

46P*. Two loudspeakers, S_1 and S_2, are 7.0 m apart and

FIGURE 18-28 Problem 45.

oscillate in phase, each emitting sound of frequency 200 Hz uniformly in all directions. S_1 emits at a power of 1.2×10^{-3} W, and S_2 at 1.8×10^{-3} W. Consider a point P that is 4.0 m from S_1 and 3.0 m from S_2. (a) How are the phases of the two waves arriving at P related? (b) What is the intensity of sound at P with both S_1 and S_2 on? (c) What is the intensity of sound at P if S_1 is turned off (S_2 on)? (d) What is the intensity of sound at P if S_2 is turned off (S_1 on)?

47P*. A large parabolic reflector having a circular opening of radius 0.50 m is used to focus sound. If the energy is delivered from the focus to the ear of a listening detective through a tube of diameter 1.0 cm with 12% efficiency, how far away can a whispered conversation be understood? (Assume that the sound level of a whisper is 20 dB at 1.0 m from the source, considered to be a point, and that the threshold for hearing is 0 dB.)

SECTION 18-5 SOURCES OF MUSICAL SOUND

48E. A sound wave of frequency 1000 Hz propagating through air has a pressure amplitude of 10.0 Pa. What are the (a) wavelength, (b) particle displacement amplitude, and (c) maximum particle speed? (d) An organ pipe open at both ends has this frequency as a fundamental. How long is the pipe?

49E. In Fig. 18-29, a rod R is clamped at its center; a disk D at its end projects into a glass tube that has cork filings spread over its interior. A plunger P is provided at the other end of the tube. The rod is made to oscillate longitudinally at frequency f to produce sound waves inside the tube, and the location of the plunger is adjusted until a standing sound wave pattern is set up inside the tube. Once the standing wave is set up, the cork filings collect in a pattern of ridges at the displacement nodes. Show that if d is the average distance between the ridges, the speed of sound v in the gas within the tube is given by

$$v = 2fd.$$

This is *Kundt's method* for determining the speed of sound in various gases.

FIGURE 18-29 Exercise 49.

50E. A sound wave in a fluid medium is reflected at a barrier so that a standing wave is formed. The distance between nodes is 3.8 cm and the speed of propagation is 1500 m/s. Find the frequency.

51E. A 15.0-cm-long violin string, fixed at both ends, oscillates in its $n = 1$ mode. The speed of waves on the string is 250 m/s, and the speed of sound in air is 348 m/s. What are (a) the frequency and (b) the wavelength of the emitted sound wave?

52E. (a) Find the speed of waves on an 800-mg violin string 22.0 cm long if the fundamental frequency is 920 Hz. (b) What is the tension in the string? For the fundamental, what is the wavelength of (c) the waves on the string, and (d) the sound waves emitted by the string?

53E. A violin string, oscillating in its fundamental mode, generates a sound wave with wavelength λ. By what multiple must the tension be increased if the string, still oscillating in its fundamental mode, is to generate a new sound wave with a wavelength of $\lambda/2$?

54E. Organ pipe A, with both ends open, has a fundamental frequency of 300 Hz. The third harmonic of organ pipe B, with one end open, has the same frequency as the second harmonic of pipe A. How long is (a) pipe A and (b) pipe B?

55E. The water level in a vertical glass tube 1.00 m long can be adjusted to any position in the tube. A tuning fork vibrating at 686 Hz is held just over the open top end of the tube. At what positions of the water level will there be resonance?

56P. A certain violin string is 30 cm long between its fixed ends and has a mass of 2.0 g. The string sounds an A note (440 Hz) when played without fingering. (a) Where must one put one's finger to play a C (523 Hz)? (b) What is the ratio of the wavelength of the string waves required for an A note to that required for a C note? (c) What is the ratio of the wavelength of the sound wave for an A note to that for a C note?

57P. A string on a cello has length L, for which the fundamental frequency is f. (a) By what length l must the string be shortened by fingering to change the fundamental frequency to rf? (b) What is l if $L = 0.80$ m and $r = \frac{6}{5}$? (c) For $r = \frac{6}{5}$, what is the ratio of the wavelength of the new sound wave emitted by the string to that of the wave emitted before fingering?

58P. In Fig. 18-30, S is a small loudspeaker driven by an audio oscillator and amplifier, adjustable in frequency from 1000 to 2000 Hz only. The tube D is a piece of cylindrical sheet-metal pipe 18.0 in. long and open at both ends. (a) If the speed of sound in air is 1130 ft/s at the existing temperature, at what frequencies will resonance occur in the pipe when the frequency emitted by the speaker is varied from 1000 to 2000 Hz? (b) Sketch the standing waves (using the convention of Fig. 18-12b) for each resonant frequency.

FIGURE 18-30 Problem 58.

59P. A well with vertical sides and water at the bottom resonates at 7.00 Hz and at no lower frequency. The air in the well has a density of 1.10 kg/m³ and a bulk modulus of 1.33×10^5 Pa. How deep is the well?

60P. A hand-clap on stage in an amphitheater (Fig. 18-31) sends out sound waves that scatter from terraces of width $w = 0.75$ m. The sound returns to the stage as a periodic series of pulses, one from each terrace; the parade of pulses sounds like a played note. At what frequency do the pulses return (in other words, what is the frequency of that perceived note)?

FIGURE 18-31 Problem 60.

61P. A tube 1.20 m long is closed at one end. A stretched wire is placed near the open end. The wire is 0.330 m long and has a mass of 9.60 g. It is fixed at both ends and vibrates in its fundamental mode. It sets the air column in the tube into oscillation at its fundamental frequency by resonance. Find (a) the frequency of oscillation of the air column and (b) the tension in the wire.

62P. The period of a pulsating variable star may be estimated by considering the star to be executing *radial* longitudinal pulsations in the fundamental standing wave mode; that is, the star's radius varies periodically with time, with a displacement antinode at the star's surface. (a) Would you expect the center of the star to be a displacement node or antinode? (b) By analogy with a pipe with one open end, show that the period of pulsation T is given by

$$T = \frac{4R}{v},$$

where R is the equilibrium radius of the star and v is the average sound speed. (c) Typical white dwarf stars are composed of material with a bulk modulus of 1.33×10^{22} Pa and a density of 10^{10} kg/m³. They have radii equal to 9.0×10^{-3} solar radii. What is the approximate pulsation period of a white dwarf?

63P. A 30.0-cm-long violin string with linear density 0.650 g/m is placed near a loudspeaker that is fed by an audio oscillator of variable frequency. It is found that the string is set into oscillation only at the frequencies 880 and 1320 Hz as the frequency of the oscillator is varied over the range 500–1500 Hz. What is the tension in the string?

SECTION 18-6 BEATS

64E. A tuning fork of unknown frequency makes three beats per second with a standard fork of frequency

384 Hz. The beat frequency decreases when a small piece of wax is put on a prong of the first fork. What is the frequency of this fork?

65E. The A string of a violin is a little too taut. Four beats per second are heard when the string is sounded together with a tuning fork that is oscillating accurately at concert A (440 Hz). What is the period of the violin string oscillation?

66E. You are given four tuning forks. The fork with the lowest frequency oscillates at 500 Hz. By striking two tuning forks at a time, the following beat frequencies are heard: 1, 2, 3, 5, 7, and 8 Hz. What are the possible frequencies of the other three tuning forks?

67P. Two identical piano wires have a fundamental frequency of 600 Hz when kept under the same tension. What fractional increase in the tension of one wire will lead to the occurrence of 6 beats/s when both wires oscillate simultaneously?

68P. You have five tuning forks that oscillate at different frequencies. By using the forks two at a time, what is the (a) maximum number and (b) minimum number of different beat frequencies that you can produce?

SECTION 18-7 THE DOPPLER EFFECT

69E. A source S generates circular waves on the surface of a lake; the pattern of wave crests is shown in Fig. 18-32. The speed of the waves is 5.5 m/s, and the crest-to-crest separation is 2.3 m. You are in a small boat heading directly toward S at a constant speed of 3.3 m/s with respect to the shore. What frequency of the waves do you observe?

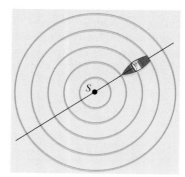

FIGURE 18-32 Exercise 69.

70E. Trooper B is chasing speeder A along a straight stretch of road. Both are moving at a speed of 100 mi/h. Trooper B, failing to catch up, sounds his siren again. Take the speed of sound in air to be 1100 ft/s and the frequency of the source to be 500 Hz. What is the Doppler shift in the frequency heard by speeder A?

71E. A whistle used to call a dog has a frequency of 30 kHz. The dog, however, ignores it. The owner of the dog, who cannot hear sounds above 20 kHz, wants to use the Doppler effect to make certain that the whistle is working. She asks a friend to blow the whistle from a moving car while the owner remains stationary and listens. (a) How fast must the car move and in what direction for the owner to hear the whistle at 20 kHz? Is the experiment practical? (b) Repeat for a whistle frequency of 22 kHz instead of 30 kHz.

72E. The 16,000-Hz whine of the turbines in the jet engines of an aircraft moving with speed 200 m/s is heard at what frequency by the pilot of a second craft trying to overtake the first at a speed of 250 m/s?

73E. An ambulance emitting a whine at 1600 Hz overtakes and passes a cyclist pedaling a bike at 8.00 ft/s. After being passed, the cyclist hears a frequency of 1590 Hz. How fast is the ambulance moving?

74E. A whistle of frequency 540 Hz moves in a circle of radius 2.00 ft at an angular speed of 15.0 rad/s. What are (a) the lowest and (b) the highest frequencies heard by a listener a long distance away at rest with respect to the center of the circle?

75E. In 1845, Buys Ballot first tested the Doppler effect for sound. He put a trumpet player on a flatcar drawn by a locomotive and another player near the tracks. If each player blows a 440-Hz note, and if there are 4.0 beats/s as they approach each other, what is the speed of the flatcar?

76E. What was the speed of the projectile in Fig. 18-22, assuming the speed of sound in the gas through which the projectile moved is 380 m/s.

77E. The speed of light in water is about three-fourths the speed of light in vacuum. A beam of high-speed electrons from a betatron emits Cerenkov radiation in water; the wave front of this light forms a cone of angle 60°. Find the speed of the electrons in the water.

78E. A bullet is fired with a speed of 2200 ft/s. Find the angle made by the shock cone with the line of motion of the bullet.

79P. Two identical tuning forks can oscillate at 440 Hz. A person is located somewhere on the line between them. Calculate the beat frequency as measured by this individual if (a) she is standing still and the tuning forks both move to the right, say, at 30.0 m/s, and (b) the tuning forks are stationary and the listener moves to the right at 30.0 m/s.

80P. A plane flies with $\frac{3}{4}$ the speed of sound. The sonic boom reaches a man on the ground exactly 1 min after the plane passed directly overhead. What is the altitude of the plane? Assume the speed of sound to be 330 m/s.

81P. A jet plane passes overhead at a height of 5000 m and a speed of Mach 1.5. (a) Find the Mach cone angle. (b) How long after the jet has passed directly overhead will the shock wave reach the ground? Use 331 m/s for the speed of sound.

82P. Figure 18-33 shows a transmitter and receiver of waves contained in a single instrument. It is used to measure the speed u of a target object (idealized as a flat plate) that is moving directly toward the unit, by analyzing the waves reflected from the target. (a) Show that the frequency f_r of the reflected waves at the receiver is related

FIGURE 18-33 Problem 82.

to their source frequency f_s by

$$f_r = f_s \left(\frac{v + u}{v - u} \right),$$

where v is the speed of the waves. (b) In a great many practical situations, $u \ll v$. In this case, show that the equation above becomes

$$\frac{f_r - f_s}{f_s} \approx \frac{2u}{v}.$$

83P. A stationary motion detector sends sound waves of 0.150 MHz toward a truck approaching at a speed of 45.0 m/s. What is the frequency of the waves reflected back to the detector?

84P. An acoustic burglar alarm consists of a source emitting waves of frequency 28.0 kHz. What will be the beat frequency of waves reflected from an intruder walking at an average speed of 0.950 m/s directly away from the alarm?

85P. A siren emitting a sound of frequency 1000 Hz moves away from you toward the face of a cliff at a speed of 10 m/s. Take the speed of sound in air as 330 m/s. (a) What is the frequency of the sound you hear coming directly from the siren? (b) What is the frequency of the sound you hear reflected off the cliff? (c) What is the beat frequency between the two sounds? Is it perceptible (it must be less than 20 Hz)?

86P. A person on a railroad car blows a trumpet sounding at 440 Hz. The car is moving toward a wall at 20.0 m/s. Calculate (a) the frequency of the sound as received at the wall and (b) the frequency of the reflected sound arriving back at the source.

87P. A French submarine and a U.S. submarine move head-on during maneuvers in motionless water in the North Atlantic (Fig. 18-34). The French sub moves at 50.0 km/h, and the U.S. sub at 70.0 km/h. The French sub sends out a sonar signal (sound wave in water) at 1000 Hz. Sonar waves travel at 5470 km/h. (a) What is the signal's frequency as detected by the U.S. sub? (b) What frequency is detected by the French sub in the signal reflected back to it by the U.S. sub?

FIGURE 18-34 Problem 87.

88P. A source of sound waves of frequency 1200 Hz moves to the right with a speed of 98.0 ft/s relative to the air. Ahead of it is a reflecting surface moving to the left with a speed of 216 ft/s relative to the air. Take the speed of sound in air to be 1080 ft/s and find (a) the wavelength of the sound emitted toward the reflector by the source, (b) the number of wave fronts per second arriving at the reflecting surface, (c) the speed of the reflected waves, (d) the wavelength of the reflected waves, and (e) the number of reflected wave fronts per second arriving at the source.

89P. In a discussion of Doppler shifts of ultrasonic waves used in medical diagnosis, the authors remark: "For every millimeter per second that a structure in the body moves, the frequency of the incident ultrasonic wave is shifted approximately 1.30 Hz/MHz." What speed of the ultrasonic waves in tissue do you deduce from this statement?

90P. A bat is flitting about in a cave, navigating via ultrasonic bleeps. Assume that the sound emission frequency of the bat is 39,000 Hz. During one fast swoop directly toward a flat wall surface, the bat is moving at $\frac{1}{40}$ of the speed of sound in air. What frequency does the bat hear reflected off the wall?

91P. A submarine, near the water surface, moves north at speed 75.0 km/h in a northbound current of speed 30.0 km/h, where both speeds are relative to the ocean floor. The sub emits a sonar signal (sound wave), of frequency $f = 1000$ Hz and speed 5470 km/h, that is detected by a destroyer north of the sub. What is the difference between the detected frequency and f if the destroyer (a) drifts with the current at 30.0 km/h and (b) is stationary relative to the ocean floor?

92P. A 2000-Hz siren and a civil defense official are both at rest with respect to the Earth. What frequency does the official hear if the wind is blowing at 12 m/s (a) from source to observer and (b) from observer to source?

93P. Two trains are traveling toward each other at 100 ft/s relative to the ground. One train is blowing a whistle at 500 Hz. (a) What frequency will be heard on the other train in still air? (b) What frequency will be heard on the other train if the wind is blowing at 100 ft/s toward the whistle and away from the listener? (c) What frequency will be heard if the wind direction is reversed?

94P. A girl is sitting near the open window of a train that is moving at a velocity of 10.00 m/s to the east. The girl's uncle stands near the tracks and watches the train move away. The locomotive whistle emits sound at frequency 500.0 Hz. The air is still. (a) What frequency does the uncle hear? (b) What frequency does the girl hear? A wind begins to blow from the east at 10.00 m/s. (c) What frequency does the uncle now hear? (d) What frequency does the girl now hear?

SECTION 18-8 THE DOPPLER EFFECT FOR LIGHT

95E. The central spot in Fig. 18-35a is a galaxy in the constellation Corona Borealis; the galaxy is 1.3×10^8 light-

FIGURE 18-35 Problem 95.

years distant. In Fig. 18-35b the central streak is the spectrum of light (distribution in wavelength) from the galaxy. The two heaviest vertical lines show the presence of calcium. The horizontal arrow shows that these calcium lines occur at longer wavelengths than those for terrestrial light sources containing calcium, the length of the arrow representing the wavelength shift. Previous measurements indicate that this galaxy is receding from us at 2.2×10^4 km/s. Calculate the fractional shift in the wavelengths of the calcium lines shown.

96E. Certain characteristic wavelengths in the light from a galaxy in the constellation Virgo are observed to be 0.4% longer than the corresponding light from terrestrial sources. What is the radial speed of this galaxy with respect to the Earth? Is it approaching or receding?

97E. In the "red shift" of radiation from a distant galaxy, the light H_y, known to have a wavelength of 434 nm when observed in the laboratory, appears to have a wavelength of 462 nm. (a) What is the speed of the galaxy in the line of sight relative to the Earth? (b) Is the galaxy approaching or receding?

98E. Assuming that Eq. 18-55 holds, find how fast you would have to go through a red light to have it appear green. Take 620 nm as the wavelength of red light and 540 nm as the wavelength of green light.

99P. The period of rotation of the sun at its equator is 24.7 d; its radius is 7.00×10^5 km. What Doppler wavelength shift is expected for light with wavelength 550 nm emitted from the edge of the sun's disk?

100P. An Earth satellite, transmitting on a frequency of 40 MHz (exactly), passes directly over a radio receiving station at an altitude of 400 km and at a speed of 3.0×10^4 km/h. Plot the change in frequency attributable to the Doppler effect as a function of time, counting $t = 0$ as the instant the satellite is over the station. (*Hint:* The speed u in the Doppler formula is not the actual speed of the satellite but its component in the direction of the station. Neglect the curvature of the Earth and of the satellite orbit.)

101P. Microwaves, which travel at the speed of light, are reflected from a distant airplane approaching the wave source. It is found that when the reflected waves are beat against the waves radiating from the source the beat frequency is 990 Hz. If the microwaves are 0.100 m in wavelength, what is the approach speed of the airplane?

CONCERT HALL ACOUSTICS: SCIENCE OR ART?

John S. Rigden
American Institute of Physics

On the evening of October 18, 1976, orchestra members of the New York Philharmonic tuned their instruments and readied themselves for a most unusual concert. The setting was Lincoln Center's Avery Fisher Hall, a new name for a concert hall with a troubled past.

Philharmonic Hall had opened in 1962 to high expectations (Fig. 1). One of America's outstanding acousticians, Leo L. Beranek, had worked with the architect who designed the hall. But from the beginning, there were problems: performers on stage could not hear other parts of the orchestra; the bass sounded weak to people in the audience; echoes from the back wall could be heard; patrons could not hear all that the orchestra was playing. Over the period 1964–1972, many attempts were made to improve the situation, but these attempts failed.

In December 1974, Cyril M. Harris, an acoustician from Columbia University, was asked to redesign the hall. Harris was initially most reluctant, but then set down his conditions: first, the inside of the hall would be demolished right back to the girders; second, acoustics would take precedence over aesthetics; third, Philip Johnson would be the architect. These conditions were granted and work began in May 1976, immediately after the Philharmonic's season-ending performance.

Since the hall was to be ready for the start of the fall season, work proceeded through the summer of 1976 at a feverish pace. Harris, wearing a hardhat, watched each step. All was ready for the October 19 beginning of the new concert season, but on October 18 a special concert was performed in the newly designed hall. It was an unusual concert because members of the audience were there by special invitation: they consisted of the construction workers, contractors, subcontractors, and architects who had worked through the summer with such devotion. Music critics were not invited, but they came. When the orchestra finished the fourth movement of Mahler's *Ninth Symphony,* the critics were most favorably impressed; the musicians were delighted; and the construction workers were happy and proud.

FIGURE 1 The auditorium of Philharmonic Hall at Lincoln Center for the Performing Arts. This view, from the loge, shows the orchestra level and the stage. Philharmonic Hall was completed in 1962 and at 10:00 a.m. on Monday, May 28, 1962, the young assistant conductor of the New York Philharmonic Orchestra, Seiji Ozawa, commanded the attention of 106 men and women of the orchestra. When his baton came down, the brass and woodwind sections of the orchestra played the opening chords of Brahms' *Third Symphony.*

John S. Rigden received his Ph.D. from Johns Hopkins University in 1960. After postdoctoral work at Harvard University, he held academic positions at Eastern Nazarene College, Middlebury College, and the University of Missouri–St. Louis. He is currently on leave from the American Institute of Physics, where he is Director of Physics Programs, to the National Academy of Sciences, where he is Director of Development of the National Science Education Standards Project. He is the past editor of the <u>American Journal of Physics</u> (1978–1988) and is the author of <u>Physics and the Sound of Music</u> (Wiley, 1977; second edition, 1985). More recently, he has written the definitive biography of the great American physicist I. I. Rabi: <u>Rabi: Scientist and Citizen</u> (Basic Books, 1987).

Tuning the Performance to the Concert Hall

Isaac Stern is a world-class violinist. Yet, when he comes on the stage of a fine concert hall, the concert hall itself becomes an additional instrument he uses to enhance his musical performance. A concert hall is not a passive enclosure in which musical performances occur; rather, it is an active participant in communicating the interpretations of an artist to an attentive patron.

Sound waves, which link a performer and a patron, carry energy. If the source of sound is a violin string oscillating with fundamental and higher harmonic frequencies, pressure undulations with the same frequencies present in the vibrating string are carried by the surrounding air away from the violin. These longitudinal pressure waves transmit acoustic energy away from the source.

A sound wave approaching an open window reaches the window, passes through, and carries energy from the source side of the window to the region beyond. An open window is a perfect absorber; it absorbs all the acoustical energy incident on it. The situation is quite different when a sound source is surrounded by reflecting surfaces as is the case in a concert hall: a sound wave strikes a wall and is reflected back into the room; it propagates across the room until it strikes another surface and is again reflected; and so on. If the source of sound delivers energy at a fixed rate into the room, the intensity of the sound builds up rapidly within the enclosed space until it approaches an equilibrium intensity level. A fraction of the incident energy is absorbed by the surface during each reflection; thus the rate at which the sound level approaches its equilibrium value depends on the nature of the reflecting surfaces. The equilibrium intensity level is reached when the rate of energy absorption by all exposed surfaces equals the rate at which the sound source delivers energy to the enclosure.

A listener in an enclosure hears first the sound that comes directly from the source. Next, after a time interval called the initial-time-delay gap, waves of once-reflected sound reach the listener. Still later, wave after wave of twice-reflected waves pass over the listener. In this fashion, the sound-intensity level increases until the listener is immersed in sound, coming from all directions, at the equilibrium level. The sum of all the reflected sound is called the reverberant sound. At the equilibrium sound level, the ratio of the loudness of the reverberant sound to that of the direct sound determines the fullness of the perceived tone and this fullness of tone is one characteristic of a fine concert hall. It is reverberant sound, coupled with the absence of interfering noise, that distinguishes musical performances experienced in a fine concert hall from those heard in a town park.

Just as sound *builds up to* an equilibrium level, so it *decays from* an equilibrium level. If the source of sound is terminated, the direct sound is the first to go and a listener perceives an abrupt drop in the sound level. Then the rate of decay slows somewhat until the last once-reflected sound waves reach the listener. The sound level trails off exponentially as weaker and weaker multiply reflected waves come successively to a listener's ears.

Reverberation Time

The time required for the sound level to build up to or to decay from its equilibrium value is called the reverberation time and it is the most important acoustical characteristic of a concert hall. Specifically, the reverberation time is defined as the time required for the sound intensity (watts per meter^2) to build up or decrease by a factor of one million. If the reverberation time is too short, musical notes are heard isolated one from the other and the music is perceived as thin. If, on the other hand, the reverberation time is too long,

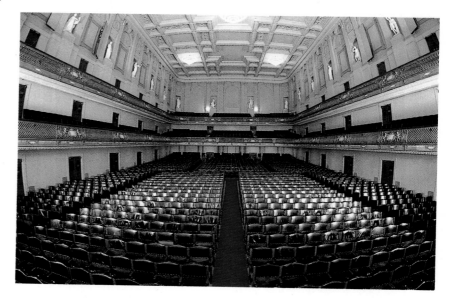

FIGURE 2 Symphony Hall in Boston. This concert hall, which opened in 1900, was planned by the Harvard physicist Wallace C. Sabine, a pioneer in the science of acoustics. Sabine developed the quantitative foundations, including an empirical formula for calculating reverberation time, which are today the heart of acoustic design. Note all the irregularities in the walls and ceilings. The conductor Bruno Walter said that Symphony Hall ''is the most noble of American concert halls.''

the sounds from earlier notes clash with the notes being played. Typically, the most desirable reverberation time for symphonic music is about 2 s: Symphony Hall in Boston, one of the finest concert halls in the world, has a reverberation time of 1.8 s when it is fully occupied (Fig. 2); the Musikvereinssaal in Vienna, another excellent concert hall, has a reverberation time of 2.05 s (fully occupied).

The reverberation time depends on the volume of the concert hall and the nature of the reflecting surfaces. The larger the volume, the longer it takes sound, traveling at approximately 345 m/s, to traverse the distances between reflecting walls and the longer it takes the reverberant sound field to build up to its equilibrium level. The volume of Symphony Hall is 61,496 m^3. The volume of Carnegie Hall in New York City is larger, 79,610 m^3, yet its reverberation time, 1.7 s, is less than that of Boston's Symphony Hall. The difference is due to the character of the reflecting surfaces. When surfaces exposed to sound waves are highly absorbent, the rate of energy absorption by all surfaces quickly becomes equal to the rate of energy production by all sources; therefore the reverberation time is smaller. Thus we can understand what Isaac Stern means when he says that Carnegie is better in rehearsal than with an audience. Since the absorbing properties of one person are equivalent to 0.5 m^2 of open window, the reverberation time of a concert hall is longer when it is unoccupied. (This is why reverberation times are measured when the hall is fully occupied. Furthermore, this is why most concert halls have lockers where patrons can hang their winter coats, which are very absorbent.)

Keeping Noise Outside the Concert Hall

Concert halls are typically located in the middle of large metropolitan areas where sound waves—noise!—are ubiquitous. A concert hall must shield patrons inside from all outside noise: jet planes overhead, city buses and ambulances on adjacent streets, and subways running underground. Sound waves playing over the external structure of a concert hall discover each and every access route to the inside. Even a keyhole can transmit appreciable amounts of sound. The walls themselves transmit sound. Sound waves of all frequencies are incident on the external walls of a concert hall and they drive the walls at the wave frequencies; the vibrating walls, in turn, become the source of sound waves inside the concert hall. The more massive the walls are, the greater the damping of the transmitted wave. As you might predict, massive walls are more likely to vibrate with low frequencies than with high frequencies; thus as sound waves of all frequencies are incident on the outside walls of a structure, it is the low frequencies to which the walls respond more readily and it is the low tones that are transmitted more readily into the interior of the structure.

Concert halls can be heated and cooled. This means there must be machinery to generate the warm or cool air, fans to push the air, and ducts to transport the air. Mechanical equipment and moving air are sources of noise. The machinery must be placed in a separate building; the ducts must be lined with a sound-absorbing material. Sharp bends, rough joints, or dampers in the ducts must be avoided as they can set the passing air into whirling, turbulent motion. Turbulent air is especially noisy.

One acoustic consultant has told of a potentially disastrous situation involving air-conditioning ducts. When he visited the construction site, he noticed that the ducts had been lowered a few centimeters so that, for support purposes, they rested on the girders. Such contact between the ducts and girders of the concert hall would have coupled the noise of the ducts to the structure itself. Fortunately, the acoustician was able to make a correction before the finished ceiling concealed a problem from view.

Assuming that all ambient sounds, from both external and internal sources, are at a level of a farmhouse bedroom on a quiet night, a conductor can use the full dynamic range of an orchestra, from the loudest *forte fortissimo* to the softest *pianissimo* and even the faintest melodic strain can be heard distinctly. Yet, even if the reverberation time of a quiet concert hall is good, the listening experience of the patron will be marred if the acoustic design of the hall is faulty; furthermore, performers will be unhappy, visiting conductors will fall short of expectations, and orchestra directors will eschew scheduling their orchestras for future performances in a problematic hall.

Concert halls acclaimed by critics, musicians, and audiences can have different shapes (rectangular, fan-shaped, horseshoe-shaped), but they all share certain qualities. Outstanding halls have reverberation times in the 1.7–2.0-s range. Almost as important as reverberation time is a quality, called intimacy, determined by the initial-time-delay gap. The bigger this gap, the less intimate the musical environment. Initial-time-delay gaps range from 10 to 70 ms, but for the best concert halls they are less than 40 ms.

Standing Waves

Standing waves can be established within a concert hall. For example, sound waves can be reflected back and forth between parallel walls. The lowest frequency standing wave, the *fundamental,* has a pressure antinode at each wall and a node in between; thus the wavelength is equal to twice the distance between the two reflecting walls. Multiples of the fundamental frequency have additional nodes and antinodes. The nodal–antinodal configuration of standing waves can, if a standing wave is prominent, produce both acoustic dead spots and

FIGURE 3   Avery Fisher Hall. Note the irregularities built into the ceiling which reflect sound waves in all directions; thus, they provide an excellent surface for diffusing sound waves. Sound diffusion equalizes the sound intensity throughout the concert hall and eliminates troublesome dead spots, focusing, and echoes.

acoustic hot spots—a condition most undesirable for a concert hall, where sound uniformity is the objective. For this reason, architects design concert halls so that dimensions are not simple multiples of each other and smooth parallel walls are avoided. Irregularities in the walls and ceiling not only diminish the chance of prominent standing waves being established, they also scatter the sound waves in many directions and contribute to the desired diffuseness of sound (Fig. 3).

Concert halls are three-dimensional structures and sound waves can be reflected in all directions; consequently, there are thousands of normal modes. A single tone sounded by a violin may well excite hundreds of the normal modes of the concert hall; vocalists may exploit the normal modes of a concert hall to attain greater dynamic range for certain notes. Thus standing waves are an important feature of a concert hall; the

sound energy briefly captured in the many normal modes of oscillation makes an important contribution to the reverberant sound.

Science and Art

In spite of our growing body of acoustical knowledge, the first concert in a newly completed concert hall is awaited with both anticipation and uncertainty. This raises the question: is acoustics a science or an art? The many myths about concert hall acoustics lend credence to this question. There is the prevalent idea that the acoustical quality of concert halls improves with age (it does not); there is a tendency to identify prominent features such as decorative gold gilt or statuary in great concert halls as the origin of fine acoustics (they are not); there is the mystical idea that the ancients were in possession of secret acoustical principles, principles unknown to moderns, and that if we could rediscover these lost principles,

our concert halls would be vastly improved (the ancients had no secret acoustical insights).

The aura of mystery surrounding the subject of acoustics is further enhanced by the fact that acoustics is, in one sense, one of the oldest branches of physics. After all, acoustics played an important role in the design and location of the open-air theaters used by the Greeks and Romans. In another sense the science of acoustics is young: it was not until around 1900 that acoustical considerations were systematically applied to the design of concert halls. Even now, the advice of the acoustician is not always followed.

The effect of a concert hall on music goes beyond those moving moments when it effectively links an inspired performer with an eager audience. Enclosures and music have had an effect on each other over centuries. More specifically, the *design* of enclosures and the *composition* of music have had a dramatic influence on each other: existing auditoriums have affected the types of music composed and new musical compositions have made acoustical demands on enclosed spaces. For example, the long reverberation times of medieval cathedrals, 5–10 s, required that music be played or sung very slowly. Because of the long reverberation times in these cathedrals, the spoken word was difficult to understand and chanting replaced speaking. Giovanni Gabrieli, the organist at St. Mark's Basilica in Venice around 1600, wrote music in a slow tempo suitable for enclosures with a long reverberation time.

At the time Gabrieli was performing in Venice, new developments were occurring in Italy. Small chapels, rectangular in shape with high ceilings, were built adjacent to large cathedrals. Characteristic of the time, the walls were ornate, rich in sculptural detail. These decorative surfaces were efficient diffusers of sound at all audible frequencies. The small rever-

FIGURE 4   The auditorium of Avery Fisher Hall. During the summer of 1976, the inside of Philharmonic Hall (see Fig. 1) was transformed into the one shown here. Avery Fisher Hall has the classic rectangular shape, like Symphony Hall in Boston (see Fig. 2). This rectangular design is a direct descendant of the small chapels built during the 17th century. Both the walls and the ceiling are amply irregular.

beration times—less than 1.5 s—of these small chapels encouraged the development of new musical forms and ushered in the Baroque period. George Frederick Handel wrote much of his music for such environments. These smaller, more intimate environments considerably enhanced the listening experience (Fig. 4).

With improved acoustical environments, musical performances became more popular. Larger structures were needed to meet the demand of an eager public. Music of the classical period—the symphonies of Josef Haydn, Wolfgang Amadeus Mozart, and Ludwig van Beethoven—is best performed in larger concert halls with reverberation times from 1.5 to 1.7 s. The romantic period followed the classical period and the music of Johannes Brahms, Peter Ilyitch Tchaikovsky, Maurice Ravel, and Richard Strauss thrives in even larger acoustical environments with larger reverberation times—1.8–2.2 s.

The future design of concert halls will exploit not only new materials and methods of construction but also the use of electronic enhancement. If we can use the past as a guide, we can be sure that as concert halls change, composers will create new musical styles that are not only matched to but also enhanced by the new enclosures.

TEMPERATURE | 19

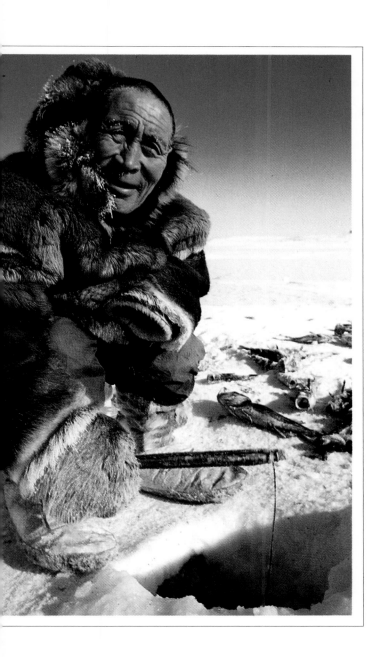

An Inuit is fishing through a hole he has cut in the ice on a lake in northern Canada. Were it not for a curious and unique thermal property of water, there would be no fish in that lake for him to catch. In fact, there might be no animal or plant life in any waters where there is an extended freeze. What is this thermal property of water that allows aquatic life in cold regions?

19-1 THERMODYNAMICS: A NEW SUBJECT

With this chapter, we leave the subject of mechanics and begin a new subject—*thermodynamics*. Mechanics deals with the mechanical (or external) energies of systems and is governed by Newton's laws. Thermodynamics deals with the internal energy of systems and is governed by a new set of laws, which you will come to know in this and the next few chapters. To give the flavor of things, some "mechanics words"—as we might call them— are force, kinetic energy, acceleration, Galileo, and Newton's second law. Some "thermodynamics words" are temperature, heat, internal energy, entropy, Kelvin, and the second law of thermodynamics.

The central concept of thermodynamics is temperature. This word is so familiar that most of us— because of our built-in sense of hot and cold—tend to be overconfident in our understanding of it. Our "temperature sense" is in fact not always reliable. On a cold winter day, for example, an iron railing seems much colder to the touch than does a wooden fence post, yet both are at the same temperature. This difference in our sense perception comes about because iron conducts heat away from our fingers much more readily than wood does.

Because of its fundamental importance, we begin the study of thermodynamics by developing the concept of temperature from its foundations, without relying in any way on our temperature sense.

19-2 TEMPERATURE

Temperature is one of the seven SI base standards. Physicists measure temperature on the **Kelvin scale.** Although the temperature of a body can apparently be raised without limit, it cannot be lowered without limit, and the limiting low temperature is taken as the zero of the Kelvin scale. Room temperature is about 290 kelvins, or 290 K as we write it, above this *absolute zero*. Figure 19-1 shows the wide range over which temperatures are determined.

When the universe began, some 10–20 billion years ago, the temperature was about 10^{39} K. As the universe expanded, it cooled down and has now reached an average temperature of about 3 K. We are a little warmer than that because we happen to live near a star. Without our sun, however, we too would be at 3 K (or rather, we could not exist).

Physicists around the world are trying to see how close they can come to the absolute zero of temperature. It turns out that the absolute zero is much like the speed c of light in that both are limits that can be approached for material bodies but never attained. As of 1992, physicists have achieved the following in the laboratory:

Fastest electron speed $0.9999999994c$

Lowest temperature   0.000000002 K

You may think that, in each case, that is surely close enough. However, new phenomena keep appearing the closer we get to these unreachable goals. As it turns out, every additional decimal place, either in electron speed or in temperature, must be fought for against ever-increasing experimental difficulties (and expenses).

Given the broad range over which temperature can vary, the very fact of our existence seems the grandest of miracles. If the temperature of the Earth were just a little lower, we would all freeze to death.

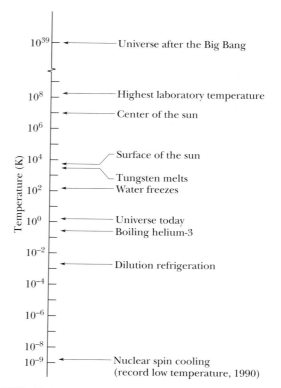

FIGURE 19-1   Some temperatures on the Kelvin scale. Note that $T = 0$ corresponds to $10^{-\infty}$ and cannot be plotted on this logarithmic plot.

And if it were just a little higher, the atoms of our bodies would be moving around in such an agitated fashion that molecules would break apart, making life impossible. As far as temperature is concerned, we hover between fire and ice, in the tiniest of ecological niches.

19-3 THE ZEROTH LAW OF THERMODYNAMICS

The properties of many bodies change as you alter their thermal environment, perhaps by moving them from a refrigerator to a warm oven. To give a few examples: as its temperature increases, the volume of a liquid increases, a metal rod grows a little longer, and the electrical resistance of a wire increases, as does the pressure exerted by a confined gas. We can use any one of these properties as the basis of an instrument that will help us to pin down the concept of temperature.

Figure 19-2 shows such an instrument. Any resourceful engineer could design and construct it, using any one of the properties listed above. The instrument is fitted with a digital readout display and has the following properties: if you heat it with a Bunsen burner, the displayed number starts to increase; if you then put it into a refrigerator, the displayed number starts to decrease. The instrument is not calibrated in any way, and the numbers have (as yet) no physical meaning. The device is a *thermoscope* but not (as yet) a *thermometer*.

Suppose that, as in Fig. 19-3a, you put the thermoscope (which we shall call body T) into intimate contact with another body (body A). The entire system is confined within a thick-walled insulating box. The numbers displayed by the thermoscope roll by

until, eventually, they come to rest (let us say the reading is "137.04") and no further change takes place. In fact, every measurable property of body T (the thermoscope) and of body A assumes a stable value, and we say that the two bodies are in *thermal equilibrium* with each other.

Now let us put body T in intimate contact with a second body (body B) as in Fig. 19-3b. Let us say that the two bodies (B and T) come to thermal equilibrium *at the same reading of the thermoscope.*

Finally, as in Fig. 19-3c, let us bring bodies A and B into intimate contact. Will they be found to be in thermal equilibrium with each other? They will. This answer, which perhaps seems obvious, is in fact *not* obvious and can be learned only from experiment.

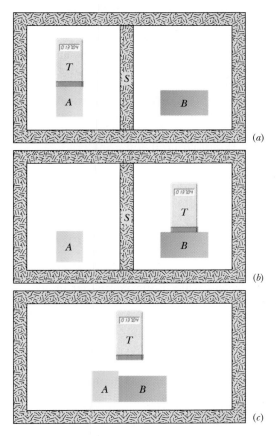

FIGURE 19-3 (a) Body T (a thermoscope) and body A are in thermal equilibrium. (Body S is a thermally insulating screen.) (b) Body T and body B are also in thermal equilibrium, at the same reading of the thermoscope. (c) If (a) and (b) are true, the zeroth law of thermodynamics states that body A and body B will also be in thermal equilibrium.

Thermal element

FIGURE 19-2 A thermoscope. The numbers increase when the device is heated and decrease when it is cooled. The thermally sensitive element could be—among many possibilities—a coil of wire whose electrical resistance is measured and displayed.

The experimental facts shown in Fig. 19-3 are summed up in the **zeroth law of thermodynamics:**

> If bodies A and B are each in thermal equilibrium with a third body T, then they are in thermal equilibrium with each other.

In less formal language, the message of the zeroth law is: "Every body has a property called **temperature.** When two bodies are found to be in thermal equilibrium, their temperatures are equal." We can now make our thermoscope (body T) into a thermometer, confident that its readings will have physical meaning. All that we have to do is calibrate it.

We use the zeroth law constantly in the laboratory. If we want to know whether the liquids in two beakers are at the same temperature, we measure the temperature of each with a thermometer. We do not need to bring the two liquids into intimate contact and observe whether or not they are in thermal equilibrium; we are quite sure that they are if their temperatures are equal.

The zeroth law, which has been called a logical afterthought, came to light only in the 1930s, long after the first and the second laws of thermodynamics had been discovered and numbered. Because the concept of temperature is fundamental to those two laws, the law that establishes temperature as a valid concept should have the lowest number. Hence the zero.

19-4 MEASURING TEMPERATURE

Let us see how we define and measure temperatures on the Kelvin scale. Equivalently, let us see how to calibrate a thermoscope so as to make it into a usable thermometer.

The Triple Point of Water

The first step in setting up a temperature scale is to pick out some reproducible thermal phenomenon and, quite arbitrarily, assign a certain Kelvin temperature to its thermal environment. That is, we select a *standard fixed point.* We could, for example, select the freezing point or the boiling point of water but, for various technical reasons, we do not. We select the **triple point of water.**

Liquid water, solid ice, and water vapor can coexist, in thermal equilibrium, at only one set of values of pressure and temperature. Figure 19-4

FIGURE 19-4 A triple-point cell, in which solid ice, liquid water, and water vapor coexist in thermal equilibrium. By international agreement, the temperature of this mixture has been defined to be 273.16 K. The bulb of a constant-volume gas thermometer is shown inserted into the well of the cell.

shows a triple-point cell, in which this so-called triple point of water can be achieved in the laboratory. By international agreement (in 1967), the triple point of water has been assigned a value of 273.16 K as the standard fixed-point temperature for the calibration of thermometers. That is,

$$T_3 = 273.16 \text{ K} \qquad \begin{array}{l}\text{(triple-point} \\ \text{temperature),}\end{array} \qquad (19\text{-}1)$$

in which the subscript 3 reminds us of the triple point.

Note that we do not use a degree mark in reporting Kelvin temperatures. It is 300 K (not 300°K) and it is pronounced "300 kelvins" (not "300 degrees Kelvin"). The usual SI prefixes apply. Thus 0.0035 K is 3.5 mK. No distinction in nomenclature is made between temperatures and temperature differences. Thus we can say, "the boiling point of sulfur is 717.8 K" and "the temperature of this water bath was raised by 8.5 K."

The Constant-Volume Gas Thermometer

So far, we have not discussed the particular physical property on which, by international agreement, we shall base our thermometer. Should it be the length of a metal rod, the electrical resistance of a wire, the pressure exerted by a confined gas, or something else? The choice is important because different choices lead to different temperatures for—to give

FIGURE 19-5 A constant-volume gas thermometer, its bulb immersed in a bath whose temperature T is to be measured. The pressure exerted by the gas is $p_0 + \rho g h$, where p_0 is the atmospheric pressure (read from a barometer) and h is the level difference of the manometer.

FIGURE 19-6 Temperatures calculated from Eq. 19-5 for a constant-volume gas thermometer whose bulb is immersed in boiling water. Different gases were used within the bulb, each at a variety of densities (which gave varying values for p_3). Note that all readings converge in the limit of zero density to a temperature of 373.125 K.

one example—the boiling point of water. For reasons that will emerge below, the standard thermometer, against which all other thermometers are to be calibrated, is based on the pressure exerted by a gas confined to a fixed volume.

Figure 19-5 shows such a **(constant-volume) gas thermometer.** It consists of a gas-filled bulb made of glass, quartz, or platinum (depending on the temperature range over which the thermometer is to be used) connected by a capillary tube to a mercury manometer. By raising and lowering reservoir R, the mercury level on the left can always be brought to the zero of the manometer scale, thus assuring that the volume of the confined gas remains constant. The temperature of any body in thermal contact with the bulb is then defined to be

$$T = Cp, \qquad (19\text{-}2)$$

in which p is the pressure exerted by the gas and C is a constant. The pressure is calculated from the relation

$$p = p_0 + \rho g h, \qquad (19\text{-}3)$$

in which p_0 is the atmospheric pressure, ρ is the density of the mercury in the manometer, and h is the

measured level difference of mercury in the two arms of the capillary tube.

With the bulb of the gas thermometer immersed in a triple-point cell, as in Fig. 19-4, we have

$$T_3 = Cp_3, \qquad (19\text{-}4)$$

in which p_3 is the pressure reading under this condition. Eliminating C between Eqs. 19-2 and 19-4 leads to

$$T = T_3 \left(\frac{p}{p_3} \right)$$

$$= (273.16 \text{ K}) \left(\frac{p}{p_3} \right) \qquad \text{(provisional).} \quad (19\text{-}5)$$

Equation 19-5 is not yet our final definition of a temperature measured with a gas thermometer. We have said nothing about what gas (or how much gas) we are to place in the thermometer. If our thermometer were used to measure some temperature, such as the boiling point of water, we would find that different choices lead to slightly different measured temperatures. However, as we use smaller and smaller amounts of gas to fill the bulb, the readings converge nicely to a single temperature, no matter what gas we use. Figure 19-6 shows this comforting convergence.*

---

*For pressure units, we shall use units introduced in Section 16-3. The SI unit for pressure is the newton per square meter, which is called the pascal (Pa). The pascal is related to other common pressure units by

$$1 \text{ atm} = 1.01 \times 10^5 \text{ Pa} = 760 \text{ torr} = 14.7 \text{ lb/in.}^2.$$

Thus we write, as our final recipe for measuring temperature with a gas thermometer,

$$T = (273.16 \text{ K}) \left(\lim_{m \to 0} \frac{p}{p_3} \right). \quad (19\text{-}6)$$

This instructs us to fill the bulb with an arbitrary mass m of *any* gas (for example, nitrogen) and to measure p_3 (using a triple-point cell) and p, the gas pressure at the temperature being measured. We next calculate the ratio p/p_3. Then we repeat both measurements with a smaller amount of gas in the bulb, again calculating this ratio. We keep on this way, using smaller and smaller amounts of gas, until we can extrapolate to the ratio p/p_3 that we would find if we had approximately no gas in the bulb. We calculate the temperature by substituting that extrapolated ratio into Eq. 19-6. Temperature defined in this way is the *ideal gas temperature.*

If temperature is to be a truly fundamental physical quantity, one in which the laws of thermodynamics may be expressed, it is absolutely necessary that its definition be independent of the properties of specific materials. It would not do, for example, to have such a basic quantity as temperature depend on the expansivity of mercury, the electrical resistivity of platinum, or any other such "handbook" property. We chose the gas thermometer as our standard instrument precisely because no such specific proper-

ties of materials are involved in its operation. You can use *any* gas and you always get the same result.

SAMPLE PROBLEM 19-1

The bulb of a gas thermometer is filled with nitrogen to a pressure of 120 kPa. What provisional value (see Fig. 19-6) would this thermometer yield for the boiling point of water and what is its error?

SOLUTION From Fig. 19-6, the curve for nitrogen shows that, at 120 kPa, the provisional temperature for the boiling point of water would be about 373.44 K. The actual boiling point of water (found by extrapolation on Fig. 19-6; see also Table 19-1) is 373.125 K. Thus using the provisional temperature leads to an error of 0.315 K, or 315 mK.

19-5 THE INTERNATIONAL TEMPERATURE SCALE

Measuring a temperature precisely with a gas thermometer is not an easy task and may require months of careful work. In practice, the gas thermometer is used only to establish certain "primary" fixed points. These points are then used to calibrate other more convenient secondary thermometers, such as a liquid-in-glass thermometer. The common household thermometers are of this type: a liquid, usually mercury, is enclosed in a glass bulb at one end of a thin tube. When the liquid is heated, it expands up the tube. The height it reaches corresponds to the temperature, which is read from a scale alongside the tube.

For practical use, as in the calibration of industrial or scientific thermometers, the *International Temperature Scale* has been adopted. This scale consists of a set of recipes for providing—in practice—the best possible approximations to the Kelvin scale. A set of primary fixed points is adopted and a set of instruments is specified to be used for interpolating between these fixed-point temperatures and for extrapolating beyond the highest temperature or below the lowest.

Table 19-1 gives the primary fixed points. The boiling point of water (sometimes called the *steam point*) is included, but the freezing point (sometimes called the *ice point*) is not. The *melting* points given in the table can also be called *freezing* points, because melting and freezing of any substance occur at the same temperature.

TABLE 19-1
PRIMARY FIXED POINTS ON THE INTERNATIONAL TEMPERATURE SCALE

SUBSTANCE	FIXED-POINT STATE	TEMPERATURE (K)
Hydrogen	Triple point	13.81
Hydrogen	Boiling point[a]	17.042
Hydrogen	Boiling point	20.28
Neon	Boiling point	27.102
Oxygen	Triple point	54.361
Argon	Triple point	83.798
Oxygen	Boiling point	90.188
Water	Boiling point	373.125
Tin	Melting point	505.074
Zinc	Melting point	692.664
Silver	Melting point	1235.08
Gold	Melting point	1337.58

[a] This boiling point is for a pressure of 25/76 atm. All other boiling or melting points are for a pressure of 1 atm.

19-6 THE CELSIUS AND FAHRENHEIT SCALES

So far, we have discussed only the Kelvin scale, used in basic scientific work. In nearly all countries of the world, the Celsius scale (formerly called the centigrade scale) is the scale of choice for popular and commercial use and much scientific use. The size of the degree is the same on the Celsius and Kelvin scales, but the zero of the former is shifted to a more convenient value. If T_C represents a Celsius temperature, then

$$T_C = T - 273.15°. \qquad (19\text{-}7)$$

In expressing temperatures on the Celsius scale, the degree symbol is commonly used. Thus we write 20.00°C, but 293.15 K.

The Fahrenheit scale, used in the United States, employs a smaller degree than the Celsius scale and a different zero of temperature. You can easily verify both differences by examining an ordinary room thermometer on which both scales are marked. The relation between the Celsius and the Fahrenheit scales is

$$T_F = \tfrac{9}{5}T_C + 32°, \qquad (19\text{-}8)$$

where T_F is Fahrenheit temperature. Transferring between these two scales can be done easily by remembering a few corresponding points (such as the freezing and boiling points of water; see Table 19-2) and by making use of the fact that 9 degrees on the Fahrenheit scale equals 5 degrees on the Celsius scale. Figure 19-7 compares the Kelvin, Celsius, and Fahrenheit scales.

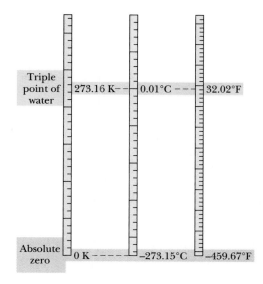

FIGURE 19-7 The Kelvin, Celsius, and Fahrenheit temperature scales compared.

TABLE 19-2
SOME CORRESPONDING TEMPERATURES

TEMPERATURE	°C	°F
Boiling point of water[a]	100	212
Normal body temperature	37.0	98.6
Accepted comfort level	20	68
Freezing point of water[a]	0	32
Zero of Fahrenheit scale	≈ − 18	0
Scales coincide	− 40	− 40

[a] Strictly, the boiling point of water on the Celsius scale is 99.975°C (see Eq. 19-7 and Table 19-1), and the freezing point is 0.00°C. Thus there is slightly less than 100 Celsius degrees between those two points.

SAMPLE PROBLEM 19-2

Suppose that you come across old scientific notes that describe a temperature scale called Z on which the boiling point of water is at 65.0°Z and the freezing point is at − 14.0°Z.

a. What temperature change ΔT on the Z scale would correspond to a change of 53.0°F?

SOLUTION To find a conversion factor between the two scales, we can use the boiling and freezing points of water. On the Z scale, the temperature difference between those two points is 65.0°Z − (− 14.0°Z), or 79.0°Z. On the Fahrenheit scale, it is 212°F − 32.0°F, or 180°F. Thus a change of 79.0°Z equals a change of 180°F. For a change of 53.0°F, we can now write

$$\Delta T = 53.0°\text{F} = 53.0°\text{F} \left(\frac{79.0°\text{Z}}{180°\text{F}} \right)$$

$$= 23.3°\text{Z}. \qquad \text{(Answer)}$$

b. To what temperature on the Fahrenheit scale would a temperature $T = -98.0°Z$ correspond?

SOLUTION The freezing point of water is − 14.0°Z, so the difference between T and the freezing point is 84.0°Z. To convert this difference to Fahrenheit degrees, we write

$$\Delta T = 84.0°\text{Z}\left(\frac{180°\text{F}}{79.0°\text{Z}}\right) = 191°\text{F}.$$

Thus T is 191°F below the freezing point of water and is, on the Fahrenheit scale,

$$T = 32.0°\text{F} - 191°\text{F} = -159°\text{F}. \quad \text{(Answer)}$$

PROBLEM SOLVING

TACTIC 1: TEMPERATURE CHANGES

Between the boiling and freezing points of water, there are (approximately) 100 kelvins and 100 Celsius degrees. Thus a kelvin is the same size as a Celsius degree. From this or from Eq. 19-7, we then know that any temperature change is the same number whether expressed in kelvins or Celsius degrees. For example, a temperature change of 10 K is equivalent to a change of 10°C.

Between the boiling and freezing points of water, there are 180 Fahrenheit degrees. Thus a Fahrenheit degree must be $\frac{9}{5}$ ($= 180°\text{F}/100$ K) the size of a kelvin or a Celsius degree. From this or from Eq. 19-8, we then know that any temperature change expressed in Fahrenheit degrees must be $\frac{9}{5}$ times that same temperature change expressed in either kelvins or Celsius degrees. For example, in Fahrenheit degrees, a temperature change of 10 K is $(9/5)(10$ K$)$, or 18°F.

You should take care not to confuse a temperature with a temperature *change*. A temperature of 10 K is certainly not the same as one of 10°C or 18°F but, as we have just seen, a temperature *change* of 10 K is the same as one of 10°C or 18°F.

19-7 THERMAL EXPANSION

Some Applications

You can often loosen a tight metal jar lid by holding it under a stream of hot water. The metal lid expands a little relative to the glass jar as the temperature of the lid rises. Such **thermal expansion** is not always desirable, as Fig. 19-8 suggests. We have all seen the expansion slots that must be placed in the roadways of bridges. Pipes at refineries often include an expansion loop, so that the pipe will not buckle as the temperature rises. Dental materials used for fillings must be matched in their expansion properties to those of tooth enamel. In aircraft manufacture,

FIGURE 19-8 Railroad tracks in Asbury Park, New Jersey, distorted because of thermal expansion on a very hot July day.

rivets and other fasteners are often designed to be cooled in dry ice before insertion and then allowed to expand to a tight fit.

Thermometers and thermostats may be based on the differences in expansion between the components of a *bimetal strip* (see Fig. 19-9). In a thermometer of a familiar type, the bimetal strip is

Brass | Steel

$T = T_0$ $T > T_0$

FIGURE 19-9 A bimetal strip, consisting of a strip of brass and a strip of steel welded together, at temperature T_0. The strip bends as shown at temperatures above this reference temperature. Below the reference temperature the strip bends the other way. Many thermostats operate on this principle, making and breaking an electrical contact as the temperature rises and falls.

FIGURE 19-10 A thermometer based on a bimetal strip. The strip is formed into a helix, which coils or uncoils as the temperature changes.

TABLE 19-3

SOME COEFFICIENTS OF LINEAR EXPANSION[a]

SUBSTANCE	α $(10^{-6}/°C)$
Ice (at 0°C)	51
Lead	29
Aluminum	23
Brass	19
Copper	17
Steel	11
Glass (ordinary)	9
Glass (Pyrex)	3.2
Invar[b]	0.7
Fused quartz	0.5

[a] Room temperature values except for the listing for ice.

[b] This alloy was designed to have a low coefficient of expansion. The word is a shortened form of "invariable."

coiled into a helix that winds and unwinds as the temperature changes (Fig. 19-10). And the familiar liquid-in-glass thermometers are based on the fact that liquids such as mercury or alcohol expand to a different (greater) extent than do their glass containers.

Thermal Expansion: Quantitative

If the temperature of a metal rod of length L is raised by an amount ΔT, its length is found to increase by an amount

$$\Delta L = L\alpha \, \Delta T, \qquad (19\text{-}9)$$

in which α is a constant called the **coefficient of linear expansion.** The value of α depends on the material and on the temperature range of interest. We can rewrite Eq. 19-9 as

$$\alpha = \frac{\Delta L/L}{\Delta T}, \qquad (19\text{-}10)$$

which shows us that α is the fractional increase in length per unit change in temperature. Although α varies somewhat with temperature, for most practical purposes at ordinary temperatures it can be taken as a constant. Table 19-3 shows some coefficients of linear expansion.

The thermal expansion of a solid is like a (three-dimensional) photographic enlargement. Figure 19-11b shows the (exaggerated) expansion of a steel rule after its temperature is increased from that of Fig. 19-11a. Equation 19-9 applies to every linear di-

(a)

Circle

Circular hole

(b)

FIGURE 19-11 The same steel rule at two different temperatures. When it expands, every dimension is increased in the same proportion. The scale, the numbers, the thickness, and the diameters of the circle and circular hole are all increased by the same factor. (The expansion has been exaggerated for clarity.)

mension of the rule, including its edge, thickness, diagonals, and the diameters of the circle etched on it and the circular hole cut in it. If the disk cut from that hole originally fits snugly in the hole, it will continue to fit snugly if it undergoes the same temperature increase as the rule.

Thermal Expansion of Liquids

If all dimensions of a solid expand with temperature, the volume of that solid must also expand. For liquids, volume expansion is the only meaningful expansion parameter. If the temperature of a solid or liquid whose volume is V is increased by an amount ΔT, the increase in volume is found to be

$$\Delta V = V\beta\,\Delta T, \tag{19-11}$$

where β is the **coefficient of volume expansion** of the solid or liquid. The coefficients of volume expansion and linear expansion for a solid are related by

$$\beta = 3\alpha. \tag{19-12}$$

The most common liquid, water, does not behave like other liquids. Figure 19-12a shows how its *specific volume* (the volume per unit mass) changes with temperature. Above about 4°C, water expands as the temperature rises, as we would expect. Between 0 and about 4°C, however, water *contracts* with increasing temperature (see Fig. 19-12b). At about 4°C, the specific volume of water passes through a minimum, which is to say that the density (the inverse of specific volume) passes through a maximum. At all other temperatures, the density of water is less than this maximum value.

This behavior of water is the reason that lakes freeze from the top down rather than from the bottom up. As the water on the surface is cooled from, say, 10°C toward the freezing point, it becomes denser than the lower water and sinks to the bottom. Below 4°C, however, further cooling makes the water then on the surface *less* dense than the lower water, so it stays on the surface until it freezes. If lakes froze from the bottom up, the ice so formed would tend not to melt completely during the summer, being insulated by the water above. After a few years, many bodies of open water in the temperate

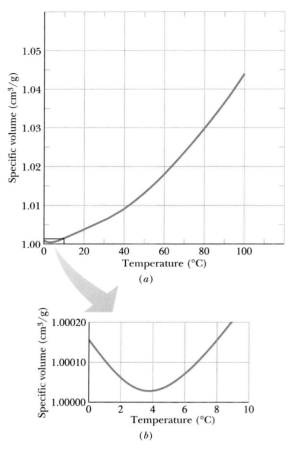

FIGURE 19-12 (*a*) The specific volume of water as a function of temperature. (*b*) An enlargement of the curve near 4°C, showing a minimum specific volume (that is, a maximum density).

zones of the Earth would be frozen solid all year round. And aquatic life as we know it could not exist. Who would guess that so much depends on the (scarcely noticeable) behavior of water in the lower left-hand corner of Fig. 19-12a?

Thermal Expansion: An Atomic View

Let us see if we can understand why a solid expands if you increase its temperature. The atoms of a crystalline solid are held together in a three-dimensional repeating lattice by springlike interatomic forces. The individual atoms vibrate about these lattice sites with an amplitude that increases

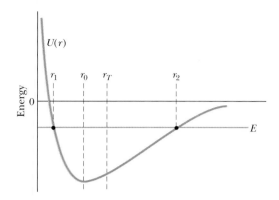

FIGURE 19-13 The potential energy $U(r)$ for two atoms separated by a distance r. As the mechanical energy E is increased (corresponding to an increase in temperature), the atoms become able to move farther apart. For a solid with a symmetrical potential energy curve there would be no thermal expansion.

with temperature. If the solid as a whole expands, the average distance between neighboring atoms must have increased.

Figure 19-13 shows the potential energy curve $U(r)$ for a pair of neighboring atoms in a lattice, r being their separation. The potential energy has a minimum at $r = r_0$, the lattice spacing that the solid would have at a temperature approaching the absolute zero. More important, the curve is not symmetrical, rising more steeply when the atoms are pushed together ($r < r_0$) than when they are pulled apart ($r > r_0$). The interatomic "springs" evidently do not obey Hooke's law.

It is this lack of symmetry of the potential energy function that accounts for the thermal expansion of solids. The horizontal line marked E shows the mechanical energy of the pair of atoms at an arbitrary temperature T. At this temperature, the interatomic separation can oscillate between r_1 and r_2, with an average value of r_T. Note that r_T is greater than r_0. Moreover, r_T must increase (shift right) as the energy E (and thus the temperature) is raised, because r_2 shifts farther to the right than r_1 shifts to the left. In other words, the average lattice spacing r_T—and thus the dimensions of the solid—increases with temperature. A solid with a symmetrical potential energy curve would not expand with temperature; r_T in Fig. 19-13 would be the same at all temperatures (provided, of course, that the solid remains a solid and does not melt or vaporize).

SAMPLE PROBLEM 19-3

Steel railroad rails are laid when the temperature is 0°C. What gap should be left between rail sections so that they just touch when the temperature rises to 42°C? The sections are 12.0 m long.

SOLUTION From Table 19-3, the coefficient of linear expansion for steel is $11 \times 10^{-6}/°C$. From Eq. 19-9 we then have

$$\Delta L = L\alpha \, \Delta T = (12.0 \text{ m})(11 \times 10^{-6}/°C)(42°C)$$

$$= 5.5 \times 10^{-3} \text{ m} = 5.5 \text{ mm.} \qquad \text{(Answer)}$$

SAMPLE PROBLEM 19-4

A steel wire whose length L is 130 cm and whose diameter d is 1.1 mm is heated to an average temperature of 830°C and stretched taut between two rigid supports. What tension develops in the wire as it cools to 20°C?

SOLUTION First let us calculate the amount by which the wire would shrink if it were allowed to cool freely. From Eq. 19-9 we have

$$\Delta L = L\alpha \, \Delta T = (1.3 \text{ m})$$

$$\times (11 \times 10^{-6}/°C)(830°C - 20°C)$$

$$= 1.16 \times 10^{-2} \text{ m} = 1.16 \text{ cm.}$$

If bridges were not built in sections, separated by expansion joints as shown here, the roadway over them might buckle due to expansion on especially hot days, or break apart via contraction on especially cold days.

However, the wire is not permitted to shrink. We therefore calculate what force would be required to stretch the wire by this amount. From Eq. 13-29 we have

$$F = \frac{\Delta L\, EA}{L} = \frac{\Delta L\, E(\pi/4)\, d^2}{L},$$

in which E is Young's modulus for steel (see Table 13-1) and A is the cross-sectional area of the wire. Substitution yields

$$F = (1.16 \times 10^{-2}\ \text{m})(200 \times 10^9\ \text{N/m}^2)(\pi/4)$$

$$\times \frac{(1.1 \times 10^{-3}\ \text{m})^2}{1.3\ \text{m}}$$

$$= 1700\ \text{N}. \qquad \text{(Answer)}$$

Can you show that this answer is independent of the length of the wire?

Sometimes the bulging brick walls of old buildings are supported by running a steel stress rod from outside wall to outside wall, through the building. The rod is then heated, and nuts on the outside walls are tightened. As the rod cools, stress develops in the rod, helping to keep the walls from bulging outward.

SAMPLE PROBLEM 19-5

On a hot day in Las Vegas, an oil trucker loaded 9785 gal of diesel fuel. He encountered cold weather on the way to Payson, Utah, where the temperature was 41°F lower than in Las Vegas, and where he delivered his entire load. How many gallons did he deliver? The coefficient of volume expansion for diesel fuel is $9.5 \times 10^{-4}/°\text{C}$, and the coefficient of linear expansion for his steel truck tank is $11 \times 10^{-6}/°\text{C}$.

SOLUTION From Eq. 19-11,

$$\Delta V = V\beta\, \Delta T = (9785\ \text{gal})$$

$$\times (9.5 \times 10^{-4}/°\text{C})(41°\text{F})(5°\text{C}/9°\text{F})$$

$$= 212\ \text{gal}.$$

Thus the amount delivered was

$$V_{\text{del}} = V - \Delta V = 9785\ \text{gal} - 212\ \text{gal}$$

$$= 9573\ \text{gal} \approx 9600\ \text{gal}. \qquad \text{(Answer)}$$

Note that the thermal expansion of the steel tank has nothing to do with the problem. Question: Who paid for the "missing" diesel fuel?

PROBLEM SOLVING

TACTIC 2: UNITS FOR TEMPERATURE CHANGES
The coefficient of linear expansion α is defined to be the fractional change in length (a dimensionless number) per unit change in temperature. In Table 19-3, that unit change in temperature is expressed in Celsius degrees (°C). Since any temperature *change* expressed in Celsius degrees is numerically the same when expressed in kelvins, the values of α in Table 19-3 could just as well be expressed in kelvins.

For example, the value of α for steel can be written as either $11 \times 10^{-6}/°\text{C}$ or $11 \times 10^{-6}/\text{K}$. This means that we could have substituted either expression for α in Sample Problems 19-3 and 19-4. We could have made a similar substitution for β in Sample Problem 19-5.

You will be seeing similar situations in the next several chapters. Quantities involving a unit change in temperature can be expressed with the equivalent notation of °C or K.

REVIEW & SUMMARY

Temperature; Thermometers
Temperature is a macroscopic quantity related to our sense of hot and cold. It is measured by a thermometer, which contains a working substance with a measurable property, such as length or pressure, that changes in a regular way as the substance becomes hotter or colder.

Zeroth Law of Thermodynamics
When a thermometer and some other object are placed in contact with each other, they eventually reach thermal equilibrium. The reading of the thermometer is then taken to be the temperature of the other object. The process provides consistent and useful temperature measurements because of the **zeroth law of thermodynamics:** if bodies A and B are each in thermal equilibrium with a third body C (the thermometer), then A and B are in thermal equilibrium with each other.

The Kelvin Temperature Scale
Temperature is measured, in the SI system, on the **Kelvin scale.** The scale is established by first defining the numerical value of the temperature at which water can exist with all three phases in equilibrium (the **triple point**) to be

273.16 K. Other temperatures are then defined in terms of a **constant-volume gas thermometer,** in which temperature is proportional to the pressure of a constant-volume sample of a gas. Since different gases give consistent results only at very low densities, the *ideal gas temperature,* as measured with a gas thermometer, is defined as

$$T = (273.16 \text{ K}) \left(\lim_{m \to 0} \frac{p}{p_3} \right). \tag{19-6}$$

Here T is the measured Kelvin temperature, p_3 and p are the pressures of the gas at the triple-point temperature and the measured temperature, respectively, and m is the mass of the gas in the thermometer.

International Scale

Several "fixed points" on the Kelvin scale have been measured as a basis for the *International Temperature Scale;* they are given in Table 19-1.

Celsius and Fahrenheit Scales

Besides the Kelvin scale, two other temperature scales are in common use: the Celsius scale, defined by

$$T_C = T - 273.15°, \tag{19-7}$$

and the Fahrenheit scale, defined by

$$T_F = \tfrac{9}{5} T_C + 32°. \tag{19-8}$$

Thermal Expansion

All objects change size with changes in temperature. The change ΔL in any linear dimension L is given by

$$\Delta L = L\alpha \, \Delta T, \tag{19-9}$$

in which α is the **coefficient of linear expansion.** The change ΔV in the volume V of a solid or liquid is

$$\Delta V = V\beta \, \Delta T. \tag{19-11}$$

Here $\beta = 3\alpha$ is the **coefficient of volume expansion** of the material.

QUESTIONS

1. Is temperature a microscopic or macroscopic concept?

2. Are there physical quantities other than temperature that tend to equalize if two different systems are joined?

3. A piece of ice and a warmer thermometer are suspended in an insulated evacuated enclosure so that they are not in contact. Why does the thermometer reading decrease for a time?

4. Let p_3 be the pressure in the bulb of a constant-volume gas thermometer when the bulb is at the triple-point temperature of 273.16 K and p the pressure when the bulb is at room temperature. Given three constant-volume gas thermometers: for A the gas is oxygen and $p_3 = 20$ cm Hg; for B the gas is also oxygen but $p_3 = 40$ cm Hg; for C the gas is hydrogen and $p_3 = 30$ cm Hg. The measured values of p for the three thermometers are p_A, p_B, and p_C. (a) An approximate value of the room temperature T can be obtained with each of the thermometers using

$$T_A = 273.16 \text{ K } \frac{p_A}{20 \text{ cm Hg}}; \qquad T_B = 273.16 \text{ K } \frac{p_B}{40 \text{ cm Hg}};$$

$$T_C = 273.16 \text{ K } \frac{p_C}{30 \text{ cm Hg}}.$$

Mark each of the following statements true or false. (1) With the method described, all three thermometers will give the same value of T. (2) The two oxygen thermometers will agree with each other but not with the hydrogen thermometer. (3) Each of the three will give a different value of T. (b) In the event that there is disagreement among the three thermometers, explain how you would change the method of using them to cause all three to give the same value of T.

5. A student, when told that the temperature at the center of the sun was thought to be about 1.5×10^7 degrees, asked whether that was on the Celsius or the Kelvin scale. How would you answer? How would you reply if he had asked whether it was on the Celsius or the Fahrenheit scale?

6. The Editor-in-Chief of a well-known business magazine, discussing possible warming effects associated with the increasing concentration of carbon dioxide in the Earth's atmosphere, wrote: "The polar regions might be three times warmer than now, ..." What do you suppose he meant? (From "Warmth and Temperature: A Comedy of Errors," by Albert A. Bartlett, *The Physics Teacher,* November 1984.)

7. Although the absolute zero of temperature seems to be experimentally unattainable, temperatures as low as 0.000000002 K have been achieved in the laboratory. Isn't this low enough for all practical purposes? Why would

physicists (as indeed they do) strive to obtain still lower temperatures?

8. Can a temperature be assigned to a vacuum?

9. Does our "temperature sense" have a built-in sense of direction; that is, does hotter necessarily mean higher temperature, or is this just an arbitrary convention? Anders Celsius (for whom the scale is named) originally chose the boiling point of water as 0°C and the freezing point as 100°C.

10. Many medicine labels inform the user to store below 86°F. Why 86? (*Hint:* Change to Celsius.)

11. How would you suggest measuring the temperature of (a) the sun, (b) the Earth's upper atmosphere, (c) an insect, (d) the moon, (e) the ocean floor, and (f) liquid helium?

12. Is one gas any better than another for purposes of a standard constant-volume gas thermometer? What properties are desirable in a gas for such purposes?

13. State some objections to using water-in-glass as a thermometer. Is mercury-in-glass an improvement?

14. Explain why the column of mercury first descends and then rises when a mercury-in-glass thermometer is put in a flame.

15. What do the Celsius and Fahrenheit temperature scales have in common?

16. What are the dimensions of α, the coefficient of linear expansion? Does the value of α depend on the unit of length used? When degrees Fahrenheit are used instead of degrees Celsius as the unit of temperature change, how does the numerical value of α change?

17. A metal ball can pass through a metal ring. When the ball is heated, however, it gets stuck in the ring. What would happen if the ring, rather than the ball, were heated?

18. Two strips, one of iron and one of zinc, are riveted together side by side to form a straight bar that curves when heated. Why is it that the iron is on the inside of the curve?

19. Explain how the period of a pendulum clock can be kept constant with temperature by attaching vertical tubes of mercury to the bottom of the pendulum.

20. Why should a chimney be freestanding, that is, not part of the structural support of the house?

21. Water expands when it freezes. Can we define a coefficient of volume expansion for the freezing process?

22. Explain why the apparent expansion, when heated, of a liquid in a glass bulb does not give the true expansion of the liquid.

EXERCISES & PROBLEMS

SECTION 19-4 MEASURING TEMPERATURE

1E. To measure temperatures, physicists and astronomers often use the variation of intensity of electromagnetic radiation emitted by an object. The wavelength at which the intensity is greatest is given by the equation

$$\lambda_{max} T = 0.2898 \text{ cm} \cdot \text{K},$$

where λ_{max} is the wavelength of greatest intensity and T is the temperature of the object in kelvins. In 1965, microwave radiation peaking at $\lambda_{max} = 0.107$ cm was discovered coming in all directions from space. To what temperature does this correspond? The interpretation of this *background radiation* is that it is left over from the Big Bang some 15 billion years ago, when the universe began rapidly expanding and cooling.

2E. A *thermocouple* is formed from two different metals, joined at two points in such a way that a small voltage is produced when the two junctions are at different temperatures. In a particular iron–constantan thermocouple, with one junction held at 0.000°C, the output voltage varies linearly from 0.000 to 28.0 mV as the temperature of the other junction is raised from 0.000 to 510°C. Find the temperature of the variable junction when the thermocouple output is 10.2 mV.

3E. The amplification or *gain* of a transistor amplifier may depend on the temperature. The gain for a certain amplifier at room temperature (20.0°C) is 30.0, whereas at 55.0°C it is 35.2. What would the gain be at 30.0°C if the gain depends linearly on temperature, over a limited range?

4E. If the temperature of a gas at the steam point is 373.15 K, what is the limiting value of the ratio of the pressure of the gas at the steam point to its pressure at the triple point of water? (Assume the volume of the gas is the same at both temperatures.)

5E. A *resistance thermometer* is a thermometer whose electrical resistance changes with temperature. We are free to define temperatures measured by such a thermometer in kelvins (K) to be directly proportional to the resistance R, measured in ohms (Ω). A certain resistance thermometer is found to have a resistance R of 90.35 Ω when its bulb is placed in water at the triple-point temperature (273.16 K). What temperature is indicated by the thermometer if the bulb is placed in an environment such that its resistance is 96.28 Ω?

6P. Two constant-volume gas thermometers are assembled, one using nitrogen as the working gas and the other using hydrogen. Both contain enough gas so that $p_3 = 80$ mm Hg. What is the difference between the pressures in the two thermometers if both are inserted into a water bath at the boiling point? Which pressure is the higher of the two?

7P. A particular gas thermometer is constructed of two gas-containing bulbs, each of which is put into a water bath, as shown in Fig. 19-14. The pressure difference between the two bulbs is measured by a mercury manometer as shown. Appropriate reservoirs, not shown in the diagram, maintain constant gas volume in the two bulbs. There is no difference in pressure when both baths are at the triple point of water. The pressure difference is 120 mm Hg when one bath is at the triple point and the other is at the boiling point of water. Finally, the pressure difference is 90.0 mm Hg when one bath is at the triple point and the other is at an unknown temperature to be measured. What is the unknown temperature?

8P. A *thermistor* is a semiconductor device with a temperature-dependent electrical resistance. It is com-

FIGURE 19-14 Problem 7.

monly used in medical thermometers and to sense overheating in electronic equipment. Over a limited range of temperature, the resistance is given by

$$R = R_a e^{B(1/T - 1/T_a)},$$

where R is the resistance of the thermistor at temperature T and R_a is the resistance at temperature T_a; B is a constant that depends on the particular semiconductor used. For one type of thermistor, $B = 4689$ K and the resistance at 273 K is 1.00×10^4 Ω (ohms). What temperature is the thermistor measuring when its resistance is 100 Ω?

SECTION 19-6 THE CELSIUS AND FAHRENHEIT SCALES

9E. On his 44th birthday, singer Tom Rush remarked: "I prefer to call it 5 Celsius." Did Tom work that out correctly? If not, what should his "age in Celsius" be?

10E. At what temperature is the Fahrenheit scale reading equal to (a) twice that of the Celsius and (b) half that of the Celsius?

11E. If your doctor tells you that your temperature is 310 degrees above absolute zero, should you worry? Explain your answer.

12E. (a) In 1964, the temperature in the Siberian village of Oymyakon reached a value of $-71°C$. What temperature is this on the Fahrenheit scale? (b) The highest officially recorded temperature in the continental United States was $134°F$ in Death Valley, California. What is this temperature on the Celsius scale?

13E. (a) The temperature of the surface of the sun is about 6000 K. Express this on the Fahrenheit scale. (b) Express normal human body temperature, $98.6°F$, on the Celsius scale. (c) In the continental United States, the lowest officially recorded temperature is $-70°F$ at Rogers Pass, Montana. Express this on the Celsius scale. (d) Express the normal boiling point of oxygen, $-183°C$, on the Fahrenheit scale. (e) At what Celsius temperature would you find a room to be uncomfortably warm?

14E. At what temperature do the following pairs of scales give the same reading: (a) Fahrenheit and Celsius (verify the listing in Table 19-2), (b) Fahrenheit and Kelvin, and (c) Celsius and Kelvin?

15P. Suppose that on a temperature scale X, water boils at $-53.5°X$ and freezes at $-170°X$. What would a temperature of 340 K be on the X scale?

16P. In the interval between the freezing point of water and $700.0°C$, a platinum resistance thermometer of definite specifications is used for interpolating temperatures on the International Temperature Scale. The Celsius temperature T_C is given by a formula for the variation of resistance with temperature:

$$R = R_0(1 + AT_C + BT_C^2).$$

R_0, A, and B are constants determined by measurements at the ice point, the steam point, and the melting point of zinc. (a) If R equals 10.000 Ω (ohms) at the ice point, 13.946 Ω at the steam point, and 24.172 Ω at zinc's melting point, find R_0, A, and B. (b) Plot R versus T_C in the temperature range from 0 to $700.0°C$.

17P. It is an everyday observation that hot and cold objects cool down or warm up to the temperature of their surroundings. If the temperature difference ΔT between an object and its surroundings ($\Delta T = T_{obj} - T_{sur}$) is not too great, the rate of cooling or warming of the object is proportional, approximately, to this temperature difference; that is,

$$\frac{d\,\Delta T}{dt} = -A(\Delta T),$$

where A is a constant. The minus sign appears because ΔT decreases with time if ΔT is positive and increases if ΔT is negative. This is known as *Newton's law of cooling*. (a) On what factors does A depend? What are its dimensions? (b) If at some instant $t = 0$ the temperature difference is ΔT_0, show that it is

$$\Delta T = \Delta T_0 e^{-At}$$

at a time t later.

18P. The heater of a house breaks down one day when the outside temperature is $7.0°C$. As a result, the inside temperature drops from 22 to $18°C$ in 1.0 h. The owner fixes the heater and adds insulation to the house. Now she finds that, on a similar day, the house takes twice as long to drop from 22 to $18°C$ when the heater is not operating. What is the ratio of the constant A in Newton's law of cooling (see Problem 17) after the insulation is added to the value before?

19P. A mercury-in-glass thermometer is placed in boiling water for a few minutes and then removed. The temperature readings at various times after removal are as shown in the table. Plot A as a function of time, assuming Newton's law of cooling to apply (see Problem 17). To what extent are you justified in assuming that Newton's law of cooling applies here?

t (s)	T (°C)	t (s)	T (°C)
0.0	98.4	100	50.3
5.0	76.1	150	43.7
10	71.1	200	38.8
15	67.7	300	32.7
20	66.4	500	27.8
25	65.1	700	26.5
30	63.9	1000	26.1
40	61.6	1400	26.0
50	59.4	2000	26.0
70	55.4	3000	26.0

SECTION 19-7 THERMAL EXPANSION

20E. A steel rod has a length of exactly 20 cm at $30°C$. How much longer is it at $50°C$?

21E. An aluminum flagpole is 33 m high. By how much does its length increase as the temperature increases by $15°C$?

22E. The Pyrex glass mirror in the telescope at the Mount Palomar Observatory has a diameter of 200 in. The temperature ranges from -10 to $50°C$ on Mount Palomar. Determine the maximum change in the diameter of the mirror.

23E. A circular hole in an aluminum plate is 2.725 cm in diameter at $0.000°C$. What is its diameter when the temperature of the plate is raised to $100.0°C$?

24E. An aluminum-alloy rod has a length of 10.000 cm at $20.000°C$ and a length of 10.015 cm at the boiling point of water. (a) What is the length of the rod at the freezing point of water? (b) What is the temperature if the length of the rod is 10.009 cm?

25E. (a) Express the coefficient of linear expansion of aluminum using the Fahrenheit temperature scale.

(b) Use your answer to calculate the change in length of a 20-ft aluminum rod if it is heated from 40 to 95°F.

26E. Soon after the Earth formed, heat released by the decay of radioactive elements raised the average internal temperature from 300 to 3000 K, at about which value it remains today. Assuming an average coefficient of volume expansion of 3.0×10^{-5} K^{-1}, by how much has the radius of the Earth increased since its formation?

27E. A rod is measured to be exactly 20.05 cm long using a steel ruler at a room temperature of 20°C. Both the rod and the ruler are placed in an oven at 270°C, where the rod now measures 20.11 cm using the same ruler. What is the coefficient of thermal expansion for the material of which the rod is made?

28E. The Stanford linear accelerator contains hundreds of brass disks tightly fitted into a steel tube. The system was assembled by cooling the disks in dry ice (at −57.00°C) to enable them to slide into the close-fitting tube. If the diameter of a disk is 80.00 mm at 43.00°C, what is its diameter in the dry ice?

29E. A glass window is exactly 20 cm by 30 cm at 10°C. By how much has its area increased when its temperature is 40°C?

30E. A brass cube has an edge length of 30 cm. What is the increase in its surface area when it is heated from 20 to 75°C?

31E. Find the change in volume of an aluminum sphere of 10-cm radius when it is heated from 0 to 100°C.

32E. What is the volume of a lead ball at 30°C if its volume at 60°C is 50 cm^3?

33E. By how much does the volume of an aluminum cube 5.00 cm on an edge increase when it is heated from 10.0 to 60.0°C?

34E. Imagine an aluminum cup of 100-cm^3 capacity filled with glycerin at 22°C. How much glycerin, if any, will spill out of the cup if the temperature of the cup and glycerin is raised to 28°C? (The coefficient of volume expansion of glycerin is 5.1×10^{-4}/°C.)

35E. A steel rod at 25.0°C is bolted securely at both ends and then cooled. At what temperature will it rupture? See Table 13-1.

36P. A steel rod is 3.000 cm in diameter at 25°C. A brass ring has an interior diameter of 2.992 cm at 25°C. At what common temperature will the ring just slide onto the rod?

37P. The area A of a rectangular plate is ab. Its coefficient of linear expansion is α. After a temperature rise ΔT, side a is longer by Δa and side b is longer by Δb. Show that if we neglect the small quantity $\Delta a\,\Delta b/ab$ (see Fig. 19-15), then $\Delta A = 2\alpha A\,\Delta T$.

38P. Prove that, if we neglect extremely small quantities, the change in volume of a solid upon expansion through a temperature rise ΔT is given by $\Delta V = 3\alpha V\,\Delta T$, where α is the coefficient of linear expansion. (See Eqs. 19-11 and 19-12.)

39P. Density is mass divided by volume. If the volume V is temperature dependent, so is the density ρ. Show that a

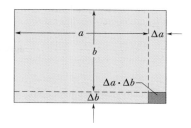

FIGURE 19-15 Problem 37.

small change in density $\Delta\rho$ with a change in temperature ΔT is given by

$$\Delta\rho = -\beta\rho\,\Delta T,$$

where β is the coefficient of volume expansion. Explain the minus sign.

40P. When the temperature of a metal cylinder is raised from 0.0°C to 100°C, its length increases by 0.23%. (a) Find the percent change in density. (b) What is the metal?

41P. Show that when the temperature of a liquid in a barometer changes by ΔT, and the pressure is constant, the height h changes by $\Delta h = \beta h\,\Delta T$, where β is the coefficient of volume expansion. Neglect the expansion of the glass tube.

42P. When the temperature of a copper penny is raised by 100°C, its diameter increases by 0.18%. To two significant figures, give the percent increase in (a) the area of a face, (b) the thickness, (c) the volume, and (d) the mass of the penny. (e) Calculate its coefficient of linear expansion.

43P. A clock pendulum made of Invar (see Table 19-3) has a period of 0.50 s and is accurate at 20°C. If the clock is used in a climate where the temperature averages 30°C, what correction (approximately) is necessary at the end of 30 days to the time given by the clock?

44P. A pendulum clock with a pendulum made of brass is designed to keep accurate time at 20°C. What will be the error, in seconds per hour, if the clock operates at 0.0°C?

45P. The timing of a certain electric watch is governed by a small tuning fork. The frequency of the fork is inversely proportional to the square root of the length of the fork. What is the fractional gain or loss in time for a quartz tuning fork 8.00 mm long at (a) −40.0°F and (b) +120°F if it keeps perfect time at 25.0°F?

46P. (a) Show that if the lengths of two rods of different solids are inversely proportional to their respective coefficients of linear expansion at the same initial temperature, the difference in length between them will be constant at all temperatures. (b) What should be the lengths of a steel and a brass rod at 0.00°C so that at all temperatures their difference in length is 0.30 m?

47P. As a result of a temperature rise of 32°C, a bar with a crack at its center buckles upward, as shown in Fig. 19-16. If the fixed distance $L_0 = 3.77$ m and the coefficient of linear expansion is $25 \times 10^{-6}/°C$, find x, the distance to which the center rises.

FIGURE 19-16   Problem 47.

48P.   In a certain experiment, it was necessary to be able to move a small radioactive source at selected, extremely slow speeds. This was accomplished by fastening the source to one end of an aluminum rod and heating the central section of the rod in a controlled way. If the effective heated section of the rod in Fig. 19-17 is 2.00 cm, at what constant rate must the temperature of the rod be changed if the source is to move at a constant speed of 100 nm/s?

FIGURE 19-17   Problem 48.

49P.   A 1.28-m-long vertical glass tube is half-filled with a liquid at 20°C. How much will the height of the liquid column change when the tube is heated to 30°C? Take $\alpha_{glass} = 1.0 \times 10^{-5}/°C$ and $\beta_{liquid} = 4.0 \times 10^{-5}/°C$.
50P.   A composite bar of length $L = L_1 + L_2$ is made from a bar of material 1 and length L_1 attached to a bar of material 2 and length L_2, as shown in Fig. 19-18. (a) Show that the effective coefficient of linear expansion α for this bar is given by $\alpha = (\alpha_1 L_1 + \alpha_2 L_2)/L$. (b) Using steel and brass, design such a composite bar whose length is 52.4 cm and whose effective coefficient of linear expansion is $13.0 \times 10^{-6}/°C$.

FIGURE 19-18   Problem 50.

51P.   A thick aluminum rod and a thin steel wire are attached in parallel, as shown in Fig. 19-19. The temperature is 10.0°C. Both the rod and the wire are 85.0 cm long and neither is under stress. The system is heated to 120°C.

Calculate the resulting stress in the wire, assuming that the rod expands freely.

FIGURE 19-19   Problem 51.

52P*.   Three equal-length straight rods, of aluminum, Invar, and steel, all at 20.0°C, form an equilateral triangle with hinge pins at the vertices. At what temperature will the angle opposite the Invar rod be 59.95°? See Appendix G for needed trigonometric formulas.
53P*.   Two rods of different materials but having the same lengths L and cross-sectional areas A are arranged end-to-end between fixed, rigid supports, as shown in Fig. 19-20a. The temperature is T and there is no initial stress. The rods are heated, so that their temperature increases by ΔT. (a) Show that the rod interface is displaced upon heating by an amount

$$\Delta L = \left(\frac{\alpha_1 E_1 - \alpha_2 E_2}{E_1 + E_2} \right) L \, \Delta T$$

FIGURE 19-20   Problem 53.

(see Fig. 19-20b), where α_1, α_2 are the coefficients of linear expansion and E_1, E_2 are Young's moduli of the materials. Ignore changes in cross-sectional areas. (b) Find the stress at the interface after heating.
54P*.   An aluminum cube 20.0 cm on an edge floats on mercury. How much farther will the block sink when the temperature rises from 270 to 320 K? (The coefficient of volume expansion of mercury is $1.80 \times 10^{-4}/°C$.)

HEAT AND THE FIRST LAW OF THERMODYNAMICS

An object with a black surface usually heats up more than one with a white surface when both are in sunlight. Such is true of the robes worn by Bedouins in the Sinai desert: black robes heat up more than white robes. Why then would a Bedouin ever wear a black robe? Wouldn't that actually decrease the chance of survival in the harsh desert environment?

20-1 HEAT

If you take a can of cola from the refrigerator and leave it on the kitchen table, its temperature will rise —rapidly at first and then increasingly slowly—until the temperature of the can equals that of the room. In the same way, the temperature of a cup of hot coffee, left to stand, will fall until it also reaches room temperature.

In generalizing this situation, we describe the cola or the coffee as a *system* (temperature T_S) and the relevant parts of the kitchen as the *environment* (temperature T_E) of that system. Our observation is that if T_S is not equal to T_E, then T_S will change until the two temperatures are equal (Fig. 20-1).

Such a change in temperature is due to the transfer of a form of energy between the system and its environment. This energy is *internal energy* (or

thermal energy), which is the collective kinetic and potential energies associated with the random motions of the atoms, molecules, and other microscopic bodies within an object. The transferred internal energy is called **heat** and is symbolized Q. Heat is *positive* when internal energy is transferred to a system from its environment (we say that heat is absorbed). And heat is *negative* when internal energy is transferred from a system to its environment (we say that heat is released or lost).

In the situation of Fig. 20-1a, in which $T_S > T_E$, internal energy is transferred from the system to the environment, so Q is negative. In Fig. 20-1b, in which $T_S = T_E$, there is no such transfer, Q is zero, and heat is neither released nor absorbed. In Fig. 20-1c, in which $T_S < T_E$, the transfer is to the system from the environment, so Q is positive.

We are led then to this definition of heat:

Heat is the energy that is transferred between a system and its environment because of a temperature difference that exists between them.

If you suspect that heat is being absorbed or released by a system, your key question is: "Where is the temperature difference?"

Energy can also be transferred between a system and its environment by means of *work* (symbol W), which we always associate with a force acting on a system during a displacement of the system. If you suspect that work is being done on or by a system, your key question is: "Where is the force and how does its point of application move?"

Both heat and work represent energy transfers between a system and its environment. Heat and work, unlike temperature, pressure, and volume, are not intrinsic properties of a system. They have meaning only as they describe the transfer of energy into or out of a system, adding to or subtracting from the system's store of *internal energy*. Thus it is proper to say: "During the last 3 min, 15 J of heat was transferred to the system from its environment" or "During the last minute, 12 J of work was done on the system by its environment." It is without meaning to say: "This system contains 450 J of heat" or "This system contains 385 J of work." The bookkeeping relation for heat, work, and internal (or

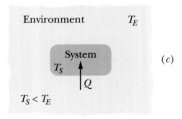

FIGURE 20-1 (*a*) If the temperature of the system exceeds that of its environment, heat is lost by the system to the environment until thermal equilibrium is established, as in (*b*). (*c*) If the temperature of the system is below that of the environment, heat is absorbed by the system until thermal equilibrium is established.

thermal) energy is summed up by the first law of thermodynamics, the subject of this chapter.

In popular usage, the word "heat" is often used where "temperature" is intended. For example, when we say: "It is a hot day," we are referring to temperature and not to heat. Do not confuse these two totally different quantities.

20-2 MEASURING HEAT: UNITS

Before scientists realized that heat is transferred energy, heat was measured in terms of its ability to raise the temperature of water. Thus the **calorie** (cal) was defined as the amount of heat that would raise the temperature of 1 g of water from 14.5 to 15.5°C. In the British system, the corresponding unit of heat was the **British thermal unit** (Btu), defined as the amount of heat that would raise the temperature of 1 lb of water from 63 to 64°F.

In 1948, the scientific community decided that, since heat (like work) is transferred energy, the SI unit for heat should be the same as that for energy, namely, the **joule.** The calorie is now defined to be 4.1860 J (exactly) with no reference to the heating of water. The "calorie" used in nutrition, sometimes called the Calorie (Cal), is really a kilocalorie.

The relations among the various heat units are

$$1 \text{ J} = 0.2389 \text{ cal} = 9.481 \times 10^{-4} \text{ Btu,}$$
$$1 \text{ Btu} = 1055 \text{ J} = 252.0 \text{ cal,}$$
$$1 \text{ cal} = 3.969 \times 10^{-3} \text{ Btu} = 4.186 \text{ J,} \quad (20\text{-}1)$$
$$1 \text{ Cal} = 10^3 \text{ cal} = 3.969 \text{ Btu} = 4186 \text{ J.}$$

In scientific work, heat is increasingly being expressed in joules, with the calorie and the Btu being gradually phased out. However, the calorie continues to be used in some areas of chemistry and the Btu in some aspects of engineering practice.

20-3 THE ABSORPTION OF HEAT BY SOLIDS AND LIQUIDS

Heat Capacity

The **heat capacity** C of an object (for example, a Pyrex coffee pot, an iron skillet, or a marble slab) is the proportionality constant between an amount of

heat and the change in temperature that this heat produces in the object. Thus

$$Q = C(T_f - T_i), \quad (20\text{-}2)$$

in which T_i and T_f are the initial and final temperatures of the object. The heat capacity C of a marble slab used in a bun warmer might, for example, be 179 cal/°C, which we can also write as 179 cal/K or as 747 J/K.

The word "capacity" in this context is really misleading in that it suggests an analogy to the capacity of a bucket to hold water. *The analogy is false,* and you should not think of the object as "containing" heat or being limited in its ability to absorb heat. The heat transfer can proceed without limit as long as the necessary temperature difference is maintained. The object may, of course, melt or vaporize during the process.

Specific Heat

Two objects made of the same material, say marble, will have heat capacities proportional to their masses. It is therefore convenient to define a "heat capacity per unit mass" or **specific heat** c that refers not to an object but to a unit mass of the material of which the object is made. Equation 20-2 then becomes

$$Q = cm(T_f - T_i). \quad (20\text{-}3)$$

Through experiment we would find that, although the heat capacity of the particular marble slab mentioned above is 179 cal/°C (or 747 J/K), the specific heat of marble itself (in that slab or in any other construction) is 0.21 cal/g·°C (or 880 J/kg·K).

From the way the calorie and the British thermal unit were initially defined, the specific heat of water is

$$c = 1 \text{ cal/g·°C} = 1 \text{ Btu/lb·°F}$$
$$= 4190 \text{ J/kg·K.} \quad (20\text{-}4)$$

Table 20-1 shows the specific heat of some substances at room temperature. Note that the value for water is relatively high. The value of the specific heat of any substance actually depends on temperature, but the values in Table 20-1 apply reasonably well in a range of temperatures near room temperature.

TABLE 20-1

SPECIFIC HEATS OF SOME SUBSTANCES AT ROOM TEMPERATURE

SUBSTANCE	SPECIFIC HEAT		MOLAR SPECIFIC HEAT
	cal/g·K	J/kg·K	J/mol·K
Elemental Solids			
Lead	0.0305	128	26.5
Tungsten	0.0321	134	24.8
Silver	0.0564	236	25.5
Copper	0.0923	386	24.5
Aluminum	0.215	900	24.4
Other Solids			
Brass	0.092	380	
Granite	0.19	790	
Glass	0.20	840	
Ice ($-10°C$)	0.530	2220	
Liquids			
Mercury	0.033	140	
Ethyl alcohol	0.58	2430	
Seawater	0.93	3900	
Water	1.00	4190	

Molar Specific Heat

In many instances the most convenient unit for specifying the amount of a substance is the mole (mol), where

$$1 \text{ mol} = 6.02 \times 10^{23} \text{ elementary units}$$

of the substance (or of *any* substance). Thus 1 mol of aluminum means 6.02×10^{23} atoms (the atom being the elementary unit), and 1 mol of aluminum oxide means 6.02×10^{23} molecules of the oxide (because the molecule is the elementary unit of a compound).

When quantities are expressed in moles, the specific heat must also involve moles (rather than a mass unit); it is then called a **molar specific heat.** Table 20-1 shows the values for some elemental solids (each consisting of a single element) at room temperature.

Note that the molar specific heats of all the elements listed in Table 20-1 have about the same value, namely, 25 J/mol·K. As Fig. 20-2 suggests, the molar specific heats of all solids vary with temperature, approaching 25 J/mol·K at high enough

FIGURE 20-2 The molar specific heats of three elements as a function of temperature. At high enough temperatures, all solids approach the same limiting value. For lead and aluminum, that value is already essentially reached at room temperature; for carbon it is not.

temperatures. Some substances, such as carbon and beryllium, do not reach this limiting value until temperatures well above room temperature. Other substances may melt or vaporize before they reach this limit.

In this photograph there is evidence for the three phases of water: Ice cubes cool the liquid water in the ice-tea drink and also the glass; the cold glass then causes water vapor in the surrounding air to condense, forming water drops on the exterior of the glass.

When we compare two substances on a molar basis, we are comparing samples that contain the same number of elementary units. The fact that, at high enough temperatures, all solid elements have the same molar specific heat tells us that all kinds of atoms—whether they be aluminum, copper, uranium, or anything else—absorb heat in the same way.

An Important Point

In assigning a specific heat to any substance, it is important to know not only how much heat was absorbed but the conditions under which that transfer took place. For solids and liquids, we usually assume that the sample was under constant pressure (usually atmospheric) during the heat transfer. It is also conceivable that the sample could have been held at constant volume while the heat was absorbed. This means that the thermal expansion of the sample was prevented, by applying external pressure. For solids and liquids, this is very hard to arrange experimentally but the effect can be calculated, and it turns out that the specific heats under constant pressure and constant volume differ usually by no more than a few percent. Gases, as you will see, have quite different values for their specific heats under constant-pressure conditions and under constant-volume conditions.

Heats of Transformation

When heat is absorbed by a solid or liquid, the temperature of the sample does not necessarily rise. Instead, the sample may change from one *phase* or state (that is, solid, liquid, or gas) to another. Thus ice may melt and water may boil, absorbing heat in each case without a temperature change. In the reverse processes (water freezing, steam condensing), heat is released by the sample, again at a constant temperature.

The amount of heat per unit mass that must be transferred when a sample completely undergoes a phase change is called the **heat of transformation** L. So when a sample of mass m completely undergoes a phase change, the total heat transferred is

$$Q = Lm, \qquad (20\text{-}5)$$

When the phase change is from a liquid phase to a gas phase (then the sample must absorb heat) or from gas to liquid (then the sample must release heat), the heat of transformation is called the **heat of vaporization** L_V. For water at its normal boiling or condensation temperature,

$$L_V = 539 \text{ cal/g} = 40.7 \text{ kJ/mol}$$
$$= 2260 \text{ kJ/kg}. \qquad (20\text{-}6)$$

When the phase change is from a solid phase to a liquid phase (then the sample must absorb heat) or from liquid to solid (then the sample must release heat), the heat of transformation is called the **heat of fusion** L_F. For water at its normal freezing or melting temperature,

$$L_F = 79.5 \text{ cal/g} = 6.01 \text{ kJ/mol}$$
$$= 333 \text{ kJ/kg}. \qquad (20\text{-}7)$$

Table 20-2 shows the heats of transformation for some substances.

TABLE 20-2
SOME HEATS OF TRANSFORMATION

| | MELTING | | BOILING | |
| | MELTING POINT (K) | HEAT OF FUSION L_F (kJ/kg) | BOILING POINT (K) | HEAT OF VAPORIZATION L_V (kJ/kg) |
SUBSTANCE				
Hydrogen	14.0	58.0	20.3	455
Oxygen	54.8	13.9	90.2	213
Mercury	234	11.4	630	296
Water	273	333	373	2256
Lead	601	23.2	2017	858
Silver	1235	105	2323	2336
Copper	1356	207	2868	4730

SAMPLE PROBLEM 20-1

A candy bar has a marked nutritional value of 350 Cal. How many kilowatt-hours of energy will it deliver to the body as it is digested?

SOLUTION The Calorie in this case is a kilocalorie so that

$$\text{energy} = (350 \times 10^3 \text{ cal})(4.19 \text{ J/cal})$$

$$= (1.466 \times 10^6 \text{ J})(1 \text{ W}\cdot\text{s/J})$$

$$\times (1 \text{ h}/3600 \text{ s})(1 \text{ kW}/1000 \text{ W})$$

$$= 0.407 \text{ kW}\cdot\text{h}. \qquad \text{(Answer)}$$

This amount of energy would keep a 100-W light bulb burning for 4.1 h. To burn up this much energy by exercise, a person would have to jog about 3 or 4 mi.

A generous daily human diet corresponds to about 3.5 kW·h per day, which represents the absolute maximum amount of work that a human can do in one day. In an industrialized country, this amount of energy can be purchased for perhaps 35 cents.

SAMPLE PROBLEM 20-2

a. How much heat is needed to take ice of mass $m = 720$ g at $-10°C$ to a liquid state at $15°C$?

SOLUTION To answer, we must consider three steps. Step 1 is to raise the temperature of the ice from $-10°C$ to the melting point at $0°C$. We use Eq. 20-3, with the specific heat c_{ice} of ice given in Table 20-1. For this step, the initial temperature T_i is $-10°C$ and the final temperature T_f is $0°C$. We find

$$Q_1 = c_{ice}m(T_f - T_i)$$

$$= (2220 \text{ J/kg}\cdot\text{K})(0.720 \text{ kg})[0°C - (-10°C)]$$

$$= 15,984 \text{ J} \approx 15.98 \text{ kJ}.$$

Step 2 is to melt the ice (there can be no change in temperature until this step is completed). We now use Eqs. 20-5 and 20-7, finding

$$Q_2 = L_F m = (333 \text{ kJ/kg})(0.720 \text{ kg}) \approx 239.8 \text{ kJ}.$$

Step 3 is to raise the temperature of the now liquid water from $0°C$ to $15°C$. We again use Eq. 20-3, but now with the specific heat c_{liq} of liquid water given in Table 20-1. In this step the initial temperature T_i is $0°C$ and the final temperature T_f is $15°C$. We find

$$Q_3 = c_{liq}m(T_f - T_i)$$

$$= (4190 \text{ J/kg}\cdot\text{K})(0.720 \text{ kg})(15°C - 0°C)$$

$$= 45,252 \text{ J} \approx 45.25 \text{ kJ}.$$

The total heat Q_{tot} required is the sum of the heat required in the three steps:

$$Q_{tot} = Q_1 + Q_2 + Q_3$$

$$= 15.98 \text{ kJ} + 239.8 \text{ kJ} + 45.25 \text{ kJ}$$

$$\approx 300 \text{ kJ}. \qquad \text{(Answer)}$$

Note that the heat required to melt the ice is much larger than the heat required to raise the temperature of either the ice or the liquid water.

b. Suppose that we supply the ice with a total heat of only 210 kJ. What then are the final state and temperature of the water?

SOLUTION From step 1, we know that 15.98 kJ is needed to raise the temperature of the ice to the melting point. The remaining heat Q_{rem} is then 210 kJ $-$ 15.98 kJ, or about 194 kJ. From step 2, we can see that this amount of heat is insufficient to melt all the ice. We can find the mass m of ice that is melted by the heat Q_{rem} by using Eqs. 20-5 and 20-7:

$$m = \frac{Q_{rem}}{L_F} = \frac{194 \text{ kJ}}{333 \text{ kJ/kg}} = 0.583 \text{ kg} \approx 580 \text{ g}.$$

Thus the mass of the ice that remains is 720 g $-$ 580 g, or 140 g. Since not all the ice is melted, the temperature of the ice–liquid water must be $0°C$. Hence we have

$$580 \text{ g water, } 140 \text{ g ice, at } 0°C. \qquad \text{(Answer)}$$

SAMPLE PROBLEM 20-3

A copper slug whose mass m_c is 75 g is heated in a laboratory oven to a temperature T of $312°C$. The slug is then dropped into a glass beaker containing a mass $m_w = 220$ g of water. The effective heat capacity C_b of the beaker is 45 cal/K. The initial temperature T_i of the water and the beaker is $12°C$. What is the final temperature T_f of the slug, the beaker, and the water?

SOLUTION Let us take as our system the *water + beaker + copper slug*. No heat enters or leaves this system, so the algebraic sum of the internal heat transfers that occur must be zero. There are three such transfers:

$$\text{for the water:} \quad Q_w = m_w c_w(T_f - T_i);$$

$$\text{for the beaker:} \quad Q_b = C_b(T_f - T_i);$$

$$\text{for the copper:} \quad Q_c = m_c c_c(T_f - T).$$

The temperature difference is written—in all three cases—as the final temperature minus the initial temperature. We do this even though we realize that Q_w

and Q_b are positive (indicating that heat is added to the initially cool water and beaker) and that Q_c is negative (indicating that heat is released by the initially hot copper slug).

From what we have said above, we must have

$$Q_w + Q_b + Q_c = 0. \qquad (20\text{-}8)$$

Substituting the heat transfer expressions above into Eq. 20-8 yields

$$m_w c_w(T_f - T_i) + C_b(T_f - T_i)$$
$$+ m_c c_c(T_f - T) = 0. \qquad (20\text{-}9)$$

We see that temperatures enter Eq. 20-9 only as differences. Thus because the intervals on the Celsius and the Kelvin scales are identical, we can use either of these scales in this equation. Solving Eq. 20-9 for T_f and substituting, we have

$$T_f = \frac{m_c c_c T + C_b T_i + m_w c_w T_i}{m_w c_w + C_b + m_c c_c}.$$

The numerator is

$$(75 \text{ g})(0.092 \text{ cal/g}\cdot\text{K})(312°\text{C}) + (45 \text{ cal/K})(12°\text{C})$$
$$+ (220 \text{ g})(1.00 \text{ cal/g}\cdot\text{K})(12°\text{C})$$
$$= 5332.8 \text{ cal},$$

and the denominator is

$$(220 \text{ g})(1.00 \text{ cal/g}\cdot\text{K}) + 45 \text{ cal/K}$$
$$+ (75 \text{ g})(0.092 \text{ cal/g}\cdot\text{K})$$
$$= 271.9 \text{ cal/°C}.$$

So, we then have

$$T_f = \frac{5332.8 \text{ cal}}{271.9 \text{ cal/°C}} = 19.6°\text{C} \approx 20°\text{C}. \qquad \text{(Answer)}$$

From the given data you can show that

$$Q_w \approx 1670 \text{ cal}, \quad Q_b \approx 342 \text{ cal}, \quad Q_c \approx -2020 \text{ cal}.$$

Apart from calculation errors due to rounding, the algebraic sum of these three heat transfers is indeed zero, as Eq. 20-8 requires.

20-4 A CLOSER LOOK AT HEAT AND WORK

Here we look in some detail at how heat and work are exchanged between a system and its environment. Let us take as our system a gas confined to a cylinder with a movable piston, as in Fig. 20-3. The pressure of the confined gas is balanced by loading lead shot onto the top of the piston. The walls of the cylinder are made of insulating material that does not allow any heat transfer, but the bottom of the

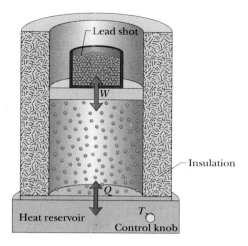

FIGURE 20-3 A gas is confined to a cylinder with a movable piston. Heat can be added to, or withdrawn from, the gas by regulating the temperature T of the adjustable heat reservoir.

cylinder rests on a heat reservoir (a hot plate, if you will) whose temperature T you can control by turning a knob.

The system starts from an *initial state i*, described by a pressure p_i, a volume V_i, and a temperature T_i. You want to change the system to a *final state f*, described by a pressure p_f, a volume V_f, and a temperature T_f. The procedure by which you change the system from its initial state to its final state is called a *thermodynamic process*. During such a process, heat may be transferred into the system from the reservoir or vice versa, and work is done by the system to raise or lower the loaded piston. That work is positive if the gas causes the piston to rise, and negative if the gas allows the piston to fall. We assume that all such changes occur slowly, with the result that the system is always in (approximate) thermodynamic equilibrium.

Suppose that you remove a little lead shot from the piston of Fig. 20-3, allowing the gas to push the piston and remaining shot upward through a differential displacement $d\mathbf{s}$ with an upward force $\mathbf{F}$. Since the displacement is tiny, we can assume that $\mathbf{F}$ is constant during the displacement. Then $\mathbf{F}$ has a magnitude that is equal to pA, where p is the pressure of the gas and A is the face area of the piston. The differential work dW done by the gas during the displacement is

$$dW = \mathbf{F}\cdot d\mathbf{s} = (pA)(ds) = (p)(A\,ds)$$
$$= p\,dV, \qquad (20\text{-}10)$$

in which dV is the differential change in the volume of the gas owing to the movement of the piston. When you have removed enough shot to change the volume of the gas from V_i to V_f, the total work done by the gas is

$$W = \int dW = \int_{V_i}^{V_f} p \, dV. \qquad (20\text{-}11)$$

To evaluate the integral in Eq. 20-11 directly, we would need to know how the pressure varies with a change in volume for the actual process by which the system changes from state i to state f.

There are actually many ways to take the gas from state i to state f. One way is shown in Fig. 20-4a, which is a plot of the pressure of the gas versus its volume and which is called a p–V diagram. The curve gives the variation of p with V. The integral of Eq. 20-11 (and thus the work W done by the gas) is represented by the shaded area under the curve between points i and f. That work is positive, owing to the fact that the gas increased its volume by forcing the piston upward.

Another way to get from state i to state f is shown in Fig. 20-4b: there the change takes place in two steps—the first from state i to state a, and the second from state a to state f.

Step ia of this process is carried out at constant pressure, which means that you leave undisturbed the lead shot that rides on top of the piston in Fig. 20-3. You cause the volume to increase (from V_i to V_f) by slowly turning up the temperature control knob, raising the temperature of the gas to some higher value T_a. During this process, work is done by the expanding gas (to lift the loaded piston) and heat is added to the system from the heat reservoir (in response to the arbitrarily small temperature differences that you create as you turn up the temperature). This heat is positive because it is added to the system.

Step af of the process of Fig. 20-4b is carried out at constant volume, so you must wedge the piston, preventing it from moving. And then you must use the control knob to reduce the temperature from T_a to its final, lower value T_f. Because the reservoir is then cooler than the gas, heat is lost by the gas to the reservoir. Thus the heat involved in this step is negative.

For the overall process iaf, the work W, which is positive and is carried out only during step ia, is represented by the shaded area under the curve. Heat is transferred during both steps ia and af, being positive during the first step and negative (but smaller in magnitude) during the second.

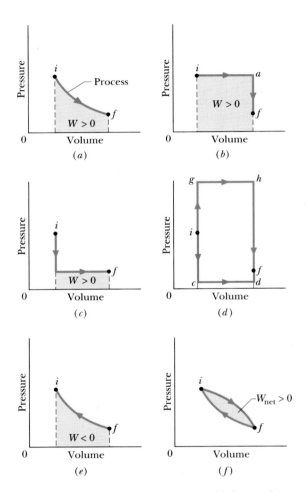

FIGURE 20-4 (*a*) The system of Fig. 20-3 goes from an *initial state i* to a *final state f* by means of a *thermodynamic process*. The area marked *W* represents the work done *by* the system during this process. The work is positive, because the process proceeds to the right on the graph. (*b*) Another process for moving between the same two states; the work is now greater than that in (*a*). (*c*) Still another process, requiring less (positive) work. (*d*) The work can be made as small as you like (path *icdf*) or as large as you like (path *ighf*). (*e*) When the volume is reduced (by some external force), the work done *by* the system is negative. (*f*) The net work done by the system during a (closed) cycle is represented by the enclosed area, which is the difference in the areas beneath the two curves that make up the cycle.

Figure 20-4*c* shows a process in which the two steps above are carried out in reverse order. The work *W* in this case is smaller than for Fig. 20-4*b*, as is the net heat absorbed. Figure 20-4*d* suggests that you can make the work done as small as you want (by following a path like *icdf*) or as large as you want (by following a path like *ighf*).

To sum up: a system can be taken from a given initial state to a given final state by an infinite number of processes. In general, the work *W* and the heat *Q* will have different values for each of these processes. We say that heat and work are *path-dependent* quantities.

Figure 20-4*e* shows an example in which negative work is done by a system, as some external force compresses the system, reducing its volume. The absolute value of the work done is still equal to the area beneath the curve, but because the volume is *compressed,* that work is negative.

Figure 20-4*f* shows a *thermodynamic cycle* in which the system is taken from some initial state *i* to some other state *f* and then back to *i*. The net work done by the system during the cycle is the sum of the positive work done during the expansion and the negative work done during the compression. In Fig. 20-4*f*, the net work happens to be positive, because the area under the expansion curve is more than the area under the compression curve.

20-5 THE FIRST LAW OF THERMODYNAMICS

You have just seen that when a system changes from a given initial state to a given final state, both *W* and *Q* depend on the nature of the process. Experimentally, however, we find a surprising thing. *The quantity $Q - W$ is the same for all processes.* It depends only on the initial and final states and does not depend at all on how the system gets from one to the other. All other combinations of *Q* and *W*, including *Q* alone, *W* alone, $Q + W$, and $Q - 2W$, are *path dependent;* only the quantity $Q - W$ is not.

$Q - W$ must represent a change in some intrinsic property of the system. We call this property the *internal energy* E_{int} and we write

$$\Delta E_{int} = E_{int,f} - E_{int,i}$$
$$= Q - W \qquad \text{(first law).} \quad (20\text{-}12)$$

Equation 20-12 is the **first law of thermodynamics.** If the thermodynamic system undergoes only a differential change, we can write the first law as*

$$dE_{int} = dQ - dW \qquad \text{(first law).} \quad (20\text{-}13)$$

We saw in Chapter 19 that the essential message of the zeroth law of thermodynamics is: "every thermodynamic system that is in thermal equilibrium has an important physical property called its *temperature T*." The essential message of the first law is: "every thermodynamic system that is in thermal equilibrium has an important physical property called its *internal energy* E_{int}."

In Chapter 8, we discussed the principle of energy conservation as it applies to isolated systems, that is, to systems in which no energy enters or leaves the system. The first law of thermodynamics is an extension of that principle to systems that are *not* isolated. In such cases, energy may be transferred through the system boundary either as work *W* or as heat *Q*. In stating the first law of thermodynamics, we assume that there are no changes in the kinetic energy or the potential energy of the system as a whole; that is, $\Delta K = \Delta U = 0$.

Before this chapter, the term *work* and the symbol *W* always meant the work done *on* a system. But starting with Eq. 20-10 and continuing through the next several chapters about thermodynamics, we focus on the work done *by* a system, such as the gas in Fig. 20-3.

The work done *on* a system is always the negative of the work done *by* the system. So if we rewrite Eq. 20-12 in terms of the work done *on* the system, we have $\Delta E_{int} = Q + W$. This tells us the following: the internal energy tends to increase if heat is absorbed by the system or if positive work is done on the system. Conversely, the internal energy tends to decrease if heat is lost by the system or if negative work is done on the system.

---

*Here *dQ* and *dW*, unlike dE_{int}, are not true differentials. That is, there are no such functions as $Q(p, V)$ and $W(p, V)$ that depend only on the state of the system. *dQ* and *dW* are called *inexact differentials* and are usually represented by the symbols đ*Q* and đ*W*. For our purposes, we can treat them simply as infinitesimally small energy transfers.

20-6 SOME SPECIAL CASES OF THE FIRST LAW OF THERMODYNAMICS

Here we look at four different thermodynamic processes, in each of which a certain restriction is imposed on the system. We then see what consequences follow when we apply the first law of thermodynamics to the process.

1. *Adiabatic processes.* During an adiabatic process the system is so well insulated that *no transfer of heat* occurs between it and its environment. Putting $Q = 0$ in the first law (Eq. 20-12) leads to

$$\Delta E_{\text{int}} = -W \qquad \text{(adiabatic process).} \qquad (20\text{-}14)$$

This tells us that if work is done *by* the system (that is, if W is positive), there must be a decrease in the internal energy of the system. Conversely, if work is done *on* the system (that is, if W is negative), there must be an increase in the internal energy of the system.

For a gas, an increase in internal energy means an increase in temperature, and conversely. The temperature of a bicycle pump increases as it is used, due to the adiabatic compression of air within the pump.

Figure 20-5 shows an idealized adiabatic process. Heat cannot enter or leave the system because of the insulation. Thus the only interaction permitted between the system and its environment is the performance of work. If we remove shot from the

piston and allow the gas to expand, the work done by the system (the gas) is positive and the temperature of the gas decreases. If, instead, we add shot to the piston and compress the gas, the work done by the system is negative and the temperature of the gas increases.

There is a second way of ensuring that heat transfer does not take place during a thermodynamic process: carry it out very rapidly so that there is no time for appreciable heat transfer. Thus the compressions and expansions of air as a sound wave passes through are adiabatic. There is simply no time for heat to transfer back and forth in synchronism with the rapidly oscillating sound wave. The compressions and expansions of steam in the cylinder of a steam engine, or of the hot gases in the cylinders of an internal combustion engine, are also essentially adiabatic for the same reason.

2. *Constant-volume processes.* If the volume of a system (such as a gas) is held constant, that system can do no work. Putting $W = 0$ into the first law (Eq. 20-12) yields

$$\Delta E_{\text{int}} = Q \qquad \text{(constant-volume process).} \qquad (20\text{-}15)$$

Thus if heat is added to a system (that is, if Q is positive), the internal energy of the system increases. Conversely, if heat is removed during the process (that is, if Q is negative), the internal energy of the system must decrease.

3. *Cyclical processes.* There are processes in which, after certain interchanges of heat and work, the system is restored to its initial state. In that case, no intrinsic property of the system—including its internal energy—can possibly change. Putting $\Delta E_{\text{int}} = 0$ in the first law (Eq. 20-12) yields

$$Q = W \qquad \text{(cyclical process).} \qquad (20\text{-}16)$$

Thus the net work done during the process must exactly equal the net amount of heat transferred; the store of internal energy of the system remains unchanged. Cyclical processes form a closed loop on a pressure–volume plot, as in Fig. 20-4*f*. We shall discuss such processes in some detail in Chapter 22.

4. *Free expansion processes.* These are adiabatic processes in which no work is done on or by the system. Thus $Q = W = 0$ and the first law requires that

$$\Delta E_{\text{int}} = 0 \qquad \text{(free expansion).} \qquad (20\text{-}17)$$

Figure 20-6 shows how such an expansion can

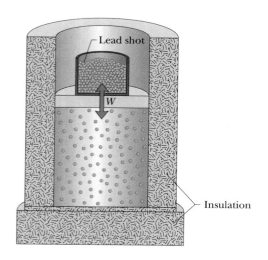

FIGURE 20-5 An adiabatic expansion can be carried out by slowly removing lead shot from the top of the piston. Adding lead shot reverses the process at any stage.

FIGURE 20-6 The initial stage of a free-expansion process. After the stopcock is opened, the gas eventually reaches an equilibrium final state, filling both chambers.

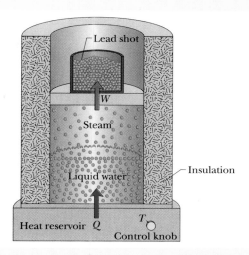

FIGURE 20-7 Sample Problem 20-4. Water boiling at constant pressure. Heat is added from the reservoir until the liquid water has changed completely into steam. Work is done by the expanding gas as it lifts the loaded piston.

be carried out. A gas is confined by a closed stopcock to half of an insulated double chamber, the other half being evacuated. The process consists of opening the stopcock and waiting until equilibrium is established, with the gas now filling both halves of the double chamber. No heat is transferred because of the insulation. No work is done because the expanding gas rushes into an evacuated space, its motion unopposed by any counteracting pressure.

A free expansion differs from all other processes that we have considered so far in that there is no way to carry it out slowly. Thus although the system is in thermal equilibrium in its initial and its final states, it is *not* in equilibrium during the process. At intermediate states, the temperature, pressure, and volume do not have unique values. Thus we cannot plot the course of the expansion on a pressure–volume diagram. All that can be done is to plot the initial and final states.

Table 20-3 summarizes the characteristics of the processes of this section.

TABLE 20-3
THE FIRST LAW OF THERMODYNAMICS:
FOUR SPECIAL CASES

The Law:	$\Delta E_{int} = Q - W$	(Eq. 20-12)

PROCESS	RESTRICTION	CONSEQUENCE
Adiabatic	$Q = 0$	$\Delta E_{int} = -W$
Constant volume	$W = 0$	$\Delta E_{int} = Q$
Closed cycle	$\Delta E_{int} = 0$	$Q = W$
Free expansion	$Q = W = 0$	$\Delta E_{int} = 0$

SAMPLE PROBLEM 20-4

Let 1.00 kg of liquid water at 100°C be converted to steam at 100°C by boiling at standard atmospheric pressure; see Fig. 20-7. The volume changes from an initial value of 1.00×10^{-3} m³ as a liquid to 1.671 m³ as steam.

a. How much work is done by the system during this process?

SOLUTION The work is given by Eq. 20-11. Because the pressure is constant (at 1.00 atm or 1.01×10^5 Pa) during the boiling process, we can take p outside the integral, obtaining

$$W = \int_{V_i}^{V_f} p \, dV = p \int_{V_i}^{V_f} dV = p(V_f - V_i)$$

$$= (1.01 \times 10^5 \text{ Pa})(1.671 \text{ m}^3 - 1.00 \times 10^{-3} \text{ m}^3)$$

$$= 1.69 \times 10^5 \text{ J} = 169 \text{ kJ}. \qquad \text{(Answer)}$$

The result is positive, indicating that work is done *by* the system on its environment, in lifting the weighted piston of Fig. 20-7.

b. How much heat must be added to the system during the process?

SOLUTION Since there is no temperature change, but

only a phase change, we use Eqs. 20-5 and 20-6:

$$Q = L_V m = (2260 \text{ kJ/kg})(1.00 \text{ kg})$$

$$= 2260 \text{ kJ}. \qquad \text{(Answer)}$$

The result is positive, which indicates that heat is *added* to the system, as we expect.

c. What is the change in the internal energy of the system during the boiling process?

SOLUTION   We find this from the first law (Eq. 20-12):

$$\Delta E_{\text{int}} = Q - W = 2260 \text{ kJ} - 169 \text{ kJ}$$

$$\approx 2090 \text{ kJ} = 2.09 \text{ MJ}. \qquad \text{(Answer)}$$

This quantity is positive, indicating that the internal energy of the system has increased during the boiling process. This energy represents the internal work done in overcoming the strong attraction that the H_2O molecules have for each other in the liquid state.

We see that, when water is boiled, about 7.5% ($= 169 \text{ kJ}/2260 \text{ kJ}$) of the added heat goes into external work in pushing back the atmosphere. The rest goes into internal energy that is added to the system.

20-7   THE TRANSFER OF HEAT

We have discussed the transfer of heat between a system and its environment but we have not yet described how that transfer takes place. There are three transfer mechanisms: conduction, convection, and radiation.

Conduction

If you leave a poker in a fire for any length of time, its handle will get hot. Energy is transferred from the fire to the handle by **conduction** along the length of the metal shaft. The vibration amplitudes of the atoms and electrons of the metal at the hot end of the shaft take on relatively large values, reflecting the elevated temperature of their environment. These increased vibrational amplitudes are passed along the shaft, from atom to atom, during collisions between adjacent atoms. In this way, a region of rising temperature extends itself along the shaft to your hand.

Consider a slab of face area A and thickness L, whose faces are maintained at temperatures T_H and T_C as in Fig. 20-8. Let Q be the heat that is transferred through the slab, from its hot face to its cold face, in time t. Experiment shows that the rate of

heat transfer H (the amount in time t) is given by

$$H = \frac{Q}{t} = kA\frac{T_H - T_C}{L}, \qquad \text{(20-18)}$$

in which k, called the *thermal conductivity*, is a constant that depends on the material of which the slab is made. Large values of k define good heat conductors, and conversely.

Thermal Resistance to Conduction (*R*-Value)

If you are interested in insulating your house or in keeping cola cans cold on a picnic, you are more concerned with poor heat conductors than with good ones. For this reason, the concept of *thermal resistance R* has been introduced into engineering practice. The *R*-value of a slab of thickness L is defined as

$$R = \frac{L}{k}. \qquad \text{(20-19)}$$

Thus the lower the thermal conductivity of the material of which a slab is made, the higher the *R*-value of the slab. Note that R is a property attributed to a slab of a specified thickness, not to a material. The commonly used unit for R (which, in this country at least, is almost never stated) is the square foot-degree Fahrenheit-hour per British thermal unit ($\text{ft}^2 \cdot {}^\circ\text{F} \cdot \text{h/Btu}$). (Now you know why it is rarely stated.)

Combining Eqs. 20-18 and 20-19 leads to

$$H = A\frac{T_H - T_C}{R}, \qquad \text{(20-20)}$$

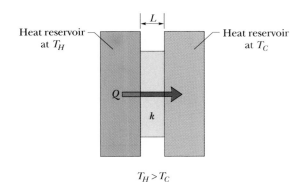

FIGURE 20-8 Thermal conduction. Heat is transferred from a reservoir at temperature T_H to a cooler reservoir at temperature T_C through a conducting slab of thickness L and thermal conductivity k.

which allows one to calculate the rate of heat flow through a slab if its *R*-value, its area, and the temperature difference between its faces are known.

In reasonably severe climates, it is recommended that the ceilings of single-family dwellings be insulated to the level of *R*-30. From Eq. 20-20 we see that this means that an average square foot of such a ceiling would lose heat by conduction at a rate of $\frac{1}{30}$ Btu/h for every 1°F difference in temperature between the two faces of the ceiling.

Table 20-4 shows the thermal conductivities of various materials and the *R*-values calculated for 1-in. slabs of those materials. The use of *R*-values is normally restricted to commercial insulating materials, but values for a wide range of materials are shown for comparison.

Inspection of the table shows why the Sierra Club recommends that hikers use a stainless steel (rather than an aluminum) cup for hot coffee. Stagnant air has an *R*-value as great as that of any of the commercial building materials shown. In fact, many such materials owe their effectiveness to their ability to entrap isolated pockets of air.

TABLE 20-4
SOME THERMAL CONDUCTIVITIES AND *R*-VALUES[a]

	CONDUCTIVITY, *k* (W/m·K)	*R*-VALUE (ft²·°F·h/Btu)
Metals		
Stainless steel	14	0.010
Lead	35	0.0041
Aluminum	235	0.00061
Copper	401	0.00036
Silver	428	0.00034
Gases		
Air (dry)	0.026	5.5
Helium	0.15	0.96
Hydrogen	0.18	0.80
Building Materials		
Polyurethane foam	0.024	5.9
Rock wool	0.043	3.3
Fiberglass	0.048	3.0
White pine	0.11	1.3
Window glass	1.0	0.14

[a] Conductivities change somewhat with temperature. The given values are at room temperature. Note that the values of *k* are given in SI units and those of *R* in the customary British units. The *R*-values are for a slab of 1 in. thickness.

In cold climates, double or triple pane windows are often installed to reduce heat loss. The glass itself is not a particularly good insulator. The air (or other gas) between the panes would insulate well if it were stagnant (it is not, because the temperature difference across the width of the air layer causes the air to circulate and thereby transport heat from the warmer pane to the cooler one). The insulating value of a window comes almost entirely from the thin *boundary layer* (see Section 16-12) of stagnant air that clings to each surface of a pane. Doubling the number of panes doubles the number of such insulating boundary layers.

You can further deduce from the table that, to build a slab insulating to *R*-30, you could use either a 5.1-in. thickness of polyurethane foam, 23 in. of white pine, 18 ft of window glass, or 1.4 mi of silver!

Conduction Through a Composite Slab

Figure 20-9 shows a composite slab, consisting of two materials having different thicknesses L_1 and L_2, and different thermal conductivities, k_1 and k_2. The temperatures of the outer surfaces of the slab are T_H and T_C. Each face of the slab has area *A*. Let us derive an expression for the rate of heat transfer through the slab under the assumption that the transfer is a *steady-state* process; that is, the temperatures everywhere in the slab and the rate of heat transfer have been established and no longer change with time.

In the steady state, the rates of heat transfer through the two materials are the same. This is the same as saying that the heat conducted through one material in a certain time must be the same as that

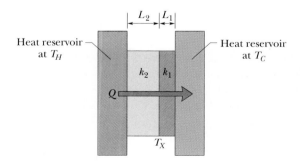

FIGURE 20-9 Heat is transferred through a composite slab made up of two different materials with different thicknesses and different thermal conductivities. The temperature at the interface of the two materials is T_X.

conducted through the other material in the same amount of time. Were this not true, temperatures in the slab would be changing and we would not have a steady-state situation. Letting T_X be the temperature of the interface between the two materials, we can now use Eq. 20-18 to write

$$H = \frac{k_2 A (T_H - T_X)}{L_2} = \frac{k_1 A (T_X - T_C)}{L_1}. \quad (20\text{-}21)$$

Solving Eq. 20-21 for T_X yields, after a little algebra,

$$T_X = \frac{k_1 L_2 T_C + k_2 L_1 T_H}{k_1 L_2 + k_2 L_1}. \quad (20\text{-}22)$$

Substituting this expression for T_X into either equality of Eq. 20-21 yields

$$H = \frac{A (T_H - T_C)}{(L_1/k_1) + (L_2/k_2)}. \quad (20\text{-}23)$$

Equation 20-19 reminds us that $L/k = R$.

We can extend Eq. 20-23 to any number of materials in the form

$$H = \frac{A (T_H - T_C)}{\Sigma (L/k)} = \frac{A (T_H - T_C)}{\Sigma R}. \quad (20\text{-}24)$$

The summation sign in the denominators tells us to add the values for all the materials.

Convection

When you look at the flame of a candle or a match, you are watching heat energy being transported upward by **convection.** Heat transfer by convection

Ever-changing convection patterns can be seen in hot black coffee: A fine mist, which scatters light and appears to be white, lies just over the regions of rising hot liquid. Narrow black lines, which lack the mist, show where liquid that has cooled on the surface is descending.

occurs when a fluid, such as air or water, is in contact with an object whose temperature is higher than that of the fluid. The temperature of the fluid that is in contact with the hot object increases, and (in most cases) the fluid expands. Being less dense than the surrounding cooler fluid, it rises because of buoyant forces (see Fig. 20-10). The surrounding cooler fluid falls to take the place of the rising warmer fluid, and a convective circulation is set up.

Convection is part of many natural processes. Atmospheric convection plays a fundamental role in determining global climate patterns and daily weather variations. Glider pilots and birds alike seek thermals (currents of warm air) that, rising from the warmer earth beneath, keep them aloft. Huge energy transfers take place within the oceans by the same process. Finally, energy is transported to the surface of the sun from the nuclear furnace at its core by enormous convection cells, the tops of which can be seen as a granulation of the solar surface.

Radiation

Energy is carried from the sun to us by electromagnetic waves that travel freely through the near vacuum of the intervening space. If you stand near a bonfire or an open fireplace, you are warmed by the

FIGURE 20-10 A Dartmouth football rally is illuminated by a fierce bonfire. As heated air and hot gases from the fire rise, cooler air flows into the base of the fire.

FIGURE 20-11 A false-color thermogram reveals the rate at which energy is radiated by houses along a street. The rates, from largest to smallest, are color coded as white, red, pink, blue, and black. Note how you can tell where there is insulation in the walls, heavy curtains over windows, and higher air temperatures at the ceilings on the second floors.

same process. All objects emit such electromagnetic radiation simply because their temperature is above absolute zero (see Fig. 20-11), and all objects absorb some of the radiation that falls on them from other objects. The average temperature of our Earth, for example, stays at about 300 K because at that temperature the Earth radiates energy into space at the same rate that it receives radiant energy from the sun (see Fig. 20-12). If the temperature of the Earth were—by some miracle—to change suddenly from 300 K to 280 K or to 320 K, it would quickly either warm up or cool down to 300 K, restoring its nice thermal balance.

When you stand in bright sunlight, it warms you because your skin and clothing absorb the light. Research has shown that a black Bedouin robe absorbs much more sunlight than a white Bedouin robe, to the extent that its temperature can be 6°C higher than that of the white robe. Why, then, would someone who must avoid overheating so as to survive in the harsh desert wear a black robe?

The answer is that the hotter black robe warms the air inside the robe. That air then rises and leaves through the porous fabric, while external air is drawn into the robe through its open bottom (Fig. 20-13). Thus the black fabric enhances air circulation under the robe, and that keeps the Bedouin from getting any hotter than a person in a white robe. In fact, it may even make him feel more comfortable: he has a continuous breeze blowing past his body.

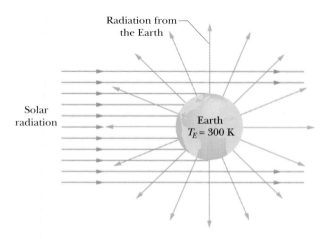

FIGURE 20-12 Solar radiation is intercepted by the Earth and is (largely) absorbed. The temperature T_E of the Earth adjusts itself to a value at which the Earth's heat loss by radiation is just equal to the solar heat it absorbs.

FIGURE 20-13 Convection up through the hotter black robe is more vigorous than that up through the cooler white robe. (After "Why Do Bedouins Wear Black Robes in Hot Deserts?" by A. Shkolnik, C. R. Taylor, V. Finch, and A. Borut, *Nature*, Vol. 283, January 24, 1980, pp. 373–374.)

SAMPLE PROBLEM 20-5

A composite slab (see Fig. 20-9) whose area A is 26 ft² is made up of 2.0 in. of rock wool and 0.75 in. of white pine. The temperature difference between the faces of the slab is 65°F. What is the rate of heat transfer through the slab?

SOLUTION The R-values given in Table 20-4 are for 1-in. slabs. Thus the R-value for the rock wool is 3.3×2.0 or 6.6 ft²·°F·h/Btu. For the wood it is 1.3×0.75 or 0.98, in the same units. The composite slab thus has an R-value of $6.6 + 0.98$ or 7.58 ft²·°F·h/Btu. Substitution into Eq. 20-24 yields

$$H = \frac{A(T_H - T_C)}{\Sigma R} = \frac{(26 \text{ ft}^2)(65°F)}{7.58 \text{ ft}^2 \cdot °F \cdot h/Btu}$$

$$= 223 \text{ Btu/h} \approx 220 \text{ Btu/h } (= 65 \text{ W}). \quad \text{(Answer)}$$

Thus at this temperature difference, each such insulating slab would transmit heat continuously to the outdoors at the rate of 65 W.

SAMPLE PROBLEM 20-6

Figure 20-14 shows the cross section of a wall made of white pine of thickness L_a and brick of thickness L_d $(= 2.0L_a)$, sandwiching two layers of unknown material with identical thicknesses and thermal conductivities. The thermal conductivity of the pine is k_a and that of the brick is k_d $(= 5.0k_a)$. The face area A of the wall is unknown. Heat conduction through the wall has reached the steady state, with the only known interface

FIGURE 20-14 Sample Problem 20-6. A wall of four layers through which there is steady-state heat transfer.

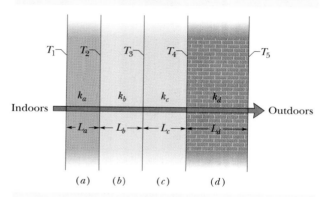

(a) (b) (c) (d)

temperatures being $T_1 = 25°C$, $T_2 = 20°C$, and $T_5 = -10°C$.

a. What is interface temperature T_4?

SOLUTION We cannot find T_4 by simply applying Eq. 20-18 layer by layer, starting with the pine and working our way rightward, because we do not know enough about the intermediate layers. However, since the heat conduction has reached the steady state, we know that the rate of conduction H_a through the pine must equal the rate of conduction H_d through the brick. From Eq. 20-18 and Fig. 20-14, we can write these rates as

$$H_a = k_a A \frac{T_1 - T_2}{L_a} \quad \text{and} \quad H_d = k_d A \frac{T_4 - T_5}{L_d}.$$

Setting $H_a = H_d$ and solving for T_4 yield

$$T_4 = \frac{k_a L_d}{k_d L_a}(T_1 - T_2) + T_5.$$

Letting $L_d = 2.0L_a$ and $k_d = 5.0k_a$, and inserting the known temperatures, we find

$$T_4 = \frac{k_a(2.0L_a)}{(5.0k_a)L_a}(25°C - 20°C) + (-10°C)$$

$$= -8.0°C. \quad \text{(Answer)}$$

b. What is interface temperature T_3?

SOLUTION Now that we know T_4, we can find T_3, even though we know little about the intermediate layers. (In fact, at this point you might be able to guess the answer.) Since the heat conduction process is in the steady state, the rate of conduction H_b through layer b is equal to the rate of conduction H_c through layer c. Then, from Eq. 20-18,

$$k_b A \frac{T_2 - T_3}{L_b} = k_c A \frac{T_3 - T_4}{L_c}.$$

Because the thermal conductivities k_b and k_c of the layers are equal and so are their thicknesses L_b and L_c, we have

$$T_2 - T_3 = T_3 - T_4,$$

which gives us

$$T_3 = \frac{T_2 + T_4}{2} = \frac{20°C + (-8.0°C)}{2}$$

$$= 6.0°C. \quad \text{(Answer)}$$

This shows us that, since the intermediate layers have identical thermal conductivities, a point midway across them has a temperature that is midway between the temperatures of their outside surfaces.

REVIEW & SUMMARY

Heat

Heat Q is the energy that is transferred between a system and its environment because of a temperature difference between them. It can be measured in **joules** (J), **calories** (cal), **kilocalories** (Cal or kcal), or **British thermal units** (Btu), with

$$1 \text{ Cal} = 10^3 \text{ cal} = 3.969 \text{ Btu} = 4186 \text{ J.} \quad (20\text{-}1)$$

Heat Capacity and Specific Heat

If heat Q is added to an object with mass m, the temperature change $T_f - T_i$ is related to Q by

$$Q = C(T_f - T_i), \quad (20\text{-}2)$$

in which C is the **heat capacity** of the object. The **specific heat** c of a material (the heat capacity per unit mass; see Table 20-1) is defined by

$$Q = cm(T_f - T_i). \quad (20\text{-}3)$$

The **molar specific heat** (the heat capacity per mole, or per 6.02×10^{23} elementary units of a material) shows an interesting regularity that helps us to understand the mechanisms involved in heat absorption. Measured specific heats depend on the conditions of measurement (under constant pressure or constant volume, for example), and these must be specified clearly.

Heat of Transformation

Heat supplied to a material may change the material's physical state, for example, from solid to liquid or from liquid to gas. The amount of heat required per unit mass for a particular material is its **heat of transformation** L; see Table 20-2. Thus

$$Q = Lm. \quad (20\text{-}5)$$

The **heat of vaporization** L_V is the amount of energy per unit mass that must be added to vaporize a liquid or that must be removed to condense a gas. The **heat of fusion** L_F is the amount of energy per unit mass that must be added to melt a solid or that must be removed to freeze a liquid.

Work Associated with Volume Change

A system may also exchange energy with its surroundings through work. The amount of work W done *by* a system as it expands or contracts from an initial volume V_i to a final volume V_f may be computed with

$$W = \int dW = \int_{V_i}^{V_f} p \, dV. \quad (20\text{-}11)$$

The integration is necessary because the presssure p may vary during the volume change. The work W may also be computed as the area under the curve of p versus V representing the change (see Fig. 20-4).

First Law of Thermodynamics

The principle of conservation of energy for a sample of material exchanging energy with its surroundings by means of work and heat is expressed in the **first law of thermodynamics**, which may assume either of the forms

$$\Delta E_{\text{int}} = E_{\text{int}, f} - E_{\text{int}, i} = Q - W \quad \text{(first law)} \quad (20\text{-}12)$$

or

$$dE_{\text{int}} = dQ - dW \quad \text{(first law).} \quad (20\text{-}13)$$

E_{int} represents the internal energy of the material, which depends only on its state (temperature, pressure, and volume). Q represents the heat exchanged by the system with its surroundings; Q is positive if the system gains heat and negative if the system loses heat. W is the work done *by* the system; W is positive if the system expands against some external force exerted by the surroundings, and negative if the system contracts because of some external force.

Q and W Are Path Dependent; E_{int} Is Not

During a thermodynamic process, the heat exchanged between a system and its surroundings and the work done by the system depend on the details of that process: they are path dependent. However, the change in the internal energy of the system does not depend on the process; it depends only on the initial and final states of the system.

Applications of the First Law

The first law of thermodynamics finds application in several special cases:

$$\text{adiabatic processes:} \quad Q = 0, \quad \Delta E_{\text{int}} = -W$$
$$\text{constant-volume processes:} \quad W = 0, \quad \Delta E_{\text{int}} = Q$$
$$\text{cyclical processes:} \quad \Delta E_{\text{int}} = 0, \quad Q = W$$
$$\text{free expansion processes:} \quad Q = W = \Delta E_{\text{int}} = 0.$$

Conduction, Convection, and Radiation

The rate H at which heat is conducted through a slab whose faces are maintained at temperatures T_H and T_C is

$$H = \frac{Q}{t} = kA \frac{T_H - T_C}{L}, \quad (20\text{-}18)$$

in which A and L are the face area and length of the slab, and k is the thermal conductivity of the material (see Fig. 20-8). For commercial building and insulating materials formed into slabs of thickness L, we often use the thermal resistance R (the R-value with $R = L/k$).

Convection occurs when temperature differences cause motion within a fluid which transfers heat. *Radiation* is heat transfer via the emission of electromagnetic energy. All objects radiate energy, the amount increasing with increasing temperature.

QUESTIONS

1. Temperature and heat are often confused, as in, "bake in an oven at moderate heat." By example, distinguish between these two concepts as carefully as you can.

2. Give an example of a process in which no heat is transferred to or from a system but the temperature of the system changes.

3. Can heat be considered a form of stored (or potential) energy? Would such an interpretation contradict the concept of heat as energy in the process of transfer because of a temperature difference?

4. Heat can be added to a substance without causing the temperature of the substance to rise. Does this contradict the concept of heat as energy in the process of transfer because of a temperature difference?

5. Why must heat be supplied to melt ice when, after all, the temperature doesn't change?

6. Explain the fact that the presence of a large body of water nearby, such as a sea or ocean, tends to moderate the temperature extremes of the climate on adjacent land.

7. An electric fan not only does not cool the air it circulates but heats it slightly. How then can it cool you?

8. Both heat conduction and wave propagation involve the transfer of energy. Is there any difference in principle between these two phenomena? Explain.

9. When a hot object warms a cool one, are their temperature changes equal in magnitude? Give examples.

10. A block of wood and a block of metal are at the *same* temperature. When the blocks feel cold, the metal feels colder than the wood; when the blocks feel hot, the metal feels hotter than the wood. Explain. At what temperature will the blocks feel equally cold or hot?

11. How can you best use a spoon to cool a cup of coffee? Stirring—which involves doing work—would seem to heat the coffee rather than to cool it.

12. How does a layer of snow protect plants during cold weather? During freezing spells, citrus growers in Florida often spray their fruit with water, hoping that it will freeze. How does that help?

13. Explain the wind-chill effect that radio and TV meteorologists mention during cold weather.

14. You put your hand in a hot oven to remove a casserole and burn your fingers on the hot dish. However, the air in the oven is at the same temperature as the casserole dish but it does not burn your fingers. Why not?

15. Why is thicker insulation used in an attic than in the walls of a house?

16. Is ice always at 0°C? Can it be colder? Can it be warmer? What about an ice–water mixture?

17. Explain why your finger sticks to a metal ice tray just taken from the refrigerator.

18. The water in a kettle makes quite a bubbling noise while it is being heated to boiling. Once it starts boiling, however, it does so quietly. What is the explanation? (*Hint:* Think of the fate of a bubble of vapor rising from the bottom of the kettle before the water is uniformly heated.)

19. On a winter day the temperature of the inside surface of a house wall is much lower than room temperature, and that of the wall's outside surface is much higher than the outdoor temperature. Explain.

20. The physiological mechanisms that maintain a person's internal temperature operate within a limited range of external temperature. Explain how this range can be extended at each extreme by the use of clothes.

21. What requirements for thermal conductivity, specific heat, and coefficient of expansion should be satisfied by a material to be used in cooking utensils?

22. Suppose that, for some strange reason, the single-pane glass windows of a house are replaced with sheets of aluminum of the same thickness as the glass. How is the rate at which heat is conducted through these window areas affected?

23. Is the temperature of an isolated system (no interaction with the environment) conserved? Explain.

24. Is heat the same as internal energy? If not, give an example in which a system's internal energy changes without a flow of heat across the system's boundary.

25. Can you tell whether the internal energy of a body was acquired by heat transfer or by performance of work?

26. If only the pressure and volume of a system are given, is the temperature always uniquely determined?

27. Discuss the process by which water freezes, from the point of view of the first law of thermodynamics. Remember that ice occupies a greater volume than an equal mass of water.

28. A thermos bottle contains coffee. The thermos bottle is vigorously shaken. Consider the coffee as the system. (a) Does its temperature rise? (b) Has heat been added to it? (c) Has work been done on it? (d) Has its internal energy changed?

29. We have seen that "energy conservation" is a universal law of nature. At the same time national leaders urge "energy conservation" upon us (for example, driving slower). Explain the two quite different meanings of these words.

30. Can heat be transferred through matter by radiation? If so, give an example. If not, explain why.

31. Why does stainless steel cookware often have a layer of copper or aluminum on the bottom?

32. Consider that heat can be transferred by convection and radiation, as well as by conduction, and explain why a

thermos bottle is double-walled, evacuated, and silvered (like a mirror).

33. A lake freezes first at its upper surface. Is convection involved? What about conduction and radiation?

34. You put two uncovered pails of water, one containing hot water and one containing an equal amount of warm water, outside in below-freezing weather. The pail with the hot water may actually develop ice first. Why? What would happen if you covered the pails?

EXERCISES & PROBLEMS

SECTION 20-3   THE ABSORPTION OF HEAT BY SOLIDS AND LIQUIDS

1E. It is possible to melt ice by rubbing one block of it against another. How much work, in joules, would you have to do to get 1.00 g of ice to melt?

2E. A certain substance has a mass per mole of 50 g/mol. When 314 J of heat is added to a 30.0-g sample of this material, its temperature rises from 25.0 to 45.0°C. (a) What is the specific heat of this substance? (b) How many moles of the substance are present? (c) What is the molar specific heat of the substance?

3E. In a certain solar house, energy from the sun is stored in barrels filled with water. In a particular winter stretch of five cloudy days, 1.00×10^6 kcal is needed to maintain the inside of the house at 22.0°C. Assuming that the water in the barrels is at 50.0°C and that the water has a density of 1.00×10^3 kg/m³, what volume of water is required?

4E. A diet doctor encourages dieting by drinking ice water. His theory is that the body must burn off enough fat to raise the temperature of the water from 0.00°C to the body temperature of 37.0°C. How many liters of ice water would have to be consumed to burn off 454 g (about 1 lb) of fat, assuming that this requires 3500 Cal? Why is it not advisable to follow this diet? One liter = 10^3 cm³. The density of water is 1.00 g/cm³.

5E. Icebergs in the North Atlantic present hazards to shipping (see Fig. 20-15), causing the lengths of shipping routes to increase by about 30% during the iceberg season. Attempts to destroy icebergs include planting explosives, bombing, torpedoing, shelling, ramming, and coating with black soot. Suppose that direct melting of the iceberg, by placing heat sources in the ice, is tried. How much heat is required to melt 10% of a 200,000-metric-ton iceberg?

6E. How much water remains unfrozen after 50.2 kJ of heat has been extracted from 260 g of liquid water initially at its freezing point?

7E. Calculate the minimum amount of heat, in joules, required to completely melt 130 g of silver initially at 15.0°C. Assume that the specific heat throughout the heating is that given in Table 20-1.

8E. A room is lighted by four 100-W incandescent light bulbs. (The power of 100 W is the rate at which a bulb converts electrical energy into heat and visible light.) Assuming that 90% of the energy is converted to heat, how much heat is added to the room in 1.00 h?

9E. What quantity of butter (6.0 Cal/g = 6000 cal/g) would supply the energy needed for a 160-lb man to ascend to the top of Mt. Everest, elevation 29,000 ft, from sea level?

10E. An energetic athlete dissipates all the energy in a diet of 4000 Cal/day. If he were to release this energy at a steady rate, how would this conversion of energy compare with that of a 100-W bulb? (The power of 100 W is the rate at which a bulb converts electrical energy into heat and visible light.)

11E. If the heat necessary to raise the temperature of a mass m of water from 68°F to 78°F were somehow converted to translational kinetic energy of that water, what would be the speed of the water?

12E. Power is supplied at the rate of 0.400 hp for 2.00 min in drilling a hole in a 1.60-lb copper block. (a) How much heat in Btu is generated? (b) What is the rise in

FIGURE 20-15 Exercise 5. Calving of an iceberg from a glacier creates the danger to shipping.

temperature of the copper if only 75.0% of the power warms the copper? (Use 1 ft·lb = 1.285 × 10⁻³ Btu.)

13E. An object of mass 6.00 kg falls through a height of 50.0 m and, by means of a mechanical linkage, rotates a paddle wheel that stirs 0.600 kg of water. The water is initially at 15.0°C. What is the maximum possible temperature rise of the water?

14E. (a) Compute the possible increase in temperature for water dropping over Niagara Falls, 162 ft high. (b) What factors would tend to prevent this possible increase?

15E. A small electric immersion heater is used to boil 100 g of water for a cup of instant coffee. The heater is labeled 200 watts, which means that it converts electrical energy to heat at this rate. Calculate the time required to bring this water from 23°C to the boiling point, ignoring any heat losses.

16E. A pickup truck whose mass is 2200 kg is speeding along the highway at 65.0 mi/h. (a) If you could use all this kinetic energy to vaporize water already at 100°C, how much water could you vaporize? (b) If you had to buy this amount of energy from your local utility company at 12¢/kW·h, how much would it cost you? Guess at the answers before you figure them out; you may be surprised.

17E. A 150-g copper bowl contains 220 g of water, both at 20.0°C. A very hot 300-g copper cylinder is dropped into the water, causing the water to boil, with 5.00 g being converted to steam. The final temperature of the system is 100°C. (a) How much heat was transferred to the water? (b) How much to the bowl? (c) What was the original temperature of the cylinder?

18P. Calculate the specific heat of a metal from the following data. A container made of the metal has a mass of 3.6 kg and contains 14 kg of water. A 1.8-kg piece of the metal initially at a temperature of 180°C is dropped into the water. The container and water initially have a temperature of 16.0°C and the final temperature of the entire system is 18.0°C.

19P. A thermometer of mass 0.0550 kg and of specific heat 0.837 kJ/kg·K reads 15.0°C. It is then completely immersed in 0.300 kg of water and it comes to the same final temperature as the water. If the thermometer reads 44.4°C, what was the temperature of the water before insertion of the thermometer?

20P. How long does it take a 2.0 × 10⁵-Btu/h water heater to raise the temperature of 40 gal of water from 70 to 100°F?

21P. An athlete needs to lose weight and decides to do it by "pumping iron." (a) How many times must an 80.0-kg weight be lifted a distance of 1.00 m in order to burn off 1 lb of fat, assuming that it takes 3500 Cal to burn off that much fat? (b) If the weight is lifted once every 2.00 s, how long does it take?

22P. A 1500-kg Buick moving at 90 km/h brakes to rest, at uniform deceleration and without skidding, over a dis-

tance of 80 m. At what average rate is thermal energy produced in the brake system?

23P. A chef, upon awaking one morning to find his stove out of order, decides to boil the water for his wife's coffee by shaking it in a thermos flask. Suppose that he uses 500 cm³ of tap water at 59°F, and that the water falls 1.0 ft each shake, the chef making 30 shakes each minute. Neglecting any loss of thermal energy by the flask, how long must he shake the flask before the water boils?

24P. A block of ice, at its melting point and of initial mass 50.0 kg, slides along a horizontal surface, starting at a speed of 5.38 m/s and finally coming to rest after traveling 28.3 m. Compute the mass of ice melted as a result of the friction between the block and the surface. (Assume that all the heat generated owing to friction goes into the block of ice.)

25P. The specific heat of a substance varies with temperature according to $c = 0.20 + 0.14T + 0.023T^2$, with T in °C and c in cal/g·K. Find the heat required to raise the temperature of 2.0 g of this substance from 5.0 to 15°C.

26P. In a solar water heater, energy from the sun is gathered by a rooftop collector, which circulate water through tubes in the collector. The solar radiation enters the collector through a transparent cover and warms the water in the tubes; this water is pumped into a holding tank. Assuming that the efficiency of the overall system is 20% (that is, 80% of the incident solar energy is lost from the system), what collector area is necessary to raise the temperature of 200 L of water in the tank from 20 to 40°C in 1.0 h? The intensity of incident sunlight is 700 W/m².

27P. An insulated thermos contains 130 cm³ of hot coffee, at a temperature of 80.0°C. You put in a 12.0-g ice cube at its melting point to cool the coffee. By how many degrees will your coffee have cooled once the ice has melted? Treat the coffee as though it were pure water.

28P. What mass of steam at 100°C must be mixed with 150 g of ice at its melting point, in a thermally insulated container, to produce liquid water at 50°C?

29P. A person makes a quantity of iced tea by mixing 500 g of hot tea (essentially water) with an equal mass of ice at its melting point. If the initial hot tea is at a temperature of (a) 90°C and (b) 70°C, what is the temperature and mass of the remaining ice when the tea and ice reach the same temperature?

30P. (a) Two 50-g ice cubes are dropped into 200 g of water in a glass. If the water were initially at a temperature of 25°C, and if the ice came directly from a freezer at −15°C, what will be the temperature of the drink when the ice and water reach the same temperature? (b) Suppose that only one ice cube had been used in (a); what would be the final temperature of the drink? Neglect the heat capacity of the glass.

31P. A 20.0-g copper ring has a diameter of exactly 1.00000 in. at its temperature of 0.000°C. An aluminum sphere has a diameter of exactly 1.00200 in. at its temperature of 100.0°C. The sphere is placed on top of the ring

(Fig. 20-16), and the two are allowed to come to thermal equilibrium, no heat being lost to the surroundings. The sphere just passes through the ring at the equilibrium temperature. What is the mass of the sphere?

FIGURE 20-16 Problem 31.

32P. A *flow calorimeter* is a device used to measure the specific heat of a liquid. Heat is added at a known rate to a stream of the liquid as it passes through the calorimeter at a known rate. Measurement of the resulting temperature difference between the inflow and the outflow points of the liquid stream enables us to compute the specific heat of the liquid. Suppose a liquid of density 0.85 g/cm³ flows through a calorimeter at the rate of 8.0 cm³/s. Heat is added at the rate of 250 W by means of an electric heating coil, and a temperature difference of 15°C is established in steady-state conditions between the inflow and the outflow points. Find the specific heat of the liquid.

33P. By means of a heating coil, energy is transferred at a constant rate to a substance in a thermally insulated container. The temperature of the substance is measured as a function of time. (a) Show how we can deduce from this information the way in which the heat capacity of the body depends on the temperature. (b) Suppose that in a certain temperature range the temperature T is proportional to t^3, where t is the time. How does the heat capacity depend on T in this range?

34P*. Two metal blocks are insulated from their surroundings. The first block, which has mass $m_1 = 3.16$ kg and is at temperature $T_1 = 17.0°C$, has a specific heat four times that of the second block. This second block is at temperature $T_2 = 47.0°C$, and its coefficient of linear expansion is $15.0 \times 10^{-6}/°C$. When the two blocks are brought together and allowed to come to thermal equilibrium, the area of one face of the second block is found to have decreased by 0.0300%. Find the mass of the second block.

SECTION 20-6 SOME SPECIAL CASES OF THE FIRST LAW OF THERMODYNAMICS

35E. A sample of gas expands from 1.0 to 4.0 m³ while its pressure decreases from 40 to 10 Pa. How much work is done by the gas if its pressure changes with volume according to each of the three processes shown in the p–V diagram in Fig. 20-17?

FIGURE 20-17 Exercise 35.

36E. Suppose that a sample of gas expands from 1.0 to 4.0 m³ along the path B in the p–V diagram shown in Fig. 20-18. It is then compressed back to 1.0 m³ along either

FIGURE 20-18 Exercise 36.

path A or path C. Compute the net work done by the gas for the complete cycle in each case.

37E. Consider that 200 J of work is done on a system and 70.0 cal of heat is extracted from the system. In the sense of the first law of thermodynamics, what are the values (including algebraic signs) of (a) W, (b) Q, and (c) ΔE_{int}?

38E. A thermodynamic system is taken from an initial state A to another state B and back again to A, via state C, as shown by path $ABCA$ in the p–V diagram of Fig. 20-19a.

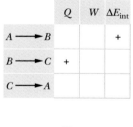

	Q	W	ΔE_{int}
$A \longrightarrow B$			+
$B \longrightarrow C$	+		
$C \longrightarrow A$			

(a)   (b)

FIGURE 20-19 Exercise 38.

(a) Complete the table in Fig. 20-19*b* by filling in either + or − for the sign of each thermodynamic quantity associated with each process. (b) Calculate the numerical value of the work done by the system for the complete cycle *ABCA*.

39E. Gas within a chamber passes through the cycle shown in Fig. 20-20. Determine the net heat added to the system during process *CA* if the heat Q_{AB} added during process *AB* is 20.0 J, no heat is transferred during process *BC*, and the net work done during the cycle is 15.0 J.

FIGURE 20-20 Exercise 39.

40E. Gas within a chamber undergoes the processes shown in the *p–V* diagram of Fig. 20-21. Calculate the net heat added to the system during one complete cycle.

41P. Figure 20-22*a* shows a cylinder containing gas and

FIGURE 20-21 Exercise 40.

(a)   *(b)*

FIGURE 20-22 Problem 41.

closed by a movable piston. The cylinder is kept submerged in an ice–water mixture. The piston is *quickly* pushed down from position 1 to position 2. The piston is held at position 2 until the gas is again at the temperature of the ice–water mixture and then is *slowly* raised back to position 1. Figure 20-22*b* is a *p–V* diagram for the process. If 100 g of ice is melted during the cycle, how much work has been done *on* the gas?

42P. When a system is taken from state *i* to state *f* along path *iaf* in Fig. 20-23, $Q = 50$ cal and $W = 20$ cal. Along path *ibf*, $Q = 36$ cal. (a) What is *W* along path *ibf*? (b) If $W = -13$ cal for the curved return path *fi*, what is *Q* for this path? (c) Take $\Delta E_{int,i} = 10$ cal. What is $\Delta E_{int,f}$? (d) If $\Delta E_{int,b} = 22$ cal, what are the values of *Q* for process *ib* and process *bf*?

FIGURE 20-23 Problem 42.

43P*. A cylinder has a well-fitted 2.0-kg metal piston whose cross-sectional area is 2.0 cm² (Fig. 20-24). The cylinder contains water and steam at constant temperature. The piston is observed to fall slowly at a rate of 0.30 cm/s because heat flows out of the cylinder through the cylinder walls. As this happens, some steam condenses in the chamber. The density of the steam inside the chamber is 6.0×10^{-4} g/cm³ and the atmospheric pressure is 1.0 atm. (a) Calculate the rate of condensation of steam. (b) At what rate is heat leaving the chamber? (c) What is the rate of change of internal energy of the steam and water inside the chamber?

FIGURE 20-24 Problem 43.

SECTION 20-7 THE TRANSFER OF HEAT

44E. The average rate at which heat is conducted through the surface of the Earth in North America is 54.0 mW/m², and the average thermal conductivity of the near-surface rocks is 2.50 W/m·K. Assuming a surface temperature of 10.0°C, what should be the temperature at a depth of 35.0 km (near the base of the crust)? Ignore the heat generated by the presence of radioactive elements.

45E. The thermal conductivity of Pyrex glass at 0°C is 2.9×10^{-3} cal/cm·°C·s. (a) Express this in W/m·K and in Btu/ft·°F·h. (b) What is the R-value for a $\frac{1}{4}$-in. sheet of such glass?

46E. (a) Calculate the rate at which body heat is conducted through the clothing of a skier in a steady-state process, given the following data: the body surface area is 1.8 m² and the clothing is 1.0 cm thick; the skin surface temperature is 33°C, whereas the outer surface of the clothing is at 1.0°C; the thermal conductivity of the clothing is 0.040 W/m·K. (b) How would the answer to (a) change if, after a fall, the skier's clothes became soaked with water? Assume that the thermal conductivity of water is 0.60 W/m·K.

47E. Consider the slab shown in Fig. 20-8. Suppose that $L = 25.0$ cm, $A = 90.0$ cm², and the material is copper. If $T_H = 125°C$, $T_C = 10.0°C$, and a steady state is reached, find the rate of heat transfer through the slab.

48E. A cylindrical copper rod of length 1.2 m and cross-sectional area 4.8 cm² is insulated to prevent heat loss through its surface. The ends are maintained at a temperature difference of 100°C by having one end in a water–ice mixture and the other in boiling water and steam. (a) Find the rate at which heat is conducted along the rod. (b) Find the rate at which ice melts at the cold end.

49E. Show that the temperature T_X at the interface of a compound slab (see Fig. 20-10) is given by

$$T_X = \frac{R_1 T_H + R_2 T_C}{R_1 + R_2}.$$

50E. Show that in a compound slab such as that in Fig. 20-25a the temperature change across each portion is inversely proportional to the thermal conductivity.

51E. Four square pieces of insulation of two different materials, all with the same thickness and area A, are available to cover an opening of area $2A$. This can be done in either of the two ways shown in Fig. 20-26. Which arrangement, (a) or (b), would give the lower heat flow if $k_2 \neq k_1$?

(a)

(b)

FIGURE 20-25 Exercise 50 and Problem 52.

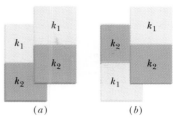

FIGURE 20-26 Exercise 51.

52P. Two identical rectangular rods of metal are welded end to end as shown in Fig. 20-25a and 10 J of heat is conducted (in a steady-state process) through the rods in 2.0 min. How long would it take for 10 J to be conducted through the rods if they are welded together as shown in Fig. 20-25b?

53P. Compute the rate of heat conduction through the following two storm doors 2.0 m high and 0.75 m wide. (a) One door is made with aluminum panels 1.5 mm thick and a 3.0-mm-thick glass pane that covers 75% of its surface (the structural frame has a negligible area). (b) The second door is made entirely of white pine averaging 2.5 cm in thickness. Take the temperature drop across each door to be 33°C, and see Table 20-4.

54P. An idealized representation of the air temperature as a function of distance from a single-pane window on a calm, winter day is shown in Fig. 20-27. The window dimensions are 60 cm × 60 cm × 0.50 cm. Assume that heat is conducted along a path that is perpendicular to the window, from points 8.0 cm from the window on one side to points 8.0 cm from it on the other side. (a) At what rate is heat conducted through the window area? (*Hint:* The temperature drop across the window glass is very small.) (b) Estimate the difference in temperature between the inner and outer glass surfaces.

55P. A large cylindrical water tank with a bottom 1.7 m in diameter is made of iron boilerplate 5.2 mm thick. As the water is being heated, the gas burner underneath is able to maintain a temperature difference of 2.3°C between the top and bottom surfaces of the bottom plate. How much

FIGURE 20-27 Problem 54.

FIGURE 20-28 Problem 57.

heat is conducted through that plate in 5.0 min? (Iron has a thermal conductivity of 67 W/m·K.)

56P. (a) What is the rate of heat loss in watts per square meter through a glass window 3.0 mm thick if the outside temperature is $-20°F$ and the inside temperature is $+72°F$? (b) A storm window is installed having the same thickness of glass but with an air gap of 7.5 cm between the two windows. What will be the corresponding rate of heat loss assuming that conduction is the only important heat-loss mechanism?

57P. A tank of water has been outdoors in cold weather and a 5.0-cm-thick slab of ice has formed on its surface (Fig. 20-28). The air above the ice is at $-10°C$. Calculate the rate of formation of ice (in centimeters per hour) on the bottom surface of the ice slab. Take the

thermal conductivity and density of ice to be 0.0040 cal/s·cm·°C and 0.92 g/cm³. Assume that heat is not transferred through the walls or bottom of the tank.

58P. Ice has formed on a shallow pond and a steady state has been reached, with the air above the ice at $-5.0°C$ and the bottom of the pond at 4.0°C. If the total depth of ice + water is 1.4 m, how thick is the ice? (Assume that the thermal conductivities of ice and water are 0.40 and 0.12 cal/m·°C·s, respectively.)

59P. Three metal rods, made of copper, aluminum, and brass, are each 6.00 cm long and 1.00 cm in diameter. These rods are placed end to end, with the aluminum between the other two. The free ends of the copper and brass rods are maintained at the boiling point and the freezing point of water, respectively. Find the steady-state temperatures of the copper–aluminum junction and the aluminum–brass junction. The thermal conductivity of brass is 109 W/m·K.

60P*. A wall assembly consists of a 20 ft × 12 ft frame made of 16 two-by-four vertical studs, each 12 ft long and set with their center lines 16 in. apart. The outside of the wall is faced with $\frac{1}{4}$-in. plywood sheet ($R = 0.30$) and $\frac{3}{4}$-in. white pine siding ($R = 0.98$). The inside is faced with $\frac{1}{4}$-in. plasterboard ($R = 0.47$), and the space between the studs is filled with polyurethane foam ($R = 5.9$ for a 1-in. layer). A "two-by-four" stud is actually 1.75 in. × 3.75 in. in size; assume that they are made of wood for which $R = 1.3$ for a 1-in. slab. (a) At what rate is heat transferred through this wall for a 30°F temperature difference? (b) What is the R-value for the assembled wall? (c) What fraction of the wall area contains studs, as opposed to foam? (d) What fraction of the heat transfer is through the studs, as opposed to the foam?

ADDITIONAL PROBLEMS

61. A sample of gas undergoes a transition from an initial state a to a final state b by three different paths (processes), as shown in the p–V diagram in Fig. 20-29. The

FIGURE 20-29 Problem 61.

heat added to the gas in process 1 is $10p_iV_i$. In terms of p_iV_i, what are (a) the heat added to the gas in process 2 and (b) the change in internal energy that the gas undergoes in process 3?

62. How many 20-g ice cubes, whose initial temperature is $-10°C$, must be added to 1.0 L of hot tea, whose initial temperature is 90°C, in order that the final mixture have a temperature of 10°C? Assume all the ice is melted in the final mixture and that the specific heat of tea is the same as that of water.

63. A sample of gas expands from an initial pressure and volume of 10 Pa and 1.0 m³ to a final volume of 2.0 m³. During the expansion, the pressure and volume are related by the equation $p = aV^2$, where $a = 10$ N/m⁸. Determine the work done by the gas during this expansion.

BOILING AND THE LEIDENFROST EFFECT

Jearl Walker
Cleveland State University

How does water boil? As commonplace as the event is, you may not have noticed all its curious features. Some of the features are important in industrial applications, while others appear to be the basis for certain dangerous stunts once performed by daredevils in carnival sideshows.

Arrange for a pan of tap water to be heated from below by a flame or electric heat source. As the water warms, air molecules are driven out of solution in the water, collecting as tiny bubbles in crevices along the bottom of the pan. The air bubbles gradually inflate, and then they begin to pinch off from the crevices and rise to the top surface of the water. As they leave, more air bubbles form in the crevices and pinch off, until the supply of air in the water is depleted. The formation of air bubbles is a sign that the water is heating but has nothing to do with boiling.

Water that is directly exposed to the atmosphere boils at what is sometimes called its normal boiling temperature T_S. For example, T_S is about 100°C when the air pressure is 1 atm. Since the water at the bottom of your pan is not directly exposed to the atmosphere, it remains liquid even when it *superheats* above T_S by as much as a few degrees. During this process, the water is constantly mixed by convection as hot water rises and cooler water descends.

If you continue to increase the pan's temperature, the bottom layer of water begins to vaporize, with water molecules gathering in small vapor bubbles in the now dry crevices. This phase of boiling is signaled by pops, pings, and eventually buzzing. The water almost sings its displeasure at being heated. Every time a vapor bubble expands upward into slightly cooler water, the bubble suddenly collapses because the vapor within it condenses. Each collapse sends out a sound wave, the ping you hear. Once the temperature of the bulk water increases, the bubbles may not collapse until after they pinch off from the crevices and ascend part of the way to the top surface of the water. This phase of boiling is labeled "isolated vapor bubbles" in Fig. 1.

If you still increase the pan's temperature, the clamor of collapsing bubbles first grows louder and then disappears. The noise begins to soften when the bulk liquid is sufficiently hot that the vapor bubbles reach the top surface of the water. There they pop open with a light splash. The water is now in full boil.

If your heat source is a kitchen stove, the story stops at this point. However, with a laboratory burner you can continue to increase the pan's temperature. The vapor bubbles next become so abundant and pinch off from their crevices so frequently that they coalesce, forming columns of vapor that violently and chaotically churn upward, sometimes meeting previously detached "slugs" of vapor.

Jearl Walker is professor of physics at Cleveland State University. He received a B.S. in physics from M.I.T. and a Ph.D. in physics from the University of Maryland. From 1977 to 1990 he conducted "The Amateur Scientist" department of Scientific American. His book The Flying Circus of Physics with Answers is published in 10 languages.

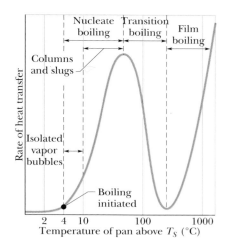

FIGURE 1 Boiling curve for water.

The production of vapor bubbles and columns is called *nucleate boiling* because the formation and growth of the bubbles depend on crevices serving as *nucleating sites* (sites of formation). Whenever you increase the pan's temperature, the rate at which heat is transferred to the water increases. If you continue to raise the pan's temperature past the stage of columns and slugs, the boiling enters a new phase called the *transition regime*. Then each increase in the pan's temperature reduces the rate at which heat is transferred to the water. The decrease is not paradoxical. In the transition regime, much of the bottom of the pan is covered by a layer of vapor. Since water vapor conducts heat about an order of magnitude more poorly than does liquid water, the transfer of heat to the water is diminished. The hotter the pan becomes, the less direct contact the water has with it and the worse the transfer of heat becomes. This situation can be dangerous in a *heat exchanger*, whose purpose is to transfer heat from a heated object. If the water in the heat exchanger is allowed to enter the transition regime, the object may destructively overheat because of diminished transfer of heat from it.

Suppose you continue to increase the temperature of the pan. Eventually, the whole of the bottom surface is covered with vapor. Then heat is slowly transferred to the liquid above the vapor by radiation and gradual conduction. This phase is called *film boiling*.

Although you cannot obtain film boiling in a pan of water on a kitchen stove, it is still commonplace in the kitchen. My grandmother once demonstrated how it serves to indicate when her skillet is hot enough for pancake batter. After she heated the empty skillet for a while, she sprinkled a few drops of water into it. The drops sizzled away within seconds. Their rapid disappearance warned her that the skillet was insufficiently hot for the batter. After further heat-

ing the skillet, she repeated her test with a few more water drops. This time they beaded up and danced over the metal, lasting well over a minute before they disappeared. The skillet was then hot enough for the batter.

To study her demonstration, I arranged for a flat metal plate to be heated by a laboratory burner. While monitoring the temperature of the plate with a thermocouple, I carefully released a drop of distilled water from a syringe held just above the plate. The drop fell into a dent I had made in the plate with a ball-peen hammer. The syringe allowed me to release drops of uniform size. Once a drop was released, I timed how long it survived on the plate. Afterward, I plotted the survival times of the drops versus the plate temperature (Fig. 2). The graph has a curious peak. When the plate temperature was between 100 and about 200°C, each drop spread over the plate in a thin layer and rapidly vaporized. When the plate temperature was about 200°C, a drop deposited on the plate beaded up and survived for over a minute. At even higher plate temperatures, the water beads did not survive quite as long. Similar experiments with tap water generated a graph with a flatter peak, probably because suspended particles of impurities in the drops

breached the vapor layer, conducting heat into the drops.

The fact that a water drop is long-lived when deposited on metal that is much hotter than the boiling temperature of water was first reported by Hermann Boerhaave in 1732. It was not investigated extensively until 1756 when Johann Gottlieb Leidenfrost published "A Tract About Some Qualities of Common Water." Because Leidenfrost's work was not translated from the Latin until 1965, it was not widely read. Still, his name is now associated with the phenomenon. In addition, the temperature corresponding to the peak in a graph such as I made is called the Leidenfrost point.

Leidenfrost conducted his experiments with an iron spoon that was heated red-hot in a fireplace. After placing a drop of water into the spoon, he timed its duration by the swings of a pendulum. He noted that the drop seemed to suck the light and heat from the spoon, leaving a spot duller than the rest of the spoon. The first drop deposited in the spoon lasted 30 s while the next drop lasted only 10 s. Additional drops lasted only a few seconds.

Leidenfrost misunderstood his demonstrations because he did not realize that the longer-lasting drops were actually boiling. Let me explain in terms of my experiments. When the temperature of the plate is less than the Leidenfrost point, the water spreads over the plate and rapidly conducts heat from it, resulting in complete vaporization within seconds. When the temperature is at or above the Leidenfrost point, the bottom surface of a drop deposited on the plate almost immediately vaporizes. The gas pressure from this vapor layer prevents the rest of the drop from touching the plate (Fig. 3). The layer thus protects and supports the drop for the next minute or so. The layer is constantly replenished as additional water vaporizes from the bottom surface of the drop because of heat radiated and conducted through

FIGURE 2   Drop lifetimes on a hot plate.

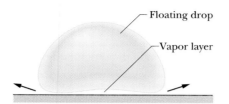

FIGURE 3 A Leidenfrost drop in cross section.

the layer from the plate. Although the layer is less than 0.1 mm thick near its outer boundary and only about 0.2 mm thick at its center, it dramatically slows the vaporization of the drop.

After reading the translation of Leidenfrost's research, I happened upon a description of a curious stunt that was performed in the sideshows of carnivals around the turn of the century. Reportedly, a performer was able to dip wet fingers into molten lead. Assuming that the stunt involved no trickery, I conjectured that it must depend on the Leidenfrost effect. As soon as the performer's wet flesh touched the hot liquid metal, part of the water vaporized, coating the fingers with a vapor layer. If the dip was brief, the flesh would not be heated significantly.

I could not resist the temptation to test my explanation. With a laboratory burner I melted down a sizable slab of lead in a crucible. I heated the lead until its temperature was over 400°C, well above its melting temperature of 328°C. After wetting a finger in tap water, I prepared to touch the top surface of the molten lead. I must confess that I had an assistant standing ready with first-aid materials. I must also confess that my first several attempts failed because by brain refused to allow this ridiculous experiment, always directing my finger to miss the lead.

When I finally overcame my fears and briefly touched the lead, I was amazed. I felt no heat. Just as I had guessed, part of the water on the finger vaporized, forming a protective

layer. Since the contact was brief, radiation and conduction of heat through the vapor layer were insufficient to raise perceptibly the temperature of my flesh. I grew braver. After wetting my hand, I dipped all my fingers into the lead, touching the bottom of the container (see opening photo). The contact with the lead was still too brief to result in a burn. Apparently, the Leidenfrost effect, or more exactly, the immediate presence of film boiling, protected my fingers.

I still questioned my explanation. Could I possibly touch the lead with a dry finger without suffering a burn? Leaving aside all rational thought, I tried it, immediately realizing my folly when pain raced through the finger. Later, I tested a dry weiner, forcing it into the molten lead for several seconds. The skin of the weiner quickly blackened. It lacked the protection of film boiling just as my dry finger had.

I must caution that dipping fingers into molten lead presents several serious dangers. If the lead is only slightly above its melting point, the loss of heat from it when the water is vaporized may solidify the lead around the fingers. If I were to pull the resulting glove of hot, solid lead up from the container, it will be in contact with my fingers so long that my fingers are certain to be badly burned. I must also contend with the possibility of splashing and spillage. In addition, there is the acute danger of having too much water on the fingers. When the surplus water rapidly vaporizes, it can blow molten lead over the surroundings and, most seriously, into the eyes. I have been scarred on my arms and face from such explosive vaporizations. *You should never repeat this demonstration.*

Film boiling can also be seen when liquid nitrogen is spilled. The drops and globs bead up as they skate over the floor. The liquid is at a temperature of about −200°C. When the spilled liquid nears the floor, its bottom surface vaporizes. The vapor

layer then provides support for the rest of the liquid, allowing the liquid to survive for a surprisingly long time.

I was told of a stunt where a performer poured liquid nitrogen into his mouth without being hurt by its extreme cold. The liquid immediately underwent film boiling on its bottom surface and thus did not directly touch the tongue. Foolishly, I repeated this demonstration. For several dozen times the stunt went smoothly and dramatically. With a large glob of liquid nitrogen in my mouth, I concentrated on not swallowing while I breathed outward. The moisture in my cold breath condensed, creating a terrific plume that extended about a meter from my mouth. However, on my last attempt the liquid thermally contracted two of my front teeth so severely that the enamel ruptured into a "road map" of fissures. My dentist convinced me to drop this demonstration.

The Leidenfrost effect may also play a role in another foolhardy demonstration: walking over hot coals. At times the news media have carried reports of a performer striding over red-hot coals with much hoopla and mystic nonsense, perhaps claiming that protection from a bad burn is afforded by "mind over matter." Actually, physics protects the feet when the walk is successful. Particularly important is the fact that although the surface of the coals is quite hot, it contains surprisingly little energy. If the performer walks at a moderate pace, a footfall is so brief that the foot conducts little energy from the coals. Of course, a slower walk invites a burn because the longer contact allows heat to be conducted to the foot from the interior of the coals.

If the feet are wet prior to the walk, the liquid might also help protect them. To wet the feet a performer might walk over wet grass just before reaching the hot coals. Instead, the feet might just be sweaty because of the heat from the coals or the excitement of the performance. Once the performer is on the coals, some of the heat vaporizes the liquid on the feet, leaving less heat to be conducted to the flesh. In addition, there may be points of contact where the liquid undergoes film boiling, thereby providing brief protection from the coals.

I have walked over hot coals on five occasions. For four of the walks I was fearful enough that my feet were sweaty. However, on the fifth walk I took my safety so much for granted that my feet were dry. The burns I suffered then were extensive and terribly painful. My feet did not heal for weeks.

My failure may have been due to a lack of film boiling on the feet, but I had also neglected an additional safety factor. On the other days I had taken the precaution of clutching an earlier edition of this book to my chest during the walks so as to bolster my belief in physics. Alas, I forgot the book on the day when I was so badly burned.

I have long argued that degree-granting programs should employ "fire-walking" as a last exam. The chairperson of the program should wait on the far side of a bed of red-hot coals while a degree candidate is forced to walk over the coals. If the candidate's belief in physics is strong enough that the feet are left undamaged, the chairperson hands the candidate a graduation certificate. The test would be far more revealing than traditional final exams.

THE KINETIC THEORY OF GASES | 21

Suppose that you return to your chilly dwelling after snowshoeing through the woods on a cold winter day. Your first thought is to light a stove. But why, exactly, would you do that? Is it because the stove will increase the store of internal (thermal) energy of the air in the cabin, until eventually the air will have enough of that internal energy to keep you comfortable? As logical as this reasoning sounds, it is flawed, because the air's store of internal energy will not be changed by the stove. How can that be? And if it is so, why would you bother to light the stove?

21-1 A NEW WAY TO LOOK AT GASES

Classical thermodynamics—the subject of the last two chapters—has nothing to say about atoms. When we apply its laws to a gas, we are concerned only with such macroscopic variables as pressure, volume, and temperature. Although we know that a gas is made up of atoms or molecules, the laws of classical thermodynamics take no account of that fact.

However, the pressure exerted by a gas must surely be related to the steady drumbeat of its molecules on the walls of its container. The ability of a gas to take on the volume of its container must surely be due to the freedom of motion of its molecules. And the temperature and internal energy of a gas must surely be related to the kinetic energy of these molecules. Perhaps we can learn something about gases by approaching the subject from this direction. The name we give to this molecular approach is the **kinetic theory of gases.** It is the subject of this chapter.

21-2 AVOGADRO'S NUMBER

When our thinking is slanted toward molecules, it makes sense to measure the sizes of our samples in moles. If we do so, we can be certain that we are comparing samples that contain the same number of molecules. The *mole* is one of the seven SI base units and is defined as follows:

> One mole is the number of atoms in a 12-g sample of carbon-12.

We speak of a "mole of helium" or "a mole of water," meaning a certain number of the elementary units of the substance. But we could just as well speak of a mole of tennis balls, where, of course, the elementary unit would be a tennis ball.

The obvious question now is: "Just how many atoms or molecules are there in a mole?" The answer is determined experimentally and, as you saw in Chapter 20, is

$$N_A = 6.02 \times 10^{23} \text{ mol}^{-1} \quad \begin{array}{c}\text{(Avogadro's}\\ \text{number).}\end{array} \quad (21\text{-}1)$$

This number is called **Avogadro's number** after Italian scientist Amedeo Avogadro (1776–1856), who first made the important suggestion that all gases contain the same number of molecules or atoms when they occupy the same volume under the same conditions of temperature and pressure.

The number of moles n contained in a sample of any substance can be found from

$$n = \frac{N}{N_A}, \quad (21\text{-}2)$$

in which N is the number of molecules in the sample. The number of moles in a sample can also be found from the mass M_{sam} of the sample and either the *molar mass M* (the mass of 1 mole of that substance) or the mass m of one molecule:

$$n = \frac{M_{sam}}{M} = \frac{M_{sam}}{mN_A}. \quad (21\text{-}3)$$

The enormously large value of Avogadro's number suggests how tiny and how numerous atoms must be. A mole of air, for example, can easily fit into a suitcase. Yet, if these molecules were spread uniformly over the surface of the Earth, there would be about 120,000 of them in every square centimeter. A second example: one mole of tennis balls would fill a volume equal to that of seven moons!

Measuring Avogadro's Number

It is not possible to see individual gas molecules as they move rapidly about, colliding with each other and with the walls of their container. Yet it *is* possible to see the direct effects of such motion and to deduce a value for N_A from these observations.

In 1827, English botanist Robert Brown observed that, when he looked through his microscope at pollen grains and other small objects suspended in water, they "danced around" in a random way. Figure 21-1 shows an example of such *Brownian motion* for a single particle. The motion occurs because the suspended particles are continually bombarded on all sides by molecules of the fluid. The numbers of molecules striking opposite sides of a particle in any short time interval—being determined by chance—will not be exactly equal. Because of these fluctuations, a randomly directed unbalanced force will act on the suspended particle, its "Brownian dance."

It is as if a bowling ball, floating in a gravity-free space, were bombarded from all sides by a swarm of randomly directed, rapidly moving Ping-Pong balls. The bowling ball would jiggle around slightly, and in

FIGURE 21-1 The motion of a tiny particle, suspended in water and viewed through a microscope. The short line segments connect its position at 30-s intervals. (The path of the particle is a good example of a *fractal*, which is a curve for which any small section resembles the curve as a whole. For example, if the path of the particle for any of the short line segments shown had been explored at, say, 0.1-s intervals, the resulting plot would be very similar to the present plot.)

a random way, exhibiting a kind of Brownian motion. By watching the erratic motion of the bowling ball, you could deduce something about the Ping-Pong balls, even if you could not see them.

It is possible to deduce the value of Avogadro's number from measurements of the Brownian motion. There are also many other ways to get at this important constant. In the early part of this century, phenomena ranging from radioactivity to the blueness of the sky were employed; Albert Einstein himself proposed four independent methods. All of these methods give us approximately the same value for Avogadro's number. It was this firm and broad experimental base for N_A that went far to convince doubters—and there were many—of the usefulness of the concept that our familiar tangible world is made up of atoms.

PROBLEM SOLVING

TACTIC 1: AVOGADRO'S NUMBER OF WHAT?
In Eq. 21-1, Avogadro's number is expressed in terms

of mol^{-1}, which is the inverse mole, or 1/mol. We could instead explicitly state the elementary unit involved in a given situation. For example, if the elementary unit is an atom, we might write $N_A = 6.02 \times 10^{23}$ atoms/mole. If the elementary unit is a molecule, then we might write $N_A = 6.02 \times 10^{23}$ molecules/mole. And if we were ever interested in a *great* many tennis balls, we might write $N_A = 6.02 \times 10^{23}$ tennis balls/mole.

21-3 IDEAL GASES

Our goal in this chapter is to explain the macroscopic properties of a gas—such as its pressure and its temperature—in terms of the behavior of the molecules that make it up. But there is an immediate problem: which gas? Should it be hydrogen or oxygen, or methane, or perhaps uranium hexafluoride? They are all different. However, experimenters have found that if we confine 1-mole samples of various gases in boxes of identical volume and hold the gases at the same temperature, then their measured pressures are nearly—though not exactly—the same. If we repeat the measurements at lower gas densities, then these small differences in the measured pressures tend to disappear. Further experiments show that, at low enough densities, all real gases tend to obey the relation

$$pV = nRT \qquad \text{(ideal gas law)}, \qquad (21\text{-}4)$$

in which p is the absolute (not gauge) pressure, n is the number of moles of gas present, and R, the **gas constant,** has the same value for all gases, namely,

$$R = 8.31 \text{ J/mol·K}. \qquad (21\text{-}5)$$

The temperature T in Eq. 21-4 must be expressed in absolute (kelvin) units. Equation 21-4 is called the **ideal gas law.** Provided the gas density is low, Eq. 21-4 holds for any type of gas, or a mixture of different types, with n being the total number of moles present.

You may well ask, "What is an *ideal gas* and what is so 'ideal' about one?"* The answer lies in the simplicity of the law (Eq. 21-4) that governs its mac-

---

*Many prefer *perfect gas* to "ideal gas"; they are the same thing.

roscopic properties. Using this law—as you will see —we can deduce many properties of the ideal gas in a simple way. Although there is no such thing in nature as a truly ideal gas, *all* gases approach the ideal state at low enough densities, that is, under conditions in which their molecules are far enough apart. Thus the ideal gas concept allows us to gain useful insights into the limiting behavior of real gases.

Work Done by an Ideal Gas at Constant Temperature

Suppose that a sample of n moles of an ideal gas, confined to a piston–cylinder arrangement, is allowed to expand from an initial volume V_i to a final volume V_f. Suppose further that the temperature T of the gas is held constant throughout the process. Let us calculate the work done by the (ideal) gas during such an **isothermal expansion.**

Our starting point is Eq. 20-11, or

$$W = \int_{V_i}^{V_f} p \, dV.$$

Because the gas is ideal, we can substitute for p from Eq. 21-4, obtaining

$$W = \int_{V_i}^{V_f} \frac{nRT}{V} \, dV. \qquad (21\text{-}6)$$

The temperature is required to be constant so that

$$W = nRT \int_{V_i}^{V_f} \frac{dV}{V} \qquad (21\text{-}7)$$

or

$$W = nRT \ln \frac{V_f}{V_i} \qquad \begin{array}{l}\text{(ideal gas,}\\ \text{isothermal process).}\end{array} \qquad (21\text{-}8)$$

Recall that the symbol ln specifies a *natural* logarithm, that is, a logarithm to base e.

For an expansion, $V_f > V_i$ by definition, so the ratio V_f/V_i in Eq. 21-8 is greater than unity. The logarithm of a quantity greater than unity is positive, and so the work W done by an ideal gas during an isothermal expansion is positive, as we expect. For a compression, we have $V_f < V_i$, so the ratio of volumes in Eq. 21-8 is less than unity. The logarithm in that equation—and hence the work W—is negative, again as we expect.

PROBLEM SOLVING

TACTIC 2: WHEN TO USE EQ. 21-8

Some students think that Eq. 21-8 will give the work done by an ideal gas during any thermodynamic process. It does not. Instead, it gives the work only when the temperature is held constant. If the temperature varies, then the symbol T in Eq. 21-6 cannot be put in front of the integral symbol as in Eq. 21-7, and thus we do not end up with Eq. 21-8. We will be discussing situations of this sort in this and the next chapter.

SAMPLE PROBLEM 21-1

A cylinder contains 12 L of oxygen at 20°C and 15 atm. The temperature is raised to 35°C, and the volume reduced to 8.5 L. What is the final pressure of the gas? Assume that the gas is ideal.

SOLUTION From Eq. 21-4 we may write

$$R = \frac{p_i V_i}{T_i} = \frac{p_f V_f}{T_f}.$$

Solving for p_f yields

$$p_f = \frac{p_i T_f V_i}{T_i V_f}. \qquad (21\text{-}9)$$

Before substituting numbers, we must make sure that we express the temperatures on the Kelvin scale. Thus

$$T_i = (273 + 20) \text{ K} = 293 \text{ K}$$

and

$$T_f = (273 + 35) \text{ K} = 308 \text{ K}.$$

Inserting the given data into Eq. 21-9 then yields

$$p_f = \frac{(15 \text{ atm})(308 \text{ K})(12 \text{ L})}{(293 \text{ K})(8.5 \text{ L})} = 22 \text{ atm.} \quad \text{(Answer)}$$

SAMPLE PROBLEM 21-2

One mole of oxygen (assume it to be an ideal gas) expands at a constant temperature T of 310 K from an initial volume V_i of 12 L to a final volume V_f of 19 L.

a. How much work is done by the expanding gas?

SOLUTION From Eq. 21-8 we have

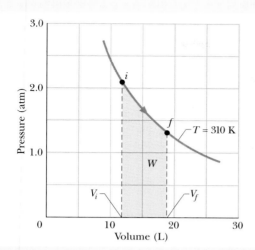

FIGURE 21-2 Sample Problem 21-2. The gold area represents the work done by 1 mol of oxygen in expanding at constant temperature T of 310 K.

$$W = nRT \ln \frac{V_f}{V_i}$$

$$= (1 \text{ mol})(8.31 \text{ J/mol·K})(310 \text{ K}) \ln \frac{19 \text{ L}}{12 \text{ L}}$$

$$= 1180 \text{ J}. \qquad \text{(Answer)}$$

The work done by the expanding gas is positive, consistent with the sign convention for work that we outlined in Section 20-4. The expansion is represented in the p–V diagram of Fig. 21-2. The curved green line there is an *isotherm*, that is, a curve giving the relation of pressure to volume for a gas when its temperature is held constant at a particular value. Thus the curve is a plot of

$$p = nRT \frac{1}{V} = (\text{a constant}) \frac{1}{V}.$$

The work done by the gas during the expansion is represented by the gold area beneath the isotherm.

b. How much work is done by the gas during an isothermal *compression* from $V_i = 19$ L to $V_f = 12$ L?

SOLUTION We proceed as in (a), finding

$$W = nRT \ln \frac{V_f}{V_i}$$

$$= (1 \text{ mol})(8.31 \text{ J/mol·K})(310 \text{ K}) \ln \frac{12 \text{ L}}{19 \text{ L}}$$

$$= -1180 \text{ J}. \qquad \text{(Answer)}$$

This result is equal in magnitude but opposite in sign to the result found in (a) for an isothermal expansion. The minus sign tells us that an external agent must have done 1180 J of work *on* the gas to compress it.

21-4 PRESSURE AND TEMPERATURE: A MOLECULAR VIEW

Here is our first kinetic theory problem. Let n moles of an ideal gas be confined in a cubical box of volume V, as in Fig. 21-3. The walls of the box are held at temperature T. What is the connection between the pressure p exerted by the gas on the walls and the speeds of the molecules?

The molecules in the box are moving in all directions and with varying speeds, bumping into each other and bouncing from the walls of the box like balls in a racquetball court. We ignore (for the time being) collisions of the molecules with one another and consider only elastic collisions with the walls.

Figure 21-3 shows a typical molecule, of mass m and velocity $\mathbf{v}$, that is about to collide with the shaded wall. Because we assume that any collision of a molecule with a wall is elastic, when this molecule collides with the shaded wall, the only component of its velocity that is changed is the x component, and that component is reversed. This means that the only change in the particle's momentum is then along the x axis, and that change is

$$(-mv_x) - (mv_x) = -2mv_x.$$

Hence the momentum Δp delivered to the wall by the molecule during the collision is $+2mv_x$. (Be-

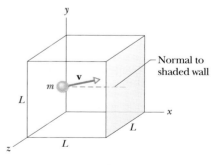

FIGURE 21-3 A cubical box of edge L, containing n moles of an ideal gas. A molecule of mass m and velocity $\mathbf{v}$ is about to collide with the shaded wall, which has area L^2. A normal to that wall is shown.

cause the symbol p represents both momentum and pressure in this book, we must be careful to note that here p represents momentum.)

The molecule of Fig. 21-3 will hit the shaded wall repeatedly. The time Δt between collisions is the time the molecule takes to travel to the opposite wall and back again (a distance of $2L$) at speed v_x. Thus Δt is equal to $2L/v_x$. (Note that this result holds even if the molecule bounces off any of the other walls along the way, because those walls are parallel to x and so cannot change v_x.) Thus the rate at which momentum is delivered to the shaded wall by this single molecule is

$$\frac{\Delta p}{\Delta t} = \frac{2mv_x}{2L/v_x} = \frac{mv_x^2}{L}.$$

From Newton's second law ($\mathbf{F} = d\mathbf{p}/dt$), the rate at which momentum is delivered to the wall is the force acting on that wall. To find this force, we must add up the contributions of all the other molecules that strike the wall, allowing for the possibility that they all have different speeds. Dividing the total force by the area of the wall (L^2) gives the pressure p on that wall, where now and in the rest of this discussion, p represents pressure. Thus

$$p = \frac{F}{L^2} = \frac{mv_{x1}^2/L + mv_{x2}^2/L + \cdots + mv_{xN}^2/L}{L^2}$$

$$= \left(\frac{m}{L^3}\right)(v_{x1}^2 + v_{x2}^2 + \cdots + v_{xN}^2), \quad (21\text{-}10)$$

where N is the number of molecules in the box.

Since $N = nN_A$, there are nN_A terms in the second parentheses of Eq. 21-10. So we can replace that quantity by $nN_A\overline{v_x^2}$, where $\overline{v_x^2}$ is the average value of the square of the x components of all the molecular speeds. Equation 21-10 then becomes

$$p = \frac{nmN_A}{L^3}\overline{v_x^2}.$$

But mN_A is the molar mass M of the gas (that is, the mass of 1 mole of the gas). Also, L^3 is the volume of the box, so

$$p = \frac{nM\overline{v_x^2}}{V}. \quad (21\text{-}11)$$

For any molecule, $v^2 = v_x^2 + v_y^2 + v_z^2$. Because there are many molecules and because they are all moving in random directions, the average values of the squares of their velocity components are equal,

so that $\overline{v_x^2} = \frac{1}{3}\overline{v^2}$. Thus Eq. 21-11 becomes

$$p = \frac{nM\overline{v^2}}{3V}. \quad (21\text{-}12)$$

The square root of $\overline{v^2}$ is a kind of average speed, called the **root-mean-square speed** of the molecules and symbolized by v_{rms}. Its name describes it rather well: you square each speed, you find the *mean* (that is, the average) of all these squared speeds, and then you take the square *root*. We can then write Eq. 21-12 as

$$p = \frac{nMv_{\text{rms}}^2}{3V}. \quad (21\text{-}13)$$

Equation 21-13 is very much in the spirit of kinetic theory. It tells us how the pressure of the gas (a purely macroscopic quantity) depends on the speed of the molecules (a purely microscopic quantity).

We can turn Eq. 21-13 around and use it to calculate v_{rms}. Combining Eq. 21-13 with the ideal gas law ($pV = nRT$) leads to

$$v_{\text{rms}} = \sqrt{\frac{3RT}{M}}. \quad (21\text{-}14)$$

Table 21-1 shows some rms speeds calculated from Eq. 21-14. The speeds are surprisingly high. For hydrogen molecules at room temperature (300 K), the rms speed is 1920 m/s or 4300 mi/h—faster than a speeding bullet! On the surface of the sun (Fig. 21-4), where the temperature is 2×10^6 K, the rms speed of hydrogen molecules would be 82 times

TABLE 21-1

SOME MOLECULAR SPEEDS AT ROOM TEMPERATURE ($T = 300$ K)[a]

GAS	MOLAR MASS[b] (g/mol)	v_{rms} (m/s)
Hydrogen	2.02	1920
Helium	4.0	1370
Water vapor	18.0	645
Nitrogen	28.0	517
Oxygen	32.0	483
Carbon dioxide	44.0	412
Sulfur dioxide	64.1	342

[a]For convenience, we often set room temperature = 300 K even though (at 27°C or 81°F) that represents a fairly warm room.

[b]Although molar masses are most often expressed in g/mol, the proper SI unit is the kg/mol.

FIGURE 21-4 A prominence (or jet) leaps outward from the high-temperature atmosphere of the sun.

larger than at room temperature were it not for the fact that at such high speeds, the molecules cannot survive the collisions among themselves. Remember too that the rms speed is only a kind of average speed; many molecules move significantly faster than this, and some are much slower.

The speed of sound in a gas is closely related to the rms speed of the molecules of that gas. In a sound wave, the disturbance is passed on from molecule to molecule by means of collisions. The wave cannot move any faster than the "average" speed of the molecules. In fact, the speed of sound must be somewhat less than this "average" molecular speed because not all molecules are moving in exactly the same direction as the wave. As examples, at room temperature, the rms speeds of hydrogen and nitrogen molecules are 1920 m/s and 517 m/s, respectively. The speeds of sound in these two gases at this temperature are 1350 m/s and 350 m/s, respectively.

A question often arises: "If molecules move so fast, why does it take as long as a minute or so before you can smell perfume if someone opens a bottle across a room? A glance ahead at Fig. 21-5 suggests that—although the molecules move very fast between collisions—a given molecule will wander only very slowly away from its release point. In practical situations, the travel time for an "alien" molecule across the air of a room is controlled by the convection currents that are always present in an ordinary room.

SAMPLE PROBLEM 21-3

Here are five pure numbers: 5, 11, 32, 67, and 89.

a. What is the average value $\bar{n}$ of these numbers?

SOLUTION We find this from

$$\bar{n} = \frac{5 + 11 + 32 + 67 + 89}{5} = 40.8 \quad \text{(Answer)}$$

b. What is the rms value n_{rms} of these numbers?

SOLUTION We find this from

$$n_{\text{rms}} = \sqrt{\frac{5^2 + 11^2 + 32^2 + 67^2 + 89^2}{5}}$$

$$= 52.1. \quad \text{(Answer)}$$

The rms value is greater than the average value because the larger numbers—being squared—are relatively more important in forming the rms value. To test this, let us replace 89 in our set of five numbers by 300. The average value of the new set of five numbers (as you can easily show) increases by a factor of 2.0. The rms value, however, increases by a factor of 2.7.

The rms values of variables occur in many branches of physics and engineering. The value 120 volts printed on an electric light bulb, for example, is an rms voltage.

SAMPLE PROBLEM 21-4

Verify the rms speed of the molecules of hydrogen gas at room temperature as given in Table 21-1.

SOLUTION The molar mass of hydrogen is 2.02 g/mol or (in SI units) 0.00202 kg/mol. Substituting into Eq. 21-14 yields

$$v_{\text{rms}} = \sqrt{\frac{3RT}{M}} = \sqrt{\frac{(3)(8.31 \text{ J/mol·K})(300 \text{ K})}{0.00202 \text{ kg/mol}}}$$

$$= 1920 \text{ m/s}, \quad \text{(Answer)}$$

as in Table 21-1.

21-5 TRANSLATIONAL KINETIC ENERGY

We again consider a single molecule as it moves around in the box of Fig. 21-3, but we now assume that its speed changes when it collides with other molecules. Its translational kinetic energy at any instant is $\frac{1}{2}mv^2$. Its *average* translational kinetic energy over the time that we watch it is

$$\overline{K} = \overline{\tfrac{1}{2}mv^2} = \tfrac{1}{2}m\overline{v^2} = \tfrac{1}{2}mv_{\mathrm{rms}}^2, \qquad (21\text{-}15)$$

in which we make the assumption that the average speed of the molecule during our observation is the same as the average speed of all the molecules at any given time. (Provided that the total energy of the gas is not changing and that we observe our molecule for long enough, this assumption is appropriate.) Substituting for v_{rms} from Eq. 21-14 leads to

$$\overline{K} = (\tfrac{1}{2}m)\frac{3RT}{M}.$$

But M/m, the molar mass divided by the mass of a molecule, is simply Avogadro's number, so

$$\overline{K} = \frac{3RT}{2N_A},$$

which we can write as

$$\overline{K} = \tfrac{3}{2}kT. \qquad (21\text{-}16)$$

The constant k, called the **Boltzmann constant,** is the ratio of the gas constant R to Avogadro's number N_A. It is sometimes called the gas constant for a single molecule (rather than for a mole), and its value is*

$$k = \frac{R}{N_A} = \frac{8.31 \text{ J/mol}\cdot\text{K}}{6.02 \times 10^{23} \text{ mol}^{-1}} = 1.38 \times 10^{-23} \text{ J/K}$$

$$= 8.62 \times 10^{-5} \text{ eV/K}. \qquad (21\text{-}17)$$

Equation 21-16 tells us what is perhaps an unexpected thing:

> At a given temperature T, all gas molecules—no matter what their mass—have the same average translational kinetic energy, namely, $\frac{3}{2}kT$. When we measure the temperature of a gas, we are measuring the average translational kinetic energy of its molecules.

Actually, nothing in our derivation limits it to molecules. We can apply it to any objects, such as tiny

*Do not confuse k (the Boltzmann constant) with K (kinetic energy) and K (the kelvin).

pollen grains or even tennis balls! We see here the basis for Brownian motion. A pollen grain, suspended in water and in thermal equilibrium with it, behaves like a very large molecule and has the same translational kinetic energy as the water molecules that surround it. Because of its very much larger mass, of course, the pollen grain has a correspondingly smaller rms speed, small enough to make its motion observable.

SAMPLE PROBLEM 21-5

What is the average translational kinetic energy (in electron-volts) of oxygen molecules in air at room temperature (= 300 K)? Of the nitrogen molecules?

SOLUTION The average translational kinetic energy depends only on the temperature and not on the nature of the molecule. For both oxygen and nitrogen molecules it is given by Eq. 21-16 as

$$\overline{K} = \tfrac{3}{2}kT = (\tfrac{3}{2})(8.62 \times 10^{-5} \text{ eV/K})(300 \text{ K})$$

$$= 0.039 \text{ eV}. \qquad \text{(Answer)}$$

Physicists find it useful to remember that the mean translational kinetic energy of *any* molecule at room temperature is about $\frac{1}{25}$ eV, which is essentially the above result.

From Table 21-1 we see that the rms speed of the molecules of oxygen (for which $M = 32.0$ g/mol) is 483 m/s. That for the molecules of nitrogen ($M = 28.0$ g/mol) is 517 m/s. Thus the lighter molecule has the larger rms speed, consistent with the fact that the two kinds of molecules have the same average translational kinetic energy.

SAMPLE PROBLEM 21-6

To enhance the effectiveness of the nuclear fission of uranium it is necessary to separate the (highly fissionable) U-235 isotope from the (less readily fissionable) U-238 isotope. One way of doing this is to form the uranium into a gas (UF_6) and allow it to diffuse repeatedly through a porous barrier. The lighter molecule will diffuse faster, the effectiveness of the barrier being determined by a *separation factor* α, defined as the ratio of the two rms speeds. What is the separation factor for the two kinds of uranium hexafluoride gas molecules?

SOLUTION From Eq. 21-14 ($v_{\mathrm{rms}} = \sqrt{3RT/M}$) we can

write

$$\alpha = \frac{v_{\text{rms, 235}}}{v_{\text{rms, 238}}} = \sqrt{\frac{M_{238}}{M_{235}}},$$

in which the M's are the molar masses of the two gas molecules. We can find these molar masses by adding six times the molar mass of fluorine (19.0 g/mol) to the molar mass of the appropriate uranium atom (235 g/mol or 238 g/mol). We get

UF_6 (uranium-238): $M_{238} = 238 + 6 \times 19.0$

$$= 352 \text{ g/mol},$$

UF_6 (uranium-235): $M_{235} = 235 + 6 \times 19.0$

$$= 349 \text{ g/mol},$$

so that

$$\alpha = \sqrt{\frac{352 \text{ g/mol}}{349 \text{ g/mol}}} = 1.0043. \quad \text{(Answer)}$$

As many as 4000 passages through the barrier are required in a practical isotope diffusion plant.

21-6 MEAN FREE PATH

Figure 21-5 shows the path of a typical molecule as it moves through a gas, changing both speed and direction abruptly as it collides elastically with other molecules. Between collisions, our typical molecule moves in a straight line at constant speed. Although the figure shows all the other molecules as stationary, they too are moving in much the same way.

One useful parameter to describe this random motion is the **mean free path** λ. As its name implies,

λ is the average distance traversed by a molecule between collisions. We expect λ to vary inversely with N/V, the number of molecules per unit volume. The large N/V is, the more collisions there should be and the smaller the mean free path. We also expect λ to vary inversely with the size of the molecules. (If the molecules were true points, they would never collide and the mean free path would be infinite.) Thus the larger the molecules, the smaller the mean free path. We can even predict that λ should vary (inversely) as the *square* of the diameter because it is the cross section of a molecule—not its diameter—that determines its effective target area.

The expression for the mean free path does, in fact, turn out to be

$$\lambda = \frac{1}{\sqrt{2}\pi d^2 \, N/V} \quad \text{(mean free path).} \quad (21\text{-}18)$$

To justify Eq. 21-18, we focus attention on a single molecule and assume—as Fig. 21-5 suggests—that our molecule is traveling with a constant speed v and that all the other molecules are at rest. Later, we shall relax this assumption.

We assume further that the molecules are spheres of diameter d. A collision will then take place if the centers of the molecules come within a distance d of each other, as in Fig. 21-6a. Another

FIGURE 21-6 (*a*) A collision occurs when the centers of two molecules come within a distance d of each other, d being the molecular diameter. (*b*) An equivalent but more convenient representation is to think of the moving molecule as having a *radius d*, and all other molecules as being points. The condition for a collision remains the same.

FIGURE 21-5 A molecule traveling through a gas, colliding with other molecules in its path. Although the other molecules are shown in stationary positions, they are also moving in a similar fashion.

way to look at the situation is to consider our single molecule to have a *radius* of d and all the other molecules to be *points*, as in Fig. 21-6b. This does not change our criterion for a collision.

As our single molecule zig-zags through the gas, it sweeps out a broken cylinder whose cross-sectional area is πd^2. If we watch this molecule for a time interval Δt, it moves a distance $v \Delta t$, where v is its assumed speed. The stretched-out length of our broken cylinder is then $v \Delta t$, and the volume of the cylinder is $(\pi d^2)(v \Delta t)$. The number of collisions that occur is then equal to the number of (point) molecules that lie within this cylinder; see Fig. 21-7.

Since N/V is the number of molecules per unit volume, the number of collisions is N/V times the volume of the cylinder, or $(N/V)(\pi d^2 v \Delta t)$. The mean free path is the length of the path (and of the cylinder) divided by this number, or

$$\lambda = \frac{\text{length of path}}{\text{number of collisions}} \approx \frac{v \Delta t}{\pi d^2 v \Delta t \, N/V}$$

$$= \frac{1}{\pi d^2 \, N/V}. \qquad (21\text{-}19)$$

This equation is only approximate because it is based on the assumption that all the molecules except one are at rest. In fact, *all* the molecules are moving; when this is taken properly into account, Eq. 21-18 results. Note that it differs from the (approximate) Eq. 21-19 only by a factor of $1/\sqrt{2}$.

We can even get a glimpse of what is "approximate" about Eq. 21-19. The v in the numerator and that in the denominator are—strictly—not the same. The v in the numerator is $\bar{v}$, the mean speed of the molecule *relative to the container*. The v in the denominator is $\overline{v}_{\text{rel}}$, the mean speed of our single molecule *relative to the other molecules*, which are mov-

ing. It is this latter average speed that determines the number of collisions. A detailed calculation, taking into account the actual speed distribution of the molecules, gives $\overline{v}_{\text{rel}} = \sqrt{2}\,\bar{v}$, which is where the factor $\sqrt{2}$ originates.

The mean free path of air molecules at sea level is about 0.1 μm. At an altitude of 100 km, the density of air has dropped to such an extent that the mean free path rises to about 16 cm. At 300 km, the mean free path is about 20 km. A problem faced by those who would study the physics and chemistry of the upper atmosphere in the laboratory is the fact that no available containers are large enough to hold gas samples that simulate upper atmospheric conditions. Yet studies of the concentrations of Freon, carbon dioxide, and ozone in the upper atmosphere are of vital public concern.

SAMPLE PROBLEM 21-7

The molecular diameters of different kinds of gas molecules can be found experimentally by measuring the rates at which different gases diffuse into each other. For oxygen, $d = 2.9 \times 10^{-10}$ m has been reported.

a. What is the mean free path for oxygen at room temperature ($T = 300$ K) and at an atmospheric pressure of 1.0 atm?

SOLUTION Let us first find N/V, the number of molecules per unit volume under these conditions. From the ideal gas law, 1.0 mol of any gas occupies a volume equal to

$$V = \frac{nRT}{p} = \frac{(1.0 \text{ mol})(8.31 \text{ J/mol} \cdot \text{K})(300 \text{ K})}{(1.0 \text{ atm})(1.01 \times 10^5 \text{ Pa/atm})}$$

$$= 2.47 \times 10^{-2} \text{ m}^3.$$

The number of molecules per unit volume is then

$$\frac{N}{V} = \frac{nN_A}{V} = \frac{(1.0 \text{ mol})(6.02 \times 10^{23} \text{ molecules/mol})}{2.47 \times 10^{-2} \text{ m}^3}$$

$$= 2.44 \times 10^{25} \text{ molecules/m}^3.$$

Equation 21-18 then gives

$$\lambda = \frac{1}{\sqrt{2}\pi d^2 \, N/V}$$

$$= \frac{1}{(\sqrt{2}\pi)(2.9 \times 10^{-10} \text{ m})^2(2.44 \times 10^{25} \text{ m}^{-3})}$$

$$= 1.1 \times 10^{-7} \text{ m}. \qquad \text{(Answer)}$$

FIGURE 21-7   In time Δt the moving molecule sweeps out a broken cylinder of length $v \Delta t$ and radius d. The cylinder is shown straightened out for convenience. The molecule takes part in a number of collisions that is equal to the number of molecules whose centers lie within the cylinder.

This is about 380 molecular diameters. On average, the molecules in such a gas are only about 11 molecular diameters apart.

b. If the average speed of an oxygen molecule is 450 m/s, what is the average collision rate?

SOLUTION We find this rate by dividing the average speed by the mean free path to get

$$\text{rate} = \frac{v}{\lambda} = \frac{450 \text{ m/s}}{1.1 \times 10^{-7} \text{ m}}$$

$$= 4.1 \times 10^9 \text{ s}^{-1}. \qquad \text{(Answer)}$$

Thus—on average—every oxygen molecule makes more than 4 billion collisions per second!

21-7 THE DISTRIBUTION OF MOLECULAR SPEEDS (OPTIONAL)

The root-mean-square speed v_{rms} gives us a general idea of the molecular speed for a gas at a given temperature. We often want to know more. For example, what fraction of the molecules have speeds greater than the rms value? Greater than twice the rms speed? To answer such questions, we need to know how the possible values of speed are distributed among the molecules. Figure 21-8a shows this distribution for oxygen molecules at room temperature; Fig. 21-8b compares it with the distribution at $T = 80$ K.

Maxwell's Distribution Law

In 1852, Scottish physicist James Clerk Maxwell first solved the problem of finding the speed distribution of gas molecules. His **Maxwell speed distribution** law is

$$P(v) = 4\pi \left(\frac{M}{2\pi RT} \right)^{3/2} v^2 e^{-Mv^2/2RT}. \qquad (21\text{-}20)$$

Here v is the molecular speed, T is the gas temperature, M is the molar mass of the gas, and R is the gas constant. It is this equation that is plotted in Fig. 21-8a,b. The quantity $P(v)$ in Eq. 21-20 and Fig. 21-8 is a *distribution function*, defined as follows:

The product $P(v)$ dv (which is a dimensionless quantity) is the fraction of molecules whose speeds lie in the range v to $v + dv$.

As Fig. 21-8a shows, this fraction is equal to the area of a strip whose height is $P(v)$ and whose width is dv. The total area under the distribution curve corresponds to the fraction of the molecules whose speeds lie between zero and infinity. All molecules

(a)

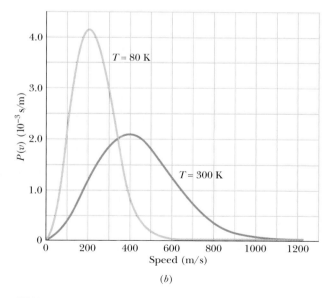

(b)

FIGURE 21-8 (a) The Maxwell speed distribution for oxygen molecules at $T = 300$ K. The three characteristic speeds are marked. (b) The curves for 300 K and 80 K. Note that the molecules move slower at the lower temperature. Because these are probability distributions, the area under each curve has a numerical value of unity.

fall into this category, so the value of this total area is unity.

Figure 21-8a also shows the root-mean-square speed v_{rms} ($= 483$ m/s) and two other measures of the oxygen speed distribution. The *most probable speed* v_P ($= 395$ m/s) is the speed at which $P(v)$ is a maximum. The *average speed* $\bar{v}$ ($= 445$ m/s) is—as its name suggests—a simple average of the molecular speeds. A small number of molecules, lying in the extreme right-hand tail of the distribution curve, can have speeds that are several times the average speed. This simple fact, as we shall demonstrate, makes possible both rain and sunshine.

Rain

The speed distribution of water molecules in, say, a pond at summertime temperatures can be represented by a curve similar to that of Fig. 21-8a. Most of the molecules do not have nearly enough kinetic energy to escape from the water through its surface. However, small numbers of very fast molecules with speeds far out in the tail of the curve can do so. It is these water molecules that "evaporate," making clouds and rain a possibility.

As the fast water molecules leave the surface, carrying energy with them, the temperature of the remaining water is maintained by heat transfer from the surroundings. Other fast molecules—produced in particularly favorable collisions—quickly take the place of those that have left, and the speed distribution is maintained.

Sunshine

Let the distribution curve of Fig. 21-8a now refer to protons in the core of the sun. The sun's energy is supplied by a nuclear fusion process that starts with the merging of two protons. However, the protons repel each other because of their electrical charges, so protons of average speed do not have enough kinetic energy to initiate this reaction. Very fast protons with speeds in the tail of the distribution curve can do so, however, and thus the sun can shine.

SAMPLE PROBLEM 21-8

A container is filled with oxygen gas maintained at 300 K. What fraction of the molecules have speeds in the range 599–601 m/s? The molar mass M of oxygen is 0.0320 kg/mol.

SOLUTION This speed interval Δv ($= 2$ m/s) is so small that we can treat it as a differential and say that the fraction frac that we seek is given very closely by $P(v)\,\Delta v$, where $P(v)$ is to be evaluated at $v = 600$ m/s, the midpoint of the interval; see the gold strip in Fig. 21-8a. Thus, using Eq. 21-20, we find

$$\text{frac} = P(v)\,\Delta v = 4\pi \left(\frac{M}{2\pi RT}\right)^{3/2} v^2 e^{-Mv^2/2RT}\,\Delta v.$$

For convenience in calculating, let us break this down into five factors, as

$$\text{frac} = (4\pi)(A)(v^2)(e^B)(\Delta v) \qquad (21\text{-}21)$$

in which A and B are

$$A = \left(\frac{M}{2\pi RT}\right)^{3/2} = \left(\frac{0.0320\ \text{kg/mol}}{(2\pi)(8.31\ \text{J/mol}\cdot\text{K})(300\ \text{K})}\right)^{3/2}$$
$$= 2.92 \times 10^{-9}\ \text{s}^3/\text{m}^3$$

and

$$B = -\frac{Mv^2}{2RT} = -\frac{(0.0320\ \text{kg/mol})(600\ \text{m/s})^2}{(2)(8.31\ \text{J/mol}\cdot\text{K})(300\ \text{K})}$$
$$= -2.31.$$

Substituting A and B into Eq. 21-21 yields

$$\text{frac} = (4\pi)(A)(v^2)(e^B)(\Delta v)$$
$$= (4\pi)(2.92 \times 10^{-9}\ \text{s}^3/\text{m}^3)$$
$$\times (600\ \text{m/s})^2(e^{-2.31})(2\ \text{m/s})$$
$$= 2.62 \times 10^{-3}. \qquad \text{(Answer)}$$

Thus, at room temperature, 0.262% of the oxygen molecules will have speeds that lie in the narrow range between 599 and 601 m/s. If the gold strip of Fig. 21-8a were drawn to the scale of this problem, it would be a very thin strip indeed.

SAMPLE PROBLEM 21-9

a. What is the average speed v of oxygen gas molecules at $T = 300$ K? The molar mass M of oxygen is 0.0320 kg/mol.

SOLUTION To find the average speed, we weight each speed v by $P(v)\,dv$, which is the fraction of the molecules whose speeds lie in the interval v to $v + dv$. We then add up (that is, integrate) these fractions over the entire range of speeds. Thus

$$\bar{v} = \int_0^\infty v P(v)\,dv. \qquad (21\text{-}22)$$

The next step is to substitute for $P(v)$ from Eq. 21-20 and evaluate the integral that results. From a table of integrals* we find

$$\bar{v} = \sqrt{\frac{8RT}{\pi M}} \quad \text{(average speed).} \quad (21\text{-}23)$$

Substituting numerical values yields

$$\bar{v} = \sqrt{\frac{(8)(8.31 \text{ J/mol} \cdot \text{K})(300 \text{ K})}{(\pi)(0.0320 \text{ kg/mol})}}$$

$$= 445 \text{ m/s.} \quad \text{(Answer)}$$

b. What is the root-mean-square speed v_{rms} of the oxygen molecules?

SOLUTION   We proceed as in (a) above except that we multiply v^2 (rather than simply v) by the weighting factor $P(v) \, dv$. This leads, after integration, to

$$\overline{v^2} = \int_0^\infty v^2 P(v) \, dv = \frac{3RT}{M}.$$

The rms speed is the square root of this quantity, or

$$v_{\text{rms}} = \sqrt{\overline{v^2}} = \sqrt{\frac{3RT}{M}} \quad \text{(rms speed).} \quad (21\text{-}24)$$

Equation 21-24 is identical to Eq. 21-14, which we derived earlier. The numerical calculation gives

$$v_{\text{rms}} = \sqrt{\frac{(3)(8.31 \text{ J/mol} \cdot \text{K})(300 \text{ K})}{(0.0320 \text{ kg/mol})}}$$

$$= 483 \text{ m/s.} \quad \text{(Answer)}$$

c. What is the most probable speed v_P?

SOLUTION   The most probable speed is the speed at which $P(v)$ of Eq. 21-20 has its maximum value. We find it by requiring that $dP/dv = 0$ and solving for v. Doing so yields (as you should show)

$$v_P = \sqrt{\frac{2RT}{M}} \quad \text{(most probable speed).} \quad (21\text{-}25)$$

Numerically, this yields

$$v_P = \sqrt{\frac{(2)(8.31 \text{ J/mol} \cdot \text{K})(300 \text{ K})}{(0.0320 \text{ kg/mol})}}$$

$$= 395 \text{ m/s.} \quad \text{(Answer)}$$

Table 21-2 summarizes the three measures of the Maxwell speed distribution.

---

*One such table is found in Section A of the familiar *CRC Handbook of Chemistry and Physics;* see integral 667 listed under "Definite Integrals."

TABLE 21-2

SPEED PARAMETERS FOR THE MAXWELL SPEED DISTRIBUTION

PARAMETER	SYMBOL	FORMULA	FOR OXYGEN AT 300 K
Most probable speed	v_P	$\sqrt{2RT/M}$	395 m/s
Average speed	$\bar{v}$	$\sqrt{8RT/\pi M}$	445 m/s
Root-mean-square speed	v_{rms}	$\sqrt{3RT/M}$	483 m/s

21-8 THE MOLAR SPECIFIC HEATS OF AN IDEAL GAS

In this section, we want to derive an expression for the internal energy E_{int} of an ideal gas. We shall then use that result to derive an expression for the molar specific heat of an ideal gas.

The Internal Energy E_{int}

Let us first assume that our ideal gas is a monatomic gas, such as helium, neon, or argon. Next, recall from Chapter 8 that internal energy is the energy associated with random motions of atoms and molecules. So let us assume that the internal energy E_{int} of our ideal gas is simply the sum of the translational kinetic energies of its molecules. The average translational kinetic energy of a single molecule depends only on the gas temperature and is given by Eq. 21-16 as $\overline{K} = \frac{3}{2}kT$. A sample of n moles of such a gas contains nN_A molecules. The internal energy E_{int} of the sample is then

$$E_{\text{int}} = (nN_A)\overline{K} = (nN_A)(\tfrac{3}{2}kT)$$

or, since $N_A k = R$, the gas constant,

$$E_{\text{int}} = \tfrac{3}{2}nRT \quad \text{(monatomic ideal gas).} \quad (21\text{-}26)$$

We see that the internal energy E_{int} is a function of the gas temperature only, being independent of other variables such as the pressure and the density.

With Eq. 21-26 in hand, we are now able to derive an expression for the molar specific heat of an ideal gas. Actually, we derive two expressions, one for the case in which the volume of the gas remains constant as heat is added to it and the other for the case in which the pressure of the gas remains constant during this process. The symbols for these two molar specific heats are C_V and C_p, respectively. (By convention, they are given capital-letter symbols, even though they are a type of specific heat.)

Molar Specific Heat at Constant Volume

Figure 21-9a shows n moles of an ideal gas at pressure p and temperature T, confined to a cylinder of fixed volume V. This *initial state i* of the gas is marked on the p–V curve of Fig. 21-9b. Suppose now that you add a small amount of heat Q to the gas, by slowly turning up the temperature of the reservoir on which the cylinder rests. The gas temperature rises a small amount to $T + \Delta T$, and its pressure to $p + \Delta p$, bringing the gas to *final state f*.

(a)

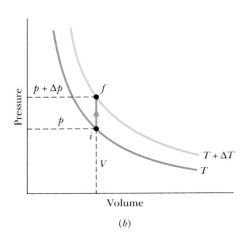

(b)

FIGURE 21-9 (a) The temperature of an ideal gas is raised from T to $T + \Delta T$ in a constant-volume process. Heat is added but no work is done. (b) The process on a p–V diagram.

The defining equation for C_V, the molar specific heat at constant volume, is, in parallel with Eq. 20-3,

$$Q = nC_V \Delta T \qquad \text{(constant volume)}. \quad (21\text{-}27)$$

Substituting this expression for Q into the first law of thermodynamics, we find

$$\Delta E_{\text{int}} + W = nC_V \Delta T.$$

With the volume held constant, $W = 0$. Then solving for C_V we obtain

$$C_V = \frac{1}{n} \frac{\Delta E_{\text{int}}}{\Delta T}. \quad (21\text{-}28)$$

From Eq. 21-26 we see that $\Delta E_{\text{int}}/\Delta T = \frac{3}{2}nR$. Substituting this result into Eq. 21-28 yields

$$C_V = \tfrac{3}{2}R = 12.5 \ \text{J/mol} \cdot \text{K}$$

$$\text{(monatomic gas)}. \quad (21\text{-}29)$$

As Table 21-3 shows, this prediction of the kinetic theory agrees very well with experiment for real monatomic gases, the case that we have assumed. The (predicted and) experimental values of C_V for diatomic and polyatomic gases are substantially higher than those for monatomic gases for reasons that will be suggested in Section 21-9.

Now that we have defined the molar specific heat at constant volume for an ideal gas, we can generalize Eq. 21-26 for the internal energy of an ideal gas:

$$E_{\text{int}} = nC_V T \qquad \text{(ideal gas)}. \quad (21\text{-}30)$$

TABLE 21-3
MOLAR SPECIFIC HEATS

MOLECULE	EXAMPLE	C_V (J/mol·K)
	Ideal	12.5
Monatomic	He	12.5
	Ar	12.6
	Ideal	20.8
Diatomic	N_2	20.7
	O_2	20.8
	Ideal	24.9
Polyatomic	NH_4	29.0
	CO_2	29.7

This equation applies not only to an ideal mon-atomic gas, but to any other ideal gas (the appropriate value of C_V must be used). Just as with Eq. 21-26, we see that the internal energy depends on the temperature of the gas but not on its pressure or density.

When a confined ideal gas undergoes a temperature change ΔT, then from either Eq. 21-28 or Eq. 21-30 we can write the resulting change in its internal energy as

$$\Delta E_{\text{int}} = nC_V \Delta T \qquad \text{(ideal gas, any process).} \qquad (21\text{-}31)$$

This equation has an important message: the change in the internal energy of a confined ideal gas does not depend on the *type of process* that produced a change in the temperature of the gas.

As examples, consider the three paths between the two isotherms in the $p-V$ diagram of Fig. 21-10. Path 1 represents a constant-volume process. Path 2 represents a constant-pressure process (that we are about to examine). And path 3 represents a process in which no heat is exchanged with the system's environment (we discuss this in Section 21-11). Although the values of heat Q and work W associated with these three paths differ, as do p_f and V_f, the values of ΔE_{int} associated with the three paths are identical because they all involve the same temperature change ΔT.

Molar Specific Heat at Constant Pressure

We now assume that the temperature of the gas is increased by the same small amount ΔT as previously, but that the necessary heat Q is added with the gas under constant pressure. A mechanism for doing this is shown in Fig. 21-11a; the $p-V$ diagram for the process is plotted in Fig. 21-11b. We can guess at once that the molar specific heat at constant pressure C_p, which we define from

$$Q = nC_p \Delta T \qquad \text{(constant pressure),} \qquad (21\text{-}32)$$

(a)

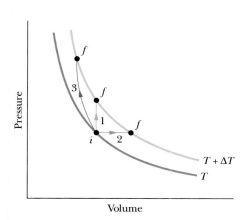

FIGURE 21-10 Three paths representing three different processes that take an ideal gas from an initial state i at temperature T to some final state f at temperature $T + \Delta T$. The change ΔE_{int} in the internal energy of the gas is the same for these three processes and for any others that result in the same change of temperature.

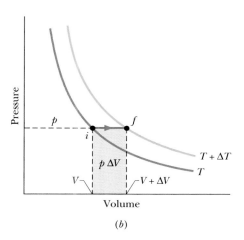

(b)

FIGURE 21-11 (a) The temperature of an ideal gas is raised from T to $T + \Delta T$ in a constant-pressure process. Heat is added and work is done in lifting the loaded piston. (b) The process on a $p-V$ diagram. The work, $p\,\Delta V$, is the sand-colored area under the line connecting the initial i and the final f states.

will be greater than the molar specific heat at constant volume: energy must be supplied not only to raise the temperature, but also to do external work, that is, to lift the weighted piston of Fig. 21-11*a*.

To relate C_p to C_V, we start with the first law of thermodynamics:

$$\Delta E_{\text{int}} = Q - W. \qquad (21\text{-}33)$$

We next replace each term in Eq. 21-33. For ΔE_{int}, we substitute from Eq. 21-31. For Q, we substitute from Eq. 21-32. To replace W, we first note that since the pressure remains constant, Eq. 20-11 tells us that $W = p \, \Delta V$. Then we note that, using the ideal gas equation ($pV = nRT$), we can write

$$W = p \, \Delta V = nR \, \Delta T.$$

Making these substitutions in Eq. 21-33, and then dividing through by $n \, \Delta T$, we find

$$C_V = C_p - R,$$

or

$$C_p - C_V = R. \qquad (21\text{-}34)$$

This prediction of kinetic theory agrees well with experiment, not only for monatomic gases but for gases in general, as long as their density is low enough so that we may treat them as ideal.

SAMPLE PROBLEM 21-10

A bubble of 5.00 mol of (monatomic) helium is submerged at a certain depth in liquid water when the water (and thus the helium) undergoes a temperature increase ΔT of 20.0°C at constant pressure. As a result, the bubble expands.

a. How much heat Q is added to the helium during the expansion and temperature increase?

SOLUTION Treating the helium as an ideal gas, we begin with Eq. 21-32 ($Q = nC_p \, \Delta T$). We then use Eqs. 21-34 and 21-29 to write

$$C_p = C_V + R = \tfrac{5}{2}R,$$

obtaining

$$
\begin{aligned}
Q = nC_p \, \Delta T &= n(\tfrac{5}{2}R) \, \Delta T \\
&= (5.00 \text{ mol})(2.5)(8.31 \text{ J/mol} \cdot \text{K})(20.0°C) \\
&= 2077.5 \text{ J} \approx 2080 \text{ J}. \qquad \text{(Answer)}
\end{aligned}
$$

b. What is the change ΔE_{int} in the internal energy of the helium during the temperature increase?

SOLUTION Even though the temperature of the helium increases at constant pressure (and *not* at constant volume), we use Eq. 21-31 to calculate the change in internal energy (for the reason given with that equation):

$$
\begin{aligned}
\Delta E_{\text{int}} = nC_V \, \Delta T \\
= (5.00 \text{ mol})(1.5)(8.31 \text{ J/mol} \cdot \text{K})(20.0°C) \\
= 1246.5 \text{ J} \approx 1250 \text{ J}. \qquad \text{(Answer)}
\end{aligned}
$$

c. How much work W is done by the helium as it expands against the surrounding water during the temperature increase?

SOLUTION Using the first law of thermodynamics, we write

$$
\begin{aligned}
W = Q - \Delta E_{\text{int}} &= 2077.5 \text{ J} - 1246.5 \text{ J} \\
&= 831 \text{ J}. \qquad \text{(Answer)}
\end{aligned}
$$

Note that during the temperature increase, only a portion (1250 J) of the heat (2080 J) that is transferred into the helium goes to increasing the internal energy of the helium and thus the temperature of the helium. The rest (831 J) is transferred out of the helium as work that the helium does during the expansion. If the water were frozen, it would not allow that expansion. Then the same temperature increase of 20.0°C would require only 1250 J of heat, because no work would be done by the helium.

21-9 THE EQUIPARTITION OF ENERGY

As Table 21-3 shows, the prediction that $C_V = \tfrac{3}{2}R$ agrees with experiment for monatomic gases but fails for diatomic and polyatomic gases. Let us try to explain the discrepancy by considering the possibility that molecules with more than one atom can store internal energy in forms other than translational motion.

Figure 21-12 shows kinetic theory models of helium (a monatomic gas), oxygen (diatomic), and methane (polyatomic). On the basis of their structure, it seems reasonable to assume that monatomic molecules—which are essentially pointlike and have only a very small rotational inertia about any axis—can store energy only in their translational motion. Diatomic and polyatomic molecules, however,

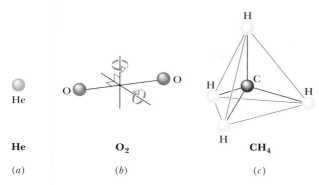

He   O₂   CH₄

(a)   (b)   (c)

FIGURE 21-12 Models of molecules as used in kinetic theory: (*a*) helium, a typical monatomic molecule; (*b*) oxygen, a typical diatomic molecule (two rotation axes are shown); and (*c*) methane, a typical polyatomic molecule.

should be able to store substantial additional amounts of energy by rotating or oscillating.

How do we take these possibilities quantitatively into account? James Clerk Maxwell first showed how to do so, by introducing the theorem of the **equipartition of energy:**

> Every kind of molecule has a certain number f of *degrees of freedom,* which are independent ways in which it can store energy. Each such degree of freedom has associated with it—on average —an energy of $\frac{1}{2}kT$ per molecule (or $\frac{1}{2}RT$ per mole).

For translational motion, there are three degrees of freedom, corresponding to the three perpendicular axes along which such motion can occur. For rotational motion, a monatomic molecule has no degrees of freedom. A diatomic molecule—the rigid dumbbell of Fig. 21-12*b*—has two rotational degrees of freedom, corresponding to the two perpendicular axes about which it can store rotational energy. Such

a molecule cannot store rotational energy about the axis connecting the nuclei of its two constituent atoms because its rotational inertia about this axis is approximately zero. A molecule with more than two atoms has six degrees of freedom, three rotational and three translational.

To extend the treatment of Section 21-8 to diatomic and polyatomic gases, it is necessary to retrace the derivations of that section in detail, replacing Eq. 21-26 ($E_{\text{int}} = \frac{3}{2}nRT$) by $E_{\text{int}} = (f/2)nRT$, where f is the number of degrees of freedom listed in Table 21-4. Doing so leads to the prediction

$$C_V = \left(\frac{f}{2}\right) R = 4.16f \text{ J/mol} \cdot \text{K}, \quad (21\text{-}35)$$

which agrees—as it must—with Eq. 21-29 for $f = 3$. As Table 21-3 shows, this prediction agrees with experiment for monatomic ($f = 3$) and diatomic ($f = 5$) molecules, but it is too low for polyatomic molecules.

SAMPLE PROBLEM 21-11

A cabin of volume V is filled with air (which we consider to be an ideal diatomic gas) at an initial low temperature T_1. After you light a wood stove, the air temperature increases to a value T_2. What is the resulting change in the store of internal energy of the air in the cabin?

SOLUTION This situation differs from other situations we have examined in that the container (the cabin) is not sealed. If it were sealed, the ideal gas law ($pV = nRT$) would tell us that as the temperature of the air in the cabin increased, the air pressure would also. However, because the cabin is not perfectly airtight, air molecules leave the cabin through various openings as the air temperature increases, so that the air pressure inside the cabin always matches the air pressure outside the cabin.

TABLE 21-4
DEGREES OF FREEDOM FOR VARIOUS MOLECULES

| | | DEGREES OF FREEDOM | | | PREDICTED MOLAR SPECIFIC HEATS | |
MOLECULE	EXAMPLE	TRANSLATIONAL	ROTATIONAL	TOTAL (f)	C_V (Eq. 21-35)	$C_p = C_V + R$
Monatomic	He	3	0	3	$\frac{3}{2}R$	$\frac{5}{2}R$
Diatomic	O_2	3	2	5	$\frac{5}{2}R$	$\frac{7}{2}R$
Polyatomic	CH_4	3	3	6	$3R$	$4R$

From Eq. 21-31, the change in the internal energy of the air in the room is

$$\Delta E_{int} = nC_V \, \Delta T.$$

Using the ideal gas law, we may substitute $\Delta(pV)/R$ for $n \, \Delta T$, finding

$$\Delta E_{int} = \frac{C_V}{R} \, \Delta(pV).$$

From this we see that since neither the pressure p nor the volume V of the air within the cabin changes,

$$\Delta E_{int} = 0, \qquad \text{(Answer)}$$

even though the temperature changes.

So why does the cabin feel more comfortable at the higher temperature? There are at least two factors involved. You have a tendency to cool because (1) you emit electromagnetic radiation (thermal radiation), and (2) you lose heat to air molecules that collide with you. If you increase the room temperature, (1) you increase the amount of thermal radiation you intercept from the surfaces within the room, replacing heat you have emitted; and (2) you increase the kinetic energy of the air molecules that collide with you, so you lose less heat to them.

21-10 A HINT OF QUANTUM THEORY

The next logical step is to see if we cannot improve the agreement of kinetic theory with experiment by taking into account the internal energy stored in the form of molecular oscillations. However, it proves unrewarding for us to push forward any further with classical kinetic theory, which is based on Newtonian mechanics.

In the middle of the last century, James Clerk Maxwell (along with others) was struggling to build a coherent picture of the structure of an atom, based on kinetic theory and on measurements of the wavelengths of light emitted by atoms. His work was to no avail, and Maxwell was led to speculate that classical theory—in this context—was in some way flawed.

Maxwell, baffled in his attempts to understand the atom, wrote that nothing remained but to adopt the attitude of "thoroughly conscious ignorance that is the prelude to every real advance in knowledge." It was only in the first quarter of the present century that—with the development of modern quantum physics—these real advances in knowledge came to pass.

Figure 21-13 suggests how the phenomenon of energy quantization (which is central to quantum

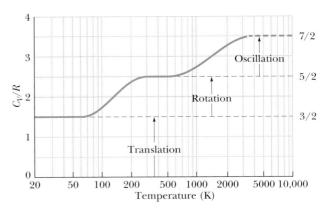

FIGURE 21-13 A plot of the ratio C_V/R versus temperature for hydrogen gas. Because rotational and oscillatory motions occur at quantized energies, only translation is possible at very low temperatures. As the temperature increases, rotational motion can be excited during collisions. At still higher temperatures, oscillatory motion can be excited.

theory) enters the picture. The ratio C_V/R for diatomic hydrogen gas is plotted against temperature, with the temperature scale made logarithmic for convenience. At very low temperatures, below about 80 K, this ratio has the value 1.5, characteristic of a *monatomic* gas. Even though hydrogen is a diatomic molecule, only its translational degrees of freedom seem to be excited. The reason is that rotational motion occurs at quantized energies and, at these low temperatures, H_2 molecules simply do not have enough kinetic energy to set each other rotating when they collide. As the temperature rises, however, rotation becomes possible so that, at more "ordinary" temperatures, the hydrogen molecule behaves like the rigid dumbbell that we have assumed in our classical theory. At still higher temperatures, oscillatory motions (which also occur only at quantized energies) become possible, but the molecule dissociates into two atoms at about 3200 K.

As you can see, quantum physics accounts in detail for this full range of phenomena. So in retrospect, it is not incorrect to say that the seeds of quantum physics lay in the kinetic theory of gases.

21-11 THE ADIABATIC EXPANSION OF AN IDEAL GAS

We saw in Section 18-2 that sound waves are propagated through air and other gases as a series of compressions and expansions that take place so rapidly that there is no time for heat transfer from one part of the medium to another. As we saw in Section 20-6, processes for which $Q = 0$ are *adiabatic processes*. We can ensure that $Q = 0$ either by carrying out the process very quickly (as in sound waves) or by doing it slowly in a heavily insulated environment. Let us see what the kinetic theory has to say about adiabatic processes.

Figure 21-14*a* shows an insulated cylinder containing an ideal gas and resting on an insulating stand. By removing weight from the piston, we can allow the gas to expand adiabatically. As the volume increases, both the pressure and the temperature drop. We shall prove below that the relation between the pressure and the volume during such an adiabatic process is

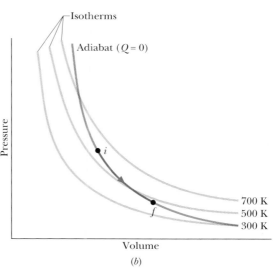

(a)

(b)

FIGURE 21-14 (*a*) The volume of an ideal gas is increased by removing weight from the piston. The process is adiabatic ($Q = 0$). (*b*) The process on a *p–V* diagram. Note that it proceeds along an adiabat.

When carbon dioxide gas in cold champagne (or soda pop) is suddenly released, it not only can shoot out a spray of liquid but can also produce a "fog." Its expansion is so rapid as to be adiabatic, and the sudden decrease in its internal energy causes it to cool, which causes water vapor in it to condense to form the tiny drops of the fog.

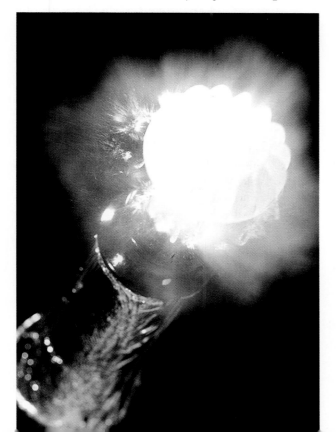

$$pV^{\gamma} = \text{a constant} \qquad \begin{matrix}\text{(adiabatic} \\ \text{process)}\end{matrix} \qquad (21\text{-}36)$$

in which $\gamma = C_p / C_V$, the ratio of the molar specific heats for the gas. Equation 21-36 tells us that on a *p–V* diagram, such as in Fig. 21-14*b*, the process occurs along a line (called an *adiabat*) that is given by $p = (\text{a constant}) / V^{\gamma}$. When the gas goes from an initial state *i* to a final state *f*, we can rewrite Eq. 21-36 as

$$p_i V_i^{\gamma} = p_f V_f^{\gamma} \qquad \text{(adiabatic process).} \qquad (21\text{-}37)$$

We can also write an equation for an adiabatic process in terms of T and V. To do so, we use the ideal gas equation ($pV = nRT$) to eliminate p from Eq. 21-36, finding

$$\left(\frac{nRT}{V}\right) V^\gamma = \text{a constant.}$$

Because n and R are constants, we can rewrite this in the alternative form

$$TV^{\gamma-1} = \text{a constant} \qquad \substack{\text{(adiabatic} \\ \text{process),}} \qquad (21\text{-}38)$$

in which the constant is different from that in Eq. 21-36. When the gas goes from an initial state i to a final state f, we can rewrite Eq. 21-38 as

$$T_i V_i^{\gamma-1} = T_f V_f^{\gamma-1} \qquad \text{(adiabatic process).} \qquad (21\text{-}39)$$

The Specific Heat Ratio γ

The ratio γ can be measured directly in various ways, without the need to measure separately the two molar specific heats C_p and C_V. A theoretical expression can also be derived for it from kinetic theory. We do so by combining Eqs. 21-35 ($C_V = \frac{1}{2}fR$) and 21-34 ($C_p - C_V = R$), to find that

$$\gamma = \frac{C_p}{C_V} = \frac{R + C_V}{C_V} = \frac{R + \frac{1}{2}fR}{\frac{1}{2}fR} = 1 + \frac{2}{f}. \qquad (21\text{-}40)$$

This result agrees well with experiment for monatomic gases (for which $f = 3$ and thus $\gamma = 1.67$) and for diatomic gases (for which $f = 5$ and $\gamma = 1.40$).

Proof of Eq. 21-36

Suppose that you remove a little lead shot from the piston of Fig. 21-14a, allowing the ideal gas to push the piston and remaining shot upward and thus to increase the volume by a differential amount dV. Since the volume change is tiny, we may assume that the pressure p of the gas on the piston is constant during the change. This assumption allows us to say that the work dW done by the gas during the volume increase is equal to $p\,dV$. From Eq. 20-13, the first law of thermodynamics can then be written as

$$dE_{\text{int}} = Q - p\,dV. \qquad (21\text{-}41)$$

Since the gas is thermally insulated (and thus the expansion is adiabatic), we substitute $Q = 0$. Then we use Eq. 21-31 to substitute $nC_V\,dT$ for dE_{int}. With these substitutions, and after some rearranging, we have

$$n\,dT = -\left(\frac{p}{C_V}\right) dV. \qquad (21\text{-}42)$$

Now from the ideal gas law ($pV = nRT$) we have

$$p\,dV + V\,dp = nR\,dT.$$

Replacing R with its equal, $C_p - C_V$, leads to

$$n\,dT = \frac{p\,dV + V\,dp}{C_p - C_V}. \qquad (21\text{-}43)$$

Equating Eqs. 21-42 and 21-43 and rearranging then lead to

$$\frac{dp}{p} + \left(\frac{C_p}{C_V}\right)\frac{dV}{V} = 0.$$

Replacing the ratio of the molar specific heats with γ and integrating (see Appendix G) yield

$$\ln p + \gamma \ln V = \text{a constant,}$$

or

$$pV^\gamma = \text{a constant,} \qquad (21\text{-}44)$$

which (compare Eq. 21-36) is what we set out to prove.

TABLE 21-5
FOUR SPECIAL PROCESSES

PATH IN FIG. 21-15	CONSTANT QUANTITY	PROCESS TYPE	SOME SPECIAL RESULTS ($\Delta E_{\text{int}} = Q - W$ AND $\Delta E_{\text{int}} = nC_V \Delta T$ FOR ALL PATHS)
1	p	Isobaric	$Q = nC_p \Delta T$; $W = p\,\Delta V$
2	T	Isothermal	$Q = W = nRT \ln(V_f/V_i)$; $\Delta E_{\text{int}} = 0$
3	$pV^\gamma, TV^{\gamma-1}$	Adiabatic	$Q = 0$; $W = -\Delta E_{\text{int}}$
4	V	Isochoric	$Q = \Delta E_{\text{int}} = nC_V \Delta T$; $W = 0$

SAMPLE PROBLEM 21-12

In Sample Problem 21-2, 1 mol of oxygen (assumed an ideal gas) expands isothermally (at 310 K) from an initial volume of 12 L to a final volume of 19 L. What would be the final temperature if the gas had expanded adiabatically to this same final volume? Oxygen (O_2) is diatomic, so that $\gamma = 1.40$.

SOLUTION From Eq. 21-39 we write

$$T_f = \frac{T_i V_i^{\gamma-1}}{V_f^{\gamma-1}} = \frac{(310\ \text{K})(12\ \text{L})^{1.40-1}}{(19\ \text{L})^{1.40-1}}$$

$$= 258\ \text{K.} \qquad\qquad \text{(Answer)}$$

The fact that the gas has cooled (from 310 to 258 K) means that its internal energy has been correspondingly reduced. The lost internal energy has gone to do the work of expansion (such as lifting the weighted piston in Fig. 21-14a).

PROBLEM SOLVING

TACTIC 3: A GRAPHICAL SUMMARY
In this chapter we have discussed four special processes that an ideal gas can undergo. An example of each is

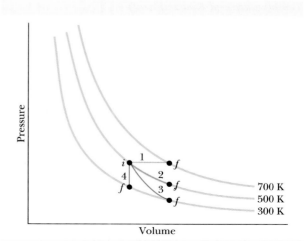

FIGURE 21-15 A p–V diagram representing four special processes through which an ideal gas may be taken.

shown in Fig. 21-15, and some associated characteristics are given in Table 21-5, including two process names (isobaric and isochoric) that we have not used but which you might see in other courses.

REVIEW & SUMMARY

Kinetic Theory of Gases
The *kinetic theory of gases* relates the *macroscopic* properties of gases (for example, pressure and temperature) to the *microscopic* properties of the gas molecules (for example, speed and kinetic energy).

Avogadro's Number
One mole of a substance contains N_A (*Avogadro's number*) elementary units (usually atoms or molecules), where N_A is found experimentally to be

$$N_A = 6.02 \times 10^{23}\ \text{mol}^{-1} \qquad \substack{\text{(Avogadro's} \\ \text{number).}} \qquad (21\text{-}1)$$

One molar mass M of any substance is the mass of one mole of the substance.

Ideal Gas
An *ideal gas* is one for which the pressure p, volume V, and temperature T are related by

$$pV = nRT \qquad \text{(ideal gas law).} \qquad (21\text{-}4)$$

Here n is the number of moles of the gas present and R ($= 8.31\ \text{J/mol}\cdot\text{K}$) is the **gas constant.**

Work in Isothermal Expansion
The work done *by* an ideal gas during an **isothermal** (constant-temperature) change from volume V_i to volume V_f is

$$W = nRT \ln \frac{V_f}{V_i} \qquad \substack{\text{(ideal gas,} \\ \text{isothermal process).}} \qquad (21\text{-}8)$$

Pressure, Temperature, and Molecular Speed
The pressure exerted by n moles of an ideal gas, in terms of the speed of its molecules, is

$$p = \frac{nMv_{\text{rms}}^2}{3V}, \qquad (21\text{-}13)$$

where $v_{\text{rms}} = \sqrt{\overline{v^2}}$ is the **root-mean-square speed** of the molecules of the gas. With Eq. 21-4 this gives

$$v_{\text{rms}} = \sqrt{\frac{3RT}{M}}. \qquad (21\text{-}14)$$

Temperature and Kinetic Energy

The average translational kinetic energy $\overline{K}$ per molecule of an ideal gas is

$$\overline{K} = \tfrac{3}{2}kT. \qquad (21\text{-}16)$$

Here k ($= R/N_A = 1.38 \times 10^{-23}$ J/K) is the **Boltzmann constant.**

Mean Free Path

The *mean free path* λ of a gas molecule is its average path length between collisions and is given by

$$\lambda = \frac{1}{\sqrt{2}\,\pi d^2\, N/V}, \qquad (21\text{-}18)$$

where N/V is the number of molecules per unit volume and d is the molecular diameter.

Maxwell Speed Distribution

The *Maxwell speed distribution* $P(v)$ is a function such that $P(v)\,dv$ gives the *fraction* of molecules with speeds between v and $v + dv$:

$$P(v) = 4\pi \left(\frac{M}{2\pi RT}\right)^{3/2} v^2 e^{-Mv^2/2RT}. \qquad (21\text{-}20)$$

Three measures of the distribution of speeds among the molecules of a gas are

$$v_P = \sqrt{\frac{2RT}{M}} \quad \text{(most probable speed)}, \qquad (21\text{-}25)$$

$$\overline{v} = \sqrt{\frac{8RT}{\pi M}} \quad \text{(average speed)}, \qquad (21\text{-}23)$$

and the rms speed defined above in Eq. 21-14.

Molar Specific Heats

The molar specific heat C_V of a gas at constant volume is

$$C_V = \frac{1}{n}\frac{Q}{\Delta T}, \qquad (21\text{-}27)$$

in which Q is heat transferred to or from the gas, and ΔT is the resulting temperature change of the gas. C_V can also be written as

$$C_V = \frac{1}{n}\frac{\Delta E_{int}}{\Delta T}, \qquad (21\text{-}28)$$

where E_{int} is the internal energy of the gas. For an ideal gas, E_{int} depends on the temperature of the gas and can be written as

$$E_{int} = nC_V T \quad \text{(ideal gas)}. \qquad (21\text{-}30)$$

Thus if a confined ideal gas undergoes a temperature change due to any process, its internal energy changes by

$$\Delta E_{int} = nC_V\,\Delta T \quad \text{(ideal gas, any process)}. \qquad (21\text{-}31)$$

The first law of thermodynamics leads to the relation

$$C_p - C_V = R, \qquad (21\text{-}34)$$

in which C_p is the molar specific heat at *constant pressure.* C_p is defined to be

$$C_p = \frac{1}{n}\frac{Q}{\Delta T}, \qquad (21\text{-}32)$$

in which Q is heat transferred to or from the gas, and ΔT is the resulting temperature change of the gas.

Equipartition Theorem

We find C_V itself by using the *equipartition of energy* theorem, which states that every *degree of freedom* of a molecule (that is, every independent way it can store energy) has associated with it—on average—an energy $\tfrac{1}{2}kT$ per molecule ($= \tfrac{1}{2}RT$ per mole). If f is the number of degrees of freedom, then $E_{int} = (f/2)nRT$ and

$$C_V = \left(\frac{f}{2}\right)R = 4.16f \text{ J/mol}\cdot\text{K}. \qquad (21\text{-}35)$$

For monatomic gases $f = 3$ (three translational degrees); for diatomic gases $f = 5$ (three translational and two rotational degrees). Table 21-3 shows that estimates of C_V computed with Eq. 21-35 for monatomic and diatomic gases agree well with experiment.

C_V and Quantum Physics

The dependence of C_V on temperature for real gases provides striking evidence for the quantum nature of rotational motion. At high temperatures, C_V is as predicted by the equipartition principle. At low temperatures, however, molecular collisions are not energetic enough to excite the quantized rotational motion. (See Fig. 21-13.)

Adiabatic Expansion

When an ideal gas undergoes an adiabatic volume change (a change for which $Q = 0$), its pressure and volume are related by

$$pV^\gamma = \text{a constant} \quad \text{(adiabatic process)}, \qquad (21\text{-}36)$$

in which γ ($= C_p/C_V$) is the ratio of molar specific heats for the gas.

QUESTIONS

1. In kinetic theory we assume that there are a large number of molecules in a gas. Real gases behave like an ideal gas at low densities. Are these statements contradictory? If not, what conclusion can you draw from them?

2. We have assumed that the collisions of gas molecules with the walls of their container are elastic. Actually, the collisions may be inelastic. Why does this make no difference as long as the walls are at the same temperature as the gas?

3. On a humid day, some say that the air is "heavy." How does the density of humid air compare with that of dry air at the same temperature and pressure?

4. Where does the average speed of air molecules in still air at room temperature fit into this sequence: 0; 2 m/s (walking speed); 30 m/s (fast car); 500 m/s (supersonic airplane); 1.1×10^4 m/s (escape speed from the Earth); 3×10^8 m/s (speed of light)?

5. Two equal-size rooms communicate through an open doorway. However, the average temperatures in the two rooms are maintained at different values. In which room is there more air?

6. Molecular motions are maintained by no outside force, yet continue indefinitely with no sign of diminishing speed. Why doesn't friction bring these tiny particles to rest, as it does other moving particles?

7. What justification is there for neglecting the changes in gravitational potential energy of molecules in a gas?

8. We have assumed that the force exerted by gas molecules on the wall of a container is steady in time. How is this justified?

9. The average velocity of the molecules in a gas must be zero if the gas as a whole and its container are not in translational motion. Explain how it can be that the *average speed* is not zero.

10. Consider a hot, stationary golf ball sitting on a tee and a cold golf ball just moving off the tee after being hit. The total kinetic energy of the molecules in the balls, relative to the tee, can be the same in the two cases. Explain how. What is the difference between the two cases?

11. Justify the fact that the pressure of a gas depends on the *square* of the speed of its particles by explaining the dependence of pressure on the collision frequency and the momentum transfer of the particles.

12. Why does the boiling temperature of a liquid increase with pressure?

13. How is the speed of sound in a gas related to the pressure and temperature of the gas?

14. Far above the Earth's surface the gas temperature is reported to be on the order of 1000 K. However, a person placed in such an environment would freeze to death rather than vaporize. Explain.

15. Why doesn't the Earth's atmosphere leak away? At the top of the atmosphere molecules will occasionally be headed out with a speed exceeding the escape speed. Isn't it just a matter of time?

16. Titan, one of Saturn's many moons, has an atmosphere, but our own moon does not. Why?

17. As heat is added to ice, it melts to become liquid, and then the liquid eventually boils to become a vapor (or gas). However, as solid carbon dioxide is heated it goes directly to the vapor state—we say it *sublimes*—without passing through a liquid state. How could liquid carbon dioxide be produced?

18. Does the concept of temperature apply to a vacuum? Consider interplanetary space, for example.

19. What direct evidence do we have for the existence of atoms? What indirect evidence is there?

20. How, if at all, would you expect the composition of the atmosphere to change with altitude?

21. We often say that we see the steam emerging from the spout of a kettle in which water is boiling. However, steam itself is a colorless gas. What is it that we really see?

22. Why does smoke rise, rather than fall, from a lighted candle? Explain in terms of molecular collisions.

23. Would a gas whose molecules were true geometric points obey the ideal gas law?

24. If you fill a saucer with water at room temperature the water, under normal conditions, will evaporate completely. It is easy to believe that some of the more energetic molecules can escape from the water surface but how can *all* of them eventually escape? Many of them—in fact the vast majority—do not have enough energy to do so.

25. Give a qualitative explanation of the connection between the mean free path of ammonia molecules in air and the time it takes to smell the ammonia when a bottle is opened across the room.

26. List effective ways of increasing the number of molecular collisions per unit time in a gas.

27. If the molecules of a gas are not spherical, what meaning can we give to d in Eq. 21-18 for the mean free path? In which gases would the molecules act the most nearly as rigid spheres?

28. The two opposite walls of a container of gas are kept at different temperatures. The air between the panes of glass in a storm window is a familiar example. Describe in terms of kinetic theory the mechanism of heat conduction through the gas.

29. A gas can transmit only those sound waves whose wavelength is long compared with the mean free path. Can you explain this? Describe a situation for which this limitation would be important.

30. Justify qualitatively the statement that, in a mixture of molecules of different kinds of complete equilibrium, each kind of molecule has the same Maxwellian distribution in speed that it would have if the other kinds were not present.

31. Is it possible for a gas to consist of molecules that all have the same speed?

32. What observation is good evidence that not all molecules of a body are moving with the same speed at a given temperature?

33. The fraction of molecules within a given range Δv of the rms speed decreases as the temperature of a gas rises. Explain.

34. (a) Do half the molecules in a gas in thermal equilibrium have speeds greater than v_P? Than $\overline{v}$? Than v_{rms}? (b) Which speed, v_P, $\overline{v}$, or v_{rms}, corresponds to a molecule having average kinetic energy?

35. Keeping in mind that the internal energy of a body consists of kinetic energy and potential energy of its particles, how would you distinguish between the internal energy of a body and its temperature?

36. The gases in two identical containers are at 1 atm pressure and room temperature. One contains helium gas (monatomic, molar mass 4 g/mol) and the other contains an equal number of moles of argon gas (monatomic, molar mass 40 g/mol). If 1 cal of heat added to the helium gas increases its temperature by a certain amount, what amount of heat must be added to the argon gas to increase its temperature by the same amount?

37. Explain how we might keep a gas at a constant temperature during a thermodynamic process.

38. Why is it more common to excite radiation from gaseous atoms by use of electrical discharge than by thermal methods?

39. A certain quantity of an ideal gas is compressed to half its initial volume. The process may be adiabatic, isothermal, or isobaric (at constant pressure). For which process is the greatest amount of mechanical work required?

40. Explain why the temperature of a gas drops in an adiabatic expansion.

41. If hot air rises, why is it cooler at the top of a mountain than near sea level? (*Hint:* Air is a poor conductor of heat.)

42. A sealed rubber balloon contains a very light gas. The balloon is released and it rises high into the atmosphere. Describe and explain the temperature of the gas and the size of the balloon.

EXERCISES & PROBLEMS

SECTION 21-2 AVOGADRO'S NUMBER

1E. Gold has a molar mass of 197 g/mol. Consider a 2.50-g sample of pure gold. (a) Calculate the number of moles of gold present. (b) How many atoms of gold are in the sample?

2E. Find the mass in kilograms of 7.50×10^{24} atoms of arsenic, which has a molar mass of 74.9 g/mol.

3P. If the water molecules in 1.00 g of water were distributed uniformly over the surface of the Earth, how many such molecules would there be on 1.00 cm^2 of the Earth's surface?

4P. Consider this sentence: A _____ of water contains about as many molecules as there are _____ s of water in all the oceans. What single word best fits both blank spaces: drop, teaspoon, tablespoon, cup, quart, barrel, or ton? The oceans cover 75% of the Earth's surface and have an average depth of about 5 km. (After Edward M. Purcell.)

5P. A distinguished scientist has written: "There are enough molecules in the ink that makes one letter of this sentence to provide not only one for every inhabitant of the Earth, but one for every creature if each star of our galaxy had a planet as populous as the Earth." Check this statement. Assume the ink sample (molar mass = 18 g/mol) to have a mass of 1 μg, the population of the Earth to be 5×10^9, and the number of stars in our galaxy to be about 10^{11}.

SECTION 21-3 IDEAL GASES

6E. (a) What is the volume occupied by 1.00 mol of an ideal gas at standard conditions, that is, a pressure of 1.00 atm ($= 1.01 \times 10^5$ Pa) and temperature of 0°C ($= 273$ K)? (b) Show that the number of molecules per cubic centimeter (the *Loschmidt number*) at standard conditions is 2.69×10^{19}.

7E. Compute (a) the number of moles and (b) the number of molecules in a gas contained in a volume of 1.00 cm^3 at a pressure of 100 Pa and a temperature of 220 K.

8E. The best vacuum that can be attained in the laboratory corresponds to a pressure of about 1.00×10^{-18} atm, or 1.01×10^{-13} Pa. How many molecules are there per cubic centimeter in such a vacuum at 293 K?

9E. A quantity of ideal gas at 10.0°C and a pressure of 100 kPa occupies a volume of 2.50 m³. (a) How many moles of the gas are present? (b) If the pressure is now raised to 300 kPa and the temperature is raised to 30.0°C, how much volume will the gas occupy? Assume no leaks.

10E. Oxygen gas having a volume of 1000 cm³ at 40.0°C and a pressure of 1.01×10^5 Pa expands until its volume is 1500 cm³ and its pressure is 1.06×10^5 Pa. Find (a) the number of moles of oxygen in the system and (b) its final temperature.

11E. An automobile tire has a volume of 1000 in.³ and contains air at a gauge pressure of 24.0 lb/in.² when the temperature is 0.00°C. What is the gauge pressure of the air in the tires when its temperature rises to 27.0°C and its volume increases to 1020 in.³? (*Hint:* It is not necessary to convert from British units to SI units; why? Use $P_{atm} = 14.7$ lb/in.².)

12E. Calculate the work done by an external agent during an isothermal compression of 1.00 mol of oxygen from a volume of 22.4 L at 0°C and 1.00 atm pressure to 16.8 L.

13P. (a) What is the number of molecules per cubic meter in air at 20°C and at a pressure of 1.0 atm $(= 1.01 \times 10^5$ Pa)? (b) What is the mass of this 1 m³ of air? Assume that 75% of the molecules are nitrogen (N_2) and 25% oxygen (O_2).

14P. Pressure p, volume V, and temperature T for a certain material are related by

$$p = \frac{AT - BT^2}{V}.$$

Find an expression for the work done by the material if the temperature changes from T_1 to T_2 while the pressure remains constant.

15P. Air that occupies 0.14 m³ at 1.03×10^5 Pa gauge pressure is expanded isothermally to atmospheric pressure and then cooled at constant pressure until it reaches its initial volume. Compute the work done by the air.

16P. Consider a given mass of an ideal gas. Compare curves representing constant-pressure, constant-volume, and isothermal processes on (a) a $p–V$ diagram, (b) a $p–T$ diagram, and (c) a $V–T$ diagram. (d) How do these curves depend on the mass of gas chosen?

17P. A container encloses two ideal gases. Two moles of the first gas are present, with molar mass M_1. The second gas has molar mass $M_2 = 3M_1$, and 0.5 mol of this gas is present. What fraction of the total pressure on the container wall is attributable to the second gas? (The kinetic theory explanation of pressure leads to the experimentally discovered law of partial pressures for a mixture of gases that do not react chemically: *the total pressure exerted by the mixture is equal to the sum of the pressures that the several gases would exert separately if each were to occupy the vessel alone.*)

18P. A weather balloon is loosely inflated with helium at a pressure of 1.0 atm (= 76 cm Hg) and a temperature of 20°C. The gas volume is 2.2 m³. At an elevation of 20,000 ft, the atmospheric pressure is down to 38 cm Hg and the helium has expanded, being under no restraint from the confining bag. At this elevation the gas temperature is −48°C. What is the gas volume now?

19P. An air bubble of 20 cm³ volume is at the bottom of a lake 40 m deep where the temperature is 4.0°C. The bubble rises to the surface, which is at a temperature of 20°C. Take the temperature of the bubble to be the same as that of the surrounding water and find its volume just before it reaches the surface.

20P. A pipe of length $L = 25.0$ m that is open at one end contains air at atmospheric pressure. It is thrust vertically into a freshwater lake until the water rises halfway up in the pipe, as shown in Fig. 21-16. What is the depth h of the lower end of the pipe? Assume that the temperature is the same everywhere and does not change.

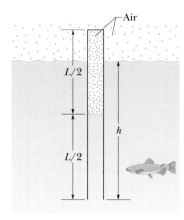

FIGURE 21-16 Problem 20.

21P. The envelope and basket of a hot-air balloon have a combined weight of 550 lb, and the envelope has a capacity of 77,000 ft³. When it is fully inflated, what should be the temperature of the enclosed air to give the balloon a lifting capacity of 600 lb (in addition to its own weight)? Assume that the surrounding air, at 20.0°C, has a weight density of 7.56×10^{-2} lb/ft³.

22P. A steel tank contains 300 g of ammonia gas (NH_3) at an absolute pressure of 1.35×10^6 Pa and temperature of 77°C. (a) What is the volume of the tank? (b) The tank is checked later when the temperature has dropped to 22°C and the absolute pressure has fallen to 8.7×10^5 Pa. How many grams of gas leaked out of the tank?

23P. Container A in Fig. 21-17 holds an ideal gas at a pressure of 5.0×10^5 Pa and a temperature of 300 K. It is connected by a thin tube to container B with four times the volume of A. Container B holds the same ideal gas at a pressure of 1.0×10^5 Pa and a temperature of 400 K. The connecting valve is opened, and equilibrium is achieved at

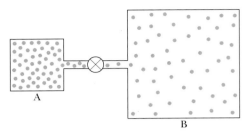

FIGURE 21-17 Problem 23.

a common pressure while the temperature of each container is kept constant at its initial value. What is the final pressure in the system?

SECTION 21-4 PRESSURE AND TEMPERATURE: A MOLECULAR VIEW

24E. Calculate the root-mean-square speed of helium atoms at 1000 K. The molar mass of helium is 4.00 g/mol.

25E. The lowest possible temperature in outer space is 2.7 K. What is the root-mean-square speed of hydrogen molecules at this temperature? (Use Table 21-1.)

26E. Find the rms speed of argon atoms at 313 K. The molar mass of argon is 39.9 g/mol.

27E. The sun is a huge ball of hot ideal gas. The temperature and pressure in the sun's atmosphere are 2.00×10^6 K and 0.0300 Pa. Calculate the rms speed of free electrons (mass $= 9.11 \times 10^{-31}$ kg) there.

28E. (a) Compute the root-mean-square speed of a nitrogen molecule at 20.0°C. (b) At what temperatures will the root-mean-square speed be half that value and twice that value?

29E. At what temperature do the atoms of helium gas have the same rms speed as the molecules of hydrogen gas at 20.0°C?

30P. At 273 K and 1.00×10^{-2} atm the density of a gas is 1.24×10^{-5} g/cm³. (a) Find v_{rms} for the gas molecules. (b) Find the molar mass of the gas and identify it.

31P. The mass of the H_2 molecule is 3.3×10^{-24} g. If 10^{23} hydrogen molecules per second strike 2.0 cm² of wall at an angle of 55° with the normal when moving with a speed of 1.0×10^5 cm/s, what pressure do they exert on the wall?

SECTION 21-5 TRANSLATIONAL KINETIC ENERGY

32E. What is the average translational kinetic energy of individual nitrogen molecules at 1600 K (a) in joules and (b) in electron-volts?

33E. (a) Determine the average value in electron-volts of the translational kinetic energy of the particles of an ideal gas at 0.00°C and at 100°C. (b) What is the translational

kinetic energy per mole of an ideal gas at these temperatures, in joules?

34E. At what temperature is the average translational kinetic energy of a molecule equal to 1.00 eV?

35E. Oxygen (O_2) gas at 273 K and 1.0 atm pressure is confined to a cubical container 10 cm on a side. Calculate the ratio of (1) the change in gravitational potential energy of an oxygen molecule falling the height of the box to (2) its average translational kinetic energy.

36P. Show that the ideal gas equation, Eq. 21-4, can be written in the alternative forms: (a) $p = \rho RT/M$, where ρ is the mass density of the gas and M the molar mass; (b) $pV = NkT$, where N is the number of gas particles (atoms or molecules).

37P. Water standing in the open at 32.0°C evaporates because of the escape of some of the surface molecules. The heat of vaporization (539 cal/g) is approximately equal to ϵn, where ϵ is the average energy of the escaping molecules and n is the number of molecules per gram. (a) Find ϵ. (b) What is the ratio of ϵ to the average kinetic energy of H_2O molecules, assuming that the kinetic energy is related to temperature in the same way as it is for gases?

38P. *Avogadro's law* states that under the same condition of temperature and pressure, equal volumes of gas contain equal numbers of molecules. Is this law equivalent to the ideal gas law?

SECTION 21-6 MEAN FREE PATH

39E. The mean free path of nitrogen molecules at 0.0°C and 1.0 atm is 0.80×10^{-5} cm. At this temperature and pressure there are 2.7×10^{19} molecules/cm³. What is the molecular diameter?

40E. At 2500 km above the Earth's surface the density of the atmosphere is about 1 molecule/cm³. (a) What mean free path is predicted by Eq. 21-18 and (b) what is its significance under these conditions? Assume a molecular diameter of 2.0×10^{-8} cm.

41E. What is the mean free path for 15 spherical jelly beans in a bag that is vigorously shaken? Take the volume of the bag to be 1.0 L and the diameter of a jelly bean to be 1.0 cm.

42E. Derive an expression, in terms of N/V, $\bar{v}$, and d, for the collision frequency of a gas atom or molecule.

43P. In a certain particle accelerator the protons travel around a circular path of diameter 23.0 m in a chamber at 1.00×10^{-6} mm Hg pressure and 295 K temperature. (a) Calculate the number of gas molecules per cubic centimeter at this pressure. (b) What is the mean free path of the gas molecules under these conditions if the molecular diameter is 2.00×10^{-8} cm?

44P. At what frequency would the wavelength of sound in air be equal to the mean free path in oxygen at 1.0 atm pressure and 0.0°C? Take the diameter of the oxygen molecule to be 3.0×10^{-8} cm.

45P. (a) What is the molar volume (the volume per mole) of an ideal gas at standard conditions (0.00°C, 1.00 atm)? (b) Calculate the ratio of the root-mean-square speed of helium atoms to that of neon atoms under these conditions. (c) What would be the mean free path of helium atoms under these conditions? Assume the atomic diameter d to be 1.00×10^{-8} cm. (d) What would be the mean free path of neon atoms under these conditions? Assume the same atomic diameter as for helium. (e) Comment on the results of parts (c) and (d) in view of the fact that the helium atoms are traveling faster than the neon atoms.

46P. The mean free path λ of the molecules of a gas may be determined from certain measurements (for example, from measurement of the viscosity of the gas). At 20°C and 75 cm Hg pressure such measurements yield values of λ_{Ar} (argon) $= 9.9 \times 10^{-6}$ cm and λ_{N_2} (nitrogen) $= 27.5 \times 10^{-6}$ cm. (a) Find the ratio of the effective diameter of argon to that of nitrogen. (b) What would be the value of the mean free path of argon at 20°C and 15 cm Hg? (c) What would be the value of the mean free path of argon at -40°C and 75 cm Hg?

47P. Show that about 10^{13} air molecules are needed to cover the period that closes this sentence. Show that there are about 10^{24} collisions of air molecules with that period each second.

SECTION 21-7 THE DISTRIBUTION OF MOLECULAR SPEEDS

48E. The speeds of a group of ten molecules are 2.0, 3.0, 4.0, . . . , 11 km/s. (a) What is the average speed of the group? (b) What is the root-mean-square speed for the group?

49E. You are given the following group of particles (N_i represents the number of particles that have a speed v_i):

N_i	v_i (cm/s)
2	1.0
4	2.0
6	3.0
8	4.0
2	5.0

(a) Compute the average speed $\bar{v}$. (b) Compute the root-mean-square speed v_{rms}. (c) Of the five speeds shown, which is the most probable speed v_P?

50E. (a) Ten particles are moving with the following speeds: four at 200 m/s, two at 500 m/s, and four at 600 m/s. Calculate the average and root-mean-square speeds. Is $v_{rms} > \bar{v}$? (b) Make up your own speed distribution for the 10 particles and show that $v_{rms} \geq \bar{v}$ for your distribution. (c) Under what condition (if any) does $v_{rms} = \bar{v}$?

51E. Consider the distribution of speeds shown in Fig. 21-18. (a) List v_{rms}, $\bar{v}$, and v_P in the order of increasing speed. (b) How does this distribution compare with the Maxwellian distribution?

FIGURE 21-18 Exercise 51.

52E. It is found that the most probable speed of molecules in a gas at equilibrium temperature T_2 is the same as the rms speed of the molecules in this gas when its equilibrium temperature is T_1. Calculate T_2/T_1.

53P. (a) Compute the temperatures at which the rms speed is equal to the speed of escape from the surface of the Earth for molecular hydrogen and for molecular oxygen. (b) Do the same for the moon, assuming the gravitational acceleration on its surface to be 0.16g. (c) The temperature high in the Earth's upper atmosphere is about 1000 K. Would you expect to find much hydrogen there? Much oxygen?

54P. A molecule of hydrogen (diameter 1.0×10^{-8} cm) escapes from a furnace ($T = 4000$ K) with the root-mean-square speed into a chamber containing atoms of cold argon (diameter 3.0×10^{-8} cm) at a density of 4.0×10^{19} atoms/cm³. (a) What is the speed of the hydrogen molecule? (b) If the H_2 molecule and an argon atom collide, what is the closest their centers can be, considering each as spherical? (c) What is the initial number of collisions per second experienced by the hydrogen molecule?

55P. Two containers are at the same temperature. The first contains gas at pressure p_1 whose molecules have mass m_1 with root-mean-square speed v_{rms1}. The second contains molecules of mass m_2 at pressure $2p_1$ that have an average speed $\bar{v}_2 = 2v_{rms1}$. Find the ratio m_1/m_2 of the masses of their molecules.

56P. For the hypothetical speed distribution for N gas particles shown in Fig. 21-19 [$P(v) = Cv^2$ for $0 < v \leq v_0$; $P(v) = 0$ for $v > v_0$], find (a) an expression for C in terms of N and v_0, (b) the average speed of the particles, and (c) the rms speed of the particles.

FIGURE 21-19 Problem 56.

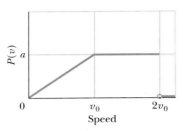

FIGURE 21-20 Problem 57.

57P. A hypothetical sample of N gas particles has the speed distribution shown in Fig. 21-20, where $P(v) = 0$ for $v > 2v_0$. (a) Express a in terms of N and v_0. (b) How many of the particles have speeds between $1.5v_0$ and $2.0v_0$? (c) Express the average speed of the particles in terms of v_0. (d) Find v_{rms}.

SECTION 21-8 THE MOLAR SPECIFIC HEATS OF AN IDEAL GAS

58E. (a) What is the internal energy of 1.0 mol of an ideal monatomic gas at 273 K? (b) Does it depend on volume or pressure?

59E. One mole of an ideal gas undergoes an isothermal expansion. Find the heat added to the gas in terms of the initial and final volumes and the temperature. (*Hint:* Use the first law of thermodynamics.)

60E. The mass of a helium atom is 6.66×10^{-27} kg. Compute the specific heat at constant volume for (monatomic) helium gas (in $J/kg \cdot K$) from the molar specific heat at constant volume.

61P. Let 20.9 J of heat be added to a particular ideal gas. As a result, its volume changes from 50.0 to 100 cm³ while the pressure remains constant at 1.00 atm. (a) By how much did the internal energy of the gas change? (b) If the quantity of gas present is 2.00×10^{-3} mol, find the molar specific heat at constant pressure. (c) Find the molar specific heat at constant volume.

62P. A quantity of ideal monatomic gas consists of n moles initially at temperature T_1. The pressure and volume are then slowly doubled in such a manner as to trace out a straight line on a p–V diagram. In terms of n, R, and T_1, what are (a) W, (b) ΔE_{int}, and (c) Q? (d) If one were to define a molar specific heat for this process, what would be its value?

63P. A container holds a mixture of three nonreacting gases: n_1 moles of the first gas with molar specific heat at constant volume C_1, and so on. Find the molar specific heat at constant volume of the mixture, in terms of the molar specific heats and quantities of the three separate gases.

64P. The mass of a gas molecule can be computed from the specific heat at constant volume c_V. Take $c_V = 0.075$ cal/g·°C for argon and calculate (a) the mass of an argon atom and (b) the molar mass of argon.

SECTION 21-9 THE EQUIPARTITION OF ENERGY

65E. Ninety joules of heat is lost by an ideal diatomic gas that has molecular rotation but not oscillation. In which kind of process, at constant volume or at constant pressure, is the decrease in the internal energy of the gas larger?

66E. One mole of oxygen (O_2) is heated at constant pressure starting at 0°C. How much heat must be added to the gas to double its volume? (The molecules rotate but do not oscillate.)

67E. Suppose 12.0 g of oxygen (O_2) is heated at constant atmospheric pressure from 25.0 to 125°C. (a) How many moles of oxygen are present? (See Table 21-1.) (b) How much heat is transferred to the oxygen? (The molecules rotate but do not oscillate.) (c) What fraction of the heat is used to raise the internal energy of the oxygen?

68P. Suppose 4.00 mol of an ideal diatomic gas, with molecular rotation but not oscillation, experiences a temperature increase of 60.0 K under constant-pressure conditions. (a) How much heat was added to the gas? (b) By how much did the internal energy of the gas increase? (c) How much work was done by the gas? (d) By how much did the internal translational kinetic energy of the gas increase?

69P. The molar mass of iodine is 127 g/mol. A standing wave in a tube filled with iodine gas at 400 K has nodes that are 6.77 cm apart when the frequency is 1000 Hz. Is iodine gas monatomic or diatomic?

SECTION 21-11 THE ADIABATIC EXPANSION OF AN IDEAL GAS

70E. A mass of gas occupies a volume of 4.3 L at a pressure of 1.2 atm and a temperature of 310 K. It is compressed adiabatically to a volume of 0.76 L. Determine (a) the final pressure and (b) the final temperature, assuming it to be an ideal gas for which $\gamma = 1.4$. (*Hint:* It is not necessary to make any unit conversions.)

71E. (a) One liter of gas with $\gamma = 1.3$ is at 273 K and 1.0 atm pressure. It is suddenly (adiabatically) compressed to half its original volume. Find its final pressure and temperature. (b) The gas is now cooled back to 273 K at constant pressure. What is its final volume?

72E. Let n moles of an ideal gas expand adiabatically from an initial temperature T_1 to a final temperature T_2. Prove that the work done by the gas is $nC_V(T_1 - T_2)$, where C_V is the molar specific heat at constant volume. (*Hint:* Use the first law of thermodynamics.)

73E. We know that $pV^\gamma = $ a constant for an adiabatic process. Evaluate "a constant" for an adiabatic process involving exactly 2.0 mol of an ideal gas passing through the state having exactly $p = 1.0$ atm and $T = 300$ K. Assume a diatomic gas whose molecules have rotation but not oscillation.

74E. For adiabatic processes in an ideal gas, show that (a) the bulk modulus is given by

$$B = -V\frac{dp}{dV} = \gamma p,$$

and therefore (b) the speed of sound is

$$v_s = \sqrt{\frac{\gamma p}{\rho}} = \sqrt{\frac{\gamma RT}{M}}.$$

See Eqs. 18-2 and 18-3.

75E. Air at 0.000°C and 1.00 atm pressure has a density of 1.29×10^{-3} g/cm³, and the speed of sound is 331 m/s at that temperature. Compute the ratio γ of the molar specific heats of air. (*Hint:* See Exercise 74.)

76E. The speed of sound in different gases at a certain temperature depends on the molar mass of the gases. Show that $v_1/v_2 = \sqrt{M_2/M_1}$ (constant T), where v_1 is the speed of sound in a gas of molar mass M_1 and v_2 is the speed of sound in a gas of molar mass M_2. (*Hint:* See Exercise 74.)

77P. Use the result of Exercise 73 to show that the speed of sound in air increases by about 0.61 m/s for each Celsius degree rise in temperature near 0°C.

78P. From the knowledge that C_V, the molar specific heat at constant volume, for a gas in a container is $5.0R$, calculate the ratio of the speed of sound in that gas to the rms speed of its molecules at temperature T. (*Hint:* See Exercise 74.)

79P. (a) An ideal gas initially at pressure p_0 undergoes a free expansion (adiabatic, no external work) until its final volume is 3.00 times its initial volume. What is the pressure of the gas after the free expansion? (b) The gas is then slowly and adiabatically compressed back to its original volume. The pressure after compression is $(3.00)^{1/3}p_0$. Determine whether the gas is monatomic, diatomic, or polyatomic. (c) How does the average kinetic energy per molecule in this final state compare with that in the initial state?

80P. An ideal gas experiences an adiabatic compression from $p = 1.0$ atm, $V = 1.0 \times 10^6$ L, $T = 0.0$°C to $p = 1.0 \times 10^5$ atm, $V = 1.0 \times 10^3$ L. (a) Is this a monatomic, a diatomic, or a polyatomic gas? (b) What is the final temperature? (c) How many moles of the gas are present? (d) What is the total translational kinetic energy per mole before and after the compression? (e) What is the ratio of the squares of the rms speeds before and after the compression?

81P. A quantity of ideal gas occcupies an initial volume V_0 at pressure p_0 and temperature T_0. It expands to volume V_1 (a) at constant pressure, (b) at constant temperature, (c) adiabatically. Graph each case on a p–V diagram. In which case is Q greatest? Least? In which case is W greatest? Least? In which case is ΔE_{int} greatest? Least?

82P. C_V for a certain ideal gas is 6.00 cal/mol·K. The temperature of 3.0 mol of the gas is raised 50 K by each of three different processes: at constant volume, at constant pressure, and by an adiabatic compression. Complete the table that follows, showing for each process the heat added (or subtracted), the work done by the gas, the change in internal energy of the gas, and the change in total translational kinetic energy of the gas.

PROCESS	HEAT ADDED	WORK DONE BY GAS	CHANGE IN INTERNAL ENERGY	CHANGE IN KINETIC ENERGY
Constant volume	————	————	————	————
Constant pressure	————	————	————	————
Adiabatic	————	————	————	————

83P. A heat engine carries 1.00 mol of an ideal monatomic gas around the cycle shown in Fig. 21-21. Process $1 \rightarrow 2$ takes place at constant volume, process $2 \rightarrow 3$ is adiabatic, and process $3 \rightarrow 1$ takes place at a constant pressure. (a) Compute the heat Q, the change in internal energy ΔE_{int}, and the work done W, for each of the three processes and for the cycle as a whole. (b) If the initial pressure at point 1 is 1.00 atm, find the pressure and the volume at points 2 and 3. Use 1.00 atm $= 1.013 \times 10^5$ Pa and $R = 8.314$ J/mol·K.

FIGURE 21-21 Problem 83.

84P. In a motorcycle engine, after combustion occurs in the top of the cylinder, the piston is forced down (toward the crankshaft) as the mixture of gaseous products undergoes an adiabatic expansion. Find the average power involved in this expansion when the engine is running at 4000 rpm, assuming that the gauge pressure immediately after combustion is 15 atm, the initial volume is 50 cm³, and the volume of the mixture at the bottom of the stroke is 250 cm³. Assume that the gases are diatomic and that the time involved in the expansion is one-half that of the total cycle. Express your answer in watts and horsepower.

ADDITIONAL PROBLEMS

85. A sample of an ideal gas is taken through the cyclic process shown on the p–V diagram in Fig. 21-22. The temperature of the gas at point a is 200 K. (a) How many moles of gas are in the sample? What are (b) the temperature of the gas at point b, (c) the temperature of the gas at point c, and (d) the net heat added to the gas during the cycle?

FIGURE 21-22 Problem 85.

86. An ideal gas initially at 300 K is compressed at a constant pressure of 25 N/m² from a volume of 3.0 m³ to a volume of 1.8 m³. In the process 75 J of heat is lost by the gas. What are (a) the change in internal energy of the gas and (b) the final temperature of the gas?

87. One mole of an ideal diatomic gas undergoes a transition from a to c along the diagonal path in Fig. 21-23. The temperature of the gas at point a is 1200 K. During the transition, (a) what is the change in internal energy of the gas, and (b) how much heat is added to the gas? (c) How much heat must be added to the gas if it goes from a to c along the indirect path abc?

FIGURE 21-23 Problem 87.

88. A sample of ideal gas expands from an initial pressure and volume of 32 atm and 1.0 L to a final volume of 4.0 L. The initial temperature of the gas is 300 K. What are the final pressure and temperature of the gas and how much work is done by the gas during the expansion, if the expansion is (a) isothermal, (b) adiabatic and the gas is monatomic, and (c) adiabatic and the gas is diatomic?

ARE MANUFACTURED EMISSIONS OF CO_2 WARMING OUR CLIMATE?

Barbara Levi
Physics Today

Barbara Goss Levi is a senior associate editor of Physics Today. She received her B.A. from Carleton College (1965) and her Ph.D. from Stanford University (1971) in particle physics. In addition to teaching (at Georgia Tech and Rutgers) and writing about physics, she has worked at Princeton University on such problems as energy conservation and arms control. Dr. Levi is a fellow of the American Physical Society and of the American Association for the Advancement of Science. She was co-editor of Global Warming: Physics and Facts, published in 1992 by the American Institute of Physics.

In the past few years stories about "global warming" have filled the news. Scientists have warned that the world's climate may grow warmer if the activities of our civilization continue to spew carbon dioxide and certain other gases into the atmosphere. The possible consequences might include the flooding of many coastal cities, the disruption of current weather patterns, and the failure of many agricultural products and ecological species. But there is no firm evidence that global warming has yet begun. So the question is whether our nation and others should undertake measures now to cut our production of carbon dioxide—just in case. It sounds prudent, but many of the measures may affect you personally. The carbon dioxide that civilization has added to the atmosphere comes from the burning of fossil fuels—coal, oil, and gas. These fuels stoke our electrical power plants and power our automobiles. Will you one day have to choose between your car or your climate?

Before humans came along, Earth's atmosphere already contained some carbon dioxide, and this carbon dioxide, along with water vapor and a few other gases in the atmosphere, has made our planet more comfortably warm than it would be without them. But our civilization is adding roughly 22 billion tons of carbon dioxide into the air every year, and much of that carbon dioxide will stay there for 50–200 years. As a result, most scientists concur that the Earth will grow warmer. However, they don't know for sure how high the mercury will rise nor how fast it will climb. The global temperature seems to have risen between 0.3 and 0.6 C°

over the last 100 years, but no one can prove beyond doubt that carbon dioxide was the cause.

The climate is such a complex system that making firm predictions requires very sophisticated computer models. Nevertheless, it is rather easy to understand the basic mechanism by which carbon dioxide now warms our planet. In this essay we try to understand by a simple model exactly what factors and equations determine the temperature of the Earth as we know it. That will help us in turn to understand how an increase in carbon dioxide is likely to affect this temperature.

The Earth's temperature is largely determined by the radiation it receives from the sun. The sun, like all warm bodies, such as smoldering logs or glowing light bulbs, radiates heat in the form of electromagnetic radiation. For our purposes here, you need to know only that electromagnetic radiation is a form of wave motion and that the waves carry energy. The intensity of radiation I emitted by any body is very strongly dependent on the temperature of that body, according to a relation known as the Stefan–Boltzmann law:

$$I = \epsilon \sigma T^4, \tag{1}$$

where I is the power (in watts) radiated from a 1-m² area of any object that is at a temperature T (K). Two constants appear in this equation: σ is known as the Stefan–Boltzmann constant (5.67×10^{-8} W/m²·K⁴) and ϵ is the emissivity of the radiating body, that is, its tendency to give off radiation. For a perfect radiator, $\epsilon = 1$, and for other bodies $\epsilon < 1$. Note the strong dependence of the radiated power on temperature: a body with twice the temperature will radiate 16 times more energy over the same time interval.

Problem 1

Use the Stefan–Boltzmann law to estimate the intensity of the solar radiation emitted by the sun, assuming that it is a perfect radiator and that its surface temperature is 6000 K. Then find the solar intensity at the Earth's surface, remembering that the solar energy will be spread over a spherical surface whose radius is equal to the mean distance from the sun to the Earth (1.495×10^{11} m). The sun's radius is about 6.96×10^8 m.

The rate of solar energy reaching the Earth per unit area is known as the solar constant S and its measured value is approximately 1360 W/m². (Why do you think that the answer you calculated was somewhat larger than the measured value?) If that energy continually streamed into the Earth, and the Earth radiated no energy back, our planet would continue to get warmer and warmer. For the Earth (or any object) to remain at an equilibrium temperature, the rate of energy absorbed by the Earth must be exactly balanced by the rate of energy radiated outward by the Earth. This principle of energy balance determines the temperature of the Earth.

The solar constant tells us the power of radiation falling on each unit of area. To find the radiation intercepted by the Earth we must multiply the solar constant by the area of the two-dimensional projection of Earth's surface. This projection is a circle whose area equals πR_E^2. Not all this solar power is absorbed by the Earth: measurements indicate that about 30% of the incident sunlight is reflected back to space. This reflectivity is called the albedo α and it is expressed in terms of the fraction of sunlight (0.3) that is reflected. A fraction $(1 - \alpha)$ is absorbed by the Earth.

The power radiated by the Earth is the intensity given by the Stefan–Boltzmann law multiplied by the surface area of the Earth, $4\pi R_E^2$. For this calculation we assume that the emissivity of the Earth is 1. Equating the incoming solar power to the power radiated by the Earth, we get

$$\pi R_E^2 (1 - \alpha) S = 4\pi R_E^2 \sigma T_E^4 \quad (2)$$

or

$$(1 - \alpha) S/4\sigma = T_E^4. \quad (3)$$

Solving for T_E, we get

$$T_E = [(1 - \alpha) S/4\sigma]^{1/4}$$
$$= 255 \text{ K } (-18°\text{C}). \quad (4)$$

This temperature is in fact just about the temperature that satellites have measured at the outer edge of the atmosphere. It sounds pretty chilly! But remember that in this calculation we have ignored the atmospheric gases that surround the Earth. The actual average global temperature at the surface of the Earth is a much more comfortable $T_s = 288$ K (15°C), or 33 C° warmer. The Earth's surface is kept at this more habitable temperature by its blanket of atmospheric gases and particles (Fig. 1).

Actually, there are only certain gases within the atmosphere that help keep the surface warm. Those gases, which we call "greenhouse gases," have two key properties: they largely transmit radiation at very short wavelengths, such as the solar radiation, and they strongly absorb radiation at longer wavelengths, such as those emitted by the Earth. (The radiation emitted by the Earth—or by any other body at about room temperature—is called "thermal energy.") The curves in Fig. 2a show how the intensity of radiation emitted from a perfect radiator varies with wavelength. The purple curve corresponds to a body (like the sun) at a temperature of 6000 K, and the red curve represents a body at 255 K (like the Earth). The curves in Figs. 2b and 2c show the wavelengths at which the main greenhouse gases—carbon dioxide and water—absorb this radiation. Although both these gases absorb radiation at several wavelengths emitted by the Earth, only water vapor appreciably absorbs some of the radiation from the sun. Thus most solar radiation passes straight through to the Earth but a lot of the

FIGURE 1 Data taken by satellites to study Earth's climate. The colors here show how much long-wavelength radiation is trapped by clouds—an effect that tends to warm the Earth. Note the high intensities over the Indian Ocean, where there are deep cirrus clouds. Clouds also reflect radiation, tending to cool the Earth. In the normal atmosphere the net effect of clouds is to cool the climate.

FIGURE 2   (a) Purple and red curves show the intensity of radiation emitted at each wavelength for perfect radiators at 6000 K (purple curve) and 255 K (red curve). The purple curve approximately represents the incoming solar radiation, and the red curve approximates the Earth's outgoing radiation. (b, c) Peaks in these curves show the wavelengths at which molecules of water and carbon dioxide in the atmosphere strongly absorb radiation. Note that these gases absorb more of the Earth's radiation than the solar radiation. Adapted from J. P. Pleixoto and H. O. Oort, *Physics of Climate,* American Institute of Physics, 1992, Fig. 6.2.

Earth's radiation gets trapped by the atmosphere.

To get an intuitive idea of how these gases affect the temperature, imagine that Earth initially did not have this blanket of greenhouse gases so that its surface was at the temperature of 255 K we calculated earlier. Suppose some greenhouse gases are now suddenly added to the Earth's atmosphere. At first Earth's surface will continue to radiate the amount of energy dictated by its temperature according to the Stefan–Boltzmann law: that intensity just balances the inward flux from the sun. But now the greenhouse gases will absorb some of this energy. These gases will reradiate the energy, some of it back to Earth's surface. The surface now will receive more energy than it is giving off and must warm. As it warms it will give off more radiant energy. The surface will continue to warm until it reaches the temperature at which the energy fluxes balance.

We can calculate the temperature at which the Earth will reach this balance by constructing a simple model. See Fig. 3. The model has two layers: the atmosphere and Earth's surface. We make the simplifying assumption that the sun's radiation S' passes unattenuated through the greenhouse gases to the surface, but that Earth's thermal radiation, denoted E, is completely trapped by the greenhouse gases in the atmosphere. These atmospheric gases absorb Earth's radiation and they reradiate the energy uniformly in all directions, as shown by the arrows labeled A. You can see from the diagram that the net effect is that Earth's surface receives not only the radiation from the sun but also the power reradiated by the atmosphere.

In Fig. 3 the incoming solar radiation is denoted by S', and its value is just the unreflected solar energy $(1 - \alpha)S$ divided by 4. (The factor of 4 comes from the fact that the radiation emitted by the Earth is proportional to its total surface area while the radiation that Earth receives is proportional to its projected area: the ratio of those two areas is 4.)

Energy conservation requires that the energy flows in and out be balanced at each of the two layers.

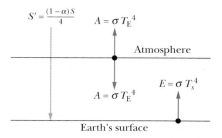

FIGURE 3   Simple model of Earth's radiation balance relates the incoming radiation from the sun S' to the radiation emitted by Earth's surface E and the radiation emitted by the atmosphere A. At the top of the atmosphere the outgoing radiation A must equal the incoming radiation S'. Similarly, at the Earth's surface E must equal $S' + A$.

Thus we get the following two equations:

top of the atmosphere: $S' = A$;   (5)

surface of the Earth: $S' + A = E$.   (6)

If we substitute the value of A given by Eq. 5 into Eq. 6 and recall that the power radiated by Earth can be written in terms of the Stefan–Boltzmann law, we get

$$E = 2S'$$

or

$$T_s^4 = 2(1 - \alpha)S/4\sigma, \quad (7)$$

$$T_s = 303 \text{ K } (30°\text{C}). \quad (8)$$

Our answer is larger than the observed surface temperature of 15°C, but we have ignored some very important effects in our simple model. Can you think of some of them? One is that the greenhouse gases actually absorb some of the solar radiation and transmit some of the Earth's thermal radiation.

Problem 2

In the two-layer model, consider the case in which the atmosphere absorbs a fraction

$a < 1$ of the Earth's radiation. In this case, the radiation leaving the top of the atmosphere will include a faction of the radiation from the Earth's surface that has not been absorbed by the atmosphere. What value of absorptivity a would the atmosphere have to have for the surface temperature to have the observed value of 288 K?

Another effect ignored in the simple model is the energy carried away from the surface by evaporation as well as radiation. A third is that the atmosphere is not a single layer at one temperature but is stratified, with its temperature gradually decreasing with altitude, up to about 10 km. Yet a fourth excluded factor is convection: as air near the surface of the Earth is warmed, it rises, carrying some of the heat with it to higher altitudes. Furthermore a realistic model would have to consider the variations of solar intensity with latitude, the subsequent convection currents driven by the temperature differences between equator and poles, surface topology, the effects of clouds, and the interactions between the oceans,

atmosphere, land, and ice masses. (Figure 4 shows a more realistic diagram of energy flows within the Earth−atmosphere system.) Those scientists who have tackled this problem have spent decades to develop very detailed computer models that require the most advanced computers we have.

Problem 3

In the early 1980s a team of five scientists warned that a nuclear war might set off a "nuclear winter," or period of dramatically cold weather. This deep cold would result if the fires ignited by nuclear weapons sent enough soot into the atmosphere essentially to block out the sun. Add a third layer—soot—above the atmosphere in the simple model. Assume that the soot absorbs all the incoming sunlight and lets out all the thermal radiation from the Earth and atmosphere. Calculate the effective surface temperature of the Earth. (The question of how severe a "nuclear winter" might be is still debated and requires far more sophisticated treatment than our simple model!)

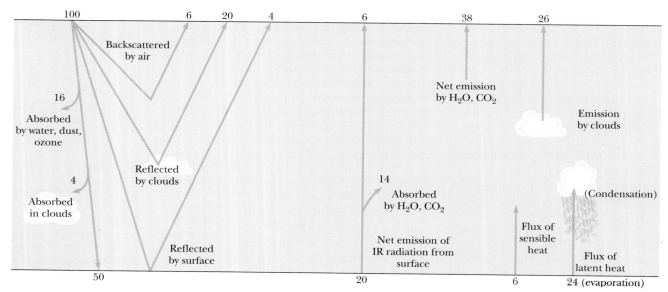

FIGURE 4 Fluxes of energy in the atmosphere are more complex than those in the simple model. The numbers are the energy fluxes expressed as a percentage of the incoming radiation, which is 100. Adapted from J. P. Pleixoto and H. O. Oort, *Physics of Climate,* American Institute of Physics, 1992, Fig. 6.3.

The carbon dioxide emitted by human activities has already increased the atmospheric concentrations by 25%, to 350 parts per million, by volume. If we continue to produce carbon dioxide and other greenhouse gases at current rates, the concentrations may reach a level in the next 50 years or so that is twice that of the pre-industrial era. The computerized climate models now developed calculate that this doubling of CO_2 will increase Earth's temperature by somewhere between 1.5 and 4.5 C°. The rise might be greater in some portions of the globe than in others and might everywhere be accompanied by other climate effects, such as altered patterns of rainfall or increased incidences of hurricanes, as well as a rise in the sea levels. No one knows whether the pace of climate change might outstrip the ability of natural ecosystems or human institutions to adapt. And yet, major efforts to cut down on CO_2 emissions are expected to be quite costly. Nevertheless, several panels of scientists have begun to call for prudent measures to curb emissions of carbon dioxide or other gases such as chlorofluorocarbons (which already pose a threat to the ozone layer), methane, and nitrous oxides. If their recommendations are followed, you may not have to give up driving your car, but you certainly might have to buy one that gets many more miles to the gallon—or runs on something other than fossil fuel!

Answers

1. 7.35×10^7 W/m²; 1590 W/m² (20% greater than the measured value).

2. $2 - 2S'/\sigma T_s^4 = 0.78$.

3. $[(1 - \alpha)S/8\sigma]^{1/4} = 214$ K.

ENTROPY AND THE SECOND LAW OF THERMODYNAMICS

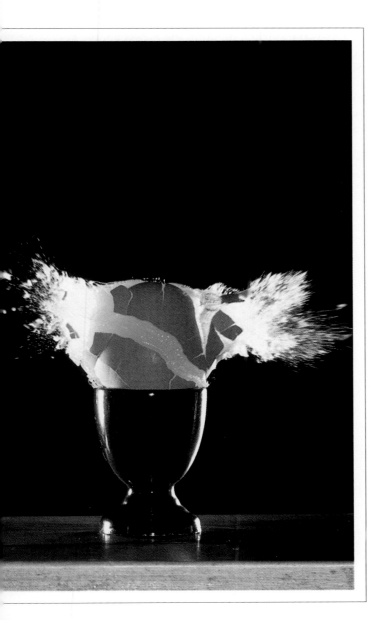

An anonymous graffito on a wall of the Pecan Street Cafe in Austin, Texas, once read: "Time is God's way of keeping things from happening all at once." Time also has direction—some things happen in a certain sequence and could never happen on their own in a reverse sequence. As an example, an accidentally dropped egg splatters in a cup. The reverse process, a splattered egg reforming into a whole egg and jumping up to an outstretched hand, will never happen on its own. But why not? Why can't that process be reversed, like a videotape run backward? What in the world gives direction to time?

22-1 SOME THINGS THAT DON'T HAPPEN

A quarter resting on a tabletop simply never—entirely by itself—rises into the air, gets so hot that you cannot touch it, or flattens itself out until it is as large as a saucer. We are not surprised at any of these nonevents and we account for them by saying: "It takes *energy* to lift the coin, to heat it, or to hammer it out flat. All these things would violate the principle of conservation of energy."

Here are three more things that never happen, but the difference is that you cannot account for their not happening in this same easy way. (1) Coffee, resting quietly in your cup, never spontaneously cools down and starts to swirl around. (2) One end of a spoon resting on a table never spontaneously gets hot while the other end cools down. (3) The molecules of air in a room never all move to one corner and stay there. Unlike the example of the coin, *these* nonevents do *not* require energy. The coffee could presumably get the kinetic energy for its swirling by cooling down. The hot end of the spoon could presumably get its energy from the cool end. And the molecules of air would not have to change their kinetic energies, just their positions.

Note one thing, however. The *reverses* of these three nonevents occur quite naturally and spontaneously. Coffee, swirling in your cup, will eventually stop swirling, its rotational energy changing into thermal energy and thus heating the coffee a little. Temperature differences set up between two ends of a spoon will tend to equalize. Air molecules will rush from one corner and fill the room uniformly.

The world is full of events that happen in one way but never in the opposite way. We are so used to this that we take these events for granted when they happen in the "right" direction, but we would be astonished beyond all belief if they happened the other way around.

The direction in which natural events happen is governed by the **second law of thermodynamics,** the subject of this chapter. The second law can be expressed in several equivalent forms, two of which in-

volve simple statements about heat and work. We shall explore these in the next few sections and then consider a third formulation of the law in terms of a new and useful concept—*entropy.*

22-2 ENGINES

If you stir a cup of room-temperature coffee, your work on the spoon results in a transfer of energy to the coffee. The coffee then has kinetic energy—it swirls. As the swirling dies out, the excess energy becomes internal (or thermal) energy of the coffee. Since the temperature of the coffee is then (slightly) higher than that of the room, the coffee transfers its excess energy to the room as heat. In this process, you have changed *work into heat* completely and rather easily.

The reverse process—changing *heat into work*—is quite another matter. Here, as a challenge, is a formulation of the second law of thermodynamics:

(a)

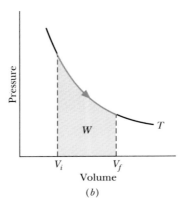

(b)

FIGURE 22-1 (*a*) An ideal gas expands isothermally, absorbing heat *Q* and doing work *W*. (*b*) The gas follows an isotherm on a *p*–*V* diagram for the expansion. Although all the heat is transformed into work, there is no violation of the second law of thermodynamics because other changes have occurred. The system is not restored to its original state at the end of the process.

SECOND LAW OF THERMODYNAMICS (FIRST FORM): It is not possible to change heat completely into work, with no other change taking place.

Rewards beyond your wildest dreams await you if you can build a device that violates this law. Although we won't succeed in doing so, let us give it a try.

Figure 22-1*a* shows a cylinder containing an ideal gas and resting on a heat reservoir at temperature *T*. By removing weight gradually from the piston, we can permit the gas to expand. The gas remains at constant temperature while doing so, absorbing heat *Q* from the reservoir. The system (the gas) follows the isotherm shown in Fig. 22-1*b* and—in lifting the weight—does work *W* as indicated by the gold-colored area in that figure. The internal energy E_{int}—which, for an ideal gas, depends only on the temperature—does not change during this isothermal expansion. From the first law of thermodynamics, $E_{int} = Q - W$, the work *W* is thus exactly equal to the heat *Q* extracted from the reservoir. Have we not turned heat completely into work?

We have indeed done so, but we have not met the essential requirement *"with no other change taking place."* Changes have taken place; the gas in the cylinder is not in the same state as it was when we started. Its volume has changed, for example, and so has its pressure. To meet our challenge, we must somehow restore the gas to its original condition. This means that the piston–cylinder arrangement must operate in a cycle, returning the gas to its original state at the end of the cycle. A device that changes heat into work while operating in a cycle is called a **heat engine** or, more simply, an **engine.**

Figure 22-2*a* suggests a generalized scheme of operation for an engine. During every cycle, energy is extracted as heat Q_H from a reservoir at temperature T_H, a portion is diverted to do useful work *W*, and the rest is discharged (lost) as heat Q_C to a reservoir at a lower temperature T_C.

Because an engine operates in a cycle, the internal energy E_{int} of the system, that is, of the gas in the cylinder, returns to its original value at the end of the cycle. Thus $\Delta E_{int} = 0$ and from the first law of thermodynamics ($\Delta E_{int} = Q - W$), the net work done per cycle by the system must equal the net heat transferred per cycle. We write this as

$$|W| = |Q_H| - |Q_C|. \qquad (22\text{-}1)$$

We have chosen here to deal with the (positive) absolute values of *Q* and *W*, which we write as $|Q|$ and $|W|$, respectively. Thus for both $Q = +10\,J$ and $Q = -10\,J$ we have $|Q| = 10\,J$ and similarly for work. We must always be clear about whether heat is being added to the system (so that *Q* is positive) or taken from the system (so that *Q* is negative) and whether positive or negative work is being done by the system. (And in later sections of this chapter, when we discuss the work done *on a system,* we must remember that such work is always the negative of the work done *by that system.*)

The purpose of an engine is to transform as much of the extracted heat Q_H into work as possible. We measure its success in doing so by its **thermal efficiency** *e*, defined as the ratio of the work it does per cycle—what you get—to the heat it absorbs per cycle—what you pay for. Using Eq. 22-1, we have

$$e = \frac{|W|}{|Q_H|} = \frac{|Q_H| - |Q_C|}{|Q_H|}. \qquad (22\text{-}2)$$

Equation 22-2 shows that the efficiency of an engine can be unity, or 100%, only if $Q_C = 0$, that is, if no heat is delivered to the low-temperature reservoir. Figure 22-2*b* is a diagram of such a "perfect" engine. From accumulated experience to date, physicists have concluded that it is impossible to build such an engine—which has never stopped hopeful inventors from trying. So another way of expressing

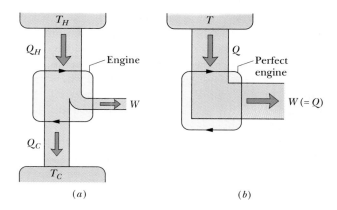

(a)   *(b)*

FIGURE 22-2 We indicate an engine by encircling it with arrows pointing in a clockwise direction. (*a*) In a real engine, heat extracted from a reservoir is converted partially into work, the rest being discharged into a reservoir of lower temperature. (*b*) In a perfect engine, all the extracted heat is converted into work. Nobody has ever built such an engine.

the second law of thermodynamics (first form) is: *there are no perfect engines.*

Real Engines

We need to be clear about the relation between the schematic engine of Fig. 22-2*a* and engines in the real world. Consider, for example, the nuclear power plant shown in Fig. 22-3. The high-temperature reservoir, marked T_H in Fig. 22-2*a*, is the nuclear reactor chamber, from which heat is removed and taken to the steam generator by circulating water. The low-temperature reservoir, marked T_C in Fig. 22-2*a*, is the steam condenser, which is cooled by river water that is pumped through it. The *working substance,* which corresponds to the ideal gas in the cylinder of Fig. 22-1*a*, is still other water that absorbs heat and becomes steam in the steam generator, passes through the turbine where it does work on the turbine, and then passes through the condenser to lose heat and become water again in its original condition. Do not confuse the *working substance* with the *fuel.* The purpose of the latter, which may be uranium fuel pellets, coal, or oil, is to maintain the temperature of the high-temperature reservoir in a real engine.

The connection between the engine in your car and the schematic engine of Fig. 22-2*a* is complicated by the fact that your car engine is an *internal combustion engine.* The high-temperature reservoir is provided—inside the cylinders—by the combustion of the fuel–air mixture. The low-temperature reservoir is the outside air into which the exhaust gases are vented. The fuel is the gasoline, and the working substance is the mixture of air and burned fuel. A far more powerful internal combustion engine is shown in Fig. 22-4.

FIGURE 22-3 The North Anna nuclear power plant near Charlottesville, Virginia, generates electrical energy at the rate of 900 MW. At the same time, by deliberate design, it discards energy into the nearby river at the rate of about 2100 MW. This plant—and all others like it, nuclear or not—throws away more energy than it delivers in useful form to the power grid. It is a real example of the engine of Fig. 22-2*a*.

FIGURE 22-4 This RD-170 engine, built in 1980, is the world's most powerful rocket engine. It is capable of producing 190 MW.

SAMPLE PROBLEM 22-1

Figure 22-5 is a *p–V* diagram for an idealized version of a small *Stirling engine,* named for the Reverend Robert Stirling of the Church of Scotland, who proposed the scheme in 1816. The engine uses $n = 8.1 \times 10^{-3}$ mol of an ideal gas, operates between hot and cold heat reservoirs of temperatures $T_H = 95°C$ and $T_C = 24°C$, and runs at the rate of 0.70 cycle per second. A cycle consists of an isothermal expansion (*ab*), an isothermal compression (*cd*), and two constant-volume processes (*bc* and *da*).

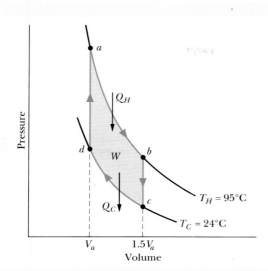

FIGURE 22-5 Sample Problem 22-1. A $p-V$ diagram for an idealized Stirling engine using an ideal gas. The net work done by the system during the cycle is represented by the gold region. Q_H represents heat added to the system during the isothermal expansion ab. Q_C represents heat lost by the system during the isothermal contraction cd. (During the constant-volume process bc, thermal energy is stored in a section of the engine, often a metal mesh. The same amount of energy is taken out of storage during the constant-volume process da.) The engine cycles clockwise (as in Fig. 22-2a).

a. What is the engine's net work per cycle?

SOLUTION To find the net work W done by the engine in one cycle, we first find the work done by the gas of the engine during each of the four processes. From Eq. 21-8, the work W_{ab} done during the isothermal expansion ab from volume V_a to volume $1.5V_a$ (at temperature T_H) is

$$W_{ab} = nRT_H \ln \frac{1.5V_a}{V_a} = nRT_H \ln 1.5. \quad (22\text{-}3)$$

Similarly, the work W_{cd} done during the isothermal compression cd from volume $1.5V_a$ to volume V_a (at temperature T_C) is

$$W_{cd} = nRT_C \ln \frac{V_a}{1.5V_a} = -nRT_C \ln 1.5,$$

where we have used the fact that $\ln(1/A) = -\ln A$. During the constant-volume processes bc and da, we know from Chapter 21 that no work is done; that is, $W_{bc} = W_{da} = 0$.

Thus the net work W done by the engine during

one cycle is

$$W = W_{ab} + W_{bc} + W_{cd} + W_{da}$$
$$= nRT_H \ln 1.5 + 0 - nRT_C \ln 1.5 + 0$$
$$= nR \ln 1.5(T_H - T_C). \quad (22\text{-}4)$$

Inserting the known data, we then have

$$W = (8.1 \times 10^{-3}\,\text{mol})(8.31\,\text{J/mol}\cdot\text{K})$$
$$\times (\ln 1.5)(95°\text{C} - 24°\text{C})$$
$$= 1.937\,\text{J} \approx 1.9\,\text{J}. \quad \text{(Answer)}$$

b. What is the power of the engine?

SOLUTION From Chapter 7, we know that power P is given by $P = W/t$. Here W is the work per cycle, and t is the time of a cycle. Since the engine cycles at the rate of $0.70\,\text{s}^{-1}$, a cycle takes $1/0.70\,\text{s}^{-1}$, or $1.429\,\text{s}$. Thus

$$P = \frac{W}{t} = \frac{1.937\,\text{J}}{1.429\,\text{s}} \approx 1.4\,\text{W}. \quad \text{(Answer)}$$

c. What is the net heat transfer into the gas during a cycle?

SOLUTION Since the gas returns to its original state at the end of each cycle, the net change ΔE_{int} in its internal energy during a cycle is zero. From the first law of thermodynamics, we then write

$$\Delta E_{\text{int}} = Q - W,$$

which, with $W = 1.937\,\text{J}$ and $\Delta E_{\text{int}} = 0$, gives us

$$Q = W = 1.937\,\text{J} \approx 1.9\,\text{J}. \quad \text{(Answer)}$$

d. What is the efficiency e of the engine?

SOLUTION In order to use Eq. 22-2 ($e = |W|/|Q_H|$), we first must find the heat transfer Q_H from the hot reservoir to the gas of the engine during the isothermal expansion ab. Recall that in an isothermal process, the heat transfer is equal to the work done by the system (see Table 21-5). Using Eq. 22-3, we write

$$Q_H = W_{ab} = nRT_H \ln 1.5,$$

and then, with Eq. 22-4, we can use Eq. 22-2 to write

$$e = \frac{|W|}{|Q_H|} = \frac{nR(T_H - T_C)\ln 1.5}{nRT_H \ln 1.5}$$
$$= \frac{T_H - T_C}{T_H}. \quad (22\text{-}5)$$

Note that this answer is independent of the value of n and of the actual pressures and volumes assumed by the gas.

Now we must be careful: although the *temperature difference* $T_H - T_C$ in the numerator can be expressed in either Celsius degrees or kelvins, the *temperature* T_H in the denominator must be expressed in kelvins—it is 368 K. Taking this precaution, we have

$$e = \frac{(95°C - 24°C)}{368 \text{ K}} = 0.1929 \approx 19\%. \quad \text{(Answer)}$$

Although this engine is reasonably efficient, its power is quite small. More sophisticated, real Stirling engines are currently being tested as alternatives to conventional car engines.

SAMPLE PROBLEM 22-2

An automobile engine, whose thermal efficiency e is 22.0%, operates at 95.0 cycles per second and does work at the rate of 120 hp.

a. How much work (in joules) does the engine do per cycle?

SOLUTION The work per cycle is

$$W = \frac{(120 \text{ hp})(746 \text{ W/hp})(1 \text{ J/W·s})}{95.0 \text{ s}^{-1}}$$

$$= 942 \text{ J}. \quad \text{(Answer)}$$

Do not confuse the symbols W for work and W for the watt, a unit of power.

b. How much heat does the engine absorb (extract from the "reservoir") per cycle?

SOLUTION From Eq. 22-2, we have

$$Q_H = \frac{W}{e} = \frac{942 \text{ J}}{0.220} = 4282 \text{ J}. \quad \text{(Answer)}$$

c. How much heat is discarded by the engine per cycle, and lost to the low-temperature reservoir?

SOLUTION From Eq. 22-1,

$$|Q_C| = |Q_H| - |W| = 4282 \text{ J} - 942 \text{ J} = 3340 \text{ J}.$$

Heat discarded *by* the engine is a negative quantity, so

$$Q_C = -3340 \text{ J}. \quad \text{(Answer)}$$

We see that this engine extracts 4282 J of heat per cycle, which must be paid for at the gas pump, does 942 J of work, and discards 3340 J of heat to the exhaust. The engine discards 3340/942 or 3.6 times more energy than it converts to useful purposes.

PROBLEM SOLVING

TACTIC 1: THE LANGUAGE OF THERMODYNAMICS
A rich, but sometimes misleading, language is used in scientific and engineering studies of thermodynamics. And regardless of your major, you will need to understand what that language means. You may see statements that say heat is absorbed, extracted, rejected, discharged, discarded, withdrawn, delivered, gained, lost, or expelled, or that it flows from one body to another (as if it were a liquid). You may also see statements that describe a body as *having* heat (as if heat can be held or possessed), or that its heat is increased or decreased. You should always keep in mind what is meant by the term *heat*:

Heat is energy that is transferred from one body to another body owing to a difference in the temperatures of the bodies.

When we identify one of the bodies as being our system of interest, any such transfer of energy into the system is positive heat Q, and any such transfer out of the system is negative Q.

The term *work* also requires close attention. You may see statements that say work is produced or generated, or combined with heat or changed from heat. Here is what is meant by the term *work*:

Work is energy that is transferred from one body to another body owing to a force that acts between them.

When we identify one of the bodies as being our system of interest, any such transfer of energy out of the system is said to be either positive work W done *by* the system or negative work W done *on* the system. And any such transfer of energy into the system is said to be negative work done *by* the system and positive work done *on* the system. (The preposition that is used is important.) Obviously, this can be confusing: whenever you see the term *work*, you should read carefully to determine the intent.

22-3 REFRIGERATORS

Heat is transferred naturally from a hot place to a cool place, as from the sun to the Earth. There is never any "natural" net transfer in the other direction. We express this observation in another formulation of the second law of thermodynamics:

SECOND LAW OF THERMODYNAMICS (SECOND FORM): It is not possible for heat to be transferred from one body to another body that is at a higher temperature with no other change taking place.

A device that transfers energy as heat from a cold place to a warm place is called a **refrigerator.** Figure 22-6a shows the heat and work transfers that occur. Heat Q_C is extracted from a low-temperature reservoir and some work W is done *on the system* by an external agent; the energies transferred as heat and as work are combined and discharged as heat Q_H to a high-temperature reservoir. In your household refrigerator, the low-temperature reservoir is the cold chamber in which the food is stored. The high-temperature reservoir is the room in which the unit is housed. Work, which shows up on a utility bill, is done by the motor that drives the unit. In an air conditioner, the low-temperature reservoir is the room to be cooled, the high-temperature reservoir is the outside air, where the condenser coils are located, and again the work is done by the motor that drives the unit.

The purpose of both a refrigerator and an air conditioner is to transfer energy as heat from the low-temperature reservoir to the high-temperature reservoir, doing as little work on the system as possible. We rate such units by their **coefficient of per-** formance K, defined as

$$K = \frac{|Q_C|}{|W|} = \frac{|Q_C|}{|Q_H| - |Q_C|}. \quad (22\text{-}6)$$

Design engineers, and those who pay utility bills, want the coefficient of performance of a refrigerator to be as high as possible. A value of 5 is typical for a household refrigerator, and a value in the range 2–3 is typical for a room air conditioner.

Figure 22-6b shows a "perfect" refrigerator— one that cools without the expenditure of work; it would have a coefficient of performance of infinity. Long experience has shown that it is impossible to build such a device. So another way to express the second law of thermodynamics (second form) is: *there are no perfect refrigerators.*

The second of the nonevents that we proposed in Section 22-1, namely, the spontaneous arising of a temperature difference between the two ends of a spoon, is equivalent to a perfect refrigerator. In the spoon, energy would be transferred spontaneously as heat from a cool place to a warm place, a violation of the second form of the second law.

SAMPLE PROBLEM 22-3

A household refrigerator, whose coefficient of performance K is 4.70, extracts heat from the cold chamber at the rate of 250 J per cycle.

a. How much work per cycle is required to operate the refrigerator?

SOLUTION From Eq. 22-6 we have

$$|W| = \frac{|Q_C|}{K} = \frac{250 \text{ J}}{4.70} = 53 \text{ J}.$$

This 53 J of energy is transferred into the system. Thus the work done *on* the refrigerating system is

$$W = +53 \text{ J} \quad \text{(on the system).} \quad \text{(Answer)}$$

We could also say, in this situation, that the work done *by* the system is -53 J. It is the former, though, that is important to the person who pays for the operation of the refrigerator.

b. How much heat per cycle is discharged to the room, which forms the high-temperature reservoir of the refrigerator?

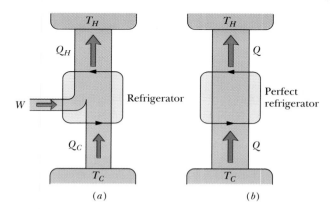

FIGURE 22-6 We indicate a refrigerator by encircling it with arrows pointing counterclockwise. (*a*) In a real refrigerator, heat is extracted from a low-temperature reservoir, some work is done, and the energy equivalent of the two is discharged as heat into a reservoir of higher temperature. (*b*) In a perfect refrigerator, no work is required. Nobody has ever built such a refrigerator.

SOLUTION Equation 22-1, which is the first law of thermodynamics for a cyclic device, holds for refrigerators as well as for engines. We then have

$$|Q_H| = |W| + |Q_C|$$

$$= 53 \text{ J} + 250 \text{ J} = 303 \text{ J}. \quad \text{(Answer)}$$

We see that a refrigerator is also an efficient room heater! By paying for 53 J of work (the motor), you get 303 J of heat delivered to the room from the condenser coils at the back of the unit. If you heated the room with an electric heater, you would get only 53 J of heat for every 53 J of work that you pay for. Think about the wisdom (?) of trying to cool the kitchen on a hot day by leaving the refrigerator door open!

22-4 THE SECOND LAW OF THERMODYNAMICS

The shorter versions of the two forms of the second law of thermodynamics are:

FIRST FORM: There are no perfect engines.

SECOND FORM: There are no perfect refrigerators.

Although these statements seem quite different, we want to show that they are exactly equivalent, in the sense that a violation of either implies a violation of the other. That is, if you succeed in building a perfect engine, you can also build a perfect refrigerator and conversely.

Let us assume first that you have built a perfect engine. You can use that engine to provide the work input to a (real!) refrigerator and thus transform it into a perfect refrigerator. Figure 22-7 shows how this is done. Couple the engine and the refrigerator together as a single unit, and adjust things so that the work done per cycle by the engine is just the amount needed per cycle to operate the refrigerator. Then no external work is involved in the combined engine + refrigerator, which we call *device X*.

Consider a numerical example. In Fig. 22-7a, the perfect engine extracts 100 J from the high-temperature reservoir and converts it into 100 J of work. The refrigerator part of the combination extracts 50 J from the low-temperature reservoir, combines it with the 100 J of work delivered to it by the engine, and discharges 150 J of heat to the high-

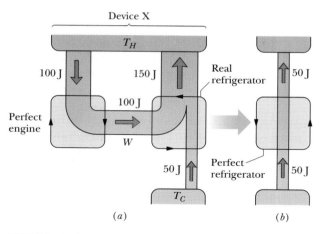

FIGURE 22-7 (*a*) The work *W* done by an engine (assumed perfect) is used to drive a real refrigerator. (*b*) The combination of these two devices (which we call device X) acts like a perfect refrigerator.

temperature reservoir. As Fig. 22-7b shows, the overall effect of device X is to extract 50 J of heat from the low-temperature reservoir and transfer it to the high-temperature reservoir, with no external work being needed. Device X is a perfect refrigerator! Thus if you can build a perfect engine, you can also build a perfect refrigerator.

As an exercise you should be able to show that, if we can build a perfect refrigerator, we can use it to transform a real engine into a perfect engine. The two formulations of the second law of thermodynamics do indeed state the same law. If you violate either, you automatically violate the other.

22-5 AN IDEAL ENGINE

There are no perfect engines. That is, no real engine can have an efficiency of 100%. A question remains: If not 100%, how high can the efficiency of a real engine be? To answer this question, we must probe into the detailed workings of an engine.

In studying gases, we avoided the complexities of real gases by introducing a useful idea: an *ideal gas*. Its usefulness lies in the fact that the ideal gas represents the limiting behavior of real gases. In studying engines, we follow this same path. We avoid the complexities of real engines by introducing another useful idea: the *ideal engine*. An ideal engine —in ways that we shall explore—represents the limiting behavior of real engines.

Our ideal engine consists of a piston–cylinder arrangement containing an ideal gas. A heat reservoir at temperature T_H, another heat reservoir at temperature T_C, and an insulating stand are also provided. The ideal gas constitutes the *system* to which we shall apply the laws of thermodynamics. The cylinder with its weighted piston, the insulating stand, and the two thermal reservoirs constitute the *environment* of that system.

We assume first that our ideal engine has no friction, no fluid turbulence, and no unwanted heat transfers. These are all the obvious things that an engineer would strive to eliminate. Beyond that, however, we assume that all the processes that make up the operating cycle of the engine—all expansions, compressions, and changes in temperature and pressure—are carried out extremely slowly; that is, we assume them to be *quasistatic* processes. By doing so we ensure that the system will be essentially in thermal equilibrium at all times and that we can plot the status of the system on a p–V diagram.

A process carried out in this way is called a *reversible process*, the test being that the process can be made to proceed in the opposite direction by making only a *tiny* change—strictly, a *differential* change—in the external conditions. Thus if we are slowly removing weight from a loaded piston, permitting the gas to expand, we can—at any stage—decide to *add* rather than subtract the weight increments, thus turning the expansion into a compression.

Because all its processes are reversible, the cycle as a whole is also reversible. This means that—at will—the engine can be run backward as an ideal refrigerator, the heat and work transfers changing in sign but not in magnitude. Our ideal engine is a *reversible* engine; indeed, that is what is ideal about it.

22-6 THE CARNOT CYCLE

It remains to describe the cycle to which we shall subject the ideal gas that forms the working substance of our ideal, reversible engine. We choose a **Carnot cycle,*** which consists of two isothermal and two adiabatic processes. Figure 22-8 suggests the mechanics of this cycle; Fig. 22-9 shows the cycle on a p–V diagram. The following four steps make up the cycle:

---

*Named for the French engineer and scientist N. L. Sadi Carnot (pronounced "car-no") who first proposed the concept in 1824.

STEP 1. Start with the cylinder on the high-temperature reservoir, so that the system, which is the ideal gas, is in the state represented by point a in Fig. 22-9. Gradually, remove some weight from the piston, allowing the system to expand slowly to point b at constant temperature T_H. During this process, heat Q_H is absorbed by the system from the high-temperature reservoir. Because this process is isothermal, the internal energy of the system does not change and all the absorbed heat is changed into the positive work done by the system during the expansion.

STEP 2. Put the cylinder on the insulating stand and, by removing more weight from the piston, allow the system to further expand slowly to point c in Fig. 22-9. This expansion is adiabatic because no heat enters or leaves the system. The system does positive work in lifting the piston farther and the temperature of the system drops to T_C, because the energy to do the work must come from the internal energy of the system.

STEP 3. Put the cylinder on the colder heat reservoir. By gradually adding weight to the piston, compress the gas slowly to point d at constant temperature T_C. During this process, heat Q_C is transferred from the gas to the reservoir. Because the compression is isothermal, Q_C is equal to the negative work done by the gas as the piston and its load descend.

STEP 4. Put the cylinder on the insulating stand and, by adding still more weight, compress the gas slowly back to its initial point a of Fig. 22-9, thus completing the cycle. The compression is adiabatic because no heat enters or leaves the system. Negative work is done by the gas, and the temperature of the gas increases to T_H.

The special property of the Carnot engine is that—as we shall show—its thermal efficiency can be written

$$e_{\text{Car}} = \frac{T_H - T_C}{T_H} \qquad \text{(Carnot engine).} \quad (22\text{-}7)$$

Thus the efficiency of a Carnot engine depends *only* on the temperatures of the two reservoirs between which it operates. The importance of the Carnot engine is that—as you will see in the next section—no real engine working between two temperatures can have an efficiency greater than a Carnot engine working between those same temperatures. It is in this sense that the Carnot engine represents the limiting behavior of real engines. If, as a practicing engineer, you are striving to increase the efficiency of a (very real!) engine, it is useful to know that there is a fundamental limit, dictated by the laws of thermodynamics, beyond which you cannot possibly go.

You cannot exceed the efficiency of a ideal Carnot engine even with another ideal engine. Note, as

an example, that the expression in Eq. 22-5 for the efficiency of an ideal Stirling engine matches that in Eq. 22-7 for a Carnot engine. (We return to this match of expressions in the next section.)

A Carnot engine—because it is reversible—can be operated backward as a Carnot refrigerator. As we shall prove below, its coefficient of performance is given by

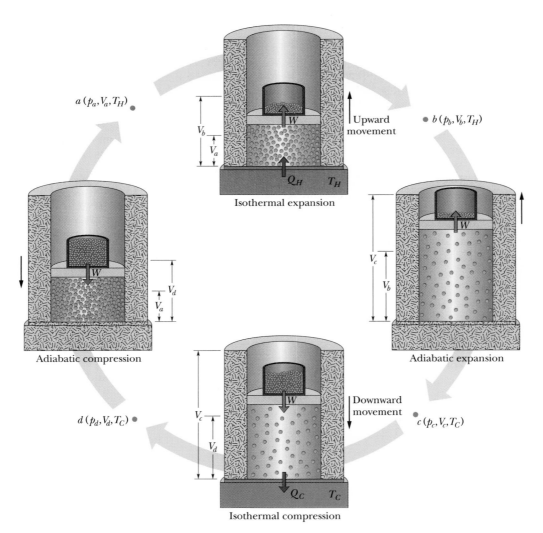

FIGURE 22-8 A Carnot cycle. The points a, b, c, and d correspond to the points so labeled in Fig. 22-9. The cylinder–piston arrangements show intermediate steps in the processes that connect adjacent points of the cycle. At top, during step 1, the decreasing weight on the piston and the heat transfer from the high-temperature reservoir allow the ideal gas in the cylinder to expand isothermally from volume V_a to volume V_b. At right, during step 2, an even lesser weight and the thermal insulation allow the gas to expand adiabatically to the largest volume V_c. At bottom, during step 3, the increasing weight and the heat loss to the low-temperature reservoir cause the gas to be compressed isothermally to volume V_d. At left, during step 4, an even greater weight and the thermal insulation cause the gas to be compressed adiabatically to the smallest volume V_a.

$$K_{\text{Car}} = \frac{T_C}{T_H - T_C} \qquad \text{(Carnot refrigerator)}. \qquad (22\text{-}8)$$

Equation 22-8 tells us that the coefficient of performance of a Carnot refrigerator increases as $T_H \to T_C$. Curiously, the less we need the refrigerator, the better it performs!

Proof of Eqs. 22-7 and 22-8

Along the isothermal path *ab* in Fig. 22-9, the temperature remains constant. Because the gas is ideal, its internal energy, which depends only on the temperature, also remains constant. From the first law of thermodynamics then, $\Delta E_{\text{int}} = 0$, and so the heat transferred from the high-temperature reservoir must equal the work done by the expanding gas. From Eq. 21-8 we then have

$$|Q_H| = |W_H| = nRT_H \ln \frac{V_b}{V_a}.$$

Similarly, for the isothermal process *cd* in Fig. 22-9, we can write

$$|Q_C| = |W_C| = nRT_C \ln \frac{V_c}{V_d}.$$

Dividing these two equations yields

$$\frac{|Q_H|}{|Q_C|} = \frac{T_H}{T_C} \frac{\ln(V_b/V_a)}{\ln(V_c/V_d)}. \qquad (22\text{-}9)$$

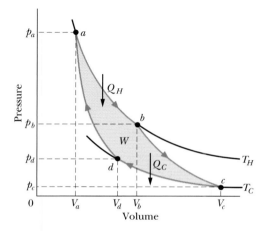

FIGURE 22-9 A *p*–*V* diagram for the Carnot cycle illustrated in Fig. 22-8. The working substance is taken to be an ideal gas.

Equation 21-38 allows us to write, for the two adiabatic processes *bc* and *da*,

$$T_H V_b^{\gamma-1} = T_C V_c^{\gamma-1} \quad \text{and} \quad T_H V_a^{\gamma-1} = T_C V_d^{\gamma-1}.$$

Dividing these two equations results in

$$\frac{V_b^{\gamma-1}}{V_a^{\gamma-1}} = \frac{V_c^{\gamma-1}}{V_d^{\gamma-1}}$$

or

$$\frac{V_b}{V_a} = \frac{V_c}{V_d}. \qquad (22\text{-}10)$$

Combining Eqs. 22-9 and 22-10 yields

$$\frac{|Q_H|}{|Q_C|} = \frac{T_H}{T_C}. \qquad (22\text{-}11)$$

Combining this result with Eq. 22-2 leads at once to Eq. 22-7; combining it with Eq. 22-3 leads to Eq. 22-8. These are the equations whose proof we sought.

22-7 THE EFFICIENCIES OF REAL ENGINES

The importance of the Carnot engine is summed up in this theorem:

> No real engine operating between two specified temperatures can have a greater efficiency than that of a Carnot engine operating between those same two temperatures.

To prove this statement, let us assume that an inventor, working in her garage, has constructed an engine, engine X, whose efficiency e_X—she claims—is greater than e_{Car}, the efficiency of a Carnot engine. That is,

$$e_X > e_{\text{Car}} \qquad \text{(a claim)}. \qquad (22\text{-}12)$$

Let us couple engine X to a Carnot engine operating backward as a Carnot refrigerator, as in Fig. 22-10a. We adjust the Carnot refrigerator so that the work it requires per cycle is just that provided by engine X.

If Eq. 22-12 is true then, from the definition of efficiency (see Eq. 22-2), we must have

$$\frac{|W|}{|Q'_H|} > \frac{|W|}{|Q_H|},$$

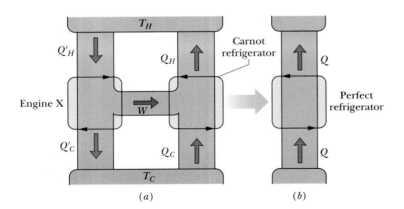

FIGURE 22-10   (*a*) Engine X drives a Carnot refrigerator. (*b*) If, as claimed, engine X is more efficient than a Carnot engine, then the combination shown in (*a*) is equivalent to the perfect refrigerator shown here. This violates the second law of thermodynamics, so we conclude that engine X *cannot* be more efficient than a Carnot engine.

where the right side of the inequality is the efficiency of the Carnot engine when it operates as an engine. This inequality requires that

$$|Q_H| > |Q'_H|. \qquad (22\text{-}13)$$

Because the work done by engine X is equal to the work done on the Carnot refrigerator, we have, from Eq. 22-1,

$$|Q_H| - |Q_C| = |Q'_H| - |Q'_C|,$$

which we can write as

$$|Q_H| - |Q'_H| = |Q_C| - |Q'_C| = Q. \qquad (22\text{-}14)$$

Because of Eq. 22-13, the quantity Q in Eq. 22-14 must be positive.

Comparison of Eq. 22-10 with Fig. 22-7 shows that the net effect of engine X and the Carnot refrigerator, working as a combination, is to transfer heat Q from a cold-temperature reservoir to a hot-temperature reservoir, without the requirement of work. Thus the combination acts like the perfect refrigerator of Fig. 22-10*b*, whose existence is a violation of the second law of thermodynamics.

Something must be wrong with one of our assumptions; it can be only Eq. 22-12. We conclude that no real engine can have an efficiency greater than that of a Carnot engine working between the same two temperatures. At most, it can have an efficiency equal to this Carnot efficiency.

A Note on the Proof

In the proof we considered the Carnot engine of Section 22-6, that is, one that has an ideal gas as its working substance. Could it be that a Carnot engine with some other working substance (say ammonia or ethanol), or some other reversible engine using a

different cycle is more efficient than the Carnot engine we considered?

It is a forlorn hope. It can be shown that *all* reversible engines working between a given pair of temperatures have *exactly* the same efficiency as a Carnot engine employing an ideal gas and working between those same temperatures. Thus the Carnot efficiency given by Eq. 22-7 can be extended to all reversible engines, regardless of the cycle they employ. (As an example, note that it is the same as Eq. 22-5 for a reversible Stirling engine.) And the requirement of an ideal gas as the working substance is not essential. The proof follows along the lines of the proof we have just given above for engine X.

SAMPLE PROBLEM 22-4

The turbine in a steam power plant takes steam from a boiler at 520°C and exhausts it into a condenser at 100°C. What is its maximum possible efficiency?

SOLUTION   Its maximum efficiency is the efficiency of a Carnot engine operating between the same two temperatures. From Eq. 22-7 then,

$$e_{max} = \frac{T_H - T_C}{T_H} = \frac{793 \text{ K} - 373 \text{ K}}{793 \text{ K}}$$

$$= 0.53 \text{ or } 53\%. \qquad \text{(Answer)}$$

Because of friction, turbulence, and unwanted thermal losses, actual efficiencies of about 40% may be realized for such a steam engine. Note that the theoretical maximum efficiency depends only on the two temperatures involved, not on the pressures or other factors.

The theoretical efficiency of an ordinary automobile engine is about 56%, but practical considerations reduce this to about 25%.

SAMPLE PROBLEM 22-5

An inventor claims to have developed an engine that, during a certain time interval, takes in 110 MJ of heat at 415 K, rejects 50 MJ of heat at 212 K while it manages to do 16.7 kW·h of work. Would you invest money in this project?

SOLUTION From Eq. 22-2, the claimed efficiency of this device is

$$e = \frac{|W|}{|Q_H|} = \frac{(16.7 \text{ kW·h})(3.60 \text{ MJ/kW·h})}{110 \text{ MJ}}$$

$$= 0.55 \text{ or } 55\%.$$

From Eq. 22-7 the maximum theoretical efficiency for the two given temperatures is

$$e = \frac{T_H - T_C}{T_H} = \frac{415 \text{ K} - 212 \text{ K}}{415 \text{ K}} = 0.49 \text{ or } 49\%.$$

The claimed efficiency is greater than the theoretical maximum. Best advice: don't invest.

SAMPLE PROBLEM 22-6

A *heat pump* is a device that—acting as a refrigerator—can heat a house by drawing heat from the outside, doing some work, and discharging heat inside the house. Suppose the outside temperature is − 10°C and the interior is to be kept at 22°C. It is necessary to deliver heat to the interior at the rate of 16 kW to make up for normal heat losses. At what minimum rate must energy be supplied to the heat pump?

SOLUTION From Eq. 22-8 the maximum coefficient of performance of the heat pump, acting as a refrigerator, is

$$K = \frac{T_C}{T_H - T_C} = \frac{(273 - 10) \text{ K}}{(273 + 22) \text{ K} - (273 - 10) \text{ K}}$$

$$= 8.22.$$

We can recast Eq. 22-6 as

$$K = \frac{|Q_C|}{|W|} = \frac{|Q_H| - |W|}{|W|}.$$

Solving for $|W|$ and dividing by time to express the result in terms of power, we obtain

$$\frac{|W|}{t} = \frac{|Q_H|/t}{K + 1} = \frac{16 \text{ kW}}{8.22 + 1} = 1.7 \text{ kW}. \quad \text{(Answer)}$$

Herein lies the "magic" of the heat pump. By using the heat pump as a refrigerator to cool the great outdoors, you can deliver 16 kW to the interior of the house but you need pay for only the 1.7 kW it takes to run the pump. Actually, the 1.7 kW is a theoretical minimum requirement because it is based on ideal performance. In practice, a greater power input would be required but there would still be a very considerable saving over, say, heating the house directly with electric heaters. In that case, you would have to pay directly for every kilowatt of heat transfer. When the outside temperature is greater than the inside temperature, the heat pump can be used as an air conditioner. Still operating as a refrigerator, it now pumps heat from inside the house to the great outdoors. Again, work must be done (and paid for) but the energy removed as heat from the house interior exceeds the energy equivalent of the work done. Another thermodynamic bargain!

22-8 ENTROPY: A NEW VARIABLE

Each of the three laws of thermodynamics is associated with a specific thermodynamic variable. For the zeroth law (see Chapter 19), the variable is the temperature T. For the first law (see Chapter 20), it is the internal energy E_{int}. For the second law, the variable is one you have not met before; it is called the **entropy** S.

It is our plan to define entropy in this section and, in later sections, to express the second law of thermodynamics in terms of that variable. We start by considering the Carnot cycle of Fig. 22-9, for which (from Eq. 22-11)

$$\frac{|Q_H|}{T_H} = \frac{|Q_C|}{T_C}.$$

We now discard the absolute value notation, recognizing in the process that whether the Carnot cycle is carried out clockwise, as an engine, or counterclockwise, as a refrigerator, Q_H and Q_C have opposite algebraic signs. With this understanding, we can write the above equation as

$$\frac{Q_H}{T_H} + \frac{Q_C}{T_C} = 0. \quad (22\text{-}15)$$

Because Q_H and Q_C are the only heat transfers in the closed cycle of Fig. 22-9, Eq. 22-15 tells us that the algebraic sum of the quantity Q/T, taken around the cycle, is zero.

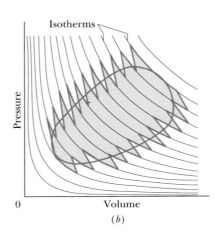

(a)   (b)

FIGURE 22-11 (a) An arbitrary reversible cycle, plotted on a p–V diagram with a family of isotherms in the background. (b) The cycle is represented as an assembly of adjacent Carnot cycles.

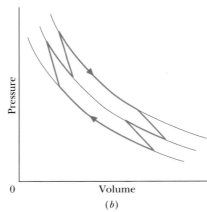

(a)   (b)

FIGURE 22-12 (a) Two of the cycles of Fig. 22-11 and (b) their equivalent.

We now wish to generalize Eq. 22-15, writing it in a form that applies not only to a Carnot cycle but to any reversible cycle. Figure 22-11a shows such a generalized cycle, superimposed on a family of isotherms. We can approximate this arbitrary cycle as closely as we wish by connecting adjacent isotherms with short, suitably chosen, adiabatic lines, as in Fig. 22-12b. In this way, we form an assembly of long, thin Carnot cycles. Convince yourself that traversing the individual Carnot cycles in Fig. 22-11b in sequence is exactly equivalent—in terms of heat transferred and work done—to traversing the jagged series of isotherms and adiabats that approximate the actual cycle. This is so because adjacent Carnot cycles (such as those in Fig. 22-12) have a common isotherm, and the two traversals, in opposite directions, of each isotherm cancel each other as far as heat transferred and work done are concerned.

We extend Eq. 22-15 by writing for the assembly of Carnot cycles, and hence for the zig-zag

isotherm–adiabat sequence in Fig. 22-11b,

$$\sum \frac{Q}{T} = 0.$$

In the limit of infinitesimal temperature differences between pairs of isotherms in Fig. 22-11b, the zig-zag sequence becomes the generalized cycle of Fig. 22-11a, and the equation above becomes

$$\oint \frac{dQ}{T} = 0 \qquad \text{(reversible cycle).} \qquad (22\text{-}16)$$

The circle on the integral sign indicates that the integral is evaluated for a complete traversal of the reversible cycle, starting and ending at any arbitrarily selected point.

We have established that the temperature T and the internal energy E_{int} are intrinsic properties of a system. One test of the validity of such a **state variable** is that, if we take the system through a complete, reversible cycle, the algebraic sum of the changes that occur in the variable must be zero. If this were not so, the variable would not return to its initial value and could not be an intrinsic property of the system alone. Thus the test of any proposed state variable X is that

$$\oint dX = 0 \qquad \text{(reversible cycle)}. \qquad (22\text{-}17)$$

Comparing Eqs. 22-16 and 22-17 shows that dQ/T must represent a differential change in some state variable that we have not previously encountered. We call this new variable the *entropy S* of the system, and we write, from Eq. 22-17,

$$dS = \frac{dQ}{T} \quad \text{and} \quad \oint dS = 0. \qquad (22\text{-}18)$$

The SI unit for entropy is the unit for heat divided by the unit for temperature, that is, the joule per kelvin.

Note that heat Q and work W are *not* state variables because $\oint dQ \neq 0$ and $\oint dW \neq 0$, as you can easily verify for the special case of a Carnot cycle.

We shall prove below that the property of a state variable X represented by Eq. 22-17 is exactly equivalent to saying that $\int dX$ between any two states has the same value for all reversible paths connecting those states. Suppose, for example, that a system is changed from an initial state i to a final state f, in which its temperature is higher by 15 K. That same temperature difference would hold no matter which of the infinite number of possible paths the system took in moving from state i to state f. The same is true for any other state variable. The pressure difference would be the same for all paths; so would the volume difference. Because entropy is also a state variable, the same must hold true for the entropy difference. Thus we can write, for the entropy difference between any two states i and f,

$$S_f - S_i = \int_i^f dS = \int_i^f \frac{dQ}{T}$$

$$\text{(reversible path)}, \qquad (22\text{-}19)$$

where the integral is carried out over any reversible path connecting these two states.

Equation 22-19 defines the *difference in entropy* between two states, rather than the actual entropy of a state. However, our concern is always with such differences. An arbitrary constant can be added to the absolute entropy (as it can be added to the absolute

internal energy) of any state without altering any of our conclusions.

Entropy Changes in Reversible Processes

Equation 22-19 holds for *reversible processes only*, that is, for processes carried out without friction and so slowly that the process can be reversed at any stage by making an *infinitesimal* change in the environment of the system. If heat is exchanged between the system and its environment in such a process, Eq. 22-19 tells us that the entropy of the system will change. If heat is *added* to the system, the increment dQ in Eq. 22-19 (according to our sign convention for heat) is *positive* so that the resulting entropy change for the system will be an *increase*. At the same time, the entropy of the environment of the system, that is, of the heat reservoir from which the heat was withdrawn and transferred to the system, will decrease, and by the same amount as the entropy of the system will increase. This must be so because every time a millijoule of heat enters the system at a given temperature, a millijoule of heat must leave the reservoir, at that same temperature.

The converse is also true. If heat is removed from the system, the entropy of the system will decrease and the entropy of the environment will increase by the same amount. Hence we conclude the following:

> For reversible processes, the entropy of the *system* may increase, decrease, or remain unchanged. The entropy change for the *environment* of the system will always be equal in magnitude but opposite in sign to the entropy change of the system. Thus, in a reversible process, the entropy of the *system + environment* remains constant.

In the next section, we shall look closely at entropy changes in *irreversible* processes and then pull our conclusions together as a new formulation of the second law of thermodynamics. First, though, here is the promised proof.

Proof of Path Independence for Eq. 22-19

We can write Eq. 22-18 (see Fig. 22-13) as

$$\int_{a}^{b} dS + \int_{b}^{a} dS = 0,$$
$$\text{path 1} \qquad \text{path 2}$$

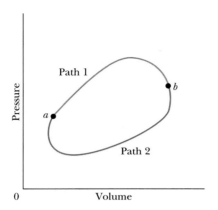

FIGURE 22-13 Paths 1 and 2 are independent paths connecting points a and b.

where a and b are arbitrary points and the integrals are taken over paths 1 and 2 connecting these points. Because the cycle is reversible, we can write this equation as

$$\int_{a}^{b} dS - \int_{a}^{b} dS = 0. \qquad (22\text{-}20)$$
$$\text{path 1} \qquad \text{path 2}$$

In Eq. 22-20, we have simply decided to traverse path 2 in the opposite direction, that is, from a to b rather than from b to a. We do this by changing the order of the limits in the second integral, which requires that we also change the sign of the integral, thus obtaining Eq. 22-20. We can then write Eq. 22-20 as

$$\int_{a}^{b} dS = \int_{a}^{b} dS. \qquad (22\text{-}21)$$
$$\text{path 1} \qquad \text{path 2}$$

This is what we set out to prove. Equations 22-19 and 22-21 both tell us that if the difference in entropy between two states (either i and f or a and b) does not depend on the path, then we can compute that difference over any (reversible) path.

■ SAMPLE PROBLEM 22-7

A lump of ice whose mass m is 235 g melts (reversibly) to water, the temperature remaining at 0°C throughout the process.

a. What is the entropy change for the ice? The heat of fusion of ice is 333 kJ/kg.

SOLUTION The requirement that we melt the ice reversibly means that we must put the ice in contact with a heat reservoir whose temperature exceeds 0°C by only a differential amount. (If we then lower the reservoir temperature to a differential amount *below* 0°C, the melted ice will start to freeze.) Because the process is reversible, we can use Eq. 22-19, obtaining

$$S_{\text{water}} - S_{\text{ice}} = \int \frac{dQ}{T} = \frac{1}{T} \int dQ = \frac{Q}{T}.$$

But

$$Q = mL_F = (0.235 \text{ kg})(333 \text{ kJ/kg}) = 7.83 \times 10^4 \text{ J}.$$

Thus

$$S_{\text{water}} - S_{\text{ice}} = \frac{Q}{T} = \frac{7.83 \times 10^4 \text{ J}}{273 \text{ K}}$$

$$= 287 \text{ J/K}. \qquad \text{(Answer)}$$

b. What is the entropy change of the environment?

SOLUTION In this case, the environment is the heat reservoir from which the heat is transferred to melt the ice. Every unit of heat that *enters* the ice must have *left* the reservoir, the temperature of both ice and reservoir being approximately the same. Therefore the entropy change of the reservoir is equal in magnitude but opposite in sign to that of the ice, or

$$\Delta S_{\text{reservoir}} = -287 \text{ J/K}. \qquad \text{(Answer)}$$

The entropy change for the *ice + reservoir* system is thus zero, as it must be for a reversible process.

In practice, the melting of ice is likely to be irreversible, as when you toss an ice cube into a glass of water at room temperature. The temperature difference between the ice and the reservoir, the water in this case, is not a differential amount but is about 20°C. The process proceeds in only one direction—the ice melts—and cannot be reversed at any stage by making only a differential change in the water temperature. You cannot use Eq. 22-19 in such a case and the calculations we have made above are not valid then.

22-9 ENTROPY CHANGES FOR IRREVERSIBLE PROCESSES

Equation 22-19 tells us how to calculate the entropy change of a system that undergoes a reversible process. Strictly speaking, there are no such processes in the real world. Friction and unwanted heat transfers are always present and the differences in pressure and temperature between a system and its environment are usually not infinitesimal. Every actual thermodynamic process is—to a greater or lesser extent—irreversible.

How are we to calculate the entropy change between the initial and final states in such cases? We take advantage of the fact that the difference in entropy—or of any other state variable, such as temperature or internal energy—between two equilibrium states does not depend on how the system passes from one state to the other:

To find the entropy change for an *irreversible* process between two equilibrium states, find a *reversible* process connecting those same states and calculate the entropy change for *that* process, using Eq. 22-19.

Consider two examples.

1. *Free expansion.* As in Section 20-6 (see Fig. 20-6), let an ideal gas increase its volume by expanding into an evacuated space. Because no work is done against the vacuum, $W = 0$. Because the system is enclosed by insulating walls, $Q = 0$. From the first law of thermodynamics, $\Delta E_{\text{int}} = Q - W$, it follows that $\Delta E_{\text{int}} = 0$ or

$$E_{\text{int},f} = E_{\text{int},i}, \qquad (22\text{-}22)$$

where i and f refer to the initial and final equilibrium states. Because the gas is ideal, the internal energy E_{int} depends on temperature only, so Eq. 22-22 tells us that $T_f = T_i$.

If we try to use Eq. 22-19 to calculate the entropy difference $S_f - S_i$ for a free expansion, we have an immediate problem. The fact that $Q = 0$ might lead us to predict from Eq. 22-19 that $\Delta S = 0$, which as you will see is incorrect. Following the procedure given above, we must find a reversible process—*any* reversible process—that connects the initial and the final states and apply Eq. 22-19—not to the free expansion but to *that* process.

Because $T_f = T_i$ for the free expansion of an ideal gas, a convenient reversible process is an isothermal expansion such as that carried out between points *a* and *b* of the Carnot cycle of Fig. 22-9. The isothermal expansion involves quite a different set of operations from the free expansion, the two processes having in common *only* the fact that they have

the same initial and final states. Applying Eq. 22-19 to the isothermal expansion yields

$$S_f - S_i = \int_{V_i}^{V_f} \frac{dQ}{T} = \frac{1}{T} \int_{V_i}^{V_f} dQ.$$

We have here removed the temperature T from the integral because the temperature remains constant in an isothermal process. In an isothermal process Q is equal to W (and so $dQ = dW$). Thus we may write this equation as

$$S_f - S_i = \frac{1}{T} \int_{V_i}^{V_f} dW. \qquad (22\text{-}23)$$

The integral in Eq. 22-23 is just the work done in an isothermal process, which (see Eq. 21-8) is

$$W = nRT \ln \frac{V_f}{V_i}. \qquad (22\text{-}24)$$

Substituting Eq. 22-24 into Eq. 22-23 leads to

$$S_f - S_i = nR \ln \frac{V_f}{V_i} \qquad \begin{array}{l}\text{(free}\\ \text{expansion)}.\end{array} \qquad (22\text{-}25)$$

Although we calculated this entropy change specifically for a reversible isothermal expansion between two states, it holds for *any* process connecting those same states, including a free expansion.

Because $V_f > V_i$, Eq. 22-25 tells us that the entropy of the system increases during a free expansion. We note that the entropy of the *environment* of the system does not change during a free expansion because the expansion takes place inside a rigid insulating box, as we considered in Fig. 20-6. Thus the entropy of the *system + environment* increases during a free expansion.

2. *Irreversible heat transfer.* Figure 22-14*a* shows two metal blocks, each of mass m and specific heat c, that are thermally insulated from each other within an insulating box. The blocks are alike in every way except that one is at a higher temperature than the other. If we remove the insulating barrier that separates the blocks and put them in thermal contact (as in Fig. 22-14*b*), they will eventually reach a common temperature T. Like the free expansion, this process is irreversible because we totally lose control of it once we put the two blocks in thermal contact with each other.

To find the entropy change between the initial state of Fig. 22-14*a* and the final state of Fig. 22-14*b*, we must once more find a *reversible* process connect-

(a)

(b)

FIGURE 22-14 (a) The initial state: two metal blocks are at different temperatures in individual insulating enclosures. (b) The final state: the insulating wall between the blocks is removed and they are allowed to come to thermal equilibrium at temperature T.

ing these states and calculate the entropy change by applying Eq. 22-19 to *that* process.

We can carry out a reversible process using a temperature reservoir of large heat capacity whose temperature is under our control, perhaps by turning a knob. We first adjust the reservoir temperature to $T + \Delta T$, the temperature of the hotter block, and we put that block in thermal contact with the reservoir. We then slowly (reversibly) lower the reservoir temperature from $T + \Delta T$ to T, extracting heat from the hot block as we do so. Because heat leaves the hot block, the entropy of that block decreases, the entropy change being

$$\Delta S_H = \int_i^f \frac{dQ}{T} = \int_{T+\Delta T}^{T} \frac{mc\,dT}{T} = mc \int_{T+\Delta T}^{T} \frac{dT}{T}$$

$$= mc \ln \frac{T}{T + \Delta T}. \qquad (22\text{-}26)$$

Here we have replaced dQ, the heat extracted from the hot block as its temperature changes by dT, by $mc\,dT$. Because the quantity whose logarithm is to be taken in Eq. 22-26 is less than unity, the logarithm is a negative number, verifying that the entropy decreases during this process.

To continue the reversible process, we now adjust the reservoir temperature to $T - \Delta T$, the temperature of the cooler block, and put *that* block in thermal contact with the reservoir. We then slowly (reversibly) raise the temperature of the reservoir from $T - \Delta T$ to T, adding heat to the cooler block

as we do so. Because heat is added to it, the entropy of the cooler block increases, the entropy change being

$$\Delta S_C = mc \ln \frac{T}{T - \Delta T}. \qquad (22\text{-}27)$$

The two blocks are now in their final equilibrium state and the reversible process is completed. The entropy change for the system is found by adding Eqs. 22-26 and 22-27, to obtain

$$S_f - S_i = \Delta S_H + \Delta S_C$$

$$= mc \ln \frac{T}{T + \Delta T} + mc \ln \frac{T}{T - \Delta T}$$

or

$$S_f - S_i = mc \ln \frac{T^2}{T^2 - \Delta T^2}$$

(irreversible heat transfer). (22-28)

In obtaining Eq. 22-28, we use the facts that $\ln a + \ln b = \ln ab$ and $(a + b)(a - b) = a^2 - b^2$.

The quantity whose logarithm is taken in Eq. 22-28 is greater than unity so that the logarithm is positive. This means that $S_f > S_i$. Thus the entropy of the system increases during this irreversible heat transfer. Because the system is thermally isolated from its surroundings, the entropy of the *system + environment* also increases during this irreversible process, just as it did for the (irreversible) free expansion.

SAMPLE PROBLEM 22-8

One mole of an ideal gas expands to twice its original volume in a free expansion, as in Fig. 20-6. What is the change in entropy of the gas? Of the environment?

SOLUTION The change in entropy of the gas is given by Eq. 22-25:

$$S_f - S_i = nR \ln \frac{V_f}{V_i}$$

$$= (1.00 \text{ mol})(8.31 \text{ J/mol·K})(\ln 2)$$

$$= 5.76 \text{ J/K}. \qquad \text{(Answer)}$$

This is an entropy increase. The entropy of the environment of the free expansion does not change because the expanding gas is thermally isolated from its surroundings. Thus the entropy change for the *system + environment* is $+5.76\,\mathrm{J/K}$.

SAMPLE PROBLEM 22-9

Two blocks of copper, the mass m of each being 850 g, are put into thermal contact in an insulated box, as in Fig. 22-14*b*. The initial temperatures of the two blocks are 325 K and 285 K and the constant specific heat c of copper is $0.386\,\mathrm{J/g \cdot K}$.

a. What is the final equilibrium temperature T of the two blocks?

SOLUTION The heat lost by the hotter block must be absorbed by the cooler one, so

$$mc(325\text{ K} - T) = mc(T - 285\text{ K}).$$

Canceling the factors mc and solving for T yield

$$T = \tfrac{1}{2}(325\text{ K} + 285\text{ K}) = 305\text{ K}. \quad \text{(Answer)}$$

b. What is the change in entropy for the two blocks?

SOLUTION The quantity ΔT that appears in Eq. 22-28 is

$$\Delta T = 325\text{ K} - 305\text{ K} = 20\text{ K}.$$

As a check, we note that $305\text{ K} - 285\text{ K} = 20\text{ K}$ as well. From Eq. 22-28 we then have

$$
\begin{aligned}
S_f - S_i &= mc \ln \frac{T^2}{T^2 - \Delta T^2} \\
&= (0.850\text{ kg})(386\text{ J/kg} \cdot \text{K}) \\
&\quad \times \ln \frac{(305\text{ K})^2}{(305\text{ K})^2 - (20\text{ K})^2} \\
&= 1.41\text{ J/K}. \quad \text{(Answer)}
\end{aligned}
$$

As for the free expansion, the entropy change for this irreversible process is an increase. Because the process is carried out in an insulated box, the entropy change of the environment is zero so that the net entropy change for the *system + environment* is $+1.41$ J/K, an increase.

22-10 ENTROPY AND THE SECOND LAW OF THERMODYNAMICS

We are now ready to formulate the second law of thermodynamics in terms of entropy:

SECOND LAW OF THERMODYNAMICS (THIRD FORM): In any thermodynamic process that proceeds from one equilibrium state to another, the entropy of the *system + environment* either remains unchanged or increases.

There is no way in which you can make the entropy of the *system + environment* decrease. It is true that the entropy of a *system* can be made to decrease, but that decrease must always be accompanied by an equal or greater increase in the entropy of the system's environment.

The third nonevent proposed in Section 22-1, the spontaneous movement of the air in the room to one corner, is what we may call a *free compression,* the opposite of a free expansion. We saw in the preceding section that a free expansion is always accompanied by an increase of entropy for the *system + environment*. A free compression would result in an entropy decrease and would thus be a violation of the entropy form of the second law.

Let us now make sure that the entropy form of the second law is consistent with the two forms that were previously presented.

1. *There are no perfect engines.* This is the short version of the first form of the second law. Because an engine operates in a cycle, the entropy change for the system, that is, for the gas that is the working substance, must be zero for one cycle of operation. We therefore need be concerned only with entropy changes of the environment. For a perfect engine, the environment is the single heat reservoir of Fig. 22-2*b* and the entropy change is a decrease, because heat is withdrawn from the reservoir. Thus a perfect engine (which is a violation of the first form) generates an entropy decrease, a violation of the entropy form of the second law.

2. *There are no perfect refrigerators.* This is the short version of the second form of the second law. Again, the entropy change for the gas in a perfect refrigerator is zero for one cycle of operation, and we need concern ourselves only with the entropy change for the environment. In this case, the environment is the two reservoirs of Fig. 22-6*b* and the entropy change is

$$\Delta S = \frac{Q}{T_H} - \frac{Q}{T_C}.$$

Because $T_H > T_C$, this entropy change is negative. So the perfect refrigerator (which violates the first form) also is a violation of the entropy form of the second law.

Entropy, Internal Energy, and Spontaneous Processes

A spontaneous process is one that occurs without an external cause and is due only to the internal nature of the system undergoing the process. For example, the sun shines spontaneously due to internal nuclear processes, and not due to any external cause.

How can we decide if some proposed process can occur spontaneously in the real world? Even if the process obeys every conservation principle of physics (such as the conservation of energy), it still must meet one more requirement in order to occur spontaneously: it must always result in an increase of entropy.

Figure 22-15 sums up the processes that can occur spontaneously for an isolated system based on entropy and internal energy considerations. The central box represents such a system, in an initial state S with definite values of internal energy, measured on the horizontal axis, and entropy, measured on the vertical axis. The only transitions that can occur spontaneously for this system are to states lying directly above the initial state. For these transitions, the energy remains constant and the entropy increases, as the first and second laws of thermodynamics require.

Transitions to states directly below S cannot occur because they represent entropy decreases; such decreases are forbidden—for isolated systems—by the second law. Transitions to states that lie to the right or to the left of S in Fig. 22-15 cannot occur because they represent changes in energy; such changes are forbidden—for isolated systems—by the first law.

Although entropy and energy considerations reveal *whether* a process may occur spontaneously, they do not tell us *when* that spontaneous process will take place. Some such processes, like the free expansion of a gas into an evacuated chamber, take place without delay. Other spontaneous processes, such as the radioactive decay of long-lived uranium isotopes, take billions of years.

22-11 WHAT IS ENTROPY ALL ABOUT?

Because our treatment of entropy has been fairly formal, you may not yet have developed a physical feeling for this important concept. Let us then summarize what we have discussed and try to make the concept more physical by extending its meaning in directions we have not yet explored.

Entropy: The Arrow of Time

Entropy S, like energy E_{int} and temperature T, is simply one of the several physical properties of a system that can be measured in the laboratory and to which a number and a unit can be assigned. (Indeed, there are those who believe that—at the deepest level—entropy is the *simplest* of these three properties to understand in physical terms.)

Consider a system, isolated from its environment, that can exist in two states—call them A and B—that have the same energy. If the system is in state A, will it move spontaneously to state B as time goes on? If it is in state B, will it move spontaneously to state A? Entropy provides the answer. The second law of thermodynamics tells us:

The only changes that are possible for an isolated system are those in which the entropy of

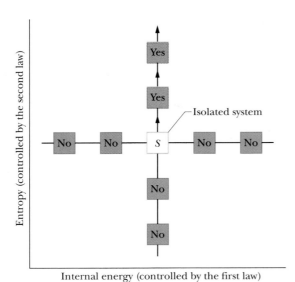

FIGURE 22-15 The central block represents the initial state S of an isolated thermodynamic system. The other blocks represent conceivable final states. Only transitions to the states vertically above S are consistent with both the first and second laws of thermodynamics. Transitions to all other states are forbidden by one or the other of these two laws.

(a)

(b)

FIGURE 22-16 Sometimes striking order appears naturally. (a) For example, Giant's Causeway in Northern Ireland consists of tall stone columns, many of which are hexagonal in cross section and appear to have been designed. The columns formed when hot magma seeped out of the ground and cooled. (b) Order can also be seen in the *sorted* *circles* of gravel and stones that occur naturally on an island north of Norway. Just how the columns and circles are produced is still debated, but we can be certain that, as the entropy of the magma and stones decreased, the entropy of their environment increased even more.

the system either increases or remains the same. Changes in which the entropy decreases will not occur.

The entropy remains the same only for reversible processes. No process in nature is truly reversible so that, in essentially all cases, we look for an entropy *increase* when we see a process occur spontaneously.

You may find that the entropy of a particular system *does* decrease in a spontaneous process, but you may be sure that a (greater) increase of entropy is going on simultaneously somewhere in the environment of the system. The ordered systems of Fig. 22-16 occurred with such an environmental increase in entropy.

The first and second laws of thermodynamics have been summed up in this way:

> The energy of the universe remains constant; the entropy of the universe always increases.

Energy obeys a conservation law; entropy does not.

Entropy: A Measure of Atomic Disorder

Entropy is also associated with the *disorder* of a system, a notion that has captured some popular attention. The claim that the disorder of the universe (or of our small isolated corner of it) always increases as time goes on is one that most of us have no trouble in accepting. If disorder and entropy *both* increase as time goes on, then perhaps they are related. In fact, they are; the trick is to define *disorder* in a quantitative and useful way.

The formal treatment of this aspect of entropy is the subject of *statistical thermodynamics,* which we do not cover in this book. We can discuss a few qualitative examples, however, that show ways in which the disorder of a system—as it is formally defined—can be increased.

1. Suddenly double the volume of a thermally insulated container of gas, as in a free expansion. We have seen earlier that such an expansion involves an increase in entropy. The atomic disorder of the gas *also* increases because there are now more positions in space that can be assumed by the individual atoms of the gas.

2. Raise the temperature of a gas in a container of fixed volume by adding heat. Even if we add heat to the gas in a reversible manner, Eq. 22-19 tells us that the entropy of the gas must increase. The disorder *also* increases because there is now a greater range of velocities that can be assumed by the atoms of the gas.

3. Let the swirling motion of the coffee in a cup gradually dissipate. We have seen that this spontaneous process results in an entropy increase. It also results in an increase in atomic disorder, because the initial swirling motion of the coffee is a relatively ordered state.

Examples of this sort could be added without limit. Suffice it to say that disorder can be defined quantitatively and can be related to entropy in a formal manner. The person who pointed the way was Ludwig Boltzmann (1844–1906), whose constant k we have used. The equation

$$S = k \log \mathcal{W},$$

which he first put forward, is the central equation of statistical thermodynamics and relates entropy to disorder quantitatively. It is engraved on his tombstone. Here S is the entropy of the system and $\mathcal{W}$ (which is usually a very large number) is a measure of the disorder of the system. We may define $\mathcal{W}$ as the number of different ways the atoms of a system can be arranged without changing the external macroscopic properties of the system.

Speaking loosely, $\mathcal{W}$ (and hence S) is much greater for a scrambled egg than for a whole egg. That is, there are more ways to construct a scrambled egg from its constituents than there are ways to construct a whole egg. If you drop a whole egg, it will scramble itself spontaneously; it never happens the other way!

A Nice Application of Entropy

Entropy considerations lie at the heart of an important method of producing very low temperatures, called *adiabatic demagnetization*. A sample of a solid such as a chrome-alum salt, some of whose atoms are equivalent to tiny magnets, is placed in an insulating enclosure at the lowest attainable temperature (perhaps a few millikelvins). A strong magnet is placed near the sample so that, as Fig. 22-17a shows, tiny atomic magnets in the salt line up, forming a very ordered state. The magnet is then wheeled away so that it no longer can affect the sample. Due to thermal agitation, the atomic magnets now assume random orientations, as in Fig. 22-17b. The disorder (and thus the entropy) associated with the atomic alignments has clearly increased.

The system is isolated, so no heat can leave or enter. The process of removing the magnet and

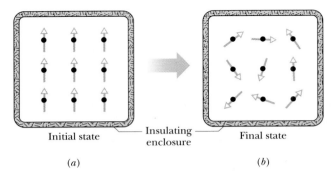

FIGURE 22-17 A method for generating low temperatures. (*a*) Atomic magnets in a chrome–alum salt are aligned by a large magnet. (*b*) When the magnet is removed, the disorder (entropy) associated with the orientations of the atoms *increases*. Because there can be no change in entropy in this thermally isolated reversible process, the disorder (entropy) associated with the temperature of the specimen must *decrease* by the same amount.

(consequent) randomizing of the orientations of the atomic magnets is very closely reversible. So from the second law, we expect that the system should have no change in entropy. However, we saw that the randomizing involves an entropy *increase*.

Where is the compensating entropy *decrease*? It can be manifested only in decreased oscillations of the structure of the specimen. In other words, the temperature of the specimen must decrease — and it does. This technique has been used with great success to reduce temperatures to record low values.

22-12 THE NATURE OF PHYSICAL LAW: AN ASIDE*

So far in this book we have encountered a number of physical laws, among them being: Newton's laws of motion; Newton's law of gravitation, the laws of conservation of energy, linear momentum, and angular momentum; and the first and second laws of thermodynamics. As we leave the subject of thermodynamics, it is perhaps well to back off and consider the nature of physical laws in general.

A law of physics is simply a statement — in word form, as for the second law of thermodynamics, or in equation form, as for Newton's law of gravitation —

---

*See Richard Feynman, *The Character of Physical Law*, MIT Press, Cambridge, MA, 1965.

that summarizes the results of experiment and observation for a certain range of physical phenomena. Because "truth" is such an abstract word, loaded with philosophical and ethical overtones, physicists rarely ask about a law, "Is this law true?" The question almost always is the much more specific and answerable, "Do the results predicted by this law agree with experiment?"

Physicists are continually pushing the limits of the variables in which a law is expressed, probing to see whether the law remains valid. Does the law hold at high temperatures and at high speeds? A law of physics is not an eternal truth but is a claim to be tested and probed. The law may survive these probings unchanged or it may have to be modified in some way. In any case, it must always await the challenge of new experimental data. As Einstein said, "No number of experiments can prove me right; a single experiment can prove me wrong."

For example, Newton's laws of motion—whose importance should be clear to all—fail to agree with experiment when tested on particles whose speeds are an appreciable fraction of the speed of light. The failure is not sudden but gradual, the discrepancy between law and experiment growing larger and larger as particle speeds get closer and closer to the speed of light.

As it happens, another law, Einstein's special theory of relativity, turns out to agree with experiment over the full range of observable particle speeds. Where does this leave Newton's laws? It leaves them as an extremely useful special case of the more comprehensive law. The usefulness of Newton's laws remains undiminished in the vast and important region in which they agree with experiment.

What about the second law of thermodynamics? Its present status is that, if we confine ourselves to processes that start and end in states of thermal equilibrium, no exceptions to it have ever been found. If an inventor claims to have built an engine that violates this law, the crushing weight of decades of experience amassed by gifted scientists and engineers the world around weighs heavily against the claim. The lone inventor may be right but must be able to prove the claim. Still, no physicist would deny that the second law of thermodynamics may one day be reinterpreted and seen as a special case of a more comprehensive law.

Perhaps the most impressive thing about physical laws is the very fact that they exist and are so simple in form. It is a source of wonderment to the discerning thinker that such vast realms of experience can be summarized in a single sentence or a single equation. Einstein put it well once again when he remarked that "the most incomprehensible thing about the universe is that it is comprehensible."

REVIEW & SUMMARY

Engines

An *engine* is a device that cyclically accepts heat Q_H from a high-temperature reservoir (at Kelvin temperature T_H) and does work W. There is always an accompanying discharge of heat Q_C at a lower temperature T_C. The **thermal efficiency** e of an engine is

$$e = \frac{|W|}{|Q_H|} = \frac{|Q_H| - |Q_C|}{|Q_H|}. \qquad (22\text{-}2)$$

The second law of thermodynamics denies the existence of perfect engines, those with $e = 1$. *It is not possible to change heat completely into work with no other changes taking place.*

Refrigerators

A *refrigerator* is a device that cyclically removes heat from a cold place and, by doing work, delivers it to a warmer place. Its effectiveness is measured by a **coefficient of per-**formance K defined as

$$K = \frac{|Q_C|}{|W|} = \frac{|Q_C|}{|Q_H| - |Q_C|}. \qquad (22\text{-}6)$$

The second law of thermodynamics denies the existence of perfect refrigerators, those with $K = \infty$. *It is not possible for heat to be transferred from one body to another at a higher temperature, with no other changes taking place.*

Reversible Process

A *reversible process* is an idealized transformation from an initial state i to a final state f that can be reversed by means of a differential change in the external conditions.

The Carnot Cycle

The *Carnot cycle* describes the action of an ideal (that is, reversible) heat engine. As Figs. 22-8 and 22-9 show, it consists of two *adiabatic processes* (for which $Q = 0$) alternating with two *isothermal processes* (for which T = constant). The efficiency of any engine using a Carnot cycle is

$$e_{\text{Car}} = \frac{T_H - T_C}{T_H} \qquad \text{(Carnot engine)} \qquad (22\text{-}7)$$

and the performance coefficient of a Carnot refrigerator is

$$K_{\text{Car}} = \frac{T_C}{T_H - T_C} \qquad \text{(Carnot refrigerator)}. \qquad (22\text{-}8)$$

Two Forms of the Second Law of Thermodynamics

No real engine (or refrigerator) operating between two temperatures can have a greater efficiency (or coefficient of performance) than that of a Carnot engine (or refrigerator) operating between the same temperatures.

Entropy

Entropy S, like pressure and volume, is a property (a **state variable**) of a system in equilibrium. The change in S for a system that goes *reversibly* from state i to state f is *defined* to be

$$S_f - S_i = \int_i^f dS = \int_i^f \frac{dQ}{T} \qquad \begin{array}{l}\text{(reversible}\\\text{path)},\end{array} \qquad (22\text{-}19)$$

where dQ is an increment of heat transferred at temperature T. The entropy change $S_f - S_i$ depends only on the initial and final states and not in any way on the nature of the reversible path connecting them. The SI unit for entropy is the joule per kelvin.

Entropy Changes for Irreversible Processes

The entropy change for an *irreversible* process may be evaluated by (1) finding a *reversible* path that connects the states i and f and (2) calculating $S_f - S_i$ for *this* path, using Eq. 22-19. The result is the entropy change for the irreversible process.

Entropy Form of the Second Law

The importance of entropy is that *in any thermodynamic process that proceeds from one equilibrium state to another, the entropy of the* system + environment *either remains unchanged or increases.* This statement of the second law of thermodynamics is equivalent to the "engine" and "refrigerator" forms given above.

Physical Law

A law of physics is a statement that summarizes the results of experiment and observation for a certain range of physical phenomena. It is always subject to further observation and, sometimes, modification as new evidence becomes available. When an older version of a law is replaced with a newer one, the older one often remains as a useful special case of the more general law.

QUESTIONS

1. Is a human being a heat engine? Explain.

2. Couldn't we define the efficiency of an engine as $e = W/Q_C$ rather than as $e = W/Q_H$? Why don't we?

3. The efficiencies of nuclear power plants are lower than those of fossil-fuel plants. Why?

4. Can a given amount of mechanical energy be converted completely into heat? If so, give an example.

5. An inventor suggested a house might be heated in the following manner: a system resembling a refrigerator draws heat from the Earth and rejects heat to the house. He claimed that the heat supplied to the house can exceed the work done by the engine of the system. What is your comment?

6. Give a qualitative explanation of how frictional forces between moving surfaces increase the temperature of those surfaces. Why does the reverse process not occur?

7. A block returns to its initial position after dissipating mechanical energy through friction. Why is this process not thermodynamically reversible?

8. Are any of the following processes reversible: (a) breaking an empty soda bottle; (b) mixing a cocktail; (c) melting an ice cube in a glass of iced tea; (d) burning a log of firewood; (e) puncturing an automobile tire; (f) finishing the "Unfinished Symphony"; (g) writing this book?

9. Give some examples of natural processes that are nearly reversible.

10. Can we calculate the work done during an irreversible process in terms of an area on a $p–V$ diagram? Is any work done?

11. Suggest a reversible process whereby heat can be added to a system. Why would adding heat by means of a Bunsen burner not be a reversible process?

12. To carry out a Carnot cycle, we need not start at point a in Fig. 22-9 but may equally well start at point b, c, or d, or any intermediate point. Explain.

13. If a Carnot engine is independent of the working substance, then perhaps real engines should be similarly independent, to a certain extent. Why then, for real engines, are we so concerned to find suitable fuels such as coal, gasoline, or fissionable material? Why not use stones as a fuel?

14. Under what conditions would an ideal heat engine be 100% efficient?

15. What factors reduce the efficiency of a heat engine from its ideal value?

16. You wish to increase the efficiency of a Carnot engine as much as possible. You can do this by increasing T_H a certain amount while keeping T_C constant, or by decreasing T_C the same amount while keeping T_H constant. Which would you do?

17. Can a kitchen be cooled by leaving the door of an electric refrigerator open? Explain.

18. Why do you get poorer gasoline mileage from your car in winter than in summer?

19. From time to time inventors will claim to have perfected a device that does useful work but consumes no (or very little) fuel. What do you think is most likely true in such cases: (a) the claimants are right, (b) the claimants are mistaken in their measurements, or (c) the claimants are swindlers? Do you think that such a claim should be examined closely by a panel of scientists and engineers? In your opinion, would the time and effort be justified?

20. We have seen that real engines always discard substantial amounts of heat to their low-temperature reservoirs. It seems a shame to throw this energy away. Why not use it to run a second engine, the low-temperature reservoir of the first engine serving as the high-temperature reservoir of the second?

21. Give examples in which the entropy of a system decreases, and explain why the second law of thermodynamics is not violated.

22. Do living things violate the second law of thermodynamics? As a chicken grows from an egg, for example, it becomes more and more ordered and organized. Increasing entropy, however, calls for disorder and decay. Is the entropy of a chicken actually decreasing as it grows?

23. Two samples of a gas initially at the same temperature and pressure are compressed from volume V to volume $V/2$, one isothermally and the other adiabatically. In which sample is the final pressure greater? Does the entropy of the gas change in either process?

24. Suppose that we had chosen to represent the state of a system by means of its entropy and its temperature (on the Kelvin scale) rather than its pressure and volume. (a) What would a Carnot cycle look like on a $T–S$ diagram? (b) What physical significance, if any, can be attached to the area under a curve on a $T–S$ diagram?

25. Does a change in entropy occur in purely mechanical motions?

26. Show that the total entropy increases when kinetic energy is dissipated by friction between sliding surfaces. Describe the increase in disorder.

27. Two pieces of molding clay of equal mass are moving in opposite directions with equal speeds. They strike and stick together. Treat the two pieces as a single system and state whether each of the following quantities is positive, negative, or zero for this process: ΔE_{int}, W, Q, and ΔS. Justify your answers.

28. Heat is transferred from the sun to the Earth. Show that the entropy of the Earth–sun system increases during this process.

29. Is it true that the energy of the universe is steadily growing less available? If so, why?

30. Discuss the following comment of the physicists Panofsky and Phillips: "From the standpoint of formal physics there is only one concept which is asymmetric in the time, namely, entropy. But this makes it reasonable to assume that the second law of thermodynamics can be used to ascertain the sense of time independently in any frame of reference; that is, we shall take the positive direction of time to be that of statistically increasing disorder, or increasing entropy." (See, in this connection, "The Arrow of Time," by David Layzer, *Scientific American*, December 1975.)

31. Explain the statement: "Cosmic rays continually *decrease* the entropy of the Earth, on which they fall." Why does this statement not contradict the second law of thermodynamics?

EXERCISES & PROBLEMS

SECTION 22-2 ENGINES

1E. A heat engine absorbs 52 kcal of heat and exhausts 36 kcal of heat each cycle. Calculate (a) the efficiency and (b) the work done in kilocalories per cycle.

2E. A car engine delivers 8.2 kJ of work per cycle. (a) Before a tune-up, the efficiency is 25%. Calculate, per cycle, the heat absorbed from the combustion of fuel and the heat exhausted to the atmosphere. (b) After a tune-up, the efficiency is 31%. What are the new values of the quantities calculated in (a)?

3E. Calculate the efficiency of a fossil-fuel power plant that consumes 380 metric tons of coal each hour to produce useful work at the rate of 750 MW. The heat of combustion of 1.0 kg of coal is 28 MJ.

4P. One mole of a monatomic ideal gas is caused to go through the cycle shown in Fig. 22-18. Process bc is an adiabatic expansion; $p_b = 10.0$ atm, $V_b = 1.00 \times 10^{-3}$ m³, and $V_c = 8.00V_b$. Calculate (a) the heat added to the gas, (b) the heat leaving the gas, (c) the net work done by the gas, and (d) the efficiency of the cycle.

FIGURE 22-18 Problem 4.

5P. One mole of a monatomic ideal gas initially at a volume of 10 L and a temperature of 300 K is heated at constant volume to a temperature of 600 K, allowed to expand isothermally to its initial pressure, and finally compressed isobarically (that is, at constant pressure) to its original volume, pressure, and temperature. (a) Compute the heat input to the system during one cycle. (b) What is the net work done by the gas during one cycle? (c) What is the efficiency of this cycle?

SECTION 22-3 REFRIGERATORS

6E. A refrigerator does 150 J of work to remove 560 J of heat from its cold compartment. (a) What is the refrigerator's coefficient of performance? (b) How much heat per cycle is exhausted to the kitchen?

7E. To make ice, a freezer extracts 42 kcal of heat at −12°C during each cycle. The freezer has a coefficient of performance of 5.7. The room temperature is 26°C. (a) How much heat per cycle is delivered to the room? (b) How much work per cycle is required to run the freezer?

SECTION 22-7 THE EFFICIENCIES OF REAL ENGINES

8E. An ideal-gas heat engine operates in a Carnot cycle between 235 and 115°C. It absorbs 6.30×10^4 cal per cycle at the higher temperature. (a) What is the efficiency of the engine? (b) How much work per cycle is this engine capable of performing?

9E. How much work must be done to extract 1.0 J of heat (a) from a reservoir at 7.0°C and transfer it to one at 27°C by means of a refrigerator using a Carnot cycle; (b) from a reservoir at −73°C to one at 27°C; (c) from a reservoir at −173°C to one at 27°C; and (d) from a reservoir at −223°C to one at 27°C?

10E. In a Carnot cycle, the isothermal expansion of an ideal gas takes place at 400 K, and the isothermal compression at 300 K. During the expansion, 500 cal of heat is transferred to the gas. Determine (a) the work performed by the gas during the isothermal expansion, (b) the heat rejected from the gas during the isothermal compression, and (c) the work done on the gas during the isothermal compression.

11E. In a hypothetical nuclear fusion reactor, the fuel is deuterium (D) gas at a temperature of about 7×10^8 K. If this gas could be used to operate an ideal heat engine with $T_C = 100°C$, what would be its efficiency?

12E. A Carnot engine has an efficiency of 22%. It operates between heat reservoirs differing in temperature by 75°C. What are the temperatures of the reservoirs?

13E. Apparatus that liquefies helium is in a room at 300 K. If the helium in the apparatus is at 4.0 K, what is the minimum ratio of heat delivered to the room to the heat removed from the helium?

14E. An ideal, reversible air conditioner takes heat from a room at 70°F and transfers it to the outdoors, which is at 96°F. For each joule of electrical energy required to operate the air conditioner, how many joules of heat are removed from the room?

15E. For the Carnot cycle illustrated in Fig. 22-9, show that the work done by the gas during process bc (step 2) has the same absolute value as the work done on the gas during process da (step 4).

16E. (a) For a Carnot (ideal) refrigerator, show that

$$|W| = |Q_C| \frac{T_H - T_C}{T_C}.$$

(b) In a real refrigerator the low-temperature coils are at a temperature of $-13°C$, and the compressed gas in the condenser has a temperature of 26°C. What is the theoretical coefficient of performance?

17E. (a) A Carnot engine operates between a hot reservoir at 320 K and a cold reservoir at 260 K. If it absorbs 500 J of heat per cycle at the hot reservoir, how much work per cycle does it deliver? (b) If the same engine, working in reverse, functions as a refrigerator between the same two reservoirs, how much work per cycle must be supplied to remove 1000 J of heat from the cold reservoir?

18E. A combination mercury–steam turbine takes saturated mercury vapor from a boiler at 876°F and exhausts it to heat a steam boiler at 460°F. The steam turbine receives steam at this temperature and exhausts it to a condenser that is at 100°F. What is the maximum efficiency of the combination?

19E. In a heat pump, heat from the outdoors at $-5.0°C$ is transferred to a room at 17°C, energy being supplied by an electric motor. How many joules of heat will be delivered to the room for each joule of electric energy consumed? Assume an ideal heat pump.

20P. A heat pump is used to heat a building. The outside temperature is $-5.0°C$, and the temperature inside the building is to be maintained at 22°C. The coefficient of performance is 3.8, and the heat pump delivers 1.8 Mcal of heat to the building each hour. At what rate must work be done to run the heat pump?

21P. A Carnot engine has a power of 500 W. It operates between heat reservoirs at 100 and 60.0°C. Calculate (a) the rate of heat input and (b) the rate of exhaust heat output in kilocalories per second.

22P. The motor in a refrigerator has a power of 200 W. If the freezing compartment is at 270 K and the outside air is at 300 K, assuming ideal efficiency, what is the maximum amount of heat that can be extracted from the freezing compartment in 10.0 min?

23P. Find the relation between the efficiency of a reversible ideal heat engine and the coefficient of performance of the reversible refrigerator obtained by running the engine backward.

24P. (a) Show that when a Carnot cycle is plotted on a temperature (Kelvin scale) versus entropy ($T-S$) diagram, the result is a rectangle. For the Carnot cycle shown in Fig. 22-19, calculate (b) the heat gained and (c) the work done by the system.

FIGURE 22-19 Problem 24.

25P. In a two-stage Carnot heat engine, a quantity of heat Q_1 is absorbed at a temperature T_1, work W_1 is done, and a quantity of heat Q_2 is expelled at a lower temperature T_2 by the first stage. The second stage absorbs the heat expelled by the first, does work W_2, and expels a quantity of heat Q_3 at a lower temperature T_3. Prove that the efficiency of the combination engine is $(T_1 - T_3)/T_1$.

26P. (a) Accurately plot a Carnot cycle on a $p-V$ diagram for 1.00 mL of an ideal gas. (See Fig. 22-9.) Let point a correspond to $p = 1.00$ atm, $T = 300$ K, and let point b correspond to $p = 0.500$ atm, $T = 300$ K; take the low-temperature reservoir to be at 100 K. Let $\gamma = 1.50$. (b) Graphically compute the work done in this cycle. (c) Compute the work analytically.

27P. A Carnot engine works between temperatures T_1

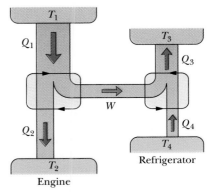

FIGURE 22-20 Problem 27.

and T_2. It drives a Carnot refrigerator that works between two different temperatures T_3 and T_4 (Fig. 22-20). Find the ratio $|Q_3|/|Q_1|$ in terms of the four temperatures.

28P. An air conditioner operating between 93 and 70°F is rated at 4000 Btu/h cooling capacity. Its coefficient of performance is 27% of that of a Carnot refrigerator operating between the same two temperatures. What is the required horsepower of the air conditioner motor?

29P. Suppose that a deep shaft were drilled in the Earth's crust near one of the poles where the surface temperature is $-40°C$, to a depth where the temperature is 800°C. (a) What is the theoretical limit to the efficiency of an engine operating between these temperatures? (b) If all the heat released into the low-temperature reservoir were used to melt ice that was initially at $-40°C$, at what rate could liquid water at 0°C be produced by a power plant having an output of 100 MW? The specific heat of ice is 0.50 cal/g·°C; its heat of fusion is 80 cal/g. (Note that the engine can operate only between 0 and 800°C in this case. Energy exhausted at $-40°C$ cannot be used to raise the temperature of anything above $-40°C$.)

30P. One mole of an ideal monatomic gas is used as the working substance of an engine that operates on the cycle shown in Fig. 22-21. Assume that $p = 2p_0$, $V = 2V_0$, $p_0 = 1.01 \times 10^5$ Pa, and $V_0 = 0.0225$ m³. Calculate (a) the work done per cycle, (b) the heat added per cycle during the expansion stroke abc, and (c) the engine efficiency. (d) What is the Carnot efficiency of an engine operating between the highest and lowest temperatures that occur in the cycle? How does this compare to the efficiency calculated in (c)?

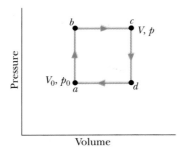

FIGURE 22-21 Problem 30.

31P. Compute the efficiency of the cycle shown in Problem 83 in Chapter 21. In this case, the heat input will not be at a fixed temperature, as it is in the Carnot cycle.

32P*. A gasoline internal combustion engine can be approximated by the cycle shown in Fig. 22-22. Assume an ideal gas and use a compression ratio of 4:1 ($V_4 = 4V_1$). Assume that $p_2 = 3p_1$. (a) Determine the pressure and

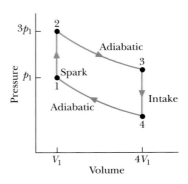

FIGURE 22-22 Problem 32.

temperature at each of the vertex points of the p–V diagram in terms of p_1, T_1, and the ratio of specific heats of the gas. (b) What is the efficiency of the cycle?

SECTION 22-8 ENTROPY: A NEW VARIABLE

33E. In Fig. 22-14, suppose that the change in entropy of the system in passing from state a to state b along path 1 is $+0.602$ cal/K. What is the entropy change in passing (a) from state a to b along path 2 and (b) from state b to a along path 2?

34E. Find (a) the heat absorbed and (b) the change in entropy of a 1.0-kg block of copper whose temperature is increased reversibly from 25 to 100°C. The specific heat of copper is 9.2×10^{-2} cal/g·°C.

35E. Heat can be removed from water at 0.0°C and atmospheric pressure without causing the water to freeze, if done with little disturbance of the water. Suppose a water drop is cooled in this way until its temperature is that of the surrounding air, which is at $-5.0°C$. The drop then suddenly freezes and transfers heat to the air until the drop is again at $-5.0°C$. What is the entropy change per gram of water during this freezing and heat transfer?

36P. An inventor claims to have invented four engines, each of which operates between heat reservoirs at 400 and 300 K. Data on each engine, per cycle of operation, are as follows: Engine (a): $Q_H = 200$ J, $Q_C = -175$ J, $W = 40$ J; engine (b): $Q_H = 500$ J, $Q_C = -200$ J, $W = 400$ J; engine (c): $Q_H = 600$ J, $Q_C = -200$ J, $W = 400$ J; engine (d): $Q_H = 100$ J, $Q_C = -90$ J, $W = 10$ J. Of the first and second laws of thermodynamics, which (if either) does each engine violate?

37P. At very low temperatures, the molar specific heat C_V for many solids is (approximately) proportional to T^3; that is, $C_V = AT^3$, where A depends on the particular substance. For aluminum, $A = 7.53 \times 10^{-6}$ cal/mol·K⁴. Find the entropy change of 4.00 mol of aluminum when its temperature is raised from 5.00 to 10.0 K.

SECTION 22-9 ENTROPY CHANGES FOR IRREVERSIBLE PROCESSES

38E. An ideal gas undergoes a reversible isothermal expansion at 77.0°C, increasing its volume from 1.30 to 3.40 L. The entropy change of the gas is 22.0 J/K. How many moles of gas are present?

39E. An ideal gas undergoes a reversible isothermal expansion at 132°C. The entropy of the gas increases by 46.0 J/K. How much heat was absorbed?

40E. One mole of an ideal gas expands isothermally at 360 K until its volume is doubled. What is the increase of entropy of the gas?

41E. Suppose that the same amount of heat, say, 260 J, is transferred by conduction from a heat reservoir at a temperature of 400 K to another reservoir, the temperature of which is (a) 100 K, (b) 200 K, (c) 300 K, and (d) 360 K. Calculate the changes in entropy and discuss the trend.

42E. A brass rod is in thermal contact with a heat reservoir at 130°C at one end and a heat reservoir at 24.0°C at the other end. (a) Compute the total change in entropy arising from the conduction of 1200 cal of heat through the rod. (b) Does the entropy of the rod change in the process?

43E. In a specific heat experiment, 200 g of aluminum ($c = 0.215$ cal/g·°C) at 100°C is mixed with 50.0 g of water at 20.0°C. (a) Calculate the equilibrium temperature. Find the entropy change (b) of the aluminum and (c) of the water. (d) Calculate the entropy change of the system. (*Hint:* See Eqs. 22-26 and 22-27.)

44P. A 10-g ice cube at -10°C is placed in a lake whose temperature is 15°C. Calculate the change in entropy of the system as the ice cube comes to thermal equilibrium with the lake. The specific heat of ice is 0.50 cal/g·°C. (*Hint:* Will the ice cube affect the temperature of the lake?)

45P. An 8.0-g ice cube at -10°C is dropped into a thermos flask containing 100 cm³ of water at 20°C. What is the change in entropy of the system when a final equilibrium state is reached? The specific heat of ice is 0.50 cal/g·°C.

46P. Four moles of an ideal gas are expanded from volume V_1 to volume $V_2 = 2V_1$. (a) If the expansion is isothermal at temperature $T = 400$ K, find the work done by the expanding gas. (b) Find the change in entropy, if any. (c) If the expansion is reversibly adiabatic instead of isothermal, what is the entropy change?

47P. A 50-g block of copper having a temperature of 400 K is placed in an insulating box with a 100-g block of lead having a temperature of 200 K. (a) What is the equilibrium temperature of this two-block system? (b) What is the change in the internal energy of the two-block system as it goes from the initial condition to the equilibrium condition? (c) What is the change in the entropy of the two-block system? (See Table 20-1.)

48P. A mole of a monatomic ideal gas is taken from an initial state of pressure p and volume V to a final state of pressure $2p$ and volume $2V$ by two different processes. (I)

It expands isothermally until its volume is doubled, and then its pressure is increased at constant volume to the final state. (II) It is compressed isothermally until its pressure is doubled, and then its volume is increased at constant pressure to the final state. Show the path of each process on a p–V diagram. For each process calculate, in terms of p and V: (a) the heat absorbed by the gas in each part of the process; (b) the work done by the gas in each part of the process; (c) the change in internal energy of the gas, $E_{\text{int},f} - E_{\text{int},i}$; and (d) the change in entropy of the gas, $S_f - S_i$.

49P. An ideal diatomic gas is caused to pass through the cycle shown on the p–V diagram in Fig. 22-23, where $V_2 = 3.00V_1$. Determine, in terms of p_1, V_1, T_1, and R: (a) p_2, p_3, and T_3 and (b) W, Q, ΔE_{int}, and ΔS per mole for all three processes.

FIGURE 22-23 Problem 49.

50P. An object of constant heat capacity C is heated from an initial temperature T_i to a final temperature T_f, by being placed in contact with a heat reservoir at T_f. (a) Represent the process on a graph of C/T versus T, and show graphically that the total change in entropy ΔS (object plus reservoir) is positive. (b) Show how the use of heat reservoirs at intermediate temperatures would allow the process to be carried out in a way that makes ΔS as small as desired.

51P. A mixture of 1773 g of water and 227 g of ice at 0.00°C is, in a reversible process, brought to a final equilibrium state where the water–ice ratio, by mass, is 1:1 at 0.00°C. (a) Calculate the entropy change of the system during this process. (The heat of fusion for water is 79.5 cal/g.) (b) The system is then returned to the first equilibrium state, but in an irreversible way (by using a Bunsen burner, for instance). Calculate the entropy change of the system during this process. (c) Is your answer consistent with the second law of thermodynamics?

52P. A round silver rod 15 cm long, with a diameter of 1.0 cm, has its ends in contact with heat reservoirs at 60 and 20°C, and steady-state heat flow has been established. What will be the initial rate of change of the entropy of the rod if (a) the hot end is suddenly insulated from the 60°C reservoir, or (b) the entire rod is suddenly insulated?

FIGURE 22-24   Problem 53.

53P. One mole of an ideal monatomic gas is caused to go through the cycle shown in Fig. 22-24. (a) How much work is done in expanding the gas from a to c along path abc? (b) What are the changes in internal energy and entropy in going from b to c? (c) What are the changes in internal energy and entropy in going through one complete cycle? Express all answers in terms of the pressure p_0, volume V_0, and temperature T_0 at point a in the diagram.

ADDITIONAL PROBLEMS

54. An ideal monatomic gas, at an initial temperature T_0 (in kelvins) expands from an initial volume V_0 to a volume $2V_0$ by one of the five processes indicated in the graph of temperature versus volume in Fig. 22-25. In which process is the expansion (a) isothermal, (b) isobaric (constant pressure), and (c) adiabatic? Explain your answers. (d) In which processes does the entropy of the gas decrease?

FIGURE 22-25   Problem 54.

FIGURE 22-26   Problem 56.

55. A Carnot engine whose low-temperature reservoir is at 17°C has an efficiency of 40%. By how much should the temperature of the high-temperature reservoir be increased to increase the efficiency to 50%?

56. One mole of an ideal gas is used as the working substance of an engine that operates on the cycle shown in Fig. 22-26. BC and DA are reversible adiabatic processes. (a) Is the gas monatomic, diatomic, or polyatomic? (b) What is the efficiency of the engine?

57. One mole of an ideal monatomic gas, at an initial pressure of 5.00 kN/m² and initial temperature of 600 K, expands from an initial volume of $V_i = 1.00$ m³ to a final volume of $V_f = 2.00$ m³. During the expansion, the pressure p and the volume V of the gas are related by the following:

$$p = 5.00 \, e^{(V_i - V)/a},$$

where p is in kN/m², V_i and V are in m³, and $a = 1.00$ m³. What are (a) the final pressure and (b) the final temperature of the gas? (c) How much work is done by the gas during the expansion? (d) What is the change in entropy of the gas during the expansion? (*Hint:* Use two simple reversible processes to find the entropy change.)

58. Two moles of an ideal monatomic gas undergo the reversible process shown in the graph of temperature versus entropy in Fig. 22-27. (a) How much heat is absorbed by the gas? (b) What is the change in internal energy of the gas? (c) How much work is done by the gas?

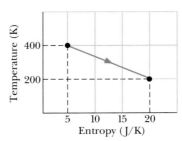

FIGURE 22-27   Problem 58.

APPENDIX A

THE INTERNATIONAL SYSTEM OF UNITS (SI)*

1. THE SI BASE UNITS

QUANTITY	NAME	SYMBOL	DEFINITION
length	meter	m	". . . the length of the path traveled by light in vacuum in 1/299,792,458 of a second." (1983)
mass	kilogram	kg	". . . this prototype [a certain platinum–iridium cylinder] shall henceforth be considered to be the unit of mass." (1889)
time	second	s	". . . the duration of 9,192,631,770 periods of the radiation corresponding to the transition between the two hyperfine levels of the ground state of the cesium-133 atom." (1967)
electric current	ampere	A	". . . that constant current which, if maintained in two straight parallel conductors of infinite length, of negligible circular cross section, and placed 1 meter apart in vacuum, would produce between these conductors a force equal to 2×10^{-7} newton per meter of length." (1946)
thermodynamic temperature	kelvin	K	". . . the fraction 1/273.16 of the thermodynamic temperature of the triple point of water." (1967)
amount of substance	mole	mol	". . . the amount of substance of a system which contains as many elementary entities as there are atoms in 0.012 kilogram of carbon-12." (1971)
luminous intensity	candela	cd	". . . the luminous intensity, in the perpendicular direction, of a surface of 1/600,000 square meter of a blackbody at the temperature of freezing platinum under a pressure of 101.325 newtons per square meter." (1967)

*Adapted from "The International System of Units (SI)," National Bureau of Standards Special Publication 330, 1972 edition. The definitions above were adopted by the General Conference of Weights and Measures, an international body, on the dates shown. In this book we do not use the candela.

2. SOME SI DERIVED UNITS

QUANTITY	NAME OF UNIT	SYMBOL	
area	square meter	m^2	
volume	cubic meter	m^3	
frequency	hertz	Hz	s^{-1}
mass density (density)	kilogram per cubic meter	kg/m^3	
speed, velocity	meter per second	m/s	
angular velocity	radian per second	rad/s	
acceleration	meter per second per second	m/s^2	
angular acceleration	radian per second per second	rad/s^2	
force	newton	N	$kg \cdot m/s^2$
pressure	pascal	Pa	N/m^2
work, energy, quantity of heat	joule	J	$N \cdot m$
power	watt	W	J/s
quantity of electric charge	coulomb	C	$A \cdot s$
potential difference, electromotive force	volt	V	W/A
electric field strength	volt per meter (or newton per coulomb)	V/m	N/C
electric resistance	ohm	Ω	V/A
capacitance	farad	F	$A \cdot s/V$
magnetic flux	weber	Wb	$V \cdot s$
inductance	henry	H	$V \cdot s/A$
magnetic flux density	tesla	T	Wb/m^2
magnetic field strength	ampere per meter	A/m	
entropy	joule per kelvin	J/K	
specific heat	joule per kilogram kelvin	$J/(kg \cdot K)$	
thermal conductivity	watt per meter kelvin	$W/(m \cdot K)$	
radiant intensity	watt per steradian	W/sr	

3. THE SI SUPPLEMENTARY UNITS

QUANTITY	NAME OF UNIT	SYMBOL
plane angle	radian	rad
solid angle	steradian	sr

APPENDIX B
SOME FUNDAMENTAL CONSTANTS OF PHYSICS*

CONSTANT	SYMBOL	COMPUTATIONAL VALUE	BEST (1986) VALUE VALUE[a]	UNCERTAINTY[b]
Speed of light in a vacuum	c	3.00×10^8 m/s	2.99792458	exact
Elementary charge	e	1.60×10^{-19} C	1.60217738	0.30
Electron mass	m_e	9.11×10^{-31} kg	9.1093897	0.59
Proton mass	m_p	1.67×10^{-27} kg	1.6726230	0.59
Ratio of proton mass to electron mass	m_p/m_e	1840	1836.152701	0.020
Neutron mass	m_n	1.68×10^{-27} kg	1.6749286	0.59
Muon mass	m_μ	1.88×10^{-28} kg	1.8835326	0.61
Electron mass[c]	m_e	5.49×10^{-4} u	5.48579902	0.023
Proton mass[c]	m_p	1.0073 u	1.007276470	0.012
Neutron mass[c]	m_n	1.0087 u	1.008664704	0.014
Hydrogen atom mass[c]	m_{1_H}	1.0078 u	1.007825035	0.011
Deuterium atom mass[c]	m_{2_H}	2.0141 u	2.0141019	0.053
Helium atom mass[c]	$m_{4_{He}}$	4.0026 u	4.0026032	0.067
Electron charge-to-mass ratio	e/m_e	1.76×10^{11} C/kg	1.75881961	0.30
Permittivity constant	ϵ_0	8.85×10^{-12} F/m	8.85418781762	exact
Permeability constant	μ_0	1.26×10^{-6} H/m	1.25663706143	exact
Planck constant	h	6.63×10^{-34} J·s	6.6260754	0.60
Electron Compton wavelength	λ_C	2.43×10^{-12} m	2.42631058	0.089
Universal gas constant	R	8.31 J/mol·K	8.314510	8.4
Avogadro constant	N_A	6.02×10^{23} mol^{-1}	6.0221367	0.59
Boltzmann constant	k	1.38×10^{-23} J/K	1.380657	11
Molar volume of ideal gas at STP[d]	V_m	2.24×10^{-2} m³/mol	2.241409	8.4
Faraday constant	F	9.65×10^4 C/mol	9.6485309	0.30
Stefan–Boltzmann constant	σ	5.67×10^{-8} W/m²·K⁴	5.67050	34
Rydberg constant	R	1.10×10^7 m^{-1}	1.0973731534	0.0012
Gravitational constant	G	6.67×10^{-11} m³/s²·kg	6.67260	100
Bohr radius	r_B	5.29×10^{-11} m	5.29177249	0.045
Electron magnetic moment	μ_e	9.28×10^{-24} J/T	9.2847700	0.34
Proton magnetic moment	μ_p	1.41×10^{-26} J/T	1.41060761	0.34
Bohr magneton	μ_B	9.27×10^{-24} J/T	9.2740154	0.34
Nuclear magneton	μ_N	5.05×10^{-27} J/T	5.0507865	0.34

[a] Values given in this column should be given the same unit and power of 10 as the computational value.   [b] Parts per million.   [c] Masses are given in unified atomic mass units (u), where 1 u = $1.6605402 \times 10^{-27}$ kg.   [d] STP means standard temperature and pressure: 0°C and 1.0 atm (0.1 MPa).

*The values in this table were largely selected from a longer list in *Symbols, Units and Nomenclature in Physics* (IUPAP), prepared by E. Richard Cohen and Pierre Giacomo, 1986.

APPENDIX C

SOME ASTRONOMICAL DATA

SOME DISTANCES FROM THE EARTH

To the moon*	3.82×10^8 m
To the sun*	1.50×10^{11} m
To the nearest star (Proxima Centauri)	4.04×10^{16} m
To the center of our galaxy	2.2×10^{20} m
To the Andromeda Galaxy	2.1×10^{22} m
To the edge of the observable universe	$\sim 10^{26}$ m

*Mean distance.

THE SUN, THE EARTH, AND THE MOON

PROPERTY	UNIT	SUN	EARTH	MOON
Mass	kg	1.99×10^{30}	5.98×10^{24}	7.36×10^{22}
Mean radius	m	6.96×10^8	6.37×10^6	1.74×10^6
Mean density	kg/m³	1410	5520	3340
Free-fall acceleration at the surface	m/s²	274	9.81	1.67
Escape velocity	km/s	618	11.2	2.38
Period of rotation[a]	—	37 d at poles[b] 26 d at equator[b]	23 h 56 min	27.3 d
Radiation power[c]	W	3.90×10^{26}		

[a] Measured with respect to the distant stars.

[b] The sun, a ball of gas, does not rotate as a rigid body.

[c] Just outside the Earth's atmosphere solar energy is received, assuming normal incidence, at the rate of 1340 W/m².

SOME PROPERTIES OF THE PLANETS

	MERCURY	VENUS	EARTH	MARS	JUPITER	SATURN	URANUS	NEPTUNE	PLUTO
Mean distance from sun, 10^6 km	57.9	108	150	228	778	1430	2870	4500	5900
Period of revolution, y	0.241	0.615	1.00	1.88	11.9	29.5	84.0	165	248
Period of rotation,[a] d	58.7	−243[b]	0.997	1.03	0.409	0.426	−0.451[b]	0.658	6.39
Orbital speed, km/s	47.9	35.0	29.8	24.1	13.1	9.64	6.81	5.43	4.74
Inclination of axis to orbit	< 28°	≈ 3°	23.4°	25.0°	3.08°	26.7°	97.9°	29.6°	57.5°
Inclination of orbit to Earth's orbit	7.00°	3.39°		1.85°	1.30°	2.49°	0.77°	1.77°	17.2°
Eccentricity of orbit	0.206	0.0068	0.0167	0.0934	0.0485	0.0556	0.0472	0.0086	0.250
Equatorial diameter, km	4880	12,100	12,800	6790	143,000	120,000	51,800	49,500	2300
Mass (Earth = 1)	0.0558	0.815	1.000	0.107	318	95.1	14.5	17.2	0.002
Density (water = 1)	5.60	5.20	5.52	3.95	1.31	0.704	1.21	1.67	2.03
Surface value of g,[c] m/s^2	3.78	8.60	9.78	3.72	22.9	9.05	7.77	11.0	0.5
Escape velocity,[c] km/s	4.3	10.3	11.2	5.0	59.5	35.6	21.2	23.6	1.1
Known satellites	0	0	1	2	16 + ring	18 + rings	15 + rings	8 + rings	1

[a] Measured with respect to the distant stars.

[b] Venus and Uranus rotate opposite their orbital motion.

[c] Gravitational acceleration measured at the planet's equator.

APPENDIX D

PROPERTIES OF THE ELEMENTS

All physical properties are for a pressure of 1 atm unless otherwise specified.

ELEMENT	SYMBOL	ATOMIC NUMBER, Z	MOLAR MASS, g/mol	DENSITY, g/cm³ AT 20°C	MELTING POINT, °C	BOILING POINT, °C	SPECIFIC HEAT, J/(g·°C) AT 25°C
Actinium	Ac	89	(227)	10.06	1323	(3473)	0.092
Aluminum	Al	13	26.9815	2.699	660	2450	0.900
Americium	Am	95	(243)	13.67	1541	—	—
Antimony	Sb	51	121.75	6.691	630.5	1380	0.205
Argon	Ar	18	39.948	1.6626×10^{-3}	−189.4	−185.8	0.523
Arsenic	As	33	74.9216	5.78	817 (28 atm)	613	0.331
Astatine	At	85	(210)	—	(302)	—	—
Barium	Ba	56	137.34	3.594	729	1640	0.205
Berkelium	Bk	97	(247)	14.79	—	—	—
Beryllium	Be	4	9.0122	1.848	1287	2770	1.83
Bismuth	Bi	83	208.980	9.747	271.37	1560	0.122
Boron	B	5	10.811	2.34	2030	—	1.11
Bromine	Br	35	79.909	3.12 (liquid)	−7.2	58	0.293
Cadmium	Cd	48	112.40	8.65	321.03	765	0.226
Calcium	Ca	20	40.08	1.55	838	1440	0.624
Californium	Cf	98	(251)	—	—	—	—
Carbon	C	6	12.01115	2.26	3727	4830	0.691
Cerium	Ce	58	140.12	6.768	804	3470	0.188
Cesium	Cs	55	132.905	1.873	28.40	690	0.243
Chlorine	Cl	17	35.453	3.214×10^{-3} (0°C)	−101	−34.7	0.486
Chromium	Cr	24	51.996	7.19	1857	2665	0.448
Cobalt	Co	27	58.9332	8.85	1495	2900	0.423
Copper	Cu	29	63.54	8.96	1083.40	2595	0.385
Curium	Cm	96	(247)	13.3	—	—	—
Dysprosium	Dy	66	162.50	8.55	1409	2330	0.172
Einsteinium	Es	99	(254)	—	—	—	—
Erbium	Er	68	167.26	9.15	1522	2630	0.167
Europium	Eu	63	151.96	5.243	817	1490	0.163
Fermium	Fm	100	(237)	—	—	—	—
Fluorine	F	9	18.9984	1.696×10^{-3} (0°C)	−219.6	−188.2	0.753
Francium	Fr	87	(223)	—	(27)	—	—
Gadolinium	Gd	64	157.25	7.90	1312	2730	0.234
Gallium	Ga	31	69.72	5.907	29.75	2237	0.377
Germanium	Ge	32	72.59	5.323	937.25	2830	0.322
Gold	Au	79	196.967	19.32	1064.43	2970	0.131
Hafnium	Hf	72	178.49	13.31	2227	5400	0.144
Helium	He	2	4.0026	0.1664×10^{-3}	−269.7	−268.9	5.23
Holmium	Ho	67	164.930	8.79	1470	2330	0.165

continued on next page

ELEMENT	SYMBOL	ATOMIC NUMBER, Z	MOLAR MASS, g/mol	DENSITY, g/cm³ AT 20°C	MELTING POINT, °C	BOILING POINT, °C	SPECIFIC HEAT, J/(g·°C) AT 25°C
Hydrogen	H	1	1.00797	0.08375×10^{-3}	-259.19	-252.7	14.4
Indium	In	49	114.82	7.31	156.634	2000	0.233
Iodine	I	53	126.9044	4.93	113.7	183	0.218
Iridium	Ir	77	192.2	22.5	2447	(5300)	0.130
Iron	Fe	26	55.847	7.874	1536.5	3000	0.447
Krypton	Kr	36	83.80	3.488×10^{-3}	-157.37	-152	0.247
Lanthanum	La	57	138.91	6.189	920	3470	0.195
Lawrencium	Lr	103	(257)	—	—	—	—
Lead	Pb	82	207.19	11.35	327.45	1725	0.129
Lithium	Li	3	6.939	0.534	180.55	1300	3.58
Lutetium	Lu	71	174.97	9.849	1663	1930	0.155
Magnesium	Mg	12	24.312	1.738	650	1107	1.03
Manganese	Mn	25	54.9380	7.44	1244	2150	0.481
Mendelevium	Md	101	(256)	—	—	—	—
Mercury	Hg	80	200.59	13.55	-38.87	357	0.138
Molybdenum	Mo	42	95.94	10.22	2617	5560	0.251
Neodymium	Nd	60	144.24	7.007	1016	3180	0.188
Neon	Ne	10	20.183	0.8387×10^{-3}	-248.597	-246.0	1.03
Neptunium	Np	93	(237)	20.25	637	—	1.26
Nickel	Ni	28	58.71	8.902	1453	2730	0.444
Niobium	Nb	41	92.906	8.57	2468	4927	0.264
Nitrogen	N	7	14.0067	1.1649×10^{-3}	-210	-195.8	1.03
Nobelium	No	102	(255)	—	—	—	—
Osmium	Os	76	190.2	22.59	3027	5500	0.130
Oxygen	O	8	15.9994	1.3318×10^{-3}	-218.80	-183.0	0.913
Palladium	Pd	46	106.4	12.02	1552	3980	0.243
Phosphorus	P	15	30.9738	1.83	44.25	280	0.741
Platinum	Pt	78	195.09	21.45	1769	4530	0.134
Plutonium	Pu	94	(244)	19.8	640	3235	0.130
Polonium	Po	84	(210)	9.32	254	—	—
Potassium	K	19	39.102	0.862	63.20	760	0.758
Praseodymium	Pr	59	140.907	6.773	931	3020	0.197
Promethium	Pm	61	(145)	7.22	(1027)	—	—
Protactinium	Pa	91	(231)	15.37 (estimated)	(1230)	—	—
Radium	Ra	88	(226)	5.0	700	—	—
Radon	Rn	86	(222)	9.96×10^{-3} (0°C)	(-71)	-61.8	0.092
Rhenium	Re	75	186.2	21.02	3180	5900	0.134
Rhodium	Rh	45	102.905	12.41	1963	4500	0.243
Rubidium	Rb	37	85.47	1.532	39.49	688	0.364
Ruthenium	Ru	44	101.107	12.37	2250	4900	0.239
Samarium	Sm	62	150.35	7.52	1072	1630	0.197
Scandium	Sc	21	44.956	2.99	1539	2730	0.569
Selenium	Se	34	78.96	4.79	221	685	0.318
Silicon	Si	14	28.086	2.33	1412	2680	0.712
Silver	Ag	47	107.870	10.49	960.8	2210	0.234
Sodium	Na	11	22.9898	0.9712	97.85	892	1.23

continued on next page

ELEMENT	SYMBOL	ATOMIC NUMBER, Z	MOLAR MASS, g/mol	DENSITY, g/cm^3 AT 20°C	MELTING POINT, °C	BOILING POINT, °C	SPECIFIC HEAT, J/(g·°C) AT 25°C
Strontium	Sr	38	87.62	2.54	768	1380	0.737
Sulfur	S	16	32.064	2.07	119.0	444.6	0.707
Tantalum	Ta	73	180.948	16.6	3014	5425	0.138
Technetium	Tc	43	(99)	11.46	2200	—	0.209
Tellurium	Te	52	127.60	6.24	449.5	990	0.201
Terbium	Tb	65	158.924	8.229	1357	2530	0.180
Thallium	Tl	81	204.37	11.85	304	1457	0.130
Thorium	Th	90	(232)	11.72	1755	(3850)	0.117
Thulium	Tm	69	168.934	9.32	1545	1720	0.159
Tin	Sn	50	118.69	7.2984	231.868	2270	0.226
Titanium	Ti	22	47.90	4.54	1670	3260	0.523
Tungsten	W	74	183.85	19.3	3380	5930	0.134
Unnilpentium	Unp	105	262	—	—	—	—
Unnilquadium	Unq	104	261	—	—	—	—
Uranium	U	92	(238)	18.95	1132	3818	0.117
Vanadium	V	23	50.942	6.11	1902	3400	0.490
Xenon	Xe	54	131.30	5.495×10^{-3}	−111.79	−108	0.159
Ytterbium	Yb	70	173.04	6.965	824	1530	0.155
Yttrium	Y	39	88.905	4.469	1526	3030	0.297
Zinc	Zn	30	65.37	7.133	419.58	906	0.389
Zirconium	Zr	40	91.22	6.506	1852	3580	0.276

The values in parentheses in the column of molar masses are the mass numbers of the longest-lived isotopes of those elements that are radioactive. Melting points and boiling points in parentheses are uncertain.

The data for gases are valid only when these are in their usual molecular state, such as H_2, He, O_2, Ne, etc. The specific heats of the gases are the values at constant pressure.

Source: Adapted from Wehr, Richards, Adair, *Physics of the Atom,* 4th ed., Addison-Wesley, Reading, MA, 1984, and from J. Emsley, *The Elements,* 2nd ed., Clarendon Press, Oxford, 1991.

PERIODIC TABLE OF THE ELEMENTS

APPENDIX F

CONVERSION FACTORS

Conversion factors may be read directly from these tables. For example, 1 degree = 2.778×10^{-3} revolutions, so $16.7° = 16.7 \times 2.778 \times 10^{-3}$ rev. The SI quantities are fully capitalized. Adapted in part from G. Shortley and D. Williams, *Elements of Physics,* Prentice-Hall, Englewood Cliffs, NJ, 1971.

PLANE ANGLE

	°	′	″	RADIAN	rev
1 degree =	1	60	3600	1.745×10^{-2}	2.778×10^{-3}
1 minute =	1.667×10^{-2}	1	60	2.909×10^{-4}	4.630×10^{-5}
1 second =	2.778×10^{-4}	1.667×10^{-2}	1	4.848×10^{-6}	7.716×10^{-7}
1 RADIAN =	57.30	3438	2.063×10^5	1	0.1592
1 revolution =	360	2.16×10^4	1.296×10^6	6.283	1

SOLID ANGLE

1 sphere = 4π steradians = 12.57 steradians

LENGTH

	cm	METER	km	in.	ft	mi
1 centimeter =	1	10^{-2}	10^{-5}	0.3937	3.281×10^{-2}	6.214×10^{-6}
1 METER =	100	1	10^{-3}	39.37	3.281	6.214×10^{-4}
1 kilometer =	10^5	1000	1	3.937×10^4	3281	0.6214
1 inch =	2.540	2.540×10^{-2}	2.540×10^{-5}	1	8.333×10^{-2}	1.578×10^{-5}
1 foot =	30.48	0.3048	3.048×10^{-4}	12	1	1.894×10^{-4}
1 mile =	1.609×10^5	1609	1.609	6.336×10^4	5280	1

1 angström = 10^{-10} m
1 nautical mile = 1852 m
  = 1.151 miles = 6076 ft

1 fermi = 10^{-15} m
1 light-year = 9.460×10^{12} km
1 parsec = 3.084×10^{13} km

1 fathom = 6 ft
1 Bohr radius = 5.292×10^{-11} m
1 yard = 3 ft

1 rod = 16.5 ft
1 mil = 10^{-3} in.
1 nm = 10^{-9} m

AREA

	METER²	cm²	ft²	in.²
1 SQUARE METER =	1	10^4	10.76	1550
1 square centimeter =	10^{-4}	1	1.076×10^{-3}	0.1550
1 square foot =	9.290×10^{-2}	929.0	1	144
1 square inch =	6.452×10^{-4}	6.452	6.944×10^{-3}	1

1 square mile = 2.788×10^7 ft²
  = 640 acres
1 barn = 10^{-28} m²

1 acre = 43,560 ft²
1 hectare = 10^4 m² = 2.471 acres

VOLUME

	METER³	cm³	L	ft³	in.³
1 CUBIC METER =	1	10^6	1000	35.31	6.102×10^4
1 cubic centimeter =	10^{-6}	1	1.000×10^{-3}	3.531×10^{-5}	6.102×10^{-2}
1 liter =	1.000×10^{-3}	1000	1	3.531×10^{-2}	61.02
1 cubic foot =	2.832×10^{-2}	2.832×10^4	28.32	1	1728
1 cubic inch =	1.639×10^{-5}	16.39	1.639×10^{-2}	5.787×10^{-4}	1

1 U.S. fluid gallon = 4 U.S. fluid quarts = 8 U.S. pints = 128 U.S. fluid ounces = 231 in.³
1 British imperial gallon = 277.4 in.³ = 1.201 U.S. fluid gallons

MASS

Quantities in the colored areas are not mass units but are often used as such. When we write, for example, 1 kg "=" 2.205 lb, this means that a kilogram is a *mass* that *weighs* 2.205 pounds at a location where g has the standard value of 9.80665 m/s².

	g	KILOGRAM	slug	u	oz	lb	ton
1 gram =	1	0.001	6.852×10^{-5}	6.022×10^{23}	3.527×10^{-2}	2.205×10^{-3}	1.102×10^{-6}
1 KILOGRAM =	1000	1	6.852×10^{-2}	6.022×10^{26}	35.27	2.205	1.102×10^{-3}
1 slug =	1.459×10^4	14.59	1	8.786×10^{27}	514.8	32.17	1.609×10^{-2}
1 atomic mass unit =	1.661×10^{-24}	1.661×10^{-27}	1.138×10^{-28}	1	5.857×10^{-26}	3.662×10^{-27}	1.830×10^{-30}
1 ounce =	28.35	2.835×10^{-2}	1.943×10^{-3}	1.718×10^{25}	1	6.250×10^{-2}	3.125×10^{-5}
1 pound =	453.6	0.4536	3.108×10^{-2}	2.732×10^{26}	16	1	0.0005
1 ton =	9.072×10^5	907.2	62.16	5.463×10^{29}	3.2×10^4	2000	1

1 metric ton = 1000 kg

DENSITY

Quantities in the colored areas are weight densities and, as such, are dimensionally different from mass densities. See note for mass table.

	slug/ft³	KILOGRAM/ METER³	g/cm³	lb/ft³	lb/in.³
1 slug per foot³ =	1	515.4	0.5154	32.17	1.862×10^{-2}
1 KILOGRAM per METER³ =	1.940×10^{-3}	1	0.001	6.243×10^{-2}	3.613×10^{-5}
1 gram per centimeter =	1.940	1000	1	62.43	3.613×10^{-2}
1 pound per foot³ =	3.108×10^{-2}	16.02	1.602×10^{-2}	1	5.787×10^{-4}
1 pound per inch³ =	53.71	2.768×10^4	27.68	1728	1

TIME

	y	d	h	min	SECOND
1 year =	1	365.25	8.766×10^3	5.259×10^5	3.156×10^7
1 day =	2.738×10^{-3}	1	24	1440	8.640×10^4
1 hour =	1.141×10^{-4}	4.167×10^{-2}	1	60	3600
1 minute =	1.901×10^{-6}	6.944×10^{-4}	1.667×10^{-2}	1	60
1 SECOND =	3.169×10^{-8}	1.157×10^{-5}	2.778×10^{-4}	1.667×10^{-2}	1

SPEED

	ft/s	km/h	METER/ SECOND	mi/h	cm/s
1 foot per second =	1	1.097	0.3048	0.6818	30.48
1 kilometer per hour =	0.9113	1	0.2778	0.6214	27.78
1 METER per SECOND =	3.281	3.6	1	2.237	100
1 mile per hour =	1.467	1.609	0.4470	1	44.70
1 centimeter per second =	3.281×10^{-2}	3.6×10^{-2}	0.01	2.237×10^{-2}	1

1 knot = 1 nautical mi/h = 1.688 ft/s   1 mi/min = 88.00 ft/s = 60.00 mi/h

FORCE

Force units in the colored areas are now little used. To clarify: 1 gram-force (= 1 gf) is the force of gravity that would act on an object whose mass is 1 gram at a location where g has the standard value of 9.80665 m/s^2.

	dyne	NEWTON	lb	pdl	gf	kgf
1 dyne =	1	10^{-5}	2.248×10^{-6}	7.233×10^{-5}	1.020×10^{-3}	1.020×10^{-6}
1 NEWTON =	10^5	1	0.2248	7.233	102.0	0.1020
1 pound =	4.448×10^5	4.448	1	32.17	453.6	0.4536
1 poundal =	1.383×10^4	0.1383	3.108×10^{-2}	1	14.10	1.410×10^{-2}
1 gram-force =	980.7	9.807×10^{-3}	2.205×10^{-3}	7.093×10^{-2}	1	0.001
1 kilogram-force =	9.807×10^5	9.807	2.205	70.93	1000	1

PRESSURE

	atm	dyne/cm^2	inch of water	cm Hg	PASCAL	lb/in.^{2}	lb/ft^2
1 atmosphere =	1	1.013×10^6	406.8	76	1.013×10^5	14.70	2116
1 dyne per centimeter^2 =	9.869×10^{-7}	1	4.015×10^{-4}	7.501×10^{-5}	0.1	1.405×10^{-5}	2.089×10^{-3}
1 inch of water[a] at 4°C =	2.458×10^{-3}	2491	1	0.1868	249.1	3.613×10^{-2}	5.202
1 centimeter of mercury[a] at 0°C =	1.316×10^{-2}	1.333×10^4	5.353	1	1333	0.1934	27.85
1 PASCAL =	9.869×10^{-6}	10	4.015×10^{-3}	7.501×10^{-4}	1	1.450×10^{-4}	2.089×10^{-2}
1 pound per inch^2 =	6.805×10^{-2}	6.895×10^4	27.68	5.171	6.895×10^3	1	144
1 pound per foot^2 =	4.725×10^{-4}	478.8	0.1922	3.591×10^{-2}	47.88	6.944×10^{-3}	1

[a] Where the acceleration of gravity has the standard value of 9.80665 m/s^2.

1 bar = 10^6 dyne/cm^2 = 0.1 MPa   1 millibar = 10^3 dyne/cm^2 = 10^2 Pa   1 torr = 1 mm Hg

ENERGY, WORK, HEAT

Quantities in the colored areas are not properly energy units but are included for convenience. They arise from the relativistic mass−energy equivalence formula $E = mc^2$ and represent the energy released if a kilogram or unified atomic mass unit (u) is completely converted to energy (bottom two rows) or the mass that would be completely converted to one unit of energy (rightmost two columns).

	Btu	erg	ft·lb	hp·h	JOULE	cal	kW·h	eV	MeV	kg	u
1 British thermal unit =	1	1.055×10^{10}	777.9	3.929×10^{-4}	1055	252.0	2.930×10^{-4}	6.585×10^{21}	6.585×10^{15}	1.174×10^{-14}	7.070×10^{12}
1 erg =	9.481×10^{-11}	1	7.376×10^{-8}	3.725×10^{-14}	10^{-7}	2.389×10^{-8}	2.778×10^{-14}	6.242×10^{11}	6.242×10^{5}	1.113×10^{-24}	670.2
1 foot-pound =	1.285×10^{-3}	1.356×10^{7}	1	5.051×10^{-7}	1.356	0.3238	3.766×10^{-7}	8.464×10^{18}	8.464×10^{12}	1.509×10^{-17}	9.037×10^{9}
1 horsepower-hour =	2545	2.685×10^{13}	1.980×10^{6}	1	2.685×10^{6}	6.413×10^{5}	0.7457	1.676×10^{25}	1.676×10^{19}	2.988×10^{-11}	1.799×10^{16}
1 JOULE =	9.481×10^{-4}	10^{7}	0.7376	3.725×10^{-7}	1	0.2389	2.778×10^{-7}	6.242×10^{18}	6.242×10^{12}	1.113×10^{-17}	6.702×10^{9}
1 calorie =	3.969×10^{-3}	4.186×10^{7}	3.088	1.560×10^{-6}	4.186	1	1.163×10^{-6}	2.613×10^{19}	2.613×10^{13}	4.660×10^{-17}	2.806×10^{10}
1 kilowatt-hour =	3413	3.600×10^{13}	2.655×10^{6}	1.341	3.600×10^{6}	8.600×10^{5}	1	2.247×10^{25}	2.247×10^{19}	4.007×10^{-11}	2.413×10^{16}
1 electron-volt =	1.519×10^{-22}	1.602×10^{-12}	1.182×10^{-19}	5.967×10^{-26}	1.602×10^{-19}	3.827×10^{-20}	4.450×10^{-26}	1	10^{-6}	1.783×10^{-36}	1.074×10^{-9}
1 million electron-volts =	1.519×10^{-16}	1.602×10^{-6}	1.182×10^{-13}	5.967×10^{-20}	1.602×10^{-13}	3.827×10^{-14}	4.450×10^{-20}	10^{6}	1	1.783×10^{-30}	1.074×10^{-3}
1 kilogram =	8.521×10^{13}	8.987×10^{23}	6.629×10^{16}	3.348×10^{10}	8.987×10^{16}	2.146×10^{16}	2.497×10^{10}	5.610×10^{35}	5.610×10^{29}	1	6.022×10^{26}
1 unified atomic mass unit =	1.415×10^{-13}	1.492×10^{-3}	1.101×10^{-10}	5.559×10^{-17}	1.492×10^{-10}	3.564×10^{-11}	4.146×10^{-17}	9.320×10^{8}	932.0	1.661×10^{-27}	1

POWER

	Btu/h	ft·lb/s	hp	cal/s	kW	WATT
1 British thermal unit per hour =	1	0.2161	3.929×10^{-4}	6.998×10^{-2}	2.930×10^{-4}	0.2930
1 foot-pound per second =	4.628	1	1.818×10^{-3}	0.3239	1.356×10^{-3}	1.356
1 horsepower =	2545	550	1	178.1	0.7457	745.7
1 calorie per second =	14.29	3.088	5.615×10^{-3}	1	4.186×10^{-3}	4.186
1 kilowatt =	3413	737.6	1.341	238.9	1	1000
1 WATT =	3.413	0.7376	1.341×10^{-3}	0.2389	0.001	1

MAGNETIC FLUX

	maxwell	WEBER
1 maxwell =	1	10^{-8}
1 WEBER =	10^8	1

MAGNETIC FIELD

	gauss	TESLA	milligauss
1 gauss =	1	10^{-4}	1000
1 TESLA =	10^4	1	10^7
1 milligauss =	0.001	10^{-7}	1

1 tesla = 1 weber/meter^2

APPENDIX G
MATHEMATICAL FORMULAS

GEOMETRY

Circle of radius r: circumference $= 2\pi r$; area $= \pi r^2$.

Sphere of radius r: area $= 4\pi r^2$; volume $= \frac{4}{3}\pi r^3$.

Right circular cylinder of radius r and height h:
area $= 2\pi r^2 + 2\pi rh$; volume $= \pi r^2 h$.

Triangle of base a and altitude h: area $= \frac{1}{2}ah$.

QUADRATIC FORMULA

If $ax^2 + bx + c = 0$, then $x = \dfrac{-b \pm \sqrt{b^2 - 4ac}}{2a}$.

TRIGONOMETRIC FUNCTIONS OF ANGLE θ

$\sin\theta = \dfrac{y}{r}$ $\quad \cos\theta = \dfrac{x}{r}$

$\tan\theta = \dfrac{y}{x}$ $\quad \cot\theta = \dfrac{x}{y}$

$\sec\theta = \dfrac{r}{x}$ $\quad \csc\theta = \dfrac{r}{y}$

PYTHAGOREAN THEOREM

In this right triangle,

$a^2 + b^2 = c^2$

TRIANGLES

Angles are A, B, C

Opposite sides are a, b, c

Angles $A + B + C = 180°$

$\dfrac{\sin A}{a} = \dfrac{\sin B}{b} = \dfrac{\sin C}{c}$

$c^2 = a^2 + b^2 - 2ab\cos C$

Exterior angle $D = A + C$

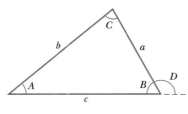

MATHEMATICAL SIGNS AND SYMBOLS

$=$ equals

$\approx$ equals approximately

$\sim$ is the order of magnitude of

$\neq$ is not equal to

$\equiv$ is identical to, is defined as

$>$ is greater than ($\gg$ is much greater than)

$<$ is less than ($\ll$ is much less than)

$\geq$ is greater than or equal to (or, is no less than)

$\leq$ is less than or equal to (or, is no more than)

$\pm$ plus or minus

$\propto$ is proportional to

Σ the sum of

$\bar{x}$ the average value of x

TRIGONOMETRIC IDENTITIES

$\sin(90° - \theta) = \cos\theta$

$\cos(90° - \theta) = \sin\theta$

$\sin\theta / \cos\theta = \tan\theta$

$\sin^2\theta + \cos^2\theta = 1$

$\sec^2\theta - \tan^2\theta = 1$

$\csc^2\theta - \cot^2\theta = 1$

$\sin 2\theta = 2\sin\theta\cos\theta$

$\cos 2\theta = \cos^2\theta - \sin^2\theta = 2\cos^2\theta - 1 = 1 - 2\sin^2\theta$

$\sin(\alpha \pm \beta) = \sin\alpha\cos\beta \pm \cos\alpha\sin\beta$

$\cos(\alpha \pm \beta) = \cos\alpha\cos\beta \mp \sin\alpha\sin\beta$

$\tan(\alpha \pm \beta) = \dfrac{\tan\alpha \pm \tan\beta}{1 \mp \tan\alpha\tan\beta}$

$\sin\alpha \pm \sin\beta = 2\sin\frac{1}{2}(\alpha \pm \beta)\cos\frac{1}{2}(\alpha \mp \beta)$

$\cos\alpha + \cos\beta = 2\cos\frac{1}{2}(\alpha + \beta)\cos\frac{1}{2}(\alpha - \beta)$

$\cos\alpha - \cos\beta = -2\sin\frac{1}{2}(\alpha + \beta)\sin\frac{1}{2}(\alpha - \beta)$

A15

BINOMIAL THEOREM

$$(1 \pm x)^n = 1 \pm \frac{nx}{1!} + \frac{n(n-1)x^2}{2!} + \cdots \qquad (x^2 < 1)$$

$$(1 \pm x)^{-n} = 1 \mp \frac{nx}{1!} + \frac{n(n+1)x^2}{2!} + \cdots \qquad (x^2 < 1)$$

EXPONENTIAL EXPANSION

$$e^x = 1 + x + \frac{x^2}{2!} = \frac{x^3}{3!} + \cdots$$

LOGARITHMIC EXPANSION

$$\ln(1 + x) = x - \tfrac{1}{2}x^2 + \tfrac{1}{3}x^3 - \cdots \qquad (|x| < 1)$$

TRIGONOMETRIC EXPANSIONS
(θ in radians)

$$\sin \theta = \theta - \frac{\theta^3}{3!} + \frac{\theta^5}{5!} - \cdots$$

$$\cos \theta = 1 - \frac{\theta^2}{2!} + \frac{\theta^4}{4!} - \cdots$$

$$\tan \theta = \theta + \frac{\theta^3}{3} + \frac{2\theta^5}{15} + \cdots$$

PRODUCTS OF VECTORS

Let $\mathbf{i}$, $\mathbf{j}$, and $\mathbf{k}$ be unit vectors in the x, y, and z directions. Then

$$\mathbf{i} \cdot \mathbf{i} = \mathbf{j} \cdot \mathbf{j} = \mathbf{k} \cdot \mathbf{k} = 1, \qquad \mathbf{i} \cdot \mathbf{j} = \mathbf{j} \cdot \mathbf{k} = \mathbf{k} \cdot \mathbf{i} = 0,$$

$$\mathbf{i} \times \mathbf{i} = \mathbf{j} \times \mathbf{j} = \mathbf{k} \times \mathbf{k} = 0,$$

$$\mathbf{i} \times \mathbf{j} = \mathbf{k}, \qquad \mathbf{j} \times \mathbf{k} = \mathbf{i}, \qquad \mathbf{k} \times \mathbf{i} = \mathbf{j}.$$

Any vector $\mathbf{a}$ with components a_x, a_y, and a_z along the x, y, and z axes can be written

$$\mathbf{a} = a_x\mathbf{i} + a_y\mathbf{j} + a_z\mathbf{k}.$$

Let $\mathbf{a}$, $\mathbf{b}$, and $\mathbf{c}$ be arbitrary vectors with magnitudes a, b, and c. Then

$$\mathbf{a} \times (\mathbf{b} + \mathbf{c}) = (\mathbf{a} \times \mathbf{b}) + (\mathbf{a} \times \mathbf{c})$$

$$(s\mathbf{a}) \times \mathbf{b} = \mathbf{a} \times (s\mathbf{b}) = s(\mathbf{a} \times \mathbf{b}) \qquad (s = \text{a scalar}).$$

Let θ be the smaller of the two angles between $\mathbf{a}$ and $\mathbf{b}$. Then

$$\mathbf{a} \cdot \mathbf{b} = \mathbf{b} \cdot \mathbf{a} = a_xb_x + a_yb_y + a_zb_z = ab \cos \theta$$

$$\mathbf{a} \times \mathbf{b} = -\mathbf{b} \times \mathbf{a} = \begin{vmatrix} \mathbf{i} & \mathbf{j} & \mathbf{k} \\ a_x & a_y & a_z \\ b_x & b_y & b_z \end{vmatrix}$$

$$= (a_yb_z - b_ya_z)\mathbf{i}$$

$$+ (a_zb_x - b_za_x)\mathbf{j} + (a_xb_y - b_xa_y)\mathbf{k}$$

$$|\mathbf{a} \times \mathbf{b}| = ab \sin \theta$$

$$\mathbf{a} \cdot (\mathbf{b} \times \mathbf{c}) = \mathbf{b} \cdot (\mathbf{c} \times \mathbf{a}) = \mathbf{c} \cdot (\mathbf{a} \times \mathbf{b})$$

$$\mathbf{a} \times (\mathbf{b} \times \mathbf{c}) = (\mathbf{a} \cdot \mathbf{c})\mathbf{b} - (\mathbf{a} \cdot \mathbf{b})\mathbf{c}$$

DERIVATIVES AND INTEGRALS

In what follows, the letters u and v stand for any functions of x, and a and m are constants. To each of the indefinite integrals should be added an arbitrary constant of integra-tion. The *Handbook of Chemistry and Physics* (CRC Press Inc.) gives a more extensive tabulation.

1. $\dfrac{dx}{dx} = 1$

2. $\dfrac{d}{dx}(au) = a\dfrac{du}{dx}$

3. $\dfrac{d}{dx}(u+v) = \dfrac{du}{dx} + \dfrac{dv}{dx}$

4. $\dfrac{d}{dx}x^m = mx^{m-1}$

5. $\dfrac{d}{dx}\ln x = \dfrac{1}{x}$

6. $\dfrac{d}{dx}(uv) = u\dfrac{dv}{dx} + v\dfrac{du}{dx}$

7. $\dfrac{d}{dx}e^x = e^x$

8. $\dfrac{d}{dx}\sin x = \cos x$

9. $\dfrac{d}{dx}\cos x = -\sin x$

10. $\dfrac{d}{dx}\tan x = \sec^2 x$

11. $\dfrac{d}{dx}\cot x = -\csc^2 x$

12. $\dfrac{d}{dx}\sec x = \tan x \sec x$

13. $\dfrac{d}{dx}\csc x = -\cot x \csc x$

14. $\dfrac{d}{dx}e^u = e^u\dfrac{du}{dx}$

15. $\dfrac{d}{dx}\sin u = \cos u\dfrac{du}{dx}$

16. $\dfrac{d}{dx}\cos u = -\sin u\dfrac{du}{dx}$

1. $\displaystyle\int dx = x$

2. $\displaystyle\int au\,dx = a\int u\,dx$

3. $\displaystyle\int (u+v)\,dx = \int u\,dx + \int v\,dx$

4. $\displaystyle\int x^m\,dx = \dfrac{x^{m+1}}{m+1} \quad (m \neq -1)$

5. $\displaystyle\int \dfrac{dx}{x} = \ln|x|$

6. $\displaystyle\int u\dfrac{dv}{dx}\,dx = uv - \int v\dfrac{du}{dx}\,dx$

7. $\displaystyle\int e^x\,dx = e^x$

8. $\displaystyle\int \sin x\,dx = -\cos x$

9. $\displaystyle\int \cos x\,dx = \sin x$

10. $\displaystyle\int \tan x\,dx = \ln|\sec x|$

11. $\displaystyle\int \sin^2 x\,dx = \tfrac{1}{2}x - \tfrac{1}{4}\sin 2x$

12. $\displaystyle\int e^{-ax}\,dx = -\dfrac{1}{a}e^{-ax}$

13. $\displaystyle\int xe^{-ax}\,dx = -\dfrac{1}{a^2}(ax+1)e^{-ax}$

14. $\displaystyle\int x^2 e^{-ax}\,dx = -\dfrac{1}{a^3}(a^2x^2 + 2ax + 2)e^{-ax}$

15. $\displaystyle\int_0^\infty x^n e^{-ax}\,dx = \dfrac{n!}{a^{n+1}}$

16. $\displaystyle\int_0^\infty x^{2n} e^{-ax^2}\,dx = \dfrac{1\cdot3\cdot5\,\cdots\,(2n-1)}{2^{n+1}a^n}\sqrt{\dfrac{\pi}{a}}$

17. $\displaystyle\int \dfrac{dx}{\sqrt{x^2+a^2}} = \ln(x+\sqrt{x^2+a^2})$

APPENDIX H

WINNERS OF THE NOBEL PRIZE IN PHYSICS*

1901	Wilhelm Konrad Röntgen	1845–1923	for the discovery of x rays
1902	Hendrik Antoon Lorentz	1853–1928	for their researches into the influence of magnetism upon radiation phenomena
	Pieter Zeeman	1865–1943	
1903	Antoine Henri Becquerel	1852–1908	for his discovery of spontaneous radioactivity
	Pierre Curie	1859–1906	for their joint researches on the radiation phenomena discovered by Becquerel
	Marie Sklowdowska-Curie	1867–1934	
1904	Lord Rayleigh (John William Strutt)	1842–1919	for his investigations of the densities of the most important gases and for his discovery of argon
1905	Philipp Eduard Anton von Lenard	1862–1947	for his work on cathode rays
1906	Joseph John Thomson	1856–1940	for his theoretical and experimental investigations on the conduction of electricity by gases
1907	Albert Abraham Michelson	1852–1931	for his optical precision instruments and metrological investigations carried out with their aid
1908	Gabriel Lippmann	1845–1921	for his method of reproducing colors photographically based on the phenomena of interference
1909	Guglielmo Marconi	1874–1937	for their contributions to the development of wireless telegraphy
	Carl Ferdinand Braun	1850–1918	
1910	Johannes Diderik van der Waals	1837–1932	for his work on the equation of state for gases and liquids
1911	Wilhelm Wien	1864–1928	for his discoveries regarding the laws governing the radiation of heat
1912	Nils Gustaf Dalén	1869–1937	for his invention of automatic regulators for use in conjunction with gas accumulators for illuminating lighthouses and buoys
1913	Heike Kamerlingh Onnes	1853–1926	for his investigations of the properties of matter at low temperatures which led, among other things, to the production of liquid helium
1914	Max von Laue	1879–1960	for his discovery of the diffraction of Röntgen rays by crystals
1915	William Henry Bragg	1862–1942	for their services in the analysis of crystal structure by means of x rays
	William Lawrence Bragg	1890–1971	
1917	Charles Glover Barkla	1877–1944	for his discovery of the characteristic x rays of the elements
1918	Max Planck	1858–1947	for his discovery of energy quanta
1919	Johannes Stark	1874–1957	for his discovery of the Doppler effect in canal rays and the splitting of spectral lines in electric fields
1920	Charles-Édouard Guillaume	1861–1938	for the service he rendered to precision measurements in physics by his discovery of anomalies in nickel steel alloys
1921	Albert Einstein	1879–1955	for his services to theoretical physics, and especially for his discovery of the law of the photoelectric effect

*See *Nobel Lectures, Physics,* 1901–1970, Elsevier Publishing Company, for biographies of the awardees and for lectures given by them on receiving the prize.

1922	Niels Bohr	1885–1962	for the investigation of the structure of atoms, and of the radiation emanating from them
1923	Robert Andrews Millikan	1868–1953	for his work on the elementary charge of electricity and on the photoelectric effect
1924	Karl Manne Georg Siegbahn	1888–1979	for his discoveries and research in the field of x-ray spectroscopy
1925	James Franck	1882–1964	for their discovery of the laws governing the impact of an electron upon an atom
	Gustav Hertz	1887–1975	
1926	Jean Baptiste Perrin	1870–1942	for his work on the discontinuous structure of matter, and especially for his discovery of sedimentation equilibrium
1927	Arthur Holly Compton	1892–1962	for his discovery of the effect named after him
	Charles Thomson Rees Wilson	1869–1959	for his method of making the paths of electrically charged particles visible by condensation of vapor
1928	Owen Willans Richardson	1879–1959	for his work on the thermionic phenomenon and especially for the discovery of the law named after him
1929	Prince Louis Victor de Broglie	1892–1987	for his discovery of the wave nature of electrons
1930	Sir Chandrasekhara Venkata Raman	1888–1970	for his work on the scattering of light and for the discovery of the effect named after him
1932	Werner Heisenberg	1901–1976	for the creation of quantum mechanics, the application of which has, among other things, led to the discovery of the allotropic forms of hydrogen
1933	Erwin Schrödinger	1887–1961	for the discovery of new productive forms of atomic theory
	Paul Adrien Maurice Dirac	1902–1984	
1935	James Chadwick	1891–1974	for his discovery of the neutron
1936	Victor Franz Hess	1883–1964	for the discovery of cosmic radiation
	Carl David Anderson	1905–1991	for his discovery of the positron
1937	Clinton Joseph Davisson	1881–1958	for their experimental discovery of the diffraction of electrons by crystals
	George Paget Thomson	1892–1975	
1938	Enrico Fermi	1901–1954	for his demonstrations of the existence of new radioactive elements produced by neutron irradiation, and for his related discovery of nuclear reactions brought about by slow neutrons
1939	Ernest Orlando Lawrence	1901–1958	for the invention and development of the cyclotron and for results obtained with it, especially for artificial radioactive elements
1943	Otto Stern	1888–1969	for his contribution to the development of the molecular-ray method and his discovery of the magnetic moment of the proton
1944	Isidor Isaac Rabi	1898–1988	for his resonance method for recording the magnetic properties of atomic nuclei
1945	Wolfgang Pauli	1900–1958	for the discovery of the Exclusion Principle (also called Pauli Principle)
1946	Percy Williams Bridgman	1882–1961	for the invention of an apparatus to produce extremely high pressures, and for the discoveries he made therewith in the field of high-pressure physics
1947	Sir Edward Victor Appleton	1892–1965	for his investigations of the physics of the upper atmosphere, especially for the discovery of the so-called Appleton layer
1948	Patrick Maynard Stuart Blackett	1897–1974	for his development of the Wilson cloud-chamber method, and his discoveries therewith in nuclear physics and cosmic radiation
1949	Hideki Yukawa	1907–1981	for his prediction of the existence of mesons on the basis of theoretical work on nuclear forces

1950	Cecil Frank Powell	1903–1969	for his development of the photographic method of studying nuclear processes and his discoveries regarding mesons made with this method
1951	Sir John Douglas Cockcroft	1897–1967	for their pioneer work on the transmutation of atomic nuclei by artificially accelerated atomic particles
	Ernest Thomas Sinton Walton	1903–	
1952	Felix Bloch	1905–1983	for their development of new nuclear-magnetic precision methods and discoveries in connection therewith
	Edward Mills Purcell	1912–	
1953	Frits Zernike	1888–1966	for his demonstration of the phase-contrast method, especially for his invention of the phase-contrast microscope
1954	Max Born	1882–1970	for his fundamental research in quantum mechanics, especially for his statistical interpretation of the wave function
	Walther Bothe	1891–1957	for the coincidence method and his discoveries made therewith
1955	Willis Eugene Lamb	1913–	for his discoveries concerning the fine structure of the hydrogen spectrum
	Polykarp Kusch	1911–	for his precision determination of the magnetic moment of the electron
1956	William Shockley	1910–1989	for their researches on semiconductors and their discovery of the transistor effect
	John Bardeen	1908–1991	
	Walter Houser Brattain	1902–1987	
1957	Chen Ning Yang	1922–	for their penetrating investigation of the parity laws which has led to important discoveries regarding the elementary particles
	Tsung Dao Lee	1926–	
1958	Pavel Aleksejevič Čerenkov	1904–1990	for the discovery and interpretation of the Cerenkov effect
	Il' ja Michajlovič Frank	1908–1990	
	Igor' Evgen' evič Tamm	1895–1971	
1959	Emilio Gino Segrè	1905–1989	for their discovery of the antiproton
	Owen Chamberlain	1920–	
1960	Donald Arthur Glaser	1926–	for the invention of the bubble chamber
1961	Robert Hofstadter	1915–1990	for his pioneering studies of electron scattering in atomic nuclei and for his thereby achieved discoveries concerning the structure of the nucleons
	Rudolf Ludwig Mössbauer	1929–	for his researches concerning the resonance absorption of γ rays and his discovery in this connection of the effect which bears his name
1962	Lev Davidovič Landau	1908–1968	for his pioneering theories of condensed matter, especially liquid helium
1963	Eugene P. Wigner	1902–	for his contributions to the theory of the atomic nucleus and the elementary particles, particularly through the discovery and application of fundamental symmetry principles
	Maria Goeppert Mayer	1906–1972	for their discoveries concerning nuclear shell structure
	J. Hans D. Jensen	1907–1973	
1964	Charles H. Townes	1915–	for fundamental work in the field of quantum electronics which has led to the construction of oscillators and amplifiers based on the maser–laser principle
	Nikolai G. Basov	1922–	
	Alexander M. Prochorov	1916–	
1965	Sin-itiro Tomonaga	1906–1979	for their fundamental work in quantum electrodynamics, with deep-ploughing consequences for the physics of elementary particles
	Julian Schwinger	1918–	
	Richard P. Feynman	1918–1988	
1966	Alfred Kastler	1902–1984	for the discovery and development of optical methods for studying Hertzian resonance in atoms
1967	Hans Albrecht Bethe	1906–	for his contributions to the theory of nuclear reactions, especially his discoveries concerning the energy production in stars

1968	Luis W. Alvarez	1911–1988	for his decisive contribution to elementary particle physics, in particular the discovery of a large number of resonance states, made possible through his development of the techniques of using the hydrogen bubble chamber and its data analysis
1969	Murray Gell-Mann	1929–	for his contributions and discoveries concerning the classification of elementary particles and their interactions
1970	Hannes Alfvén	1908–	for fundamental work and discoveries in magneto-hydrodynamics with fruitful applications in different parts of plasma physics
	Louis Néel	1904–	for fundamental work and discoveries concerning antiferromagnetism and ferrimagnetism which have led to important applications in solid state physics
1971	Dennis Gabor	1900–1979	for his discovery of the principles of holography
1972	John Bardeen	1908–1991	for their development of a theory of superconductivity
	Leon N. Cooper	1930–	
	J. Robert Schrieffer	1931–	
1973	Leo Esaki	1925–	for his discovery of tunneling in semiconductors
	Ivar Giaever	1929–	for his discovery of tunneling in superconductors
	Brian D. Josephson	1940–	for his theoretical prediction of the properties of a supercurrent through a tunnel barrier
1974	Antony Hewish	1924–	for the discovery of pulsars
	Sir Martin Ryle	1918–1984	for his pioneering work in radioastronomy
1975	Aage Bohr	1922–	for the discovery of the connection between collective motion and particle motion and the development of the theory of the structure of the atomic nucleus based on this connection
	Ben Mottelson	1926–	
	James Rainwater	1917–1986	
1976	Burton Richter	1931–	for their (independent) discovery of an important fundamental particle
	Samuel Chao Chung Ting	1936–	
1977	Philip Warren Anderson	1923–	for their fundamental theoretical investigations of the electronic structure of magnetic and disordered systems
	Nevill Francis Mott	1905–	
	John Hasbrouck Van Vleck	1899–1980	
1978	Peter L. Kapitza	1894–1984	for his basic inventions and discoveries in low-temperature physics
	Arno A. Penzias	1933–	for their discovery of cosmic microwave background radiation
	Robert Woodrow Wilson	1936–	
1979	Sheldon Lee Glashow	1932–	for their unified model of the action of the weak and electromagnetic forces and for their prediction of the existence of neutral currents
	Abdus Salam	1926–	
	Steven Weinberg	1933–	
1980	James W. Cronin	1931–	for the discovery of violations of fundamental symmetry principles in the decay of neutral K mesons
	Val L. Fitch	1923–	
1981	Nicolaas Bloembergen	1920–	for their contribution to the development of laser spectroscopy
	Arthur Leonard Schawlow	1921–	
	Kai M. Siegbahn	1918–	for his contribution to high-resolution electron spectroscopy
1982	Kenneth Geddes Wilson	1936–	for his method of analyzing the critical phenomena inherent in the changes of matter under the influence of pressure and temperature
1983	Subrehmanyan Chandrasekhar	1910–	for his theoretical studies of the structure and evolution of stars
	William A. Fowler	1911–	for his studies of the formation of the chemical elements in the universe
1984	Carlo Rubbia	1934–	for their decisive contributions to the Large Project, which led to the discovery of the field particles W and Z, communicators of the weak interaction
	Simon van der Meer	1925–	

1985	Klaus von Klitzing	1943–	for his discovery of the quantized Hall resistance
1986	Ernst Ruska	1906–1988	for his invention of the electron microscope
	Gerd Binnig	1947–	for their invention of the scanning tunneling microscope
	Heinrich Rohrer	1933–	
1987	Karl Alex Müller	1927–	for their discovery of a new class of superconductors
	J. George Bednorz	1950–	
1988	Leon M. Lederman	1922–	for the first use of a neutrino beam and the discovery of the muon neutrino
	Melvin Schwartz	1932–	
	Jack Steinberger	1921–	
1989	Norman Ramsey	1915–	for their work that led to the development of atomic clocks and precision timing
	Hans Dehmelt	1922–	
	Wolfgang Paul	1913–	
1990	Jerome I. Friedman	1930–	for demonstrating that protons and neutrons consist of quarks
	Henry W. Kendall	1926–	
	Richard E. Taylor	1929–	
1991	Pierre de Gennes	1932–	for studies of order phenomena, such as in liquid crystals and polymers
1992	George Charpak	1924–	for his invention of fast electronic detectors for high-energy particles

ANSWERS TO ODD-NUMBERED EXERCISES AND PROBLEMS

Chapter 1

3. (a) 186 mi. (b) 3.0×10^8 mm. **5.** (a) 10^9.
(b) 10^{-4}. (c) 9.1×10^5. **7.** 32.2 km. **9.** 0.020 km³.
11. (a) 250 ft². (b) 23.3 m². (c) 3060 ft³.
(d) 86.6 m³. **13.** 8×10^2 km. **15.** (a) 11 m²/L.
(b) 1.13×10^4 m⁻¹. (c) 2.17×10^{-3} gal/ft².
17. (a) $d_{sun}/d_{moon} = 400$. (b) $V_{sun}/V_{moon} = 6.4 \times 10^7$.
(c) 3.5×10^3 km. **19.** (a) 31 m. (b) 21 m.
(c) Lake Ontario. **21.** 52.6 min; 5.2%. **23.** 720 days.
25. (a) Yes. (b) 8.6 s. **27.** 0.12 AU/min. **29.** 3.3 ft.
31. 2 h. **33.** 6.0×10^{26}. **35.** 9.0×10^{49}.
37. (a) 10^3 km/m³. (b) 158 kg/s. **39.** 0.260 kg.

Chapter 2

1. (a) Lewis: 10.0 m/s; Rodgers: 5.41 m/s.
(b) 1 h 10 min. **3.** 310 ft. **5.** 2 cm/y.
7. 6.71×10^8 mi/h, 9.84×10^8 ft/s, 1.00 ly/y.
9. (a) 5.7 ft/s. (b) 7.0 ft/s.
11. (a) 45 mi/h (72 km/h). (b) 43 mi/h (69 km/h).
(c) 44 mi/h (71 km/h). (d) 0. **13.** (a) 28.5 cm/s.
(b) 18.0 cm/s. (c) 40.5 cm/s. (d) 28.1 m/s.
(e) 30.3 cm/s. **15.** (a) Mathematically, an infinite
number. (b) 60 km. **17.** (a) $4\,s > t > 2\,s$.
(b) $3\,s > t > 0$. (c) $6\,s > t > 3\,s$. (d) $t = 3\,s$.
19. 100 m. **21.** (a) 20 m/s², in the direction opposite to
its initial velocity. **23.** (a) The signs of v and a are:
OA: $+$, 0; AB: $+$, $-$; BC: 0, 0; CD: $-$, $+$. (b) No.
(c) No. **27.** (a) No. (b) m²/s²; m/s².
29. (a) $t = 1.2$ s. (b) $t = 0$. (c) $t > 0$, $t < 0$.
31. (a) $\bar{v} = 14$ m/s, $\bar{a} = 18$ m/s².
(b) $v(2) = 24$ m/s, $v(1) = 6$ m/s, $a(2) = 24$ m/s²,
$a(1) = 12$ m/s². **33.** (a) L/T^2, ft/s²; L/T^3, ft/s³.
(b) 2.0 s. (c) 24 ft. (d) -16 ft.
(e) 3, 0, -9, -24 ft/s. (f) 0, -6, -12, -18 ft/s².
35. (a) 1.6 m/s. (b) 18 m/s. **37.** (a) 3.1×10^6 s.
(b) 4.6×10^{13} m. **39.** (a) 0.10 m.
41. (a) 8.3 m/s²; 0.85g. (b) 3.2 s; $\approx 8T$.
43. (a) 5.00 s. (b) 61.5 m. **45.** (a) 2.6 s.
47. (a) 5.0 m/s². (b) 4.0 s. (c) 6.0 s. (d) 90 m.
49. (a) 5.00 m/s. (b) 1.67 m/s². (c) 7.50 m.
51. (a) 0.74 s. (b) -20 ft/s². **53.** (a) 32.9 m/s.
(b) 49.1 s. (c) 11.7 m/s. **55.** (a) 34.7 ft. (b) 41.6 s.
57. Collide. **59.** (a) 29.4 m. (b) 2.45 s. **61.** 183 m/s.

63. (a) 1.54 s. (b) 27.1 m/s. **67.** (a) 3.70 m/s.
(b) 1.74 m/s. (c) 0.154 m. **69.** 4.0 m/s.
73. (a) $v = (v_0^2 + 2gh)^{1/2}$, downward.
(b) $t = [(v_0^2 + 2gh)^{1/2} - v_0]/g$.
(c) same; $t = [(v_0^2 + 2gh)^{1/2} + v_0]/g$, more. **75.** Four
times as high. **77.** 1650 m/s², upward. **79.** (a) 38.1 m.
(b) 9.02 m/s. (c) 14.5 m/s, up. **81.** 96g. **83.** (a) 17 s.
(b) 290 m. **85.** ≈ 0.3 s. **87.** (a) 76 m above the
ground. (b) 4.1 s. **89.** 2.34 m.

Chapter 3

1. The displacements should be (a) parallel,
(b) antiparallel, (c) perpendicular.
3. (a) 370 m, 36° north of east.
(b) Displacement magnitude = 370 m;
distance walked = 425 m. **5.** 81 km, 40° north of east.
7. (a) 38 units at 320°. (b) 130 units at 1.2°.
(c) 62 units at 130°. **9.** $a_x = -2.5$; $a_y = -6.9$.
11. $r_x = 13$ m; $r_y = 7.5$ m.
13. (a) 14 cm, 45° left of straight down.
(b) 20 cm, vertically up. (c) Zero. **15.** 4.74 km.
17. 168 cm, 32.5° above the floor. **19.** (a) 21.0 ft.
(b) No; yes; yes. (c) $14\mathbf{i} + 12\mathbf{j} + 10\mathbf{k}$, a possible answer.
(d) 26.1 ft. **21.** $r_x = 12$, $r_y = -5.8$, $r_z = -2.8$.
23. (a) $8\mathbf{i} + 2\mathbf{j}$, 8.2, 14°.
(b) $2\mathbf{i} - 6\mathbf{j}$, 6.3, $-72°$ relative to $\mathbf{i}$. **25.** (a) 5.0, $-37°$.
(b) 10, 53°. (c) 11, 27°. (d) 11, 80°.
(e) 11, 260°. The angles are relative to $+x$. Last two
vectors are in opposite directions.
27. (a) $r_x = 1.59$, $r_y = 12.1$. (b) 12.2. (c) 82.5°.
29. 3390 ft, horizontally.
31. (a) -2.83 m, -2.83 m; $+5.00$ m, 0 m; 3.00 m, 5.20 m.
(b) 5.17 m, 2.37 m. (c) 5.69 m, 24.6° north of east.
(d) 5.69 m, 24.6° south of west. **35.** (b) 11,200 km.
37. (a) 10 m, north. (b) 7.5 m, south. **39.** No.
43. (a) 30. (b) 52. **45.** (a) 0. (b) -16. (c) -9.
49. (a) $11\mathbf{i} + 5\mathbf{j} - 7\mathbf{k}$. (b) 120°. **51.** (a) $2\mathbf{k}$. (b) 26.
(c) 46. **53.** (a) 2.97. (b) $1.51\mathbf{i} + 2.67\mathbf{j} - 1.36\mathbf{k}$.
(c) 48. **55.** 70.5°. **57.** (b) $a^2b \sin \phi$.

Chapter 4

1. (a) $-(5.0\text{ m})\mathbf{i} + (8.0\text{ m})\mathbf{j}$. (b) 9.4 m, 122° from $+x$.
3. (a) $-(7.0\text{ m})\mathbf{i} + (12\text{ m})\mathbf{j}$. (b) xy plane.
5. (a) 671 mi, 63.4° south of east. (b) 298 mi/h,
63.4° south of east; angle must be same as in (a).

(c) 400 mi/h. **7.** (a) 6.79 km/h. (b) 6.96°.
9. (a) $(3 \text{ m/s})\mathbf{i} - (8t \text{ m/s})\mathbf{j}$. (b) $(3 \text{ m/s})\mathbf{i} - (16 \text{ m/s})\mathbf{j}$.
(c) 16 m/s, $-79°$ to $+x$. **11.** (a) $8t\mathbf{j} + \mathbf{k}$. (b) $8\mathbf{j}$.
13. $-(2.10 \text{ m/s}^2)\mathbf{i} + (2.81 \text{ m/s}^2)\mathbf{j}$.
15. (a) $-(1.5 \text{ m/s})\mathbf{j}$. (b) $(4.5 \text{ m})\mathbf{i} - (2.25 \text{ m})\mathbf{j}$.
17. (a) 18 cm. (b) 1.9 m. **19.** (a) 5.4×10^{-13} m.
(b) Decreases. **21.** (a) 0.50 s. (b) 10 ft/s.
23. (a) 0.21 s; 0.21 s. (b) 8.1 in. (c) 24 in.
25. (a) 16.9 m; 8.21 m. (b) 27.6 m; 7.26 m.
(c) 40.2 m; 0. **27.** (a) 1.15 s. (b) 12.0 m.
(c) 19.2 m/s; 4.80 m/s. (d) No. **29.** (b) 27°.
31. (a) 194 m/s. (b) 38°. **33.** 1.9 in.
35. Not accidental; horizontal launch speed about 20%
of world-class sprint speed. **37.** (a) 11 m. (b) 23 m.
(c) 17 m/s, 63° below horizontal. **39.** (a) 73 ft.
(b) 7.6°. (c) 1.0 s. **41.** 23 ft/s.
43. Approximately 40 m.
45. 30 m above the release point. **47.** The third.
49. (a) 202 ft/s. (b) 806 m. (c) 161 m/s; -171 m/s.
51. 78.5°. **53.** 25 m.
55. Between the angles 31° and 63° above the horizontal.
57. (a) 310 s. (b) 105 km. (c) 139 km.
59. (a) 4.0 m/s². (b) Toward center of circle.
61. (a) 22 m. (b) 15 s. **63.** (a) 4.6×10^{12} m.
(b) 2.8 d. **65.** (a) 7.3 km. (b) Less than 80 km/h.
67. (a) 19 m/s. (b) 35 rev/min. **69.** (a) 0.034 m/s².
(b) 84 min. **71.** (a) 4.2 m, 45°; 5.5 m, 68°; 6.0 m, 90°.
(b) 4.2 m, 135°. (c) 0.85 m/s, 135°.
(d) 0.94 m/s, 90°; 0.94 m/s, 180°. (e) 0.30 m/s², 180°;
0.30 m/s², 270°. Angles measured counterclockwise from $+x$. **73.** (a) 5 km/h, upstream.
(b) 1 km/h, downstream.
75. Wind blows from the west at 53 mi/h. **77.** 48 s.
79. (a) $(80 \text{ km/h})\mathbf{i} - (60 \text{ km/h})\mathbf{j}$.
(b) $\mathbf{v}$ happens to be along the line of sight.
(c) Answers do not change. **81.** $(0.96 \text{ m/s})\mathbf{j}$.
83. 80 m/s. **85.** 185 km/h, 22° south of west.
87. 87° from the direction of motion of the car.
89. (a) 47° east of north. (b) 6 min 35 s. **91.** $0.83c$.

Chapter 5

1. (a) $F_x = 1.88$ N, $F_y = 0.684$ N.
(b) $(1.88 \text{ N})\mathbf{i} + (0.684 \text{ N})\mathbf{j}$.
3. (a) $-(6.26 \text{ N})\mathbf{i} - (3.23 \text{ N})\mathbf{j}$.
(b) 7.0 N, 207° relative to $+x$.
5. $(3 \text{ N})\mathbf{i} + (-11 \text{ N})\mathbf{j} + (4 \text{ N})\mathbf{k}$.
7. (a) $(-32 \text{ N})\mathbf{i} + (-21 \text{ N})\mathbf{j}$. (b) 38 N, 213° from $+x$.
9. (a) $(0.86 \text{ m/s}^2)\mathbf{i} + (-0.16 \text{ m/s}^2)\mathbf{j}$.
(b) 0.88 m/s², $-11°$ relative to $+x$.
11. (a) Mass = 44 slug; weight = 1400 lb.
(b) Mass = 421 kg; weight = 4100 N. **13.** (a) 740 N.
(b) 290 N. (c) Zero. (d) 75 kg at each location.

15. (a) 147 N, downward. (b) 147 N, upward.
(c) 147 N. **17.** (a) 54 N. (b) 152 N. **19.** 1.18×10^4 N.
21. 1.2×10^5 N. **23.** 16 N. **25.** 8.0 cm/s².
27. (a) The 4-kg mass. (b) 6.5 m/s². (c) 13 N.
29. 1.2×10^6 N. **31.** 69 lb. **33.** 1.5 mm.
35. 10 m/s². **37.** (a) 110 lb, up. (b) 110 lb, down.
39. (a) 0.62 m/s². (b) 0.13 m/s². (c) 2.6 m.
41. (a) 0.74 m/s². (b) 7.3 m/s².
43. (a) $5\mathbf{i} + 4.3\mathbf{j}$, m/s. (b) $15\mathbf{i} + 6.4\mathbf{j}$, m.
45. (a) 65 N. (b) 49 N. **47.** (a) 220 kN. (b) 50 kN.
49. (a) 0.970 m/s². (b) $T_1 = 11.6$ N, $T_2 = 34.9$ N.
51. (a) 7020 lb. (b) 5460 lb. **53.** (a) 5.1 m/s.
55. (a) 3260 N. (b) 2.7×10^3 kg. (c) -1.2 m/s².
57. (a) 1.23; 2.46; 3.69; 4.92, N. (b) 6.15 N. (c) 0.25 N.
59. (a) Allow a downward acceleration with
magnitude ≥ 4.2 ft/s². (b) 13 ft/s or greater.
61. (a) 7.3 kg. (b) 89 N. **63.** (a) 4.9 m/s².
(b) 2.0 m/s², upward. (c) 120 N. **65.** (a) 120 m/s².
(b) $12g$. (c) 1.4×10^8 N. (d) 4.2 y.
67. (a) 2.18 m/s². (b) 116 N. (c) 21.0 m/s².
69. 4.6 m/s². (b) 2.6 m/s². **71.** (a) 9.4 km.
(b) 61 km. **73.** (a) -466 N. (b) -527 N.
(c) -931 N, -1050 N.
(d) First two cases: -931 N. Third case: -1860 N.
Fourth case: -1980 N.

Chapter 6

1. (a) 200 N. (b) 120 N. **3.** (a) 110 N. (b) 130 N.
(c) No. (d) 46 N. (e) 17 N. **5.** (a) (i) 245 N, 100 N.
(ii) 195 N, 86.6 N. (iii) 158 N, 50.0 N.
(b) (i) At rest. (ii) Slides. (iii) At rest. **7.** 9.3 m/s².
9. (a) 90 N. (b) 70 N. (c) 0.89 m/s². **11.** (a) No.
(b) $(12 \text{ N})\mathbf{i} + (5 \text{ N})\mathbf{j}$. **13.** 20°. **15.** (a) 0.13 N.
(b) 0.12. **19.** (a) 56 N. (b) 59 N. (c) 1100 N.
21. (a) $v_0^2/(4g \sin \theta)$. (b) No. **23.** 0.53.
25. (a) 11 N, rightward. (b) 0.14 m/s².
27. (a) 2.0 m/s², down the plane. (b) 4.0 m.
(c) It stays there. **29.** (a) 8.6 N. (b) 46 N. (c) 39 N.
31. (a) Zero. (b) 3.9 m/s² down the incline.
(c) 1.0 m/s² down the incline. **33.** (a) 13 N.
(b) 1.6 m/s². **35.** (a) 1.05 N, in tension.
(b) 3.62 m/s² down the plane.
(c) Answers are the same except that the rod is under
compression.
37. (a) 6.1 m/s², rightward. (b) 0.98 m/s², rightward.
39. $g(\sin \theta - \sqrt{2}\mu_k \cos \theta)$. **41.** (a) 19°. (b) 3300 N.
43. 6200 N. **45.** 2.3. **47.** 10 m/s. **49.** 68 ft.
51. (a) 11°. (b) 0.19. **53.** (a) 0.96 m/s. (b) 0.021.
55. (a) 2.2×10^6 m/s.
(b) 9.1×10^{22} m/s², toward the nucleus.
(c) 8.3×10^{-8} N. **57.** 178 km/h. **59.** 0.12, 0.23.
61. 874 N. **63.** $\sqrt{gR}$. **65.** 1.52 km.

67. (a) 5.1 m/s^2, radially inward. (b) 4.8 N. (c) 10 N.
69. (a) 0.0338 N. (b) 9.77 N. **71.** (a) 5.8′. (b) Zero.
(c) Zero.

Chapter 7

1. (a) 2.7×10^7 ft·lb. (b) 150 ft·lb. **3.** (a) 314 J.
(b) -155 J. (c) Zero. (d) 158 J. **5.** (a) 230 N.
(b) -400 J. (c) 400 J. (d) Zero. (e) Zero.
7. (a) $c = 4$ m. (b) $c < 4$ m. (c) $c > 4$ m.
9. (a) 215 lb. (b) 10,100 ft·lb. (c) 48 ft.
(d) 10,300 ft·lb. **11.** (a) 2200 J. (b) -1500 J.
13. 25 J. **17.** -6 J. **19.** 1250 J. **21.** 1.8×10^{13} J.
23. (a) 3610 J. (b) 1900 J. (c) 1.1×10^{10} J.
25. 47 keV. **27.** 7.9 J. **29.** (a) 48 km/h.
(b) 9.0×10^4 J. **31.** (a) 1×10^5 megatons TNT.
(b) 1×10^7 bombs. **33.** 530 J. **35.** (a) 1.2×10^4 J.
(b) -1.1×10^4 J. (c) 1100 J. (d) 5.4 m/s.
37. (a) 797 N. (b) Zero. (c) -1550 J. (d) Zero.
(e) 1550 J. (f) F varies during displacement.
39. 270 kW. **41.** 235 kW. **43.** 17 kW.
45. (a) 1.8×10^5 ft·lb. (b) 0.55 hp.
47. (a) 0.83, 2.5, 4.2 J. (b) 5.0 W. **49.** 90 hp.
51. (a) 79.37 keV. (b) 3.12 MeV. (c) 10.9 MeV.

Chapter 8

1. 89 N/cm. **3.** (a) 200 J. (b) 170 J. (c) 13 m/s.
5. (a) 4.0×10^4 J. (b) 4.0×10^4 J. **7.** (a) v_0.
(b) $\sqrt{v_0^2 + gh}$. (c) $\sqrt{v_0^2 + 2gh}$. (d) $(v_0^2/2g) + h$.
9. 56 m/s. **11.** (a) 7.84 N/cm. (b) 62.7 J.
(c) 80.0 cm. **13.** (a) mgL. (b) $\sqrt{2gL}$. **15.** (a) 2.8 m/s.
(b) 2.7 m/s. **17.** (a) 35 cm. (b) 1.7 m/s.
19. (a) 1.2 J. (b) 11 m/s. (c) No. (d) No.
21. (a) 25 kJ. (b) 7.8 kJ. (c) 160 m. **23.** (a) 4.8 m/s.
(b) 2.4 m/s. **25.** 10 cm. **27.** 1.25 cm. **29.** (a) 19 J.
(b) 6.4 m/s. (c) 11 J, 6.4 m/s.
31. It comes close to breaking, but does not break.
33. (a) $2\sqrt{gL}$. (b) $5mg$. (c) 71°. **35.** $mgL/32$.
39. (a) $1.12(A/B)^{1/6}$. (b) Repulsive. (c) Attractive.
41. (a) Turning point on left, none on right.
(b) Turning points on both left and right.
(c) -1.2×10^{-19} J. (d) 2.2×10^{-19} J.
(e) $\approx 1 \times 10^{-9}$ N, on each, directed toward the other.
(f) $r < 0.2$ nm. (g) $r > 0.2$ nm. (h) $r = 0.2$ nm.
43. (a) 7.9×10^4 J. (b) 1.8 W. **45.** (a) 2700 MJ.
(b) 2700 MW. (c) 240 M$. **47.** (a) -0.74 J.
(b) -0.53 J. **49.** (a) 0.77 mi. (b) 71 kW. **51.** 690 W.
53. 5.5×10^6 N. **55.** 24 W. **57.** (b) 3.4.
59. (a) -3800 J. (b) 3.1×10^4 N. **61.** 54%.
63. -12 J. **65.** (a) 1.5 MJ. (b) 0.51 MJ. (c) 1.0 MJ.
(d) 63 m/s. **67.** 0.191. **69.** (a) 44 m/s. (b) 0.036.
73. (a) 5.0 in. (b) 8.7 ft/s. **75.** (a) 560 J. (b) 150 J.
(c) 5.5 m/s. **77.** 1.2 m.
81. In the center of the flat part. **83.** (a) 24 ft/s.

(b) 3.0 ft. (c) 9.0 ft. (d) 49 ft. **85.** 180 W.
87. (a) 2.1×10^6 kg. (b) $\sqrt{100 + 1.5t}$ m/s.
(c) $[(1.5 \times 10^6)/\sqrt{100 + 1.5t}]$ N. (d) 6.7 km.
89. (a) 110 rev/min. (b) 19 W. **91.** (a) 216 J.
(b) 1180 N. (c) 432 J, twice answer to part (a).
93. (a) 1.1×10^{17} J. (b) 1.2 kg. **95.** 1.10 kg.
97. 270 times the equatorial circumference of the Earth.
99. 2×10^5 kg. **101.** (a) 2.46×10^{15} s^{-1}. (b) Emitted.

Chapter 9

1. (a) 4700 km. (b) $0.72R_E$.
3. (a) $x_{cm} = 1.1$ m; $y_{cm} = 1.3$ m.
(b) Shifts toward topmost particle.
5. $x_{cm} = -0.25$ m; $y_{cm} = 0$.
7. Within the iron, at midheight and midwidth, and 2.7
cm from midlength.
9. $x_{cm} = y_{cm} = 20$ cm; $z_{cm} = 16$ cm. **11.** 36.8 m.
13. 6.2 m. **15.** (a) Down; $mv/(m + M)$.
(b) Balloon again stationary. **17.** (a) L. (b) Zero.
19. 58 kg. **21.** (a) 25 mm from each body.
(b) 26 mm from lighter body, along an interconnecting
line. (c) Down. (d) -1.6×10^{-2} m/s^2.
23. 8100 slug·ft/s, in the direction of motion.
25. (a) 52.0 km/h. (b) 28.8 km/h. **27.** A proton.
29. (a) 30°. (b) (-0.572 kg·m/s)**j**. **31.** (a) 6.4 J.
(b) $P_i = 0.80$ kg·m/s, 30° above the horizontal;
$P_f = 0.80$ kg·m/s, 30° below the horizontal.
33. 9.8×10^{-3} ft/s, backward. **35.** 4400 km/h.
37. $wv_{rel}/(W + w)$.
39. 14 m/s, 135° from the other pieces.
41. (a) 0.54 m/s. (b) 0 m/s. (c) 1.1 m/s.
43. (a) 721 m/s. (b) 937 m/s. **45.** (a) $0.200v_{rel}$.
(b) $0.210v_{rel}$. (c) $0.209v_{rel}$. **47.** (a) 8.0×10^4 N.
(b) 27 kg/s. **49.** (a) 1.57×10^6 N. (b) 1.35×10^5 kg.
(c) 2.08 km/s. **51.** 2.2×10^{-3}. **55.** 6.1 s.
57. (a) 2.3×10^4 N. (b) 4.2×10^6 W.
59. 2.7 m/s. **61.** (a) -500 J. (b) 1700 N.

Chapter 10

1. 400 N·s. **3.** 2.5 m/s. **5.** (a) $2mv/\Delta t$. (b) 570 N.
7. 6400 lb. **9.** 67 m/s.
11. (a) 2.3 N·s, in initial direction of flight.
(b) 2.3 N·s, opposite initial direction of flight.
(c) 1400 N, in initial direction of flight. (d) 58 J.
13. 10 m/s. **15.** 216. **17.** 29. **19.** $2\mu v$. **21.** 990 N.
23. (a) 1.8 N·s, to the left. (b) 180 N, to the right.
27. 8 m/s. **29.** (a) 1.9 m/s, to the right. (b) Yes.
(c) No, total kinetic energy would have increased.
31. 0.22%. **33.** (a) 99 g. (b) 1.9 m/s.
35. (a) 2.47 m/s. (b) 1.23 m/s. **37.** 100 g. **39.** $m_1/3$.
41. ≈ 2 mm/y. **43.** 1.81 m/s. **45.** 310 m/s.
47. 2.7 m/s. (b) 1400 m/s. **49.** 190 tons. **51.** $\frac{1}{6}mv^2$.
53. 13 tons. **55.** 25 cm. **57.** (a) 62.5 km/h. (b) 0.75.

59. $\sqrt{2E\dfrac{M+m}{mM}}$.

61. (a) 30° from the incoming proton's direction.
(b) 250 m/s and 430 m/s. **63.** (a) 41°. (b) 4.76 m/s.
(c) No. **65.** $v = V/4$.
67. (a) 117° from the final direction of *B*. (b) No.
69. 120°. **71.** (a) 1.9 m/s, 30° to initial direction.
(b) No. **73.** (a) 3.4 m/s, deflected by 17° to the right.
(b) 0.95 MJ. **75.** (a) 117 MeV.
(b) Equal and opposite momenta. (c) π^-.
77. (a) 4.94 MeV. (b) Zero. (c) 4.85 MeV.
(d) 0.09 MeV.

Chapter 11

1. (a) 1.50 rad. (b) 85.9°. (c) 1.49 m.
3. (a) 5.5×10^{15} s. (b) 26. **5.** (a) 2 rad. (b) 0.
(c) 130 rad/s. (d) 32 rad/s². (e) No.
7. (a) 0.105 rad/s. (b) 1.75×10^{-3} rad/s.
(c) 1.45×10^{-4} rad/s. **9.** 11 rad/s. **11.** (a) 30 s.
(b) 1800 rad. **13.** (a) 9000 rev/min². (b) 420 rev.
15. (a) -1.25 rad/s². (b) 250 rad. (c) 39.8 rev.
17. (a) 140 rad. (b) 14 s. **19.** 8.0 s. **21.** (a) 340 s.
(b) -4.5×10^{-3} rad/s². (c) 98 s. **23.** (a) 1.0 rev/s².
(b) 4.8 s. (c) 9.6 s. (d) 48 rev.
25. (b) -2.3×10^{-9} rad/s². (c) 2600 y. (d) 24 ms.
27. (a) 3.5 rad/s. (b) 21 in./s. (c) 10 in./s.
29. (a) 20.9 rad/s. (b) 12.5 m/s. (c) 800 rev/min².
(d) 600 rev. **31.** (a) 2.0×10^{-7} rad/s. (b) 30 km/s.
(c) 5.9 mm/s², toward the sun.
33. (a) 2.50×10^{-3} rad/s. (b) 20.2 m/s². (c) 0.
35. (a) 7.3×10^{-5} rad/s. (b) 350 m/s.
(c) 7.3×10^{-5} rad/s, 460 m/s. **37.** (a) 40.2 cm/s².
(b) 2.36×10^3 m/s². (c) 83.2 m.
39. (a) 3.8×10^3 rad/s. (b) 190 m/s. **41.** 16 s.
43. (a) 73 cm/s². (b) 0.075. (c) 0.11. **45.** 12.3 kg·m².
47. First cylinder: 1100 J; second cylinder: 9700 J.
49. (a) 1300 g·cm², (b) 550 g·cm². (c) 1900 g·cm².
(d) $A + B$. **51.** (a) $5ml^2 + \frac{8}{3}Ml^2$. (b) $(\frac{5}{2}m + \frac{4}{3}M)l^2\omega^2$.
53. (a) 9.71×10^{37} kg·m². (b) 2.57×10^{29} J.
(c) 1.9×10^9 y. **57.** $\frac{1}{3}M(a^2 + b^2)$. **59.** (a) 49 MJ.
(b) 100 min. **61.** 140 N·m.
63. (a) $r_1F_1 \sin\theta_1 - r_2F_2 \sin\theta_2$. (b) -3.8 N·m.
65. 1.28 kg·m². **67.** 9.7 rad/s², counterclockwise.
69. (a) 155 kg·m². (b) 64.4 kg. **71.** 130 N.
73. (a) 6.00 cm/s². (b) 4.87 N. (c) 4.54 N.
(d) 1.20 rad/s². (e) 0.0138 kg·m². **75.** (a) $2\theta/t^2$.
(b) $2R\theta/t^2$. (c) $T_1 = M(g - 2R\theta/t^2)$;
$T_2 = Mg - (2\theta/t^2)(MR + I/R)$.
77. (a) $3g(1 - \cos\theta)$. (b) $\frac{3}{2}g \sin\theta$. (c) 41.8°.
79. 292 ft·lb (396 N·m). **81.** (a) $ml^2\omega^2/6$.
(b) $l^2\omega^2/6g$. **83.** $\sqrt{9g/4l}$. **85.** (a) 4.8×10^5 N.
(b) 1.1×10^4 N·m. (c) 1.3×10^6 J.

87. (a) -7.66 rad/s². (b) -11.7 N·m.
(c) -4.60×10^4 J. (d) 624 rev.
(e) The work done by friction, -4.60×10^4 J.

Chapter 12

1. 1.00. **3.** (a) 59.3 rad/s. (b) -9.31 rad/s².
(c) 70.7 m. **5.** (a) 990 J. (b) 3000 J. (c) 1.1×10^5 J.
7. (a) 44.8 ft·lb. (b) 11.2 ft. (c) No.
9. (a) 0 m/s, 0 m/s². (b) 22 m/s, 1500 m/s².
(c) -22 m/s, 1500 m/s². (d) Center: 22 m/s, 0 m/s²;
top: 44 m/s, 1500 m/s²; bottom: 0 m/s, 1500 m/s².
11. 48 m. **13.** (a) 2.7R. (b) (50/7)mg. **15.** (a) 1.13 s.
(b) 13.6 m. **17.** 70 rev/s.
21. (a) 10 N·m, parallel to *yz* plane, at 53° to +*y*.
(b) 22 N·m, $-x$.
23. (a) $(6.0\ N·m)\mathbf{i} - (3.0\ N·m)\mathbf{j} - (6.0\ N·m)\mathbf{k}$.
(b) $(26\ N·m)\mathbf{i} + (3.0\ N·m)\mathbf{j} - (18\ N·m)\mathbf{k}$.
(c) $(32\ N·m)\mathbf{i} - (24\ N·m)\mathbf{k}$. (d) 0.
25. (a) $(50\ N·m)\mathbf{k}$. (b) 90°. **27.** 9.8 kg·m²/s.
31. 2.5×10^{11} kg·m²/s. **33.** *mvd*, about any origin.
35. (a) 3.15×10^{43} kg·m²/s. (b) 0.616.
37. 4.5 N·m, parallel to *xy* plane at $-63°$ from +*x*.
39. (a) 0. (b) 0.
(c) $30t^3$ kg·m²/s, $90t^2$ N·m, both in $-z$ direction.
(d) $30t^3$ kg·m²/s, $90t^2$ N·m, both in $+z$ direction.
41. (a) $\frac{1}{2}mgt^2v_0 \cos\theta_0$. (b) $mgtv_0 \cos\theta_0$.
(c) $mgtv_0 \cos\theta_0$. **43.** (a) -1.47 N·m. (b) 20.4 rad.
(c) -29.9 J. (d) 19.9 W. **45.** (a) 12.2 kg·m².
(b) 308 kg·m²/s, down. **49.** (a) 1.2 s. (b) 8.6 m.
(c) 5.2 rev. (d) 6.1 m/s. (e) No.
51. (a) 3.6 rev/s. (b) 3.0.
(c) Work done by man in moving weights inward.
53. (a) 267 rev/min. (b) 2/3. **55.** 3.0 min.
57. 12.7 rad/s, clockwise viewed from above.
59. (a) $\frac{7}{12}ML^2$. (b) $\frac{7}{12}ML^2\omega_0$; down. (c) $\frac{14}{5}\omega_0$.
(d) $\frac{21}{40}mL^2\omega_0^2$. **61.** (a) $(mRv - I\omega_0)/(I + mR^2)$.
(b) No, energy transferred to internal energy of
cockroach.
63. The day would be longer by about 0.8 s.
65. (a) 0.148 rad/s. (b) 0.0123. (c) 181°.
67. 0.43 rev/min.

Chapter 13

1. (a) Two. (b) Seven. **3.** 8.7 N. **5.** (a) $-27\mathbf{i} + 2\mathbf{j}$ N.
(b) 176° counterclockwise from +*x* direction.
7. 7920 N. **9.** 0.29. **11.** (a) 2770 N. (b) 3890 N.
13. (a) 1160 N, down. (b) 1740 N, up.
(c) Left, stretched; right, compressed.
15. One-fourth the beam length from the free end.
17. (a) 3*W*. (b) 4*W*. **19.** (a) Bottom: 2*W*; sides: *W*.
(b) $\sqrt{2}W$. **21.** (a) 49 N. (b) 28 N. (c) 57 N. (d) 29°.
25. (a) 408 N. (b) $F_h = 245$ N (right).

(c) $F_v = 163$ N (up). **27.** $W\dfrac{\sqrt{2rh - h^2}}{r - h}$. **29.** (a) 42 N.

(b) 66 N. **31.** (a) 43.3 lb. (b) 21.7 lb. (c) 12.5 lb.

33. (a) 6630 N. (b) $F_h = 5740$ N. (c) $F_v = 5960$ N.

35. (a) $Wx/(L \sin \theta)$. (b) $Wx/(L \tan \theta)$.

(c) $W(1 - x/L)$. **37.** (a) -180 lb, 60 lb.

(b) 180 lb, 60 lb. (c) 180 lb, 210 lb.

(d) -180 lb, -60 lb. **39.** 0.34.

41. Bars BC, CD, and DA are in tension due to forces **T**;
diagonals AC and BD are in compression due to forces
$\sqrt{2}$**T**. **43.** (a) 445 N. (b) 0.50. (c) 315 N.

45. (a) 3.9 m/s². (b) 2000 N on each rear wheel;
3500 N on each front wheel.

(c) 790 N on each rear wheel; 1410 N on each front
wheel. **47.** (a) 1.9×10^{-3}. (b) 1.3×10^7 N/m².

(c) 6.9×10^9 N/m². **49.** 3.1 cm. **51.** 2.4×10^9 N/m².

53. (a) 1.8×10^7 N. (b) 1.4×10^7 N. (c) 16.

55. (a) 867 N. (b) 143 N. (c) 0.165.

Chapter 14

1. (a) 0.50 s. (b) 2.0 Hz. (c) 18 cm. **3.** (a) 245 N/m.
(b) 0.284 s. **5.** 708 N/m. **7.** $f > 500$ Hz.

9. (a) 100 N/m. (b) 0.45 s.

11. (a) 6.28×10^5 rad/s. (b) 1.59 mm.

13. (a) 1.0 mm. (b) 0.75 m/s. (c) 570 m/s².

15. (a) 1.29×10^5 N/m. (b) 2.68 Hz. **17.** (a) 4.0 s.

(b) $\pi/2$ rad/s. (c) 0.37 cm. (d) $(0.37 \text{ cm})\cos (\pi/2)t$.

(e) $(-0.58 \text{ cm/s})\sin (\pi/2)t$. (f) 0.58 cm/s.

(g) 0.91 cm/s². (h) Zero. (i) 0.58 cm/s.

19. (b) 12.47 kg. (c) 54.43 kg. **21.** 1.6 kg.

23. (a) 1.6 Hz. (b) 1.0 m/s, 0. (c) 10 m/s², ± 10 cm.

(d) $-(10 \text{ N/m})x$. **25.** 22 cm. **27.** (a) 25 cm.

(b) 2.2 Hz. **29.** (a) 0.500 m. (b) -0.251 m.

(c) 3.06 m/s. **31.** (a) $0.183A$. (b) Same direction.

37. (a) $k_1 = (n + 1)k/n$; $k_2 = (n + 1)k$.

(b) $f_1 = \sqrt{(n + 1)/n}\, f$; $f_2 = \sqrt{n + 1}\, f$. **39.** 3.7×10^{-2} J.

41. (a) 130 N/m. (b) 0.62 s. (c) 1.6 Hz. (d) 5.0 cm.

(e) 0.17 J. (f) 0.51 m/s. **43.** (a) 7.25×10^6 N/m.

(b) 49,400. **45.** (a) $mv/(m + M)$. (b) $mv/\sqrt{k(m + M)}$.

47. (a) $-(80 \text{ N})\cos[(2000 \text{ rad/s})t - \pi/3 \text{ rad}]$.

(b) 3.1 ms. (c) 4.0 m/s. (d) 0.080 J. **49.** (a) 16.7 cm.

(b) 1.23%. **51.** (a) 0.735 kg·m². (b) 0.024 N·m.

(c) 0.181 rad/s. **53.** 0.079 kg·m². **55.** 9.79 in.

57. 99 cm. **61.** $T/\sqrt{2}$. **63.** 5.6 cm. **65.** (a) 0.869 s.

(b) $r = R/2$. **69.** (a) $2\pi\sqrt{\dfrac{L^2 + 12x^2}{12gx}}$. (b) 0.289.

71. (a) 0.35 Hz. (b) 0.39 Hz. (c) Zero. **73.** 14.0°.

75. $2\pi\sqrt{m/3k}$. **77.** (a) 2.0 s. (b) 19.8 N·m/rad.

79. $0.29L$. **81.** 0.39. **83.** (a) 0.102 kg/s. (b) 0.137 J.

85. $k = 490$ N/cm; $b = 1100$ kg/s. **87.** 1.9 in.

Chapter 15

1. 19 m. **3.** 2.16. **5.** 1/2. **7.** 3.4×10^5 km.

9. 3.7×10^{-5} N, increasing y. **11.** $M = m$.

13. 3.2×10^{-7} N. **15.** $\dfrac{GmM}{d^2}\left(1 - \dfrac{1}{8(1 - R/2d)^2}\right)$.

17. (a) 1.62 m/s². (b) 4.9 s. **19.** 0.016 lb.

21. $3.4 \times 10^{-3}g$. (b) $6.1 \times 10^{-4}g$. (c) $1.4 \times 10^{-11}g$.

23. 9.78 m/s². **25.** (b) 1.9 h. **27.** (b) 3.2 m.

31. (a) $G(M_1 + M_2)m/a^2$. (b) $GM_1 m/b^2$. (c) 0.

35. (b) 2.0×10^8 N/m². (c) 360 km.

37. (a) -4.4×10^{-11} J. (b) -2.9×10^{-11} J.

(c) 2.9×10^{-11} J. **39.** 1/2. **41.** 220 km/s.

44. $-Gm(M_E/R + M_m/r)$. **49.** 2.6×10^4 km.

51. (a) 82 km/s. (b) 1.8×10^4 km/s.

53. (a) $\dfrac{GMmx}{(x^2 + R^2)^{3/2}}$. (b) $v^2 = 2GM\left(\dfrac{1}{R} - \dfrac{1}{\sqrt{R^2 + x^2}}\right)$.

55. 1.87 y. **57.** 5.93×10^{24} kg. **59.** 0.35 lunar months.

61. 3.9 y. **63.** 5.01×10^9 m or 7.20 solar radii.

65. 3.58×10^4 km. **67.** 81°. **73.** (a) 42.1 km/s.

(b) 12.3 km/s. (c) 16.6 km/s.

75. 1.6 cm/s, to the west along the equator.

77. (a) $-GM_E m/r$. (b) $-2GM_E m/r$. (c) It falls radially
to Earth. **81.** (a) No. (b) Same. (c) Yes.

83. (a) $T \propto r^{3/2}$. (b) $K \propto r^{-1}$. (c) $L \propto r^{1/2}$.

(d) $v \propto r^{-1/2}$. **85.** (a) 0. (b) 1.8×10^{32} J.

(c) 1.8×10^{32} J. (d) 0.99 km/s.

Chapter 16

1. 1000 kg/m³. **3.** 1.1×10^5 Pa or 1.1 atm.

5. 2.9×10^4 N. **7.** 6.0 lb/in.².

9. 1.90×10^4 Pa (143 mm Hg). **11.** 5.4×10^4 Pa.

13. 0.52 m. **15.** (a) 6.06×10^9 N. (b) 20 atm.

17. 0.412 cm. **19.** 2.0. **21.** 44 km. **23.** (a) $\rho g W D^2/2$.

(b) $\rho g W D^3/6$. (c) $D/3$. **27.** -3.9×10^{-3} atm.

29. (a) fA/a. (b) 20 lb. **31.** 1070 g. **33.** 1.5 g/cm³.

35. 600 kg/m³. **37.** (a) 670 kg/m³. (b) 740 kg/m³.

39. 390 kg. **41.** (a) 1.2 kg. (b) 1300 kg/m³.

43. 0.126 m³. **45.** Five. **47.** (a) 1.80 m³. (b) 4.75 m³.

49. 2.79 g/cm³. **51.** (b) 3.17 s. **53.** 4.0 m.

55. 28 ft/s. **57.** 43 cm/s. **59.** (a) 2.40 m/s.

(b) 245 Pa. **61.** (a) 12 ft/s. (b) 13 lb/in.².

63. 0.72 ft·lb/ft³. **65.** (a) 2. (b) $R_1/R_2 = \frac{1}{2}$.

(c) Drain it until $h_2 = h_1/4$. **67.** 116 m/s.

71. 110 m/s. **73.** 1.11×10^4 N.

75. $\frac{1}{2}\rho v^2 A$, where ρ is the density of air. **77.** 1.4 cm.

79. (a) 74 N. (b) 150 m³.

83. (a) $v = 4.1$ m/s, $v' = 21$ m/s. (b) 8.0×10^{-3} m³/s.

85. (a) 830 W. (b) 1100 W.

Chapter 17

1. (a) 75 Hz. (b) 13 ms. **3.** (a) 1.7 s. (b) 2.0 m/s.

(c) 3.3 m. (d) 15 cm. **5.** (a) 0.68 s. (b) 1.47 Hz.

(c) 2.06 m/s. **7.** (c) 200 cm/s; $-x$ direction.

9. (a) 5 cm/s. (b) Toward increasing x.

11. (a) 2.0 mm; 96 Hz; 30 m/s; 31 cm. (b) 1.2 m/s.
13. (a) 6.0 cm. (b) 100 cm. (c) 2.0 Hz.
(d) 200 cm/s. (e) $-x$ direction. (f) 75 cm/s.
(g) -2.0 cm. **17.** 129 m/s. **19.** 135 N.
23. (a) 15 m/s. (b) 0.036 N.
25. $y = (0.12 \text{ mm}) \sin[(141 \text{ m}^{-1})x + (628 \text{ s}^{-1})t]$.
27. (a) 5.0 cm. (b) 40 cm. (c) 12 m/s. (d) 0.033 s.
(e) 9.4 m/s. (f) $5.0 \sin(16x + 190t + 0.79)$, with
x in m, y in cm, t in s.
29. (a) $v_1 = 28.6$ m/s; $v_2 = 22.1$ m/s.
(b) $M_1 = 188$ g; $M_2 = 313$ g.
31. (a) $\sqrt{k(\Delta l)(l + \Delta l)/m}$. **33.** (a) $P_2 = 2P_1$.
(b) $P_2 = P_1/4$. **35.** (a) 3.65 m/s. (b) 12.2 N.
(c) Zero. (d) 44.3 W. (e) Zero. (f) Zero.
(g) 0.50 cm. **37.** 82.8°, 1.45 rad. **41.** 5.0 cm.
43. 10 cm. **45.** (a) 82.0 m/s. (b) 16.8 m.
(c) 4.88 Hz. **47.** (a) -0.0390 m.
(b) $y = 0.15 \sin(0.79x + 13t)$. (c) -0.14 m.
49. (a) 66.1 m/s. (b) 26.4 Hz. **51.** (a) 144 m/s.
(b) 3.00 m; 1.50 m. (c) 48.0 Hz; 96.0 Hz.
53. First and second harmonics of A match fourth and
eighth harmonics of B, respectively.
55. (a) 0.25 cm, 120 cm/s. (b) 3.0 cm. (c) Zero.
61. (a) 1.3 m. (b) $y = (0.002) \sin(9.4x) \cos(3800t)$, with
x and y in m, t in s.

Chapter 18

1. (a) $\approx 6\%$. **3.** The radio listener, by about 0.85 s.
5. 7.9×10^{10} Pa.
7. If only the length is uncertain, it must be known to
within 10^{-4} cm. If only the time is imprecise, the
uncertainty must be no more than 1 part in 10^5.
9. 43.5 m. **11.** 40.7 m. **13.** 100 kHz.
15. (a) 2.29, 0.229, 22.9 kHz. (b) 1.14, 0.114, 11.4 kHz.
17. (a) 6.0 m/s.
(b) $y = 0.30 \sin(\pi x/12 + 50\pi t)$, with x and y
in cm and t in s. **19.** 4.12 rad.
21. (a) $343(1 + 2m)$ Hz, with m being an integer
from 0 to 28.
(b) $686m$ Hz, with m being an integer from 1 to 29.
23. (a) Eight. (b) Eight. **25.** 64.7 Hz, 129 Hz.
27. (a) 0.080 W/m². (b) 0.013 W/m².
29. 35.7 nm. **31.** (a) 1000. (b) 32.
33. (a) 39.7 μW/m². (b) 166 nm. (c) 0.923 Pa.
35. (a) 59.7. (b) 2.81×10^{-4}. **37.** (a) $I \propto r^{-1}$.
(b) $s_m \propto r^{-1/2}$. **39.** (a) 5000. (b) 71. (c) 71.
41. 171 m. **43.** 3.16 km. **45.** (a) 5200 Hz.
(b) Amplitude$_{SAD}$/Amplitude$_{SBD}$ = 2. **47.** 350 m.
51. (a) 833 Hz. (b) 0.418 m. **53.** By a factor of 4.
55. Water filled to a height of $\frac{7}{8}, \frac{5}{8}, \frac{3}{8}, \frac{1}{8}$ m.
57. (a) $L\left(1 - \dfrac{1}{r}\right)$. (b) 13 cm. (c) $\frac{5}{6}$. **59.** 12.4 m.

61. (a) 71.5 Hz. (b) 64.8 N. **63.** 45.3 N. **65.** 2.25 ms.
67. 0.02. **69.** 3.8 Hz.
71. (a) 380 mi/h, away from owner.
(b) 77 mi/h, away from owner. **73.** 15.1 ft/s.
75. 3.1 m/s. **77.** 2.6×10^8 m/s. **79.** (a) 77.6 Hz.
(b) 77.0 Hz. **81.** (a) 42°. (b) 11 s. **83.** 0.189 MHz.
85. (a) 970 Hz. (b) 1030 Hz. (c) 60 Hz. No.
87. (a) 1.02 kHz. (b) 1.04 kHz. **89.** 1540 m/s.
91. (a) 8.29 Hz. (b) 13.9 Hz. **93.** (a) 598 Hz.
(b) 608 Hz. (c) 589 Hz. **95.** 0.073.
97. (a) 1.94×10^4 km/s. (b) Receding.
99. ± 3.78 pm. **101.** 49.5 m/s.

Chapter 19

1. 2.71 K. **3.** 31.5. **5.** 291.1 K. **7.** 348 K.
9. 7 Celsius. **11.** No; 310 K = 98.6°F.
13. (a) 10,000°F. (b) 37.0°C. (c) -57°C.
(d) -297°F. (e) 25°C = 77°F, for example.
15. -91.9°X. **17.** (a) Dimensions are inverse time.
21. 1.1 cm. **23.** 2.731 cm. **25.** (a) 13×10^{-6}/°F.
(b) 0.17 in. **27.** 23×10^{-6}/°C. **29.** 0.32 cm².
31. 29 cm³. **33.** 0.432 cm³. **35.** -157°C. **43.** 9.1 s; the
clock is running slowly. **45.** (a) $+9.0 \times 10^{-6}$.
(b) -1.3×10^{-5}. **47.** 7.5 cm. **49.** Increases by 0.1 mm.
51. 2.64×10^8 Pa. **53.** (b) $E_1 E_2 (\alpha_1 + \alpha_2) \Delta T/(E_1 + E_2)$.

Chapter 20

1. 333 J. **3.** 35.7 m³. **5.** 6.7×10^{12} J. **7.** 42.7 kJ.
9. 250 g. **11.** 220 m/s. **13.** 1.17°C. **15.** 160 s.
17. (a) 20,300 cal. (b) 1110 cal. (c) 873°C.
19. 45.4°C. **21.** (a) 18,700. (b) 10.4 h. **23.** 2.8 days.
25. 82 cal. **27.** 13.5°C. **29.** (a) 5.3°C, no ice remaining.
(b) 0°C, 60 g of ice left. **31.** 8.72 g.
33. (b) $C \propto T^{-2/3}$. **35.** A: 120 J; B: 75 J; C: 30 J.
37. (a) -200 J. (b) -293 J. (c) -93 J. **39.** -5.0 J.
41. 33.3 kJ. **43.** (a) 0.36 mg/s. (b) 0.81 J/s.
(c) -0.69 J/s. **45.** (a) 1.2 W/m·K; 0.70 Btu/ft·°F·h.
(b) 0.030 ft²·°F·h/Btu. **47.** 1660 J/s.
51. Arrangement (b). **53.** (a) 2.0 MW. (b) 220 W.
55. 2.0×10^7 J. **57.** 0.40 cm/h. **59.** Cu–Al, 84.3°C;
Al–brass, 57.6°C.

Chapter 21

1. (a) 0.0127. (b) 7.65×10^{21}. **3.** 6560.
5. Number of molecules in the ink $\approx 3 \times 10^{16}$. Number
of people $\approx 5 \times 10^{20}$. Statement is wrong, by a factor of
about 20,000. **7.** (a) 5.47×10^{-8} mol.
(b) 3.29×10^{16}. **9.** (a) 106. (b) 0.892 m³.
11. 27.0 lb/in.². **13.** (a) 2.5×10^{25}. (b) 1.2 kg.
15. 5700 N·m. **17.** 1/5. **19.** 100 cm³. **21.** 198°F.
23. 2.0×10^5 Pa. **25.** 180 m/s. **27.** 9.53×10^6 m/s.
29. 307°C. **31.** 1.9×10^4 dyne/cm².

33. (a) 0.0353 eV; 0.0483 eV. (b) 3400 J; 4650 J.
35. 9.1×10^{-6}. **37.** (a) 6.75×10^{-20} J. (b) 10.7.
39. 0.32 nm. **41.** 15 cm. **43.** (a) 3.26×10^{10}.
(b) 173 m. **45.** (a) 22.5 L. (b) 2.25.
(c) 8.4×10^{-5} cm. (d) Same as (c). **49.** (a) 3.2 cm/s.
(b) 3.4 cm/s. (c) 4.0 cm/s. **51.** (a) $\bar{v}$, v_{rms}, v_P.
(b) v_P, $\bar{v}$, v_{rms}. **53.** (a) 1.0×10^4 K; 1.6×10^5 K.
(b) 440 K; 7000 K. **55.** 4.7. **57.** $2N/3v_0$. (b) $N/3$.
(c) $1.22\, v_0$. (d) $1.31\, v_0$. **59.** $RT \ln(V_f/V_i)$.
61. (a) 15.9 J. (b) 34.4 J/mol·K. (c) 26.1 J/mol·K.
63. $(n_1 C_1 + n_2 C_2 + n_3 C_3)/(n_1 + n_2 + n_3)$.
65. Constant volume. **67.** (a) 0.375 mol. (b) 1090 J.
(c) 0.714. **69.** Diatomic. **71.** (a) 2.5 atm, 340 K.
(b) 0.40 L. **73.** 1500 N·m$^{2.2}$. **75.** 1.40. **79.** (a) $p_0/3$.
(b) Polyatomic (ideal). (c) $K_f/K_i = 1.44$.
81. Q, W and ΔE_{int} are all greatest for (a) and least for (c).
83. (a) In joules, in the order Q, ΔE_{int}, W: $1 \to 2$: 3740,
3740, 0; $2 \to 3$: 0, -1810, 1810; $3 \to 1$: -3220, -1930,
-1290; Cycle: 520, 0, 520. (b) $V_2 = 0.0246$ m^3;
$p_2 = 2.00$ atm; $V_3 = 0.0373$ m^3; $p_3 = 1.00$ atm.

Chapter 22

1. (a) 31%. (b) 16 kcal. **3.** 25%. **5.** (a) 7200 J.
(b) 960 J. (c) 13%. **7.** (a) 49 kcal. (b) 31 kJ.
9. (a) 0.071 J. (b) 0.50 J. (c) 2.0 J. (d) 5.0 J.
11. 99.999947%. **13.** 75. **17.** (a) 94 J. (b) 230 J.
19. 13 J. **21.** (a) 1.11 kcal/s. (b) 0.995 kcal/s.

23. $e = \dfrac{1}{K+1}$. **27.** $[1 - (T_2/T_1)]/[1 - (T_4/T_3)]$.

29. (a) 78%. (b) 81 kg/s. **31.** 0.139.
33. (a) $+0.602$ cal/K. (b) -0.602 cal/K.
35. -0.30 cal/g·K. **37.** 8.79×10^{-3} cal/K.
39. 4450 cal. **41.** (a) 1.95 J/K. (b) 0.650 J/K.
(c) 0.217 J/K. (d) 0.072 J/K. **43.** (a) 57°C.
(b) -5.27 cal/K. (c) $+5.95$ cal/K. (d) $+0.68$/K.
45. $+0.15$ cal/K. **47.** (a) 320 K. (b) Zero.
(c) $+0.41$ cal/K. **49.** (a) $p_1/3$; $p_1/3^{1.4}$; $T_1/3^{0.4}$.
(b) In the order W, Q, ΔE_{int}, ΔS: $1 \to 2$: $1.10RT_1$,
$1.10RT_1$, 0, $1.10R$; $2 \to 3$: 0, $-0.889RT_1$, $-0.889RT_1$,
$-1.10R$; $3 \to 1$: $-0.889RT_1$, 0, $0.889RT_1$, 0.
51. (a) -225 cal/K. (b) $+225$ cal/K. **53.** (a) $3p_0 V$.
(b) $6RT_0$; $\frac{3}{2}R \ln 2$. (c) Both are zero.

Chapter 23

1. (a) 8.99×10^9 N. (b) 8990 N. **3.** 1.39 m.
5. (a) 4.9×10^{-7} kg. (b) 7.1×10^{-11} C. **7.** $\frac{3}{8}$**F**.
9. (a) $q_1 = 9q_2$. (b) $q_1 = -25q_2$. **11.** 1.2×10^{-5} C and
3.8×10^{-5} C. **13.** 14 cm from the positive charge, 24
cm from the negative charge.
15. (a) A charge $-4q/9$ must be located on the line
segment joining the two positive charges, a distance $L/3$
from the $+q$ charge. **17.** (a) $Q = -2\sqrt{2}q$. (b) No.

19. (b) $\pm 2.4 \times 10^{-8}$ C. **21.** (a) $\dfrac{L}{2}\left(1 + \dfrac{1}{4\pi\epsilon_0} \dfrac{qQ}{Wh^2}\right)$.

(b) $\sqrt{\dfrac{3}{4\pi\epsilon_0} \dfrac{qQ}{W}}$. **23.** 3.8 N. **25.** 0.19 MC.

27. (a) 8.99×10^{-19} N. (b) 625. **29.** 11.9 cm.
31. 1.3 days. **33.** 1.3×10^7 C. **35.** 1.7×10^8 N.
37. (a) Positron. (b) Electron. **39.** (a) 510 N.
(b) 7.7×10^{28} m/s^2. **41.** (a) $(Gh/2\pi c^3)^{1/2}$.
(b) 1.61×10^{-35} m.

Chapter 24

1. (a) 6.4×10^{-18} N. (b) 20 N/C.
3. To the right in the figure. **7.** 0.111 nC. **9.** 56 pC.
11. (a) 6.4×10^5 N/C, toward the negative charge.
(b) 1.0×10^{-13} N, toward the positive charge.

13. $\dfrac{1}{4\pi\epsilon_0} \dfrac{3q}{d^2}$, directly toward $-2q$.

15. (a) $1.7a$ to the right of the $+2q$ charge.
17. 50 cm from q_1 and 100 cm from q_2. **19.** 9:30.
21. $E = q/\pi\epsilon_0 a^2$, along bisector, away from triangle.
23. 6.88×10^{-28} C·m. **29.** $R/\sqrt{2}$.

31. $\dfrac{1}{4\pi\epsilon_0} \dfrac{4q}{\pi R^2}$, toward increasing y. **37.** (a) 0.10 μC.

(b) 1.3×10^{17}. (c) 5.0×10^{-6}. **39.** 3.51×10^{15} m/s^2.
41. (a) 4.8×10^{-13} N. (b) 4.8×10^{-13} N.
43. (a) 1.5×10^3 N/C. (b) 2.4×10^{-16} N, up.
(c) 1.6×10^{-26} N. (d) 1.5×10^{10}.
45. (a) 2.46×10^{17} m/s^2. (b) 0.122 ns. (c) 1.83 mm.
47. (a) 7.12 cm. (b) 28.5 ns. (c) 11.2%. **49.** $-5e$.
51. (a) 0.245 N, 11.3° clockwise from the $+x$ axis.
(b) $x = 108$ m; $y = -21.6$ m.
53. (a) $-(2.1 \times 10^{13}$ m/s$^2)$**j**.
(b) $(1.5 \times 10^5$ m/s$)$**i** $- (2.8 \times 10^6$ m/s$)$**j**.

55. (a) $2\pi\sqrt{\dfrac{l}{|g - qE/m|}}$. (b) $2\pi\sqrt{\dfrac{l}{g + qE/m}}$.

57. (a) 9.30×10^{-15} C·m. (b) 2.05×10^{-11} J.
59. $2pE \cos\theta_0$.

Chapter 25

1. (a) 693 kg/s. (b) 693 kg/s. (c) 347 kg/s.
(d) 347 kg/s. (e) 575 kg/s. **3.** (a) Zero.
(b) -3.92 N·m^2/C. (c) Zero. (d) Zero for each field.
5. (a) Enclose $2q$ and $-2q$, or enclose all four charges.
(b) Enclose $2q$ and q. (c) Not possible.
7. 2.0×10^5 N·m^2/C. **9.** $q/6\epsilon_0$. **11.** 3.54 μC.
13. Through each of the three faces meeting at q: zero;
through each of the three other faces: $q/24\epsilon_0$.
15. 2.0 μC/m^2. **17.** (a) 4.5×10^{-7} C/m^2.
(b) 5.1×10^4 N/C. **19.** (a) -3.0×10^{-6} C.
(b) $+1.3 \times 10^{-5}$ C. **21.** 5.0 μC/m. **23.** $E = \lambda/2\pi\epsilon_0 r$.
(b) Zero. **25.** 3.8×10^{-8} C/m^2.
27. (a) $E = q/2\pi\epsilon_0 Lr$, radially inward.

(b) $-q$ on both inner and outer surfaces.
(c) $E = q/2\pi\epsilon_0 Lr$, radially outward.   **29.** 270 eV.
31. (a) $E = \sigma/\epsilon_0$, to the left.   (b) $E = 0$.

(c) $E = \sigma/\epsilon_0$, to the right.   **33.** $E = \dfrac{\sigma}{2\epsilon_0\sqrt{z^2 + R^2}}$.

35. 0.44 mm.   **37.** $\pm 4.9 \times 10^{-10}$ C.   **39.** (a) $\rho x/\epsilon_0$.
(b) $\rho d/2\epsilon_0$.   **41.** (a) -750 N·m²/C.   (b) -6.64 nC.
43. 2.50×10^4 N/C.   (b) 1.35×10^4 N/C.
47. (a) $E = q/4\pi\epsilon_0 r^2$, radially outward.   (b) Same as (a).
(c) No.   (d) Yes, charges are induced on the surfaces.
(e) Yes.   (f) No.   (g) No.   **51.** -1.04 nC.   **53.** $q/2\pi a^2$.

Chapter 26
1. 1.2 GeV.   **3.** (a) 3.0×10^{10} J.   (b) 7.7 km/s.
(c) 9.0×10^4 kg.   **5.** (a) -2.46 V.   (b) -2.46 V.
(c) Zero.   **7.** 2.90 kV.   **9.** 8.8 mm.
11. (a) $-qr^2/(8\pi\epsilon_0 R^3)$.   (b) $q/(8\pi\epsilon_0 R)$.   (c) Center.
13. (b) Because $V = 0$ point is chosen differently.
(c) $q/(8\pi\epsilon_0 R)$.   (d) Potential differences are
independent of the choice for the $V = 0$ point.
15. (a) -4500 V.   (b) -4500 V.   **17.** 843 V.
19. 2.8×10^5.   **23.** (a) 3.3 nC.   (b) 12 nC/m².
25. 200 mV.   **27.** (a) 38 s.   (b) 280 days.   **29.** None.
31. (a) None.

(b) 41 cm from $+q$, between the charges.   **35.** $\dfrac{-8}{4\pi\epsilon_0}\dfrac{e}{d}$.

37. $\dfrac{-1}{4\pi\epsilon_0}\dfrac{Q}{R}$.   **39.** (a) $\dfrac{-5}{4\pi\epsilon_0}\dfrac{Q}{R}$.   (b) $\dfrac{-5}{4\pi\epsilon_0}\dfrac{Q}{(z^2 + R^2)^{1/2}}$.

41. $\dfrac{-Q/L}{4\pi\epsilon_0}\ln\left(\dfrac{L}{d} + 1\right)$.

43. In V/m, ab: -6.0; bc: zero; ce: 3.0; ef: 15; fg: zero; gh:
-3.0.   **49.** (a) $\dfrac{\lambda}{4\pi\epsilon_0}\ln\left(\dfrac{L + x}{x}\right)$.   (b) $\dfrac{\lambda}{4\pi\epsilon_0}\dfrac{L}{x(L + x)}$.

(c) Zero.   **51.** (a) $qd/2\pi\epsilon_0 a(a + d)$.   **53.** -1.9 J.
55. (a) 0.484 MeV.   (b) Zero.   **57.** -1.2×10^{-6} J.
59. (a) $+6.0 \times 10^4$ V.   (b) -7.8×10^5 V.   (c) 2.5 J.
(d) Increase.   (e) Same.   (f) Same.

61. $W = \dfrac{qQ}{8\pi\epsilon_0}\left(\dfrac{1}{r_1} - \dfrac{1}{r_2}\right)$.   **63.** 1.8×10^{-10} J.

65. (a) 0.225 J.   (b) 22.5 m/s².
(c) A: 7.75 m/s; B: 3.87 m/s.   **67.** (a) 25 fm.
(b) Twice as much.   **69.** $\sqrt{2eV/m}$.
71. 23 km/s.   **73.** 400 V.   **75.** 2.5×10^{-8} C.
79. (a) -180 V.   (b) 3000 V; -9000 V.

81. $r < R_1$: $E = 0$; $V = \dfrac{1}{4\pi\epsilon_0}\left(\dfrac{q_1}{R_1} + \dfrac{q_2}{R_2}\right)$. $R_1 < r < R_2$:

$E = \dfrac{1}{4\pi\epsilon_0}\dfrac{q_1}{r^2}$; $V = \dfrac{1}{4\pi\epsilon_0}\left(\dfrac{q_1}{r} + \dfrac{q_2}{R_2}\right)$. $r > R_2$:

$E = \dfrac{1}{4\pi\epsilon_0}\left(\dfrac{q_1 + q_2}{r^2}\right)$; $V = \dfrac{1}{4\pi\epsilon_0}\left(\dfrac{q_1 + q_2}{r}\right)$.   **83.** 9.52 kW.

Chapter 27
1. 7.5 pC.   **3.** 3.0 mC.   **5.** (a) 140 pF.   (b) 17 nC.
7. 0.55 pF.   **9.** 4×10^{-7} C.   **11.** $5.05\pi\epsilon_0 R$.   **15.** 9090.
17. 7.33 μF.   **19.** (a) 2.40 μF.   (b) 0.480 mC on both.
(c) $V_2 = 120$ V; $V_1 = 80$ V.   **21.** (a) $d/3$.   (b) $3d$.
25. (a) Five in a series.   (b) Three arrays as in (a) in
parallel. There are other possibilities.   **27.** 43 pF.

29. $q_1 = \dfrac{C_1 C_2 + C_1 C_3}{C_1 C_2 + C_1 C_3 + C_2 C_3} C_1 V_0$;

$q_2 = q_3 = \dfrac{C_2 C_3}{C_1 C_2 + C_1 C_3 + C_2 C_3} C_1 V_0$.

31. First case: 50.0 V; second case: zero.
33. (a) 3.05 MJ.   (b) 0.847 kW·h.   **35.** (a) 0.204 μJ.
(b) No.   **37.** 0.27 J.   **39.** 4.9%.   **41.** 10.4 ¢.
43. (a) 2.0 J.   **45.** (a) $q_1 = q_2 = 0.33$ mC, $q_3 = 0.40$ mC.
(b) $V_1 = 33$ V; $V_2 = 67$ V; $V_3 = 100$ V.   (c) $U_1 = 5.4$ mJ;
$U_2 = 11$ mJ; $U_3 = 20$ mJ.   **53.** Pyrex.   **55.** (a) 6.2 cm.
(b) 280 pF.   **57.** 0.63 m².   **59.** (a) 2.85 m³.
(b) 1.01×10^4.   **61.** (a) $\epsilon_0 A/(d - b)$.   (b) $d/(d - b)$.

(c) $-q^2 b/2\epsilon_0 A$; sucked in.   **65.** $\dfrac{\epsilon_0 A}{4d}\left(\kappa_1 + \dfrac{2\kappa_2\kappa_3}{\kappa_2 + \kappa_3}\right)$.

67. (a) 13.4 pF.   (b) 1.15 nC.   (c) 1.13×10^4 N/C.
(d) 4.33×10^3 N/C.   **69.** (a) 89 pF.   (b) 120 pF.
(c) 11 nC; 11 nC.   (d) 10 kV/m.   (e) 2.1 kV/m.
(f) 88 V.   (g) 0.17 μJ.

Chapter 28
1. (a) 1200 C.   (b) 7.5×10^{21}.   **3.** 5.6 ms.
5. (a) 6.4 A/m², north.   **7.** 0.38 mm.
9. 0.67 A, toward the negative terminal.
11. (a) 0.654 μA/m².   (b) 83.4 MA.   **13.** 13 min.
15. (a) $J_0 A/3$.   (b) $2J_0 A/3$.   **17.** 2.0×10^{-8} Ω·m.
19. 2.4 Ω.   **21.** 2.0×10^6 $(\Omega \cdot$m$)^{-1}$.   **23.** 57°C.
25. (a) 0.38 mV.   (b) Negative.   (c) 3 min 58 s.
27. 54 Ω.   **29.** 2.9 mm.   **31.** (a) 2.39, iron being larger.
(b) No.   **33.** (a) Silver.   (b) 51.6 nΩ.   **35.** 2000 K.
37. (a) 38.3 mA.   (b) 109 A/m².   (c) 1.28 cm/s.
(d) 227 V/m.   **39.** (a) 1.73 cm/s.   (b) 3.24 pA/m².
41. (a) 0.43%; 0.0017%; 0.0034%.   **45.** 560 W.
47. 0.20 hp.   **49.** 0.135 W.   **51.** (a) 4.9 MA/m².
(b) 83 mV/m.   (c) 25 V.   (d) 640 W.
53. New length $= 1.369L$; new area $= 0.730A$.
55. (a) 5.85 m.   (b) 10.4 m.
57. (a) \$4.46 for a 31-day month.   (b) 144 Ω.
(c) 0.833 A.   **59.** (a) 9.4×10^{13} s^{-1}.   (b) 240 W.
61. 710 cal/g.   **63.** (a) 8.6%.   (b) Smaller.

Chapter 29
1. (a) 1.9×10^{-18} J ($= 12$ eV).   (b) 6.5 W.
3. (a) \$320.   (b) 9.6¢.   **5.** (a) Counterclockwise.
(b) Battery 1.   (c) B.

7. (c) Third plot gives rate of energy dissipation by R.
9. (a) 14 V.   (b) 100 W.   (c) 600 W.   (d) 10 V; 100 W.
11. (a) 50 V.   (b) 48 V.   (c) B is the negative terminal.
13. 2.5 V.   **15.** (a) 990 Ω.   (b) 9.4×10^{-4} W.
17. 8.0 Ω.   **19.** The cable.   **21.** (a) 1000 Ω.
(b) 300 mV.   (c) 2.3×10^{-3}.
23. (a) 1.32×10^7 A/m² in each.
(b) $V_A = 8.90$ V; $V_B = 51.1$ V.   (c) A: copper; B: iron.
25. Silicon: 85.0 Ω; iron: 915 Ω.   **27.** 4.00 Ω, in parallel.
29. $i_1 = 50$ mA; $i_2 = 60$ mA; $V_{ab} = 9.0$ V.
31. (a) 6.67 Ω.   (b) 6.67 Ω.   (c) Zero.   **33.** (a) R_2.
(b) R_1.   **35.** 3d.   **37.** 7.5 V.   **39.** Nine.
41. (a) $2\mathcal{E}/(2r + R)$, series; $2\mathcal{E}/(r + 2R)$, parallel.
(b) Series.   (c) Parallel.
43. (a) Left branch, 0.67 A, down; center branch, 0.33 A, up; right branch, 0.33 A, up.   (b) 3.3 V.   **45.** (a) 120 Ω.
(b) $i_1 = 50$ mA; $i_2 = i_3 = 20$ mA; $i_4 = 10$ mA.
47. (a) 19.5 Ω.   (b) 0.   (c) ∞.   (d) 82.3 W, 57.6 W.
49. (a) 2.50 Ω.   (b) 3.13 Ω.
51. $100R(\mathcal{E}x/R_0)^2/(100R/R_0 + 10x - x^2)^2$, x in cm.
53. (a) 13.5 kΩ.   (b) 1500 Ω.   (c) 167 Ω.   (d) 1480 Ω.
55. 0.45 A.   **57.** (a) 12.5 V.   (b) 50 A.   **59.** 0.9%.
65. (a) 2.52 s.   (b) 21.6 μC.   (c) 3.40 s.   **67.** (a) 0.41τ.
(b) 1.1τ.   **69.** (a) 2.17 s.   (b) 39.6 mV.   **71.** (a) 10^{-3} C.
(b) 10^{-3} A.   (c) $V_C = 10^3 e^{-t}$, $V_R = -10^3 e^{-t}$, volts.
73. 0.72 MΩ.   **77.** Decreases by 13 μC.

Chapter 30

1. M/QT.   **3.** (a) 9.56×10^{-14} N; zero.   (b) 0.267°.
5. (a) $(6.2 \times 10^{-14}$ N$)\mathbf{k}$.   (b) $-(6.2 \times 10^{-14}$ N$)\mathbf{k}$.
7. (a) East.   (b) 6.28×10^{14} m/s².   (c) 2.98 mm.   **9.** 2.
11. (a) 3.75 km/s.   **13.** 680 kV/m.
17. (b) 2.84×10^{-3}.   **19.** 21 μT.   **21.** 1.6×10^{-8} T.
23. (a) 1.11×10^7 m/s.   (b) 0.316 mm.
25. (a) 2.60×10^6 m/s.   (b) 0.109 μs.   (c) 0.140 MeV.
(d) 70 kV.   **29.** (a) $K_p = K_d = \frac{1}{2}K_\alpha$.
(b) $R_d = R_\alpha = 14$ cm.   **33.** (a) 495 mT.   (b) 22.7 mA.
(c) 8.17 MJ.   **35.** (a) 0.36 ns.   (b) 0.17 mm.
(c) 1.5 mm.   **37.** (a) 2.9998×10^8 m/s.   **39.** (a) 22 cm.
(b) 21 MHz.   **41.** Neutron moves tangent to original path; proton moves in a circular orbit of radius 25 cm.
43. Case (b).   **45.** 20.1 N.
47. $-(2.5 \times 10^{-3}$ N$)\mathbf{j} + (0.75 \times 10^{-3}$ N$)\mathbf{k}$.
51. (a) 3.3×10^8 A.   (b) 1.0×10^{17} W.   (c) Totally unrealistic.   **53.** (a) 0; 1.38 mN; 1.38 mN.
55. (a) 20 min.   (b) 5.9×10^{-2} N·m.
59. $2\pi aiB \sin \theta$, normal to the plane of the loop (up).
61. 2.45 A.   **63.** 2.08 GA.   **65.** (a) 0.30 J/T.
(b) 0.024 N·m.
67. (a) $(8.0 \times 10^{-4}$ N·m$)(-1.2\mathbf{i} - 0.90\mathbf{j} + 1.0\mathbf{k})$.
(b) -6.0×10^{-4} J.

Chapter 31

1. 7.7 mT.   **3.** (a) 3.3×10^{-6} T.   (b) Yes.
5. (a) $(0.24$ nT$)\mathbf{i}$.   (b) Zero.   (c) $-(43$ pT$)\mathbf{k}$.
(d) $(0.14$ nT$)\mathbf{k}$.   **7.** (a) 16 A.   (b) West to east.
9. (a) 3.2×10^{-16} N, parallel to the current.
(b) 3.2×10^{-16} N, radially outward if $\mathbf{v}$ is parallel to the current.   (c) Zero.   **11.** (a) Zero.
(b) $\mu_0 i/4R$, into the page.   (c) Same as (b).
13. $\dfrac{\mu_0 i}{4} \left(\dfrac{1}{R_1} - \dfrac{1}{R_2} \right)$, into the page.   **15.** 2 rad.
25. 200 μT, into the page.   **27.** (a) It is impossible to have other than $B = 0$ midway between them.
(b) 30 A.   **29.** At all points between the wires, on a line parallel to them, at a distance $d/4$ from the wire carrying current i.
35. $0.338\mu_0 i^2/a$, toward the center of the square.
37. (b) To the right.   **39.** (b) 2.3 km/s.   **41.** $+5\mu_0 i_0$.
43. 4.5×10^{-6} T·m.   **47.** (a) $\mu_0 ir/2\pi c^2$.   (b) $\mu_0 i/2\pi r$.
(c) $\dfrac{\mu_0 i}{2\pi(a^2 - b^2)} \left(\dfrac{a^2 - r^2}{r} \right)$.   (d) Zero.
49. $3i_0/8$, into the page.   **53.** 5.71 mT.   **55.** 108 m.
61. 0.272 A.   **63.** 0.47 A·m².   **65.** $8\mu_0 Ni/5\sqrt{5}R$.
67. (b) ia^2.   **71.** (a) 79 μT.   (b) 1.1×10^{-6} N·m.
73. (a) $(\mu_0 i/2R)(1 + 1/\pi)$, out of page.
(b) $(\mu_0 i/2\pi R)\sqrt{1 + \pi^2}$, 18° out of page.

Chapter 32

1. 57 μWb.   **3.** 1.5 mV.   **5.** (a) 31 mV.
(b) Right to left.   **7.** $A^2 B^2/R\,\Delta t$.   **9.** (b) 58 mA.
11. 1.2 mV.   **13.** 1.15 μWb.
15. 51 mV; clockwise when viewed along the direction of $\mathbf{B}$.
17. (b) No.   **19.** (a) 21.7 V.   (b) Counterclockwise.
21. (a) 13 μWb/m.   (b) 17%.   (c) Zero flux.
23. (a) 48.1 mV.   (b) 2.67 mA.
25. $BiLt/m$, away from generator.   **27.** (a) 85.2 T·m².
(b) 56.8 V.   (c) Linearly.
29. (b) Design it so that $Nab = (5/2\pi)$ m².
31. 268 W.   **33.** 15.5 μC.   **35.** (a) 0.598 μV.
(b) Counterclockwise.   **37.** (a) $\dfrac{\mu_0 ia}{2\pi} \ln \left(\dfrac{2r + b}{2r - b} \right)$.
(b) $2\mu_0 iabv/\pi R(4r^2 - b^2)$.
39. (a) $3.4(2 + \theta)$ mΩ, θ in rad.   (b) 4.3θ mWb, θ in rad.
(c) 2.0 rad.   (d) 2.2 A.
41. 1: -1.07 mV; 2: -2.40 mV; 3: 1.33 mV.
43. At a: 4.4×10^7 m/s², to the right. At b: zero. At c: 4.4×10^7 m/s², to the left.   **45.** (a) 1st, 2nd, 5th, 6th.
(b) 1st, 4th, 5th, 8th.   (c) 1st, 5th.

Chapter 33

1. 0.1 μWb.   **3.** (a) 800.   (b) 2.5×10^{-4} H.
5. (b) So that the changing magnetic field of one does

not induce current in the other. (c) $L_{eq} = \sum_{j=1}^{N} L_j$.

7. (a) $\mu_0 i / W$. (b) $\pi \mu_0 R^2 / W$. **9.** (a) Decreasing.
(b) 0.68 mH. **11.** (a) 0.10 H. (b) 1.3 V.
13. (a) 16 kV. (b) 3.1 kV. (c) 23 kV. **15.** 6.91 τ_L.
17. 1.54 s. **19.** (a) 8.45 ns. (b) 7.37 mA.
21. $(42 + 20t)$ V. **23.** 12.0 A/s.
25. (a) $i_1 = i_2 = 3.33$ A. (b) $i_1 = 4.55$ A; $i_2 = 2.73$ A.
(c) $i_1 = 0$; $i_2 = 1.82$ A. (d) $i_1 = i_2 = 0$. **27.** (a) 1.5 s.
29. (a) 13.9 H. (b) 120 mA. **31.** (a) 10 A. (b) 100 J.
33. 25.6 ms. **35.** (a) 18.7 J. (b) 5.10 J. (c) 13.6 J.
39. 5.58 A. **41.** 3×10^{36} J. **43.** (a) 1.3 mT.
(b) 0.63 J/m³. **45.** (a) 1.0 J/m³.
(b) 4.8×10^{-15} J/m³. **47.** (a) 1.67 mH. (b) 6.00 mWb.
49. (b) Have the turns of one solenoid wrapped in
direction opposite to those in other solenoid.
51. Magnetic field exists only within cross section of
solenoid.
53. (a) $\dfrac{\mu_0 N l}{2\pi} \ln\left(1 + \dfrac{b}{a}\right)$. (b) 13 μH.

Chapter 34

5. (b) In the direction of the angular momentum vector.
7. $+3$ Wb. **9.** $(\mu_0 iL / \pi)\ln 3$. **11.** 13 MWb, outward.
15. 1660 km. **17.** 61.1 μT; 84.2°. **19.** 20.8 mJ/T.
21. Yes. **23.** (a) 3.7 K. (b) 1.3 K.
27. $\Delta\mu = e^2 r^2 B / 4m$. **29.** (a) 3.0 μT.
(b) 5.6×10^{-10} eV. **31.** (a) 8.9 A·m². (b) 13 N·m.

Chapter 35

1. 9.14 nF. **3.** 45.2 mA. **5.** (a) 6.00 μs. (b) 167 kHz.
(c) 3.00 μs. **7.** (a) 89 rad/s. (b) 70 ms. (c) 25 μF.
9. 38 μH. **11.** 7.0×10^{-4} s. **15.** (a) 3.0 nC.
(b) 1.7 mA. (c) 4.5 nJ. **17.** (a) 3.60 mH.
(b) 1.33 kHz. (c) 0.188 ms.
19. 600, 710, 1100, 1300 Hz. **21.** (a) $Q/\sqrt{3}$. (b) 0.152.
25. (a) 1.98 μJ. (b) 5.56 μC. (c) 12.6 mA.
(d) $-46.9°$. (c) $+46.9°$. **27.** (a) 356 μs.
(b) 2.50 mH. (c) 3.20 mJ. **29.** (a) Zero. (b) $2i(t)$.
31. 8.66 mΩ. **33.** $(L/R)\ln 2$. **35.** (b) 2.10×10^{-3}.
37. 1.84 kHz.
39. 1.13 kHz; 1.45 kHz; 1.78 kHz; 2.30 kHz.

Chapter 36

1. 377 rad/s. **3.** (a) 955 mA. (b) 119 mA.
5. (a) 4.60 kHz. (b) 26.6 nF.
(c) $X_L = 2.60$ kΩ; $X_C = 0.650$ kΩ. **7.** (a) 0.65 kHz.
(b) 24 Ω. **9.** (a) 39.1 mA. (b) Zero. (c) 33.8 mA.
(d) Supplying energy. **11.** (a) 6.73 ms. (b) 2.24 ms.
(c) Capacitor. (d) 59.0 μF.
13. (a) $X_C = 0$; $X_L = 86.7$ Ω; $Z = 182$ Ω; $I = 198$ mA;
$\phi = 28.5°$.

15. (a) $X_C = 37.9$ Ω; $X_L = 86.7$ Ω; $Z = 167$ Ω;
$I = 216$ mA; $\phi = 17.1°$. **19.** 89 Ω.
21. (a) 224 rad/s. (b) 6.00 A. (c) 228 rad/s; 219 rad/s.
(d) 0.039. **23.** (a) 45.0°. (b) 70.7 Ω. **29.** 141 V.
31. Zero; 9.00 W; 3.14 W; 1.82 W. **33.** 177 Ω.
35. 7.61 A. **41.** (a) 117 μF. (b) Zero.
(c) 90.0 W; zero. (d) 0°; 90°. (e) 1; 0. **43.** (a) 2.59 A.
(b) 38.8 V; 159 V, 224 V, 64.2 V; 75.0 V.
(c) 100 W for R; zero for L and C. **45.** (a) 2.4 V.
(b) 3.2 mA; 0.16 A. **47.** (a) 1.9 V; 5.8 W.
(b) 19 V; 0.58 kW. (c) 0.19 kV; 58 kW.

Chapter 37

3. At $r = 27.5$ mm and $r = 110$ mm.
7. Change the potential difference between the plates at
the rate of 1.0 MV/s.
11. (a) 0.63 μT. (b) 2.3×10^{12} V/m·s. **13.** (a) 2.0 A.
(b) 2.3×10^{11} V/m·s. (c) 0.50 A. (d) 0.63 μT·m.
15. (a) 7.60 μA. (b) 859 kV·m/s. (c) 3.39 mm.
(d) 5.16 pT.

Chapter 38

1. (a) 4.7×10^{-3} Hz. (b) 3 min 32 s.
3. (a) 4.5×10^{24} Hz.
(b) 1.0×10^4 km, or 1.6 Earth radii.
7. (a) It would steadily increase. (b) The summed
discrepancies between the apparent times of eclipse and
those observed from x; the radius of the Earth's orbit.
9. 5.0×10^{-21} H. **11.** 1.07 pT.
17. 4.8×10^{-29} W/m². **19.** 4.51×10^{-10}. **21.** 89 cm.
23. 1.2 MW/m². **25.** 820 m. **27.** 1.03 kV/m; 3.43 μT.
29. (a) $\pm EBa^2/\mu_0$ for faces parallel to the xy plane; zero
through each of the other four faces. (b) Zero.
31. (a) 83 W/m². (b) 1.7 MW. **33.** (a) 3.5 μW/m².
(b) 0.78 μW. (c) 1.5×10^{-17} W/m². (d) 110 nV/m.
(e) 0.25 fT. **35.** 1.0×10^7 N/m². **37.** (a) 6.0×10^8 N.
(b) $F_g = 3.6 \times 10^{22}$ N. **39.** (a) 100 MHz.
(b) 1.0 μT along the z axis.
(c) 2.1 m⁻¹; 6.3×10^8 rad/s. (d) 120 W/m².
(e) 8.0×10^{-7} N; 4.0×10^{-7} N/m². **41.** 491 nm.
45. 1.9 mm/s. **47.** (b) 580 nm. **49.** (a) 1.9 V/m.
(b) 1.7×10^{-11} N/m². **51.** $\frac{1}{8}$. **53.** 47%.
55. 20° or 70°. **57.** 19 W/m². **59.** (a) Two sheets.
(b) Five sheets.

Chapter 39

1. (a) 38.0°. (b) 52.9°. **3.** 1.26. **5.** 1.07 m.
9. (a) 1.8 m. (b) 1.2 m.
11. 351 cm beneath the mirror surface.
13. (b) $(6.05 \times 10^{-3})°$. **15.** 1.41. **17.** 1.22. **19.** 41.2°.
21. (a) Cover the center of each face with an opaque
disk of radius 4.5 mm. (b) About 0.63.

23. (a) $\sqrt{1 + \sin^2 \theta}$. (b) $\sqrt{2}$. **25.** (b) 0.170.
27. (b) 51.6 ns. **29.** 49.0°. **31.** (a) 1.60. (b) 58.0°.
33. 180°. **35.** 9.10 m. **37.** a and c. **39.** (a) 7. (b) 5.
(c) 1 to 3, depending on position of O and your
perspective. **43.** New illumination is 10/9 of the old.
51. (a) 2.00. (b) None.
53. $i = \dfrac{(2 - n)r}{2(n - 1)}$, to the right of the right side of the
sphere. **55.** (b) Separate the lenses by a distance $f_2 - f_1$,
where f_2 is the focal length of the converging lens.
57. 45 mm; 90 mm. **59.** (a) $+40$ cm. (b) At infinity.
63. (a) 40 cm, real. (b) 80 cm, real. (c) 240 cm, real.
(d) -40 cm, virtual. (e) -80 cm, virtual.
(f) -240 cm, virtual. **65.** An X means that the quantity
cannot be found from the given data:
(a) $+$, X, X, $+20$, X, -1.0, yes, no.
(b) Converging, X, X, -10, X, $+2.0$, no, yes.
(c) Converging, $+$, X, X, -10, X, no, yes.
(d) Diverging, $-$, X, X, -3.3, X, no, yes.
(e) Converging, $+30$, -15, $+1.5$, no, yes.
(f) Diverging, -30, -7.5, $+0.75$, no, yes.
(g) Diverging, -120, -9.2, $+0.92$, no, yes.
(h) Diverging, -10, X, X, -5, X, $+$, no.
(i) Converging, $+3.3$, X, X, $+5$, X, no.
67. (a) 0.60 m on the side of the lens away from the
mirror. (b) Real. (c) Upright. (d) $+0.20$.
69. (a) Converging. (b) 26.7 cm. (c) 8.89 cm.
73. 22 cm. **77.** 2.1 mm. **79.** (a) 2.35 cm.
(b) Decrease. **81.** (a) 5.3 cm. (b) 3.0 mm. **83.** -75.

Chapter 40

1. (a) 5.09×10^{14} Hz. (b) 388 nm.
(c) 1.97×10^8 m/s. **5.** 2.1×10^8 m/s. **7.** The time is
longer for the pipeline containing air, by about 1.55 ns.
9. 22°. Refraction reduces θ. **11.** (a) 3.60 μm.
(b) Intermediate, close to fully constructive
interference. **13.** (a) 1.55 μm. (b) 4.65 μm.
15. (a) 0.216 rad. (b) 12.4°.
17. The distance D must also be doubled. **19.** 33 μm.
21. (a) 0.010 rad. (b) 5.0 mm. **23.** 0.15°.
25. 23.1 Hz. **27.** 8.75λ. **31.** 24 μm. **33.** 8.0 μm.
37. Zero. **39.** 2.65. **41.** $y = 17 \sin(\omega t + 13°)$.
43. $I = \frac{1}{9}I_m[1 + 8 \cos^2(\pi d \sin \theta/\lambda)]$, I_m = intensity of
central maximum.
45. $L = (m + \frac{1}{2})\lambda/2$, for $m = 0, 1, 2, \ldots$.
47. Fully constructive interference. **49.** 131 nm.
51. 492 nm. **53.** 120 nm.
55. (a) and (c). **57.** 673 nm. **59.** (a) 169 nm.
(c) Blue-violet will be sharply reduced. **61.** 840 nm.
63. (a) $\lambda/2n_2$. (b) $\lambda/4n_2$. (c) $\lambda/2n_2$. **65.** 141.
67. (a) 1800 nm. (b) 8. **69.** 1.89 μm. **71.** 1.00025.

73. (a) 34. (b) 46. **77.** (a) π rad. (b) Dark.
(c) $2h \sin \theta = m\lambda$ (minima); $2h \sin \theta = (m + \frac{1}{2})\lambda$
(maxima). **79.** 5.2 μm. **81.** $I = I_m \cos^2(2\pi x/\lambda)$.

Chapter 41

1. 690 nm. **3.** 60.4 μm. **5.** (a) 2.5 mm.
(b) 2.2×10^{-4} rad. **7.** (a) 70 cm. (b) 1.0 mm.
9. 41.2 m from perpendicular to speaker. **11.** 160°.
15. (d) 53°; 10°; 5.1°. **19.** (a) 1.3×10^{-4} rad.
(b) 21 m. **21.** 30 m. **23.** (a) 1.1×10^4 km.
(b) 11 km. **25.** 53 m. **27.** (a) 19 cm. (b) Larger.
29. 4.7 cm. **31.** (a) 0.347°. (b) 0.97°. **33.** (a) Red.
(b) 100 μm. **35.** Five. **39.** $\lambda D/d$. **41.** (a) 5.05 μm.
(b) 20.2 μm. **43.** (a) 3330 nm.
(b) 0; $\pm 10.2°$; $\pm 20.7°$; $\pm 32.0°$; $\pm 45.0°$; $\pm 62.2°$.
45. All wavelengths shorter than 635 nm. **47.** 13,600.
49. (a) 6.0 μm. (b) 1.5 μm.
(c) $m = 0, 1, 2, 3, 5, 6, 7, 9$. **51.** 1100. **61.** (a) 56 pm.
(b) None. **63.** (a) 23,100. (b) 28.7°.
67. (a) 2400 nm. (b) 800 nm. (c) $m = 0, 1, 2$.
71. 2.9°. **73.** 26 pm; 39 pm. **75.** 39.8 pm.
77. Yes; $m = 3$ for $\lambda = 130$ pm; $m = 4$ for $\lambda = 97.2$ pm.
79. (a) $a_0/\sqrt{2}$; $a_0/\sqrt{5}$; $a_0/\sqrt{10}$; $a_0/\sqrt{13}$; $a_0/\sqrt{17}$.

Chapter 42

1. (a) 3×10^{-18}. (b) 2×10^{-12}. (c) 8.2×10^{-8}.
(d) 6.4×10^{-6}. (e) 1.1×10^{-6}. (f) 3.7×10^{-5}.
(g) 9.9×10^{-5}. (h) 0.10. **3.** $0.75c$. **5.** $0.99c$.
7. 55 m. **9.** 1.32 m. **11.** 1.53 cm. **13.** (a) 87.4 m.
(b) 394 ns. **15.** (a) 2.21×10^{-12}. (b) 5.25 d.
17. $x' = 138$ km; $t' = -374$ μs.
19. $t'_1 = 0$; $t'_2 = -2.5$ μs.
23. (a) S' must move toward S, along their common axis,
at a speed of $0.480c$.
(b) The "red" flash (suitably Doppler shifted).
(c) 4.39 μs. **25.** $0.81c$. **27.** $0.95c$.
29. $0.588c$, recession. **31.** (a) 34,000 mi/h.
(b) 6.4×10^{-10}. **33.** 22.9 MHz. **35.** $+2.97$ nm.
37. (a) $\tau_0/\sqrt{(1 - v^2/c^2)}$. **39.** (a) $0.134c$. (b) 4.65 keV.
(c) 1.1%. **41.** (a) 0.9988; 20.6. (b) 0.145; 1.01.
(c) 0.073; 1.0027.
43. (a) 5.71 GeV; 6.65 GeV; 6.58 GeV/c.
(b) 3.11 MeV; 3.62 MeV; 3.59 MeV/c. **45.** 18 smu/y.
47. (a) $0.943c$. (b) $0.866c$. **49.** (a) 1.41. (b) $0.707c$.
(c) $0.414\ mc^2$. **51.** (a) The photon. (b) The proton.
(c) The proton. (d) The photon.
53. (c) $207m_e$; the particle is a muon.
55. (a) $0.948c$. (b) 226 MeV. (c) 314 MeV/c.
57. (a) 0.776 mm. (b) 16.0 mm. (c) 0.335 ns; no.
59. 660 km. **61.** (a) 534. (b) 0.99999825. (c) 2.23 T.

Chapter 43

3. 2.1 μm; infrared. **5.** (a) 35.4 keV.
(b) 8.57×10^{18} Hz.
(c) 35.4 keV/$c = 1.89 \times 10^{-23}$ kg·m/s.
7. (a) 1.24×10^{20} Hz. (b) 2.43 pm.
(c) 2.73×10^{-22} kg·m/s $= 0.511$ MeV/c.
9. (a) The infrared bulb. (b) 6.0×10^{20}.
11. 4.7×10^{26}. **13.** (a) 2.96×10^{20} photons/s.
(b) 48,600 km. (c) 280 m.
(d) 5.89×10^{18} m^{-2}·s^{-1}; 1.96×10^{10} m^{-3}.
15. 233 nm. **17.** 10 eV. **19.** 676 km/s. **21.** (a) 1.3 V.
(b) 680 km/s. **23.** (a) 382 nm. (b) 1.82 eV.
27. (a) 3.1 keV. (b) 14 keV. **29.** (a) 2.7 pm.
(b) 6.05 pm. **31.** (a) $+4.8$ pm. (b) -41 keV.
(c) 41 keV. **33.** (a) 8.1×10^{-9}%. (b) 4.9×10^{-4}%.
(c) 8.8%. (d) 66%. **35.** 2.65 fm. **38.** 44°.
43. 9.99 μm. **45.** 91 K. **47.** (a) 0.97 mm; microwave.
(b) 9.9 μm; infrared. (c) 1.6 μm; infrared.
(d) 0.26 nm; x ray.
(e) 2.9×10^{-41} m; very high energy gamma ray.
53. $4\Delta T/T$; 0.0130. **55.** 2.6 eV. **57.** -80.7 keV.
59. (a) 121.5 nm. (b) 91.2 nm.
61. 656 nm; 486 nm; 434 nm. **63.** (a) 12.7 eV.
(b) 12.7 eV ($4 \to 1$); 2.55 eV ($4 \to 2$); 0.66 eV ($4 \to 3$);
12.1 eV ($3 \to 1$); 1.89 eV ($3 \to 2$); 10.2 eV ($2 \to 1$).
65. (a) 30.5 nm; 291 nm; 1050 nm.
(b) 8.25×10^{14} Hz; 3.65×10^{14} Hz; 2.06×10^{14} Hz.
69. 4.1 m/s. **75.** (a) 3×10^{74}. (b) No. **77.** (b) n^2.
(c) n. (d) $1/n$. (e) $1/n^3$. (f) $1/n$. (g) $1/n^4$.
(h)$1/n^4$. (i) $1/n^2$. (j) $1/n^2$. (k) $1/n^2$.

Chapter 44

1. (a) 1.7×10^{-35} m. **3.** 7.75 pm.
5. (a) 3.3×10^{-24} kg·m/s for each.
(b) 38 eV for the electron; 6.2 keV for the photon.
7. (a) 38.8 meV. (b) 146 pm. **9.** (a) 73 pm; 3.4 nm.
(b) Yes. **11.** (a) 1.24 keV; 1.50 eV.
(b) 1.24 GeV; 1.24 GeV. **13.** 0.025 fm. **15.** A neutron.
17. 9.70 kV (relativistic calculation); 9.79 kV (classical
calculation). **19.** 11.5°, 23.6°, 36.9°, 53.1°.
21. (a) 20.5 meV. (b) 37.7 eV. **23.** (a) 1900 MeV.
(b) No. **25.** 90.5 eV.
27. 18.1, 36.2, 54.3, 66.3, 72.4 μeV. **29.** (a) 0.196.
(b) 0.608. (c) 0.196. **31.** 0.323. **33.** 0.439.
37. Proton: 9.2×10^{-6}; deuteron: 7.6×10^{-8}.
39. 10^{104} y. **41.** 7×10^{-23} kg·m/s. **43.** 1.2 m.
45. 0.41 μeV; $E_2 = -3.4$ eV. **47.** (a) 124 keV.
(b) 40.5 keV.

Chapter 45

3. 3.64×10^{-34} J·s. **5.** 24.1°.

7. $n > 3$; $m_\ell = +3, +2, +1, 0, -1, -2, -3$;
$m_s = +\frac{1}{2}, -\frac{1}{2}$. **9.** 50. **17.** 5.54 nm^{-1}. **19.** 0.0054.
21. (b) 16.4 nm$^{-3/2}$. **25.** 0.981 nm^{-1}; 3.61 nm^{-1}.
27. 0.0019. **29.** 54.7°; 125°. **31.** (a) 58 μeV.
(b) 14 GHz. (c) 2.1 cm; short radio wave region.
33. 5.35 cm. **35.** 19 mT. **37.** All statements are true.
39. (a) $(2, 0, 0, \pm\frac{1}{2})$.
(b) $n = 2$, $\ell = 1$; $m_\ell = 1, 0, -1$; $m_s = \pm\frac{1}{2}$. **45.** 12.4 kV.
47. 49.6 pm; 99.2 pm. **49.** (a) 24.8 pm.
(b) and (c) remain unchanged. **51.** 6.44 keV; 10.2 eV.
53. $\frac{9}{16}$. **55.** (a) 69.5 kV. (b) 17.9 pm. (c) 21.4 pm;
18.5 pm. **57.** 282 pm.
59. (b) 24%; 15%; 11%; 7.9%; 6.5%; 4.7%; 3.5%; 2.5%;
2.0%; 1.5%. **61.** 10,000 K. **63.** 4.4×10^{17}.
65. 2.0×10^{16} s^{-1}. **67.** 4.8 km. **69.** 1.8 pm.
71. (a) 7.33 μm. (b) 707 kW/m^2. (c) 24.9 GW/m^2.
73. (a) 53 GPa. (b) 1.2×10^8 K.

Chapter 46

1. 3460 atm. **3.** (a) 2.7×10^{25} m^{-3}.
(b) 8.43×10^{28} m^{-3}. (c) 3100.
(d) 3.3 nm for oxygen, 0.228 nm for the electrons.
7. 1.92×10^{28} m^{-3}·eV^{-1}. **9.** (a) 6.81 eV.
(b) 1.77×10^{28} m^{-3}·eV^{-1}. (c) 1.59×10^{28} m^{-3}·eV^{-1}.
11. 5.53 eV. **13.** $T \gg 10^5$ K. **15.** 3.
19. (a) 5.86×10^{28} m^{-3}. (b) 5.52 eV. (c) 1390 km/s.
(d) 0.522 nm. **21.** 137 MeV. **25.** 200°C.
29. (a) 19.8 kJ. (b) 3 min 18 s. **31.** (a) n-type.
(b) 5×10^{21} m^{-3}. (c) 2.5×10^5.
33. (a) Pure: 4.78×10^{-10}; doped: 0.0141. (b) 0.824.
37. (b) 2.49×10^8. **39.** 4.20 eV.

Chapter 47

1. 15.8 fm. **3.** (a) 0.39 MeV. (b) 4.61 MeV.
5. (a) Six. (b) Eight. **9.** (a) 1150 MeV.
(b) 4.8 MeV/nucleon; 12 MeV/proton.
12. (a) 3.0×10^{17} kg/m^3 for each.
(b) 1.3×10^{25} C/m^3 for ^{55}Mn; 1.1×10^{25} C/m^3 for
^{209}Bi. **15.** 4×10^{-22} s. **17.** $K \approx 30$ MeV.
21. (a) 19.8 MeV, 6.26 MeV, 2.22 MeV. (b) 28.3 MeV.
(c) 7.07 MeV. **23.** 1.6×10^{25} MeV. **25.** 7.92 MeV.
27. 280 d. **29.** (a) 7.6×10^{16} s^{-1}. (b) 4.9×10^{16} s^{-1}.
31. (a) 4.8×10^{-18} s^{-1}. (b) 4.6×10^9 y.
33. 5.3×10^{22}. **35.** 265 mg. **37.** 209 d. **39.** 87.8 mg.
41. 730 cm^2. **45.** (a) 3.66×10^7 s^{-1}. (b) $t \gg 3.82$ d.
(c) 3.66×10^7 s^{-1}. (d) 6.42 ng.
47. Pu: 5.5×10^{-9}; Cm: zero. **49.** (a) 4.25 MeV.
(b) -24.1 MeV. (c) 28.3 MeV.
51. $Q_3 = -9.50$ MeV; $Q_4 = 4.66$ MeV;
$Q_5 = -1.30$ MeV. **53.** 1.40 MeV. **55.** 0.782 MeV.
59. (b) 0.961 MeV. **61.** 78.4 eV. **63.** 1600 y.
65. 1.7 mg. **67.** 1.02 mg. **69.** (a) 18 mJ.

(b) 0.29 rem. **71.** (a) 6.3×10^{18}. (b) 2.5×10^{11}.
(c) 0.20 J. (d) 0.23 rad. (e) 3.0 rem.
73. 3.87×10^{10} K. **75.** (a) 25.35 MeV. (b) 12.8 MeV.
(c) 25.0 MeV. **77.** (a) 3.85 MeV, 7.95 MeV.
(b) 3.98 MeV, 7.33 MeV.

Chapter 48

1. (a) 2.6×10^{24}. (b) 8.2×10^{13} J. (c) 2.6×10^4 y.
3. 3.1×10^{10} s^{-1}. **7.** $+5.00$ MeV.
9. (a) 16 fissions/day. (b) 4.3×10^8. **11.** (a) 10.
(b) 231 MeV. **13.** (a) 252 MeV.
(b) Typical fission energy = 200 MeV. **15.** 463 kg.
17. Yes. **19.** (a) 1.15 MeV. (b) 3.2 kg.
21. (a) 44 kton. **25.** 1.7×10^k.
27. (b) 1.0, 0.89, 0.28, 0.019. (c) 8. **29.** (a) 75 kW.
(b) 5800 kg. **33.** (a) 30 MeV. (b) 6 MeV.
(c) 170 keV. **35.** (a) 170 kV. **37.** 1.41 MeV.
41. (a) 3.1×10^{31} photons/m^3. (b) 1.2×10^6 times.

43. (a) 4.0×10^{27} MeV. (b) 5.1×10^{26} MeV.
45. (a) 1.83×10^{38} s^{-1}. (b) 8.25×10^{28} s^{-1}.
49. (a) 6.3×10^{14} J/kg. (b) 6.2×10^{11} kg/s.
(c) 4.3×10^9 kg/s. (d) 15×10^9 y. **51.** (a) 24.9 MeV.
(b) 8.65 Mton. **53.** $K_\alpha = 3.5$ MeV; $K_n = 14.1$ MeV.
57. (a) 230 kJ. (b) 0.11 lb. (c) 23 MW.

Chapter 49

1. 6.03×10^{-29} kg. **3.** 2.4×10^{-43}. **5.** 1.08×10^{42} J.
7. 27 cm/s. **9.** (a) 1.90×10^{-18} kg·m/s. (b) 9.90 m.
13. (a) Strangeness. (b) Charge. (c) Energy.
15. $Q = 0$; $B = -1$; $S = 0$. **17.** (a) Energy.
(b) Angular momentum. (c) Charge. **19.** 338 MeV.
23. (a) $\overline{uud}$. (b) $\overline{udd}$. **25.** (a) sud. (b) uss.
29. Σ^0; 7530 km/s. **31.** 665 nm. **33.** (b) 0.934.
(c) 1.65×10^{10} ly. **35.** (a) 256 μeV. (b) 4.84 nm.
37. (a) 121 m/s. (b) 246 y. **39.** (a) 2.6 K. (b) 29 nm.

PHOTO CREDITS

Chapter 11
Opener: Guido Alberto Rossi/Image Bank. Page 286: Rick Rickman/Duomo. Page 293: Art Tilley/FPG International. Page 295: Roger Ressmeyer/Starlight. Page 305: Fabricius-Taylor/AllStock. Page 306: Zimmerman/FPG International. Page 312: Courtesy Lick Observatory. Page 316: Courtesy Lawrence Livermore Laboratory, University of California.

Chapter 12
Opener: Courtesy Ringling Brothers and Barnum & Bailey Circus. Page 320: Richard Megna/Fundamental Photographs. Page 321: Courtesy Alice Halliday. Page 322: Elisabeth Weiland/Photo Researchers. Page 327: Doug Lee/Peter Arnold. Page 332: Kerstgens/Sipa Press. Page 335: Steven E. Sutton/Duomo. Page 336: (top) Courtesy NASA; (bottom) E. Sander/Gamma-Liaison. Page 339: copyright © Estate of Harold Edgerton, courtesy of Palm Press, Inc.

ESSAY 2
Page E2-1: (top left and bottom) Susan Cook: (top right) courtesy Kenneth Laws. Pages E2-2–E2-4: Martha Swope.

Chapter 13
Opener: Jose Azel/Woodfin Camp & Associates. Page 354: (left) Fred Hirschmann/Allstock; (right) Andy Levin/Photo Researchers. Page 355: Randy G. Taylor/Leo de Wys. Page 356: Richard Negri/Gamma-Liaison. Page 358: Chris Dalmas/Gamma-Liaison. Page 364: David Bitters/The Picture Cube. Page 368: Courtesy Micro-Measurements Division, Measurements Group, Inc., Raleigh, North Carolina. Page 372: Hideo Kurihara/Tony Stone Worldwide.

Chapter 14
Opener: Tom van Dyke/Sygma. Page 382: Kent Knudson/FPG International. Page 398: Bettmann Archive. Page 404: Courtesy NASA.

Chapter 15
Opener: Courtesy Lund Observatory. Page 412: Courtesy NASA. Page 421: Kim Gordon/AstroStock. Page 422: Courtesy Lockheed Missile and Space Co., Inc. Pages 425 and 438: Courtesy Finley-Holiday Film Corporation. Page 429: Courtesy NASA.

ESSAY 3
Pages E3-1 and E3-3: Courtesy NASA.

Chapter 16
Opener: Steven Frink. Page 453: T. Orban/Sygma. Page 454: John Amos/Photo Researchers. Page 455: (left) Will McIntyre/Photo Researchers; (right) courtesy D. H. Peregrine, University of Bristol, England. Page 456: Courtesy Volvo North America Corporation. Page 463: Courtesy San Francisco Maritime National Historical Park Museum. Page 466: T. Campion/Sygma. Page 470: Kick Stewart/Allsport.

ESSAY 4
Page E4-1: Courtesy Peter Brancazio.

Chapter 17
Opener: John Visser/Bruce Coleman. Page 493: Richard Megna/Fundamental Photographs. Page 494: Courtesy T. D. Rossing, Northern Illinois University.

Chapter 18
Opener: Stephen Dalton/Animals Animals. Page 504: (left) Howard Sochurack/Stock Market; (right) courtesy John Foster, IBM Corporation. Page 510: Ben Rose/Image Bank. Page 513: Bob Gruen/Star File Photos. Page 514: John Eastcott/Yva Momatiuk/DRK Photo. Page 521: Philippe Plailly/Science Photo Library/Photo Researchers. Page 532: Courtesy Mt. Wilson and Palomar Observatories.

ESSAY 5
Page E5-1: (top) Courtesy John Rigden; (center) courtesy Lincoln Center. Page E5-2: Lincoln Russell/Stock, Boston. Pages E5-4 and E5-5: Courtesy Lincoln Center, photos by Norman McGrath.

Chapter 19
Opener: Jim Brandenburg/Minden Pictures. Page 540: AP/Wide World Photos. Page 543: Richard Choy/Peter Arnold.

Chapter 20
Opener: Tom Owen Edmunds/Image Bank. Page 554: Obremski/Image Bank. Page 564: (left) Fritz Goro; (right) Peter Arnold/Peter Arnold. Page 565: Courtesy Daedalus Enterprises. Page 569: Mark Newman.

ESSAY 6
Page E6-1: Jeff R. Werner.

Chapter 21
Opener: Bryan and Cherry Alexander Photography. Page 581: Courtesy NASA. Page 593: Tom Branch/Photo Researchers.

ESSAY 7
Page E7-1: Courtesy Barbara Levi. Page E7-2: Courtesy NASA/Langley Research Center.

Chapter 22
Opener: Steven Dalton/Photo Researchers. Page 608: (top) Richard Ustinich/Image Bank; (bottom) V. Kiselev/Sovfoto. Page 625: (left) Carry Wolinski/Stock, Boston; (right) Courtesy Professor Hallet, University of Washington, Seattle.

Chapter 23
Opener: Michael Watson. Page 636: Fundamental Photographs. Page 637: Courtesy Saran Wrap, Dow Brands, Household Products Division. Page 638: (left) Courtesy Xerox Corporation; (right) Johann Gabriel Doppelmayr, Neuentdeckte Phaenomena von bewündernswurdigen Würckungen der Natur, Nuremberg 1744. Page 646: Courtesy Lawrence Berkeley Laboratory, University of California.

1019: (left) Frans Lanting/Minden Pictures; (right) Wayne Sorce. Page 1027: Courtesy Matthew G. Wheeler. Page 1042: Piergiorgio Scharandis/Black Star.

ESSAY 12
Page E12-1: Courtesy Suzanne Nagel.

Chapter 40
Opener: E. R. Degginger. Page 1053: Runk/Schoenberger/Grant Heilman. Page 1054: From *Atlas of Optical Phenomena* by M. Cagnet et al., Springer-Verlag, Prentice Hall, 1962. Pages 1055 and 1062: Richard Megna/Fundamental Photographs. Page 1064: Courtesy Dr. Helen Ghiradella, Department of Biological Sciences, SUNY, Albany. Page 1073: Courtesy Bausch & Lomb.

ESSAY 13
Page E13-1: Courtesy Elsa Garmire.

Chapter 41
Opener: Courtesy The Art Institute of Chicago. Page 1076: Ken Kay/Fundamental Photographs. Pages 1077 and 1083: From *Atlas of Optical Phenomena* by Cagnet, Francon, Thierr, Springer-Verlag, Berlin, 1962. Page 1084: (left) AP/Wide World Photos; (right) Cath Ellis/Science Photo Library/Photo Researchers. Pages 1086 and 1089: From *Atlas of Optical Phenomena* by Cagnet, Francon, Thierr, Springer-Verlag, Berlin, 1962. Page 1091: Department of Physics, Imperial College/Science Photo Library/Photo Researchers. Page 1092: Peter L. Chapman/Stock, Boston. Page 1101: Courtesy Robert Greenler.

ESSAY 14
Pages E14-1–4: Courtesy Tung Jeong.

Chapter 42
Opener: T. Tracy/FPG International. Page 1106: Courtesy Hebrew University of Jerusalem, Israel.

ESSAY 15
Page E15-1: (top) Photo by Vittorio Giannella, courtesy Joseph Ford; (bottom) Reproduced with permission from ONERA, courtesy Joseph Ford.

Chapter 43
Opener: Courtesy Jearl Walker. Page 1143: Courtesy AIP Neils Bohr Library.

Chapter 44
Opener: Lawrence Berkeley Laboratory/Science Photo Library/Photo Researchers. Page 1159: (top left) Courtesy Riber Division of Instruments, Inc.; (top right and bottom) From PSSC film ''Matter Waves,'' courtesy Education Development Center, Newton, Massachusetts. Page 1167: Courtesy IBM Corporation, Research Division, Almadon

Research Center, San Jose, CA. Page 1173: Courtesy A. Tonomura, J. Endo, T. Matsuda, and T. Kawasaki/Advanced Research Laboratory, Hitachi, Ltd., Kokubinju, Tokyo; H. Ezawa, Department of Physics, Gakushuin University, Mejiro, Tokyo.

ESSAY 16
Pages E16-1 and E16-4: Courtesy Ivar Giaever.

Chapter 45
Opener: Kurt Coste/Tony Stone Worldwide. Page 1180: Courtesy Professor Hatsu Uyeda, Kyoto University. Page 1190: CNRI/Science Photo Library/Photo Researchers. Page 1195: (left) Courtesy AIP Neils Bohr Library; (right) Scott Camazine/Photo Researchers. Page 1197: (left) Will & Deni McIntyre/Photo Researchers; (right) Tony Stone Worldwide.

Chapter 46
Opener: Courtesy IBM Corporation, Research Division. Page 1217: Alfred Pasieka/Peter Arnold. Page 1224: Courtesy AT&T Bell Laboratories.

ESSAY 17
Pages E17-1–2: Courtesy Patricia Cladis.

Chapter 47
Opener: Elscint/Science Photo Library/Photo Researchers. Page 1245: (inset) R. Perry/Sygma; (bottom) George Rockwin/Bruce Coleman.

Chapter 48
Opener: Courtesy U.S. Department of Energy. Page 1266: Ivleva/Magnum. Page 1267: Courtesy Chicago Historical Society. Page 1272: Courtesy Anglo-Australian Telescope Board. Page 1275: (left) Courtesy Los Alamos National Laboratory, New Mexico; (right) Roger Ressmeyer/Starlight. Page 1279: Courtesy Martin Marietta Energy Systems, Inc., for the Department of Energy.

Chapter 49
Opener: Courtesy NASA. Page 1284: David Parker/Science Photo Library/Photo Researchers. Page 1287: Courtesy Lawrence Berkeley Laboratory. Page 1298: Courtesy AIP Neils Bohr Library. Page 1299: Courtesy AT&T Bell Laboratories.

Table of Contents
Jerry Yulsman/Image Bank
Lois Greenfield/Bruce Coleman
Stephen Dalton/Animals Animals
Stephen Dalton/Photo Researchers
Johnny Johnson/Earth Scenes
Courtesy The Art Institute of Chicago

INDEX

AN OUTLINE OF TOPICS